Ebook Set up and Use [illegible]

Use of the ebook is subject to the single user licence at the back of this book.

Electronic access to your price book is now provided as an ebook on the VitalSource® Bookshelf platform. You can access it online or offline on your PC/Mac, smartphone or tablet. You can browse and search the content across all the books you've got, make notes and highlights and share these notes with other users.

Setting up

1. Create a VitalSource Bookshelf account at https://online.vitalsource.com/user/new (or log into your existing account if you already have one).

2. Retrieve the code by scratching off the security-protected label inside the front cover of this book. Log in to Bookshelf and click the **Redeem** menu at the top right of the screen. and Enter the code in the **Redeem code** box and press **Redeem**. Once the code has been redeemed your Spon's Price Book will download and appear in your **library**. N.B. the code in the scratch-off panel can only be used once, and has to be redeemed before end December 2017.

When you have created a Bookshelf account and redeemed the code you will be able to access the ebook online or offline on your smartphone, tablet or PC/Mac. Your notes and highlights will automatically stay in sync no matter where you make them.

Use ONLINE

1. Log in to your Bookshelf account at https://online.vitalsource.com).
2. Double-click on the title in your **library** to open the ebook.

Use OFFLINE

Download BookShelf to your PC, Mac, iOS device, Android device or Kindle Fire, and log in to your Bookshelf account to access your ebook, as follows:

On you[illegible]
Go to [illegible] [illegible]d follow the instructions [illegible] [illegible]ce Bookshelf app to your PC or Mac. [illegible] the VitalSource Bookshelf icon that appears on your desktop and log into your Bookshelf account. Select **All Titles** from the menu on the left – you should see your price book on your Bookshelf. If your Price Book does not appear, select **Update Booklist** from the **Account** menu. Double-click the price book to open it.

On your iPhone/iPod Touch/iPad
Download the free VitalSource Bookshelf App available via the iTunes App Store. Open the Bookshelf app and log into your Bookshelf account. Select **All Titles** - you should see your price book on your Bookshelf. Select the price book to open it. You can find more information at https://support.vitalsource.com/hc/en-us/categories/200134217-Bookshelf-for-iOS

On your Android™ smartphone or tablet
Download the free VitalSource Bookshelf App available via Google Play. Open the Bookshelf app and log into your Bookshelf account. You should see your price book on your Bookshelf. Select the price book to open it. You can find more information at https://support.vitalsource.com/hc/en-us/categories/200139976-Bookshelf-for-Android-and-Kindle-Fire

On your Kindle Fire
Download the free VitalSource Bookshelf App from Amazon. Open the Bookshelf app and log into your Bookshelf account. Select All Titles – you should see your price book on your Bookshelf. Select the price book to open it. You can find more information at https://support.vitalsource.com/hc/en-us/categories/200139976-Bookshelf-for-Android-and-Kindle-Fire

Support

If you have any questions about downloading Bookshelf, creating your account, or accessing and using your ebook edition, please visit http://support.vitalsource.com/

For questions or comments on content, please contact us on sponsonline@tandf.co.uk

---

All materials in *Spon's Online*, which is supplied with *Spon's Mechanical & Electrical Services Price Book 2017*, are copyright protected. No such materials may be used, displayed, modified, adapted, distributed, transmitted, transferred, published or otherwise reproduced in any form or by any means now or hereafter developed other than strictly in accordance with the terms of the Electronic Version Single-User Licence Agreement at the end of this book. The ebook may be used by more than one user if additional access codes are purchased from Electronic Sales, Spon Press, 3 Park Square, Milton Park, Abingdon, Oxon OX14 4RN Tel: +44 (0) 20 7017 6000

© 2017 COPYRIGHT ALL RIGHTS RESERVED

Free Updates

with three easy steps...

1. Register today on www.pricebooks.co.uk/updates
2. We'll alert you by email when new updates are posted on our website
3. Then go to www.pricebooks.co.uk/updates
   and download the update.

All four Spon Price Books – *Architects' and Builders'*, *Civil Engineering and Highway Works*, *External Works and Landscape* and *Mechanical and Electrical Services* – are supported by an updating service. Two or three updates are loaded on our website during the year, typically in November, February and May. Each gives details of changes in prices of materials, wage rates and other significant items, with regional price level adjustments for Northern Ireland, Scotland and Wales and regions of England. The updates terminate with the publication of the next annual edition.

As a purchaser of a Spon Price Book you are entitled to this updating service for this 2017 edition – free of charge. Simply register via the website www.pricebooks.co.uk/updates and we will send you an email when each update becomes available.

If you haven't got internet access or if you've some questions about the updates please write to us at Spon Price Book Updates, Spon Press Marketing Department, 3 Park Square, Milton Park, Abingdon, Oxfordshire, OX14 4RN.

Find out more about Spon books
Visit www.pricebooks.co.uk for more details.

M
Elec

CAMBRIDGESHIRE
LIBRARIES
REFERENCE
AND
INFORMATION
SERVICE

............................LIBRARY

10010010466991

Spon's Mechanical and Electrical Services Price Book

Edited by

AECOM

2017

Forty-eighth edition

CRC Press
Taylor & Francis Group

First edition 1968
Forty-eighth edition published 2017
by CRC Press
2 Park Square, Milton Park, Abingdon, Oxon, OX14 4RN

and by CRC Press
Taylor & Francis, 6000 Broken Sound Parkway, NW, Suite 300, Boca Raton, FL 33487

CRC Press is an imprint of the Taylor & Francis Group, an informa business

© 2017 Taylor & Francis

The right of AECOM Ltd to be identified as the Author of this Work has been asserted by them in accordance with the Copyright, Designs and Patents Act 1988

All rights reserved. No part of this book may be reprinted or reproduced or utilized in any form or by any electronic, mechanical, or other means, now known or hereafter invented, including photocopying and recording, or in any information storage or retrieval system, without permission in writing from the publishers.

Trademark notice: Product or corporate names may be trademarks or registered trademarks, and are used only for identification and explanation without intent to infringe.

The publisher makes no representation, express or implied, with regard to the accuracy of the information contained in this book and cannot accept legal responsibility or liability for any errors or omissions that may be made.

British Library Cataloguing in Publication Data
A catalogue record for this book is available from the British Library

ISBN13: 978-1-4987-8616-4
Ebook: 978-1-4987-8649-2

ISSN: 0305-4543

Typeset in Arial by Taylor & Francis Books

Cambridgeshire Libraries	
10010010466991	
Askews & Holts	31-Oct-2016
697 REF	£160.00

MIX
Paper from responsible sources
FSC www.fsc.org FSC® C013056

Printed and bound in Great Britain by
TJ International Ltd, Padstow, Cornwall

Contents

Preface to the Forty-Eighth Edition

The forty-eighth edition of *Spon's Mechanical and Electrical Services Price Book* continues to cover the widest range and depth of engineering services, reflecting the many alternative systems and products that are commonly used in the industry as well as current industry trends.

The Construction industry is still experiencing growth, with MEP tender pricing seeing a 5–7% increase during 2016.

Key influences

- Increase in staffing costs
- Weakening pound against the Euro & US Dollar
- Commodity prices have fallen
- Reduced supplier discounts and increased fixed price allowances

2016 Wage increases with both Mechanical and Electrical trades have been agreed.

Before referring to prices or other information in the book, readers are advised to study the `Directions' which precede each section of the Materials Costs/Measured Work Prices. As before, no allowance has been made in any of the sections for Value Added Tax.

The order of the book reflects the order of the estimating process, from broad outline costs through to detailed unit rate items.

The approximate estimating section has been revised to provide up to date key data in terms of square metre rates, all-in rates for key elements and selected specialist activities and elemental analyses on a comprehensive range of building types.

The prime purpose of the Materials Costs/Measured Work Prices part is to provide industry average prices for mechanical and electrical services, giving a reasonably accurate indication of their likely cost. Supplementary information is included which will enable readers to make adjustments to suit their own requirements. It cannot be emphasized too strongly that it is not intended that these prices are used in the preparation of an actual tender without adjustment for the circumstances of the particular project in terms of productivity, locality, project size and current market conditions. Adjustments should be made to standard rates for time, location, local conditions, site constraints and any other factor likely to affect the costs of a specific scheme. Readers are referred to the build up of the gang rates, where allowances are included for supervision, labour related insurances, and where the percentage allowances for overhead, profit and preliminaries are defined.

Readers are reminded of the service available on the Spon's website detailing significant changes to the published information. www.pricebooks.co.uk/updates

As with previous editions the Editors invite the views of readers, critical or otherwise, which might usefully be considered when preparing future editions of this work.

Whilst every effort is made to ensure the accuracy of the information given in this publication, neither the Editors nor Publishers in any way accept liability for loss of any kind resulting from the use made by any person of such information.

In conclusion, the Editors record their appreciation of the indispensable assistance received from the many individuals and organizations in compiling this book.

AECOM Ltd
Aldgate Tower
2 Leman Street
London E1 8FAS

Telephone: 020 7061 7000
Facsimile: 020 7061 7061

Special Acknowledgements

The Editors wish to record their appreciation of the special assistance given by the following organizations in the compilation of this edition.

redstone

160 Centennial Business Park
Centennial Avenue
Elstree
Hertfordshire
WD6 3SH
Tel: 0845 201 0000
Email: salesenquiries@redstone.com
www.redstone.com

HOTCHKISS

Hampden Park Industrial Estate
Eastbourne
East Sussex
BN22 9AX
Tel: 01323 501234
Fax: 01323 508752
Email: info@Hotchkiss.co.uk
www.Hotchkiss.co.uk

Abbey

Abbey Thermal Insulation Ltd
23–24 Riverside House
Lower Southend Road
Wickford
Essex
SS11 8BB
Tel: 01268 572116
Fax: 01268 572117
Email: general@abbeythermal.com

MJL

Michael J Lonsdale
22–24 Cowper Street
London
EC2A 4AP
Tel: 0845 241 6622
Fax: 0845 241 633

T.Clarke
BUILDING SERVICES GROUP

Stanhope House
116–118 Walworth Road
London
SE17 1JY
Tel: 020 7358 5000
Email: info@tclarke.co.uk
www.tclarke.co.uk

TROX® TECHNIK
The art of handling air

Trox UK Ltd
Caxton Way
Thetford
Norfolk
IP24 3SQ
Tel: 01842 754545
Fax: 01842 763051
www.troxuk.co.uk

Gardner&Co

Gardner & Co Ltd
1–5 Bermondsey Street
London
SE1 2ER
Tel: 020 7403 5894
Fax: 020 7378 0316
www.gardnerco.net

debbas

Debbas
F48 Business & Technology Centre (BTC)
Bessemer Drive
Stevenage
SG1 2DX
Tel: 01438 791 032
www.debbas.com

SPON'S PRICEBOOKS 2017 From AECOM

Spon's Architects' and Builders' Price Book 2017

Editor: AECOM

To NRM1 and NRM2, and with more plant prices separated out in the measured works section. With further developments:

- The in situ concrete section and plasterboard linings and partitions sections have been heavily revised and developed.
- A laboratory, a car park and an updated London Office have been added as new Cost Models
- And new items are added: Foamglas insulation, Carlite waterproof concrete, and an expanded range of stone flooring.

Hbk & VitalSource® ebook 824pp approx.
978-1-4987-8611-9 £160
VitalSource® ebook
978-1-4987-8643-0 £160
(inc. sales tax where appropriate)

Spon's Civil Engineering and Highway Works Price Book 2017

Editor:AECOM

This year some item descriptions have been revised to reflect prices in the global market. Plus a revised structure for manhole construction; new prices for pre-cast manholes with monolithic precast concrete bases; updated rail supply and installation rates and a few more added items; revised piling rates and descriptions; new highway pipe rates; reduced steel prices.

Hbk & VitalSource® ebook 688pp approx.
978-1-4987-8612-6 £180
VitalSource® ebook
978-1-4987-8646-1 £180
(inc. sales tax where appropriate)

Spon's External Works and Landscape Price Book 2017

Editor:AECOM

This NRM edition includes a number of new and significantly developed items: acoustic fencing, architectural metalwork, block and beam systems, culverts, deep concrete manholes to 3 m, headwalls, pre-cast steps and new retaining walls. Green roofs and sportsfields have now been broken down into details, rather than presented as a lump sum.

Hbk & VitalSource® ebook 624pp approx.
978-1-4987-8615-7 £150
VitalSource® ebook
978-1-4987-8648-5 £150
(inc. sales tax where appropriate)

Spon's Mechanical and Electrical Services Price Book 2017

Editor:AECOM

This NRM edition includes a number of new and significantly developed items: WiFi equipment, FP 600 cable, wireless fire alarms; and a much enhanced clock section.

Hbk & VitalSource® ebook 840pp approx.
978-1-4987-8616-4 £160
VitalSource® ebook
978-1-4987-8650-8 £160
(inc. sales tax where appropriate)

Receive our VitalSource® ebook free when you order any hard copy Spon 2017 Price Book

Visit www.pricebooks.co.uk

To order:
Tel: 01235 400524 Fax: 01235 400525
Post: Taylor & Francis Customer Services, Bookpoint Ltd, 200 Milton Park, Abingdon, Oxon, OX14 4SB, UK
Email: book.orders@tandf.co.uk
A complete listing of all our books is on www.crcpress.com

CRC Press
Taylor & Francis Group

Acknowledgements

The Editors wish to record their appreciation of the assistance given by many individuals and organizations in the compilation of this edition.

Manufacturers, Distributors and Subcontractors who have contributed this year include:

Actionair
Joseph Wilson Ind. Estate
South Street
Whitstable
Kent CT5 3DU
Tel: 01227 276100
Fax: 01227 264262
Email: sales@actionair.co.uk
www.actionair.co.uk
Dampers

Aquilar Limited
Dial Post Court
Horsham Road
Rusper
West Sussex RH12 4QX
Tel: 08707 940310
Fax: 08707 940320
www.aquilar.co.uk
Leak Detection

Balmoral Tanks
Balmoral Park
Loirston
Aberdeen AB12 3GY
Tel: 01224 859000
Fax: 01224 859123
www.balmoral-group.com
GRP Water Storage Tanks

Biddle Air Systems Ltd
St Mary's Road, Nuneaton
Warwickshire
CV11 5AU
Tel: 024 7638 4233
Fax: 024 7637 3621
Email: info@biddle-air.co.uk
Air Curtains

Braithwaite Engineers Ltd
Neptune Works
Uskway
Newport
South Wales NP9 2UY
Tel: 01633 262141
Fax: 01633 250631
www.braithwaite.co.uk
Sectional Steel Water Storage Tanks

Broadcrown Limited
Alliance Works
Airfield Industrial Estate
Hixon
Staffs ST18 0PF
Tel: 01889 272200
Fax: 01889 272220
www.broadcrown.co.uk
Generators

Chloride Power Protection
Unit C, George Curl Way
Southampton SO18 2RY
Tel: 023 8061 0311
Fax: 023 8061 0852
www.chloridepower.com
Static UPS Systems

Cooper Lighting and Security
London Project Office
Suite 8, King Harold Court
Sun Street
Waltham Abbey
Essex EN9 1ER
Tel: 01302 303303
Fax: 01392 367155
www.cooper-ls.com
Emergency Lighting and Luminaires

Dewey Waters Limited
Cox's Green
Wrington
Bristol BS40 5QS
Tel: 01934 862601
Fax: 01934 862604
www.deweywaters.co.uk
Tanks

DMS Flow Measurement and Controls Limited
X-Cel House
Chrysalis Way
Langley Bridge
Eastwood
Nottinghamshire
NG16 3RY
Tel: 01773 534555
Fax: 01773 534666
www.dmsltd.com
Chilled Water Plant

Dunham-Bush Limited
8 Downley Road
Havant
Hampshire PO9 2JD
Tel: 02392 477700
Fax: 02392 450396
www.dunham-bush.com
Convectors and Heaters

Flakt Woods Limited
Axial Way
Colchester CO4 5ZD
Tel: 01206 222555
Fax: 01206 222777
Fans

Frontline Security Solutions Ltd
Reflex House
The Vale
Chalfont St Peter
Bucks
SL9 9RZ
Tel: 01753 482 248
Access Control and Security Detection Alarm

Hattersley, Newman, Hender Ltd
Burscough Road
Ormskirk
Lancashire L39 2XG
Tel: 01695 577199
Fax: 01695 578775
Email: uksales@hattersley-valves.co.uk
www.hattersley.com
Valves

Hawker Siddeley Switchgear Ltd Technology Centre
Falcon Works
Loughborough
Leicester LE11 1EX
Tel: 01509 611511
Fax: 01509 612009
Email: sales@hss-ltd.com
HV Supply, Cables and HV Switchgear and Transformers

Hitec Power Protection Limited
Unit B21a
Holly Farm Business Park
Honiley
Kenilworth
Warwickshire CV8 1NP
Tel: 01926 484535
Fax: 01926 484336
www.hitecups.co.uk
Uninterruptible Power Supply (Rotary/Diesel)

Honeywell CS Limited
Honeywell House
Anchor Boulevard
Crossways Business Park
Dartford
Kent DA2 6QH
Tel: 01322 484800
Fax: 01322 484898
www.honeywell.com
Control Components

Hoval Limited
Northgate
Newark
Notts NG24 1JN
Tel: 01636 672711
Fax: 01636 673532
www.hoval.co.uk
Boilers

HRS Hevac Ltd
10–12 Caxton Way
Watford Business Park
Watford
Herts WD18 8JY
Tel: 01923 232335
Fax: 01923 230266
www.hrshevac.co.uk
Heat Exchangers

Hudevad
Bridge House
Bridge Street
Walton on Thames
Surrey KT12 1AL
Tel: 01932 247835
Fax: 01932 247694
www.hudevad.co.uk
Radiators

Hydrotec (UK) Limited
Hydrotec House
5 Mannor Courtyard
Hughenden Avenue
High Wycombe HP13 5RE
Tel: 01494 796040
Fax: 01494 796049
www.hydrotec.com
Chemical Treatment

IAC
IEC House
Moorside Road
Winchester
Hampshire SO23 7US
Tel: 01962 873000
Fax: 01962 873102
www.industrialacoustics.com
Attenuators

IC Service & Maintenance Ltd
Unit K3, Temple Court
Knights Place
Knight Road
Strood
Kent ME2 2LT
Tel: 01634 290300
Fax: 01634 290700
www.icservice.biz
Fire Detection & Alarm

Kampmann
Benson Environmental Limited
47 Central Avenue
West Molesey
Surrey KT8 2QZ
Tel: 020 8783 0033
Fax: 020 8783 0140
www.diffusionenv.com
Trench Heating

Kiddie Fire Protection Services
Enterprise House
Jasmine Grove
London SE20 8JW
Tel: 020 8659 7235
Fax: 020 8659 7237
www.kfp.co.uk
Fire Protection Equipment

Metcraft Ltd
Harwood Industrial Estate
Littlehampton
West Sussex BN17 7BB
Tel: 01903 714226
Fax: 01903 723206
www.metcraft.co.uk
Oil Storage Tanks

Mitsubishi Electric Europe BV
Unit 8, Electra Park
Bidder Street
Canning Town
London E16 4ES
Tel (switchboard): 020 7511 5664
www.mitsubishi-lifts.co.uk
Lifts and Escalators

Osma Underfloor Heating
18 Apple Lane
Sowton Trade City
Exeter
Devon EX2 5GL
Tel: 01392 444122
Fax: 01392 444135
www.osmaufh.co.uk
Underfloor Heating

Pullen Pumps Limited
158 Beddington Lane
Croydon CR9 4PT
Tel: 020 8684 9521
Fax: 020 8689 8892
www.pullenpumps.co.uk
Pumps, Booster Sets

Rycroft
Duncombe Road
Bradford BD8 9TB
Tel: 01274 490911
Fax: 01274 498580
www.rycroft.com
Storage Cylinders

Safelincs Ltd
Farlesthorpe Road Industrial Estate
Alford
Lincolnshire LN13 9PS
Telephone: 01507 462176
Email: support@safelincs.co.uk
Fire Protection Equipment

Saint-Gobain PAM UK Ltd
Lows Lane
Stanton-By-Dale
Ilkeston
Derbyshire DE7 4RU
Tel: 0115 930 5000
Fax: 0115 932 9513
www.saint-gobain-pam.co.uk
Cast Iron Pipework

Schneider Electric Limited
120 New Cavendish Street
London W1W 6XX
Tel: 0870 608 8608
www.schneider-electric.com
Energy Management

Senior Hargreaves
Lord Street
Bury
Lancashire
BL9 0RG
Tel: 01204 596 015
Fax: 01204 598 751
Email: jane.holland@sanberlabservices.co.uk
www.sanberlabservices.co.uk
Plastic Ductwork

SF Limited
Pottington Business Park
Barnstaple
Devon EX31 1LZ
Tel: 01271 326633
Fax: 01271 334303
Flues

Simmtronic Limited
Waterside
Charlton Mead Lane
Hoddesdon
Hertfordshire EN11 0QR
Tel: 01992 456869
Fax: 01992 445132
www.simmtronic.com
Lighting Controls

Socomec Limited
Knowl Piece
Wilbury Way
Hitchin
Hertfordshire SG4 0TY
Tel: 01462 440033
Fax: 01462 431143
www.socomec.com
Automatic Transfer Switches

Spirax-Sarco Ltd
Charlton House
Cheltenham
Gloucestershire GL53 8ER
Tel: 01242 521361
Fax: 01242 573342
www.spiraxsarco.com
Traps and Valves

Tyco Limited
Unit 6, West Point Enterprize Park
Clarence Avenue
Trafford Park
Manchester M17 1QS
Tel: 0161 875 0400
Fax: 0161 875 0491
www.tyco.com
Fire Protection

Utile Engineering Company Ltd
Irthlingborough
Northants NN9 5UG
Tel: 01933 650216
Fax: 01933 652738
www.utileengineering.com
Gas Boosters

Woods of Colchester
Tufnell Way
Colchester
Essex CO4 5AR
Tel: 01206 544122
Fax: 01206 574434
Air Distribution, Fans, Anti-Vibration Mountings

ESSENTIAL READING FROM TAYLOR AND FRANCIS

Spon's Asia Pacific Construction Costs Handbook, Fifth Edition

LANGDON & SEAH

In the last few years, the global economic outlook has continued to be shrouded in uncertainty and volatility following the financial crisis in the Euro zone. While the US and Europe are going through a difficult period, investors are focusing more keenly on Asia. This fifth edition provides overarching construction cost data for 16 countries: Brunei, Cambodia, China, Hong Kong, India, Indonesia, Japan, Malaysia, Myanmar, Philippines, Singapore, South Korea, Sri Lanka, Taiwan, Thailand and Vietnam.

May 2015: 234X156 mm: 452 pp
Hb: 978-1-4822-4358-1: £160.00

To Order: Tel: +44 (0) 1235 400524 Fax: +44 (0) 1235 400525
or Post: Taylor and Francis Customer Services,
Bookpoint Ltd, Unit T1, 200 Milton Park, Abingdon, Oxon, OX14 4TA UK
Email: book.orders@tandf.co.uk

For a complete listing of all our titles visit:
www.tandf.co.uk

Taylor & Francis
Taylor & Francis Group

It is one thing to
imagine a better world.
It's another to deliver it.

**Understanding change,
unlocking potential, creating
brilliant new communities.**

The Tate Modern extension
takes an iconic building and
adds to it. Cost management
provided by AECOM.

AECOM
Built to deliver a better world

aecom.com

SPON'S PRICEBOOKS 2017 From AECOM

Spon's Architects' and Builders' Price Book 2017

Editor: AECOM

To NRM1 and NRM2, and with more plant prices separated out in the measured works section. With further developments:

- The in situ concrete section and plasterboard linings and partitions sections have been heavily revised and developed.
- A laboratory, a car park and an updated London Office have been added as new Cost Models
- And new items are added: Foamglas insulation, Carlite waterproof concrete, and an expanded range of stone flooring.

Hbk & VitalSource® ebook 824pp approx.
978-1-4987-8611-9 £160
VitalSource® ebook
978-1-4987-8643-0 £160
(inc. sales tax where appropriate)

Spon's Civil Engineering and Highway Works Price Book 2017

Editor:AECOM

This year some item descriptions have been revised to reflect prices in the global market. Plus a revised structure for manhole construction; new prices for pre-cast manholes with monolithic precast concrete bases; updated rail supply and installation rates and a few more added items; revised piling rates and descriptions; new highway pipe rates; reduced steel prices.

Hbk & VitalSource® ebook 688pp approx.
978-1-4987-8612-6 £180
VitalSource® ebook
978-1-4987-8646-1 £180
(inc. sales tax where appropriate)

Spon's External Works and Landscape Price Book 2017

Editor:AECOM

This NRM edition includes a number of new and significantly developed items: acoustic fencing, architectural metalwork, block and beam systems, culverts, deep concrete manholes to 3 m, headwalls, pre-cast steps and new retaining walls. Green roofs and sportsfields have now been broken down into details, rather than presented as a lump sum.

Hbk & VitalSource® ebook 624pp approx.
978-1-4987-8615-7 £150
VitalSource® ebook
978-1-4987-8648-5 £150
(inc. sales tax where appropriate)

Spon's Mechanical and Electrical Services Price Book 2017

Editor:AECOM

This NRM edition includes a number of new and significantly developed items: WiFi equipment, FP 600 cable, wireless fire alarms; and a much enhanced clock section.

Hbk & VitalSource® ebook 840pp approx.
978-1-4987-8616-4 £160
VitalSource® ebook
978-1-4987-8650-8 £160
(inc. sales tax where appropriate)

Receive our VitalSource® ebook free when you order any hard copy Spon 2017 Price Book

Visit www.pricebooks.co.uk

To order:
Tel: 01235 400524 Fax: 01235 400525
Post: Taylor & Francis Customer Services, Bookpoint Ltd, 200 Milton Park, Abingdon, Oxon, OX14 4SB, UK
Email: book.orders@tandf.co.uk
A complete listing of all our books is on www.crcpress.com

CRC Press
Taylor & Francis Group

ESSENTIAL READING FROM TAYLOR AND FRANCIS

BIM and Quantity Surveying

PITTARD & SELL

The sudden arrival of Building Information Modelling (BIM) as a key part of the building industry is redefining the roles and working practices of its stakeholders. Many clients, designers, contractors, quantity surveyors, and building managers are still finding their feet in an industry where BIM compliance can bring great rewards.

This guide is designed to help quantity surveying practitioners and students understand what BIM means for them, and how they should prepare to work successfully on BIM compliant projects. The case studies show how firms at the forefront of this technology have integrated core quantity surveying responsibilities like cost estimating, tendering, and development appraisal into high profile BIM projects. In addition to this, the implications for project management, facilities management, contract administration and dispute resolution are also explored through case studies, making this a highly valuable guide for those in a range of construction project management roles.

Featuring a chapter describing how the role of the quantity surveyor is likely to permanently shift as a result of this development, as well as descriptions of tools used, this covers both the organisational and practical aspects of a crucial topic.

December 2015: 234 x 156 mm: 258 pp
Pb: 978-0-415-87043-6; £24.99

To Order: Tel: +44 (0) 1235 400524 Fax: +44 (0) 1235 400525
or Post: Taylor and Francis Customer Services,
Bookpoint Ltd, Unit T1, 200 Milton Park, Abingdon, Oxon, OX14 4TA UK
Email: book.orders@tandf.co.uk

For a complete listing of all our titles visit:
www.tandf.co.uk

Taylor & Francis
Taylor & Francis Group

ESSENTIAL READING FROM TAYLOR AND FRANCIS

NRM1 Cost Management Handbook

David P Benge

The definitive guide to measurement and estimating using NRM1, written by the author of NRM1

The 'RICS New rules of measurement: Order of cost estimating and cost planning of capital building works' (referred to as NRM1) is the cornerstone of good cost management of capital building works projects - enabling more effective and accurate cost advice to be given to clients and other project team members, while facilitating better cost control.

The NRM1 Cost Management Handbook is the essential guide to how to successfully interpret and apply these rules, including explanations of how to:

- quantify building works and prepare order of cost estimates and cost plans
- use the rules as a toolkit for risk management and procurement
- analyse actual costs for the purpose of collecting benchmark data and preparing cost analyses
- capture historical cost data for future order of cost estimates and elemental cost plans
- employ the rules to aid communication
- manage the complete 'cost management cycle'
- use the elemental breakdown and cost structures, together with the coding system developed for NRM1, to effectively integrate cost management with Building Information Modelling (BIM).

March 2014: 246 x 174: 640pp
Pb: 978-0-415-72077-9: £41.99

To Order: Tel: +44 (0) 1235 400524 Fax: +44 (0) 1235 400525
or Post: Taylor and Francis Customer Services,
Bookpoint Ltd, Unit T1, 200 Milton Park, Abingdon, Oxon, OX14 4TA UK
Email: book.orders@tandf.co.uk

For a complete listing of all our titles visit:
www.tandf.co.uk

Taylor & Francis
Taylor & Francis Group

ESSENTIAL READING FROM TAYLOR AND FRANCIS

Estimator's Pocket Book

Duncan Cartlidge

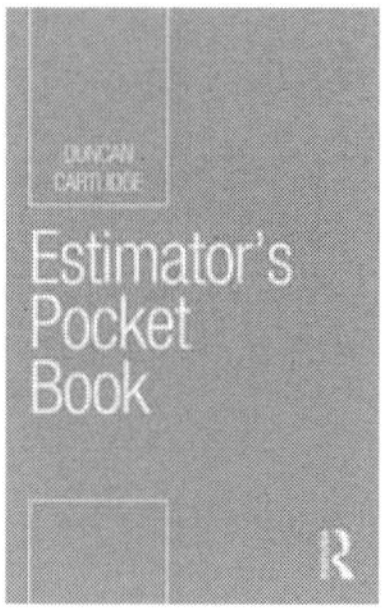

The Estimator's Pocket Book is a concise and practical reference covering the main pricing approaches, as well as useful information such as how to process sub-contractor quotations, tender settlement and adjudication. It is fully up-to-date with NRM2 throughout, features a look ahead to NRM3 and describes the implications of BIM for estimators.

It includes instructions on how to handle:

- the NRM order of cost estimate;
- unit-rate pricing for different trades;
- pro rata pricing and dayworks
- builders' quantities;
- approximate quantities.

Worked examples show how each of these techniques should be carried out in clear, easy-to-follow steps. This is the indispensible estimating reference for all quantity surveyors, cost managers, project managers and anybody else with estimating responsibilities. Particular attention is given to NRM2, but the overall focus is on the core estimating skills needed in practice.

May 2013 186x123: 310pp
Pb: 978-0-415-52711-8: £21.99

To Order: Tel: +44 (0) 1235 400524 Fax: +44 (0) 1235 400525
or Post: Taylor and Francis Customer Services,
Bookpoint Ltd, Unit T1, 200 Milton Park, Abingdon, Oxon, OX14 4TA UK
Email: book.orders@tandf.co.uk

For a complete listing of all our titles visit:
www.tandf.co.uk

Taylor & Francis
Taylor & Francis Group

ESSENTIAL READING FROM TAYLOR AND FRANCIS

Building Services Handbook, 8th Edition

GREENO & HALL

This eighth edition of Hall and Greeno's leading textbook has been reviewed and updated in relation to the latest building and water regulations, new technology, and new legislation, and even more design calculations have been added. In addition, topics such as: alternative sources of natural energy, solar, ground source, heat pumps, renewable energy sources, geothermal methods, and wind power, are now covered.

Building Services Handbook summarises the application of all common elements of building services practice, technique and procedure, to provide an essential information resource for students as well as practitioners working in building services, building management and the facilities administration and maintenance sectors of the construction industry. Information is presented in the highly illustrated and accessible style of the best-selling companion title Building Construction Handbook.

The comprehensive reference for all construction and building services students, Building Services Handbook is ideal for a wide range of courses including NVQ and BTEC National through Higher National Certificate and Diploma to Foundation and three-year Degree level. The clear illustrations and complementary references to industry Standards combine essential guidance with a resource base for further reading and development of specific topics.

September 2015: 234 x 156 mm: 768 pp
Pb: 978-1-138-80563-7; £32.99

To Order: Tel: +44 (0) 1235 400524 Fax: +44 (0) 1235 400525
or Post: Taylor and Francis Customer Services,
Bookpoint Ltd, Unit T1, 200 Milton Park, Abingdon, Oxon, OX14 4TA UK
Email: book.orders@tandf.co.uk

For a complete listing of all our titles visit:
www.tandf.co.uk

Taylor & Francis
Taylor & Francis Group

PART 1

Engineering Features

This section on Engineering Features, deals with current issues and/or technical advancements within the industry. These shall be complimented by cost models and/or itemized prices for items that form part of such.

The intention is that the book shall develop to provide more than just a schedule of prices to assist the user in the preparation and evaluation of costs.

ESSENTIAL READING FROM TAYLOR AND FRANCIS

Building Construction Handbook, 11th Edition

CHUDLEY & GREENO

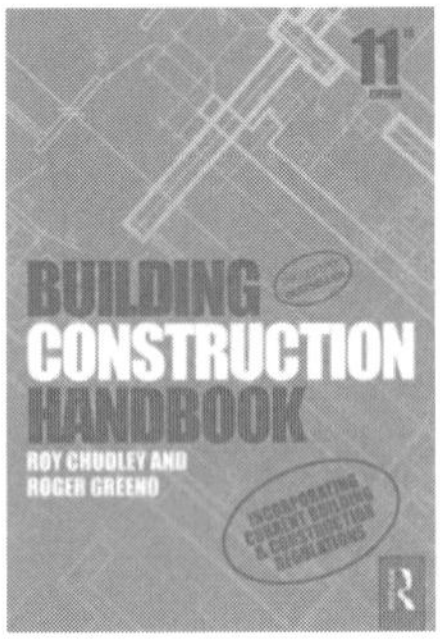

The Building Construction Handbook is THE authoritative reference for all construction students and professionals. Its detailed drawings clearly illustrate the construction of building elements, and have been an invaluable guide for builders since 1988. The principles and processes of construction are explained with the concepts of design included where appropriate. Extensive coverage of building construction practice, techniques, and regulations representing both traditional procedures and modern developments are included to provide the most comprehensive and easy to understand guide to building construction.

This new edition has been updated to reflect recent changes to the building regulations, as well as new material on the latest technologies used in domestic construction.

Building Construction Handbook is the essential, easy-to-use resource for undergraduate and vocational students on a wide range of courses including NVQ and BTEC National, through to Higher National Certificate and Diploma, to Foundation and three-year Degree level. It is also a useful practical reference for building designers, contractors and others engaged in the construction industry.

April 2016: 234 x 156 mm: 1012 pp
Pb: 978-1-138-90709-6: £28.99

To Order: Tel: +44 (0) 1235 400524 Fax: +44 (0) 1235 400525
or Post: Taylor and Francis Customer Services,
Bookpoint Ltd, Unit T1, 200 Milton Park, Abingdon, Oxon, OX14 4TA UK
Email: book.orders@tandf.co.uk

For a complete listing of all our titles visit:
www.tandf.co.uk

Taylor & Francis
Taylor & Francis Group

Building Information Modelling (BIM)

With the UK Government mandating the use of Building Information Modelling (BIM) on all its construction projects by 2016, BIM is currently the hot topic of the construction industry. Some say it will revolutionize building design and construction where many processes will become highly-integrated, efficient and effective. Others say they have 'seen it all before' and argue that BIM is not much more than a new set of design tools that construction must adapt to and work with. However the majority agree that BIM is introducing new ways of working for design and construction teams bringing considerable benefits for all involved in construction – and most of all for clients.

So what is BIM?

BIM exploits the potential in computer-based modelling technologies to provide a new way of designing buildings and managing design and construction processes. BIM goes well beyond the effective application of 3D CAD. It provides a major step-change in the ability of design and construction teams to structure and exchange information around shared, computer-based models of a building project. This of course can bring great benefits, including better design coordination, reductions in design costs and improved communications throughout the design and construction process. But the real benefits of BIM go further. By working together to develop sophisticated, information-rich and coordinated computer models, the different design and construction disciplines can 'prototype' projects before they are built. Designs can be developed and tested 'virtually' so that the performance and cost is optimized. They can be coordinated so that many potential problems are either designed-out or avoided altogether. Ultimately, the benefits not only lie in more efficient and effective design and construction processes, but better and more certain project outcomes – better buildings that are more fit-for-purpose and meet their brief requirements and design intent.

How does BIM work?

BIM is based on digital models of a building that store information about the project (relating to architecture, engineering, construction and so on) in a way that enables it to be shared across and between different design and construction disciplines. Not only that, 3D dynamic modelling software can be used to develop and manipulate these digital models to refine the design, and also to test and validate its potential performance across a range of criteria, including buildability, energy performance-in-use, whole life costing etc. The potential for all key project information to be stored and manipulated on a computer is what sets BIM apart from more conventional approaches, and BIM-based design solutions differ from their traditional counterparts in that they:

- Are created and developed on digital databases which enable collaboration and effective data exchange between different disciplines;
- Allow change to be managed through these databases, so that changes in one part of the database are reflected in (and coordinated through) changes in other parts; and
- Capture and preserve information for reuse by all members of the design and construction team, including facilities management (FM), and user operation and management.

Conventionally, a good deal of design and construction work is document based. Information is communicated and stored via a variety of drawings and reports that, despite being stored and distributed in digital form, are essentially 'unstructured' and thus of limited use. Not only is this information unstructured, it is also held in a variety of forms and locations that are not formally coordinated (information on individual building components, for example, are contained on drawings, specifications, bills of quantity descriptions, etc.). Such an approach has considerable potential for data conflicts and redundancy as well as risks to data integrity and security. Conversely, by providing an intelligent, digital structure for project information – and ultimately a means by which the information can all be held centrally as a 'single' model – BIM opens up a wide range of possibilities for improvement. These include better ways of generating, exchanging, storing and reusing project information that greatly improve communications between different design and construction disciplines through the life of the asset.

While the idea of a single, central project model is something of a 'holy grail' for BIM, in practice digital information is held in a number of project models that are used for different purposes – architectural 'rendering', structural analysis, costing, and so on. But the key feature of BIM working is that these models are coordinated and work from shared databases. Thus BIM is as much a process of generating, sharing and managing project information throughout the lifecycle of the asset, as it is the digital model itself. That is why BIM is sometimes referred to as BIM (M) – Building Information Modelling and Management.

Of course, digitally structuring information can take many forms, from simple temporary models created for a specific purpose (i.e. a room schedule with key room dimensions in spreadsheet form) through to shared 3D whole-building models containing architectural, structural, servicing and other data all in the same place ('Full BIM').

And the greater the degree of information-sharing and collaboration in the development of the models, the more accurate and complete the models will be and the greater are the benefits of using BIM. As shared models are developed, problems – such as clashes between structure and services – tend to get ironed out, and designs become more consistent and coordinated. As all project information comes to be based on a single model, there is considerably less scope for misinterpretation and consequential change, disruption and re-work. And all of this helps greatly to improve the predictability and certainty of project outcomes.

BIM benefits

Generally, BIM is recognized as providing a wide range of valuable benefits including:

- **Design:** improved coordination of design and deliverables between disciplines; improved project understanding through visualization; improved design management and control, including change control; and improved understanding of design changes and implications through parametric modelling.
- **Compliance:** ability to perform simulation and analysis for regulatory compliance; and ability to simulate and optimize energy and wider sustainability performance.
- **Costing/economics:** ability to perform cost analysis as the design develops, and to check for adherence to budget/cost targets; ability to understand cost impacts of design changes; and improved accuracy of cost estimates.
- **Construction:** reduction of construction risks through identification of constructability issues early in the design process; early detection and avoidance of clashes; ability to model impact of design changes on schedule and programme; and ability to integrate contractor/subcontractor design input directly to the model.
- **Operation and management:** creation of an FM database directly from the project (as built) model; ability to perform FM costing and procurement from the model; and ability to update the model with real-time information on actual performance through the life of the building.

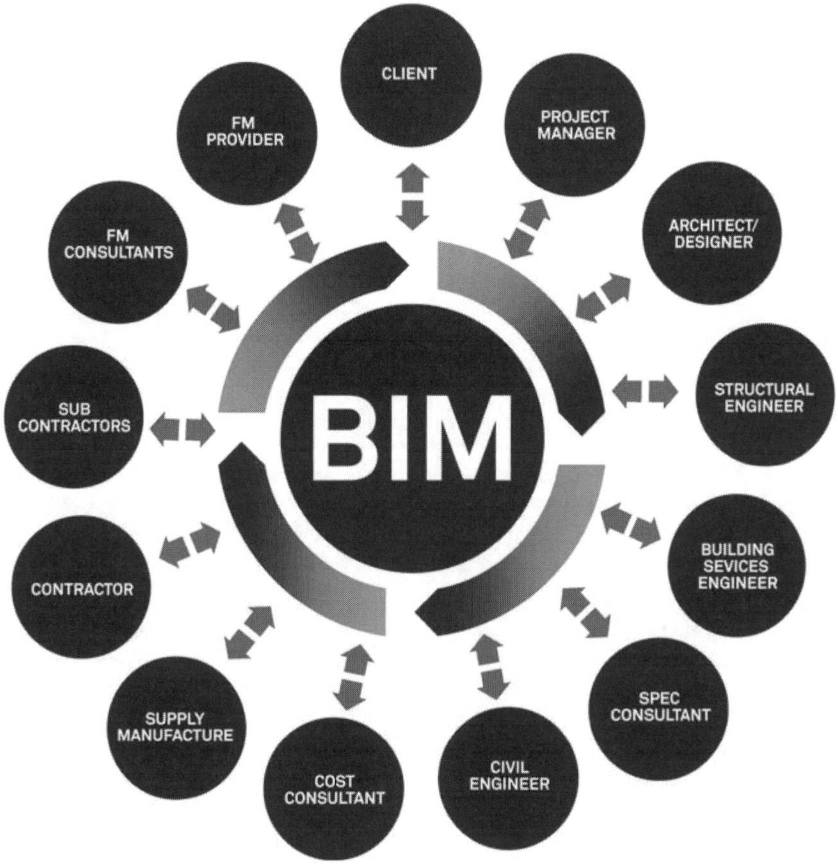

Figure 1: Information sharing (BIM)
Source: BuildingSMART

Ongoing BIM issues and risks

Primarily because BIM is a relatively new approach that has not yet achieved widespread adoption in the construction sector, there are a number of uncertainties and potential risks in its adoption, including:

- **Legal, contractual and insurance issues:** Fundamentally, there is a lack of precedent to provide clear guidance on issues such as ownership of intellectual property rights (IPR) in project models and information. Standard forms of building contract – including design and consultant appointments – do not currently make specific provision for BIM. The implications for design liability and associated insurance arrangements of the greater integration of design, construction and operation information envisaged under BIM have not yet been fully worked through.
- **Current practice:** Related to the issues of liability and insurance is the question of the standards of professional practice (and associated duties of care) that are expected across the sector. New modelling capabilities are blurring the boundaries between who does what in design and construction teams; 'best practice' in BIM is not yet clearly defined and different disciplines are developing their own methods and standards; and training requirements are not yet clear.

- **IT and software:** While a critical requirement of BIM is interoperability of systems and data, not all software and associated IT platforms currently used for BIM deploy IFC-compliant information standards, for example, and manufacturers have not yet agreed on a single standard. The variety of systems in use can impose heavy training burdens on firms who need to operate with some or all of them.
- **Awareness and expectations:** Paradoxically, BIM suffers from relatively low levels of client and sector awareness combined with very high expectations typical of new information technology developments. Because the ultimate potential of BIM is so considerable there is a tendency for some proponents to believe that all the possible benefits are available today, and to over-promise on what can currently be achieved.

A balanced view

The benefits outweigh the risks to a very significant degree and the sheer transformative affect of BIM should not be underestimated. More efficient and effective design processes; greatly improved information quality and co-ordination between design and construction; and better prototyping prior to construction are all made possible through BIM.

Potentially these bring enormous benefits, not only by improving the efficiency of design and construction, but by improving its effectiveness also, ultimately providing greater certainty of project outcomes and better buildings. More than that, the potential for project models to support the management of facilities in the post-construction phase is considerable and could lead to more effective operation of buildings through the whole life cycle. One of these benefits is to provide new learning about what really works in design and construction which can be fed back to inform and improve these processes.

Conversely, the downside risks, though real, are generally resolvable and do not create insurmountable barriers to BIM adoption. Indeed, as BIM use becomes more widespread, it may be expected that, amongst other things, issues associated with IPR, insurances and best practice will all become clearer and more easily managed through improved contracts and management arrangements.

In conclusion

The most important decision clients will face is the selection and appointment of the design and construction team. Finding the right team of people with a positive approach to using BIM, to sharing information and above all to learning from the experience is key. In this regard, clients should expect all their consultants, contractors and specialists to be familiar with BIM and its requirements; to be positively engaged in its adoption; and to be actively developing ways in which processes can be made more value-adding and effective. Many projects have seen the adoption of BIM for clash detection and coordination purposes with a BIM consultant being appointed to manage the process. Greater understanding from the design teams and demand from Clients will see more traditionally based tasks being achieved through BIM in the subsequent years to come.

Building Management Systems (BMS)

A **Building Management System (BMS)** is a computer based control system installed in buildings that controls and monitors the building's mechanical and electrical equipment in the most efficient way.

Benefits

- Switches plant on and off automatically according to time, time of day and environmental conditions (e.g. senses conditions and activates devices to correct settings)
- Optimizes plant operation and services
- Monitors plant status and environmental conditions and thus improves standards of operation and maintenance
- Allows remote access, control and monitoring (e.g. using graphical information and alarms)

Types of BMS points

- Is the physical connection of a BMS point to a controller or outstation I/O terminals? The points schedule is critical when designing the system
- Input – gathers data/monitor; Output – provide command/control
- Analogue points – has a variable value 0% to 100% (e.g. temperature sensor, modulating control valves)
- Digital points – binary I/O (0 or 1) (e.g. on/off control, run/off/fault status)
- Network integration points (NI) – allows to communicate or integrate to stand alone controller such as chiller, meters, boilers, generators

BMS Components – hardware and software/programming

Hardware

- **Primary network/BMS backbone**
  - Head end equipment (desktop computer, laptop, monitors, keyboard, dot matrix and colour printer, UPS supply, etc.)
  - Ethernet Network Cabling (normally Cat 5e or Cat 6 cables)
  - Ethernet switch (located inside the control panels)
- **Secondary network**
  - Control enclosure (CE) or Control panel (CP) or BMS outstation
  - Motor control centre (MCC)
  - Unitary controller for FCU/VAV/Chilled beams
  - Power cables and isolators (from MCC panel to mechanical equipments)
  - Control cables (from MCC/CE panel to field devices)
  - Field devices (control valves, meter, sensors, VSD/Inverter, DOL starter, star-delta starter, actuators)

Software/programming

- **BMS programme protocol** – Refers to the communication of BMS from PC software to the different types of controller/sensors *(e.g. Modbus, LonWorks, BACnet)*
- **Graphics** – Provide true representation of the installed plant. Colour graphics shall incorporate automatic updating of real time field data
- **Software System Integration** (*e.g. energy management system, chillers, boiler, CHP, generator, water treatment plant, closed control units, blind control, lighting control, security and access controls)*

BMS Pricing – *Quick Estimate for a typical medium size office building:*

Shell and Core:

1.	Head end equipment (Workstation) *Extra cost:* • *Additional remote workstation*	£10,000–£20,000 *£5,000–£10,000*
2.	Network connection (BMS backbone – Copper)	£15,000–£30,000
3.	Main plant – cost per BMS points *Factors that affect the cost:* • *UPS and touch pad display for each control panel* • *Panels enclosure – fire-rated or GRP enclosure*	£525–£840 per pt
4.	Control enclosure (CE) for future tenants. Additional I/O modules will be part of fitting out cost *Factors that affect the cost:* • *UPS and touch pad display for each control panel* • *Panels enclosure – fire-rated or GRP enclosure*	£5,000–£10,000 per panel
5.	Landlord FCUs – depending on points per/FCU controller *Factors that affect the cost:* • *Local controller for cellular office, meeting rooms, etc.*	£420–£735 per unit
6.	Software, graphics, commissioning, engineering	Included in BMS pt rate
7.	Trade contract preliminaries	20% to 25%

Category A, B or C Office Fit-out:

1.	Tenant plants – cost per points utilizing existing MCC/CE panel but require additional I/O modules	£315–£525 per pt
2.	Energy meters	
	a. LTHW/CHW meters	£750–£1,500
	b. Electricity meters (meters by Electrical TC)	£210–£315
	Factors that affect the cost: • *Size of heat meter or type of meter such as ultrasonic flow meter*	
3.	Intelligent unitary controller (IUC) – depends on points per/IUC controller and the air conditioning solution such as 4-pipe FCU, 2-pipe w/electric heating, VAV, chilled beams, chilled ceiling *Factors that affect the cost:* • *Local controller for cellular office/meeting room* • *For chilled beam and chilled ceiling, controls depends on the zoning* • *Depend also on zone controls*	£400–£735 per unit
4.	Software, graphics, commissioning *Extra over cost:* • *Integration to other standalone system (e.g. lighting control, AV, fire alarm, security, central battery)*	Included in pt rate
5.	Trade contract preliminaries	20% to 25%

ESSENTIAL READING FROM TAYLOR AND FRANCIS

Watts Pocket Handbook, 29th Edition
RUSHTON

Back in print for the first time in years, the Watts Pocket Handbook renews its commitment to share industry knowledge by providing technical and legal information across a comprehensive spread of property and construction topics.

Compiled by the Watts Technical Director, the Handbook provides specialist information and guidance on a vast selection of construction related subjects including:

- Contracts and procurement
- Insurance
- Materials and defects
- Environmental and sustainability issues

Watts Pocket Handbook remains the must-have reference book for professionals and students engaged in construction, building surveying, service engineering, property development and much more.

April 2016; 186 x 123 mm; 324 pp
Pb: 978-1-138-66595-8; £19.99

To Order: Tel: +44 (0) 1235 400524 Fax: +44 (0) 1235 400525
or Post: Taylor and Francis Customer Services,
Bookpoint Ltd, Unit T1, 200 Milton Park, Abingdon, Oxon, OX14 4TA UK
Email: book.orders@tandf.co.uk

For a complete listing of all our titles visit:
www.tandf.co.uk

Taylor & Francis
Taylor & Francis Group

Renewable Energy Options

This article focuses on building-integrated options rather than large-scale utility solutions such as wind farms, which are addressed separately, and provides an analysis of where they may be best installed.

The legislative background, imperatives and incentives

In recognition of the causes and effects of global climate change, the Kyoto protocol was signed by the UK and other nations in 1992, with a commitment to reduce the emission of greenhouse gases relative to 1990 as the base year.

The first phase of European Union Emission Trading Scheme (EU ETS) covered the power sector and high-energy users such as oil refineries, metal processing, mineral and paper pulp industries. From 1 January 2005, all such companies in the EU had to limit their CO_2 emissions to allocated levels in line with Kyoto. The EU ETS is now in its third phase, running from 2013 to 2020 and covers the 27 EU Member States, as well as Iceland, Liechtenstein, Norway and Croatia. The main change from previous phases being that additional sectors and gases are now included. Key principles of the EU ETS are that participating organizations can:

- Meet the targets by reducing their own emissions (e.g. by implementing energy efficiency measures, using renewable energy sources), or
- Exceed the targets and sell or bank their excess emission allowances, or
- Fail to meet the targets and buy emission allowances from other participants.

The EU ETS is designated as a 'cap and trade' system, where participation is mandatory for the sectors covered and it accepts credits from emission-saving projects carried out under the Kyoto Protocol's Clean Development Mechanism (CDM) and Joint Implementation instrument (JI).

In the UK, the Utilities Act (2000) requires power suppliers to provide some electricity from renewables, starting at 3% in 2003 and rising to15% by 2015. In a similar way to the EU ETS, generating companies receive and can trade Renewables Obligation Certificates (ROCs) for the qualifying electricity they generate.

The focus is on Greenhouse Gases (GHG) and Carbon Dioxide (CO_2) in particular as the main direct contributor to the greenhouse effect. The goals set by the UK government are:

- 34% GHG emissions reduction by 2020 (below 1990 baseline).
- Around 30% of electricity from renewables by 2020.
- 80% GHG emissions reduction by 2050 (below 1990 baseline).

Four years after the introduction of ROCs, it was estimated that less than 3% of UK electricity was being generated from renewable sources. A 'step change' in policy was required, and the Office of the Deputy Prime Minister (ODPM) published 'Planning Policy Statement 22 (PPS 22): Renewable Energy' in order to promote renewable energy through the UK's regional and local planning authorities. PPS 22 was replaced by the National Planning Policy Framework, published in March 2012 by the Department for Communities and Local Government (DCLG), which encourages the use of renewable resources.

Local planning & building regulations

More than 100 local authorities embraced PPS 22 by adopting pro-renewables planning policies typically requiring a percentage (e.g. 10%) of a development's electricity or thermal energy needs to be derived from renewable sources. Government is developing further guidance to support local authorities addressing the sustainability of planned developments and to ensure a level playing field across the country.

Government is also committed to successive improvements in national new-build standards through changes to the Building Regulations, Part L, Conservation of Fuel and Power. In October 2010, new regulations introduced a 25% improvement on 2006 standards.

Energy performance certificates (EPCs)

A Recast of the 2002 EU Directive on the Energy Performance of Buildings (EPBD) was published in 2010. It requires that EPCs be produced for buildings constructed, sold, or rented out to new tenants. For buildings occupied by public authorities and frequently visited by the public, the EPC must be displayed in a prominent place clearly visible to the public. These provisions enable prospective buyers and tenants to be informed of a building's energy performance, ensure the public sector leads by example, and raise public awareness. The Recast was transposed in the UK by national Governments, for example in England & Wales through the Energy Performance of Buildings (England & Wales) Regulations 2012 and other regulations. Other requirements of the Recast include: adopting a method to calculate the energy performance of buildings, setting minimum energy performance requirements and for new buildings to be Nearly Zero Energy Buildings by 2020.

Assessing the regulated carbon emissions associated with new buildings is now an important part of the design and building permitting process with the regulatory approach set out in Part L of the Building Regulations (Conservation of Fuel & Power), the associated National Calculation Methodology (NCM) Modelling Guide and the Standard Assessment Procedure (SAP). On-site renewable energy sources are taken into account and there are limits on design flexibility to discourage inappropriate trade-offs such as buildings with poor insulation standards offset by renewable energy systems.

Technology options and applications

- **Wind generators** – In a suitable location, wind energy can be an effective source of renewable power. Without grant, an installed cost range of £3500 to £5500 per kW of generator capacity may be achieved for small building-mounted turbines. A common arrangement was for a turbine with three blades on a horizontal axis, all mounted on a tower or, for small generators in inner city areas, on a building. Such arrangements typically compare poorly against other renewable options, as they are highly dependent on wind speed at the turbine, obstructions (e.g. nearby buildings and trees), turbulences, the elevation of the turbine above ground, and mitigating other impacts such as aesthetics consideration for planning permission, noise, and vibrations. With suitable conditions, average site wind speeds of 4 m/s can produce useful amounts of energy from a small generator up to about 3 kW, but larger generators require at least 7 m/s. A small increase in average site wind speed will typically result a large increase in output. There will be a need for inverters, synchronizing equipment, and metering for a grid connection.

  Larger, stand-alone turbines typically compare more favourably than smaller building-mounted turbines. Third party provision through an Energy Service Company (ESCo) can be successful for larger (stand-alone) installations located within or close to the host building's site, especially in industrial settings where there may be less aesthetic or noise issues than inner city locations. The ESCo provides funding, installs and operates the plant and the client signs up for the renewable electrical energy at an agreed price for a period of time.

- **Building integrated photovoltaics (BIPV)** – Photovoltaic materials, commonly known as solar cells, generate direct current electrical power when exposed to light. Solar cells are constructed from semiconducting materials that absorb solar radiation; electrons are displaced within the material, thus starting a flow of current through an external connected circuit. PVs are available in a number of forms including monocrystalline, polycrystalline, amorphous silicon (thin film) or hybrid panels that are mounted on or integrated into the roofs or facades of buildings. Conversion efficiency of solar energy to electrical power is improving with advances in technology and ranges from 10% to 20%. In practice, allowing for UK weather conditions, an installation of 7 m^2 of monocrystalline modules (south facing at 30° from horizontal) typically produces 1000 watts peak (1 kWp), yielding about 800 kWh in a year. Installed costs range from £1300 to £2750/kWp.

- **Ground source heat pumps (heating & cooling)** – At a particular depth (about 10 m in the UK), the ground temperature remains substantially constant throughout the year. Heat (and coolth) may be extracted through either an 'open' system – discharging ground water to river or sewer after passing it through a heat exchanger, or a 'closed' system – circulating a fluid (often water) through a heat exchanger and (typically) vertical pipes extending below the ground water table. An electrically driven heat pump is then used to raise the fluid temperature via the refrigeration cycle, and low temperature hot water is delivered to the building.

  Most inner city ground heating and cooling systems consist of a cluster of pipes inserted into vertical holes typically 50 to 100 metres deep depending on space and ground conditions. Horizontal systems can be used where site circumstances allow. Costs for the drilling operation vary according to location, site accessibility, and ground conditions. Geological investigations are recommended to confirm ground conditions, reduce risks, and improve design and cost certainty.

  Such systems may achieve a Coefficient of Performance (COP = heat output/ electrical energy input) of between 3 and 4, achieving good savings of energy compared with conventional fossil fuels based systems. Installed costs are in the range £650 to £1800/kW depending on system type (vertical or horizontal), its size and complexity.

  Note that there is some debate on the status of ground source heat pumps as a renewable source of energy as it requires an external source of power which may not be renewable, typically electricity from the grid.

- **Solar water heating** – The basic principle is to collect heat from the sun via a fluid which is circulated in a roof solar panel or 'collector'. The heated fluid is then used to preheat hot water for space heating or domestic hot water, either in a separate tank or a twin coil hot water cylinder. Purpose-designed 'evacuated tube collectors' were developed to increase performance against the typical 'flat plate collectors'. A typical residential 'evacuated tube collectors' system has a cost ranging from £750 to £1100/m^2 depending on pipe runs and complexity. Such system may produce approximately 500 to 800 kWh/m^2 per year. Commercial systems are larger and more complex, and may achieve similar performance, providing there is sufficient hot water demand. Low-density residential, retail and leisure developments with washrooms and showers may also be suitable applications providing adequate demand for hot water.

- **Biomass boilers** – Wood chips or pellets derived from waste or farmed coppices or forests are available commercially and are considered carbon neutral, having absorbed carbon dioxide during growth. With a suitable fuel storage hopper and automatic screw drive and controls, biomass boilers can replace conventional boilers with little technical or aesthetic impact. However, they do depend on a viable source of fuel, and there are requirements for fuel deliveries access (in particular for inner city or remote locations), fuel storage, ash removal/disposal, as well as periodic de-coking. In individual dwellings, space may be a problem because a biomass boiler does not integrate readily into a typical modern kitchen. However, communal systems (serving multiple dwellings/flats) may be a viable domestic application. Biomass boilers are available in a wide range of domestic and commercial sizes. For a large installation, biomass boilers are more likely to form part of a modular system rather than to displace conventional boilers entirely. There is a cost premium for the biomass storage and feed system, and the cost of the fuel is currently comparable with other solid fuels. Installed costs of a biomass boiler range from £220 to £400/kW.

- **Biomass combined heat & power (CHP)** – Conventional CHP installations consist of either an internal combustion engine or a gas turbine driving an alternator, with maximum recovery of heat, particularly from the exhaust system. For best efficiency, there needs to be a convenient and constant requirement for the heat energy output and the generated electricity should also be utilized locally, with any excess exported to the grid.

  Considering the cost implications for biomass storage and handling as described for biomass boilers above, biomass CHP would only be viable in specific circumstances, with installed system costs in the order of £2600 to £3700/kW (electrical). Note that, at the time of writing, the authors are not aware of any small scale biomass CHP system successfully operated in the UK over any significant period.

ESSENTIAL READING FROM TAYLOR AND FRANCIS

Faber & Kell's Heating & Air-conditioning of Buildings

11th Edition

Doug Oughton & Ant Wilson

For over 70 years, Faber & Kell's has been the definitive reference text in its field. It provides an understanding of the principles of heating and air-conditioning of buildings in a concise manner, illustrating practical information with simple, easy-to-use diagrams, now in full-colour.

This new-look 11th edition has been re-organised for ease of use and includes fully updated chapters on sustainability and renewable energy sources, as well as information on the new Building Regulations Parts F and L. As well as extensive updates to regulations and codes, it now includes an introduction that explains the role of the building services engineer in the construction process. Its coverage of design calculations, advice on using the latest technologies, building management systems, operation and maintenance makes this an essential reference for all building services professionals.

January 2015: 246 x 189: 968pp
Pb: 978-0-415-52265-6: £105.00

To Order: Tel: +44 (0) 1235 400524 Fax: +44 (0) 1235 400525
or Post: Taylor and Francis Customer Services,
Bookpoint Ltd, Unit T1, 200 Milton Park, Abingdon, Oxon, OX14 4TA UK
Email: book.orders@tandf.co.uk

For a complete listing of all our titles visit:
www.tandf.co.uk

Taylor & Francis
Taylor & Francis Group

Grey Water Recycling and Rainwater Harvesting

The potential for grey water recycling and rainwater harvesting for both domestic residential and for various types of commercial building, considering the circumstances in which the systems offer benefits, both as stand-alone installations and combined.

Water usage trends

Water usage in the UK has increased dramatically over last century or so and it is still accelerating. The current average per capita usage is estimated to be at least 150 litres per day, and the population is predicted to rise from 60 million now to 65 million by 2017 and to 75 million by 2031, with an attendant increase in loading on water supply and drainage infrastructures.

Even at the current levels of consumption, it is clear from recent experience that long, dry summers can expose the drier regions of the UK to water shortages and restrictions. The predicted effects of climate change include reduced summer rainfall, more extreme weather patterns, and an increase in the frequency of exceptionally warm dry summers. This is likely to result in a corresponding increase in demand to satisfy more irrigation of gardens, parks, additional usage of sports facilities and other open spaces, together with additional needs for agriculture. The net effect, therefore, at least in the drier regions of the UK, is for increased demand coincident with a reduction in water resource, thereby increasing the risk of shortages.

Water applications and reuse opportunities

Average domestic water utilization can be summarized as follows, as a percentage of total usage (Source: – Three Valleys Water)

- Wash hand basin 8%
- Toilet 35%
- Dishwasher 4%
- Washing machine 12%
- Shower 5%
- Kitchen sink 15%
- Bath 15%
- External use 6%

Water for drinking and cooking makes up less than 20% of the total, and more than a third of the total is used for toilet flushing. The demand for garden watering, although still relatively small, is increasing year by year and coincides with summer shortages, thereby exacerbating the problem.

In many types of building it is feasible to collect rainwater from the roof area and to store it, after suitable filtration, in order to meet the demand for toilet flushing, cleaning, washing machines and outdoor use – thereby saving in many cases a third of the water demand. The other, often complementary recycling approach, is to collect and disinfect 'grey water' – the waste water from baths, showers and washbasins. Hotels, leisure centres, care homes and apartment blocks generate large volumes of waste water and therefore present a greater opportunity for recycling. With intelligent design, even offices can make worthwhile water savings by recycling grey water, not necessarily to flush all of the toilets in the building but perhaps just those in one or two primary cores, with the grey water plant and distribution pipework dimensioned accordingly.

Intuitively, rainwater harvesting and grey water recycling seem like 'the right things to do' and rainwater harvesting is already common practice in many counties in Northern continental Europe. In Germany, for example, some 60,000 to 80,000 systems are being installed every year – compared with perhaps 2,000 systems in the UK. Grey water recycling systems, which are less widespread than rainwater harvesting, have been developed over the last 20 years and the 'state of the art' is to use biological and UV (ultra-violet) disinfection rather than chemicals, and to reduce the associated energy use through advanced technology membrane micro filters.

In assessing the environmental credentials of new developments, the sustainable benefits are recognized by the Building Research Establishment Environmental Assessment Method (BREEAM) whereby additional points can be gained for efficient systems – those designed to achieve enough water savings to satisfy at least 50% of the relevant demand. Furthermore, rainwater harvesting systems from several manufacturers are included in the 'Energy Technology Product List' and thereby qualify for Enhanced Capital Allowances (ECAs). Claims are allowed not only for the equipment, but also to directly associated project costs including:

- Transportation – the cost of getting equipment to the site.
- Installation – cranage (to lift heavy equipment into place), project management costs and labour, plus any necessary modifications to the site or existing equipment.
- Professional Fees – if they are directly related to the acquisition and installation of the equipment.

Rainwater harvesting

A typical domestic rainwater harvesting system can be installed at reasonable cost if properly designed and installed at the same time as building the house. The collection tank can either be buried or installed in a basement area. Rainwater enters the drainage system through sealed gullies and passes through a pre-filter to remove leaves and other debris before passing into the collection tank. A submersible pump, under the control of the monitoring and sensing panel, delivers recycled rainwater on demand. The non-potable distribution pipework to the washing machine, cleaner's tap, outside tap and toilets etc. could be either a boosted system or configured for a header tank in the loft, with mains supply back-up, monitors and sensors located there instead of at the control panel.

Calculating the collection tank size brings into play the concept of system efficiency – relating the water volume saved to the annual demand. In favourable conditions – ample rainfall and large roof collection area – it would be possible in theory to achieve almost 100%. In practice, systems commonly achieve 50 to 70% efficiency, with enough storage to meet demand for typically one week, though this is subject to several variables. As well as reducing the demand for drinking quality mains supply water, rainwater harvesting tanks act as an effective storm water attenuator, thereby reducing the drainage burden and the risk of local flooding which is a benefit to the wider community. Many urban buildings are located where conditions are unfavourable for rainwater harvesting – low rainfall and small roof collection area. In these circumstances it may still be worth considering water savings through grey water recycling.

Grey water recycling

In a grey water recycling system, waste water from baths, showers and washbasins is collected by conventional fittings and pipework, to enter a pretreatment sedimentation tank which removes the larger dirt particles. This is followed by the aerobic treatment tank in which cleaning bacteria ensure that all biodegradable substances are broken down. The water then passes onto a third tank, where an ultra-filtration membrane removes all particles larger than 0.00005 mm, (this includes viruses and bacteria) effectively disinfecting the recycled grey water. The clean water is then stored in the fourth tank from where it is pumped on demand under the control of monitors and sensors in the control panel. Recycled grey water may then be used for toilet and urinal flushing, for laundry and general cleaning, and for outdoor use such as vehicle washing and garden irrigation – a substantial water saving for premises such as hotels. If the tank becomes depleted, the distribution is switched automatically to the mains water back-up supply. If there is insufficient plant room space, then the tanks may be buried but with adequate arrangements for maintenance access. Overflow soakaways are recommended where feasible but are not an inherent part of the rainwater harvesting and grey water recycling systems.

Combined rainwater/grey water systems

Rainwater can be integrated into a grey water scheme with very little added complication other than increased tank size. In situations where adequate rainwater can be readily collected and diverted to pretreatment, then heavy demands such as garden irrigation can be met more easily than with grey water alone. An additional benefit is that they reduce the risk of flooding by keeping collected storm water on site instead of passing it immediately into the drains.

Indicative system cost and payback considerations

Rainwater harvesting scenario: Office building in Leeds having a roof area of 2000 m² and accommodating 435 people over 3 floors. Local annual rainfall is 875 mm and the application is for toilet and urinal flushing. An underground collection tank of 25,000 litres has been specified to give 4.5 storage days.

Rainwater harvesting cost breakdown		Cost £
System tanks and filters and controls		21,000
Mains water back-up and distribution pump arrangement		5,000
Non-potable distribution pipework		1,000
Connections to drainage		1,000
Civil works and tank installation (assumption normal ground conditions)		8,000
System installation & commissioning		2,000
	TOTAL COST	38,000

Rainwater harvesting payback considerations

Annual water saving: 1100 m³ @ average cost £2.30/m³ = £2,530

Annual maintenance and system energy cost = £800

Indicative payback period = 38,000/1,730 = 22 years

Grey water recycling scenario: Urban leisure hotel building with 200 bedrooms offers little opportunity for rainwater harvesting but has a grey water demand of up to 12000 litres per day for toilet and urinal flushing, plus a laundry. The grey water recycling plant is located in a basement plant room.

Grey water recycling cost breakdown		Cost £
System tanks and controls		37,000
Mains water back-up and distribution pump arrangement		4,000
Non-potable distribution pipework		4,000
Grey water waste collection pipework		5,000
Connections to drainage		1,000
System installation & commissioning		6,000
	TOTAL COST	57,000

Grey water recycling payback considerations

Annual water saving: 4260 m³ @ average cost £2.30/m³ = £9,798

Annual maintenance and system energy cost = £2,200

Indicative payback period = 57,000/7,598 = 7.5 years

Exclusions

- Site organization and management costs other than specialist contractor's allowances
- Contingency/design reserve
- Main contractor's overhead and profit or management fee
- Professional fees
- Tax allowances
- Value added tax

Conclusions

There is a justified and growing interest in saving and recycling water by way of both grey water recycling and rainwater harvesting. The financial incentive at today's water cost is not great for small or inefficient systems but water costs are predicted to rise and demand to increase – not least as a result of population growth.

The payback periods for the above scenarios are not intended to compare to potential payback periods of rainwater harvesting against grey water recycling but rather to illustrate the importance of choosing 'horses for courses'. The office building with its relatively small roof area and limited demand for toilet flushing results in a fairly inefficient system. Burying the collection tank also adds a cost so that payback exceeds 20 years. Payback periods of less than 10 years are feasible for buildings with large roofs and a large demand for toilet flushing or for other uses. Therefore sports stadia, exhibition halls, supermarkets, schools and similar structures are likely to be suitable.

The hotel scenario is good application for grey water recycling. Many hotels and residential developments will generate more than enough grey water to meet the demand for toilet flushing etc. and in these circumstances large quantities of water can be saved and recycled with attractive payback periods.

Ground Water Cooling

The use of ground water cooling systems, considering the technical and cost implications of this renewable energy technology.

The application of ground water cooling systems is quickly becoming an established technology in the UK with numerous installations having been completed for a wide range of building types, both new build and existing (refurbished).

Buildings in the UK are significant users of energy, accounting for 60% of UK carbon emissions in relation to their construction and occupation. The drivers for considering renewable technologies such as ground water cooling are well documented and can briefly be summarized as follows:

- Government set targets – The Energy White Paper, published in 2003, setting a target of producing 10% of UK electricity from renewable sources by 2010 and the aspiration of doubling this by 2020.
- The proposed revision to the Building Regulations Part L 2006, in raising the overall energy efficiency of non-domestic buildings, through the reduction in carbon emissions, by 27%. Local Government policy for sustainable development. In the case of London, major new developments (i.e. City of London schemes over 30,000 m^2) are required to demonstrate how they will generate a proportion of the site's delivered energy requirements from on-site renewable sources where feasible. The GLA's expectation is that, overall, large developments will contribute 10% of their energy requirement using renewables, although the actual requirement will vary from site to site. Local authorities are also likely to set lower targets for buildings which fall below the GLA's renewables threshold.
- Company policies of building developers and end users to minimize detrimental impact to the environment.

The ground as a heat source/sink

The thermal capacity of the ground can provide an efficient means of tempering the internal climate of buildings. Whereas the annual swing in mean air temperature in the UK is around 20 K, the temperature of the ground is far more stable. At the modest depth of 2 m, the swing in temperature reduces to 8 K, while at a depth of 50 m the temperature of the ground is stable at 11–13°C. This stability and ambient temperature therefore makes ground water a useful source of renewable energy for heating and cooling systems in buildings.

Furthermore, former industrial cities like Nottingham, Birmingham, Liverpool and London have a particular problem with rising ground water as they no longer need to abstract water from below ground for use in manufacturing. The use of ground water for cooling is therefore encouraged by the Environment Agency in areas with rising ground water as a means of combating this problem.

System types

Ground water cooling systems may be defined as either open or closed loop.

Open loop systems

Open loop systems generally involve the direct abstraction and use of ground water, typically from aquifers (porous water bearing rock). Water is abstracted via one or more boreholes and passed through a heat exchanger and is returned via a separate borehole or boreholes, discharged to foul water drainage or released into a suitable available source such as a river. Typical ground water supply temperatures are in the range 6–10°C and typical re-injection temperatures 12–18°C.

All open loop abstractions from ground water require an abstraction licence. If the water is being returned to the ground then it is designated as non-consumptive use and also requires an environmental permit to discharge the water back into the aquifer. Where it is a simple abstraction, it is regarded as a consumptive use and the abstraction charges are higher.

Open loop systems fed by ground water at 8°C, can typically cool water to 12°C on the secondary side of the heat exchanger to serve conventional cooling systems.

Open loop systems are thermally efficient but over time can suffer from blockages caused by silt, and corrosion due to dissolved salts. As a result, additional cost may be incurred in having to provide filtration or water treatment, before the water can be used in the building.

There is also a risk that minerals can precipitate out of the ground water, especially where the water is iron and/or manganese rich and there is either and/or a change in pressure or exposure of the water to air. This could require cleaning of the pipework and the borehole lining or some form of filtration as indicated above. Several water supply boreholes suffer from this problem, which can cause a reduction in borehole yield

Abstraction licence and discharge consent needs to be obtained for each installation, and this together with the maintenance and durability issues can significantly affect whole life operating costs, making this system less attractive.

Closed loop systems

Closed loop systems do not rely on the direct abstraction of water, but instead comprise a continuous pipework loop buried in the ground. Water circulates in the pipework and provides the means of heat transfer with the ground. Since ground water is not being directly used, closed loop systems therefore suffer fewer of the operational problems of open loop systems, being designed to be virtually maintenance free, but do not contribute to the control of ground water levels.

There are two types of closed loop system:

Vertical Boreholes – Vertical loops are inserted as U tubes into pre-drilled boreholes, typically less than 150 mm in diameter. These are backfilled with a high conductivity grout to seal the bore, prevent any cross contamination and to ensure good thermal conductivity between the pipe wall and surrounding ground. Vertical boreholes have the highest performance and means of heat rejection, but also have the highest cost due to associated drilling and excavation requirements.

Alterative systems to the U tubes are available on the market with larger diameter propriety pipework systems with flow and return pipes. The larger pipework can have better heat transfer properties, reduced resistance and therefore achieve better efficiencies or less pipework and therefore less ground area requirement than a traditional U tube system which may outweigh the additional cost per m length of the pipework.

As an alternative to having a separate borehole housing the pipe loop, it can also be integrated with the piling, where the loop is encased within the structural piles. This obviously saves on the costs of drilling and excavation since these would be carried out as part of the piling installation. The feasibility of this option would depend on marrying up the piling layout with the load requirement, and hence the number of loops, for the building.

Horizontal Loops – These are single (or pairs) of pipes laid in 2 m deep trenches, which are backfilled with fine aggregate. These obviously require a greater physical area than vertical loops but are cheaper to install. As they are located closer to the surface where ground temperatures are less stable, efficiency is lower compared to open systems. Alternatively, coiled pipework can also be used where excavation is more straightforward and a large amount of land is available. Although performance may be reduced with this system as the pipe overlaps itself, it does represent a cost effective way of maximising the length of pipe installed and hence overall system capacity.

The case for heat pumps

Instead of using the ground water source directly in the building, referred to as passive cooling, when coupled to a reverse cycle heat pump, substantially increased cooling loads can be achieved.

Heat is extracted from the building and transferred by the heat pump into the water circulating through the loop. As it circulates, it gives up heat to the cooler earth, with the cooler water returning to the heat pump to pick up more heat. In heating mode the cycle is reversed, with the heat being extracted from the earth and being delivered to the HVAC system.

The use of heat pumps provides greater flexibility for heating and cooling applications within the building than passive systems. Ground source heat pumps are inherently more efficient than air source heat pumps, their energy requirement is therefore lower and their associated CO_2 emissions are also reduced, so they are well suited for connection to a ground water source.

Closed loop systems can typically achieve outputs of 50 W/m (of bore length), although this will vary with geology and borehole construction. When coupled to a reverse cycle heat pump, 1m of vertical borehole will typically deliver 140 kWh of useful heating and 110 kWh of cooling per annum, although this will depend on hours run and length of heating and cooling seasons.

Key factors affecting cost

- The cost is obviously dependant on the type of system used. Deciding on what system is best suited to a particular project is dependant on the peak cooling and heating loads of the building and its likely load profile. This in turn determines the performance required from the ground loop, in terms of area of coverage in the case of the horizontal looped system, and in the case of vertical boreholes, the depth and number or bores. The cost of the system is therefore a function of the building load.
- In the case of vertical boreholes, drilling costs are significant factor, as specific ground conditions can be variable, and there are potential problems in drilling through sand layers, pebble beds, gravels and clay, which may mean additional costs through having to drill additional holes or the provision of sleeving etc. The costs of excavation obviously make the vertical borehole solution significantly more expensive than the equivalent horizontal loop.
- The thermal efficiency of the building is also a factor. The higher load associated with a thermally inefficient building obviously results in the requirement for a greater number of boreholes or greater area of horizontal loop coverage, however in the case of boreholes the associated cost differential between a thermally inefficient building and a thermally efficient one is substantially greater than the equivalent increase in the cost of conventional plant. Reducing the energy consumption of the building is cheaper than producing the energy from renewables and the use of renewable energy only becomes cost effective, and indeed should only be considered, when a building is energy efficient.
- With open loop systems, the principal risk in terms of operation is that the user is not in control of the quantity or quality of the water being taken out of the ground, this being dependant on the local ground conditions. Reduced performance due to blockage (silting etc.) may lead to the system not delivering the design duties whilst bacteriological contamination may lead to the expensive water treatment or the system being taken temporarily out of operation. In order to mitigate the above risk, it may be decided to provide additional means of heat rejection and heating by mechanical means as a back up to the borehole system, in the event of operational problems. This obviously carries a significant cost. If this additional plant were not provided, then there are space savings to be had over conventional systems due to the absence of heating, heat rejection and possibly refrigeration plant.
- With the design of open loop systems close analysis can be undertaken of existing boreholes in the area and also detailed desktop studies can be undertaken by a hydrogeologist to predict the yield for the proposed borehole. As noted above the delivered yield of the borehole is subject to the local conditions in the exact position where the borehole delivered and as such the delivered yield will only be validated after the borehole is completed and tested. As such there is a risk that you can get less (you could get more) than was predicted requiring additional boreholes (and cost) to be delivered to get the required yield.

- Open loop systems may lend themselves particularly well to certain applications increasing their cost effectiveness, i.e. in the case of a leisure centre, the removal of heat from the air conditioned parts of the centre and the supply of fresh water to the swimming pool.
- In terms of the requirements for abstraction and disposal of the water for open loop systems, there are risks associated with the future availability and cost of the necessary licenses; particularly in areas of high forecast energy consumption, such as the South East of England, which needs to be borne in mind when selecting a suitable system.
- Closed loop systems and non-consumptive open loop systems generally require balanced heating and cooling loads (if there is moving ground water this is not as critical) in order to not cause the ground to heat up or cool down over time. If the heating and cooling loads (kWh) over the year are not balanced this can cause the ground to heat up or cool down over time causing a reduction in the efficiency of the system over time which will have a detrimental effect to the seasonal COP's and operational running costs of the system over time.

Whilst open loop systems would suit certain applications or end user clients, for commercial buildings the risks associated with this system tend to mean that closed loop applications are the system of choice. When coupled to a reversible heat pump, the borehole acts simply as a heat sink or heat source so the problems associated with open loop systems do not arise.

Typical costs

Table 1 gives details of the typical borehole cost to an existing site in Central London, using one 140 m deep borehole working on the open loop principle, providing heat rejection for the 600 kW of cooling provided to the building. The borehole passes through rubble, river gravel terraces, clay and finally chalk, and is lined above the chalk level to prevent the hole collapsing. The breakdown includes all costs associated with the provision of a working borehole up to the well head, including the manhole chamber and manhole. The costs of any plant or equipment from the well head are not included.

Heat is drawn out of the cooling circuit and the water is discharged into the Thames at an elevated temperature. In this instance, although the boreholes are more expensive than the dry air cooler alternative, the operating cost is significantly reduced as the system can operate at around three times the efficiency of conventional dry air coolers, so the payback period is a reasonable one. Additionally, the borehole system does not generate any noise, does not require rooftop space and does not require as much maintenance.

This is representative of a typical cost of providing a borehole for an open loop scheme within the London basin. There are obviously economies of scale to be had in drilling more than one well at the same time, with two wells saving approximately 10% of the comparative cost of two separate wells and four wells typically saving 15%.

Table 2 provides a summary of the typical range of costs that could expected for the different types of system based on current prices.

Table 1: Breakdown of the Cost of a Typical Open Loop Borehole System

Description	**Cost £**
General Items	
• Mobilization, insurances, demobilization on completion	21,000
• Fencing around working area for the duration of drilling and testing	2,000
• Modifications to existing LV panel and installation of new power supplies for borehole installation	15,000
Trial Hole	
• Allowance for breakout access to nearest walkway (Existing borehole on site used for trial purposes, hence no drilling costs included)	3,000
Construct Borehole	
• Drilling, using temporary casing where required, permanent casing and grouting	32,000
Borehole Cap and Chamber	
• Cap borehole with PN16 flange, construct manhole chamber in roadway, rising main, header pipework, valves, flow meter	12,000
• Permanent pump	14,000
Samples	
• Water samples	1,000
Acidization	
• Mobilization, set up and removal of equipment for acidization of borehole, carry out acidization	12,000
Development and Test Pumping	
• Mobilize pumping equipment and materials and remove on completion of testing	4,000
• Calibration test, pretest monitoring, step testing	4,000
• Constant rate testing and monitoring	20,000
• Waste removal and disposal	3,000
Reinstatement	
• Reinstatement and making good	2,000
Total	**145,000**

Table 2 Summary of the Range of Costs for Different Systems

	Range			
System	**Small – 4 kWth**	**Medium – 50 kWth**	**Large – 400 kWth**	**Notes**
Heat pump (per unit)	£ 3,000–5,000	£ 30,000–42,000	£ 145,000–175,000	
Slinky pipe (per installation) including excavation	£ 3,000–4,000	£ 42,000–52,000	£ 360,000–390,000 [(1)]	[(1)] Based on 90 nr 50 m lengths
Vertical, closed (per installation) using structural piles	N/A	£ 42,000–63,000	Not available	Based on 50 nr piles. Includes borehole cap and header pipework but excludes connection to pump room and heat pumps
Vertical, closed (per installation) including excavation	£ 2,000–3,000	£ 63,000–85,000	£ 370,000–400,000	Includes borehole cap and header pipework but excludes connection to pump room and heat pumps
Vertical, open (per installation) including excavation	£ 2,000–3,000	£ 45,000–65,000	£ 335,000–370,000	Excludes connection to pump room and heat exchangers

Fuel Cells

The application of fuel cell technology within buildings

Fuel cells are electrochemical devices that convert the chemical energy in fuel into electrical energy directly, without combustion, with high electrical efficiency and low pollutant emissions. They represent a new type of power generation technology that offers modularity, efficient operation across a wide range of load conditions, and opportunities for integration into co-generation systems. With the publication of the energy white paper in February this year, the Government confirmed it's commitment to the development of fuel cells as a key technology in the UK's future energy system, as the move is made away from a carbon based economy.

There are currently very few fuel cells available commercially, and those that are available are not financially viable. Demand has therefore been limited to niche applications, where the end user is willing to pay the premium for what they consider to be the associated key benefits. Indeed, the UK currently has only one fuel cell in regular commercial operation. However, fuel cell technology has made significant progress in recent years, with prices predicted to approach those of the principal competition in the near future.

Fuel cell technology

A fuel cell is composed of an anode (a negative electrode that repels electrons), an electrolyte membrane in the centre, and a cathode (a positive electrode that attracts electrons). As hydrogen flows into the cell on the anode side, a platinum coating on the anode facilitates the separation of the hydrogen gas into electrons and protons. The electrolyte membrane only allows the protons to pass through to the cathode side of the fuel cell. The electrons cannot pass through this membrane and flow through an external circuit to form an electric current.

As oxygen flows into the fuel cell cathode, another platinum coating helps the oxygen, protons, and electrons combine to produce pure water and heat.

The voltage from a single cell is about 0.7 volts, just enough for a light bulb. However by stacking the cells, higher outputs are achieved, with the number of cells in the stack determining the total voltage, and the surface area of each cell determining the total current. Multiplying the two together yields the total electrical power generated.

In a fuel cell the conversion process from chemical energy to electricity is direct. In contrast, conventional energy conversion processes first transform chemical energy to heat through combustion and then convert heat to electricity through some form of power cycle (e.g. gas turbine or internal combustion engine) together with a generator. The fuel cell is therefore not limited by the Carnot efficiency limits of an internal combustion engine in converting fuel to power, resulting in efficiencies 2 to 3 times greater.

Fuel cell systems

In addition to the fuel cell itself, the system comprises the following subsystems:

- A fuel processor – This allows the cell to operate with available hydrocarbon fuels, by cleaning the fuel and converting (or reforming) it as required.
- A power conditioner – This regulates the DC electricity output of the cell to meet the application, and to power the fuel cell auxiliary systems.
- An air management system – This delivers air at the required temperature, pressure and humidity to the fuel stack and fuel processor.
- A thermal management system – This heats or cools the various process streams entering and leaving the fuel cell and fuel processor, as required.
- A water management system – Pure water is required for fuel processing in all fuel cell systems, and for dehumidification in the PEMFC.

The overall electrical conversion efficiency of a fuel cell system (defined as the electrical power out divided by the chemical energy into the system, taking into account the individual efficiencies of the subsystems) ranges from 35–55%. Taking into account the thermal energy available from the system, the overall or cogeneration efficiency is 75–90%.

Also, unlike most conventional generating systems (which operate most efficiently near full load, and then suffer declining efficiency as load decreases), fuel cell systems can maintain high efficiency at loads as low as 20% of full load.

Fuel cell systems also offer the following potential benefits:

- At operating temperature, they respond quickly to load changes, the limiting factor usually being the response time of the auxiliary systems.
- They are modular and can be built in a wide range of outputs. This also allows them to be located close to the point of electricity use, facilitating cogeneration systems.
- Noise levels are comparable with residential or light commercial air conditioning systems.
- Commercially available systems are designed to operate unattended and manufactured as packaged units.
- Since the fuel cell stack has no moving parts, other than the replacement of the stack at 3–5 year intervals there is little on-site maintenance. The maintenance requirements are well established for the auxiliary system plant.
- Fuel cell stacks fuelled by hydrogen produce only water, therefore the fuel processor is the primary source of emissions, and these are significantly lower than emissions from conventional combustion systems.
- Since fuel cell technology generates 50% more electricity than the conventional equivalent without directly burning any fuel, CO_2 emissions are significantly reduced in the production of the source fuel.
- Potentially zero carbon emissions when using hydrogen produced from renewable energy sources.
- The facilitation of embedded generation, where electricity is generated close to the point of use, minimizing transmission losses.
- The fast response times of fuel cells offer potential for use in UPS systems, replacing batteries and standby generators.

Types of fuel cell

There are four main types of fuel cell technology that are applicable for building systems, classed in terms of the electrolyte they use. The chemical reactions involved in each cell are very different.

Phosphoric Acid Fuel Cells (PAFCs) are the dominant current technology for large stationary applications and have been available commercially for some time. There is less potential for PAFC unit cost reduction than for some other fuel cell systems, and this technology may be superseded in time by the other technologies.

The Solid Oxide Fuel Cell (SOFC) offers significant flexibility due to its large power range and wide fuel compatibility. SOFCs represent one of the most promising technologies for stationary applications. There are difficulties when operating at high temperatures with the stability of the materials, however, significant further development and cost reduction is anticipated with this type.

The relative complexity of Molten Carbonate Fuel Cells (MCFCs) has tended to limit developments to large scale stationary applications, although the technology is still very much in the development stages.

The quick start-up times and size range make Polymer Electrolyte Membrane Fuel Cells (PEMFCs) suitable for small to medium sized stationary applications. They have a high power density and can vary output quickly, making them well suited for transport applications as well as UPS systems. The development efforts in the transport sector suggest there will continue to be substantial cost reductions over both the short and long term.

All four technologies remain the subject of extensive research and development programmes to reduce initial costs and improve reliability through improvements in materials, optimization of operating conditions and advances in manufacturing.

The market for fuel cells

The stationary applications market for fuel cells can be sectaries as follows:

- Distributed generation/CHP – For large scale applications, there are no drivers specifically advantageous to fuel cells, with economics (and specifically initial cost) therefore being the main consideration. So, until cost competitive and thoroughly proven and reliable fuel cells are available, their use is likely to be limited to niche applications such as environmentally sensitive areas.
- Domestic and small scale CHP – The drivers for the use of fuel cells in this emerging market are better value for customers than separate gas and electricity purchase, reduction in domestic CO_2 emissions, and potential reduction in electricity transmissioncosts. However, the barriers of resistance to distributed generation, high capital costs and competition from Stirling engines needs to be overcome.
- Small generator sets and remote power – The drivers for the use of fuel cells are high reliability, low noise and low refuelling frequencies, which cannot be met by existing technologies. Since cost is often not the primary consideration, fuel cells will find early markets in this sector. Existing PEMFC systems are close to meeting the requirements in terms of cost, size and performance. Small SOFCs have potential in this market, but require further development.

Costs

Fuel cell technologies are still significantly more expensive than the existing technologies typically and, dependent on the fuel cell type, between £3,000–£5,000/kW. To extend fuel cell application beyond niche markets, their cost needs to reduce significantly. The successful and widespread commercial application of fuel cells is dependent on the projected cost reductions indicated, with electricity generated from fuel cells being competitive with current centralized and distributed power generation.

Conclusions

Despite significant growth in recent years, fuel cells are still at a relatively early stage of commercial development, with prohibitively high capital costs preventing them from competing with the incumbent technology in the market place. However, costs are forecast to reduce significantly over the next five years as the technology moves from niche applications, and into mass production.

However, in order for these projected cost reductions to be achieved, customers need to be convinced that the end product is not only cost competitive but also thoroughly proven, and Government support represents a key part in achieving this.

The Governments of Canada, USA, Japan and Germany have all been active in supporting development of the fuel cell sector through integrated strategies, however the UK has been slow in this respect, and support has to date been small in comparison. It is clear that without Government intervention, fuel cell applications may struggle to reach the cost and performance requirements of the emerging fuel cell market.

ESSENTIAL READING FROM TAYLOR AND FRANCIS

Cost Studies of Buildings

Allan Ashworth & Srinath Perera

This practical guide to cost studies of buildings has been updated and revised throughout for the 6th edition. New developments in RICS New Rules of Measurement (NRM) are incorporated throughout the book, in addition to new material on e-business, the internet, social media, building information modelling, sustainability, building resilience and carbon estimating.

This trusted and easy to use guide to the cost management role:

Focuses on the importance of costs of constructing projects during the different phases of the construction process

Features learning outcomes and self-assessment questions for each chapter

Addresses the requirements of international readers

From introductory data on the construction industry and the history of construction economics, to recommended methods for cost analysis and post-contract cost control, Cost Studies of Buildings is an ideal companion for anyone learning about cost management.

July 2015: 246x174: 544pp Pb:
978-1-138-01735-1: £42.99

To Order: Tel: +44 (0) 1235 400524 Fax: +44 (0) 1235 400525
or Post: Taylor and Francis Customer Services,
Bookpoint Ltd, Unit T1, 200 Milton Park, Abingdon, Oxon, OX14 4TA UK
Email: book.orders@tandf.co.uk

For a complete listing of all our titles visit:
www.tandf.co.uk

Taylor & Francis
Taylor & Francis Group

Biomass Energy

The potential for biomass energy systems, with regards to the adequacy of the fuel supply and the viability of various system types at different scales.

Biomass heating and combined heat and power (CHP) systems have become a major component of the low-carbon strategy for many projects, as they can provide a large renewable energy component at a relatively low initial cost. Work by the Carbon Trust has demonstrated that both large and small biomass systems were viable even before recent increases in gas and fuel oil prices, so it is no surprise that recent research by South Bank University into the renewables strategies to large London projects has found that 25% feature biomass or biofuel systems.

These proposals are not without risk, however. Although the technology is well established, few schemes are in operation in the UK and the long-term success depends more on the effectiveness of the local supply chain than the quality of the design and installation.

How the biomass market works

Biomass is defined as living or recently dead biological material that can be used as an energy source. Biomass is generally used to provide heat, generate electricity or drive CHP engines. The biomass family includes biofuels, which are being specified in city centre schemes, but which provide lower energy outputs and could transfer farmland away from food production.

In the UK, much of the focus in biomass development is on the better utilization of waste materials such as timber and the use of set-aside land for low-intensity energy crops such as willow, rather than expansion of the biofuels sector. There are a variety of drivers behind the development of a biomass strategy. In addition to carbon neutrality, another policy goal is the promotion of the UK's energy security through the development of independent energy sources. A third objective is to address energy poverty, particularly for off-grid energy users, who are most vulnerable to the effects of high long-term costs of fuel oil and bottled gas.

Biomass' position in the zero-carbon hierarchy is a little ambiguous in that its production, transport and combustion all produce carbon emissions, albeit most is offset during a plant's growth cycle. The key to neutrality is that the growing and combustion cycles need to occur over a short period, so that combustion emissions are genuinely offset. Biomass strategy is also concerned about minimizing waste and use of landfill, and the ash produced by combustion can be used as a fertilizer.

Dramatic increases in fossil fuel prices have swung considerations decisively in favour of technologies such as biomass. Research by the Carbon Trust has demonstrated that, with oil at $50 (£25) a barrel, rates of return of more than 10% could be achieved with both small and large heating installations. CHP and electricity-only schemes have more complex viability issues linked to renewable incentives, but with oil currently trading at over ($100) £50 a barrel and a plentiful supply of source material, it is argued that biomass input prices will not rise and so the sector should become increasingly competitive.

The main sources of biomass in the UK include:

- Forestry crops, including the waste products of tree surgery industry
- Industrial waste, particularly timber, paper and card: timber pallets account for 30% of this waste stream by weight
- Woody energy crops, particularly those grown through 'short rotation' methods such as willow coppicing
- Wastes and residues taken from food, agriculture and manufacturing

Biomass is an emerging UK energy sector. Most suppliers are small and there remains a high level of commercial risk associated with finding appropriate, reliable sources of biomass. This is particularly the case for larger-scale schemes such as those proposed for Greater London, which will have sourced biomass either from multiple UK suppliers or from overseas. Many have adopted biofuels as an alternative.

The UK's only large-scale biomass CHP in Slough has a throughput of 180,000 tonnes of biomass per year requiring the total production of more than 20 individual suppliers – not a recipe for easy management or product consistency. However, Carbon Trust research has identified significant potential capacity in waste wood (5 million to 6 million tonnes) and short-rotation coppicing, which could create the conditions for wider adoption of small and large-scale biomass.

Biomass technologies

A wide range of technologies have been developed for processing various forms of biomass, including anaerobic digesters and gasifiers. However, the main biomass technology is solid fuel combustion, as a heat source, CHP unit or energy source for electricity generation.

Solid fuel units use either wood chippings or wood pellets. Wood chippings are largely unprocessed and need few material inputs, other than seasoning, chipping and transport. Wood pellets are formed from compressed sawdust. As a result they have a lower moisture content than wood chippings and consistent dimensions, so are easier to handle but are about twice as expensive.

Solid fuel burners operate in the same way as other fossil fuel-based heat sources, with the following key differences:

- Biomass heat output can be controlled but not instantaneously, so systems cannot respond to rapid load changes. Solutions to provide more flexibility include provision of peak capacity from gas-fired systems, or the use of thermal stores that capture excess heat energy during off-peak periods, enabling extended operation of the biomass system itself
- Heat output cannot be throttled back by as much as gas-fired systems, so for heat-only installations it may be necessary to have an alternative summer system for water heating, such as a solar collector
- Biomass feedstock is bulky and needs a mechanized feed system as well as extensive storage
- Biomass systems are large, and the combustion unit, feed hopper and fuel store take up substantial floor area. A large unit with an output of 500 kW has a footprint of 7.5 m × 2 m
- Biomass systems need maintenance related to fuel deliveries, combustion efficiency, ash removal, adding to the lifetime cost
- Fuel stores need to be physically isolated from the boiler and the rest of the building in order to minimize fire risk. The fuel store needs to be sized to provide for at least 100 hours of operation, which is approximately 100 m^2 for a 500 kW boiler. The space taken up by storage and delivery access may compromise other aspects of site planning
- Fire-protection measures include anti-blowback arrangements on conveyors and fire dampers, together with the specification of elements such as flues for higher operating temperatures
- Collocation of the fuel source and burner at ground level require larger, free-standing flues

As a result of these issues, which drive up initial costs, affect development efficiency and add to management overheads, take-up of biomass has initially been mostly at the small-scale, heat-only end of the market, based on locally sourced feedstock. In such systems the initial cost premium of the biomass boiler can be offset against long-term savings in fuel costs.

Sourcing biomass

Compared with solar or wind power installations, the initial costs of biomass systems are low, the technology is well established and energy output is dependable. As a result, the real challenge for successful operation of a biomass system is associated with the reliable sourcing of feedstock.

Heat-only systems themselves cost between £150 and £750 per kW (excluding costs of storage), depending on scale and technology adopted. This compares with a typical cost of £50 to £300 per kW for a gas-fired boiler – which does not require further investment in fuel or thermal storage bunkers.

As a high proportion of lifetime cost is associated with the operation of a system, availability of good-quality, locally sourced feedstock is essential for long-term viability – particularly in areas where incentivization through policies like the Merton rule is driving up demand.

Research funded by BioRegional in connection with medium-scale biomass systems in the South-east shows that considerable feedstock is already in the system but far more is required to respond to emerging requirements.

Fortunately, the scale of the UK's untapped resource is considerable, with 5 million to 6 million tonnes of waste wood going to landfill annually, and 680,000 ha of set-aside land that could be used for energy crops without affecting agricultural output.

Based on these figures, it is estimated that 15% of the UK's building-related energy load could be supported without recourse to imported material. However, the supply chain is fragmented in terms of producers, processors and distributors – presenting potential biomass users with a range of complexities that gas users simply do not need to worry about. These include:

- Ensuring quality. Guaranteeing biomass quality is important for the assurance of performance and reliability. Variation in moisture content affects combustion, while inconsistent woodchip size or differences in sawdust content can result in malfunction. The presence of contaminants in waste wood causes problems too. High-profile schemes including the 180,000-tonne generator in Slough have had to shutdown because of variations in fuel quality. Use of pellets reduces to risk, but they are more expensive and require more energy for processing and transport
- Functioning markets. The scale of trade in biomass compares unfavourably with gas or oil, in that there are no standard contracts, fixed-price deals or opportunities for hedging which enable major users to manage their energy cost risk
- Security of supply. The potential for competing uses could lead to price inflation. Biofuels carry the greatest such risk, but many biomass streams have alternative uses. Lack of capacity in the marketplace is another security issue, with no mechanism to encourage strategic stockpiling for improved response to crop failure or fluctuations in demand
- Installation and maintenance infrastructure. The different technologies used in biomass systems creates maintenance requirements not yet met by a readily available pool of skilled system engineers

Optimum uses of biomass technology

Biomass is a high-grade, locally available source of energy that can be used at a range of scales to support domestic and commercial use. Following increases in fossil fuel prices, one of the main barriers to adoption is, now, the capability of the supply chain.

The Carbon Trust's biomass sector review, completed before the large energy price rises in recent years, drew the following key conclusions about the most effective application of the technologies:

- Returns on CHP and electricity-generating systems depend heavily on government incentives such as renewable obligations certificates. Under the present arrangements, large CHP systems provide the best returns
- Heat-only systems are very responsive to changes in fuel prices
- Small-scale heat-only plants produce the best returns, because the cost of the displaced fuel (typically fuel oil) is more expensive
- Small-scale electricity and large-scale heat-only installations produce very poor returns
- There is little difference in the impact of fuel type in the returns generated by projects

The study also concluded that heat installations at all scales had the greater potential for carbon saving, based on a finite supply of biomass. This is because heat-generating processes have the greatest efficiency and, in the case of small-scale systems in isolated, off-grid dwellings, displace fuels such as oil that have the greatest carbon intensity. Ninety percent of the UK's existing biomass resource of 5.6 m tonnes per annum could be used in displacing carbon-intensive off-grid heating, saving 2.5 m tonnes of carbon emissions.

Small-scale systems are well established in Europe and the existing local supply chain suits the demand pattern. In addition, since the target market is in rural, off-grid locations, affected dwellings are less likely to suffer space constraints related to storage. As fuel costs continue to rise, the benefits of avoiding fuel poverty, combined with the effective reduction of carbon emissions from existing buildings, mean smaller systems are likely to offer the best mid-term use of the existing biomass supply base, with large-scale systems being developed as the supply chain matures and expands.

Large-scale systems also offer the opportunity to generate significant returns, but the barriers that developers or operators face are significant, particularly if there is an electricity supply component, which requires a supply agreement. However, while developers are required to delivery renewable energy on site, biomass in the form of biofuels, has the great attraction of being able to provide a scale of renewable energy generation that other systems such as ground source heating or photovoltaics simply cannot compete with.

Whether biomass plant should be used on commercial schemes in urban locations is potentially a policy issue. Sizing of both CHP and heat-only systems should be determined by the heat load, which for city-centre schemes may not be that large – affecting the potential for the CHP component. The costs of a district heating element on these schemes may also be prohibitively high, and considerations of biomass transport and storage also make it harder to get city-centre schemes to stack up.

It may be a more appropriate policy to encourage industrial users or large scale regenerators to take first call on the expanding biomass resource, rather than commercial schemes. The launch of a 45 MW biomass power station in Scotland illustrates this trend. Data shows that 50% of the market potential for industrial applications of CHP could utilize 100% of the UK's available biomass resource. The issues that city-centre biomass schemes face in connection with storage, transport, emissions and supply chain management might be better addressed by industrial users or their energy suppliers in low-cost locations rather than by developers in prime city-centre sites.

Indicative costs

System	Indicative load kW	Capital cost (£/kWh)
Gas-fired boiler	50 400	90 50
Biomass-fired boiler	50 500	530 265
Biomass-fired CHP	1,000	480

Allowance for stand-alone boiler house and fuel store £30,000–60,000 for 50 kWh system indicative costs exclude flues and plant room installation

Fuel costs (typical bulk prices at small commercial scale – February 2012)

Wood chip: 2.9 p/kWh

Wood pellet: 4.2 p/kWh

Fuel oil: 6.0 p/kWh

Natural gas: 4.8 p/kWh

Bottled LPG: 7.6 p/kWh

Capital Allowances

Introduction

Capital allowances provide tax relief by prescribing a statutory rate of depreciation for tax purposes in place of that used for accounting purposes. They are utilized by government to provide an incentive to invest in capital equipment, including assets within commercial property, by allowing the majority of taxpayers a deduction from taxable profits for certain types of capital expenditure, thereby reducing or deferring tax liabilities.

The capital allowances most commonly applicable to real estate are those given for capital expenditure on existing commercial buildings in disadvantaged areas, and plant and machinery in all buildings other than residential dwellings. Relief for certain expenditure on industrial buildings and hotels was withdrawn from April 2011, although the ability to claim plant and machinery remains.

Enterprise Zone Allowances are also available for capital expenditure within designated areas only where there is a focus on high value manufacturing. Enhanced rates of allowances are available on certain types of energy and water saving plant and machinery assets, whilst reduced rates apply to 'integral features' and items with an expected economic life of more than 25 years.

The Act

The primary legislation is contained in the Capital Allowances Act 2001. Major changes to the system were announced by the Government in 2007 and there have been further changes in subsequent Finance Acts.

The Act is arranged in 12 Parts (plus two addenda) and was published with an accompanying set of Explanatory Notes.

Plant and machinery

The Finance Act 1994 introduced major changes to the availability of Capital Allowances on real estate. A definition was introduced which precludes expenditure on the provision of a building from qualifying for plant and machinery, with prescribed exceptions.

List A in Section 21 of the 2001 Act sets out those assets treated as parts of buildings:

- *Walls, floors, ceilings, doors, gates, shutters, windows and stairs.*
- *Mains services, and systems, for water, electricity and gas.*
- *Waste disposal systems.*
- *Sewerage and drainage systems.*
- *Shafts or other structures in which lifts, hoists, escalators and moving walkways are installed.*
- *Fire safety systems.*

Similarly, List B in Section 22 identifies excluded structures and other assets.

Both sections are, however, subject to Section 23. This section sets out expenditure, which although being part of a building, may still be expenditure on the provision of Plant and Machinery.

List C in Section 23 is reproduced below:

Sections 21 and 22 do not affect the question whether expenditure on any item in List C is expenditure on the provision of Plant or Machinery

1. Machinery (including devices for providing motive power) not within any other item in this list.
2. Gas and sewerage systems provided mainly –
   a. to meet the particular requirements of the qualifying activity, or
   b. to serve particular plant or machinery used for the purposes of the qualifying activity.
3. Omitted.
4. Manufacturing or processing equipment; storage equipment (including cold rooms); display equipment; and counters, checkouts and similar equipment.
5. Cookers, washing machines, dishwashers, refrigerators and similar equipment; washbasins, sinks, baths, showers, sanitary ware and similar equipment; and furniture and furnishings.
6. Hoists.
7. Sound insulation provided mainly to meet the particular requirements of the qualifying activity.
8. Computer, telecommunication and surveillance systems (including their wiring or other links).
9. Refrigeration or cooling equipment.
10. Fire alarm systems; sprinkler and other equipment for extinguishing or containing fires.
11. Burglar alarm systems.
12. Strong rooms in bank or building society premises; safes.
13. Partition walls, where moveable and intended to be moved in the course of the qualifying activity.
14. Decorative assets provided for the enjoyment of the public in hotel, restaurant or similar trades.
15. Advertising hoardings; signs, displays and similar assets.
16. Swimming pools (including diving boards, slides & structures on which such boards or slides are mounted).
17. Any glasshouse constructed so that the required environment (namely, air, heat, light, irrigation and temperature) for the growing of plants is provided automatically by means of devices forming an integral part of its structure.
18. Cold stores.
19. Caravans provided mainly for holiday lettings.
20. Buildings provided for testing aircraft engines run within the buildings.
21. Moveable buildings intended to be moved in the course of the qualifying activity.
22. The alteration of land for the purpose only of installing Plant or Machinery.
23. The provision of dry docks.
24. The provision of any jetty or similar structure provided mainly to carry Plant or Machinery.
25. The provision of pipelines or underground ducts or tunnels with a primary purpose of carrying utility conduits.
26. The provision of towers to support floodlights.
27. The provision of –
   a. any reservoir incorporated into a water treatment works, or
   b. any service reservoir of treated water for supply within any housing estate or other particular locality.
28. The provision of –
   a. silos provided for temporary storage, or
   b. storage tanks.
29. The provision of slurry pits or silage clamps.
30. The provision of fish tanks or fish ponds.
31. The provision of rails, sleepers and ballast for a railway or tramway.
32. The provision of structures and other assets for providing the setting for any ride at an amusement park or exhibition.
33. The provision of fixed zoo cages.

Capital allowances on plant and machinery are given in the form of writing down allowances at the rate of 18% per annum on a reducing balance basis. For every £100 of qualifying expenditure £18 is claimable in year 1, £14.76 in year 2 and so on until either all the allowances have been claimed or the asset is sold.

ESSENTIAL READING FROM TAYLOR AND FRANCIS

Quantity Surveyor's Pocket Book

2nd Edition

D Cartlidge

This second edition of the Quantity Surveyor's Pocket Book is fully updated in line with NRM1, NRM2 and JCT(11), and remains a must-have guide for students and qualified practitioners. Its focussed coverage of the data, techniques, and skills essential to the quantity surveying role make it an invaluable companion for everything from initial cost advice to the final account stage.

Key features include:

- the structure of the construction industry
- cost forecasting and feasibility studies
- measurement and quantification, with NRM2 and SMM7 examples
- estimating and bidding
- whole life costs
- contract selection
- final account procedure.

June 2012: 186x123: 440pp
Pb: 978-0-415-50110-1: £22.99

To Order: Tel: +44 (0) 1235 400524 Fax: +44 (0) 1235 400525
or Post: Taylor and Francis Customer Services,
Bookpoint Ltd, Unit T1, 200 Milton Park, Abingdon, Oxon, OX14 4TA UK
Email: book.orders@tandf.co.uk

For a complete listing of all our titles visit:
www.tandf.co.uk

Taylor & Francis
Taylor & Francis Group

Integral features

The category of qualifying expenditure on 'integral features' was introduced with effect from April 2008. The following items are integral features:

- An electrical system (including a lighting system)
- A cold water system
- A space or water heating system, a powered system of ventilation, air cooling or air purification, and any floor or ceiling comprised in such a system
- A lift, an escalator or a moving walkway
- External solar shading

A reduced writing down allowance of 8% per annum is available on integral features.

Thermal insulation

For many years the addition of thermal insulation to an existing industrial building has been treated as qualifying for plant and machinery allowances. From April 2008 this has been extended to include all commercial buildings but not residential buildings.

A reduced writing down allowance of 8% per annum is available on thermal insulation.

Long-life assets

A reduced writing down allowance of 8% per annum is available on long-life assets. Allowances were given at the rate of 6% before April 2008.

A long-life asset is defined as plant and machinery that can reasonably be expected to have a useful economic life of at least 25 years. The useful economic life is taken as the period from first use until it is likely to cease to be used as a fixed asset of any business. It is important to note that this likely to be a shorter period than an item's physical life.

Plant and machinery provided for use in a building used wholly or mainly as dwelling house, showroom, hotel, office or retail shop or similar premises, or for purposes ancillary to such use, cannot be long-life assets.

In contrast plant and machinery assets in buildings such as factories, cinemas, hospitals and so on are all potentially long-life assets.

Case law

The fact that an item appears in List C does not automatically mean that it will qualify for capital allowances. It only means that it may potentially qualify.

Guidance about the meaning of plant has to be found in case law. The cases go back a long way, beginning in 1887. The current state of the law on the meaning of plant derives from the decision in the case of *Wimpy International Ltd and Associated Restaurants Ltd v Warland* in the late 1980s.

The Judge in that case said that there were three tests to be applied when considering whether or not an item is plant.

1. Is the item stock in trade? If the answer yes, then the item is not plant.
2. Is the item used for carrying on the business? In order to pass the business use test the item must be employed in carrying on the business; it is not enough for the asset to be simply used in the business. For example, product display lighting in a retail store may be plant but general lighting in a warehouse would fail the test.

3. Is the item the business premises or part of the business premises? An item cannot be plant if it fails the premises test, i.e. if the business use is as the premises (or part of the premises) or place on which the business is conducted. The meaning of part of the premises in this context should not be confused with the law of real property. The Inland Revenue's internal manuals suggest there are four general factors to be considered, each of which is a question of fact and degree:
   - Does the item appear visually to retain a separate identity
   - With what degree of permanence has it been attached to the building
   - To what extent is the structure complete without it
   - To what extent is it intended to be permanent or alternatively is it likely to be replaced within a short period

There is obviously a core list of items that will usually qualify in the majority of cases. However, many other still need to be looked at on a case-by-case basis. For example, decorative assets in a hotel restaurant may be plant but similar assets in an office reception area would almost certainly not be.

One of the benefits of the integral features rules, apart from simplification, is that items that did not qualify by applying these rules, such as general lighting in a warehouse or an office building, will now qualify albeit at a reduced rate.

Refurbishment schemes

Building refurbishment projects will typically be a mixture of capital costs and revenue expenses, unless the works are so extensive that they are more appropriately classified a redevelopment. A straightforward repair or a 'like for like' replacement of part of an asset would be a revenue expense, meaning that the entire amount can be deducted from taxable profits in the same year.

Where capital expenditure is incurred that is incidental to the installation of plant or machinery then Section 25 of the 2001 Act allows it to be treated as part of the expenditure on the qualifying item. Incidental expenditure will often include parts of the building that would be otherwise disallowed, as shown in the Lists reproduced above. For example, the cost of forming a lift shaft inside an existing building would be deemed to be part of the expenditure on the provision of the lift.

The extent of the application of Section 25 was reviewed for the first time by the Special Commissioners in December 2007 and by the First Tier Tribunal (Tax Chamber) in December 2009, in the case of JD Wetherspoon. The key areas of expenditure considered were overheads and preliminaries where it was held that such costs could be allocated on a pro-rata basis; decorative timber panelling which was found to be part of the premises and so ineligible for allowances; toilet lighting which was considered to provide an attractive ambience and qualified for allowances; and incidental building alterations of which enclosing walls to toilets and kitchens and floor finishes did not qualify but tiled splash backs, toilet cubicles and drainage did qualify along with the related sanitary fittings and kitchen equipment.

Annual investment allowance

The annual investment allowance is available to all businesses of any size and allows a deduction for the whole of the first £500,000 from 19 March 2014 (£250,000 before 19 March 2014) of qualifying expenditure on plant and machinery, including integral features and long-life assets. The annual investment allowance will return to £25,000.00 from January 2016.

The enhanced capital allowances scheme

The scheme is one of a series of measures introduced to ensure that the UK meets its target for reducing greenhouse gases under the Kyoto Protocol. 100% first year allowances are available on products included on the Energy Technology List published on the website at www.eca.gov.uk and other technologies supported by the scheme. All businesses will be able to claim the enhanced allowances, but only investments in new and unused Machinery and Plant can qualify.

There are currently 15 technologies with multiple sub-technologies currently covered by the scheme:

- Air-to-air energy recovery
- Automatic monitoring and targeting (AMT)
- Boiler equipment
- Combined heat and power (CHP)
- Compressed air equipment
- Heat pumps
- Heating ventilation and air conditioning (HVAC) equipment
- High speed hand air dryers
- Lighting
- Motors and drives
- Pipework insulation
- Radiant and warm air heaters
- Refrigeration equipment
- Solar thermal systems
- Uninterruptible power supplies

The Finance Act 2003 introduced a new category of environmentally beneficial plant and machinery qualifying for 100% first-year allowances. The Water Technology List includes 14 technologies:

- Cleaning in place equipment
- Efficient showers
- Efficient taps
- Efficient toilets
- Efficient washing machines
- Flow controllers
- Grey water recovery and reuse equipment
- Leakage detection equipment
- Meters and monitoring equipment
- Rainwater harvesting equipment.
- Small scale slurry and sludge dewatering equipment
- Vehicle wash water reclaim units
- Water efficient industrial cleaning equipment
- Water management equipment for mechanical seals

Buildings and structures and long-life assets as defined above cannot qualify under the scheme. However, following the introduction of the integral features rules, lighting in any non-residential building may potentially qualify for enhanced capital allowances if it meets the relevant criteria.

A limited payable ECA tax credit equal to 19% of the loss surrendered was also introduced for UK companies in April 2008.

From April 2012 expenditure on plant and machinery for which tariff payments are received under the renewable energy schemes introduced by the Department of Energy and Climate Change (Feed-in Tariffs or Renewable Heat Incentives) will not be entitled to enhanced capital allowances.

Enterprise zones

The creation of 11 new enterprise zones was announced in the 2011 Budget. Additional zones have since been added bringing the number to 24 in total. Originally introduced in the early 1980s as a stimulus to commercial development and investment, they had virtually faded from the real estate psyche.

The original zones benefited from a 100% first year allowance on capital expenditure incurred on the construction (or the purchase within two years of first use) of any commercial building within a designated enterprise zone, within 10 years of the site being so designated. Like other allowances given under the industrial buildings code the building has a life of 25 years for tax purposes.

The majority of these enterprise zones had reached the end of their 10-year life by 1993. However, in certain very limited circumstances it may still be possible to claim these allowances up to 20 years after the site was first designated.

Enterprise zones benefit from a number of reliefs, including a 100% first year allowance for new and unused non-leased plant and machinery assets, where there is a focus on high-value manufacturing.

Flat conversion allowances

Tax relief is available on capital expenditure incurred on or after 11 May 2001 on the renovation or conversion of vacant or underused space above shops and other commercial premises to provide flats for rent.

In order to qualify the property must have been built before 1980 and the expenditure incurred on, or in connection with:

- Converting part of a qualifying building into a qualifying flat.
- Renovating an existing flat in a qualifying building if the flat is, or will be a qualifying flat.
- Repairs incidental to conversion or renovation of a qualifying flat.
- The cost of providing access to the flat(s).

The property must not have more than four storeys above the ground floor and it must appear that, when the property was constructed, the floors above the ground floor were primarily for residential use. The ground floor must be authorized for business use at the time of the conversion work and for the period during which the flat is held for letting. Each new flat must be a self-contained dwelling, with external access separate from the ground-floor premises. It must have no more than 4 rooms, excluding kitchen and bathroom. None of the flats can be 'high value' flats, as defined in the legislation. The new flats must be available for letting as a dwelling for a period of not more than 5 years.

An initial allowance of 100% is available or, alternatively, a lower amount may be claimed, in which case the balance may be claimed at a rate of 25% per annum in subsequent a years. The allowances may be recovered if the flat is sold or ceases to be let within 7 years.

At Budget 2011 the Government announced that the relief would be abolished from April 2013.

Business premises renovation allowance

The business premises renovation allowance (BPRA) was first announced in December 2003. The idea behind the scheme is to bring long-term vacant properties back into productive use by providing 100% capital allowances for the cost of renovating and converting unused premises in disadvantaged areas. The legislation was included in Finance Act 2005 and was finally implemented on 11 April 2007 following EU state aid approval.

The legislation is identical in many respects to that for flat conversion allowances. The scheme will apply to properties within the areas specified in the Assisted Areas Order 2007 and Northern Ireland.

BPRA is available to both individuals and companies who own or lease business property that has been unused for 12 months or more. Allowances will be available to a person who incurs qualifying capital expenditure on the renovation of business premises.

An announcement to extend the scheme by a further five years to 2017 was made within the 2011 Budget, along with a further 11 new designated Enterprise Zones.

Legislation was introduced in Finance Bill 2014 to clarify the scope of expenditure qualifying for relief to actual costs of construction and building work and for certain specified activities such as architectural and surveying services. The changes will have effect for qualifying expenditure incurred on or after 1 April 2014 for businesses within the charge to corporation tax, and 6 April 2014 for businesses within the charge to income tax.

Other capital allowances

Other types of allowances include those available for capital expenditure on Mineral Extraction, Research and Development, Know-How, Patents, Dredging and Assured Tenancy.

Enhanced Capital Allowances (ECAs), the Energy Technology List (ETL) and the Water Technology List (WTL)

Background

This bulletin is intended to raise awareness of the existence, location and relevance of the ETL and the WTL which are often referred to for simplicity as 'The Technology Lists'. The lists comprise the technologies and products which qualify for the UK Government's Enhanced Capital Allowance (ECA) scheme. Also included in the lists are the energy-saving and water-saving performance criteria for each product or technology.

What are ECAs?

Enhanced Capital Allowances are a form of tax relief enabling a business to claim 100% first-year capital allowances on their spending on qualifying plant and machinery instead of the normal reducing balance writing down allowances of 8% for integral features and 18% for main pool plant. There are two building services-related schemes for ECAs:

- Energy-saving plant and machinery
- Water conservation plant and machinery

Businesses can write off the whole of the capital cost of their investment in these technologies against their taxable profits of the period during which they make the investment. This can deliver a helpful cash flow boost and a shortened payback period via both the energy, or water, saved and the allowances claimed. Alternatively a loss making business can surrender the ECA value of the loss to obtain a 19% tax credit.

ECAs are only available to investors of qualifying plant and machinery and not developers. Developers are 'traders' and they incur revenue expenditure and not capital expenditure and so cannot claim capital allowances. However the purchaser of a newly constructed development for investment can still make use of ECAs provided they are sold unused and the purchaser is a taxpayer. This can be a valuable selling point for developers if highlighted in the marketing literature.

Interest in ECAs has been highlighted by the changes to the existing capital allowances regime which have reduced the value of allowances for many qualifying items. The introduction of Energy Performance Certificates and the Carbon Reduction Commitment Energy Saving Scheme will also promote the types of technologies that could qualify for ECAs.

Who manages the technology lists and where are they?

The lists are part of the ECA scheme which was developed by the Treasury, HM Revenue & and the Department of Energy & Climate Change (DECC). The ETL is managed and maintained by the Carbon Trust and can be found at http://etl.decc.gov.uk/. The WTL is managed by Business Link and the Department for Environment Food & Rural Affairs (DEFRA) and can be found at http://wtl.defra.gov.uk.

What technologies and products are included in the ETL? (as at 2013)

- Air-to-air energy recovery
- Automatic monitoring and targeting
- Boiler equipment
- Combined heat and power (CHP)
- Compact heat exchangers
- Compressed air equipment
- Heat pumps
- Heating ventilation and air conditioning (HVAC) equipment
- High speed hand air dryers
- Lighting (high efficiency lighting units, lighting controls and LEDs)
- Motors and drives
- Pipework insulation
- Radiant and warm air heaters
- Refrigeration equipment
- Solar thermal systems
- Thermal screens
- Uninterruptible power supplies

And the WTL?

- Cleaning in place equipment
- Efficient showers
- Efficient taps
- Efficient toilets
- Efficient washing machines
- Flow controllers
- Leakage detection equipment
- Meters and monitoring equipment
- Rainwater harvesting equipment
- Small scale slurry and sludge dewatering equipment
- Vehicle wash water reclaim units
- Water efficient industrial cleaning equipment
- Water management equipment for mechanical seals
- Water reuse

Eligibility criteria

It is not possible to summarize the eligibility criteria here because each technology has different definitions and considerations. Most of the above technologies have specific products listed that are eligible, but others are defined by detailed performance criteria. The key technologies worthy of note which operate on a Performance criteria are Lighting, CHP, Pipework Insulation, Automatic Monitoring & Targeting equipment and Water Reuse. Qualification is generally extremely strict. To give a brief example; Lighting doesn't merely concentrate on the lamp type used but also has very strict requirements for the light fitting itself, having to meet high Light Output Ratios, colour rendering requirements and use high frequency control equipment. CHP and AMT equipment is also heavily guided.

Scope of ECA claims

For products on the Technology Lists, claims will be considered for the cost of the equipment itself, and other costs directly involved in installing it. These include:

- Transportation – the cost of getting equipment to the site.
- Installation – cranage (to lift heavy equipment into place), project management costs and labour, plus any necessary modifications to the site or existing equipment.
- Preliminary costs and oncosts – if they are directly related to the acquisition and installation of the equipment.
- Professional Fees – if they are directly related to the acquisition and installation of the equipment.

Making a claim for ECAs

If investing in eligible technology claimants should submit their ECA claims as part of their normal Income or Corporation Tax return. The best advice we can give clients on relevant projects is to highlight the fact that their investment may be eligible, and introduce Davis Langdon Banking Tax & Finance to the client as early as possible in the design process so that they can offer a view on the potential for worthwhile tax relief before, during and after the investment programme.

Follow-up

For further general advice on any aspect of this Bulletin please contact Andy White of AECOM Fiscal Incentives.

LED Lighting

LED lighting – is it the 'best thing since sliced bread'?

Background

Current LED technology is becoming a viable alternative to other energy efficient lighting solutions in a variety of applications which include external lighting, industrial and office environments. With careful selection, LED systems can deliver the qualitative design requirements without compromise to the overall visual comfort and ambiance of the space. LED characteristics of high light output, good light colour quality and rendering and engineered product efficiency are beginning to become comparable to fluorescent lighting systems, however, there is an indeterminate risk regarding the longevity and stability of LED operating characteristics simply because the latest LEDs have not been around long enough. They are relatively untested on the marketplace when compared to a proven technology of fluorescent.

Currently available types of LED luminaries

The initial concept of LED lighting was developed for electronic devices, signal lighting, and decorative external lighting. In continued development, high output 'white' LED have been designed and engineered to replace traditional lighting equipment for more energy efficient systems.

Current market LEDs are available with lumen efficacy up to 120 lumens, provide high lumen maintenance, long life, and provide instant dimmable light. LEDs are also noted for their benefits of no infrared and ultra-violet radiation, no mercury content, and a high degree of recyclable content for a sustainable lighting solution.

Recent LED installations have included office environments, restaurants, hotels, roadways and car parks, and feature lighting. With improvements in high colour rendering white LEDs, street lighting installations have the side benefit of dramatically improving the resolution of CCTV coverage to areas such as car parks and building elevations.

- Colour temperature range (2,500–10,000 K)
- Colour rendering (65–95r CIE Ra)
- RGB vivid colours (millions)
- Dimmable
- Instant light
- Compact & robust (can integrate into fabric of built environment)
- Retrofit & new build applications
- Most of content recyclable & no mercury
- No IR or UV radiation

Warning though!

It should be noted the marketplace is inundated with varying qualities of LEDs. There are no agreed standardized formats leading to a myriad of different styles, types and colours available. As LED technology is pioneered and developed by semiconductor electronic manufacturers, the industry can no longer rely on the traditional lamp manufacturers to set relatively clear guidance to qualitative aspects of their products. With this lack of clarity, some suppliers are taking advantage by flogging inferior quality LEDs with misleading and unreliable data.

The LED industry is also hindered by its own high speed of development. Month by month new developments are introduced. This is leading to hesitation in commitment to the technology from lighting manufacturers, designers, and clients leading to questions like, 'will a replacement LED module be available in the event of failure?'.

Currently, a great deal of trust and faith is put in manufacturer's ability to provide proper documentation and support for their products. It is important to use the manufacturers that can provide sufficient data and that they have a reliable reputation in the marketplace. A manufacturer should be able to advise on issues of thermal management, colour consistency and appearance (binning), colour rendering, dimming capability, true lumen output and efficacy, operational lumen depreciation, and ideally provide independent tested photometric reports, evidence of electrical safety and thermal testing.

Finally, it is important to get the manufacturer to prove their products and design. For a true LED replacement lighting system, the design scheme should not compromise on the qualitative demands defined in good practice design guidances. For example, the market place has various 'replacement tubes' available where on the face of it you simply install a LED 'tube' – this is not the case. The 'tube' is often simply a linear strip of LEDs within a glass tube. The retrofitted luminaire may require modifications, and the photometric performance may not be directly comparable. It should also be remembered that all diffusers are specially designed for a fluorescent tube and will probably adversely affect the light distribution and will be completely different to that of a normal tube. So caution, when dealing with these, if requested by a client.

Cost implications

Manufacturers and suppliers make great hay regarding the energy savings that can be made against more traditional light sources.

The manufacturers' cost calculations should be carefully examined as in some instances they base their calculations on several incorrect assumptions such as they may state 365 twelve hour days but, even after correcting these incorrect assumptions, the payback and subsequent energy cost savings are dramatic.

Payback

As the LED luminaires are currently more expensive than more traditional sources of illumination the payback period is of particular importance as actual energy savings will only be realized after the LED luminaires if you like have paid for itself.

Therefore based on the following material costs only as it is assumed the labour content will be similar:

Power cost (£)/kWh	0.12
Days in use/year	260
Hours in use/day	12

	Typical LED Downlight	Compact Fluorescent Downlight	LV Halogen (IRC) Downlight
Input power (Watts)	16	28	35
Lifetime (hours)	50,000	12,000	4,000
Replacement lamp cost including labour (£)	–	15.00	15.00
Annual energy cost per lamp (£)	6.04	10.57	13.21
Total energy cost for 25 luminaires per year (£)	151.01	264.28	330.28
Luminaire unit cost (£)	130.00	70.00	80.00

Supply only cost of 25 luminaires (£)	3,250.00	1,750.00	2,000.00
Allowance for 4nr emergency battery packs (£)	320.00	320.00	320.00
Lamps (£)	Included	375.00	375.00
Total cost of luminaries/emergencies/ lamps (£)	3,570.00	2,445.00	2,695.00

Extra over cost of LED luminaires and emergency luminaires compared to traditional luminaires is (£)		1,125.00	875.00
Yearly cost of energy and yearly re-lamping allowance (compact fluorescent @ every 4 years; halogen @ every 1.5 year) (£)	151.01	358.03	580.28
The calculated yearly energy and re-lamping saving of using LED luminaires when compared to traditional luminaires is (£)		207.02	429.27
Therefore time taken in years to 'pay' for the additional cost of the LED luminaires based on the energy and re-lamping costs of 'traditional' luminaires in **years** is		**5.43**	**2.04**

Power cost (£)/kWh	0.12
Days in use/year	260
Hours in use/day	12

	Recessed LED Office Luminaire	Recessed Fluorescent Office Luminaire
Input power (Watts)	37.5	42
Typical lifetime (hours)	50,000	12,000
Replacement lamp cost including labour (£)	–	20.00
Annual energy cost per lamp (£)	14.15	15.85
Total energy cost for 25 luminaires per year (£)	353.93	396.28
Luminaire unit cost including lamps (£)	225.00	165.00
Supply only cost of 25 luminaires including lamps (£)	5,625.00	4,125.00
Allowance for 4nr emergency battery packs (£)	520.00	400.00
Total cost of luminaries/emergencies/lamps (£)	6,145.00	4,525.00

Extra over cost of LED luminaires and emergency luminaires compared to fluorescent luminaires is (£)		1,620.00
Yearly cost of energy and yearly re-lamping allowance (fluorescent @ every 4 years) (£)	353.93	521.28
The calculated yearly energy and re-lamping saving of using LED luminaires when compared to fluorescent luminaires is (£)		167.35
Therefore time taken in years to 'pay' for the additional cost of the LED luminaires based on the energy and re-lamping costs of fluorescent luminaires in **years** is		**9.68**

Notes and conclusions

1. LED luminaires do save energy.
2. LED luminaires are generally twice the price of a 'traditional' downlight but LED luminaire prices are becoming more competitive as the technology develops.
3. Some LED luminaires may not be converted to emergency so 'additional' luminaires may be required.
4. LED replacement lamps may have completely different light emitting characteristics. And, the luminaire may require modification to accept the retrofit LED lamp.

Getting the Connection

Details the process involved in the provision of a new electrical supply to a site.

The provision of an electricity connection has both a physical and contractual element. Physically, it involves the design, planning and construction of electrical infrastructure (cables, switchgear, civil works), whilst contractually, it requires legal agreements to be drawn up and agreed (construction, connection, adoption).

Planning

The Planning Stage, typically RIBA Stage C, is where the site's developer (the Developer) should be formulating their plans for the scheme and, in doing so, consult the local distribution network operator's (the Host)[1] long-term development statement, which will identify potential connection point opportunities. This information is normally readily available from the Host and should provide an early indication of whether the Host's network may need to be reinforced before the development can be connected.

As the Planning Stage progresses, the Developer should discuss its proposals with the Host. Relatively simple connections for single building supplies are straight forward, however schemes of a more complex nature may require some form of feasibility study to be carried out to assess connection options and provide indicative costs for the contestable[2] and non-contestable[3] work elements.

Design

The Design Stage, typically RIBA Stage D, is the point at which the Developer submits its formal connection request to the Host. It is important that this is completed in accordance with the specific procedures of the Host, since if it does not include all supporting information required by the Host's application process, there is likely to be a delay in the processing of a firm offer. The Host is expected to provide a documented application process to assist the applicant make a complete application.

The convention is to request a Section 16/16 A connection[4] where the terms are standard and non-negotiable. The alternative is a Section 22 connection[2] offer, where the terms are fully negotiable. If the Section 22 route is chosen then caution is required, as disputes over Section 22 agreements cannot be referred to the regulator for determination after the connection agreement has been signed; a post contract dispute will have to be pursued as a civil action. It is always advised that a Section 16/16 A offer is requested before considering the Section 22 option.

On receipt of the Host's firm offer[5], the Developer generally has 90 days to accept its terms and to undertake its own review of the offer to ensure it meets its requirements. If for any reason the Developer and Host are unable to reach agreement of the terms, it is recommended that the Developer seeks specialist advice. It should be noted that in extreme circumstances, it may be necessary to refer an issue to regulator for determination but, be warned, this can be a lengthy process, taking upto 16 weeks to conclude.

1. *A distribution network operator (DNO) is a company that is responsible for the design, construction operation and maintenance of a public electricity distribution network. The host DNO is the electricity distribution network to which the development site will directly connect.*
2. *Contestable is work in providing the connection that can be carried out by an accredited independent party.*
3. *Non-contestable is work that can only be carried by the Host.*
4. *The Electricity Act 1989.*
5. *Unless the offer becomes interactive, i.e. another connection scheme is vying for network capacity. Where the connection request becomes interactive the Developer has 30 days to accept the offer. The Host will inform the Developer if the connection request is interactive.*

Competition

One of the key decisions affecting the way in which the connection process proceeds is whether the Developer wishes to introduce competition into the procurement process by appointing a third party to design and construct the contestable connection works. In these circumstances, the Developer can requisition a non-contestable quotation from the Host or ask the third party to do so on its behalf. The Host is obliged to provide information within standardized time frames against which its performance is monitored by the industry regulator (Ofgem).

Contestable works are those that may be carried out either by the Host or by an approved contractor, on the Developer's behalf. This contrasts with non-contestable works, which can only be carried out by the Host. Broadly speaking, the Host will make the connection to its network for the new supply and undertake any upstream network reinforcement works. All works downstream from the point of connection to the Host network into the site and to each building are contestable works. Therefore, the extent of the contestable works can vary significantly depending on where the point of connection is designated by the Host. Alternatively, the Developer can request an independent licensed network operator (IDNO) to adopt assets constructed by the third party.

If the Developer decides to contract with a third party to construct the contestable works, it is the Developer's responsibility to ensure that the construction works meet the Host's network adoption requirements.

Network reinforcement

Reinforcement works may be required to increase the capacity of the network to enable the connection to meet a site's projected demand. In terms of the capacity made available by reinforcement, the following possible scenarios arise:

- Where reinforcement works are necessary for sole use by the Developer, the Developer is charged the full cost of the works. In some circumstances the most economic method of reinforcement may introduce spare network capacity in excess of the Developer's requirements. In such circumstances the Developer can receive a rebate where this spare capacity is absorbed by subsequent developments. However, the possibility of a rebate is time constrained and the original development will only qualify for a rebate for up to 5 years after the connection is completed.
- Where reinforcement works are necessary but also cater for the Host's future network requirements, the Developer is charged for a proportion of the cost of the works in the form of a capital contribution. This is calculated using cost apportionment factors for security (the ratio of capacity requested to that which is made available) and fault level (the ratio of fault level contribution of connection to that which is made available).

The charges levied for reinforcement works are attributable to those reinforcement works undertaken at one voltage level above the connection voltage only. That is, if connection voltage is 400 V then costs are chargeable for reinforcement works undertaken at the next highest voltage, which is usually 11 kV, and not for works conducted at the next highest voltage, which is usually 33 kV[6]. These deeper network reinforcement costs are recovered as part of the system charges built into the electricity supply tariffs.

Costs

The cost of the Developer's connection depends on the nature and extent of the works to be undertaken. The distance between the site and the Host's network, the size of the customer demand in relation to available capacity (and hence the potential need for reinforcement), customer-specific timescales and to an extent market value of raw material and labour, are all significant factors that will affect the cost.

6. *Some DNOs use voltages of 6.6 kV and 22 kV.*

As part of the firm offer received from the Host, the Developer is provided with a charging statement, which includes the charges to be levied for the following items:

- Assessment and Design – to identify and design the most appropriate point on the existing network for the connection.
- Design Approval – to ensure design of a connection meets the safety and operation requirements of the Host.
- Non-contestable Works/Reinforcement – to include circuits and plant forming part of the connection that can be undertaken by the Host only, including land rights issues and consents.
- Contestable Connection Works – to include circuits and plant forming part of the connection that can be undertaken by approved contractors or the Host.
- Inspection of Works – to ensure that works are being constructed in accordance with the design requirements of the Host.
- Commissioning of Works – to include circuit outages and testing that will ensure that connection is safe to be energized.

In some circumstances, in addition to the cost of the physical connection works there may be chargeable costs associated with operation, maintenance repair and replacement of the new or modified connection. These are known as O&M costs and are chargeable as capitalized up-front costs where the Developer requests a solution that is in excess of the minimum necessary to provide the connection. For example, where extra resilience is requested for a connection, over and above that which the Host is obligated to provide, then O&M costs can be levied but only for the extra resilience element of the connection, not the total cost of the connection.

ESSENTIAL READING FROM TAYLOR AND FRANCIS

BIM and Quantity Surveying

PITTARD & SELL

The sudden arrival of Building Information Modelling (BIM) as a key part of the building industry is redefining the roles and working practices of its stakeholders. Many clients, designers, contractors, quantity surveyors, and building managers are still finding their feet in an industry where BIM compliance can bring great rewards.

This guide is designed to help quantity surveying practitioners and students understand what BIM means for them, and how they should prepare to work successfully on BIM compliant projects. The case studies show how firms at the forefront of this technology have integrated core quantity surveying responsibilities like cost estimating, tendering, and development appraisal into high profile BIM projects. In addition to this, the implications for project management, facilities management, contract administration and dispute resolution are also explored through case studies, making this a highly valuable guide for those in a range of construction project management roles.

Featuring a chapter describing how the role of the quantity surveyor is likely to permanently shift as a result of this development, as well as descriptions of tools used, this covers both the organisational and practical aspects of a crucial topic.

December 2015: 234 x 156 mm: 258 pp

Pb: 978-0-415-87043-6; £24.99

To Order: Tel: +44 (0) 1235 400524 Fax: +44 (0) 1235 400525

or Post: Taylor and Francis Customer Services,

Bookpoint Ltd, Unit T1, 200 Milton Park, Abingdon, Oxon, OX14 4TA UK

Email: book.orders@tandf.co.uk

For a complete listing of all our titles visit:

www.tandf.co.uk

Taylor & Francis

Taylor & Francis Group

Feed-In Tariffs (FITs)

To ensure that everyone plays their part towards achieving the Renewable Energy Strategy's (RES) target of 15% for renewable energy by 2020, the need to develop new methods of renewable energy in all sectors is apparent.

On 1 February 2010, the Department for Energy and Climate Change (DECC) announced their intention to implement tariff levels following the Energy Act 2008 introduction of Feed-in Tariffs (FITs). The FIT scheme is intended to encourage the adoption of small-scale low carbon technologies up to a maximum of 5 MW through tariff payments made on both generation and export of produced renewable energy.

Feed-in Tariffs (FITs) are effectively payments which are made for every kilowatt per hour of renewable electricity (kWhr) that is generated. Homeowners/businesses will have the opportunity to benefit financially from the scheme either by consuming it on-site, earning a generation tariff, or by selling spare capacity to the grid, generating an export tariff.

Amendment 2 to the Feed-in Tariffs Order 2010 came into effect on 1 August 2012. The schemes policy and tariff rates are set by the Government with the scheme being administered by energy suppliers and Ofgem.

Feed-in Tariffs will be paid by electricity suppliers. Generators between 50 kW and 5 MW should apply to Ofgem for accreditation. Micro-generators can obtain accreditation via the Microgeneration Certification Scheme (MCS), before then applying to a supplier for a FIT agreement. All generation will be metered and FIT payments will be made in accordance with the Electricity Act 1989. The consumer will be responsible for the capital of the plant; the necessary access/connections to the electricity and transmission systems and organization of the payment receipts.

Since the scheme was introduced on 1 April 2010 and up to December 2012 a total of 358,658 renewable installations have been registered, with 1,657 MW of total installed capacity also being registered.

FITs review

The first review of the Feed-in Tariff scheme was scheduled to be carried out in 2012 however it was brought forward by a year due to the surge in uptake which placed a huge strain on the FITs budget. The surge was due to the sharp fall in the cost of solar PV installations and also the creation of large-scale solar farms, which went against the ethos of the scheme which was to promote small scale installations.

August 2011 saw the introduction of lower rates for large-scale solar installations. October 2011 the Department of Energy and Climate Control (DECC) launched a consultation into cuts for smaller scale solar installations. The consultation resulted in the decision to reduce FITs to approximately half the original tariff, however this was brought to the High Court and after appeals were heard the ruling found the cuts to tariffs were illegal.

February 2012 saw the DECC announce wide ranging reforms to the FIT scheme which they hope will provide a better framework for consumers and communities. There are five key areas that the DECC has identified as needing refinements;

- The tariff for micro-CHP installations will be increased to recognize the benefits this technology could bring and to encourage its development.
- A tariff of 21 p/kWh to take effect from 1 April 2012 for domestic-size solar panels with an eligibility date on or after 3 March 2012.
- Installations of solar PV panels on or after 1 April 2012 will be required to produce and Energy Performance Certificate rating of 'D' or above to qualify for a full FIT.

- 'Multi-installation' tariff rates set at 80% of the standard tariffs will be introduced for solar PV installations where a single individual or organization is already receiving FITs for other solar PV installations.
- Individuals or organizations with 25 or fewer installations will still be eligible for the individual rate. DECC is consulting on a proposal that social housing, community projects and distributed energy schemes be exempt from multi-installation tariff rates.

Following the FITs review the Department of Energy and Climate Change (DECC) announced the first scheduled reduction in tariff rates for all PV installations with an eligibility date on or after 1 November 2012. The new tariff rates will be published by Ofgem by 31 August 2015. Tariff rates will be continuously reviewed as set out in the degression mechanism as follows:

- Degression will take place on a quarterly basis, with tariffs changing on the first day of the first month of the quarter for new installations with an eligibility date from that date.
- Tariffs will be published at least two months before the start of the new quarter, and will be based on deployment of new PV generating capacity in the previous quarter.
- The baseline degression rate will be 3.5% per quarter.
- Degression will be skipped if deployment is below a floor threshold.
- Deployment statistics will be published on a monthly basis by DECC.

Eligibility

Small-scale (up to a maximum capacity of 5 MW) low-carbon electricity technologies eligible for FITs are:

- Wind
- Solar Photovoltaics (PV)
- Hydro
- Anaerobic Digestion
- Domestic scale micro Combined Heat & Power (CHP), eligible up to 2 kW.

As of 1 April 2010, micro generators (<50 kW) in Anaerobic Digestion, Hydro, Solar PV and Wind will be ineligible for support under the Renewable Obligation (RO) due to FIT inclusion. Refurbished/renovated installations will be ineligible. Small generators that applied for accreditation on or after the 15 July 2009 and before 1 April 2010 will be eligible to transfer to FITs.

Generating stations whose electricity is sold under a Non-Fossil Fuel Obligation (NFFO) arrangement will be ineligible for FITs; however will remain eligible to receive RO support.

Any existing systems installed before 15 July 2009 will only qualify for the FIT scheme if they are under 50 kW and registered with the RO.

From 1 December 2012, new provisions have been introduced for schools and some community organizations.

Tariffs

The Feed-in Tariffs (FITs) provide three main financial benefits to the consumer:

- A 'Generation' tariff (up to 21.65 p/kWh depending on type and size) based on the total generation and the energy type.
- An 'export' tariff (4.50 p/kWh) for any surplus energy produced exported into the grid and due to producing a percentage of the energy used.
- On-site usage of generated electricity, therefore reducing the amount purchased from an electricity supplier.

Energy Source	Scale	Effective Dates	Applicable Rates	Tariff (p/kWh)
Anaerobic Digestion	≤250 kW	to 31 March 2016		10.13
Anaerobic Digestion	>250 kW–500 kW	to 31 March 2016		9.36
Anaerobic Digestion	>500 kW–5MW	to 31 March 2016		8.68
Hydro generating station	≤15 kW	to 31 March 2016		17.17
Hydro generating station	>15 kW–100 kW	to 31 March 2016		16.03
Hydro generating station	>100 kW–500 kW	from 15 March 2016		12.67
Hydro generating station	>500 kW–2 MW	to 31 March 2016		9.90
Hydro generating station	>2 MW–5 MW	to 31 March 2016		2.70
Combined Heat & Power (CHP)	≤2 kW	from 15 March 2016	Limited to 30,000 units	13.45
Solar photovoltaic	≤4 kW	to 30 September 2015	Higher	12.92
Solar photovoltaic	≤4 kW	to 30 September 2015	Middle	11.63
Solar photovoltaic	≤4 kW	to 30 September 2015	Lower	5.94
Solar photovoltaic	>4 kW–10 kW	to 30 September 2015	Higher	11.71
Solar photovoltaic	>4 kW–10 kW	to 30 September 2015	Middle	10.54
Solar photovoltaic	>4 kW–10 kW	to 30 September 2015	Lower	5.94
Solar photovoltaic	>10 kW–50 kW	to 30 September 2015	Higher	11.71
Solar photovoltaic	>10 kW–50 kW	to 30 September 2015	Middle	10.54
Solar photovoltaic	>10 kW–50 kW	to 30 September 2015	Lower	5.94
Solar photovoltaic	>50 kW–100 kW	to 30 September 2015	Higher	9.63
Solar photovoltaic	>50 kW–100 kW	to 30 September 2015	Middle	8.67
Solar photovoltaic	>50 kW–100 kW	to 30 September 2015	Lower	5.94
Solar photovoltaic	>100 kW–150 kW	to 30 September 2015	Higher	9.63
Solar photovoltaic	>100 kW–150 kW	to 30 September 2015	Middle	8.67
Solar photovoltaic	>100 kW–150 kW	to 30 September 2015	Lower	5.94
Solar photovoltaic	>150 kW–250 kW	to 30 September 2015	Higher	9.21
Solar photovoltaic	>150 kW–250 kW	to 30 September 2015	Middle	8.29
Solar photovoltaic	>150 kW–250 kW	to 30 September 2015	Lower	5.94
Solar photovoltaic	>250 kW–5 MW	to 30 September 2015		5.94
Wind	≤1.5 kW	to 31 March 2016		14.45
Wind	>1.5 kW–15 kW	to 31 March 2016		14.45
Wind	>15 kW–100 kW	to 31 March 2016		14.45
Wind	>100 kW–500 kW	to 31 March 2016		12.05
Wind	>500 kW–1.5 MW	to 31 March 2016		6.54
Wind	>1.5 MW–5 MW	to 31 March 2016		2.77

ESSENTIAL READING FROM TAYLOR AND FRANCIS

NRM1 Cost Management Handbook

David P Benge

The definitive guide to measurement and estimating using NRM1, written by the author of NRM1

The 'RICS New rules of measurement: Order of cost estimating and cost planning of capital building works' (referred to as NRM1) is the cornerstone of good cost management of capital building works projects - enabling more effective and accurate cost advice to be given to clients and other project team members, while facilitating better cost control.

The NRM1 Cost Management Handbook is the essential guide to how to successfully interpret and apply these rules, including explanations of how to:

- quantify building works and prepare order of cost estimates and cost plans
- use the rules as a toolkit for risk management and procurement
- analyse actual costs for the purpose of collecting benchmark data and preparing cost analyses
- capture historical cost data for future order of cost estimates and elemental cost plans
- employ the rules to aid communication
- manage the complete 'cost management cycle'
- use the elemental breakdown and cost structures, together with the coding system developed for NRM1, to effectively integrate cost management with Building Information Modelling (BIM).

March 2014: 246 x 174: 640pp
Pb: 978-0-415-72077-9: £41.99

To Order: Tel: +44 (0) 1235 400524 Fax: +44 (0) 1235 400525
or Post: Taylor and Francis Customer Services,
Bookpoint Ltd, Unit T1, 200 Milton Park, Abingdon, Oxon, OX14 4TA UK
Email: book.orders@tandf.co.uk

For a complete listing of all our titles visit:
www.tandf.co.uk

Taylor & Francis
Taylor & Francis Group

Carbon Trading

What is carbon trading?

Carbon trading is currently the central pillar of the Kyoto Protocol and other international agreements aimed at slowing climate change.

Emissions' trading (also known as cap and trade) is a market-based approach used to control pollution by providing economic incentives for achieving reductions in the emissions of pollutants. Carbon emissions' trading is a form of emissions trading that specifically targets carbon dioxide.

A central authority (usually a governmental body) sets a limit or cap on the amount of a pollutant that can be emitted. The limit or cap is allocated or sold to firms in the form of emissions permits which represent the right to emit or discharge a specific volume of the specified pollutant. Firms are required to hold a number of permits (or credits) equivalent to their emission levels. The total number of permits cannot exceed the cap, limiting total emissions to that level. Firms that need to increase their emission permits must buy permits from those who require fewer permits. The transfer of permits is referred to as a trade. In effect, the buyer is paying a charge for polluting, while the seller is being rewarded for having reduced emissions. Thus, in theory, those who can reduce emissions most cheaply do so, achieving the pollution reduction at the lowest cost to society.

Carbon trading in the UK

New rules set out by the government will pitch some of the UK's largest organizations against each other in a drive to cut carbon dioxide (CO_2) emissions.

For the first time large non-energy intensive organizations, which account for about 10% of UK CO_2 emissions, will be legally bound to closely monitor and report their emissions from energy use in preparation for carbon trading.

The Government scheme, known as the CRC Energy Efficiency Scheme, will include businesses will be ranked according to reductions in energy use and improvements in energy efficiency alongside public sector organizations such as NHS trusts, local authorities and government departments.

Analysis for the Environment Agency suggests that the scheme could reduce CO_2 emissions by up to 11.6 million tonnes per year by 2020, the equivalent of taking four million cars off the road. It is also expected to save organizations money through reduced energy bills, benefiting the economy by at least one billion pounds by 2020.

During the introductory phase in 2012, allowances are being sold at a fixed price of £12 per tonne of CO_2. A further 15,000 organizations that use less than 6,000 MWh, but still have at least one half hourly electricity meter, will be obliged to register and declare their electricity use.

How does the EU Emission Trading System (EU ETS) work?

Each EU member state must develop a National Allocation Plan (NAP) approved by the European Commission. This sets an overall cap on the total emissions allowed from all the installations covered by the system. This is converted into allowances (one allowance equals one tonne of CO_2), which are then distributed by EU member states to installations covered by the system.

At the end of each year, installations are required to surrender allowances to account for their actual emissions. They may use all or part of their allocation. Installations can emit more than their allocation by buying allowances from the market. Similarly, an installation that emits less than its allocation can sell its surplus allowances. The environmental outcome is not affected because the amount of allowances allocated is fixed.

CRC scheme

The Carbon Reduction Commitment Energy Efficiency Scheme (CRC) was introduced in 2010 following years of design and many rounds of consultation. Registration and the making of Information Disclosures took considerable effort for many organizations and most hoped that things would settle down from 30 September 2010. But before a month was out the Comprehensive Spending Review heralded further change, which was consulted on over a 4 week period in the run up to Christmas 2010. And in January 2011 a further five Government discussions papers were issued in an attempt to engage business in further simplification measures.

Qualification for the scheme is based on electricity supply across organizations and groups of undertakings, rather than at an individual site basis. Organizations that are a single entity, that is, they are not part of a group, will need to establish qualification for CRC across the sites that are part of that organization.

Organizations that are part of a group will be grouped under the highest parent undertaking.

Organizations qualified as participants for Phase 1 if, during the 2008 calendar year, they had at least one half-hourly electricity meter (HHM), settled on the half hourly market and if they consumed at least 6,000 MWh (megawatt hours) through all half-hourly meters.

If an organization consumed less than 6,000 MWh of electricity, they did not need to register for Phase 1.

A megawatt hour (MWh) is an amount of electricity. It does not mean watts per hour. A megawatt hour is the equivalent of a million-watt device being used for a period of one hour.

To put the CRC energy use into perspective, 100 computers running constantly 24 hours a day, for 273 years would equal 6,000 MWh. There are a 1,000 kilowatt hours (kWh) in a megawatt hour.

Organizations that qualified for Phase 1 may not for Phase 2. This will be especially important for organizations that were only just above the Introductory Phase threshold of 6000 MWh electricity usage in 2008, and so qualify, but have made progress in energy reduction projects since then. If an organization consumes more than 6,000 MWh during the Phase 2 qualification year (1 April 2012 to 31 March 2013) then they will qualify for the scheme. This same principal applies to all planned future phases.

Phase 1 registration and information disclosures are closed. The Environment Agency is currently contacting those organizations that it thinks are responsible for half-hourly electricity meters that were not declared in those registrations and information disclosures.

Participants in Phase 1 (those who made full registrations) should be tracking and recording ALL their energy consumptions from 1 April 2010.

Participants should also be conducting internal audits, keeping records of them and be prepared for an Environment Agency audit. There is planned 20–25% coverage of random audits each year.

Year 1 (Qualification year) full energy reports needed to be submitted summer 2011 – Participants in Phase 1 needed to report their full year 1 energy consumptions (1 April 2010 to 31 March 2011) in the period 1 April 2011 to 31 July 2011. The report had to include ALL energy used, not just the electricity through the half hourly meters that were declared during registration. CO_2 credits are not required and do not need to be paid for in the qualification period. At the end of the 2nd year (registration period), CO_2 credits will be purchased based on the 2nd year's carbon emissions.

Participants in Phase 1 have to buy their first tranche of emissions allowances in summer 2012 and they must buy enough to cover their actual emissions in year 2 (1 April 2011 to 31 March 2012). In other words they will be buying in arrears. Prices will be fixed at £12/tonne, equivalent to approximately 10% of the total cost of the energy itself.

There is no revenue recycling so CRC is now a flat tax at £12/tonne of CO_2 or roughly 10% of the energy bill itself.

The league table will still be published, based on the same performance metrics as before – absolute emissions, relative emissions and early action metrics – with the early action metrics tapering away over the first three years.

CRC Phase 1 has been extended and now runs until 31 March 2014, in other words it has four reporting years rather than the previous three.

The Phase 2 qualification year is 1 April 2012 to 31 March 2013. Phase 2 registration occurs in the period 2013/2014. The first reporting year in Phase 2 will be 1 April 2013 to 31 March 2014 and the fifth reporting year of Phase 2 will end 31 March 2018. The third phase will begin on 1 April 2018.

Simplifying the CRC energy efficiency scheme

In 2011, the Government set out its proposals for the future of the scheme. Responding to feedback received, the government proposed a number of simplifications to the CRC scheme which, subject to further analysis and formal consultation, will be implemented from Phase 2 onwards (where registration for Phase 2 begins in April 2013).

Amongst other simplifications the Government is proposing to:

- **Reduce the number of fuels covered by the scheme**
  Under the current scheme, businesses have to report on the emissions from 29 different fuels and we propose reducing this to four: electricity, gas, kerosene and diesel for heating. This will cover over approximately 95% of emissions captured under the CRC come while significantly reducing the administration burden of the scheme.
- **Move to fixed price allowance sales**
  Instead of establishing an emissions cap and holding annual auctions as proposed by the previous Administration we propose from the start of phase 2 in 2014 there should be two sales per year where the price of allowances is fixed. This would remove the need for businesses to come up with auctioning strategies and give price certainty to help investment decisions. A lower price in the first sale will incentivize good energy management and reward those who successfully forecast energy use.
- **Simplify the organizational rules**
  Abolish the need for large businesses to participate in groups which do not reflect their natural structure.
- **Make qualification processes easier**
  To make qualification a one step process instead of two. Previously businesses had to firstly prove they had a qualifying electricity meter and then declare they used a particular amount of electricity. This would be abolished in favor of participants just having to prove they use a certain amount of electricity from the qualifying meter.
- **Reducing overlap with other schemes**
  Any CCA or EU ETS site would be automatically exempt from the CRC scheme.

Carbon floor

What is a carbon floor price and why is it needed?

Creating a carbon floor price in the UK essentially requires our industries to pay a top up if the market price for carbon falls below a certain level.

A carbon floor price is a regulatory/taxation policy that states that polluters must pay a minimum amount of money for the right to pollute. This is likely to take the form of a tax that requires those who qualify to make a payment to the Treasury. It is expected to replace the existing Climate Change Levy, which is a downstream tax on energy use rather than a direct upstream tax on greenhouse gas pollution.

Roughly half of Europe's emissions are covered by a European regulation that caps emissions (the EU Emissions Trading Scheme) requiring them to submit sufficient permits to cover their emissions. Permits, known as allowances, can be freely traded and the price someone is willing to pay to acquire them determines the price of pollution. At present because there are too many allowances available in the market compared to the demand prices are relatively low – at around €15 per tonne of carbon dioxide equivalent. This low price is not necessarily enough to dissuade polluters from continuing to emit and does not provide an attractive enough return for would-be investors in low carbon solutions. There is also the risk that it could fall even lower. This lack of price certainty is seen as a potentially important barrier to investment.

A carbon floor price is therefore primarily designed to attract low carbon investment into a country by making the price of pollution higher and increasing the rewards for low carbon projects. As explained below it is not in and of itself an environmental policy and in terms of value for money it must be assessed as an industrial policy.

The UK could decide to impose an immediately effective floor price, which takes the price of pollution all the way up to this projected level of cost. This would however have a very significant impact on the competitiveness of UK industry relative to competitors in the rest of Europe. It is much more likely that the floor price will initially be set either at today's market levels (around £13–14 per tonne) or slightly above, incorporating the cost of the Climate Change Levy which is currently equivalent to around £4–6 per tonne. A price escalator may also be built-in to steadily increase the floor over time.

Renewable Obligation Certificates (ROCs)

The Renewable Obligation (RO) is the main support scheme for renewable electricity projects in the UK, placing an obligation on licensed UK suppliers of electricity to increase their proportion of electricity production from renewable sources, or result in a penalty.

Since its introduction in 2002, it has succeeded in tripling the level of renewable electricity in the UK from 1.8% to 5.4% and is currently worth around £1 billion/year in support to the renewable electricity industry. The RO applies to all powered plant with a power capacity greater than 5 MW. The target started at 3% and is presently at 11.4% rising incrementally to 15.4% by 2015. It is likely to be extended to 20% by 2020.

In April 2009, the introduction of banding under the Renewables Obligation Order 2009 meant different technologies receive different levels of support, providing a greater incentive to those that are further from the market.

The RO was extended from its current end date of 2027 to 2037, in April 2010, for new projects with a view to providing greater long-term certainty for investors and an increase in support for offshore wind projects.

The RO is administered by the Office of the Gas and Electricity Markets (Ofgem) and suppliers of electricity have to prove they have met this obligation, producing Renewable Obligation Certificates (ROCs) to renewable electricity generators at the end of each year.

A Renewable Obligation Certificate (ROC) is a green certificate that is issued by Ofgem to an accredited generator for eligible renewable electricity generated within the UK and supplied to customers in the UK by a licensed supplier. A ROC is issued for each megawatt hour (MWh) of eligible renewable output generated.

Failing to meet the obligation results in 'buy-out' fines being paid to Ofgem on the shortfall of every MWh sold that was not renewable. Ofgem then distributes the funds to all electricity supply companies possessing ROCs, the amount received being in proportion to the number of ROCs held (at the end of the year). If a supplier meets part or all of its RO, but other companies do not, the supplier who has ROCs will be rewarded with a share of the fines.

Previously, 1 ROC was issued for each megawatt hour (MWh) of eligible generation, regardless of technology. Since April 2009, the reforms introduced means that new generators joining the RO now receive different numbers of ROCs, depending on their costs and potential for large-scale deployment. For example, onshore wind continues to receive 1 ROC/MWh, whereas offshore wind and energy crops currently receive 2 ROCs/MWh.

Obligation periods are valid for a year, beginning on the 1 April to 31 March. Supply companies have until the 31 September following the period to submit sufficient ROCs to cover their obligation, or submit sufficient payment to Ofgem to cover their shortfall.

Buy-out price

Suppliers can meet all, or part of their obligations by making a buy-out payment. The buy-out price set by Ofgem for the compliance period of 2015–2016 is £44.33 per Renewables Obligation Certificate (ROC). The buy-out price sets the rate which suppliers must pay if they fail to meet their obligations under the scheme and is adjusted annually in accordance with the Retail Prices Index (RPI).

Buy-out fund redistribution

At the end of the year, the funds made to Ofgem are distributed to all the electricity suppliers possessing ROCs, with the amount received in proportion to the number of ROCs held. If a supplier meets all or part of its RO, it will be rewarded with a share of the buy-out fines.

Pricing

Due to ROCs having the potential to save the supplier from having to commit to a buy-out payment, it increases the price of the electricity. When the renewable generator sells the electricity to a supplier it is not uncommon for the ROC to be sold in addition ultimately forcing the cost of electricity upwards. Also, due to the fact that ROCs entitles suppliers to a share of the 'buy-out' fund at the end of year, increases its value. Electrical suppliers can benefit financially by participating in the RO system due to the renewable targets set by the Government likely to be under-fulfilled and the fact that the RO is not over-subscribed will result in the ROCs and their recycled values being worth more than the £42.02 per MWh.

e-ROC

The most efficient method of buying and selling Renewable Obligation Certificates (ROCs) is through the e-ROC online auctions. They offer renewable generators access to the whole supplier market in the UK, delivering high ROC prices for low fees. The average price of ROCs sold through the auctions 22 May 2015 was £42.44 and with the fees set at only 60p per ROC (subject to a minimum fee of £50), indicates a profitable return for those in participation. Auctions are operated by NFPAS, a subsidiary of the Non-Fossil Purchasing Agency Limited (NFPA) and are usually held four times a year. NFPAS runs regular e-ROC online auctions for the sale of Renewable Obligation Certificates (ROCs).

Eligibility

The reforms stated in Renewable Obligations Order 2009 introduced the concept of 'banding' for the Renewable Obligation Certificates (ROCs). The aim of ROC banding is to establish the number of ROCs per MWh that can be obtained according to the type of technology that is used to generate the renewable electricity.

There are 28 renewable technologies covered by ROCs Banding, resulting in an increasingly complex regulatory environment for technology providers, project developers and finance providers to navigate.

Band	Renewable Technology	Level of Banding (ROCs/MWh)
Established 1	Landfill Gas	0.25
Established 2	Sewage Gas Co-Firing of Non-Energy Crops (Regular) Biomass	0.5
Reference	Onshore Wind Hydro-Electric Co-Firing of Energy Crops Co-Firing of Biomass with CHP Energy from Waste with CHP Geo Pressure Pre-Banded Gasification Pre-Banded Pyrolysis Standard Gasification Standard Pyrolysis	1
Post-Demonstration	Offshore Wind Dedicated Regular Biomass Co-Firing of Energy Crops with CHP	1.5

Band	Renewable Technology	Level of Banding (ROCs/MWh)
Emerging	Wave Tidal Steam Advanced Gasification Advanced Pyrolysis Anaerobic Digestion Dedicated Energy Crops Dedicated Energy Crops with CHP Dedicated Regular Biomass with HP Solar Photovoltaic Geothermal Tidal Lagoons Tidal Barrages	2

ESSENTIAL READING FROM TAYLOR AND FRANCIS

Estimator's Pocket Book

Duncan Cartlidge

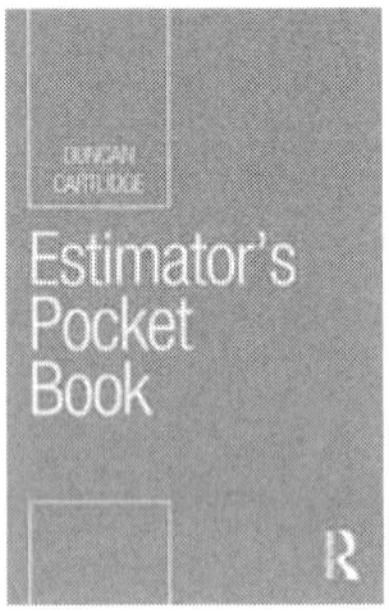

The Estimator's Pocket Book is a concise and practical reference covering the main pricing approaches, as well as useful information such as how to process sub-contractor quotations, tender settlement and adjudication. It is fully up-to-date with NRM2 throughout, features a look ahead to NRM3 and describes the implications of BIM for estimators.

It includes instructions on how to handle:

- the NRM order of cost estimate;
- unit-rate pricing for different trades;
- pro rata pricing and dayworks
- builders' quantities;
- approximate quantities.

Worked examples show how each of these techniques should be carried out in clear, easy-to-follow steps. This is the indispensible estimating reference for all quantity surveyors, cost managers, project managers and anybody else with estimating responsibilities. Particular attention is given to NRM2, but the overall focus is on the core estimating skills needed in practice.

May 2013 186x123: 310pp
Pb: 978-0-415-52711-8: £21.99

To Order: Tel: +44 (0) 1235 400524 Fax: +44 (0) 1235 400525
or Post: Taylor and Francis Customer Services,
Bookpoint Ltd, Unit T1, 200 Milton Park, Abingdon, Oxon, OX14 4TA UK
Email: book.orders@tandf.co.uk

For a complete listing of all our titles visit:
www.tandf.co.uk

Taylor & Francis
Taylor & Francis Group

RICS Ska Rating

The RICS SKA Rating is an assessment method that focuses on promoting environmental and sustainable good practice in the fit-out and refurbishment of existing buildings. The RICS SKA Rating is now seven years old and has grown from an idea into a credible, widely adopted environmental assessment method with 45,087 users and 470 certified projects on the system.

Existing non-domestic building stock represents 18% of UK carbon emissions and up to 40 fit-outs can take place over a building's life. Fit-out and refurbishment activity represents roughly 10% of UK construction spend and until SKA was launched, it was something of a sustainability blind spot.

SKA provides a set of good practice measures that can be implemented on a project and has been widely adopted by property occupiers in both the office and retail sectors including Bank of China, Yell, BBC, GE, Westpac Banking, Derwent London, Nationwide and Lush.

From a standing start, SKA has become a widely recognized assessment method that is mentioned in the same breath as BREEAM and LEED and is referenced in documents such as the UK Government's Low Carbon Construction Action Plan and the British Council for Offices 'Guide to Fit-out'.

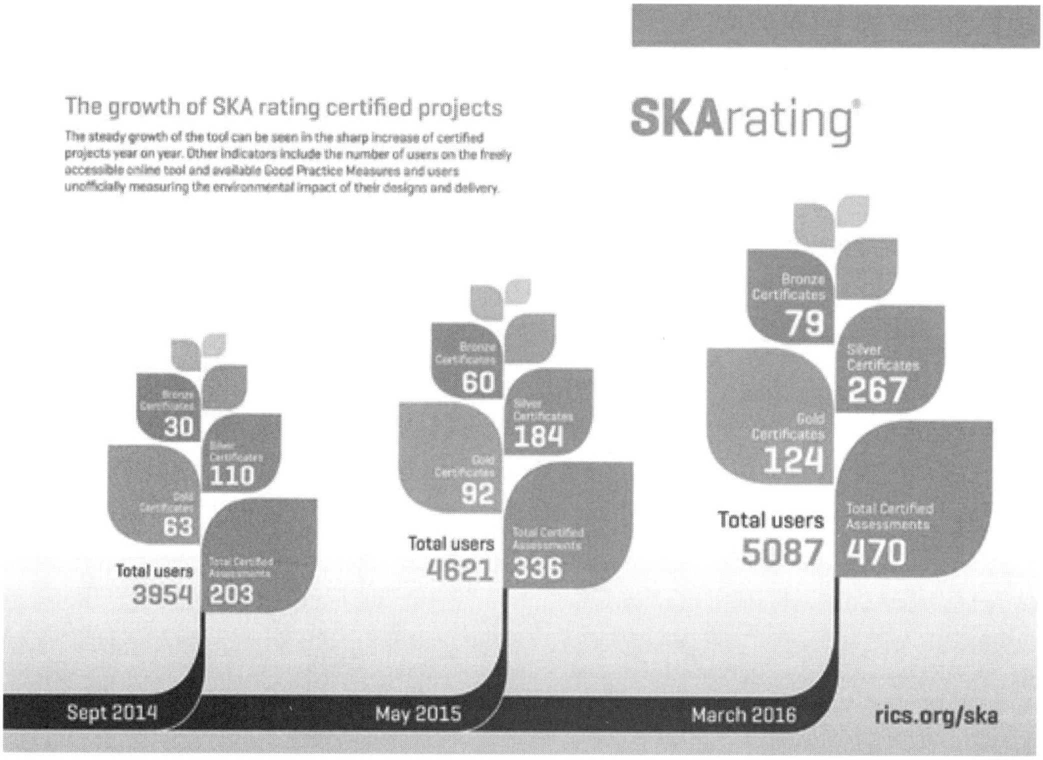

Figure 1: Diagram showing the growth of SKA rating certified projects over last 3 years

The key to SKA's growth is that it has been designed to be much cheaper and simpler to implement than other environmental assessment schemes. SKA is entirely online with the guidance, assessment methodology and even the certification generated through the tool. The aim is to promote good practice in fit-out projects, so the online tool and the associated good practice guidance can be accessed by anyone (ska-tool.rics.org). Also, anyone can train as an assessor and RICS provides a one day Foundation course either undertaken via face-to-face training or via the RICS Online Academy. Then there are separate online modules for the Offices, Retail and new Higher Education Schemes.

There are currently 177 accredited SKA assessors and the numbers are steadily growing. Qualified SKA assessors can assess projects and can generate certificates once they have demonstrated that the project meets the SKA criteria.

SKA covers the following topics and issues:

- Reducing energy and water use by selecting efficient equipment and promoting metering etc.
- Selecting materials with lower environmental impact.
- Reducing construction site impacts, including waste.
- Reducing pollutants such as refrigerant leakage.
- Promoting health and wellbeing in the working environment, including improving internal air quality, daylight, etc.
- Rewarding more sustainable project delivery, such as registration to the Considerate Constructors Scheme and seasonal commissioning.

These issues are addressed by providing a long list of good practice measures that are rewarded if the project demonstrates implementation. For example, SKA aims to reduce energy use in the completed fit-out by promoting the use of energy efficient equipment.

The installation of more efficient equipment means that the fit-out can use less energy, providing it is well managed and the use (or the intensity of use) has not changed. A British Council of Shopping Centres' report on Low Carbon Fit-out (Cutting Carbon, Cutting Costs: Achieving Performance in Retail fit-outs 2013) includes a case study that demonstrates the potential savings. Nationwide moved its premises in Oxford from one location to another on the same street (Queen Street) in Oxford. The fit-out was specifically designed to be low carbon and achieved a SKA Gold rating.

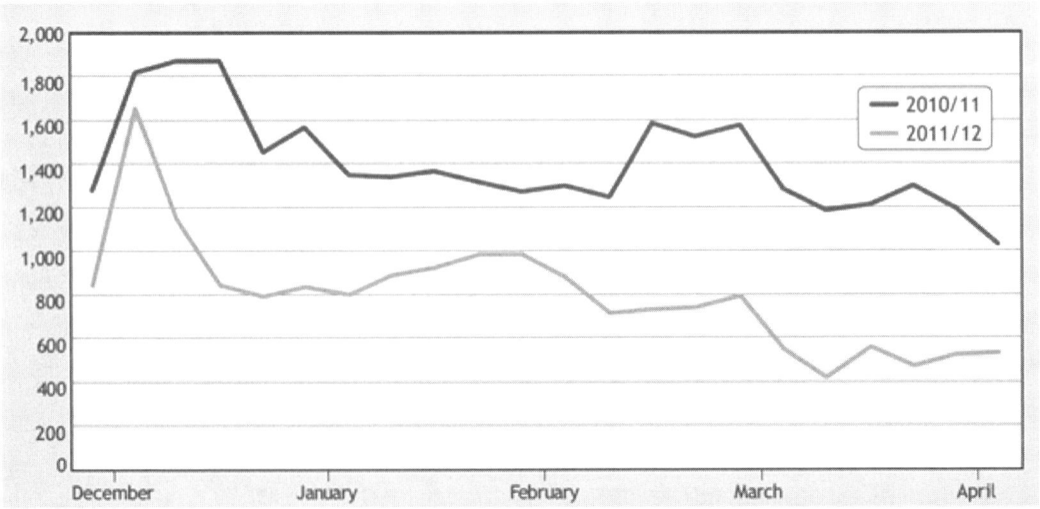

Figure 2: Comparison of energy use before and after fit-out of a Nationwide branch

Fit-out projects are typically of a short duration and vary widely in scope from a quick refresh to a minor refurbishment or a comprehensive fit-out programme. Other environmental assessment tools tend to employ a 'one size fits all' approach and they assess the whole building or rely upon the basebuild to complete the assessment. SKA has a flexible scope as it assesses the project rather than the whole building. This means that the project team can get rewarded for implementing a good practice fit-out, even if the building has poor transport links or if it has inefficient central air conditioning plant.

SKA broke new ground when the offices scheme was launched in 2009 and again in 2011 when the retail version was launched. The aim is to increase the adoption of SKA, particularly in retail and to continue to push out sustainable good practice into other sectors.

Higher education

RICS have been working with the Association of Directors of Estates (AUDE) to develop SKA Higher Education. The scheme is based on SKA rating methodology and has been evolved to reflect the specific requirements of a Higher Education estate including:

- lecture theatres
- laboratories
- and other specialist educational uses

Over 30 universities have been using the SKA online tool for fit-outs of their office areas. So the launch of the scheme will allow other parts of the estate to be formally assessed and certified.

Product labelling

Product selection is a key part of SKA. This includes everything from procuring energy efficient technologies through to using furniture that have low environmental impacts during manufacture and high levels of recycled material. Grigoriou Interiors, a SKA development partner, has now launched a label and directory for products that are proven to be SKA-compliant. This will allow designers to find and specify products much more quickly and it will make the assessment process even simpler than it is now. For more information see: http://www.rics.org/uk/knowledge/ska-rating-/ska-rating-product-compliance-label/.

Conclusion

SKA has been a great success story to date and has helped to promote sustainable good practice in fit-out and refurbishment projects. It can provide tangible and reputational benefits to occupiers and other property stakeholders and the ultimate aim of SKA is to get it widely adopted to the point that it is embedded in all projects as part of the established process.

ESSENTIAL READING FROM TAYLOR AND FRANCIS

Building Services Handbook, 8th Edition

GREENO & HALL

This eighth edition of Hall and Greeno's leading textbook has been reviewed and updated in relation to the latest building and water regulations, new technology, and new legislation, and even more design calculations have been added. In addition, topics such as: alternative sources of natural energy, solar, ground source, heat pumps, renewable energy sources, geothermal methods, and wind power, are now covered.

Building Services Handbook summarises the application of all common elements of building services practice, technique and procedure, to provide an essential information resource for students as well as practitioners working in building services, building management and the facilities administration and maintenance sectors of the construction industry. Information is presented in the highly illustrated and accessible style of the best-selling companion title Building Construction Handbook.

The comprehensive reference for all construction and building services students, Building Services Handbook is ideal for a wide range of courses including NVQ and BTEC National through Higher National Certificate and Diploma to Foundation and three-year Degree level. The clear illustrations and complementary references to industry Standards combine essential guidance with a resource base for further reading and development of specific topics.

September 2015: 234 x 156 mm: 768 pp
Pb: 978-1-138-80563-7; £32.99

To Order: Tel: +44 (0) 1235 400524 Fax: +44 (0) 1235 400525
or Post: Taylor and Francis Customer Services,
Bookpoint Ltd, Unit T1, 200 Milton Park, Abingdon, Oxon, OX14 4TA UK
Email: book.orders@tandf.co.uk

For a complete listing of all our titles visit:
www.tandf.co.uk

Taylor & Francis
Taylor & Francis Group

The Aggregates Levy

The Aggregates Levy came into operation on 1 April 2002 in the UK, except for Northern Ireland where it has been phased in over five years from 2003.

It was introduced to ensure that the external costs associated with the exploitation of aggregates are reflected in the price of aggregate, and to encourage the use of recycled aggregate. There continues to be strong evidence that the levy is achieving its environmental objectives, with sales of primary aggregate down and production of recycled aggregate up. The Government expects that the rates of the levy will at least keep pace with inflation over time, although it accepts that the levy is still bedding in.

The rate of the levy remains at £2.00 per tonne from 1 April 2014 and is levied on anyone considered to be responsible for commercially exploiting 'virgin' aggregates in the UK and should naturally be passed by price increase to the ultimate user.

All materials falling within the definition of 'Aggregates' are subject to the levy unless specifically exempted.

It does not apply to clay, soil, vegetable or other organic matter.

The intention is that it will:

- Encourage the use of alternative materials that would otherwise be disposed of to landfill sites
- Promote development of new recycling processes, such as using waste tyres and glass
- Promote greater efficiency in the use of virgin aggregates
- Reduce noise and vibration, dust and other emissions to air, visual intrusion, loss of amenity and damage to wildlife habitats

Definitions

'Aggregates' means any rock, gravel or sand which is extracted or dredged in the UK for aggregates use. It includes whatever substances are for the time being incorporated in it or naturally occur mixed with it.

'Exploitation' is defined as involving any one or a combination of any of the following:

- Being removed from its original site
- Becoming subject to a contract or other agreement to supply to any person
- Being used for construction purposes
- Being mixed with any material or substance other than water, except in permitted circumstances

Incidence

It is a tax on primary aggregates production – i.e. 'virgin' aggregates won from a source and used in a location within the UK territorial boundaries (land or sea). The tax is not levied on aggregates which are exported or on aggregates imported from outside the UK territorial boundaries.

It is levied at the point of sale.

Exemption from tax

An 'aggregate' is exempt from the levy if it is:

- Material which has previously been used for construction purposes
- Aggregate that has already been subject to a charge to the Aggregates Levy
- Aggregate which was previously removed from its originating site before the start date of the levy
- Aggregate which is being returned to the land from which it was won
- Aggregate won from a farm land or forest where used on that farm or forest
- Rock which has not been subjected to an industrial crushing process
- Aggregate won by being removed from the ground on the site of any building or proposed building in the course of excavations carried out in connection with the modification or erection of the building and exclusively for the purpose of laying foundations or of laying any pipe or cable
- Aggregate won by being removed from the bed of any river, canal or watercourse or channel in or approach to any port or harbour (natural or artificial), in the course of carrying out any dredging exclusively for the purpose of creating, restoring, improving or maintaining that body of water
- Aggregate won by being removed from the ground along the line of any highway or proposed highway in the course of excavations for improving, maintaining or constructing the highway otherwise than purely to extract the aggregate
- Drill cuttings from petroleum operations on land and on the seabed
- Aggregate resulting from works carried out in exercise of powers under the New Road and Street Works Act 1991, the Roads (Northern Ireland) Order 1993 or the Street Works (Northern Ireland) Order 1995
- Aggregate removed for the purpose of cutting of rock to produce dimension stone, or the production of lime or cement from limestone.
- Aggregate arising as a waste material during the processing of the following industrial minerals:
    - ball clay
    - barytes
    - calcite
    - china clay
    - coal, lignite, slate or shale
    - feldspar
    - flint
    - fluorspar
    - fuller's earth
    - gems and semi-precious stones
    - gypsum
    - any metal or the ore of any metal
    - muscovite
    - perlite
    - potash
    - pumice
    - rock phosphates
    - sodium chloride
    - talc
    - vermiculite

However, the levy is still chargeable on any aggregates arising as the spoil or waste from or the by-products of the above exempt processes. This includes quarry overburden.

Anything that consists 'wholly or mainly' of the following is exempt from the levy (note that 'wholly' is defined as 100% but 'mainly' as more than 50%, thus exempting any contained aggregates amounting to less than 50% of the original volumes:

- clay, soil, vegetable or other organic matter
- coal, slate or shale
- china clay waste and ball clay waste

Relief from the levy either in the form of credit or repayment is obtainable where:

- it is subsequently exported from the UK in the form of aggregate
- it is used in an exempt process
- where it is used in a prescribed industrial or agricultural process
- it is waste aggregate disposed of by dumping or otherwise, e.g. sent to landfill or returned to the originating site

The Aggregates Levy Credit Scheme (ALCS) for Northern Ireland was suspended with effect from 1 December 2010 following a ruling by the European General Court.

A new exemption for aggregate obtained as a by-product of railway, tramway and monorail improvement, maintenance and construction was introduced in 2007.

Discounts

From 1 July 2005 the standard added water percentage discounts listed below can be used. Alternatively a more exact percentage can be agreed and this must be done for dust dampening of aggregates.

- washed sand 7%
- washed gravel 3.5%
- washed rock/aggregate 4%

Impact

The British Aggregates Association suggested that the additional cost imposed by quarries is more likely to be in the order of £3.40 per tonne on mainstream products, applying an above average rate on these in order that by-products and low grade waste products can be held at competitive rates, as well as making some allowance for administration and increased finance charges.

With many gravel aggregates costing in the region of £16.00 to £18.00 per tonne, there is a significant impact on construction costs.

Avoidance

An alternative to using new aggregates in filling operations is to crush and screen rubble which may become available during the process of demolition and site clearance as well as removal of obstacles during the excavation processes.

Example: Assuming that the material would be suitable for fill material under buildings or roads, a simple cost comparison would be as follows (note that for the purpose of the exercise, the material is taken to be 1.80 tonne per m³ and the total quantity involved less than 1,000 m³):

Importing fill material:	**£/m³**	**£/tonne**
Cost of 'new' aggregates delivered to site	31.39	17.44
Addition for Aggregates Tax	3.60	2.00
Total cost of importing fill materials	34.99	19.44
Disposing of site material:	**£/m³**	**£/tonne**
Cost of removing materials from site	17.00	9.44
Crushing site materials:	**£/m³**	**£/tonne**
Transportation of material from excavations or demolition to stockpiles	1.50	0.83
Transportation of material from temporary stockpiles to the crushing plant	2.00	1.11
Establishing plant and equipment on site; removing on completion	2.00	1.11
Maintain and operate plant	9.00	5.00
Crushing hard materials on site	13.00	7.22
Screening material on site	2.00	1.11
Total cost of crushing site materials	29.50	16.38

From the above it can be seen that potentially there is a great benefit in crushing site materials for filling rather than importing fill materials.

Setting the cost of crushing against the import price would produce a saving of £5.49 per m³. If the site materials were otherwise intended to be removed from the site, then the cost benefit increases by the saved disposal cost to £22.49 per m³.

Even if there is no call for any or all of the crushed material on site, it ought to be regarded as a useful asset and either sold on in crushed form or else sold with the prospects of crushing elsewhere.

Specimen unit rates	**Unit³**	**£**
Establishing plant and equipment on site; removing on completion		
crushing plant	trip	1,200.00
screening plant	trip	600.00
Maintain and operate plant		
crushing plant	week	7,200.00
screening plant	week	1,800.00
Transportation of material from excavations or demolition places to temporary stockpiles	m³	3.00
Transportation of material from temporary stockpiles to the crushing plant	m³	2.40
Breaking up material on site using impact breakers		
mass concrete	m³	14.00
reinforced concrete	m³	16.00
brickwork	m³	6.00

Crushing material on site		
mass concrete not exceeding 1000 m^3	m^3	13.00
mass concrete 1000–5000 m^3	m^3	12.00
mass concrete over 5000 m^3	m^3	11.00
reinforced concrete not exceeding 1000 m^3	m^3	15.00
reinforced concrete 1000–5000 m^3	m^3	14.00
reinforced concrete over 5000 m^3	m^3	13.00
brickwork not exceeding 1000 m^3	m^3	12.00
brickwork 1000–5000 m^3	m^3	11.00
brickwork over 5000 m^3	m^3	10.00
Screening material on site	m^3	2.00

More detailed information can be found on the HMRC website (www.hmrc.gov.uk) in Notice AGL 1 Aggregates Levy published in May 2009.

ESSENTIAL READING FROM TAYLOR AND FRANCIS

Building Construction Handbook, 11th Edition

CHUDLEY & GREENO

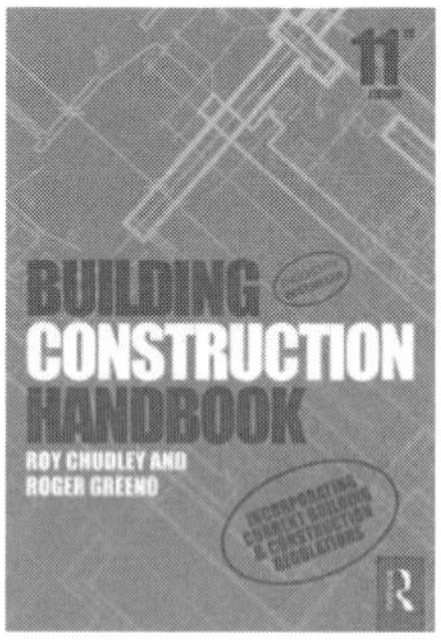

The Building Construction Handbook is THE authoritative reference for all construction students and professionals. Its detailed drawings clearly illustrate the construction of building elements, and have been an invaluable guide for builders since 1988. The principles and processes of construction are explained with the concepts of design included where appropriate. Extensive coverage of building construction practice, techniques, and regulations representing both traditional procedures and modern developments are included to provide the most comprehensive and easy to understand guide to building construction.

This new edition has been updated to reflect recent changes to the building regulations, as well as new material on the latest technologies used in domestic construction.

Building Construction Handbook is the essential, easy-to-use resource for undergraduate and vocational students on a wide range of courses including NVQ and BTEC National, through to Higher National Certificate and Diploma, to Foundation and three-year Degree level. It is also a useful practical reference for building designers, contractors and others engaged in the construction industry.

April 2016: 234 x 156 mm: 1012 pp
Pb: 978-1-138-90709-6: £28.99

To Order: Tel: +44 (0) 1235 400524 Fax: +44 (0) 1235 400525
or Post: Taylor and Francis Customer Services,
Bookpoint Ltd, Unit T1, 200 Milton Park, Abingdon, Oxon, OX14 4TA UK
Email: book.orders@tandf.co.uk

For a complete listing of all our titles visit:
www.tandf.co.uk

Taylor & Francis
Taylor & Francis Group

Value Added Tax

Introduction

Value Added Tax (VAT) is a tax on the consumption of goods and services. The UK adopted VAT when it joined the European Community in 1973. The principal source of European law in relation to VAT is Council Directive 2006/112/EC, a recast of Directive 77/388/EEC which is currently restated and consolidated in the UK through the VAT Act 1994 and various Statutory Instruments, as amended by subsequent Finance Acts.

VAT Notice 708: Buildings and construction (June 2007) gives an interpretation of the law in connection with construction works from the point of view of HM Revenue & Customs. VAT tribunals and court decisions since the date of this publication will affect the application of the law in certain instances. The Notice is available on HM Revenue & Customs website at www.hmrc.gov.uk.

The scope of VAT

VAT is payable on:

- Supplies of goods and services made in the UK
- By a taxable person
- In the course or furtherance of business
- Which are not specifically exempted or zero-rated

Rates of VAT

There are three rates of VAT:

- A standard rate, currently 20% since January 2011
- A reduced rate, currently 5%
- A zero rate

Additionally some supplies are exempt from VAT and others are outside the scope of VAT.

Recovery of VAT

When a taxpayer makes taxable supplies he must account for VAT at the appropriate rate of either 20% or 5%. This VAT then has to be passed to HM Revenue & Customs and will normally be charged to the taxpayer's customers.

As a VAT registered person, the taxpayer can reclaim from HM Revenue & Customs as much of the VAT incurred on their purchases as relates to the standard-rated, reduced-rated and zero-rated onward supplies they make. A person cannot however reclaim VAT that relates to any non-business activities (but see below) or to any exempt supplies they make.

At predetermined intervals the taxpayer will pay to HM Revenue & Customs the excess of VAT collected over the VAT they can reclaim. However if the VAT reclaimed is more than the VAT collected, the taxpayer can reclaim the difference from HM Revenue & Customs.

Example

X Ltd constructs a block of flats. It sells long leases to buyers for a premium. X Ltd has constructed a new building designed as a dwelling and will have granted a long lease. This sale of a long lease is VAT zero-rated. This means any VAT incurred in connection with the development which X Ltd will have paid (e.g. payments for consultants and certain preliminary services) will be reclaimable. For reasons detailed below the contractor employed by X Ltd will not have charged VAT on his construction services.

Use for business and non-business activities

Where VAT relates partly to business use and partly to non-business use then the basic rule is that it must be apportioned so that only the business element is potentially recoverable. In some cases VAT on land, buildings and certain construction services purchased for both business and non-business use could be recovered in full by applying what is known as 'Lennartz' accounting to reclaim VAT relating to the non-business use and account for VAT on the non-business use over a maximum period of 10 years. Following an ECJ case restricting the scope of this approach, its application to immovable property was removed completely in January 2011 when UK law was amended to comply with EU Directive 2009/162/EU.

Taxable persons

A taxable person is an individual, firm, company etc. who is required to be registered for VAT. A person who makes taxable supplies above certain value limits is required to be registered. The current registration limit is £81,000 for 2015–16. The threshold is exceeded if at the end of any month the value of taxable supplies in the period of one year then ending is over the limit, or at any time, if there are reasonable grounds for believing that the value of the taxable supplies in the period of 30 days than beginning will exceed £81,000.

A person who makes taxable supplies below these limits is entitled to be registered on a voluntary basis if they wish, for example in order to recover VAT incurred in relation to those taxable supplies.

In addition, a person who is not registered for VAT in the UK but acquires goods from another EC member state, or make distance sales in the UK, above certain value limits may be required to register for VAT in the UK.

VAT exempt supplies

If a supply is exempt from VAT this means that no tax is payable – but equally the person making the exempt supply cannot normally recover any of the VAT on their own costs relating to that supply.

Generally property transactions such as leasing of land and buildings are exempt unless a landlord chooses to standard-rate its supplies by a process known as opting to tax. This means that VAT is added to rental income and also that VAT incurred, on say, an expensive refurbishment, is recoverable.

Supplies outside the scope of VAT

Supplies are outside the scope of VAT if they are:

- Made by someone who is not a taxable person
- Made outside the UK
- Not made in the course or furtherance of business

In course or furtherance of business

VAT must be accounted for on all taxable supplies made in the course or furtherance of business with the corresponding recovery of VAT on expenditure incurred.

If a taxpayer also carries out non-business activities then VAT incurred in relation to such supplies is generally not recoverable.

In VAT terms, business means any activity continuously performed which is mainly concerned with making supplies for a consideration. This includes:

- Any one carrying on a trade, vocation or profession
- The provision of membership benefits by clubs, associations and similar bodies in return for a subscription or other consideration
- Admission to premises for a charge

It may also include the activities of other bodies including charities and non-profit making organizations.

Examples of non-business activities are:

- Providing free services or information
- Maintaining some museums or particular historic sites
- Publishing religious or political views

Construction services

In general the provision of construction services by a contractor will be VAT standard rated at 20%, however, there are a number of exceptions for construction services provided in relation to certain residential and charitable use buildings.

The supply of building materials is VAT standard rated at 20%, however, where these materials are supplied as part of the construction services the VAT liability of those materials follows that of the construction services supplied.

Zero-rated construction services

The following construction services are VAT zero-rated including the supply of related building materials.

The construction of new dwellings

The supply of services in the course of the construction of a building designed for use as a dwelling or number of dwellings is zero-rated other than the services of an architect, surveyor or any other person acting as a consultant or in a supervisory capacity.

The following conditions must be satisfied in order for the works to qualify for zero-rating:

1. The work must not amount to the conversion, reconstruction or alteration of an existing building;
2. The work must not be an enlargement of, or extension to, an existing building except to the extent that the enlargement or extension creates an additional dwelling or dwellings;
3. The building must be designed as a dwelling or number of dwellings. Each dwelling must consist of self-contained living accommodation with no provision for direct internal access from the dwelling to any other dwelling or part of a dwelling;
4. Statutory planning consent must have been granted for the construction of the dwelling, and construction carried out in accordance with that consent;
5. Separate use or disposal of the dwelling must not be prohibited by the terms of any covenant, statutory planning consent or similar provision.

The construction of a garage at the same time as the dwelling can also be zero-rated as can the demolition of any existing building on the site of the new dwelling.

A building only ceases to be an existing building (see points 1. and 2. above) when it is:

1. Demolished completely to ground level; or when
2. The part remaining above ground level consists of no more than a single facade (or a double facade on a corner site) the retention of which is a condition or requirement of statutory planning consent or similar permission.

The construction of a new building for 'relevant residential or charitable' use

The supply of services in the course of the construction of a building designed for use as a relevant residential or charitable building is zero-rated other than the services of an architect, surveyor or any other person acting as a consultant or in a supervisory capacity.

A 'relevant residential' use building means:

1. A home or other institution providing residential accommodation for children;
2. A home or other institution providing residential accommodation with personal care for persons in need of personal care by reason of old age, disablement, past or present dependence on alcohol or drugs or past or present mental disorder;
3. A hospice;
4. Residential accommodation for students or school pupils;
5. Residential accommodation for members of any of the armed forces;
6. A monastery, nunnery, or similar establishment; or
7. An institution which is the sole or main residence of at least 90% of its residents.

A 'relevant residential' purpose building does not include use as a hospital, a prison or similar institution or as a hotel, inn or similar establishment.

A 'relevant charitable' use means use by a charity:

1. Otherwise than in the course or furtherance of a business; or
2. As a village hall or similarly in providing social or recreational facilities for a local community.

Non-qualifying use which is not expected to exceed 10% of the time the building is normally available for use can be ignored. The calculation of business use can be based on time, floor area or head count subject to approval being acquired from HM Revenue & Customs.

The construction services can only be zero-rated if a certificate is given by the end user to the contractor carrying out the works confirming that the building is to be used for a qualifying purpose i.e. for a 'relevant residential or charitable' purpose. It follows that such services can only be zero-rated when supplied to the end user and, unlike supplies relating to dwellings, supplies by subcontractors cannot be zero-rated.

The construction of an annex used for a 'relevant charitable' purpose

Construction services provided in the course of construction of an annexe for use entirely or partly for a 'relevant charitable' purpose can be zero-rated.

In order to qualify the annexe must:

1. Be capable of functioning independently from the existing building;
2. Have its own main entrance; and
3. Be covered by a qualifying use certificate.

The conversion of a non-residential building into dwellings or the conversion of a building from non-residential use to 'relevant residential' use where the supply is to a 'relevant' housing association

The supply to a 'relevant' housing association in the course of conversion of a non-residential building or non-residential part of a building into:

1. A building or part of a building designed as a dwelling or number of dwellings; or
2. A building or part of a building for use solely for a relevant residential purpose,

of any services related to the conversion other than the services of an architect, surveyor or any person acting as a consultant or in a supervisory capacity are zero-rated.

A 'relevant' housing association is defined as:

1. A private registered provider of social housing;
2. A registered social landlord within the meaning of Part I of the Housing Act 1996 (Welsh registered social landlords);
3. A registered social landlord within the meaning of the Housing (Scotland) Act 2001 (Scottish registered social landlords); or
4. A registered housing association within the meaning of Part II of the Housing (Northern Ireland) Order 1992 (Northern Irish registered housing associations).

If the building is to be used for a 'relevant residential' purpose the housing association should issue a qualifying use certificate to the contractor completing the works.

The construction of a permanent park for residential caravans

The supply in the course of the construction of any civil engineering work 'necessary for' the development of a permanent park for residential caravans of any services related to the construction can be VAT zero-rated. This includes access roads, paths, drainage, sewerage and the installation of mains water, power and gas supplies.

Certain building alterations for 'disabled' persons

Certain goods and services supplied to a 'disabled' person, or a charity making these items and services available to 'disabled' persons can be zero-rated. The recipient of these goods or services needs to give the supplier an appropriate written declaration that they are entitled to benefit from zero rating.

The following services (amongst others) are zero-rated:

1. The installation of specialist lifts and hoists and their repair and maintenance;
2. The construction of ramps, widening doorways or passageways including any preparatory work and making good work;
3. The provision, extension and adaptation of a bathroom, washroom or lavatory; and
4. Emergency alarm call systems.

Approved alterations to protected buildings

The zero rate for approved alterations to protected buildings was withdrawn from 1 October 2012, other than for projects where a contract was entered into or where listed building consent (or equivalent approval for listed places of worship) had been applied for before 21 March 2012.

Provided the application was in place before 21 March 2012, zero rating will continue under the transitional rules until 30 September 2015.

All other projects will be subject to the standard rate of VAT on or after 1 October 2012.

Sale of reconstructed buildings

Since 1 October 2012 a protected building shall not be regarded as substantially reconstructed unless, when the reconstruction is completed, the reconstructed building incorporates no more of the original building than the external walls, together with other external features of architectural or historical interest. Transitional arrangements protect contracts entered into before 21 March 2012 for the first grant of a major interest in the protected building made on or before 20 March 2013.

DIY builders and converters

Private individuals who decide to construct their own home are able to reclaim VAT they pay on goods they use to construct their home by use of a special refund mechanism made by way of an application to HM Revenue & Customs. This also applies to services provided in the conversion of an existing non-residential building to form a new dwelling.

The scheme is meant to ensure that private individuals do not suffer the burden of VAT if they decide to construct their own home.

Charities may also qualify for a refund on the purchase of materials incorporated into a building used for non-business purposes where they provide their own free labour for the construction of a 'relevant charitable' use building.

Reduced-rated construction services

The following construction services are subject to the reduced rate of VAT of 5%, including the supply of related building materials.

A changed number of dwellings conversion

In order to qualify for the 5% rate there must be a different number of 'single household dwellings' within a building than there were before commencement of the conversion works. A 'single household dwelling' is defined as a dwelling that is designed for occupation by a single household.

These conversions can be from 'relevant residential' purpose buildings, non-residential buildings and houses in multiple occupation.

A house in multiple occupation conversion

This relates to construction services provided in the course of converting a 'single household dwelling', a number of 'single household dwellings', a non-residential building or a 'relevant residential' purpose building into a house for multiple occupation such as a bed sit accommodation.

A special residential conversion

A special residential conversion involves the conversion of a 'single household dwelling', a house in multiple occupation or a non-residential building into a 'relevant residential' purpose building such as student accommodation or a care home.

Renovation of derelict dwellings

The provision of renovation services in connection with a dwelling or 'relevant residential' purpose building that has been empty for two or more years prior to the date of commencement of construction works can be carried out at a reduced rate of VAT of 5%.

Installation of energy saving materials

The supply and installation of certain energy saving materials including insulation, draught stripping, central heating and hot water controls and solar panels in a residential building or a building used for a relevant charitable purpose.

Buildings that are used by charities for non-business purposes, and/or as village halls, were removed from the scope of the reduced rate for the supply of energy saving materials under legislation introduced in Finance Bill 2013.

Grant-funded of heating equipment or connection of a gas supply

The grant funded supply and installation of heating appliances, connection of a mains gas supply, supply, installation, maintenance and repair of central heating systems, and supply and installation of renewable source heating systems, to qualifying persons. A qualifying person is someone aged 60 or over or is in receipt of various specified benefits.

Installation of security goods

The grant funded supply and installation of security goods to a qualifying person.

Housing alterations for the elderly

Certain home adaptations that support the needs of elderly people were reduced rated with effect from 1 July 2007.

Building contracts

Design and build contracts

If a contractor provides a design and build service relating to works to which the reduced or zero rate of VAT is applicable then any design costs incurred by the contractor will follow the VAT liability of the principal supply of construction services.

Management contracts

A management contractor acts as a main contractor for VAT purposes and the VAT liability of his services will follow that of the construction services provided. If the management contractor only provides advice without engaging trade contractors his services will be VAT standard rated.

Construction management and project management

The project manager or construction manager is appointed by the client to plan, manage and coordinate a construction project. This will involve establishing competitive bids for all the elements of the work and the appointment of trade contractors. The trade contractors are engaged directly by the client for their services.

The VAT liability of the trade contractors will be determined by the nature of the construction services they provide and the building being constructed.

The fees of the construction manager or project manager will be VAT standard rated. If the construction manager also provides some construction services these works may be zero or reduced rated if the works qualify.

Liquidated and ascertained damages

Liquidated damages are outside of the scope of VAT as compensation. The employer should not reduce the VAT amount due on a payment under a building contract on account of a deduction of damages. In contrast an agreed reduction in the contract price will reduce the VAT amount.

Similarly, in certain circumstances HM Revenue & Customs may agree that a claim by a contractor under a JCT or other form of contract is also compensation payment and outside the scope of VAT.

ESSENTIAL READING FROM TAYLOR AND FRANCIS

Watts Pocket Handbook, 29th Edition

RUSHTON

Back in print for the first time in years, the Watts Pocket Handbook renews its commitment to share industry knowledge by providing technical and legal information across a comprehensive spread of property and construction topics.

Compiled by the Watts Technical Director, the Handbook provides specialist information and guidance on a vast selection of construction related subjects including:

- Contracts and procurement
- Insurance
- Materials and defects
- Environmental and sustainability issues

Watts Pocket Handbook remains the must-have reference book for professionals and students engaged in construction, building surveying, service engineering, property development and much more.

April 2016; 186 x 123 mm; 324 pp
Pb: 978-1-138-66595-8; £19.99

To Order: Tel: +44 (0) 1235 400524 Fax: +44 (0) 1235 400525
or Post: Taylor and Francis Customer Services,
Bookpoint Ltd, Unit T1, 200 Milton Park, Abingdon, Oxon, OX14 4TA UK
Email: book.orders@tandf.co.uk

For a complete listing of all our titles visit:
www.tandf.co.uk

Taylor & Francis
Taylor & Francis Group

Typical Engineering Details

In addition to the Engineering Features, Typical Engineering Details are included. These are indicative schematics to assist in the compilation of costing exercises. The user should note that these are only examples and cannot be construed to reflect the design for each and every situation. They are merely provided to assist the user with gaining an understanding of the Engineering concepts and elements making up such.

ELECTRICAL

- Urban Network Mainly Underground
- Urban Network Mainly Underground with Reinforcement
- Urban Network Mainly Underground with Substation Reinforcement
- Typical Simple 11 kV Network Connection For LV Intakes Up To 1000 kVA
- Typical 11 kV Network Connections For HV Intakes 1000 kVA To 6000 kVA
- Static UPS System – Simplified Single Line Schematic For a Single Module
- Typical Data Transmission (Structured Cabling)
- Typical Networked Lighting Control System
- Typical Standby Power System, Single Line Schematic
- Typical Fire Detection and Alarm Schematic
- Typical Block Diagram – Access Control System (ACS)
- Typical Block Diagram – Intruder Detection System (IDS)
- Typical Block Diagram – Digital CCTV

MECHANICAL

- Fan Coil Unit System
- Displacement System
- Chilled Ceiling System (Passive System)
- Chilled Beam System (Passive System)
- Variable Air Volume (VAV)
- Variable Refrigerant Volume System (VRV)
- Reverse cycle heat pump

Urban network mainly underground

Details: Connection to small housing development 10 houses, 60 m of LV cable from local 11 kV/400 V substation route in footpath and verge, 10 m of service cable to each plot in verge

Supply Capacity: 200 kVA

Connection Voltage: LV

3 phase supply

Breakdown of detailed cost information

	Labour	Plant	Materials	Overheads	Total
Mains Cable	£300	£100	£ 1,200	£320	£1,920
Service Cable	£300	£100	£500	£180	£1,080
Jointing	£1,200	£400	£1,500	£300	£3400
Termination	Incl.	Incl.	Incl.	Incl.	Incl.
Trench/Reinstate	£1,800	£600	£1,500	£780	£4,680
Substation	–	–	–	–	–
HV Trench & Joint	–	–	–	–	–
HV Cable Install	–	–	–	–	–
Total Calculated Price	£3,600	£1,200	£4,700	£1,580	£11,080
Total Non-contestable Elements and Associated Charges					£3,000
Grand Total Calculated Price excl. VAT					£14,080

Urban network mainly underground with reinforcement

Details: Connection to small housing development 10 houses, 60 m of LV cable from local 11 kV/400 V substation route in footpath and verge, 10 m of service cable to each plot in unmade ground. Scheme includes reinforcement of LV distribution board at substation

Supply Capacity: 200 kVA

Connection Voltage: LV

Single phase supply

Breakdown of detailed cost information

	Labour	Plant	Materials	Overheads	Total
Mains Cable	£300	£100	£ 1,200	£320	£1,920
Service Cable	£300	£100	£500	£180	£1,080
Jointing	£1,200	£400	£1,500	£300	£3400
Termination	Incl.	Incl.	Incl.	Incl.	Incl.
Trench/Reinstate	£1,800	£600	£1,500	£780	£4,680
Substation	–	–	–	–	–
HV Trench & Joint	–	–	–	–	–
HV Cable Install	–	–	–	–	–
Total Calculated Price	£3,600	£1,200	£4,700	£1,580	£11,080
Total Non-contestable Elements and Associated Charges					£13,000
Grand Total Calculated Price excl. VAT					£24,080

Urban network mainly underground with substation reinforcement

Details: Connection to small housing development 10 houses, 60 m of LV cable from local 11 kV/400 V substation route in footpath and verge, 10 m of service cable to each plot in verge. Scheme includes reinforcement of LV distribution board and new substation and 20 m of HV cable

Supply Capacity: 200 kVA

Connection Voltage: LV

Single phase

Breakdown of detailed cost information

	Labour	Plant	Materials	Overheads	Total
Mains Cable	£300	£100	£ 1,500	£380	£2,280
Service Cable	£300	£100	£500	£180	£1,080
Jointing	£1,200	£400	£1,500	£300	£3400
Termination	Incl.	Incl.	Incl.	Incl.	Incl.
Trench/Reinstate	£1,800	£600	£1,500	£780	£4,680
Substation	£10,000	£3,000	£80,000	£18,600	£111,600
HV Trench & Joint	£1,800	£600	£1,000	£680	£4,080
HV Cable Install	£300	£100	£1,125	£305	£1,830
Total Calculated Price	£15,700	£4,900	£87,125	£21,225	£128,950
Total Non-contestable Elements and Associated Charges					£15,000
Grand Total Calculated Price excl. VAT					£143,950

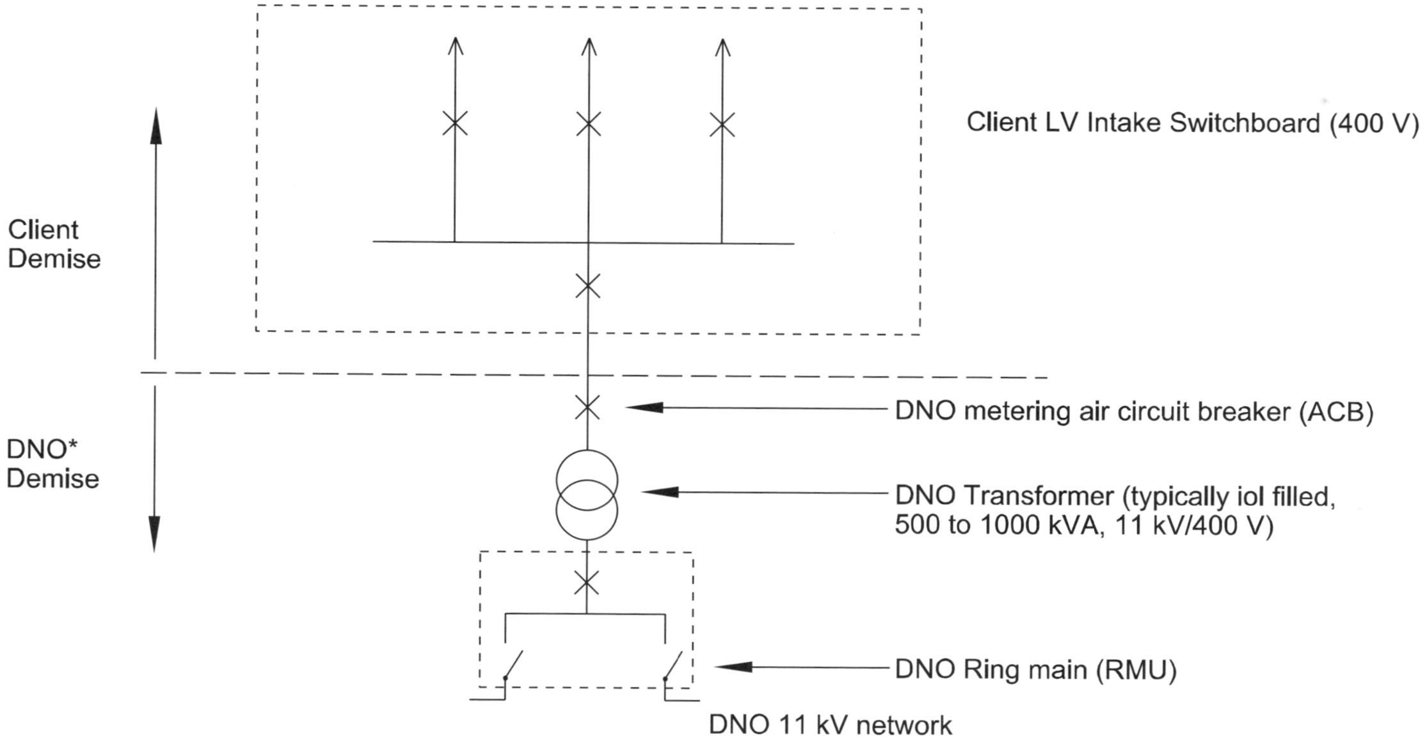

Note: *DNO - Distribution Network Operator

Typical Simple 11 kV Network Connection for LV Intakes up to 1000 kVA

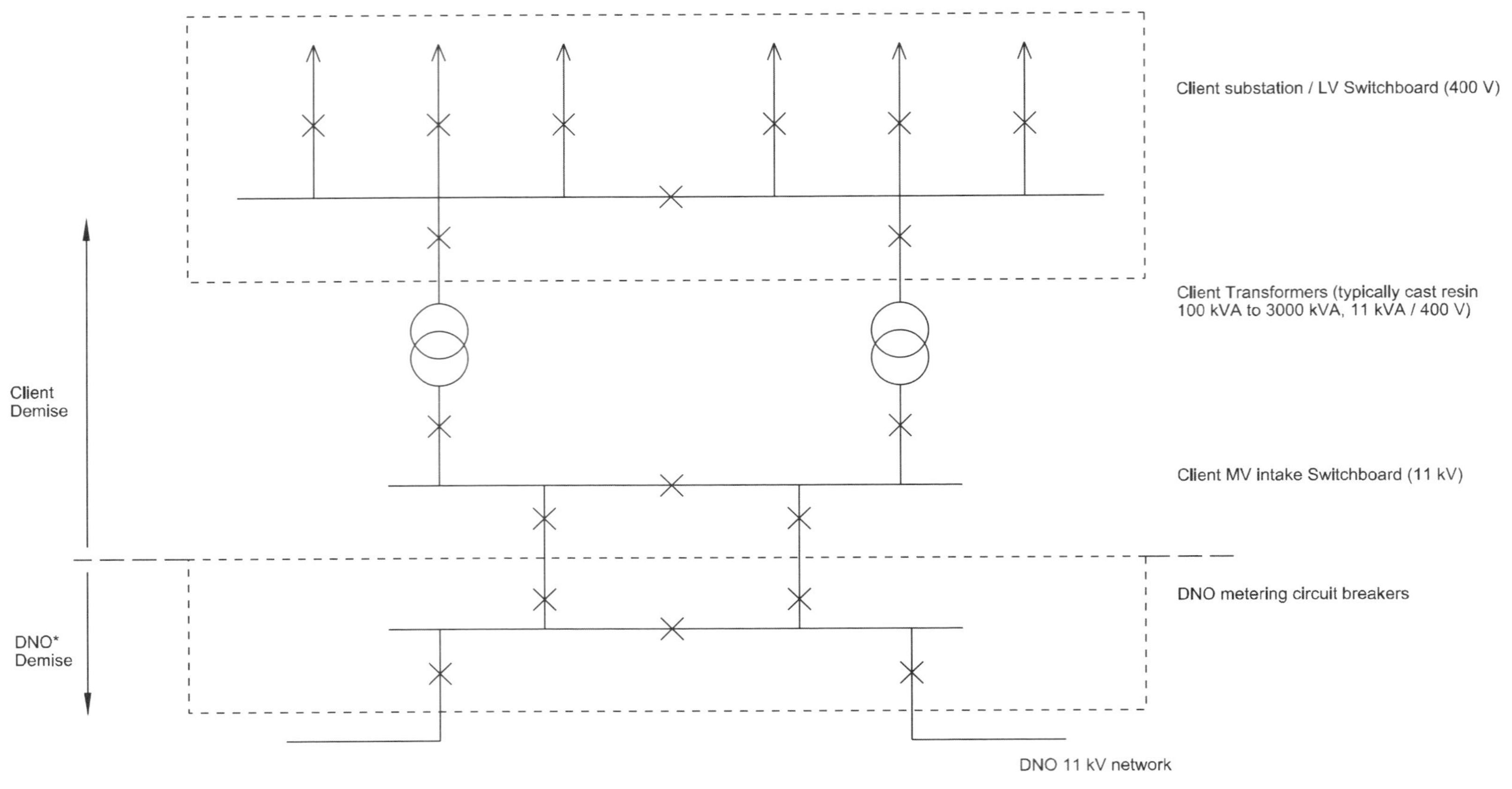

Note: *DNO - Distribution Network Operator

Typical 11 kV Network Connection for MV Intakes 1000 kVA up to 6000 kVA

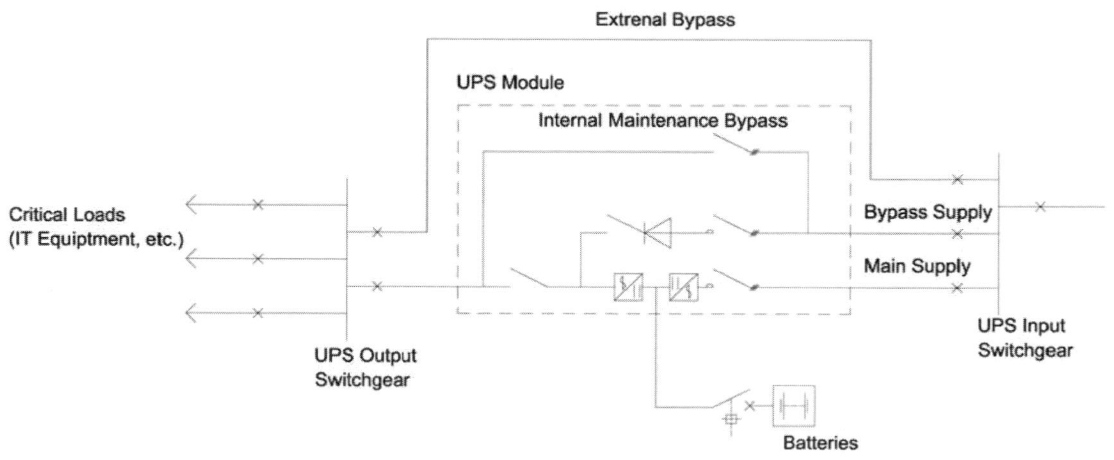

Static UPS System – Simplified Schematic For Single Module

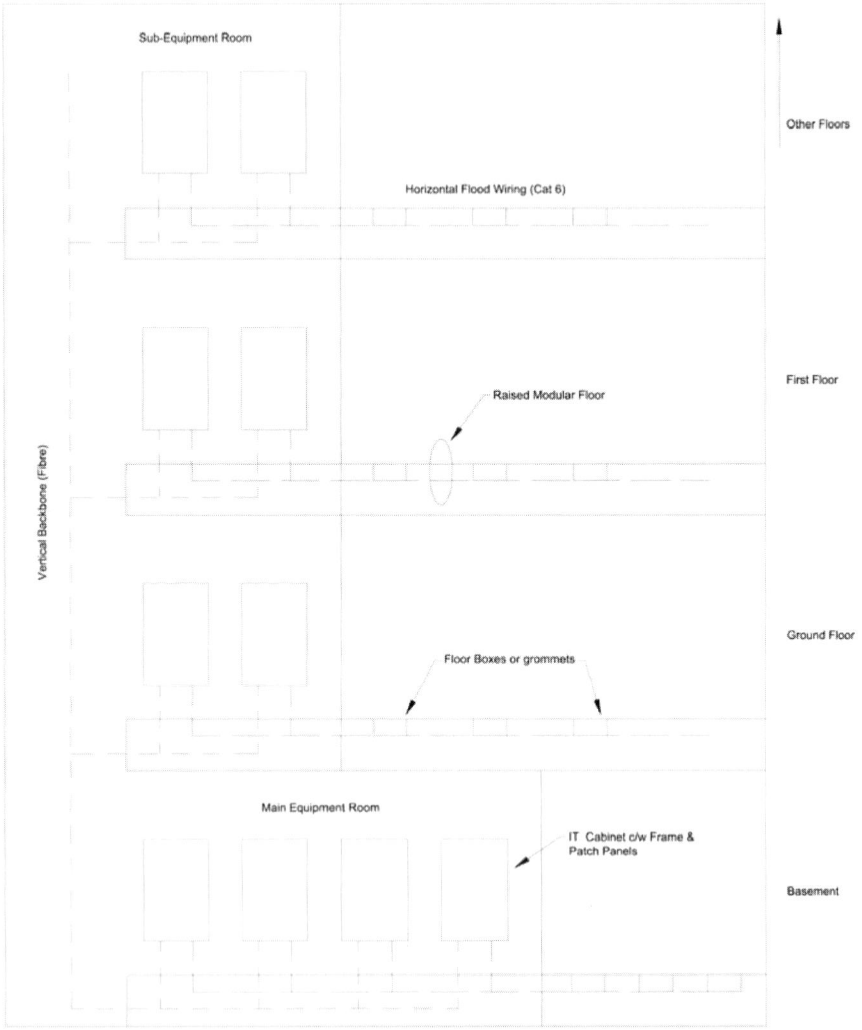

Typical Data Transmission (Structured Cabling)

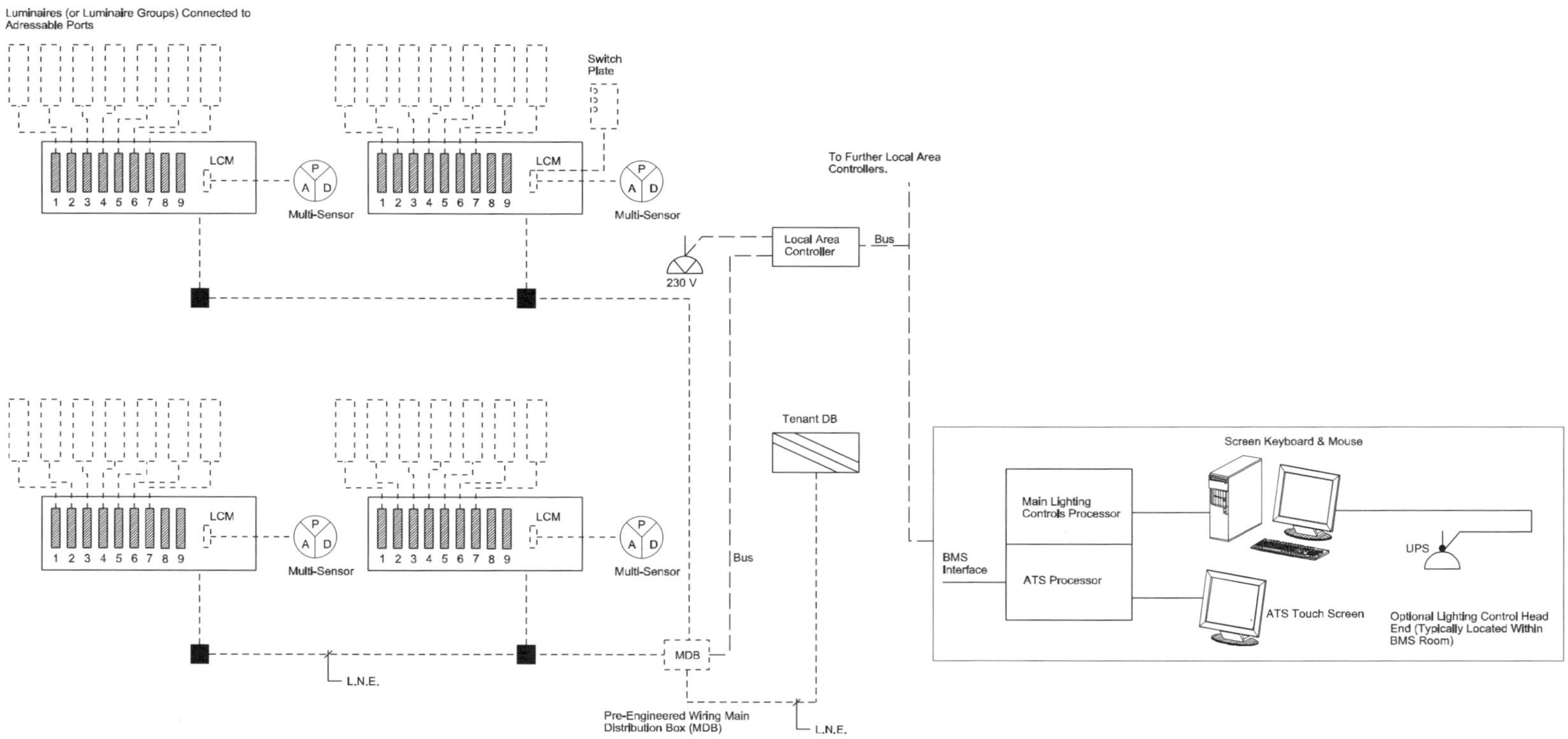

Typical Networked Lighting Control System

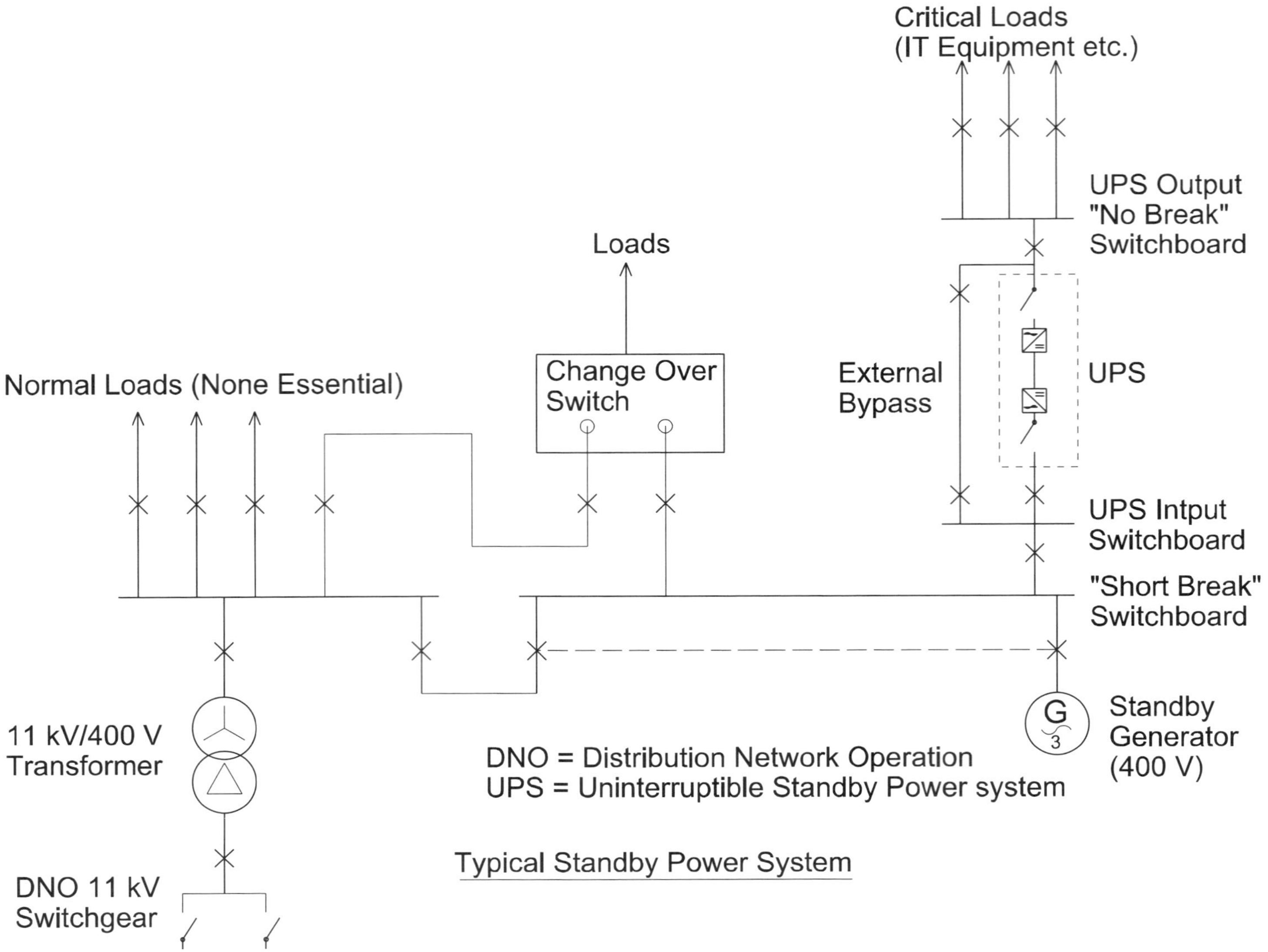

Typical Standby Power System

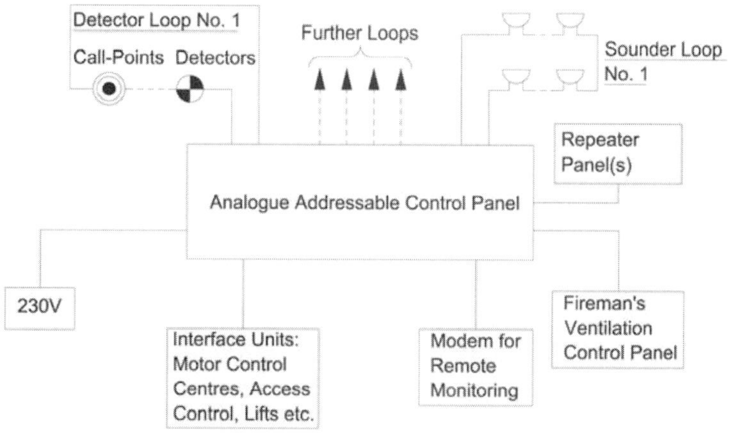

Typical Fire Detection and Alarm Schematic

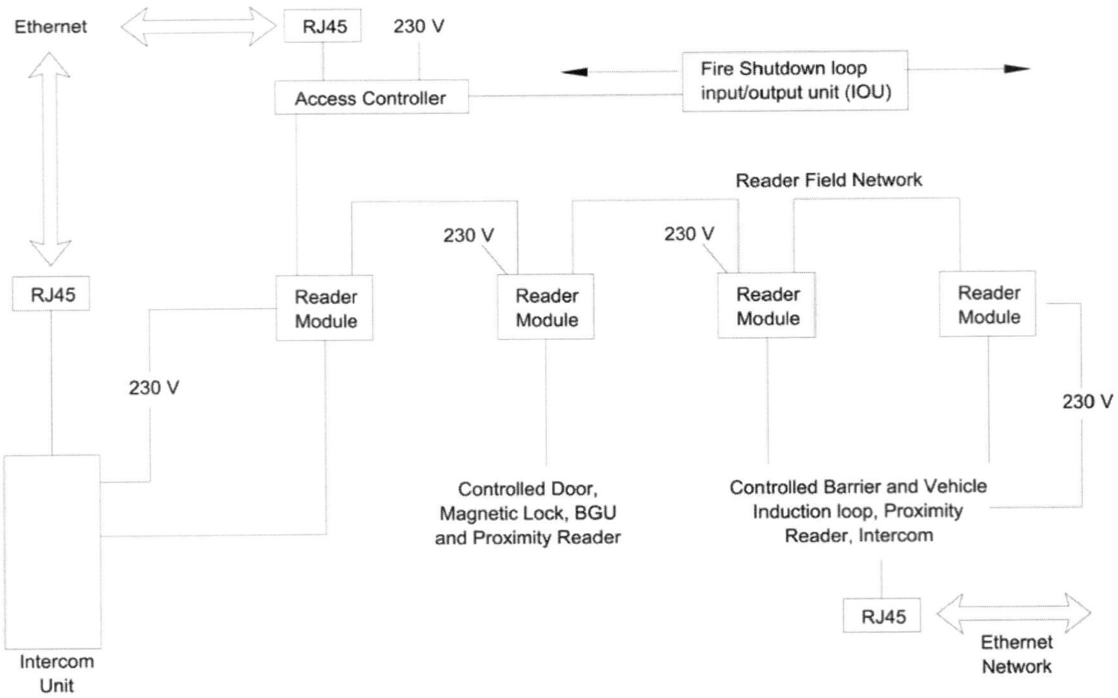

Typical Block Diagram – Access Control System

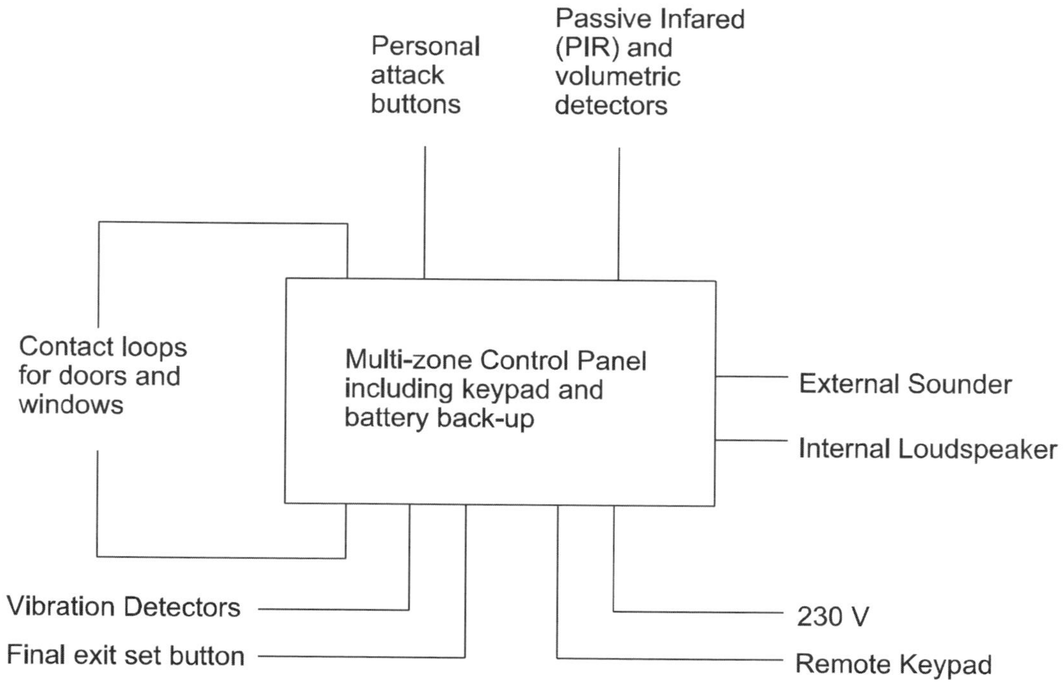

Typical Block Diagram – Intruder Detection System (IDS)

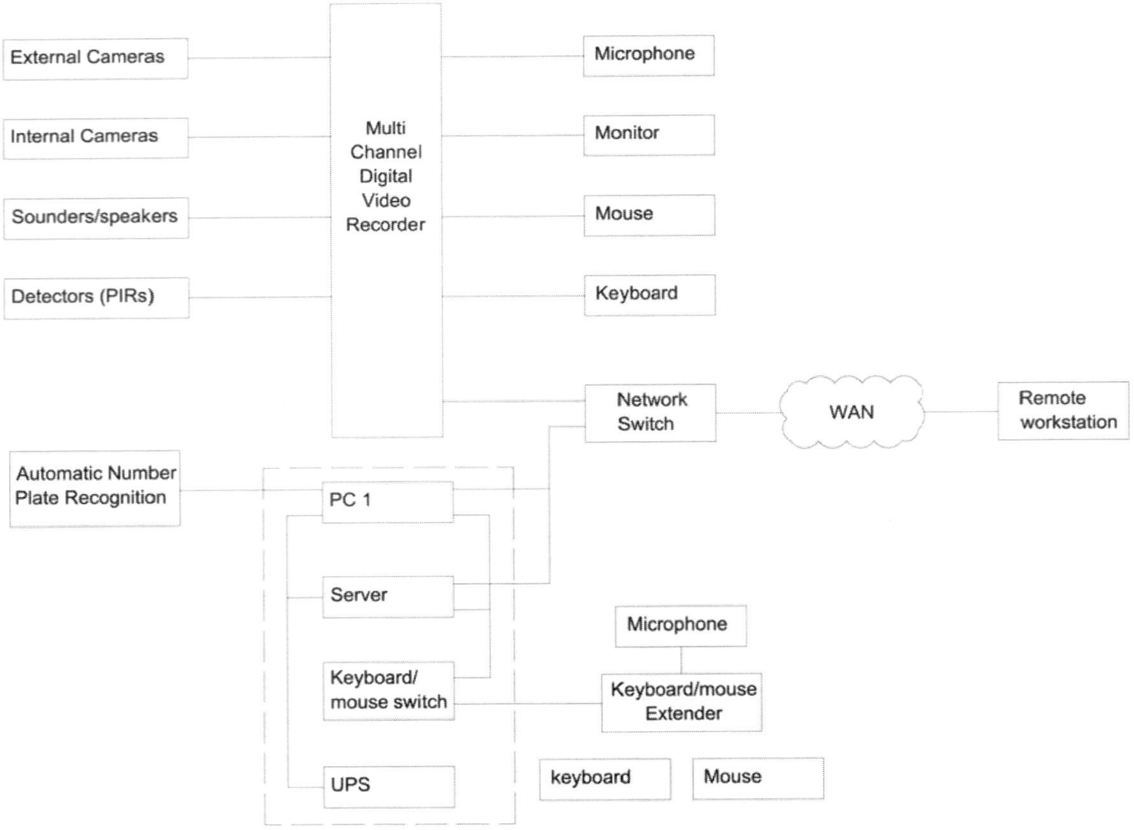

Typical Block Diagram – Digital CCTV

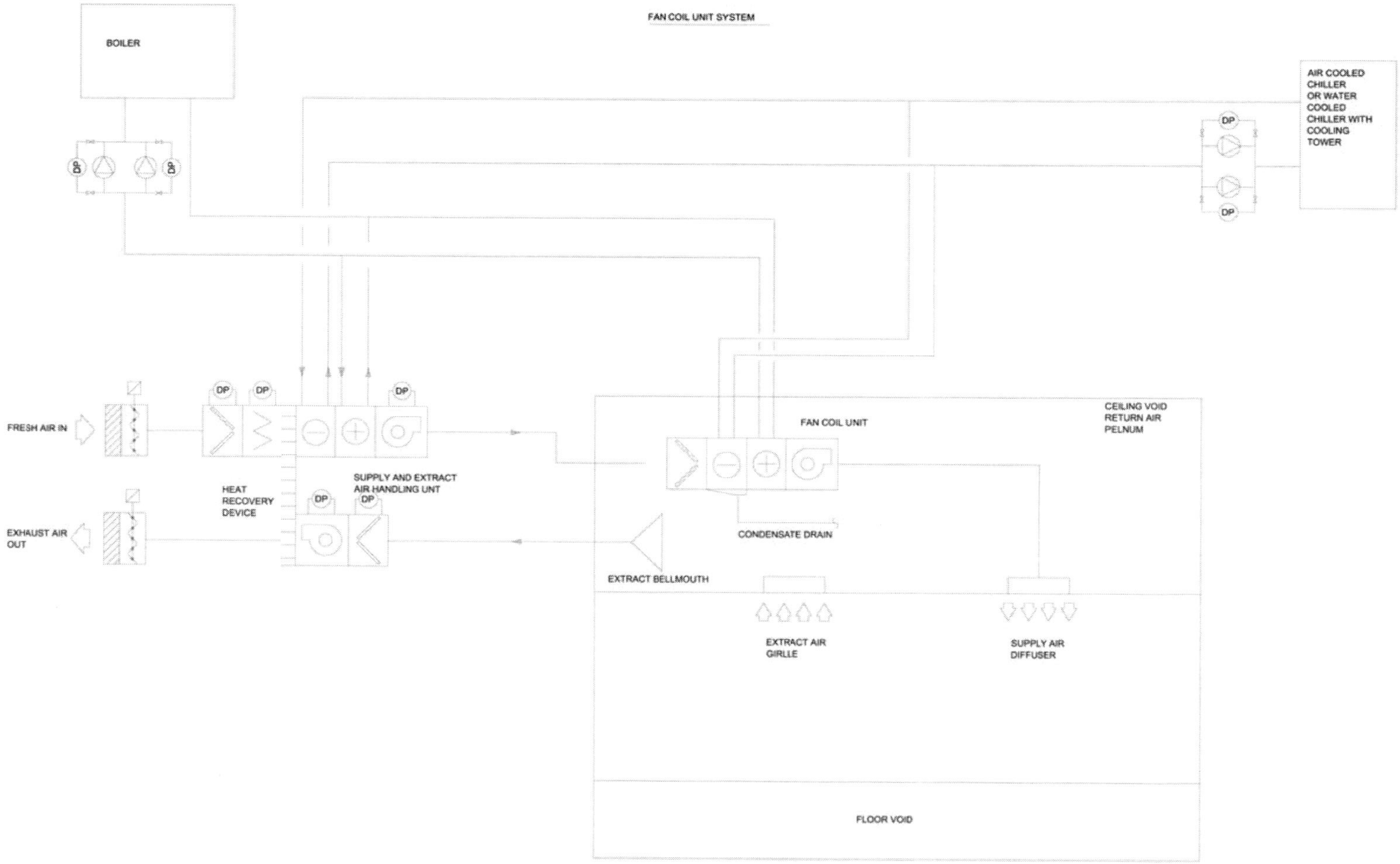
FAN COIL UNIT SYSTEM
BOILER
DP
DP
AIR COOLED CHILLER OR WATER COOLED CHILLER WITH COOLING TOWER
DP
DP
FRESH AIR IN
DP
DP
DP
HEAT RECOVERY DEVICE
SUPPLY AND EXTRACT AIR HANDLING UNT
DP
DP
EXHAUST AIR OUT
CEILING VOID RETURN AIR PELNUM
FAN COIL UNIT
CONDENSATE DRAIN
EXTRACT BELLMOUTH
EXTRACT AIR GIRLLE
SUPPLY AIR DIFFUSER
FLOOR VOID

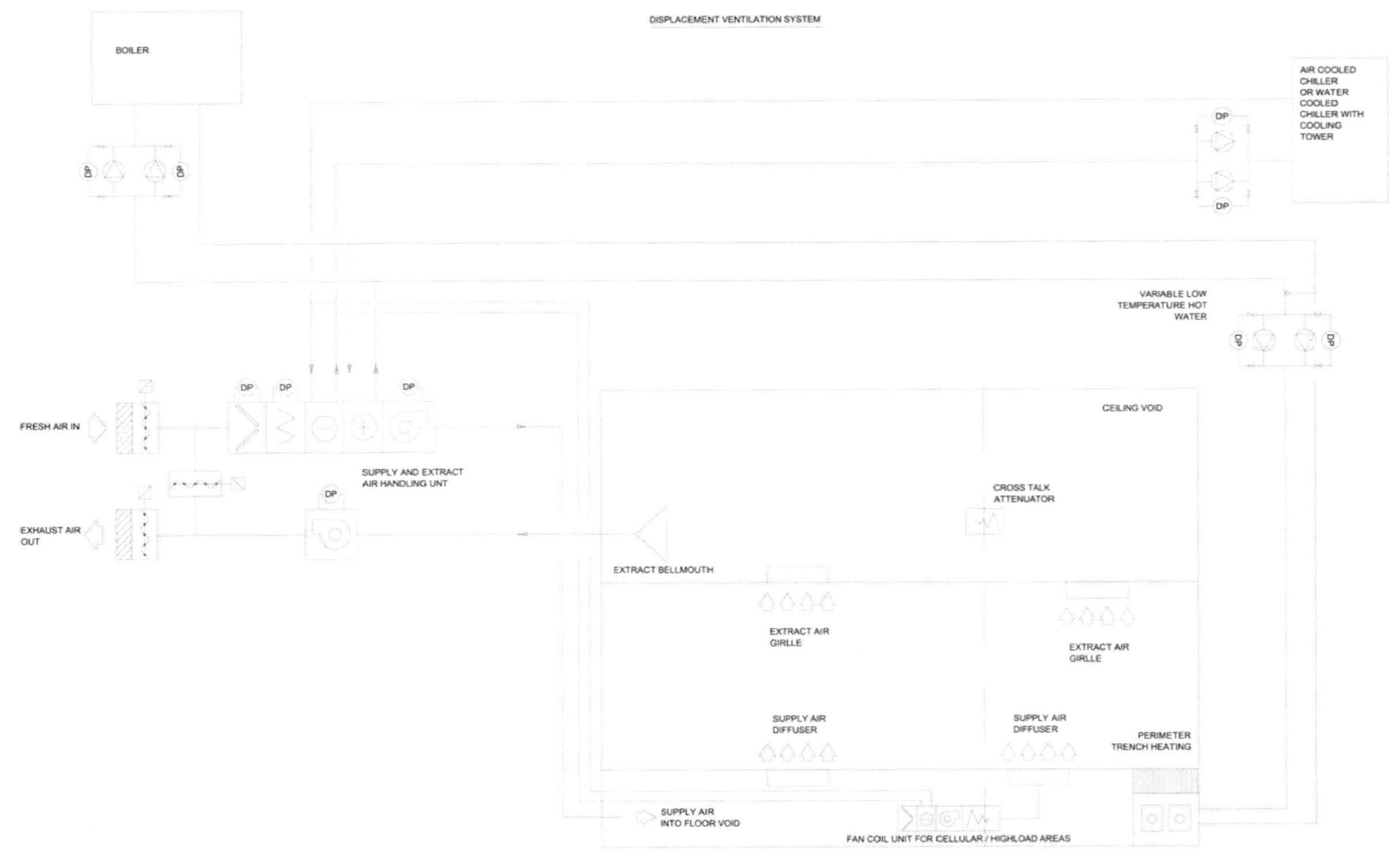
DISPLACEMENT VENTILATION SYSTEM
BOILER
AIR COOLED CHILLER OR WATER COOLED CHILLER WITH COOLING TOWER
DP
VARIABLE LOW TEMPERATURE HOT WATER
FRESH AIR IN
EXHAUST AIR OUT
SUPPLY AND EXTRACT AIR HANDLING UNT
CEILING VOID
CROSS TALK ATTENUATOR
EXTRACT BELLMOUTH
EXTRACT AIR GIRLLE
EXTRACT AIR GIRLLE
SUPPLY AIR DIFFUSER
SUPPLY AIR DIFFUSER
PERIMETER TRENCH HEATING
SUPPLY AIR INTO FLOOR VOID
FAN COIL UNIT FOR CELLULAR / HIGHLOAD AREAS

CHILLED CEILING SYSTEM

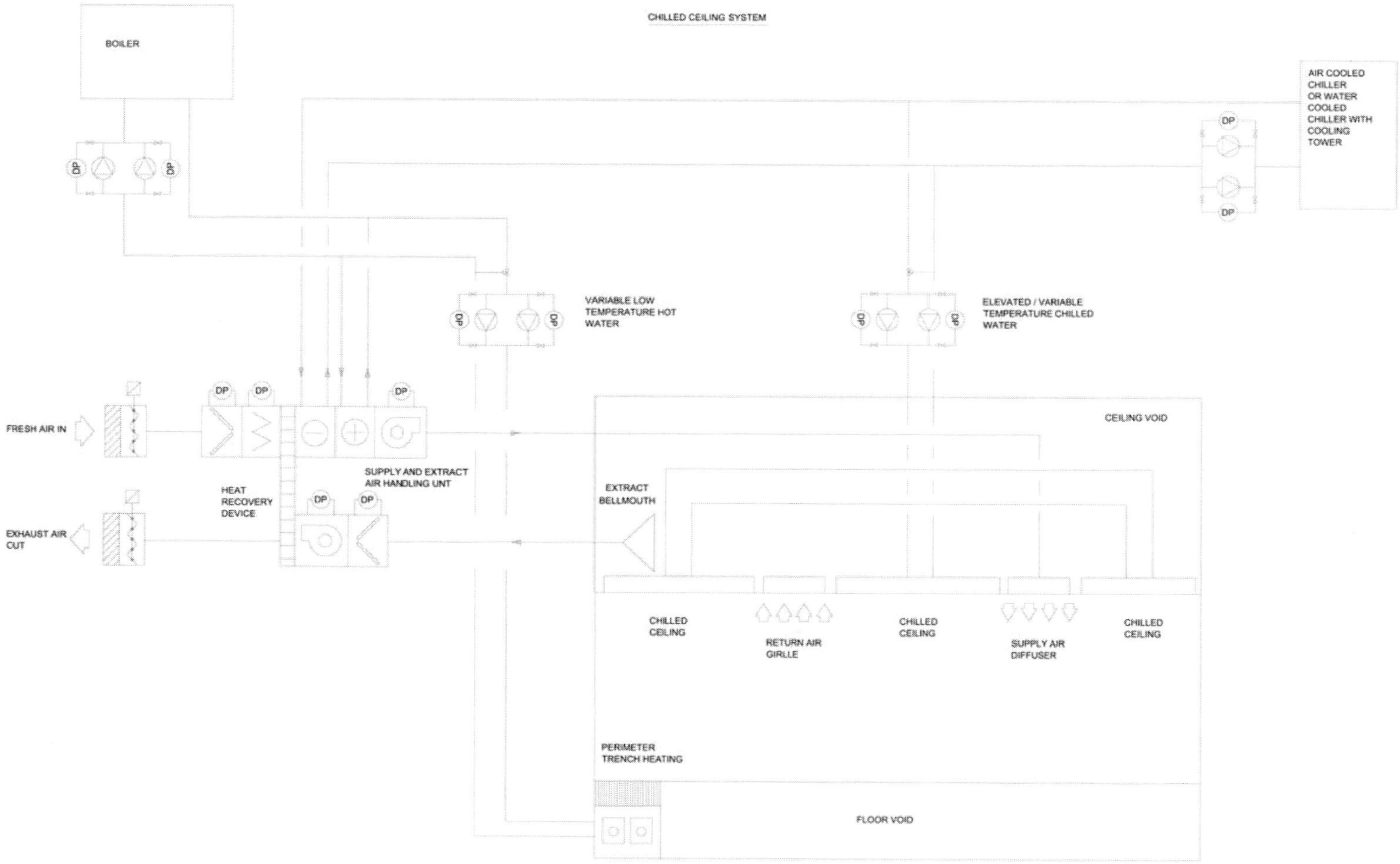

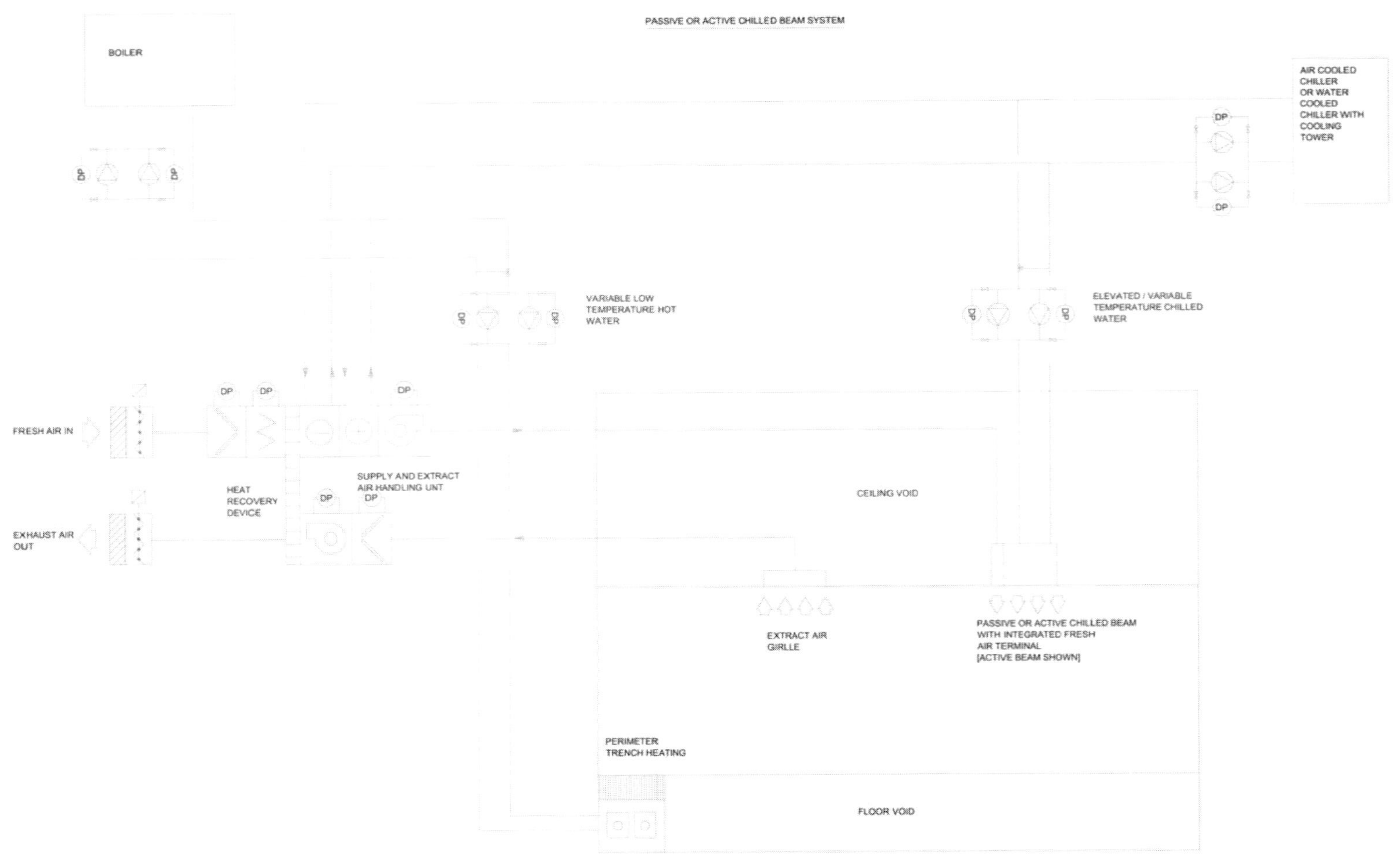
PASSIVE OR ACTIVE CHILLED BEAM SYSTEM
BOILER
AIR COOLED CHILLER OR WATER COOLED CHILLER WITH COOLING TOWER
DP
VARIABLE LOW TEMPERATURE HOT WATER
ELEVATED / VARIABLE TEMPERATURE CHILLED WATER
FRESH AIR IN
EXHAUST AIR OUT
HEAT RECOVERY DEVICE
SUPPLY AND EXTRACT AIR HANDLING UNT
CEILING VOID
EXTRACT AIR GIRLLE
PASSIVE OR ACTIVE CHILLED BEAM WITH INTEGRATED FRESH AIR TERMINAL [ACTIVE BEAM SHOWN]
PERIMETER TRENCH HEATING
FLOOR VOID

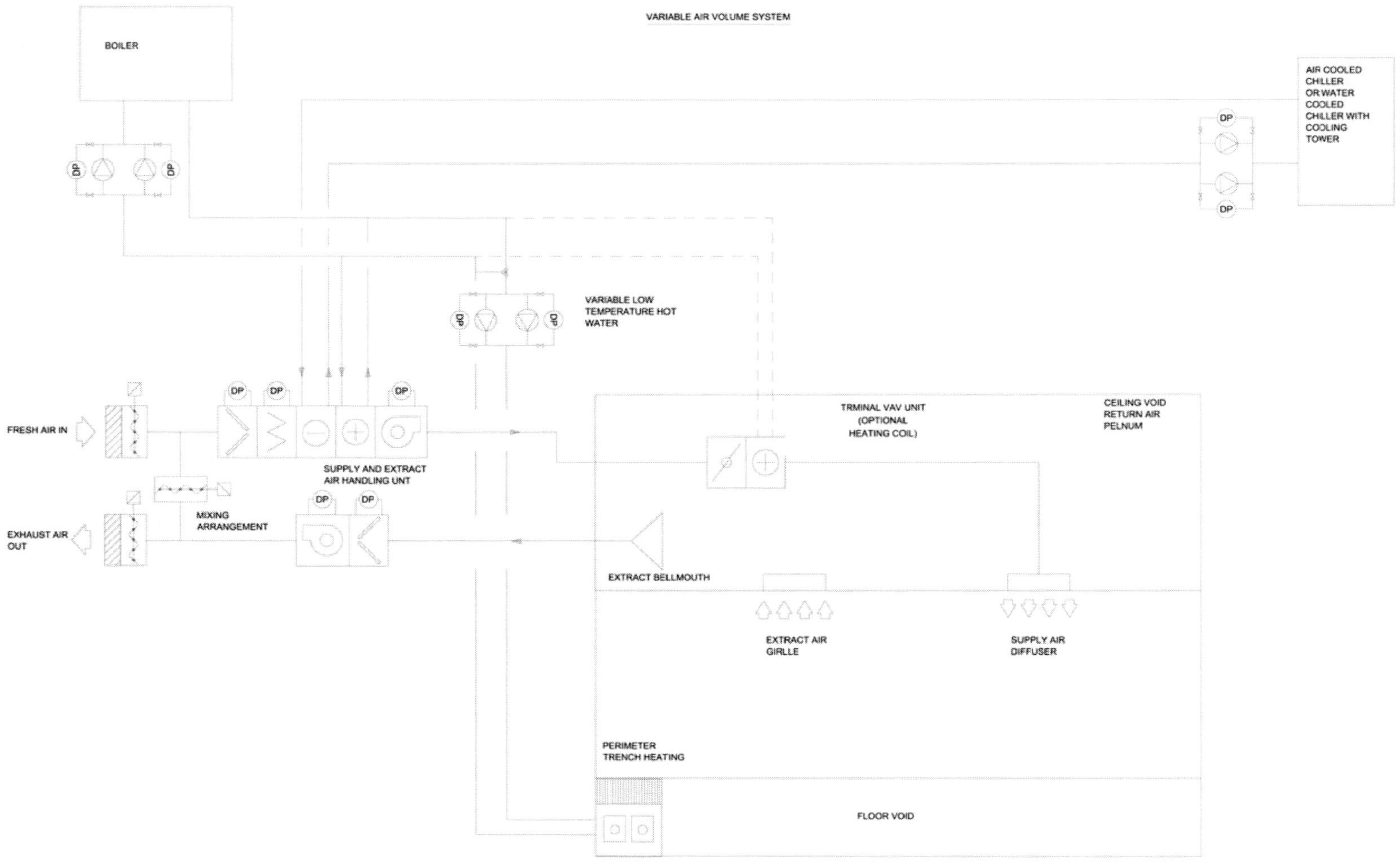
VARIABLE AIR VOLUME SYSTEM
BOILER
DP
AIR COOLED CHILLER OR WATER COOLED CHILLER WITH COOLING TOWER
VARIABLE LOW TEMPERATURE HOT WATER
FRESH AIR IN
SUPPLY AND EXTRACT AIR HANDLING UNT
MIXING ARRANGEMENT
EXHAUST AIR OUT
TRMINAL VAV UNIT (OPTIONAL HEATING COIL)
CEILING VOID RETURN AIR PELNUM
EXTRACT BELLMOUTH
EXTRACT AIR GIRLLE
SUPPLY AIR DIFFUSER
PERIMETER TRENCH HEATING
FLOOR VOID

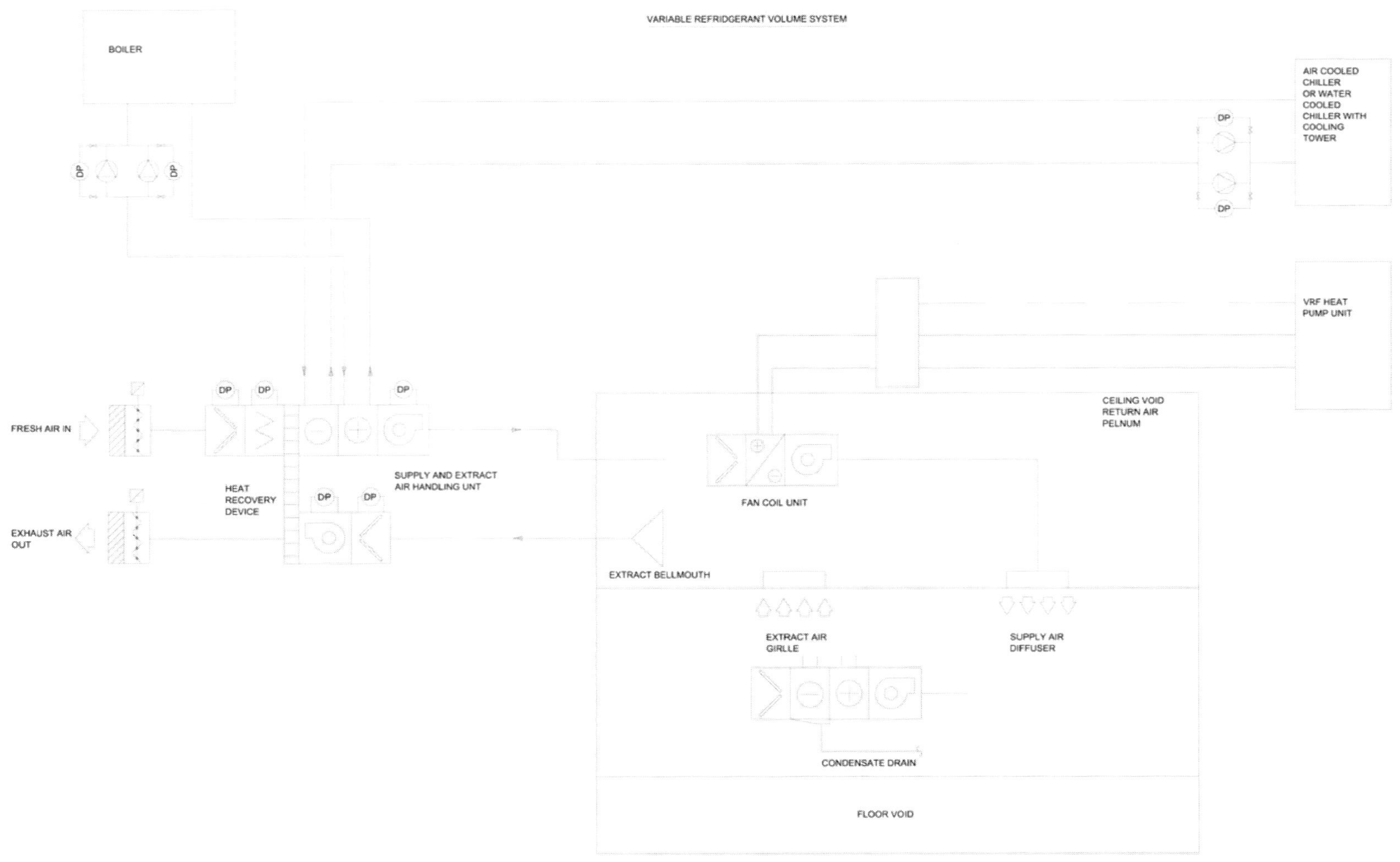

VARIABLE REFRIDGERANT VOLUME SYSTEM
BOILER
DP
DP
AIR COOLED CHILLER OR WATER COOLED CHILLER WITH COOLING TOWER
DP
DP
VRF HEAT PUMP UNIT
DP
DP
DP
FRESH AIR IN
HEAT RECOVERY DEVICE
SUPPLY AND EXTRACT AIR HANDLING UNT
DP
DP
EXHAUST AIR OUT
CEILING VOID RETURN AIR PELNUM
FAN COIL UNIT
EXTRACT BELLMOUTH
EXTRACT AIR GIRLLE
SUPPLY AIR DIFFUSER
CONDENSATE DRAIN
FLOOR VOID

REVERSE CYCLE HEAT PUMP

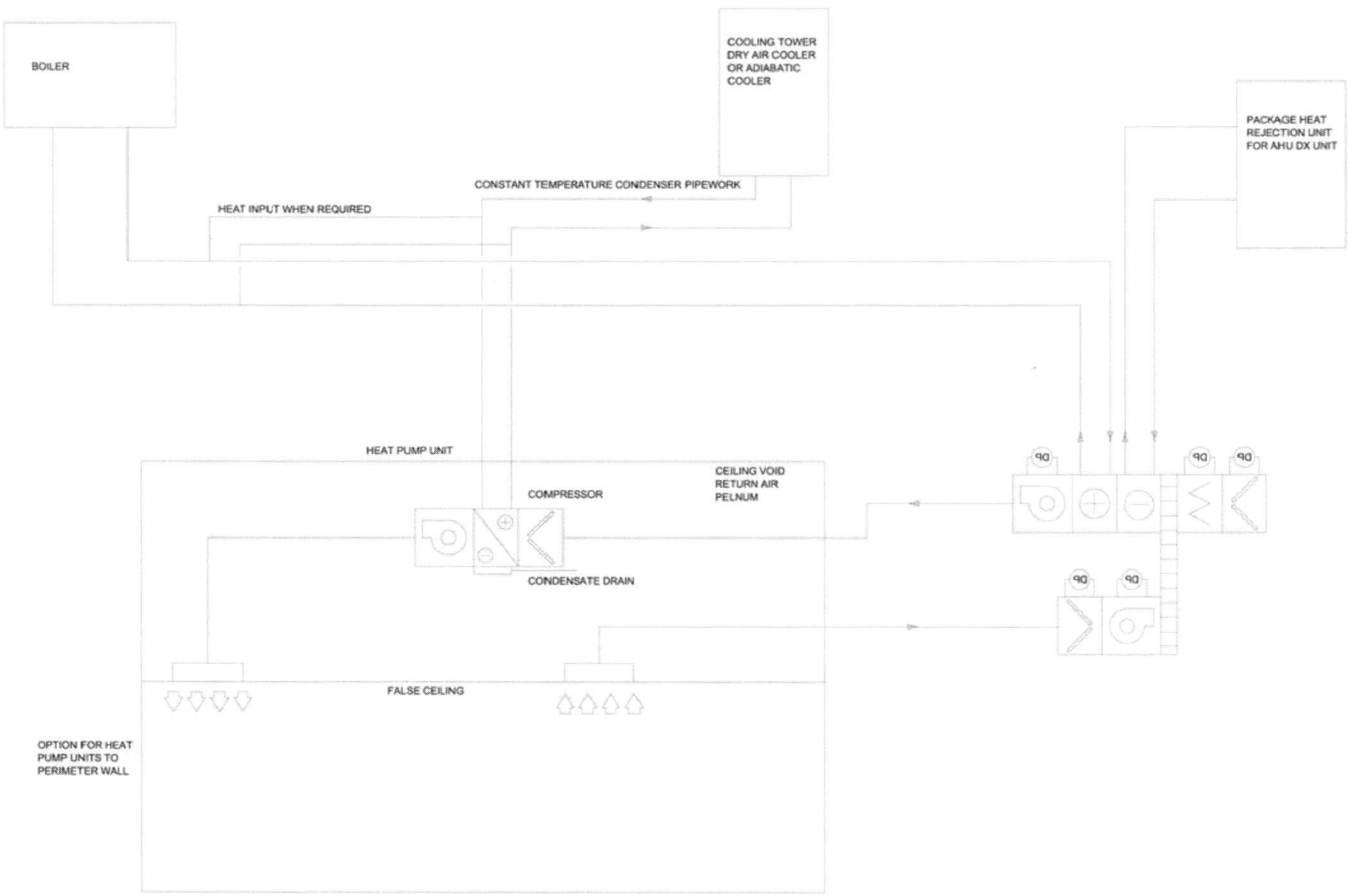

ESSENTIAL READING FROM TAYLOR AND FRANCIS

Faber & Kell's Heating & Air-conditioning of Buildings

11th Edition

Doug Oughton & Ant Wilson

For over 70 years, Faber & Kell's has been the definitive reference text in its field. It provides an understanding of the principles of heating and air-conditioning of buildings in a concise manner, illustrating practical information with simple, easy-to-use diagrams, now in full-colour.

This new-look 11th edition has been re-organised for ease of use and includes fully updated chapters on sustainability and renewable energy sources, as well as information on the new Building Regulations Parts F and L. As well as extensive updates to regulations and codes, it now includes an introduction that explains the role of the building services engineer in the construction process. Its coverage of design calculations, advice on using the latest technologies, building management systems, operation and maintenance makes this an essential reference for all building services professionals.

January 2015: 246 x 189: 968pp
Pb: 978-0-415-52265-6: £105.00

To Order: Tel: +44 (0) 1235 400524 Fax: +44 (0) 1235 400525
or Post: Taylor and Francis Customer Services,
Bookpoint Ltd, Unit T1, 200 Milton Park, Abingdon, Oxon, OX14 4TA UK
Email: book.orders@tandf.co.uk

For a complete listing of all our titles visit:
www.tandf.co.uk

Taylor & Francis
Taylor & Francis Group

ESSENTIAL READING FROM TAYLOR AND FRANCIS

Spon's Asia Pacific Construction Costs Handbook, Fifth Edition

LANGDON & SEAH

In the last few years, the global economic outlook has continued to be shrouded in uncertainty and volatility following the financial crisis in the Euro zone. While the US and Europe are going through a difficult period, investors are focusing more keenly on Asia. This fifth edition provides overarching construction cost data for 16 countries: Brunei, Cambodia, China, Hong Kong, India, Indonesia, Japan, Malaysia, Myanmar, Philippines, Singapore, South Korea, Sri Lanka, Taiwan, Thailand and Vietnam.

May 2015: 234X156 mm: 452 pp
Hb: 978-1-4822-4358-1: £160.00

To Order: Tel: +44 (0) 1235 400524 Fax: +44 (0) 1235 400525
or Post: Taylor and Francis Customer Services,
Bookpoint Ltd, Unit T1, 200 Milton Park, Abingdon, Oxon, OX14 4TA UK
Email: book.orders@tandf.co.uk

For a complete listing of all our titles visit:
www.tandf.co.uk

Taylor & Francis
Taylor & Francis Group

It is one thing to
imagine a better world.
It's another to deliver it.

**Understanding change,
unlocking potential, creating
brilliant new communities.**

The Tate Modern extension
takes an iconic building and
adds to it. Cost management
provided by AECOM.

AECOM
Built to deliver a better world

aecom.com

SPON'S PRICEBOOKS 2017 From AECOM

Spon's Architects' and Builders' Price Book 2017

Editor: AECOM

To NRM1 and NRM2, and with more plant prices separated out in the measured works section. With further developments:

- The in situ concrete section and plasterboard linings and partitions sections have been heavily revised and developed.
- A laboratory, a car park and an updated London Office have been added as new Cost Models
- And new items are added: Foamglas insulation, Carlite waterproof concrete, and an expanded range of stone flooring.

Hbk & VitalSource® ebook 824pp approx.
978-1-4987-8611-9 £160
VitalSource® ebook
978-1-4987-8643-0 £160
(inc. sales tax where appropriate)

Spon's Civil Engineering and Highway Works Price Book 2017

Editor:AECOM

This year some item descriptions have been revised to reflect prices in the global market. Plus a revised structure for manhole construction; new prices for pre-cast manholes with monolithic precast concrete bases; updated rail supply and installation rates and a few more added items; revised piling rates and descriptions; new highway pipe rates; reduced steel prices.

Hbk & VitalSource® ebook 688pp approx.
978-1-4987-8612-6 £180
VitalSource® ebook
978-1-4987-8646-1 £180
(inc. sales tax where appropriate)

Spon's External Works and Landscape Price Book 2017

Editor:AECOM

This NRM edition includes a number of new and significantly developed items: acoustic fencing, architectural metalwork, block and beam systems, culverts, deep concrete manholes to 3 m, headwalls, pre-cast steps and new retaining walls. Green roofs and sportsfields have now been broken down into details, rather than presented as a lump sum.

Hbk & VitalSource® ebook 624pp approx.
978-1-4987-8615-7 £150
VitalSource® ebook
978-1-4987-8648-5 £150
(inc. sales tax where appropriate)

Spon's Mechanical and Electrical Services Price Book 2017

Editor:AECOM

This NRM edition includes a number of new and significantly developed items: WiFi equipment, FP 600 cable, wireless fire alarms; and a much enhanced clock section.

Hbk & VitalSource® ebook 840pp approx.
978-1-4987-8616-4 £160
VitalSource® ebook
978-1-4987-8650-8 £160
(inc. sales tax where appropriate)

Receive our VitalSource® ebook free when you order any hard copy Spon 2017 Price Book

Visit www.pricebooks.co.uk

To order:
Tel: 01235 400524 Fax: 01235 400525
Post: Taylor & Francis Customer Services, Bookpoint Ltd, 200 Milton Park, Abingdon, Oxon, OX14 4SB, UK
Email: book.orders@tandf.co.uk
A complete listing of all our books is on www.crcpress.com

CRC Press
Taylor & Francis Group

PART 2

Approximate Estimating

ESSENTIAL READING FROM TAYLOR AND FRANCIS

Cost Studies of Buildings

Allan Ashworth & Srinath Perera

This practical guide to cost studies of buildings has been updated and revised throughout for the 6th edition. New developments in RICS New Rules of Measurement (NRM) are incorporated throughout the book, in addition to new material on e-business, the internet, social media, building information modelling, sustainability, building resilience and carbon estimating.

This trusted and easy to use guide to the cost management role:

Focuses on the importance of costs of constructing projects during the different phases of the construction process

Features learning outcomes and self-assessment questions for each chapter

Addresses the requirements of international readers

From introductory data on the construction industry and the history of construction economics, to recommended methods for cost analysis and post-contract cost control, Cost Studies of Buildings is an ideal companion for anyone learning about cost management.

July 2015: 246x174: 544pp Pb:
978-1-138-01735-1: £42.99

To Order: Tel: +44 (0) 1235 400524 Fax: +44 (0) 1235 400525
or Post: Taylor and Francis Customer Services,
Bookpoint Ltd, Unit T1, 200 Milton Park, Abingdon, Oxon, OX14 4TA UK
Email: book.orders@tandf.co.uk

For a complete listing of all our titles visit:
www.tandf.co.uk

Taylor & Francis
Taylor & Francis Group

DIRECTIONS

The prices shown in this section of the book are average prices on a fixed price basis for typical buildings tendered during the second quarter of 2016. Unless otherwise noted, they exclude external services and professional fees.

The information in this section has been arranged to follow more closely the order in which estimates may be developed, in accordance with RIBA work stages.

a) Cost Indices and Regional Variations – These provide information regarding the adjustments to be made to estimates taking into account current pricing levels for different locations in the UK.

b) Feasibility Costs – These provide a range of data (based on a rate per square metre) for all-in engineering costs, excluding lifts, associated with a wide variety of building types. These would typically be used at work stage A/B (feasibility) of a project.

c) Elemental Rates – The outline costs for offices have been developed further to provide rates for the alternative solutions for each of the services elements. These would typically be used at work stage C, outline proposal.

   Where applicable, costs have been identified as Shell and Core and Fit-Out to reflect projects where the choice of procurement has dictated that the project is divided into two distinctive contractual parts.

   Such detail would typically be required at work stage D, detailed proposals.

d) All-in Rates – These are provided for a number of items and complete parts of a system i.e. boiler plant, ductwork, pipework, electrical switchgear and small power distribution, together with lifts and escalators. Refer to the relevant section for further guidance notes.

e) Elemental Costs – These are provided for a diverse range of building types; offices, laboratory, shopping mall, airport terminal building, supermarket, performing arts centre, sports hall, luxury hotel, hospital and secondary school. Also included is a separate analysis of a building management system for an office block. In each case, a full analysis of engineering services costs is given to show the division between all elements and their relative costs to the total building area. A regional variation factor has been applied to bring these analyses to a common London base.

Prices should be applied to the total floor area of all storeys of the building under consideration. The area should be measured between the external walls without deduction for internal walls and staircases/lift shafts i.e. GIA (Gross Internal Area).

Although prices are reviewed in the light of recent tenders it has only been possible to provide a range of prices for each building type. This should serve to emphasize that these can only be average prices for typical requirements and that such prices can vary widely depending on variations in size, location, phasing, specification, site conditions, procurement route, programme, market conditions and net to gross area efficiencies. Rates per square metre should not therefore be used indiscriminately and each case needs to be assessed on its own merits.

The prices do not include for incidental builder's work nor for profit and attendance by a Main Contractor where the work is executed as a subcontract: they do however include for preliminaries, profit and overheads for the services contractor. Capital contributions to statutory authorities and public undertakings and the cost of work carried out by them have been excluded.

Where services works are procured indirectly, i.e. ductwork via a mechanical Subcontractor, the reader should make due allowance for the addition of a further level of profit etc.

COST INDICES

The following tables reflect the major changes in cost to contractors but do not necessarily reflect changes in tender levels. In addition to changes in labour and materials costs, tenders are affected by other factors such as the degree of competition in the particular industry, the area where the work is to be carried out, the availability of labour and the prevailing economic conditions. This has meant in recent years that, when there has been an abundance of work, tender levels have tended to increase at a greater rate than can be accounted for solely by increases in basic labour and material costs and, conversely, when there is a shortage of work this has tended to result in keener tenders. Allowances for these factors are impossible to assess on a general basis and can only be based on experience and knowledge of the particular circumstances.

In compiling the tables the cost of labour has been calculated on the basis of a notional gang as set out elsewhere in the book. The proportion of labour to materials has been assumed as follows:

Mechanical Services – 30:70, Electrical Services – 50:50, (1976 = 100)

Mechanical Services

Year	*First Quarter*	*Second Quarter*	*Third Quarter*	*Fourth Quarter*
2003	443	446	447	456
2004	458	464	467	482
2005	486	488	492	508
2006	513	522	527	533
2007	535	541	546	555
2008	555	560	567	584
2009	582	578	577	577
2010	576	583	585	592
2011	600	606	607	605
2012	607	607	605	607
2013	613	618	617	619
2014	621	621	620	630
2015	628	629	625	636
2016	P 636	P 638	F 639	F 649
2017	F 650	F 651	F 651	F 663

Electrical Services

Year	*First Quarter*	*Second Quarter*	*Third Quarter*	*Fourth Quarter*
2003	530	533	533	541
2004	571	574	576	589
2005	607	608	607	615
2006	631	636	641	649
2007	665	666	668	676
2008	693	697	700	710
2009	720	721	724	737
2010	760	766	766	768
2011	777	780	778	783
2012	776	776	775	789
2013	789	788	786	797

COST INDICES

Electrical Services				
Year	*First Quarter*	*Second Quarter*	*Third Quarter*	*Fourth Quarter*
2014	797	795	796	799
2015	812	811	809	808
2016	P 827	P 827	F 827	F 828
2017	F 848	F 848	F 848	F 849

(P = Provisional)
(F = Forecast)

COST INDICES

Regional variations

Prices throughout this Book apply to work in the London area (see Directions at the beginning of the Mechanical Installations and Electrical Installations sections). However, prices for mechanical and electrical services installations will of course vary from region to region, largely as a result of differing labour costs but also depending on the degree of accessibility, urbanization and local market conditions.

The following table of regional factors is intended to provide readers with indicative adjustments that may be made to the prices in the Book for locations outside of London. The figures are of necessity averages for regions and further adjustments should be considered for city centre or very isolated locations, or other known local factors.

Greater London	1.03	Yorkshire & Humberside	0.86
South East	0.96 (Excl GL)	North West	0.88
South West	0.90	Northern	0.84
East Midlands	0.87	Scotland	0.89
West Midlands	0.87	Wales	0.86
East Anglia	0.91	Northern Ireland	0.74

RIBA STAGE A/B PREPARATION

Item	Unit	Range £
Typical square metre rates for Engineering Services		
The following examples indicate the range of rates within each building type for engineering services, excluding lifts etc., utilities services and professional fees. Based on Gross Internal Area (GIA).		
Industrial Buildings		
Factories		
Owner occupation: Includes for rainwater, soil/waste, LTHW heating via HL radiant heaters, BMS, LV installations, lighting, fire alarms, security, earthing	m²	115.00 to 140.00
Owner occupation: Includes for rainwater, soil/waste, sprinklers, LTHW heating via HL gas fired heaters, local air conditioning, BMS, HV/LV installations, lighting, fire alarms, security, earthing	m²	170.00 to 210.00
Warehouses		
High bay for owner occupation: Includes for rainwater, soil/waste, LTHW heating via HL gas fired heaters, BMS, HV/LV installations, lighting, fire alarms, security, earthing	m²	95.00 to 115.00
High bay for owner occupation: Includes for rainwater, soil/waste, sprinklers, LTHW heating via HL radiant heaters, local air conditioning, BMS, HV/LV installations, lighting, fire alarms, security, earthing	m²	190.00 to 230.00
Distribution Centres		
High bay for letting: Includes for rainwater, soil/waste, LTHW heating via HL gas fired heaters, BMS, HV/LV installations, lighting, fire alarms, security, earthing	m²	115.00 to 140.00
High bay for owner occupation: Includes for rainwater, soil/waste, sprinklers, LTHW heating via HL radiant heaters, local air conditioning, BMS, HV/LV installations, lighting, fire alarms, security	m²	200.00 to 240.00
Office Buildings 5,000 m² to 15,000 m²		
Offices for Letting		
Shell & Core and Cat A non-air conditioned: Includes for rainwater, soil/waste, cold water, hot water via local electrical heaters, LTHW heating via radiator heaters, toilet extract, controls, LV installations, lighting, small power (landlords), fire alarms, earthing, security & IT wireways	m²	270.00 to 330.00
Shell & Core and Cat A non-air conditioned: Includes for rainwater, soil/waste, cold water, hot water, LTHW heating via perimeter heaters, toilet extract, controls, LV installations, lighting, small power (landlords), fire alarms, earthing, security wireways, IT wireways	m²	280.00 to 340.00
Shell & Core and Cat A air conditioned: Includes for rainwater, soil/waste, VRV 3 pipe heat pumps, toilet extract, BMS, LV installations, lighting, small power (landlords), fire alarms, earthing, security & IT wireways	m²	390.00 to 480.00
Shell & Core and Cat A air conditioned: Includes for rainwater, soil/waste, cold water, hot water via local electrical heaters, LTHW heating via perimeter heaters, 2 pipe, toilet extract, BMS, LV installations, lighting, small power (landlords), fire alarms, earthing, security & IT wireways	m²	430.00 to 530.00

RIBA STAGE A/B PREPARATION

Item	Unit	Range £		
Typical square metre rates for Engineering Services – cont				
Office Buildings 5,000 m² to 15,000 m² – cont				
Offices for Owner Occupation				
Non-air conditioned: Includes for rainwater, soil/waste, cold water, hot water via local electrical heaters, dry risers, LTHW heating via radiator heaters, toilet extract, controls, LV installations, lighting, small power (landlords), fire alarms, earthing, security, IT wireways	m²	280.00	to	340.00
Non-air conditioned: Includes for rainwater, soil/waste, cold water, hot water, dry risers, LTHW heating via perimeter heaters, toilet extract, BMS, LV installations, lighting, small power (landlords), fire alarms, earthing, security, IT wireways	m²	300.00	to	370.00
Air conditioned: Includes for rainwater, soil/waste, cold water, hot water via electrical heaters, dry risers, LTHW heating via perimeter heating, 4 pipe air conditioning, toilet extract, kitchen extract, sprinkler protection, BMS, LV installations, life safety standby generators, lighting, small power, fire alarms, earthing, security, IT wireways small power, fire alarms L1/P1, earthing, security and IT wireways	m²	485.00	to	590.00
Health and Welfare Facilities				
District General Hospitals				
Natural ventilation: Includes for rainwater, soil/waste, cold water, hot water, dry risers, medical gases, LTHW heating via perimeter heating, toilet extract, kitchen extract, BMS, LV installations, standby generation, lighting, small power, fire alarms, earthing/lightning protection, nurse call systems, security, IT wireways	m²	430.00	to	530.00
Natural ventilation: Includes for rainwater, soil/waste, cold water, hot water, dry risers, LTHW heating via perimeter heaters, localized VAV air conditioning, kitchen/toilet extract, BMS, LV installations, standby generation, lighting, small power, fire alarms, earthing/lightning protection, nurse call systems, security, IT wireways	m²	610.00	to	750.00
Private Hospitals				
Air conditioned: Includes for rainwater, soil/waste, cold water, hot water, dry risers, medical gases, LTHW heating, 2 pipe air conditioning, toilet extract, kitchen extract, BMS, LV installations, standby generation, lighting, small power, fire alarms, earthing, nurse call systems, security, IT wireways	m²	680.00	to	830.00
Air conditioned: Includes for rainwater, soil/waste, cold water, hot water, dry risers, medical gases, LTHW heating, 4 pipe air conditioning, kitchen/toilet extract, BMS, LV installations, standby generation, lighting, small power, fire alarms, earthing, nurse call system, security, IT wireways	m²	730.00	to	890.00
Day Care Unit				
Natural ventilation: Includes for rainwater, soil/waste, cold water, hot water via local electrical heaters, medical gases, LTHW heating, toilet extract, kitchen extract, BMS, LV installations, lighting, small power, fire alarms, earthing, security, IT wireways	m²	430.00	to	530.00
Comfort cooled: Includes for rainwater, soil/waste, cold water, hot water, medical gases, LTHW heating, DX air conditioning, kitchen/toilet extract, BMS, LV installations, lighting, small power, fire alarms, earthing, security, IT wireways	m²	460.00	to	560.00

RIBA STAGE A/B PREPARATION

Item	Unit	Range £
Entertainment and Recreation Buildings		
Non-Performing		
Natural ventilation: Includes for rainwater, soil/waste, cold water, central hot water, dry risers, LTHW heating, toilet extract, kitchen extract, controls, LV installations, lighting, small power, fire alarms, earthing, security, IT wireways	m²	310.00 to 380.00
Comfort cooled: Includes for rainwater, soil/waste, cold water, hot water via electrical heaters, sprinklers/dry risers, LTHW heating, DX air conditioning, kitchen/toilet extract, BMS, LV installations, lighting, small power, fire alarms, earthing, security, IT wireways	m²	530.00 to 650.00
Performing Arts (with Theatre)		
Natural ventilation: Includes for rainwater, soil/waste, cold water, central hot water, sprinklers/dry risers, LTHW heating, toilet extract, kitchen extract, controls, LV installations, lighting, small power, fire alarms, earthing, security, IT wireways	m²	495.00 to 610.00
Comfort cooled: Includes for rainwater, soil/waste, cold water, central hot water, sprinklers/dry risers, LTHW heating, DX air conditioning, kitchen/toilet extract, BMS, LV installations, lighting including enhanced dimming/scene setting, small power, fire alarms, earthing, security, IT wireways including production, audio and video recording, EPOS system	m²	670.00 to 810.00
Sports Halls		
Natural ventilation: Includes for rainwater, soil/waste, cold water, hot water gas fired heaters, LTHW heating, toilet extract, BMS, LV installations, lighting, small power, fire alarms, earthing, security	m²	225.00 to 275.00
Comfort cooled: Includes for rainwater, soil/waste, cold water, hot water via LTHW heat exchangers, LTHW heating, air conditioning via AHU's, toilet extract, BMS, LV installations, lighting, small power, fire alarms, earthing, security	m²	320.00 to 390.00
Multi-Purpose Leisure Centre		
Natural ventilation: Includes for rainwater, soil/waste, cold water, hot water gas fired heaters, LTHW heating, toilet extract, BMS, LV installations, lighting, small power, fire alarms, earthing, security, IT wireways	m²	315.00 to 385.00
Comfort cooled: Includes for rainwater, soil/waste, cold water, hot water LTHW heat exchangers, LTHW heating, air conditioning via AHU, pool hall supply/extract, kitchen/toilet extract, BMS, LV installations, lighting, small power, fire alarms, earthing, security, IT wireways	m²	550.00 to 670.00
Retail Buildings		
Open Arcade		
Natural ventilation: Includes for rainwater, soil/waste, cold water, hot water, sprinklers/dry risers, LTHW heating, toilet extract, smoke extract, BMS, LV installations, life safety standby generators, lighting, small power, fire alarms, public address, earthing/lightning protection, security, IT wireways	m²	325.00 to 395.00
Enclosed Shopping Mall		
Air conditioned: Includes for rainwater, soil/waste, cold water, hot water, sprinklers/dry risers, LTHW heating, air conditioning via AHUs, toilet extract, smoke extract, BMS, LV installations, life safety standby generators, lighting, small power, fire alarms, public address, earthing/lightning protection, CCTV/security, IT wireways people counting systems	m²	470.00 to 570.00

RIBA STAGE A/B PREPARATION

Item	Unit	Range £
Typical square metre rates for Engineering Services – cont		
Retail Buildings – cont		
Department Stores		
Air conditioned: Includes for sanitaryware, soil/waste, cold water, hot water, sprinklers/dry risers, LTHW heating, air conditioning via AHUs, toilet extract, smoke extract, BMS, LV installations, life safety standby generators, lighting, small power, fire alarms, public address, earthing, lightning protection, CCTV/ security, IT installation wireways	m²	370.00 to 450.00
Supermarkets		
Air conditioned: Includes for rainwater, soil/waste, cold water, hot water, sprinklers, LTHW heating, air conditioning via AHUs, toilet extract, BMS, LV installations, lighting, small power, fire alarms, earthing, lightning protection, security, IT wireways, refrigeration	m²	540.00 to 660.00
Educational Buildings		
Secondary Schools (Academy)		
Natural Ventilation: Includes for rainwater, soil/waste, cold water, central hot water, LTHW heating, toilet extract, BMS, LV installations, lighting, small power, fire alarms, earthing, security, IT wireways	m²	310.00 to 380.00
Natural vent with comfort cooling to selected areas (BB93 compliant): Includes for rainwater, soil/waste, cold water, central hot water, LTHW heating, DX air conditioning, general supply/extract, toilet extract, BMS, LV installations, lighting, small power, fire alarms, earthing, security, IT wireways	m²	370.00 to 450.00
Scientific Buildings		
Educational Research		
Comfort cooled: Includes for rainwater, soil/waste, cold water, central hot water, dry risers, compressed air, medical gasses, LTHW heating, 4 pipe air conditioning, toilet extract, fume, BMS, LV installations, lighting, small power, fire alarms, earthing, lightning protection, security, IT wireways	m²	690.00 to 850.00
Air conditioned: Includes for rainwater, soil/waste, laboratory waste, cold water, central hot water, specialist water, dry risers, compressed air, medical gasses, steam, LTHW heating, VAV air conditioning, Comm room cooling, toilet extract, fume extract, BMS, LV installations, UPS, standby generators, lighting, small power, fire alarms, earthing, lightning protection, security, IT wireways	m²	1225.00 to 1500.00
Commercial Research		
Air conditioned; Includes for rainwater, soil/waste, laboratory waste, cold water, central hot water, specialist water, dry risers, compressed air, medical gasses, steam, LTHW heating, VAV air conditioning, Comm room cooling, toilet extract, fume extract, BMS, LV installations, UPS, standby generators, lighting, small power, fire alarms, earthing, lightning protection, security, IT wireways	m²	1350.00 to 1650.00

RIBA STAGE A/B PREPARATION

Item	Unit	Range £
Residential Facilities		
Hotels		
1 to 3 Star: Includes for rainwater, soil/waste, cold water, hot water, dry risers, LTHW heating via radiators, toilet/bathroom extract, kitchen extract, BMS, LV installations, lighting, small power, fire alarms, earthing, lightning protection, security, IT wireways	m²	420.00 to 520.00
4 to 5 Star: Includes for rainwater, soil/waste, cold water, hot water, sprinklers, dry risers, 4 pipe air conditioning, kitchen extract, toilet/bathroom extract, BMS, LV installations, life safety standby generators, lighting, small power, fire alarms, earthing, security, IT wireways	m²	750.00 to 910.00

RIBA STAGE C/D/E DESIGN

Item	Unit	Range £		
Elemental Rates for Alternative Engineering Services Solutions – Offices				
The following examples of building types indicate the range of rates for alternative design solutions for each of the engineering services elements based on Gross Internal Area for the Shell and Core and Net Internal Area for the Fit-Out. Fit-Out is assumed to be to Cat A Standard. Consideration should be made for the size of the building, which may affect the economies of scale for rates, i.e. the larger the building the lower the rates.				
Mechanical and Electrical Services				
Shell & Core				
5.1 Sanitary Installations				
Building up to 3,000 m²	m² GIA	9.55	to	11.70
Building over 3,000 m² up to 15,000 m² (low rise)	m² GIA	7.95	to	9.70
5.3 Disposal Installations				
Building up to 3,000 m²	m² GIA	21.00	to	26.00
Building over 3,000 m² up to 15,000 m²	m² GIA	18.90	to	23.00
5.4 Water Installations				
Building up to 3,000 m²	m² GIA	21.00	to	25.00
Building over 3,000 m² up to 15,000 m²	m² GIA	21.00	to	25.00
5.6 Space Heating and Air Conditioning				
LPHW Heating Installation; including gas installations				
Building up to 3,000 m²	m² GIA	42.50	to	52.00
Building over 3,000 m² up to 15,000 m²	m² GIA	45.00	to	55.00
Comfort cooling				
2 pipe fan coil for building up to 3,000 m²	m² GIA	66.00	to	81.00
2 pipe fan coil for building over 3,000 m² up to 15,000 m²	m² GIA	61.00	to	74.00
2 pipe variable refrigerant volume (VRV) for building up to 3,000 m²	m² GIA	48.00	to	58.00
Full air conditioning				
4 pipe fan coil for building up to 3,000 m²	m² GIA	94.00	to	115.00
4 pipe fan coil for building over 3,000 m² up to 15,000 m²	m² GIA	94.00	to	115.00
4 pipe variable refrigerant volume for building up to 3,000 m²	m² GIA	82.00	to	100.00
Ventilated (active) chilled beams for building over 3,000 m² up to 15,000 m²	m² GIA	94.00	to	115.00
Chilled beam exposed services for building over 3,000 m² to 15,000 m²	m² GIA	94.00	to	115.00
Concealed passive chilled beams for building over 3,000 m² up to 15,000 m²	m² GIA	94.00	to	115.00
Chilled ceiling for building over 3,000 m² to 15,000 m²	m² GIA	94.00	to	115.00
Chilled ceiling/perimeter beams for building over 3,000 m² up to 15,000 m²	m² GIA	94.00	to	115.00
Displacement for building over 3,000 m² up to 15,000 m²	m² GIA	94.00	to	115.00
5.7 Ventilation Systems (excluding smoke extract)				
Building up to 3,000 m²	m² GIA	36.50	to	45.00
Building over 3,000 m² up to 15,000 m²	m² GIA	34.00	to	42.00
5.8 Electrical Installations				
LV Installations				
Standby generators (life safety only)				
Buildings over 3,000 m² up to 15,000 m²	m² GIA	18.90	to	23.00
LV Distribution				
Buildings up to 3,000 m²	m² GIA	28.50	to	35.00
Buildings over 3,000 m² up to 15,000 m²	m² GIA	37.00	to	45.00

RIBA STAGE C/D/E DESIGN

Item	Unit	Range £		
Lighting Installations (including lighting controls and luminaires)				
Buildings up to 3,000 m²	m² GIA	18.90	to	23.00
Buildings over 3,000 m² up to 15,000 m²	m² GIA	18.90	to	23.00
Small Power				
Buildings up to 3,000 m²	m² GIA	5.70	to	6.90
Buildings over 3,000 m² to up to 15,000 m²	m² GIA	5.20	to	6.30
Electrical Installations for Mechanical Plant				
Buildings up to 3,000 m²	m² GIA	7.30	to	8.90
Buildings over 3,000 m² to up to 15,000 m²	m² GIA	8.30	to	10.10
5.11 Protective Installations				
Fire Protection over 3,000 m² to 15,000 m²				
Dry risers	m² GIA	4.25	to	5.20
Sprinkler installation	m² GIA	19.30	to	23.50
Protective Installations				
Earthing				
Buildings up to 3,000 m²	m² GIA	2.65	to	3.25
Buildings over 3,000 m² to 15,000 m²	m² GIA	2.65	to	3.25
Lightning Protection				
Buildings up to 3,000 m²	m² GIA	3.80	to	4.60
Buildings over 3,000 m² to 15,000 m²	m² GIA	2.60	to	3.20
5.12 Communications, Security and Control Installations				
Fire Alarms (single stage)				
Buildings up to 3,000 m²	m² GIA	9.40	to	11.50
Buildings over 3,000 m² to 15,000 m²	m² GIA	11.90	to	14.50
Fire Alarms (phased evacuation)				
Buildings over 3,000 m² to 15,000 m²	m² GIA	14.00	to	17.20
IT (Wireways only)				
Buildings up to 3,000 m²	m² GIA	2.55	to	3.15
Buildings over 3,000 m² to 15,000 m²	m² GIA	2.55	to	3.15
Security				
Buildings up to 3,000 m²	m² GIA	7.30	to	8.90
Buildings over 3,000 m² to 15,000 m²	m² GIA	8.30	to	10.20
BMS Controls; including MCC panels and control cabling				
Full air conditioning	m² GIA	18.80	to	23.00
Fit-Out				
5.6 Space Heating and Air Conditioning				
LPHW Heating Installation; including gas installations				
Building up to 3,000 m²	m² NIA	66.00	to	80.00
Building over 3,000 m² to 15,000 m²	m² NIA	53.00	to	65.00
Comfort cooling				
2 pipe fan coil for building up to 3,000 m²	m² NIA	95.00	to	115.00
2 pipe fan coil for building over 3,000 m² to 15,000 m²	m² NIA	92.00	to	110.00
2 pipe variable refrigerant volume (VRV) for building up to 3,000 m²	m² NIA	72.00	to	88.00
Full air conditioning				
4 pipe fan coil for building up to 3,000 m²	m² NIA	145.00	to	180.00
4 pipe fan coil for building over 3,000 m² to 15,000 m²	m² NIA	125.00	to	155.00
4 pipe variable refrigerant volume for building up to 3,000 m²	m² NIA	115.00	to	140.00
Ventilated (active) chilled beams for building over 3,000 m² to 15,000 m²	m² NIA	140.00	to	175.00
Chilled beam exposed services for building over 3,000 m² to 15,000 m²	m² NIA	220.00	to	265.00
Concealed passive chilled beams for building over 3,000 m² to 15,000 m²	m² NIA	125.00	to	150.00
Chilled ceiling for building over 3,000 m² to 15,000 m²	m² NIA	220.00	to	265.00

RIBA STAGE C/D/E DESIGN

Item	Unit	Range £		
Elemental Rates for Alternative Engineering Services Solutions – Offices – cont				
Fit-Out – cont				
Chilled ceiling/perimeter beams for building over 3,000 m² to 15,000 m²	m² NIA	235.00	to	290.00
Displacement for building over 3,000 m² to 15,000 m²	m² NIA	84.00	to	100.00
5.8 Electrical Installations				
Lighting Installations (including lighting controls and luminaires)				
Buildings up to 3,000 m²	m² NIA	52.00	to	64.00
Buildings over 3,000 m² to 15,000 m²	m² NIA	65.00	to	79.00
Electrical Installations for Mechanical Plant				
Buildings up to 3,000 m²	m² NIA	7.80	to	9.55
Buildings over 3,000 m² to 15,000 m²	m² NIA	7.80	to	9.55
5.11 Protective Installations				
Fire Protection over 3,000 m² to 15,000 m²				
Sprinkler installation	m² NIA	29.00	to	35.00
Earthing				
Buildings up to 3,000 m²	m² NIA	2.60	to	3.20
Buildings over 3,000 m² to 15,000 m²	m² NIA	2.60	to	3.20
5.12 Communications, Security and Control Installations				
Fire Alarms (single stage)				
Buildings up to 3,000 m²	m² NIA	11.90	to	14.50
Buildings over 3,000 m² to 15,000 m²	m² NIA	11.90	to	14.50
Fire Alarms (phased evacuation)				
Buildings over 3,000 m² to 15,000 m²	m² NIA	16.10	to	19.50
BMS Controls; including MCC panels and control cabling				
Full air conditioning	m² NIA	19.30	to	23.50

RIBA STAGE C/D/E DESIGN

Item	Unit	Range £		
Elemental Rates for Alternative Engineering Services Solutions – Hotels				
The following examples of building types indicate the range of rates for alternative design solutions for each of the engineering services elements based on Gross Internal Area for the Shell and Core and Net Internal Area for the Fit-Out. Fit-Out is assumed to be to Cat A Standard. Consideration should be made for the size of the building, which may affect the economies of scale for rates, i.e. the larger the building the lower the rates.				
5 Services				
Shell & Core				
5.1 Sanitary Installations				
2 to 3 star	m² GIA	26.00	to	32.00
4 to 5 star	m² GIA	38.00	to	46.00
5.3 Disposal Installations				
2 to 3 star	m² GIA	29.00	to	35.00
4 to 5 star	m² GIA	30.00	to	36.00
5.4 Water Installations				
2 to 3 star	m² GIA	36.00	to	44.00
4 to 5 star	m² GIA	49.00	to	60.00
5.6 Space Heating and Air Conditioning				
LPHW Heating Installation; including gas installations				
2 to 3 star	m² GIA	40.00	to	49.00
4 to 5 star	m² GIA	41.00	to	50.00
Air Conditioning; including ventilation				
2 to 3 star – 4 pipe Fan Coil	m² GIA	235.00	to	290.00
4 to 5 star – 4 pipe Fan Coil	m² GIA	240.00	to	295.00
2 to 3 star – 3 pipe variable refrigerant volume	m² GIA	145.00	to	180.00
4 to 5 star – 3 pipe variable refrigerant volume	m² GIA	145.00	to	180.00
5.8 Electrical Installations				
LV Installations				
Standby generators (life safety only)				
2 to 3 star	m² GIA	15.70	to	19.10
4 to 5 star	m² GIA	16.10	to	19.70
LV distribution				
2 to 3 star	m² GIA	34.00	to	42.00
4 to 5 star	m² GIA	45.00	to	55.00
Lighting Installations				
2 to 3 star	m² GIA	23.00	to	28.50
4 to 5 star	m² GIA	35.00	to	43.00
Small Power				
2 to 3 star	m² GIA	8.10	to	9.90
4 to 5 star	m² GIA	13.60	to	16.60
Electrical installations for mechanical plant				
2 to 3 star	m² GIA	7.00	to	8.60
4 to 5 star	m² GIA	7.00	to	8.60

RIBA STAGE C/D/E DESIGN

Item	Unit	Range £		
Elemental Rates for Alternative Engineering Services Solutions – Hotels – cont				
Shell & Core – cont				
5.11 Protective Installations				
Fire Protection				
2 to 3 star – Dry risers	m² GIA	10.90	to	13.30
4 to 5 star – Dry risers	m² GIA	10.90	to	13.30
2 to 3 star – Sprinkler installation	m² GIA	23.50	to	29.00
4 to 5 star – Sprinkler installation	m² GIA	26.50	to	32.00
Earthing				
2 to 3 star	m² GIA	1.60	to	2.00
4 to 5 star	m² GIA	1.60	to	2.00
Lightning Protection				
2 to 3 star	m² GIA	1.60	to	2.00
4 to 5 star	m² GIA	1.60	to	2.00
5.12 Communications, Security and Control Installations				
Fire Alarms				
2 to 3 star	m² GIA	15.70	to	19.10
4 to 5 star	m² GIA	16.10	to	19.70
IT				
2 to 3 star	m² GIA	13.10	to	16.10
4 to 5 star	m² GIA	13.10	to	16.10
Security				
2 to 3 star	m² GIA	20.00	to	25.00
4 to 5 star	m² GIA	21.00	to	25.00
BMS Controls; including MCC panels and control cabling				
2 to 3 star	m² GIA	11.60	to	14.20
4 to 5 star	m² GIA	29.00	to	35.00

RIBA STAGE C/D/E DESIGN

Item	Unit	Range £		
Elemental Rates for Alternative Engineering Services Solutions – Apartments				
The following examples of building types indicate the range of rates for alternative design solutions for each of the engineering services elements based on Gross Internal Area for the Shell and Core and Net Internal Area for the Fit-Out. Fit-Out is assumed to be to Cat A Standard. Consideration should be made for the size of the building, which may affect the economies of scale for rates, i.e. the larger the building the lower the rates.				
5 Services				
Shell & Core				
5.1 Sanitary Installations				
Affordable	m² GIA	0.35	to	0.45
Private	m² GIA	1.60	to	1.90
5.3 Disposal Installations				
Affordable	m² GIA	18.70	to	23.00
Private	m² GIA	21.00	to	25.00
5.4 Water Installations				
Affordable	m² GIA	18.30	to	22.00
Private	m² GIA	36.00	to	44.00
5.5 Heat Source				
Affordable	m² GIA	8.30	to	10.10
Private	m² GIA	10.30	to	12.70
5.6 Space Heating and Air Conditioning				
Affordable	m² GIA	22.50	to	27.50
Private	m² GIA	72.00	to	88.00
5.7 Ventilation Systems				
Affordable	m² GIA	15.50	to	18.90
Private	m² GIA	21.00	to	25.00
5.8 Electrical Installations				
Affordable	m² GIA	40.50	to	49.50
Private	m² GIA	62.00	to	76.00
5.9 Fuel Installations (Gas)				
Affordable	m² GIA	2.05	to	2.50
Private	m² GIA	3.15	to	3.85
5.11 Protective Installations				
Affordable	m² GIA	19.80	to	24.00
Private	m² GIA	28.50	to	35.00
5.12 Communications, Security and Control Installations				
Affordable	m² GIA	32.00	to	39.00
Private	m² GIA	43.00	to	52.00
5.13 Special Installations				
Affordable	m² GIA	8.10	to	9.90
Private	m² GIA	35.50	to	43.50

RIBA STAGE C/D/E DESIGN

Item	Unit	Range £		
Elemental Rates for Alternative Engineering Services Solutions – Apartments – cont				
Fit-Out				
5.1 Sanitary Installations				
Affordable	m² NIA	32.00	to	39.00
Private	m² NIA	90.00	to	110.00
5.3 Disposal Installations				
Affordable	m² NIA	9.40	to	11.50
Private	m² NIA	22.00	to	27.00
5.4 Water Installations				
Affordable	m² NIA	26.00	to	32.00
Private	m² NIA	49.00	to	60.00
5.6 Space Heating and Air Conditioning				
Affordable	m² NIA	72.00	to	88.00
Private	m² NIA	190.00	to	230.00
5.7 Ventilation Systems				
Affordable (whole house vent)	m² NIA	34.00	to	41.00
Private (whole house vent)	m² NIA	49.00	to	60.00
5.8 Electrical Installations				
Affordable	m² NIA	56.00	to	68.00
Private	m² NIA	100.00	to	125.00
5.9 Fuel Installations (Gas)				
Affordable	m² NIA			
Private	m² NIA	13.90	to	17.10
5.11 Protective Installations				
Affordable	m² NIA	21.00	to	26.00
Private	m² NIA	26.00	to	32.00
5.12 Communications, Security and Control Installations				
Affordable	m² NIA	17.60	to	21.50
Private	m² NIA	91.00	to	110.00
5.13 Special Installations				
Affordable	m² NIA	10.10	to	12.30
Private	m² NIA	28.00	to	34.00
Note: The range in cost differs due to the vast diversity in services strategies available. The lower end of the scale reflects no comfort cooling, radiator heating in lieu of underfloor, etc. The high end of the scale is based on central plant installations with good quality apartments, which includes comfort cooling, sprinklers, home network installations, video entry, higher quality of sanitaryware and good quality LED lighting.				

APPROXIMATE ESTIMATING RATES – 5 SERVICES

Item	Unit	Range £
5.3 Disposal Installations		
ABOVE GROUND DRAINAGE		
Soil and waste	point	365.00 to 445.00
5.4 Water Installations		
WATER INSTALLATIONS		
Cold Water	point	350.00 to 430.00
Hot Water	point	420.00 to 510.00
PIPEWORK		
HOT AND COLD WATER		
Excludes insulation, valves and ancillaries, etc.		
Light gauge copper tube to EN1057 R250 (TX) formerly BS 2871 Part 1 Table X with joints as described including allowance for waste, fittings and supports assuming average runs with capillary joints up to 54mm and bronze welded thereafter		
Horizontal High Level Distribution		
15 mm	m	33.00 to 41.00
22 mm	m	34.00 to 42.00
28 mm	m	43.00 to 53.00
35 mm	m	49.00 to 60.00
42 mm	m	57.00 to 69.00
54 mm	m	72.00 to 88.00
67 mm	m	105.00 to 130.00
76 mm	m	105.00 to 130.00
108 mm	m	155.00 to 190.00
Risers		
15 mm	m	19.00 to 23.00
22 mm	m	22.00 to 27.00
28 mm	m	33.00 to 41.00
35 mm	m	37.00 to 45.50
42 mm	m	45.00 to 55.00
54 mm	m	52.00 to 64.00
67 mm	m	91.00 to 110.00
76 mm	m	105.00 to 130.00
108 mm	m	130.00 to 160.00
Toilet Areas, etc., at Low Level		
15 mm	m	57.00 to 69.00
22 mm	m	70.00 to 86.00
28 mm	m	94.00 to 115.00
LTHW AND CHILLED WATER		
Excludes insulation, valves and ancillaries, etc.		
Black heavy weight mild steel tube to BS 1387 with joints in the running length, allowance for waste, fittings and supports assuming average runs		
Horizontal Distribution – Basements, etc.		
15 mm	m	41.00 to 50.00
20 mm	m	44.00 to 54.00
25 mm	m	49.50 to 61.00

APPROXIMATE ESTIMATING RATES – 5 SERVICES

Item	Unit	Range £
5.4 Water Installations – cont		
LTHW AND CHILLED WATER – cont		
32 mm	m	57.00 to 69.00
40 mm	m	65.00 to 79.00
50 mm	m	80.00 to 98.00
65 mm	m	86.00 to 105.00
80 mm	m	115.00 to 140.00
100 mm	m	145.00 to 180.00
125 mm	m	210.00 to 250.00
150 mm	m	280.00 to 340.00
200 mm	m	380.00 to 460.00
250 mm	m	460.00 to 560.00
300 mm	m	540.00 to 670.00
Risers		
15 mm	m	24.00 to 30.00
20 mm	m	29.00 to 35.00
25 mm	m	32.00 to 39.00
32 mm	m	36.00 to 44.00
40 mm	m	41.00 to 50.00
50 mm	m	51.00 to 62.00
65 mm	m	63.00 to 77.00
80 mm	m	86.00 to 105.00
100 mm	m	105.00 to 130.00
125 mm	m	130.00 to 160.00
150 mm	m	170.00 to 210.00
200 mm	m	270.00 to 330.00
250 mm	m	355.00 to 435.00
300 mm	m	400.00 to 490.00
On Floor Distribution		
15 mm	m	39.00 to 47.00
20 mm	m	44.00 to 54.00
25 mm	m	48.00 to 58.00
32 mm	m	55.00 to 67.00
40 mm	m	63.00 to 77.00
50 mm	m	74.00 to 90.00
Plantroom Areas, etc.		
15 mm	m	41.00 to 50.00
20 mm	m	45.00 to 55.00
25 mm	m	52.00 to 63.00
32 mm	m	60.00 to 74.00
40 mm	m	68.00 to 83.00
50 mm	m	80.00 to 98.00
65 mm	m	89.00 to 110.00
80 mm	m	125.00 to 150.00
100 mm	m	145.00 to 180.00
125 mm	m	220.00 to 265.00
150 mm	m	290.00 to 360.00
200 mm	m	385.00 to 470.00
250 mm	m	490.00 to 600.00
300 mm	m	570.00 to 700.00

APPROXIMATE ESTIMATING RATES – 5 SERVICES

Item	Unit	Range £		
5.5 Heat Source				
HEAT SOURCE				
Gas fired boilers including gas train and controls	kW	34.00	to	42.00
Gas fired boilers including gas train, controls, flue, plantroom pipework, valves and insulation, pumps and pressurization unit	kW	93.00	to	110.00
5.6 Space Heating and Air Conditioning				
CHILLED WATER				
Air cooled R134a refrigerant chiller including control panel, anti-vibration mountings	kW	130.00	to	160.00
Air cooled R134a refrigerant chiller including control panel, anti-vibration mountings, plantroom pipework, valves, insulation, pumps and pressurization units	kW	220.00	to	270.00
Water cooled R134a refrigerant chiller including control panel, anti-vibration mountings	kW	78.00	to	96.00
Water cooled R134a refrigerant chiller including control panel, anti-vibration mountings, plantroom pipework, valves, insulation, pumps and pressurization units	kW	165.00	to	200.00
Absorption steam medium chiller including control panel, anti-vibration mountings, plantroom pipework, valves, insulation, pumps and pressurization units	kW	305.00	to	370.00
HEAT REJECTION				
Open circuit, forced draft cooling tower	kW	72.00	to	88.00
Closed circuit, forced draft cooling tower	kW	65.00	to	79.00
Dry Air	kW	36.00	to	44.00
PUMPS				
Pumps including flexible connections, anti-vibration mountings	kPa	36.00	to	44.00
Pumps including flexible connections, anti-vibration mountings, plantroom pipework, valves, insulation and accessories	kPa	90.00	to	110.00
DUCTWORK				
The rates below allow for ductwork and for all other labour and material in fabrication, fittings, supports and jointing to equipment, stop and capped ends, elbows, bends, diminishing and transition pieces, regular and reducing couplings, volume control dampers, branch diffuser and 'snap on' grille connections, ties, 'Ys', crossover spigots, etc., turning vanes, regulating dampers, access doors and openings, hand-holes, test holes and covers, blanking plates, flanges, stiffeners, tie rods and all supports and brackets fixed to structure.				
Rectangular galvanized mild steel ductwork as HVCA DW 144 up to 1000 mm longest side	m²	60.00	to	74.00
Rectangular galvanized mild steel ductwork as HVCA DW 144 up to 2500 mm longest side	m²	65.00	to	79.00
Rectangular galvanized mild steel ductwork as HVCA DW 144 up to 3000 mm longest side and above	m²	89.00	to	110.00
Circular galvanized mild steel ductwork as HVCA DW 144	m²	65.00	to	79.00
Flat oval galvanized mild steel ductwork as HVCA DW 144 up to 545 mm wide	m²	65.00	to	79.00
Flat oval galvanized mild steel ductwork as HVCA DW 144 up to 880 mm wide	m²	72.00	to	88.00
Flat oval galvanized mild steel ductwork as HVCA DW 144 up to 1785 mm wide	m²	81.00	to	99.00

APPROXIMATE ESTIMATING RATES – 5 SERVICES

Item	Unit	Range £
5.6 Space Heating and Air Conditioning – cont		
PACKAGED AIR HANDLING UNITS		
Air handling unit including LPHW preheater coil, pre-filter panel, LPHW heater coils, chilled water coil, filter panels, inverter drive, motorized volume control dampers, sound attenuation, flexible connections to ductwork and all anti-vibration mountings	m^3/s	7200.00 to 8900.00
5.7 Ventilation Systems		
EXTRACT FANS		
Extract fan including inverter drive, sound attenuation, flexible connections to ductwork and all anti-vibration mountings	m^3/s	1425.00 to 1750.00
5.8 Electrical Installations		
HV/LV INSTALLATIONS		
The cost of HV/LV equipment will vary according to the electricity supplier's requirements, the duty required and the actual location of the site. For estimating purposes, the items indicated below are typical of the equipment required in a HV substation incorporated into a building.		
RING MAIN UNIT		
Ring Main Unit, 11 kV including electrical terminations	point	15500.00 to 19000.00
TRANSFORMERS		
Oil filled transformers, 11 kV to 415 kV including electrical terminations	kVA	17.10 to 21.00
Cast resin transformers, 11 kV to 415 kV including electrical terminations	kVA	18.00 to 22.00
Midal filled transformers, 11 kV to 415 kV including electrical terminations	kVA	19.10 to 23.00
HV SWITCHGEAR		
Cubicle section HV switchpanel, Form 4 type 6 including air circuit breakers, meters and electrical terminations	section	23000.00 to 28000.00
LV SWITCHGEAR		
LV switchpanel, Form 3 including all isolators, fuses, meters and electrical terminations	isolator	2250.00 to 2750.00
LV switchpanel, Form 4 type 5 including all isolators, fuses, meters and electrical terminations	isolator	3550.00 to 4350.00
EXTERNAL PACKAGED SUBSTATION		
Extra over cost for prefabricated packaged substation housing, excludes base and protective security fencing	each	23000.00 to 28000.00
STANDBY GENERATING SETS		
Diesel powered including control panel, flue, oil day tank and attenuation		
Approximate installed cost, LV	kVA	285.00 to 350.00
Approximate installed cost, HV	kVA	320.00 to 390.00
UNINTERRUPTIBLE POWER SUPPLY		
Rotary UPS including control panel and choke transformer (excludes distribution)		
Approximate installed cost (range 1000 kVA to 2500 kVA)	kVA	455.00 to 560.00
Static UPS including control panel, automatic bypass, DC isolator and batteries for 30 minutes standby (excludes distribution)		
Approximate installed cost (range 500 kVA to 1000 kVA)	kVA	245.00 to 300.00

APPROXIMATE ESTIMATING RATES – 5 SERVICES

Item	Unit	Range £		
SMALL POWER				
Approximate prices for wiring of power points of length not exceeding 20 m, including accessories, wireways but excluding distribution boards				
13 amp Accessories				
Wired in PVC insulated twin and earth cable in ring main circuit				
Domestic properties	point	56.00	to	68.00
Commercial properties	point	77.00	to	94.00
Industrial properties	point	77.00	to	94.00
Wired in PVC insulated twin and earth cable in radial circuit				
Domestic properties	point	77.00	to	94.00
Commercial properties	point	91.00	to	110.00
Industrial properties	point	91.00	to	110.00
Wired in LSF insulated single cable in ring main circuit				
Commercial properties	point	91.00	to	110.00
Industrial properties	point	91.00	to	110.00
Wired in LSF insulated single cable in radial circuit				
Commercial properties	point	110.00	to	140.00
Industrial properties	point	110.00	to	140.00
45 amp wired in PVC insulated twin and earth cable				
Domestic properties	point	110.00	to	140.00
Low Voltage Power Circuits				
Three phase four wire radial circuit feeding an individual load, wired in LSF insulated single cable including all wireways, isolator, not exceeding 10 metres; in commercial properties				
1.5	point	185.00	to	225.00
2.5	point	200.00	to	250.00
4	point	210.00	to	260.00
6	point	230.00	to	280.00
10	point	275.00	to	335.00
16	point	305.00	to	375.00
Three phase four core radial circuit feeding an individual load item, wired in LSF/SWA/XLPE insulated cable including terimations, isolator, clipped to surface, not exceeding 10 metres; in commercial properties				
1.5	point	140.00	to	170.00
2.5	point	150.00	to	190.00
4	point	165.00	to	200.00
6	point	185.00	to	225.00
10	point	275.00	to	335.00
16	point	350.00	to	430.00
LIGHTING				
Approximate prices for wiring of lighting points including rose, wireways but excluding distribution boards, luminaires and switches				
Final Circuits				
Wired in PVC insulated twin and earth cable				
Domestic properties	point	40.00	to	49.00
Commercial properties	point	50.00	to	61.00
Industrial properties	point	50.00	to	61.00

APPROXIMATE ESTIMATING RATES – 5 SERVICES

Item	Unit	Range £
5.8 Electrical Installations – cont		
Final Circuits – cont		
Wired in LSF insulated single cable		
Commercial properties	point	66.00 to 80.00
Industrial properties	point	66.00 to 80.00
ELECTRICAL WORKS IN CONNECTION WITH MECHANICAL SERVICES		
The cost of electrical connections to mechanical services equipment will vary depending on the type of building and complexity of the equipment.		
Typical rate for power wiring, isolators and associated wireways.	m^2	7.00 to 8.60
FIRE ALARMS		
Cost per point for two core MICC insulated wired system including all terminations, supports and wireways		
Call point	point	255.00 to 310.00
Smoke detector	point	220.00 to 270.00
Smoke/heat detector	point	255.00 to 310.00
Heat detector	point	235.00 to 290.00
Heat detector and sounder	point	215.00 to 260.00
Input/output/relay units	point	305.00 to 375.00
Alarm sounder	point	235.00 to 290.00
Alarm sounder/beacon	point	285.00 to 350.00
Speakers/voice sounders	point	285.00 to 350.00
Speakers/voice sounders (weatherproof)	point	305.00 to 375.00
Beacon/strobe	point	220.00 to 270.00
Beacon/strobe (weatherproof)	point	320.00 to 390.00
Door release units	point	320.00 to 390.00
Beam detector	point	910.00 to 1100.00
For costs for zone control panel, battery chargers and batteries, see 'Prices for Measured Work' section.		
EXTERNAL LIGHTING		
Estate Road Lighting		
Post type road lighting lantern 70 watt CDM-T 3000 K complete with 5 m high column with hinged lockable door, control gear and cut-out including 2.5 mm two core butyl cable internal wiring, interconnectionns and earthing fed by 16 mm^2 four core XLPE/SWA/LSF cable and terminations. Approximate installed price per metre road length (based on 300 metres run) including time switch but excluding builder's work in connection		
Columns erected at 30 m intervals along road (cost per m of road)	m	70.00 to 86.00
Bollard Lighting		
Bollard lighting fitting 26 watt TC-D 3500k including control gear, all internal wiring, interconnectionns, earthing and 25 metres of 2.5 mm^2 three core XLPE/SWA/LSF cable		
Approximate installed price excluding builder's work in connection	each	1000.00 to 1225.00

APPROXIMATE ESTIMATING RATES – 5 SERVICES

Item	Unit	Range £		
Outdoor Flood Lighting				
Wall mounted outdoor flood light fitting complete with tungsten halogen lamp, mounting bracket, wire guard and all internal wiring and containment, fixed to brickwork or concrete and connected				
Installed price 500 watt	point	205.00	to	250.00
Installed price 1000 watt	point	245.00	to	300.00
Pedestal mounted outdoor flood light fitting complete with 1000 watt MBF/U lamp, control gear, contained in weatherproof steel box, all internal wiring and containment, interconnections and earthing, fixed to brickwork or concrete and connected				
Approximate installed price excluding builder's work in connection	each	1000.00	to	1225.00
BUILDING MANAGEMENT INSTALLATIONS				
Category A Fit-Out				
Option 1 – 185 Nr four pipe fan coil – 740 points				
1.0 Field Equipment				
Network devices; Valves/actuators; Sensing devices	point	70.00	to	70.00
2.0 Cabling				
Power – from local isolator to DDC controller; Control – from DDC controller to field equipment	point	40.00	to	40.00
3.0 Programming				
Software – central facility; Software – network devices; Graphics	point	22.00	to	22.00
4.0 On site testing and commissioning				
Equipment; Programming/graphics; Power and control cabling	point	23.00	to	23.00
Total Option 1 – Four pipefan coil	point	155.00	to	155.00
Cost/FCU	each	630.00	to	630.00
Option 2 – 185 Nr two pipe fan coil system with electric heating – 740 points				
1.0 Field Equipment				
Network devices; Valves/actuators/thyristors; Sensing devices	point	91.00	to	91.00
2.0 Cabling				
Power – from local isolator to DDC controller; Control – from DDC controller to field equipment	point	42.00	to	42.00
3.0 Programming				
Software – central facility; Software – network devices; Graphics	point	23.00	to	23.00
4.0 On site testing and commissioning				
Equipment; Programming/graphics; Power and control cabling	point	24.00	to	24.00
Total Option 2 – Two pipe fan coil with electric heating	point	180.00	to	180.00
Cost/FCU	each	710.00	to	710.00
Option 3 – 180 Nr chilled beams with perimeter heating – 567 points				
1.0 Field Equipment				
Network devices; Valves/actuators; Sensing devices	point	87.00	to	87.00
2.0 Cabling				
Power – from local isolator to DDC controller; Control – from DDC controller to field equipment	point	59.00	to	59.00

APPROXIMATE ESTIMATING RATES – 5 SERVICES

Item	Unit	Range £
5.8 Electrical Installations – cont		
Option 3 – 180 Nr chilled beams with perimeter heating – 567 points – cont		
3.0 Programming		
Cost/Point	point	29.00 to 29.00
4.0 On site testing and commissioning		
Equipment; Programming/Graphics; Power and control cabling	point	31.00 to 31.00
Total Option 3 – Chilled beams with perimeter heating	point	205.00 to 205.00
Cost/Chilled Beam	each	660.00 to 660.00
Shell & Core Only		
Main Plant – Cost/BMS Points	each	680.00 to 680.00
Landlord FCUs – Cost/Terminal Units	each	680.00 to 680.00
Trade Contract Preliminaries	%	26.00 to 26.00
Notes: The following are included in points, rates		
– DDC Controllers/Control Enclosures/Control Panels		
– Motor Control Centre (MCC)		
– Field Devices		
– Control and Power Cabling from DDC Controllers/MCC		
– Programming		
– On Site Testing and Commissioning		
5.10 Lift and Conveyor Systems		
LIFT INSTALLATIONS		
The cost of lift installations will vary depending upon a variety of circumstances. The following prices assume a car height of 2.2 metres, manufacturer's standard car finish, brushed stainless steel 2 panel centre opening doors to BSEN81 Part 1 & 2 and Lift Regulations 1997.		
Passenger Lifts, machine room above		
Electrically operated AC drive serving 2 levels with directional collective controls and a speed of 1.0 m/s		
8 Person	item	60000.00 to 73000.00
10 Person	item	64000.00 to 78000.00
13 Person	item	68000.00 to 83000.00
17 Person	item	77000.00 to 93000.00
21 Person	item	86000.00 to 105000.00
26 Person	item	96000.00 to 120000.00
Electrically operated AC drive serving 4 levels and a speed of 1.0 m/s		
8 Person	item	70000.00 to 86000.00
10 Person	item	74000.00 to 90000.00
13 Person	item	78000.00 to 96000.00
17 Person	item	87000.00 to 105000.00
21 Person	item	97000.00 to 120000.00
26 Person	item	115000.00 to 140000.00
Electrically operated AC drive serving 6 levels and a speed of 1.0 m/s		
8 Person	item	79000.00 to 97000.00
10 Person	item	83000.00 to 100000.00
13 Person	item	87000.00 to 105000.00

APPROXIMATE ESTIMATING RATES – 5 SERVICES

Item	Unit	Range £
17 Person	item	97000.00 to 120000.00
21 Person	item	115000.00 to 140000.00
26 Person	item	125000.00 to 150000.00
Electrically operated AC drive serving 8 levels and a speed of 1.0 m/s		
8 Person	item	88000.00 to 110000.00
10 Person	item	93000.00 to 110000.00
13 Person	item	97000.00 to 120000.00
17 Person	item	115000.00 to 140000.00
21 Person	item	120000.00 to 145000.00
26 Person	item	140000.00 to 170000.00
Electrically operated AC drive serving 10 levels and a speed of 1.0 m/s		
8 Person	item	97000.00 to 120000.00
10 Person	item	100000.00 to 125000.00
13 Person	item	100000.00 to 125000.00
17 Person	item	120000.00 to 145000.00
21 Person	item	135000.00 to 160000.00
26 Person	item	155000.00 to 190000.00
Electrically operated AC drive serving 12 levels and a speed of 1.0 m/s		
8 Person	item	100000.00 to 125000.00
10 Person	item	115000.00 to 140000.00
13 Person	item	115000.00 to 140000.00
17 Person	item	130000.00 to 160000.00
21 Person	item	145000.00 to 180000.00
26 Person	item	165000.00 to 200000.00
Electrically operated AC drive serving 14 levels and a speed of 1.0 m/s		
8 Person	item	120000.00 to 145000.00
10 Person	item	120000.00 to 145000.00
13 Person	item	130000.00 to 160000.00
17 Person	item	145000.00 to 180000.00
21 Person	item	155000.00 to 190000.00
26 Person	item	180000.00 to 220000.00
Add to above for:		
Increase speed from 1.0 m/s to 1.6 m/s		
8 Person	item	4050.00 to 4950.00
10 Person	item	4200.00 to 5200.00
13 Person	item	4300.00 to 5300.00
17 Person	item	4300.00 to 5300.00
21 Person	item	4300.00 to 5300.00
26 Person	item	4300.00 to 5300.00
Increase speed from 1.6 m/s to 2.0 m/s		
8 Person	item	5700.00 to 6900.00
10 Person	item	6100.00 to 7500.00
13 Person	item	6700.00 to 8100.00
17 Person	item	3000.00 to 3600.00
21 Person	item	3050.00 to 3750.00
26 Person	item	3050.00 to 3750.00
Increase speed from 2.0 m/s to 2.5 m/s		
8 Person	item	2075.00 to 2500.00
10 Person	item	2075.00 to 2500.00
13 Person	item	2475.00 to 3000.00

APPROXIMATE ESTIMATING RATES – 5 SERVICES

Item	Unit	Range £
5.10 Lift and Conveyor Systems – cont		
Add to above for: – cont		
17 Person	item	2475.00 to 3000.00
21 Person	item	3000.00 to 3600.00
26 Person	item	3000.00 to 3600.00
Enhanced finish to car – Centre mirror, flat ceiling, carpet		
8 Person	item	2900.00 to 3500.00
10 Person	item	3100.00 to 3800.00
13 Person	item	3100.00 to 3800.00
17 Person	item	3500.00 to 4300.00
21 Person	item	4100.00 to 5000.00
26 Person	item	4700.00 to 5800.00
Bottom motor room		
8 Person	item	7100.00 to 8600.00
10 Person	item	7100.00 to 8600.00
13 Person	item	7100.00 to 8600.00
17 Person	item	8700.00 to 10500.00
21 Person	item	8700.00 to 10500.00
26 Person	item	8800.00 to 11000.00
Fire fighting control		
8 Person	item	5700.00 to 6900.00
10 Person	item	5700.00 to 6900.00
13 Person	item	5700.00 to 6900.00
17 Person	item	5700.00 to 6900.00
21 Person	item	5700.00 to 6900.00
26 Person	item	5700.00 to 6900.00
Glass back		
8 Person	item	2550.00 to 3100.00
10 Person	item	2900.00 to 3500.00
13 Person	item	3400.00 to 4200.00
17 Person	item	4100.00 to 5000.00
21 Person	item	4100.00 to 5000.00
26 Person	item	4100.00 to 5000.00
Glass doors		
8 Person	item	20000.00 to 24000.00
10 Person	item	20000.00 to 24000.00
13 Person	item	21500.00 to 26000.00
17 Person	item	23000.00 to 28500.00
21 Person	item	21500.00 to 26000.00
26 Person	item	21500.00 to 26000.00
Painting to entire pit		
8 Person	item	2050.00 to 2500.00
10 Person	item	2050.00 to 2500.00
13 Person	item	2050.00 to 2500.00
17 Person	item	2050.00 to 2500.00
21 Person	item	2050.00 to 2500.00
26 Person	item	2050.00 to 2500.00

APPROXIMATE ESTIMATING RATES – 5 SERVICES

Item	Unit	Range £
Dual seal shaft		
8 Person	item	4100.00 to 5000.00
10 Person	item	4100.00 to 5000.00
13 Person	item	4100.00 to 5000.00
17 Person	item	4950.00 to 6100.00
21 Person	item	4950.00 to 6100.00
26 Person	item	4950.00 to 6100.00
Dust sealing machine room		
8 Person	item	810.00 to 990.00
10 Person	item	810.00 to 990.00
13 Person	item	1300.00 to 1575.00
17 Person	item	1300.00 to 1575.00
21 Person	item	1300.00 to 1575.00
26 Person	item	1300.00 to 1575.00
Intercom to reception desk and security room		
8 Person	item	405.00 to 495.00
10 Person	item	405.00 to 495.00
13 Person	item	405.00 to 495.00
17 Person	item	405.00 to 495.00
21 Person	item	405.00 to 495.00
26 Person	item	405.00 to 495.00
Heating, cooling and ventilation to machine room		
8 Person	item	1025.00 to 1275.00
10 Person	item	1025.00 to 1275.00
13 Person	item	1025.00 to 1275.00
17 Person	item	1025.00 to 1275.00
21 Person	item	1025.00 to 1275.00
26 Person	item	1025.00 to 1275.00
Shaft lighting/small power		
8 Person	item	3550.00 to 4350.00
10 Person	item	3550.00 to 4350.00
13 Person	item	3550.00 to 4350.00
17 Person	item	3550.00 to 4350.00
21 Person	item	3550.00 to 4350.00
26 Person	item	3550.00 to 4350.00
Motor room lighting/small power		
8 Person	item	1300.00 to 1575.00
10 Person	item	1300.00 to 1575.00
13 Person	item	1625.00 to 1975.00
17 Person	item	1625.00 to 1975.00
21 Person	item	1625.00 to 1975.00
26 Person	item	1625.00 to 1975.00
Lifting beams		
8 Person	item	1550.00 to 1875.00
10 Person	item	1550.00 to 1875.00
13 Person	item	1550.00 to 1875.00
17 Person	item	1725.00 to 2125.00
21 Person	item	1725.00 to 2125.00
26 Person	item	1725.00 to 2125.00

APPROXIMATE ESTIMATING RATES – 5 SERVICES

Item	Unit	Range £
5.10 Lift and Conveyor Systems – cont		
Add to above for: – cont		
10 mm Equipotential bonding of all entrance metalwork		
8 Person	item	810.00 to 990.00
10 Person	item	810.00 to 990.00
13 Person	item	810.00 to 990.00
17 Person	item	810.00 to 990.00
21 Person	item	810.00 to 990.00
26 Person	item	810.00 to 990.00
Shaft secondary steelwork		
8 Person	item	5900.00 to 7300.00
10 Person	item	6100.00 to 7400.00
13 Person	item	6300.00 to 7700.00
17 Person	item	6400.00 to 7900.00
21 Person	item	6400.00 to 7900.00
26 Person	item	6400.00 to 7900.00
Independent insurance inspection		
8 Person	item	1950.00 to 2375.00
10 Person	item	1950.00 to 2375.00
13 Person	item	1950.00 to 2375.00
17 Person	item	1950.00 to 2375.00
21 Person	item	1950.00 to 2375.00
26 Person	item	1950.00 to 2375.00
12 Month warranty service		
8 Person	item	1050.00 to 1275.00
10 Person	item	1050.00 to 1275.00
13 Person	item	1050.00 to 1275.00
17 Person	item	1050.00 to 1275.00
21 Person	item	1050.00 to 1275.00
26 Person	item	1050.00 to 1275.00
Passenger Lifts, machine room less		
Electrically operated AC drive serving 2 levels with directional collective controls and a speed of 1.0 m/s		
8 Person	item	54000.00 to 65000.00
10 Person	item	60000.00 to 74000.00
13 Person	item	63000.00 to 77000.00
17 Person	item	76000.00 to 92000.00
21 Person	item	83000.00 to 100000.00
26 Person	item	92000.00 to 110000.00
Electrically operated AC drive serving 4 levels and a speed of 1.0 m/s		
8 Person	item	62000.00 to 76000.00
10 Person	item	68000.00 to 83000.00
13 Person	item	72000.00 to 88000.00
17 Person	item	86000.00 to 105000.00
21 Person	item	93000.00 to 110000.00
26 Person	item	100000.00 to 125000.00

APPROXIMATE ESTIMATING RATES – 5 SERVICES

Item	Unit	Range £
Electrically operated AC drive serving 6 levels and a speed of 1.0 m/s		
8 Person	item	72000.00 to 88000.00
10 Person	item	77000.00 to 93000.00
13 Person	item	81000.00 to 99000.00
17 Person	item	93000.00 to 110000.00
21 Person	item	100000.00 to 125000.00
26 Person	item	110000.00 to 130000.00
Electrically operated AC drive serving 8 levels and a speed of 1.0 m/s		
8 Person	item	80000.00 to 98000.00
10 Person	item	85000.00 to 100000.00
13 Person	item	90000.00 to 110000.00
17 Person	item	100000.00 to 125000.00
21 Person	item	110000.00 to 130000.00
26 Person	item	120000.00 to 150000.00
Electrically operated AC drive serving 10 levels and a speed of 1.0 m/s		
8 Person	item	89000.00 to 110000.00
10 Person	item	92000.00 to 110000.00
13 Person	item	100000.00 to 120000.00
17 Person	item	110000.00 to 130000.00
21 Person	item	120000.00 to 150000.00
26 Person	item	140000.00 to 170000.00
Electrically operated AC drive serving 12 levels and a speed of 1.0 m/s		
8 Person	item	100000.00 to 120000.00
10 Person	item	100000.00 to 120000.00
13 Person	item	110000.00 to 130000.00
17 Person	item	120000.00 to 150000.00
21 Person	item	130000.00 to 160000.00
26 Person	item	145000.00 to 180000.00
Electrically operated AC drive serving 14 levels and a speed of 1.0 m/s		
8 Person	item	110000.00 to 130000.00
10 Person	item	110000.00 to 130000.00
13 Person	item	115000.00 to 140000.00
17 Person	item	140000.00 to 170000.00
21 Person	item	145000.00 to 180000.00
26 Person	item	165000.00 to 200000.00
Add to above for:		
Increase speed from 1.0 m/s to 1.6 m/s		
8 Person	item	2650.00 to 3250.00
10 Person	item	2650.00 to 3250.00
13 Person	item	2800.00 to 3400.00
17 Person	item	4300.00 to 5200.00
21 Person	item	4950.00 to 6100.00
26 Person	item	6300.00 to 7700.00
Enhanced finish to car – Centre mirror, flat ceiling, carpet		
8 Person	item	2900.00 to 3500.00
10 Person	item	3050.00 to 3750.00
13 Person	item	3050.00 to 3750.00
17 Person	item	3500.00 to 4300.00
21 Person	item	4050.00 to 4950.00
26 Person	item	4800.00 to 5800.00

APPROXIMATE ESTIMATING RATES – 5 SERVICES

Item	Unit	Range £
5.10 Lift and Conveyor Systems – cont		
Add to above for: – cont		
Fire fighting control		
8 Person	item	5600.00 to 6800.00
10 Person	item	5600.00 to 6800.00
13 Person	item	5600.00 to 6800.00
17 Person	item	5600.00 to 6800.00
21 Person	item	5600.00 to 6800.00
26 Person	item	5600.00 to 6800.00
Painting to entire pit		
8 Person	item	2025.00 to 2475.00
10 Person	item	2025.00 to 2475.00
13 Person	item	2025.00 to 2475.00
17 Person	item	2025.00 to 2475.00
21 Person	item	2025.00 to 2475.00
26 Person	item	2025.00 to 2475.00
Dual seal shaft		
8 Person	item	4100.00 to 5000.00
10 Person	item	4100.00 to 5000.00
13 Person	item	4100.00 to 5000.00
17 Person	item	4950.00 to 6100.00
21 Person	item	4950.00 to 6100.00
26 Person	item	4950.00 to 6100.00
Shaft lighting/small power		
8 Person	item	3550.00 to 4350.00
10 Person	item	3550.00 to 4350.00
13 Person	item	3550.00 to 4350.00
17 Person	item	3550.00 to 4350.00
21 Person	item	3550.00 to 4350.00
26 Person	item	3550.00 to 4350.00
Intercom to reception desk and security room		
8 Person	item	405.00 to 495.00
10 Person	item	405.00 to 495.00
13 Person	item	405.00 to 495.00
17 Person	item	405.00 to 495.00
21 Person	item	405.00 to 495.00
26 Person	item	405.00 to 495.00
Lifting beams		
8 Person	item	1525.00 to 1875.00
10 Person	item	1525.00 to 1875.00
13 Person	item	1525.00 to 1875.00
17 Person	item	1725.00 to 2125.00
21 Person	item	1725.00 to 2125.00
26 Person	item	1725.00 to 2125.00
10 mm Equipotential bonding of all entrance metalwork		
8 Person	item	810.00 to 990.00
10 Person	item	810.00 to 990.00
13 Person	item	810.00 to 990.00
17 Person	item	810.00 to 990.00
21 Person	item	810.00 to 990.00
26 Person	item	810.00 to 990.00

APPROXIMATE ESTIMATING RATES – 5 SERVICES

Item	Unit	Range £
Shaft secondary steelwork		
8 Person	item	5800.00 to 7000.00
10 Person	item	5900.00 to 7300.00
13 Person	item	6100.00 to 7500.00
17 Person	item	6300.00 to 7700.00
21 Person	item	6300.00 to 7700.00
26 Person	item	6300.00 to 7700.00
Independent insurance inspection		
8 Person	item	1875.00 to 2275.00
10 Person	item	1875.00 to 2275.00
13 Person	item	1875.00 to 2275.00
17 Person	item	1875.00 to 2275.00
21 Person	item	1875.00 to 2275.00
26 Person	item	1875.00 to 2275.00
12 Month warranty service		
8 Person	item	990.00 to 1200.00
10 Person	item	990.00 to 1200.00
13 Person	item	990.00 to 1200.00
17 Person	item	990.00 to 1200.00
21 Person	item	990.00 to 1200.00
26 Person	item	990.00 to 1200.00
Goods Lifts, machine room above		
Electrically operated two speed serving 2 levels to take 1000 kg load, prime coated internal finish and a speed of 1.0 m/s		
2000 kg	item	100000.00 to 125000.00
2250 kg	item	110000.00 to 130000.00
2500 kg	item	110000.00 to 130000.00
3000 kg	item	120000.00 to 150000.00
Electrically operated two speed serving 4 levels and a speed of 1.0 m/s		
2000 kg	item	110000.00 to 130000.00
2250 kg	item	120000.00 to 150000.00
2500 kg	item	125000.00 to 150000.00
3000 kg	item	140000.00 to 170000.00
Electrically operated two speed serving 6 levels and a speed of 1.0 m/s		
2000 kg	item	125000.00 to 150000.00
2250 kg	item	140000.00 to 170000.00
2500 kg	item	140000.00 to 170000.00
3000 kg	item	155000.00 to 190000.00
Electrically operated two speed serving 8 levels and a speed of 1.0 m/s		
2000 kg	item	140000.00 to 170000.00
2250 kg	item	145000.00 to 180000.00
2500 kg	item	155000.00 to 190000.00
3000 kg	item	175000.00 to 215000.00
Electrically operated two speed serving 10 levels and a speed of 1.0 m/s		
2000 kg	item	155000.00 to 190000.00
2250 kg	item	165000.00 to 200000.00
2500 kg	item	165000.00 to 200000.00
3000 kg	item	190000.00 to 230000.00

APPROXIMATE ESTIMATING RATES – 5 SERVICES

Item	Unit	Range £
5.10 Lift and Conveyor Systems – cont		
Goods Lifts, machine room above – cont		
Electrically operated two speed serving 12 levels and a speed of 1.0 m/s		
2000 kg	item	165000.00 to 200000.00
2250 kg	item	175000.00 to 215000.00
2500 kg	item	175000.00 to 215000.00
3000 kg	item	210000.00 to 250000.00
Electrically operated two speed serving 14 levels and a speed of 1.0 m/s		
2000 kg	item	180000.00 to 220000.00
2250 kg	item	195000.00 to 240000.00
2500 kg	item	195000.00 to 240000.00
3000 kg	item	220000.00 to 270000.00
Add to above for:		
Increase speed of travel from 1.0 m/s to 1.6 m/s		
2000 kg	item	1300.00 to 1575.00
Enhanced finish to car – Centre mirror, flat ceiling, carpet		
2000 kg	item	3550.00 to 4350.00
2250 kg	item	3550.00 to 4350.00
2500 kg	item	3550.00 to 4350.00
Bottom motor room		
2000 kg	item	8700.00 to 10500.00
Painting to entire pit		
2000 kg	item	2025.00 to 2475.00
2250 kg	item	2025.00 to 2475.00
2500 kg	item	2025.00 to 2475.00
3000 kg	item	2025.00 to 2475.00
Dual seal shaft		
2000 kg	item	4100.00 to 5000.00
2250 kg	item	4100.00 to 5000.00
2500 kg	item	4100.00 to 5000.00
3000 kg	item	4100.00 to 5000.00
Intercom to reception desk and security room		
2000 kg	item	410.00 to 500.00
2250 kg	item	410.00 to 500.00
2500 kg	item	410.00 to 500.00
3000 kg	item	410.00 to 500.00
Heating, cooling and ventilation to machine room		
2000 kg	item	1125.00 to 1375.00
2250 kg	item	1125.00 to 1375.00
2500 kg	item	1125.00 to 1375.00
3000 kg	item	1125.00 to 1375.00
Lifting beams		
2000 kg	item	1500.00 to 1850.00
2250 kg	item	1500.00 to 1850.00
2500 kg	item	1500.00 to 1850.00
3000 kg	item	1500.00 to 1850.00

APPROXIMATE ESTIMATING RATES – 5 SERVICES

Item	Unit	Range £
10 mm Equipotential bonding of all entrance metalwork		
2000 kg	item	810.00 to 990.00
2250 kg	item	810.00 to 990.00
2500 kg	item	810.00 to 990.00
3000 kg	item	810.00 to 990.00
Independent insurance inspection		
2000 kg	item	1925.00 to 2375.00
2250 kg	item	1925.00 to 2375.00
2500 kg	item	1925.00 to 2375.00
3000 kg	item	1925.00 to 2375.00
12 Month warranty service		
2000 kg	item	1050.00 to 1275.00
2250 kg	item	1225.00 to 1475.00
2500 kg	item	1225.00 to 1475.00
3000 kg	item	1225.00 to 1475.00
Goods Lift, machine room less		
Electrically operated two speed serving 2 levels to take 1000 kg load, prime coated internal finish and a speed of 1.0 m/s		
2000 kg	item	89000.00 to 110000.00
2250 kg	item	93000.00 to 110000.00
2500 kg	item	100000.00 to 125000.00
Electrically operated two speed serving 4 levels and a speed of 1.0 m/s		
2000 kg	item	100000.00 to 125000.00
2250 kg	item	105000.00 to 130000.00
2500 kg	item	115000.00 to 140000.00
Electrically operated two speed serving 6 levels and a speed of 1.0 m/s		
2000 kg	item	115000.00 to 140000.00
2250 kg	item	115000.00 to 140000.00
2500 kg	item	130000.00 to 160000.00
Electrically operated two speed serving 8 levels and a speed of 1.0 m/s		
2000 kg	item	120000.00 to 150000.00
2250 kg	item	130000.00 to 160000.00
2500 kg	item	140000.00 to 170000.00
Electrically operated two speed serving 10 levels and a speed of 1.0 m/s		
2000 kg	item	135000.00 to 165000.00
2250 kg	item	140000.00 to 170000.00
2500 kg	item	145000.00 to 180000.00
Electrically operated two speed serving 12 levels and a speed of 1.0 m/s		
2000 kg	item	120000.00 to 150000.00
2250 kg	item	155000.00 to 190000.00
2500 kg	item	165000.00 to 200000.00
Electrically operated two speed serving 14 levels and a speed of 1.0 m/s		
2000 kg	item	155000.00 to 190000.00
2250 kg	item	165000.00 to 200000.00
2500 kg	item	175000.00 to 215000.00

APPROXIMATE ESTIMATING RATES – 5 SERVICES

Item	Unit	Range £
5.10 Lift and Conveyor Systems – cont		
Add to above for:		
Increase speed of travel from 1.0 m/s to 1.6 m/s		
2000 kg	item	4950.00 to 6100.00
2250 kg	item	7900.00 to 9700.00
Enhanced finish to car – Centre mirror, flat ceiling, carpet		
2000 kg	item	4200.00 to 5100.00
2250 kg	item	5500.00 to 6700.00
2500 kg	item	8100.00 to 9900.00
Painting to entire pit		
2000 kg	item	2050.00 to 2500.00
2250 kg	item	2050.00 to 2500.00
2500 kg	item	2050.00 to 2500.00
Dual seal shaft		
2000 kg	item	4100.00 to 5000.00
2250 kg	item	4100.00 to 5000.00
2500 kg	item	4100.00 to 5000.00
Intercom to reception desk and security room		
2000 kg	item	410.00 to 500.00
2250 kg	item	410.00 to 500.00
2500 kg	item	410.00 to 500.00
Lifting beams		
2000 kg	item	1550.00 to 1875.00
2250 kg	item	1550.00 to 1875.00
2500 kg	item	1550.00 to 1875.00
10 mm Equipotential bonding of all entrance metalwork		
2000 kg	item	810.00 to 990.00
2250 kg	item	810.00 to 990.00
2500 kg	item	810.00 to 990.00
Independent insurance inspection		
2000 kg	item	1950.00 to 2375.00
2250 kg	item	1950.00 to 2375.00
2500 kg	item	1950.00 to 2375.00
12 Month warranty service		
2000 kg	item	1050.00 to 1275.00
2250 kg	item	1050.00 to 1275.00
2500 kg	item	1050.00 to 1275.00
ESCALATOR INSTALLATIONS		
30Ø Pitch escalator with a rise of 3 to 6 metres with standard balustrades		
1000 mm step width	item	81000.00 to 99000.00
Add to above for:		
Balustrade lighting	item	2500.00 to 3100.00
Skirting lighting	item	10000.00 to 12500.00
Emergency stop button pedestals	item	4500.00 to 5500.00
Truss cladding – stainless steel	item	25500.00 to 31000.00
Truss cladding – spray painted steel	item	22500.00 to 27500.00

APPROXIMATE ESTIMATING RATES – 5 SERVICES

Item	Unit	Range £
5.11 Protective Installations		
SPRINKLER INSTALLATION		
Recommended maximum area coverage per sprinkler head:		
Extra light hazard, 21 m² of floor area		
Ordinary hazard, 12 m² of floor area		
Extra high hazard, 9 m² of floor area		
Equipment		
Sprinkler equipment installation, pipework, valve sets, booster pumps and water storage	item	72000.00 to 88000.00
Price per sprinkler head; including pipework, valves and supports	point	190.00 to 230.00
HOSE REELS AND DRY RISERS		
Wall mounted concealed hose reel with 36 metre hose including approximately 15 metres of pipework and isolating valve:		
Price per hose reel	point	1800.00 to 2200.00
100 mm dry riser main including 2 way breeching valve and box, 65 mm landing valve, complete with padlock and leather strap and automatic air vent and drain valve:		
Price per landing	point	1525.00 to 1875.00
5.12 Communications, Security and Control Installations		
ACCESS CONTROL SYSTEMS		
Door mounted access control unit inclusive of door furniture, lock plus software; including up to 50 metres of cable and termination; including documentation testing and commissioning		
Internal single leaf door	point	1075.00 to 1325.00
Internal double door	point	1225.00 to 1475.00
External single leaf door	point	1300.00 to 1575.00
External double leaf door	point	1500.00 to 1850.00
Management control PC with printer software and commissioning up to 1000 users	point	15500.00 to 19000.00
CCTV INSTALLATIONS		
CCTV equipment inclusive of 50 m of cable including testing and commissioning		
Internal camera with bracket	point	900.00 to 1100.00
Internal camera with housing	point	1100.00 to 1350.00
Internal PTZ camera with bracket	point	1700.00 to 2075.00
External fixed camera with housing	point	1225.00 to 1475.00
External PTZ camera dome	point	2375.00 to 2900.00
External PTZ camera dome with power	point	3000.00 to 3600.00

APPROXIMATE ESTIMATING RATES – 5 SERVICES

Item	Unit	Range £		
5.12 Communications, Security and Control Installations – cont				
IT INSTALLATIONS				
DATA CABLING				
Complete channel link including patch leads, cable, panels, testing and documentation (excludes cabinets and/or frames, patch cords, backbone/ harness connectivity as well as containment)				
Low Level				
Cat 5e (up to 5,000 outlets)	point	49.50	to	61.00
Cat 5e (5,000 to 15,000 outlets)	point	39.50	to	48.50
Cat 6 (up to 5,000 outlets)	point	59.00	to	73.00
Cat 6 (5,000 to 15,000 outlets)	point	55.00	to	67.00
Cat 6a (up to 5,000 outlets)	point	75.00	to	91.00
Cat 6a (5,000 to 15,000 outlets)	point	69.00	to	85.00
Cat 7 (up to 5,000 outlets)	point	89.00	to	110.00
Cat 7 (5,000 to 15,000 outlets)	point	85.00	to	100.00
Note: LSZH cable based on average of 50 metres false floor low level installation assuming 1 workstation in 2.5 m × 2.5 m (to 3.2 m × 3.2 m) density, with 4 data points per workstation. Not applicable for installations with less than 250 No. outlets.				
High Level				
Cat 5e (up to 500 outlets)	point	59.00	to	73.00
Cat 5e (over 500 outlets)	point	49.50	to	61.00
Cat 6 (up to 500 outlets)	point	69.00	to	85.00
Cat 6 (over 500 outlets)	point	65.00	to	79.00
Cat 6a (up to 500 outlets)	point	85.00	to	100.00
Cat 6a (over 500 outlets)	point	79.00	to	97.00
Cat 7 (up to 500 outlets)	point	99.00	to	120.00
Cat 7 (over 500 outlets)	point	95.00	to	115.00
Note: High level at 10 × 10 m grid.				

BUILDING MODELS – ELEMENTAL COST SUMMARIES

Item	Unit	Range £		
AIRPORT TERMINAL BUILDING				
New build airport terminal building, premium quality, located in the South East, handling both domestic and international flights with a gross internal floor area (GIA) of 25,000 m². These costs exclude baggage handling, check-in systems, pre-check in and boarding security systems, vertical transportation and services to aircraft stands. Heat source is via district mains (excluded). Costs assume the works are executed under landside access/logistics environment				
5 Services				
5.1 Sanitary Installations	m²	5.00	to	7.00
5.3 Disposal Installations				
rainwater	m²	6.00	to	7.20
soil and waste	m²	7.00	to	8.40
condensate	m²	1.00	to	2.00
5.4 Water Installations	m²	17.00	to	20.00
5.6 Space Heating and Air Conditioning				
LTHW heating system	m²	74.00	to	81.00
chilled water system	m²	71.00	to	78.00
supply and extract air conditioning system	m²	99.00	to	110.00
allowance for service to communications room	m²	9.00	to	10.80
5.7 Ventilation Systems				
mechanical ventilation to baggage handling and plantrooms	m²	23.00	to	26.00
toilet extract ventilation	m²	8.00	to	9.60
smoke extract installation	m²	9.00	to	10.80
kitchen extract system	m²	3.00	to	3.90
5.8 Electrical Installations				
HV/LV Switchgear	m²	61.00	to	70.00
standby generator	m²	38.00	to	42.00
mains and sub-mains installation	m²	42.00	to	48.00
small power installation	m²	15.00	to	18.00
lighting and luminaries	m²	95.00	to	105.00
emergency llighting installation	m²	11.00	to	13.20
power to mechanical services	m²	6.00	to	7.20
5.9 Fuel Installations (gas)	m²	2.00	to	3.00
5.11 Protective Installations				
lightning protection	m²	3.00	to	3.90
earthing and bonding	m²	3.00	to	5.10
sprinkler installations	m²	32.00	to	37.00
dry riser and hose reel installation	m²	6.00	to	7.20
fire suppression installation to comunications room	m²	3.00	to	3.90
5.12 Communications, Security and Control Installations				
fire and smoke detection and alarm system	m²	29.00	to	32.00
voice/public address system	m²	19.00	to	22.00
intruder detection	m²	9.00	to	10.80
security, CCTV and access control	m²	32.00	to	37.00
wireways for telephones, data and structure cable	m²	14.00	to	16.10
structured cable installation	m²	34.00	to	37.50
flight information display system	m²	23.00	to	27.50
BMS installation	m²	52.00	to	58.00
Total Cost/m² (based on GIFA of 25,000 m²)	m²	860.00	to	1025.00

BUILDING MODELS – ELEMENTAL COST SUMMARIES

Item	Unit	Range £		
SHOPPING MALL (TENANT'S FIT-OUT EXCLUDED)				
Natural ventilation shopping mall with approximately 33,000 m² two storey retail area and a 13,000 m² above ground, mechanically ventilated, covered car park, situated in a town centre in South East England. Total M&E services value approximately £10,500,000				
RETAIL BUILDING				
5 Services				
5.1 Sanitary Installations	m²	1.05	to	1.05
5.3 Disposal Installations				
rainwater	m²	6.45	to	6.45
soil, waste and vent	m²	6.45	to	6.45
5.4 Water Installations				
cold water installation	m²	6.45	to	6.45
hot water installation	m²	5.40	to	5.40
5.5 Heat Source	m²	N/A		
5.6 Space Heating and Air Conditioning				
condenser water system	m²	32.00	to	32.00
LTHW installation	m²	4.50	to	4.50
air conditioning system	m²	32.00	to	32.00
over door heaters at entrances	m²	1.05	to	1.05
5.7 Ventilation Systems				
public toilet ventilation	m²	1.05	to	1.05
plant room ventilation	m²	4.50	to	4.50
supply and extract systems to shop units	m²	14.00	to	14.00
toilet extract systems to shop units	m²	2.15	to	2.15
smoke ventilation to Mall area	m²	11.00	to	11.00
service corridor ventilation	m²	2.15	to	2.15
other miscellaneous ventilation	m²	18.50	to	18.50
5.8 Electrical Installations				
LV distribution	m²	24.00	to	24.00
standby power	m²	4.50	to	4.50
general lighting	m²	66.00	to	66.00
external lighting	m²	5.50	to	5.50
emergency lighting	m²	13.30	to	13.30
small power	m²	10.80	to	10.80
mechanical services power supplies	m²	3.50	to	3.50
general earthing	m²	1.05	to	1.05
UPS for security and CCTV equipment	m²	1.05	to	1.05
5.9 Fuel Installations				
gas supply and boilers	m²	2.10	to	2.10
gas supplies to anchor (major) stores	m²	1.05	to	1.05
5.11 Protective Installations				
lightning protection	m²	1.05	to	1.05
sprinkler installations	m²	15.00	to	15.00
dry risers	m²	N/A		
hose reel installation	m²	N/A		

BUILDING MODELS – ELEMENTAL COST SUMMARIES

Item	Unit	Range £		
5.12 Communications, Security and Control Installations				
fire alarm installation	m²	8.75	to	8.75
public address/voice alarm	m²	5.50	to	5.50
security installation	m²	11.80	to	11.80
general containment	m²	13.00	to	13.00
5.13 Special Installations				
BMS/Controls	m²	18.50	to	18.50
Total Cost/m² (based on GIFA of 33,000 m²)	m²	355.00	to	355.00
CAR PARK – 13,000M²				
5 Services				
5.3 Disposal Intallations				
car park drainage	m²	5.25	to	5.25
5.7 Ventilation Systems				
car park ventilation (impulse fans)	m²	32.00	to	32.00
5.8 Electrical Installations				
LV distribution	m²	10.80	to	10.80
standby power	m²	N/A		
general lighting	m²	22.00	to	22.00
external lighting	m²	N/A		
emergency lighting	m²	5.50	to	5.50
small power	m²	3.25	to	3.25
mechanical services power supplies	m²	5.30	to	5.30
general earthing	m²	1.05	to	1.05
ramp frost protection	m²	2.10	to	2.10
5.11 Protective Installations				
sprinkler installation	m²	21.50	to	21.50
dry riser and hose reel installation	m²	N/A		
fire alarm installtions	m²	21.50	to	21.50
5.12 Communications, Security and Control Installations				
security installations	m²	9.60	to	9.60
BMS/Controls	m²	6.40	to	6.40
5.13 Special Installations				
entry/exit barriers, pay stations	m²	6.30	to	6.30
Total Cost/m² (based on GIFA of 13,000 m²)	m²	150.00	to	150.00

BUILDING MODELS – ELEMENTAL COST SUMMARIES

Item	Unit	Range £
SUPERMARKET		
Supermarket located in the South East with a total gross floor area of 4,000 m², including a sales area of 2,350 m². The building is on one level and incorporates a main sales, coffee shop, bakery, offices and amenities areas and warehouse. Total M&E services value is approximately £1,000,000		
5 Services		
5.1 Sanitary Installations	m²	1.55 to 1.55
5.3 Disposal Installations	m²	3.75 to 3.75
5.4 Water Installations		
hot and cold water services	m²	27.00 to 27.00
5.6 Space Heating and Air Conditioning		
heating & ventilation with cooling via DX units	m²	10.80 to 10.80
5.7 Ventilation Systems		
supply and extract sytems	m²	2.75 to 2.75
5.8 Electrical Installations		
panels/boards	m²	30.50 to 30.50
containment	m²	1.65 to 1.65
general lighting	m²	20.00 to 20.00
small power	m²	11.00 to 11.00
mechanical serviceds wiring	m²	2.25 to 2.25
5.9 Fuel Installations	m²	N/A
5.11 Protective Installations		
lightning protection	m²	1.25 to 1.25
5.12 Communications, Security and Control Installations		
fire alarms, detection and public address	m²	2.75 to 2.75
CCTV	m²	2.75 to 2.75
intruder alarm, detection and store security	m²	2.75 to 2.75
telecom and structured cabling	m²	0.75 to 0.75
BMS	m²	7.50 to 7.50
data cabinet	m²	2.75 to 2.75
controls wiring	m²	4.00 to 4.00
5.13 Special Installations		
refrigeration		
installation	m²	30.00 to 30.00
plant	m²	30.00 to 30.00
cold store	m²	11.00 to 11.00
cabinets	m²	81.00 to 81.00
Total Cost/m² (based on GIFA of 4,000 m²)	m²	290.00 to 290.00

BUILDING MODELS – ELEMENTAL COST SUMMARIES

Item	Unit	Range £		
OFFICE BUILDING				
Speculative 15 storey office in Central London for multiple tenant occupancy with a gross floor area of 19,300 m². A four pipe fan coil system, with roof mounted cooling towers, gas fired boilers and basement mounted water cooled chillers.				
SHELL & CORE – 19,300 m²				
5 Services				
5.1 Sanitary Installations	m²	7.00	to	9.10
5.3 Disposal Installations				
rainwater/soil and waste	m²	14.00	to	18.20
condensate	m²	2.00	to	2.60
5.4 Water Installations				
hot and cold water service	m²	14.00	to	18.20
5.5 Heat Source	m²	8.00	to	10.40
5.6 Space Heating and Air Conditioning				
LTHW, plant and distribution	m²	14.00	to	18.20
chilled water, plant and distribution	m²	48.00	to	62.00
air, plant and distribution	m²	37.00	to	48.00
5.7 Ventilation Systems				
toilet extract ventilation	m²	5.00	to	6.50
basement extract	m²	14.00	to	18.20
miscellaneous ventilation systems	m²	16.00	to	21.00
5.8 Electrical Installations				
life safety generator, fuel and flue	m²	9.00	to	11.70
HV/LV supply/distribution	m²	48.00	to	62.00
general lighting	m²	22.00	to	28.50
general power	m²	5.00	to	6.50
electrical services for mechanical equipment	m²	4.00	to	5.20
5.9 Fuel Installatiuons (gas)	m²	1.00	to	1.30
5.11 Protective Installations				
wet risers	m²	8.00	to	10.40
sprinklers	m²	16.00	to	21.00
earting and bonding	m²	2.00	to	2.60
lightning protection	m²	2.00	to	2.60
5.12 Communications, Security and Control Installations				
fire and voice alarms	m²	12.00	to	15.60
voice and data (wireways)	m²	1.00	to	1.30
security (wireways)	m²	1.00	to	1.30
disbled/refuge alarms	m²	2.00	to	2.60
CCTV/Access control/intruder detection	m²	3.00	to	3.90
BMS	m²	19.00	to	25.00
Total Cost/m² (based on GIFA of 19,300 m²)	m²	335.00	to	435.00

BUILDING MODELS – ELEMENTAL COST SUMMARIES

Item	Unit	Range £		
OFFICE BUILDING – cont				
CATEGORY 'A' FIT-OUT – 12,500M² NIA				
5 Services				
5.6 Space Heating and Air Conditioning				
4 pipe fan coil units	m²	24.00	to	31.00
LTHW heating	m²	22.00	to	28.50
chilled water	m²	35.00	to	45.50
condensate	m²	55.00	to	72.00
5.8 Electrical Installations				
tenant distribution boards	m²	3.00	to	3.00
lighting installation, lighting control and emergency	m²	63.00	to	82.00
supply only floor boxes	m²	3.00	to	3.90
earthing and bonding	m²	1.00	to	1.30
electrical services in connection with mechanical	m²	2.00	to	2.60
tenant distribution board	m²	N/A		
5.11 Protective Installations				
sprinkler installation	m²	18.00	to	23.00
5.12 Communications, Security and Control Installations				
fire and voice alarms	m²	10.00	to	13.00
BMS	m²	18.00	to	23.00
Total Cost/m² (based on NIFA of 12,500 m²)	m²	220.00	to	290.00

BUILDING MODELS – ELEMENTAL COST SUMMARIES

Item	Unit	Range £		
BUSINESS PARK				
New build office in South East within M25 part of a speculative business park with a gross floor area of 10,000 m². A full air displacement system with roof mounted air cooled chillers, gas fired boilers and air handling plant. Total M&E services value approximately £2,500,000 (shell & core) and £1,000,000 (Cat A Fit-out)				
SHELL & CORE – 10,000 m² GIA				
5 Services				
5.1 Sanitary Installations	m²	7.00	to	9.10
5.3 Disposal Installations				
rainwater	m²	2.00	to	3.00
soil and waste	m²	9.00	to	10.80
5.4 Water Installations				
cold water services	m²	9.65	to	11.60
hot water services	m²	2.00	to	4.00
5.5 Heat Source	m²	6.00	to	7.80
5.6 Space Heating and Air Conditioning				
LTHW heating; plantroom and risers	m²	12.00	to	15.60
chilled water; plantroom and risers	m²	30.00	to	36.00
ductwork; plantroom and risers	m²	59.00	to	68.00
5.7 Ventilation Systems				
toilet and miscellaneous ventilation	m²	12.00	to	15.60
5.8 Electrical Installations				
LV supply/distribution	m²	24.00	to	29.00
general lighting	m²	20.00	to	24.00
general power	m²	4.00	to	5.60
electrical services in connection with mechanical services	m²	2.00	to	4.00
5.9 Fuel Installations (gas)	m²	1.00	to	3.00
5.11 Protective Installations				
earthing and bonding	m²	2.00	to	4.00
lightning protection	m²	2.00	to	4.00
dry risers	m²	1.00	to	3.00
5.12 Communications, Security and Control Installations				
fire alarms	m²	9.00	to	10.80
security (wireways)	m²	1.00	to	3.00
data and voice (wireways)	m²	1.00	to	3.00
BMS	m²	20.00	to	24.00
Total Cost/m² (based on GIFA of 10,000 m²)	m²	235.00	to	280.00

BUILDING MODELS – ELEMENTAL COST SUMMARIES

Item	Unit	Range £
BUSINESS PARK – cont		
CATEGORY 'A' FIT-OUT – 8,000 m² NIA		
5 Services		
5.6 Space Heating and Air Conditioning		
LTHW heating and perimeter heaters	m²	37.00 to 44.50
floor swirl diffusers and supply ductwork	m²	23.00 to 27.50
5.8 Electrical Installations		
distribution boards	m²	1.00 to 3.00
general lightining, recessed including lighting controls	m²	50.00 to 60.00
5.11 Protective Installations		
earthing and bonding	m²	1.00 to 2.00
5.12 Communications, Security and Control Installations		
fire alarms	m²	7.00 to 9.10
BMS	m²	5.00 to 7.00
Total Cost/m² (based on NIA of 8,000 m²)	m²	120.00 to 150.00

BUILDING MODELS – ELEMENTAL COST SUMMARIES

Item	Unit	Range £		
PERFORMING ARTS CENTRE (MEDIUM SPECIFICATION)				
Performing Arts centre with a Gross Internal Area (GIA) of approximately 8,000 m², based on a medium specification with cooling to the Auditorium. The development comprises dance studios and a theatre auditorium in the outer London area. The theatre would require all the necessary stage lighting, machinery and equipment installed in a modern professional theatre (these are excluded from the model, as assumed to be FF&E, but the containment and power wiring is included).				
5 Services				
5.1 Sanitary Installations	m²	9.00	to	10.80
5.3 Disposal Installations				
soil, waste and rainwater	m²	11.00	to	13.20
5.4 Water Installations				
cold water services	m²	14.00	to	16.80
hot water services	m²	11.00	to	13.20
5.5 Heat Source	m²	11.00	to	13.20
5.6 Space Heating and Air Conditioning				
heating	m²	66.00	to	73.00
chilled water system	m²	28.00	to	31.00
supply and extract air systems	m²	100.00	to	115.00
5.7 Ventilation Systems				
ventilation and extract systems to toilets, kitchen and workshop	m²	15.00	to	18.00
5.8 Electrical Installations				
LV supply/distribution	m²	38.00	to	42.00
general lighting	m²	80.00	to	90.00
small power	m²	26.00	to	28.50
power to mechanical plant	m²	7.00	to	9.10
5.9 Fuel Installations (gas)	m²	3.00	to	3.90
5.11 Protective Installations				
lightning protection	m²	3.00	to	3.90
5.12 Communications, Security and Control Installations				
fire alarms and detection	m²	21.00	to	25.00
voice and data complete installation (excluding active equipment)	m²	26.00	to	31.00
security; access; control; disabled alarms; staff paging	m²	24.00	to	29.00
BMS	m²	38.00	to	42.00
5.13 Special Installations				
theatre systems includes for containment and power wiring	m²	19.00	to	21.00
Total Cost/m² (based on GIFA of 8,000 m²)	m²	550.00	to	660.00

BUILDING MODELS – ELEMENTAL COST SUMMARIES

Item	Unit	Range £		
SPORTS HALL				
Single storey sports hall, located in the South East, with a gross internal area of 1,200 m² (40 m × 30 m). Total M&E services value approximately £220,000				
5 Services				
5.1 Sanitary Installations	m²	10.50	to	10.50
5.3 Disposal Installations				
rainwater	m²	3.05	to	3.05
soil and waste	m²	6.50	to	6.50
5.4 Water Installations				
hot and cold water services	m²	14.50	to	14.50
5.5 Heat Source				
boilers, flues, pumps and controls	m²	15.00	to	15.00
5.6 Space Heating and Air Conditioning				
warm air heating to sports hall area	m²	18.00	to	18.00
radiator heating to ancillary areas	m²	21.00	to	21.00
5.7 Ventilation Systems				
ventilation to changing, fitness and sports hall areas	m²	20.00	to	20.00
5.8 Electrical Installations				
main swithcgear and sub-mains	m²	12.00	to	12.00
small power	m²	11.00	to	11.00
lighting and luminaires to sports areas	m²	25.00	to	25.00
lighting and luminaires to ancillary areas	m²	30.00	to	30.00
5.9 Fuel Installations (included in 5.5)	m²	N/A		
5.11 Protective and Lightning Protection				
lightning protection	m²	3.50	to	3.50
5.12 Communications, Security and Control Installations				
fire, smoke detection and alarm system, intruder detection	m²	12.00	to	12.00
CCTV installation	m²	14.00	to	14.00
public address and music systems	m²	6.50	to	6.50
wireways for voice and data	m²	2.50	to	2.50
Total Cost/m² (based on GIFA of 1,200 m²)	m²	225.00	to	225.00

BUILDING MODELS – ELEMENTAL COST SUMMARIES

Item	Unit	Range £		
STADIUM – NEW				
A three storey stadium, located in Greater London with a gross internal area of 85,000 m² and incorporating 60,000 spectator seats. Total M&E services value approximately £28,000,000				
5 Services				
5.1 Sanitary Installations	m²	10.80	to	10.80
5.3 Disposal Installations				
rainwater	m²	4.50	to	4.50
above ground drainage	m²	11.80	to	11.80
5.4 Water Installations				
hot and cold water	m²	21.00	to	21.00
5.5 Heat Source	m²	9.75	to	9.75
5.6 Space Heating and Air Conditioning				
heating	m²	6.50	to	6.50
cooling	m²	15.00	to	15.00
5.7 Ventilation Systems	m²	60.00	to	60.00
5.8 Electrical Installations				
HV/LV Supply	m²	10.80	to	10.80
LV distribution	m²	25.50	to	25.50
general lighting	m²	63.00	to	63.00
small power	m²	20.00	to	20.00
earting and bonding	m²	1.05	to	1.05
power supply to mechanical equipmemt	m²	1.05	to	1.05
pitch lighting	m²	11.00	to	11.00
5.9 Fuel Installations (gas)	m²	5.00	to	5.00
5.11 Protective Installations				
lightning protection	m²	1.05	to	1.05
hydrants	m²	2.15	to	2.15
5.12 Communications, Security and Control Installations				
wireways for data, TV, telecom and PA	m²	8.65	to	8.65
public address	m²	16.30	to	16.30
security	m²	15.20	to	15.20
data voice installations	m²	32.50	to	32.50
fire alarms	m²	8.75	to	8.75
disabled/refuse alarm/call systems	m²	3.25	to	3.25
BMS	m²	15.20	to	15.20
Total Cost/m² (based on GIFA of 85,000 m²)	m²	380.00	to	380.00
Total Cost/seat (based on 60,000 seats)	£/seat	540.00	to	540.00

BUILDING MODELS – ELEMENTAL COST SUMMARIES

Item	Unit	Range £
HOTELS		
200 Bedroom, four star hotel, situated in Central London, with a gross internal floor area of 16,500 m². The development comprises a ten storey building with large suites on each guest floor, together with banqueting, meeting rooms and leisure facilities.		
5 Services		
5.1 Sanitary Installations	m²	45.00 to 55.00
5.3 Disposal Installations		
rainwater	m²	4.00 to 4.90
soil and waste	m²	23.00 to 28.00
5.4 Water Installations		
hot and cold water services	m²	49.00 to 60.00
5.5 Heat Source		
condensing boiler, CHP and flues	m²	11.00 to 13.50
5.6 Space Heating and Air Conditioning		
air conditioning and space heating system; chillers, pumps, CHW and LTHW pipework, insulation, 4 pipe FCU, ductwork, grilles and diffusers, to guest rooms, public areas, meeting and banquet rooms	m²	135.00 to 170.00
5.7 Ventilation Systems		
general bathroom extract from guest suites, ventilation to kitchens and bathrooms, etc.	m²	34.00 to 42.00
smoke extract	m²	10.00 to 12.20
5.8 Electrical Installations		
HV/LV Installation, standby power, lighting, emergency lighting and small power to guest rooms and public areas, including earthing and bonding	m²	145.00 to 180.00
5.9 Fuel Installations	m²	2.70 to 3.30
gas	m²	2.00 to 2.50
5.11 Protective Installations		
dry risers and sprinkler installation	m²	28.00 to 34.00
lightning protection	m²	2.00 to 3.00
5.12 Communications, Security and Control Installations		
fire/smoke detection and fire alarm system	m²	10.00 to 12.20
security/access control	m²	23.00 to 28.00
integrated sound and AV system	m²	20.00 to 24.50
telephone and data and TV installation	m²	15.00 to 18.30
containment	m²	14.00 to 17.10
wire ways for voice and data (no hotel management and head end equipment)	m²	9.00 to 11.00
BMS	m²	30.00 to 36.00
Total Cost/m² (based on GIFA of 16,500 m²)	m²	610.00 to 750.00

BUILDING MODELS – ELEMENTAL COST SUMMARIES

Item	Unit	Range £
PRIVATE HOSPITAL		
New build project building. The works consist of a new 80 bed hospital of approximately 15,000 m², eight storey with a plant room. All heat is provided from existing steam boiler plant, medical gases are also served from existing plant. The project includes the provision of additional standby electrical generation to serve the wider site requirements. This hospital has six operating theatres, ITU/HDU department, pathology facilities, diagnostic imaging, out patient facilities and physiotherapy.		
5 Services		
5.1 Sanitary Installations	m²	19.00 to 21.00
5.3 Disposal Installations		
rainwater	m²	3.00 to 5.10
soil and waste	m²	24.00 to 26.00
specialist drainage (above ground)	m²	2.00 to 4.00
5.4 Water Installations		
hot and cold water services	m²	47.00 to 53.00
5.5 Heat Source (included in 5.6)	m²	
5.6 Space Heating and Air Conditioning		
LPHW heating	m²	42.00 to 47.00
chilled water	m²	38.00 to 42.00
steam and condensate	m²	24.00 to 26.00
ventilation, comfort cooling and air conditioning	m²	130.00 to 145.00
5.7 Ventilation Systems (included in 5.6)	m²	
5.8 Electrical Installations		
HV distribution	m²	4.00 to 6.00
LV supply/distribution	m²	33.00 to 37.00
standby power	m²	30.00 to 34.00
UPS	m²	19.00 to 21.00
general lighting	m²	52.00 to 52.00
general power	m²	38.00 to 42.00
emergency lighting	m²	19.00 to 21.00
theatre lighting	m²	14.00 to 16.10
specialist lighting	m²	14.00 to 16.10
external lighting	m²	2.00 to 4.00
electrical supplies for mechanical equipment	m²	9.00 to 10.80
5.9 Fuel Installations		
gas installations	m²	4.00 to 6.00
oil installations	m²	7.00 to 9.10
5.11 Protective Installations		
dry risers	m²	2.00 to 3.00
lightning protection	m²	1.00 to 3.00

BUILDING MODELS – ELEMENTAL COST SUMMARIES

Item	Unit	Range £
PRIVATE HOSPITAL – cont		
5 Services – cont		
5.12 Communications, Security and Control Installations		
fire alarms and detection	m²	19.00 to 21.00
voice and data	m²	11.00 to 13.20
security and CCTV	m²	14.00 to 16.10
nurse call and cardiac alarm system	m²	19.00 to 21.00
personnel paging	m²	4.00 to 6.00
hospital radio (entertainment)	m²	23.00 to 27.50
BMS	m²	36.00 to 39.50
5.13 Special Installations		
pneumatic tube conveying system	m²	5.00 to 7.00
group 1 equipment	m²	35.00 to 38.50
Total Cost/m² (based on GIFA of 15,000 m²)	m²	750.00 to 900.00

BUILDING MODELS – ELEMENTAL COST SUMMARIES

Item	Unit	Range £		
SCHOOL				
New build secondary school (Academy) located in Southern England, with a gross internal floor area of 10,000 m². Total M&E services value approximately £4,000,000 The building comprises a three storey teaching block, including provision for music, drama, catering, sports hall, science laboratories, food technology, workshops and reception/admin (BB93 compliant). Excludes IT cabling and sprinkler protection.				
5 Services				
5.1 Sanitary Installations				
toilet cores and changing facilities and lab sinks	m²	14.00	to	16.10
5.3 Disposal Installations				
rainwater installations	m²	4.00	to	6.00
soil and waste	m²	11.00	to	13.20
5.4 Water Installations				
potable hot and cold water services	m²	26.00	to	31.00
non-potable hot and cold water services to labs and art rooms	m²	9.00	to	10.80
5.5 Heat Source				
gas fired boiler installation	m²	11.00	to	13.20
5.6 Space Heating and Air Conditioning				
LTHW heating system (primary)	m²	33.00	to	36.50
LTHW heating system (secondary)	m²	7.00	to	9.10
DX cooling system to ICT server rooms	m²	4.00	to	6.00
mechanical supply and extract ventilation including DX type cooling to Music, Drama, Kitchen/Dining and Sports Hall	m²	52.00	to	57.00
5.7 Ventilation Systems				
toilet extract systems	m²	7.00	to	9.10
changing area extract systems	m²	4.00	to	6.00
extract ventilation from design/food technology and science labs	m²	9.00	to	10.80
5.8 Electrical Installations				
mains and sub-mains distribution	m²	33.00	to	36.50
lighting and luminaires including emergency fittings	m²	61.00	to	67.00
small power installation	m²	28.00	to	32.00
earthig and bonding	m²	1.00	to	2.00
5.9 Fuel Installations (gas)	m²	7.00	to	9.10
5.11 Protective Installations				
lightning protection	m²	3.00	to	3.90
5.12 Communications, Security and Control Installations				
containment for telephone, IT data, AV and security systems	m²	4.00	to	6.00
fire, smoke detection and alarm system	m²	17.00	to	20.00
security installations including CCTV, access control and intruder alarm	m²	19.00	to	21.00
disabled toilet, refuge and induction loop systems	m²	4.00	to	6.00
BMS – to plant	m²	19.00	to	21.00
BMS – to opening vents/windows	m²	9.00	to	10.80
Total Cost/m² (based on GIFA of 10,000 m²)	m²	395.00	to	475.00

BUILDING MODELS – ELEMENTAL COST SUMMARIES

Item	Unit	Range £		
AFFORDABLE RESIDENTIAL DEVELOPMENT				
A 8 storey, 117 affordable residential development with a gross internal area of 11,400 m² and a net internal area of 8,400 m², situated within the Central London area. The development has no car park or communal facilities and achieves a net to gross efficiency of 74%. Based upon radiator LTHW heating within each apartment, with plate heat exchanger, whole house ventilation, plastic cold and hot water services pipework, sprinkler protection to apartments. Excludes remote metering. Based upon 2 bed units with 1 bathroom.				
SHELL & CORE				
5 Services				
5.1 Sanitary Installations	m²	1.00	to	2.00
5.3 Disposal Installations	m²	28.00	to	36.50
5.4 Water Installations	m²	26.00	to	31.00
5.5 Heat Source	m²	10.00	to	13.00
5.6 Space Heating and Air Conditioning	m²	19.00	to	23.00
5.7 Ventilation Services	m²	13.00	to	16.90
5.8 Electrical Installations	m²	42.00	to	50.00
5.9 Fuel Installations – Gas	m²	1.00	to	3.00
5.11 Protective Installations	m²	22.00	to	26.00
5.12 Communications, Security and Control Installations	m²	36.00	to	43.00
5.13 Special Installations	m²	12.00	to	14.40
Total Cost/m² (based on GIFA of 11,400 m²)	m²	210.00	to	298.30
FITTING OUT				
5 Services				
5.1 Sanitary Installations	m²	27.00	to	32.50
5.3 Disposal Installations	m²	9.00	to	10.80
5.4 Water Installations	m²	23.00	to	27.50
5.5 Heat Source	m²	32.00	to	38.50
5.6 Space Heating and Air Conditioning	m²	40.00	to	48.00
5.7 Ventilating Services	m²	63.00	to	76.00
5.8 Electrical Installation	m²	59.00	to	71.00
5.11 Protective Installations	m²	17.00	to	20.00
5.12 Communications, Security and Control Installations	m²	27.00	to	32.50
5.13 Special Installations	m²	6.00	to	7.80
Total Cost/m² (based on a NIA of 8,400 m²)	m²	303.00	to	364.00
Total cost per apartment – Shell & Core and Fitting Out	£/apt	42000.00	to	55000.00

BUILDING MODELS – ELEMENTAL COST SUMMARIES

Item	Unit	Range £
PRIVATE RESIDENTIAL DEVELOPMENT		
A 24 storey, 203 apartment private residential development with a gross internal area of 50,000 m² and a net internal area of 33,000 m², situated within the London's prime residential area. The development has limited car park facilities, residents communal areas such as gym, cinema room, self-stimulated, etc. Based upon LTHW underfloor heating, 4 pipe fan coils to principle rooms, with plate heat and cooling exchangers, whole house ventilation with summer house facility, copper hot and cold water services pipework, sprinkler protection to apartments, lighting control, TV/data outlets. Flood wiring for apartment home automation and sound system only. Based upon a combination of 2/3 bedroom apartments with family bathroom and ensuite.		
SHELL & CORE		
5 Services		
5.1 Sanitary Installations	m²	1.00 to 2.00
5.3 Disposal Installations	m²	21.00 to 27.00
5.4 Water Installations	m²	18.00 to 23.00
5.5 Heat Source	m²	9.00 to 11.70
5.6 Space Heating and Air Conditioning	m²	70.00 to 84.00
5.7 Ventilation Systems	m²	17.00 to 20.00
5.8 Electrical Installations	m²	61.00 to 76.00
5.9 Fuel Installations (gas)	m²	1.00 to 2.00
5.11 Protective Installations	m²	20.00 to 26.00
5.12 Communications, Security and Control Installations	m²	40.00 to 52.00
5.13 Special Installations	m²	32.00 to 38.50
Total Cost/m² (based on GIFA of 50,000 m²)	m²	290.00 to 380.00
FITTING OUT		
5 Services		
5.1 Sanitary Installations	m²	83.00 to 100.00
5.3 Disposal Installations	m²	20.00 to 24.00
5.4 Water Installations	m²	32.00 to 38.50
5.5 Heat Source	m²	40.00 to 52.00
5.6 Space Heating and Air Conditioning	m²	140.00 to 170.00
5.7 Ventilation Systems	m²	39.00 to 47.00
5.8 Electrical Installations	m²	95.00 to 120.00
5.11 Protective Installations	m²	23.00 to 30.00
5.12 Communications, Security and Control Installations	m²	110.00 to 140.00
5.13 Special Installations	m²	15.00 to 25.50
Total Cost/m² (based on NIFA of 33,000 m²)	m²	600.00 to 780.00
Total cost per apartment – Shell & Core and Fit-Out	£/apt	75000.00 to 98000.00

BUILDING MODELS – ELEMENTAL COST SUMMARIES

Item	Unit	Range £		
DISTRIBUTION CENTRE				
Distribution centre located in London with a total gross floor area of 75,000 m², including a refrigerated cold box of 17,500 m². Total M&E services value approximately £15,000,000. The building is on one level and also incorporates an office area, vehicle recovery unit, gate house and plantrooms.				
5 Services				
5.3 Disposal Installations				
soil and waste	m²	2.10	to	2.10
rainwater	m²	6.30	to	6.30
5.4 Water Installations				
hot and cold water services	m²	2.20	to	2.20
5.6 Space Heating and Air Conditioning				
heating with ventilation to offices, displacement system to main warehouse	m²	23.00	to	23.00
5.7 Ventilation Systems				
smoke extract system	m²	5.40	to	5.40
5.8 Electrical Installations				
generator	m²	16.30	to	16.30
main HV installation	m²	21.00	to	21.00
MV distribution	m²	13.70	to	13.70
lighting installation	m²	11.60	to	11.60
small power installation	m²	13.10	to	13.10
5.9 Fuel Installations				
gas mains services to plantroom	m²	0.65	to	0.65
5.11 Protective Installations				
sprinklers including racking protection	m²	45.00	to	45.00
earthing/lightning protection	m²	0.65	to	0.65
5.12 Communications, Security and Control Installations				
fire alarms, detection and public address	m²	12.10	to	12.10
CCTV	m²	7.50	to	7.50
BMS	m²	5.50	to	5.50
5.13 Special Installations				
refrigeration	m²	38.00	to	38.00
Total Cost/m² (based on GIFA of 75,000 m²)	m²	225.00	to	225.00

BUILDING MODELS – ELEMENTAL COST SUMMARIES

Item	Unit	Range £		
DATA CENTRE				
New build data centre located in the London area/proximity to M25. Net Technical Area (NTA) provided at 2,000 m² with typically other areas of 250 m² office space, 250 m² ancillary space and 1,000 m² internal plant. Total GIA 3,500 m². Total M&E services value approximately £16,500,000				
Power and cooling to Technical Space @ 1,500 w/m². Cost/m² against Nett Technical Area – NTA.				
5 Services				
5.3 Disposal Installations				
soil and waste	m²	14.00	to	14.00
rainwater	m²	10.00	to	10.00
condensate	m²	10.00	to	10.00
5.4 Water Installations				
hot and cold water services	m²	25.00	to	25.00
5.6 Space Heating and Air Conditioning				
chilled water plant with redundancy of N+1 to provide 1,000 w/m² net technical space cooling	m²	385.00	to	385.00
chilled water distribution to data centre to free standing cooling units and distribution to ancillary office and workshop/build areas	m²	395.00	to	395.00
freestanding cooling units with redundancy of N+20% to technical space and switchrooms. Based on single coil cooling units 30% of units with humidification	m²	355.00	to	355.00
floor grilles	m²	74.00	to	74.00
5.7 Ventilation Systems				
supply and extract ventilation systems to data centre, switchrooms and ancillary spaces including dedicated gas extract and hot aisle extract system	m²	440.00	to	440.00
5.8 Electrical Installations				
main HV installations including transformers with redundant capacity of N+1	m²	670.00	to	670.00
LV distribution including cabling and busbar installations to provide fulll system – system dual supplies/redundancy including supplies to mechanical services, office and ancillary areas	m²	800.00	to	800.00
Generator Installation				
standby rated containerized generators with redundant capacity of N+1 and including synchronization panel, 72 hours bulk fuel store and controls	m²	950.00	to	950.00
Uninterruptible Power Supplies				
static UPS to provide 2 × (N+1) system redundancy with 10 minute battery autonomy	m²	1125.00	to	1125.00
LV Switchgear				
Incoming LV switchgear, UPS input and output boards for electrical and mechanical services systems	m²	890.00	to	890.00
Power Distribution Units (PDUs)				
PDUs to provide a redundant capacity of 2 Nr PDU to include static transfer switch and isolating transformer	m²	810.00	to	810.00
Lighting Installation				
lighting to technical, workshop and plant areas including office/ancillary spaces and external areas	m²	110.00	to	110.00

BUILDING MODELS – ELEMENTAL COST SUMMARIES

Item	Unit	Range £		
DATA CENTRE – cont				
5 Services – cont				
5.9 Fuel Installation (assumed not required)				
5.11 Protective Installations				
gaseous suppression to technical areas and switchrooms	m^2	250.00	to	250.00
lightning protection	m^2	15.00	to	15.00
earthing and clean earth	m^2	30.00	to	30.00
leak detection	m^2	15.00	to	15.00
5.12 Communications, Security and Control Installations				
very early smoke detection alarm (VESDA) to technical areas	m^2	145.00	to	145.00
CCTV installations	m^2	175.00	to	175.00
access control installations	m^2	145.00	to	145.00
BMS	m^2	375.00	to	375.00
PLC/Electrical monitoring system	m^2	255.00	to	255.00
Total Cost/m² (based on NTA of 2,000 m²)	m^2	8400.00	to	8400.00

ESSENTIAL READING FROM TAYLOR AND FRANCIS

Spon's Asia Pacific Construction Costs Handbook, Fifth Edition

LANGDON & SEAH

In the last few years, the global economic outlook has continued to be shrouded in uncertainty and volatility following the financial crisis in the Euro zone. While the US and Europe are going through a difficult period, investors are focusing more keenly on Asia. This fifth edition provides overarching construction cost data for 16 countries: Brunei, Cambodia, China, Hong Kong, India, Indonesia, Japan, Malaysia, Myanmar, Philippines, Singapore, South Korea, Sri Lanka, Taiwan, Thailand and Vietnam.

May 2015: 234X156 mm: 452 pp
Hb: 978-1-4822-4358-1: £160.00

To Order: Tel: +44 (0) 1235 400524 Fax: +44 (0) 1235 400525
or Post: Taylor and Francis Customer Services,
Bookpoint Ltd, Unit T1, 200 Milton Park, Abingdon, Oxon, OX14 4TA UK
Email: book.orders@tandf.co.uk

For a complete listing of all our titles visit:
www.tandf.co.uk

Taylor & Francis
Taylor & Francis Group

ESSENTIAL READING FROM TAYLOR AND FRANCIS

BIM and Quantity Surveying

PITTARD & SELL

The sudden arrival of Building Information Modelling (BIM) as a key part of the building industry is redefining the roles and working practices of its stakeholders. Many clients, designers, contractors, quantity surveyors, and building managers are still finding their feet in an industry where BIM compliance can bring great rewards.

This guide is designed to help quantity surveying practitioners and students understand what BIM means for them, and how they should prepare to work successfully on BIM compliant projects. The case studies show how firms at the forefront of this technology have integrated core quantity surveying responsibilities like cost estimating, tendering, and development appraisal into high profile BIM projects. In addition to this, the implications for project management, facilities management, contract administration and dispute resolution are also explored through case studies, making this a highly valuable guide for those in a range of construction project management roles.

Featuring a chapter describing how the role of the quantity surveyor is likely to permanently shift as a result of this development, as well as descriptions of tools used, this covers both the organisational and practical aspects of a crucial topic.

December 2015: 234 x 156 mm: 258 pp
Pb: 978-0-415-87043-6; £24.99

To Order: Tel: +44 (0) 1235 400524 Fax: +44 (0) 1235 400525
or Post: Taylor and Francis Customer Services,
Bookpoint Ltd, Unit T1, 200 Milton Park, Abingdon, Oxon, OX14 4TA UK
Email: book.orders@tandf.co.uk

For a complete listing of all our titles visit:
www.tandf.co.uk

Taylor & Francis
Taylor & Francis Group

ESSENTIAL READING FROM TAYLOR AND FRANCIS

NRM1 Cost Management Handbook

David P Benge

The definitive guide to measurement and estimating using NRM1, written by the author of NRM1

The 'RICS New rules of measurement: Order of cost estimating and cost planning of capital building works' (referred to as NRM1) is the cornerstone of good cost management of capital building works projects - enabling more effective and accurate cost advice to be given to clients and other project team members, while facilitating better cost control.

The NRM1 Cost Management Handbook is the essential guide to how to successfully interpret and apply these rules, including explanations of how to:

- quantify building works and prepare order of cost estimates and cost plans
- use the rules as a toolkit for risk management and procurement
- analyse actual costs for the purpose of collecting benchmark data and preparing cost analyses
- capture historical cost data for future order of cost estimates and elemental cost plans
- employ the rules to aid communication
- manage the complete 'cost management cycle'
- use the elemental breakdown and cost structures, together with the coding system developed for NRM1, to effectively integrate cost management with Building Information Modelling (BIM).

March 2014: 246 x 174: 640pp
Pb: 978-0-415-72077-9: £41.99

To Order: Tel: +44 (0) 1235 400524 Fax: +44 (0) 1235 400525
or Post: Taylor and Francis Customer Services,
Bookpoint Ltd, Unit T1, 200 Milton Park, Abingdon, Oxon, OX14 4TA UK
Email: book.orders@tandf.co.uk

For a complete listing of all our titles visit:
www.tandf.co.uk

Taylor & Francis
Taylor & Francis Group

ESSENTIAL READING FROM TAYLOR AND FRANCIS

Estimator's Pocket Book

Duncan Cartlidge

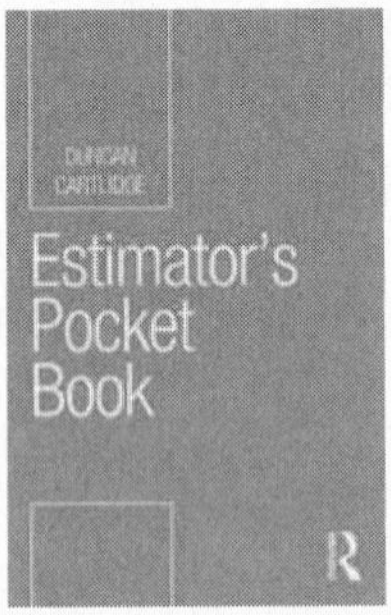

The Estimator's Pocket Book is a concise and practical reference covering the main pricing approaches, as well as useful information such as how to process sub-contractor quotations, tender settlement and adjudication. It is fully up-to-date with NRM2 throughout, features a look ahead to NRM3 and describes the implications of BIM for estimators.

It includes instructions on how to handle:

- the NRM order of cost estimate;
- unit-rate pricing for different trades;
- pro rata pricing and dayworks
- builders' quantities;
- approximate quantities.

Worked examples show how each of these techniques should be carried out in clear, easy-to-follow steps. This is the indispensible estimating reference for all quantity surveyors, cost managers, project managers and anybody else with estimating responsibilities. Particular attention is given to NRM2, but the overall focus is on the core estimating skills needed in practice.

May 2013 186x123: 310pp
Pb: 978-0-415-52711-8: £21.99

To Order: Tel: +44 (0) 1235 400524 Fax: +44 (0) 1235 400525
or Post: Taylor and Francis Customer Services,
Bookpoint Ltd, Unit T1, 200 Milton Park, Abingdon, Oxon, OX14 4TA UK
Email: book.orders@tandf.co.uk

For a complete listing of all our titles visit:
www.tandf.co.uk

Taylor & Francis
Taylor & Francis Group

ESSENTIAL READING FROM TAYLOR AND FRANCIS

Building Services Handbook, 8th Edition

GREENO & HALL

This eighth edition of Hall and Greeno's leading textbook has been reviewed and updated in relation to the latest building and water regulations, new technology, and new legislation, and even more design calculations have been added. In addition, topics such as: alternative sources of natural energy, solar, ground source, heat pumps, renewable energy sources, geothermal methods, and wind power, are now covered.

Building Services Handbook summarises the application of all common elements of building services practice, technique and procedure, to provide an essential information resource for students as well as practitioners working in building services, building management and the facilities administration and maintenance sectors of the construction industry. Information is presented in the highly illustrated and accessible style of the best-selling companion title Building Construction Handbook.

The comprehensive reference for all construction and building services students, Building Services Handbook is ideal for a wide range of courses including NVQ and BTEC National through Higher National Certificate and Diploma to Foundation and three-year Degree level. The clear illustrations and complementary references to industry Standards combine essential guidance with a resource base for further reading and development of specific topics.

September 2015: 234 x 156 mm: 768 pp
Pb: 978-1-138-80563-7; £32.99

To Order: Tel: +44 (0) 1235 400524 Fax: +44 (0) 1235 400525
or Post: Taylor and Francis Customer Services,
Bookpoint Ltd, Unit T1, 200 Milton Park, Abingdon, Oxon, OX14 4TA UK
Email: book.orders@tandf.co.uk

For a complete listing of all our titles visit:
www.tandf.co.uk

Taylor & Francis
Taylor & Francis Group

ESSENTIAL READING FROM TAYLOR AND FRANCIS

Building Construction Handbook, 11th Edition

CHUDLEY & GREENO

The Building Construction Handbook is THE authoritative reference for all construction students and professionals. Its detailed drawings clearly illustrate the construction of building elements, and have been an invaluable guide for builders since 1988. The principles and processes of construction are explained with the concepts of design included where appropriate. Extensive coverage of building construction practice, techniques, and regulations representing both traditional procedures and modern developments are included to provide the most comprehensive and easy to understand guide to building construction.

This new edition has been updated to reflect recent changes to the building regulations, as well as new material on the latest technologies used in domestic construction.

Building Construction Handbook is the essential, easy-to-use resource for undergraduate and vocational students on a wide range of courses including NVQ and BTEC National, through to Higher National Certificate and Diploma, to Foundation and three-year Degree level. It is also a useful practical reference for building designers, contractors and others engaged in the construction industry.

April 2016: 234 x 156 mm: 1012 pp
Pb: 978-1-138-90709-6: £28.99

To Order: Tel: +44 (0) 1235 400524 Fax: +44 (0) 1235 400525
or Post: Taylor and Francis Customer Services,
Bookpoint Ltd, Unit T1, 200 Milton Park, Abingdon, Oxon, OX14 4TA UK
Email: book.orders@tandf.co.uk

For a complete listing of all our titles visit:
www.tandf.co.uk

Taylor & Francis
Taylor & Francis Group

PART 3

Material Costs/Measured Work Prices

ESSENTIAL READING FROM TAYLOR AND FRANCIS

Quantity Surveyor's Pocket Book

2nd Edition

D Cartlidge

This second edition of the Quantity Surveyor's Pocket Book is fully updated in line with NRM1, NRM2 and JCT(11), and remains a must-have guide for students and qualified practitioners. Its focussed coverage of the data, techniques, and skills essential to the quantity surveying role make it an invaluable companion for everything from initial cost advice to the final account stage.

Key features include:

- the structure of the construction industry
- cost forecasting and feasibility studies
- measurement and quantification, with NRM2 and SMM7 examples
- estimating and bidding
- whole life costs
- contract selection
- final account procedure.

June 2012: 186x123: 440pp
Pb: 978-0-415-50110-1: £22.99

To Order: Tel: +44 (0) 1235 400524 Fax: +44 (0) 1235 400525
or Post: Taylor and Francis Customer Services,
Bookpoint Ltd, Unit T1, 200 Milton Park, Abingdon, Oxon, OX14 4TA UK
Email: book.orders@tandf.co.uk

For a complete listing of all our titles visit:
www.tandf.co.uk

Taylor & Francis
Taylor & Francis Group

Material Costs/Measured Work Prices

DIRECTIONS

The following explanations are given for each of the column headings and letter codes.

Unit	Prices for each unit are given as singular (i.e. 1 metre, 1 nr) unless stated otherwise.
Net price	Industry tender prices, plus nominal allowance for fixings (unless measured separately), waste and applicable trade discounts.
Material cost	Net price plus percentage allowance for overheads (7%), profit (5%) and preliminaries (10%).
Labour norms	In man-hours for each operation.
Labour cost	Labour constant multiplied by the appropriate all-in man-hour cost based on gang rate (see also relevant Rates of Wages Section) plus percentage allowance for overheads, profit and preliminaries.
Measured work Price (total rate)	Material cost plus Labour cost.

MATERIAL COSTS

The Material Costs given are based at Second Quarter 2016 but exclude any charges in respect of VAT. The average rate of copper during this quarter is US$4,630/UK£3,305 per tonne. Users of the book are advised to register on the Spon's website www.pricebooks.co.uk/updates to receive the free quarterly updates – alerts will then be provided by e-mail as changes arise.

MEASURED WORK PRICES

These prices are intended to apply to new work in the London area. The prices are for reasonable quantities of work and the user should make suitable adjustments if the quantities are especially small or especially large. Adjustments may also be required for locality (e.g. outside London – refer to cost indices in approximate estimating section for details of adjustment factors) and for the market conditions (e.g. volume of work secured or being tendered) at the time of use.

MECHANICAL INSTALLATIONS

The labour rate has been based on average gang rates per man hour effective from 5 October 2016. To this rate has been added 10% and 7% to cover preliminary items, site and head office overheads together with 5% for profit, resulting in an inclusive rate of £30.55 per man hour. The rate has been calculated on a working year of 2,016 hours; a detailed build-up of the rate is given at the end of these directions.

The rates and allowances will be subject to increases as covered by the National Agreements. Future changes will be published in the free Spon's quarterly update by registering on their website.

DIRECTIONS

DUCTWORK INSTALLATIONS

The labour rate basis is as per Mechanical above and to this rate has been added 10% plus 7% to cover site and head office overheads only (factory overhead is included in the material rate) and preliminary items together with 5% for profit, resulting in an inclusive rate of £29.25 per man hour. The rate has been calculated on a working year of 2,016 hours; a detailed build-up of the rate is given at the end of these directions.

In calculating the 'Measured Work Prices' the following assumptions have been made:

(a) That the work is carried out as a subcontract under the Standard Form of Building Contract.
(b) That, unless otherwise stated, the work is being carried out in open areas at a height which would not require more than simple scaffolding.
(c) That the building in which the work is being carried out is no more than six storey's high.

Where these assumptions are not valid, as for example where work is carried out in ducts and similar confined spaces or in multi-storey structures when additional time is needed to get to and from upper floors, then an appropriate adjustment must be made to the prices. Such adjustment will normally be to the labour element only.

Note: The rates do not include for any uplift applied if the ductwork package is procured via the Mechanical Subcontractor

DIRECTIONS

LABOUR RATE – MECHANICAL & PUBLIC HEALTH

The annual cost of a notional twelve man gang of 12 men

	FOREMAN	SENIOR CRAFTSMAN (+2 Welding skill)	SENIOR CRAFTSMAN	CRAFTSMAN	INSTALLER	MATE (Over 18)	SUB TOTALS
	1 NR	1 NR	2 NR	4 NR	2 NR	2 NR	
Hourly Rate from 6 October 2014 – assumed +3.50%	**16.64**	**14.33**	**13.77**	**12.63**	**11.43**	**9.64**	
Working hours per annum per man	1,680.00	1,680.00	1,680.00	1,680.00	1,680.00	1,680.00	
× Hourly rate × nr of men = £ per annum	**27,959.90**	**24,082.38**	**46,252.08**	**84,853.44**	**38,392.70**	**32,376.46**	**253,916.96**
Overtime Rate	23.45	20.18	19.41	17.80	16.11	13.57	
Overtime hours per annum per man	336.00	336.00	336.00	336.00	336.00	336.00	
× Hourly rate × nr of men = £ per annum	**7,880.24**	**6,781.32**	**13,041.00**	**23,925.89**	**10,829.25**	**9,118.27**	**71,575.96**
Total	**35,840.15**	**30,863.70**	**59,293.08**	**108,779.33**	**49,221.95**	**41,494.72**	**325,492.93**
Incentive schemes 0.00%	**0.00**	**0.00**	**0.00**	**0.00**	**0.00**	**0.00**	**0.00**
Daily Travel Time Allowance (15–20 miles each way)	10.33	10.33	10.33	10.33	10.33	10.33	
Days per annum per man	224.00	224.00	224.00	224.00	224.00	224.00	
× nr of men = £ per annum	**2,313.76**	**2,313.76**	**4,627.53**	**9,255.05**	**4,627.53**	**4,627.53**	**27,765.16**
Daily Travel Fare (15–20 miles each way)	14.60	14.60	14.60	14.60	14.60	14.60	
Days per annum per man	224.00	224.00	224.00	224.00	224.00	224.00	
× nr of men = £ per annum	**3,270.40**	**3,270.40**	**6,540.80**	**13,081.60**	**6,540.80**	**6,540.80**	**39,244.50**
Employer's Pension contributions at 05/10/16							
£ Contributions/annum	**1,460.67**	**1,306.21**	**2,414.37**	**4,828.73**	**2,005.33**	**1,692.10**	**13,707.42**
National Insurance Contributions							
Gross pay – subject to NI	46,903.69	41,365.02	79,705.87	148,247.32	68,240.97	59,444.35	
% of NI Contributions	13.80	13.80	13.80	13.80	13.80	13.80	
£ Contributions/annum	**4,932.85**	**4,168.51**	**7,919.69**	**14,298.69**	**6,337.54**	**5,123.60**	**42,780.89**
Holiday Credit and Welfare contributions							
Number of weeks	52	52	52	52	52	52	
Total weekly £ contribution each	87.33	81.82	73.90	68.66	63.12	54.77	
× nr of men = £ Contributions/annum	**4,541.33**	**4,254.64**	**7,685.50**	**14,281.28**	**6,564.96**	**5,696.31**	**43,024.02**
Holiday Top-up Funding including overtime	18.04	12.74	14.99	13.70	12.36	10.43	
Cost	**938.05**	**662.52**	**1,558.97**	**2,850.06**	**1,285.73**	**1,084.99**	**8,380.31**

	SUBTOTAL		**500,395.52**
	TRAINING (INCLUDING ANY TRADE REGISTRATIONS) – SAY	1.00%	**5,003.96**
	SEVERANCE PAY AND SUNDRY COSTS – SAY	1.50%	**7,580.99**
	EMPLOYER'S LIABILITY AND THIRD PARTY INSURANCE – SAY	2.00%	**10,259.61**
	ANNUAL COST OF NOTIONAL GANG		**523,240.08**
MEN ACTUALLY WORKING = 10.5	THEREFORE ANNUAL COST PERPRODUCTIVE MAN		**49,832.39**
AVERAGE NR OF HOURS WORKEDPER MAN = 2016	THEREFORE ALL-IN MAN HOURS		**24.72**
	PRELIMINARY ITEMS – SAY	10%	**2.47**
	SITE AND HEAD OFFICE OVERHEADS – SAY	7%	**1.90**
	PROFIT – SAY	5%	**1.45**
	THEREFORE INCLUSIVE MAN HOUR RATE		**30.55**

DIRECTIONS

Notes:

1) The following assumptions have been made in the above calculations:
    a) Increase in Hourly Rate from 5 October 2016 is assumed +3.50%.
    b) The working week of 37.5 hours i.e. the normal working week as defined by the National Agreement.
    c) The actual hours worked are five days of 9 hours each.
    d) A working year of 2016 hours.
    e) Five days in the year are lost through sickness or similar reason.
2) The incentive scheme addition of 5% is intended to reflect bonus schemes typically in use.
3) National insurance contributions are those effective from 6 April 2016. The National Insurance concession will be withdrawn for the construction industry on 30 October 2014 therefore NI will be paid for 52 weeks.
4) Weekly Holiday Credit/Welfare Stamp values are those effective from 5 October 2016.
5) Overtime rates are based on Premium Rate 1.
6) Fares with Oyster Card (New Malden to Waterloo + Zone 1 – Anytime fare) current at June 2016 (TfL – 0845 748 4950 or 0845 330 9876).

DIRECTIONS

LABOUR RATE – DUCTWORK

The annual cost of notional eight man gang of 8 men

		FOREMAN	SENIOR CRAFTSMAN	CRAFTSMAN	INSTALLER	SUB TOTALS
		1 NR	1 NR	4 NR	2 NR	
Hourly Rate from 5 October 2015		**16.64**	**13.77**	**12.63**	**11.43**	
Working hours per annum per man		1,680.00	1,680.00	1,680.00	1,680.00	
× Hourly rate × nr of men = £ per annum		**27,959.90**	**23,126.04**	**84,853.44**	**38,392.70**	**174,332.09**
Overtime Rate		23.45	19.41	17.80	16.11	
Overtime hours per annum per man		336.00	336.00	336.00	336.00	
× hourly rate × nr of men = £ per annum		**7,880.24**	**6,520.50**	**23,925.89**	**10,829.25**	**49,155.88**
Total		**35,840.15**	**29,646.54**	**108,779.33**	**49,221.95**	**223,487.96**
Incentive schemes	**0.00%**	**0.00**	**0.00**	**0.00**	**0.00**	**0.00**
Daily Travel Time Allowance (15–20 miles each way)		10.33	10.33	10.33	10.33	
Days per annum per man		224.00	224.00	224.00	224.00	
× nr of men = £ per annum		**2,313.76**	**2,313.76**	**9,255.05**	**4,627.53**	**18,510.11**
Daily Travel Fare (15–20 miles each way)		14.60	14.60	14.60	14.60	**14.60**
Days per annum per man		224.00	224.00	224.00	224.00	
× nr of men = £ per annum		**3,270.40**	**3,270.40**	**13,081.60**	**6,540.80**	**26,163.20**
Employer's Pension Contributions from 5 October 2016						
£ Contributions/annum		**1,460.67**	**1,207.18**	**4,828.73**	**2,005.33**	**9,501.92**
National Insurance Contributions:						
Gross pay – subject to NI		46,903.69	39,852.94	148,247.32	68,240.97	
% of NI Contributions		13.8	13.8	13.8	13.8	
£ Contributions/annum		**4,932.85**	**3,959.85**	**14,298.69**	**6,337.54**	**29,528.92**
Holiday Credit and Welfare contributions:						
Number of weeks		52	52	52	52	
Total weekly £ contribution each		87.33	73.90	68.66	63.12	
× nr of men = £ Contributions/annum		**4,541.33**	**3,842.75**	**14,281.28**	**6,564.96**	**29,230.32**
Holiday Top-up Funding including overtime		**18.04**	**14.99**	**13.70**	**12.36**	
Cost		**938.05**	**779.48**	**2,850.06**	**1,285.73**	**5,853.32**

	SUBTOTAL		**342,276.76**
	TRAINING (INCLUDING ANY TRADE REGISTRATIONS) – SAY	**1.00%**	**3,422.76**
	SEVERANCE PAY AND SUNDRY COSTS – SAY	**1.50%**	**5,185.48**
	EMPLOYER'S LIABILITY AND THIRD PARTY INSURANCE – SAY	**2.00%**	**7,017.68**
	ANNUAL COST OF NOTIONAL GANG		**357,901.68**
MEN ACTUALLY WORKING = 7.5	THEREFORE ANNUAL COST PERPRODUCTIVE MAN		**47,720.22**
AVERAGE NR OF HOURS WORKEDPER MAN = 2016	THEREFORE ALL-IN MAN HOURS		**23.67**
	PRELIMINARY ITEMS – SAY	**10%**	**2.37**
	SITE AND HEAD OFFICE OVERHEADS – SAY	**7%**	**1.82**
	PROFIT – SAY	**5%**	**1.39**
	THEREFORE INCLUSIVE MAN HOUR RATE		**29.25**

DIRECTIONS

Notes:

1) The following assumptions have been made in the above calculations:
   a) The working week of 37.5 hours i.e. the normal working week as defined by the National Agreement.
   b) The actual hours worked are five days of 9 hours each.
   c) A working year of 2016 hours.
   d) Five days in the year are lost through sickness or similar reason.
2) The incentive scheme addition of 5% is intended to reflect bonus schemes typically in use.
3) National insurance contributions are those effective from 6 April 2016. The National Insurance concession will be withdrawn for the construction industry on 30 October 2012 therefore NI will be paid for 52 weeks.
4) Weekly Holiday Credit/Welfare Stamp values are those effective from 6 October 2014.
5) Rates are based from 5 October 2014 assumed +3.50%.
6) Fares with Oyster Card (New Malden to Waterloo + Zone 1 – Anytime fare) current at June 2016 (TfL).
7) Overtime rates are based on Premium Rate 1.

33 DRAINAGE ABOVE GROUND

Item	Net Price £	Material £	Labour hours	Labour £	Unit	**Total rate £**
RAINWATER PIPEWORK/GUTTERS						
PVC-u gutters: push fit joints; fixed with brackets to backgrounds; BS 4576 BS EN 607						
Half round gutter, with brackets measured separately						
75 mm	3.75	4.61	0.69	25.89	m	**30.50**
100 mm	7.69	9.45	0.64	24.01	m	**33.46**
150 mm	8.68	10.67	0.82	30.78	m	**41.45**
Brackets: including fixing to backgrounds. For minimum fixing distances, refer to the Tables and Memoranda at the rear of the book						
75 mm; Fascia	1.47	1.81	0.15	5.63	nr	**7.44**
100 mm; Jointing	2.33	2.87	0.16	6.01	nr	**8.88**
100 mm; Support	0.95	1.17	0.16	6.01	nr	**7.18**
150 mm; Fascia	2.84	3.49	0.16	6.01	nr	**9.50**
Bracket supports: including fixing to backgrounds. For minimum fixing distances, refer to the Tables and Memoranda at the rear of the book						
Side rafter	3.73	4.58	0.16	6.01	nr	**10.59**
Top rafter	3.73	4.58	0.16	6.01	nr	**10.59**
Rise and fall	4.03	4.96	0.16	6.01	nr	**10.97**
Extra over fittings half round PVC-u gutter						
Union						
75 mm	2.18	2.68	0.19	7.13	nr	**9.81**
100 mm	2.33	2.87	0.24	9.01	nr	**11.88**
150 mm	8.34	10.25	0.28	10.50	nr	**20.75**
Rainwater pipe outlets						
Running: 75 × 53 mm dia.	5.02	6.16	0.12	4.50	nr	**10.66**
Running: 100 × 68 mm dia.	3.43	4.21	0.12	4.50	nr	**8.71**
Running: 150 × 110 mm dia.	16.05	19.72	0.12	4.50	nr	**24.22**
Stop end: 100 × 68 mm dia.	3.84	4.71	0.12	4.50	nr	**9.21**
Internal stop ends: short						
75 mm	2.18	3.92	0.09	3.38	nr	**7.30**
100 mm	1.10	2.59	0.09	3.38	nr	**5.97**
150 mm	5.79	7.11	0.09	3.38	nr	**10.49**
External stop ends: short						
75 mm	4.42	5.43	0.09	3.38	nr	**8.81**
100 mm	5.99	7.36	0.09	3.38	nr	**10.74**
150 mm	5.79	7.11	0.09	3.38	nr	**10.49**
Angles						
75 mm; 45°	5.93	7.29	0.20	7.51	nr	**14.80**
75 mm; 90°	5.93	7.29	0.20	7.51	nr	**14.80**
100 mm; 90°	4.09	5.03	0.20	7.51	nr	**12.54**

33 DRAINAGE ABOVE GROUND

Item	Net Price £	Material £	Labour hours	Labour £	Unit	**Total rate £**
RAINWATER PIPEWORK/GUTTERS – cont						
Extra over fittings half round PVC-u gutter – cont						
Angles – cont						
100 mm; 120°	4.53	5.57	0.20	7.51	nr	**13.08**
100 mm; 135°	4.53	5.57	0.20	7.51	nr	**13.08**
100 mm; Prefabricated to special angle	25.55	31.38	0.23	8.64	nr	**40.02**
100 mm; Prefabricated to raked angle	27.65	33.97	0.23	8.64	nr	**42.61**
150 mm; 90°	14.67	18.02	0.20	7.51	nr	**25.53**
Gutter adaptors						
100 mm; Stainless steel clip	2.11	2.59	0.16	6.01	nr	**8.60**
100 mm; Cast iron spigot	5.78	7.10	0.23	8.64	nr	**15.74**
100 mm; Cast iron socket	5.78	7.10	0.23	8.64	nr	**15.74**
100 mm; Cast iron 'ogee' spigot	5.85	7.18	0.23	8.64	nr	**15.82**
100 mm; Cast iron 'ogee' socket	5.85	7.18	0.23	8.64	nr	**15.82**
100 mm; Half round to Square PVC-u	10.48	12.87	0.23	8.64	nr	**21.51**
100 mm; Gutter overshoot guard	12.96	15.92	0.58	21.77	nr	**37.69**
Square gutter, with brackets measured separately						
120 mm square	3.54	4.35	0.82	30.78	m	**35.13**
Brackets: including fixing to backgrounds. For minimum fixing distances, refer to the Tables and Memoranda at the rear of the book						
Jointing	2.44	2.99	0.16	6.01	nr	**9.00**
Support	1.07	1.31	0.16	6.01	nr	**7.32**
Bracket support: including fixing to backgrounds. For minimum fixing distances, refer to the Tables and Memoranda at the rear of the book						
Side rafter	3.73	4.58	0.16	6.01	nr	**10.59**
Top rafter	3.73	4.58	0.16	6.01	nr	**10.59**
Rise and fall	4.03	4.96	0.16	6.01	nr	**10.97**
Extra over fittings square PVC-u gutter						
Rainwater pipe outlets						
Running: 62 mm square	3.45	4.24	0.12	4.50	nr	**8.74**
Stop end: 62 mm square	3.90	4.79	0.12	4.50	nr	**9.29**
Stop ends: short						
External	1.79	2.19	0.09	3.38	nr	**5.57**
Angles						
90°	4.42	5.43	0.20	7.51	nr	**12.94**
120°	11.08	13.61	0.20	7.51	nr	**21.12**
135°	5.18	6.36	0.20	7.51	nr	**13.87**
Prefabricated to special angle	26.20	32.18	0.23	8.64	nr	**40.82**
Prefabricated to raked angle	32.30	39.68	0.23	8.64	nr	**48.32**
Gutter adaptors						
Cast iron	16.82	20.66	0.23	8.64	nr	**29.30**

33 DRAINAGE ABOVE GROUND

Item	Net Price £	Material £	Labour hours	Labour £	Unit	Total rate £
High capacity square gutter, with brackets measured separately						
137 mm	8.08	9.92	0.82	30.78	m	**40.70**
Brackets: including fixing to backgrounds. For minimum fixing distances, refer to the Tables and Memoranda at the rear of the book						
Jointing	7.68	9.44	0.16	6.01	nr	**15.45**
Support	3.20	3.93	0.16	6.01	nr	**9.94**
Overslung	2.98	3.66	0.16	6.01	nr	**9.67**
Bracket supports: including fixing to backgrounds. For minimum fixing distances, refer to the Tables and Memoranda at the rear of the book						
Side rafter	4.58	5.63	0.16	6.01	nr	**11.64**
Top rafter	4.58	5.63	0.16	6.01	nr	**11.64**
Rise and fall	7.35	9.03	0.16	6.01	nr	**15.04**
Extra over fittings high capacity square PVC-u						
Rainwater pipe outlets						
Running: 75 mm square	12.58	15.46	0.12	4.50	nr	**19.96**
Running: 82 mm dia.	12.58	15.46	0.12	4.50	nr	**19.96**
Running: 110 mm dia.	11.01	13.52	0.12	4.50	nr	**18.02**
Screwed outlet adaptor						
75 mm square pipe	8.00	9.83	0.23	8.64	nr	**18.47**
Stop ends: short						
External	4.42	5.43	0.09	3.38	nr	**8.81**
Angles						
90°	11.02	13.53	0.20	7.51	nr	**21.04**
135°	12.96	15.92	0.20	7.51	nr	**23.43**
Prefabricated to special angle	37.43	45.98	0.23	8.64	nr	**54.62**
Prefabricated to raked internal angle	65.17	80.06	0.23	8.64	nr	**88.70**
Prefabricated to raked external angle	65.17	80.06	0.23	8.64	nr	**88.70**
Deep eliptical gutter, with brackets measured separately						
137 mm	4.38	5.38	0.82	30.78	m	**36.16**
Brackets: including fixing to backgrounds. For minimum fixing distances, refer to the Tables and Memoranda at the rear of the book						
Jointing	3.46	4.25	0.16	6.01	nr	**10.26**
Support	1.48	1.82	0.16	6.01	nr	**7.83**
Bracket support: including fixing to backgrounds. For minimum fixing distances, refer to the Tables and Memoranda at the rear of the book						
Side rafter	4.80	5.90	0.16	6.01	nr	**11.91**
Top rafter	4.80	5.90	0.16	6.01	nr	**11.91**
Rise and fall	7.72	9.48	0.16	6.01	nr	**15.49**

33 DRAINAGE ABOVE GROUND

Item	Net Price £	Material £	Labour hours	Labour £	Unit	Total rate £
RAINWATER PIPEWORK/GUTTERS – cont						
Extra over fittings deep eliptical PVC-u gutter						
Rainwater pipe outlets						
Running: 68 mm dia.	5.51	6.77	0.12	4.50	nr	**11.27**
Running: 82 mm dia.	5.24	6.44	0.12	4.50	nr	**10.94**
Stop end: 68 mm dia.	5.60	6.88	0.12	4.50	nr	**11.38**
Stop ends: short						
External	2.69	3.31	0.09	3.38	nr	**6.69**
Angles						
90°	5.57	6.85	0.20	7.51	nr	**14.36**
135°	6.61	8.12	0.20	7.51	nr	**15.63**
Prefabricated to special angle	19.33	23.75	0.23	8.64	nr	**32.39**
Gutter adaptors						
Stainless steel clip	4.74	5.83	0.16	6.01	nr	**11.84**
Marley deepflow	3.83	4.70	0.23	8.64	nr	**13.34**
Ogee profile PVC-u gutter, with brackets measured separately						
122 mm	4.61	5.66	0.82	30.78	m	**36.44**
Brackets: including fixing to backgrounds. For minimum fixing distances, refer to the Tables and Memoranda at the rear of the book						
Jointing	3.66	4.49	0.16	6.01	nr	**10.50**
Support	1.41	1.73	0.16	6.01	nr	**7.74**
Overslung	1.41	1.73	0.16	6.01	nr	**7.74**
Extra over fittings Ogee profile PVC-u gutter						
Rainwater pipe outlets						
Running: 68 mm dia.	5.24	6.44	0.12	4.50	nr	**10.94**
Stop ends: short						
Internal/External: left or right hand	2.65	3.26	0.09	3.38	nr	**6.64**
Angles						
90°: internal or external	5.30	6.51	0.20	7.51	nr	**14.02**
135°: internal or external	5.30	6.51	0.20	7.51	nr	**14.02**
PVC-u rainwater pipe: dry push fit joints; fixed with brackets to backgrounds; BS 4576/ BS EN 607						
Pipe: circular, with brackets measured separately						
53 mm	5.95	7.31	0.61	22.90	m	**30.21**
68 mm	6.20	7.61	0.61	22.90	m	**30.51**

33 DRAINAGE ABOVE GROUND

Item	Net Price £	Material £	Labour hours	Labour £	Unit	Total rate £
Pipe clip: including fixing to backgrounds. For minimum fixing distances, refer to the Tables and Memoranda at the rear of the book						
68 mm	1.39	1.71	0.16	6.01	nr	**7.72**
Pipe clip adjustable: including fixing to backgrounds. For minimum fixing distances, refer to the Tables and Memoranda at the rear of the book						
53 mm	2.12	2.60	0.16	6.01	nr	**8.61**
68 mm	2.97	3.64	0.16	6.01	nr	**9.65**
Pipe clip drive in: including fixing to backgrounds. For minimum fixing distances, refer to the Tables and Memoranda at the rear of the book						
68 mm	3.33	4.10	0.16	6.01	nr	**10.11**
Extra over fittings circular pipework PVC-u						
Pipe coupler: PVC-u to PVC-u						
68 mm	1.73	2.12	0.12	4.50	nr	**6.62**
Pipe coupler: PVC-u to Cast Iron						
68 mm: to 3" cast iron	6.79	8.34	0.17	6.37	nr	**14.71**
68 mm: to 3 ¾" cast iron	23.32	28.64	0.17	6.37	nr	**35.01**
Access pipe: single socket						
68 mm	10.84	13.31	0.15	5.63	nr	**18.94**
Bend: short radius						
53 mm: 67.5°	2.52	3.10	0.20	7.51	nr	**10.61**
68 mm: 92.5°	3.71	4.56	0.20	7.51	nr	**12.07**
68 mm: 112.5°	3.71	4.56	0.20	7.51	nr	**12.07**
Bend: long radius						
68 mm: 112°	3.51	4.32	0.20	7.51	nr	**11.83**
Branch						
68 mm: 92°	22.78	27.98	0.23	8.64	nr	**36.62**
68 mm: 112°	22.85	28.07	0.23	8.64	nr	**36.71**
Double branch						
68 mm: 112°	45.08	55.38	0.24	9.01	nr	**64.39**
Shoe						
53 mm	3.74	4.60	0.12	4.50	nr	**9.10**
68 mm	5.51	6.77	0.12	4.50	nr	**11.27**
Rainwater head: including fixing to backgrounds						
68 mm	10.84	13.31	0.29	10.89	nr	**24.20**
Pipe: square, with brackets measured separately						
62 mm	3.79	4.65	0.45	16.89	m	**21.54**
75 mm	7.22	8.87	0.45	16.89	m	**25.76**
Pipe clip: including fixing to backgrounds. For minimum fixing distances, refer to the Tables and Memoranda at the rear of the book						
62 mm	1.28	1.58	0.16	6.01	nr	**7.59**
75 mm	2.82	3.46	0.16	6.01	nr	**9.47**

33 DRAINAGE ABOVE GROUND

Item	Net Price £	Material £	Labour hours	Labour £	Unit	Total rate £
RAINWATER PIPEWORK/GUTTERS – cont						
Pipe: square, with brackets measured separately – cont						
Pipe clip adjustable: including fixing to backgrounds. For minimum fixing distances, refer to the Tables and Memoranda at the rear of the book						
62 mm	3.93	4.83	0.16	6.01	nr	**10.84**
Extra over fittings square pipework PVC-u						
Pipe coupler: PVC-u to PVC-u						
62 mm	1.86	2.29	0.20	7.51	nr	**9.80**
75 mm	2.89	3.55	0.20	7.51	nr	**11.06**
Square to circular adaptor: single socket						
62 mm to 68 mm	3.63	4.46	0.20	7.51	nr	**11.97**
Square to circular adaptor: single socket						
75 mm to 62 mm	4.69	5.76	0.20	7.51	nr	**13.27**
Access pipe						
62 mm	21.68	26.64	0.16	6.01	nr	**32.65**
75 mm	22.88	28.11	0.16	6.01	nr	**34.12**
Bends						
62 mm: 92.5°	3.35	4.12	0.20	7.51	nr	**11.63**
62 mm: 112.5°	2.44	2.99	0.20	7.51	nr	**10.50**
75 mm: 112.5°	6.12	7.52	0.20	7.51	nr	**15.03**
Bends: prefabricated special angle						
62 mm	21.78	26.75	0.23	8.64	nr	**35.39**
75 mm	30.26	37.17	0.23	8.64	nr	**45.81**
Offset						
62 mm	5.25	6.45	0.20	7.51	nr	**13.96**
75 mm	14.45	17.76	0.20	7.51	nr	**25.27**
Offset: prefabricated special angle						
62 mm	21.78	26.75	0.23	8.64	nr	**35.39**
Shoe						
62 mm	3.08	3.78	0.12	4.50	nr	**8.28**
75 mm	5.28	6.49	0.12	4.50	nr	**10.99**
Branch						
62 mm	7.61	9.35	0.23	8.64	nr	**17.99**
75 mm	37.14	45.62	0.23	8.64	nr	**54.26**
Double branch						
62 mm	40.87	50.21	0.24	9.01	nr	**59.22**
Rainwater head						
62 mm	34.27	42.11	0.29	10.89	nr	**53.00**
75 mm	36.05	44.29	3.45	129.42	nr	**173.71**

33 DRAINAGE ABOVE GROUND

Item	Net Price £	Material £	Labour hours	Labour £	Unit	Total rate £
PVC-u rainwater pipe: solvent welded joints; fixed with brackets to backgrounds; BS 4576/ BS EN 607						
Pipe: circular, with brackets measured separately						
82 mm	11.69	14.36	0.35	13.14	m	**27.50**
Pipe clip: galvanized; including fixing to backgrounds. For minimum fixing distances, refer to the Tables and Memoranda at the rear of the book						
82 mm	4.75	5.84	0.58	21.77	nr	**27.61**
Pipe clip: galvanized plastic coated; including fixing to backgrounds. For minimum fixing distances, refer to the Tables and Memoranda at the rear of the book						
82 mm	6.58	8.09	0.58	21.77	nr	**29.86**
Pipe clip: PVC-u including fixing to backgrounds. For minimum fixing distances, refer to the Tables and Memoranda at the rear of the book						
82 mm	3.53	4.34	0.58	21.77	nr	**26.11**
Pipe clip: PVC-u adjustable: including fixing to backgrounds. For minimum fixing distances, refer to the Tables and Memoranda at the rear of the book						
82 mm	6.47	7.95	0.58	21.77	nr	**29.72**
Extra over fittings circular pipework PVC-u						
Pipe coupler: PVC-u to PVC-u						
82 mm	5.93	7.29	0.21	7.89	nr	**15.18**
Access pipe						
82 mm	36.50	44.85	0.23	8.64	nr	**53.49**
Bend						
82 mm: 92, 112.5 and 135°	14.90	18.30	0.29	10.89	nr	**29.19**
Shoe						
82 mm	9.20	11.30	0.29	10.89	nr	**22.19**
110 mm	11.54	14.17	0.32	12.01	nr	**26.18**
Branch						
82 mm: 92, 112.5 and 135°	22.20	27.27	0.35	13.14	nr	**40.41**
Rainwater head						
82 mm	19.48	23.93	0.58	21.77	nr	**45.70**
110 mm	17.74	21.80	0.58	21.77	nr	**43.57**
Roof outlets: 178 dia.; Flat						
50 mm	19.47	23.92	1.15	43.16	nr	**67.08**
82 mm	19.47	23.92	1.15	43.16	nr	**67.08**
Roof outlets: 178 mm dia.; Domed						
50 mm	19.47	23.92	1.15	43.16	nr	**67.08**
82 mm	19.47	23.92	1.15	43.16	nr	**67.08**

33 DRAINAGE ABOVE GROUND

Item	Net Price £	Material £	Labour hours	Labour £	Unit	Total rate £
RAINWATER PIPEWORK/GUTTERS – cont						
Extra over fittings circular pipework PVC-u – cont						
Roof outlets: 406 mm dia.; Flat						
82 mm	38.05	46.75	1.15	43.16	nr	**89.91**
110 mm	38.05	46.75	1.15	43.16	nr	**89.91**
Roof outlets: 406 mm dia.; Domed						
82 mm	38.05	46.75	1.15	43.16	nr	**89.91**
110 mm	38.05	46.75	1.15	43.16	nr	**89.91**
Roof outlets: 406 mm dia.; Inverted						
82 mm	83.75	102.89	1.15	43.16	nr	**146.05**
110 mm	83.75	102.89	1.15	43.16	nr	**146.05**
Roof outlets: 406 mm dia.; Vent Pipe						
82 mm	55.01	67.58	1.15	43.16	nr	**110.74**
110 mm	55.01	67.58	1.15	43.16	nr	**110.74**
Balcony outlets: screed						
82 mm	32.56	40.01	1.15	43.16	nr	**83.17**
Balcony outlets: asphalt						
82 mm	32.38	39.77	1.15	43.16	nr	**82.93**
Adaptors						
82 mm × 62 mm square pipe	4.36	5.36	0.21	7.89	nr	**13.25**
82 mm × 68 mm circular pipe	4.36	5.36	0.21	7.89	nr	**13.25**
For 110 mm dia. pipework and fittings refer to 33 Drainage Above Ground						
Cast iron gutters: mastic and bolted joints; BS 460; fixed with brackets to backgrounds						
Half round gutter, with brackets measured separately						
100 mm	15.02	18.45	0.85	31.90	m	**50.35**
115 mm	15.67	19.25	0.97	36.40	m	**55.65**
125 mm	18.33	22.52	0.97	36.40	m	**58.92**
150 mm	31.33	38.49	1.12	42.04	m	**80.53**
Brackets; fixed to backgrounds. For minimum fixing distances, refer to the Tables and Memoranda at the rear of the book						
Fascia						
100 mm	3.52	4.33	0.16	6.01	nr	**10.34**
115 mm	3.52	4.33	0.16	6.01	nr	**10.34**
125 mm	3.52	4.33	0.16	6.01	nr	**10.34**
150 mm	4.46	5.48	0.16	6.01	nr	**11.49**
Rise and fall						
100 mm	7.00	8.60	0.39	14.63	nr	**23.23**
115 mm	7.00	8.60	0.39	14.63	nr	**23.23**
125 mm	7.19	8.83	0.39	14.63	nr	**23.46**
150 mm	7.32	8.99	0.39	14.63	nr	**23.62**

33 DRAINAGE ABOVE GROUND

Item	Net Price £	Material £	Labour hours	Labour £	Unit	**Total rate £**
Top rafter						
100 mm	4.31	5.29	0.16	6.01	nr	**11.30**
115 mm	4.31	5.29	0.16	6.01	nr	**11.30**
125 mm	5.83	7.16	0.16	6.01	nr	**13.17**
150 mm	7.93	9.74	0.16	6.01	nr	**15.75**
Side rafter						
100 mm	4.31	5.29	0.16	6.01	nr	**11.30**
115 mm	4.31	5.29	0.16	6.01	nr	**11.30**
125 mm	5.83	7.16	0.16	6.01	nr	**13.17**
150 mm	7.93	9.74	0.16	6.01	nr	**15.75**

33 DRAINAGE ABOVE GROUND

Item	Net Price £	Material £	Labour hours	Labour £	Unit	**Total rate £**
RAINWATER PIPEWORK/GUTTERS – cont						
Extra over fittings half round gutter cast iron BS 460						
Union						
100 mm	8.26	10.14	0.39	14.63	nr	**24.77**
115 mm	10.31	12.66	0.48	18.01	nr	**30.67**
125 mm	11.65	14.31	0.48	18.01	nr	**32.32**
150 mm	13.07	16.05	0.55	20.64	nr	**36.69**
Stop end; internal						
100 mm	4.21	5.18	0.12	4.50	nr	**9.68**
115 mm	5.45	6.70	0.15	5.63	nr	**12.33**
125 mm	5.45	6.70	0.15	5.63	nr	**12.33**
150 mm	7.57	9.30	0.20	7.51	nr	**16.81**
Stop end; external						
100 mm	4.21	5.18	0.12	4.50	nr	**9.68**
115 mm	5.35	6.57	0.15	5.63	nr	**12.20**
125 mm	5.45	6.70	0.15	5.63	nr	**12.33**
150 mm	7.57	9.30	0.20	7.51	nr	**16.81**
90° angle; single socket						
100 mm	12.53	15.39	0.39	14.63	nr	**30.02**
115 mm	12.90	15.84	0.43	16.14	nr	**31.98**
125 mm	15.20	18.67	0.43	16.14	nr	**34.81**
150 mm	27.78	34.13	0.50	18.77	nr	**52.90**
90° angle; double socket						
100 mm	15.18	18.65	0.39	14.63	nr	**33.28**
115 mm	16.11	19.79	0.43	16.14	nr	**35.93**
125 mm	20.85	25.61	0.43	16.14	nr	**41.75**
135° angle; single socket						
100 mm	12.80	15.73	0.39	14.63	nr	**30.36**
115 mm	12.92	15.88	0.43	16.14	nr	**32.02**
125 mm	19.08	23.44	0.43	16.14	nr	**39.58**
150 mm	28.32	34.79	0.50	18.77	nr	**53.56**
Running outlet						
65 mm outlet						
100 mm	12.23	15.03	0.39	14.63	nr	**29.66**
115 mm	13.31	16.35	0.43	16.14	nr	**32.49**
125 mm	15.20	18.67	0.43	16.14	nr	**34.81**
75 mm outlet						
100 mm	12.23	15.03	0.39	14.63	nr	**29.66**
115 mm	13.31	16.35	0.43	16.14	nr	**32.49**
125 mm	15.20	18.67	0.43	16.14	nr	**34.81**
150 mm	26.33	32.35	0.50	18.77	nr	**51.12**
100 mm outlet						
150 mm	26.33	32.35	0.50	18.77	nr	**51.12**
Stop end outlet; socket						
65 mm outlet						
100 mm	14.33	17.61	0.39	14.63	nr	**32.24**
115 mm	16.07	19.74	0.43	16.14	nr	**35.88**

33 DRAINAGE ABOVE GROUND

Item	Net Price £	Material £	Labour hours	Labour £	Unit	**Total rate £**
75 mm outlet						
125 mm	14.33	17.61	0.43	16.14	nr	**33.75**
150 mm	30.12	69.35	0.50	18.77	nr	**88.12**
100 mm outlet						
150 mm	26.33	32.35	0.50	18.77	nr	**51.12**
Stop end outlet; spigot						
65 mm outlet						
100 mm	14.33	17.61	0.39	14.63	nr	**32.24**
115 mm	16.07	19.74	0.43	16.14	nr	**35.88**
75 mm outlet						
125 mm	14.33	17.61	0.43	16.14	nr	**33.75**
150 mm	30.12	37.00	0.50	18.77	nr	**55.77**
100 mm outlet						
150 mm	30.12	37.00	0.50	18.77	nr	**55.77**
Half round; 3 mm thick double beaded gutter, with brackets measured separately						
100 mm	15.30	18.80	0.85	31.90	m	**50.70**
115 mm	15.97	19.61	0.85	31.90	m	**51.51**
125 mm	18.32	22.50	0.97	36.40	m	**58.90**
Brackets; fixed to backgrounds. For minimum fixing distances, refer to the Tables and Memoranda at the rear of the book						
Fascia						
100 mm	3.52	4.33	0.16	6.01	nr	**10.34**
115 mm	3.52	4.33	0.16	6.01	nr	**10.34**
125 mm	3.52	4.33	0.16	6.01	nr	**10.34**
Extra over fittings half round 3 mm thick Gutter BS 460						
Union						
100 mm	8.26	10.14	0.38	14.26	nr	**24.40**
115 mm	10.08	12.38	0.38	14.26	nr	**26.64**
125 mm	11.65	14.31	0.43	16.14	nr	**30.45**
Stop end; internal						
100 mm	4.21	5.18	0.12	4.50	nr	**9.68**
115 mm	5.45	6.70	0.12	4.50	nr	**11.20**
125 mm	5.48	6.73	0.15	5.63	nr	**12.36**
Stop end; external						
100 mm	4.21	5.18	0.12	4.50	nr	**9.68**
115 mm	5.45	6.70	0.12	4.50	nr	**11.20**
125 mm	5.48	6.73	0.15	5.63	nr	**12.36**
90° angle; single socket						
100 mm	13.01	15.98	0.38	14.26	nr	**30.24**
115 mm	13.20	16.21	0.38	14.26	nr	**30.47**
125 mm	15.20	18.67	0.43	16.14	nr	**34.81**
135° angle; single socket						
100 mm	12.80	15.73	0.38	14.26	nr	**29.99**
115 mm	12.90	15.84	0.38	14.26	nr	**30.10**
125 mm	16.08	19.75	0.43	16.14	nr	**35.89**

33 DRAINAGE ABOVE GROUND

Item	Net Price £	Material £	Labour hours	Labour £	Unit	**Total rate £**
RAINWATER PIPEWORK/GUTTERS – cont						
Extra over fittings half round 3 mm thick Gutter BS 460 – cont						
Running outlet						
65 mm outlet						
100 mm	13.12	16.12	0.38	14.26	nr	**30.38**
115 mm	13.56	16.66	0.38	14.26	nr	**30.92**
125 mm	17.96	22.06	0.43	16.14	nr	**38.20**
75 mm outlet						
115 mm	13.56	16.66	0.38	14.26	nr	**30.92**
125 mm	15.75	19.35	0.43	16.14	nr	**35.49**
Stop end outlet; socket						
65 mm outlet						
100 mm	14.33	17.61	0.38	14.26	nr	**31.87**
115 mm	16.07	19.74	0.38	14.26	nr	**34.00**
125 mm	17.96	22.06	0.43	16.14	nr	**38.20**
75 mm outlet						
125 mm	18.28	22.46	0.43	16.14	nr	**38.60**
Stop end outlet; spigot						
65 mm outlet						
100 mm	14.33	17.61	0.38	14.26	nr	**31.87**
115 mm	16.07	19.74	0.38	14.26	nr	**34.00**
125 mm	17.96	22.06	0.43	16.14	nr	**38.20**
Deep half round gutter, with brackets measured separately						
100 × 75 mm	25.16	30.91	0.85	31.90	m	**62.81**
125 × 75 mm	32.52	39.95	0.97	36.40	m	**76.35**
Brackets; fixed to backgrounds. For minimum fixing distances, refer to the Tables and Memoranda at the rear of the book						
Fascia						
100 × 75 mm	11.88	14.60	0.16	6.01	nr	**20.61**
125 × 75 mm	14.64	17.99	0.16	6.01	nr	**24.00**
Extra over fittings deep half round gutter BS 460						
Union						
100 × 75 mm	13.82	16.98	0.38	14.26	nr	**31.24**
125 × 75 mm	14.64	17.99	0.43	16.14	nr	**34.13**
Stop end; internal						
100 × 75 mm	12.10	14.87	0.12	4.50	nr	**19.37**
125 × 75 mm	14.93	18.34	0.15	5.63	nr	**23.97**
Stop end; external						
100 × 75 mm	12.10	14.87	0.12	4.50	nr	**19.37**
125 × 75 mm	14.93	18.34	0.15	5.63	nr	**23.97**
90° angle; single socket						
100 × 75 mm	34.50	42.39	0.38	14.26	nr	**56.65**
125 × 75 mm	43.79	53.79	0.43	16.14	nr	**69.93**

33 DRAINAGE ABOVE GROUND

Item	Net Price £	Material £	Labour hours	Labour £	Unit	**Total rate £**
135° angle; single socket						
100 × 75 mm	34.50	42.39	0.38	14.26	nr	**56.65**
125 × 75 mm	43.79	53.79	0.43	16.14	nr	**69.93**
Running outlet						
65 mm outlet						
100 × 75 mm	34.50	42.39	0.38	14.26	nr	**56.65**
125 × 75 mm	43.79	53.79	0.43	16.14	nr	**69.93**
75 mm outlet						
100 × 75 mm	34.50	42.39	0.38	14.26	nr	**56.65**
125 × 75 mm	31.78	39.04	0.43	16.14	nr	**55.18**
Stop end outlet; socket						
65 mm outlet						
100 × 75 mm	46.60	57.25	0.38	14.26	nr	**71.51**
75 mm outlet						
100 × 75 mm	46.60	57.25	0.38	14.26	nr	**71.51**
125 × 75 mm	46.60	57.25	0.43	16.14	nr	**73.39**
Stop end outlet; spigot						
65 mm outlet						
100 × 75 mm	46.60	57.25	0.38	14.26	nr	**71.51**
75 mm outlet						
100 × 75 mm	46.60	57.25	0.38	14.26	nr	**71.51**
125 × 75 mm	46.60	57.25	0.43	16.14	nr	**73.39**
Ogee gutter, with brackets measured separately						
100 mm	16.76	20.59	0.85	31.90	m	**52.49**
115 mm	18.44	22.65	0.97	36.40	m	**59.05**
125 mm	19.34	23.76	0.97	36.40	m	**60.16**
Brackets; fixed to backgrounds. For minimum fixing distances, refer to the Tables and Memoranda at the rear of the book						
Fascia						
100 mm	3.83	4.70	0.16	6.01	nr	**10.71**
115 mm	3.83	4.70	0.16	6.01	nr	**10.71**
125 mm	4.32	5.30	0.16	6.01	nr	**11.31**
Extra over fittings Ogee cast iron gutter BS 460						
Union						
100 mm	8.27	10.16	0.38	14.26	nr	**24.42**
115 mm	10.08	12.38	0.43	16.14	nr	**28.52**
125 mm	11.65	14.31	0.43	16.14	nr	**30.45**
Stop end; internal						
100 mm	4.31	5.29	0.12	4.50	nr	**9.79**
115 mm	5.57	6.85	0.15	5.63	nr	**12.48**
125 mm	5.57	6.85	0.15	5.63	nr	**12.48**
Stop end; external						
100 mm	4.31	5.29	0.12	4.50	nr	**9.79**
115 mm	5.57	6.85	0.15	5.63	nr	**12.48**
125 mm	5.57	6.85	0.15	5.63	nr	**12.48**

33 DRAINAGE ABOVE GROUND

Item	Net Price £	Material £	Labour hours	Labour £	Unit	Total rate £
RAINWATER PIPEWORK/GUTTERS – cont						
Extra over fittings Ogee cast iron gutter BS 460 – cont						
90° angle; internal						
100 mm	13.07	16.05	0.38	14.26	nr	**30.31**
115 mm	14.17	17.41	0.43	16.14	nr	**33.55**
125 mm	15.47	19.01	0.43	16.14	nr	**35.15**
90° angle; external						
100 mm	13.07	16.05	0.38	14.26	nr	**30.31**
115 mm	14.17	17.41	0.43	16.14	nr	**33.55**
125 mm	15.47	19.01	0.43	16.14	nr	**35.15**
135° angle; internal						
100 mm	13.58	16.68	0.38	14.26	nr	**30.94**
115 mm	14.46	17.77	0.43	16.14	nr	**33.91**
125 mm	19.05	23.40	0.43	16.14	nr	**39.54**
135° angle; external						
100 mm	13.58	16.68	0.38	14.26	nr	**30.94**
115 mm	14.46	17.77	0.43	16.14	nr	**33.91**
125 mm	19.05	23.40	0.43	16.14	nr	**39.54**
Running outlet						
65 mm outlet						
100 mm	13.32	16.36	0.38	14.26	nr	**30.62**
115 mm	14.18	17.42	0.43	16.14	nr	**33.56**
125 mm	15.47	19.01	0.43	16.14	nr	**35.15**
75 mm outlet						
125 mm	15.47	19.01	0.43	16.14	nr	**35.15**
Stop end outlet; socket						
65 mm outlet						
100 mm	21.03	25.84	0.38	14.26	nr	**40.10**
115 mm	21.03	25.84	0.43	16.14	nr	**41.98**
125 mm	21.03	25.84	0.43	16.14	nr	**41.98**
75 mm outlet						
125 mm	21.03	25.84	0.43	16.14	nr	**41.98**
Stop end outlet; spigot						
65 mm outlet						
100 mm	21.03	25.84	0.38	14.26	nr	**40.10**
115 mm	21.03	25.84	0.43	16.14	nr	**41.98**
125 mm	21.03	25.84	0.43	16.14	nr	**41.98**
75 mm outlet						
125 mm	21.03	25.84	0.43	16.14	nr	**41.98**

33 DRAINAGE ABOVE GROUND

Item	Net Price £	Material £	Labour hours	Labour £	Unit	**Total rate £**
Notts Ogee gutter, with brackets measured separately						
115 mm	29.78	36.58	0.85	31.90	m	**68.48**
Brackets; fixed to backgrounds. For minimum fixing distances, refer to the Tables and Memoranda at the rear of the book						
Fascia						
115 mm	11.61	14.26	0.16	6.01	nr	**20.27**
Extra over fittings Notts Ogee cast iron gutter BS 460						
Union						
115 mm	14.10	17.32	0.38	14.26	nr	**31.58**
Stop end; internal						
115 mm	11.61	14.26	0.16	6.01	nr	**20.27**
Stop end; external						
115 mm	11.61	14.26	0.16	6.01	nr	**20.27**
90° angle; internal						
115 mm	33.69	41.39	0.43	16.14	nr	**57.53**
90° angle; external						
115 mm	33.69	41.39	0.43	16.14	nr	**57.53**
135° angle; internal						
115 mm	34.32	42.16	0.43	16.14	nr	**58.30**
135° angle; external						
115 mm	34.32	42.16	0.43	16.14	nr	**58.30**
Running outlet						
65 mm outlet						
115 mm	40.38	49.60	0.43	16.14	nr	**65.74**
75 mm outlet						
115 mm	40.38	49.60	0.43	16.14	nr	**65.74**
Stop end outlet; socket						
65 mm outlet						
115 mm	51.99	63.87	0.43	16.14	nr	**80.01**
Stop end outlet; spigot						
65 mm outlet						
115 mm	51.99	63.87	0.43	16.14	nr	**80.01**
No 46 moulded gutter, with brackets measured separately						
100 × 75 mm	28.79	35.36	0.85	31.90	m	**67.26**
125 × 100 mm	42.22	51.87	0.97	36.40	m	**88.27**
Brackets; fixed to backgrounds. For minimum fixing distances, refer to the Tables and Memoranda at the rear of the book						
Fascia						
100 × 75 mm	6.43	7.90	0.16	6.01	nr	**13.91**
125 × 100 mm	6.43	7.90	0.16	6.01	nr	**13.91**
Extra over fittings						

33 DRAINAGE ABOVE GROUND

Item	Net Price £	Material £	Labour hours	Labour £	Unit	Total rate £
RAINWATER PIPEWORK/GUTTERS – cont						
No 46 moulded gutter, with brackets measured separately – cont						
Union						
100 × 75 mm	13.56	16.66	0.38	14.26	nr	**30.92**
125 × 100 mm	15.72	19.31	0.43	16.14	nr	**35.45**
Stop end; internal						
100 × 75 mm	12.15	14.93	0.12	4.50	nr	**19.43**
125 × 100 mm	15.72	19.31	0.15	5.63	nr	**24.94**
Stop end; external						
100 × 75 mm	12.15	14.93	0.12	4.50	nr	**19.43**
125 × 100 mm	15.72	19.31	0.15	5.63	nr	**24.94**
90° angle; internal						
100 × 75 mm	31.81	39.08	0.38	14.26	nr	**53.34**
125 × 100 mm	45.72	56.16	0.43	16.14	nr	**72.30**
90° angle; external						
100 × 75 mm	31.81	39.08	0.38	14.26	nr	**53.34**
125 × 100 mm	45.72	56.16	0.43	16.14	nr	**72.30**
135° angle; internal						
100 × 75 mm	32.42	39.83	0.38	14.26	nr	**54.09**
125 × 100 mm	45.72	56.16	0.43	16.14	nr	**72.30**
135° angle; external						
100 × 75 mm	32.42	39.83	0.38	14.26	nr	**54.09**
125 × 100 mm	45.72	56.16	0.43	16.14	nr	**72.30**
Running outlet						
65 mm outlet						
100 × 75 mm	32.42	39.83	0.38	14.26	nr	**54.09**
125 × 100 mm	45.72	56.16	0.43	16.14	nr	**72.30**
75 mm outlet						
100 × 75 mm	32.42	39.83	0.38	14.26	nr	**54.09**
125 × 100 mm	45.72	56.16	0.43	16.14	nr	**72.30**
100 mm outlet						
100 × 75 mm	32.42	39.83	0.38	14.26	nr	**54.09**
125 × 100 mm	45.72	56.16	0.43	16.14	nr	**72.30**
100 × 75 mm outlet						
125 × 100 mm	32.42	39.83	0.43	16.14	nr	**55.97**
Stop end outlet; socket						
65 mm outlet						
100 × 75 mm	61.44	75.47	0.38	14.26	nr	**89.73**
75 mm outlet						
125 × 100 mm	61.44	75.47	0.43	16.14	nr	**91.61**
Stop end outlet; spigot						
65 mm outlet						
100 × 75 mm	61.44	75.47	0.38	14.26	nr	**89.73**
75 mm outlet						
125 × 100 mm	61.44	75.47	0.43	16.14	nr	**91.61**

33 DRAINAGE ABOVE GROUND

Item	Net Price £	Material £	Labour hours	Labour £	Unit	**Total rate £**
Box gutter, with brackets measured separately						
100 × 75 mm	46.69	57.36	0.85	31.90	m	**89.26**
Brackets; fixed to backgrounds. For minimum fixing distances, refer to the Tables and Memoranda at the rear of the book						
Fascia						
100 × 75 mm	7.40	9.09	0.16	6.01	nr	**15.10**
Extra over fittings Box cast iron gutter BS 460						
Union						
100 × 75 mm	9.82	12.06	0.38	14.26	nr	**26.32**
Stop end; external						
100 × 75 mm	7.40	9.09	0.12	4.50	nr	**13.59**
90° angle						
100 × 75 mm	37.31	45.83	0.38	14.26	nr	**60.09**
135° angle						
100 × 75 mm	37.31	45.83	0.38	14.26	nr	**60.09**
Running outlet						
65 mm outlet						
100 × 75 mm	37.31	45.83	0.38	14.26	nr	**60.09**
75 mm outlet						
100 × 75 mm	37.31	45.83	0.38	14.26	nr	**60.09**
100 × 75 mm outlet						
100 × 75 mm	37.31	45.83	0.38	14.26	nr	**60.09**
Cast iron rainwater pipe; dry joints; BS 460; fixed to backgrounds: Circular						
Plain socket pipe, with brackets measured separately						
65 mm	27.10	33.30	0.69	25.89	m	**59.19**
75 mm	27.10	33.30	0.69	25.89	m	**59.19**
100 mm	36.99	45.44	0.69	25.89	m	**71.33**
Bracket; fixed to backgrounds. For minimum fixing distances, refer to the Tables and Memoranda at the rear of the book						
65 mm	9.90	12.16	0.29	10.89	nr	**23.05**
75 mm	9.94	12.21	0.29	10.89	nr	**23.10**
100 mm	10.07	12.37	0.29	10.89	nr	**23.26**
Eared socket pipe, with wall spacers measured separately						
65 mm	28.96	35.57	0.62	23.27	m	**58.84**
75 mm	28.96	35.57	0.62	23.27	m	**58.84**
100 mm	38.88	47.76	0.62	23.27	m	**71.03**
Wall spacer plate; eared pipework						
65 mm	6.39	7.85	0.16	6.01	nr	**13.86**
75 mm	6.51	8.00	0.16	6.01	nr	**14.01**
100 mm	10.07	12.37	0.16	6.01	nr	**18.38**

33 DRAINAGE ABOVE GROUND

Item	Net Price £	Material £	Labour hours	Labour £	Unit	**Total rate £**
RAINWATER PIPEWORK/GUTTERS – cont						
Extra over fittings circular cast iron pipework BS 460						
Loose sockets						
Plain socket						
65 mm	7.83	9.62	0.23	8.64	nr	**18.26**
75 mm	7.83	9.62	0.23	8.64	nr	**18.26**
100 mm	12.11	14.88	0.23	8.64	nr	**23.52**
Eared socket						
65 mm	11.33	13.92	0.29	10.89	nr	**24.81**
75 mm	11.33	13.92	0.29	10.89	nr	**24.81**
100 mm	15.29	18.78	0.29	10.89	nr	**29.67**
Shoe; front projection						
Plain socket						
65 mm	24.37	29.94	0.23	8.64	nr	**38.58**
75 mm	24.37	29.94	0.23	8.64	nr	**38.58**
100 mm	32.85	40.35	0.23	8.64	nr	**48.99**
Eared socket						
65 mm	28.24	34.69	0.29	10.89	nr	**45.58**
75 mm	28.24	34.69	0.29	10.89	nr	**45.58**
100 mm	37.48	46.04	0.29	10.89	nr	**56.93**
Access pipe						
65 mm	43.98	54.03	0.23	8.64	nr	**62.67**
75 mm	46.18	56.73	0.23	8.64	nr	**65.37**
100 mm	81.91	100.62	0.23	8.64	nr	**109.26**
100 mm; eared	92.48	113.61	0.29	10.89	nr	**124.50**
Bends; any°						
65 mm	17.60	21.62	0.23	8.64	nr	**30.26**
75 mm	20.97	25.76	0.23	8.64	nr	**34.40**
100 mm	29.63	36.40	0.23	8.64	nr	**45.04**
Branch						
92.5°						
65 mm	33.95	41.71	0.29	10.89	nr	**52.60**
75 mm	37.44	45.99	0.29	10.89	nr	**56.88**
100 mm	43.64	53.61	0.29	10.89	nr	**64.50**
112.5°						
65 mm	33.95	41.71	0.29	10.89	nr	**52.60**
75 mm	37.44	45.99	0.29	10.89	nr	**56.88**
135°						
65 mm	33.95	41.71	0.29	10.89	nr	**52.60**
75 mm	37.44	45.99	0.29	10.89	nr	**56.88**

33 DRAINAGE ABOVE GROUND

Item	Net Price £	Material £	Labour hours	Labour £	Unit	Total rate £
Offsets						
75 to 150 mm projection						
65 mm	26.45	32.50	0.25	9.39	nr	**41.89**
75 mm	26.45	32.50	0.25	9.39	nr	**41.89**
100 mm	49.89	61.29	0.25	9.39	nr	**70.68**
225 mm projection						
65 mm	26.96	33.12	0.25	9.39	nr	**42.51**
75 mm	26.96	33.12	0.25	9.39	nr	**42.51**
100 mm	49.89	61.29	0.25	9.39	nr	**70.68**
305 mm projection						
65 mm	36.06	44.30	0.25	9.39	nr	**53.69**
75 mm	37.86	46.51	0.25	9.39	nr	**55.90**
100 mm	61.61	75.68	0.25	9.39	nr	**85.07**
380 mm projection						
65 mm	71.96	88.40	0.25	9.39	nr	**97.79**
75 mm	71.96	88.40	0.25	9.39	nr	**97.79**
100 mm	98.23	120.68	0.25	9.39	nr	**130.07**
455 mm projection						
65 mm	84.24	103.49	0.25	9.39	nr	**112.88**
75 mm	84.24	103.49	0.25	9.39	nr	**112.88**
100 mm	122.17	150.09	0.25	9.39	nr	**159.48**
Cast iron rainwater pipe; dry joints; BS 460; fixed to backgrounds: Rectangular						
Plain socket						
100 × 75 mm	76.58	94.08	1.04	39.03	m	**133.11**
Bracket; fixed to backgrounds. For minimum fixing distances, refer to the Tables and Memoranda at the rear of the book						
100 × 75 mm; build in holdabat	43.23	53.11	0.35	13.14	nr	**66.25**
100 × 75 mm; trefoil earband	34.44	42.30	0.29	10.89	nr	**53.19**
100 × 75 mm; plain earband	33.31	40.92	0.29	10.89	nr	**51.81**
Eared socket, with wall spacers measured separately						
100 × 75 mm	77.84	95.62	1.16	43.53	m	**139.15**
Wall spacer plate; eared pipework						
100 × 75	10.07	12.37	0.16	6.01	nr	**18.38**
Extra over fittings rectangular cast iron pipework BS 460						
Loose socket						
100 × 75 mm; plain	32.48	39.90	0.23	8.64	nr	**48.54**
100 × 75 mm; eared	53.82	66.12	0.29	10.89	nr	**77.01**
Shoe; front						
100 × 75 mm; plain	86.15	105.84	0.23	8.64	nr	**114.48**
100 × 75 mm; eared	105.27	129.33	0.29	10.89	nr	**140.22**
Shoe; side						
100 × 75 mm; plain	104.52	128.40	0.23	8.64	nr	**137.04**
100 × 75 mm; eared	130.68	160.55	0.29	10.89	nr	**171.44**

33 DRAINAGE ABOVE GROUND

Item	Net Price £	Material £	Labour hours	Labour £	Unit	Total rate £
RAINWATER PIPEWORK/GUTTERS – cont						
Extra over fittings rectangular cast iron pipework BS 460 – cont						
Bends; side; any°						
100 × 75 mm; plain	79.28	97.40	0.25	9.39	nr	**106.79**
100 × 75 mm; 135°; plain	80.98	99.49	0.25	9.39	nr	**108.88**
Bends; side; any°						
100 × 75 mm; eared	100.24	123.14	0.25	9.39	nr	**132.53**
Bends; front; any°						
100 × 75 mm; plain	75.07	92.22	0.25	9.39	nr	**101.61**
100 × 75 mm; eared	85.07	104.51	0.25	9.39	nr	**113.90**
Offset; side						
Plain socket						
75 mm projection	103.61	127.28	0.25	9.39	nr	**136.67**
115 mm projection	107.76	132.38	0.25	9.39	nr	**141.77**
225 mm projection	134.26	164.93	0.25	9.39	nr	**174.32**
305 mm projection	154.73	190.08	0.25	9.39	nr	**199.47**
Offset; front						
Plain socket						
75 mm projection	78.83	96.84	0.25	9.39	nr	**106.23**
150 mm projection	87.12	107.03	0.25	9.39	nr	**116.42**
225 mm projection	107.85	132.49	0.25	9.39	nr	**141.88**
305 mm projection	128.31	157.63	0.25	9.39	nr	**167.02**
Eared socket						
75 mm projection	100.84	123.88	0.25	9.39	nr	**133.27**
150 mm projection	108.74	133.59	0.25	9.39	nr	**142.98**
225 mm projection	128.63	158.02	0.25	9.39	nr	**167.41**
305 mm projection	149.65	183.84	0.25	9.39	nr	**193.23**
Offset; plinth						
115 mm projection; plain	81.36	99.95	0.25	9.39	nr	**109.34**
115 mm projection; eared	104.78	128.72	0.25	9.39	nr	**138.11**
Rainwater heads						
Flat hopper						
210 × 160 × 185 mm; 65 mm outlet	61.90	76.04	0.40	15.02	nr	**91.06**
210 × 160 × 185 mm; 75 mm outlet	61.90	76.04	0.40	15.02	nr	**91.06**
250 × 215 × 215 mm; 100 mm outlet	73.37	90.13	0.40	15.02	nr	**105.15**
Flat rectangular						
225 × 125 × 125 mm; 65 mm outlet	86.06	105.72	0.40	15.02	nr	**120.74**
225 × 125 × 125 mm; 75 mm outlet	86.06	105.72	0.40	15.02	nr	**120.74**
280 × 150 × 130 mm; 100 mm outlet	118.82	145.97	0.40	15.02	nr	**160.99**

33 DRAINAGE ABOVE GROUND

Item	Net Price £	Material £	Labour hours	Labour £	Unit	**Total rate £**
Rectangular						
250 × 180 × 175 mm; 75 mm outlet	80.25	98.58	0.40	15.02	nr	**113.60**
250 × 180 × 175 mm; 100 mm outlet	80.25	98.58	0.40	15.02	nr	**113.60**
300 × 250 × 200 mm; 65 mm outlet	111.77	137.31	0.40	15.02	nr	**152.33**
300 × 250 × 200 mm; 75 mm outlet	111.77	137.31	0.40	15.02	nr	**152.33**
300 × 250 × 200 mm; 100 mm outlet	111.77	137.31	0.40	15.02	nr	**152.33**
300 × 250 × 200 mm; 100 × 75 mm outlet	111.77	137.31	0.40	15.02	nr	**152.33**
Castellated rectangular						
250 × 180 × 175 mm; 65 mm outlet	80.25	98.58	0.40	15.02	nr	**113.60**

33 DRAINAGE ABOVE GROUND

Item	Net Price £	Material £	Labour hours	Labour £	Unit	Total rate £
DISPOSAL SYSTEMS						
Pricing note:° angles are only indicated where material prices differ						
PVC-u overflow pipe; solvent welded joints; fixed with clips to backgrounds						
Pipe, with brackets measured separately						
19 mm	1.31	1.61	0.21	7.89	m	**9.50**
Fixings						
Pipe clip: including fixing to backgrounds. For minimum fixing distances, refer to the Tables and Memoranda at the rear of the book						
19 mm	0.51	0.63	0.18	6.76	nr	**7.39**
Extra over fittings overflow pipework PVC-u						
Straight coupler						
19 mm	1.39	1.71	0.17	6.37	nr	**8.08**
Bend						
19 mm: 91.25°	1.64	2.02	0.17	6.37	nr	**8.39**
19 mm: 135°	1.65	2.03	0.17	6.37	nr	**8.40**
Tee						
19 mm	1.79	2.19	0.18	6.76	nr	**8.95**
Reverse nut connector						
19 mm	0.65	0.80	0.15	5.63	nr	**6.43**
BSP adaptor: solvent welded socket to threaded socket						
19 mm × ¾"	2.39	2.94	0.14	5.26	nr	**8.20**
Straight tank connector						
19 mm	2.17	2.67	0.21	7.89	nr	**10.56**
32 mm	1.07	1.31	0.28	10.50	nr	**11.81**
40 mm	1.07	1.31	0.30	11.26	nr	**12.57**
Bent tank connector						
19 mm	2.56	3.15	0.21	7.89	nr	**11.04**
Tundish						
19 mm	33.95	41.71	0.38	14.26	nr	**55.97**
MuPVC waste pipe; solvent welded joints; fixed with clips to backgrounds; BS 5255						
Pipe, with brackets measured separately						
32 mm	2.07	2.54	0.23	8.64	m	**11.18**
40 mm	2.56	3.15	0.23	8.64	m	**11.79**
50 mm	3.87	4.76	0.26	9.75	m	**14.51**
Fixings						
Pipe clip: including fixing to backgrounds. For minimum fixing distances, refer to the Tables and Memoranda at the rear of the book						
32 mm	0.43	0.53	0.13	4.87	nr	**5.40**
40 mm	0.49	0.60	0.13	4.87	nr	**5.47**
50 mm	0.65	0.80	0.13	4.87	nr	**5.67**

33 DRAINAGE ABOVE GROUND

Item	Net Price £	Material £	Labour hours	Labour £	Unit	**Total rate £**
Pipe clip: expansion: including fixing to backgrounds. For minimum fixing distances, refer to the Tables and Memoranda at the rear of the book						
32 mm	0.51	0.63	0.13	4.87	nr	**5.50**
40 mm	0.53	0.65	0.13	4.87	nr	**5.52**
50 mm	1.25	1.53	0.13	4.87	nr	**6.40**
Pipe clip: metal; including fixing to backgrounds. For minimum fixing distances, refer to the Tables and Memoranda at the rear of the book						
32 mm	1.85	2.27	0.13	4.87	nr	**7.14**
40 mm	2.21	2.72	0.13	4.87	nr	**7.59**
50 mm	2.81	3.45	0.13	4.87	nr	**8.32**
Extra over fittings waste pipework MuPVC						
Screwed access plug						
32 mm	1.20	1.47	0.18	6.76	nr	**8.23**
40 mm	1.20	1.47	0.18	6.76	nr	**8.23**
50 mm	1.73	2.12	0.25	9.39	nr	**11.51**
Straight coupling						
32 mm	1.29	1.59	0.27	10.13	nr	**11.72**
40 mm	1.29	1.59	0.27	10.13	nr	**11.72**
50 mm	2.36	2.90	0.27	10.13	nr	**13.03**
Expansion coupling						
32 mm	2.28	2.80	0.27	10.13	nr	**12.93**
40 mm	2.74	3.37	0.27	10.13	nr	**13.50**
50 mm	3.73	4.58	0.27	10.13	nr	**14.71**
MuPVC to copper coupling						
32 mm	2.28	2.80	0.27	10.13	nr	**12.93**
40 mm	2.74	3.37	0.27	10.13	nr	**13.50**
50 mm	3.73	4.58	0.27	10.13	nr	**14.71**
Spigot and socket coupling						
32 mm	2.28	2.80	0.27	10.13	nr	**12.93**
40 mm	2.74	3.37	0.27	10.13	nr	**13.50**
50 mm	3.73	4.58	0.27	10.13	nr	**14.71**
Union						
32 mm	5.42	6.66	0.28	10.50	nr	**17.16**
40 mm	7.10	8.73	0.28	10.50	nr	**19.23**
50 mm	8.09	9.94	0.28	10.50	nr	**20.44**
Reducer: socket						
32 × 19 mm	1.96	2.40	0.27	10.13	nr	**12.53**
40 × 32 mm	1.29	1.59	0.27	10.13	nr	**11.72**
50 × 32 mm	1.86	2.29	0.27	10.13	nr	**12.42**
50 × 40 mm	2.28	2.80	0.27	10.13	nr	**12.93**
Reducer: level invert						
40 × 32 mm	1.61	1.97	0.27	10.13	nr	**12.10**
50 × 32 mm	2.00	2.46	0.27	10.13	nr	**12.59**
50 × 40 mm	2.00	2.46	0.27	10.13	nr	**12.59**

33 DRAINAGE ABOVE GROUND

Item	Net Price £	Material £	Labour hours	Labour £	Unit	Total rate £
DISPOSAL SYSTEMS – cont						
Extra over fittings waste pipework MuPVC – cont						
Swept bend						
32 mm	1.32	1.62	0.27	10.13	nr	**11.75**
32 mm: 165°	1.37	1.68	0.27	10.13	nr	**11.81**
40 mm	1.47	1.81	0.27	10.13	nr	**11.94**
40 mm: 165°	2.59	3.18	0.27	10.13	nr	**13.31**
50 mm	2.56	3.15	0.30	11.26	nr	**14.41**
50 mm: 165°	3.41	4.19	0.30	11.26	nr	**15.45**
Knuckle bend						
32 mm	1.20	1.47	0.27	10.13	nr	**11.60**
40 mm	1.33	1.64	0.27	10.13	nr	**11.77**
Spigot and socket bend						
32 mm	2.16	2.66	0.27	10.13	nr	**12.79**
32 mm: 150°	2.23	2.74	0.27	10.13	nr	**12.87**
40 mm	2.47	3.03	0.27	10.13	nr	**13.16**
50 mm	3.52	4.33	0.30	11.26	nr	**15.59**
Swept tee						
32 mm: 91.25°	1.77	2.17	0.31	11.63	nr	**13.80**
32 mm: 135°	2.11	2.59	0.31	11.63	nr	**14.22**
40 mm: 91.25°	2.24	2.75	0.31	11.63	nr	**14.38**
40 mm: 135°	2.79	3.42	0.31	11.63	nr	**15.05**
50 mm	4.38	5.38	0.31	11.63	nr	**17.01**
Swept cross						
40 mm: 91.25°	5.44	6.68	0.31	11.63	nr	**18.31**
50 mm: 91.25°	5.70	7.00	0.43	16.14	nr	**23.14**
50 mm: 135°	7.20	8.84	0.31	11.63	nr	**20.47**
Male iron adaptor						
32 mm	1.97	2.42	0.28	10.50	nr	**12.92**
40 mm	2.30	2.82	0.28	10.50	nr	**13.32**
Female iron adaptor						
32 mm	2.30	2.82	0.28	10.50	nr	**13.32**
40 mm	2.30	2.82	0.28	10.50	nr	**13.32**
50 mm	3.33	4.10	0.31	11.63	nr	**15.73**
Reverse nut adaptor						
32 mm	2.94	3.61	0.20	7.51	nr	**11.12**
40 mm	2.94	3.61	0.20	7.51	nr	**11.12**
Automatic air admittance valve						
32 mm	16.52	20.30	0.27	10.13	nr	**30.43**
40 mm	16.52	20.30	0.28	10.50	nr	**30.80**
50 mm	16.52	20.30	0.31	11.63	nr	**31.93**
MuPVC to metal adpator: including heat shrunk joint to metal						
50 mm	7.11	8.74	0.38	14.26	nr	**23.00**
Caulking bush: including joint to metal						
32 mm	3.10	3.81	0.31	11.63	nr	**15.44**
40 mm	3.10	3.81	0.31	11.63	nr	**15.44**
50 mm	3.10	3.81	0.32	12.01	nr	**15.82**

33 DRAINAGE ABOVE GROUND

Item	Net Price £	Material £	Labour hours	Labour £	Unit	**Total rate £**
Weathering apron						
50 mm	2.73	3.35	0.65	24.40	nr	**27.75**
Vent cowl						
50 mm	3.11	3.82	0.19	7.13	nr	**10.95**
ABS waste pipe; solvent welded joints; fixed with clips to backgrounds; BS 5255						
Pipe, with brackets measured separately						
32 mm	1.89	2.32	0.23	8.64	m	**10.96**
40 mm	2.36	2.90	0.23	8.64	m	**11.54**
50 mm	2.98	3.66	0.26	9.75	m	**13.41**
Fixings						
Pipe clip: including fixing to backgrounds. For minimum fixing distances, refer to the Tables and Memoranda at the rear of the book						
32 mm	0.32	0.39	0.17	6.37	nr	**6.76**
40 mm	0.41	0.50	0.17	6.37	nr	**6.87**
50 mm	1.20	1.47	0.17	6.37	nr	**7.84**
Pipe clip: expansion: including fixing to backgrounds. For minimum fixing distances, refer to the Tables and Memoranda at the rear of the book						
32 mm	0.32	0.39	0.17	6.37	nr	**6.76**
40 mm	0.41	0.50	0.17	6.37	nr	**6.87**
50 mm	1.20	1.47	0.17	6.37	nr	**7.84**
Pipe clip: metal; including fixing to backgrounds. For minimum fixing distances, refer to the Tables and Memoranda at the rear of the book						
32 mm	1.82	2.24	0.17	6.37	nr	**8.61**
40 mm	2.17	2.67	0.17	6.37	nr	**9.04**
50 mm	2.77	3.40	0.17	6.37	nr	**9.77**
Extra over fittings waste pipework ABS						
Screwed access plug						
32 mm	1.25	1.53	0.18	6.76	nr	**8.29**
40 mm	1.25	1.53	0.18	6.76	nr	**8.29**
50 mm	2.58	3.17	0.25	9.39	nr	**12.56**
Straight coupling						
32 mm	1.25	1.53	0.27	10.13	nr	**11.66**
40 mm	1.25	1.53	0.27	10.13	nr	**11.66**
50 mm	2.58	3.17	0.27	10.13	nr	**13.30**
Expansion coupling						
32 mm	2.58	3.17	0.27	10.13	nr	**13.30**
40 mm	2.58	3.17	0.27	10.13	nr	**13.30**
50 mm	3.95	4.85	0.27	10.13	nr	**14.98**
ABS to Copper coupling						
32 mm	2.44	2.99	0.27	10.13	nr	**13.12**
40 mm	2.44	2.99	0.27	10.13	nr	**13.12**
50 mm	3.74	4.60	0.27	10.13	nr	**14.73**

33 DRAINAGE ABOVE GROUND

Item	Net Price £	Material £	Labour hours	Labour £	Unit	Total rate £
DISPOSAL SYSTEMS – cont						
Extra over fittings waste pipework ABS – cont						
Reducer: socket						
40 × 32 mm	1.25	1.53	0.27	10.13	nr	**11.66**
50 × 32 mm	2.84	3.49	0.27	10.13	nr	**13.62**
50 × 40 mm	2.84	3.49	0.27	10.13	nr	**13.62**
Swept bend						
32 mm	1.25	1.53	0.27	10.13	nr	**11.66**
40 mm	1.25	1.53	0.27	10.13	nr	**11.66**
50 mm	2.58	3.17	0.30	11.26	nr	**14.43**
Knuckle bend						
32 mm	1.25	1.53	0.27	10.13	nr	**11.66**
40 mm	1.25	1.53	0.27	10.13	nr	**11.66**
Swept tee						
32 mm	1.80	2.22	0.31	11.63	nr	**13.85**
40 mm	1.80	2.22	0.31	11.63	nr	**13.85**
50 mm	4.72	5.80	0.31	11.63	nr	**17.43**
Swept cross						
40 mm	6.02	7.39	0.23	8.64	nr	**16.03**
50 mm	6.87	8.44	0.43	16.14	nr	**24.58**
Male iron adaptor						
32 mm	2.58	3.17	0.28	10.50	nr	**13.67**
40 mm	2.58	3.17	0.28	10.50	nr	**13.67**
Female iron adapator						
32 mm	2.58	3.17	0.28	10.50	nr	**13.67**
40 mm	2.58	3.17	0.28	10.50	nr	**13.67**
50 mm	3.87	4.76	0.31	11.63	nr	**16.39**
Tank connectors						
32 mm	1.79	2.19	0.29	10.89	nr	**13.08**
40 mm	1.96	2.40	0.29	10.89	nr	**13.29**
Caulking bush: including joint to pipework						
50 mm	3.05	3.75	0.50	18.77	nr	**22.52**
Polypropylene waste pipe; push fit joints; fixed with clips to backgrounds; BS 5254. Pipe, with brackets measured separately						
32 mm	1.10	1.35	0.21	7.89	m	**9.24**
40 mm	1.36	1.67	0.21	7.89	m	**9.56**
50 mm	2.26	2.77	0.38	14.26	m	**17.03**
Fixings						
Pipe clip: saddle; including fixing to backgrounds. For minimum fixing distances, refer to the Tables and Memoranda at the rear of the book						
32 mm	0.37	0.45	0.17	6.37	nr	**6.82**
40 mm	0.37	0.45	0.17	6.37	nr	**6.82**

33 DRAINAGE ABOVE GROUND

Item	Net Price £	Material £	Labour hours	Labour £	Unit	**Total rate £**
Pipe clip: including fixing to backgrounds. For minimum fixing distances, refer to the Tables and Memoranda at the rear of the book						
50 mm	0.89	1.09	0.17	6.37	nr	**7.46**
Extra over fittings waste pipework polypropylene						
Screwed access plug						
32 mm	1.05	1.29	0.16	6.01	nr	**7.30**
40 mm	1.05	1.29	0.16	6.01	nr	**7.30**
50 mm	1.80	2.22	0.20	7.51	nr	**9.73**
Straight coupling						
32 mm	1.05	1.29	0.19	7.13	nr	**8.42**
40 mm	1.05	1.29	0.19	7.13	nr	**8.42**
50 mm	1.80	2.22	0.20	7.51	nr	**9.73**
Universal waste pipe coupler						
32 mm dia.	1.78	2.18	0.20	7.51	nr	**9.69**
40 mm dia.	2.01	2.47	0.20	7.51	nr	**9.98**
Reducer						
40 × 32 mm	1.80	2.22	0.19	7.13	nr	**9.35**
50 × 32 mm	1.88	2.31	0.19	7.13	nr	**9.44**
50 × 40 mm	1.94	2.38	0.20	7.51	nr	**9.89**
Swept bend						
32 mm	1.05	1.29	0.19	7.13	nr	**8.42**
40 mm	1.05	1.29	0.19	7.13	nr	**8.42**
50 mm	1.80	2.22	0.20	7.51	nr	**9.73**
Knuckle bend						
32 mm	1.05	1.29	0.19	7.13	nr	**8.42**
40 mm	1.05	1.29	0.19	7.13	nr	**8.42**
50 mm	1.80	2.22	0.20	7.51	nr	**9.73**
Spigot and socket bend						
32 mm	1.05	1.29	0.19	7.13	nr	**8.42**
40 mm	1.05	1.29	0.19	7.13	nr	**8.42**
Swept tee						
32 mm	1.15	1.42	0.22	8.25	nr	**9.67**
40 mm	1.15	1.42	0.22	8.25	nr	**9.67**
50 mm	1.94	2.38	0.23	8.64	nr	**11.02**
Male iron adaptor						
32 mm	1.05	1.29	0.13	4.87	nr	**6.16**
40 mm	1.05	1.29	0.19	7.13	nr	**8.42**
50 mm	1.80	2.22	0.15	5.63	nr	**7.85**
Tank connector						
32 mm	1.05	1.29	0.24	9.01	nr	**10.30**
40 mm	1.05	1.29	0.24	9.01	nr	**10.30**
50 mm	1.80	2.22	0.35	13.14	nr	**15.36**

33 DRAINAGE ABOVE GROUND

Item	Net Price £	Material £	Labour hours	Labour £	Unit	Total rate £
DISPOSAL SYSTEMS – cont						
Polypropylene traps; including fixing to appliance and connection to pipework; BS 3943						
Tubular P trap; 75 mm seal						
32 mm dia.	4.25	5.22	0.20	7.51	nr	**12.73**
40 mm dia.	4.91	6.03	0.20	7.51	nr	**13.54**
Tubular S trap; 75 mm seal						
32 mm dia.	5.36	6.58	0.20	7.51	nr	**14.09**
40 mm dia.	6.32	7.76	0.20	7.51	nr	**15.27**
Running tubular P trap; 75 mm seal						
32 mm dia.	6.53	8.02	0.20	7.51	nr	**15.53**
40 mm dia.	7.13	8.76	0.20	7.51	nr	**16.27**
Running tubular S trap; 75 mm seal						
32 mm dia.	7.85	9.64	0.20	7.51	nr	**17.15**
40 mm dia.	8.46	10.40	0.20	7.51	nr	**17.91**
Spigot and socket bend; converter from P to S Trap						
32 mm	1.67	2.05	0.20	7.51	nr	**9.56**
40 mm	1.79	2.19	0.21	7.89	nr	**10.08**
Bottle P trap; 75 mm seal						
32 mm dia.	4.74	5.83	0.20	7.51	nr	**13.34**
40 mm dia.	5.67	6.96	0.20	7.51	nr	**14.47**
Bottle S trap; 75 mm seal						
32 mm dia.	5.71	7.01	0.20	7.51	nr	**14.52**
40 mm dia.	6.95	8.54	0.25	9.39	nr	**17.93**
Bottle P trap; resealing; 75 mm seal						
32 mm dia.	5.90	7.25	0.20	7.51	nr	**14.76**
40 mm dia.	6.89	8.46	0.25	9.39	nr	**17.85**
Bottle S trap; resealing; 75 mm seal						
32 mm dia.	6.75	8.30	0.20	7.51	nr	**15.81**
40 mm dia.	7.83	9.62	0.25	9.39	nr	**19.01**
Bath trap, low level; 38 mm seal						
40 mm dia.	5.91	7.26	0.25	9.39	nr	**16.65**
Bath trap, low level; 38 mm seal complete with overflow hose						
40 mm dia.	9.14	11.22	0.25	9.39	nr	**20.61**
Bath trap; 75 mm seal complete with overlow hose						
40 mm dia.	9.10	11.18	0.25	9.39	nr	**20.57**
Bath trap; 75 mm seal complete with overflow hose and overflow outlet						
40 mm dia.	15.54	19.09	0.20	7.51	nr	**26.60**
Bath trap; 75 mm seal complete with overflow hose, overflow outlet and ABS chrome waste						
40 mm dia.	21.48	26.39	0.20	7.51	nr	**33.90**
Washing machine trap; 75 mm seal including stand pipe						
40 mm dia.	12.36	15.18	0.25	9.39	nr	**24.57**

33 DRAINAGE ABOVE GROUND

Item	Net Price £	Material £	Labour hours	Labour £	Unit	Total rate £
Washing machine standpipe						
40 mm dia.	6.45	7.93	0.25	9.39	nr	**17.32**
Plastic unslotted chrome plated basin/sink waste including plug						
32 mm	7.77	9.54	0.34	12.77	nr	**22.31**
40 mm	10.39	12.77	0.34	12.77	nr	**25.54**
Plastic slotted chrome plated basin/sink waste including plug						
32 mm	6.16	7.57	0.34	12.77	nr	**20.34**
40 mm	10.32	12.67	0.34	12.77	nr	**25.44**
Bath overflow outlet; plastic; white						
42 mm	5.41	6.65	0.37	13.88	nr	**20.53**
Bath overlow outlet; plastic; chrome plated						
42 mm	7.15	8.79	0.37	13.88	nr	**22.67**
Combined cistern and bath overflow outlet; plastic; white						
42 mm	11.11	13.65	0.39	14.63	nr	**28.28**
Combined cistern and bath overlow outlet; plastic; chrome plated						
42 mm	11.10	13.64	0.39	14.63	nr	**28.27**
Cistern overflow outlet; plastic; white						
42 mm	8.95	10.99	0.15	5.63	nr	**16.62**
Cistern overlow outlet; plastic; chrome plated						
42 mm	8.07	9.91	0.15	5.63	nr	**15.54**
PVC-u soil and waste pipe; solvent welded joints; fixed with clips to backgrounds; BS 4514/BS EN 607						
Pipe, with brackets measured separately						
82 mm	9.40	11.55	0.35	13.14	m	**24.69**
110 mm	9.58	11.77	0.41	15.39	m	**27.16**
160 mm	24.84	30.51	0.51	19.14	m	**49.65**
Fixings						
Galvanized steel pipe clip: including fixing to backgrounds. For minimum fixing distances, refer to the Tables and Memoranda at the rear of the book						
82 mm	3.90	4.79	0.18	6.76	nr	**11.55**
110 mm	4.05	4.98	0.18	6.76	nr	**11.74**
160 mm	9.75	11.98	0.18	6.76	nr	**18.74**
Plastic coated steel pipe clip: including fixing to backgrounds. For minimum fixing distances, refer to the Tables and Memoranda at the rear of the book						
82 mm	5.46	6.71	0.18	6.76	nr	**13.47**
110 mm	7.43	9.12	0.18	6.76	nr	**15.88**
160 mm	9.39	11.54	0.18	6.76	nr	**18.30**

33 DRAINAGE ABOVE GROUND

Item	Net Price £	Material £	Labour hours	Labour £	Unit	**Total rate £**
DISPOSAL SYSTEMS – cont						
Fixings – cont						
Plastic pipe clip: including fixing to backgrounds. For minimum fixing distances, refer to the Tables and Memoranda at the rear of the book						
82 mm	2.89	3.55	0.18	6.76	nr	**10.31**
110 mm	5.56	6.84	0.18	6.76	nr	**13.60**
Plastic coated steel pipe clip: adjustable; including fixing to backgrounds. For minimum fixing distances, refer to the Tables and Memoranda at the rear of the book						
82 mm	4.42	5.43	0.20	7.51	nr	**12.94**
110 mm	5.51	6.77	0.20	7.51	nr	**14.28**
Galvanized steel pipe clip: drive in; including fixing to backgrounds. For minimum fixing distances, refer to the Tables and Memoranda at the rear of the book						
110 mm	8.42	10.34	0.22	8.25	nr	**18.59**
Extra over fittings solvent welded pipework PVC-u						
Straight coupling						
82 mm	4.99	6.13	0.21	7.89	nr	**14.02**
110 mm	6.23	7.65	0.22	8.25	nr	**15.90**
160 mm	17.98	22.09	0.24	9.01	nr	**31.10**
Expansion coupling						
82 mm	7.46	9.17	0.21	7.89	nr	**17.06**
110 mm	7.64	9.39	0.22	8.25	nr	**17.64**
160 mm	22.97	28.21	0.24	9.01	nr	**37.22**
Slip coupling; double ring socket						
82 mm	15.50	19.04	0.21	7.89	nr	**26.93**
110 mm	19.42	23.86	0.22	8.25	nr	**32.11**
160 mm	30.83	37.87	0.24	9.01	nr	**46.88**
Puddle flanges						
110 mm	151.48	186.09	0.45	16.89	nr	**202.98**
160 mm	244.87	300.83	0.55	20.64	nr	**321.47**
Socket reducer						
82 to 50 mm	7.59	9.32	0.18	6.76	nr	**16.08**
110 to 50 mm	9.45	11.61	0.18	6.76	nr	**18.37**
110 to 82 mm	9.75	11.98	0.22	8.25	nr	**20.23**
160 to 110 mm	19.75	24.27	0.26	9.75	nr	**34.02**
Socket plugs						
82 mm	5.88	7.22	0.15	5.63	nr	**12.85**
110 mm	8.60	10.56	0.20	7.51	nr	**18.07**
160 mm	15.82	19.44	0.27	10.13	nr	**29.57**
Access door; including cutting into pipe						
82 mm	16.66	20.46	0.28	10.50	nr	**30.96**
110 mm	16.66	20.46	0.34	12.77	nr	**33.23**
160 mm	29.78	36.58	0.46	17.26	nr	**53.84**

33 DRAINAGE ABOVE GROUND

Item	Net Price £	Material £	Labour hours	Labour £	Unit	**Total rate £**
Screwed access cap						
82 mm	11.80	14.50	0.15	5.63	nr	**20.13**
110 mm	13.91	17.08	0.20	7.51	nr	**24.59**
160 mm	26.18	32.16	0.27	10.13	nr	**42.29**
Access pipe: spigot and socket						
110 mm	20.57	25.27	0.22	8.25	nr	**33.52**
Access pipe: double socket						
110 mm	20.57	25.27	0.22	8.25	nr	**33.52**
Swept bend						
82 mm	12.52	15.38	0.29	10.89	nr	**26.27**
110 mm	14.65	18.00	0.32	12.01	nr	**30.01**
160 mm	36.50	44.85	0.49	18.39	nr	**63.24**
Bend; special angle						
82 mm	24.12	29.63	0.29	10.89	nr	**40.52**
110 mm	28.75	35.32	0.32	12.01	nr	**47.33**
160 mm	48.55	59.64	0.49	18.39	nr	**78.03**
Spigot and socket bend						
82 mm	12.11	14.88	0.26	9.75	nr	**24.63**
110 mm	14.20	17.44	0.32	12.01	nr	**29.45**
110 mm: 135°	15.94	19.58	0.32	12.01	nr	**31.59**
160 mm: 135°	35.30	43.36	0.44	16.51	nr	**59.87**
Variable bend: single socket						
110 mm	27.48	33.76	0.33	12.38	nr	**46.14**
Variable bend: double socket						
110 mm	27.41	33.67	0.33	12.38	nr	**46.05**
Access bend						
110 mm	40.66	49.95	0.33	12.38	nr	**62.33**
Single branch: two bosses						
82 mm	17.49	21.48	0.35	13.14	nr	**34.62**
82 mm: 104°	18.66	22.92	0.35	13.14	nr	**36.06**
110 mm	19.39	23.82	0.42	15.76	nr	**39.58**
110 mm: 135°	20.21	24.83	0.42	15.76	nr	**40.59**
160 mm	41.16	50.57	0.50	18.77	nr	**69.34**
160 mm: 135°	42.10	51.72	0.50	18.77	nr	**70.49**
Single branch; four bosses						
110 mm	23.60	28.99	0.42	15.76	nr	**44.75**
Single access branch						
82 mm	56.72	69.68	0.35	13.14	nr	**82.82**
110 mm	33.19	40.77	0.42	15.76	nr	**56.53**
Unequal single branch						
160 × 160 × 110 mm	46.48	57.10	0.50	18.77	nr	**75.87**
160 × 160 × 110 mm: 135°	49.93	61.34	0.50	18.77	nr	**80.11**
Double branch						
110 mm	50.09	61.54	0.42	15.76	nr	**77.30**
110 mm: 135°	47.90	58.84	0.42	15.76	nr	**74.60**
Corner branch						
110 mm	84.62	103.96	0.42	15.76	nr	**119.72**
Unequal double branch						
160 × 160 × 110 mm	86.56	106.34	0.50	18.77	nr	**125.11**

33 DRAINAGE ABOVE GROUND

Item	Net Price £	Material £	Labour hours	Labour £	Unit	Total rate £
DISPOSAL SYSTEMS – cont						
Extra over fittings solvent welded pipework PVC-u – cont						
Single boss pipe; single socket						
110 × 110 × 32 mm	5.24	6.44	0.24	9.01	nr	**15.45**
110 × 110 × 40 mm	5.24	6.44	0.24	9.01	nr	**15.45**
110 × 110 × 50 mm	5.52	6.78	0.24	9.01	nr	**15.79**
Single boss pipe; triple socket						
110 × 110 × 40 mm	8.50	10.45	0.24	9.01	nr	**19.46**
Waste boss; including cutting into pipe						
82 to 32 mm	6.85	8.41	0.29	10.89	nr	**19.30**
82 to 40 mm	6.85	8.41	0.29	10.89	nr	**19.30**
110 to 32 mm	6.85	8.41	0.29	10.89	nr	**19.30**
110 to 40 mm	6.85	8.41	0.29	10.89	nr	**19.30**
110 to 50 mm	7.10	8.73	0.29	10.89	nr	**19.62**
160 to 32 mm	9.69	11.91	0.30	11.26	nr	**23.17**
160 to 40 mm	9.69	11.91	0.35	13.14	nr	**25.05**
160 to 50 mm	9.69	11.91	0.40	15.02	nr	**26.93**
Self-locking waste boss; including cutting into pipe						
110 to 32 mm	9.16	11.26	0.30	11.26	nr	**22.52**
110 to 40 mm	9.56	11.75	0.30	11.26	nr	**23.01**
110 to 50 mm	12.35	15.17	0.30	11.26	nr	**26.43**
Adaptor saddle; including cutting to pipe						
82 to 32 mm	4.56	5.61	0.29	10.89	nr	**16.50**
110 to 40 mm	5.63	6.92	0.29	10.89	nr	**17.81**
160 to 50 mm	10.20	12.53	0.29	10.89	nr	**23.42**
Branch boss adaptor						
32 mm	2.72	3.34	0.26	9.75	nr	**13.09**
40 mm	2.72	3.34	0.26	9.75	nr	**13.09**
50 mm	3.88	4.77	0.26	9.75	nr	**14.52**
Branch boss adaptor bend						
32 mm	3.73	4.58	0.26	9.75	nr	**14.33**
40 mm	4.06	4.99	0.26	9.75	nr	**14.74**
50 mm	4.87	5.99	0.26	9.75	nr	**15.74**
Automatic air admittance valve						
82 to 110 mm	42.73	52.49	0.19	7.13	nr	**59.62**
PVC-u to metal adpator: including heat shrunk joint to metal						
110 mm	11.89	14.61	0.57	21.39	nr	**36.00**
Caulking bush: including joint to pipework						
82 mm	11.71	14.39	0.46	17.26	nr	**31.65**
110 mm	11.71	14.39	0.46	17.26	nr	**31.65**
Vent cowl						
82 mm	3.53	4.34	0.13	4.87	nr	**9.21**
110 mm	3.56	4.38	0.13	4.87	nr	**9.25**
160 mm	9.33	11.47	0.13	4.87	nr	**16.34**

33 DRAINAGE ABOVE GROUND

Item	Net Price £	Material £	Labour hours	Labour £	Unit	**Total rate £**
Weathering apron; to lead slates						
82 mm	3.53	4.34	1.15	43.16	nr	**47.50**
110 mm	4.05	4.98	1.15	43.16	nr	**48.14**
160 mm	12.17	14.95	1.15	43.16	nr	**58.11**
Weathering apron; to asphalt						
82 mm	14.88	18.28	1.10	41.28	nr	**59.56**
110 mm	14.88	18.28	1.10	41.28	nr	**59.56**
Weathering slate; flat; 406 × 406 mm						
82 mm	41.94	51.52	1.04	39.03	nr	**90.55**
110 mm	41.94	51.52	1.04	39.03	nr	**90.55**
Weathering slate; flat; 457 × 457 mm						
82 mm	43.02	52.85	1.04	39.03	nr	**91.88**
110 mm	43.02	52.85	1.04	39.03	nr	**91.88**
Weathering slate; angled; 610 × 610 mm						
82 mm	58.15	71.44	1.04	39.03	nr	**110.47**
110 mm	58.15	71.44	1.04	39.03	nr	**110.47**
PVC-u soil and waste pipe; ring seal joints; fixed with clips to backgrounds; BS 4514/ BS EN 607						
Pipe, with brackets measured separately						
82 mm dia.	8.29	10.18	0.35	13.14	m	**23.32**
110 mm dia.	8.35	10.26	0.41	15.39	m	**25.65**
160 mm dia.	29.95	36.79	0.51	19.14	m	**55.93**
Fixings						
Galvanized steel pipe clip: including fixing to backgrounds. For minimum fixing distances, refer to the Tables and Memoranda at the rear of the book						
82 mm	3.90	4.79	0.18	6.76	nr	**11.55**
110 mm	4.05	4.98	0.18	6.76	nr	**11.74**
160 mm	9.75	11.98	0.18	6.76	nr	**18.74**
Plastic coated steel pipe clip: including fixing to backgrounds. For minimum fixing distances, refer to the Tables and Memoranda at the rear of the book						
82 mm	5.46	6.71	0.18	6.76	nr	**13.47**
110 mm	7.43	9.12	0.18	6.76	nr	**15.88**
160 mm	9.39	11.54	0.18	6.76	nr	**18.30**
Plastic pipe clip: including fixing to backgrounds. For minimum fixing distances, refer to the Tables and Memoranda at the rear of the book						
82 mm	2.89	3.55	0.18	6.76	nr	**10.31**
110 mm	5.56	6.84	0.18	6.76	nr	**13.60**
Plastic coated steel pipe clip: adjustable; including fixing to backgrounds. For minimum fixing distances, refer to the Tables and Memoranda at the rear of the book						
82 mm	4.42	5.43	0.20	7.51	nr	**12.94**
110 mm	5.51	6.77	0.20	7.51	nr	**14.28**

33 DRAINAGE ABOVE GROUND

Item	Net Price £	Material £	Labour hours	Labour £	Unit	Total rate £
DISPOSAL SYSTEMS – cont						
Fixings – cont						
Galvanized steel pipe clip: drive in; including fixing to backgrounds. For minimum fixing distances, refer to the Tables and Memoranda at the rear of the book						
110 mm	8.42	10.34	0.22	8.25	nr	**18.59**
Extra over fittings ring seal pipework PVC-u						
Straight coupling						
82 mm	5.43	6.67	0.21	7.89	nr	**14.56**
110 mm	6.09	7.49	0.22	8.25	nr	**15.74**
160 mm	20.57	25.27	0.24	9.01	nr	**34.28**
Straight coupling; double socket						
82 mm	9.58	11.77	0.21	7.89	nr	**19.66**
110 mm	8.70	10.69	0.22	8.25	nr	**18.94**
160 mm	20.57	25.27	0.24	9.01	nr	**34.28**
Reducer; socket						
82 to 50 mm	8.94	10.98	0.15	5.63	nr	**16.61**
110 to 50 mm	12.47	15.32	0.15	5.63	nr	**20.95**
110 to 82 mm	12.47	15.32	0.19	7.13	nr	**22.45**
160 to 110 mm	17.95	22.05	0.31	11.63	nr	**33.68**
Access cap						
82 mm	9.73	11.95	0.15	5.63	nr	**17.58**
110 mm	11.46	14.08	0.17	6.37	nr	**20.45**
Access cap; pressure plug						
160 mm	26.94	33.10	0.33	12.38	nr	**45.48**
Access pipe						
82 mm	23.88	29.34	0.22	8.25	nr	**37.59**
110 mm	23.82	29.26	0.22	8.25	nr	**37.51**
160 mm	50.04	61.48	0.24	9.01	nr	**70.49**
Bend						
82 mm	12.14	14.91	0.29	10.89	nr	**25.80**
82 mm; adjustable radius	27.38	33.63	0.29	10.89	nr	**44.52**
110 mm	13.06	16.04	0.32	12.01	nr	**28.05**
110 mm; adjustable radius	24.38	29.95	0.32	12.01	nr	**41.96**
160 mm	39.22	48.18	0.49	18.39	nr	**66.57**
160 mm; adjustable radius	51.45	63.21	0.49	18.39	nr	**81.60**
Bend; spigot and socket						
110 mm	13.19	16.20	0.32	12.01	nr	**28.21**
Bend; offset						
82 mm	10.69	13.14	0.21	7.89	nr	**21.03**
110 mm	17.15	21.07	0.32	12.01	nr	**33.08**
160 mm	37.83	46.47	0.32	12.01	nr	**58.48**
Bend; access						
110 mm	35.98	44.20	0.33	12.38	nr	**56.58**

33 DRAINAGE ABOVE GROUND

Item	Net Price £	Material £	Labour hours	Labour £	Unit	**Total rate £**
Single branch						
82 mm	19.06	23.41	0.35	13.14	nr	**36.55**
110 mm	17.62	21.65	0.42	15.76	nr	**37.41**
110 mm; 45°	18.13	22.27	0.31	11.63	nr	**33.90**
160 mm	45.45	55.84	0.50	18.77	nr	**74.61**
Single branch; access						
82 mm	28.57	35.10	0.35	13.14	nr	**48.24**
110 mm	41.36	50.81	0.42	15.76	nr	**66.57**
Unequal single branch						
160 × 160 × 110 mm	46.45	57.07	0.50	18.77	nr	**75.84**
160 × 160 × 110 mm; 45°	49.93	61.34	0.50	18.77	nr	**80.11**
Double branch; 4 bosses						
110 mm	46.15	56.70	0.49	18.39	nr	**75.09**
Corner branch; 2 bosses						
110 mm	84.61	103.94	0.49	18.39	nr	**122.33**
Multi-branch; 4 bosses						
110 mm	25.26	31.03	0.52	19.52	nr	**50.55**
Boss Branch						
110 × 32 mm	5.22	6.42	0.34	12.77	nr	**19.19**
110 × 40 mm	5.22	6.42	0.34	12.77	nr	**19.19**
Strap on boss						
110 × 32 mm	5.82	7.15	0.30	11.26	nr	**18.41**
110 × 40 mm	5.82	7.15	0.30	11.26	nr	**18.41**
110 × 50 mm	5.87	7.21	0.30	11.26	nr	**18.47**
Patch boss						
82 × 32 mm	6.85	8.41	0.31	11.63	nr	**20.04**
82 × 40 mm	6.85	8.41	0.31	11.63	nr	**20.04**
82 × 50 mm	6.85	8.41	0.31	11.63	nr	**20.04**
Boss pipe; collar 4 boss						
110 mm	8.50	10.45	0.35	13.14	nr	**23.59**
Boss adaptor; rubber; push fit						
32 mm	3.18	3.91	0.26	9.75	nr	**13.66**
40 mm	3.18	3.91	0.26	9.75	nr	**13.66**
50 mm	3.60	4.42	0.26	9.75	nr	**14.17**
WC connector; cap and seal; solvent socket						
110 mm	8.86	10.89	0.23	8.64	nr	**19.53**
110 mm; 90°	14.24	17.49	0.27	10.13	nr	**27.62**
Vent terminal						
82 mm	3.53	4.34	0.13	4.87	nr	**9.21**
110 mm	3.56	4.38	0.13	4.87	nr	**9.25**
160 mm	9.35	11.49	0.13	4.87	nr	**16.36**
Weathering slate; inclined; 610 × 610 mm						
82 mm	58.15	71.44	1.04	39.03	nr	**110.47**
110 mm	58.15	71.44	1.04	39.03	nr	**110.47**
Weathering slate; inclined; 450 × 450 mm						
82 mm	43.02	52.85	1.04	39.03	nr	**91.88**
110 mm	43.02	52.85	1.04	39.03	nr	**91.88**
Weathering slate; flat; 400 × 400 mm						
82 mm	41.97	51.56	1.04	39.03	nr	**90.59**
110 mm	41.97	51.56	1.04	39.03	nr	**90.59**

33 DRAINAGE ABOVE GROUND

Item	Net Price £	Material £	Labour hours	Labour £	Unit	Total rate £
DISPOSAL SYSTEMS – cont						
Extra over fittings ring seal pipework PVC-u – cont						
Air admittance valve						
82 mm	42.73	52.49	0.19	7.13	nr	**59.62**
110 mm	42.73	52.49	0.19	7.13	nr	**59.62**
Cast iron pipe; nitrile rubber gasket joint with continuity clip BSEN 877; fixed vertically to backgrounds						
Pipe, with brackets and jointing couplings measured separately						
50 mm	21.07	25.88	0.31	11.63	m	**37.51**
75 mm	24.56	30.18	0.34	12.77	m	**42.95**
100 mm	27.29	33.53	0.37	13.88	m	**47.41**
150 mm	56.98	70.00	0.60	22.52	m	**92.52**
Fixings						
Brackets; fixed to backgrounds. For minimum fixing distances, refer to the Tables and Memoranda at the rear of the book						
50 mm	7.85	9.64	0.15	5.63	nr	**15.27**
75 mm	8.34	10.25	0.18	6.76	nr	**17.01**
100 mm	8.55	10.50	0.18	6.76	nr	**17.26**
150 mm	15.83	19.45	0.20	7.51	nr	**26.96**
Extra over fittings nitrile gasket cast iron pipework BS 416/6087, with jointing couplings measured separately						
Standard coupling						
50 mm	10.19	12.52	0.17	6.37	nr	**18.89**
75 mm	11.17	13.72	0.17	6.37	nr	**20.09**
100 mm	14.60	17.93	0.17	6.37	nr	**24.30**
150 mm	29.15	35.82	0.17	6.37	nr	**42.19**
Conversion coupling						
65 × 75 mm	11.85	14.55	0.60	22.53	nr	**37.08**
70 × 75 mm	11.85	14.55	0.60	22.53	nr	**37.08**
90 × 100 mm	15.26	18.74	0.67	25.15	nr	**43.89**
Access pipe; round door						
50 mm	29.86	36.69	0.41	15.39	nr	**52.08**
75 mm	42.93	52.74	0.46	17.26	nr	**70.00**
100 mm	45.13	55.44	0.67	25.18	nr	**80.62**
150 mm	75.07	92.22	0.83	31.16	nr	**123.38**
Access pipe; square door						
100 mm	88.92	109.24	0.67	25.18	nr	**134.42**
150 mm	136.12	167.22	0.83	31.16	nr	**198.38**
Taper reducer						
75 mm	24.54	30.15	0.60	22.53	nr	**52.68**
100 mm	31.15	38.27	0.67	25.15	nr	**63.42**
150 mm	60.65	74.51	0.83	31.16	nr	**105.67**

33 DRAINAGE ABOVE GROUND

Item	Net Price £	Material £	Labour hours	Labour £	Unit	**Total rate £**
Blank cap						
50 mm	6.19	7.60	0.24	9.01	nr	**16.61**
75 mm	7.74	9.51	0.26	9.75	nr	**19.26**
100 mm	7.50	9.21	0.32	12.01	nr	**21.22**
150 mm	10.85	13.32	0.40	15.02	nr	**28.34**
Blank cap; 50 mm screwed tapping						
75 mm	15.25	18.73	0.26	9.75	nr	**28.48**
100 mm	17.70	21.75	0.32	12.01	nr	**33.76**
150 mm	19.66	24.15	0.40	15.02	nr	**39.17**
Universal connector						
50 × 56/48/40 mm	9.21	11.32	0.33	12.38	nr	**23.70**
Change piece; BS416						
100 mm	19.66	24.15	0.47	17.64	nr	**41.79**
WC connector						
100 mm	31.62	38.85	0.49	18.39	nr	**57.24**
Boss pipe; 2" BSPT socket						
50 mm	25.25	31.02	0.58	21.77	nr	**52.79**
75 mm	36.95	45.39	0.65	24.40	nr	**69.79**
100 mm	44.14	54.22	0.79	29.64	nr	**83.86**
150 mm	68.76	84.47	0.86	32.28	nr	**116.75**
Boss pipe; 2" BSPT socket; 135°						
100 mm	53.17	65.32	0.79	29.64	nr	**94.96**
Boss pipe; 2 × 2" BSPT socket; opposed						
75 mm	48.94	60.12	0.65	24.40	nr	**84.52**
100 mm	54.20	66.58	0.79	29.64	nr	**96.22**
Boss pipe; 2 × 2" BSPT socket; in line						
100 mm	57.63	70.80	0.79	29.64	nr	**100.44**
Boss pipe; 2 × 2" BSPT socket; 90°						
100 mm	57.07	70.11	0.79	29.64	nr	**99.75**
Bend; short radius						
50 mm	17.85	21.92	0.50	18.77	nr	**40.69**
75 mm	20.20	24.81	0.60	22.53	nr	**47.34**
100 mm	24.70	30.35	0.67	25.15	nr	**55.50**
100 mm; 11°	21.29	26.16	0.67	25.15	nr	**51.31**
100 mm; 67°	24.70	30.35	0.67	25.15	nr	**55.50**
150 mm	44.14	54.22	0.83	31.16	nr	**85.38**
Access bend; short radius						
50 mm	43.98	54.03	0.50	18.77	nr	**72.80**
75 mm	47.72	58.62	0.60	22.53	nr	**81.15**
100 mm	52.27	64.22	0.67	25.15	nr	**89.37**
100 mm; 45°	52.27	64.22	0.67	25.15	nr	**89.37**
150 mm	74.20	91.15	0.83	31.16	nr	**122.31**
150 mm; 45°	74.20	91.15	0.83	31.16	nr	**122.31**
Long radius bend						
75 mm	33.70	41.40	0.60	22.53	nr	**63.93**
100 mm	40.01	49.15	0.67	25.15	nr	**74.30**
100 mm; 5°	24.70	30.35	0.67	25.15	nr	**55.50**
150 mm	87.22	107.15	0.83	31.16	nr	**138.31**
150 mm; 22.5°	91.43	112.32	0.83	31.16	nr	**143.48**

33 DRAINAGE ABOVE GROUND

Item	Net Price £	Material £	Labour hours	Labour £	Unit	Total rate £
DISPOSAL SYSTEMS – cont						
Fixings – cont						
Access bend; long radius						
75 mm	59.81	73.48	0.60	22.53	nr	**96.01**
100 mm	67.59	83.03	0.67	25.15	nr	**108.18**
150 mm	119.04	146.24	0.83	31.16	nr	**177.40**
Long tail bend						
100 × 250 mm long	31.91	39.20	0.70	26.28	nr	**65.48**
100 × 815 mm long	101.45	124.64	0.70	26.28	nr	**150.92**
Offset						
75 mm projection						
75 mm	26.35	32.37	0.53	19.89	nr	**52.26**
100 mm	26.00	31.94	0.66	24.77	nr	**56.71**
115 mm projection						
75 mm	28.95	35.56	0.53	19.89	nr	**55.45**
100 mm	32.46	39.88	0.66	24.77	nr	**64.65**
150 mm projection						
75 mm	34.02	41.79	0.53	19.89	nr	**61.68**
100 mm	34.02	41.79	0.66	24.77	nr	**66.56**
225 mm projection						
100 mm	37.19	45.69	0.66	24.77	nr	**70.46**
300 mm projection						
100 mm	40.01	49.15	0.66	24.77	nr	**73.92**
Branch; equal and unequal						
50 mm	26.84	32.97	0.78	29.27	nr	**62.24**
75 mm	30.39	37.34	0.85	31.92	nr	**69.26**
100 mm	38.17	46.89	1.00	37.53	nr	**84.42**
150 mm	94.66	116.29	1.20	45.07	nr	**161.36**
150 × 100 mm; 87.5°	72.43	88.98	1.21	45.22	nr	**134.20**
150 × 100 mm; 45°	106.24	130.51	1.21	45.22	nr	**175.73**
Branch; 2” BSPT screwed socket						
100 mm	51.15	62.84	1.00	37.53	nr	**100.37**
Branch; long tail						
100 × 915 mm long	105.77	129.94	1.00	37.53	nr	**167.47**
Access branch; equal and unequal						
50 mm	58.48	71.84	0.78	29.27	nr	**101.11**
75 mm	58.48	71.84	0.85	31.92	nr	**103.76**
100 mm	65.75	80.78	1.02	38.29	nr	**119.07**
150 mm	139.97	171.95	1.20	45.07	nr	**217.02**
150 × 100 mm; 87.5°	117.64	144.52	1.20	45.07	nr	**189.59**
150 × 100 mm; 45°	127.51	156.65	1.20	45.07	nr	**201.72**
Parallel branch						
100 mm	40.01	49.15	1.00	37.53	nr	**86.68**
Double branch						
75 mm	45.13	55.44	0.95	35.65	nr	**91.09**
100 mm	47.23	58.02	1.30	48.80	nr	**106.82**
150 × 100 mm	133.00	163.39	1.56	58.55	nr	**221.94**

33 DRAINAGE ABOVE GROUND

Item	Net Price £	Material £	Labour hours	Labour £	Unit	**Total rate £**
Double access branch						
100 mm	74.79	91.88	1.43	53.69	nr	**145.57**
Corner branch						
100 mm	67.16	82.51	1.30	48.80	nr	**131.31**
Puddle flange; grey epoxy coated						
100 mm	44.90	55.16	1.00	37.53	nr	**92.69**
Roof vent connector; asphalt						
75 mm	54.65	67.14	0.90	33.78	nr	**100.92**
100 mm	42.67	52.42	0.97	36.40	nr	**88.82**
P trap						
100 mm	39.59	48.64	1.00	37.53	nr	**86.17**
P trap with access						
50 mm	60.60	74.44	0.77	28.90	nr	**103.34**
75 mm	60.60	74.44	0.90	33.78	nr	**108.22**
100 mm	67.17	82.52	1.16	43.53	nr	**126.05**
150 mm	117.27	144.07	1.77	66.42	nr	**210.49**
Bellmouth gully inlet						
100 mm	57.31	70.40	1.08	40.53	nr	**110.93**
Balcony gully inlet						
100 mm	173.04	212.58	1.08	40.53	nr	**253.11**
Roof outlet						
Flat grate						
75 mm	106.82	131.23	0.83	31.15	nr	**162.38**
100 mm	150.27	184.61	1.08	40.53	nr	**225.14**
Dome grate						
75 mm	106.82	131.23	0.83	31.15	nr	**162.38**
100 mm	168.55	207.06	1.08	40.53	nr	**247.59**
Top hat						
100 mm	228.20	280.34	1.08	40.53	nr	**320.87**
Cast iron pipe; EPDM rubber gasket joint with continuity clip; BS EN877; fixed to backgrounds						
Pipe, with brackets and jointing couplings measured separately						
50 mm	21.68	26.64	0.31	11.63	m	**38.27**
70 mm	25.35	31.14	0.34	12.77	m	**43.91**
100 mm	30.18	37.08	0.37	13.88	m	**50.96**
125 mm	48.41	59.47	0.65	24.40	m	**83.87**
150 mm	59.79	73.45	0.70	26.28	m	**99.73**
200 mm	99.90	122.72	1.14	42.79	m	**165.51**
250 mm	139.59	171.49	1.25	46.91	m	**218.40**
300 mm	173.82	213.54	1.53	57.42	m	**270.96**

33 DRAINAGE ABOVE GROUND

Item	Net Price £	Material £	Labour hours	Labour £	Unit	Total rate £
DISPOSAL SYSTEMS – cont						
Fixings						
Brackets; fixed to backgrounds. For minimum fixing distances, refer to the Tables and Memoranda at the rear of the book						
Ductile iron						
50 mm	9.21	11.32	0.10	3.76	nr	**15.08**
70 mm	9.21	11.32	0.10	3.76	nr	**15.08**
100 mm	10.64	13.07	0.15	5.63	nr	**18.70**
150 mm	19.72	24.22	0.20	7.51	nr	**31.73**
200 mm	73.13	89.84	0.25	9.39	nr	**99.23**
Mild steel; vertical						
125 mm	17.94	22.04	0.15	5.63	nr	**27.67**
Mild steel; stand off						
250 mm	39.08	48.01	0.25	9.39	nr	**57.40**
300 mm	43.05	52.89	0.25	9.39	nr	**62.28**
Stack support; rubber seal						
70 mm	36.55	44.90	0.55	20.64	nr	**65.54**
100 mm	40.66	49.95	0.65	24.40	nr	**74.35**
125 mm	45.10	55.41	0.74	27.77	nr	**83.18**
150 mm	64.36	79.07	0.86	32.28	nr	**111.35**
Wall spacer plate; cast iron (eared sockets)						
50 mm	7.84	9.63	0.10	3.76	nr	**13.39**
70 mm	7.84	9.63	0.10	3.76	nr	**13.39**
100 mm	7.84	9.63	0.10	3.76	nr	**13.39**
Extra over fittings EPDM rubber jointed cast iron pipework BS EN 877, with jointing couplings measured separately						
Coupling						
50 mm	9.85	12.10	0.10	3.76	nr	**15.86**
70 mm	10.84	13.31	0.10	3.76	nr	**17.07**
100 mm	14.12	17.35	0.10	3.76	nr	**21.11**
125 mm	17.53	21.54	0.15	5.63	nr	**27.17**
150 mm	28.29	34.76	0.30	11.26	nr	**46.02**
200 mm	63.27	77.73	0.35	13.14	nr	**90.87**
250 mm	90.63	111.34	0.40	15.02	nr	**126.36**
300 mm	104.87	128.83	0.50	18.77	nr	**147.60**
Plain socket						
50 mm	24.74	30.40	0.10	3.76	nr	**34.16**
70 mm	24.74	30.40	0.10	3.76	nr	**34.16**
100 mm	28.41	34.90	0.10	3.76	nr	**38.66**
150 mm	47.93	58.88	0.10	3.76	nr	**62.64**
Eared socket						
50 mm	25.48	31.30	0.25	9.39	nr	**40.69**
70 mm	25.48	31.30	0.25	9.39	nr	**40.69**
100 mm	30.88	37.94	0.25	9.39	nr	**47.33**
150 mm	50.64	62.21	0.25	9.39	nr	**71.60**

33 DRAINAGE ABOVE GROUND

Item	Net Price £	Material £	Labour hours	Labour £	Unit	**Total rate £**
Slip socket						
50 mm	32.05	39.37	0.25	9.39	nr	**48.76**
70 mm	32.05	39.37	0.25	9.39	nr	**48.76**
100 mm	37.44	45.99	0.25	9.39	nr	**55.38**
150 mm	56.35	69.23	0.25	9.39	nr	**78.62**
Stack support pipe						
70 mm	23.66	29.06	0.74	27.77	nr	**56.83**
100 mm	32.49	39.91	0.88	33.02	nr	**72.93**
125 mm	34.75	42.69	1.00	37.53	nr	**80.22**
150 mm	47.09	57.86	1.19	44.66	nr	**102.52**
Access pipe; round door						
50 mm	45.42	55.80	0.27	10.13	nr	**65.93**
70 mm	48.06	59.04	0.30	11.26	nr	**70.30**
100 mm	52.83	64.90	0.32	12.01	nr	**76.91**
150 mm	95.61	117.45	0.71	26.65	nr	**144.10**
Access pipe; square door						
100 mm	102.15	125.50	0.32	12.01	nr	**137.51**
125 mm	106.30	130.59	0.67	25.15	nr	**155.74**
150 mm	159.93	196.48	0.71	26.65	nr	**223.13**
200 mm	317.70	390.30	1.21	45.41	nr	**435.71**
250 mm	499.86	614.08	1.31	49.16	nr	**663.24**
300 mm	623.55	766.03	1.41	52.92	nr	**818.95**
Taper reducer						
70 mm	26.22	32.21	0.30	11.26	nr	**43.47**
100 mm	30.83	37.87	0.32	12.01	nr	**49.88**
125 mm	31.00	38.08	0.64	24.01	nr	**62.09**
150 mm	59.19	72.71	0.67	25.15	nr	**97.86**
200 mm	96.13	118.09	1.15	43.16	nr	**161.25**
250 mm	198.56	243.94	1.25	46.91	nr	**290.85**
300 mm	272.96	335.33	1.37	51.41	nr	**386.74**
Blank cap						
50 mm	6.87	8.44	0.24	9.01	nr	**17.45**
70 mm	7.25	8.90	0.26	9.75	nr	**18.65**
100 mm	8.42	10.34	0.32	12.01	nr	**22.35**
125 mm	11.92	14.65	0.35	13.14	nr	**27.79**
150 mm	12.18	14.96	0.40	15.02	nr	**29.98**
200 mm	53.78	66.07	0.60	22.52	nr	**88.59**
250 mm	109.93	135.05	0.65	24.40	nr	**159.45**
300 mm	156.67	192.46	0.72	27.03	nr	**219.49**
Blank cap; 50 mm screwed tapping						
70 mm	16.15	19.84	0.26	9.75	nr	**29.59**
100 mm	17.45	21.44	0.32	12.01	nr	**33.45**
150 mm	20.97	25.76	0.40	15.02	nr	**40.78**
Universal connector; EPDM rubber						
50 × 56/48/40 mm	14.01	17.21	0.30	11.26	nr	**28.47**
Blank end; push fit						
100 × 38/32 mm	15.65	19.23	0.32	12.01	nr	**31.24**
Boss pipe; 2" BSPT socket						
50 mm	38.41	47.19	0.27	10.13	nr	**57.32**
75 mm	38.41	47.19	0.30	11.26	nr	**58.45**
100 mm	46.94	57.67	0.32	12.01	nr	**69.68**
150 mm	76.53	94.02	0.71	26.65	nr	**120.67**

33 DRAINAGE ABOVE GROUND

Item	Net Price £	Material £	Labour hours	Labour £	Unit	Total rate £
DISPOSAL SYSTEMS – cont						
Extra over fittings EPDM rubber jointed cast iron pipework BS EN 877, with jointing couplings measured separately – cont						
Boss pipe; 2 × 2" BSPT socket; opposed						
100 mm	60.63	74.49	0.32	12.01	nr	**86.50**
Boss pipe; 2 × 2" BSPT socket; 90°						
100 mm	60.63	74.49	0.32	12.01	nr	**86.50**
Manifold connector						
100 mm	93.63	115.03	0.64	24.01	nr	**139.04**
150 mm	130.45	160.26	1.00	37.53	nr	**197.79**
Bend; short radius						
50 mm	17.05	20.95	0.27	10.13	nr	**31.08**
70 mm	19.19	23.57	0.30	11.26	nr	**34.83**
100 mm	22.71	27.90	0.32	12.01	nr	**39.91**
125 mm	40.29	49.50	0.62	23.27	nr	**72.77**
150 mm	40.80	50.13	0.67	25.15	nr	**75.28**
200 mm; 45°	121.46	149.22	1.21	45.41	nr	**194.63**
250 mm; 45°	237.00	291.15	1.31	49.16	nr	**340.31**
300 mm; 45°	333.39	409.57	1.43	53.68	nr	**463.25**
Access bend; short radius						
70 mm	37.20	45.70	0.30	11.26	nr	**56.96**
100 mm	54.33	66.75	0.32	12.01	nr	**78.76**
150 mm	95.60	117.44	0.67	25.15	nr	**142.59**
Bend; long radius bend						
100 mm; 88°	57.74	70.94	0.32	12.01	nr	**82.95**
100 mm; 22°	42.58	52.31	0.32	12.01	nr	**64.32**
150 mm; 88°	165.50	203.32	0.67	25.15	nr	**228.47**
Access bend; long radius						
100 mm	70.28	86.34	0.32	12.01	nr	**98.35**
150 mm	170.73	209.74	0.32	12.01	nr	**221.75**
Bend; long tail						
100 mm	39.81	48.91	0.32	12.01	nr	**60.92**
Bend; long tail double						
70 mm	61.70	75.80	0.30	11.26	nr	**87.06**
100 mm	68.05	83.60	0.32	12.01	nr	**95.61**
Offset; 75 mm projection						
100 mm	34.92	42.90	0.32	12.01	nr	**54.91**
Offset; 130 mm projection						
50 mm	28.88	35.48	0.27	10.13	nr	**45.61**
70 mm	43.83	53.84	0.30	11.26	nr	**65.10**
100 mm	57.56	70.72	0.32	12.01	nr	**82.73**
125 mm	72.95	89.62	0.67	25.15	nr	**114.77**
Branch; equal and unequal						
50 mm	27.35	33.60	0.37	13.88	nr	**47.48**
70 mm	28.87	35.47	0.40	15.02	nr	**50.49**
100 mm	39.62	48.68	0.42	15.76	nr	**64.44**
125 mm	79.39	97.53	0.76	28.53	nr	**126.06**
150 mm	86.38	106.11	0.97	36.40	nr	**142.51**
200 mm	221.59	272.22	1.51	56.67	nr	**328.89**
250 mm	283.19	347.90	1.63	61.18	nr	**409.08**

33 DRAINAGE ABOVE GROUND

Item	Net Price £	Material £	Labour hours	Labour £	Unit	**Total rate £**
300 mm	467.28	574.06	1.77	66.42	nr	**640.48**
Branch; radius; equal and unequal						
70 mm	35.17	43.21	0.40	15.02	nr	**58.23**
100 mm	46.31	56.89	0.42	15.76	nr	**72.65**
150 mm	100.15	123.04	1.37	51.41	nr	**174.45**
200 mm	281.82	346.22	1.51	56.67	nr	**402.89**
Branch; long tail						
100 mm	129.18	158.70	0.52	19.52	nr	**178.22**
Access branch; radius; equal and unequal						
70 mm	51.58	63.37	0.40	15.02	nr	**78.39**
100 mm	69.93	85.91	0.42	15.76	nr	**101.67**
150 mm	166.94	205.09	0.97	36.40	nr	**241.49**
Double branch; equal and unequal						
100 mm	39.00	47.91	0.52	19.52	nr	**67.43**
100 mm; 70°	58.18	71.47	0.52	19.52	nr	**90.99**
150 mm	166.94	205.09	1.37	51.41	nr	**256.50**
200 mm	285.51	350.75	1.51	56.67	nr	**407.42**
Double branch; radius; equal and unequal						
100 mm	50.30	61.79	0.52	19.52	nr	**81.31**
150 mm	205.93	252.99	1.37	51.41	nr	**304.40**
Corner branch						
100 mm	101.75	125.00	0.52	19.52	nr	**144.52**
150 mm	112.49	138.19	0.52	19.52	nr	**157.71**
Corner branch; long tail						
100 mm	151.79	186.47	0.52	19.52	nr	**205.99**
Roof vent connector; asphalt						
100 mm	64.01	78.63	0.32	12.01	nr	**90.64**
Movement connector						
100 mm	81.31	99.89	0.32	12.01	nr	**111.90**
150 mm	150.51	184.90	0.67	25.15	nr	**210.05**
Expansion plugs						
70 mm	18.09	22.23	0.32	12.01	nr	**34.24**
100 mm	22.54	27.69	0.39	14.63	nr	**42.32**
150 mm	40.46	49.71	0.55	20.64	nr	**70.35**
P trap						
100 mm dia.	42.17	51.81	0.32	12.01	nr	**63.82**
P trap with access						
50 mm	64.40	79.12	0.27	10.13	nr	**89.25**
70 mm	64.40	79.12	0.30	11.26	nr	**90.38**
100 mm	69.74	85.68	0.32	12.01	nr	**97.69**
150 mm	124.70	153.19	0.67	25.15	nr	**178.34**
Branch trap						
100 mm	152.84	187.76	0.42	15.76	nr	**203.52**
Stench trap						
100 mm	297.21	365.13	0.42	15.76	nr	**380.89**
Balcony gully inlet						
100 mm	168.69	207.24	1.00	37.53	nr	**244.77**

33 DRAINAGE ABOVE GROUND

Item	Net Price £	Material £	Labour hours	Labour £	Unit	Total rate £
DISPOSAL SYSTEMS – cont						
Roof outlet						
Flat grate						
70 mm	104.13	127.92	1.00	37.53	nr	**165.45**
100 mm	146.49	179.96	1.00	37.53	nr	**217.49**
Dome grate						
70 mm	104.13	127.92	1.00	37.53	nr	**165.45**
100 mm	164.31	201.85	1.00	37.53	nr	**239.38**
Top hat						
100 mm	222.46	273.29	1.00	37.53	nr	**310.82**
Floor drains; for cast iron pipework BS 416 and BS EN877						
Adjustable clamp plate body						
100 mm; 165 mm nickel bronze grate and frame	99.17	121.83	0.50	18.77	nr	**140.60**
100 mm; 165 mm nickel bronze rodding eye	115.40	141.77	0.50	18.77	nr	**160.54**
100 mm; 150 × 150 mm nickel bronze grate and frame	111.27	136.70	0.50	18.77	nr	**155.47**
100 mm; 150 × 150 mm nickel bronze rodding eye	115.40	141.77	0.50	18.77	nr	**160.54**
Deck plate body						
100 mm; 165 mm nickel bronze grate and frame	99.17	121.83	0.50	18.77	nr	**140.60**
100 mm; 165 mm nickel bronze rodding eye	115.40	141.77	0.50	18.77	nr	**160.54**
100 mm; 150 × 150 mm nickel bronze grate and frame	111.27	136.70	0.50	18.77	nr	**155.47**
100 mm; 150 × 150 mm nickel bronze rodding eye	115.40	141.77	0.50	18.77	nr	**160.54**
Extra for						
100 mm; Srewed extension piece	38.67	47.50	0.30	11.26	nr	**58.76**
100 mm; Grating extension piece; screwed or spigot	29.34	36.05	0.30	11.26	nr	**47.31**
100 mm; Brewary trap	1067.25	1311.11	2.00	75.06	nr	**1386.17**

38 PIPED SUPPLY SYSTEMS

Item	Net Price £	Material £	Labour hours	Labour £	Unit	**Total rate £**
COLD WATER PIPELINES: COPPER PIPEWORK						
Copper pipe; capillary or compression joints in the running length; EN1057 R250 (TX) formerly BS 2871 Table X						
Fixed vertically or at low level, with brackets measured separately						
12 mm dia.	1.59	1.95	0.39	14.63	m	**16.58**
15 mm dia.	1.78	2.18	0.40	15.02	m	**17.20**
22 mm dia.	3.58	4.40	0.47	17.64	m	**22.04**
28 mm dia.	4.54	5.58	0.51	19.14	m	**24.72**
35 mm dia.	10.76	13.22	0.58	21.77	m	**34.99**
42 mm dia.	13.11	16.11	0.66	24.77	m	**40.88**
54 mm dia.	16.85	20.70	0.72	27.03	m	**47.73**
67 mm dia.	22.04	27.08	0.75	28.14	m	**55.22**
76 mm dia.	31.17	38.29	0.76	28.53	m	**66.82**
108 mm dia.	44.84	55.08	0.78	29.27	m	**84.35**
133 mm dia.	58.18	71.47	1.05	39.41	m	**110.88**
159 mm dia.	92.21	113.28	1.15	43.16	m	**156.44**
Fixed horizontally at high level or suspended, with brackets measured separately						
12 mm dia.	1.59	1.95	0.45	16.89	m	**18.84**
15 mm dia.	1.78	2.18	0.46	17.26	m	**19.44**
22 mm dia.	3.58	4.40	0.54	20.28	m	**24.68**
28 mm dia.	4.54	11.17	0.59	22.13	m	**33.30**
35 mm dia.	10.76	13.22	0.67	25.15	m	**38.37**
42 mm dia.	13.11	16.11	0.76	28.53	m	**44.64**
54 mm dia.	16.85	20.70	0.83	31.15	m	**51.85**
67 mm dia.	22.04	27.08	0.86	32.28	m	**59.36**
76 mm dia.	31.17	38.29	0.87	32.65	m	**70.94**
108 mm dia.	44.84	55.08	0.90	33.78	m	**88.86**
133 mm dia.	58.18	71.47	1.21	45.41	m	**116.88**
159 mm dia.	92.21	113.28	1.32	49.55	m	**162.83**
Copper pipe; capillary or compression joints in the running length; EN1057 R250 (TY) formerly BS 2871 Table Y						
Fixed vertically or at low level with brackets measured separately (Refer to Copper Pipe Table X Section)						
12 mm dia.	2.05	2.52	0.41	15.39	m	**17.91**
15 mm dia.	2.93	3.60	0.43	16.14	m	**19.74**
22 mm dia.	5.16	6.34	0.50	18.77	m	**25.11**
28 mm dia.	6.66	8.18	0.54	20.28	m	**28.46**
35 mm dia.	9.72	11.94	0.62	23.27	m	**35.21**
42 mm dia.	11.79	14.48	0.71	26.65	m	**41.13**
54 mm dia.	20.10	24.70	0.78	29.27	m	**53.97**
67 mm dia.	26.69	32.79	0.82	30.78	m	**63.57**
76 mm dia.	39.00	47.91	0.60	22.52	m	**70.43**
108 mm dia.	54.16	66.54	0.88	33.02	m	**99.56**

38 PIPED SUPPLY SYSTEMS

Item	Net Price £	Material £	Labour hours	Labour £	Unit	Total rate £
COLD WATER PIPELINES: COPPER PIPEWORK – cont						
Copper pipe; capillary or compression joints in the running length; EN1057 R250 (TX) formerly BS 2871 Table X						
Plastic coated gas and cold water service pipe for corrosive environments, fixed vertically or at low level with brackets measured separtely						
15 mm dia. (white)	4.10	5.04	0.59	22.13	m	**27.17**
22 mm dia. (white)	7.26	8.91	0.68	25.53	m	**34.44**
28 mm dia. (white)	7.49	9.20	0.74	27.77	m	**36.97**
FIXINGS						
Saddle band						
6 mm dia.	0.06	0.08	0.05	1.92	nr	**2.00**
8 mm dia.	0.06	0.08	0.07	2.57	nr	**2.65**
10 mm dia.	0.08	0.10	0.09	3.20	nr	**3.30**
12 mm dia.	0.08	0.10	0.11	4.02	nr	**4.12**
15 mm dia.	0.08	0.10	0.13	4.87	nr	**4.97**
22 mm dia.	0.16	0.20	0.13	4.87	nr	**5.07**
28 mm dia.	0.09	0.12	0.16	6.01	nr	**6.13**
35 mm dia.	0.29	0.36	0.18	6.76	nr	**7.12**
42 mm dia.	0.61	0.75	0.21	7.89	nr	**8.64**
54 mm dia.	0.82	1.01	0.21	7.89	nr	**8.90**
Single spacing clip						
15 mm dia.	0.17	0.21	0.14	5.26	nr	**5.47**
22 mm dia.	0.17	0.21	0.15	5.63	nr	**5.84**
28 mm dia.	0.42	0.51	0.17	6.37	nr	**6.88**
Two piece spacing clip						
8 mm dia. bottom	0.15	0.19	0.11	4.13	nr	**4.32**
8 mm dia. top	0.15	0.19	0.11	4.13	nr	**4.32**
12 mm dia. bottom	0.15	0.19	0.13	4.87	nr	**5.06**
12 mm dia. top	0.15	0.19	0.13	4.87	nr	**5.06**
15 mm dia. bottom	0.16	0.20	0.13	4.87	nr	**5.07**
15 mm dia. top	0.16	0.20	0.13	4.87	nr	**5.07**
22 mm dia. bottom	0.16	0.20	0.14	5.26	nr	**5.46**
22 mm dia. top	0.17	0.21	0.14	5.26	nr	**5.47**
28 mm dia. bottom	0.16	0.20	0.16	6.01	nr	**6.21**
28 mm dia. top	0.28	0.35	0.16	6.01	nr	**6.36**
35 mm dia. bottom	0.17	0.21	0.21	7.89	nr	**8.10**
35 mm dia. top	0.41	0.50	0.21	7.89	nr	**8.39**
Single pipe bracket						
15 mm dia.	1.43	1.75	0.14	5.26	nr	**7.01**
22 mm dia.	1.45	1.78	0.14	5.26	nr	**7.04**
28 mm dia.	1.99	2.45	0.17	6.37	nr	**8.82**
Single pipe ring						
15 mm dia.	2.50	3.07	0.26	9.75	nr	**12.82**
22 mm dia.	2.71	3.33	0.26	9.75	nr	**13.08**
28 mm dia.	3.41	4.19	0.31	11.63	nr	**15.82**

38 PIPED SUPPLY SYSTEMS

Item	Net Price £	Material £	Labour hours	Labour £	Unit	**Total rate £**
35 mm dia.	4.80	5.90	0.32	12.01	nr	**17.91**
42 mm dia.	5.56	6.84	0.32	12.01	nr	**18.85**
54 mm dia.	7.13	8.76	0.34	12.77	nr	**21.53**
67 mm dia.	11.96	14.69	0.35	13.14	nr	**27.83**
76 mm dia.	24.56	30.18	0.42	15.76	nr	**45.94**
108 mm dia.	45.20	55.52	0.42	15.76	nr	**71.28**
Double pipe ring						
15 mm dia.	3.43	4.21	0.26	9.75	nr	**13.96**
22 mm dia.	4.29	5.27	0.26	9.75	nr	**15.02**
28 mm dia.	6.77	8.32	0.31	11.63	nr	**19.95**
35 mm dia.	9.02	11.08	0.32	12.01	nr	**23.09**
42 mm dia.	10.14	12.45	0.32	12.01	nr	**24.46**
54 mm dia.	10.45	12.84	0.34	12.77	nr	**25.61**
67 mm dia.	13.18	16.19	0.35	13.14	nr	**29.33**
76 mm dia.	19.99	24.56	0.42	15.76	nr	**40.32**
108 mm dia.	27.31	33.55	0.42	15.76	nr	**49.31**
Wall bracket						
15 mm dia.	2.84	3.49	0.05	1.88	nr	**5.37**
22 mm dia.	3.72	4.57	0.05	1.88	nr	**6.45**
28 mm dia.	4.44	5.45	0.05	1.88	nr	**7.33**
35 mm dia.	6.66	8.18	0.05	1.88	nr	**10.06**
42 mm dia.	8.81	10.83	0.05	1.88	nr	**12.71**
54 mm dia.	13.73	16.86	0.05	1.88	nr	**18.74**
Hospital bracket						
15 mm dia.	4.88	6.00	0.26	9.75	nr	**15.75**
22 mm dia.	5.77	7.09	0.26	9.75	nr	**16.84**
28 mm dia.	6.77	8.32	0.31	11.63	nr	**19.95**
35 mm dia.	7.16	8.80	0.32	12.01	nr	**20.81**
42 mm dia.	10.25	12.59	0.32	12.01	nr	**24.60**
54 mm dia.	13.71	16.84	0.34	12.77	nr	**29.61**
Screw on backplate, female						
All sizes 15 mm to 54 mm × 10 mm	1.56	1.92	0.10	3.76	nr	**5.68**
Screw on backplate, male						
All sizes 15 mm to 54 mm × 10 mm	2.07	2.54	0.10	3.76	nr	**6.30**
Pipe joist clips, single						
15 mm dia.	0.99	1.22	0.08	2.99	nr	**4.21**
22 mm dia.	0.99	1.22	0.08	2.99	nr	**4.21**
Pipe joist clips, double						
15 mm dia.	1.38	1.69	0.08	2.99	nr	**4.68**
22 mm dia.	0.75	0.92	0.08	2.99	nr	**3.91**
Extra over channel sections for fabricated hangers and brackets						
Galvanized steel; including inserts, bolts, nuts, washers; fixed to backgrounds						
41 × 21 mm	6.02	7.39	0.29	10.89	m	**18.28**
41 × 41 mm	7.22	8.87	0.29	10.89	m	**19.76**

38 PIPED SUPPLY SYSTEMS

Item	Net Price £	Material £	Labour hours	Labour £	Unit	Total rate £
COLD WATER PIPELINES: COPPER PIPEWORK – cont						
Extra over channel sections for fabricated hangers and brackets – cont						
Threaded rods; metric thread; including nuts, washers etc.						
10 mm dia. × 600 mm long for ring clips up to 54 mm	1.97	2.42	0.18	6.76	nr	**9.18**
12 mm dia. × 600 mm long for ring clips from 54 mm	3.04	3.74	0.18	6.76	nr	**10.50**
Extra over copper pipes; capillary fittings; BS 864						
Stop end						
15 mm dia.	1.44	1.76	0.13	4.87	nr	**6.63**
22 mm dia.	2.68	3.30	0.14	5.26	nr	**8.56**
28 mm dia.	4.80	5.90	0.17	6.37	nr	**12.27**
35 mm dia.	10.58	13.00	0.19	7.13	nr	**20.13**
42 mm dia.	18.22	22.39	0.22	8.25	nr	**30.64**
54 mm dia.	25.43	31.24	0.23	8.64	nr	**39.88**
Straight coupling; copper to copper						
6 mm dia.	1.51	1.86	0.23	8.64	nr	**10.50**
8 mm dia.	1.54	1.89	0.23	8.64	nr	**10.53**
10 mm dia.	0.79	0.97	0.23	8.64	nr	**9.61**
15 mm dia.	0.17	0.21	0.23	8.64	nr	**8.85**
22 mm dia.	0.46	0.57	0.26	9.75	nr	**10.32**
28 mm dia.	1.33	1.64	0.30	11.26	nr	**12.90**
35 mm dia.	4.32	5.30	0.34	12.77	nr	**18.07**
42 mm dia.	7.22	8.87	0.38	14.26	nr	**23.13**
54 mm dia.	13.32	16.36	0.42	15.76	nr	**32.12**
67 mm dia.	39.56	48.60	0.53	19.89	nr	**68.49**
Adaptor coupling; imperial to metric						
½" × 15 mm dia.	3.13	3.84	0.27	10.13	nr	**13.97**
¾" × 22 mm dia.	2.75	3.38	0.31	11.63	nr	**15.01**
1" × 28 mm dia.	5.39	6.63	0.36	13.51	nr	**20.14**
1 ¼" × 35 mm dia.	9.01	11.07	0.41	15.39	nr	**26.46**
1 ½" × 42 mm dia.	11.46	14.08	0.46	17.26	nr	**31.34**
Reducing coupling						
15 × 10 mm dia.	3.29	4.04	0.23	8.64	nr	**12.68**
22 × 10 mm dia.	4.83	5.93	0.26	9.75	nr	**15.68**
22 × 15 mm dia.	5.16	6.34	0.27	10.13	nr	**16.47**
28 × 15 mm dia.	5.95	7.31	0.28	10.50	nr	**17.81**
28 × 22 mm dia.	5.99	7.36	0.30	11.26	nr	**18.62**
35 × 28 mm dia.	8.58	10.54	0.34	12.77	nr	**23.31**
42 × 35 mm dia.	12.58	15.46	0.38	14.26	nr	**29.72**
54 × 35 mm dia.	22.07	27.11	0.42	15.76	nr	**42.87**
54 × 42 mm dia.	24.08	29.58	0.42	15.76	nr	**45.34**

38 PIPED SUPPLY SYSTEMS

Item	Net Price £	Material £	Labour hours	Labour £	Unit	**Total rate £**
Straight female connector						
15 mm × ½" dia.	3.27	4.02	0.27	10.13	nr	**14.15**
22 mm × ¾" dia.	4.73	5.81	0.31	11.63	nr	**17.44**
28 mm × 1" dia.	8.95	10.99	0.36	13.51	nr	**24.50**
35 mm × 1 ¼" dia.	15.47	19.01	0.41	15.39	nr	**34.40**
42 mm × 1 ½" dia.	20.09	24.69	0.46	17.26	nr	**41.95**
54 mm × 2" dia.	31.85	39.12	0.52	19.52	nr	**58.64**
Straight male connector						
15 mm × ½" dia.	2.78	3.41	0.27	10.13	nr	**13.54**
22 mm × ¾" dia.	4.97	6.10	0.31	11.63	nr	**17.73**
28 mm × 1" dia.	8.03	9.87	0.36	13.51	nr	**23.38**
35 mm × 1 ¼" dia.	14.11	17.34	0.41	15.39	nr	**32.73**
42 mm × 1 ½" dia.	18.16	22.31	0.46	17.26	nr	**39.57**
54 mm × 2" dia.	27.57	33.87	0.52	19.52	nr	**53.39**
67 mm × 2 ½" dia.	44.04	54.11	0.63	23.65	nr	**77.76**
Female reducing connector						
15 mm × ¾" dia.	8.12	9.97	0.27	10.13	nr	**20.10**
Male reducing connector						
15 mm × ¾" dia.	7.26	8.91	0.27	10.13	nr	**19.04**
22 mm × 1" dia.	11.06	13.59	0.31	11.63	nr	**25.22**
Flanged connector						
28 mm dia.	51.92	63.79	0.36	13.51	nr	**77.30**
35 mm dia.	65.74	80.77	0.41	15.39	nr	**96.16**
42 mm dia.	78.57	96.53	0.46	17.26	nr	**113.79**
54 mm dia.	118.77	145.91	0.52	19.52	nr	**165.43**
67 mm dia.	139.67	171.58	0.61	22.90	nr	**194.48**
Tank connector						
15 mm × ½" dia.	6.95	8.54	0.25	9.39	nr	**17.93**
22 mm × ¾" dia.	10.60	13.02	0.28	10.50	nr	**23.52**
28 mm × 1" dia.	13.93	17.11	0.32	12.01	nr	**29.12**
35 mm × 1 ¼" dia.	17.87	21.96	0.37	13.88	nr	**35.84**
42 mm × 1 ½" dia.	23.42	28.77	0.43	16.14	nr	**44.91**
54 mm × 2" dia.	35.81	44.00	0.46	17.26	nr	**61.26**
Tank connector with long thread						
15 mm × ½" dia.	9.01	11.07	0.30	11.26	nr	**22.33**
22 mm × ¾" dia.	12.83	15.76	0.33	12.38	nr	**28.14**
28 mm × 1" dia.	15.87	19.50	0.39	14.63	nr	**34.13**
Reducer						
15 × 10 mm dia.	1.11	1.37	0.23	8.64	nr	**10.01**
22 × 15 mm dia.	0.78	0.96	0.26	9.75	nr	**10.71**
28 × 15 mm dia.	2.93	3.60	0.28	10.50	nr	**14.10**
28 × 22 mm dia.	2.25	2.76	0.30	11.26	nr	**14.02**
35 × 22 mm dia.	8.32	10.22	0.34	12.77	nr	**22.99**
42 × 22 mm dia.	15.03	18.47	0.36	13.51	nr	**31.98**
42 × 35 mm dia.	11.62	14.28	0.38	14.26	nr	**28.54**
54 × 35 mm dia.	24.41	29.99	0.40	15.02	nr	**45.01**
54 × 42 mm dia.	21.05	25.86	0.42	15.76	nr	**41.62**
67 × 54 mm dia.	28.60	35.13	0.53	19.89	nr	**55.02**

38 PIPED SUPPLY SYSTEMS

Item	Net Price £	Material £	Labour hours	Labour £	Unit	Total rate £
COLD WATER PIPELINES: COPPER PIPEWORK – cont						
Extra over copper pipes; capillary fittings; BS 864 – cont						
Adaptor; copper to female iron						
15 mm × ½" dia.	5.63	6.92	0.27	10.13	nr	**17.05**
22 mm × ¾" dia.	8.59	10.55	0.31	11.63	nr	**22.18**
28 mm × 1" dia.	12.10	14.87	0.36	13.51	nr	**28.38**
35 mm × 1 ¼" dia.	21.89	26.89	0.41	15.39	nr	**42.28**
42 mm × 1 ½" dia.	27.57	33.87	0.46	17.26	nr	**51.13**
54 mm × 2" dia.	33.18	40.76	0.52	19.52	nr	**60.28**
Adaptor; copper to male iron						
15 mm × ½" dia.	5.75	7.07	0.27	10.13	nr	**17.20**
22 mm × ¾" dia.	7.36	9.04	0.31	11.63	nr	**20.67**
28 mm × 1" dia.	12.29	15.10	0.36	13.51	nr	**28.61**
35 mm × 1 ¼" dia.	17.88	21.97	0.41	15.39	nr	**37.36**
42 mm × 1 ½" dia.	24.71	30.36	0.46	17.26	nr	**47.62**
54 mm × 2" dia.	33.18	40.76	0.52	19.52	nr	**60.28**
Union coupling						
15 mm dia.	7.79	9.57	0.41	15.39	nr	**24.96**
22 mm dia.	12.46	15.31	0.45	16.89	nr	**32.20**
28 mm dia.	18.16	22.31	0.51	19.14	nr	**41.45**
35 mm dia.	23.84	29.28	0.64	24.01	nr	**53.29**
42 mm dia.	34.83	42.79	0.68	25.51	nr	**68.30**
54 mm dia.	66.27	81.42	0.78	29.27	nr	**110.69**
67 mm dia.	112.23	137.88	0.96	36.04	nr	**173.92**
Elbow						
15 mm dia.	0.30	0.37	0.23	8.64	nr	**9.01**
22 mm dia.	0.80	0.99	0.26	9.75	nr	**10.74**
28 mm dia.	2.12	2.60	0.31	11.63	nr	**14.23**
35 mm dia.	9.26	11.37	0.35	13.14	nr	**24.51**
42 mm dia.	15.30	18.80	0.41	15.39	nr	**34.19**
54 mm dia.	31.61	38.83	0.44	16.51	nr	**55.34**
67 mm dia.	82.03	100.78	0.54	20.28	nr	**121.06**
Backplate elbow						
15 mm dia.	5.84	7.17	0.51	19.14	nr	**26.31**
22 mm dia.	12.55	15.41	0.54	20.28	nr	**35.69**
Overflow bend						
22 mm dia.	17.67	21.70	0.26	9.75	nr	**31.45**
Return bend						
15 mm dia.	8.75	10.75	0.23	8.64	nr	**19.39**
22 mm dia.	17.20	21.13	0.26	9.75	nr	**30.88**
28 mm dia.	21.97	26.98	0.31	11.63	nr	**38.61**
Obtuse elbow						
15 mm dia.	1.12	1.38	0.23	8.64	nr	**10.02**
22 mm dia.	2.34	2.88	0.26	9.75	nr	**12.63**
28 mm dia.	4.49	5.51	0.31	11.63	nr	**17.14**
35 mm dia.	13.98	17.18	0.36	13.51	nr	**30.69**
42 mm dia.	24.88	30.57	0.41	15.39	nr	**45.96**
54 mm dia.	44.99	55.27	0.44	16.51	nr	**71.78**
67 mm dia.	81.64	100.30	0.54	20.28	nr	**120.58**

38 PIPED SUPPLY SYSTEMS

Item	Net Price £	Material £	Labour hours	Labour £	Unit	**Total rate £**
Straight tap connector						
15 mm × ½" dia.	1.50	1.85	0.13	4.87	nr	**6.72**
22 mm × ¾" dia.	2.24	2.75	0.14	5.26	nr	**8.01**
Bent tap connector						
15 mm × ½" dia.	2.23	2.74	0.13	4.87	nr	**7.61**
22 mm × ¾" dia.	6.81	8.37	0.14	5.26	nr	**13.63**
Bent male union connector						
15 mm × ½" dia.	11.39	14.00	0.41	15.39	nr	**29.39**
22 mm × ¾" dia.	14.79	18.16	0.45	16.89	nr	**35.05**
28 mm × 1" dia.	21.17	26.01	0.51	19.14	nr	**45.15**
35 mm × 1 ¼" dia.	34.54	42.43	0.64	24.01	nr	**66.44**
42 mm × 1 ½" dia.	56.14	68.96	0.68	25.51	nr	**94.47**
54 mm × 2" dia.	88.70	108.97	0.78	29.27	nr	**138.24**
Bent female union connector						
15 mm dia.	11.39	14.00	0.41	15.39	nr	**29.39**
22 mm × ¾" dia.	14.79	18.16	0.45	16.89	nr	**35.05**
28 mm × 1" dia.	21.17	26.01	0.51	19.14	nr	**45.15**
35 mm × 1 ¼" dia.	34.54	42.43	0.64	24.01	nr	**66.44**
42 mm × 1 ½" dia.	56.14	68.96	0.68	25.51	nr	**94.47**
54 mm × 2" dia.	88.70	108.97	0.78	29.27	nr	**138.24**
Straight union adaptor						
15 mm × ¾" dia.	4.87	5.99	0.41	15.39	nr	**21.38**
22 mm × 1" dia.	6.91	8.48	0.45	16.89	nr	**25.37**
28 mm × 1 ¼" dia.	11.17	13.72	0.51	19.14	nr	**32.86**
35 mm × 1 ½" dia.	17.20	21.13	0.64	24.01	nr	**45.14**
42 mm × 2" dia.	21.72	26.68	0.68	25.51	nr	**52.19**
54 mm × 2 ½" dia.	33.53	41.19	0.78	29.27	nr	**70.46**
Straight male union connector						
15 mm × ½" dia.	9.70	11.92	0.41	15.39	nr	**27.31**
22 mm × ¾" dia.	12.58	15.46	0.45	16.89	nr	**32.35**
28 mm × 1" dia.	18.74	23.03	0.51	19.14	nr	**42.17**
35 mm × 1 ¼" dia.	27.00	33.17	0.64	24.01	nr	**57.18**
42 mm × 1 ½" dia.	42.43	52.12	0.68	25.51	nr	**77.63**
54 mm × 2" dia.	60.96	74.89	0.78	29.27	nr	**104.16**
Straight female union connector						
15 mm × ½" dia.	9.70	11.92	0.41	15.39	nr	**27.31**
22 mm × ¾" dia.	12.58	15.46	0.45	16.89	nr	**32.35**
28 mm × 1" dia.	18.74	23.03	0.51	19.14	nr	**42.17**
35 mm × 1 ¼" dia.	27.00	33.17	0.64	24.01	nr	**57.18**
42 mm × 1 ½" dia.	42.43	52.12	0.68	25.51	nr	**77.63**
54 mm × 2" dia.	60.96	74.89	0.78	29.27	nr	**104.16**
Male nipple						
¾ × ½" dia.	2.29	2.81	0.24	8.90	nr	**11.71**
1 × ¾" dia.	2.75	3.38	0.32	12.01	nr	**15.39**
1 ¼ × 1" dia.	3.01	3.70	0.37	13.88	nr	**17.58**
1 ½ × 1 ¼" dia.	11.21	13.78	0.42	15.76	nr	**29.54**
2 × 1 ½" dia.	22.94	28.18	0.46	17.26	nr	**45.44**
2 ½ × 2" dia.	29.88	36.71	0.56	21.02	nr	**57.73**

38 PIPED SUPPLY SYSTEMS

Item	Net Price £	Material £	Labour hours	Labour £	Unit	**Total rate £**
COLD WATER PIPELINES: COPPER PIPEWORK – cont						
Extra over copper pipes; capillary fittings; BS 864 – cont						
Female nipple						
¾ × ½" dia.	3.55	4.36	0.19	7.13	nr	**11.49**
1 × ¾" dia.	5.55	6.81	0.32	12.01	nr	**18.82**
1 ¼ × 1" dia.	7.57	9.30	0.37	13.89	nr	**23.19**
1 ½ × 1 ¼" dia.	11.21	13.78	0.42	15.76	nr	**29.54**
2 × 1 ½" dia.	22.94	28.18	0.46	17.26	nr	**45.44**
2 ½ × 2" dia.	29.88	36.71	0.56	21.02	nr	**57.73**
Equal tee						
10 mm dia.	3.04	3.74	0.25	9.39	nr	**13.13**
15 mm dia.	0.30	0.37	0.36	13.51	nr	**13.88**
22 mm dia.	0.80	0.99	0.39	14.63	nr	**15.62**
28 mm dia.	5.92	7.28	0.43	16.14	nr	**23.42**
35 mm dia.	15.08	18.52	0.57	21.39	nr	**39.91**
42 mm dia.	24.19	29.72	0.60	22.52	nr	**52.24**
54 mm dia.	48.78	59.92	0.65	24.40	nr	**84.32**
67 mm dia.	66.36	81.52	0.78	29.27	nr	**110.79**
Female tee, reducing branch FI						
15 × 15 mm × ¼" dia.	7.29	8.96	0.36	13.51	nr	**22.47**
22 × 22 mm × ½" dia.	8.94	10.98	0.39	14.66	nr	**25.64**
28 × 28 mm × ¾" dia.	17.50	21.50	0.43	16.14	nr	**37.64**
35 × 35 mm × ¾" dia.	25.25	31.02	0.47	17.64	nr	**48.66**
42 × 42 mm × ½" dia.	30.34	37.27	0.60	22.52	nr	**59.79**
Backplate tee						
15 × 15 mm × ½" dia.	13.82	16.98	0.62	23.27	nr	**40.25**
Heater tee						
½ × ½" × 15 mm dia.	12.42	15.26	0.36	13.51	nr	**28.77**
Union heater tee						
½ × ½" × 15 mm dia.	12.42	15.26	0.36	13.51	nr	**28.77**
Sweep tee – equal						
15 mm dia.	9.87	12.13	0.36	13.51	nr	**25.64**
22 mm dia.	12.70	15.60	0.39	14.63	nr	**30.23**
28 mm dia.	21.38	26.26	0.43	16.14	nr	**42.40**
35 mm dia.	30.32	37.24	0.57	21.39	nr	**58.63**
42 mm dia.	44.96	55.23	0.60	22.52	nr	**77.75**
54 mm dia.	49.79	61.16	0.65	24.40	nr	**85.56**
67 mm dia.	67.88	83.39	0.78	29.27	nr	**112.66**
Sweep tee – reducing						
22 × 22 × 15 mm dia.	10.65	13.08	0.39	14.63	nr	**27.71**
28 × 28 × 22 mm dia.	18.09	22.23	0.43	16.14	nr	**38.37**
35 × 35 × 22 mm dia.	30.32	37.24	0.57	21.39	nr	**58.63**
Sweep tee – double						
15 mm dia.	11.17	13.72	0.36	13.51	nr	**27.23**
22 mm dia.	15.21	18.69	0.39	14.63	nr	**33.32**
28 mm dia.	23.10	28.38	0.43	16.14	nr	**44.52**

38 PIPED SUPPLY SYSTEMS

Item	Net Price £	Material £	Labour hours	Labour £	Unit	Total rate £
Cross						
15 mm dia.	14.78	18.15	0.48	18.01	nr	**36.16**
22 mm dia.	16.50	20.28	0.53	19.89	nr	**40.17**
28 mm dia.	23.69	29.11	0.61	22.90	nr	**52.01**
Extra over copper pipes; high duty capillary fittings; BS 864						
Stop end						
15 mm dia.	6.92	8.50	0.16	6.01	nr	**14.51**
Straight coupling; copper to copper						
15 mm dia.	3.16	3.89	0.27	10.13	nr	**14.02**
22 mm dia.	5.08	6.24	0.32	12.01	nr	**18.25**
28 mm dia.	6.78	8.33	0.37	13.88	nr	**22.21**
35 mm dia.	12.67	15.56	0.43	16.14	nr	**31.70**
42 mm dia.	13.85	17.01	0.50	18.77	nr	**35.78**
54 mm dia.	20.38	25.03	0.54	20.28	nr	**45.31**
Reducing coupling						
15 × 12 mm dia.	5.97	7.33	0.27	10.13	nr	**17.46**
22 × 15 mm dia.	6.92	8.50	0.32	12.01	nr	**20.51**
28 × 22 mm dia.	9.53	11.71	0.37	13.88	nr	**25.59**
Straight female connector						
15 mm × ½" dia.	7.79	9.57	0.32	12.01	nr	**21.58**
22 mm × ¾" dia.	8.79	10.79	0.36	13.51	nr	**24.30**
28 mm × 1" dia.	12.95	15.91	0.42	15.76	nr	**31.67**
Straight male connector						
15 mm × ½" dia.	7.57	9.30	0.32	12.01	nr	**21.31**
22 mm × ¾" dia.	8.79	10.79	0.36	13.51	nr	**24.30**
28 mm × 1" dia.	12.95	15.91	0.42	15.76	nr	**31.67**
42 mm × 1 ½" dia.	25.27	31.05	0.53	19.89	nr	**50.94**
54 mm × 2" dia.	41.06	50.44	0.62	23.27	nr	**73.71**
Reducer						
15 × 12 mm dia.	3.92	4.82	0.27	10.13	nr	**14.95**
22 × 15 mm dia.	3.82	4.69	0.32	12.01	nr	**16.70**
28 × 22 mm dia.	6.92	8.50	0.37	13.88	nr	**22.38**
35 × 28 mm dia.	8.79	10.79	0.43	16.14	nr	**26.93**
42 × 35 mm dia.	11.31	13.89	0.50	18.77	nr	**32.66**
54 × 42 mm dia.	18.24	22.41	0.39	14.63	nr	**37.04**
Straight union adaptor						
15 mm × ¾" dia.	6.33	7.78	0.27	10.13	nr	**17.91**
22 mm × 1" dia.	8.56	10.52	0.32	12.01	nr	**22.53**
28 mm × 1 ¼" dia.	11.31	13.89	0.37	13.88	nr	**27.77**
35 mm × 1 ½" dia.	20.50	25.18	0.43	16.14	nr	**41.32**
42 mm × 2" dia.	25.97	31.90	0.50	18.77	nr	**50.67**
Bent union adaptor						
15 mm × ¾" dia.	16.45	20.21	0.27	10.13	nr	**30.34**
22 mm × 1" dia.	22.19	27.26	0.32	12.01	nr	**39.27**
28 mm × 1 ¼" dia.	29.88	36.71	0.37	13.88	nr	**50.59**
Adaptor; male copper to FI						
15 mm × ½" dia.	7.41	9.10	0.27	10.13	nr	**19.23**
22 mm × ¾" dia.	12.70	15.60	0.32	12.01	nr	**27.61**

38 PIPED SUPPLY SYSTEMS

Item	Net Price £	Material £	Labour hours	Labour £	Unit	Total rate £
COLD WATER PIPELINES: COPPER PIPEWORK – cont						
Extra over copper pipes – cont						
Union coupling						
15 mm dia.	14.26	17.51	0.54	20.28	nr	**37.79**
22 mm dia.	18.24	22.41	0.60	22.52	nr	**44.93**
28 mm dia.	25.32	31.10	0.68	25.51	nr	**56.61**
35 mm dia.	44.22	54.33	0.83	31.15	nr	**85.48**
42 mm dia.	52.07	63.97	0.89	33.40	nr	**97.37**
Elbow						
15 mm dia.	9.18	11.28	0.27	10.13	nr	**21.41**
22 mm dia.	9.81	12.05	0.32	12.01	nr	**24.06**
28 mm dia.	14.57	17.90	0.37	13.88	nr	**31.78**
35 mm dia.	22.77	27.97	0.43	16.14	nr	**44.11**
42 mm dia.	28.37	34.85	0.50	18.77	nr	**53.62**
54 mm dia.	49.35	60.63	0.52	19.52	nr	**80.15**
Return bend						
28 mm dia.	21.97	26.98	0.37	13.88	nr	**40.86**
35 mm dia.	27.07	33.25	0.43	16.14	nr	**49.39**
Bent male union connector						
15 mm × ½" dia.	21.28	26.15	0.54	20.28	nr	**46.43**
22 mm × ¾" dia.	28.66	35.21	0.60	22.52	nr	**57.73**
28 mm × 1" dia.	52.07	63.97	0.68	25.51	nr	**89.48**
Composite flange						
35 mm dia.	4.30	5.28	0.38	14.26	nr	**19.54**
42 mm dia.	5.57	6.85	0.41	15.39	nr	**22.24**
54 mm dia.	7.12	8.75	0.43	16.14	nr	**24.89**
Equal tee						
15 mm dia.	10.55	12.96	0.44	16.51	nr	**29.47**
22 mm dia.	13.26	16.29	0.47	17.64	nr	**33.93**
28 mm dia.	17.48	21.47	0.53	19.89	nr	**41.36**
35 mm dia.	29.88	36.71	0.70	26.28	nr	**62.99**
42 mm dia.	38.04	46.74	0.84	31.52	nr	**78.26**
54 mm dia.	59.90	73.58	0.79	29.64	nr	**103.22**
Reducing tee						
15 × 12 mm dia.	14.45	17.76	0.44	16.51	nr	**34.27**
22 × 15 mm dia.	17.07	20.97	0.47	17.64	nr	**38.61**
28 × 22 mm dia.	24.39	29.97	0.53	19.89	nr	**49.86**
35 × 28 mm dia.	38.66	47.49	0.73	27.39	nr	**74.88**
42 × 28 mm dia.	49.50	60.81	0.84	31.52	nr	**92.33**
54 × 28 mm dia.	78.16	96.02	1.01	37.92	nr	**133.94**
Extra over copper pipes; compression fittings; BS 864						
Stop end						
15 mm dia.	1.62	2.00	0.10	3.76	nr	**5.76**
22 mm dia.	1.92	2.36	0.12	4.50	nr	**6.86**
28 mm dia.	2.21	2.72	0.15	5.63	nr	**8.35**

38 PIPED SUPPLY SYSTEMS

Item	Net Price £	Material £	Labour hours	Labour £	Unit	**Total rate £**
Straight connector; copper to copper						
15 mm dia.	3.36	4.13	0.18	6.76	nr	**10.89**
22 mm dia.	4.45	5.47	0.21	7.89	nr	**13.36**
28 mm dia.	5.91	7.26	0.24	9.01	nr	**16.27**
Straight connector; copper to imperial copper						
22 mm dia.	5.54	6.80	0.21	7.89	nr	**14.69**
Male coupling; copper to MI (BSP)						
15 mm dia.	0.95	1.17	0.19	7.13	nr	**8.30**
22 mm dia.	1.50	1.85	0.23	8.64	nr	**10.49**
28 mm dia.	3.49	4.28	0.26	9.75	nr	**14.03**
Male coupling with long thread and backnut						
15 mm dia.	6.14	7.54	0.19	7.13	nr	**14.67**
22 mm dia.	7.80	9.59	0.23	8.64	nr	**18.23**
Female coupling; copper to FI (BSP)						
15 mm dia.	1.17	1.44	0.19	7.13	nr	**8.57**
22 mm dia.	1.68	2.07	0.23	8.64	nr	**10.71**
28 mm dia.	4.88	6.00	0.27	10.13	nr	**16.13**
Elbow						
15 mm dia.	1.25	1.53	0.18	6.76	nr	**8.29**
22 mm dia.	2.13	2.61	0.21	7.89	nr	**10.50**
28 mm dia.	7.21	8.86	0.24	9.01	nr	**17.87**
Male elbow; copper to FI (BSP)						
15 mm × ½" dia.	2.40	2.95	0.19	7.13	nr	**10.08**
22 mm × ¾" dia.	3.12	3.83	0.23	8.64	nr	**12.47**
28 mm × 1" dia.	7.53	9.25	0.27	10.13	nr	**19.38**
Female elbow; copper to FI (BSP)						
15 mm × ½" dia.	3.69	4.54	0.19	7.13	nr	**11.67**
22 mm × ¾" dia.	5.33	6.55	0.23	8.64	nr	**15.19**
28 mm × 1" dia.	9.40	11.55	0.27	10.13	nr	**21.68**
Backplate elbow						
15 mm × ½" dia.	5.33	6.55	0.50	18.77	nr	**25.32**
Tank coupling; long thread						
22 mm dia.	7.80	9.59	0.46	17.26	nr	**26.85**
Tee equal						
15 mm dia.	1.77	2.17	0.28	10.50	nr	**12.67**
22 mm dia.	2.98	3.66	0.30	11.26	nr	**14.92**
28 mm dia.	13.58	16.68	0.34	12.77	nr	**29.45**
Tee reducing						
22 mm dia.	7.68	9.44	0.30	11.26	nr	**20.70**
Backplate tee						
15 mm dia.	18.75	23.04	0.62	23.27	nr	**46.31**
Extra over fittings; silver brazed welded joints						
Reducer						
76 × 67 mm dia.	30.97	38.04	1.40	52.54	nr	**90.58**
108 × 76 mm dia.	64.89	79.72	1.80	67.56	nr	**147.28**
133 × 108 mm dia.	129.39	158.96	2.20	82.57	nr	**241.53**
159 × 133 mm dia.	165.49	203.30	2.60	97.58	nr	**300.88**

38 PIPED SUPPLY SYSTEMS

Item	Net Price £	Material £	Labour hours	Labour £	Unit	Total rate £
COLD WATER PIPELINES: COPPER PIPEWORK – cont						
Extra over fittings – cont						
90° elbow						
76 mm dia.	74.80	91.90	1.60	60.05	nr	**151.95**
108 mm dia.	137.70	169.17	2.00	75.06	nr	**244.23**
133 mm dia.	291.08	357.59	2.40	90.07	nr	**447.66**
159 mm dia.	361.36	443.93	2.80	105.08	nr	**549.01**
45° elbow						
76 mm dia.	70.08	86.09	1.60	60.05	nr	**146.14**
108 mm dia.	109.91	135.02	2.00	75.06	nr	**210.08**
133 mm dia.	281.06	345.28	2.40	90.07	nr	**435.35**
159 mm dia.	404.55	496.99	2.80	105.08	nr	**602.07**
Equal tee						
76 mm dia.	91.10	111.92	2.40	90.07	nr	**201.99**
108 mm dia.	141.49	173.82	3.00	112.59	nr	**286.41**
133 mm dia.	333.24	409.38	3.60	135.11	nr	**544.49**
159 mm dia.	397.43	488.24	4.20	157.63	nr	**645.87**
Extra over copper pipes; dezincification resistant compression fittings; BS 864						
Stop end						
15 mm dia.	2.32	2.85	0.10	3.76	nr	**6.61**
22 mm dia.	3.36	4.13	0.13	4.87	nr	**9.00**
28 mm dia.	7.19	8.83	0.15	5.63	nr	**14.46**
35 mm dia.	11.27	13.85	0.18	6.76	nr	**20.61**
42 mm dia.	18.75	23.04	0.20	7.51	nr	**30.55**
Straight coupling; copper to copper						
15 mm dia.	1.85	2.27	0.18	6.76	nr	**9.03**
22 mm dia.	3.04	3.74	0.21	7.89	nr	**11.63**
28 mm dia.	6.87	8.44	0.24	9.01	nr	**17.45**
35 mm dia.	14.55	17.87	0.29	10.89	nr	**28.76**
42 mm dia.	19.11	23.48	0.33	12.38	nr	**35.86**
54 mm dia.	28.58	35.11	0.38	14.26	nr	**49.37**
Straight swivel connector; copper to imperial copper						
22 mm dia.	6.84	8.40	0.20	7.51	nr	**15.91**
Male coupling; copper to MI (BSP)						
15 mm × ½" dia.	1.66	2.04	0.19	7.13	nr	**9.17**
22 mm × ¾" dia.	2.52	3.10	0.23	8.64	nr	**11.74**
28 mm × 1" dia.	4.87	5.99	0.26	9.75	nr	**15.74**
35 mm × 1 ¼" dia.	11.05	13.58	0.32	12.01	nr	**25.59**
42 mm × 1 ½" dia.	16.57	20.36	0.37	13.88	nr	**34.24**
54 mm × 2" dia.	24.48	30.07	0.57	21.39	nr	**51.46**
Male coupling with long thread and backnuts						
22 mm dia.	9.40	11.55	0.23	8.64	nr	**20.19**
28 mm dia.	10.41	12.79	0.24	9.01	nr	**21.80**

38 PIPED SUPPLY SYSTEMS

Item	Net Price £	Material £	Labour hours	Labour £	Unit	**Total rate £**
Female coupling; copper to FI (BSP)						
15 mm × ½" dia.	1.99	2.45	0.19	7.13	nr	**9.58**
22 mm × ¾" dia.	2.91	3.57	0.23	8.64	nr	**12.21**
28 mm × 1" dia.	6.30	7.74	0.27	10.13	nr	**17.87**
35 mm × 1 ¼" dia.	13.27	16.31	0.32	12.01	nr	**28.32**
42 mm × 1 ½" dia.	17.84	21.91	0.37	13.88	nr	**35.79**
54 mm × 2" dia.	26.17	32.15	0.42	15.76	nr	**47.91**
Elbow						
15 mm dia.	2.24	2.75	0.18	6.76	nr	**9.51**
22 mm dia.	3.56	4.38	0.21	7.89	nr	**12.27**
28 mm dia.	8.87	10.90	0.24	9.01	nr	**19.91**
35 mm dia.	19.62	24.11	0.29	10.89	nr	**35.00**
42 mm dia.	26.58	32.65	0.33	12.38	nr	**45.03**
54 mm dia.	45.73	56.17	0.38	14.26	nr	**70.43**
Male elbow; copper to MI (BSP)						
15 mm × ½" dia.	3.87	4.76	0.19	7.13	nr	**11.89**
22 mm × ¾" dia.	4.33	5.32	0.23	8.64	nr	**13.96**
28 mm × 1" dia.	8.12	9.97	0.27	10.13	nr	**20.10**
Female elbow; copper to FI (BSP)						
15 mm × ½" dia.	4.15	5.10	0.19	7.13	nr	**12.23**
22 mm × ¾" dia.	5.98	7.35	0.23	8.64	nr	**15.99**
28 mm × 1" dia.	9.91	12.17	0.27	10.13	nr	**22.30**
Backplate elbow						
15 mm × ½" dia.	6.04	7.42	0.50	18.77	nr	**26.19**
Straight tap connector						
15 mm dia.	3.43	4.21	0.13	4.87	nr	**9.08**
22 mm dia.	7.44	9.14	0.15	5.63	nr	**14.77**
Tank coupling						
15 mm dia.	4.72	5.80	0.19	7.13	nr	**12.93**
22 mm dia.	5.23	6.43	0.23	8.64	nr	**15.07**
28 mm dia.	11.03	13.56	0.27	10.13	nr	**23.69**
35 mm dia.	19.49	23.94	0.32	12.01	nr	**35.95**
42 mm dia.	31.69	38.93	0.37	13.88	nr	**52.81**
54 mm dia.	40.82	50.15	0.31	11.63	nr	**61.78**
Tee equal						
15 mm dia.	3.14	3.85	0.28	10.50	nr	**14.35**
22 mm dia.	5.18	6.36	0.30	11.26	nr	**17.62**
28 mm dia.	14.14	17.37	0.34	12.77	nr	**30.14**
35 mm dia.	25.52	31.35	0.43	16.14	nr	**47.49**
42 mm dia.	40.14	49.31	0.46	17.26	nr	**66.57**
54 mm dia.	64.46	79.19	0.54	20.28	nr	**99.47**
Tee reducing						
22 mm dia.	8.29	10.18	0.30	11.26	nr	**21.44**
28 mm dia.	13.66	16.78	0.34	12.77	nr	**29.55**
35 mm dia.	24.94	30.64	0.43	16.14	nr	**46.78**
42 mm dia.	38.56	47.38	0.46	17.26	nr	**64.64**
54 mm dia.	64.46	79.19	0.54	20.28	nr	**99.47**

38 PIPED SUPPLY SYSTEMS

Item	Net Price £	Material £	Labour hours	Labour £	Unit	Total rate £
COLD WATER PIPELINES: COPPER PIPEWORK – cont						
Extra over copper pipes; bronze one piece brazing flanges; metric, including jointing ring and bolts						
Bronze flange; PN6						
15 mm dia.	29.30	35.99	0.27	10.13	nr	**46.12**
22 mm dia.	34.92	42.90	0.32	12.01	nr	**54.91**
28 mm dia.	38.35	47.11	0.36	13.51	nr	**60.62**
35 mm dia.	45.83	56.30	0.47	17.64	nr	**73.94**
42 mm dia.	49.99	61.41	0.54	20.28	nr	**81.69**
54 mm dia.	58.00	71.25	0.63	23.65	nr	**94.90**
67 mm dia.	68.37	83.99	0.77	28.90	nr	**112.89**
76 mm dia.	82.39	101.22	0.93	34.90	nr	**136.12**
108 mm dia.	108.86	133.74	1.14	42.79	nr	**176.53**
133 mm dia.	149.16	183.25	1.41	52.92	nr	**236.17**
159 mm dia.	215.18	264.35	1.74	65.31	nr	**329.66**
Bronze flange; PN10						
15 mm dia.	33.55	41.21	0.27	10.13	nr	**51.34**
22 mm dia.	35.71	43.87	0.32	12.01	nr	**55.88**
28 mm dia.	39.28	48.26	0.38	14.26	nr	**62.52**
35 mm dia.	47.47	58.32	0.47	17.64	nr	**75.96**
42 mm dia.	53.22	65.38	0.54	20.28	nr	**85.66**
54 mm dia.	58.24	71.55	0.63	23.65	nr	**95.20**
67 mm dia.	68.42	84.05	0.77	28.90	nr	**112.95**
76 mm dia.	82.39	101.22	0.93	34.90	nr	**136.12**
108 mm dia.	104.14	127.93	1.14	42.79	nr	**170.72**
133 mm dia.	139.05	170.82	1.41	52.92	nr	**223.74**
159 mm dia.	176.23	216.50	1.74	65.31	nr	**281.81**
Bronze flange; PN16						
15 mm dia.	36.90	45.33	0.27	10.13	nr	**55.46**
22 mm dia.	39.29	48.27	0.32	12.01	nr	**60.28**
28 mm dia.	43.20	53.07	0.38	14.26	nr	**67.33**
35 mm dia.	52.22	64.16	0.47	17.64	nr	**81.80**
42 mm dia.	58.55	71.92	0.54	20.28	nr	**92.20**
54 mm dia.	64.06	78.70	0.63	23.65	nr	**102.35**
67 mm dia.	75.25	92.44	0.77	28.90	nr	**121.34**
76 mm dia.	90.62	111.33	0.93	34.90	nr	**146.23**
108 mm dia.	114.54	140.71	1.14	42.79	nr	**183.50**
133 mm dia.	152.94	187.89	1.41	52.92	nr	**240.81**
159 mm dia.	193.84	238.13	1.74	65.31	nr	**303.44**
Extra over copper pipes; bronze blank flanges; metric, including jointing ring and bolts						
Gunmetal blank flange; PN6						
15 mm dia.	24.59	30.21	0.27	10.13	nr	**40.34**
22 mm dia.	31.39	38.57	0.27	10.13	nr	**48.70**
28 mm dia.	32.26	39.63	0.27	10.13	nr	**49.76**

38 PIPED SUPPLY SYSTEMS

Item	Net Price £	Material £	Labour hours	Labour £	Unit	**Total rate £**
35 mm dia.	52.79	64.85	0.32	12.01	nr	**76.86**
42 mm dia.	71.21	87.49	0.32	12.01	nr	**99.50**
54 mm dia.	78.52	96.46	0.34	12.77	nr	**109.23**
67 mm dia.	96.61	118.68	0.36	13.51	nr	**132.19**
76 mm dia.	124.45	152.89	0.37	13.88	nr	**166.77**
108 mm dia.	197.74	242.93	0.41	15.39	nr	**258.32**
133 mm dia.	233.27	286.58	0.58	21.77	nr	**308.35**
159 mm dia.	291.57	358.20	0.61	22.90	nr	**381.10**
Gunmetal blank flange; PN10						
15 mm dia.	29.74	36.54	0.27	10.13	nr	**46.67**
22 mm dia.	38.49	47.28	0.27	10.13	nr	**57.41**
28 mm dia.	42.62	52.36	0.27	10.13	nr	**62.49**
35 mm dia.	52.79	64.85	0.32	12.01	nr	**76.86**
42 mm dia.	98.93	121.54	0.32	12.01	nr	**133.55**
54 mm dia.	112.79	138.56	0.34	12.77	nr	**151.33**
67 mm dia.	121.00	148.65	0.46	17.26	nr	**165.91**
76 mm dia.	160.79	197.53	0.47	17.64	nr	**215.17**
108 mm dia.	188.71	231.83	0.51	19.14	nr	**250.97**
133 mm dia.	250.68	307.97	0.58	21.77	nr	**329.74**
159 mm dia.	341.41	419.42	0.71	26.65	nr	**446.07**
Gunmetal blank flange; PN16						
15 mm dia.	29.74	36.54	0.27	10.13	nr	**46.67**
22 mm dia.	39.07	48.00	0.27	10.13	nr	**58.13**
28 mm dia.	42.62	52.36	0.27	10.13	nr	**62.49**
35 mm dia.	52.79	64.85	0.32	12.01	nr	**76.86**
42 mm dia.	98.93	121.54	0.32	12.01	nr	**133.55**
54 mm dia.	112.79	138.56	0.34	12.77	nr	**151.33**
67 mm dia.	140.54	172.65	0.46	17.26	nr	**189.91**
76 mm dia.	160.79	197.53	0.47	17.64	nr	**215.17**
108 mm dia.	188.71	231.83	0.51	19.14	nr	**250.97**
133 mm dia.	433.41	532.44	0.58	21.77	nr	**554.21**
159 mm dia.	517.69	635.99	0.71	26.65	nr	**662.64**
Extra over copper pipes; bronze screwed flanges; metric, including jointing ring and bolts						
Gunmetal screwed flange; 6 BSP						
15 mm dia.	24.59	30.21	0.35	13.14	nr	**43.35**
22 mm dia.	28.44	34.93	0.47	17.64	nr	**52.57**
28 mm dia.	29.64	36.41	0.52	19.52	nr	**55.93**
35 mm dia.	39.95	49.08	0.62	23.27	nr	**72.35**
42 mm dia.	47.96	58.92	0.70	26.28	nr	**85.20**
54 mm dia.	65.17	80.06	0.84	31.52	nr	**111.58**
67 mm dia.	81.79	100.47	1.03	38.66	nr	**139.13**
76 mm dia.	98.72	121.28	1.22	45.79	nr	**167.07**
108 mm dia.	156.35	192.08	1.41	52.92	nr	**245.00**
133 mm dia.	185.34	227.69	1.75	65.68	nr	**293.37**
159 mm dia.	235.94	289.85	2.21	82.95	nr	**372.80**

38 PIPED SUPPLY SYSTEMS

Item	Net Price £	Material £	Labour hours	Labour £	Unit	Total rate £
COLD WATER PIPELINES: COPPER PIPEWORK – cont						
Extra over copper pipes – cont						
Gunmetal screwed flange; 10 BSP						
15 mm dia.	29.79	36.59	0.35	13.14	nr	**49.73**
22 mm dia.	34.69	42.62	0.47	17.64	nr	**60.26**
28 mm dia.	38.30	47.05	0.52	19.52	nr	**66.57**
35 mm dia.	54.59	67.06	0.62	23.27	nr	**90.33**
42 mm dia.	66.65	81.88	0.70	26.28	nr	**108.16**
54 mm dia.	94.78	116.43	0.84	31.52	nr	**147.95**
67 mm dia.	110.96	136.31	1.03	38.66	nr	**174.97**
76 mm dia.	125.26	153.88	1.22	45.79	nr	**199.67**
108 mm dia.	166.28	204.28	1.41	52.92	nr	**257.20**
133 mm dia.	201.52	247.57	1.75	65.68	nr	**313.25**
159 mm dia.	357.62	439.34	2.21	82.95	nr	**522.29**
Gunmetal screwed flange; 16 BSP						
15 mm dia.	23.40	28.75	0.35	13.14	nr	**41.89**
22 mm dia.	29.59	36.35	0.47	17.64	nr	**53.99**
28 mm dia.	34.25	42.07	0.52	19.52	nr	**61.59**
35 mm dia.	52.34	64.30	0.62	23.27	nr	**87.57**
42 mm dia.	62.99	77.39	0.70	26.28	nr	**103.67**
54 mm dia.	81.03	99.55	0.84	31.52	nr	**131.07**
67 mm dia.	117.12	143.88	1.03	38.66	nr	**182.54**
76 mm dia.	134.30	164.99	1.22	45.79	nr	**210.78**
108 mm dia.	157.15	193.06	1.41	52.92	nr	**245.98**
133 mm dia.	254.63	312.82	1.75	65.68	nr	**378.50**
159 mm dia.	314.98	386.96	2.21	82.95	nr	**469.91**
Extra over copper pipes; labour						
Made bend						
15 mm dia.	–	–	0.26	9.75	nr	**9.75**
22 mm dia.	–	–	0.28	10.50	nr	**10.50**
28 mm dia.	–	–	0.31	11.63	nr	**11.63**
35 mm dia.	–	–	0.42	15.76	nr	**15.76**
42 mm dia.	–	–	0.51	19.14	nr	**19.14**
54 mm dia.	–	–	0.58	21.77	nr	**21.77**
67 mm dia.	–	–	0.69	25.89	nr	**25.89**
76 mm dia.	–	–	0.80	30.02	nr	**30.02**
Bronze butt weld						
15 mm dia.	–	–	0.25	9.39	nr	**9.39**
22 mm dia.	–	–	0.31	11.63	nr	**11.63**
28 mm dia.	–	–	0.37	13.88	nr	**13.88**
35 mm dia.	–	–	0.49	18.39	nr	**18.39**
42 mm dia.	–	–	0.58	21.77	nr	**21.77**
54 mm dia.	–	–	0.72	27.03	nr	**27.03**
67 mm dia.	–	–	0.88	33.02	nr	**33.02**
76 mm dia.	–	–	1.08	40.53	nr	**40.53**
108 mm dia.	–	–	1.37	51.41	nr	**51.41**
133 mm dia.	–	–	1.73	64.92	nr	**64.92**
159 mm dia.	–	–	2.03	76.19	nr	**76.19**

38 PIPED SUPPLY SYSTEMS

Item	Net Price £	Material £	Labour hours	Labour £	Unit	**Total rate £**
PRESS FIT (copper fittings); Mechanical press fit joints; butyl rubber O ring						
Coupler						
15 mm dia.	1.05	1.29	0.36	13.51	nr	**14.80**
22 mm dia.	1.66	2.04	0.36	13.51	nr	**15.55**
28 mm dia.	3.40	4.18	0.44	16.51	nr	**20.69**
35 mm dia.	4.28	5.26	0.44	16.51	nr	**21.77**
42 mm dia.	7.70	9.46	0.52	19.52	nr	**28.98**
54 mm dia.	9.82	12.06	0.60	22.52	nr	**34.58**
Stop end						
22 mm dia.	2.65	3.26	0.18	6.76	nr	**10.02**
28 mm dia.	4.20	5.16	0.22	8.25	nr	**13.41**
35 mm dia.	7.20	8.84	0.22	8.25	nr	**17.09**
42 mm dia.	10.83	13.30	0.26	9.75	nr	**23.05**
54 mm dia.	13.03	16.01	0.30	11.26	nr	**27.27**
Reducer						
22 × 15 mm dia.	1.22	1.50	0.36	13.51	nr	**15.01**
28 × 15 mm dia.	3.23	3.97	0.40	15.02	nr	**18.99**
28 × 22 mm dia.	3.35	4.12	0.40	15.02	nr	**19.14**
35 × 22 mm dia.	4.01	4.92	0.40	15.02	nr	**19.94**
35 × 28 mm dia.	4.44	5.45	0.44	16.51	nr	**21.96**
42 × 22 mm dia.	6.94	8.53	0.44	16.51	nr	**25.04**
42 × 28 mm dia.	6.61	8.12	0.48	18.01	nr	**26.13**
42 × 35 mm dia.	6.61	8.12	0.48	18.01	nr	**26.13**
54 × 35 mm dia.	8.95	10.99	0.52	19.52	nr	**30.51**
54 × 42 mm dia.	8.95	10.99	0.56	21.02	nr	**32.01**
90° elbow						
15 mm dia.	1.14	1.40	0.36	13.51	nr	**14.91**
22 mm dia.	1.91	2.34	0.36	13.51	nr	**15.85**
28 mm dia.	4.10	5.04	0.44	16.51	nr	**21.55**
35 mm dia.	8.46	10.40	0.44	16.51	nr	**26.91**
42 mm dia.	15.89	19.52	0.52	19.52	nr	**39.04**
54 mm dia.	22.06	27.10	0.60	22.52	nr	**49.62**
45° elbow						
15 mm dia.	1.40	1.72	0.36	13.51	nr	**15.23**
22 mm dia.	1.94	2.38	0.36	13.51	nr	**15.89**
28 mm dia.	5.81	7.14	0.44	16.51	nr	**23.65**
35 mm dia.	8.29	10.18	0.44	16.51	nr	**26.69**
42 mm dia.	13.79	16.94	0.52	19.52	nr	**36.46**
54 mm dia.	19.61	24.09	0.60	22.52	nr	**46.61**
Equal tee						
15 mm dia.	1.81	2.23	0.54	20.28	nr	**22.51**
22 mm dia.	3.29	4.04	0.54	20.28	nr	**24.32**
28 mm dia.	5.92	7.28	0.66	24.77	nr	**32.05**
35 mm dia.	10.23	12.57	0.66	24.77	nr	**37.34**
42 mm dia.	20.20	24.81	0.78	29.27	nr	**54.08**
54 mm dia.	25.22	30.99	0.90	33.78	nr	**64.77**

38 PIPED SUPPLY SYSTEMS

Item	Net Price £	Material £	Labour hours	Labour £	Unit	Total rate £
COLD WATER PIPELINES: COPPER PIPEWORK – cont						
Reducing tee						
22 × 15 mm dia.	2.68	3.30	0.54	20.28	nr	**23.58**
28 × 15 mm dia.	5.20	6.38	0.62	23.27	nr	**29.65**
28 × 22 mm dia.	6.99	8.59	0.62	23.27	nr	**31.86**
35 × 22 mm dia.	9.11	11.19	0.62	23.27	nr	**34.46**
35 × 28 mm dia.	10.14	12.45	0.62	23.27	nr	**35.72**
42 × 28 mm dia.	18.35	22.54	0.70	26.28	nr	**48.82**
42 × 35 mm dia.	18.35	22.54	0.70	26.28	nr	**48.82**
54 × 35 mm dia.	31.00	38.08	0.82	30.78	nr	**68.86**
54 × 42 mm dia.	31.00	38.08	0.82	30.78	nr	**68.86**
Male iron connector; BSP thread						
15 mm dia.	3.82	4.69	0.18	6.76	nr	**11.45**
22 mm dia.	5.68	6.98	0.18	6.76	nr	**13.74**
28 mm dia.	7.59	9.32	0.22	8.25	nr	**17.57**
35 mm dia.	13.73	16.86	0.22	8.25	nr	**25.11**
42 mm dia.	18.40	22.61	0.26	9.75	nr	**32.36**
54 mm dia.	35.52	43.64	0.30	11.26	nr	**54.90**
90° elbow; male iron BSP thread						
15 mm dia.	6.20	7.61	0.36	13.51	nr	**21.12**
22 mm dia.	9.70	11.92	0.36	13.51	nr	**25.43**
28 mm dia.	14.86	18.26	0.44	16.51	nr	**34.77**
35 mm dia.	19.32	23.73	0.44	16.51	nr	**40.24**
42 mm dia.	25.20	30.95	0.52	19.52	nr	**50.47**
54 mm dia.	36.81	45.22	0.60	22.52	nr	**67.74**
Female iron connector; BSP thread						
15 mm dia.	4.37	5.37	0.18	6.76	nr	**12.13**
22 mm dia.	5.77	7.09	0.18	6.76	nr	**13.85**
28 mm dia.	7.78	9.56	0.22	8.25	nr	**17.81**
35 mm dia.	15.21	18.69	0.22	8.25	nr	**26.94**
42 mm dia.	21.73	26.69	0.26	9.75	nr	**36.44**
54 mm dia.	37.26	45.77	0.30	11.26	nr	**57.03**
90° elbow; female iron BSP thread						
15 mm dia.	5.23	6.43	0.36	13.51	nr	**19.94**
22 mm dia.	7.70	9.46	0.36	13.51	nr	**22.97**
28 mm dia.	12.70	15.60	0.44	16.51	nr	**32.11**
35 mm dia.	16.42	20.17	0.44	16.51	nr	**36.68**
42 mm dia.	22.35	27.46	0.52	19.52	nr	**46.98**
54 mm dia.	32.87	40.38	0.60	22.52	nr	**62.90**

38 PIPED SUPPLY SYSTEMS

Item	Net Price £	Material £	Labour hours	Labour £	Unit	Total rate £
COLD WATER PIPELINES: STAINLESS STEEL PIPEWORK						
Stainless steel pipes; capillary or compression joints; BS 4127, vertical or at low level, with brackets measured separately						
Grade 304; satin finish						
15 mm dia.	3.81	4.68	0.41	15.39	m	**20.07**
22 mm dia.	5.36	6.58	0.51	19.15	m	**25.73**
28 mm dia.	7.29	8.96	0.58	21.77	m	**30.73**
35 mm dia.	11.01	13.52	0.65	24.40	m	**37.92**
42 mm dia.	13.98	17.18	0.71	26.66	m	**43.84**
54 mm dia.	19.48	23.93	0.80	30.02	m	**53.95**
Grade 316 satin finish						
15 mm dia.	4.90	6.02	0.61	22.90	m	**28.92**
22 mm dia.	9.17	11.27	0.76	28.54	m	**39.81**
28 mm dia.	10.88	13.37	0.87	32.67	m	**46.04**
35 mm dia.	19.72	24.22	0.98	36.79	m	**61.01**
42 mm dia.	25.53	31.36	1.06	39.81	m	**71.17**
54 mm dia.	29.75	36.55	1.16	43.53	m	**80.08**
FIXINGS						
Single pipe ring						
15 mm dia.	11.88	14.60	0.26	9.75	nr	**24.35**
22 mm dia.	13.84	17.00	0.26	9.75	nr	**26.75**
28 mm dia.	14.50	17.81	0.31	11.63	nr	**29.44**
35 mm dia.	16.47	20.23	0.32	12.01	nr	**32.24**
42 mm dia.	18.71	22.98	0.32	12.01	nr	**34.99**
54 mm dia.	21.36	26.24	0.34	12.77	nr	**39.01**
Screw on backplate, female						
All sizes 15 mm to 54 mm dia.	10.71	13.16	0.10	3.76	nr	**16.92**
Screw on backplate, male						
All sizes 15 mm to 54 mm dia.	12.20	14.98	0.10	3.76	nr	**18.74**
Stainless steel threaded rods; metric thread; including nuts, washers etc.						
10 mm dia. × 600 mm long	12.87	15.81	0.18	6.76	nr	**22.57**
Extra over stainless steel pipes; capillary fittings						
Straight coupling						
15 mm dia.	1.40	1.72	0.25	9.39	nr	**11.11**
22 mm dia.	2.27	2.79	0.28	10.50	nr	**13.29**
28 mm dia.	2.99	3.68	0.33	12.38	nr	**16.06**
35 mm dia.	6.90	8.47	0.37	13.89	nr	**22.36**
42 mm dia.	7.96	9.78	0.42	15.76	nr	**25.54**
54 mm dia.	11.97	14.70	0.45	16.89	nr	**31.59**
45° bend						
15 mm dia.	7.53	9.25	0.25	9.39	nr	**18.64**
22 mm dia.	9.89	12.15	0.30	11.13	nr	**23.28**
28 mm dia.	12.15	14.93	0.33	12.38	nr	**27.31**

38 PIPED SUPPLY SYSTEMS

Item	Net Price £	Material £	Labour hours	Labour £	Unit	**Total rate £**
COLD WATER PIPELINES: STAINLESS STEEL PIPEWORK – cont						
Extra over stainless steel pipes – cont						
45° bend – cont						
35 mm dia.	14.39	17.68	0.37	13.89	nr	**31.57**
42 mm dia.	18.52	22.75	0.42	15.76	nr	**38.51**
54 mm dia.	22.91	28.14	0.45	16.89	nr	**45.03**
90° bend						
15 mm dia.	3.90	4.79	0.28	10.50	nr	**15.29**
22 mm dia.	5.25	6.45	0.28	10.50	nr	**16.95**
28 mm dia.	7.41	9.10	0.33	12.38	nr	**21.48**
35 mm dia.	18.06	22.19	0.37	13.89	nr	**36.08**
42 mm dia.	24.87	30.56	0.42	15.76	nr	**46.32**
54 mm dia.	33.71	41.41	0.45	16.89	nr	**58.30**
Reducer						
22 × 15 mm dia.	8.98	11.04	0.28	10.50	nr	**21.54**
28 × 22 mm dia.	10.01	12.30	0.33	12.38	nr	**24.68**
35 × 28 mm dia.	12.26	15.06	0.37	13.89	nr	**28.95**
42 × 35 mm dia.	13.21	16.23	0.42	15.76	nr	**31.99**
54 × 42 mm dia.	39.15	48.10	0.48	18.05	nr	**66.15**
Tap connector						
15 mm dia.	18.94	23.27	0.13	4.87	nr	**28.14**
22 mm dia.	25.03	30.75	0.14	5.26	nr	**36.01**
28 mm dia.	34.75	42.69	0.17	6.37	nr	**49.06**
Tank connector						
15 mm dia.	24.46	30.05	0.13	4.87	nr	**34.92**
22 mm dia.	36.40	44.72	0.13	4.87	nr	**49.59**
28 mm dia.	44.83	55.07	0.15	5.63	nr	**60.70**
35 mm dia.	64.81	79.62	0.18	6.76	nr	**86.38**
42 mm dia.	85.58	105.14	0.21	7.89	nr	**113.03**
54 mm dia.	129.58	159.19	0.24	9.01	nr	**168.20**
Tee equal						
15 mm dia.	6.97	8.56	0.37	13.89	nr	**22.45**
22 mm dia.	8.68	10.67	0.40	15.02	nr	**25.69**
28 mm dia.	10.49	12.88	0.45	16.89	nr	**29.77**
35 mm dia.	25.19	30.94	0.59	22.16	nr	**53.10**
42 mm dia.	31.09	38.20	0.62	23.28	nr	**61.48**
54 mm dia.	62.78	77.12	0.67	25.15	nr	**102.27**
Unequal tee						
22 × 15 mm dia.	14.14	17.37	0.37	13.89	nr	**31.26**
28 × 15 mm dia.	15.92	19.56	0.45	16.89	nr	**36.45**
28 × 22 mm dia.	15.92	19.56	0.45	16.89	nr	**36.45**
35 × 22 mm dia.	27.84	34.20	0.59	22.16	nr	**56.36**
35 × 28 mm dia.	27.84	34.20	0.59	22.16	nr	**56.36**
42 × 28 mm dia.	34.21	42.03	0.62	23.28	nr	**65.31**
42 × 35 mm dia.	34.21	42.03	0.62	23.28	nr	**65.31**
54 × 35 mm dia.	70.83	87.01	0.67	25.15	nr	**112.16**
54 × 42 mm dia.	70.83	87.01	0.67	25.15	nr	**112.16**

38 PIPED SUPPLY SYSTEMS

Item	Net Price £	Material £	Labour hours	Labour £	Unit	**Total rate £**
Union, conical seat						
15 mm dia.	31.36	38.52	0.25	9.39	nr	**47.91**
22 mm dia.	49.39	60.68	0.28	10.50	nr	**71.18**
28 mm dia.	63.85	78.43	0.33	12.38	nr	**90.81**
35 mm dia.	83.83	102.98	0.37	13.89	nr	**116.87**
42 mm dia.	105.72	129.87	0.42	15.76	nr	**145.63**
54 mm dia.	139.88	171.84	0.45	16.89	nr	**188.73**
Union, flat seat						
15 mm dia.	32.75	40.24	0.25	9.39	nr	**49.63**
22 mm dia.	51.00	62.65	0.28	10.50	nr	**73.15**
28 mm dia.	65.93	81.00	0.33	12.38	nr	**93.38**
35 mm dia.	86.14	105.82	0.37	13.89	nr	**119.71**
42 mm dia.	108.53	133.33	0.42	15.76	nr	**149.09**
54 mm dia.	145.52	178.77	0.45	16.89	nr	**195.66**
Extra over stainless steel pipes; compression fittings						
Straight coupling						
15 mm dia.	27.27	33.51	0.18	6.76	nr	**40.27**
22 mm dia.	51.94	63.81	0.22	8.25	nr	**72.06**
28 mm dia.	69.92	85.90	0.25	9.39	nr	**95.29**
35 mm dia.	107.94	132.60	0.30	11.26	nr	**143.86**
42 mm dia.	125.99	154.78	0.40	15.02	nr	**169.80**
90° bend						
15 mm dia.	34.37	42.22	0.18	6.76	nr	**48.98**
22 mm dia.	68.27	83.87	0.22	8.25	nr	**92.12**
28 mm dia.	93.11	114.39	0.25	9.39	nr	**123.78**
35 mm dia.	188.54	231.62	0.33	12.38	nr	**244.00**
42 mm dia.	275.52	338.48	0.35	13.14	nr	**351.62**
Reducer						
22 × 15 mm dia.	49.47	60.77	0.28	10.50	nr	**71.27**
28 × 22 mm dia.	67.75	83.23	0.28	10.50	nr	**93.73**
35 × 28 mm dia.	99.00	121.62	0.30	11.26	nr	**132.88**
42 × 35 mm dia.	131.71	161.81	0.37	13.89	nr	**175.70**
Stud coupling						
15 mm dia.	28.33	34.81	0.42	15.76	nr	**50.57**
22 mm dia.	47.93	58.88	0.25	9.39	nr	**68.27**
28 mm dia.	66.55	81.75	0.25	9.39	nr	**91.14**
35 mm dia.	106.59	130.95	0.37	13.89	nr	**144.84**
42 mm dia.	125.99	154.78	0.42	15.76	nr	**170.54**
Equal tee						
15 mm dia.	48.39	59.45	0.37	13.89	nr	**73.34**
22 mm dia.	99.98	122.83	0.40	15.02	nr	**137.85**
28 mm dia.	136.74	167.99	0.45	16.89	nr	**184.88**
35 mm dia.	271.90	334.03	0.59	22.16	nr	**356.19**
42 mm dia.	377.02	463.17	0.62	23.28	nr	**486.45**
Running tee						
15 mm dia.	59.58	73.20	0.37	13.89	nr	**87.09**
22 mm dia.	107.33	131.86	0.40	15.02	nr	**146.88**
28 mm dia.	181.71	223.23	0.59	22.16	nr	**245.39**

38 PIPED SUPPLY SYSTEMS

Item	Net Price £	Material £	Labour hours	Labour £	Unit	Total rate £
COLD WATER PIPELINES: STAINLESS STEEL PIPEWORK – cont						
PRESS FIT (stainless steel); Press fit jointing system; butyl rubber O ring mechanical joint						
Pipework						
15 mm dia.	4.69	5.76	0.46	17.26	m	**23.02**
22 mm dia.	7.49	9.20	0.48	18.01	m	**27.21**
28 mm dia.	9.24	11.35	0.52	19.52	m	**30.87**
35 mm dia.	13.61	16.72	0.56	21.02	m	**37.74**
42 mm dia.	16.74	20.57	0.58	21.77	m	**42.34**
54 mm dia.	21.32	26.19	0.66	24.77	m	**50.96**
FIXINGS						
For stainless steel pipes						
Refer to fixings for stainless steel pipes; capillary or compression joints; BS 4127						
Extra over stainless steel pipes; Press fit jointing system						
Coupling						
15 mm dia.	5.51	6.77	0.36	13.51	nr	**20.28**
22 mm dia.	6.93	8.52	0.36	13.51	nr	**22.03**
28 mm dia.	7.80	9.59	0.44	16.51	nr	**26.10**
35 mm dia.	9.70	11.92	0.44	16.51	nr	**28.43**
42 mm dia.	13.24	16.26	0.52	19.52	nr	**35.78**
54 mm dia.	15.92	19.56	0.60	22.52	nr	**42.08**
Stop end						
22 mm dia.	5.24	6.44	0.18	6.76	nr	**13.20**
28 mm dia.	6.09	7.49	0.22	8.25	nr	**15.74**
35 mm dia.	9.97	12.24	0.22	8.25	nr	**20.49**
42 mm dia.	14.00	17.20	0.26	9.75	nr	**26.95**
54 mm dia.	16.20	19.90	0.30	11.26	nr	**31.16**
Reducer						
22 × 15 mm dia.	6.57	8.07	0.36	13.51	nr	**21.58**
28 × 15 mm dia.	7.43	9.12	0.40	15.02	nr	**24.14**
28 × 22 mm dia.	7.70	9.46	0.40	15.02	nr	**24.48**
35 × 22 mm dia.	9.39	11.54	0.40	15.02	nr	**26.56**
35 × 28 mm dia.	11.63	14.29	0.44	16.51	nr	**30.80**
42 × 35 mm dia.	12.27	15.08	0.48	18.01	nr	**33.09**
54 × 42 mm dia.	14.02	17.22	0.56	21.02	nr	**38.24**
90° bend						
15 mm dia.	7.87	9.67	0.36	13.51	nr	**23.18**
22 mm dia.	11.01	13.52	0.36	13.51	nr	**27.03**
28 mm dia.	13.89	17.06	0.44	16.51	nr	**33.57**
35 mm dia.	21.86	26.86	0.44	16.51	nr	**43.37**
42 mm dia.	36.50	44.85	0.52	19.52	nr	**64.37**
54 mm dia.	50.42	61.94	0.60	22.52	nr	**84.46**

38 PIPED SUPPLY SYSTEMS

Item	Net Price £	Material £	Labour hours	Labour £	Unit	Total rate £
45° bend						
15 mm dia.	10.69	13.14	0.36	13.51	nr	**26.65**
22 mm dia.	13.29	16.33	0.36	13.51	nr	**29.84**
28 mm dia.	15.46	18.99	0.44	16.51	nr	**35.50**
35 mm dia.	18.14	22.28	0.44	16.51	nr	**38.79**
42 mm dia.	29.20	35.87	0.52	19.52	nr	**55.39**
54 mm dia.	37.94	46.61	0.60	22.52	nr	**69.13**
Equal tee						
15 mm dia.	12.91	15.86	0.54	20.28	nr	**36.14**
22 mm dia.	15.85	19.47	0.54	20.28	nr	**39.75**
28 mm dia.	18.51	22.74	0.66	24.77	nr	**47.51**
35 mm dia.	23.45	28.81	0.66	24.77	nr	**53.58**
42 mm dia.	33.29	40.90	0.78	29.27	nr	**70.17**
54 mm dia.	39.84	48.94	0.90	33.78	nr	**82.72**
Reducing tee						
22 × 15 mm dia.	13.55	16.64	0.54	20.28	nr	**36.92**
28 × 15 mm dia.	16.42	20.17	0.62	23.27	nr	**43.44**
28 × 22 mm dia.	17.78	21.84	0.62	23.27	nr	**45.11**
35 × 22 mm dia.	21.08	25.89	0.62	23.27	nr	**49.16**
35 × 28 mm dia.	21.99	27.02	0.62	23.27	nr	**50.29**
42 × 28 mm dia.	31.28	38.43	0.70	26.28	nr	**64.71**
42 × 35 mm dia.	32.22	39.59	0.70	26.28	nr	**65.87**
54 × 35 mm dia.	36.38	44.69	0.82	30.78	nr	**75.47**
54 × 42 mm dia.	37.42	45.97	0.82	30.78	nr	**76.75**
FIXINGS						
For stainless steel pipes						
Refer to fixings for stainless steel pipes; capillary or compression joints; BS 4127						

38 PIPED SUPPLY SYSTEMS

Item	Net Price £	Material £	Labour hours	Labour £	Unit	Total rate £
COLD WATER PIPELINES: MEDIUM DENSITY POLYETHYLENE – BLUE PIPEWORK						
Note: MDPE is sized on Outside Diameter i.e. OD not ID						
Pipes for water distribution; laid underground; electrofusion joints in the running length; BS 6572						
Coiled service pipe						
20 mm dia.	0.87	1.07	0.37	13.89	m	**14.96**
25 mm dia.	1.13	1.39	0.41	15.39	m	**16.78**
32 mm dia.	1.89	2.32	0.47	17.64	m	**19.96**
50 mm dia.	4.52	5.55	0.53	19.90	m	**25.45**
63 mm dia.	7.10	8.73	0.60	22.53	m	**31.26**
Mains service pipe						
90 mm dia.	10.71	13.16	0.90	33.78	m	**46.94**
110 mm dia.	16.03	19.70	1.10	41.28	m	**60.98**
125 mm dia.	20.29	24.93	1.20	45.07	m	**70.00**
160 mm dia.	32.33	39.72	1.48	55.55	m	**95.27**
180 mm dia.	42.11	51.73	1.50	56.35	m	**108.08**
225 mm dia.	64.00	78.62	1.77	66.42	m	**145.04**
250 mm dia.	80.79	99.25	1.75	65.73	m	**164.98**
315 mm dia.	124.74	153.25	1.90	71.32	m	**224.57**
Extra over fittings; MDPE blue; electrofusion joints						
Coupler						
20 mm dia.	7.29	8.96	0.36	13.51	nr	**22.47**
25 mm dia.	7.29	8.96	0.40	15.02	nr	**23.98**
32 mm dia.	7.29	8.96	0.44	16.51	nr	**25.47**
40 mm dia.	10.75	13.21	0.48	18.01	nr	**31.22**
50 mm dia.	10.34	12.70	0.52	19.52	nr	**32.22**
63 mm dia.	13.50	16.58	0.58	21.77	nr	**38.35**
90 mm dia.	19.85	24.38	0.67	25.15	nr	**49.53**
110 mm dia.	31.92	39.22	0.74	27.77	nr	**66.99**
125 mm dia.	36.01	44.24	0.83	31.15	nr	**75.39**
160 mm dia.	57.28	70.37	1.00	37.53	nr	**107.90**
180 mm dia.	67.36	82.75	1.25	46.91	nr	**129.66**
225 mm dia.	107.66	132.26	1.35	50.66	nr	**182.92**
250 mm dia.	157.63	193.65	1.50	56.30	nr	**249.95**
315 mm dia.	259.95	319.35	1.80	67.56	nr	**386.91**
Extra over fittings; MDPE blue; butt fused joints						
Cap						
25 mm dia.	13.83	16.99	0.20	7.51	nr	**24.50**
32 mm dia.	13.83	16.99	0.22	8.25	nr	**25.24**
40 mm dia.	14.66	18.01	0.24	9.01	nr	**27.02**

38 PIPED SUPPLY SYSTEMS

Item	Net Price £	Material £	Labour hours	Labour £	Unit	**Total rate £**
50 mm dia.	21.37	26.25	0.26	9.75	nr	**36.00**
63 mm dia.	24.52	30.12	0.32	12.01	nr	**42.13**
90 mm dia.	40.43	49.67	0.37	13.88	nr	**63.55**
110 mm dia.	81.40	100.00	0.40	15.02	nr	**115.02**
125 mm dia.	64.98	79.83	0.46	17.26	nr	**97.09**
160 mm dia.	74.30	91.28	0.50	18.77	nr	**110.05**
180 mm dia.	124.27	152.67	0.60	22.52	nr	**175.19**
225 mm dia.	146.78	180.32	0.68	25.51	nr	**205.83**
250 mm dia.	215.46	264.69	0.75	28.14	nr	**292.83**
315 mm dia.	277.82	341.30	0.90	33.78	nr	**375.08**
Reducer						
63 × 32 mm dia.	18.68	22.95	0.54	20.28	nr	**43.23**
63 × 50 mm dia.	21.85	26.84	0.60	22.52	nr	**49.36**
90 × 63 mm dia.	27.65	33.97	0.67	25.15	nr	**59.12**
110 × 90 mm dia.	38.09	46.80	0.74	27.77	nr	**74.57**
125 × 90 mm dia.	55.40	68.06	0.83	31.15	nr	**99.21**
125 × 110 mm dia.	61.04	74.99	1.00	37.53	nr	**112.52**
160 × 110 mm dia.	93.78	115.21	1.10	41.28	nr	**156.49**
180 × 125 mm dia.	101.85	125.12	1.25	46.91	nr	**172.03**
225 × 160 mm dia.	166.09	204.05	1.40	52.54	nr	**256.59**
250 × 180 mm dia.	127.61	156.76	1.80	67.56	nr	**224.32**
315 × 250 mm dia.	146.79	180.33	2.40	90.07	nr	**270.40**
Bend; 45°						
50 mm dia.	28.51	35.03	0.50	18.77	nr	**53.80**
63 mm dia.	34.48	42.36	0.58	21.77	nr	**64.13**
90 mm dia.	53.40	65.60	0.67	25.15	nr	**90.75**
110 mm dia.	78.29	96.18	0.74	27.77	nr	**123.95**
125 mm dia.	87.51	107.51	0.83	31.15	nr	**138.66**
160 mm dia.	162.18	199.24	1.00	37.53	nr	**236.77**
180 mm dia.	184.62	226.81	1.25	46.91	nr	**273.72**
225 mm dia.	235.32	289.09	1.40	52.53	nr	**341.62**
250 mm dia.	254.63	312.82	1.80	67.56	nr	**380.38**
315 mm dia.	317.25	389.74	2.40	90.07	nr	**479.81**
Bend; 90°						
50 mm dia.	28.51	35.03	0.50	18.77	nr	**53.80**
63 mm dia.	34.48	42.36	0.58	21.77	nr	**64.13**
90 mm dia.	53.40	65.60	0.67	25.15	nr	**90.75**
110 mm dia.	78.29	96.18	0.74	27.77	nr	**123.95**
125 mm dia.	87.51	107.51	0.83	31.15	nr	**138.66**
160 mm dia.	162.18	199.24	1.00	37.53	nr	**236.77**
180 mm dia.	314.98	386.96	1.25	46.91	nr	**433.87**
225 mm dia.	398.27	489.28	1.40	52.54	nr	**541.82**
250 mm dia.	435.63	535.17	1.80	67.56	nr	**602.73**
315 mm dia.	545.36	669.97	2.40	90.07	nr	**760.04**
Equal tee						
50 mm dia.	31.16	38.28	0.70	26.28	nr	**64.56**
63 mm dia.	34.01	41.78	0.75	28.14	nr	**69.92**
90 mm dia.	62.13	76.32	0.87	32.65	nr	**108.97**
110 mm dia.	92.65	113.82	1.00	37.53	nr	**151.35**
125 mm dia.	119.61	146.94	1.08	40.53	nr	**187.47**
160 mm dia.	196.67	241.60	1.35	50.66	nr	**292.26**
180 mm dia.	201.24	247.22	1.63	61.18	nr	**308.40**

38 PIPED SUPPLY SYSTEMS

Item	Net Price £	Material £	Labour hours	Labour £	Unit	Total rate £
COLD WATER PIPELINES: MEDIUM DENSITY POLYETHYLENE – BLUE PIPEWORK – cont						
Extra over fittings – cont						
Equal tee – cont						
225 mm dia.	243.91	299.64	1.90	71.32	nr	**370.96**
250 mm dia.	336.53	413.43	2.70	101.33	nr	**514.76**
315 mm dia.	838.95	1030.65	3.60	135.11	nr	**1165.76**
Extra over plastic fittings, compression joints						
Straight connector						
20 mm dia.	3.22	3.96	0.38	14.26	nr	**18.22**
25 mm dia.	3.38	4.15	0.45	16.89	nr	**21.04**
32 mm dia.	8.02	9.85	0.50	18.77	nr	**28.62**
50 mm dia.	18.47	22.69	0.68	25.53	nr	**48.22**
63 mm dia.	27.81	34.17	0.85	31.92	nr	**66.09**
Reducing connector						
25 mm dia.	6.62	8.14	0.38	14.26	nr	**22.40**
32 mm dia.	10.69	13.14	0.45	16.89	nr	**30.03**
50 mm dia.	29.62	36.39	0.50	18.77	nr	**55.16**
63 mm dia.	41.33	50.78	0.62	23.28	nr	**74.06**
Straight connector; polyethylene to MI						
20 mm dia.	2.92	3.59	0.31	11.63	nr	**15.22**
25 mm dia.	4.97	6.10	0.35	13.14	nr	**19.24**
32 mm dia.	5.38	6.60	0.40	15.02	nr	**21.62**
50 mm dia.	13.74	16.88	0.55	20.65	nr	**37.53**
63 mm dia.	19.36	23.78	0.65	24.40	nr	**48.18**
Straight connector; polyethylene to FI						
20 mm dia.	3.92	4.82	0.31	11.63	nr	**16.45**
25 mm dia.	4.25	5.22	0.35	13.14	nr	**18.36**
32 mm dia.	5.08	6.24	0.40	15.02	nr	**21.26**
50 mm dia.	16.14	19.82	0.55	20.65	nr	**40.47**
63 mm dia.	22.61	27.77	0.75	28.16	nr	**55.93**
Elbow						
20 mm dia.	4.30	5.28	0.38	14.26	nr	**19.54**
25 mm dia.	6.35	7.80	0.45	16.89	nr	**24.69**
32 mm dia.	9.25	11.36	0.50	18.77	nr	**30.13**
50 mm dia.	21.46	26.37	0.68	25.53	nr	**51.90**
63 mm dia.	29.17	35.84	0.80	30.02	nr	**65.86**
Elbow; polyethylene to MI						
25 mm dia.	5.47	6.72	0.35	13.14	nr	**19.86**
Elbow; polyethylene to FI						
20 mm dia.	3.90	4.79	0.31	11.63	nr	**16.42**
25 mm dia.	5.31	6.52	0.35	13.14	nr	**19.66**
32 mm dia.	7.95	9.77	0.42	15.76	nr	**25.53**
50 mm dia.	18.89	23.20	0.50	18.77	nr	**41.97**
63 mm dia.	24.77	30.43	0.55	20.65	nr	**51.08**

38 PIPED SUPPLY SYSTEMS

Item	Net Price £	Material £	Labour hours	Labour £	Unit	**Total rate £**
Tank coupling						
25 mm dia.	8.19	10.06	0.42	15.76	nr	**25.82**
Equal tee						
20 mm dia.	5.78	7.10	0.53	19.90	nr	**27.00**
25 mm dia.	9.04	11.11	0.55	20.65	nr	**31.76**
32 mm dia.	11.32	13.90	0.64	24.03	nr	**37.93**
50 mm dia.	26.40	32.43	0.75	28.16	nr	**60.59**
63 mm dia.	40.90	50.24	0.87	32.67	nr	**82.91**
Equal tee; FI branch						
20 mm dia.	5.60	6.88	0.45	16.89	nr	**23.77**
25 mm dia.	8.92	10.96	0.50	18.77	nr	**29.73**
32 mm dia.	10.86	13.35	0.60	22.53	nr	**35.88**
50 mm dia.	25.01	30.72	0.68	25.53	nr	**56.25**
63 mm dia.	35.11	43.13	0.81	30.42	nr	**73.55**
Equal tee; MI branch						
25 mm dia.	8.76	10.76	0.50	18.77	nr	**29.53**

38 PIPED SUPPLY SYSTEMS

Item	Net Price £	Material £	Labour hours	Labour £	Unit	Total rate £
COLD WATER PIPELINES: ABS PIPEWORK						
Pipes; solvent welded joints in the running length, brackets measured separately						
Class C (9 bar pressure)						
1" dia.	5.79	7.11	0.30	11.26	m	**18.37**
1 ¼" dia.	9.74	11.97	0.33	12.38	m	**24.35**
1 ½" dia.	12.37	15.19	0.36	13.51	m	**28.70**
2" dia.	16.66	20.46	0.39	14.63	m	**35.09**
3" dia.	34.32	42.16	0.46	17.26	m	**59.42**
4" dia.	56.57	69.50	0.53	19.89	m	**89.39**
6" dia.	111.87	137.43	0.76	28.53	m	**165.96**
8" dia.	192.10	236.00	0.97	36.40	m	**272.40**
Class E (15 bar pressure)						
½" dia.	4.42	5.43	0.24	9.01	m	**14.44**
¾" dia.	6.80	8.36	0.27	10.13	m	**18.49**
1" dia.	8.96	11.00	0.30	11.26	m	**22.26**
1 ¼" dia.	13.38	16.43	0.33	12.38	m	**28.81**
1 ½" dia.	17.64	21.67	0.36	13.51	m	**35.18**
2" dia.	22.10	27.15	0.39	14.63	m	**41.78**
3" dia.	44.42	54.57	0.49	18.39	m	**72.96**
4" dia.	71.46	87.79	0.57	21.39	m	**109.18**
FIXINGS						
Refer to steel pipes; galvanized iron. For minimum fixing dimensions, refer to the Tables and Memoranda at the rear of the book						
Extra over fittings; solvent welded joints						
Cap						
½" dia.	1.89	2.32	0.16	6.01	nr	**8.33**
¾" dia.	2.18	2.68	0.19	7.13	nr	**9.81**
1" dia.	2.51	3.09	0.22	8.25	nr	**11.34**
1 ¼" dia.	4.19	5.15	0.25	9.39	nr	**14.54**
1 ½" dia.	6.47	7.95	0.28	10.50	nr	**18.45**
2" dia.	8.18	10.05	0.31	11.63	nr	**21.68**
3" dia.	24.59	30.21	0.36	13.51	nr	**43.72**
4" dia.	37.61	46.20	0.44	16.51	nr	**62.71**
Elbow 90°						
½" dia.	2.64	3.24	0.29	10.89	nr	**14.13**
¾" dia.	3.16	3.89	0.34	12.77	nr	**16.66**
1" dia.	4.42	5.43	0.40	15.02	nr	**20.45**
1 ¼" dia.	7.47	9.18	0.45	16.89	nr	**26.07**
1 ½" dia.	9.70	11.92	0.51	19.14	nr	**31.06**
2" dia.	14.76	18.13	0.56	21.02	nr	**39.15**
3" dia.	42.41	52.10	0.65	24.40	nr	**76.50**
4" dia.	63.33	77.81	0.80	30.02	nr	**107.83**
6" dia.	254.93	313.18	1.21	45.41	nr	**358.59**
8" dia.	389.07	477.97	1.45	54.42	nr	**532.39**

38 PIPED SUPPLY SYSTEMS

Item	Net Price £	Material £	Labour hours	Labour £	Unit	**Total rate £**
Elbow 45°						
½" dia.	5.10	6.27	0.29	10.89	nr	**17.16**
¾" dia.	5.18	6.36	0.34	12.77	nr	**19.13**
1" dia.	6.47	7.95	0.40	15.02	nr	**22.97**
1 ¼" dia.	9.46	11.62	0.45	16.89	nr	**28.51**
1 ½" dia.	11.73	14.41	0.51	19.14	nr	**33.55**
2" dia.	16.29	20.01	0.56	21.02	nr	**41.03**
3" dia.	38.34	47.10	0.65	24.40	nr	**71.50**
4" dia.	79.50	97.67	0.80	30.02	nr	**127.69**
6" dia.	164.80	202.46	1.21	45.41	nr	**247.87**
8" dia.	354.63	435.67	1.45	54.42	nr	**490.09**
Reducing bush						
¾" × ½" dia.	1.95	2.39	0.42	15.76	nr	**18.15**
1" × ½" dia.	2.51	3.09	0.45	16.89	nr	**19.98**
1" × ¾" dia.	2.51	3.09	0.45	16.89	nr	**19.98**
1 ¼" × 1" dia.	3.38	4.15	0.48	18.01	nr	**22.16**
1 ½" × ¾" dia.	4.42	5.43	0.51	19.14	nr	**24.57**
1 ½" × 1" dia.	4.42	5.43	0.51	19.14	nr	**24.57**
1 ½" × 1 ¼" dia.	4.42	5.43	0.51	19.14	nr	**24.57**
2" × 1" dia.	5.79	7.11	0.56	21.02	nr	**28.13**
2" × 1 ¼" dia.	5.79	7.11	0.56	21.02	nr	**28.13**
2" × 1 ½" dia.	5.79	7.11	0.56	21.02	nr	**28.13**
3" × 1 ½" dia.	16.29	20.01	0.65	24.40	nr	**44.41**
3" × 2" dia.	16.29	20.01	0.65	24.40	nr	**44.41**
4" × 3" dia.	22.45	27.58	0.80	30.02	nr	**57.60**
6" × 4" dia.	69.17	84.98	1.21	45.41	nr	**130.39**
Union						
½" dia.	10.50	12.89	0.34	12.77	nr	**25.66**
¾" dia.	11.32	13.90	0.39	14.63	nr	**28.53**
1" dia.	15.25	18.73	0.43	16.14	nr	**34.87**
1 ¼" dia.	18.70	22.97	0.50	18.77	nr	**41.74**
1 ½" dia.	25.75	31.64	0.57	21.39	nr	**53.03**
2" dia.	33.58	41.25	0.62	23.27	nr	**64.52**
Sockets						
½" dia.	1.95	2.39	0.34	12.77	nr	**15.16**
¾" dia.	2.18	2.68	0.39	14.63	nr	**17.31**
1" dia.	2.51	3.09	0.43	16.14	nr	**19.23**
1 ¼" dia.	4.42	5.43	0.50	18.77	nr	**24.20**
1 ½" dia.	5.32	6.53	0.57	21.39	nr	**27.92**
2" dia.	7.47	9.18	0.62	23.27	nr	**32.45**
3" dia.	30.03	36.90	0.70	26.28	nr	**63.18**
4" dia.	42.63	52.37	0.70	26.28	nr	**78.65**
6" dia.	106.50	130.84	1.26	47.28	nr	**178.12**
8" dia.	212.73	261.33	1.55	58.17	nr	**319.50**
Barrel nipple						
½" dia.	3.67	4.50	0.34	12.77	nr	**17.27**
¾" dia.	4.78	5.87	0.39	14.63	nr	**20.50**
1" dia.	6.19	7.60	0.43	16.14	nr	**23.74**
1 ¼" dia.	8.58	10.54	0.50	18.77	nr	**29.31**
1 ½" dia.	10.11	12.42	0.57	21.39	nr	**33.81**
2" dia.	12.25	15.05	0.62	23.27	nr	**38.32**
3" dia.	32.55	39.98	0.70	26.28	nr	**66.26**

38 PIPED SUPPLY SYSTEMS

Item	Net Price £	Material £	Labour hours	Labour £	Unit	**Total rate £**
COLD WATER PIPELINES: ABS PIPEWORK – cont						
Extra over fittings – cont						
Tee, 90°						
½" dia.	3.00	3.69	0.41	15.39	nr	**19.08**
¾" dia.	4.19	5.15	0.47	17.64	nr	**22.79**
1" dia.	5.79	7.11	0.55	20.64	nr	**27.75**
1 ¼" dia.	8.32	10.22	0.64	24.01	nr	**34.23**
1 ½" dia.	12.25	15.05	0.71	26.65	nr	**41.70**
2" dia.	18.70	22.97	0.78	29.27	nr	**52.24**
3" dia.	54.52	66.98	0.91	34.16	nr	**101.14**
4" dia.	80.03	98.32	1.12	42.04	nr	**140.36**
6" dia.	279.67	343.57	1.69	63.43	nr	**407.00**
8" dia.	436.07	535.71	2.03	76.19	nr	**611.90**
Full face flange						
½" dia.	32.58	40.03	0.10	3.76	nr	**43.79**
¾" dia.	33.34	40.96	0.13	4.87	nr	**45.83**
1" dia.	36.12	44.37	0.15	5.63	nr	**50.00**
1 ¼" dia.	40.14	49.31	0.18	6.76	nr	**56.07**
1 ½" dia.	48.27	59.30	0.21	7.89	nr	**67.19**
2" dia.	65.34	80.27	0.29	10.89	nr	**91.16**
3" dia.	112.04	137.64	0.37	13.88	nr	**151.52**
4" dia.	146.84	180.39	0.41	15.39	nr	**195.78**

38 PIPED SUPPLY SYSTEMS

Item	Net Price £	Material £	Labour hours	Labour £	Unit	**Total rate £**
COLD WATER PIPELINES: PVC-u PIPEWORK						
Pipes; solvent welded joints in the running length, brackets Measured separately						
Class C (9 bar pressure)						
2" dia.	13.74	16.88	0.41	15.39	m	**32.27**
3" dia.	26.33	32.35	0.47	17.64	m	**49.99**
4" dia.	46.70	57.37	0.50	18.77	m	**76.14**
6" dia.	101.07	124.16	1.76	66.06	m	**190.22**
Class D (12 bar pressure)						
1 ¼" dia.	8.00	9.83	0.41	15.39	m	**25.22**
1 ½" dia.	11.00	13.51	0.42	15.76	m	**29.27**
2" dia.	17.06	20.96	0.45	16.89	m	**37.85**
3" dia.	36.54	44.89	0.48	18.01	m	**62.90**
4" dia.	61.18	75.16	0.53	19.89	m	**95.05**
6" dia.	113.53	139.47	0.58	21.77	m	**161.24**
Class E (15 bar pressure)						
½" dia.	3.91	4.80	0.38	14.26	m	**19.06**
¾" dia.	5.60	6.88	0.40	15.02	m	**21.90**
1" dia.	6.52	8.01	0.41	15.39	m	**23.40**
1 ¼" dia.	9.56	11.75	0.41	15.39	m	**27.14**
1 ½" dia.	12.44	15.28	0.42	15.76	m	**31.04**
2" dia.	19.42	23.86	0.45	16.89	m	**40.75**
3" dia.	42.08	51.69	0.47	17.64	m	**69.33**
4" dia.	69.10	84.89	0.50	18.77	m	**103.66**
6" dia.	149.68	183.89	0.53	19.89	m	**203.78**
Class 7						
½" dia.	6.93	8.52	0.32	12.01	m	**20.53**
¾" dia.	9.71	11.93	0.33	12.38	m	**24.31**
1" dia.	14.80	18.19	0.40	15.02	m	**33.21**
1 ¼" dia.	20.34	24.99	0.40	15.02	m	**40.01**
1 ½" dia.	25.19	30.94	0.41	15.39	m	**46.33**
2" dia.	41.87	51.44	0.43	16.14	m	**67.58**
FIXINGS						
Refer to steel pipes; galvanized iron. For minimum fixing dimensions, refer to the Tables and Memoranda at the rear of the book						
Extra over fittings; solvent welded joints						
End cap						
½" dia.	1.34	1.65	0.17	6.37	nr	**8.02**
¾" dia.	1.56	1.92	0.19	7.13	nr	**9.05**
1" dia.	1.75	2.15	0.22	8.25	nr	**10.40**
1 ¼" dia.	2.74	3.37	0.25	9.39	nr	**12.76**
1 ½" dia.	4.60	5.65	0.28	10.50	nr	**16.15**
2" dia.	5.63	6.92	0.31	11.63	nr	**18.55**
3" dia.	17.28	21.23	0.36	13.51	nr	**34.74**
4" dia.	26.68	32.78	0.44	16.51	nr	**49.29**
6" dia.	64.46	79.19	0.67	25.15	nr	**104.34**

38 PIPED SUPPLY SYSTEMS

Item	Net Price £	Material £	Labour hours	Labour £	Unit	**Total rate £**
COLD WATER PIPELINES: PVC-u PIPEWORK – cont						
FIXINGS – cont						
Socket						
½" dia.	1.41	1.73	0.31	11.63	nr	**13.36**
¾" dia.	1.56	1.92	0.35	13.14	nr	**15.06**
1" dia.	1.82	2.24	0.42	15.76	nr	**18.00**
1 ¼" dia.	3.28	4.03	0.45	16.89	nr	**20.92**
1 ½" dia.	3.85	4.73	0.51	19.14	nr	**23.87**
2" dia.	5.47	6.72	0.56	21.02	nr	**27.74**
3" dia.	20.88	25.65	0.65	24.40	nr	**50.05**
4" dia.	30.27	37.19	0.80	30.02	nr	**67.21**
6" dia.	75.92	93.27	1.21	45.41	nr	**138.68**
Reducing socket						
¾ × ½" dia.	1.65	2.03	0.31	11.63	nr	**13.66**
1 × ¾" dia.	2.08	2.55	0.35	13.14	nr	**15.69**
1 ¼ × 1" dia.	3.96	4.86	0.42	15.76	nr	**20.62**
1 ½ × 1 ¼" dia.	4.42	5.43	0.45	16.89	nr	**22.32**
2 × 1 ½" dia.	6.67	8.19	0.51	19.14	nr	**27.33**
3 × 2" dia.	20.31	24.95	0.56	21.02	nr	**45.97**
4 × 3" dia.	30.07	36.94	0.65	24.40	nr	**61.34**
6 × 4" dia.	109.57	134.61	0.80	30.02	nr	**164.63**
8 × 6" dia.	169.71	208.49	1.21	45.41	nr	**253.90**
Elbow, 90°						
½" dia.	1.87	2.30	0.31	11.63	nr	**13.93**
¾" dia.	2.24	2.75	0.35	13.14	nr	**15.89**
1" dia.	3.12	3.83	0.42	15.76	nr	**19.59**
1 ¼" dia.	5.47	6.72	0.45	16.89	nr	**23.61**
1 ½" dia.	7.03	8.64	0.45	16.89	nr	**25.53**
2" dia.	10.42	12.80	0.56	21.02	nr	**33.82**
3" dia.	30.07	36.94	0.65	24.40	nr	**61.34**
4" dia.	45.28	55.63	0.80	30.02	nr	**85.65**
6" dia.	179.29	220.26	1.21	45.41	nr	**265.67**
Elbow 45°						
½" dia.	3.56	4.38	0.31	11.63	nr	**16.01**
¾" dia.	3.80	4.67	0.35	13.14	nr	**17.81**
1" dia.	4.60	5.65	0.45	16.89	nr	**22.54**
1 ¼" dia.	6.59	8.10	0.45	16.89	nr	**24.99**
1 ½" dia.	8.26	10.14	0.51	19.14	nr	**29.28**
2" dia.	11.64	14.30	0.56	21.02	nr	**35.32**
3" dia.	27.44	33.70	0.65	24.40	nr	**58.10**
4" dia.	56.38	69.26	0.80	30.02	nr	**99.28**
6" dia.	116.32	142.89	1.21	45.41	nr	**188.30**
Bend 90° (long radius)						
3" dia.	83.82	102.97	0.65	24.40	nr	**127.37**
4" dia.	169.32	208.00	0.80	30.02	nr	**238.02**
6" dia.	372.08	457.10	1.21	45.41	nr	**502.51**

38 PIPED SUPPLY SYSTEMS

Item	Net Price £	Material £	Labour hours	Labour £	Unit	**Total rate £**
Bend 45° (long radius)						
1 ½" dia.	19.90	24.44	0.51	19.14	nr	**43.58**
2" dia.	32.53	39.96	0.56	21.02	nr	**60.98**
3" dia.	69.52	85.41	0.65	24.40	nr	**109.81**
4" dia.	135.30	166.22	0.80	30.02	nr	**196.24**
Socket union						
½" dia.	7.23	8.88	0.34	12.77	nr	**21.65**
¾" dia.	8.26	10.14	0.39	14.63	nr	**24.77**
1" dia.	10.73	13.18	0.45	16.89	nr	**30.07**
1 ¼" dia.	13.36	16.41	0.50	18.77	nr	**35.18**
1 ½" dia.	18.30	22.48	0.57	21.39	nr	**43.87**
2" dia.	23.67	29.07	0.62	23.27	nr	**52.34**
3" dia.	88.13	108.27	0.70	26.28	nr	**134.55**
4" dia.	119.32	146.58	0.89	33.40	nr	**179.98**
Saddle plain						
2" × 1 ¼" dia.	18.60	22.85	0.42	15.76	nr	**38.61**
3" × 1 ½" dia.	26.13	32.10	0.48	18.01	nr	**50.11**
4" × 2" dia.	29.44	36.16	0.68	25.51	nr	**61.67**
6" × 2" dia.	34.57	42.47	0.91	34.16	nr	**76.63**
Straight tank connector						
½" dia.	4.80	5.90	0.13	4.87	nr	**10.77**
¾" dia.	5.43	6.67	0.14	5.26	nr	**11.93**
1" dia.	11.55	14.19	0.14	5.26	nr	**19.45**
1 ¼" dia.	29.32	36.02	0.16	6.01	nr	**42.03**
1 ½" dia.	32.14	39.48	0.18	6.76	nr	**46.24**
2" dia.	38.51	47.31	0.24	9.01	nr	**56.32**
3" dia.	39.49	48.51	0.29	10.89	nr	**59.40**
Equal tee						
½" dia.	2.17	2.67	0.44	16.51	nr	**19.18**
¾" dia.	2.74	3.37	0.48	18.01	nr	**21.38**
1" dia.	4.13	5.07	0.54	20.28	nr	**25.35**
1 ¼" dia.	5.84	7.17	0.70	26.28	nr	**33.45**
1 ½" dia.	8.42	10.34	0.74	27.77	nr	**38.11**
2" dia.	13.36	16.41	0.80	30.02	nr	**46.43**
3" dia.	38.70	47.54	1.04	39.03	nr	**86.57**
4" dia.	56.75	69.72	1.28	48.04	nr	**117.76**
6" dia.	197.70	242.88	1.93	72.43	nr	**315.31**

38 PIPED SUPPLY SYSTEMS

Item	Net Price £	Material £	Labour hours	Labour £	Unit	Total rate £
COLD WATER PIPELINES: PVC-C PIPEWORK						
Pipes; solvent welded in the running length, brackets measured separately						
Pipe; 3 m long; PN25						
16 × 2.0 mm	4.37	5.37	0.20	7.51	m	**12.88**
20 × 2.3 mm	6.60	8.11	0.20	7.51	m	**15.62**
25 × 2.8 mm	8.55	10.50	0.20	7.51	m	**18.01**
32 × 3.6 mm	12.70	15.60	0.20	7.51	m	**23.11**
Pipe; 5 m long; PN25						
40 × 4.5 mm	15.48	19.02	0.20	7.51	m	**26.53**
50 × 5.6 mm	23.32	28.64	0.20	7.51	m	**36.15**
63 × 7.0 mm	36.01	44.24	0.20	7.51	m	**51.75**
FIXINGS						
Refer to steel pipes; galvanized iron. For minimum fixing dimensions, refer to the Tables and Memoranda at the rear of the book						
Extra over fittings; solvent welded joints						
Straight coupling; PN25						
16 mm	0.68	0.84	0.20	7.51	nr	**8.35**
20 mm	0.95	1.17	0.20	7.51	nr	**8.68**
25 mm	1.18	1.45	0.20	7.51	nr	**8.96**
32 mm	3.62	4.45	0.20	7.51	nr	**11.96**
40 mm	4.66	5.72	0.20	7.51	nr	**13.23**
50 mm	6.24	7.66	0.20	7.51	nr	**15.17**
63 mm	11.01	13.52	0.20	7.51	nr	**21.03**
Elbow; 90°; PN25						
16 mm	1.09	1.34	0.20	7.51	nr	**8.85**
20 mm	1.67	2.05	0.20	7.51	nr	**9.56**
25 mm	2.08	2.55	0.20	7.51	nr	**10.06**
32 mm	4.34	5.33	0.20	7.51	nr	**12.84**
40 mm	6.68	8.21	0.20	7.51	nr	**15.72**
50 mm	9.26	11.37	0.20	7.51	nr	**18.88**
63 mm	15.81	19.43	0.20	7.51	nr	**26.94**
Elbow; 45°; PN25						
20 mm	1.67	2.05	0.20	7.51	nr	**9.56**
25 mm	2.08	2.55	0.20	7.51	nr	**10.06**
32 mm	4.34	5.33	0.20	7.51	nr	**12.84**
40 mm	6.68	8.21	0.20	7.51	nr	**15.72**
50 mm	9.26	11.37	0.20	7.51	nr	**18.88**
63 mm	15.81	19.43	0.20	7.51	nr	**26.94**
Reducer fitting; single stage reduction						
20/16 mm	1.18	1.45	0.20	7.51	nr	**8.96**
25/20 mm	1.42	1.74	0.20	7.51	nr	**9.25**
32/25 mm	2.85	3.50	0.20	7.51	nr	**11.01**
40/32 mm	3.75	4.61	0.20	7.51	nr	**12.12**
50/40 mm	4.34	5.33	0.20	7.51	nr	**12.84**
63/50 mm	6.57	8.07	0.20	7.51	nr	**15.58**

38 PIPED SUPPLY SYSTEMS

Item	Net Price £	Material £	Labour hours	Labour £	Unit	**Total rate £**
Equal tee; 90°; PN25						
16 mm	1.82	2.24	0.20	7.51	nr	**9.75**
20 mm	2.49	3.06	0.20	7.51	nr	**10.57**
25 mm	3.15	3.87	0.20	7.51	nr	**11.38**
32 mm	5.15	6.33	0.20	7.51	nr	**13.84**
40 mm	8.90	10.93	0.20	7.51	nr	**18.44**
50 mm	13.32	16.36	0.20	7.51	nr	**23.87**
63 mm	22.48	27.61	0.20	7.51	nr	**35.12**
Cap; PN25						
20 mm	1.25	1.53	0.20	7.51	nr	**9.04**
25 mm	1.67	2.05	0.20	7.51	nr	**9.56**
32 mm	2.42	2.97	0.20	7.51	nr	**10.48**
40 mm	3.33	4.10	0.20	7.51	nr	**11.61**
50 mm	4.66	5.72	0.20	7.51	nr	**13.23**
63 mm	7.42	9.11	0.20	7.51	nr	**16.62**

38 PIPED SUPPLY SYSTEMS

Item	Net Price £	Material £	Labour hours	Labour £	Unit	Total rate £
COLD WATER PIPELINES: SCREWED STEEL PIPEWORK						
Galvanized steel pipes; screwed and socketed joints; BS 1387: 1985						
Galvanized; medium, fixed vertically, with brackets measured separately, screwed joints are within the running length, but any flanges are additional						
10 mm dia.	4.64	5.70	0.51	19.14	m	**24.84**
15 mm dia.	4.18	5.13	0.52	19.52	m	**24.65**
20 mm dia.	4.73	5.81	0.55	20.64	m	**26.45**
25 mm dia.	6.61	8.12	0.60	22.52	m	**30.64**
32 mm dia.	8.17	10.04	0.67	25.15	m	**35.19**
40 mm dia.	9.50	11.67	0.75	28.14	m	**39.81**
50 mm dia.	13.32	16.36	0.85	31.90	m	**48.26**
65 mm dia.	18.06	22.19	0.93	34.90	m	**57.09**
80 mm dia.	23.39	28.74	1.07	40.16	m	**68.90**
100 mm dia.	33.10	40.67	1.46	54.79	m	**95.46**
125 mm dia.	52.63	64.66	1.72	64.55	m	**129.21**
150 mm dia.	61.11	75.08	1.96	73.56	m	**148.64**
Galvanized; heavy, fixed vertically, with brackets measured separately, screwed joints are within the running length, but any flanges are additional						
15 mm dia.	4.96	6.09	0.52	19.52	m	**25.61**
20 mm dia.	5.62	6.91	0.55	20.64	m	**27.55**
25 mm dia.	8.02	9.85	0.60	22.52	m	**32.37**
32 mm dia.	9.94	12.21	0.67	25.15	m	**37.36**
40 mm dia.	11.60	14.25	0.75	28.14	m	**42.39**
50 mm dia.	16.08	19.75	0.85	31.90	m	**51.65**
65 mm dia.	21.84	26.83	0.93	34.90	m	**61.73**
80 mm dia.	27.74	34.08	1.07	40.16	m	**74.24**
100 mm dia.	38.66	47.49	1.46	54.79	m	**102.28**
125 mm dia.	56.03	68.84	1.72	64.55	m	**133.39**
150 mm dia.	65.53	80.50	1.96	73.56	m	**154.06**
Galvanized; medium, fixed horizontaly or suspended at high level, with brackets measured separately, screwed joints are within the running length, but any flanges are additional						
10 mm dia.	4.64	5.70	0.51	19.14	m	**24.84**
15 mm dia.	4.18	5.13	0.52	19.52	m	**24.65**
20 mm dia.	4.73	5.81	0.55	20.64	m	**26.45**
25 mm dia.	6.61	8.12	0.60	22.52	m	**30.64**
32 mm dia.	8.17	10.04	0.67	25.15	m	**35.19**
40 mm dia.	9.50	11.67	0.75	28.14	m	**39.81**
50 mm dia.	13.32	16.36	0.85	31.90	m	**48.26**
65 mm dia.	18.06	22.19	0.93	34.90	m	**57.09**
80 mm dia.	23.39	28.74	1.07	40.16	m	**68.90**
100 mm dia.	33.10	40.67	1.46	54.79	m	**95.46**

38 PIPED SUPPLY SYSTEMS

Item	Net Price £	Material £	Labour hours	Labour £	Unit	**Total rate £**
125 mm dia.	52.63	64.66	1.72	64.55	m	**129.21**
150 mm dia.	61.11	75.08	1.96	73.56	m	**148.64**
Galvanized; heavy, fixed horizontaly or suspended at high level, with brackets measured separately, screwed joints are within the running length, but any flanges are additional						
15 mm dia.	4.96	6.09	0.52	19.52	m	**25.61**
20 mm dia.	5.62	6.91	0.55	20.64	m	**27.55**
25 mm dia.	8.02	9.85	0.60	22.52	m	**32.37**
32 mm dia.	9.94	12.21	0.67	25.15	m	**37.36**
40 mm dia.	11.60	14.25	0.75	28.14	m	**42.39**
50 mm dia.	16.08	19.75	0.85	31.90	m	**51.65**
65 mm dia.	21.84	26.83	0.93	34.90	m	**61.73**
80 mm dia.	27.74	34.08	1.07	40.16	m	**74.24**
100 mm dia.	38.66	47.49	1.46	54.79	m	**102.28**
125 mm dia.	56.03	68.84	1.72	64.55	m	**133.39**
150 mm dia.	65.53	80.50	1.96	73.56	m	**154.06**
FIXINGS						
For steel pipes; galvanized iron. For minimum fixing dimensions, refer to the Tables and Memoranda at the rear of the book						
Single pipe bracket, screw on, galvanized iron; screwed to wood						
15 mm dia.	1.21	1.49	0.14	5.26	nr	**6.75**
20 mm dia.	1.35	1.66	0.14	5.26	nr	**6.92**
25 mm dia.	1.58	1.94	0.17	6.37	nr	**8.31**
32 mm dia.	2.15	2.65	0.19	7.13	nr	**9.78**
40 mm dia.	3.19	3.92	0.22	8.25	nr	**12.17**
50 mm dia.	4.24	5.21	0.22	8.25	nr	**13.46**
65 mm dia.	5.00	6.14	0.28	10.50	nr	**16.64**
80 mm dia.	7.84	9.63	0.32	12.01	nr	**21.64**
100 mm dia.	11.33	13.92	0.35	13.14	nr	**27.06**
Single pipe bracket, screw on, galvanized iron; plugged and screwed						
15 mm dia.	1.21	1.49	0.25	9.39	nr	**10.88**
20 mm dia.	1.35	1.66	0.25	9.39	nr	**11.05**
25 mm dia.	1.58	1.94	0.30	11.26	nr	**13.20**
32 mm dia.	2.15	2.65	0.32	12.01	nr	**14.66**
40 mm dia.	3.19	3.92	0.32	12.01	nr	**15.93**
50 mm dia.	4.24	5.21	0.32	12.01	nr	**17.22**
65 mm dia.	5.00	6.14	0.35	13.14	nr	**19.28**
80 mm dia.	7.84	9.63	0.42	15.76	nr	**25.39**
100 mm dia.	11.33	13.92	0.42	15.76	nr	**29.68**
Single pipe bracket for building in, galvanized iron						
15 mm dia.	1.28	1.58	0.10	3.76	nr	**5.34**
20 mm dia.	1.44	1.76	0.11	4.13	nr	**5.89**

38 PIPED SUPPLY SYSTEMS

Item	Net Price £	Material £	Labour hours	Labour £	Unit	Total rate £
COLD WATER PIPELINES: SCREWED STEEL PIPEWORK – cont						
FIXINGS – cont						
Single pipe bracket for building in, galvanized iron – cont						
25 mm dia.	1.58	1.94	0.12	4.50	nr	**6.44**
32 mm dia.	1.60	1.96	0.14	5.26	nr	**7.22**
40 mm dia.	2.06	2.53	0.15	5.63	nr	**8.16**
50 mm dia.	2.60	3.19	0.16	6.01	nr	**9.20**
Pipe ring, single socket, galvanized iron						
15 mm dia.	1.28	1.58	0.10	3.76	nr	**5.34**
20 mm dia.	1.44	1.76	0.11	4.13	nr	**5.89**
25 mm dia.	1.58	1.94	0.12	4.50	nr	**6.44**
32 mm dia.	1.60	1.96	0.15	5.63	nr	**7.59**
40 mm dia.	2.06	2.53	0.15	5.63	nr	**8.16**
50 mm dia.	2.60	3.19	0.16	6.01	nr	**9.20**
65 mm dia.	3.78	4.64	0.30	11.26	nr	**15.90**
80 mm dia.	4.54	5.58	0.35	13.14	nr	**18.72**
100 mm dia.	7.15	8.79	0.40	15.02	nr	**23.81**
125 mm dia.	14.45	17.76	0.60	22.53	nr	**40.29**
150 mm dia.	17.46	21.45	0.77	28.90	nr	**50.35**
Pipe ring, double socket, galvanized iron						
15 mm dia.	11.38	13.98	0.10	3.76	nr	**17.74**
20 mm dia.	13.09	16.09	0.11	4.13	nr	**20.22**
25 mm dia.	14.43	17.72	0.12	4.50	nr	**22.22**
32 mm dia.	15.90	19.53	0.14	5.26	nr	**24.79**
40 mm dia.	20.09	24.69	0.15	5.63	nr	**30.32**
50 mm dia.	22.85	28.07	0.16	6.01	nr	**34.08**
Screw on backplate (Male), galvanized iron; plugged and screwed						
All sizes 15 mm to 50 mm × M12	1.01	1.24	0.10	3.76	nr	**5.00**
Screw on backplate (Female), galvanized iron; plugged and screwed						
All sizes 15 mm to 50 mm × M12	1.01	1.24	0.10	3.76	nr	**5.00**
Extra over channel sections for fabricated hangers and brackets						
Galvanized steel; including inserts, bolts, nuts, washers; fixed to backgrounds						
41 × 21 mm	6.02	7.39	0.29	10.89	m	**18.28**
41 × 41 mm	7.22	8.87	0.29	10.89	m	**19.76**
Threaded rods; metric thread; including nuts, washers etc.						
10 mm dia. × 600 mm long	1.97	2.42	0.18	6.76	nr	**9.18**
12 mm dia. × 600 mm long	3.04	3.74	0.18	6.76	nr	**10.50**

38 PIPED SUPPLY SYSTEMS

Item	Net Price £	Material £	Labour hours	Labour £	Unit	**Total rate £**
Extra over steel flanges, screwed and drilled; metric; BS 4504						
Screwed flanges; PN6						
15 mm dia.	15.91	19.54	0.35	13.14	nr	**32.68**
20 mm dia.	15.91	19.54	0.47	17.64	nr	**37.18**
25 mm dia.	15.91	19.54	0.53	19.89	nr	**39.43**
32 mm dia.	15.91	19.54	0.62	23.27	nr	**42.81**
40 mm dia.	15.91	19.54	0.70	26.28	nr	**45.82**
50 mm dia.	16.97	20.84	0.84	31.52	nr	**52.36**
65 mm dia.	23.60	28.99	1.03	38.66	nr	**67.65**
80 mm dia.	33.32	40.93	1.23	46.17	nr	**87.10**
100 mm dia.	39.40	48.40	1.41	52.92	nr	**101.32**
125 mm dia.	72.28	88.80	1.77	66.42	nr	**155.22**
150 mm dia.	72.28	88.80	2.21	82.95	nr	**171.75**
Screwed flanges; PN16						
15 mm dia.	20.23	24.85	0.35	13.14	nr	**37.99**
20 mm dia.	20.23	24.85	0.47	17.64	nr	**42.49**
25 mm dia.	20.23	24.85	0.53	19.89	nr	**44.74**
32 mm dia.	21.61	26.54	0.62	23.27	nr	**49.81**
40 mm dia.	21.61	26.54	0.70	26.28	nr	**52.82**
50 mm dia.	26.36	32.38	0.84	31.52	nr	**63.90**
65 mm dia.	32.93	40.46	1.03	38.66	nr	**79.12**
80 mm dia.	40.10	49.27	1.23	46.17	nr	**95.44**
100 mm dia.	44.97	55.24	1.41	52.92	nr	**108.16**
125 mm dia.	78.71	96.69	1.77	66.42	nr	**163.11**
150 mm dia.	77.45	95.15	2.21	82.95	nr	**178.10**
Extra over steel flanges, screwed and drilled; imperial; BS 10						
Screwed flanges; table E						
½" dia.	27.19	33.40	0.35	13.14	nr	**46.54**
¾" dia.	27.19	33.40	0.47	17.64	nr	**51.04**
1" dia.	27.19	33.40	0.53	19.89	nr	**53.29**
1 ¼" dia.	27.19	33.40	0.62	23.27	nr	**56.67**
1 ½" dia.	27.19	33.40	0.70	26.28	nr	**59.68**
2" dia.	27.19	33.40	0.84	31.52	nr	**64.92**
2 ½" dia.	32.36	39.75	1.03	38.66	nr	**78.41**
3" dia.	38.99	47.90	1.23	46.17	nr	**94.07**
4" dia.	49.60	60.93	1.41	52.92	nr	**113.85**
5" dia.	105.14	129.16	1.77	66.42	nr	**195.58**
Extra over steel flange connections						
Bolted connection between pair of flanges; including gasket, bolts, nuts and washers						
50 mm dia.	46.92	57.65	0.53	19.89	nr	**77.54**
65 mm dia.	59.10	72.61	0.53	19.89	nr	**92.50**
80 mm dia.	67.79	83.28	0.53	19.89	nr	**103.17**
100 mm dia.	80.92	99.41	0.53	19.89	nr	**119.30**
125 mm dia.	152.83	187.75	0.61	22.90	nr	**210.65**
150 mm dia.	157.19	193.11	0.90	33.78	nr	**226.89**

38 PIPED SUPPLY SYSTEMS

Item	Net Price £	Material £	Labour hours	Labour £	Unit	Total rate £
COLD WATER PIPELINES: SCREWED STEEL PIPEWORK – cont						
Extra over heavy steel tubular fittings; BS 1387						
Long screw connection with socket and backnut						
15 mm dia.	6.27	7.71	0.63	23.65	nr	**31.36**
20 mm dia.	7.87	9.67	0.84	31.52	nr	**41.19**
25 mm dia.	10.30	12.65	0.95	35.65	nr	**48.30**
32 mm dia.	13.56	16.66	1.11	41.65	nr	**58.31**
40 mm dia.	16.52	20.30	1.28	48.04	nr	**68.34**
50 mm dia.	24.36	29.93	1.53	57.42	nr	**87.35**
65 mm dia.	55.95	68.73	1.87	70.18	nr	**138.91**
80 mm dia.	72.37	88.90	2.21	82.95	nr	**171.85**
100 mm dia.	82.09	100.85	3.05	114.47	nr	**215.32**
Running nipple						
15 mm dia.	1.58	1.94	0.50	18.77	nr	**20.71**
20 mm dia.	1.96	2.40	0.68	25.51	nr	**27.91**
25 mm dia.	2.10	2.58	0.77	28.90	nr	**31.48**
32 mm dia.	3.38	4.15	0.90	33.78	nr	**37.93**
40 mm dia.	4.57	5.62	1.03	38.66	nr	**44.28**
50 mm dia.	6.96	8.55	1.23	46.17	nr	**54.72**
65 mm dia.	14.96	18.38	1.50	56.30	nr	**74.68**
80 mm dia.	23.33	28.66	1.78	66.80	nr	**95.46**
100 mm dia.	36.56	44.92	2.38	89.32	nr	**134.24**
Barrel nipple						
15 mm dia.	1.31	1.61	0.50	18.77	nr	**20.38**
20 mm dia.	1.98	2.44	0.68	25.51	nr	**27.95**
25 mm dia.	2.22	2.73	0.77	28.90	nr	**31.63**
32 mm dia.	3.68	4.53	0.90	33.78	nr	**38.31**
40 mm dia.	4.11	5.05	1.03	38.66	nr	**43.71**
50 mm dia.	5.88	7.22	1.23	46.17	nr	**53.39**
65 mm dia.	12.54	15.40	1.50	56.30	nr	**71.70**
80 mm dia.	17.52	21.52	1.78	66.80	nr	**88.32**
100 mm dia.	31.66	38.89	2.38	89.32	nr	**128.21**
125 mm dia.	58.83	72.27	2.87	107.72	nr	**179.99**
150 mm dia.	92.67	113.84	3.39	127.23	nr	**241.07**
Close taper nipple						
15 mm dia.	1.86	2.29	0.50	18.77	nr	**21.06**
20 mm dia.	2.41	2.96	0.68	25.51	nr	**28.47**
25 mm dia.	3.16	3.89	0.77	28.90	nr	**32.79**
32 mm dia.	4.73	5.81	0.90	33.78	nr	**39.59**
40 mm dia.	5.85	7.18	1.03	38.66	nr	**45.84**
50 mm dia.	8.99	11.05	1.23	46.17	nr	**57.22**
65 mm dia.	14.20	17.44	1.50	56.30	nr	**73.74**
80 mm dia.	23.27	28.59	1.78	66.80	nr	**95.39**
100 mm dia.	44.25	54.36	2.38	89.32	nr	**143.68**

38 PIPED SUPPLY SYSTEMS

Item	Net Price £	Material £	Labour hours	Labour £	Unit	**Total rate £**
90° bend with socket						
15 mm dia.	5.14	6.31	0.64	24.01	nr	**30.32**
20 mm dia.	6.91	8.48	0.85	31.90	nr	**40.38**
25 mm dia.	10.58	13.00	0.97	36.40	nr	**49.40**
32 mm dia.	15.16	18.63	1.12	42.04	nr	**60.67**
40 mm dia.	18.52	22.75	1.29	48.42	nr	**71.17**
50 mm dia.	28.81	35.40	1.55	58.17	nr	**93.57**
65 mm dia.	58.09	71.37	1.89	70.94	nr	**142.31**
80 mm dia.	86.22	105.92	2.24	84.06	nr	**189.98**
100 mm dia.	152.89	187.82	3.09	115.97	nr	**303.79**
125 mm dia.	374.49	460.06	3.92	147.13	nr	**607.19**
150 mm dia.	562.23	690.70	4.74	177.90	nr	**868.60**
Extra over heavy steel fittings; BS 1740						
Plug						
15 mm dia.	1.43	1.75	0.28	10.50	nr	**12.25**
20 mm dia.	2.21	2.72	0.38	14.26	nr	**16.98**
25 mm dia.	3.88	4.77	0.44	16.51	nr	**21.28**
32 mm dia.	6.03	7.41	0.51	19.14	nr	**26.55**
40 mm dia.	6.66	8.18	0.59	22.13	nr	**30.31**
50 mm dia.	9.52	11.70	0.70	26.28	nr	**37.98**
65 mm dia.	22.78	27.98	0.85	31.90	nr	**59.88**
80 mm dia.	42.62	52.36	1.00	37.53	nr	**89.89**
100 mm dia.	81.84	100.54	1.44	54.04	nr	**154.58**
Socket						
15 mm dia.	1.58	1.94	0.64	24.01	nr	**25.95**
20 mm dia.	1.78	2.18	0.85	31.90	nr	**34.08**
25 mm dia.	2.53	3.11	0.97	36.40	nr	**39.51**
32 mm dia.	3.65	4.48	1.12	42.04	nr	**46.52**
40 mm dia.	4.43	5.44	1.29	48.42	nr	**53.86**
50 mm dia.	6.83	8.39	1.55	58.17	nr	**66.56**
65 mm dia.	13.55	16.64	1.89	70.94	nr	**87.58**
80 mm dia.	17.53	21.54	2.24	84.06	nr	**105.60**
100 mm dia.	32.99	40.53	3.09	115.97	nr	**156.50**
150 mm dia.	78.75	96.75	4.74	177.90	nr	**274.65**
Elbow, female/female						
15 mm dia.	9.06	11.13	0.64	24.01	nr	**35.14**
20 mm dia.	11.80	14.50	0.85	31.90	nr	**46.40**
25 mm dia.	16.03	19.70	0.97	36.40	nr	**56.10**
32 mm dia.	29.84	36.66	1.12	42.04	nr	**78.70**
40 mm dia.	35.58	43.71	1.29	48.42	nr	**92.13**
50 mm dia.	58.33	71.66	1.55	58.17	nr	**129.83**
65 mm dia.	142.51	175.08	1.89	70.94	nr	**246.02**
80 mm dia.	169.93	208.76	2.24	84.06	nr	**292.82**
100 mm dia.	294.27	361.51	3.09	115.97	nr	**477.48**
Equal tee						
15 mm dia.	11.26	13.83	0.91	34.16	nr	**47.99**
20 mm dia.	13.07	16.05	1.22	45.79	nr	**61.84**
25 mm dia.	19.27	23.68	1.40	52.54	nr	**76.22**
32 mm dia.	39.82	48.92	1.62	60.80	nr	**109.72**
40 mm dia.	43.33	53.24	1.86	69.80	nr	**123.04**
50 mm dia.	70.47	86.57	2.21	82.95	nr	**169.52**
65 mm dia.	170.41	209.35	2.72	102.09	nr	**311.44**

38 PIPED SUPPLY SYSTEMS

Item	Net Price £	Material £	Labour hours	Labour £	Unit	Total rate £
COLD WATER PIPELINES: SCREWED STEEL PIPEWORK – cont						
Extra over heavy steel tubular fittings – cont						
Equal tee – cont						
80 mm dia.	182.88	224.67	3.21	120.48	nr	**345.15**
100 mm dia.	294.31	361.56	4.44	166.63	nr	**528.19**
Extra over malleable iron fittings; BS 143						
Cap						
15 mm dia.	3.53	4.34	0.32	12.01	nr	**16.35**
20 mm dia.	3.69	4.54	0.43	16.14	nr	**20.68**
25 mm dia.	6.67	8.19	0.49	18.39	nr	**26.58**
32 mm dia.	12.66	15.55	0.58	21.77	nr	**37.32**
40 mm dia.	13.60	16.71	0.66	24.77	nr	**41.48**
50 mm dia.	21.61	26.54	0.78	29.27	nr	**55.81**
65 mm dia.	37.15	45.64	0.96	36.04	nr	**81.68**
80 mm dia.	61.27	75.27	1.13	42.41	nr	**117.68**
100 mm dia.	104.46	128.33	1.70	63.81	nr	**192.14**
Plain plug, hollow						
15 mm dia.	1.71	2.10	0.28	10.50	nr	**12.60**
20 mm dia.	2.66	3.27	0.38	14.26	nr	**17.53**
25 mm dia.	4.67	5.73	0.44	16.51	nr	**22.24**
32 mm dia.	7.23	8.88	0.51	19.14	nr	**28.02**
40 mm dia.	7.99	9.82	0.59	22.13	nr	**31.95**
50 mm dia.	11.44	14.05	0.70	26.28	nr	**40.33**
65 mm dia.	27.34	33.59	0.85	31.90	nr	**65.49**
80 mm dia.	51.15	62.84	1.00	37.53	nr	**100.37**
100 mm dia.	98.22	120.67	1.44	54.04	nr	**174.71**
Plain plug, solid						
15 mm dia.	2.13	2.61	0.29	10.89	nr	**13.50**
20 mm dia.	2.27	2.79	0.38	14.26	nr	**17.05**
25 mm dia.	3.02	3.71	0.44	16.51	nr	**20.22**
32 mm dia.	4.62	5.68	0.51	19.14	nr	**24.82**
40 mm dia.	6.23	7.65	0.59	22.13	nr	**29.78**
50 mm dia.	8.18	10.05	0.70	26.28	nr	**36.33**
Elbow, male/female						
15 mm dia.	1.12	1.38	0.64	24.01	nr	**25.39**
20 mm dia.	1.48	1.82	0.85	31.90	nr	**33.72**
25 mm dia.	2.47	3.03	0.97	36.40	nr	**39.43**
32 mm dia.	5.62	6.91	1.12	42.04	nr	**48.95**
40 mm dia.	7.75	9.52	1.29	48.42	nr	**57.94**
50 mm dia.	9.99	12.27	1.55	58.17	nr	**70.44**
65 mm dia.	22.27	27.36	1.89	70.94	nr	**98.30**
80 mm dia.	30.44	37.39	2.24	84.06	nr	**121.45**
100 mm dia.	53.22	65.38	3.09	115.97	nr	**181.35**
Elbow						
15 mm dia.	0.99	1.22	0.64	24.01	nr	**25.23**
20 mm dia.	1.37	1.68	0.85	31.90	nr	**33.58**
25 mm dia.	2.11	2.59	0.97	36.40	nr	**38.99**

38 PIPED SUPPLY SYSTEMS

Item	Net Price £	Material £	Labour hours	Labour £	Unit	Total rate £
32 mm dia.	4.39	5.40	1.12	42.04	nr	**47.44**
40 mm dia.	6.58	8.09	1.29	48.42	nr	**56.51**
50 mm dia.	7.71	9.47	1.55	58.17	nr	**67.64**
65 mm dia.	17.20	21.13	1.89	70.94	nr	**92.07**
80 mm dia.	25.25	31.02	2.24	84.06	nr	**115.08**
100 mm dia.	43.36	53.27	3.09	115.97	nr	**169.24**
125 mm dia.	104.03	127.81	4.44	166.63	nr	**294.44**
150 mm dia.	193.68	237.94	5.79	217.30	nr	**455.24**
45° elbow						
15 mm dia.	2.56	3.15	0.64	24.01	nr	**27.16**
20 mm dia.	3.16	3.89	0.85	31.90	nr	**35.79**
25 mm dia.	4.35	5.34	0.97	36.40	nr	**41.74**
32 mm dia.	10.13	12.44	1.12	42.04	nr	**54.48**
40 mm dia.	11.92	14.65	1.29	48.42	nr	**63.07**
50 mm dia.	16.34	20.08	1.55	58.17	nr	**78.25**
65 mm dia.	22.97	28.21	1.89	70.94	nr	**99.15**
80 mm dia.	34.53	42.42	2.24	84.06	nr	**126.48**
100 mm dia.	66.58	81.80	3.09	115.97	nr	**197.77**
150 mm dia.	202.63	248.93	5.79	217.30	nr	**466.23**
Bend, male/female						
15 mm dia.	1.96	2.40	0.64	24.01	nr	**26.41**
20 mm dia.	3.22	3.96	0.85	31.90	nr	**35.86**
25 mm dia.	4.52	5.55	0.97	36.40	nr	**41.95**
32 mm dia.	7.65	9.40	1.12	42.04	nr	**51.44**
40 mm dia.	11.21	13.78	1.29	48.42	nr	**62.20**
50 mm dia.	21.08	25.89	1.55	58.17	nr	**84.06**
65 mm dia.	32.25	39.62	1.89	70.94	nr	**110.56**
80 mm dia.	49.66	61.00	2.24	84.06	nr	**145.06**
100 mm dia.	122.99	151.09	3.09	115.97	nr	**267.06**
Bend, male						
15 mm dia.	4.49	5.51	0.64	24.01	nr	**29.52**
20 mm dia.	5.04	6.19	0.85	31.90	nr	**38.09**
25 mm dia.	7.40	9.09	0.97	36.40	nr	**45.49**
32 mm dia.	16.36	20.10	1.12	42.04	nr	**62.14**
40 mm dia.	22.95	28.19	1.29	48.42	nr	**76.61**
50 mm dia.	34.82	42.78	1.55	58.17	nr	**100.95**
Bend, female						
15 mm dia.	2.01	2.47	0.64	24.01	nr	**26.48**
20 mm dia.	2.87	3.53	0.85	31.90	nr	**35.43**
25 mm dia.	4.03	4.96	0.97	36.40	nr	**41.36**
32 mm dia.	7.85	9.64	1.12	42.04	nr	**51.68**
40 mm dia.	9.35	11.49	1.29	48.42	nr	**59.91**
50 mm dia.	14.72	18.08	1.55	58.17	nr	**76.25**
65 mm dia.	32.25	39.62	1.89	70.94	nr	**110.56**
80 mm dia.	47.82	58.75	2.24	84.06	nr	**142.81**
100 mm dia.	100.34	123.27	3.09	115.97	nr	**239.24**
125 mm dia.	202.44	248.69	4.44	166.63	nr	**415.32**
150 mm dia.	444.77	546.40	5.79	217.30	nr	**763.70**

38 PIPED SUPPLY SYSTEMS

Item	Net Price £	Material £	Labour hours	Labour £	Unit	Total rate £
COLD WATER PIPELINES: SCREWED STEEL PIPEWORK – cont						
Extra over malleable iron fittings – cont						
Return bend						
15 mm dia.	10.27	12.62	0.64	24.01	nr	**36.63**
20 mm dia.	16.60	20.39	0.85	31.90	nr	**52.29**
25 mm dia.	20.71	25.44	0.97	36.40	nr	**61.84**
32 mm dia.	28.76	35.33	1.12	42.04	nr	**77.37**
40 mm dia.	38.96	47.86	1.29	48.42	nr	**96.28**
50 mm dia.	59.47	73.06	1.55	58.17	nr	**131.23**
Equal socket, parallel thread						
15 mm dia.	0.96	1.18	0.64	24.01	nr	**25.19**
20 mm dia.	1.26	1.54	0.85	31.90	nr	**33.44**
25 mm dia.	1.60	1.96	0.97	36.40	nr	**38.36**
32 mm dia.	3.55	4.36	1.12	42.04	nr	**46.40**
40 mm dia.	4.81	5.91	1.29	48.42	nr	**54.33**
50 mm dia.	6.96	8.55	1.55	58.17	nr	**66.72**
65 mm dia.	11.04	13.57	1.89	70.94	nr	**84.51**
80 mm dia.	15.65	19.23	2.24	84.06	nr	**103.29**
100 mm dia.	25.84	31.74	3.09	115.97	nr	**147.71**
Concentric reducing socket						
20 × 15 mm dia.	1.53	1.88	0.76	28.53	nr	**30.41**
25 × 15 mm dia.	2.01	2.47	0.86	32.28	nr	**34.75**
25 × 20 mm dia.	1.90	2.33	0.86	32.28	nr	**34.61**
32 × 25 mm dia.	3.65	4.48	1.01	37.92	nr	**42.40**
40 × 25 mm dia.	4.81	5.91	1.16	43.53	nr	**49.44**
40 × 32 mm dia.	5.33	6.55	1.16	43.53	nr	**50.08**
50 × 25 mm dia.	9.24	11.35	1.38	51.80	nr	**63.15**
50 × 40 mm dia.	7.47	9.18	1.38	51.80	nr	**60.98**
65 × 50 mm dia.	13.04	16.02	1.69	63.43	nr	**79.45**
80 × 50 mm dia.	16.26	19.97	2.00	75.06	nr	**95.03**
100 × 50 mm dia.	32.43	39.84	2.75	103.20	nr	**143.04**
100 × 80 mm dia.	96.98	119.14	2.75	103.20	nr	**222.34**
150 × 100 mm dia.	90.12	110.71	4.10	153.87	nr	**264.58**
Eccentric reducing socket						
20 × 15 mm dia.	3.49	4.28	0.76	28.53	nr	**32.81**
25 × 15 mm dia.	9.90	12.16	0.86	32.28	nr	**44.44**
25 × 20 mm dia.	11.25	13.82	0.86	32.28	nr	**46.10**
32 × 25 mm dia.	14.52	17.84	1.01	37.92	nr	**55.76**
40 × 25 mm dia.	16.65	20.45	1.16	43.53	nr	**63.98**
40 × 32 mm dia.	18.12	22.26	1.16	43.53	nr	**65.79**
50 × 25 mm dia.	18.18	22.33	1.18	44.29	nr	**66.62**
50 × 40 mm dia.	18.23	22.40	1.28	48.04	nr	**70.44**
65 × 50 mm dia.	18.47	22.69	1.69	63.43	nr	**86.12**
80 × 50 mm dia.	19.66	24.15	2.00	75.06	nr	**99.21**
Hexagon bush						
20 × 15 mm dia.	0.87	1.07	0.37	13.88	nr	**14.95**
25 × 15 mm dia.	1.19	1.46	0.43	16.14	nr	**17.60**
25 × 20 mm dia.	1.12	1.38	0.43	16.14	nr	**17.52**

38 PIPED SUPPLY SYSTEMS

Item	Net Price £	Material £	Labour hours	Labour £	Unit	**Total rate £**
32 × 25 mm dia.	1.50	1.85	0.51	19.14	nr	**20.99**
40 × 25 mm dia.	2.24	2.75	0.58	21.77	nr	**24.52**
40 × 32 mm dia.	2.24	2.75	0.58	21.77	nr	**24.52**
50 × 25 mm dia.	4.75	5.84	0.71	26.65	nr	**32.49**
50 × 40 mm dia.	4.43	5.44	0.71	26.65	nr	**32.09**
65 × 50 mm dia.	8.16	10.03	0.84	31.52	nr	**41.55**
80 × 50 mm dia.	12.31	15.12	1.00	37.53	nr	**52.65**
100 × 50 mm dia.	28.52	35.04	1.52	57.05	nr	**92.09**
100 × 80 mm dia.	23.74	29.17	1.52	57.05	nr	**86.22**
150 × 100 mm dia.	85.29	104.78	2.48	93.07	nr	**197.85**
Hexagon nipple						
15 mm dia.	0.94	1.16	0.28	10.50	nr	**11.66**
20 mm dia.	1.06	1.30	0.38	14.26	nr	**15.56**
25 mm dia.	1.48	1.82	0.44	16.51	nr	**18.33**
32 mm dia.	3.17	3.90	0.51	19.14	nr	**23.04**
40 mm dia.	3.65	4.48	0.59	22.13	nr	**26.61**
50 mm dia.	6.65	8.17	0.70	26.28	nr	**34.45**
65 mm dia.	11.13	13.67	0.85	31.90	nr	**45.57**
80 mm dia.	15.39	18.91	1.00	37.53	nr	**56.44**
100 mm dia.	27.31	33.55	1.44	54.04	nr	**87.59**
150 mm dia.	75.60	92.87	2.32	87.08	nr	**179.95**
Union, male/female						
15 mm dia.	4.67	5.73	0.64	24.01	nr	**29.74**
20 mm dia.	5.72	7.02	0.85	31.90	nr	**38.92**
25 mm dia.	6.65	8.17	0.97	36.40	nr	**44.57**
32 mm dia.	11.74	14.43	1.12	42.04	nr	**56.47**
40 mm dia.	15.03	18.47	1.29	48.42	nr	**66.89**
50 mm dia.	23.64	29.04	1.55	58.17	nr	**87.21**
65 mm dia.	52.90	64.98	1.89	70.94	nr	**135.92**
Union, female						
15 mm dia.	4.43	5.44	0.64	24.01	nr	**29.45**
20 mm dia.	4.86	5.97	0.85	31.90	nr	**37.87**
25 mm dia.	5.70	7.00	0.97	36.40	nr	**43.40**
32 mm dia.	9.82	12.06	1.12	42.04	nr	**54.10**
40 mm dia.	11.09	13.63	1.29	48.42	nr	**62.05**
50 mm dia.	16.53	20.31	1.55	58.17	nr	**78.48**
65 mm dia.	42.35	52.03	1.89	70.94	nr	**122.97**
80 mm dia.	55.99	68.79	2.24	84.06	nr	**152.85**
100 mm dia.	106.61	130.97	3.09	115.97	nr	**246.94**
Union elbow, male/female						
15 mm dia.	7.05	8.66	0.64	24.01	nr	**32.67**
20 mm dia.	8.84	10.86	0.85	31.90	nr	**42.76**
25 mm dia.	12.41	15.25	0.97	36.40	nr	**51.65**
Twin elbow						
15 mm dia.	6.78	8.33	0.91	34.16	nr	**42.49**
20 mm dia.	7.49	9.20	1.22	45.79	nr	**54.99**
25 mm dia.	12.15	14.93	1.39	52.16	nr	**67.09**
32 mm dia.	24.32	29.87	1.62	60.80	nr	**90.67**
40 mm dia.	30.79	37.82	1.86	69.80	nr	**107.62**
50 mm dia.	39.58	48.63	2.21	82.95	nr	**131.58**
65 mm dia.	63.96	78.57	2.72	102.09	nr	**180.66**
80 mm dia.	72.59	89.18	3.21	120.48	nr	**209.66**

38 PIPED SUPPLY SYSTEMS

Item	Net Price £	Material £	Labour hours	Labour £	Unit	Total rate £
COLD WATER PIPELINES: SCREWED STEEL PIPEWORK – cont						
Extra over malleable iron fittings – cont						
Equal tee						
15 mm dia.	1.37	1.68	0.91	34.16	nr	**35.84**
20 mm dia.	1.98	2.44	1.22	45.79	nr	**48.23**
25 mm dia.	2.85	3.50	1.39	52.16	nr	**55.66**
32 mm dia.	6.02	7.39	1.62	60.80	nr	**68.19**
40 mm dia.	8.24	10.12	1.86	69.80	nr	**79.92**
50 mm dia.	11.86	14.57	2.21	82.95	nr	**97.52**
65 mm dia.	27.82	34.18	2.72	102.09	nr	**136.27**
80 mm dia.	32.43	39.84	3.21	120.48	nr	**160.32**
100 mm dia.	58.79	72.22	4.44	166.63	nr	**238.85**
125 mm dia.	109.05	133.97	5.38	201.92	nr	**335.89**
150 mm dia.	260.77	320.35	6.31	236.82	nr	**557.17**
Tee reducing on branch						
20 × 15 mm dia.	2.03	2.50	1.22	45.79	nr	**48.29**
25 × 15 mm dia.	2.78	3.41	1.39	52.16	nr	**55.57**
25 × 20 mm dia.	3.16	3.89	1.39	52.16	nr	**56.05**
32 × 25 mm dia.	6.12	7.52	1.62	60.80	nr	**68.32**
40 × 25 mm dia.	7.75	9.52	1.86	69.80	nr	**79.32**
40 × 32 mm dia.	11.38	13.98	1.86	69.80	nr	**83.78**
50 × 25 mm dia.	10.30	12.65	2.21	82.95	nr	**95.60**
50 × 40 mm dia.	16.01	19.67	2.21	82.95	nr	**102.62**
65 × 50 mm dia.	24.70	30.35	2.72	102.09	nr	**132.44**
80 × 50 mm dia.	33.39	41.02	3.21	120.48	nr	**161.50**
100 × 50 mm dia.	55.28	67.91	4.44	166.63	nr	**234.54**
100 × 80 mm dia.	85.29	104.78	4.44	166.63	nr	**271.41**
150 × 100 mm dia.	138.34	169.95	6.28	235.68	nr	**405.63**
Equal pitcher tee						
15 mm dia.	4.73	5.81	0.91	34.16	nr	**39.97**
20 mm dia.	5.82	7.15	1.22	45.79	nr	**52.94**
25 mm dia.	8.72	10.71	1.39	52.16	nr	**62.87**
32 mm dia.	13.51	16.60	1.62	60.80	nr	**77.40**
40 mm dia.	20.90	25.67	1.86	69.80	nr	**95.47**
50 mm dia.	29.34	36.05	2.21	82.95	nr	**119.00**
65 mm dia.	41.75	51.29	2.72	102.09	nr	**153.38**
80 mm dia.	65.15	80.04	3.21	120.48	nr	**200.52**
100 mm dia.	146.57	180.06	4.44	166.63	nr	**346.69**
Cross						
15 mm dia.	3.92	4.82	1.00	37.53	nr	**42.35**
20 mm dia.	6.14	7.54	1.33	49.92	nr	**57.46**
25 mm dia.	7.78	9.56	1.51	56.67	nr	**66.23**
32 mm dia.	11.57	14.22	1.77	66.42	nr	**80.64**
40 mm dia.	15.57	19.13	2.02	75.81	nr	**94.94**
50 mm dia.	24.19	29.72	2.42	90.83	nr	**120.55**
65 mm dia.	34.53	42.42	2.97	111.46	nr	**153.88**
80 mm dia.	45.92	56.42	3.50	131.35	nr	**187.77**
100 mm dia.	94.76	116.41	4.84	181.65	nr	**298.06**

38 PIPED SUPPLY SYSTEMS

Item	Net Price £	Material £	Labour hours	Labour £	Unit	**Total rate £**
COLD WATER PIPELINES: PIPEWORK ANCILLARIES – VALVES						
Regulators						
Gunmetal; self-acting two port thermostat; single seat; screwed; normally closed; with adjustable or fixed bleed device						
25 mm dia.	690.06	847.74	1.46	54.79	nr	**902.53**
32 mm dia.	709.97	872.19	1.45	54.42	nr	**926.61**
40 mm dia.	758.88	932.28	1.55	58.18	nr	**990.46**
50 mm dia.	913.53	1122.27	1.68	63.04	nr	**1185.31**
Self-acting temperature regulator for storage calorifier; integral sensing element and pocket; screwed ends						
15 mm dia.	667.06	819.48	1.32	49.55	nr	**869.03**
25 mm dia.	732.21	899.52	1.52	57.05	nr	**956.57**
32 mm dia.	945.54	1161.59	1.79	67.18	nr	**1228.77**
40 mm dia.	1156.69	1421.00	1.99	74.68	nr	**1495.68**
50 mm dia.	1352.09	1661.05	2.26	84.82	nr	**1745.87**
Self-acting temperature regulator for storage calorifier; integral sensing element and pocket; flanged ends; bolted connection						
15 mm dia.	979.25	1203.01	0.61	22.90	nr	**1225.91**
25 mm dia.	1120.74	1376.83	0.72	27.03	nr	**1403.86**
32 mm dia.	1412.73	1735.53	0.94	35.28	nr	**1770.81**
40 mm dia.	1673.27	2055.62	1.03	38.66	nr	**2094.28**
50 mm dia.	1942.79	2386.71	1.18	44.29	nr	**2431.00**
Chrome plated thermostatic mixing valves including non-return valves and inlet swivel connections with strainers; copper compression fittings						
15 mm dia.	108.09	132.79	0.69	25.89	nr	**158.68**
Chrome plated thermostatic mixing valves including non-return valves and inlet swivel connections with angle pattern combined isolating valves and strainers; copper compression fittings						
15 mm dia.	177.37	217.90	0.69	25.89	nr	**243.79**
Gunmetal thermostatic mixing valves including non-return valves and inlet swivel connections with strainers; copper compression fittings						
15 mm dia.	224.31	275.56	0.69	25.89	nr	**301.45**
Gunmetal thermostatic mixing valves including non-return valves and inlet swivel connections with angle pattern combined isolating valves and strainers; copper compression fittings						
15 mm dia.	235.83	289.72	0.69	25.89	nr	**315.61**

38 PIPED SUPPLY SYSTEMS

Item	Net Price £	Material £	Labour hours	Labour £	Unit	**Total rate £**
COLD WATER PIPELINES: PIPEWORK ANCILLARIES – VALVES – cont						
Ball float valves						
Bronze, equilibrium; copper float; working pressure cold services up to 16 bar; flanged ends; BS 4504 Table 16/21; bolted connections						
25 mm dia.	146.38	179.82	1.04	39.03	nr	**218.85**
32 mm dia.	203.06	249.46	1.22	45.79	nr	**295.25**
40 mm dia.	275.42	338.35	1.38	51.80	nr	**390.15**
50 mm dia.	432.80	531.70	1.66	62.30	nr	**594.00**
65 mm dia.	464.13	570.18	1.93	72.43	nr	**642.61**
80 mm dia.	517.51	635.76	2.16	81.07	nr	**716.83**
Heavy, equilibrium; with long tail and backnut; copper float; screwed for iron						
25 mm dia.	134.83	165.64	1.58	59.30	nr	**224.94**
32 mm dia.	185.11	227.41	1.78	66.80	nr	**294.21**
40 mm dia.	260.97	320.60	1.90	71.32	nr	**391.92**
50 mm dia.	342.70	421.01	2.65	99.46	nr	**520.47**
Brass, ball valve; BS 1212; copper float; screwed						
15 mm dia.	7.29	8.96	0.25	9.39	nr	**18.35**
22 mm dia.	13.24	16.26	0.29	10.89	nr	**27.15**
28 mm dia.	57.71	70.90	0.35	13.14	nr	**84.04**
Gate valves						
DZR copper alloy wedge non-rising stem; capillary joint to copper						
15 mm dia.	15.32	18.82	0.84	31.52	nr	**50.34**
22 mm dia.	18.78	23.07	1.01	37.92	nr	**60.99**
28 mm dia.	25.40	31.21	1.19	44.66	nr	**75.87**
35 mm dia.	45.59	56.01	1.38	51.80	nr	**107.81**
42 mm dia.	77.46	95.16	1.62	60.80	nr	**155.96**
54 mm dia.	108.54	133.34	1.94	72.82	nr	**206.16**
Cocks; capillary joints to copper						
Stopcock; brass head with gunmetal body						
15 mm dia.	6.70	8.23	0.45	16.89	nr	**25.12**
22 mm dia.	12.50	15.36	0.46	17.26	nr	**32.62**
28 mm dia.	35.55	43.67	0.54	20.28	nr	**63.95**
Lockshield stop cocks; brass head with gunmetal body						
15 mm dia.	7.74	9.51	0.45	16.89	nr	**26.40**
22 mm dia.	7.74	9.51	0.46	17.26	nr	**26.77**
28 mm dia.	7.74	9.51	0.54	20.28	nr	**29.79**
DZR stopcock; brass head with gunmetal body						
15 mm dia.	18.35	22.54	0.45	16.89	nr	**39.43**
22 mm dia.	31.78	39.04	0.46	17.26	nr	**56.30**
28 mm dia.	52.96	65.06	0.54	20.28	nr	**85.34**

38 PIPED SUPPLY SYSTEMS

Item	Net Price £	Material £	Labour hours	Labour £	Unit	**Total rate £**
Gunmetal stopcock						
35 mm dia.	69.22	85.04	0.69	25.89	nr	**110.93**
42 mm dia.	91.93	112.94	0.71	26.65	nr	**139.59**
54 mm dia.	137.31	168.68	0.81	30.41	nr	**199.09**
Double union stopcock						
15 mm dia.	19.72	24.22	0.60	22.52	nr	**46.74**
22 mm dia.	27.73	34.06	0.60	22.52	nr	**56.58**
28 mm dia.	49.33	60.61	0.69	25.89	nr	**86.50**
Double union DZR stopcock						
15 mm dia.	32.19	39.54	0.60	22.52	nr	**62.06**
22 mm dia.	39.57	48.61	0.61	22.90	nr	**71.51**
28 mm dia.	73.15	89.87	0.69	25.89	nr	**115.76**
Double union gunmetal stopcock						
35 mm dia.	121.87	149.72	0.63	23.65	nr	**173.37**
42 mm dia.	167.26	205.47	0.67	25.15	nr	**230.62**
54 mm dia.	263.12	323.24	0.85	31.90	nr	**355.14**
Double union stopcock with easy clean cover						
15 mm dia.	34.17	41.98	0.60	22.52	nr	**64.50**
22 mm dia.	42.63	52.37	0.61	22.90	nr	**75.27**
28 mm dia.	79.25	97.36	0.69	25.89	nr	**123.25**
Combined stopcock and drain						
15 mm dia.	33.89	41.63	0.67	25.15	nr	**66.78**
22 mm dia.	41.36	50.81	0.68	25.51	nr	**76.32**
Combined DZR stopcock and drain						
15 mm dia.	44.79	55.02	0.67	25.15	nr	**80.17**
Gate valve						
DZR copper alloy wedge non-rising stem; compression joint to copper						
15 mm dia.	15.32	18.82	0.84	31.52	nr	**50.34**
22 mm dia.	18.78	23.07	1.01	37.92	nr	**60.99**
28 mm dia.	25.40	31.21	1.19	44.66	nr	**75.87**
35 mm dia.	45.59	56.01	1.38	51.80	nr	**107.81**
42 mm dia.	77.46	95.16	1.62	60.80	nr	**155.96**
54 mm dia.	108.54	133.34	1.94	72.82	nr	**206.16**
Cocks; compression joints to copper						
Stopcock; brass head gunmetal body						
15 mm dia.	6.09	7.49	0.42	15.76	nr	**23.25**
22 mm dia.	11.37	13.96	0.42	15.76	nr	**29.72**
28 mm dia.	32.32	39.70	0.45	16.89	nr	**56.59**
Lockshield stopcock; brass head gunmetal body						
15 mm dia.	6.45	7.93	0.42	15.76	nr	**23.69**
22 mm dia.	9.23	11.34	0.42	15.76	nr	**27.10**
28 mm dia.	16.43	20.18	0.45	16.89	nr	**37.07**
DZR stopcock						
15 mm dia.	15.30	18.80	0.38	14.26	nr	**33.06**
22 mm dia.	26.49	32.54	0.39	14.63	nr	**47.17**
28 mm dia.	44.12	54.20	0.40	15.02	nr	**69.22**

38 PIPED SUPPLY SYSTEMS

Item	Net Price £	Material £	Labour hours	Labour £	Unit	Total rate £
COLD WATER PIPELINES: PIPEWORK ANCILLARIES – VALVES – cont						
Cocks – cont						
DZR stopcock – cont						
35 mm dia.	69.22	85.04	0.52	19.52	nr	**104.56**
42 mm dia.	91.93	112.94	0.54	20.28	nr	**133.22**
54 mm dia.	137.31	168.68	0.63	23.65	nr	**192.33**
DZR lockshield stopcock						
15 mm dia.	6.45	7.93	0.38	14.26	nr	**22.19**
22 mm dia.	10.18	12.51	0.39	14.63	nr	**27.14**
Combined stop/draincock						
15 mm dia.	33.89	41.63	0.22	8.25	nr	**49.88**
22 mm dia.	43.61	53.57	0.45	16.89	nr	**70.46**
DZR combined stop/draincock						
15 mm dia.	44.79	55.02	0.41	15.39	nr	**70.41**
22 mm dia.	58.04	71.31	0.42	15.76	nr	**87.07**
Stopcock to polyethylene						
15 mm dia.	23.41	28.76	0.38	14.26	nr	**43.02**
20 mm dia.	29.49	36.22	0.39	14.63	nr	**50.85**
25 mm dia.	39.23	48.19	0.40	15.02	nr	**63.21**
Draw off coupling						
15 mm dia.	14.22	17.47	0.38	14.26	nr	**31.73**
DZR draw off coupling						
15 mm dia.	14.22	17.47	0.38	14.26	nr	**31.73**
22 mm dia.	16.46	20.22	0.39	14.63	nr	**34.85**
Draw off elbow						
15 mm dia.	15.61	19.17	0.38	14.26	nr	**33.43**
22 mm dia.	19.23	23.63	0.39	14.63	nr	**38.26**
Lockshield drain cock						
15 mm dia.	16.32	20.04	0.41	15.39	nr	**35.43**
Check valves						
DZR copper alloy and bronze, WRC approved cartridge double check valve; BS 6282; working pressure cold services up to 10 bar at 65°C; screwed ends						
32 mm dia.	100.92	123.98	1.38	51.80	nr	**175.78**
40 mm dia.	115.01	141.29	1.62	60.80	nr	**202.09**
50 mm dia.	188.59	231.68	1.94	72.82	nr	**304.50**

38 PIPED SUPPLY SYSTEMS

Item	Net Price £	Material £	Labour hours	Labour £	Unit	Total rate £
COLD WATER PIPELINES: PIPEWORK ANCILLARIES – PUMPS						
Packaged cold water pressure booster set; fully automatic; 3 phase supply; includes fixing in position; electrical work elsewhere.						
Pressure booster set						
0.75 l/s @ 30 m head	4210.43	5172.51	9.38	352.04	nr	**5524.55**
1.5 l/s @ 30 m head	4999.88	6142.35	9.38	352.04	nr	**6494.39**
3 l/s @ 30 m head	6108.88	7504.76	10.38	389.57	nr	**7894.33**
6 l/s @ 30 m head	13796.67	16949.21	10.38	389.57	nr	**17338.78**
12 l/s @ 30 m head	17236.43	21174.95	12.38	464.64	nr	**21639.59**
0.75 l/s @ 50 m head	4793.13	5888.36	9.38	352.04	nr	**6240.40**
1.5 l/s @ 50 m head	6108.88	7504.76	9.38	352.04	nr	**7856.80**
3 l/s @ 50 m head	6747.97	8289.88	10.38	389.57	nr	**8679.45**
6 l/s @ 50 m head	15409.41	18930.46	10.38	389.57	nr	**19320.03**
12 l/s @ 50 m head	18928.11	23253.18	12.38	464.64	nr	**23717.82**
0.75 l/s @ 70 m head	5244.23	6442.54	9.38	352.04	nr	**6794.58**
1.5 l/s @ 70 m head	6744.21	8285.27	9.38	352.04	nr	**8637.31**
3 l/s @ 70 m head	7180.28	8820.98	10.38	389.57	nr	**9210.55**
6 l/s @ 70 m head	16822.91	20666.94	10.38	389.57	nr	**21056.51**
12 l/s @ 70 m head	20450.62	25123.59	12.38	464.64	nr	**25588.23**
Automatic sump pump for clear and drainage water; single stage centrifugal pump, presure tight electric motor; single phase supply; includes fixing in position; electrical work elsewhere						
Single pump						
1 l/s @ 2.68 m total head	222.53	273.38	3.50	131.35	nr	**404.73**
1 l/s @ 4.68 m total head	244.94	300.91	3.50	131.35	nr	**432.26**
1 l/s @ 6.68 m total head	323.38	397.27	3.50	131.35	nr	**528.62**
2 l/s @ 4.38 m total head	323.38	397.27	4.00	150.12	nr	**547.39**
2 l/s @ 6.38 m total head	323.38	397.27	4.00	150.12	nr	**547.39**
2 l/s @ 8.38 m total head	416.23	511.34	4.00	150.12	nr	**661.46**
3 l/s @ 3.7 m total head	323.38	397.27	4.50	168.88	nr	**566.15**
3 l/s @ 5.7 m total head	416.23	511.34	4.50	168.88	nr	**680.22**
4 l/s @ 2.9 m total head	323.38	397.27	5.00	187.66	nr	**584.93**
4 l/s @ 4.9 m total head	416.23	511.34	5.00	187.66	nr	**699.00**
4 l/s @ 6.9 m total head	1154.26	1418.00	5.00	187.66	nr	**1605.66**
Extra for high level alarm box with single float switch, local alarm and volt free contacts for remote alarm	338.71	416.10	–	–	nr	**416.10**
Duty/standby pump unit						
1 l/s @ 2.68 m total head	422.63	519.20	5.00	187.66	nr	**706.86**
1 l/s @ 4.68 m total head	464.27	570.36	5.00	187.66	nr	**758.02**
1 l/s @ 6.68 m total head	627.56	770.96	5.00	187.66	nr	**958.62**
2 l/s @ 4.38 m total head	627.56	770.96	5.50	206.43	nr	**977.39**
2 l/s @ 6.38 m total head	627.56	770.96	5.50	206.43	nr	**977.39**
2 l/s @ 8.38 m total head	803.66	987.29	5.50	206.43	nr	**1193.72**
3 l/s @ 3.7 m total head	627.56	770.96	6.00	225.18	nr	**996.14**

38 PIPED SUPPLY SYSTEMS

Item	Net Price £	Material £	Labour hours	Labour £	Unit	Total rate £
COLD WATER PIPELINES: PIPEWORK ANCILLARIES – PUMPS – cont						
Automatic sump pump for clear and drainage water – cont						
Duty/standby pump unit – cont						
3 l/s @ 5.7 m total head	803.66	987.29	6.00	225.18	nr	**1212.47**
4 l/s @ 2.9 m total head	627.56	770.96	6.50	243.95	nr	**1014.91**
4 l/s @ 4.9 m total head	803.66	987.29	6.50	243.95	nr	**1231.24**
4 /s @ 6.9 m total head	2135.61	2623.59	7.00	262.71	nr	**2886.30**
Extra for 4nr float switches to give pump on, off and high level alarm	353.88	434.74	–	–	nr	**434.74**
Extra for dual pump control panel, internal wall mounted IP54, including volt free contacts	1685.17	2070.23	4.00	150.12	nr	**2220.35**

38 PIPED SUPPLY SYSTEMS

Item	Net Price £	Material £	Labour hours	Labour £	Unit	**Total rate £**
COLD WATER PIPELINES: PIPEWORK ANCILLARIES – TANKS						
Cisterns; fibreglass; complete with ball valve, fixing plate and fitted covers						
Rectangular						
60 litre capacity	383.00	470.52	1.33	49.92	nr	**520.44**
100 litre capacity	402.00	493.86	1.40	52.54	nr	**546.40**
150 litre capacity	430.00	528.26	1.61	60.43	nr	**588.69**
250 litre capacity	462.00	567.57	1.61	60.43	nr	**628.00**
420 litre capacity	633.00	777.64	1.99	74.68	nr	**852.32**
730 litre capacity	890.00	1093.36	3.31	124.23	nr	**1217.59**
800 litre capacity	918.00	1127.76	3.60	135.11	nr	**1262.87**
1700 litre capacity	1236.00	1518.43	13.32	499.92	nr	**2018.35**
2250 litre capacity	1497.00	1839.06	20.18	757.38	nr	**2596.44**
3400 litre capacity	1832.00	2250.61	24.50	919.51	nr	**3170.12**
4500 litre capacity	2417.00	2969.28	29.91	1122.54	nr	**4091.82**
Cisterns; polypropylene; complete with ball valve, fixing plate and cover; includes placing in position						
Rectangular						
18 litre capacity	67.94	83.46	1.00	37.53	nr	**120.99**
68 litre capacity	91.55	112.47	1.00	37.53	nr	**150.00**
91 litre capacity	92.15	113.21	1.00	37.53	nr	**150.74**
114 litre capacity	102.52	125.95	1.00	37.53	nr	**163.48**
182 litre capacity	135.39	166.33	1.00	37.53	nr	**203.86**
227 litre capacity	172.45	211.86	1.00	37.53	nr	**249.39**
Circular						
114 litre capacity	95.90	117.81	1.00	37.53	nr	**155.34**
227 litre capacity	97.79	120.13	1.00	37.53	nr	**157.66**
318 litre capacity	162.66	199.83	1.00	37.53	nr	**237.36**
455 litre capacity	186.96	229.68	1.00	37.53	nr	**267.21**
Steel sectional water storage tank; hot pressed steel tank to BS 1564 TYPE 1; 5 mm plate; pre-insulated and complete with all connections and fittings to comply with BSEN 13280; 2001 and WRAS water supply (water fittings) regulations 1999; externally flanged base and sides; cost of erection (on prepared base) is included within the net price, labour cost allows for offloading and positioning materials						
Note: Prices are based on the most economical tank size for each volume, and the cost will vary with differing tank dimensions, for the same volume						

38 PIPED SUPPLY SYSTEMS

Item	Net Price £	Material £	Labour hours	Labour £	Unit	Total rate £
COLD WATER PIPELINES: PIPEWORK ANCILLARIES – TANKS – cont						
Volume, size						
4,900 litres, 3.66 m × 1.22 m × 1.22 m (h)	5555.00	6824.32	6.00	225.18	nr	**7049.50**
20,300 litres, 3.66 m × 2.4 m × 2.4 m (h)	11835.00	14539.30	12.00	450.37	nr	**14989.67**
52,000 litres, 6.1 m × 3.6 m × 2.4 m (h)	21100.00	25921.35	19.00	713.09	nr	**26634.44**
94,000 litres, 7.3 m × 3.6 m × 3.6 m (h)	37150.00	45638.78	28.00	1050.86	nr	**46689.64**
140,000 litres, 9.7 m × 6.1 m × 2.44 m (h)	41450.00	50921.32	28.00	1050.86	nr	**51972.18**
GRP sectional water storage tank; pre-insulated and complete with all connections and fittings to comply with BSEN 13280; 2001 and WRAS water supply (water fittings) regulations 1999; externally flanged base and sides; cost of erection (on prepared base) is included within the net price, labour cost allows for offloading and positioning materials						
Note: Prices are based on the most economical tank size for each volume, and the cost will vary with differing tank dimensions, for the same volume						
Volume, size						
4,500 litres, 3 m × 1 m × 1.5 m (h)	4838.00	5943.48	5.00	187.66	nr	**6131.14**
10,000 litres, 2.5 m × 2 m × 2 m (h)	6299.00	7738.32	7.00	262.71	nr	**8001.03**
20,000 litres, 4 m × 2.5 m × 2 m (h)	8194.00	10066.33	10.00	375.31	nr	**10441.64**
30,000 litres 5 m × 3 m × 2 m (h)	10055.00	12352.57	12.00	450.37	nr	**12802.94**
40,000 litres, 5 m × 4 m × 2 m (h)	11746.00	14429.96	12.00	450.37	nr	**14880.33**
50,000 litres, 5 m × 4 m × 2.5 m (h)	14725.00	18089.66	14.00	525.43	nr	**18615.09**
60,000 litres, 6 m × 4 m × 2.5 m (h)	15624.00	19194.08	16.00	600.49	nr	**19794.57**
70,000 litres, 7 m × 4 m × 2.5 m (h)	17539.00	21546.66	16.00	600.49	nr	**22147.15**
80,000 litres, 8 m × 4 m × 2.5 m (h)	18272.00	22447.15	16.00	600.49	nr	**23047.64**
90,000 litres, 6 m × 5 m × 3 m (h)	19538.00	24002.43	16.00	600.49	nr	**24602.92**
105,000 litres, 7 m × 5 m × 3 m (h)	21857.00	26851.32	24.00	900.73	nr	**27752.05**
120,000 litres, 8 m × 5 m × 3 m (h)	22982.00	28233.39	24.00	900.73	nr	**29134.12**
135,000 litres, 9 m × 6 m × 2.5 m (h)	22550.00	27702.67	24.00	900.73	nr	**28603.40**
144,000 litres, 8 m × 6 m × 3 m (h)	25534.00	31368.52	24.00	900.73	nr	**32269.25**

38 PIPED SUPPLY SYSTEMS

Item	Net Price £	Material £	Labour hours	Labour £	Unit	**Total rate £**
CLEANING AND CHEMICAL TREATMENT						
Electromagnetic water conditioner, complete with control box; maximum inlet pressure 16 bar; electrical work elsewhere						
Connection size, nominal flow rate at 50 mbar						
20 mm dia. 0.3l/s	1645.00	2020.88	1.25	46.91	nr	**2067.79**
25 mm dia. 0.6l/s	2160.00	2653.56	1.45	54.42	nr	**2707.98**
32 mm dia. 1.2l/s	3080.00	3783.78	1.55	58.17	nr	**3841.95**
40 mm dia. 1.7l/s	3680.00	4520.88	1.65	61.93	nr	**4582.81**
50 mm dia. 3.4l/s	4820.00	5921.37	1.75	65.68	nr	**5987.05**
65 mm dia. 5.2l/s	5275.00	6480.34	1.90	71.32	nr	**6551.66**
100 mm dia. 30.5l/s	16420.00	20171.97	3.00	112.59	nr	**20284.56**
Ultraviolet water sterillising unit, complete with control unit; UV lamp housed in quartz tube; unit complete with UV intensity sensor, flushing and discharge valve and facilities for remote alarm; electrical work elsewhere						
Maximum flow rate (@ 250J/m^2 exposure), connection size						
1.28l/s, 40 mm dia.	3065.00	3765.35	1.98	74.31	nr	**3839.66**
2.00l/s, 40 mm dia.	3875.00	4760.44	1.98	74.31	nr	**4834.75**
1.28l/s, 50 mm dia.	4110.81	5050.13	1.98	74.31	nr	**5124.44**
7.4l/s, 80 mm dia.	10070.00	12371.00	3.60	135.11	nr	**12506.11**
16.8l/s 100 mm dia.	12510.00	15368.54	3.60	135.11	nr	**15503.65**
32.3l/s 100 mm dia.	14070.00	17285.00	3.60	135.11	nr	**17420.11**
Base exchange water softener complete with resin tank, brine tank and consumption data monitoring facilities						
Capacities of softeners are based on 300 ppm hardness and quoted in m^3 of softened water produced. Design flow rates are recommended for continuous use						
Simplex configuration						
Design flow rate, min–max softenend water produced						
0.97l/s, 7.7 m^3	2975.00	3654.79	10.00	375.31	nr	**4030.10**
1.25l/s, 11.7 m^3	3050.00	3746.93	12.00	450.37	nr	**4197.30**
1.8l/s, 19.8 m^3	4545.00	5583.53	12.00	450.37	nr	**6033.90**
2.23l/s, 23.4 m^3	5100.00	6265.35	12.00	450.37	nr	**6715.72**
2.92l/s, 31.6 m^3	5275.00	6480.34	15.00	562.96	nr	**7043.30**
3.75l/s, 39.8 m^3	6300.00	7739.55	15.00	562.96	nr	**8302.51**
4.72l/s, 60.3 m^3	7155.00	8789.92	18.00	675.55	nr	**9465.47**
6.66l/s, 96.6 m^3	8450.00	10380.82	20.00	750.61	nr	**11131.43**

38 PIPED SUPPLY SYSTEMS

Item	Net Price £	Material £	Labour hours	Labour £	Unit	**Total rate £**
CLEANING AND CHEMICAL TREATMENT – cont						
Capacities of softeners are based on 300 ppm hardness and quoted in m³ of softened water produced. – cont						
Duplex configuration						
Design flow rate, min–max softenend water produced						
0.97l/s, 7.7 m³	4390.00	5393.11	12.00	450.37	nr	**5843.48**
1.25l/s, 11.7 m³	4570.00	5614.25	12.00	450.37	nr	**6064.62**
1.81l/s, 19.8 m³	7665.00	9416.45	15.00	562.96	nr	**9979.41**
2.23l/s, 23.4 m³	8785.00	10792.37	15.00	562.96	nr	**11355.33**
2.29l/s, 31.6 m³	9055.00	11124.07	18.00	675.55	nr	**11799.62**
3.75l/s, 39.8 m³	11260.00	13832.91	18.00	675.55	nr	**14508.46**
4.72l/s, 60.3 m³	12850.00	15786.23	18.00	675.55	nr	**16461.78**
6.66/s, 96.6 m³	15890.00	19520.87	23.00	863.20	nr	**20384.07**
Triplex configuration						
Design flow rate, min–max softenend water produced						
3l/s, 5.8 m³–33.6 m³	4224.14	5189.35	15.00	562.96	nr	**5752.31**
3.9l/s, 11.7 m³–64.2 m³	5986.62	7354.57	15.00	562.96	nr	**7917.53**
3.9l/s, 15.5 m³–85.5 m³	6321.78	7766.30	18.00	675.55	nr	**8441.85**
4.8l/s, 23.3 m³-128.1 m³	8228.71	10108.97	18.00	675.55	nr	**10784.52**
4.8l/s, 38.8 m³-213.6 m³	9453.77	11613.96	22.00	825.68	nr	**12439.64**
5.7l/s, 11.7 m³-64.2 m³	10586.37	13005.35	22.00	825.68	nr	**13831.03**
9.6l/s, 19.4 m³-106.8 m³	11476.28	14098.61	22.00	825.68	nr	**14924.29**
13.2l/s, 31.1 m³-171.0 m³	13163.63	16171.52	27.00	1013.32	nr	**17184.84**
15.3l/s, 46.6 m³-256.2 m³	21172.75	26010.73	27.00	1013.32	nr	**27024.05**
15.3l/s, 77.7 m³-427.2 m³	26454.38	32499.20	32.00	1200.98	nr	**33700.18**
THERMAL INSULATION						
Flexible closed cell walled insulation; Class 1/Class O; adhesive joints; including around fittings						
6 mm wall thickness						
15 mm dia.	1.29	1.59	0.15	5.63	m	**7.22**
22 mm dia.	1.53	1.88	0.15	5.63	m	**7.51**
28 mm dia.	1.94	2.38	0.15	5.63	m	**8.01**
9 mm wall thickness						
15 mm dia.	1.48	1.82	0.15	5.63	m	**7.45**
22 mm dia.	1.87	2.30	0.15	5.63	m	**7.93**
28 mm dia.	1.99	2.45	0.15	5.63	m	**8.08**
35 mm dia.	2.25	2.76	0.15	5.63	m	**8.39**
42 mm dia.	2.57	3.16	0.15	5.63	m	**8.79**
54 mm dia.	2.67	3.28	0.15	5.63	m	**8.91**
13 mm wall thickness						
15 mm dia.	1.90	2.33	0.15	5.63	m	**7.96**
22 mm dia.	2.33	2.87	0.15	5.63	m	**8.50**
28 mm dia.	2.78	3.41	0.15	5.63	m	**9.04**

38 PIPED SUPPLY SYSTEMS

Item	Net Price £	Material £	Labour hours	Labour £	Unit	**Total rate £**
35 mm dia.	2.99	3.68	0.15	5.63	m	**9.31**
42 mm dia.	3.52	4.33	0.15	5.63	m	**9.96**
54 mm dia.	4.90	6.02	0.15	5.63	m	**11.65**
67 mm dia.	7.24	8.89	0.15	5.63	m	**14.52**
76 mm dia.	8.52	10.47	0.15	5.63	m	**16.10**
108 mm dia.	8.93	10.97	0.15	5.63	m	**16.60**
19 mm wall thickness						
15 mm dia.	3.09	3.80	0.15	5.63	m	**9.43**
22 mm dia.	3.76	4.62	0.15	5.63	m	**10.25**
28 mm dia.	5.10	6.27	0.15	5.63	m	**11.90**
35 mm dia.	5.91	7.26	0.15	5.63	m	**12.89**
42 mm dia.	6.98	8.58	0.15	5.63	m	**14.21**
54 mm dia.	8.82	10.84	0.15	5.63	m	**16.47**
67 mm dia.	10.58	13.00	0.15	5.63	m	**18.63**
76 mm dia.	12.22	15.02	0.22	8.25	m	**23.27**
108 mm dia.	16.67	20.48	0.22	8.25	m	**28.73**
25 mm wall thickness						
15 mm dia.	6.06	7.44	0.15	5.63	m	**13.07**
22 mm dia.	6.63	8.15	0.15	5.63	m	**13.78**
28 mm dia.	7.51	9.23	0.15	5.63	m	**14.86**
35 mm dia.	8.32	10.22	0.15	5.63	m	**15.85**
42 mm dia.	8.87	10.90	0.15	5.63	m	**16.53**
54 mm dia.	10.43	12.81	0.15	5.63	m	**18.44**
67 mm dia.	14.05	17.26	0.15	5.63	m	**22.89**
76 mm dia.	15.84	19.46	0.22	8.25	m	**27.71**
32 mm wall thickness						
15 mm dia.	7.59	9.32	0.15	5.63	m	**14.95**
22 mm dia.	8.32	10.22	0.15	5.63	m	**15.85**
28 mm dia.	9.65	11.85	0.15	5.63	m	**17.48**
35 mm dia.	9.97	12.24	0.15	5.63	m	**17.87**
42 mm dia.	11.59	14.24	0.15	5.63	m	**19.87**
54 mm dia.	14.59	17.92	0.15	5.63	m	**23.55**
76 mm dia.	21.82	26.81	0.22	8.25	m	**35.06**
Note: For mineral fibre sectional insulation; bright class O foil faced; bright class O foil taped joints; 19 mm aluminium bands rates, refer to section – Thermal Insulation						
Note: For mineral fibre sectional insulation; bright class O foil faced; bright class O foil taped joints; 22 swg plain/embossed aluminium cladding; pop riveted rates, refer to section – Thermal Insulation						
Note: For mineral fibre sectional insulation; bright class O foil faced; bright class O foil taped joints; 0.8 mm polyisobutylene sheeting; welded joints rates, refer to section – Thermal Insulation						

38 PIPED SUPPLY SYSTEMS

Item	Net Price £	Material £	Labour hours	Labour £	Unit	Total rate £
HOT WATER						
PIPELINES						
Note: For pipework prices refer to section – Cold Water						
PIPELINE ANCILLARIES						
Note: For prices for ancillaries refer to section – Cold Water						
STORAGE CYLINDERS/CALORIFIERS						
Insulated copper storage cylinders; BS 699; includes placing in position						
Grade 3 (maximum 10 m working head)						
BS size 6; 115 litre capacity; 400 mm dia.; 1050 mm high	194.22	238.60	1.50	56.35	nr	**294.95**
BS size 7; 120 litre capacity; 450 mm dia.; 900 mm high	229.62	282.09	2.00	75.06	nr	**357.15**
BS size 8; 144 litre capacity; 450 mm dia.; 1050 mm high	244.10	299.88	2.80	105.13	nr	**405.01**
Grade 4 (maximum 6 m working head)						
BS size 2; 96 litre capacity; 400 mm dia.; 900 mm high	150.38	184.74	1.50	56.35	nr	**241.09**
BS size 7; 120 litre capacity; 450 mm dia.; 900 mm high	158.03	194.15	1.50	56.35	nr	**250.50**
BS size 8; 144 litre capacity; 450 mm dia.; 1050 mm high	188.65	231.76	1.50	56.35	nr	**288.11**
BS size 9; 166 litre capacity; 450 mm dia.; 1200 mm high	275.84	338.87	1.50	56.35	nr	**395.22**
Storage cylinders; brazed copper construction; to BS 699; screwed bosses; includes placing in position						
Tested to 2.2 bar, 15 m maximum head						
144 litres	854.89	1050.23	3.00	112.71	nr	**1162.94**
160 litres	966.42	1187.25	3.00	112.71	nr	**1299.96**
200 litres	996.13	1223.74	3.76	141.09	nr	**1364.83**
255 litres	1133.65	1392.69	3.76	141.09	nr	**1533.78**
290 litres	1531.36	1881.27	3.76	141.09	nr	**2022.36**
370 litres	1754.40	2155.28	4.50	169.05	nr	**2324.33**
450 litres	2389.85	2935.93	5.00	187.66	nr	**3123.59**
Tested to 2.55 bar, 17 m maximum head						
550 litres	2587.73	3179.02	5.00	187.66	nr	**3366.68**
700 litres	3025.31	3716.59	6.02	226.10	nr	**3942.69**
800 litres	3499.14	4298.69	6.54	245.29	nr	**4543.98**
900 litres	3786.43	4651.63	8.00	300.25	nr	**4951.88**
1000 litres	3995.74	4908.77	8.00	300.25	nr	**5209.02**
1250 litres	4376.25	5376.22	13.16	493.82	nr	**5870.04**
1500 litres	6703.03	8234.68	15.15	568.65	nr	**8803.33**

38 PIPED SUPPLY SYSTEMS

Item	Net Price £	Material £	Labour hours	Labour £	Unit	Total rate £
2000 litres	8044.23	9882.34	17.24	647.07	nr	**10529.41**
3000 litres	11299.38	13881.28	24.39	915.38	nr	**14796.66**
Indirect cylinders; copper; bolted top; up to 5 tappings for connections; BS 1586; includes placing in position						
Grade 3, tested to 1.45 bar, 10 m maximum head						
74 litre capacity	391.97	481.53	1.50	56.35	nr	**537.88**
96 litre capacity	399.05	490.23	1.50	56.35	nr	**546.58**
114 litre capacity	409.74	503.37	1.50	56.35	nr	**559.72**
117 litre capacity	425.29	522.47	2.00	75.06	nr	**597.53**
140 litre capacity	438.23	538.37	2.50	93.83	nr	**632.20**
162 litre capacity	612.84	752.87	3.00	112.71	nr	**865.58**
190 litre capacity	669.89	822.96	3.51	131.68	nr	**954.64**
245 litre capacity	783.89	963.01	3.80	142.71	nr	**1105.72**
280 litre capacity	1389.58	1707.10	4.00	150.12	nr	**1857.22**
360 litre capacity	1503.62	1847.20	4.50	169.05	nr	**2016.25**
440 litre capacity	1745.91	2144.85	4.50	169.05	nr	**2313.90**
Grade 2, tested to 2.2 bar, 15 m maximum head						
117 litre capacity	566.51	695.96	2.00	75.06	nr	**771.02**
140 litre capacity	616.43	757.28	2.50	93.83	nr	**851.11**
162 litre capacity	705.48	866.68	2.80	105.13	nr	**971.81**
190 litre capacity	819.55	1006.81	3.00	112.71	nr	**1119.52**
245 litre capacity	990.56	1216.91	4.00	150.12	nr	**1367.03**
280 litre capacity	1582.00	1943.49	4.00	150.12	nr	**2093.61**
360 litre capacity	1745.91	2144.85	4.50	169.05	nr	**2313.90**
440 litre capacity	2066.56	2538.77	4.50	169.05	nr	**2707.82**
Grade 1, tested 3.65 bar, 25 m maximum head						
190 litre capacity	1218.61	1497.06	3.00	112.71	nr	**1609.77**
245 litre capacity	1386.00	1702.70	3.00	112.71	nr	**1815.41**
280 litre capacity	1959.71	2407.50	4.00	150.12	nr	**2557.62**
360 litre capacity	2479.95	3046.62	4.50	169.05	nr	**3215.67**
440 litre capacity	3010.78	3698.74	4.50	169.05	nr	**3867.79**
Indirect cylinders, including manhole; BS 853						
Grade 3, tested to 1.5 bar, 10 m maximum head						
550 litre capacity	2534.35	3113.45	5.21	195.47	nr	**3308.92**
700 litre capacity	2805.91	3447.06	6.02	226.10	nr	**3673.16**
800 litre capacity	3258.47	4003.03	6.54	245.29	nr	**4248.32**
1000 litre capacity	4073.12	5003.83	7.04	264.30	nr	**5268.13**
1500 litre capacity	4706.70	5782.18	10.00	375.31	nr	**6157.49**
2000 litre capacity	6516.99	8006.12	16.13	605.34	nr	**8611.46**
Grade 2, tested to 2.55 bar, 15 m maximum head						
550 litre capacity	2809.81	3451.85	5.21	195.47	nr	**3647.32**
700 litre capacity	3516.67	4320.22	6.02	226.10	nr	**4546.32**
800 litre capacity	3711.05	4559.03	6.54	245.29	nr	**4804.32**

38 PIPED SUPPLY SYSTEMS

Item	Net Price £	Material £	Labour hours	Labour £	Unit	Total rate £
HOT WATER – cont						
STORAGE CYLINDERS/CALORIFIERS – cont						
Indirect cylinders, including manhole – cont						
Grade 2, tested to 2.55 bar, 15 m maximum head – cont						
1000 litre capacity	4594.64	5644.52	7.04	264.30	nr	**5908.82**
1500 litre capacity	5654.94	6947.09	10.00	375.31	nr	**7322.40**
2000 litre capacity	7068.71	8683.91	16.13	605.34	nr	**9289.25**
Grade 1, tested to 4 bar, 25 m maximum head						
550 litre capacity	3269.26	4016.28	5.21	195.47	nr	**4211.75**
700 litre capacity	3711.05	4559.03	6.02	226.10	nr	**4785.13**
800 litre capacity	3976.15	4884.71	6.54	245.29	nr	**5130.00**
1000 litre capacity	5301.54	6512.94	7.04	264.30	nr	**6777.24**
1500 litre capacity	6361.79	7815.45	10.00	375.31	nr	**8190.76**
2000 litre capacity	7775.54	9552.25	16.13	605.34	nr	**10157.59**
Storage calorifiers; copper; heater battery capable of raising temperature of contents from 10°C to 65°C in one hour; static head not exceeding 1.35 bar; BS 853; includes fixing in position on cradles or legs.						
Horizontal; primary LPHW at 82°C/71°C						
400 litre capacity	3508.75	4310.50	7.04	264.30	nr	**4574.80**
1000 litre capacity	5613.96	6896.75	8.00	300.25	nr	**7197.00**
2000 litre capacity	11228.00	13793.60	14.08	528.60	nr	**14322.20**
3000 litre capacity	13859.54	17026.44	25.00	938.27	nr	**17964.71**
4000 litre capacity	16841.96	20690.34	40.00	1501.23	nr	**22191.57**
4500 litre capacity	18980.15	23317.12	50.00	1876.54	nr	**25193.66**
Vertical; primary LPHW at 82°C/71°C						
400 litre capacity	3954.37	4857.94	7.04	264.30	nr	**5122.24**
1000 litre capacity	6355.26	7807.43	8.00	300.25	nr	**8107.68**
2000 litre capacity	12105.18	14871.21	14.08	528.60	nr	**15399.81**
3000 litre capacity	15131.49	18589.03	25.00	938.27	nr	**19527.30**
4000 litre capacity	18561.22	22802.46	40.00	1501.23	nr	**24303.69**
4500 litre capacity	20982.29	25776.74	50.00	1876.54	nr	**27653.28**
Storage calorifiers; galvanized mild steel; heater battery capable of raising temperature of contents from 10°C to 65°C in one hour; static head not exceeding 1.35 bar; BS 853; includes fixing in position on cradles or legs						
Horizontal; primary LPHW at 82°C/71°C						
400 litre capacity	3508.75	4310.50	7.04	264.30	nr	**4574.80**
1000 litre capacity	5613.96	6896.75	8.00	300.25	nr	**7197.00**
2000 litre capacity	11228.00	13793.60	14.08	528.60	nr	**14322.20**

38 PIPED SUPPLY SYSTEMS

Item	Net Price £	Material £	Labour hours	Labour £	Unit	Total rate £
3000 litre capacity	13859.54	17026.44	25.00	938.27	nr	**17964.71**
4000 litre capacity	16841.96	20690.34	40.00	1501.23	nr	**22191.57**
4500 litre capacity	18980.15	23317.12	50.00	1876.54	nr	**25193.66**
Vertical; primary LPHW at 82°C/71°C						
400 litre capacity	3954.37	4857.94	7.04	264.30	nr	**5122.24**
1000 litre capacity	6355.26	7807.43	8.00	300.25	nr	**8107.68**
2000 litre capacity	12105.18	14871.21	14.08	528.60	nr	**15399.81**
3000 litre capacity	15131.49	18589.03	25.00	938.27	nr	**19527.30**
4000 litre capacity	18561.22	22802.46	40.00	1501.23	nr	**24303.69**
4500 litre capacity	20982.29	25776.74	50.00	1876.54	nr	**27653.28**
Indirect cylinders; mild steel, welded throughout, galvanized; with bolted connections; includes placing in position						
3.2 mm plate						
136 litre capacity	1400.80	1720.89	2.50	93.83	nr	**1814.72**
159 litre capacity	1508.58	1853.29	2.80	105.13	nr	**1958.42**
182 litre capacity	1853.43	2276.94	3.00	112.71	nr	**2389.65**
227 litre capacity	2284.41	2806.40	3.00	112.66	nr	**2919.06**
273 litre capacity	2664.76	3273.66	3.00	112.71	nr	**3386.37**
364 litre capacity	3103.35	3812.47	4.00	150.12	nr	**3962.59**
455 litre capacity	3205.37	3937.79	5.00	187.66	nr	**4125.45**
683 litre capacity	5187.27	6372.57	6.02	226.10	nr	**6598.67**
910 litre capacity	6024.61	7401.23	7.04	264.30	nr	**7665.53**
Local Electric Hot Water Heaters						
Unvented multipoint water heater; providing hot water for one or more outlets: Used with conventional taps or mixers: Factory fitted temperature and pressure relief valve: Externally adjustable thermostat: Elemental 'on' indicator: Fitted with 1 metre of 3 core cable: Electrical supply and connection excluded						
5 litre capacity, 2.2 kW rating	197.89	243.11	1.50	56.30	nr	**299.41**
10 litre capacity, 2.2 kW rating	220.70	271.13	1.50	56.30	nr	**327.43**
15 litre capacity, 2.2 kW rating	245.24	301.28	1.50	56.30	nr	**357.58**
30 litre capacity, 3 kW rating	437.09	536.97	2.00	75.06	nr	**612.03**
50 litre capacity, 3 kW rating	460.17	565.32	2.00	75.06	nr	**640.38**
80 litre capacity, 3 kW rating	905.49	1112.39	2.00	75.06	nr	**1187.45**
100 litre capacity, 3 kW rating	972.08	1194.20	2.00	75.06	nr	**1269.26**
Accessories						
Pressure reducing valve and expansion kit	161.13	197.95	2.00	75.06	nr	**273.01**
Thermostatic blending valve	85.75	105.35	1.00	37.53	nr	**142.88**

38 PIPED SUPPLY SYSTEMS

Item	Net Price £	Material £	Labour hours	Labour £	Unit	**Total rate £**
NATURAL GAS: PIPELINES: MEDIUM DENSITY POLYETHYLENE – YELLOW						
Pipe; laid underground; electrofusion joints in the running length; BS 6572; BGT PL2 standards						
Coiled service pipe						
20 mm dia.	1.58	1.94	0.37	13.89	m	**15.83**
25 mm dia.	2.06	2.53	0.41	15.39	m	**17.92**
32 mm dia.	3.42	4.20	0.47	17.64	m	**21.84**
63 mm dia.	13.00	15.97	0.60	22.52	m	**38.49**
90 mm dia.	17.06	20.96	0.90	33.78	m	**54.74**
Mains service pipe						
63 mm dia.	12.62	15.51	0.60	22.52	m	**38.03**
90 mm dia.	16.57	20.36	0.90	33.78	m	**54.14**
125 mm dia.	31.98	39.29	1.20	45.03	m	**84.32**
180 mm dia.	66.04	81.13	1.50	56.30	m	**137.43**
250 mm dia.	121.55	149.32	1.75	65.68	m	**215.00**
Extra over fittings, electrofusion joints						
Straight connector						
32 mm dia.	9.89	12.15	0.47	17.64	nr	**29.79**
63 mm dia.	18.58	22.83	0.58	21.77	nr	**44.60**
90 mm dia.	27.41	33.67	0.67	25.15	nr	**58.82**
125 mm dia.	51.32	63.04	0.83	31.15	nr	**94.19**
180 mm dia.	92.43	113.55	1.25	46.91	nr	**160.46**
Reducing connector						
90 × 63 mm dia.	38.27	47.02	0.67	25.15	nr	**72.17**
125 × 90 mm dia.	76.75	94.29	0.83	31.15	nr	**125.44**
180 × 125 mm dia.	140.82	173.00	1.25	46.91	nr	**219.91**
Bend; 45°						
90 mm dia.	73.97	90.87	0.67	25.15	nr	**116.02**
125 mm dia.	121.00	148.65	0.83	31.15	nr	**179.80**
180 mm dia.	274.49	337.21	1.25	46.91	nr	**384.12**
Bend; 90°						
63 mm dia.	47.92	58.87	0.58	21.77	nr	**80.64**
90 mm dia.	73.97	90.87	0.67	25.15	nr	**116.02**
125 mm dia.	121.00	148.65	0.83	31.15	nr	**179.80**
180 mm dia.	274.49	337.21	1.25	46.91	nr	**384.12**
Extra over malleable iron fittings, compression joints						
Straight connector						
20 mm dia.	14.12	17.35	0.38	14.26	nr	**31.61**
25 mm dia.	15.38	18.89	0.45	16.89	nr	**35.78**
32 mm dia.	17.25	21.19	0.50	18.77	nr	**39.96**
63 mm dia.	34.65	42.57	0.85	31.92	nr	**74.49**
Straight connector; polyethylene to MI						
20 mm dia.	11.99	14.73	0.31	11.63	nr	**26.36**
25 mm dia.	13.08	16.07	0.35	13.14	nr	**29.21**
32 mm dia.	14.61	17.94	0.40	15.02	nr	**32.96**
63 mm dia.	24.47	30.06	0.65	24.40	nr	**54.46**

38 PIPED SUPPLY SYSTEMS

Item	Net Price £	Material £	Labour hours	Labour £	Unit	Total rate £
Straight connector; polyethylene to FI						
20 mm dia.	11.99	14.73	0.31	11.63	nr	**26.36**
25 mm dia.	14.61	17.94	0.35	13.14	nr	**31.08**
32 mm dia.	14.61	17.94	0.40	15.02	nr	**32.96**
63 mm dia.	24.47	30.06	0.75	28.16	nr	**58.22**
Elbow						
20 mm dia.	18.35	22.54	0.38	14.26	nr	**36.80**
25 mm dia.	20.02	24.59	0.45	16.89	nr	**41.48**
32 mm dia.	22.43	27.55	0.50	18.77	nr	**46.32**
63 mm dia.	45.06	55.36	0.80	30.02	nr	**85.38**
Equal tee						
20 mm dia.	24.47	30.06	0.53	19.90	nr	**49.96**
25 mm dia.	28.61	35.14	0.55	20.65	nr	**55.79**
32 mm dia.	36.03	44.27	0.64	24.03	nr	**68.30**
GAS BOOSTERS						
Complete skid mounted gas booster set, including AV mounts, flexible connections, low pressure switch,control panel and NRV (for run/standby unit); 3 phase supply; in accordance with IGE/UP/2; includes delivery, offloading and positioning						
Single unit						
Flow, pressure range						
0–200 m³/hour, 0.1–2.6 kPa	3688.64	4531.50	10.00	375.31	nr	**4906.81**
0–200 m³/hour, 0.1–4.0 kPa	4208.83	5170.55	10.00	375.31	nr	**5545.86**
0–200 m³/hour, 0.1—7 kPa	4897.88	6017.05	10.00	375.31	nr	**6392.36**
0–200 m³/hour, 0.1–9.5 kPa	5100.56	6266.04	10.00	375.31	nr	**6641.35**
0–200 m³/hour 0.1–11.0 kPa	5698.46	7000.56	10.00	375.31	nr	**7375.87**
0–400 m³/hour, 0.1–4.0 kPa	4647.93	5709.98	10.00	375.31	nr	**6085.29**
0–1000 m³/hour, 0.1–7.4 kPa	6438.18	7909.30	10.00	375.31	nr	**8284.61**
50–1000 m³/hour, 0.1–16.0 kPa	12028.56	14777.09	20.00	750.61	nr	**15527.70**
50–1000 m³/hour, 0.1–24.5 kPa	13599.23	16706.65	20.00	750.61	nr	**17457.26**
50–1000 m³/hour, 0.1–31.0 kPa	15467.21	19001.47	20.00	750.61	nr	**19752.08**
50–1000 m³/hour, 0.1–41.0 kPa	17071.71	20972.59	20.00	750.61	nr	**21723.20**
50–1000 m³/hour, 0.1–51.0 kPa	17713.44	21760.96	20.00	750.61	nr	**22511.57**
100–1800 m³/hour, 3.5–23.5 kPa	18176.27	22329.55	20.00	750.61	nr	**23080.16**
100–1800 m³/hour, 4.5–27.0 kPa	19605.09	24084.86	20.00	750.61	nr	**24835.47**
100–1800 m³/hour, 6.0–32.5 kPa	21712.87	26674.26	20.00	750.61	nr	**27424.87**
100–1800 m³/hour, 7.2–39.0 kPa	24979.25	30687.01	20.00	750.61	nr	**31437.62**
100–1800 m³/hour, 9.0–42.0 kPa	26256.09	32255.61	20.00	750.61	nr	**33006.22**

38 PIPED SUPPLY SYSTEMS

Item	Net Price £	Material £	Labour hours	Labour £	Unit	Total rate £
NATURAL GAS: PIPELINES: MEDIUM DENSITY POLYETHYLENE – YELLOW – cont						
GAS BOOSTERS – cont						
Run/Standby unit						
Flow, pressure range						
0–200 m³/hour, 0.1–2.6 kPa	18972.08	23307.20	16.00	600.49	nr	**23907.69**
0–200 m³/hour, 0.1—4.0 kPa	19549.90	24017.05	16.00	600.49	nr	**24617.54**
0–200 m³/hour, 0.1–7 kPa	20002.69	24573.31	16.00	600.49	nr	**25173.80**
0–200 m³/hour, 0.1–9.5 kPa	20498.34	25182.21	16.00	600.49	nr	**25782.70**
0–200 m³/hour 0.1—1.0 kPa	20837.94	25599.41	16.00	600.49	nr	**26199.90**
0–400 m³/hour, 0.1–4.0 kPa	22979.55	28230.37	16.00	600.49	nr	**28830.86**
0–1000 m³/hour, 0.1–7.4 kPa	29291.19	35984.22	25.00	938.27	nr	**36922.49**
50–1000 m³/hour, 0.1–16.0 kPa	40146.11	49319.50	25.00	938.27	nr	**50257.77**
50–1000 m³/hour, 0.1–24.5 kPa	45386.97	55757.89	25.00	938.27	nr	**56696.16**
50–1000 m³/hour, 0.1–31.0 kPa	50615.55	62181.20	25.00	938.27	nr	**63119.47**
50–1000 m³/hour, 0.1–41.0 kPa	55859.46	68623.35	25.00	938.27	nr	**69561.62**
50–1000 m³/hour, 0.1–51.0 kPa	56110.33	68931.54	25.00	938.27	nr	**69869.81**
100–1800 m³/hour, 3.5–23.5 kPa	58378.64	71718.16	25.00	938.27	nr	**72656.43**
100–1800 m³/hour, 4.5–27.0 kPa	59979.19	73684.43	25.00	938.27	nr	**74622.70**
100–1800 m³/hour, 6.0–32.5 kPa	63722.21	78282.74	25.00	938.27	nr	**79221.01**
100–1800 m³/hour, 7.2–39.0 kPa	67730.11	83206.44	25.00	938.27	nr	**84144.71**
100–1800 m³/hour, 9.0–42.0 kPa	71193.43	87461.13	25.00	938.27	nr	**88399.40**

38 PIPED SUPPLY SYSTEMS

Item	Net Price £	Material £	Labour hours	Labour £	Unit	**Total rate £**
NATURAL GAS: PIPELINES: SCREWED STEEL						
For prices for steel pipework refer to section – Low Temperature Hot Water Heating						
PIPE IN PIPE						
Note: For pipe in pipe, a sleeve size two pipe sizes bigger than actual pipe size has been allowed. All rates refer to actual pipe size.						
Black steel pipes – Screwed and socketed joints; BS 1387: 1985 upto 50 mm pipe size. Butt welded joints; BS 1387: 1985 65 mm pipe size and above.						
Pipe dia.						
25 mm	15.02	18.45	1.73	64.92	m	**83.37**
32 mm	19.95	24.51	1.95	73.18	m	**97.69**
40 mm	25.66	31.52	2.16	81.07	m	**112.59**
50 mm	33.68	41.38	2.44	91.57	m	**132.95**
65 mm	46.19	56.74	2.95	110.71	m	**167.45**
80 mm	55.82	68.58	3.42	128.35	m	**196.93**
100 mm	69.73	85.66	4.00	150.12	m	**235.78**
Extra over black steel pipes – Screwed pipework; black malleable iron fittings; BS 143. Welded pipework; butt welded steel fittings; BS 1965						
Bend, 90°						
25 mm	8.66	10.64	2.91	109.21	m	**119.85**
32 mm	13.21	16.23	3.45	129.49	m	**145.72**
40 mm	18.49	22.71	5.34	200.41	m	**223.12**
50 mm	20.84	25.60	6.53	245.07	m	**270.67**
65 mm	28.16	34.60	8.84	331.77	m	**366.37**
80 mm	47.99	58.96	10.73	402.71	m	**461.67**
100 mm	63.72	78.28	12.76	478.89	m	**557.17**
Bend, 45°						
25 mm	10.96	13.46	2.91	109.21	m	**122.67**
32 mm	14.20	17.44	3.45	129.49	m	**146.93**
40 mm	15.87	19.50	5.34	200.41	m	**219.91**
50 mm	18.93	23.26	6.53	245.07	m	**268.33**
65 mm	24.71	30.36	8.84	331.77	m	**362.13**
80 mm	42.29	51.95	10.73	402.71	m	**454.66**
100 mm	54.16	66.54	12.76	478.89	m	**545.43**
Equal tee						
25 mm	72.96	89.63	4.18	156.88	m	**246.51**
32 mm	74.90	92.01	4.94	185.41	m	**277.42**
40 mm	92.15	113.21	7.28	273.22	m	**386.43**
50 mm	94.55	116.15	8.48	318.25	m	**434.40**
65 mm	125.88	154.64	11.47	430.48	m	**585.12**
80 mm	215.97	265.31	14.23	534.06	m	**799.37**
100 mm	244.42	300.27	17.92	672.56	m	**972.83**

38 PIPED SUPPLY SYSTEMS

Item	Net Price £	Material £	Labour hours	Labour £	Unit	**Total rate £**
NATURAL GAS: PIPELINES: SCREWED STEEL – cont						
Copper pipe; capillary or compression joints in the running length; EN1057 R250 (TX) formerly BS 2871 Table X						
Plastic coated gas service pipe for corrosive environments, fixed vertically or at low level with brackets measured separtely						
15 mm dia. (yellow)	8.50	10.45	0.85	31.92	m	**42.37**
22 mm dia. (yellow)	16.75	20.58	0.96	36.06	m	**56.64**
28 mm dia. (yellow)	21.27	26.13	1.06	39.77	m	**65.90**
Copper pipe; capillary or compression joints in the running length; EN1057 R250 (TY) formerly BS 2871 Table Y						
Plastic coated gas and cold water service pipe for corrosive environments, fixed vertically or at low level with brackets measured separately (Refer to Copper Pipe Table X Section)						
15 mm dia. (yellow)	9.82	12.06	0.61	22.90	m	**34.96**
22 mm dia. (yellow)	17.85	21.92	0.69	25.89	m	**47.81**
FIXINGS						
Refer to Section – Cold Water						
Extra over copper pipes; capillary fittings; BS 864						
Refer to Section – Cold Water						

38 PIPED SUPPLY SYSTEMS

Item	Net Price £	Material £	Labour hours	Labour £	Unit	**Total rate £**
FUEL OIL STORAGE/DISTRIBUTION						
PIPELINES						
For pipework prices refer to Section – Low Temperature Hot Water Heating						
TANKS						
Fuel storage tanks; mild steel; with all necessary screwed bosses; oil resistant joint rings; includes placing in position. Rectangular shape						
1360 litre (300 gallon) capacity; 2 mm plate	377.83	464.16	12.03	451.50	nr	**915.66**
2730 litre (600 gallon) capacity; 2.5 mm plate	504.88	620.25	18.60	698.07	nr	**1318.32**
4550 litre (1000 gallon) capacity; 3 mm plate	1058.50	1300.36	25.00	938.27	nr	**2238.63**
Fuel storage tanks; 5 mm plate mild steel to BS 799 type J; complete with raised neck manhole with bolted cover, screwed connections, vent and fill connections, drain valve, gauge and overfill alarm; includes placing in position; excludes pumps and control panel. Nominal capacity size						
5,600 litres, 2.5 m × 1.5 m × 1.5 m high	2470.62	3035.16	20.00	750.61	nr	**3785.77**
Extra for bund unit (internal use)	1420.80	1745.46	30.00	1125.93	nr	**2871.39**
Extra for external use with bund (watertight)	899.83	1105.44	2.00	75.06	nr	**1180.50**
10,200 litres, 3.05 m × 1.83 m × 1.83 m high	3070.51	3772.13	30.00	1125.93	nr	**4898.06**
Extra for bund unit (internal use)	2068.05	2540.60	40.00	1501.23	nr	**4041.83**
Extra for external use with bund (watertight)	1105.06	1357.57	2.00	75.06	nr	**1432.63**
15,000 litres, 3.75 m × 2 m × 2 m high	3883.52	4770.91	40.00	1501.23	nr	**6272.14**
Extra for bund unit (internal use)	2778.46	3413.34	55.00	2064.18	nr	**5477.52**
Extra for external use with bund (watertight)	1294.50	1590.29	2.00	75.06	nr	**1665.35**
20,000 litres, 4 m × 2.5 m × 2 m high	5122.78	6293.33	50.00	1876.54	nr	**8169.87**
Extra for bund unit (internal use)	3370.48	4140.63	65.00	2439.50	nr	**6580.13**
Extra for external use with bund (watertight)	1531.30	1881.20	2.00	75.06	nr	**1956.26**
Extra for BMS output (all tank sizes)	536.74	659.39	–	–	nr	**659.39**

38 PIPED SUPPLY SYSTEMS

Item	Net Price £	Material £	Labour hours	Labour £	Unit	Total rate £
FUEL OIL STORAGE/DISTRIBUTION – cont						
Fuel storage tanks; plastic; with all necessary screwed bosses; oil resistant joint rings; includes placing in position						
Cylindrical; horizontal						
1250 litre (285 gallon) capacity	305.70	375.55	3.73	139.99	nr	**515.54**
1350 litre (300 gallon) capacity	312.31	383.67	4.30	161.39	nr	**545.06**
2500 litre (550 gallon) capacity	480.94	590.84	4.88	183.14	nr	**773.98**
Cylindrical; vertical						
1365 litre (300 gallon) capacity	193.90	238.20	3.73	139.99	nr	**378.19**
2600 litre (570 gallon) capacity	295.11	362.54	4.88	183.14	nr	**545.68**
3635 litre (800 gallon) capacity	460.08	565.20	4.88	183.14	nr	**748.34**
5455 litre (1200 gallon) capacity	670.23	823.38	5.95	223.30	nr	**1046.68**
Bunded tanks						
1135 litre (250 gallon) capacity	621.60	763.63	4.30	161.39	nr	**925.02**
1590 litre (350 gallon) capacity	739.49	908.46	4.88	183.14	nr	**1091.60**
2500 litre (550 gallon) capacity	878.98	1079.83	5.95	223.30	nr	**1303.13**
5000 litre (1100 gallon) capacity	1773.71	2179.00	6.53	245.07	nr	**2424.07**

38 PIPED SUPPLY SYSTEMS

Item	Net Price £	Material £	Labour hours	Labour £	Unit	**Total rate £**
FIRE HOSE REELS: PIPEWORK ANCILLARIES						
PIPELINES						
For pipework prices refer to section – Cold Water						
PIPELINE ANCILLARIES						
For prices for ancillaries refer to section – Cold Water						
Hose reels; automatic; connection to 25 mm screwed joint; reel with 30.5 metres, 19 mm rubber hose; suitable for working pressure up to 7 bar						
Reels						
Non-swing pattern	306.11	376.06	3.75	140.74	nr	**516.80**
Recessed non-swing pattern	342.54	420.81	3.75	140.74	nr	**561.55**
Swinging pattern	360.62	443.03	3.75	140.74	nr	**583.77**
Recessed swinging pattern	360.62	443.03	3.75	140.74	nr	**583.77**
Hose reels; manual; connection to 25 mm screwed joint; reel with 30.5 metres, 19 mm rubber hose; suitable for working pressure up to 7 bar						
Reels						
Non-swing pattern	306.11	376.06	3.25	121.98	nr	**498.04**
Recessed non-swing pattern	342.54	420.81	3.25	121.98	nr	**542.79**
Swinging pattern	360.62	443.03	3.25	121.98	nr	**565.01**
Recessed swinging pattern	360.62	443.03	3.25	121.98	nr	**565.01**

38 PIPED SUPPLY SYSTEMS

Item	Net Price £	Material £	Labour hours	Labour £	Unit	Total rate £
DRY RISERS: PIPEWORK ANCILLARIES						
PIPELINES						
For pipework prices refer to section – Cold Water						
VALVES (BS 5041, parts 2 and 3)						
Bronze/gunmetal inlet breeching for pumping in with 65 mm dia. instantaneous male coupling; with cap, chain and 25 mm drain valve						
Double inlet with back pressure valve, flanged to steel	267.58	328.72	1.75	65.68	nr	**394.40**
Quadruple inlet with back pressure valve, flanged to steel	599.13	736.03	1.75	65.68	nr	**801.71**
Bronze/gunmetal gate type outlet valve with 65 mm dia. instantaneous female coupling; cap and chain; wheel head secured by padlock and leather strap						
Flanged to BS 4504 PN6 (bolted connection to counter flanges measured separately)	214.59	263.62	1.75	65.68	nr	**329.30**
Bronze/gunmetal landing type outlet valve, with 65 mm dia. instantaneous female coupling; cap and chain; wheelhead secured by padlock and leather strap; bolted connections to counter flanges measured separately						
Horizontal, flanged to BS 4504 PN6	233.71	287.11	1.50	56.30	nr	**343.41**
Oblique, flanged to BS 4504 PN6	233.71	287.11	1.50	56.30	nr	**343.41**
Air valve, screwed joints to steel						
25 mm dia.	46.45	57.07	0.55	20.64	nr	**77.71**
INLET BOXES (BS 5041, part 5)						
Steel dry riser inlet box with hinged wire glazed door suitably lettered (fixing by others)						
610 × 460 × 325 mm; double inlet	284.78	349.85	3.00	112.59	nr	**462.44**
610 × 610 × 356 mm; quadruple inlet	531.50	652.95	3.00	112.59	nr	**765.54**
OUTLET BOXES (BS 5041, part 5)						
Steel dry riser outlet box with hinged wire glazed door suitably lettered (fixing by others)						
610 × 460 × 325 mm; single outlet	278.10	341.65	3.00	112.59	nr	**454.24**

38 PIPED SUPPLY SYSTEMS

Item	Net Price £	Material £	Labour hours	Labour £	Unit	**Total rate £**
SPRINKLERS						
PIPELINES						
Prefabricated black steel pipework; screwed joints, including all coupliings, unions and the like to BS 1387:1985; includes fixing to backgrounds, with brackets measured separately						
Heavy weight						
25 mm dia.	6.14	7.54	0.47	17.64	m	**25.18**
32 mm dia.	7.62	9.37	0.53	19.89	m	**29.26**
40 mm dia.	8.88	10.91	0.58	21.77	m	**32.68**
50 mm dia.	12.33	15.15	0.63	23.65	m	**38.80**
FIXINGS						
For steel pipes; black malleable iron. For minimum fixing distances, refer to the Tables and Memoranda at the rear of the book						
Pipe ring, single socket, black malleable iron						
25 mm dia.	1.09	1.34	0.12	4.50	nr	**5.84**
32 mm dia.	1.16	1.43	0.14	5.26	nr	**6.69**
40 mm dia.	1.50	1.85	0.15	5.63	nr	**7.48**
50 mm dia.	1.89	2.32	0.16	6.01	nr	**8.33**
Extra over channel sections for fabricated hangers and brackets						
Galvanized steel; including inserts, bolts, nuts, washers; fixed to backgrounds						
41 × 21 mm	6.02	7.39	0.29	10.89	m	**18.28**
41 × 41 mm	7.22	8.87	0.29	10.89	m	**19.76**
Threaded rods; metric thread; including nuts, washers etc.						
12 mm dia. × 600 mm long	3.04	3.74	0.18	6.76	nr	**10.50**
Extra over for black malleable iron fittings; BS 143						
Plain plug, solid						
25 mm dia.	2.19	2.69	0.40	15.02	nr	**17.71**
32 mm dia.	3.15	3.87	0.44	16.51	nr	**20.38**
40 mm dia.	4.43	5.44	0.48	18.01	nr	**23.45**
50 mm dia.	5.82	7.15	0.56	21.02	nr	**28.17**
Concentric reducing socket						
32 mm dia.	2.60	3.19	0.48	18.01	nr	**21.20**
40 mm dia.	3.38	4.15	0.55	20.64	nr	**24.79**
50 mm dia.	4.74	5.83	0.60	22.52	nr	**28.35**

38 PIPED SUPPLY SYSTEMS

Item	Net Price £	Material £	Labour hours	Labour £	Unit	Total rate £
SPRINKLERS – cont						
Extra over for black malleable iron fittings – cont						
Elbow; 90° female/female						
25 mm dia.	1.52	1.87	0.44	16.51	nr	**18.38**
32 mm dia.	2.79	3.42	0.53	19.89	nr	**23.31**
40 mm dia.	4.68	5.75	0.60	22.52	nr	**28.27**
50 mm dia.	5.49	6.74	0.65	24.40	nr	**31.14**
Tee						
25 mm dia equal	2.08	2.55	0.51	19.14	nr	**21.69**
32 mm dia reducing to 25 mm dia.	3.79	4.65	0.54	20.28	nr	**24.93**
40 mm dia.	5.82	7.15	0.65	24.40	nr	**31.55**
50 mm dia.	8.37	10.28	0.78	29.27	nr	**39.55**
Cross tee						
25 mm dia equal	5.65	6.94	1.16	43.53	nr	**50.47**
32 mm dia.	8.23	10.11	1.40	52.54	nr	**62.65**
40 mm dia.	10.59	13.01	1.60	60.05	nr	**73.06**
50 mm dia.	17.21	21.15	1.68	63.04	nr	**84.19**
Prefabricated black steel pipework; welded joints, including all coupliings, unions and the like to BS 1387:1985; fixing to backgrounds						
Heavy weight						
65 mm dia.	16.11	19.79	0.65	24.40	m	**44.19**
80 mm dia.	20.50	25.18	0.70	26.28	m	**51.46**
100 mm dia.	28.60	35.13	0.85	31.90	m	**67.03**
150 mm dia.	38.70	47.54	1.15	43.16	m	**90.70**
FIXINGS						
For steel pipes; black malleable iron. For minimum fixing distances, refer to the Tables and Memoranda at the rear of the book						
Pipe ring, single socket, black malleable iron						
65 mm dia.	2.73	3.35	0.30	11.26	nr	**14.61**
80 mm dia.	3.28	4.03	0.35	13.14	nr	**17.17**
100 mm dia.	4.96	6.09	0.40	15.02	nr	**21.11**
150 mm dia.	11.23	13.80	0.77	28.90	nr	**42.70**
Extra over fittings						
Reducer (one size down)						
65 mm dia.	8.27	10.16	2.70	101.33	nr	**111.49**
80 mm dia.	10.76	13.22	2.86	107.33	nr	**120.55**
100 mm dia.	21.45	26.35	3.22	120.84	nr	**147.19**
150 mm dia.	59.60	73.22	4.20	157.63	nr	**230.85**

38 PIPED SUPPLY SYSTEMS

Item	Net Price £	Material £	Labour hours	Labour £	Unit	Total rate £
Elbow; 90°						
65 mm dia.	11.37	13.96	3.06	114.84	nr	**128.80**
80 mm dia.	16.70	20.52	3.40	127.61	nr	**148.13**
100 mm dia.	20.88	25.65	3.70	138.86	nr	**164.51**
150 mm dia.	31.32	38.47	5.20	195.16	nr	**233.63**
Branch bend						
65 mm dia.	11.57	14.22	3.60	135.11	nr	**149.33**
80 mm dia.	13.55	16.64	3.80	142.62	nr	**159.26**
100 mm dia.	20.85	25.61	5.10	191.42	nr	**217.03**
150 mm dia.	53.94	66.27	7.50	281.48	nr	**347.75**
Prefabricated black steel pipe; victaulic joints; including all couplings and the like to BS 1387: 1985; fixing to backgrounds						
Heavy weight						
65 mm dia.	13.90	17.07	0.70	26.28	m	**43.35**
80 mm dia.	18.05	22.18	0.78	29.27	m	**51.45**
100 mm dia.	26.31	32.32	0.93	34.90	m	**67.22**
150 mm dia.	52.39	64.36	1.25	46.91	m	**111.27**
FIXINGS						
For fixings refer to For steel pipes; black malleable iron.						
Extra over fittings						
Coupling						
65 mm dia.	10.17	12.49	0.26	9.75	nr	**22.24**
80 mm dia.	10.90	13.39	0.26	9.75	nr	**23.14**
100 mm dia.	14.28	17.55	0.32	12.01	nr	**29.56**
150 mm dia.	25.15	30.90	0.35	13.14	nr	**44.04**
Reducer						
65 mm dia.	8.27	10.16	0.48	18.01	nr	**28.17**
80 mm dia.	10.76	13.22	0.43	16.14	nr	**29.36**
100 mm dia.	21.45	26.35	0.46	17.26	nr	**43.61**
150 mm dia.	32.18	39.53	0.45	16.89	nr	**56.42**
Elbow; any°						
65 mm dia.	11.37	13.96	0.56	21.02	nr	**34.98**
80 mm dia.	16.70	20.52	0.63	23.65	nr	**44.17**
100 mm dia.	32.22	39.59	0.71	26.65	nr	**66.24**
150 mm dia.	48.32	59.36	0.80	30.02	nr	**89.38**
Equal tee						
65 mm dia.	17.61	21.63	0.74	27.77	nr	**49.40**
80 mm dia.	21.22	26.07	0.83	31.15	nr	**57.22**
100 mm dia.	38.89	47.77	0.94	35.28	nr	**83.05**
150 mm dia.	58.32	71.64	1.05	39.41	nr	**111.05**

38 PIPED SUPPLY SYSTEMS

Item	Net Price £	Material £	Labour hours	Labour £	Unit	Total rate £
PIPELINE ANCILLARIES						
SPRINKLER HEADS						
Sprinkler heads; brass body; frangible glass bulb; manufactured to standard operating temperature of 57–141°C; quick response; RTI<50						
Conventional pattern; 15 mm dia.	4.53	5.57	0.15	5.63	nr	**11.20**
Sidewall pattern; 15 mm dia.	6.67	8.19	0.15	5.63	nr	**13.82**
Conventional pattern; 15 mm dia.; satin chrome plated	5.32	6.53	0.15	5.63	nr	**12.16**
Sidewall pattern; 15 mm dia.; satin chrome plated	6.92	8.50	0.15	5.63	nr	**14.13**
Fully concealed; fusible link; 15 mm dia.	16.64	20.44	0.15	5.63	nr	**26.07**
VALVES						
Wet system alarm valves; including internal non-return valve; working pressure up to 12.5 bar; BS4504 PN16 flanged ends; bolted connections						
100 mm dia.	1464.49	1799.12	25.00	938.27	nr	**2737.39**
150 mm dia.	1788.91	2197.67	25.00	938.27	nr	**3135.94**
Wet system by-pass alarm valves; including internal non-return valve; working pressure up to 12.5 bar; BS4504 PN16 flanged ends; bolted connections						
100 mm dia.	2581.50	3171.38	25.00	938.27	nr	**4109.65**
150 mm dia.	3263.89	4009.69	25.00	938.27	nr	**4947.96**
Alternate system wet/dry alarm station; including butterfly valve, wet alarm valve, dry pipe differential pressure valve and pressure gauges; working pressure up to 10.5 bar; BS4505 PN16 flanged ends; bolted connections						
100 mm dia.	3199.38	3930.43	40.00	1501.23	nr	**5431.66**
150 mm dia.	3728.51	4580.48	40.00	1501.23	nr	**6081.71**
Alternate system wet/dry alarm station; including electrically operated butterfly valve, water supply accelerator set, wet alarm valve, dry pipe differential pressure valve and pressure gauges; working pressure up to 10.5 bar; BS4505 PN16 flanged ends; bolted connections						
100 mm dia.	3646.90	4480.21	45.00	1688.88	nr	**6169.09**
150 mm dia.	4181.06	5136.43	45.00	1688.88	nr	**6825.31**
ALARM/GONGS						
Water operated motor alarm and gong; stainless steel and aluminum body and gong; screwed connections						
Connection to sprinkler system and drain pipework	446.40	548.40	6.00	225.18	nr	**773.58**

38 PIPED SUPPLY SYSTEMS

Item	Net Price £	Material £	Labour hours	Labour £	Unit	Total rate £
WATER TANKS						
Note: Prices are based on the most economical tank size for each volume, and the cost will vary with differing tank dimensions, for the same volume						
Steel sectional sprinkler tank; ordinary hazard, life safety classification; two compartment tank, complete with all fittings and accessories to comply with LPCB type A requirements; cost of erection (on prepared supports) is included within net price, labour cost allows for offloading and positioning of materials						
Volume, size						
70 m³, 6.1 m × 4.88 m × 2.44 m (h)	25451.50	31267.17	24.00	900.73	nr	**32167.90**
105 m³, 7.3 m × 6.1 m × 2.44 m (h)	29865.89	36690.24	28.00	1050.86	nr	**37741.10**
168 m³, 9.76 m × 4.88 mx 3.66 m (h)	36900.09	45331.77	28.00	1050.86	nr	**46382.63**
211 m³, 9.76 m × 6.1 m × 3.66 m (h)	40738.12	50046.78	32.00	1200.98	nr	**51247.76**
Steel sectional sprinkler tank; ordinary hazard, property protection classification; single compartment tank, complete with all fittings and accessories to comply with LPCB type A requirements; cost of erection (on prepared supports) is included within net price, labour cost allows for offloading and positioning of materials						
Volume, size						
70 m³, 6.1 m × 4.88 m × 2.44 m (h)	17945.73	22046.33	24.00	900.73	nr	**22947.06**
105 m³, 7.3 m × 6.1 m × 2.44 m (h)	22170.19	27236.08	28.00	1050.86	nr	**28286.94**
168 m³, 9.76 m × 4.88 m × 3.66 m (h)	28647.68	35193.68	28.00	1050.86	nr	**36244.54**
211 m³, 9.76 m × 6.1 m × 3.66 m (h)	32145.13	39490.29	32.00	1200.98	nr	**40691.27**
GRP sectional sprinkler tank; ordinary hazard, life safety classification; two compartment tank, complete with all fittings and accessories to comply with LPCB type A requirements; cost of erection (on prepared supports) is included within net price, labour cost allows for offloading and positioning of materials						
Volume, size						
55 m³, 6 m × 4 m × 3 m (h)	29142.59	35801.67	14.00	525.43	nr	**36327.10**
70 m³, 6 m × 5 m × 3 m (h)	32583.54	40028.88	16.00	600.49	nr	**40629.37**
80 m³, 8 m × 4 m × 3 m (h)	33768.20	41484.23	16.00	600.49	nr	**42084.72**
105 m³, 10 m × 4 m × 3 m (h)	38211.35	46942.64	24.00	900.73	nr	**47843.37**
125 m³, 10 m × 5 m × 3 m (h)	43093.93	52940.89	24.00	900.73	nr	**53841.62**
140 m³, 8 m × 7 m × 3 m (h)	44878.65	55133.42	24.00	900.73	nr	**56034.15**
135 m³, 9 m × 6 m × 3 m (h)	46235.49	56800.30	24.00	900.73	nr	**57701.03**
160 m³, 13 m × 5 m × 3 m (h)	50762.15	62361.31	24.00	900.73	nr	**63262.04**
185 m³, 12 m × 6 m × 3 m (h)	53696.84	65966.56	24.00	900.73	nr	**66867.29**

38 PIPED SUPPLY SYSTEMS

Item	Net Price £	Material £	Labour hours	Labour £	Unit	**Total rate £**
WATER TANKS – cont						
GRP sectional sprinkler tank; ordinary hazard, property protection classification; single compartment tank, complete with all fittings and accessories to comply with LPCB type A requirements; cost of erection (on prepared supports) is within net price, labour cost allows for offloading and positioning of materials						
Volume, size						
55 m³, 6 m × 4 m × 3 m (h)	22420.06	27543.04	14.00	525.43	nr	**28068.47**
70 m³, 6 m × 5 m × 3 m (h)	25329.05	31116.74	16.00	600.49	nr	**31717.23**
80 m³, 8 m × 4 m × 3 m (h)	27035.37	33212.95	16.00	600.49	nr	**33813.44**
105 m³, 10 m × 4 m × 3 m (h)	31672.54	38909.71	24.00	900.73	nr	**39810.44**
125 m³, 10 m × 5 m × 3 m (h)	35815.03	43998.77	24.00	900.73	nr	**44899.50**
140 m³, 8 m × 7 m × 3 m (h)	37076.79	45548.83	24.00	900.73	nr	**46449.56**
135 m³, 9 m × 6 m × 3 m (h)	38955.49	47856.82	24.00	900.73	nr	**48757.55**
160 m³, 13 m × 5 m × 3 m (h)	43672.13	53651.21	24.00	900.73	nr	**54551.94**
185 m³, 12 m × 6 m × 3 m (h)	45878.29	56361.48	24.00	900.73	nr	**57262.21**

38 PIPED SUPPLY SYSTEMS

Item	Net Price £	Material £	Labour hours	Labour £	Unit	Total rate £
FIRE EXTINGUISHERS AND HYDRANTS						
EXTINGUISHERS						
Fire extinguishers; hand held; BS 5423; placed in position						
Water type; cartridge operated; for Class A fires						
Water type, 9 litre capacity; 55 gm CO_2 cartridge; Class A fires (fire rating 13A)	84.29	103.55	1.00	37.53	nr	**141.08**
Foam type, 9 litre capacity; 75 gm CO_2 cartridge; Class A & B fires (fire rating 13A:183B)	99.95	122.79	1.00	37.53	nr	**160.32**
Dry powder type; cartridge operated; for Class A, B & C fires and electrical equipment fires						
Dry powder type, 1 kg capacity; 12 gm CO_2 cartridge; Class A, B & C fires (fire rating 5A:34B)	35.64	43.78	1.00	37.53	nr	**81.31**
Dry powder type, 2 kg capacity; 28 gm CO_2 cartridge; Class A, B & C fires (fire rating 13A:55B)	48.00	58.97	1.00	37.53	nr	**96.50**
Dry powder type, 4 kg capacity; 90 gm CO_2 cartridge; Class A, B & C fires (fire rating 21A:183B)	85.50	105.04	1.00	37.53	nr	**142.57**
Dry powder type, 9 kg capacity; 190 gm CO_2 cartridge; Class A, B & C fires (fire rating 43A:233B)	114.55	140.72	1.00	37.53	nr	**178.25**
Dry powder type; stored pressure type; for Class A, B & C fires and electrical equipment fires						
Dry powder type, 1 kg capacity; Class A, B & C fires (fire rating 5A:34B)	32.83	40.33	1.00	37.53	nr	**77.86**
Dry powder type, 2 kg capacity; Class A, B & C fires (fire rating 13A:55B)	38.98	47.89	1.00	37.53	nr	**85.42**
Dry powder type, 4 kg capacity; Class A, B & C fires (fire rating 21A:183B)	72.81	89.45	1.00	37.53	nr	**126.98**
Dry powder type, 9 kg capacity; Class A, B & C fires (fire rating 43A:233B)	94.33	115.89	1.00	37.53	nr	**153.42**
Carbon dioxide type; for Class B fires and electrical equipment fires						
CO_2 type with hose and horn, 2 kg capacity, Class B fires (fire rating 34B)	90.18	110.79	1.00	37.53	nr	**148.32**
CO_2 type with hose and horn, 5 kg capacity, Class B fires (fire rating 55B)	141.30	173.59	1.00	37.53	nr	**211.12**
Glass fibre blanket, in GRP container						
1100 × 1100 mm	28.62	35.16	0.50	18.77	nr	**53.93**
1200 × 1200 mm	31.36	38.52	0.50	18.77	nr	**57.29**
1800 × 1200 mm	41.83	51.39	0.50	18.77	nr	**70.16**

38 PIPED SUPPLY SYSTEMS

Item	Net Price £	Material £	Labour hours	Labour £	Unit	Total rate £
FIRE EXTINGUISHERS AND HYDRANTS – cont						
HYDRANTS						
Fire hydrants; bolted connections						
Underground hydrants, complete with frost plug to BS 750						
sluice valve pattern type 1	235.33	289.11	4.50	168.88	nr	**457.99**
screw down pattern type 2	170.93	209.99	4.50	168.88	nr	**378.87**
Stand pipe for underground hydrant; screwed base; light alloy						
single outlet	135.35	166.28	1.00	37.53	nr	**203.81**
double outlet	198.04	243.30	1.00	37.53	nr	**280.83**
64 mm dia. bronze/gunmetal outlet valves						
oblique flanged landing valve	148.14	181.99	1.00	37.53	nr	**219.52**
oblique screwed landing valve	148.14	181.99	1.00	37.53	nr	**219.52**
Cast iron surface box; fixing by others						
400 × 200 × 100 mm	129.72	159.36	1.00	37.53	nr	**196.89**
500 × 200 × 150 mm	174.78	214.71	1.00	37.53	nr	**252.24**
frost plug	31.48	38.67	0.25	9.39	nr	**48.06**

38 MECHANICAL/COOLING/HEATING SYSTEMS

Item	Net Price £	Material £	Labour hours	Labour £	Unit	Total rate £
GAS/OIL FIRED BOILERS – DOMESTIC						
Domestic water boilers; stove enamelled casing; electric controls; placing in position; assembling and connecting; electrical work elsewhere						
Gas fired; floor standing; connected to conventional flue						
9–12 kW	629.01	772.74	8.59	322.38	nr	**1095.12**
12–15 kW	661.71	812.91	8.59	322.38	nr	**1135.29**
15–18 kW	704.79	865.83	8.88	333.27	nr	**1199.10**
18–21 kW	818.81	1005.91	9.92	372.31	nr	**1378.22**
21–23 kW	907.74	1115.16	10.66	400.07	nr	**1515.23**
23–29 kW	1177.33	1446.35	11.81	443.25	nr	**1889.60**
29–37 kW	1392.49	1710.67	11.81	443.25	nr	**2153.92**
37–41 kW	1448.13	1779.03	12.68	475.88	nr	**2254.91**
Gas fired; wall hung; connected to conventional flue						
9–12 kW	552.84	679.16	8.59	322.38	nr	**1001.54**
12–15 kW	713.74	876.83	8.59	322.38	nr	**1199.21**
13–18 kW	905.80	1112.78	8.59	322.38	nr	**1435.16**
Gas fired; floor standing; connected to balanced flue						
9–12 kW	785.74	965.29	9.16	343.78	nr	**1309.07**
12–15 kW	817.51	1004.31	10.95	410.96	nr	**1415.27**
15–18 kW	879.99	1081.07	11.98	449.62	nr	**1530.69**
18–21 kW	1036.99	1273.94	12.78	479.64	nr	**1753.58**
21–23 kW	1196.02	1469.31	12.78	479.64	nr	**1948.95**
23–29 kW	1524.47	1872.81	15.45	579.85	nr	**2452.66**
29–37 kW	2930.64	3600.29	17.65	662.42	nr	**4262.71**
Gas fired; wall hung; connected to balanced flue						
6–9 kW	593.26	728.82	9.16	343.78	nr	**1072.60**
9–12 kW	678.88	834.00	9.16	343.78	nr	**1177.78**
12–15 kW	768.74	944.40	9.45	354.67	nr	**1299.07**
15–18 kW	921.76	1132.38	9.74	365.56	nr	**1497.94**
18–22 kW	978.41	1201.98	9.74	365.56	nr	**1567.54**
Gas fired; wall hung; connected to fan flue (including flue kit)						
6–9 kW	678.51	833.55	9.16	343.78	nr	**1177.33**
9–12 kW	756.14	928.91	9.16	343.78	nr	**1272.69**
12–15 kW	820.82	1008.38	10.95	410.96	nr	**1419.34**
15–18 kW	884.15	1086.18	11.98	449.62	nr	**1535.80**
18–23 kW	1152.02	1415.25	12.78	479.64	nr	**1894.89**
23–29 kW	1491.84	1832.72	15.45	579.85	nr	**2412.57**
29–35 kW	1831.55	2250.06	17.65	662.42	nr	**2912.48**

38 MECHANICAL/COOLING/HEATING SYSTEMS

Item	Net Price £	Material £	Labour hours	Labour £	Unit	Total rate £
GAS/OIL FIRED BOILERS – DOMESTIC – cont						
Oil fired; floor standing; connected to conventional flue						
12–15 kW	1178.10	1447.30	10.38	389.57	nr	**1836.87**
15–19 kW	1228.93	1509.74	12.20	457.87	nr	**1967.61**
21–25 kW	1402.56	1723.05	14.30	536.70	nr	**2259.75**
26–32 kW	1537.73	1889.10	15.80	592.99	nr	**2482.09**
35–50 kW	1736.15	2132.86	20.46	767.88	nr	**2900.74**
Fire place mounted natural gas fire and back boiler; cast iron water boiler; electric control box; fire output 3 kW with wood surround						
10.50 kW	343.70	422.24	8.88	333.27	nr	**755.51**

38 MECHANICAL/COOLING/HEATING SYSTEMS

Item	Net Price £	Material £	Labour hours	Labour £	Unit	Total rate £
GAS/OIL FIRED BOILERS – COMMERCIAL; FORCED DRAFT						
Commercial steel shell floor standing boilers; pressure jet burner; 6 bar max working pressure; including controls, enamelled jacket, insulation, placing in position and commissioning; electrical work elsewhere						
Natural gas; burner and boiler (high/low type), connected to conventional flue						
1151–1250 kW	17175.00	21099.49	14.00	525.43	nr	**21624.92**
1426–1500 kW	21030.00	25835.36	14.00	525.43	nr	**26360.79**
1625–1760 kW	24780.00	30442.23	14.00	525.43	nr	**30967.66**
Natural gas; burner and boiler (modulating type), connected to conventional flue						
1151–1250 kW	19260.00	23660.91	12.00	450.37	nr	**24111.28**
1426–1500 kW	22990.00	28243.21	14.00	525.43	nr	**28768.64**
1625–1760 kW	27650.00	33968.03	14.00	525.43	nr	**34493.46**
1900–2050 kW	29010.00	35638.79	14.00	525.43	nr	**36164.22**
Natural gas; low NOx burner and boiler (modulating type), connected to conventional flue						
1151–1250 kW	20320.00	24963.12	12.00	450.37	nr	**25413.49**
1426–1500 kW	25025.00	30743.21	14.00	525.43	nr	**31268.64**
1625–1760 kW	28705.00	35264.09	14.00	525.43	nr	**35789.52**
1900–2050 kW	30215.00	37119.13	14.00	525.43	nr	**37644.56**
Oil; burner and boiler (high/low type), connected to conventional flue						
1151–1250 kW	15780.00	19385.73	12.00	450.37	nr	**19836.10**
1426–1500 kW	19215.00	23605.63	14.00	525.43	nr	**24131.06**
1625–1760 kW	22890.00	28120.37	14.00	525.43	nr	**28645.80**
Oil; burner and boiler (modulating type), connected to conventional flue						
1151–1250 kW	18505.00	22733.39	12.00	450.37	nr	**23183.76**
1426–1500 kW	22850.00	28071.22	14.00	525.43	nr	**28596.65**
1625–1760 kW	26940.00	33095.79	14.00	525.43	nr	**33621.22**
1900–2050 kW	27728.00	34063.85	14.00	525.43	nr	**34589.28**
Dual fuel; burner and boiler (high/low oil and modulating gast type), connected to conventional flue						
1151–1250 kW	22400.00	27518.40	12.00	450.37	nr	**27968.77**
1426–1500 kW	26940.00	33095.79	14.00	525.43	nr	**33621.22**
1625–1760 kW	31185.00	38310.77	14.00	525.43	nr	**38836.20**
1900–2050 kW	37665.00	46271.45	14.00	525.43	nr	**46796.88**
Dual fuel; burner and boiler (high/low oil and modulating gast type), connected to conventional flue						
1151–1250 kW	22455.00	27585.97	12.00	450.37	nr	**28036.34**
1426–1500 kW	26939.00	33094.56	14.00	525.43	nr	**33619.99**
1625–1760 kW	34323.00	42165.81	14.00	525.43	nr	**42691.24**
1900–2050 kW	37665.00	46271.45	14.00	525.43	nr	**46796.88**

38 MECHANICAL/COOLING/HEATING SYSTEMS

Item	Net Price £	Material £	Labour hours	Labour £	Unit	Total rate £
GAS/OIL FIRED BOILERS – COMMERCIAL; CONDENSING						
Low Nox wall mounted condensing boiler with high efficiency modulating premix burner: Aluminium heat exchanger: Placing in position						
Maximum output						
35 kW	2000.00	2457.00	11.00	412.84	nr	**2869.84**
45 kW	2306.00	2832.92	11.00	412.84	nr	**3245.76**
60 kW	2863.00	3517.20	11.00	412.84	nr	**3930.04**
80 kW	3375.00	4146.19	11.00	412.84	nr	**4559.03**
Low Nox floor standing condensing boiler with high efficiency modulating premix burner: Stainless steel heat exchanger: Including controls: Placing in position.						
Maximum output						
50 kW	3965.00	4871.00	8.00	300.25	nr	**5171.25**
70 kW	4380.00	5380.83	8.00	300.25	nr	**5681.08**
100 kW	4740.00	5823.09	8.00	300.25	nr	**6123.34**
125 kW	4895.00	6013.51	10.00	375.31	nr	**6388.82**
150 kW	5460.00	6707.61	10.00	375.31	nr	**7082.92**
200 kW	6285.00	7721.12	10.00	375.31	nr	**8096.43**
250 kW	7205.00	8851.34	10.00	375.31	nr	**9226.65**
300 kW	8035.00	9871.00	10.00	375.31	nr	**10246.31**
350 kW	9065.00	11136.35	10.00	375.31	nr	**11511.66**
400 kW	10200.00	12530.70	10.00	375.31	nr	**12906.01**
450 kW	11325.00	13912.76	10.00	375.31	nr	**14288.07**
500 kW	12360.00	15184.26	10.00	375.31	nr	**15559.57**
650 kW	14835.00	18224.80	10.00	375.31	nr	**18600.11**
720 kW	17305.00	21259.19	10.00	375.31	nr	**21634.50**
850 kW	21375.00	26259.19	12.00	450.37	nr	**26709.56**
1000 kW	23820.00	29262.87	12.00	450.37	nr	**29713.24**
1300 kW	21390.00	26277.62	12.00	450.37	nr	**26727.99**

38 MECHANICAL/COOLING/HEATING SYSTEMS

Item	Net Price £	Material £	Labour hours	Labour £	Unit	Total rate £
GAS/OIL FIRED BOILERS – FLUE SYSTEMS						
Flues; suitable for domestic, medium sized industrial and commercial oil and gas appliances; stainless steel, twin wall, insulated; for use internally or externally						
Straight length; 120 mm long; including one locking band						
127 mm dia.	44.89	55.15	0.49	18.39	nr	**73.54**
152 mm dia.	50.30	61.79	0.51	19.14	nr	**80.93**
178 mm dia.	58.38	71.72	0.54	20.28	nr	**92.00**
203 mm dia.	66.43	81.61	0.58	21.77	nr	**103.38**
254 mm dia.	78.55	96.49	0.70	26.28	nr	**122.77**
304 mm dia.	98.42	120.91	0.74	27.77	nr	**148.68**
355 mm dia.	140.94	173.14	0.80	30.02	nr	**203.16**
Straight length; 300 mm long; including one locking band						
127 mm dia.	69.24	85.06	0.52	19.52	nr	**104.58**
152 mm dia.	78.28	96.17	0.52	19.52	nr	**115.69**
178 mm dia.	89.94	110.49	0.55	20.64	nr	**131.13**
203 mm dia.	101.79	125.04	0.64	24.01	nr	**149.05**
254 mm dia.	112.88	138.67	0.79	29.64	nr	**168.31**
304 mm dia.	135.42	166.36	0.86	32.28	nr	**198.64**
355 mm dia.	148.60	182.55	0.94	35.28	nr	**217.83**
400 mm dia.	159.02	195.35	1.03	38.66	nr	**234.01**
450 mm dia.	181.90	223.46	1.03	38.66	nr	**262.12**
500 mm dia.	195.10	239.68	1.10	41.28	nr	**280.96**
550 mm dia.	215.30	264.50	1.10	41.28	nr	**305.78**
600 mm dia.	237.56	291.85	1.10	41.28	nr	**333.13**
Straight length; 500 mm long; including one locking band						
127 mm dia.	81.38	99.97	0.55	20.64	nr	**120.61**
152 mm dia.	90.73	111.46	0.55	20.64	nr	**132.10**
178 mm dia.	102.17	125.52	0.63	23.65	nr	**149.17**
203 mm dia.	119.43	146.72	0.63	23.65	nr	**170.37**
254 mm dia.	138.73	170.43	0.86	32.28	nr	**202.71**
304 mm dia.	166.20	204.17	0.95	35.65	nr	**239.82**
355 mm dia.	186.64	229.29	1.03	38.66	nr	**267.95**
400 mm dia.	203.80	250.37	1.12	42.04	nr	**292.41**
450 mm dia.	235.36	289.14	1.12	42.04	nr	**331.18**
500 mm dia.	253.60	311.55	1.19	44.66	nr	**356.21**
550 mm dia.	279.55	343.42	1.19	44.66	nr	**388.08**
600 mm dia.	289.59	355.76	1.19	44.66	nr	**400.42**
Straight length; 1000 mm long; including one locking band						
127 mm dia.	145.48	178.72	0.62	23.27	nr	**201.99**
152 mm dia.	162.14	199.18	0.68	25.51	nr	**224.69**
178 mm dia.	182.42	224.10	0.74	27.77	nr	**251.87**
203 mm dia.	214.88	263.98	0.80	30.02	nr	**294.00**
254 mm dia.	243.52	299.17	0.87	32.65	nr	**331.82**
304 mm dia.	281.10	345.33	1.06	39.77	nr	**385.10**
355 mm dia.	322.56	396.27	1.16	43.53	nr	**439.80**

38 MECHANICAL/COOLING/HEATING SYSTEMS

Item	Net Price £	Material £	Labour hours	Labour £	Unit	**Total rate £**
GAS/OIL FIRED BOILERS – FLUE SYSTEMS – cont						
Straight length; 1000 mm long – cont						
400 mm dia.	345.43	424.36	1.26	47.28	nr	**471.64**
450 mm dia.	364.49	447.77	1.26	47.28	nr	**495.05**
500 mm dia.	395.55	485.93	1.33	49.92	nr	**535.85**
550 mm dia.	435.02	534.42	1.33	49.92	nr	**584.34**
600 mm dia.	457.02	561.45	1.33	49.92	nr	**611.37**
Adjustable length; boiler removal; internal use only; including one locking band						
127 mm dia.	66.69	81.93	0.52	19.52	nr	**101.45**
152 mm dia.	75.35	92.57	0.55	20.64	nr	**113.21**
178 mm dia.	85.40	104.92	0.59	22.13	nr	**127.05**
203 mm dia.	97.94	120.32	0.64	24.01	nr	**144.33**
254 mm dia.	145.21	178.40	0.79	29.64	nr	**208.04**
304 mm dia.	173.92	213.66	0.86	32.28	nr	**245.94**
355 mm dia.	195.04	239.61	0.99	37.15	nr	**276.76**
400 mm dia.	341.30	419.29	0.91	34.16	nr	**453.45**
450 mm dia.	365.33	448.81	0.91	34.16	nr	**482.97**
500 mm dia.	398.45	489.50	0.99	37.15	nr	**526.65**
550 mm dia.	434.68	534.01	0.99	37.15	nr	**571.16**
600 mm dia.	455.05	559.03	0.99	37.15	nr	**596.18**
Inspection length; 500 mm long; including one locking band						
127 mm dia.	171.40	210.57	0.55	20.64	nr	**231.21**
152 mm dia.	177.48	218.03	0.55	20.64	nr	**238.67**
178 mm dia.	186.94	229.66	0.63	23.65	nr	**253.31**
203 mm dia.	197.96	243.19	0.63	23.65	nr	**266.84**
254 mm dia.	255.94	314.42	0.86	32.28	nr	**346.70**
304 mm dia.	277.39	340.78	0.95	35.65	nr	**376.43**
355 mm dia.	313.21	384.78	1.03	38.66	nr	**423.44**
400 mm dia.	502.50	617.32	1.12	42.04	nr	**659.36**
450 mm dia.	516.17	634.12	1.12	42.04	nr	**676.16**
500 mm dia.	566.39	695.81	1.19	44.66	nr	**740.47**
550 mm dia.	592.26	727.59	1.19	44.66	nr	**772.25**
600 mm dia.	602.33	739.97	1.19	44.66	nr	**784.63**
Adapters						
127 mm dia.	14.34	17.62	0.49	18.39	nr	**36.01**
152 mm dia.	15.69	19.28	0.51	19.14	nr	**38.42**
178 mm dia.	16.78	20.61	0.54	20.28	nr	**40.89**
203 mm dia.	19.09	23.46	0.58	21.77	nr	**45.23**
254 mm dia.	21.01	25.81	0.70	26.28	nr	**52.09**
304 mm dia.	26.21	32.20	0.74	27.77	nr	**59.97**
355 mm dia.	31.83	39.10	0.80	30.02	nr	**69.12**
400 mm dia.	35.70	43.86	0.89	33.40	nr	**77.26**
450 mm dia.	38.14	46.85	0.89	33.40	nr	**80.25**
500 mm dia.	40.48	49.73	0.96	36.04	nr	**85.77**
550 mm dia.	47.06	57.81	0.96	36.04	nr	**93.85**
600 mm dia.	56.56	69.49	0.96	36.04	nr	**105.53**

38 MECHANICAL/COOLING/HEATING SYSTEMS

Item	Net Price £	Material £	Labour hours	Labour £	Unit	Total rate £
Fittings for flue system						
90° insulated tee; including two locking bands						
127 mm dia.	166.64	204.72	1.89	70.94	nr	**275.66**
152 mm dia.	192.33	236.28	2.04	76.56	nr	**312.84**
178 mm dia.	210.22	258.26	2.39	89.69	nr	**347.95**
203 mm dia.	246.14	302.38	2.56	96.09	nr	**398.47**
254 mm dia.	249.35	306.33	2.95	110.71	nr	**417.04**
304 mm dia.	310.08	380.93	3.41	127.98	nr	**508.91**
355 mm dia.	395.31	485.64	3.77	141.49	nr	**627.13**
400 mm dia.	520.91	639.93	4.25	159.51	nr	**799.44**
450 mm dia.	542.93	666.99	4.76	178.65	nr	**845.64**
500 mm dia.	614.37	754.75	5.12	192.16	nr	**946.91**
550 mm dia.	660.40	811.30	5.61	210.56	nr	**1021.86**
600 mm dia.	687.16	844.18	5.98	224.44	nr	**1068.62**
135° insulated tee; including two locking bands						
127 mm dia.	216.13	265.51	1.89	70.94	nr	**336.45**
152 mm dia.	233.69	287.09	2.04	76.56	nr	**363.65**
178 mm dia.	255.50	313.89	2.39	89.69	nr	**403.58**
203 mm dia.	324.27	398.37	2.56	96.09	nr	**494.46**
254 mm dia.	368.45	452.64	2.95	110.71	nr	**563.35**
304 mm dia.	277.62	341.06	3.41	127.98	nr	**469.04**
355 mm dia.	546.25	671.07	3.77	141.49	nr	**812.56**
400 mm dia.	719.06	883.36	4.25	159.51	nr	**1042.87**
450 mm dia.	778.43	956.30	4.76	178.65	nr	**1134.95**
500 mm dia.	906.86	1114.08	5.12	192.16	nr	**1306.24**
550 mm dia.	930.44	1143.04	5.61	210.56	nr	**1353.60**
600 mm dia.	981.98	1206.37	5.98	224.44	nr	**1430.81**
Wall sleeve; for 135° tee through wall						
127 mm dia.	22.12	27.17	1.89	70.94	nr	**98.11**
152 mm dia.	29.68	36.47	2.04	76.56	nr	**113.03**
178 mm dia.	31.09	38.20	2.39	89.69	nr	**127.89**
203 mm dia.	34.97	42.96	2.56	96.09	nr	**139.05**
254 mm dia.	39.00	47.91	2.95	110.71	nr	**158.62**
304 mm dia.	45.45	55.84	3.41	127.98	nr	**183.82**
355 mm dia.	50.49	62.02	3.77	141.49	nr	**203.51**
15° insulated elbow; including two locking bands						
127 mm dia.	115.64	142.06	1.57	58.92	nr	**200.98**
152 mm dia.	128.53	157.90	1.79	67.18	nr	**225.08**
178 mm dia.	137.45	168.86	2.05	76.94	nr	**245.80**
203 mm dia.	145.73	179.02	2.33	87.44	nr	**266.46**
254 mm dia.	150.10	184.40	2.45	91.95	nr	**276.35**
304 mm dia.	189.76	233.12	3.43	128.73	nr	**361.85**
355 mm dia.	253.85	311.85	4.71	176.77	nr	**488.62**
30° insulated elbow; including two locking bands						
127 mm dia.	115.64	142.06	1.44	54.04	nr	**196.10**
152 mm dia.	128.53	157.90	1.62	60.80	nr	**218.70**
178 mm dia.	137.45	168.86	1.89	70.94	nr	**239.80**
203 mm dia.	145.73	179.02	2.17	81.44	nr	**260.46**
254 mm dia.	150.10	184.40	2.16	81.07	nr	**265.47**
304 mm dia.	189.76	233.12	2.74	102.84	nr	**335.96**
355 mm dia.	253.38	311.27	3.17	118.97	nr	**430.24**

38 MECHANICAL/COOLING/HEATING SYSTEMS

Item	Net Price £	Material £	Labour hours	Labour £	Unit	Total rate £
GAS/OIL FIRED BOILERS – FLUE SYSTEMS – cont						
30° insulated elbow mm long – cont						
400 mm dia.	253.85	311.85	3.53	132.48	nr	**444.33**
450 mm dia.	294.20	361.42	3.88	145.61	nr	**507.03**
500 mm dia.	308.43	378.90	4.24	159.13	nr	**538.03**
550 mm dia.	331.02	406.65	4.61	173.02	nr	**579.67**
600 mm dia.	360.94	443.41	4.96	186.15	nr	**629.56**
45° insulated elbow; including two locking bands						
127 mm dia.	115.64	142.06	1.44	54.04	nr	**196.10**
152 mm dia.	128.53	157.90	1.51	56.67	nr	**214.57**
178 mm dia.	137.45	168.86	1.58	59.30	nr	**228.16**
203 mm dia.	145.73	179.02	1.66	62.30	nr	**241.32**
254 mm dia.	150.10	184.40	1.72	64.55	nr	**248.95**
304 mm dia.	189.76	233.12	1.80	67.56	nr	**300.68**
355 mm dia.	253.85	311.85	1.94	72.82	nr	**384.67**
400 mm dia.	323.72	397.69	2.01	75.44	nr	**473.13**
450 mm dia.	339.08	416.56	2.09	78.43	nr	**494.99**
500 mm dia.	363.72	446.83	2.16	81.07	nr	**527.90**
550 mm dia.	396.69	487.34	2.23	83.70	nr	**571.04**
600 mm dia.	406.74	499.68	2.30	86.33	nr	**586.01**
Flue supports						
Wall support, galvanized; including plate and brackets						
127 mm dia.	70.21	86.26	2.24	84.06	nr	**170.32**
152 mm dia.	77.19	94.83	2.44	91.57	nr	**186.40**
178 mm dia.	85.22	104.70	2.52	94.58	nr	**199.28**
203 mm dia.	88.74	109.02	2.77	103.96	nr	**212.98**
254 mm dia.	105.83	130.01	2.98	111.85	nr	**241.86**
304 mm dia.	119.65	146.99	3.46	129.85	nr	**276.84**
355 mm dia.	160.16	196.76	4.08	153.12	nr	**349.88**
400 mm dia.; with 300 mm support length and collar	411.39	505.40	4.80	180.15	nr	**685.55**
450 mm dia.; with 300 mm support length and collar	438.62	538.85	5.62	210.92	nr	**749.77**
500 mm dia.; with 300 mm support length and collar	479.18	588.67	6.24	234.19	nr	**822.86**
550 mm dia.; with 300 mm support length and collar	524.25	644.04	6.97	261.59	nr	**905.63**
600 mm dia.; with 300 mm support length and collar	553.42	679.88	7.49	281.11	nr	**960.99**
Ceiling/floor support						
127 mm dia.	21.74	26.71	1.86	69.80	nr	**96.51**
152 mm dia.	24.16	29.68	2.14	80.31	nr	**109.99**
178 mm dia.	28.93	35.54	1.93	72.43	nr	**107.97**
203 mm dia.	43.55	53.50	2.74	102.84	nr	**156.34**
254 mm dia.	49.69	61.05	3.21	120.48	nr	**181.53**
304 mm dia.	56.79	69.76	3.68	138.11	nr	**207.87**
355 mm dia.	67.68	83.15	4.28	160.63	nr	**243.78**

38 MECHANICAL/COOLING/HEATING SYSTEMS

Item	Net Price £	Material £	Labour hours	Labour £	Unit	**Total rate £**
400 mm dia.	86.45	106.21	4.86	182.40	nr	**288.61**
450 mm dia.	93.88	115.33	5.46	204.92	nr	**320.25**
500 mm dia.	113.98	140.03	6.04	226.68	nr	**366.71**
550 mm dia.	139.44	171.30	6.65	249.58	nr	**420.88**
600 mm dia.	200.10	245.83	7.24	271.72	nr	**517.55**
Ceiling/floor firestop spacer						
127 mm dia.	4.22	5.19	0.66	24.77	nr	**29.96**
152 mm dia.	4.71	5.79	0.69	25.89	nr	**31.68**
178 mm dia.	5.32	6.53	0.70	26.28	nr	**32.81**
203 mm dia.	6.31	7.75	0.87	32.65	nr	**40.40**
254 mm dia.	6.51	8.00	0.91	34.16	nr	**42.16**
304 mm dia.	7.83	9.62	0.95	35.65	nr	**45.27**
355 mm dia.	14.25	17.50	0.99	37.15	nr	**54.65**
Wall band; internal or external use						
127 mm dia.	25.60	31.45	1.03	38.66	nr	**70.11**
152 mm dia.	26.73	32.83	1.07	40.16	nr	**72.99**
178 mm dia.	27.78	34.13	1.11	41.65	nr	**75.78**
203 mm dia.	29.33	36.04	1.18	44.29	nr	**80.33**
254 mm dia.	30.53	37.51	1.30	48.79	nr	**86.30**
304 mm dia.	32.98	40.52	1.45	54.42	nr	**94.94**
355 mm dia.	35.09	43.11	1.65	61.93	nr	**105.04**
400 mm dia.	43.52	53.47	1.85	69.44	nr	**122.91**
450 mm dia.	46.37	56.96	2.39	89.69	nr	**146.65**
500 mm dia.	55.74	68.48	2.25	84.45	nr	**152.93**
550 mm dia.	58.38	71.72	2.45	91.95	nr	**163.67**
600 mm dia.	61.73	75.83	2.66	99.82	nr	**175.65**
Flashings and terminals						
Insulated top stub; including one locking band						
127 mm dia.	63.05	77.46	1.49	55.92	nr	**133.38**
152 mm dia.	71.29	87.58	1.90	71.32	nr	**158.90**
178 mm dia.	76.76	94.30	1.92	72.06	nr	**166.36**
203 mm dia.	81.68	100.35	2.20	82.57	nr	**182.92**
254 mm dia.	86.72	106.53	2.49	93.45	nr	**199.98**
304 mm dia.	118.55	145.63	2.79	104.71	nr	**250.34**
355 mm dia.	156.52	192.29	3.19	119.72	nr	**312.01**
400 mm dia.	150.90	185.38	3.59	134.73	nr	**320.11**
450 mm dia.	163.95	201.41	3.97	148.99	nr	**350.40**
500 mm dia.	189.66	233.00	4.38	164.39	nr	**397.39**
550 mm dia.	198.45	243.80	4.78	179.40	nr	**423.20**
600 mm dia.	205.38	252.30	5.17	194.03	nr	**446.33**
Rain cap; including one locking band						
127 mm dia.	33.71	41.41	1.49	55.92	nr	**97.33**
152 mm dia.	35.23	43.28	1.54	57.80	nr	**101.08**
178 mm dia.	38.79	47.65	1.72	64.55	nr	**112.20**
203 mm dia.	46.38	56.97	2.00	75.06	nr	**132.03**
254 mm dia.	60.99	74.93	2.49	93.45	nr	**168.38**
304 mm dia.	41.76	51.30	2.80	105.08	nr	**156.38**
355 mm dia.	110.17	135.35	3.19	119.72	nr	**255.07**

38 MECHANICAL/COOLING/HEATING SYSTEMS

Item	Net Price £	Material £	Labour hours	Labour £	Unit	Total rate £
GAS/OIL FIRED BOILERS – FLUE SYSTEMS – cont						
Flashings and terminals – cont						
Rain cap – cont						
400 mm dia.	109.91	135.02	3.45	129.49	nr	**264.51**
450 mm dia.	119.54	146.85	3.97	148.99	nr	**295.84**
500 mm dia.	129.24	158.77	4.38	164.39	nr	**323.16**
550 mm dia.	138.67	170.35	4.78	179.40	nr	**349.75**
600 mm dia.	148.26	182.13	5.17	194.03	nr	**376.16**
Round top; including one locking band						
127 mm dia.	64.45	79.18	1.49	55.92	nr	**135.10**
152 mm dia.	70.27	86.33	1.65	61.93	nr	**148.26**
178 mm dia.	80.11	98.42	1.92	72.06	nr	**170.48**
203 mm dia.	94.27	115.81	2.20	82.57	nr	**198.38**
254 mm dia.	111.46	136.93	2.49	93.45	nr	**230.38**
304 mm dia.	146.40	179.85	2.80	105.08	nr	**284.93**
355 mm dia.	195.31	239.94	3.19	119.72	nr	**359.66**
Coping cap; including one locking band						
127 mm dia.	36.10	44.35	1.49	55.92	nr	**100.27**
152 mm dia.	37.81	46.45	1.65	61.93	nr	**108.38**
178 mm dia.	41.60	51.10	1.92	72.06	nr	**123.16**
203 mm dia.	49.92	61.33	2.20	82.57	nr	**143.90**
254 mm dia.	60.99	74.93	2.49	93.45	nr	**168.38**
304 mm dia.	82.23	101.02	2.79	104.71	nr	**205.73**
355 mm dia.	110.17	135.35	3.19	119.72	nr	**255.07**
Storm collar						
127 mm dia.	6.84	8.40	0.52	19.52	nr	**27.92**
152 mm dia.	7.32	8.99	0.55	20.64	nr	**29.63**
178 mm dia.	8.12	9.97	0.57	21.39	nr	**31.36**
203 mm dia.	8.50	10.45	0.66	24.77	nr	**35.22**
254 mm dia.	10.64	13.07	0.66	24.77	nr	**37.84**
304 mm dia.	11.09	13.63	0.72	27.03	nr	**40.66**
355 mm dia.	11.86	14.57	0.77	28.90	nr	**43.47**
400 mm dia.	31.44	38.62	0.82	30.78	nr	**69.40**
450 mm dia.	34.61	42.51	0.87	32.65	nr	**75.16**
500 mm dia.	37.73	46.35	0.92	34.53	nr	**80.88**
550 mm dia.	40.89	50.23	0.98	36.78	nr	**87.01**
600 mm dia.	44.03	54.10	1.03	38.66	nr	**92.76**
Flat flashing; including storm collar and sealant						
127 mm dia.	38.43	47.21	1.49	55.92	nr	**103.13**
152 mm dia.	39.71	48.78	1.65	61.93	nr	**110.71**
178 mm dia.	41.66	51.18	1.92	72.06	nr	**123.24**
203 mm dia.	45.62	56.05	2.20	82.57	nr	**138.62**
254 mm dia.	62.35	76.60	2.49	93.45	nr	**170.05**
304 mm dia.	74.49	91.51	2.80	105.08	nr	**196.59**
355 mm dia.	117.39	144.22	3.20	120.10	nr	**264.32**
400 mm dia.	163.76	201.18	3.59	134.73	nr	**335.91**
450 mm dia.	188.13	231.12	3.97	148.99	nr	**380.11**
500 mm dia.	203.81	250.38	4.38	164.39	nr	**414.77**
550 mm dia.	216.40	265.85	4.78	179.40	nr	**445.25**
600 mm dia.	224.22	275.46	5.17	194.03	nr	**469.49**

38 MECHANICAL/COOLING/HEATING SYSTEMS

Item	Net Price £	Material £	Labour hours	Labour £	Unit	**Total rate £**
5°–30° rigid adjustable flashing; including storm collar and sealant						
127 mm dia.	65.44	80.39	1.49	55.92	nr	**136.31**
152 mm dia.	68.96	84.71	1.65	61.93	nr	**146.64**
178 mm dia.	73.33	90.09	1.92	72.06	nr	**162.15**
203 mm dia.	77.11	94.73	2.20	82.57	nr	**177.30**
254 mm dia.	81.17	99.72	2.49	93.45	nr	**193.17**
304 mm dia.	100.45	123.41	2.80	105.08	nr	**228.49**
355 mm dia.	114.25	140.35	3.19	119.72	nr	**260.07**
400 mm dia.	368.33	452.50	3.59	134.73	nr	**587.23**
450 mm dia.	429.85	528.07	3.97	148.99	nr	**677.06**
500 mm dia.	458.89	563.74	4.38	164.39	nr	**728.13**
550 mm dia.	480.20	589.92	4.77	179.01	nr	**768.93**
600 mm dia.	522.46	641.84	5.17	194.03	nr	**835.87**
Domestic and small commercial; twin walled gas vent system suitable for gas fired appliances; domestic gas boilers; small commercial boilers with internal or external flues						
152 mm long						
100 mm dia.	7.00	8.60	0.52	19.52	nr	**28.12**
125 mm dia.	8.61	10.57	0.52	19.52	nr	**30.09**
150 mm dia.	9.32	11.45	0.52	19.52	nr	**30.97**
305 mm long						
100 mm dia.	10.62	13.05	0.52	19.52	nr	**32.57**
125 mm dia.	12.47	15.32	0.52	19.52	nr	**34.84**
150 mm dia.	14.79	18.16	0.52	19.52	nr	**37.68**
457 mm long						
100 mm dia.	11.75	14.44	0.55	20.64	nr	**35.08**
125 mm dia.	13.21	16.23	0.55	20.64	nr	**36.87**
150 mm dia.	16.35	20.09	0.55	20.64	nr	**40.73**
914 mm long						
100 mm dia.	21.00	25.80	0.62	23.27	nr	**49.07**
125 mm dia.	24.48	30.07	0.62	23.27	nr	**53.34**
150 mm dia.	28.06	34.47	0.62	23.27	nr	**57.74**
1524 mm long						
100 mm dia.	30.33	37.26	0.82	30.78	nr	**68.04**
125 mm dia.	37.33	45.86	0.84	31.52	nr	**77.38**
150 mm dia.	40.07	49.22	0.84	31.52	nr	**80.74**
Adjustable length 305 mm long						
100 mm dia.	13.45	16.53	0.56	21.02	nr	**37.55**
125 mm dia.	15.10	18.55	0.56	21.02	nr	**39.57**
150 mm dia.	19.03	23.38	0.56	21.02	nr	**44.40**
Adjustable length 457 mm long						
100 mm dia.	18.13	22.27	0.56	21.02	nr	**43.29**
125 mm dia.	21.98	27.01	0.56	21.02	nr	**48.03**
150 mm dia.	24.46	30.05	0.56	21.02	nr	**51.07**

38 MECHANICAL/COOLING/HEATING SYSTEMS

Item	Net Price £	Material £	Labour hours	Labour £	Unit	Total rate £
GAS/OIL FIRED BOILERS – FLUE SYSTEMS – cont						
Domestic and small commercial – cont						
Adjustable elbow 0°–90°						
100 mm dia.	15.34	18.85	0.48	18.01	nr	**36.86**
125 mm dia.	18.13	22.27	0.48	18.01	nr	**40.28**
150 mm dia.	22.71	27.90	0.48	18.01	nr	**45.91**
Draughthood connector						
100 mm dia.	4.73	5.81	0.48	18.01	nr	**23.82**
125 mm dia.	5.33	6.55	0.48	18.01	nr	**24.56**
150 mm dia.	5.79	7.11	0.48	18.01	nr	**25.12**
Adaptor						
100 mm dia.	11.47	14.09	0.48	18.01	nr	**32.10**
125 mm dia.	11.72	14.40	0.48	18.01	nr	**32.41**
150 mm dia.	11.97	14.70	0.48	18.01	nr	**32.71**
Support plate						
100 mm dia.	8.28	10.17	0.48	18.01	nr	**28.18**
125 mm dia.	8.81	10.83	0.48	18.01	nr	**28.84**
150 mm dia.	9.43	11.58	0.48	18.01	nr	**29.59**
Wall band						
100 mm dia.	7.49	9.20	0.48	18.01	nr	**27.21**
125 mm dia.	7.98	9.81	0.48	18.01	nr	**27.82**
150 mm dia.	10.12	12.43	0.48	18.01	nr	**30.44**
Firestop						
100 mm dia.	3.24	3.98	0.48	18.01	nr	**21.99**
125 mm dia.	3.24	3.98	0.48	18.01	nr	**21.99**
150 mm dia.	3.73	4.58	0.48	18.01	nr	**22.59**
Flat flashing						
125 mm dia.	22.08	27.12	0.55	20.64	nr	**47.76**
150 mm dia.	30.71	37.73	0.55	20.64	nr	**58.37**
Adjustable flashing 5°–30°						
100 mm dia.	57.53	70.68	0.55	20.64	nr	**91.32**
125 mm dia.	89.81	110.33	0.55	20.64	nr	**130.97**
Storm collar						
100 mm dia.	4.59	5.64	0.55	20.64	nr	**26.28**
125 mm dia.	4.69	5.76	0.55	20.64	nr	**26.40**
150 mm dia.	4.80	5.90	0.55	20.64	nr	**26.54**
Gas vent terminal						
100 mm dia.	17.51	21.51	0.55	20.64	nr	**42.15**
125 mm dia.	19.24	23.64	0.55	20.64	nr	**44.28**
150 mm dia.	24.68	30.32	0.55	20.64	nr	**50.96**
Twin wall galvanized steel flue box, 125 mm dia.; fitted for gas fire, where no chimney exists						
Free standing	111.24	136.66	2.15	80.69	nr	**217.35**
Recess	111.24	136.66	2.15	80.69	nr	**217.35**
Back boiler	81.96	100.68	2.40	90.07	nr	**190.75**

38 MECHANICAL/COOLING/HEATING SYSTEMS

Item	Net Price £	Material £	Labour hours	Labour £	Unit	**Total rate £**
PACKAGED STEAM GENERATORS						
Packaged steam boilers; boiler mountings centrifugal water feed pump; insulation; and sheet steel wrap around casing; plastic coated						
Gas fired						
104 kW	12532.00	15395.56	71.00	2664.68	nr	**18060.24**
276 kW	17042.00	20936.10	86.00	3227.64	nr	**24163.74**
1384 kW	40022.00	49167.03	148.00	5554.54	nr	**54721.57**
2940 KW	57899.00	71128.92	207.00	7768.84	nr	**78897.76**
4843 kW	68308.00	83916.38	295.00	11071.55	nr	**94987.93**
Gas oil fired						
104 kW	9791.00	12028.24	86.45	3244.53	nr	**15272.77**
276 kW	14083.00	17300.97	148.22	5562.79	nr	**22863.76**
1384 kW	35791.00	43969.24	207.50	7787.62	nr	**51756.86**
2940 kW	54280.00	66682.98	207.00	7768.84	nr	**74451.82**
4843 kW	64804.00	79611.71	295.00	11071.55	nr	**90683.26**

38 MECHANICAL/COOLING/HEATING SYSTEMS

Item	Net Price £	Material £	Labour hours	Labour £	Unit	Total rate £
LOW TEMPERATURE HOT WATER HEATING; PIPELINE: SCREWED STEEL						
Black steel pipes; screwed and socketed joints; BS 1387: 1985. Fixed vertically, brackets measured separately. Screwed joints are within the running length, but any flanges are additional						
Medium weight						
10 mm dia.	4.71	5.79	0.37	13.88	m	**19.67**
15 mm dia.	2.97	3.64	0.37	13.88	m	**17.52**
20 mm dia.	3.50	4.29	0.37	13.88	m	**18.17**
25 mm dia.	5.02	6.16	0.41	15.39	m	**21.55**
32 mm dia.	6.22	7.64	0.48	18.01	m	**25.65**
40 mm dia.	7.23	8.88	0.52	19.52	m	**28.40**
50 mm dia.	10.16	12.48	0.62	23.27	m	**35.75**
65 mm dia.	13.80	16.96	0.65	24.40	m	**41.36**
80 mm dia.	17.94	22.04	1.10	41.28	m	**63.32**
100 mm dia.	25.41	31.22	1.31	49.16	m	**80.38**
125 mm dia.	32.34	39.73	1.66	62.30	m	**102.03**
150 mm dia.	37.55	46.13	1.88	70.55	m	**116.68**
Heavy weight						
15 mm dia.	3.54	4.35	0.37	13.88	m	**18.23**
20 mm dia.	4.19	5.15	0.37	13.88	m	**19.03**
25 mm dia.	6.14	7.54	0.41	15.39	m	**22.93**
32 mm dia.	7.62	9.37	0.48	18.01	m	**27.38**
40 mm dia.	8.88	10.91	0.52	19.52	m	**30.43**
50 mm dia.	12.33	15.15	0.62	23.27	m	**38.42**
65 mm dia.	16.78	20.61	0.64	24.01	m	**44.62**
80 mm dia.	21.35	26.23	1.10	41.28	m	**67.51**
100 mm dia.	29.79	36.59	1.31	49.16	m	**85.75**
125 mm dia.	34.48	42.36	1.66	62.30	m	**104.66**
150 mm dia.	40.32	49.53	1.88	70.55	m	**120.08**
200 mm dia.	63.53	78.05	2.99	112.21	m	**190.26**
250 mm dia.	79.66	97.86	3.49	130.99	m	**228.85**
300 mm dia.	94.85	116.52	3.91	146.75	m	**263.27**
Black steel pipes; screwed and socketed joints; BS 1387: 1985. Fixed at high level or suspended, brackets measured separately. Screwed joints are within the running length, but any flanges are additional						
Medium weight						
10 mm dia.	4.71	5.79	0.58	21.77	m	**27.56**
15 mm dia.	2.97	3.64	0.58	21.77	m	**25.41**
20 mm dia.	3.50	4.29	0.58	21.77	m	**26.06**
25 mm dia.	5.02	6.16	0.60	22.52	m	**28.68**
32 mm dia.	6.22	7.64	0.68	25.51	m	**33.15**
40 mm dia.	7.23	8.88	0.73	27.39	m	**36.27**
50 mm dia.	10.16	12.48	0.85	31.90	m	**44.38**

38 MECHANICAL/COOLING/HEATING SYSTEMS

Item	Net Price £	Material £	Labour hours	Labour £	Unit	**Total rate £**
65 mm dia.	13.80	16.96	0.88	33.02	m	**49.98**
80 mm dia.	17.94	22.04	1.45	54.42	m	**76.46**
100 mm dia.	25.41	31.22	1.74	65.31	m	**96.53**
125 mm dia.	32.34	39.73	2.21	82.95	m	**122.68**
150 mm dia.	37.55	46.13	2.50	93.83	m	**139.96**
Heavy weight						
15 mm dia.	3.54	4.35	0.58	21.77	m	**26.12**
20 mm dia.	4.19	5.15	0.58	21.77	m	**26.92**
25 mm dia.	6.14	7.54	0.60	22.52	m	**30.06**
32 mm dia.	7.62	9.37	0.68	25.51	m	**34.88**
40 mm dia.	8.88	10.91	0.73	27.39	m	**38.30**
50 mm dia.	12.33	15.15	0.85	31.90	m	**47.05**
65 mm dia.	16.78	20.61	0.88	33.02	m	**53.63**
80 mm dia.	21.35	26.23	1.45	54.42	m	**80.65**
100 mm dia.	29.79	36.59	1.74	65.31	m	**101.90**
125 mm dia.	34.48	42.36	2.21	82.95	m	**125.31**
150 mm dia.	40.32	49.53	2.50	93.83	m	**143.36**
200 mm dia.	63.53	78.05	2.99	112.21	m	**190.26**
250 mm dia.	79.66	97.86	3.49	130.99	m	**228.85**
300 mm dia.	94.85	116.52	3.91	146.75	m	**263.27**
FIXINGS						
For steel pipes; black malleable iron. For minimum fixing distances, refer to the Tables and Memoranda to the rear of the book						
Single pipe bracket, screw on, black malleable iron; screwed to wood						
15 mm dia.	0.77	0.95	0.14	5.26	nr	**6.21**
20 mm dia.	0.87	1.07	0.14	5.26	nr	**6.33**
25 mm dia.	1.01	1.24	0.17	6.37	nr	**7.61**
32 mm dia.	1.84	2.26	0.19	7.13	nr	**9.39**
40 mm dia.	1.75	2.15	0.22	8.25	nr	**10.40**
50 mm dia.	2.45	3.01	0.22	8.25	nr	**11.26**
65 mm dia.	3.22	3.96	0.28	10.50	nr	**14.46**
80 mm dia.	4.41	5.42	0.32	12.01	nr	**17.43**
100 mm dia.	6.44	7.91	0.35	13.14	nr	**21.05**
Single pipe bracket, screw on, black malleable iron; plugged and screwed						
15 mm dia.	0.77	0.95	0.25	9.39	nr	**10.34**
20 mm dia.	0.87	1.07	0.25	9.39	nr	**10.46**
25 mm dia.	1.01	1.24	0.30	11.26	nr	**12.50**
32 mm dia.	1.84	2.26	0.32	12.01	nr	**14.27**
40 mm dia.	1.75	2.15	0.32	12.01	nr	**14.16**
50 mm dia.	2.45	3.01	0.32	12.01	nr	**15.02**
65 mm dia.	3.22	3.96	0.35	13.14	nr	**17.10**
80 mm dia.	4.41	5.42	0.42	15.76	nr	**21.18**
100 mm dia.	6.44	7.91	0.42	15.76	nr	**23.67**

38 MECHANICAL/COOLING/HEATING SYSTEMS

Item	Net Price £	Material £	Labour hours	Labour £	Unit	Total rate £
LOW TEMPERATURE HOT WATER HEATING; PIPELINE: SCREWED STEEL – cont						
For steel pipes – cont						
Single pipe bracket for building in, black malleable iron						
15 mm dia.	2.27	2.79	0.10	3.76	nr	**6.55**
20 mm dia.	2.27	2.79	0.11	4.13	nr	**6.92**
25 mm dia.	2.27	2.79	0.12	4.50	nr	**7.29**
32 mm dia.	2.58	3.17	0.14	5.26	nr	**8.43**
40 mm dia.	2.60	3.19	0.15	5.63	nr	**8.82**
50 mm dia.	2.71	3.33	0.16	6.01	nr	**9.34**
Pipe ring, single socket, black malleable iron						
15 mm dia.	0.90	1.10	0.10	3.76	nr	**4.86**
20 mm dia.	1.00	1.23	0.11	4.13	nr	**5.36**
25 mm dia.	1.09	1.34	0.12	4.50	nr	**5.84**
32 mm dia.	1.16	1.43	0.14	5.26	nr	**6.69**
40 mm dia.	1.50	1.85	0.15	5.63	nr	**7.48**
50 mm dia.	1.89	2.32	0.16	6.01	nr	**8.33**
65 mm dia.	2.73	3.35	0.30	11.26	nr	**14.61**
80 mm dia.	3.28	4.03	0.35	13.14	nr	**17.17**
100 mm dia.	4.96	6.09	0.40	15.02	nr	**21.11**
125 mm dia.	10.03	12.33	0.60	22.53	nr	**34.86**
150 mm dia.	11.23	13.80	0.77	28.90	nr	**42.70**
200 mm dia.	31.07	38.17	0.90	33.78	nr	**71.95**
250 mm dia.	38.84	47.71	1.10	41.28	nr	**88.99**
300 mm dia.	46.61	57.26	1.25	46.91	nr	**104.17**
350 mm dia.	54.39	66.82	1.50	56.30	nr	**123.12**
400 mm dia.	64.37	79.08	1.75	65.68	nr	**144.76**
Pipe ring, double socket, black malleable iron						
15 mm dia.	1.03	1.27	0.10	3.76	nr	**5.03**
20 mm dia.	1.20	1.47	0.11	4.13	nr	**5.60**
25 mm dia.	1.34	1.65	0.12	4.50	nr	**6.15**
32 mm dia.	1.59	1.95	0.14	5.26	nr	**7.21**
40 mm dia.	1.71	2.10	0.15	5.63	nr	**7.73**
50 mm dia.	2.68	3.30	0.16	6.01	nr	**9.31**
Screw on backplate (Male), black malleable iron; plugged and screwed						
M12	0.54	0.66	0.10	3.76	nr	**4.42**
Screw on backplate (Female), black malleable iron; plugged and screwed						
M12	0.54	0.66	0.10	3.76	nr	**4.42**
Extra over channel sections for fabricated hangers and brackets						
Galvanized steel; including inserts, bolts, nuts, washers; fixed to backgrounds						
41 × 21 mm	6.02	7.39	0.29	10.89	m	**18.28**
41 × 41 mm	7.22	8.87	0.29	10.89	m	**19.76**

38 MECHANICAL/COOLING/HEATING SYSTEMS

Item	Net Price £	Material £	Labour hours	Labour £	Unit	Total rate £
Threaded rods; metric thread; including nuts, washers etc.						
12 mm dia. × 600 mm long	3.04	3.74	0.18	6.76	nr	**10.50**
Pipe roller and chair						
Roller and chair; black malleable						
Up to 50 mm dia.	16.38	20.12	0.20	7.51	nr	**27.63**
65 mm dia.	16.86	20.72	0.20	7.51	nr	**28.23**
80 mm dia.	18.31	22.49	0.20	7.51	nr	**30.00**
100 mm dia.	18.63	22.89	0.20	7.51	nr	**30.40**
125 mm dia.	26.10	32.07	0.20	7.51	nr	**39.58**
150 mm dia.	26.10	32.07	0.30	11.26	nr	**43.33**
175 mm dia.	44.04	54.11	0.30	11.26	nr	**65.37**
200 mm dia.	67.46	82.88	0.30	11.26	nr	**94.14**
250 mm dia.	88.01	108.12	0.30	11.26	nr	**119.38**
300 mm dia.	110.82	136.14	0.30	11.26	nr	**147.40**
Roller bracket; black malleable						
25 mm dia.	3.35	4.12	0.20	7.51	nr	**11.63**
32 mm dia.	3.52	4.33	0.20	7.51	nr	**11.84**
40 mm dia.	3.76	4.62	0.20	7.51	nr	**12.13**
50 mm dia.	3.97	4.87	0.20	7.51	nr	**12.38**
65 mm dia.	5.22	6.42	0.20	7.51	nr	**13.93**
80 mm dia.	7.52	9.24	0.20	7.51	nr	**16.75**
100 mm dia.	8.36	10.27	0.20	7.51	nr	**17.78**
125 mm dia.	13.80	16.96	0.20	7.51	nr	**24.47**
150 mm dia.	13.80	16.96	0.30	11.26	nr	**28.22**
175 mm dia.	30.76	37.79	0.30	11.26	nr	**49.05**
200 mm dia.	30.76	37.79	0.30	11.26	nr	**49.05**
250 mm dia.	40.89	50.23	0.30	11.26	nr	**61.49**
300 mm dia.	50.23	61.71	0.30	11.26	nr	**72.97**
350 mm dia.	80.76	99.21	0.30	11.26	nr	**110.47**
400 mm dia.	92.32	113.41	0.30	11.26	nr	**124.67**
Extra over black steel screwed pipes; black steel flanges, screwed and drilled; metric; BS 4504						
Screwed flanges; PN6						
15 mm dia.	8.54	10.49	0.35	13.14	nr	**23.63**
20 mm dia.	8.54	10.49	0.47	17.64	nr	**28.13**
25 mm dia.	8.54	10.49	0.53	19.89	nr	**30.38**
32 mm dia.	8.54	10.49	0.62	23.27	nr	**33.76**
40 mm dia.	8.54	10.49	0.70	26.28	nr	**36.77**
50 mm dia.	9.10	11.18	0.84	31.52	nr	**42.70**
65 mm dia.	12.65	15.54	1.03	38.66	nr	**54.20**
80 mm dia.	17.86	21.94	1.23	46.17	nr	**68.11**
100 mm dia.	21.11	25.93	1.41	52.92	nr	**78.85**
125 mm dia.	38.74	47.60	1.77	66.42	nr	**114.02**
150 mm dia.	38.74	47.60	2.21	82.95	nr	**130.55**

38 MECHANICAL/COOLING/HEATING SYSTEMS

Item	Net Price £	Material £	Labour hours	Labour £	Unit	Total rate £
LOW TEMPERATURE HOT WATER HEATING; PIPELINE: SCREWED STEEL – cont						
Extra over black steel screwed pipes – cont						
Screwed flanges; PN16						
15 mm dia.	10.84	13.31	0.35	13.14	nr	**26.45**
20 mm dia.	10.84	13.31	0.47	17.64	nr	**30.95**
25 mm dia.	10.84	13.31	0.53	19.89	nr	**33.20**
32 mm dia.	11.59	14.24	0.62	23.27	nr	**37.51**
40 mm dia.	11.59	14.24	0.70	26.28	nr	**40.52**
50 mm dia.	14.13	17.36	0.84	31.52	nr	**48.88**
65 mm dia.	17.65	21.68	1.03	38.66	nr	**60.34**
80 mm dia.	21.49	26.40	1.23	46.17	nr	**72.57**
100 mm dia.	24.11	29.62	1.41	52.92	nr	**82.54**
125 mm dia.	42.18	51.82	1.77	66.42	nr	**118.24**
150 mm dia.	41.50	50.98	2.21	82.95	nr	**133.93**
Extra over black steel screwed pipes; black steel flanges, screwed and drilled; imperial; BS 10						
Screwed flanges; Table E						
½" dia.	16.20	19.90	0.35	13.14	nr	**33.04**
¾" dia.	16.20	19.90	0.47	17.64	nr	**37.54**
1" dia.	16.17	19.87	0.53	19.89	nr	**39.76**
1 ¼" dia.	16.20	19.90	0.62	23.27	nr	**43.17**
1 ½" dia.	16.20	19.90	0.70	26.28	nr	**46.18**
2" dia.	16.20	19.90	0.84	31.52	nr	**51.42**
2 ½" dia.	19.27	23.68	1.03	38.66	nr	**62.34**
3" dia.	21.84	26.83	1.23	46.17	nr	**73.00**
4" dia.	29.55	36.30	1.41	52.92	nr	**89.22**
5" dia.	62.60	76.90	1.77	66.42	nr	**143.32**
6" dia.	62.60	76.90	2.21	82.95	nr	**159.85**
Extra over black steel screwed pipes; black steel flange connections						
Bolted connection between pair of flanges; including gasket, bolts, nuts and washers						
50 mm dia.	8.12	9.97	0.53	19.89	nr	**29.86**
65 mm dia.	9.61	11.80	0.53	19.89	nr	**31.69**
80 mm dia.	14.43	17.72	0.53	19.89	nr	**37.61**
100 mm dia.	15.67	19.25	0.61	22.90	nr	**42.15**
125 mm dia.	18.43	22.64	0.61	22.90	nr	**45.54**
150 mm dia.	26.33	32.35	0.90	33.78	nr	**66.13**
Extra over black steel screwed pipes; black heavy steel tubular fittings; BS 1387						

38 MECHANICAL/COOLING/HEATING SYSTEMS

Item	Net Price £	Material £	Labour hours	Labour £	Unit	**Total rate £**
Long screw connection with socket and backnut						
15 mm dia.	5.62	6.91	0.63	23.65	nr	**30.56**
20 mm dia.	7.07	8.68	0.84	31.52	nr	**40.20**
25 mm dia.	9.26	11.37	0.95	35.65	nr	**47.02**
32 mm dia.	12.42	15.26	1.11	41.65	nr	**56.91**
40 mm dia.	14.94	18.35	1.28	48.04	nr	**66.39**
50 mm dia.	22.00	27.03	1.53	57.42	nr	**84.45**
65 mm dia.	30.78	37.81	1.87	70.18	nr	**107.99**
80 mm dia.	70.37	86.45	2.21	82.95	nr	**169.40**
100 mm dia.	114.01	140.06	3.05	114.47	nr	**254.53**
Running nipple						
15 mm dia.	1.45	1.78	0.50	18.77	nr	**20.55**
20 mm dia.	1.84	2.26	0.68	25.51	nr	**27.77**
25 mm dia.	2.26	2.77	0.77	28.90	nr	**31.67**
32 mm dia.	3.16	3.89	0.90	33.78	nr	**37.67**
40 mm dia.	4.26	5.23	1.03	38.66	nr	**43.89**
50 mm dia.	6.48	7.96	1.23	46.17	nr	**54.13**
65 mm dia.	13.96	17.15	1.50	56.30	nr	**73.45**
80 mm dia.	21.78	26.75	1.78	66.80	nr	**93.55**
100 mm dia.	34.12	41.92	2.38	89.32	nr	**131.24**
Barrel nipple						
15 mm dia.	0.82	1.01	0.50	18.77	nr	**19.78**
20 mm dia.	1.31	1.61	0.68	25.51	nr	**27.12**
25 mm dia.	1.73	2.12	0.77	28.90	nr	**31.02**
32 mm dia.	2.60	3.19	0.90	33.78	nr	**36.97**
40 mm dia.	3.19	3.92	1.03	38.66	nr	**42.58**
50 mm dia.	4.56	5.61	1.23	46.17	nr	**51.78**
65 mm dia.	9.76	11.99	1.50	56.30	nr	**68.29**
80 mm dia.	13.61	16.72	1.78	66.80	nr	**83.52**
100 mm dia.	24.65	30.28	2.38	89.32	nr	**119.60**
125 mm dia.	45.74	56.20	2.87	107.72	nr	**163.92**
150 mm dia.	72.09	88.57	3.39	127.23	nr	**215.80**
Close taper nipple						
15 mm dia.	1.73	2.12	0.50	18.77	nr	**20.89**
20 mm dia.	2.26	2.77	0.68	25.51	nr	**28.28**
25 mm dia.	2.93	3.60	0.77	28.90	nr	**32.50**
32 mm dia.	4.41	5.42	0.90	33.78	nr	**39.20**
40 mm dia.	5.46	6.71	1.03	38.66	nr	**45.37**
50 mm dia.	8.41	10.33	1.23	46.17	nr	**56.50**
65 mm dia.	16.32	20.04	1.50	56.30	nr	**76.34**
80 mm dia.	20.12	24.72	1.78	66.80	nr	**91.52**
100 mm dia.	41.30	50.74	2.38	89.32	nr	**140.06**
Extra over black steel screwed pipes; black malleable iron fittings; BS 143						
Cap						
15 mm dia.	0.70	0.86	0.32	12.01	nr	**12.87**
20 mm dia.	0.80	0.99	0.43	16.14	nr	**17.13**
25 mm dia.	1.01	1.24	0.49	18.39	nr	**19.63**

38 MECHANICAL/COOLING/HEATING SYSTEMS

Item	Net Price £	Material £	Labour hours	Labour £	Unit	Total rate £
LOW TEMPERATURE HOT WATER HEATING; PIPELINE: SCREWED STEEL – cont						
Extra over black steel screwed pipes – cont						
Cap – cont						
32 mm dia.	1.76	2.16	0.58	21.77	nr	**23.93**
40 mm dia.	2.09	2.57	0.66	24.77	nr	**27.34**
50 mm dia.	4.35	5.34	0.78	29.27	nr	**34.61**
65 mm dia.	7.26	8.91	0.96	36.04	nr	**44.95**
80 mm dia.	8.21	10.09	1.13	42.41	nr	**52.50**
100 mm dia.	17.98	22.09	1.70	63.81	nr	**85.90**
Plain plug, hollow						
15 mm dia.	0.49	0.60	0.28	10.50	nr	**11.10**
20 mm dia.	0.63	0.78	0.38	14.26	nr	**15.04**
25 mm dia.	0.85	1.04	0.44	16.51	nr	**17.55**
32 mm dia.	1.19	1.46	0.51	19.14	nr	**20.60**
40 mm dia.	2.23	2.74	0.59	22.13	nr	**24.87**
50 mm dia.	3.12	3.83	0.70	26.28	nr	**30.11**
65 mm dia.	5.11	6.28	0.85	31.90	nr	**38.18**
80 mm dia.	7.97	9.79	1.00	37.53	nr	**47.32**
100 mm dia.	14.68	18.04	1.44	54.04	nr	**72.08**
Plain plug, solid						
15 mm dia.	1.52	1.87	0.28	10.50	nr	**12.37**
20 mm dia.	1.47	1.81	0.38	14.26	nr	**16.07**
25 mm dia.	2.19	2.69	0.44	16.51	nr	**19.20**
32 mm dia.	3.15	3.87	0.51	19.14	nr	**23.01**
40 mm dia.	4.43	5.44	0.59	22.13	nr	**27.57**
50 mm dia.	5.82	7.15	0.70	26.28	nr	**33.43**
90° elbow, male/female						
15 mm dia.	0.80	0.99	0.64	24.01	nr	**25.00**
20 mm dia.	1.09	1.34	0.85	31.90	nr	**33.24**
25 mm dia.	1.79	2.19	0.97	36.40	nr	**38.59**
32 mm dia.	3.29	4.04	1.12	42.04	nr	**46.08**
40 mm dia.	5.52	6.78	1.29	48.42	nr	**55.20**
50 mm dia.	7.10	8.73	1.55	58.17	nr	**66.90**
65 mm dia.	15.34	18.85	1.89	70.94	nr	**89.79**
80 mm dia.	20.96	25.75	2.24	84.06	nr	**109.81**
100 mm dia.	36.66	45.03	3.09	115.97	nr	**161.00**
90° elbow						
15 mm dia.	0.72	0.88	0.64	24.01	nr	**24.89**
20 mm dia.	0.99	1.22	0.85	31.90	nr	**33.12**
25 mm dia.	1.52	1.87	0.97	36.40	nr	**38.27**
32 mm dia.	2.79	3.42	1.12	42.04	nr	**45.46**
40 mm dia.	4.68	5.75	1.29	48.42	nr	**54.17**
50 mm dia.	5.49	6.74	1.55	58.17	nr	**64.91**
65 mm dia.	11.84	14.54	1.89	70.94	nr	**85.48**
80 mm dia.	17.40	21.38	2.24	84.06	nr	**105.44**
100 mm dia.	33.56	41.23	3.09	115.97	nr	**157.20**
125 mm dia.	81.62	100.28	4.44	166.63	nr	**266.91**
150 mm dia.	151.93	186.65	5.79	217.30	nr	**403.95**

38 MECHANICAL/COOLING/HEATING SYSTEMS

Item	Net Price £	Material £	Labour hours	Labour £	Unit	**Total rate £**
45° elbow						
15 mm dia.	1.72	2.11	0.64	24.01	nr	**26.12**
20 mm dia.	2.11	2.59	0.85	31.90	nr	**34.49**
25 mm dia.	3.15	3.87	0.97	36.40	nr	**40.27**
32 mm dia.	6.44	7.91	1.12	42.04	nr	**49.95**
40 mm dia.	7.90	9.70	1.29	48.42	nr	**58.12**
50 mm dia.	10.82	13.29	1.55	58.17	nr	**71.46**
65 mm dia.	15.83	19.45	1.89	70.94	nr	**90.39**
80 mm dia.	23.79	29.22	2.24	84.06	nr	**113.28**
100 mm dia.	52.17	64.09	3.09	115.97	nr	**180.06**
150 mm dia.	145.68	178.97	5.79	217.30	nr	**396.27**
90° bend, male/female						
15 mm dia.	1.41	1.73	0.64	24.01	nr	**25.74**
20 mm dia.	2.07	2.54	0.85	31.90	nr	**34.44**
25 mm dia.	3.05	3.75	0.97	36.40	nr	**40.15**
32 mm dia.	4.84	5.94	1.12	42.04	nr	**47.98**
40 mm dia.	7.65	9.40	1.29	48.42	nr	**57.82**
50 mm dia.	13.36	16.41	1.55	58.17	nr	**74.58**
65 mm dia.	22.23	27.31	1.89	70.94	nr	**98.25**
80 mm dia.	34.20	42.01	2.24	84.06	nr	**126.07**
100 mm dia.	84.75	104.12	3.09	115.97	nr	**220.09**
90° bend, male						
15 mm dia.	3.27	4.02	0.64	24.01	nr	**28.03**
20 mm dia.	3.65	4.48	0.85	31.90	nr	**36.38**
25 mm dia.	5.36	6.58	0.97	36.40	nr	**42.98**
32 mm dia.	11.63	14.29	1.12	42.04	nr	**56.33**
40 mm dia.	16.33	20.07	1.29	48.42	nr	**68.49**
50 mm dia.	24.78	30.44	1.55	58.17	nr	**88.61**
90° bend, female						
15 mm dia.	1.29	1.59	0.64	24.01	nr	**25.60**
20 mm dia.	1.85	2.27	0.85	31.90	nr	**34.17**
25 mm dia.	2.61	3.20	0.97	36.40	nr	**39.60**
32 mm dia.	4.94	6.07	1.12	42.04	nr	**48.11**
40 mm dia.	6.58	8.09	1.29	48.42	nr	**56.51**
50 mm dia.	9.27	11.39	1.55	58.17	nr	**69.56**
65 mm dia.	19.80	24.33	1.89	70.94	nr	**95.27**
80 mm dia.	31.54	38.74	2.24	84.06	nr	**122.80**
100 mm dia.	69.15	84.96	3.09	115.97	nr	**200.93**
125 mm dia.	163.10	200.37	4.44	166.63	nr	**367.00**
150 mm dia.	296.64	364.42	5.79	217.30	nr	**581.72**
Return bend						
15 mm dia.	7.45	9.16	0.64	24.01	nr	**33.17**
20 mm dia.	12.04	14.79	0.85	31.90	nr	**46.69**
25 mm dia.	15.02	18.45	0.97	36.40	nr	**54.85**
32 mm dia.	23.27	28.59	1.12	42.04	nr	**70.63**
40 mm dia.	27.72	34.05	1.29	48.42	nr	**82.47**
50 mm dia.	42.30	51.96	1.55	58.17	nr	**110.13**

38 MECHANICAL/COOLING/HEATING SYSTEMS

Item	Net Price £	Material £	Labour hours	Labour £	Unit	Total rate £
LOW TEMPERATURE HOT WATER HEATING; PIPELINE: SCREWED STEEL – cont						
Extra over black steel screwed pipes – cont						
Equal socket, parallel thread						
15 mm dia.	0.70	0.86	0.64	24.01	nr	**24.87**
20 mm dia.	0.85	1.04	0.85	31.90	nr	**32.94**
25 mm dia.	1.18	1.45	0.97	36.40	nr	**37.85**
32 mm dia.	2.35	2.89	1.12	42.04	nr	**44.93**
40 mm dia.	3.42	4.20	1.29	48.42	nr	**52.62**
50 mm dia.	4.95	6.08	1.55	58.17	nr	**64.25**
65 mm dia.	7.61	9.35	1.89	70.94	nr	**80.29**
80 mm dia.	10.78	13.24	2.24	84.06	nr	**97.30**
100 mm dia.	17.79	21.85	3.09	115.97	nr	**137.82**
Concentric reducing socket						
20 × 15 mm dia.	1.10	1.35	0.76	28.53	nr	**29.88**
25 × 15 mm dia.	1.29	1.59	0.85	31.90	nr	**33.49**
25 × 20 mm dia.	1.38	1.69	0.86	32.28	nr	**33.97**
32 × 25 mm dia.	2.60	3.19	1.01	37.92	nr	**41.11**
40 × 25 mm dia.	3.27	4.02	1.16	43.53	nr	**47.55**
40 × 32 mm dia.	3.38	4.15	1.16	43.53	nr	**47.68**
50 × 25 mm dia.	4.86	5.97	1.38	51.80	nr	**57.77**
50 × 40 mm dia.	4.74	5.83	1.38	51.80	nr	**57.63**
65 × 50 mm dia.	8.61	10.57	1.69	63.43	nr	**74.00**
80 × 50 mm dia.	11.21	13.78	2.00	75.06	nr	**88.84**
100 × 50 mm dia.	22.35	27.46	2.75	103.20	nr	**130.66**
100 × 80 mm dia.	30.82	37.86	2.75	103.20	nr	**141.06**
150 × 100 mm dia.	62.08	76.26	4.10	153.87	nr	**230.13**
Eccentric reducing socket						
20 × 15 mm dia.	1.97	2.42	0.73	27.39	nr	**29.81**
25 × 15 mm dia.	5.65	6.94	0.85	31.90	nr	**38.84**
25 × 20 mm dia.	6.40	7.86	0.85	31.90	nr	**39.76**
32 × 25 mm dia.	8.74	10.74	1.01	37.92	nr	**48.66**
40 × 25 mm dia.	10.00	12.28	1.16	43.53	nr	**55.81**
40 × 32 mm dia.	5.02	6.16	1.16	43.53	nr	**49.69**
50 × 25 mm dia.	6.48	7.96	1.38	51.80	nr	**59.76**
50 × 40 mm dia.	8.94	10.98	1.38	51.80	nr	**62.78**
65 × 50 mm dia.	10.40	12.78	1.69	63.43	nr	**76.21**
80 × 50 mm dia.	18.22	22.39	2.00	75.06	nr	**97.45**
Hexagon bush						
20 × 15 mm dia.	0.63	0.78	0.37	13.88	nr	**14.66**
25 × 15 mm dia.	0.76	0.93	0.43	16.14	nr	**17.07**
25 × 20 mm dia.	0.80	0.99	0.43	16.14	nr	**17.13**
32 × 25 mm dia.	1.03	1.27	0.51	19.14	nr	**20.41**
40 × 25 mm dia.	1.64	2.02	0.58	21.77	nr	**23.79**
40 × 32 mm dia.	1.50	1.85	0.58	21.77	nr	**23.62**
50 × 25 mm dia.	3.38	4.15	0.71	26.65	nr	**30.80**

38 MECHANICAL/COOLING/HEATING SYSTEMS

Item	Net Price £	Material £	Labour hours	Labour £	Unit	**Total rate £**
50 × 40 mm dia.	3.12	3.83	0.71	26.65	nr	**30.48**
65 × 50 mm dia.	5.24	6.44	0.85	31.90	nr	**38.34**
80 × 50 mm dia.	8.49	10.43	1.00	37.53	nr	**47.96**
100 × 50 mm dia.	19.64	24.13	1.52	57.05	nr	**81.18**
100 × 80 mm dia.	27.09	33.28	1.52	57.05	nr	**90.33**
150 × 100 mm dia.	58.76	72.19	2.57	96.45	nr	**168.64**
Hexagon nipple						
15 mm dia.	0.68	0.84	0.28	10.50	nr	**11.34**
20 mm dia.	0.76	0.93	0.38	14.26	nr	**15.19**
25 mm dia.	1.09	1.34	0.44	16.51	nr	**17.85**
32 mm dia.	2.23	2.74	0.51	19.14	nr	**21.88**
40 mm dia.	2.60	3.19	0.59	22.13	nr	**25.32**
50 mm dia.	4.70	5.78	0.70	26.28	nr	**32.06**
65 mm dia.	7.67	9.42	0.85	31.90	nr	**41.32**
80 mm dia.	11.08	13.61	1.00	37.53	nr	**51.14**
100 mm dia.	18.81	23.11	1.44	54.04	nr	**77.15**
150 mm dia.	41.98	51.58	2.32	87.08	nr	**138.66**
Union, male/female						
15 mm dia.	3.54	4.35	0.64	24.01	nr	**28.36**
20 mm dia.	4.32	5.30	0.85	31.90	nr	**37.20**
25 mm dia.	5.42	6.66	0.97	36.40	nr	**43.06**
32 mm dia.	8.34	10.25	1.12	42.04	nr	**52.29**
40 mm dia.	10.66	13.09	1.29	48.42	nr	**61.51**
50 mm dia.	16.80	20.64	1.55	58.17	nr	**78.81**
65 mm dia.	32.06	39.39	1.89	70.94	nr	**110.33**
80 mm dia.	44.09	54.17	2.24	84.06	nr	**138.23**
Union, female						
15 mm dia.	2.89	3.55	0.64	24.01	nr	**27.56**
20 mm dia.	3.15	3.87	0.85	31.90	nr	**35.77**
25 mm dia.	3.67	4.50	0.97	36.40	nr	**40.90**
32 mm dia.	6.92	8.50	1.12	42.04	nr	**50.54**
40 mm dia.	7.83	9.62	1.29	48.42	nr	**58.04**
50 mm dia.	12.96	15.92	1.55	58.17	nr	**74.09**
65 mm dia.	27.93	34.31	1.89	70.94	nr	**105.25**
80 mm dia.	38.58	47.40	2.24	84.06	nr	**131.46**
100 mm dia.	73.45	90.24	3.09	115.97	nr	**206.21**
Union elbow, male/female						
15 mm dia.	5.12	6.29	0.55	20.64	nr	**26.93**
20 mm dia.	6.41	7.88	0.85	31.90	nr	**39.78**
25 mm dia.	8.99	11.05	0.97	36.40	nr	**47.45**
Twin elbow						
15 mm dia.	4.14	5.08	0.91	34.16	nr	**39.24**
20 mm dia.	4.58	5.63	1.22	45.79	nr	**51.42**
25 mm dia.	7.42	9.11	1.39	52.16	nr	**61.27**
32 mm dia.	15.24	18.72	1.62	60.80	nr	**79.52**
40 mm dia.	19.30	23.71	1.86	69.80	nr	**93.51**
50 mm dia.	24.81	30.48	2.21	82.95	nr	**113.43**
65 mm dia.	44.07	54.14	2.72	102.09	nr	**156.23**
80 mm dia.	75.09	92.25	3.21	120.48	nr	**212.73**

38 MECHANICAL/COOLING/HEATING SYSTEMS

Item	Net Price £	Material £	Labour hours	Labour £	Unit	Total rate £
LOW TEMPERATURE HOT WATER HEATING; PIPELINE: SCREWED STEEL – cont						
Extra over black steel screwed pipes – cont						
Equal tee						
15 mm dia.	0.99	1.22	0.91	34.16	nr	**35.38**
20 mm dia.	1.44	1.76	1.22	45.79	nr	**47.55**
25 mm dia.	2.08	2.55	1.39	52.16	nr	**54.71**
32 mm dia.	3.79	4.65	1.62	60.80	nr	**65.45**
40 mm dia.	5.82	7.15	1.86	69.80	nr	**76.95**
50 mm dia.	8.37	10.28	2.21	82.95	nr	**93.23**
65 mm dia.	18.35	22.54	2.72	102.09	nr	**124.63**
80 mm dia.	22.35	27.46	3.21	120.48	nr	**147.94**
100 mm dia.	40.51	49.77	4.44	166.63	nr	**216.40**
125 mm dia.	117.67	144.55	5.38	201.92	nr	**346.47**
150 mm dia.	187.48	230.32	6.31	236.82	nr	**467.14**
Tee reducing on branch						
20 × 15 mm dia.	1.29	1.59	1.22	45.79	nr	**47.38**
25 × 15 mm dia.	1.79	2.19	1.39	52.16	nr	**54.35**
25 × 20 mm dia.	2.14	2.63	1.39	52.16	nr	**54.79**
32 × 25 mm dia.	3.72	4.57	1.62	60.80	nr	**65.37**
40 × 25 mm dia.	4.92	6.05	1.86	69.80	nr	**75.85**
40 × 32 mm dia.	6.45	7.93	1.86	69.80	nr	**77.73**
50 × 25 mm dia.	7.32	8.99	2.21	82.95	nr	**91.94**
50 × 40 mm dia.	10.90	13.39	2.21	82.95	nr	**96.34**
65 × 50 mm dia.	17.02	20.91	2.72	102.09	nr	**123.00**
80 × 50 mm dia.	23.01	28.27	3.21	120.48	nr	**148.75**
100 × 50 mm dia.	38.09	46.80	4.44	166.63	nr	**213.43**
100 × 80 mm dia.	58.76	72.19	4.44	166.63	nr	**238.82**
150 × 100 mm dia.	138.11	169.67	6.31	236.82	nr	**406.49**
Equal pitcher tee						
15 mm dia.	3.03	3.73	0.91	34.16	nr	**37.89**
20 mm dia.	3.76	4.62	1.22	45.79	nr	**50.41**
25 mm dia.	5.65	6.94	1.39	52.16	nr	**59.10**
32 mm dia.	8.57	10.53	1.62	60.80	nr	**71.33**
40 mm dia.	13.24	16.26	1.86	69.80	nr	**86.06**
50 mm dia.	18.60	22.85	2.21	82.95	nr	**105.80**
65 mm dia.	28.76	35.33	2.72	102.09	nr	**137.42**
80 mm dia.	44.89	55.15	3.21	120.48	nr	**175.63**
100 mm dia.	100.98	124.06	4.44	166.63	nr	**290.69**
Cross						
15 mm dia.	2.96	3.63	1.00	37.53	nr	**41.16**
20 mm dia.	4.45	5.47	1.33	49.92	nr	**55.39**
25 mm dia.	5.65	6.94	1.51	56.67	nr	**63.61**
32 mm dia.	8.23	10.11	1.77	66.42	nr	**76.53**
40 mm dia.	10.59	13.01	2.02	75.81	nr	**88.82**
50 mm dia.	17.21	21.15	2.42	90.83	nr	**111.98**
65 mm dia.	23.79	29.22	2.97	111.46	nr	**140.68**
80 mm dia.	31.64	38.87	3.50	131.35	nr	**170.22**
100 mm dia.	65.29	80.21	4.84	181.65	nr	**261.86**

38 MECHANICAL/COOLING/HEATING SYSTEMS

Item	Net Price £	Material £	Labour hours	Labour £	Unit	**Total rate £**
LOW TEMPERATURE HOT WATER HEATING; PIPELINE: BLACK WELDED STEEL						
Black steel pipes; butt welded joints; BS 1387: 1985; including protective painting. Fixed Vertical with brackets measured separately (Refer to Screwed Steel Section). Welded butt joints are within the running length, but any flanges are additional						
Medium weight						
10 mm dia.	2.79	3.42	0.37	13.88	m	**17.30**
15 mm dia.	2.97	3.64	0.37	13.88	m	**17.52**
20 mm dia.	3.50	4.29	0.37	13.88	m	**18.17**
25 mm dia.	5.02	6.16	0.41	15.39	m	**21.55**
32 mm dia.	6.22	7.64	0.48	18.01	m	**25.65**
40 mm dia.	7.23	8.88	0.52	19.52	m	**28.40**
50 mm dia.	10.16	12.48	0.62	23.27	m	**35.75**
65 mm dia.	13.80	16.96	0.64	24.01	m	**40.97**
80 mm dia.	17.94	22.04	1.10	41.28	m	**63.32**
100 mm dia.	25.41	31.22	1.31	49.16	m	**80.38**
125 mm dia.	32.34	39.73	1.66	62.30	m	**102.03**
150 mm dia.	37.55	46.13	1.88	70.55	m	**116.68**
Heavy weight						
15 mm dia.	3.54	4.35	0.37	13.88	m	**18.23**
20 mm dia.	4.19	5.15	0.37	13.88	m	**19.03**
25 mm dia.	6.14	7.54	0.41	15.39	m	**22.93**
32 mm dia.	7.62	9.37	0.48	18.01	m	**27.38**
40 mm dia.	8.88	10.91	0.52	19.52	m	**30.43**
50 mm dia.	12.33	15.15	0.62	23.27	m	**38.42**
65 mm dia.	16.78	20.61	0.64	24.01	m	**44.62**
80 mm dia.	21.35	26.23	1.10	41.28	m	**67.51**
100 mm dia.	29.79	36.59	1.31	49.16	m	**85.75**
125 mm dia.	34.48	42.36	1.66	62.30	m	**104.66**
150 mm dia.	40.32	49.53	1.88	70.55	m	**120.08**
Black steel pipes; butt welded joints; BS 1387: 1985; including protective painting. Fixed at High Level or Suspended with brackets measured separately (Refer to Screwed Steel Section). Welded butt joints are within the running length, but any flanges are additional						
Medium weight						
10 mm dia.	2.79	3.42	0.58	21.77	m	**25.19**
15 mm dia.	2.97	3.64	0.58	21.77	m	**25.41**
20 mm dia.	3.50	4.29	0.58	21.77	m	**26.06**
25 mm dia.	5.02	6.16	0.60	22.52	m	**28.68**
32 mm dia.	6.22	7.64	0.68	25.51	m	**33.15**
40 mm dia.	7.23	8.88	0.73	27.39	m	**36.27**
50 mm dia.	10.16	12.48	0.85	31.90	m	**44.38**

38 MECHANICAL/COOLING/HEATING SYSTEMS

Item	Net Price £	Material £	Labour hours	Labour £	Unit	Total rate £
LOW TEMPERATURE HOT WATER HEATING; PIPELINE: BLACK WELDED STEEL – cont						
Black steel pipes – cont						
Medium weight – cont						
65 mm dia.	13.80	16.96	0.88	33.02	m	**49.98**
80 mm dia.	17.94	22.04	1.45	54.42	m	**76.46**
100 mm dia.	25.41	31.22	1.74	65.31	m	**96.53**
125 mm dia.	32.34	39.73	2.21	82.95	m	**122.68**
150 mm dia.	37.55	46.13	2.50	93.83	m	**139.96**
Heavy weight						
15 mm dia.	3.54	4.35	0.58	21.77	m	**26.12**
20 mm dia.	4.19	5.15	0.58	21.77	m	**26.92**
25 mm dia.	6.14	7.54	0.60	22.52	m	**30.06**
32 mm dia.	7.62	9.37	0.68	25.51	m	**34.88**
40 mm dia.	8.88	10.91	0.73	27.39	m	**38.30**
50 mm dia.	12.33	15.15	0.85	31.90	m	**47.05**
65 mm dia.	16.78	20.61	0.88	33.02	m	**53.63**
80 mm dia.	21.35	26.23	1.45	54.42	m	**80.65**
100 mm dia.	29.79	36.59	1.74	65.31	m	**101.90**
125 mm dia.	34.48	42.36	2.21	82.95	m	**125.31**
150 mm dia.	40.32	49.53	2.50	93.83	m	**143.36**
FIXINGS						
Refer to steel pipes; black malleable iron. For minimum fixing distances, refer to the Tables and Memoranda to the rear of the book						
Extra over black steel butt welded pipes; black steel flanges, welded and drilled; metric; BS 4504						
Welded flanges; PN6						
15 mm dia.	3.68	4.53	0.59	22.13	nr	**26.66**
20 mm dia.	3.68	4.53	0.69	25.89	nr	**30.42**
25 mm dia.	3.68	4.53	0.84	31.52	nr	**36.05**
32 mm dia.	6.12	7.52	1.00	37.53	nr	**45.05**
40 mm dia.	6.52	8.01	1.11	41.65	nr	**49.66**
50 mm dia.	6.71	8.24	1.37	51.41	nr	**59.65**
65 mm dia.	9.87	12.13	1.54	57.80	nr	**69.93**
80 mm dia.	14.82	18.21	1.67	62.67	nr	**80.88**
100 mm dia.	16.19	19.89	2.22	83.32	nr	**103.21**
125 mm dia.	21.22	26.07	2.61	97.97	nr	**124.04**
150 mm dia.	21.43	26.32	2.99	112.21	nr	**138.53**
Welded flanges; PN16						
15 mm dia.	7.39	9.08	0.59	22.13	nr	**31.21**
20 mm dia.	7.39	9.08	0.69	25.89	nr	**34.97**
25 mm dia.	7.39	9.08	0.84	31.52	nr	**40.60**

38 MECHANICAL/COOLING/HEATING SYSTEMS

Item	Net Price £	Material £	Labour hours	Labour £	Unit	Total rate £
32 mm dia.	9.96	12.23	1.00	37.53	nr	**49.76**
40 mm dia.	9.96	12.23	1.11	41.65	nr	**53.88**
50 mm dia.	13.09	16.09	1.37	51.41	nr	**67.50**
65 mm dia.	15.11	18.56	1.54	57.80	nr	**76.36**
80 mm dia.	19.20	23.58	1.67	62.67	nr	**86.25**
100 mm dia.	19.59	24.07	2.22	83.32	nr	**107.39**
125 mm dia.	31.10	38.21	2.61	97.97	nr	**136.18**
150 mm dia.	37.13	45.61	2.99	112.21	nr	**157.82**
Blank flanges, slip on for welding; PN6						
15 mm dia.	2.32	2.85	0.48	18.01	nr	**20.86**
20 mm dia.	2.32	2.85	0.55	20.64	nr	**23.49**
25 mm dia.	2.32	2.85	0.64	24.01	nr	**26.86**
32 mm dia.	4.75	5.84	0.76	28.53	nr	**34.37**
40 mm dia.	4.75	5.84	0.84	31.52	nr	**37.36**
50 mm dia.	4.31	5.29	1.01	37.92	nr	**43.21**
65 mm dia.	8.55	10.50	1.30	48.79	nr	**59.29**
80 mm dia.	8.92	10.96	1.41	52.92	nr	**63.88**
100 mm dia.	9.18	11.28	1.78	66.80	nr	**78.08**
125 mm dia.	18.31	22.49	2.06	77.31	nr	**99.80**
150 mm dia.	17.33	21.29	2.35	88.19	nr	**109.48**
Blank flanges, slip on for welding; PN16						
15 mm dia.	2.10	2.58	0.48	18.01	nr	**20.59**
20 mm dia.	3.22	3.96	0.55	20.64	nr	**24.60**
25 mm dia.	3.92	4.82	0.64	24.01	nr	**28.83**
32 mm dia.	4.01	4.92	0.76	28.53	nr	**33.45**
40 mm dia.	5.19	6.37	0.84	31.52	nr	**37.89**
50 mm dia.	8.04	9.88	1.01	37.92	nr	**47.80**
65 mm dia.	9.87	12.13	1.30	48.79	nr	**60.92**
80 mm dia.	12.80	15.73	1.41	52.92	nr	**68.65**
100 mm dia.	15.87	19.50	1.78	66.80	nr	**86.30**
125 mm dia.	23.29	28.61	2.06	77.31	nr	**105.92**
150 mm dia.	30.04	36.91	2.35	88.19	nr	**125.10**
Extra over black steel butt welded pipes; black steel flanges, welding and drilled; imperial; BS 10						
Welded flanges; Table E						
15 mm dia.	7.58	9.31	0.59	22.13	nr	**31.44**
20 mm dia.	7.58	9.31	0.69	25.89	nr	**35.20**
25 mm dia.	7.58	9.31	0.84	31.52	nr	**40.83**
32 mm dia.	7.58	9.31	1.00	37.53	nr	**46.84**
40 mm dia.	7.79	9.57	1.11	41.65	nr	**51.22**
50 mm dia.	9.95	12.22	1.37	51.41	nr	**63.63**
65 mm dia.	12.11	14.88	1.54	57.80	nr	**72.68**
80 mm dia.	15.91	19.54	1.67	62.67	nr	**82.21**
100 mm dia.	23.50	28.88	2.22	83.32	nr	**112.20**
125 mm dia.	45.71	56.15	2.61	97.97	nr	**154.12**
150 mm dia.	45.71	56.15	2.99	112.21	nr	**168.36**

38 MECHANICAL/COOLING/HEATING SYSTEMS

Item	Net Price £	Material £	Labour hours	Labour £	Unit	Total rate £
LOW TEMPERATURE HOT WATER HEATING; PIPELINE: BLACK WELDED STEEL – cont						
Extra over black steel butt welded pipes – cont						
Blank flanges, slip on for welding; Table E						
15 mm dia.	5.21	6.40	0.48	18.01	nr	**24.41**
20 mm dia.	5.21	6.40	0.55	20.64	nr	**27.04**
25 mm dia.	5.21	6.40	0.64	24.01	nr	**30.41**
32 mm dia.	6.55	8.04	0.76	28.53	nr	**36.57**
40 mm dia.	7.13	8.76	0.84	31.52	nr	**40.28**
50 mm dia.	7.87	9.67	1.01	37.92	nr	**47.59**
65 mm dia.	9.07	11.14	1.30	48.79	nr	**59.93**
80 mm dia.	19.80	24.33	1.41	52.92	nr	**77.25**
100 mm dia.	22.49	27.63	1.78	66.80	nr	**94.43**
125 mm dia.	36.99	45.44	2.06	77.31	nr	**122.75**
150 mm dia.	47.56	58.43	2.35	88.19	nr	**146.62**
Extra over black steel butt welded pipes; black steel flange connections						
Bolted connection between pair of flanges; including gasket, bolts, nuts and washers						
50 mm dia.	8.12	9.97	0.50	18.77	nr	**28.74**
65 mm dia.	9.61	11.80	0.50	18.77	nr	**30.57**
80 mm dia.	14.43	17.72	0.50	18.77	nr	**36.49**
100 mm dia.	15.67	19.25	0.50	18.77	nr	**38.02**
125 mm dia.	18.43	22.64	0.50	18.77	nr	**41.41**
150 mm dia.	26.33	32.35	0.88	33.02	nr	**65.37**
Extra over fittings; BS 1965; butt welded						
Cap						
25 mm dia.	17.87	21.96	0.47	17.64	nr	**39.60**
32 mm dia.	17.87	21.96	0.59	22.13	nr	**44.09**
40 mm dia.	17.98	22.09	0.70	26.28	nr	**48.37**
50 mm dia.	21.01	25.81	0.99	37.15	nr	**62.96**
65 mm dia.	24.69	30.33	1.35	50.66	nr	**80.99**
80 mm dia.	25.13	30.87	1.66	62.30	nr	**93.17**
100 mm dia.	32.85	40.35	2.23	83.70	nr	**124.05**
125 mm dia.	46.21	56.77	3.03	113.73	nr	**170.50**
150 mm dia.	53.44	65.65	3.79	142.23	nr	**207.88**
Concentric reducer						
20 × 15 mm dia.	9.95	12.22	0.69	25.89	nr	**38.11**
25 × 15 mm dia.	12.42	15.26	0.87	32.65	nr	**47.91**
25 × 20 mm dia.	9.61	11.80	0.87	32.65	nr	**44.45**
32 × 25 mm dia.	13.24	16.26	1.08	40.53	nr	**56.79**
40 × 25 mm dia.	17.25	21.19	1.38	51.80	nr	**72.99**
40 × 32 mm dia.	11.82	14.52	1.38	51.80	nr	**66.32**
50 × 25 mm dia.	16.41	20.16	1.82	68.30	nr	**88.46**

38 MECHANICAL/COOLING/HEATING SYSTEMS

Item	Net Price £	Material £	Labour hours	Labour £	Unit	**Total rate £**
50 × 40 mm dia.	11.63	14.29	1.82	68.30	nr	**82.59**
65 × 50 mm dia.	16.05	19.72	2.52	94.58	nr	**114.30**
80 × 50 mm dia.	16.33	20.07	3.24	121.60	nr	**141.67**
100 × 50 mm dia.	27.04	33.22	4.08	153.12	nr	**186.34**
100 × 80 mm dia.	18.60	22.85	4.08	153.12	nr	**175.97**
125 × 80 mm dia.	40.62	49.91	4.71	176.77	nr	**226.68**
150 × 100 mm dia.	43.94	53.98	5.33	200.04	nr	**254.02**
Eccentric reducer						
20 × 15 mm dia.	14.80	18.19	0.69	25.89	nr	**44.08**
25 × 15 mm dia.	19.38	23.80	0.87	32.65	nr	**56.45**
25 × 20 mm dia.	16.26	19.97	0.87	32.65	nr	**52.62**
32 × 25 mm dia.	17.98	22.09	1.08	40.53	nr	**62.62**
40 × 25 mm dia.	22.19	27.26	1.38	51.80	nr	**79.06**
40 × 32 mm dia.	21.20	26.04	1.38	51.80	nr	**77.84**
50 × 25 mm dia.	25.28	31.06	1.82	68.30	nr	**99.36**
50 × 40 mm dia.	18.23	22.40	1.82	68.30	nr	**90.70**
65 × 50 mm dia.	21.49	26.40	2.52	94.58	nr	**120.98**
80 × 50 mm dia.	26.33	32.35	3.24	121.60	nr	**153.95**
100 × 50 mm dia.	44.59	54.78	4.08	153.12	nr	**207.90**
100 × 80 mm dia.	32.89	40.40	4.08	153.12	nr	**193.52**
125 × 80 mm dia.	87.88	107.96	4.71	176.77	nr	**284.73**
150 × 100 mm dia.	66.55	81.75	5.33	200.04	nr	**281.79**
45° elbow, long radius						
15 mm dia.	3.74	4.60	0.56	21.02	nr	**25.62**
20 mm dia.	3.79	4.65	0.75	28.14	nr	**32.79**
25 mm dia.	4.92	6.05	0.93	34.90	nr	**40.95**
32 mm dia.	5.97	7.33	1.17	43.91	nr	**51.24**
40 mm dia.	6.03	7.41	1.46	54.79	nr	**62.20**
50 mm dia.	8.27	10.16	1.97	73.93	nr	**84.09**
65 mm dia.	10.65	13.08	2.70	101.33	nr	**114.41**
80 mm dia.	9.83	12.07	3.32	124.60	nr	**136.67**
100 mm dia.	14.88	18.28	4.09	153.50	nr	**171.78**
125 mm dia.	31.64	38.87	4.94	185.41	nr	**224.28**
150 mm dia.	39.29	48.27	5.78	216.93	nr	**265.20**
90° elbow, long radius						
15 mm dia.	3.86	4.75	0.56	21.02	nr	**25.77**
20 mm dia.	3.78	4.64	0.75	28.14	nr	**32.78**
25 mm dia.	5.01	6.15	0.93	34.90	nr	**41.05**
32 mm dia.	5.99	7.36	1.17	43.91	nr	**51.27**
40 mm dia.	6.04	7.42	1.46	54.79	nr	**62.21**
50 mm dia.	8.27	10.16	1.97	73.93	nr	**84.09**
65 mm dia.	10.65	13.08	2.70	101.33	nr	**114.41**
80 mm dia.	11.57	14.22	3.32	124.60	nr	**138.82**
100 mm dia.	17.50	21.50	4.09	153.50	nr	**175.00**
125 mm dia.	36.42	44.74	4.94	185.41	nr	**230.15**
150 mm dia.	46.22	56.78	5.78	216.93	nr	**273.71**
Branch bend (based on branch and pipe sizes being the same)						
15 mm dia.	15.39	18.91	0.85	31.90	nr	**50.81**
20 mm dia.	15.25	18.73	0.85	31.90	nr	**50.63**
25 mm dia.	15.41	18.93	1.02	38.28	nr	**57.21**

38 MECHANICAL/COOLING/HEATING SYSTEMS

Item	Net Price £	Material £	Labour hours	Labour £	Unit	Total rate £
LOW TEMPERATURE HOT WATER HEATING; PIPELINE: BLACK WELDED STEEL – cont						
Extra over fittings – cont						
Branch bend (based on branch and pipe sizes being the same) – cont						
32 mm dia.	14.56	17.89	1.11	41.65	nr	**59.54**
40 mm dia.	14.37	17.65	1.36	51.04	nr	**68.69**
50 mm dia.	13.96	17.15	1.70	63.81	nr	**80.96**
65 mm dia.	20.64	25.36	1.78	66.80	nr	**92.16**
80 mm dia.	32.24	39.61	1.82	68.30	nr	**107.91**
100 mm dia.	42.05	51.66	1.87	70.18	nr	**121.84**
125 mm dia.	74.10	91.03	2.21	82.95	nr	**173.98**
150 mm dia.	113.94	139.98	2.65	99.46	nr	**239.44**
Equal tee						
15 mm dia.	36.48	44.81	0.82	30.78	nr	**75.59**
20 mm dia.	36.48	44.81	1.10	41.28	nr	**86.09**
25 mm dia.	36.48	44.81	1.35	50.66	nr	**95.47**
32 mm dia.	36.48	44.81	1.63	61.18	nr	**105.99**
40 mm dia.	36.48	44.81	2.14	80.31	nr	**125.12**
50 mm dia.	38.42	47.20	3.02	113.34	nr	**160.54**
65 mm dia.	56.04	68.85	3.61	135.49	nr	**204.34**
80 mm dia.	56.13	68.95	4.18	156.88	nr	**225.83**
100 mm dia.	69.84	85.80	5.24	196.65	nr	**282.45**
125 mm dia.	159.84	196.36	6.70	251.46	nr	**447.82**
150 mm dia.	174.60	214.49	8.45	317.14	nr	**531.63**
Extra over black steel butt welded pipes; labour						
Made bend						
15 mm dia.	–	–	0.42	15.76	nr	**15.76**
20 mm dia.	–	–	0.42	15.76	nr	**15.76**
25 mm dia.	–	–	0.50	18.77	nr	**18.77**
32 mm dia.	–	–	0.62	23.27	nr	**23.27**
40 mm dia.	–	–	0.74	27.77	nr	**27.77**
50 mm dia.	–	–	0.89	33.40	nr	**33.40**
65 mm dia.	–	–	1.05	39.41	nr	**39.41**
80 mm dia.	–	–	1.13	42.41	nr	**42.41**
100 mm dia.	–	–	2.90	108.83	nr	**108.83**
125 mm dia.	–	–	3.56	133.61	nr	**133.61**
150 mm dia.	–	–	4.18	156.88	nr	**156.88**
Splay cut end						
15 mm dia.	–	–	0.14	5.26	nr	**5.26**
20 mm dia.	–	–	0.16	6.01	nr	**6.01**
25 mm dia.	–	–	0.18	6.76	nr	**6.76**
32 mm dia.	–	–	0.25	9.39	nr	**9.39**
40 mm dia.	–	–	0.27	10.13	nr	**10.13**
50 mm dia.	–	–	0.31	11.63	nr	**11.63**
65 mm dia.	–	–	0.35	13.14	nr	**13.14**

38 MECHANICAL/COOLING/HEATING SYSTEMS

Item	Net Price £	Material £	Labour hours	Labour £	Unit	**Total rate £**
80 mm dia.	–	–	0.40	15.02	nr	**15.02**
100 mm dia.	–	–	0.48	18.01	nr	**18.01**
125 mm dia.	–	–	0.56	21.02	nr	**21.02**
150 mm dia.	–	–	0.64	24.01	nr	**24.01**
Screwed joint to fitting						
15 mm dia.	–	–	0.30	11.26	nr	**11.26**
20 mm dia.	–	–	0.40	15.02	nr	**15.02**
25 mm dia.	–	–	0.46	17.26	nr	**17.26**
32 mm dia.	–	–	0.53	19.89	nr	**19.89**
40 mm dia.	–	–	0.61	22.90	nr	**22.90**
50 mm dia.	–	–	0.73	27.39	nr	**27.39**
65 mm dia.	–	–	0.89	33.40	nr	**33.40**
80 mm dia.	–	–	1.05	39.41	nr	**39.41**
100 mm dia.	–	–	1.46	54.79	nr	**54.79**
125 mm dia.	–	–	2.10	78.82	nr	**78.82**
150 mm dia.	–	–	2.73	102.46	nr	**102.46**
Straight butt weld						
15 mm dia.	–	–	0.31	11.63	nr	**11.63**
20 mm dia.	–	–	0.42	15.76	nr	**15.76**
25 mm dia.	–	–	0.52	19.52	nr	**19.52**
32 mm dia.	–	–	0.69	25.89	nr	**25.89**
40 mm dia.	–	–	0.83	31.15	nr	**31.15**
50 mm dia.	–	–	1.22	45.79	nr	**45.79**
65 mm dia.	–	–	1.57	58.92	nr	**58.92**
80 mm dia.	–	–	1.95	73.18	nr	**73.18**
100 mm dia.	–	–	2.38	89.32	nr	**89.32**
125 mm dia.	–	–	2.83	106.22	nr	**106.22**
150 mm dia.	–	–	3.27	122.72	nr	**122.72**

38 MECHANICAL/COOLING/HEATING SYSTEMS

Item	Net Price £	Material £	Labour hours	Labour £	Unit	Total rate £
LOW TEMPERATURE HOT WATER HEATING; PIPELINE: CARBON WELDED STEEL						
Hot finished seamless carbon steel pipe; BS 806 and BS 3601; wall thickness to BS 3600; butt welded joints; including protective painting, fixed vertically or at low level, brackets measured separately (Refer to Screwed Pipework Section). Welded butt joints are within the running length, but any flanges are additional						
Pipework						
200 mm dia.	91.92	112.93	2.04	76.56	m	**189.49**
250 mm dia.	82.77	101.68	2.59	97.20	m	**198.88**
300 mm dia.	194.58	239.04	2.99	112.21	m	**351.25**
350 mm dia.	205.37	252.29	3.52	132.11	m	**384.40**
400 mm dia.	358.04	439.86	4.08	153.12	m	**592.98**
Hot finished seamless carbon steel pipe; BS 806 and BS 3601; wall thickness to BS 3600; butt welded joints; including protective painting, fixed at high level or suspended, brackets measured separately (Refer to Screwed Pipework Section). Welded butt joints are within the running length, but any flanges are additional						
Pipework						
200 mm dia.	91.92	112.93	3.70	138.86	m	**251.79**
250 mm dia.	82.77	101.68	4.73	177.52	m	**279.20**
300 mm dia.	194.58	239.04	5.65	212.05	m	**451.09**
350 mm dia.	205.37	252.29	6.68	250.70	m	**502.99**
400 mm dia.	358.04	439.86	7.70	288.99	m	**728.85**
FIXINGS						
Refer to steel pipes; black malleable iron. For minimum fixing distances, refer to the Tables and Memoranda to the rear of the book						
Extra over fittings; BS 1965 part 1; butt welded						
Cap						
200 mm dia.	75.26	92.45	3.70	138.86	nr	**231.31**
250 mm dia.	144.42	177.42	4.73	177.52	nr	**354.94**
300 mm dia.	157.97	194.06	5.65	212.05	nr	**406.11**
350 mm dia.	243.39	299.01	6.68	250.70	nr	**549.71**
400 mm dia.	281.52	345.85	7.70	288.99	nr	**634.84**

38 MECHANICAL/COOLING/HEATING SYSTEMS

Item	Net Price £	Material £	Labour hours	Labour £	Unit	Total rate £
Concentric reducer						
200 mm × 150 mm dia.	82.22	101.01	7.27	272.85	nr	**373.86**
250 mm × 150 mm dia.	119.24	146.49	9.05	339.65	nr	**486.14**
250 mm × 200 mm dia.	72.97	89.64	9.10	341.52	nr	**431.16**
300 mm × 150 mm dia.	137.73	169.20	10.75	403.45	nr	**572.65**
300 mm × 200 mm dia.	142.63	175.22	10.75	403.45	nr	**578.67**
300 mm × 250 mm dia.	126.80	155.78	11.15	418.47	nr	**574.25**
350 mm × 200 mm dia.	216.29	265.71	12.50	469.14	nr	**734.85**
350 mm × 250 mm dia.	181.34	222.78	12.70	476.65	nr	**699.43**
350 mm × 300 mm dia.	181.34	222.78	13.00	487.90	nr	**710.68**
400 mm × 250 mm dia.	265.16	325.75	14.46	542.69	nr	**868.44**
400 mm × 300 mm dia.	328.38	403.41	14.51	544.57	nr	**947.98**
400 mm × 350 mm dia.	286.22	351.62	15.16	568.96	nr	**920.58**
Eccentric reducer						
200 mm × 150 mm dia.	148.50	182.44	7.27	272.85	nr	**455.29**
250 mm × 150 mm dia.	197.87	243.09	9.05	339.65	nr	**582.74**
250 mm × 200 mm dia.	129.47	159.05	9.10	341.52	nr	**500.57**
300 mm × 150 mm dia.	237.38	291.62	10.75	403.45	nr	**695.07**
300 mm × 200 mm dia.	273.79	336.35	10.75	403.45	nr	**739.80**
300 mm × 250 mm dia.	220.39	270.75	11.15	418.47	nr	**689.22**
350 mm × 200 mm dia.	251.47	308.93	12.50	469.14	nr	**778.07**
350 mm × 250 mm dia.	209.06	256.83	12.70	476.65	nr	**733.48**
350 mm × 300 mm dia.	209.06	256.83	13.00	487.90	nr	**744.73**
400 mm × 250 mm dia.	335.76	412.48	14.46	542.69	nr	**955.17**
400 mm × 300 mm dia.	314.43	386.27	14.51	544.57	nr	**930.84**
400 mm × 350 mm dia.	387.41	475.93	15.16	568.96	nr	**1044.89**
45° elbow						
200 mm dia.	78.64	96.61	7.75	290.86	nr	**387.47**
250 mm dia.	147.70	181.45	10.05	377.19	nr	**558.64**
300 mm dia.	215.79	265.09	12.20	457.87	nr	**722.96**
350 mm dia.	174.63	214.54	14.65	549.83	nr	**764.37**
400 mm dia.	290.25	356.57	17.12	642.53	nr	**999.10**
90° elbow						
200 mm dia.	92.52	113.66	7.75	290.86	nr	**404.52**
250 mm dia.	173.79	213.50	10.05	377.19	nr	**590.69**
300 mm dia.	253.89	311.90	12.20	457.87	nr	**769.77**
350 mm dia.	349.03	428.79	14.65	549.83	nr	**978.62**
400 mm dia.	445.81	547.68	17.12	642.53	nr	**1190.21**
Equal tee						
200 mm dia.	244.42	300.27	11.25	422.23	nr	**722.50**
250 mm dia.	419.12	514.89	14.53	545.32	nr	**1060.21**
300 mm dia.	520.14	638.99	17.55	658.66	nr	**1297.65**
350 mm dia.	697.84	857.29	20.98	787.40	nr	**1644.69**
400 mm dia.	797.52	979.76	24.38	915.00	nr	**1894.76**
Extra over black steel butt welded pipes; labour						
Straight butt weld						
200 mm dia.	–	–	4.08	153.12	nr	**153.12**
250 mm dia.	–	–	5.20	195.16	nr	**195.16**
300 mm dia.	–	–	6.22	233.44	nr	**233.44**

38 MECHANICAL/COOLING/HEATING SYSTEMS

Item	Net Price £	Material £	Labour hours	Labour £	Unit	Total rate £
LOW TEMPERATURE HOT WATER HEATING; PIPELINE: CARBON WELDED STEEL – cont						
Extra over black steel butt welded pipes – cont						
Straight butt weld – cont						
350 mm dia.	–	–	7.33	275.10	nr	**275.10**
400 mm dia.	–	–	8.41	315.64	nr	**315.64**
Branch weld						
100 mm dia.	–	–	3.46	129.85	nr	**129.85**
125 mm dia.	–	–	4.23	158.76	nr	**158.76**
150 mm dia.	–	–	5.00	187.66	nr	**187.66**
Extra over black steel butt welded pipes; black steel flanges, welding and drilled; metric; BS 4504						
Welded flanges; PN16						
200 mm dia.	50.33	61.83	4.10	153.87	nr	**215.70**
250 mm dia.	72.28	88.80	5.33	200.04	nr	**288.84**
300 mm dia.	101.26	124.39	6.40	240.20	nr	**364.59**
350 mm dia.	195.84	240.59	7.43	278.86	nr	**519.45**
400 mm dia.	258.57	317.66	8.45	317.14	nr	**634.80**
Welded flanges; PN25						
200 mm dia.	167.09	205.28	4.10	153.87	nr	**359.15**
250 mm dia.	200.30	246.07	5.33	200.04	nr	**446.11**
300 mm dia.	270.86	332.76	6.40	240.20	nr	**572.96**
Blank flanges, slip on for welding; PN16						
200 mm dia.	116.36	142.95	2.70	101.33	nr	**244.28**
250 mm dia.	176.84	217.25	3.48	130.60	nr	**347.85**
300 mm dia.	188.94	232.11	4.20	157.63	nr	**389.74**
350 mm dia.	295.70	363.27	4.78	179.40	nr	**542.67**
400 mm dia.	391.00	480.34	5.35	200.78	nr	**681.12**
Blank flanges, slip on for welding; PN25						
200 mm dia.	116.36	142.95	2.70	101.33	nr	**244.28**
250 mm dia.	176.84	217.25	3.48	130.60	nr	**347.85**
300 mm dia.	188.94	232.11	4.20	157.63	nr	**389.74**
Extra over black steel butt welded pipes; black steel flange connections						
Bolted connection between pair of flanges; including gasket, bolts, nuts and washers						
200 mm dia.	97.65	119.96	3.83	143.75	nr	**263.71**
250 mm dia.	151.12	185.65	4.93	185.02	nr	**370.67**
300 mm dia.	206.06	253.14	5.90	221.43	nr	**474.57**

38 MECHANICAL/COOLING/HEATING SYSTEMS

Item	Net Price £	Material £	Labour hours	Labour £	Unit	**Total rate £**
LOW TEMPERATURE HOT WATER HEATING; PIPELINE: PRESS FIT						
Press fit jointing system; operating temperature −20°C to +120°C; operating pressure 16 bar; butyl rubber 'O' ring mechanical joint. With brackets measured separately (Refer to Screwed Steel Section)						
Carbon steel						
Pipework						
15 mm dia.	1.59	1.95	0.46	17.26	m	**19.21**
20 mm dia.	2.55	3.13	0.48	18.01	m	**21.14**
25 mm dia.	3.59	4.41	0.52	19.52	m	**23.93**
32 mm dia.	4.63	5.69	0.56	21.02	m	**26.71**
40 mm dia.	6.34	7.79	0.58	21.77	m	**29.56**
50 mm dia.	8.22	10.10	0.66	24.77	m	**34.87**
Extra over for carbon steel press fit fittings						
Coupling						
15 mm dia.	1.08	1.32	0.36	13.51	nr	**14.83**
20 mm dia.	1.31	1.61	0.36	13.51	nr	**15.12**
25 mm dia.	1.66	2.04	0.44	16.51	nr	**18.55**
32 mm dia.	2.78	3.41	0.44	16.51	nr	**19.92**
40 mm dia.	3.72	4.57	0.52	19.52	nr	**24.09**
50 mm dia.	4.38	5.38	0.60	22.52	nr	**27.90**
Reducer						
20 × 15 mm dia.	1.00	1.23	0.36	13.51	nr	**14.74**
25 × 15 mm dia.	1.31	1.61	0.40	15.02	nr	**16.63**
25 × 20 mm dia.	1.38	1.69	0.40	15.02	nr	**16.71**
32 × 20 mm dia.	1.53	1.88	0.40	15.02	nr	**16.90**
32 × 25 mm dia.	1.64	2.02	0.44	16.51	nr	**18.53**
40 × 32 mm dia.	3.56	4.38	0.48	18.01	nr	**22.39**
50 × 20 mm dia.	10.27	12.62	0.48	18.01	nr	**30.63**
50 × 25 mm dia.	10.36	12.73	0.52	19.52	nr	**32.25**
50 × 40 mm dia.	10.89	13.38	0.56	21.02	nr	**34.40**
90° elbow						
15 mm dia.	1.57	1.93	0.36	13.51	nr	**15.44**
20 mm dia.	2.05	2.52	0.36	13.51	nr	**16.03**
25 mm dia.	2.81	3.45	0.44	16.51	nr	**19.96**
32 mm dia.	7.03	8.64	0.44	16.51	nr	**25.15**
40 mm dia.	11.23	13.80	0.52	19.52	nr	**33.32**
50 mm dia.	13.43	16.50	0.60	22.52	nr	**39.02**
45° elbow						
15 mm dia.	1.86	2.29	0.36	13.51	nr	**15.80**
20 mm dia.	2.08	2.55	0.36	13.51	nr	**16.06**
25 mm dia.	2.83	3.48	0.44	16.51	nr	**19.99**
32 mm dia.	5.57	6.85	0.44	16.51	nr	**23.36**
40 mm dia.	6.98	8.58	0.52	19.52	nr	**28.10**
50 mm dia.	7.90	9.70	0.60	22.52	nr	**32.22**

38 MECHANICAL/COOLING/HEATING SYSTEMS

Item	Net Price £	Material £	Labour hours	Labour £	Unit	**Total rate £**
LOW TEMPERATURE HOT WATER HEATING; PIPELINE: PRESS FIT – cont						
Extra over for carbon steel press fit fittings – cont						
Equal tee						
15 mm dia.	2.99	3.68	0.54	20.28	nr	**23.96**
20 mm dia.	3.47	4.26	0.54	20.28	nr	**24.54**
25 mm dia.	4.63	5.69	0.66	24.77	nr	**30.46**
32 mm dia.	7.22	8.87	0.66	24.77	nr	**33.64**
40 mm dia.	10.66	13.09	0.78	29.27	nr	**42.36**
50 mm dia.	12.79	15.71	0.90	33.78	nr	**49.49**
Reducing tee						
20 × 15 mm dia.	3.40	4.18	0.54	20.28	nr	**24.46**
25 × 15 mm dia.	4.58	5.63	0.62	23.27	nr	**28.90**
25 × 20 mm dia.	4.99	6.13	0.62	23.27	nr	**29.40**
32 × 15 mm dia.	6.72	8.25	0.62	23.27	nr	**31.52**
32 × 20 mm dia.	7.28	8.95	0.62	23.27	nr	**32.22**
32 × 25 mm dia.	7.39	9.08	0.62	23.27	nr	**32.35**
40 × 20 mm dia.	9.74	11.97	0.70	26.28	nr	**38.25**
40 × 25 mm dia.	10.11	12.42	0.70	26.28	nr	**38.70**
40 × 32 mm dia.	9.87	12.13	0.70	26.28	nr	**38.41**
50 × 20 mm dia.	11.61	14.26	0.82	30.78	nr	**45.04**
50 × 25 mm dia.	11.84	14.54	0.82	30.78	nr	**45.32**
50 × 32 mm dia.	12.18	14.96	0.82	30.78	nr	**45.74**
50 × 40 mm dia.	12.78	15.70	0.82	30.78	nr	**46.48**

38 MECHANICAL/COOLING/HEATING SYSTEMS

Item	Net Price £	Material £	Labour hours	Labour £	Unit	**Total rate £**
LOW TEMPERATURE HOT WATER HEATING; PIPELINE: MECHANICAL GROOVED						
MECHANICAL GROOVED						
Mechanical grooved jointing system; working temperature not exceeding 82°C BS 5750; pipework complete with grooved joints; painted finish. With brackets measured separately (Refer to Screwed Steel Section)						
Grooved Joints						
65 mm	14.54	17.86	0.58	21.77	m	**39.63**
80 mm	15.24	18.72	0.68	25.51	m	**44.23**
100 mm	19.77	24.29	0.79	29.64	m	**53.93**
125 mm	31.04	38.14	1.02	38.28	m	**76.42**
150 mm	32.38	39.77	1.15	43.16	m	**82.93**
Extra over mechanical grooved system fittings						
Couplings						
65 mm	14.54	17.86	0.41	15.39	nr	**33.25**
80 mm	15.24	18.72	0.41	15.39	nr	**34.11**
100 mm	19.77	24.29	0.66	24.77	nr	**49.06**
125 mm	31.04	38.14	0.68	25.51	nr	**63.65**
150 mm	32.38	39.77	0.80	30.02	nr	**69.79**
Concentric reducers (one size down)						
80 mm	23.33	28.66	0.59	22.13	nr	**50.79**
100 mm	23.38	28.72	0.71	26.65	nr	**55.37**
125 mm	40.32	49.53	0.85	31.90	nr	**81.43**
150 mm	52.26	64.20	0.98	36.78	nr	**100.98**
Short radius elbow; 90°						
65 mm	26.71	32.81	0.53	19.89	nr	**52.70**
80 mm	27.27	33.51	0.61	22.90	nr	**56.41**
100 mm	36.50	44.85	0.80	30.02	nr	**74.87**
125 mm	60.19	73.94	0.90	33.78	nr	**107.72**
150 mm	78.13	95.98	0.94	35.28	nr	**131.26**
Short radius elbow; 45°						
65 mm	22.87	28.10	0.53	19.89	nr	**47.99**
80 mm	25.70	31.57	0.61	22.90	nr	**54.47**
100 mm	31.94	39.24	0.80	30.02	nr	**69.26**
125 mm	54.03	66.38	0.90	33.78	nr	**100.16**
150 mm	59.48	73.07	0.94	35.28	nr	**108.35**
Equal tee						
65 mm	48.09	59.08	0.83	31.15	nr	**90.23**
80 mm	50.96	62.60	0.93	34.90	nr	**97.50**
100 mm	56.98	70.00	1.18	44.29	nr	**114.29**
125 mm	151.11	185.64	1.37	51.41	nr	**237.05**
150 mm	140.10	172.12	1.43	53.68	nr	**225.80**

38 MECHANICAL/COOLING/HEATING SYSTEMS

Item	Net Price £	Material £	Labour hours	Labour £	Unit	Total rate £
LOW TEMPERATURE HOT WATER HEATING; PIPELINE: PLASTIC PIPEWORK						
Polypropylene PP-R 80 pipe, mechanically stabilized by fibre compound mixture in middle layer; suitable for continuous working temperatures of 0–90°C; thermally fused joints in the running length						
Pipe; 4 m long; PN 20						
20 mm dia.	2.90	3.56	0.35	13.14	m	**16.70**
25 mm dia.	4.36	5.36	0.39	14.63	m	**19.99**
32 mm dia.	4.97	6.10	0.43	16.14	m	**22.24**
40 mm dia.	6.64	8.16	0.47	17.64	m	**25.80**
50 mm dia.	9.65	11.85	0.51	19.14	m	**30.99**
63 mm dia.	15.94	19.58	0.52	19.52	m	**39.10**
75 mm dia.	20.62	25.34	0.60	22.52	m	**47.86**
90 mm dia.	31.79	39.05	0.69	25.89	m	**64.94**
110 mm dia.	47.80	58.73	0.69	25.89	m	**84.62**
125 mm dia.	51.22	62.93	0.85	31.90	m	**94.83**
FIXINGS						
Refer to steel pipes; black malleable iron. For minimum fixing distances, refer to the Tables and Memoranda to the rear of the book						
Extra over fittings; thermally fused joints						
Overbridge bow						
20 mm dia.	1.92	2.36	0.51	19.14	nr	**21.50**
25 mm dia.	3.54	4.35	0.56	21.02	nr	**25.37**
32 mm dia.	7.06	8.67	0.65	24.40	nr	**33.07**
Elbow 90°						
20 mm dia.	0.65	0.80	0.44	16.51	nr	**17.31**
25 mm dia.	0.86	1.06	0.52	19.52	nr	**20.58**
32 mm dia.	1.23	1.51	0.59	22.13	nr	**23.64**
40 mm dia.	1.94	2.38	0.66	24.77	nr	**27.15**
50 mm dia.	4.17	5.12	0.73	27.39	nr	**32.51**
63 mm dia.	6.39	7.85	0.85	31.90	nr	**39.75**
75 mm dia.	14.15	17.39	0.85	31.90	nr	**49.29**
90 mm dia.	26.14	32.11	1.04	39.03	nr	**71.14**
110 mm dia.	37.17	45.66	1.04	39.03	nr	**84.69**
125 mm dia.	57.27	70.36	1.30	48.79	nr	**119.15**
Long bend 90°						
20 mm dia.	3.44	4.22	0.48	18.01	nr	**22.23**
25 mm dia.	3.60	4.42	0.57	21.39	nr	**25.81**
32 mm dia.	4.22	5.19	0.65	24.40	nr	**29.59**
40 mm dia.	7.72	9.48	0.73	27.39	nr	**36.87**

38 MECHANICAL/COOLING/HEATING SYSTEMS

Item	Net Price £	Material £	Labour hours	Labour £	Unit	**Total rate £**
Elbow 90°, female/male						
20 mm dia.	0.66	0.81	0.44	16.51	nr	**17.32**
25 mm dia.	0.87	1.07	0.52	19.52	nr	**20.59**
32 mm dia.	1.24	1.52	0.59	22.13	nr	**23.65**
Elbow 45°						
20 mm dia.	0.66	0.81	0.44	16.51	nr	**17.32**
25 mm dia.	0.87	1.07	0.52	19.52	nr	**20.59**
32 mm dia.	1.24	1.52	0.59	22.13	nr	**23.65**
40 mm dia.	1.95	2.39	0.66	24.77	nr	**27.16**
50 mm dia.	4.17	5.12	0.73	27.39	nr	**32.51**
63 mm dia.	6.41	7.88	0.85	31.90	nr	**39.78**
75 mm dia.	14.16	17.40	0.85	31.90	nr	**49.30**
90 mm dia.	26.14	32.11	1.04	39.03	nr	**71.14**
110 mm dia.	37.18	45.67	1.04	39.03	nr	**84.70**
125 mm dia.	57.28	70.37	1.30	48.79	nr	**119.16**
Elbow 45°, female/male						
20 mm dia.	0.66	0.81	0.44	16.51	nr	**17.32**
25 mm dia.	0.87	1.07	0.52	19.52	nr	**20.59**
32 mm dia.	1.24	1.52	0.59	22.13	nr	**23.65**
T-piece 90°						
20 mm dia.	0.90	1.10	0.61	22.90	nr	**24.00**
25 mm dia.	1.22	1.50	0.72	27.03	nr	**28.53**
32 mm dia.	1.59	1.95	0.83	31.15	nr	**33.10**
40 mm dia.	2.45	3.01	0.92	34.53	nr	**37.54**
50 mm dia.	6.96	8.55	1.01	37.92	nr	**46.47**
63 mm dia.	9.98	12.26	1.11	41.65	nr	**53.91**
75 mm dia.	16.63	20.43	1.18	44.29	nr	**64.72**
90 mm dia.	30.61	37.60	1.46	54.79	nr	**92.39**
110 mm dia.	47.75	58.66	1.46	54.79	nr	**113.45**
125 mm dia.	63.45	77.95	1.82	68.30	nr	**146.25**
T-piece 90° reducing						
25 × 20 × 20 mm	1.24	1.52	0.72	27.03	nr	**28.55**
25 × 20 × 25 mm	1.24	1.52	0.72	27.03	nr	**28.55**
32 × 20 × 32 mm	1.59	1.95	0.83	31.15	nr	**33.10**
32 × 25 × 32 mm	1.59	1.95	0.83	31.15	nr	**33.10**
40 × 20 × 40 mm	2.45	3.01	0.92	34.53	nr	**37.54**
40 × 25 × 40 mm	2.45	3.01	0.92	34.53	nr	**37.54**
40 × 32 × 40 mm	2.45	3.01	0.92	34.53	nr	**37.54**
50 × 25 × 50 mm	6.96	8.55	1.01	37.92	nr	**46.47**
50 × 32 × 50 mm	6.96	8.55	1.01	37.92	nr	**46.47**
50 × 40 × 50 mm	6.96	8.55	1.01	37.92	nr	**46.47**
63 × 20 × 63 mm	9.39	11.54	1.11	41.65	nr	**53.19**
63 × 25 × 63 mm	9.39	11.54	1.11	41.65	nr	**53.19**
63 × 32 × 63 mm	9.39	11.54	1.11	41.65	nr	**53.19**
63 × 40 × 63 mm	9.39	11.54	1.01	37.92	nr	**49.46**
63 × 50 × 63 mm	9.39	11.54	1.01	37.92	nr	**49.46**
75 × 20 × 75 mm	15.26	18.74	1.18	44.29	nr	**63.03**
75 × 25 × 75 mm	15.26	18.74	1.18	44.29	nr	**63.03**
75 × 32 × 75 mm	15.26	18.74	1.18	44.29	nr	**63.03**
75 × 40 × 75 mm	15.26	18.74	1.18	44.29	nr	**63.03**

38 MECHANICAL/COOLING/HEATING SYSTEMS

Item	Net Price £	Material £	Labour hours	Labour £	Unit	Total rate £
LOW TEMPERATURE HOT WATER HEATING; PIPELINE: PLASTIC PIPEWORK – cont						
Extra over fittings – cont						
T-piece 90° reducing – cont						
75 × 50 × 75 mm	15.26	18.74	1.18	44.29	nr	**63.03**
75 × 63 × 75 mm	15.26	18.74	1.18	44.29	nr	**63.03**
90 × 63 × 90 mm	30.61	37.60	1.46	54.79	nr	**92.39**
110 × 75 × 110 mm	47.75	58.66	1.46	54.79	nr	**113.45**
110 × 90 × 110 mm	47.75	58.66	1.46	54.79	nr	**113.45**
125 × 90 × 125 mm	56.80	69.78	1.82	68.30	nr	**138.08**
125 × 110 × 125 mm	57.96	71.20	1.82	68.30	nr	**139.50**
Reducer						
25 × 20 mm	0.71	0.87	0.59	22.13	nr	**23.00**
32 × 20 mm	0.92	1.13	0.62	23.27	nr	**24.40**
32 × 25 mm	0.92	1.13	0.62	23.27	nr	**24.40**
40 × 20 mm	1.46	1.80	0.66	24.77	nr	**26.57**
40 × 25 mm	1.46	1.80	0.66	24.77	nr	**26.57**
40 × 32 mm	1.46	1.80	0.66	24.77	nr	**26.57**
50 × 20 mm	2.34	2.88	0.73	27.39	nr	**30.27**
50 × 25 mm	2.34	2.88	0.73	27.39	nr	**30.27**
50 × 32 mm	2.34	2.88	0.73	27.39	nr	**30.27**
50 × 40 mm	2.34	2.88	0.73	27.39	nr	**30.27**
63 × 40 mm	4.74	5.83	0.78	29.27	nr	**35.10**
63 × 25 mm	4.74	5.83	0.78	29.27	nr	**35.10**
63 × 32 mm	4.74	5.83	0.78	29.27	nr	**35.10**
63 × 50 mm	4.74	5.83	0.78	29.27	nr	**35.10**
75 × 50 mm	5.28	6.49	0.85	31.90	nr	**38.39**
75 × 63 mm	5.28	6.49	0.85	31.90	nr	**38.39**
90 × 63 mm	11.81	14.51	1.04	39.03	nr	**53.54**
90 × 75 mm	11.81	14.51	1.04	39.03	nr	**53.54**
110 × 90 mm	19.09	23.46	1.17	43.91	nr	**67.37**
125 × 110 mm	29.82	36.63	1.43	53.68	nr	**90.31**
Socket						
20 mm dia.	0.64	0.79	0.51	19.14	nr	**19.93**
25 mm dia.	0.71	0.87	0.56	21.02	nr	**21.89**
32 mm dia.	0.92	1.13	0.65	24.40	nr	**25.53**
40 mm dia.	1.13	1.39	0.74	27.77	nr	**29.16**
50 mm dia.	2.34	2.88	0.81	30.41	nr	**33.29**
63 mm dia.	4.74	5.83	0.86	32.28	nr	**38.11**
75 mm dia.	5.28	6.49	0.91	34.16	nr	**40.65**
90 mm dia.	13.65	16.77	0.91	34.16	nr	**50.93**
110 mm dia.	23.19	28.49	0.91	34.16	nr	**62.65**
125 mm dia.	32.31	39.69	1.30	48.79	nr	**88.48**
End cap						
20 mm dia.	0.99	1.22	0.25	9.39	nr	**10.61**
25 mm dia.	1.22	1.50	0.29	10.89	nr	**12.39**
32 mm dia.	1.51	1.86	0.33	12.38	nr	**14.24**

38 MECHANICAL/COOLING/HEATING SYSTEMS

Item	Net Price £	Material £	Labour hours	Labour £	Unit	**Total rate £**
40 mm dia.	2.41	2.96	0.36	13.51	nr	**16.47**
50 mm dia.	3.34	4.11	0.40	15.02	nr	**19.13**
63 mm dia.	5.62	6.91	0.44	16.51	nr	**23.42**
75 mm dia.	8.10	9.95	0.47	17.64	nr	**27.59**
90 mm dia.	18.34	22.53	0.57	21.39	nr	**43.92**
110 mm dia.	22.03	27.07	0.57	21.39	nr	**48.46**
125 mm dia.	33.58	41.25	0.85	31.90	nr	**73.15**
Stub flange with gasket						
32 mm dia.	22.27	27.36	0.23	8.64	nr	**36.00**
40 mm dia.	28.01	34.41	0.27	10.13	nr	**44.54**
50 mm dia.	33.89	41.63	0.38	14.26	nr	**55.89**
63 mm dia.	40.70	50.00	0.43	16.14	nr	**66.14**
75 mm dia.	47.75	58.66	0.48	18.01	nr	**76.67**
90 mm dia.	64.62	79.39	0.53	19.89	nr	**99.28**
110 mm dia.	90.49	111.16	0.53	19.89	nr	**131.05**
125 mm dia.	129.69	159.33	0.75	28.14	nr	**187.47**
Weld in saddle with female thread						
40–½"	1.62	2.00	0.36	13.51	nr	**15.51**
50–½"	1.62	2.00	0.36	13.51	nr	**15.51**
63–½"	1.62	2.00	0.40	15.02	nr	**17.02**
75–½"	1.62	2.00	0.40	15.02	nr	**17.02**
90–½"	1.62	2.00	0.46	17.26	nr	**19.26**
110–½"	1.62	2.00	0.46	17.26	nr	**19.26**
Weld in saddle with male thread						
50–½"	1.62	2.00	0.36	13.51	nr	**15.51**
63–½"	1.62	2.00	0.40	15.02	nr	**17.02**
75–½"	1.62	2.00	0.40	15.02	nr	**17.02**
90–½"	1.62	2.00	0.46	17.26	nr	**19.26**
110–½"	1.62	2.00	0.46	17.26	nr	**19.26**
Transition piece, round with female thread						
20 × ½"	3.74	4.60	0.29	10.89	nr	**15.49**
20 × ¾"	4.93	6.06	0.29	10.89	nr	**16.95**
25 × ½"	3.74	4.60	0.33	12.38	nr	**16.98**
25 × ¾"	4.93	6.06	0.33	12.38	nr	**18.44**
Transition piece, hexagon with female thread						
32 × 1"	13.94	17.13	0.36	13.51	nr	**30.64**
40 × 1 ¼"	22.05	27.09	0.36	13.51	nr	**40.60**
50 × 1 ½"	25.60	31.45	0.36	13.51	nr	**44.96**
63 × 2"	39.66	48.72	0.40	15.02	nr	**63.74**
75 × 2"	41.37	50.82	0.40	15.02	nr	**65.84**
125 × 5"	79.27	97.39	0.51	19.14	nr	**116.53**
Stop valve for surface assembly						
20 mm dia.	15.39	18.91	0.25	9.39	nr	**28.30**
25 mm dia.	15.39	18.91	0.29	10.89	nr	**29.80**
32 mm dia.	28.98	35.61	0.33	12.38	nr	**47.99**
Ball valve						
20 mm dia.	54.20	66.58	0.25	9.39	nr	**75.97**
25 mm dia.	58.07	71.34	0.29	10.89	nr	**82.23**
32 mm dia.	69.77	85.71	0.33	12.38	nr	**98.09**
40 mm dia.	89.05	109.40	0.36	13.51	nr	**122.91**
50 mm dia.	122.25	150.18	0.40	15.02	nr	**165.20**
63 mm dia.	137.84	169.33	0.44	16.51	nr	**185.84**

38 MECHANICAL/COOLING/HEATING SYSTEMS

Item	Net Price £	Material £	Labour hours	Labour £	Unit	**Total rate £**
LOW TEMPERATURE HOT WATER HEATING; PIPELINE: PLASTIC PIPEWORK – cont						
Floor or ceiling cover plates						
Plastic						
15 mm dia.	0.49	0.60	0.16	6.01	nr	**6.61**
20 mm dia.	0.56	0.69	0.22	8.25	nr	**8.94**
25 mm dia.	0.59	0.72	0.22	8.25	nr	**8.97**
32 mm dia.	0.66	0.81	0.24	9.01	nr	**9.82**
40 mm dia.	1.46	1.80	0.26	9.75	nr	**11.55**
50 mm dia.	1.62	2.00	0.26	9.75	nr	**11.75**
Chromium plated						
15 mm dia.	3.39	4.17	0.16	6.01	nr	**10.18**
20 mm dia.	3.62	4.45	0.17	6.37	nr	**10.82**
25 mm dia.	3.75	4.61	0.21	7.89	nr	**12.50**
32 mm dia.	3.84	4.71	0.22	8.25	nr	**12.96**
40 mm dia.	4.33	5.32	0.26	9.75	nr	**15.07**
50 mm dia.	5.18	6.36	0.26	9.75	nr	**16.11**

38 MECHANICAL/COOLING/HEATING SYSTEMS

Item	Net Price £	Material £	Labour hours	Labour £	Unit	Total rate £
LOW TEMPERATURE HOT WATER HEATING; PIPELINE: PIPELINE ANCILLERIES						
EXPANSION JOINTS						
Axial movement bellows expansion joints; stainless steel						
Screwed ends for steel pipework; up to 6 bar G at 100°C						
15 mm dia.	83.12	102.11	0.68	25.51	nr	**127.62**
20 mm dia.	88.79	109.07	0.81	30.41	nr	**139.48**
25 mm dia.	92.23	113.31	0.93	34.90	nr	**148.21**
32 mm dia.	97.65	119.96	1.06	39.77	nr	**159.73**
40 mm dia.	103.43	127.06	1.16	43.53	nr	**170.59**
50 mm dia.	103.04	126.59	1.19	44.66	nr	**171.25**
Screwed ends for steel pipework; aluminium and steel outer sleeves; up to 16 bar G at 120°C						
20 mm dia.	90.56	111.26	1.32	49.55	nr	**160.81**
25 mm dia.	92.23	113.31	1.52	57.05	nr	**170.36**
32 mm dia.	97.65	119.96	1.80	67.56	nr	**187.52**
40 mm dia.	103.43	127.06	2.03	76.19	nr	**203.25**
50 mm dia.	105.10	129.12	2.26	84.82	nr	**213.94**
Flanged ends for steel pipework; aluminium and steel outer sleeves; up to 16 bar G at 120°C						
20 mm dia.	80.12	98.43	0.53	19.89	nr	**118.32**
25 mm dia.	85.84	105.45	0.64	24.01	nr	**129.46**
32 mm dia.	118.27	145.30	0.74	27.77	nr	**173.07**
40 mm dia.	127.80	157.01	0.82	30.78	nr	**187.79**
50 mm dia.	151.64	186.29	0.89	33.40	nr	**219.69**
Flanged ends for steel pipework; up to 16 bar G at 120°C						
65 mm dia.	140.21	172.25	1.10	41.28	nr	**213.53**
80 mm dia.	150.69	185.13	1.31	49.16	nr	**234.29**
100 mm dia.	172.62	212.07	1.78	66.80	nr	**278.87**
150 mm dia.	264.19	324.56	3.08	115.59	nr	**440.15**
Screwed ends for non-ferrous pipework; up to 6 bar G at 100°C						
20 mm dia.	90.56	111.26	0.72	27.03	nr	**138.29**
25 mm dia.	92.23	113.31	0.84	31.52	nr	**144.83**
32 mm dia.	97.65	119.96	1.02	38.28	nr	**158.24**
40 mm dia.	103.43	127.06	1.11	41.65	nr	**168.71**
50 mm dia.	105.10	129.12	1.18	44.29	nr	**173.41**
Flanged ends for steel, copper or non-ferrous pipework; up to 16 bar G at 120°C						
65 mm dia.	180.26	221.44	0.87	32.65	nr	**254.09**
80 mm dia.	185.98	228.48	0.95	35.65	nr	**264.13**
100 mm dia.	214.60	263.63	1.15	43.16	nr	**306.79**
150 mm dia.	351.94	432.36	1.36	51.04	nr	**483.40**

38 MECHANICAL/COOLING/HEATING SYSTEMS

Item	Net Price £	Material £	Labour hours	Labour £	Unit	**Total rate £**
LOW TEMPERATURE HOT WATER HEATING; PIPELINE: PIPELINE ANCILLERIES – cont						
Angular movement bellows expansion joints; stainless steel						
Flanged ends for steel pipework; up to 16 bar G at 120°C						
50 mm dia.	207.80	255.29	0.71	26.65	nr	**281.94**
65 mm dia.	252.19	309.81	0.83	31.15	nr	**340.96**
80 mm dia.	294.69	362.03	0.91	34.16	nr	**396.19**
100 mm dia.	380.45	467.39	0.97	36.40	nr	**503.79**
125 mm dia.	502.52	617.35	1.16	43.53	nr	**660.88**
150 mm dia.	597.54	734.08	1.18	44.29	nr	**778.37**

38 MECHANICAL/COOLING/HEATING SYSTEMS

Item	Net Price £	Material £	Labour hours	Labour £	Unit	Total rate £
LOW TEMPERATURE HOT WATER HEATING; PIPELINE: VALVES						
Gate valves						
Bronze gate valve; non-rising stem; BS 5154, series B, PN 16; working pressure up to 16 from −10°C to 100°C; 7 bar for saturated steam; screwed ends to steel						
15 mm dia.	49.85	61.24	1.18	44.29	nr	**105.53**
20 mm dia.	68.10	83.66	1.24	46.54	nr	**130.20**
25 mm dia.	86.66	106.46	1.31	49.16	nr	**155.62**
32 mm dia.	118.05	145.03	1.43	53.68	nr	**198.71**
40 mm dia.	157.59	193.60	1.53	57.42	nr	**251.02**
50 mm dia.	217.44	267.12	1.63	61.18	nr	**328.30**
Bronze gate valve; non-rising stem; BS 5154, series B, PN 32; working pressure up to 14 bar for saturated steam, 32 bar from −10°C to 100°C; screwed ends to steel						
15 mm dia.	43.73	53.72	1.11	41.65	nr	**95.37**
20 mm dia.	56.73	69.69	1.28	48.04	nr	**117.73**
25 mm dia.	76.84	94.40	1.49	55.92	nr	**150.32**
32 mm dia.	106.13	130.38	1.88	70.55	nr	**200.93**
40 mm dia.	142.49	175.05	2.31	86.70	nr	**261.75**
50 mm dia.	200.21	245.96	2.80	105.08	nr	**351.04**
Bronze gate valve; non-rising stem; BS 5154, series B, PN 20; working pressure up to 9 bar for saturated steam, 20 bar from −10°C to 100°C; screwed ends to steel						
15 mm dia.	29.82	36.63	0.84	31.52	nr	**68.15**
20 mm dia.	42.38	52.06	1.01	37.92	nr	**89.98**
25 mm dia.	54.84	67.37	1.19	44.66	nr	**112.03**
32 mm dia.	78.25	96.13	1.38	51.80	nr	**147.93**
40 mm dia.	107.09	131.57	1.62	60.80	nr	**192.37**
50 mm dia.	153.03	188.00	1.94	72.82	nr	**260.82**
Bronze gate valve; non-rising stem; BS 5154, series B, PN 16; working pressure up to 7 bar for saturated steam, 16 bar from −10°C to 100°C; BS4504 flanged ends; bolted connections						
15 mm dia.	134.53	165.27	1.18	44.29	nr	**209.56**
20 mm dia.	148.10	181.94	1.24	46.54	nr	**228.48**
25 mm dia.	195.02	239.58	1.31	49.16	nr	**288.74**
32 mm dia.	244.69	300.60	1.43	53.68	nr	**354.28**
40 mm dia.	295.83	363.43	1.53	57.42	nr	**420.85**
50 mm dia.	408.98	502.44	1.63	61.18	nr	**563.62**
65 mm dia.	635.17	780.31	1.71	64.18	nr	**844.49**
80 mm dia.	898.97	1104.38	1.88	70.55	nr	**1174.93**
100 mm dia.	1633.70	2007.00	2.03	76.19	nr	**2083.19**

38 MECHANICAL/COOLING/HEATING SYSTEMS

Item	Net Price £	Material £	Labour hours	Labour £	Unit	Total rate £
LOW TEMPERATURE HOT WATER HEATING; PIPELINE: VALVES – cont						
Cast iron gate valve; bronze trim; non-rising stem; BS 5150, PN6; working pressure 6 bar from −10°C to 120°C; BS4504 flanged ends; bolted connections						
50 mm dia.	265.99	326.77	1.85	69.44	nr	**396.21**
65 mm dia.	278.02	341.54	2.00	75.06	nr	**416.60**
80 mm dia.	321.30	394.72	2.27	85.20	nr	**479.92**
100 mm dia.	406.43	499.30	2.76	103.58	nr	**602.88**
125 mm dia.	581.77	714.70	6.05	227.06	nr	**941.76**
150 mm dia.	669.43	822.39	8.03	301.37	nr	**1123.76**
200 mm dia.	1275.10	1566.46	9.17	344.15	nr	**1910.61**
250 mm dia.	1962.00	2410.32	10.72	402.33	nr	**2812.65**
300 mm dia.	2327.08	2858.81	11.75	440.98	nr	**3299.79**
Cast iron gate valve; bronze trim; non-rising stem; BS 5150, PN10; working pressure up to 8.4 bar for saturated steam, 10 bar from −10° C to 120°C; BS4504 flanged ends; bolted connections						
50 mm dia.	279.01	342.76	1.85	69.44	nr	**412.20**
65 mm dia.	346.73	425.95	2.00	75.06	nr	**501.01**
80 mm dia.	386.02	474.22	2.27	85.20	nr	**559.42**
100 mm dia.	513.88	631.30	2.76	103.58	nr	**734.88**
125 mm dia.	718.46	882.63	6.05	227.06	nr	**1109.69**
150 mm dia.	847.65	1041.34	8.03	301.37	nr	**1342.71**
200 mm dia.	1557.40	1913.27	9.17	344.15	nr	**2257.42**
250 mm dia.	2342.20	2877.39	10.72	402.33	nr	**3279.72**
300 mm dia.	2635.67	3237.92	11.75	440.98	nr	**3678.90**
350 mm dia.	3065.57	3766.06	12.67	475.51	nr	**4241.57**
Cast iron gate valve; bronze trim; non-rising stem; BS 5163 series A, PN16; working pressure for cold water services up to 16 bar; BS4504 flanged ends; bolted connections						
50 mm dia.	279.01	342.76	1.85	69.44	nr	**412.20**
65 mm dia.	346.73	425.95	2.00	75.06	nr	**501.01**
80 mm dia.	386.02	474.22	2.27	85.20	nr	**559.42**
100 mm dia.	513.88	631.30	2.76	103.58	nr	**734.88**
125 mm dia.	718.46	882.63	6.05	227.06	nr	**1109.69**
150 mm dia.	847.65	1041.34	8.03	301.37	nr	**1342.71**
Ball valves						
Malleable iron body; lever operated stainless steel ball and stem; working pressure up to 12 bar; flanged ends to BS 4504 16/11; bolted connections						
40 mm dia.	370.82	455.55	1.54	57.80	nr	**513.35**
50 mm dia.	467.27	574.05	1.64	61.55	nr	**635.60**
80 mm dia.	755.86	928.58	1.92	72.06	nr	**1000.64**
100 mm dia.	781.50	960.07	2.80	105.08	nr	**1065.15**
150 mm dia.	1444.54	1774.62	12.05	452.25	nr	**2226.87**

38 MECHANICAL/COOLING/HEATING SYSTEMS

Item	Net Price £	Material £	Labour hours	Labour £	Unit	**Total rate £**
Malleable iron body; lever operated stainless steel ball and stem; working pressure up to 16 bar; screwed ends to steel						
15 mm dia.	33.20	40.78	1.34	50.31	nr	**91.09**
20 mm dia.	44.05	54.12	1.34	50.31	nr	**104.43**
25 mm dia.	45.29	55.64	1.40	52.56	nr	**108.20**
32 mm dia.	62.40	76.66	1.46	54.86	nr	**131.52**
40 mm dia.	62.40	76.66	1.54	57.82	nr	**134.48**
50 mm dia.	74.61	91.65	1.64	61.62	nr	**153.27**
Carbon steel body; lever operated stainless steel ball and stem; Class 150; working pressure up to 19 bar; screwed ends to steel						
15 mm dia.	81.48	100.10	0.84	31.53	nr	**131.63**
20 mm dia.	91.60	112.53	1.14	42.80	nr	**155.33**
25 mm dia.	99.18	121.84	1.30	48.80	nr	**170.64**
32 mm dia.	127.60	156.75	1.42	53.29	nr	**210.04**
40 mm dia.	159.50	195.94	1.56	58.55	nr	**254.49**
50 mm dia.	199.38	244.93	1.68	63.04	nr	**307.97**
Globe valves						
Bronze; rising stem; renewable disc; BS 5154 series B, PN32; working pressure up to 14 bar for saturated steam, 32 bar from −10°C to 100°C; screwed ends to steel						
15 mm dia.	44.45	54.61	0.77	28.90	nr	**83.51**
20 mm dia.	60.77	74.66	1.03	38.66	nr	**113.32**
25 mm dia.	87.18	107.10	1.19	44.66	nr	**151.76**
32 mm dia.	122.75	150.80	1.38	51.80	nr	**202.60**
40 mm dia.	170.94	210.00	1.62	60.80	nr	**270.80**
50 mm dia.	262.38	322.33	1.61	60.43	nr	**382.76**
Bronze; needle valve; rising stem; BS 5154, series B, PN32; working pressure up to 14 bar for saturated steam, 32 bar from −10°C to 100°C; screwed ends to steel						
15 mm dia.	33.78	41.50	1.07	40.16	nr	**81.66**
20 mm dia.	57.19	70.26	1.18	44.29	nr	**114.55**
25 mm dia.	80.65	99.08	1.27	47.67	nr	**146.75**
32 mm dia.	167.87	206.23	1.35	50.66	nr	**256.89**
40 mm dia.	264.46	324.89	1.47	55.17	nr	**380.06**
50 mm dia.	334.95	411.48	1.61	60.43	nr	**471.91**
Bronze; rising stem; renewable disc; BS 5154, series B, PN16; working pressure upto 7 bar for saturated steam, 16 bar from −10°C to 100°C; BS4504 flanged ends; bolted connections						
15 mm dia.	89.80	110.32	1.16	43.53	nr	**153.85**
20 mm dia.	103.86	127.60	1.26	47.28	nr	**174.88**
25 mm dia.	182.13	223.74	1.38	51.80	nr	**275.54**
32 mm dia.	229.92	282.46	1.47	55.17	nr	**337.63**
40 mm dia.	261.34	321.06	1.56	58.55	nr	**379.61**
50 mm dia.	374.96	460.63	1.71	64.18	nr	**524.81**

38 MECHANICAL/COOLING/HEATING SYSTEMS

Item	Net Price £	Material £	Labour hours	Labour £	Unit	Total rate £
LOW TEMPERATURE HOT WATER HEATING; PIPELINE: VALVES – cont						
Globe valves – cont						
Bronze; rising stem; renewable disc; BS 2060, class 250; working pressure up to 24 bar for saturated steam, 38 bar from −10°C to 100°C; flanged ends (BS 10 table H); bolted connections						
15 mm dia.	228.69	280.95	1.16	43.53	nr	**324.48**
20 mm dia.	265.77	326.50	1.26	47.28	nr	**373.78**
25 mm dia.	365.22	448.68	1.38	51.80	nr	**500.48**
32 mm dia.	477.10	586.12	1.47	55.17	nr	**641.29**
40 mm dia.	571.33	701.88	1.56	58.55	nr	**760.43**
50 mm dia.	888.39	1091.39	1.71	64.18	nr	**1155.57**
65 mm dia.	1062.44	1305.20	1.88	70.55	nr	**1375.75**
80 mm dia.	1350.08	1658.57	2.03	76.19	nr	**1734.76**
Check valves						
Bronze; swing pattern; BS 5154 series B, PN 25; working pressure up to 10.5 bar for saturated steam, 25 bar from −10°C to 100°C; screwed ends to steel						
15 mm dia.	20.65	25.37	0.77	28.90	nr	**54.27**
20 mm dia.	24.53	30.14	1.03	38.66	nr	**68.80**
25 mm dia.	34.01	41.78	1.19	44.66	nr	**86.44**
32 mm dia.	57.62	70.79	1.38	51.80	nr	**122.59**
40 mm dia.	71.68	88.06	1.62	60.80	nr	**148.86**
50 mm dia.	109.93	135.05	1.94	72.82	nr	**207.87**
65 mm dia.	210.35	258.42	2.45	91.95	nr	**350.37**
80 mm dia.	297.43	365.39	2.83	106.22	nr	**471.61**
Bronze; vertical lift pattern; BS 5154 series B, PN32; working pressure up to 14 bar for saturated steam, 32 bar from −10°C to 100°C; screwed ends to steel						
15 mm dia.	26.21	32.20	0.96	36.04	nr	**68.24**
20 mm dia.	37.27	45.79	1.07	40.16	nr	**85.95**
25 mm dia.	54.72	67.22	1.17	43.91	nr	**111.13**
32 mm dia.	83.36	102.41	1.33	49.92	nr	**152.33**
40 mm dia.	108.62	133.44	1.41	52.92	nr	**186.36**
50 mm dia.	162.76	199.95	1.55	58.17	nr	**258.12**
65 mm dia.	459.27	564.22	1.80	67.56	nr	**631.78**
80 mm dia.	686.60	843.49	1.99	74.68	nr	**918.17**
Bronze; oblique swing pattern; BS 5154 series A, PN32; working pressure up to 14 bar for saturated steam, 32 bar from −10°C to 120°C; screwed connections to steel						
15 mm dia.	33.29	40.90	0.96	36.04	nr	**76.94**
20 mm dia.	39.55	48.58	1.07	40.16	nr	**88.74**
25 mm dia.	54.83	67.36	1.17	43.91	nr	**111.27**

38 MECHANICAL/COOLING/HEATING SYSTEMS

Item	Net Price £	Material £	Labour hours	Labour £	Unit	**Total rate £**
32 mm dia.	81.98	100.72	1.33	49.92	nr	**150.64**
40 mm dia.	105.67	129.81	1.41	52.92	nr	**182.73**
50 mm dia.	175.06	215.06	1.55	58.17	nr	**273.23**
Cast iron; swing pattern; BS 5153 PN6; working pressure up to 6 bar from −10°C to 120°C; BS 4504 flanged ends; bolted connections						
50 mm dia.	575.56	707.08	1.86	69.80	nr	**776.88**
65 mm dia.	630.17	774.16	2.00	75.06	nr	**849.22**
80 mm dia.	709.31	871.38	2.56	96.09	nr	**967.47**
100 mm dia.	908.65	1116.28	2.76	103.58	nr	**1219.86**
125 mm dia.	1355.58	1665.33	6.05	227.06	nr	**1892.39**
150 mm dia.	1524.18	1872.45	8.11	304.37	nr	**2176.82**
200 mm dia.	3370.69	4140.90	9.26	347.53	nr	**4488.43**
250 mm dia.	4213.26	5175.99	10.72	402.33	nr	**5578.32**
300 mm dia.	5055.92	6211.20	11.75	440.98	nr	**6652.18**
Cast iron; horizontal lift pattern; BS 5153 PN16; working pressure up to 13 bar for saturated steam, 16 bar from −10°C to 120°C; BS 4504 flanged ends; bolted connections						
50 mm dia.	519.75	638.52	1.86	69.80	nr	**708.32**
65 mm dia.	519.75	638.52	2.00	75.06	nr	**713.58**
80 mm dia.	643.49	790.52	2.56	96.09	nr	**886.61**
100 mm dia.	749.37	920.60	2.96	111.09	nr	**1031.69**
125 mm dia.	1118.04	1373.52	7.76	291.24	nr	**1664.76**
150 mm dia.	1257.01	1544.23	10.50	394.07	nr	**1938.30**
Cast iron; semi-lugged butterfly valve; BS5155 PN16; working pressure 16 bar from −10°C to 120°C; BS 4504 flanged ends; bolted connections						
50 mm dia.	173.59	213.26	2.20	82.57	nr	**295.83**
65 mm dia.	179.14	220.07	2.31	86.70	nr	**306.77**
80 mm dia.	211.02	259.23	2.88	108.09	nr	**367.32**
100 mm dia.	293.36	360.39	3.11	116.72	nr	**477.11**
125 mm dia.	428.17	526.01	5.02	188.40	nr	**714.41**
150 mm dia.	491.46	603.76	6.98	261.96	nr	**865.72**
200 mm dia.	784.53	963.79	8.25	309.63	nr	**1273.42**
250 mm dia.	1157.94	1422.53	10.47	392.95	nr	**1815.48**
300 mm dia.	1736.94	2133.83	11.48	430.85	nr	**2564.68**
Cast iron; semi-lugged butterfly valve; BS5155 PN20; working pressure up to 20 bar from −10°C to 120°C; BS 4504 flanged ends; bolted connections						
50 mm dia.	1547.83	1901.51	2.20	82.57	nr	**1984.08**
65 mm dia.	1601.40	1967.32	2.31	86.70	nr	**2054.02**
80 mm dia.	1709.68	2100.35	2.88	108.09	nr	**2208.44**
100 mm dia.	2081.94	2557.66	3.11	116.72	nr	**2674.38**
125 mm dia.	2672.28	3282.90	5.02	188.40	nr	**3471.30**
150 mm dia.	3032.75	3725.74	5.02	188.40	nr	**3914.14**
200 mm dia.	3488.08	4285.10	8.25	309.63	nr	**4594.73**
250 mm dia.	5500.10	6756.88	10.47	392.95	nr	**7149.83**
300 mm dia.	6664.44	8187.26	11.48	430.85	nr	**8618.11**

38 MECHANICAL/COOLING/HEATING SYSTEMS

Item	Net Price £	Material £	Labour hours	Labour £	Unit	**Total rate £**
LOW TEMPERATURE HOT WATER HEATING; PIPELINE: VALVES – cont						
Check valves – cont						
Cast iron; semi-lugged butterfly valve; BS5155 PN20; working pressure up to 30 bar from −10°C to 120°C; BS 4504 flanged ends; bolted connections						
50 mm dia.	1547.83	1901.51	2.20	82.57	nr	**1984.08**
65 mm dia.	1601.40	1967.32	2.31	86.70	nr	**2054.02**
80 mm dia.	1709.68	2100.35	2.88	108.09	nr	**2208.44**
100 mm dia.	2081.94	2557.66	3.11	116.72	nr	**2674.38**
125 mm dia.	2672.28	3282.90	5.02	188.40	nr	**3471.30**
150 mm dia.	3032.75	3725.74	5.02	188.40	nr	**3914.14**
200 mm dia.	3488.08	4285.10	8.25	309.63	nr	**4594.73**
250 mm dia.	5500.10	6756.88	10.47	392.95	nr	**7149.83**
300 mm dia.	6664.44	8187.26	11.48	430.85	nr	**8618.11**
Commissioning valves						
Bronze commissioning set; metering station; double regulating valve; BS5154 PN20 Series B; working pressure 20 bar from −10°C to 100°C; screwed ends to steel						
15 mm dia.	87.08	106.97	1.08	40.53	nr	**147.50**
20 mm dia.	139.91	171.87	1.46	54.79	nr	**226.66**
25 mm dia.	166.37	204.38	1.68	63.04	nr	**267.42**
32 mm dia.	223.07	274.04	1.95	73.18	nr	**347.22**
40 mm dia.	323.49	397.40	2.27	85.20	nr	**482.60**
50 mm dia.	488.20	599.75	2.73	102.46	nr	**702.21**
Cast iron commissioning set; metering station; double regulating valve; BS5152 PN16; working pressure 16 bar from −10°C to 90°C; flanged ends (BS 4504, Part 1, Table 16); bolted connections						
65 mm dia.	739.63	908.64	1.80	67.56	nr	**976.20**
80 mm dia.	885.62	1087.99	2.56	96.09	nr	**1184.08**
100 mm dia.	1232.93	1514.66	2.30	86.33	nr	**1600.99**
125 mm dia.	1805.93	2218.59	2.44	91.57	nr	**2310.16**
150 mm dia.	2378.97	2922.56	2.90	108.83	nr	**3031.39**
200 mm dia.	6089.24	7480.63	8.26	310.00	nr	**7790.63**
250 mm dia.	7617.54	9358.15	10.49	393.70	nr	**9751.85**
300 mm dia.	8272.82	10163.16	11.49	431.22	nr	**10594.38**
Cast iron variable orifice double regulating valve; orifice valve; BS5152 PN16; working pressure 16 bar from −10° to 90°C; flanged ends (BS 4504, Part 1, Table 16); bolted connections						
65 mm dia.	541.51	665.25	2.00	75.06	nr	**740.31**
80 mm dia.	662.81	814.26	2.56	96.09	nr	**910.35**
100 mm dia.	909.60	1117.44	2.96	111.09	nr	**1228.53**

38 MECHANICAL/COOLING/HEATING SYSTEMS

Item	Net Price £	Material £	Labour hours	Labour £	Unit	**Total rate £**
125 mm dia.	1362.12	1673.36	7.76	291.24	nr	**1964.60**
150 mm dia.	1749.61	2149.39	10.50	394.07	nr	**2543.46**
200 mm dia.	4585.54	5633.33	8.26	310.00	nr	**5943.33**
250 mm dia.	6972.29	8565.46	10.49	393.70	nr	**8959.16**
300 mm dia.	12506.88	15364.70	11.49	431.22	nr	**15795.92**
Cast iron globe valve with double regulating feature; BS5152 PN16; working pressure 16 bar from −10°C to 120°C; flanged ends (BS 4504, Part 1, Table 16); bolted connections						
65 mm dia.	645.08	792.48	2.00	75.06	nr	**867.54**
80 mm dia.	780.11	958.37	2.56	96.09	nr	**1054.46**
100 mm dia.	1027.64	1262.46	2.96	111.09	nr	**1373.55**
125 mm dia.	1432.80	1760.20	7.76	291.24	nr	**2051.44**
150 mm dia.	1813.40	2227.76	10.50	394.07	nr	**2621.83**
200 mm dia.	4694.74	5767.49	8.26	310.00	nr	**6077.49**
250 mm dia.	7103.03	8726.08	10.49	393.70	nr	**9119.78**
300 mm dia.	11159.89	13709.92	11.49	431.22	nr	**14141.14**
Bronze autoflow commissioning valve; PN25 ; working pressure 25 bar up to 100°C; screwed ends to steel						
15 mm dia.	98.73	121.29	0.82	30.78	nr	**152.07**
20 mm dia.	104.55	128.44	1.08	40.53	nr	**168.97**
25 mm dia.	130.18	159.93	1.27	47.67	nr	**207.60**
32 mm dia.	203.30	249.75	1.50	56.30	nr	**306.05**
40 mm dia.	222.62	273.49	1.76	66.06	nr	**339.55**
50 mm dia.	329.14	404.34	2.13	79.94	nr	**484.28**
Ductile iron autoflow commissioning valves; PN16; working pressure 16 bar from −10°C to 120°C; for ANSI 150 flanged ends						
65 mm dia.	866.78	1064.84	2.31	86.70	nr	**1151.54**
80 mm dia.	958.34	1177.32	2.88	108.09	nr	**1285.41**
100 mm dia.	1224.25	1503.99	3.11	116.72	nr	**1620.71**
150 mm dia.	2521.96	3098.22	6.98	261.96	nr	**3360.18**
200 mm dia.	3673.46	4512.85	8.26	310.00	nr	**4822.85**
250 mm dia.	5101.89	6267.67	10.49	393.70	nr	**6661.37**
300 mm dia.	6515.92	8004.81	11.49	431.22	nr	**8436.03**
Strainers						
Bronze strainer; Y type; PN32 ; working pressure 32 bar from −10°C to 100°C; screwed ends to steel						
15 mm dia.	39.57	48.61	0.82	30.78	nr	**79.39**
20 mm dia.	50.27	61.76	1.08	40.53	nr	**102.29**
25 mm dia.	66.67	81.90	1.27	47.67	nr	**129.57**
32 mm dia.	107.69	132.30	1.50	56.30	nr	**188.60**
40 mm dia.	139.29	171.12	1.76	66.06	nr	**237.18**
50 mm dia.	233.29	286.60	2.13	79.94	nr	**366.54**

38 MECHANICAL/COOLING/HEATING SYSTEMS

Item	Net Price £	Material £	Labour hours	Labour £	Unit	Total rate £
LOW TEMPERATURE HOT WATER HEATING; PIPELINE: VALVES – cont						
Strainers – cont						
Cast iron strainer; Y type; PN16; working pressure 16 bar from −10°C to 120°C; BS 4504 flanged ends						
65 mm dia.	246.13	302.37	2.31	86.70	nr	**389.07**
80 mm dia.	291.33	357.90	2.88	108.09	nr	**465.99**
100 mm dia.	426.26	523.66	3.11	116.72	nr	**640.38**
125 mm dia.	785.80	965.36	5.02	188.40	nr	**1153.76**
150 mm dia.	1008.18	1238.55	6.98	261.96	nr	**1500.51**
200 mm dia.	1692.19	2078.85	8.26	310.00	nr	**2388.85**
250 mm dia.	2447.00	3006.14	10.49	393.70	nr	**3399.84**
300 mm dia.	3984.89	4895.44	11.49	431.22	nr	**5326.66**
Regulators						
Gunmetal; self-acting two port thermostatic regulator; single seat; screwed ends; complete with sensing element, 2 m long capillary tube						
15 mm dia.	696.84	856.07	1.37	51.41	nr	**907.48**
20 mm dia.	719.25	883.60	1.24	46.54	nr	**930.14**
25 mm dia.	738.15	906.82	1.34	50.30	nr	**957.12**
Gunmetal; self-acting two port thermostatic regulator; double seat; flanged ends (BS 4504 PN25); with sensing element, 2 m long capillary tube; steel body						
65 mm dia.	2584.30	3174.81	1.23	49.20	nr	**3224.01**
80 mm dia.	3050.20	3747.17	1.62	64.81	nr	**3811.98**
Control valves						
Electrically operated (electrical work elsewhere)						
Pressure independant control valve (PICV)						
Brass; pressure independant control valve; rotary type; PN 25; maximum working pressure 25 bar at from −10°C to 120°C; screwed ends to steel; electrical work elsewhere						
15 mm dia.	79.80	98.04	1.08	40.53	nr	**138.57**
20 mm dia.	95.55	117.38	1.18	44.29	nr	**161.67**
25 mm dia.	140.70	172.85	1.35	50.66	nr	**223.51**
32 mm dia.	204.75	251.54	1.46	54.79	nr	**306.33**
40 mm dia.	226.80	278.63	1.53	57.42	nr	**336.05**
50 mm dia.	395.85	486.30	1.61	60.43	nr	**546.73**

38 MECHANICAL/COOLING/HEATING SYSTEMS

Item	Net Price £	Material £	Labour hours	Labour £	Unit	Total rate £
Brass; pressure independant control valve; rotary type with electronic actuator; 24 V motor; PN 25; maximum working pressure 25 bar at from −10°C to 120°C; screwed ends to steel; electrical work elsewhere						
15 mm dia.	166.95	205.10	2.06	77.31	nr	**282.41**
20 mm dia.	182.70	224.45	2.15	80.69	nr	**305.14**
25 mm dia.	227.85	279.91	2.27	85.20	nr	**365.11**
32 mm dia.	321.30	394.72	2.35	88.19	nr	**482.91**
40 mm dia.	343.35	421.81	2.47	92.70	nr	**514.51**
50 mm dia.	523.95	643.67	2.55	95.70	nr	**739.37**
Differential pressure control valves (DPCV)						
Brass; differential control valve; PN 25; maximum working pressure 25 bar at from −20°C to 120°C; screwed ends to steel						
15 mm dia.	219.45	269.60	1.99	74.68	nr	**344.28**
20 mm dia.	236.33	290.34	2.03	76.19	nr	**366.53**
25 mm dia.	262.61	322.61	2.11	79.19	nr	**401.80**
32 mm dia.	297.27	365.20	2.17	81.44	nr	**446.64**
40 mm dia.	512.38	629.45	2.26	84.82	nr	**714.27**
50 mm dia.	615.23	755.81	2.31	86.70	nr	**842.51**
Cast iron; butterfly type; two position electrically controlled 240 V motor and linkage mechanism; for low pressure hot water; maximum pressure 6 bar at 120°C; flanged ends; electrical work elsewhere						
25 mm dia.	858.00	1054.05	1.47	55.17	nr	**1109.22**
32 mm dia.	886.55	1089.12	1.52	57.05	nr	**1146.17**
40 mm dia.	1238.81	1521.88	1.61	60.43	nr	**1582.31**
50 mm dia.	1274.02	1565.13	1.71	64.18	nr	**1629.31**
65 mm dia.	1305.92	1604.33	2.51	94.21	nr	**1698.54**
80 mm dia.	1356.99	1667.06	2.69	100.96	nr	**1768.02**
100 mm dia.	1424.05	1749.45	2.81	105.46	nr	**1854.91**
125 mm dia.	1577.31	1937.72	2.94	110.34	nr	**2048.06**
150 mm dia.	1709.27	2099.84	3.33	124.97	nr	**2224.81**
200 mm dia.	2120.07	2604.50	3.67	137.74	nr	**2742.24**
Cast iron; three way 240 V motorized; for low pressure hot water; maximum pressure 6 bar 120°C; flanged ends, drilled (BS 10, Table F)						
25 mm dia.	958.94	1178.06	1.99	74.68	nr	**1252.74**
40 mm dia.	993.01	1219.91	2.13	79.94	nr	**1299.85**
50 mm dia.	1005.79	1235.61	3.21	120.48	nr	**1356.09**
65 mm dia.	1109.51	1363.04	3.23	121.23	nr	**1484.27**
80 mm dia.	1243.23	1527.31	3.50	131.35	nr	**1658.66**

38 MECHANICAL/COOLING/HEATING SYSTEMS

Item	Net Price £	Material £	Labour hours	Labour £	Unit	**Total rate £**
LOW TEMPERATURE HOT WATER HEATING; PIPELINE: VALVES – cont						
2 port control valves						
Cast iron; 2 port; motorized control valve; normally open for heating applications; PN 16; screwed ends to steel						
15 mm dia.	37.80	46.44	1.24	46.54	nr	**92.98**
22 mm dia.	57.71	70.90	1.31	49.16	nr	**120.06**
28 mm dia.	69.02	84.79	1.37	51.41	nr	**136.20**
32 mm dia.	80.64	99.07	1.45	54.42	nr	**153.49**
40 mm dia.	100.80	123.84	1.52	57.05	nr	**180.89**
50 mm dia.	126.00	154.79	1.60	60.05	nr	**214.84**
Two port normally closed motorized valve; electric actuator; spring return; domestic usage						
22 mm dia.	130.48	160.29	1.18	44.29	nr	**204.58**
28 mm dia.	176.86	217.28	1.35	50.66	nr	**267.94**
Two port on/off motorized valve; electric actuator; spring return; domestic usage						
22 mm dia.	130.48	160.29	1.18	44.29	nr	**204.58**
3 port control valves						
Cast iron; 3 port; motorized control valve with temperatyre control unit; PN 25; screwed ends to steel						
22 mm dia.	58.80	72.24	2.00	75.06	nr	**147.30**
25 mm dia.	66.67	81.90	2.13	79.94	nr	**161.84**
32 mm dia.	85.34	104.84	2.19	82.18	nr	**187.02**
40 mm dia.	106.68	131.06	2.26	84.82	nr	**215.88**
Three port motorized valve; electric actuator; spring return; domestic usage						
22 mm dia.	198.39	243.73	1.18	44.29	nr	**288.02**
4 port control valves						
Cast iron; 4 port; fixed orifice double regulating and control valve; PN 16; screwed ends to steel						
22 mm dia.	52.11	64.02	2.89	108.47	nr	**172.49**
Safety and relief valves						
Bronze relief valve; spring type; side outlet; working pressure up to 20.7 bar at 120°C; screwed ends to steel						
15 mm dia.	151.62	186.27	0.26	9.75	nr	**196.02**
20 mm dia.	166.30	204.30	0.36	13.51	nr	**217.81**
Bronze relief valve; spring type; side outlet; working pressure up to 17.2 bar at 120°C; screwed ends to steel						
25 mm dia.	227.44	279.40	0.38	14.26	nr	**293.66**
32 mm dia.	308.14	378.55	0.48	18.01	nr	**396.56**

38 MECHANICAL/COOLING/HEATING SYSTEMS

Item	Net Price £	Material £	Labour hours	Labour £	Unit	**Total rate £**
Bronze relief valve; spring type; side outlet; working pressure up to 13.8 bar at 120°C; screwed ends to steel						
40 mm dia.	402.29	494.21	0.64	24.01	nr	**518.22**
50 mm dia.	524.56	644.43	0.76	28.53	nr	**672.96**
65 mm dia.	971.24	1193.17	0.94	35.28	nr	**1228.45**
80 mm dia.	1274.25	1565.41	1.10	41.28	nr	**1606.69**
Cocks; screwed joints to steel						
Bronze gland cock; complete with malleable iron lever; working pressure up to 10 bar at 100°C; screwed ends to steel						
15 mm dia.	53.28	65.46	0.77	28.90	nr	**94.36**
20 mm dia.	75.80	93.12	1.03	38.66	nr	**131.78**
25 mm dia.	102.14	125.47	1.19	44.66	nr	**170.13**
32 mm dia.	154.35	189.62	1.38	51.80	nr	**241.42**
40 mm dia.	215.52	264.77	1.62	60.80	nr	**325.57**
50 mm dia.	328.92	404.08	1.94	72.82	nr	**476.90**
Bronze three-way plug cock; complete with malleable iron lever; working pressure up to 10 bar at 100°C; screwed ends to steel						
15 mm dia.	108.02	132.70	0.77	28.90	nr	**161.60**
20 mm dia.	125.03	153.60	1.03	38.66	nr	**192.26**
25 mm dia.	174.78	214.71	1.19	44.66	nr	**259.37**
32 mm dia.	247.89	304.53	1.38	51.80	nr	**356.33**
40 mm dia.	298.88	367.17	1.62	60.80	nr	**427.97**
Air vents; including regulating, adjusting and testing						
Automatic air vent; maximum pressure up to 7 bar at 93°C; screwed ends to steel						
15 mm dia.	13.59	16.70	0.80	30.02	nr	**46.72**
Automatic air vent; maximum pressure up to 7 bar at 93°C; lockhead isolating valve; screwed ends to steel						
15 mm dia.	13.59	16.70	0.83	31.15	nr	**47.85**
Automatic air vent; maximum pressure up to 17 bar at 200°C; flanged ends (BS10, Table H); bolted connections to counter flange (measured separately)						
15 mm dia.	488.99	600.73	0.83	31.15	nr	**631.88**

38 MECHANICAL/COOLING/HEATING SYSTEMS

Item	Net Price £	Material £	Labour hours	Labour £	Unit	**Total rate £**
LOW TEMPERATURE HOT WATER HEATING; PIPELINE: VALVES – cont						
Radiator valves						
Bronze; wheelhead or lockshield; chromium plated finish; screwed joints to steel						
Straight						
15 mm dia.	17.35	21.32	0.59	22.13	nr	**43.45**
20 mm dia.	27.00	33.17	0.73	27.39	nr	**60.56**
25 mm dia.	33.89	41.63	0.85	31.90	nr	**73.53**
Angled						
15 mm dia.	58.70	72.11	0.59	22.13	nr	**94.24**
20 mm dia.	77.41	95.10	0.73	27.39	nr	**122.49**
25 mm dia.	99.51	122.25	0.85	31.90	nr	**154.15**
Bronze; wheelhead or lockshield; chromium plated finish; compression joints to copper						
Straight						
15 mm dia.	31.08	38.18	0.59	22.13	nr	**60.31**
20 mm dia.	39.16	48.11	0.73	27.39	nr	**75.50**
25 mm dia.	49.21	60.46	0.85	31.90	nr	**92.36**
Angled						
15 mm dia.	23.45	28.81	0.59	22.13	nr	**50.94**
20 mm dia.	24.60	30.22	0.73	27.39	nr	**57.61**
25 mm dia.	30.50	37.46	0.85	31.90	nr	**69.36**
Twin entry						
8 mm dia.	31.40	38.58	0.23	8.64	nr	**47.22**
10 mm dia.	35.63	43.77	0.23	8.64	nr	**52.41**
Bronze; thermostatic head; chromium plated finish; compression joints to copper						
Straight						
15 mm dia.	12.84	15.77	0.59	22.13	nr	**37.90**
20 mm dia.	18.85	23.15	0.73	27.39	nr	**50.54**
Angled						
15 mm dia.	11.96	14.69	0.59	22.13	nr	**36.82**
20 mm dia.	17.52	21.52	0.73	27.39	nr	**48.91**

38 MECHANICAL/COOLING/HEATING SYSTEMS

Item	Net Price £	Material £	Labour hours	Labour £	Unit	Total rate £
LOW TEMPERATURE HOT WATER HEATING; PIPELINE: EQUIPMENT						
GAUGES						
Thermometers and pressure gauges						
Dial thermometer; coated steel case and dial; glass window; brass pocket; BS 5235; pocket length 100 mm; screwed end						
Back/bottom entry						
100 mm dia face	68.66	84.35	0.81	30.42	nr	**114.77**
150 mm dia face	112.53	138.24	0.81	30.42	nr	**168.66**
Dial thermometer; coated steel case and dial; glass window; brass pocket; BS 5235; pocket length 100 mm; screwed end						
100 mm dia face	101.78	125.03	0.81	30.42	nr	**155.45**
150 mm dia face	113.98	140.03	0.81	30.42	nr	**170.45**
PRESSURIZATION UNITS						
LTHW pressurization unit complete with expansion vessel(s), interconnecting pipework and all necessary isolating and drain valves; includes placing in position; electrical work elsewhere. Selection based on a final working pressure of 4 bar, a 3 m static head and system operating temperatures of 82/71° C						
System volume						
2,400 litres	2037.70	2503.32	15.00	562.96	nr	**3066.28**
6,000–20,000 litres	4659.21	5723.84	22.00	825.68	nr	**6549.52**
25,000 litres	7176.30	8816.08	22.00	825.68	nr	**9641.76**
DIRT SEPARATORS						
Dirt seperator; maximum operating temperature and pressure of 110°C and 10 bar; fitted with drain valve						
Bore size, flow rate (at 1.0 m/s velocity); threaded connections						
32 mm dia. 3.7 m³/h	95.48	117.30	2.29	85.94	nr	**203.24**
40 mm dia. 5.0 m³/h	113.66	139.63	2.45	91.95	nr	**231.58**
Bore size, flow rate (at 1.5 ms velocity); flanged connections to PN16						
50 mm dia. 13.0 m³/h	642.58	789.41	3.00	112.59	nr	**902.00**
65 mm dia. 21.0 m³/h	669.87	822.94	3.00	112.59	nr	**935.53**
80 mm dia. 29.0 m³/h	942.66	1158.06	3.84	144.11	nr	**1302.17**
100 mm dia. 49.0 m³/h	997.22	1225.09	4.44	166.63	nr	**1391.72**
125 mm dia. 74.0 m³/h	1911.07	2347.75	11.64	436.85	nr	**2784.60**
150 mm dia. 109.0 m³/h	1992.91	2448.29	15.75	591.11	nr	**3039.40**
200 mm dia. 181.0 m³/h	3003.77	3690.13	15.75	591.11	nr	**4281.24**

38 MECHANICAL/COOLING/HEATING SYSTEMS

Item	Net Price £	Material £	Labour hours	Labour £	Unit	**Total rate £**
LOW TEMPERATURE HOT WATER HEATING; PIPELINE: EQUIPMENT – cont						
DIRT SEPARATORS – cont						
250 mm dia. 288.0 m³/h	5324.04	6540.59	15.75	591.11	nr	**7131.70**
300 mm dia. 407.0 m³/h	7782.20	9560.43	17.24	647.03	nr	**10207.46**
MICROBUBBLE DEAERATORS						
Microbubble deaerator; maximum operating temperature and pressure of 110°C and 10 bar; fitted with drain valve						
Bore size, flow rate (at 1.0 m/s velocity); threaded connections						
32 mm dia. 3.7 m³/h	87.90	107.98	2.29	85.94	nr	**193.92**
40 mm dia. 5.0 m³/h	104.58	128.48	2.45	91.95	nr	**220.43**
Bore size, flow rate (at 1.5 ms velocity); flanged connections to PN16						
50 mm dia. 13.0 m³/h	642.58	789.41	3.00	112.59	nr	**902.00**
65 mm dia. 21.0 m³/h	668.59	821.36	3.00	112.59	nr	**933.95**
80 mm dia. 29.0 m³/h	942.66	1158.06	3.84	144.11	nr	**1302.17**
100 mm dia. 49.0 m³/h	997.22	1225.09	4.44	166.63	nr	**1391.72**
125 mm dia. 74.0 m³/h	1911.07	2347.75	11.64	436.85	nr	**2784.60**
150 mm dia. 109.0 m³/h	1992.91	2448.29	15.75	591.11	nr	**3039.40**
200 mm dia. 181.0 m³/h	3003.77	3690.13	15.75	591.11	nr	**4281.24**
250 mm dia. 288.0 m³/h	5324.04	6540.59	15.75	591.11	nr	**7131.70**
300 mm dia. 407.0 m³/h	7809.48	9593.94	17.24	647.03	nr	**10240.97**
COMBINED MICROBUBBLE DEAERATORS AND DIRT SEPARATORS						
Combined deaerator and dirt separators; maximum operating temperature and pressure of 110°C and 10 bar; fitted with drain valve						
Bore size, flow rate (at 1.5 ms velocity); threaded connections						
25 mm dia. 2.0 m³/h	121.24	148.94	2.75	103.20	nr	**252.14**
Bore size, flow rate (at 1.5 ms velocity); flanged connections to PN16						
50 mm dia. 13.0 m³/h	819.90	1007.24	3.60	135.11	nr	**1142.35**
65 mm dia. 21.0 m³/h	847.18	1040.76	3.60	135.11	nr	**1175.87**
80 mm dia. 29.0 m³/h	1147.25	1409.39	4.61	173.02	nr	**1582.41**
100 mm dia. 49.0 m³/h	1202.12	1476.80	5.33	200.04	nr	**1676.84**
125 mm dia. 74.0 m³/h	2292.99	2816.94	13.97	524.30	nr	**3341.24**
150 mm dia. 109.0 m³/h	2376.34	2919.34	18.90	709.33	nr	**3628.67**
200 mm dia. 181.0 m³/h	3685.75	4527.95	18.90	709.33	nr	**5237.28**
250 mm dia. 288.0 m³/h	6498.57	7983.50	18.90	709.33	nr	**8692.83**
300 mm dia. 407.0 m³/h	9966.08	12243.33	20.69	776.51	nr	**13019.84**

38 MECHANICAL/COOLING/HEATING SYSTEMS

Item	Net Price £	Material £	Labour hours	Labour £	Unit	Total rate £
LOW TEMPERATURE HOT WATER HEATING; PIPELINE: PUMPS						
Centrifugal heating and chilled water pump; belt drive; 3 phase, 1450 rpm motor; max. pressure 1000 kN/m²; max. temperature 125°C; bed plate; coupling guard; bolted connections; supply only mating flanges;includes fixing on prepared concrete base; electrical work elsewhere						
40 mm pump size; 4.0 l/s at 70 kPa max head; 0.25 kW max motor rating	1484.35	1823.52	7.59	284.85	nr	**2108.37**
40 mm pump size; 4.0 l/s at 130 kPa max head; 1.5 kW max motor rating	1857.64	2282.11	8.09	303.63	nr	**2585.74**
50 mm pump size; 8.5 l/s at 90 kPa max head; 2.2 kW max motor rating	1915.73	2353.47	8.67	325.40	nr	**2678.87**
50 mm pump size; 8.5 l/s at 190 kPa max head; 3 kW max motor rating	3329.65	4090.47	11.20	420.35	nr	**4510.82**
50 mm pump size; 8.5 l/s at 215 kPa max head; 4 kW max motor rating	2176.34	2673.64	11.70	439.11	nr	**3112.75**
65 mm pump size; 14.0 l/s at 90 kPa max head; 3 kW max motor rating	1908.70	2344.84	11.70	439.11	nr	**2783.95**
65 mm pump size; 14.0 l/s at 160 kPa max head; 4 kW max motor rating	2116.47	2600.08	11.70	439.11	nr	**3039.19**
80 mm pump size; 14.5 l/s at 210 kPa max head; 5.5 kW max motor rating	3246.90	3988.81	11.70	439.11	nr	**4427.92**
80 mm pump size; 22.0 l/s at 130 kPa max head; 5.5 kW max motor rating	3246.90	3988.81	13.64	511.92	nr	**4500.73**
80 mm pump size; 22.0 l/s at 200 kPa max head; 7.5 kW max motor rating	3400.08	4176.99	13.64	511.92	nr	**4688.91**
100 mm pump size; 22.0 l/s at 250 kPa max head; 11 kW max motor rating	4771.74	5862.09	13.64	511.92	nr	**6374.01**
100 mm pump size; 30.0 l/s at 100 kPa max head; 4.0 kW max motor rating	2671.11	3281.46	19.15	718.71	nr	**4000.17**
100 mm pump size; 36.0 l/s at 250 kPa max head; 15.0 kW max motor rating	5053.47	6208.19	19.15	718.71	nr	**6926.90**
100 mm pump size; 36.0 l/s at 550 kPa max head; 30.0 kW max motor rating	6858.27	8425.39	19.15	718.71	nr	**9144.10**

38 MECHANICAL/COOLING/HEATING SYSTEMS

Item	Net Price £	Material £	Labour hours	Labour £	Unit	Total rate £
LOW TEMPERATURE HOT WATER HEATING; PIPELINE: PUMPS – cont						
Centrifugal heating and chilled water pump; TWIN HEAD BELT DRIVE; 3 phase, 1450 rpm motor; max. pressure 1000 kN/m²; max. temperature 125°C; bed plate; coupling guard; bolted connections; supply only mating flanges; includes fixing on prepared concrete base; electrical work elsewhere						
40 mm pump size; 4.0 l/s at 70 kPa max head; 0.75 kW max motor rating	2974.00	3653.56	7.59	284.85	nr	**3938.41**
40 mm pump size; 4.0 l/s at 130 kPa max head; 1.5 kW max motor rating	3579.99	4398.02	8.09	303.63	nr	**4701.65**
50 mm pump size; 8.5 l/s at 90 kPa max head; 2.2 kW max motor rating	3657.40	4493.12	8.67	325.40	nr	**4818.52**
50 mm pump size; 8.5 l/s at 190 kPa max head; 4 kW max motor rating	4215.62	5178.89	11.20	420.35	nr	**5599.24**
65 mm pump size; 8.5 l/s at 215 kPa max head; 4 kW max motor rating	5793.16	7116.90	11.70	439.11	nr	**7556.01**
65 mm pump size; 14.0 l/s at 90 kPa max head; 3 kW max motor rating	6082.82	7472.74	11.70	439.11	nr	**7911.85**
65 mm pump size; 14.0 l/s at 160 kPa max head; 4 kW max motor rating	6386.97	7846.39	11.70	439.11	nr	**8285.50**
80 mm pump size; 14.5 l/s at 210 kPa max head; 7.5 kW max motor rating	6605.03	8114.28	13.64	511.92	nr	**8626.20**
Centrifugal heating and chilled water pump; CLOSE COUPLED; 3 phase, 1450 rpm motor; max. pressure 1000 kN/m²; max. temperature 110°C; bed plate; coupling guard; bolted connections; supply only mating flanges; includes fixing on prepared concrete base; electrical work elsewhere						
40 mm pump size; 4.0 l/s at 23 kPa max head; 0.55 kW max motor rating	838.22	1029.76	7.31	274.34	nr	**1304.10**
50 mm pump size; 4.0 l/s at 75 kPa max head; 0.75 kW max motor rating	929.15	1141.47	7.31	274.34	nr	**1415.81**
50 mm pump size; 7.0 l/s at 65 kPa max head; 0.75 kW max motor rating	929.15	1141.47	8.01	300.63	nr	**1442.10**
65 mm pump size; 10.0 l/s at 33 kPa max head; 0.75 kW max motor rating	1053.16	1293.81	8.01	300.63	nr	**1594.44**
50 mm pump size; 4.0 l/s at 120 kPa max head; 1.5 kW max motor rating	1268.06	1557.81	8.01	300.63	nr	**1858.44**
80 mm pump size; 16.0 l/s at 80 kPa max head; 2.2 kW max motor rating	1641.71	2016.84	12.35	463.50	nr	**2480.34**

38 MECHANICAL/COOLING/HEATING SYSTEMS

Item	Net Price £	Material £	Labour hours	Labour £	Unit	Total rate £
80 mm pump size; 16.0 l/s at 120 kPa max head; 4.0 kW max motor rating	1600.38	1966.06	12.35	463.50	nr	**2429.56**
100 mm pump size; 28.0 l/s at 40 kPa max head; 2.2 kW max motor rating	1621.90	1992.50	17.86	670.30	nr	**2662.80**
100 mm pump size; 28.0 l/s at 90 kPa max head; 4.0 kW max motor rating	1790.51	2199.65	17.86	670.30	nr	**2869.95**
125 mm pump size; 40.0 l/s at 50 kPa max head; 3.0 kW max motor rating	1707.85	2098.09	25.85	970.17	nr	**3068.26**
125 mm pump size; 40.0 l/s at 120 kPa max head; 7.5 kW max motor rating	2107.94	2589.60	25.85	970.17	nr	**3559.77**
150 mm pump size; 70.0 l/s at 75 kPa max head; 11 kW max motor rating	3098.25	3806.20	30.43	1142.06	nr	**4948.26**
150 mm pump size; 70.0 l/s at 120 kPa max head; 15.0 kW max motor rating	3311.54	4068.22	30.43	1142.06	nr	**5210.28**
150 mm pump size; 70.0 l/s at 150 kPa max head; 15.0 kW max motor rating	3311.54	4068.22	30.43	1142.06	nr	**5210.28**
Centrifugal heating & chilled water pump; close coupled; 3 phase, VARIABLE SPEED motor; max. system pressure 1000 kN/m²; max. temperature 110°C; bed plate; coupling guard; bolted connections; supply only mating flanges; includes fixing on prepared concrete base; electrical work elsewhere.						
40 mm pump size; 4.0 l/s at 23 kPa max head; 0.55 kW max motor rating	1512.76	1858.43	7.31	274.34	nr	**2132.77**
40 mm pump size; 4.0 l/s at 75 kPa max head; 0.75 kW max motor rating	1669.82	2051.37	7.31	274.34	nr	**2325.71**
50 mm pump size; 7.0 l/s at 65 kPa max head; 1.5 kW max motor rating	2048.43	2516.49	8.01	300.63	nr	**2817.12**
50 mm pump size; 10.0 l/s at 33 kPa max head; 1.5 kW max motor rating	2048.43	2516.49	8.01	300.63	nr	**2817.12**
50 mm pump size; 4.0 l/s at 120 kPa max head; 1.5 kW max motor rating	2048.43	2516.49	8.01	300.63	nr	**2817.12**
80 mm pump size; 16.0 l/s at 80 kPa max head; 2.2 kW max motor rating	2670.06	3280.17	12.35	463.50	nr	**3743.67**
80 mm pump size; 16.0 l/s at 120 kPa max head; 3.0 kW max motor rating	2880.03	3538.12	12.35	463.50	nr	**4001.62**
100 mm pump size; 28.0 l/s at 40 kPa max head; 2.2 kW max motor rating	2736.19	3361.41	17.86	670.30	nr	**4031.71**
100 mm pump size; 28.0 l/s at 90 kPa max head; 4.0.kW max motor rating	3263.59	4009.32	17.86	670.30	nr	**4679.62**
125 mm pump size; 40.0 l/s at 50 kPa max head; 3.0 kW max motor rating	3132.92	3848.80	25.85	970.17	nr	**4818.97**
125 mm pump size; 40.0 l/s at 120 kPa max head; 7.5 kW max motor rating	1289.31	1583.91	25.85	970.17	nr	**2554.08**
150 mm pump size; 70.0 l/s at 75 kPa max head; 7.5 kW max motor rating	5097.08	6261.76	30.43	1142.06	nr	**7403.82**

38 MECHANICAL/COOLING/HEATING SYSTEMS

Item	Net Price £	Material £	Labour hours	Labour £	Unit	Total rate £
LOW TEMPERATURE HOT WATER HEATING; PIPELINE: PUMPS – cont						
Glandless domestic heating pump; for low pressure domestic hot water heating systems; 240 volt; 50 Hz electric motor; max working pressure 1000 N/m² and max temperature of 130°C; includes fixing in position; electrical work elsewhere						
1" BSP unions – 2 speed	143.04	175.73	1.58	59.30	nr	**235.03**
1.25" BSP unions – 3 speed	210.78	258.94	1.58	59.30	nr	**318.24**
Glandless pumps; for hot water secondary supply; silent running; 3 phase; max pressure 1000 kN/m²; max temperature 130°C ; bolted connections; supply only mating flanges; including fixing in position; electrical elsewhere						
1" BSP unions – 3 speed	292.64	359.51	1.58	59.30	nr	**418.81**
Pipeline mounted circulator; for heating and chilled water; silent running; 3 phase; 1450 rpm motor; max pressure 1000 kN/m²; max temperature 120°C; bolted connections; supply only mating flanges; includes fixing in position; electrical elsewhere						
32 mm pump size; 2.0 l/s at 17 kPa max head; 0.2 kW max motor rating	585.27	719.01	6.44	241.70	nr	**960.71**
50 mm pump size; 3.0 l/s at 20 kPa max head; 0.2 kW max motor rating	585.27	719.01	6.86	257.46	nr	**976.47**
65 mm pump size; 5.0 l/s at 30 kPa max head; 0.37 kW max motor rating	925.84	1137.39	7.48	280.73	nr	**1418.12**
65 mm pump size; 8.0 l/s at 37 kPa max head; 0.75 kW max motor rating	1036.61	1273.47	7.48	280.73	nr	**1554.20**
80 mm pump size; 12.0 l/s at 42 kPa max head; 1.1 kW max motor rating	1205.26	1480.66	8.01	300.63	nr	**1781.29**
100 mm pump size; 25.0 l/s at 37 kPa max head; 2.2 kW max motor rating	1717.77	2110.28	9.11	341.90	nr	**2452.18**
Dual pipeline mounted circulator; for heating & chilled water; silent running; 3 phase; 1450 rpm motor; max pressure 1000 kN/m²; max temperature 120°C; bolted connections; supply only mating flanges; includes fixing in position; electrical work elsewhere						
40 mm pump size; 2.0 l/s at 17 kPa max head; 0.8 kW max motor rating	1079.60	1326.29	7.88	295.73	nr	**1622.02**
50 mm pump size; 3.0 l/s at 20 kPa max head; 0.2 kW max motor rating	1082.90	1330.34	8.01	300.63	nr	**1630.97**
65 mm pump size; 5.0 l/s at 30 kPa max head; 0.37 kW max motor rating	1757.44	2159.01	9.20	345.28	nr	**2504.29**

38 MECHANICAL/COOLING/HEATING SYSTEMS

Item	Net Price £	Material £	Labour hours	Labour £	Unit	**Total rate £**
65 mm pump size; 8.0 l/s at 37 kPa max head; 0.75 kW max motor rating	2031.90	2496.19	9.20	345.28	nr	**2841.47**
100 mm pump size; 12.0 l/s at 42 kPa max head; 1.1 kW max motor rating	2298.07	2823.18	9.45	354.67	nr	**3177.85**
Glandless accelerator pumps; for low and medium pressure heating services; silent running; 3 phase; 1450 rpm motor; max pressure 1000 kN/m²; max temperature 130°C; bolted connections; supply only mating flanges; includes fixing in position; electrical work elsewhere						
40 mm pump size; 4.0 l/s at 15 kPa max head; 0.35 kW max motor rating	453.65	557.31	6.94	260.46	nr	**817.77**
50 mm pump size; 6.0 l/s at 20 kPa max head; 0.45 kW max motor rating	483.46	593.93	7.35	275.85	nr	**869.78**
80 mm pump size; 13.0 l/s at 28 kPa max head; 0.58 kW max motor rating	952.65	1170.33	7.76	291.24	nr	**1461.57**

38 MECHANICAL/COOLING/HEATING SYSTEMS

Item	Net Price £	Material £	Labour hours	Labour £	Unit	Total rate £
LOW TEMPERATURE HOT WATER HEATING; PIPELINE: HEAT EXCHANGERS						
Plate heat exchanger; for use in LTHW systems; painted carbon steel frame; stainless steel plates, nitrile rubber gaskets; design pressure of 10 bar and operating temperature of 110/135°C						
Primary side; 80°C in, 69°C out; secondary side; 82°C in, 71°C out						
107 kW, 2.38 l/s	2186.65	2686.30	10.00	375.31	nr	**3061.61**
245 kW, 5.46 l/s	3360.12	4127.91	10.00	375.31	nr	**4503.22**
287 kW, 6.38 l/s	3729.91	4582.19	10.00	375.31	nr	**4957.50**
328 kW, 7.31 l/s	4069.92	4999.90	10.00	375.31	nr	**5375.21**
364 kW, 8.11 l/s	4830.10	5933.78	10.00	375.31	nr	**6309.09**
403 kW, 8.96 l/s	5070.52	6229.14	10.00	375.31	nr	**6604.45**
453 kW, 10.09 l/s	5275.45	6480.89	10.00	375.31	nr	**6856.20**
490 kW, 10.89 l/s	5468.92	6718.57	10.00	375.31	nr	**7093.88**
1000 kW, 21.7 l/s	4068.78	4998.49	12.00	450.37	nr	**5448.86**
1500 kW, 32.6 l/s	5183.86	6368.38	12.00	450.37	nr	**6818.75**
2000 kW, 43.4 l/s	6218.80	7639.80	15.00	562.96	nr	**8202.76**
2500 kW, 54.3 l/s	7644.13	9390.81	15.00	562.96	nr	**9953.77**
Note: For temperature conditions different to those above, the cost of the units can vary significantly, and so manufacturers advice should be sought.						

38 MECHANICAL/COOLING/HEATING SYSTEMS

Item	Net Price £	Material £	Labour hours	Labour £	Unit	**Total rate £**
LOW TEMPERATURE HOT WATER HEATING; PIPELINE: CALORIFIERS						
Non-storage calorifiers; mild steel; heater battery duty 116°C/90°C to BS 853, maximum test on shell 11.55 bar, tubes 26.25 bar						
Horizontal or vertical; primary water at 116°C on, 90°C off						
40 kW capacity	902.38	1108.57	3.00	112.59	nr	**1221.16**
88 kW capacity	1051.28	1291.50	5.00	187.66	nr	**1479.16**
176 kW capacity	1204.71	1479.99	7.04	264.30	nr	**1744.29**
293 kW capacity	1670.69	2052.45	9.01	338.12	nr	**2390.57**
586 kW capacity	2488.98	3057.72	22.22	834.01	nr	**3891.73**
879 kW capacity	3466.39	4258.46	28.57	1072.31	nr	**5330.77**
1465 kW capacity	5568.95	6841.45	50.00	1876.54	nr	**8717.99**
2000 kW capacity	10136.23	12452.36	60.00	2251.84	nr	**14704.20**

38 MECHANICAL/COOLING/HEATING SYSTEMS

Item	Net Price £	Material £	Labour hours	Labour £	Unit	Total rate £
LOW TEMPERATURE HOT WATER HEATING; PIPELINE: HEAT EMITTERS						
Perimeter convector heating; metal casing with standard finish; aluminium extruded grille; including backplates						
Top/sloping/flat front outlet						
60 × 200 mm	35.35	43.43	2.00	75.06	m	**118.49**
60 × 300 mm	38.45	47.24	2.00	75.06	m	**122.30**
60 × 450 mm	47.66	58.55	2.00	75.06	m	**133.61**
60 × 525 mm	50.73	62.32	2.00	75.06	m	**137.38**
60 × 600 mm	55.35	68.00	2.00	75.06	m	**143.06**
90 × 260 mm	38.45	47.24	2.00	75.06	m	**122.30**
90 × 300 mm	39.97	49.10	2.00	75.06	m	**124.16**
90 × 450 mm	49.22	60.47	2.00	75.06	m	**135.53**
90 × 525 mm	53.83	66.13	2.00	75.06	m	**141.19**
90 × 600 mm	56.90	69.90	2.00	75.06	m	**144.96**
Extra over for dampers						
Damper	16.91	20.77	0.25	9.39	nr	**30.16**
Extra over for fittings						
60 mm end caps	13.53	16.62	0.25	9.39	nr	**26.01**
90 mm end caps	21.29	26.16	0.25	9.39	nr	**35.55**
60 mm corners	28.76	35.33	0.25	9.39	nr	**44.72**
90 mm corners	42.30	51.96	0.25	9.39	nr	**61.35**

38 MECHANICAL/COOLING/HEATING SYSTEMS

Item	Net Price £	Material £	Labour hours	Labour £	Unit	Total rate £
LOW TEMPERATURE HOT WATER HEATING; PIPELINE: RADIATORS						
Radiant strip heaters						
Suitable for connection to hot water system; aluminium sheet panels with steel pipe clamped to upper surface; including insulation, sliding brackets, cover plates, end closures; weld or screwed BSP ends						
One pipe						
1500 mm long	71.50	87.83	3.11	116.72	nr	**204.55**
3000 mm long	113.08	138.92	3.11	116.72	nr	**255.64**
4500 mm long	153.39	188.44	3.11	116.72	nr	**305.16**
6000 mm long	211.46	259.78	3.11	116.72	nr	**376.50**
Two pipe						
1500 mm long	133.48	163.98	4.15	155.75	nr	**319.73**
3000 mm long	210.84	259.01	4.15	155.75	nr	**414.76**
4500 mm long	287.91	353.69	4.15	155.75	nr	**509.44**
6000 mm long	387.89	476.52	4.15	155.75	nr	**632.27**
Pressed steel panel type radiators; fixed with and including brackets; taking down once for decoration; refixing						
300 mm high; single panel						
500 mm length	19.52	23.98	2.03	76.19	nr	**100.17**
1000 mm length	39.00	47.91	2.03	76.19	nr	**124.10**
1500 mm length	51.11	62.79	2.03	76.19	nr	**138.98**
2000 mm length	57.34	70.44	2.47	92.70	nr	**163.14**
2500 mm length	63.55	78.07	2.97	111.46	nr	**189.53**
3000 mm length	76.05	93.43	3.22	120.84	nr	**214.27**
300 mm high; double panel; convector						
500 mm length	37.52	46.10	2.13	79.94	nr	**126.04**
1000 mm length	75.09	92.25	2.13	79.94	nr	**172.19**
1500 mm length	112.62	138.36	2.13	79.94	nr	**218.30**
2000 mm length	150.17	184.49	2.57	96.45	nr	**280.94**
2500 mm length	187.69	230.58	3.07	115.22	nr	**345.80**
3000 mm length	225.25	276.72	3.31	124.23	nr	**400.95**
450 mm high; single panel						
500 mm length	18.21	22.38	2.08	78.06	nr	**100.44**
1000 mm length	36.45	44.78	2.08	78.06	nr	**122.84**
1600 mm length	58.30	71.62	2.53	94.95	nr	**166.57**
2000 mm length	72.88	89.53	2.97	111.46	nr	**200.99**
2400 mm length	87.44	107.42	3.47	130.23	nr	**237.65**
3000 mm length	109.30	134.27	3.82	143.37	nr	**277.64**
450 mm high; double panel; convector						
500 mm length	33.36	40.98	2.18	81.82	nr	**122.80**
1000 mm length	66.72	81.96	2.18	81.82	nr	**163.78**
1600 mm length	122.20	150.12	2.63	98.71	nr	**248.83**
2000 mm length	205.44	252.38	3.06	114.84	nr	**367.22**
2400 mm length	246.54	302.87	3.37	126.47	nr	**429.34**
3000 mm length	308.17	378.59	3.92	147.13	nr	**525.72**

38 MECHANICAL/COOLING/HEATING SYSTEMS

Item	Net Price £	Material £	Labour hours	Labour £	Unit	Total rate £
LOW TEMPERATURE HOT WATER HEATING; PIPELINE: RADIATORS – cont						
Pressed steel panel type radiators – cont						
600 mm high; single panel						
500 mm length	24.40	29.98	2.18	81.82	nr	**111.80**
1000 mm length	48.80	59.95	2.43	91.20	nr	**151.15**
1600 mm length	78.08	95.92	3.13	117.47	nr	**213.39**
2000 mm length	97.59	119.89	3.77	141.49	nr	**261.38**
2400 mm length	117.13	143.89	4.07	152.75	nr	**296.64**
3000 mm length	146.40	179.85	5.11	191.78	nr	**371.63**
600 mm high; double panel; convector						
500 mm length	42.02	51.62	2.28	85.56	nr	**137.18**
1000 mm length	84.01	103.20	2.28	85.56	nr	**188.76**
1600 mm length	153.89	189.05	3.23	121.23	nr	**310.28**
2000 mm length	258.70	317.81	3.87	145.25	nr	**463.06**
2400 mm length	310.45	381.39	4.17	156.50	nr	**537.89**
3000 mm length	388.05	476.72	5.24	196.65	nr	**673.37**
700 mm high; single panel						
500 mm length	28.54	35.06	2.23	83.70	nr	**118.76**
1000 mm length	57.08	70.12	2.83	106.22	nr	**176.34**
1600 mm length	91.33	112.20	3.73	139.99	nr	**252.19**
2000 mm length	114.18	140.27	4.46	167.38	nr	**307.65**
2400 mm length	137.03	168.35	4.48	168.14	nr	**336.49**
3000 mm length	171.26	210.39	5.24	196.65	nr	**407.04**
700 mm high; double panel; convector						
500 mm length	54.60	67.07	2.33	87.44	nr	**154.51**
1000 mm length	146.90	180.46	3.08	115.59	nr	**296.05**
1600 mm length	235.02	288.72	3.83	143.75	nr	**432.47**
2000 mm length	293.77	360.90	4.17	156.50	nr	**517.40**
2400 mm length	352.55	433.10	4.37	164.01	nr	**597.11**
3000 mm length	440.67	541.36	4.82	180.89	nr	**722.25**
Flat panel type steel radiators; fixed with and including brackets; taking down once for decoration; refixing						
300 mm high; single panel (44 mm deep)						
500 mm length	200.93	246.84	2.03	76.19	nr	**323.03**
1000 mm length	315.59	387.70	2.03	76.19	nr	**463.89**
1500 mm length	430.25	528.56	2.03	76.19	nr	**604.75**
2000 mm length	544.91	669.42	2.47	92.70	nr	**762.12**
2400 mm length	636.64	782.11	2.97	111.46	nr	**893.57**
3000 mm length	915.97	1125.26	3.22	120.84	nr	**1246.10**
300 mm high; double panel (100 mm deep)						
500 mm length	398.58	489.66	2.03	76.19	nr	**565.85**
1000 mm length	624.62	767.35	2.03	76.19	nr	**843.54**
1500 mm length	850.67	1045.04	2.03	76.19	nr	**1121.23**
2000 mm length	1076.71	1322.74	2.47	92.70	nr	**1415.44**
2400 mm length	1257.55	1544.90	2.97	111.46	nr	**1656.36**
3000 mm length	1670.54	2052.26	3.22	120.84	nr	**2173.10**

38 MECHANICAL/COOLING/HEATING SYSTEMS

Item	Net Price £	Material £	Labour hours	Labour £	Unit	**Total rate £**
500 mm high; single panel (44 mm deep)						
500 mm length	236.96	291.10	2.13	79.94	nr	**371.04**
1000 mm length	387.66	476.24	2.13	79.94	nr	**556.18**
1500 mm length	538.36	661.37	2.13	79.94	nr	**741.31**
2000 mm length	689.05	846.50	2.57	96.45	nr	**942.95**
2400 mm length	809.61	994.60	3.07	115.22	nr	**1109.82**
3000 mm length	1132.19	1390.89	3.31	124.23	nr	**1515.12**
500 mm high; double panel (100 mm deep)						
500 mm length	464.10	570.15	2.08	78.06	nr	**648.21**
1000 mm length	755.66	928.33	2.08	78.06	nr	**1006.39**
1500 mm length	1047.20	1286.48	2.53	94.95	nr	**1381.43**
2000 mm length	1338.79	1644.70	2.97	111.46	nr	**1756.16**
2400 mm length	1572.04	1931.25	3.47	130.23	nr	**2061.48**
3000 mm length	2063.66	2535.20	3.82	143.37	nr	**2678.57**
600 mm high; single panel (44 mm deep)						
500 mm length	267.54	328.67	2.18	81.82	nr	**410.49**
1000 mm length	437.89	537.95	2.18	81.82	nr	**619.77**
1500 mm length	608.24	1285.18	2.63	98.71	nr	**1383.89**
2000 mm length	778.60	956.51	3.06	114.84	nr	**1071.35**
2400 mm length	914.88	1123.93	3.37	126.47	nr	**1250.40**
3000 mm length	1261.04	1549.19	3.92	147.13	nr	**1696.32**
600 mm high; double panel (100 mm deep)						
500 mm length	528.31	649.03	2.18	81.82	nr	**730.85**
1000 mm length	862.46	1059.53	2.43	91.20	nr	**1150.73**
1500 mm length	1196.61	1470.03	3.13	117.47	nr	**1587.50**
2000 mm length	1530.77	1880.55	3.77	141.49	nr	**2022.04**
2400 mm length	1798.09	2208.96	4.07	152.75	nr	**2361.71**
3000 mm length	2340.81	2875.69	5.11	191.78	nr	**3067.47**
700 mm high; single panel (44 mm deep)						
500 mm length	290.47	356.84	2.28	85.56	nr	**442.40**
1000 mm length	483.76	594.30	2.28	85.56	nr	**679.86**
1500 mm length	677.04	831.75	3.23	121.23	nr	**952.98**
2000 mm length	870.32	1069.18	3.87	145.25	nr	**1214.43**
2400 mm length	1025.26	1259.53	4.17	156.50	nr	**1416.03**
3000 mm length	1398.63	1718.22	5.24	196.65	nr	**1914.87**
700 mm high; double panel (100 mm deep)						
500 mm length	570.90	701.35	2.23	83.70	nr	**785.05**
1000 mm length	947.64	1164.18	2.83	106.22	nr	**1270.40**
1500 mm length	1324.38	1627.00	3.73	139.99	nr	**1766.99**
2000 mm length	1701.12	2089.83	4.46	167.38	nr	**2257.21**
2400 mm length	2002.51	2460.09	4.48	168.14	nr	**2628.23**

38 MECHANICAL/COOLING/HEATING SYSTEMS

Item	Net Price £	Material £	Labour hours	Labour £	Unit	Total rate £
LOW TEMPERATURE HOT WATER HEATING; PIPELINE: RADIATORS – cont						
Fan convector; sheet metal casing with lockable access panel; centrifugal fan; air filter; LPHW heating coil; extruded aluminium grilles; 3 speed; includes fixing in position; electrical work elsewhere						
Free standing flat top, 695 mm high, medium speed rating						
Entering air temperature, 18°C						
695 mm long, 1 row 1.94 kW, 75 l/s	679.61	834.90	2.73	102.46	nr	**937.36**
695 mm long, 2 row 2.64 kW, 75 l/s	679.61	834.90	2.73	102.46	nr	**937.36**
895 mm long, 1 row 4.02 kW, 150 l/s	765.81	940.80	2.73	102.46	nr	**1043.26**
895 mm long, 2 row 5.62 kW, 150 l/s	765.81	940.80	2.73	102.46	nr	**1043.26**
1195 mm long, 1 row 6.58 kW, 250 l/s	871.89	1071.12	3.00	112.59	nr	**1183.71**
1195 mm long, 2 row 9.27 kW, 250 l/s	871.89	1071.12	3.00	112.59	nr	**1183.71**
1495 mm long, 1 row 9.04 kW, 340 l/s	973.01	1195.34	3.26	122.35	nr	**1317.69**
1495 mm long, 2 row 12.73 kW, 340 l/s	973.01	1195.34	3.26	122.35	nr	**1317.69**
Free standing flat top, 695 mm high, medium speed rating, c/w floor plinth						
695 mm long, 1 row 1.94 kW, 75 l/s	713.59	876.65	2.73	102.46	nr	**979.11**
695 mm long, 2 row 2.64 kW, 75 l/s	713.59	876.65	2.73	102.46	nr	**979.11**
895 mm long, 1 row 4.02 kW, 150 l/s	804.09	987.83	2.73	102.46	nr	**1090.29**
895 mm long, 2 row 5.62 kW, 150 l/s	804.09	987.83	2.73	102.46	nr	**1090.29**
1195 mm long, 1 row 6.58 kW, 250 l/s	915.49	1124.68	3.00	112.59	nr	**1237.27**
1195 mm long, 2 row 9.27 kW, 250 l/s	915.49	1124.68	3.00	112.59	nr	**1237.27**
1495 mm long, 1 row 9.04 kW, 340 l/s	1021.65	1255.10	3.26	122.35	nr	**1377.45**
1495 mm long, 2 row 12.73 kW, 340 l/s	1021.65	1255.10	3.26	122.35	nr	**1377.45**
Free standing sloping top, 695 mm high, medium speed rating, c/w floor plinth						
695 mm long, 1 row 1.94 kW, 75 l/s	738.45	907.19	2.73	102.46	nr	**1009.65**
695 mm long, 2 row 2.64 kW, 75 l/s	738.45	907.19	2.73	102.46	nr	**1009.65**
895 mm long, 1 row 4.02 kW, 150 l/s	828.96	1018.37	2.73	102.46	nr	**1120.83**
895 mm long, 2 row 5.62 kW, 150 l/s	828.96	1018.37	2.73	102.46	nr	**1120.83**
1195 mm long, 1 row 6.58 kW, 250 l/s	940.35	1155.22	3.00	112.59	nr	**1267.81**
1195 mm long, 2 row 9.27 kW, 250 l/s	940.35	1155.22	3.00	112.59	nr	**1267.81**
1495 mm long, 1 row 9.04 kW, 340 l/s	1046.51	1285.64	3.26	122.35	nr	**1407.99**
1495 mm long, 2 row 12.73 kW, 340 l/s	1046.51	1285.64	3.26	122.35	nr	**1407.99**
Wall mounted high level sloping discharge						
695 mm long, 1 row 1.94 kW, 75 l/s	745.91	916.35	2.73	102.46	nr	**1018.81**
695 mm long, 2 row 2.64 kW, 75 l/s	745.91	916.35	2.73	102.46	nr	**1018.81**
895 mm long, 1 row 4.02 kW, 150 l/s	767.48	942.85	2.73	102.46	nr	**1045.31**
895 mm long, 2 row 5.62 kW, 150 l/s	767.48	942.85	2.73	102.46	nr	**1045.31**
1195 mm long, 1 row 6.58 kW, 250 l/s	934.03	1147.46	3.00	112.59	nr	**1260.05**
1195 mm long, 2 row 9.27 kW, 250 l/s	934.03	1147.46	3.00	112.59	nr	**1260.05**
1495 mm long, 1 row 9.04 kW, 340 l/s	1012.77	1244.19	3.26	122.35	nr	**1366.54**
1495 mm long, 2 row 12.73 kW, 340 l/s	1012.77	1244.19	3.26	122.35	nr	**1366.54**

38 MECHANICAL/COOLING/HEATING SYSTEMS

Item	Net Price £	Material £	Labour hours	Labour £	Unit	Total rate £
Ceiling mounted sloping inlet/outlet 665 mm wide						
895 mm long, 1 row 4.02 kW, 150 l/s	843.71	1036.50	4.15	155.75	nr	**1192.25**
895 mm long, 2 row 5.62 kW, 150 l/s	843.71	1036.50	4.15	155.75	nr	**1192.25**
1195 mm long, 1 row 6.58 kW, 250 l/s	948.15	1164.81	4.15	155.75	nr	**1320.56**
1195 mm long, 2 row 9.27 kW, 250 l/s	948.15	1164.81	4.15	155.75	nr	**1320.56**
1495 mm long, 1 row 9.04 kW, 340 l/s	1040.96	1278.82	4.15	155.75	nr	**1434.57**
1495 mm long, 2 row 12.73 kW, 340 l/s	1040.96	1278.82	4.15	155.75	nr	**1434.57**
Free standing unit, extended height 1700/ 1900/2100 mm						
895 mm long, 1 row 4.02 kW, 150 l/s	976.65	1199.81	3.11	116.72	nr	**1316.53**
895 mm long, 2 row 5.62 kW, 150 l/s	976.33	1199.43	3.11	116.72	nr	**1316.15**
1195 mm long, 1 row 6.58 kW, 250 l/s	1140.42	1401.00	3.11	116.72	nr	**1517.72**
1195 mm long, 2 row 9.27 kW, 250 l/s	1140.42	1401.00	3.11	116.72	nr	**1517.72**
1495 mm long, 1 row 9.04 kW, 340 l/s	1254.80	1541.53	3.11	116.72	nr	**1658.25**
1495 mm long, 2 row 12.73 kW, 340 l/s	1254.80	1541.53	3.11	116.72	nr	**1658.25**
LTHW trench heating; water temperatures 90°C/70°C; room air temperature 20°C; convector with copper tubes and aluminium fins within steel duct; Includes fixing within floor screed; electrical work elsewhere						
Natural convection type						
Normal capacity, 182 mm width, complete with linear, natural anodized aluminium grille (grille also costed separately below)						
92 mm deep						
1250 mm long, 234 W output	238.68	293.22	2.00	75.06	nr	**368.28**
2250 mm long, 471 W output	381.89	469.15	4.00	150.12	nr	**619.27**
3250 mm long, 709 W output	522.38	641.74	5.00	187.66	nr	**829.40**
4250 mm long, 946 W output	666.94	819.34	7.00	262.71	nr	**1082.05**
5000 mm long, 1124 W output	911.07	1119.25	8.00	300.25	nr	**1419.50**
120 mm deep						
1250 mm long, 294 W output	278.23	341.81	2.00	75.06	nr	**416.87**
2250 mm long, 471 W output	437.82	537.86	4.00	150.12	nr	**687.98**
3250 mm long, 891 W output	596.02	732.21	5.00	187.66	nr	**919.87**
4250 mm long, 1190 W output	754.23	926.57	7.00	262.71	nr	**1189.28**
5000 mm long, 1414 W output	1016.10	1248.28	8.00	300.25	nr	**1548.53**
150 mm deep						
1250 mm long, 329 W output	285.06	350.20	2.00	75.06	nr	**425.26**
2250 mm long, 664 W output	450.09	552.94	4.00	150.12	nr	**703.06**
3250 mm long, 998 W output	612.39	752.33	5.00	187.66	nr	**939.99**
4250 mm long, 1333 W output	778.78	956.73	7.00	262.71	nr	**1219.44**
5000 mm long, 1584 W output	1044.73	1283.45	8.00	300.25	nr	**1583.70**
200 mm deep						
1250 mm long, 396 W output	297.33	365.27	2.00	75.06	nr	**440.33**
2250 mm long, 799 W output	544.20	668.55	4.00	150.12	nr	**818.67**
3250 mm long, 1201 W output	647.85	795.88	5.00	187.66	nr	**983.54**

38 MECHANICAL/COOLING/HEATING SYSTEMS

Item	Net Price £	Material £	Labour hours	Labour £	Unit	Total rate £
LOW TEMPERATURE HOT WATER HEATING; PIPELINE: RADIATORS – cont						
Normal capacity, 182 mm width – cont						
200 mm deep – cont						
4250 mm long, 1603 W output	829.24	1018.72	7.00	262.71	nr	**1281.43**
5000 mm long, 1905 W output	1102.01	1353.82	8.00	300.25	nr	**1654.07**
Fan assisted type (outputs assume fan at 50%)						
Normal capacity, 182 mm width, complete with Natural anodized aluminium grille						
112 mm deep						
1250 mm long, 437 W output	630.11	774.09	2.00	75.06	nr	**849.15**
2250 mm long, 1019 W output	762.42	936.63	4.00	150.12	nr	**1086.75**
3250 mm long, 1488 W output	896.07	1100.82	5.00	187.66	nr	**1288.48**
4250 mm long, 1845 W output	1028.37	1263.35	7.00	262.71	nr	**1526.06**
5000 mm long, 2038 W output	1160.67	1425.88	8.00	300.25	nr	**1726.13**
Linear grille anodized aluminium, 170 mm width (if supplied as a separate item)	148.67	182.64	–	–	m	**182.64**
Roll up grille, natural anodized aluminium	148.67	182.64	–	–	m	**182.64**
Thermostatic valve with remote regulator (c/w valve body)	61.38	75.40	4.00	150.12	nr	**225.52**
Fan speed controller	30.00	36.85	2.00	75.06	nr	**111.91**
Note: as an alternative to thermostatic control, the system can be controlled via two port valves. Refer to valve section for valve prices						
LTHW underfloor heating; water flow and return temperatures of 60°C and 70°C; pipework at 300 mm centres; pipe fixings; flow and return manifolds and zone actuators; wiring block; insulation; includes fixing in position; excludes secondary pump, mixing valve, zone thermostats and floor finishes; electrical work elsewhere						
Note: All rates are expressed on a m² basis, for the following example areas						
Screeded floor with 15–25 mm stone/marble finish (producing 80–100 W/m²)						
250 m² area (single zone)	26.06	32.01	0.14	5.26	m²	**37.27**
1000 m² area (single zone)	25.39	31.20	0.12	4.50	m²	**35.70**
5000 m² area (multi-zone)	25.33	31.12	0.10	3.76	m²	**34.88**
Screeded floor with 10 mm carpet tile (producing 80–100 W/m²)						
250 m² area (single zone)	26.06	32.01	0.14	5.26	m²	**37.27**
1000 m² area (single zone)	25.39	31.20	0.12	4.50	m²	**35.70**
5000 m² area (multi-zone)	25.33	31.12	0.10	3.76	m²	**34.88**

38 MECHANICAL/COOLING/HEATING SYSTEMS

Item	Net Price £	Material £	Labour hours	Labour £	Unit	Total rate £
Floating timber floor with 20 mm timber finish (producing 70–80 Wm²)						
250 m² are (single zone)	32.85	40.35	0.14	5.26	m²	**45.61**
1000 m² area(single zone)	35.21	43.26	0.12	4.50	m²	**47.76**
5000 m² area (multi-zone)	33.63	41.32	0.10	3.76	m²	**45.08**
Floating timber floor with 10 mm carpet tile (producing 70–80 W/m²)						
250 m² are (single zone)	42.75	52.52	0.16	6.01	m²	**58.53**
1000 m² area(single zone)	39.64	48.70	0.12	4.50	m²	**53.20**
5000 m² area (multi-zone)	37.93	46.60	0.10	3.76	m²	**50.36**

38 MECHANICAL/COOLING/HEATING SYSTEMS

Item	Net Price £	Material £	Labour hours	Labour £	Unit	**Total rate £**
LOW TEMPERATURE HOT WATER HEATING; PIPELINE: PIPE FREEZING						
Freeze isolation of carbon steel or copper pipelines containing static water, either side of work location, freeze duration not exceeding 4 hours assuming that flow and return circuits are treated concurrently and activities undertaken during normal working hours						
Up to 4 freezes						
50 mm dia.	429.88	528.11	1.00	37.53	nr	**565.64**
65 mm dia.	429.88	528.11	1.00	37.53	nr	**565.64**
80 mm dia.	491.45	603.75	1.00	37.53	nr	**641.28**
100 mm dia.	553.03	679.40	1.00	37.53	nr	**716.93**
150 mm dia.	920.20	1130.46	1.00	37.53	nr	**1167.99**
200 mm dia.	1410.51	1732.82	1.00	37.53	nr	**1770.35**

38 MECHANICAL/COOLING/HEATING SYSTEMS

Item	Net Price £	Material £	Labour hours	Labour £	Unit	Total rate £
LOW TEMPERATURE HOT WATER HEATING; PIPELINE: ENERGY METERS						
Ultrasonic						
Energy meter for measuring energy use in LTHW systems; includes ultrasonic flow meter (with sensor and signal converter), energy calculator, pair of temperature sensors with brass pockets, and 3 m of interconnecting cable; includes fixing in position; electrical work elsewhere						
Pipe size (flanged connections to PN16); maximum flow rate						
50 mm, 36 m³/hr	1278.34	1570.44	1.80	67.56	nr	**1638.00**
65 mm, 60 m³/hr	1409.11	1731.09	2.32	87.08	nr	**1818.17**
80 mm, 100 m³/hr	1579.51	1940.43	2.56	96.09	nr	**2036.52**
125 mm, 250 m³/hr	1830.01	2248.17	3.60	135.11	nr	**2383.28**
150 mm, 360 m³/hr	1986.58	2440.52	4.80	180.15	nr	**2620.67**
200 mm, 600 m³/hr	2219.60	2726.78	6.24	234.19	nr	**2960.97**
250 mm, 1000 m³/hr	2561.27	3146.52	9.60	360.30	nr	**3506.82**
300 mm, 1500 m³/hr	3010.72	3698.67	10.80	405.33	nr	**4104.00**
350 mm, 2000 m³/hr	3627.78	4456.73	13.20	495.40	nr	**4952.13**
400 mm, 2500 m³/hr	4152.75	5101.66	15.60	585.48	nr	**5687.14**
500 mm, 3000 m³/hr	4712.70	5789.55	24.00	900.73	nr	**6690.28**
600 mm, 3500 m³/hr	5294.77	6504.62	28.00	1050.86	nr	**7555.48**

38 MECHANICAL/COOLING/HEATING SYSTEMS

Item	Net Price £	Material £	Labour hours	Labour £	Unit	Total rate £
LOW TEMPERATURE HOT WATER HEATING: CONTROL COMPONENTS MECHANICAL						
Room thermostats; light and medium duty; installed and connected						
Range 3°C to 27°C; 240 Volt						
1 amp; on/off type	27.52	33.81	0.30	11.26	nr	**45.07**
Range 0°C to +15°C; 240 Volt						
6 amp; frost thermostat	18.56	22.81	0.30	11.26	nr	**34.07**
Range 3°C to 27°C; 250 Volt						
2 amp; changeover type; dead zone	46.03	56.55	0.30	11.26	nr	**67.81**
2 amp; changeover type	21.35	26.23	0.30	11.26	nr	**37.49**
2 amp; changeover type; concealed setting	26.93	33.09	0.30	11.26	nr	**44.35**
6 amp; on/off type	16.62	20.42	0.30	11.26	nr	**31.68**
6 amp; temperature set-back	34.40	42.26	0.30	11.26	nr	**53.52**
16 amp; on/off type	25.60	31.45	0.30	11.26	nr	**42.71**
16 amp; on/off type; concealed setting	27.91	34.28	0.30	11.26	nr	**45.54**
20 amp; on/off type; concealed setting	23.77	29.20	0.30	11.26	nr	**40.46**
20 amp; indicated ‘off’ position	29.97	36.81	0.30	11.26	nr	**48.07**
20 amp; manual; double pole on/off and neon indicator	54.56	67.03	0.30	11.26	nr	**78.29**
20 amp; indicated ‘off’ position	35.71	43.87	0.30	11.26	nr	**55.13**
Range 10°C to 40°C; 240 Volt						
20 amp; changeover contacts	32.54	39.97	0.30	11.26	nr	**51.23**
2 amp; 'heating-cooling' switch	70.37	86.45	0.30	11.26	nr	**97.71**
Surface thermostats						
Cylinder thermostat						
6 amp; changeover type; with cable	19.21	23.60	0.25	9.39	nr	**32.99**
Electrical thermostats; installed and connected						
Range 5°C to 30°C; 230 Volt standard port single time						
10 amp with sensor	26.67	32.76	0.30	11.26	nr	**44.02**
Range 5°C to 30°C; 230 Volt standard port double time						
10 amp with sensor	30.23	37.14	0.30	11.26	nr	**48.40**
10 amp with sensor and on/off switch	46.11	56.65	0.30	11.26	nr	**67.91**
Radiator thermostats						
Angled valve body; thermostatic head; built in sensor						
15 mm; liquid filled	16.63	20.43	0.84	31.53	nr	**51.96**
15 mm; wax filled	16.63	20.43	0.84	31.53	nr	**51.96**

38 MECHANICAL/COOLING/HEATING SYSTEMS

Item	Net Price £	Material £	Labour hours	Labour £	Unit	**Total rate £**
Immersion thermostats; stem type; domestic water boilers; fitted; electrical work elsewhere						
Temperature range 0°C to 40°C						
Non-standard; 280 mm stem	10.12	12.43	0.25	9.39	nr	**21.82**
Temperature range 18°C to 88°C						
13 amp; 178 mm stem	6.72	8.25	0.25	9.39	nr	**17.64**
20 amp; 178 mm stem	10.45	12.84	0.25	9.39	nr	**22.23**
Non-standard; pocket clip; 280 mm stem	9.69	11.91	0.25	9.39	nr	**21.30**
Temperature range 40°C to 80°C						
13 amp; 178 mm stem	3.69	4.54	0.25	9.39	nr	**13.93**
20 amp; 178 mm stem	6.98	8.58	0.25	9.39	nr	**17.97**
Non-standard; pocket clip; 280 mm stem	10.86	13.35	0.25	9.39	nr	**22.74**
13 amp; 457 mm stem	4.37	5.37	0.25	9.39	nr	**14.76**
20 amp; 457 mm stem	7.64	9.39	0.25	9.39	nr	**18.78**
Temperature range 50°C to 100°C						
Non-standard; 1780 mm stem	9.14	11.22	0.25	9.39	nr	**20.61**
Non-standard; 280 mm stem	9.53	11.71	0.25	9.39	nr	**21.10**
Pockets for thermostats						
For 178 mm stem	11.96	14.69	0.25	9.39	nr	**24.08**
For 280 mm stem	12.16	14.94	0.25	9.39	nr	**24.33**
Immersion thermostats; stem type; industrial installations; fitted; electrical work elsewhere						
Temperature range 5°C to 105°C						
For 305 mm stem	150.22	184.55	0.50	18.77	nr	**203.32**

38 MECHANICAL/COOLING/HEATING SYSTEMS

Item	Net Price £	Material £	Labour hours	Labour £	Unit	**Total rate £**
THERMAL INSULATION						
For flexible closed cell insulation see Section – Cold Water						
Mineral fibre sectional insulation; bright class O foil faced; bright class O foil taped joints; 19 mm aluminium bands						
Concealed pipework						
20 mm thick						
15 mm dia.	3.48	4.27	0.15	5.63	m	**9.90**
20 mm dia.	3.71	4.56	0.15	5.63	m	**10.19**
25 mm dia.	3.99	4.90	0.15	5.63	m	**10.53**
32 mm dia.	4.45	5.47	0.15	5.63	m	**11.10**
40 mm dia.	4.76	5.85	0.15	5.63	m	**11.48**
50 mm dia.	5.45	6.70	0.15	5.63	m	**12.33**
Extra over for fittings concealed insulation						
Flange/union						
15 mm dia.	1.74	2.14	0.13	4.87	nr	**7.01**
20 mm dia.	1.86	2.29	0.13	4.87	nr	**7.16**
25 mm dia.	2.00	2.46	0.13	4.87	nr	**7.33**
32 mm dia.	2.23	2.74	0.13	4.87	nr	**7.61**
40 mm dia.	2.37	2.91	0.13	4.87	nr	**7.78**
50 mm dia.	2.72	3.34	0.13	4.87	nr	**8.21**
Valves						
15 mm dia.	3.48	4.27	0.15	5.63	nr	**9.90**
20 mm dia.	3.99	4.90	0.15	5.63	nr	**10.53**
25 mm dia.	3.99	4.90	0.15	5.63	nr	**10.53**
32 mm dia.	4.45	5.47	0.15	5.63	nr	**11.10**
40 mm dia.	4.76	5.85	0.15	5.63	nr	**11.48**
50 mm dia.	5.45	6.70	0.15	5.63	nr	**12.33**
Expansion bellows						
15 mm dia.	6.95	8.54	0.22	8.25	nr	**16.79**
20 mm dia.	7.43	9.12	0.22	8.25	nr	**17.37**
25 mm dia.	7.98	9.81	0.22	8.25	nr	**18.06**
32 mm dia.	8.88	10.91	0.22	8.25	nr	**19.16**
40 mm dia.	9.52	11.70	0.22	8.25	nr	**19.95**
50 mm dia.	10.87	13.36	0.22	8.25	nr	**21.61**
25 mm thick						
15 mm dia.	3.83	4.70	0.15	5.63	m	**10.33**
20 mm dia.	4.12	5.06	0.15	5.63	m	**10.69**
25 mm dia.	4.62	5.68	0.15	5.63	m	**11.31**
32 mm dia.	5.03	6.18	0.15	5.63	m	**11.81**
40 mm dia.	5.37	6.59	0.15	5.63	m	**12.22**
50 mm dia.	6.16	7.57	0.15	5.63	m	**13.20**
65 mm dia.	7.04	8.65	0.15	5.63	m	**14.28**
80 mm dia.	7.72	9.48	0.22	8.25	m	**17.73**
100 mm dia.	10.17	12.49	0.22	8.25	m	**20.74**
125 mm dia.	11.77	14.46	0.22	8.25	m	**22.71**
150 mm dia.	14.02	17.22	0.22	8.25	m	**25.47**

38 MECHANICAL/COOLING/HEATING SYSTEMS

Item	Net Price £	Material £	Labour hours	Labour £	Unit	Total rate £
200 mm dia.	19.81	24.34	0.25	9.39	m	**33.73**
250 mm dia.	23.71	29.13	0.25	9.39	m	**38.52**
300 mm dia.	25.35	31.14	0.25	9.39	m	**40.53**
Extra over for fittings concealed insulation						
Flange/union						
15 mm dia.	1.92	2.36	0.13	4.87	nr	**7.23**
20 mm dia.	2.06	2.53	0.13	4.87	nr	**7.40**
25 mm dia.	2.31	2.84	0.13	4.87	nr	**7.71**
32 mm dia.	2.53	3.11	0.13	4.87	nr	**7.98**
40 mm dia.	2.68	3.30	0.13	4.87	nr	**8.17**
50 mm dia.	3.08	3.78	0.13	4.87	nr	**8.65**
65 mm dia.	3.52	4.33	0.13	4.87	nr	**9.20**
80 mm dia.	3.87	4.76	0.18	6.76	nr	**11.52**
100 mm dia.	5.09	6.26	0.18	6.76	nr	**13.02**
125 mm dia.	5.88	7.22	0.18	6.76	nr	**13.98**
150 mm dia.	7.01	8.61	0.18	6.76	nr	**15.37**
200 mm dia.	9.90	12.16	0.22	8.25	nr	**20.41**
250 mm dia.	11.86	14.57	0.22	8.25	nr	**22.82**
300 mm dia.	12.68	15.58	0.22	8.25	nr	**23.83**
Valves						
15 mm dia.	3.83	4.70	0.15	5.63	nr	**10.33**
20 mm dia.	4.12	5.06	0.15	5.63	nr	**10.69**
25 mm dia.	4.62	5.68	0.15	5.63	nr	**11.31**
32 mm dia.	5.03	6.18	0.15	5.63	nr	**11.81**
40 mm dia.	5.37	6.59	0.15	5.63	nr	**12.22**
50 mm dia.	6.16	7.57	0.15	5.63	nr	**13.20**
65 mm dia.	7.04	8.65	0.15	5.63	nr	**14.28**
80 mm dia.	7.72	9.48	0.20	7.51	nr	**16.99**
100 mm dia.	10.17	12.49	0.20	7.51	nr	**20.00**
125 mm dia.	11.77	14.46	0.20	7.51	nr	**21.97**
150 mm dia.	14.02	17.22	0.20	7.51	nr	**24.73**
200 mm dia.	19.81	24.34	0.25	9.39	nr	**33.73**
250 mm dia.	23.71	29.13	0.25	9.39	nr	**38.52**
300 mm dia.	25.35	31.14	0.25	9.39	nr	**40.53**
Expansion bellows						
15 mm dia.	7.65	9.40	0.22	8.25	nr	**17.65**
20 mm dia.	8.26	10.14	0.22	8.25	nr	**18.39**
25 mm dia.	9.22	11.33	0.22	8.25	nr	**19.58**
32 mm dia.	10.04	12.34	0.22	8.25	nr	**20.59**
40 mm dia.	10.77	13.23	0.22	8.25	nr	**21.48**
50 mm dia.	12.34	15.16	0.22	8.25	nr	**23.41**
65 mm dia.	14.08	17.29	0.22	8.25	nr	**25.54**
80 mm dia.	15.42	18.94	0.29	10.89	nr	**29.83**
100 mm dia.	20.34	24.99	0.29	10.89	nr	**35.88**
125 mm dia.	23.53	28.91	0.29	10.89	nr	**39.80**
150 mm dia.	28.05	34.46	0.29	10.89	nr	**45.35**
200 mm dia.	39.61	48.66	0.36	13.51	nr	**62.17**
250 mm dia.	47.44	58.27	0.36	13.51	nr	**71.78**
300 mm dia.	50.72	62.31	0.36	13.51	nr	**75.82**

38 MECHANICAL/COOLING/HEATING SYSTEMS

Item	Net Price £	Material £	Labour hours	Labour £	Unit	Total rate £
THERMAL INSULATION – cont						
Extra over for fittings concealed insulation – cont						
30 mm thick						
15 mm dia.	4.98	6.12	0.15	5.63	m	**11.75**
20 mm dia.	5.33	6.55	0.15	5.63	m	**12.18**
25 mm dia.	5.64	6.93	0.15	5.63	m	**12.56**
32 mm dia.	6.14	7.54	0.15	5.63	m	**13.17**
40 mm dia.	6.51	8.00	0.15	5.63	m	**13.63**
50 mm dia.	7.45	9.16	0.15	5.63	m	**14.79**
65 mm dia.	8.43	10.35	0.15	5.63	m	**15.98**
80 mm dia.	9.20	11.30	0.22	8.25	m	**19.55**
100 mm dia.	11.89	14.61	0.22	8.25	m	**22.86**
125 mm dia.	13.71	16.84	0.22	8.25	m	**25.09**
150 mm dia.	16.09	19.77	0.22	8.25	m	**28.02**
200 mm dia.	22.48	27.61	0.25	9.39	m	**37.00**
250 mm dia.	26.74	32.85	0.25	9.39	m	**42.24**
300 mm dia.	28.34	34.82	0.25	9.39	m	**44.21**
350 mm dia.	31.15	38.27	0.25	9.39	m	**47.66**
Extra over for fittings concealed insulation						
Flange/union						
15 mm dia.	2.48	3.04	0.13	4.87	nr	**7.91**
20 mm dia.	2.67	3.28	0.13	4.87	nr	**8.15**
25 mm dia.	2.81	3.45	0.13	4.87	nr	**8.32**
32 mm dia.	3.08	3.78	0.13	4.87	nr	**8.65**
40 mm dia.	3.26	4.00	0.13	4.87	nr	**8.87**
50 mm dia.	3.73	4.58	0.13	4.87	nr	**9.45**
65 mm dia.	4.23	5.20	0.13	4.87	nr	**10.07**
80 mm dia.	4.58	5.63	0.18	6.76	nr	**12.39**
100 mm dia.	5.96	7.32	0.18	6.76	nr	**14.08**
125 mm dia.	6.83	8.39	0.18	6.76	nr	**15.15**
150 mm dia.	8.06	9.90	0.18	6.76	nr	**16.66**
200 mm dia.	11.25	13.82	0.22	8.25	nr	**22.07**
250 mm dia.	13.38	16.43	0.22	8.25	nr	**24.68**
300 mm dia.	14.18	17.42	0.22	8.25	nr	**25.67**
350 mm dia.	15.57	19.13	0.22	8.25	nr	**27.38**
Valves						
15 mm dia.	5.33	6.55	0.15	5.63	nr	**12.18**
20 mm dia.	5.33	6.55	0.15	5.63	nr	**12.18**
25 mm dia.	5.64	6.93	0.15	5.63	nr	**12.56**
32 mm dia.	6.14	7.54	0.15	5.63	nr	**13.17**
40 mm dia.	6.51	8.00	0.15	5.63	nr	**13.63**
50 mm dia.	7.45	9.16	0.15	5.63	nr	**14.79**
65 mm dia.	8.43	10.35	0.15	5.63	nr	**15.98**
80 mm dia.	9.20	11.30	0.20	7.51	nr	**18.81**
100 mm dia.	11.89	14.61	0.20	7.51	nr	**22.12**
125 mm dia.	13.71	16.84	0.20	7.51	nr	**24.35**
150 mm dia.	16.09	19.77	0.20	7.51	nr	**27.28**

38 MECHANICAL/COOLING/HEATING SYSTEMS

Item	Net Price £	Material £	Labour hours	Labour £	Unit	**Total rate £**
200 mm dia.	22.48	27.61	0.25	9.39	nr	**37.00**
250 mm dia.	26.74	32.85	0.25	9.39	nr	**42.24**
300 mm dia.	28.34	34.82	0.25	9.39	nr	**44.21**
350 mm dia.	31.15	38.27	0.25	9.39	nr	**47.66**
Expansion bellows						
15 mm dia.	9.97	12.24	0.22	8.25	nr	**20.49**
20 mm dia.	10.64	13.07	0.22	8.25	nr	**21.32**
25 mm dia.	11.28	13.86	0.22	8.25	nr	**22.11**
32 mm dia.	12.32	15.13	0.22	8.25	nr	**23.38**
40 mm dia.	13.02	15.99	0.22	8.25	nr	**24.24**
50 mm dia.	14.89	18.29	0.22	8.25	nr	**26.54**
65 mm dia.	16.85	20.70	0.22	8.25	nr	**28.95**
80 mm dia.	18.39	22.60	0.29	10.89	nr	**33.49**
100 mm dia.	23.81	29.25	0.29	10.89	nr	**40.14**
125 mm dia.	27.38	33.63	0.29	10.89	nr	**44.52**
150 mm dia.	32.16	39.51	0.29	10.89	nr	**50.40**
200 mm dia.	44.97	55.24	0.36	13.51	nr	**68.75**
250 mm dia.	53.49	65.71	0.36	13.51	nr	**79.22**
300 mm dia.	56.73	69.69	0.36	13.51	nr	**83.20**
350 mm dia.	62.28	76.51	0.36	13.51	nr	**90.02**
40 mm thick						
15 mm dia.	6.37	7.82	0.15	5.63	m	**13.45**
20 mm dia.	6.57	8.07	0.15	5.63	m	**13.70**
25 mm dia.	7.07	8.68	0.15	5.63	m	**14.31**
32 mm dia.	7.54	9.26	0.15	5.63	m	**14.89**
40 mm dia.	7.92	9.73	0.15	5.63	m	**15.36**
50 mm dia.	8.97	11.01	0.15	5.63	m	**16.64**
65 mm dia.	10.05	12.35	0.15	5.63	m	**17.98**
80 mm dia.	10.93	13.43	0.22	8.25	m	**21.68**
100 mm dia.	14.21	17.46	0.22	8.25	m	**25.71**
125 mm dia.	16.05	19.72	0.22	8.25	m	**27.97**
150 mm dia.	18.72	22.99	0.22	8.25	m	**31.24**
200 mm dia.	25.88	31.79	0.25	9.39	m	**41.18**
250 mm dia.	30.33	37.26	0.25	9.39	m	**46.65**
300 mm dia.	32.43	39.84	0.25	9.39	m	**49.23**
350 mm dia.	35.86	44.06	0.25	9.39	m	**53.45**
400 mm dia.	39.96	49.09	0.25	9.39	m	**58.48**
Extra over for fittings concealed insulation						
Flange/union						
15 mm dia.	3.20	3.93	0.13	4.87	nr	**8.80**
20 mm dia.	3.28	4.03	0.13	4.87	nr	**8.90**
25 mm dia.	3.53	4.34	0.13	4.87	nr	**9.21**
32 mm dia.	3.76	4.62	0.13	4.87	nr	**9.49**
40 mm dia.	3.96	4.86	0.13	4.87	nr	**9.73**
50 mm dia.	4.49	5.51	0.13	4.87	nr	**10.38**
65 mm dia.	5.03	6.18	0.13	4.87	nr	**11.05**
80 mm dia.	5.48	6.73	0.18	6.76	nr	**13.49**
100 mm dia.	7.10	8.73	0.18	6.76	nr	**15.49**
125 mm dia.	8.01	9.84	0.18	6.76	nr	**16.60**
150 mm dia.	9.37	11.51	0.18	6.76	nr	**18.27**

38 MECHANICAL/COOLING/HEATING SYSTEMS

Item	Net Price £	Material £	Labour hours	Labour £	Unit	Total rate £
THERMAL INSULATION – cont						
Extra over for fittings concealed insulation – cont						
Flange/union – cont						
200 mm dia.	12.95	15.91	0.22	8.25	nr	**24.16**
250 mm dia.	15.15	18.62	0.22	8.25	nr	**26.87**
300 mm dia.	16.20	19.90	0.22	8.25	nr	**28.15**
350 mm dia.	17.93	22.03	0.22	8.25	nr	**30.28**
400 mm dia.	20.00	24.57	0.22	8.25	nr	**32.82**
Valves						
15 mm dia.	6.37	7.82	0.15	5.63	nr	**13.45**
20 mm dia.	6.57	8.07	0.15	5.63	nr	**13.70**
25 mm dia.	7.07	8.68	0.15	5.63	nr	**14.31**
32 mm dia.	7.54	9.26	0.15	5.63	nr	**14.89**
40 mm dia.	7.92	9.73	0.15	5.63	nr	**15.36**
50 mm dia.	8.97	11.01	0.15	5.63	nr	**16.64**
65 mm dia.	10.05	12.35	0.15	5.63	nr	**17.98**
80 mm dia.	10.93	13.43	0.20	7.51	nr	**20.94**
100 mm dia.	14.21	17.46	0.20	7.51	nr	**24.97**
125 mm dia.	16.05	19.72	0.20	7.51	nr	**27.23**
150 mm dia.	18.72	22.99	0.20	7.51	nr	**30.50**
200 mm dia.	25.88	31.79	0.25	9.39	nr	**41.18**
250 mm dia.	30.33	37.26	0.25	9.39	nr	**46.65**
300 mm dia.	32.43	39.84	0.25	9.39	nr	**49.23**
350 mm dia.	35.86	44.06	0.25	9.39	nr	**53.45**
400 mm dia.	39.96	49.09	0.25	9.39	nr	**58.48**
Expansion bellows						
15 mm dia.	12.77	15.69	0.22	8.25	nr	**23.94**
20 mm dia.	13.13	16.13	0.22	8.25	nr	**24.38**
25 mm dia.	14.15	17.39	0.22	8.25	nr	**25.64**
32 mm dia.	15.07	18.51	0.22	8.25	nr	**26.76**
40 mm dia.	15.85	19.47	0.22	8.25	nr	**27.72**
50 mm dia.	17.93	22.03	0.22	8.25	nr	**30.28**
65 mm dia.	20.10	24.70	0.22	8.25	nr	**32.95**
80 mm dia.	21.89	26.89	0.29	10.89	nr	**37.78**
100 mm dia.	28.39	34.88	0.29	10.89	nr	**45.77**
125 mm dia.	32.09	39.43	0.29	10.89	nr	**50.32**
150 mm dia.	37.47	46.03	0.29	10.89	nr	**56.92**
200 mm dia.	51.76	63.59	0.36	13.51	nr	**77.10**
250 mm dia.	60.65	74.51	0.36	13.51	nr	**88.02**
300 mm dia.	64.85	79.66	0.36	13.51	nr	**93.17**
350 mm dia.	35.86	44.06	0.36	13.51	nr	**57.57**
400 mm dia.	79.94	98.21	0.36	13.51	nr	**111.72**
50 mm thick						
15 mm dia.	8.87	10.90	0.15	5.63	m	**16.53**
20 mm dia.	9.32	11.45	0.15	5.63	m	**17.08**
25 mm dia.	9.91	12.17	0.15	5.63	m	**17.80**
32 mm dia.	10.35	12.72	0.15	5.63	m	**18.35**
40 mm dia.	10.89	13.38	0.15	5.63	m	**19.01**
50 mm dia.	12.22	15.02	0.15	5.63	m	**20.65**
65 mm dia.	13.36	16.41	0.15	5.63	m	**22.04**

38 MECHANICAL/COOLING/HEATING SYSTEMS

Item	Net Price £	Material £	Labour hours	Labour £	Unit	**Total rate £**
80 mm dia.	14.29	17.56	0.22	8.25	m	**25.81**
100 mm dia.	18.28	22.46	0.22	8.25	m	**30.71**
125 mm dia.	20.52	25.21	0.22	8.25	m	**33.46**
150 mm dia.	23.68	29.10	0.22	8.25	m	**37.35**
200 mm dia.	32.32	39.70	0.25	9.39	m	**49.09**
250 mm dia.	37.32	45.84	0.25	9.39	m	**55.23**
300 mm dia.	39.55	48.58	0.25	9.39	m	**57.97**
350 mm dia.	43.65	53.62	0.25	9.39	m	**63.01**
400 mm dia.	48.42	59.48	0.25	9.39	m	**68.87**
Extra over for fittings concealed insulation						
Flange/union						
15 mm dia.	4.44	5.45	0.13	4.87	nr	**10.32**
20 mm dia.	4.67	5.73	0.13	4.87	nr	**10.60**
25 mm dia.	4.97	6.10	0.13	4.87	nr	**10.97**
32 mm dia.	5.20	6.38	0.13	4.87	nr	**11.25**
40 mm dia.	5.46	6.71	0.13	4.87	nr	**11.58**
50 mm dia.	6.09	7.49	0.13	4.87	nr	**12.36**
65 mm dia.	6.68	8.21	0.13	4.87	nr	**13.08**
80 mm dia.	7.15	8.79	0.18	6.76	nr	**15.55**
100 mm dia.	9.15	11.25	0.18	6.76	nr	**18.01**
125 mm dia.	10.26	12.60	0.18	6.76	nr	**19.36**
150 mm dia.	11.84	14.54	0.18	6.76	nr	**21.30**
200 mm dia.	16.16	19.86	0.22	8.25	nr	**28.11**
250 mm dia.	18.66	22.92	0.22	8.25	nr	**31.17**
300 mm dia.	19.79	24.31	0.22	8.25	nr	**32.56**
350 mm dia.	21.85	26.84	0.22	8.25	nr	**35.09**
400 mm dia.	24.22	29.76	0.22	8.25	nr	**38.01**
Valves						
15 mm dia.	8.87	10.90	0.15	5.63	nr	**16.53**
20 mm dia.	9.32	11.45	0.15	5.63	nr	**17.08**
25 mm dia.	9.91	12.17	0.15	5.63	nr	**17.80**
32 mm dia.	10.35	12.72	0.15	5.63	nr	**18.35**
40 mm dia.	10.89	13.38	0.15	5.63	nr	**19.01**
50 mm dia.	12.22	15.02	0.15	5.63	nr	**20.65**
65 mm dia.	13.36	16.41	0.15	5.63	nr	**22.04**
80 mm dia.	14.29	17.56	0.20	7.51	nr	**25.07**
100 mm dia.	18.28	22.46	0.20	7.51	nr	**29.97**
125 mm dia.	20.52	25.21	0.20	7.51	nr	**32.72**
150 mm dia.	23.68	29.10	0.20	7.51	nr	**36.61**
200 mm dia.	32.32	39.70	0.25	9.39	nr	**49.09**
250 mm dia.	37.32	45.84	0.25	9.39	nr	**55.23**
300 mm dia.	39.55	48.58	0.25	9.39	nr	**57.97**
350 mm dia.	43.65	53.62	0.25	9.39	nr	**63.01**
400 mm dia.	48.42	59.48	0.25	9.39	nr	**68.87**
Expansion bellows						
15 mm dia.	17.70	21.75	0.22	8.25	nr	**30.00**
20 mm dia.	18.65	22.91	0.22	8.25	nr	**31.16**
25 mm dia.	19.82	24.35	0.22	8.25	nr	**32.60**

38 MECHANICAL/COOLING/HEATING SYSTEMS

Item	Net Price £	Material £	Labour hours	Labour £	Unit	Total rate £
THERMAL INSULATION – cont						
Extra over for fittings concealed insulation – cont						
Expansion bellows – cont						
32 mm dia.	20.73	25.46	0.22	8.25	nr	**33.71**
40 mm dia.	21.80	26.79	0.22	8.25	nr	**35.04**
50 mm dia.	24.43	30.01	0.22	8.25	nr	**38.26**
65 mm dia.	26.69	32.79	0.22	8.25	nr	**41.04**
80 mm dia.	28.60	35.13	0.29	10.89	nr	**46.02**
100 mm dia.	36.53	44.88	0.29	10.89	nr	**55.77**
125 mm dia.	41.01	50.38	0.29	10.89	nr	**61.27**
150 mm dia.	47.35	58.17	0.29	10.89	nr	**69.06**
200 mm dia.	64.66	79.43	0.36	13.51	nr	**92.94**
250 mm dia.	74.64	91.70	0.36	13.51	nr	**105.21**
300 mm dia.	79.13	97.21	0.36	13.51	nr	**110.72**
350 mm dia.	87.32	107.27	0.36	13.51	nr	**120.78**
400 mm dia.	96.83	118.95	0.36	13.51	nr	**132.46**
Mineral fibre sectional insulation; bright class O foil faced; bright class O foil taped joints; 22 SWG plain/embossed aluminium cladding; pop riveted						
Plantroom pipework						
20 mm thick						
15 mm dia.	5.69	6.99	0.44	16.51	m	**23.50**
20 mm dia.	6.01	7.38	0.44	16.51	m	**23.89**
25 mm dia.	6.43	7.90	0.44	16.51	m	**24.41**
32 mm dia.	6.97	8.56	0.44	16.51	m	**25.07**
40 mm dia.	7.34	9.02	0.44	16.51	m	**25.53**
50 mm dia.	8.18	10.05	0.44	16.51	m	**26.56**
Extra over for fittings plantroom insulation						
Flange/union						
15 mm dia.	6.34	7.79	0.58	21.77	nr	**29.56**
20 mm dia.	6.74	8.28	0.58	21.77	nr	**30.05**
25 mm dia.	7.22	8.87	0.58	21.77	nr	**30.64**
32 mm dia.	7.89	9.69	0.58	21.77	nr	**31.46**
40 mm dia.	8.33	10.24	0.58	21.77	nr	**32.01**
50 mm dia.	9.35	11.49	0.58	21.77	nr	**33.26**
Bends						
15 mm dia.	3.11	3.82	0.44	16.51	nr	**20.33**
20 mm dia.	3.32	4.07	0.44	16.51	nr	**20.58**
25 mm dia.	3.54	4.35	0.44	16.51	nr	**20.86**
32 mm dia.	3.83	4.70	0.44	16.51	nr	**21.21**
40 mm dia.	4.02	4.93	0.44	16.51	nr	**21.44**
50 mm dia.	4.50	5.53	0.44	16.51	nr	**22.04**

38 MECHANICAL/COOLING/HEATING SYSTEMS

Item	Net Price £	Material £	Labour hours	Labour £	Unit	**Total rate £**
Tees						
15 mm dia.	1.87	2.30	0.44	16.51	nr	**18.81**
20 mm dia.	1.99	2.45	0.44	16.51	nr	**18.96**
25 mm dia.	2.12	2.60	0.44	16.51	nr	**19.11**
32 mm dia.	2.30	2.82	0.44	16.51	nr	**19.33**
40 mm dia.	2.43	2.98	0.44	16.51	nr	**19.49**
50 mm dia.	2.69	3.31	0.44	16.51	nr	**19.82**
Valves						
15 mm dia.	3.48	4.27	0.78	29.27	nr	**33.54**
20 mm dia.	3.99	4.90	0.78	29.27	nr	**34.17**
25 mm dia.	3.99	4.90	0.78	29.27	nr	**34.17**
32 mm dia.	4.45	5.47	0.78	29.27	nr	**34.74**
40 mm dia.	4.76	5.85	0.78	29.27	nr	**35.12**
50 mm dia.	5.45	6.70	0.78	29.27	nr	**35.97**
Pumps						
15 mm dia.	18.66	22.92	2.34	87.82	nr	**110.74**
20 mm dia.	19.80	24.33	2.34	87.82	nr	**112.15**
25 mm dia.	21.26	26.11	2.34	87.82	nr	**113.93**
32 mm dia.	23.19	28.49	2.34	87.82	nr	**116.31**
40 mm dia.	24.49	30.08	2.34	87.82	nr	**117.90**
50 mm dia.	27.54	33.83	2.34	87.82	nr	**121.65**
Expansion bellows						
15 mm dia.	14.91	18.31	1.05	39.41	nr	**57.72**
20 mm dia.	15.84	19.46	1.05	39.41	nr	**58.87**
25 mm dia.	17.01	20.90	1.05	39.41	nr	**60.31**
32 mm dia.	18.54	22.77	1.05	39.41	nr	**62.18**
40 mm dia.	19.58	24.06	1.05	39.41	nr	**63.47**
50 mm dia.	22.02	27.05	1.05	39.41	nr	**66.46**
25 mm thick						
15 mm dia.	6.24	7.66	0.44	16.51	m	**24.17**
20 mm dia.	6.74	8.28	0.44	16.51	m	**24.79**
25 mm dia.	7.39	9.08	0.44	16.51	m	**25.59**
32 mm dia.	7.86	9.66	0.44	16.51	m	**26.17**
40 mm dia.	8.46	10.40	0.44	16.51	m	**26.91**
50 mm dia.	9.48	11.64	0.44	16.51	m	**28.15**
65 mm dia.	10.71	13.16	0.44	16.51	m	**29.67**
80 mm dia.	11.61	14.26	0.52	19.52	m	**33.78**
100 mm dia.	14.41	17.70	0.52	19.52	m	**37.22**
125 mm dia.	16.73	20.55	0.52	19.52	m	**40.07**
150 mm dia.	19.48	23.93	0.52	19.52	m	**43.45**
200 mm dia.	26.61	32.69	0.60	22.52	m	**55.21**
250 mm dia.	31.61	38.83	0.60	22.52	m	**61.35**
300 mm dia.	35.40	43.49	0.60	22.52	m	**66.01**
Extra over for fittings plantroom insulation						
Flange/union						
15 mm dia.	6.97	8.56	0.58	21.77	nr	**30.33**
20 mm dia.	7.54	9.26	0.58	21.77	nr	**31.03**
25 mm dia.	8.32	10.22	0.58	21.77	nr	**31.99**

38 MECHANICAL/COOLING/HEATING SYSTEMS

Item	Net Price £	Material £	Labour hours	Labour £	Unit	Total rate £
THERMAL INSULATION – cont						
Extra over for fittings plantroom insulation – cont						
Flange/union – cont						
32 mm dia.	8.93	10.97	0.58	21.77	nr	**32.74**
40 mm dia.	9.58	11.77	0.58	21.77	nr	**33.54**
50 mm dia.	10.80	13.27	0.58	21.77	nr	**35.04**
65 mm dia.	12.25	15.05	0.58	21.77	nr	**36.82**
80 mm dia.	13.31	16.35	0.67	25.15	nr	**41.50**
100 mm dia.	16.85	20.70	0.67	25.15	nr	**45.85**
125 mm dia.	19.57	24.05	0.67	25.15	nr	**49.20**
150 mm dia.	22.93	28.17	0.67	25.15	nr	**53.32**
200 mm dia.	31.68	38.92	0.87	32.65	nr	**71.57**
250 mm dia.	37.75	46.38	0.87	32.65	nr	**79.03**
300 mm dia.	41.72	51.25	0.87	32.65	nr	**83.90**
Bends						
15 mm dia.	3.42	4.20	0.44	16.51	nr	**20.71**
20 mm dia.	3.71	4.56	0.44	16.51	nr	**21.07**
25 mm dia.	4.06	4.99	0.44	16.51	nr	**21.50**
32 mm dia.	4.33	5.32	0.44	16.51	nr	**21.83**
40 mm dia.	4.67	5.73	0.44	16.51	nr	**22.24**
50 mm dia.	5.22	6.42	0.44	16.51	nr	**22.93**
65 mm dia.	5.88	7.22	0.44	16.51	nr	**23.73**
80 mm dia.	6.38	7.83	0.52	19.52	nr	**27.35**
100 mm dia.	7.92	9.73	0.52	19.52	nr	**29.25**
125 mm dia.	9.22	11.33	0.52	19.52	nr	**30.85**
150 mm dia.	10.71	13.16	0.52	19.52	nr	**32.68**
200 mm dia.	14.63	17.98	0.60	22.52	nr	**40.50**
250 mm dia.	17.39	21.37	0.60	22.52	nr	**43.89**
300 mm dia.	19.46	23.91	0.60	22.52	nr	**46.43**
Tees						
15 mm dia.	2.06	2.53	0.44	16.51	nr	**19.04**
20 mm dia.	2.21	2.72	0.44	16.51	nr	**19.23**
25 mm dia.	2.44	2.99	0.44	16.51	nr	**19.50**
32 mm dia.	2.60	3.19	0.44	16.51	nr	**19.70**
40 mm dia.	2.79	3.42	0.44	16.51	nr	**19.93**
50 mm dia.	3.11	3.82	0.44	16.51	nr	**20.33**
65 mm dia.	3.53	4.34	0.44	16.51	nr	**20.85**
80 mm dia.	3.83	4.70	0.52	19.52	nr	**24.22**
100 mm dia.	4.75	5.84	0.52	19.52	nr	**25.36**
125 mm dia.	5.53	6.79	0.52	19.52	nr	**26.31**
150 mm dia.	6.43	7.90	0.52	19.52	nr	**27.42**
200 mm dia.	8.77	10.77	0.60	22.52	nr	**33.29**
250 mm dia.	10.45	12.84	0.60	22.52	nr	**35.36**
300 mm dia.	11.69	14.36	0.60	22.52	nr	**36.88**
Valves						
15 mm dia.	11.07	13.60	0.78	29.27	nr	**42.87**
20 mm dia.	11.97	14.70	0.78	29.27	nr	**43.97**
25 mm dia.	13.21	16.23	0.78	29.27	nr	**45.50**

38 MECHANICAL/COOLING/HEATING SYSTEMS

Item	Net Price £	Material £	Labour hours	Labour £	Unit	**Total rate £**
32 mm dia.	14.17	17.41	0.78	29.27	nr	**46.68**
40 mm dia.	15.21	18.69	0.78	29.27	nr	**47.96**
50 mm dia.	17.13	21.04	0.78	29.27	nr	**50.31**
65 mm dia.	19.45	23.90	0.78	29.27	nr	**53.17**
80 mm dia.	21.15	25.99	0.92	34.53	nr	**60.52**
100 mm dia.	26.75	32.86	0.92	34.53	nr	**67.39**
125 mm dia.	31.10	38.21	0.92	34.53	nr	**72.74**
150 mm dia.	36.42	44.74	0.92	34.53	nr	**79.27**
200 mm dia.	50.33	61.83	1.12	42.04	nr	**103.87**
250 mm dia.	59.97	73.67	1.12	42.04	nr	**115.71**
300 mm dia.	66.25	81.39	1.12	42.04	nr	**123.43**
Pumps						
15 mm dia.	20.50	25.18	2.34	87.82	nr	**113.00**
20 mm dia.	22.16	27.23	2.34	87.82	nr	**115.05**
25 mm dia.	24.47	30.06	2.34	87.82	nr	**117.88**
32 mm dia.	26.23	32.22	2.34	87.82	nr	**120.04**
40 mm dia.	28.17	34.61	2.34	87.82	nr	**122.43**
50 mm dia.	31.74	39.00	2.34	87.82	nr	**126.82**
65 mm dia.	36.01	44.24	2.34	87.82	nr	**132.06**
80 mm dia.	39.18	48.13	2.76	103.58	nr	**151.71**
100 mm dia.	49.56	60.89	2.76	103.58	nr	**164.47**
125 mm dia.	57.55	70.70	2.76	103.58	nr	**174.28**
150 mm dia.	67.45	82.87	2.76	103.58	nr	**186.45**
200 mm dia.	93.21	114.51	3.36	126.10	nr	**240.61**
250 mm dia.	111.05	136.43	3.36	126.10	nr	**262.53**
300 mm dia.	122.67	150.70	3.36	126.10	nr	**276.80**
Expansion bellows						
15 mm dia.	16.39	20.14	1.05	39.41	nr	**59.55**
20 mm dia.	17.72	21.77	1.05	39.41	nr	**61.18**
25 mm dia.	19.57	24.05	1.05	39.41	nr	**63.46**
32 mm dia.	21.01	25.81	1.05	39.41	nr	**65.22**
40 mm dia.	22.54	27.69	1.05	39.41	nr	**67.10**
50 mm dia.	25.40	31.21	1.05	39.41	nr	**70.62**
65 mm dia.	28.81	35.40	1.05	39.41	nr	**74.81**
80 mm dia.	31.35	38.51	1.26	47.28	nr	**85.79**
100 mm dia.	39.64	48.70	1.26	47.28	nr	**95.98**
125 mm dia.	46.02	56.53	1.26	47.28	nr	**103.81**
150 mm dia.	53.95	66.28	1.26	47.28	nr	**113.56**
200 mm dia.	74.59	91.63	1.53	57.42	nr	**149.05**
250 mm dia.	88.84	109.14	1.53	57.42	nr	**166.56**
300 mm dia.	98.13	120.55	1.53	57.42	nr	**177.97**
30 mm thick						
15 mm dia.	7.68	9.44	0.44	16.51	m	**25.95**
20 mm dia.	8.13	9.99	0.44	16.51	m	**26.50**
25 mm dia.	8.61	10.57	0.44	16.51	m	**27.08**
32 mm dia.	9.43	11.58	0.44	16.51	m	**28.09**
40 mm dia.	9.82	12.06	0.44	16.51	m	**28.57**
50 mm dia.	10.89	13.38	0.44	16.51	m	**29.89**
65 mm dia.	12.29	15.10	0.44	16.51	m	**31.61**

38 MECHANICAL/COOLING/HEATING SYSTEMS

Item	Net Price £	Material £	Labour hours	Labour £	Unit	Total rate £
THERMAL INSULATION – cont						
Extra over for fittings plantroom insulation – cont						
30 mm thick – cont						
80 mm dia.	13.43	16.50	0.52	19.52	m	**36.02**
100 mm dia.	16.51	20.29	0.52	19.52	m	**39.81**
125 mm dia.	18.85	23.15	0.52	19.52	m	**42.67**
150 mm dia.	21.98	27.01	0.52	19.52	m	**46.53**
200 mm dia.	29.54	36.29	0.60	22.52	m	**58.81**
250 mm dia.	34.94	42.92	0.60	22.52	m	**65.44**
300 mm dia.	38.47	47.26	0.60	22.52	m	**69.78**
350 mm dia.	42.83	52.62	0.60	22.52	m	**75.14**
Extra over for fittings plantroom insulation						
Flange/union						
15 mm dia.	8.74	10.74	0.58	21.77	nr	**32.51**
20 mm dia.	9.28	11.40	0.58	21.77	nr	**33.17**
25 mm dia.	9.83	12.07	0.58	21.77	nr	**33.84**
32 mm dia.	10.77	13.23	0.58	21.77	nr	**35.00**
40 mm dia.	11.26	13.83	0.58	21.77	nr	**35.60**
50 mm dia.	12.62	15.51	0.58	21.77	nr	**37.28**
65 mm dia.	14.26	17.51	0.58	21.77	nr	**39.28**
80 mm dia.	15.57	19.13	0.67	25.15	nr	**44.28**
100 mm dia.	19.45	23.90	0.67	25.15	nr	**49.05**
125 mm dia.	22.26	27.34	0.67	25.15	nr	**52.49**
150 mm dia.	26.05	32.00	0.67	25.15	nr	**57.15**
200 mm dia.	35.46	43.56	0.87	32.65	nr	**76.21**
250 mm dia.	42.04	51.65	0.87	32.65	nr	**84.30**
300 mm dia.	45.74	56.20	0.87	32.65	nr	**88.85**
350 mm dia.	50.74	62.34	0.87	32.65	nr	**94.99**
Bends						
15 mm dia.	4.23	5.20	0.44	16.51	nr	**21.71**
20 mm dia.	4.46	5.48	0.44	16.51	nr	**21.99**
25 mm dia.	4.74	5.83	0.44	16.51	nr	**22.34**
32 mm dia.	5.18	6.36	0.44	16.51	nr	**22.87**
40 mm dia.	5.39	6.63	0.44	16.51	nr	**23.14**
50 mm dia.	6.01	7.38	0.44	16.51	nr	**23.89**
65 mm dia.	6.76	8.31	0.44	16.51	nr	**24.82**
80 mm dia.	7.37	9.05	0.52	19.52	nr	**28.57**
100 mm dia.	9.07	11.14	0.52	19.52	nr	**30.66**
125 mm dia.	10.37	12.74	0.52	19.52	nr	**32.26**
150 mm dia.	12.10	14.87	0.52	19.52	nr	**34.39**
200 mm dia.	16.26	19.97	0.60	22.52	nr	**42.49**
250 mm dia.	19.24	23.64	0.60	22.52	nr	**46.16**
300 mm dia.	21.15	25.99	0.60	22.52	nr	**48.51**
350 mm dia.	23.57	28.96	0.60	22.52	nr	**51.48**
Tees						
15 mm dia.	2.54	3.12	0.44	16.51	nr	**19.63**
20 mm dia.	2.68	3.30	0.44	16.51	nr	**19.81**
25 mm dia.	2.84	3.49	0.44	16.51	nr	**20.00**

38 MECHANICAL/COOLING/HEATING SYSTEMS

Item	Net Price £	Material £	Labour hours	Labour £	Unit	**Total rate £**
32 mm dia.	3.10	3.81	0.44	16.51	nr	**20.32**
40 mm dia.	3.24	3.98	0.44	16.51	nr	**20.49**
50 mm dia.	3.59	4.41	0.44	16.51	nr	**20.92**
65 mm dia.	4.05	4.98	0.44	16.51	nr	**21.49**
80 mm dia.	4.44	5.45	0.52	19.52	nr	**24.97**
100 mm dia.	5.46	6.71	0.52	19.52	nr	**26.23**
125 mm dia.	6.24	7.66	0.52	19.52	nr	**27.18**
150 mm dia.	7.27	8.94	0.52	19.52	nr	**28.46**
200 mm dia.	9.76	11.99	0.60	22.52	nr	**34.51**
250 mm dia.	11.54	14.17	0.60	22.52	nr	**36.69**
300 mm dia.	12.70	15.60	0.60	22.52	nr	**38.12**
350 mm dia.	14.15	17.39	0.60	22.52	nr	**39.91**
Valves						
15 mm dia.	13.89	17.06	0.78	29.27	nr	**46.33**
20 mm dia.	14.74	18.11	0.78	29.27	nr	**47.38**
25 mm dia.	15.61	19.17	0.78	29.27	nr	**48.44**
32 mm dia.	17.11	21.02	0.78	29.27	nr	**50.29**
40 mm dia.	17.90	21.99	0.78	29.27	nr	**51.26**
50 mm dia.	20.04	24.62	0.78	29.27	nr	**53.89**
65 mm dia.	22.65	27.82	0.78	29.27	nr	**57.09**
80 mm dia.	24.72	30.37	0.92	34.53	nr	**64.90**
100 mm dia.	30.88	37.94	0.92	34.53	nr	**72.47**
125 mm dia.	35.37	43.45	0.92	34.53	nr	**77.98**
150 mm dia.	41.36	50.81	0.92	34.53	nr	**85.34**
200 mm dia.	56.31	69.17	1.12	42.04	nr	**111.21**
250 mm dia.	66.77	82.03	1.12	42.04	nr	**124.07**
300 mm dia.	72.63	89.23	1.12	42.04	nr	**131.27**
350 mm dia.	80.59	99.00	1.12	42.04	nr	**141.04**
Pumps						
15 mm dia.	25.71	31.58	2.34	87.82	nr	**119.40**
20 mm dia.	27.29	33.53	2.34	87.82	nr	**121.35**
25 mm dia.	28.91	35.51	2.34	87.82	nr	**123.33**
32 mm dia.	31.67	38.90	2.34	87.82	nr	**126.72**
40 mm dia.	33.13	40.70	2.34	87.82	nr	**128.52**
50 mm dia.	37.12	45.60	2.34	87.82	nr	**133.42**
65 mm dia.	41.92	51.50	2.34	87.82	nr	**139.32**
80 mm dia.	45.77	56.23	2.76	103.58	nr	**159.81**
100 mm dia.	57.18	70.25	2.76	103.58	nr	**173.83**
125 mm dia.	65.50	80.46	2.76	103.58	nr	**184.04**
150 mm dia.	76.58	94.08	2.76	103.58	nr	**197.66**
200 mm dia.	104.28	128.11	3.36	126.10	nr	**254.21**
250 mm dia.	123.66	151.91	3.36	126.10	nr	**278.01**
300 mm dia.	134.53	165.27	3.36	126.10	nr	**291.37**
350 mm dia.	149.23	183.33	3.36	126.10	nr	**309.43**
Expansion bellows						
15 mm dia.	20.58	25.28	1.05	39.41	nr	**64.69**
20 mm dia.	21.83	26.82	1.05	39.41	nr	**66.23**
25 mm dia.	23.13	28.41	1.05	39.41	nr	**67.82**
32 mm dia.	25.34	31.13	1.05	39.41	nr	**70.54**
40 mm dia.	26.50	32.55	1.05	39.41	nr	**71.96**
50 mm dia.	29.67	36.45	1.05	39.41	nr	**75.86**
65 mm dia.	33.54	41.20	1.05	39.41	nr	**80.61**

38 MECHANICAL/COOLING/HEATING SYSTEMS

Item	Net Price £	Material £	Labour hours	Labour £	Unit	Total rate £
THERMAL INSULATION – cont						
Extra over for fittings plantroom insulation – cont						
Expansion bellows – cont						
80 mm dia.	36.64	45.01	1.26	47.28	nr	**92.29**
100 mm dia.	45.75	56.21	1.26	47.28	nr	**103.49**
125 mm dia.	52.42	64.40	1.26	47.28	nr	**111.68**
150 mm dia.	61.27	75.27	1.26	47.28	nr	**122.55**
200 mm dia.	83.43	102.49	1.53	57.42	nr	**159.91**
250 mm dia.	98.92	121.53	1.53	57.42	nr	**178.95**
300 mm dia.	107.62	132.22	1.53	57.42	nr	**189.64**
350 mm dia.	119.39	146.67	1.53	57.42	nr	**204.09**
40 mm thick						
15 mm dia.	9.40	11.55	0.44	16.51	m	**28.06**
20 mm dia.	9.86	12.12	0.44	16.51	m	**28.63**
25 mm dia.	10.51	12.91	0.44	16.51	m	**29.42**
32 mm dia.	11.01	13.52	0.44	16.51	m	**30.03**
40 mm dia.	11.71	14.39	0.44	16.51	m	**30.90**
50 mm dia.	12.96	15.92	0.44	16.51	m	**32.43**
65 mm dia.	14.44	17.73	0.44	16.51	m	**34.24**
80 mm dia.	15.54	19.09	0.52	19.52	m	**38.61**
100 mm dia.	22.75	27.95	0.52	19.52	m	**47.47**
125 mm dia.	21.32	26.19	0.52	19.52	m	**45.71**
150 mm dia.	24.86	30.54	0.52	19.52	m	**50.06**
200 mm dia.	32.87	40.38	0.60	22.52	m	**62.90**
250 mm dia.	38.51	47.31	0.60	22.52	m	**69.83**
300 mm dia.	42.58	52.31	0.60	22.52	m	**74.83**
350 mm dia.	47.58	58.45	0.60	22.52	m	**80.97**
400 mm dia.	53.51	65.74	0.60	22.52	m	**88.26**
Extra over for fittings plantroom insulation						
Flange/union						
15 mm dia.	10.84	13.31	0.58	21.77	nr	**35.08**
20 mm dia.	11.32	13.90	0.58	21.77	nr	**35.67**
25 mm dia.	12.11	14.88	0.58	21.77	nr	**36.65**
32 mm dia.	12.75	15.67	0.58	21.77	nr	**37.44**
40 mm dia.	13.53	16.62	0.58	21.77	nr	**38.39**
50 mm dia.	15.07	18.51	0.58	21.77	nr	**40.28**
65 mm dia.	16.84	20.68	0.58	21.77	nr	**42.45**
80 mm dia.	18.16	22.31	0.67	25.15	nr	**47.46**
100 mm dia.	22.75	27.95	0.67	25.15	nr	**53.10**
125 mm dia.	25.47	31.29	0.67	25.15	nr	**56.44**
150 mm dia.	29.72	36.51	0.67	25.15	nr	**61.66**
200 mm dia.	39.88	48.99	0.87	32.65	nr	**81.64**
250 mm dia.	46.74	57.42	0.87	32.65	nr	**90.07**
300 mm dia.	51.15	62.84	0.87	32.65	nr	**95.49**
350 mm dia.	57.01	70.03	0.87	32.65	nr	**102.68**
400 mm dia.	63.92	78.53	0.87	32.65	nr	**111.18**

38 MECHANICAL/COOLING/HEATING SYSTEMS

Item	Net Price £	Material £	Labour hours	Labour £	Unit	**Total rate £**
Bends						
15 mm dia.	5.17	6.35	0.44	16.51	nr	**22.86**
20 mm dia.	5.42	6.66	0.44	16.51	nr	**23.17**
25 mm dia.	5.78	7.10	0.44	16.51	nr	**23.61**
32 mm dia.	6.04	7.42	0.44	16.51	nr	**23.93**
40 mm dia.	6.44	7.91	0.44	16.51	nr	**24.42**
50 mm dia.	7.12	8.75	0.44	16.51	nr	**25.26**
65 mm dia.	7.95	9.77	0.44	16.51	nr	**26.28**
80 mm dia.	8.53	10.48	0.52	19.52	nr	**30.00**
100 mm dia.	10.53	12.94	0.52	19.52	nr	**32.46**
125 mm dia.	11.72	14.40	0.52	19.52	nr	**33.92**
150 mm dia.	13.67	16.79	0.52	19.52	nr	**36.31**
200 mm dia.	18.06	22.19	0.60	22.52	nr	**44.71**
250 mm dia.	21.04	25.85	0.60	22.52	nr	**48.37**
300 mm dia.	23.43	28.78	0.60	22.52	nr	**51.30**
350 mm dia.	26.18	32.16	0.60	22.52	nr	**54.68**
400 mm dia.	29.43	36.15	0.60	22.52	nr	**58.67**
Tees						
15 mm dia.	3.10	3.81	0.44	16.51	nr	**20.32**
20 mm dia.	3.24	3.98	0.44	16.51	nr	**20.49**
25 mm dia.	3.47	4.26	0.44	16.51	nr	**20.77**
32 mm dia.	3.63	4.46	0.44	16.51	nr	**20.97**
40 mm dia.	3.87	4.76	0.44	16.51	nr	**21.27**
50 mm dia.	4.27	5.25	0.44	16.51	nr	**21.76**
65 mm dia.	4.76	5.85	0.44	16.51	nr	**22.36**
80 mm dia.	5.12	6.29	0.52	19.52	nr	**25.81**
100 mm dia.	6.32	7.76	0.52	19.52	nr	**27.28**
125 mm dia.	7.04	8.65	0.52	19.52	nr	**28.17**
150 mm dia.	8.21	10.09	0.52	19.52	nr	**29.61**
200 mm dia.	10.84	13.31	0.60	22.52	nr	**35.83**
250 mm dia.	12.72	15.62	0.60	22.52	nr	**38.14**
300 mm dia.	14.04	17.25	0.60	22.52	nr	**39.77**
350 mm dia.	15.72	19.31	0.60	22.52	nr	**41.83**
400 mm dia.	17.66	21.69	0.60	22.52	nr	**44.21**
Valves						
15 mm dia.	17.24	21.18	0.78	29.27	nr	**50.45**
20 mm dia.	17.99	22.10	0.78	29.27	nr	**51.37**
25 mm dia.	19.24	23.64	0.78	29.27	nr	**52.91**
32 mm dia.	20.25	24.87	0.78	29.27	nr	**54.14**
40 mm dia.	21.46	26.37	0.78	29.27	nr	**55.64**
50 mm dia.	23.93	29.40	0.78	29.27	nr	**58.67**
65 mm dia.	26.73	32.83	0.78	29.27	nr	**62.10**
80 mm dia.	28.85	35.44	0.92	34.53	nr	**69.97**
100 mm dia.	36.16	44.43	0.92	34.53	nr	**78.96**
125 mm dia.	40.43	49.67	0.92	34.53	nr	**84.20**
150 mm dia.	47.21	58.00	0.92	34.53	nr	**92.53**
200 mm dia.	63.33	77.81	1.12	42.04	nr	**119.85**
250 mm dia.	74.24	91.20	1.12	42.04	nr	**133.24**
300 mm dia.	81.24	99.80	1.12	42.04	nr	**141.84**
350 mm dia.	90.55	111.24	1.12	42.04	nr	**153.28**
400 mm dia.	101.54	124.74	1.12	42.04	nr	**166.78**

38 MECHANICAL/COOLING/HEATING SYSTEMS

Item	Net Price £	Material £	Labour hours	Labour £	Unit	Total rate £
THERMAL INSULATION – cont						
Extra over for fittings plantroom insulation – cont						
Pumps						
15 mm dia.	31.90	39.19	2.34	87.82	nr	**127.01**
20 mm dia.	33.30	40.91	2.34	87.82	nr	**128.73**
25 mm dia.	35.62	43.76	2.34	87.82	nr	**131.58**
32 mm dia.	37.48	46.04	2.34	87.82	nr	**133.86**
40 mm dia.	39.76	48.85	2.34	87.82	nr	**136.67**
50 mm dia.	44.32	54.44	2.34	87.82	nr	**142.26**
65 mm dia.	49.51	60.83	2.34	87.82	nr	**148.65**
80 mm dia.	53.43	65.64	2.76	103.58	nr	**169.22**
100 mm dia.	66.94	82.24	2.76	103.58	nr	**185.82**
125 mm dia.	74.88	91.99	2.76	103.58	nr	**195.57**
150 mm dia.	87.43	107.40	2.76	103.58	nr	**210.98**
200 mm dia.	117.28	144.08	3.36	126.10	nr	**270.18**
250 mm dia.	137.49	168.90	3.36	126.10	nr	**295.00**
300 mm dia.	196.72	241.67	3.36	126.10	nr	**367.77**
350 mm dia.	167.68	206.00	3.36	126.10	nr	**332.10**
400 mm dia.	188.04	231.01	3.36	126.10	nr	**357.11**
Expansion bellows						
15 mm dia.	25.52	31.35	1.05	39.41	nr	**70.76**
20 mm dia.	26.66	32.75	1.05	39.41	nr	**72.16**
25 mm dia.	28.49	35.00	1.05	39.41	nr	**74.41**
32 mm dia.	29.98	36.83	1.05	39.41	nr	**76.24**
40 mm dia.	31.82	39.09	1.05	39.41	nr	**78.50**
50 mm dia.	35.46	43.56	1.05	39.41	nr	**82.97**
65 mm dia.	39.61	48.66	1.05	39.41	nr	**88.07**
80 mm dia.	42.74	52.51	1.26	47.28	nr	**99.79**
100 mm dia.	53.55	65.78	1.26	47.28	nr	**113.06**
125 mm dia.	59.89	73.57	1.26	47.28	nr	**120.85**
150 mm dia.	69.93	85.91	1.26	47.28	nr	**133.19**
200 mm dia.	93.81	115.25	1.53	57.42	nr	**172.67**
250 mm dia.	110.00	135.13	1.53	57.42	nr	**192.55**
300 mm dia.	120.35	147.85	1.53	57.42	nr	**205.27**
350 mm dia.	134.14	164.79	1.53	57.42	nr	**222.21**
400 mm dia.	150.43	184.80	1.53	57.42	nr	**242.22**
50 mm thick						
15 mm dia.	12.26	15.06	0.44	16.51	m	**31.57**
20 mm dia.	12.92	15.88	0.44	16.51	m	**32.39**
25 mm dia.	13.68	16.81	0.44	16.51	m	**33.32**
32 mm dia.	14.39	17.68	0.44	16.51	m	**34.19**
40 mm dia.	15.05	18.49	0.44	16.51	m	**35.00**
50 mm dia.	16.60	20.39	0.44	16.51	m	**36.90**
65 mm dia.	17.85	21.92	0.44	16.51	m	**38.43**
80 mm dia.	19.16	23.54	0.52	19.52	m	**43.06**
100 mm dia.	23.68	29.10	0.52	19.52	m	**48.62**
125 mm dia.	26.20	32.18	0.52	19.52	m	**51.70**
150 mm dia.	30.10	36.98	0.52	19.52	m	**56.50**

38 MECHANICAL/COOLING/HEATING SYSTEMS

Item	Net Price £	Material £	Labour hours	Labour £	Unit	**Total rate £**
200 mm dia.	39.47	48.49	0.60	22.52	m	**71.01**
250 mm dia.	45.61	56.03	0.60	22.52	m	**78.55**
300 mm dia.	49.65	60.99	0.60	22.52	m	**83.51**
350 mm dia.	55.34	67.99	0.60	22.52	m	**90.51**
400 mm dia.	61.90	76.04	0.60	22.52	m	**98.56**
Extra over for fittings plantroom insulation						
Flange/union						
15 mm dia.	14.44	17.73	0.58	21.77	nr	**39.50**
20 mm dia.	15.24	18.72	0.58	21.77	nr	**40.49**
25 mm dia.	16.14	19.82	0.58	21.77	nr	**41.59**
32 mm dia.	16.94	20.81	0.58	21.77	nr	**42.58**
40 mm dia.	17.75	21.81	0.58	21.77	nr	**43.58**
50 mm dia.	19.70	24.20	0.58	21.77	nr	**45.97**
65 mm dia.	21.29	26.16	0.58	21.77	nr	**47.93**
80 mm dia.	22.81	28.02	0.67	25.15	nr	**53.17**
100 mm dia.	28.54	35.06	0.67	25.15	nr	**60.21**
125 mm dia.	31.73	38.98	0.67	25.15	nr	**64.13**
150 mm dia.	36.51	44.86	0.67	25.15	nr	**70.01**
200 mm dia.	48.54	59.63	0.87	32.65	nr	**92.28**
250 mm dia.	56.09	68.91	0.87	32.65	nr	**101.56**
300 mm dia.	60.52	74.35	0.87	32.65	nr	**107.00**
350 mm dia.	67.30	82.68	0.87	32.65	nr	**115.33**
400 mm dia.	75.08	92.23	0.87	32.65	nr	**124.88**
Bend						
15 mm dia.	6.75	8.30	0.44	16.51	nr	**24.81**
20 mm dia.	7.11	8.74	0.44	16.51	nr	**25.25**
25 mm dia.	7.53	9.25	0.44	16.51	nr	**25.76**
32 mm dia.	7.92	9.73	0.44	16.51	nr	**26.24**
40 mm dia.	8.26	10.14	0.44	16.51	nr	**26.65**
50 mm dia.	9.13	11.21	0.44	16.51	nr	**27.72**
65 mm dia.	9.82	12.06	0.44	16.51	nr	**28.57**
80 mm dia.	10.53	12.94	0.52	19.52	nr	**32.46**
100 mm dia.	13.01	15.98	0.52	19.52	nr	**35.50**
125 mm dia.	14.41	17.70	0.52	19.52	nr	**37.22**
150 mm dia.	16.57	20.36	0.52	19.52	nr	**39.88**
200 mm dia.	21.70	26.66	0.60	22.52	nr	**49.18**
250 mm dia.	25.09	30.83	0.60	22.52	nr	**53.35**
300 mm dia.	27.30	33.54	0.60	22.52	nr	**56.06**
350 mm dia.	30.43	37.38	0.60	22.52	nr	**59.90**
400 mm dia.	34.03	41.81	0.60	22.52	nr	**64.33**
Tee						
15 mm dia.	4.05	4.98	0.44	16.51	nr	**21.49**
20 mm dia.	4.26	5.23	0.44	16.51	nr	**21.74**
25 mm dia.	4.51	5.54	0.44	16.51	nr	**22.05**
32 mm dia.	4.75	5.84	0.44	16.51	nr	**22.35**
40 mm dia.	4.97	6.10	0.44	16.51	nr	**22.61**
50 mm dia.	5.48	6.73	0.44	16.51	nr	**23.24**
65 mm dia.	5.88	7.22	0.44	16.51	nr	**23.73**

38 MECHANICAL/COOLING/HEATING SYSTEMS

Item	Net Price £	Material £	Labour hours	Labour £	Unit	Total rate £
THERMAL INSULATION – cont						
Extra over for fittings plantroom insulation – cont						
Tee – cont						
80 mm dia.	6.32	7.76	0.52	19.52	nr	**27.28**
100 mm dia.	7.82	9.61	0.52	19.52	nr	**29.13**
125 mm dia.	8.64	10.62	0.52	19.52	nr	**30.14**
150 mm dia.	9.92	12.19	0.52	19.52	nr	**31.71**
200 mm dia.	13.01	15.98	0.60	22.52	nr	**38.50**
250 mm dia.	15.06	18.50	0.60	22.52	nr	**41.02**
300 mm dia.	16.39	20.14	0.60	22.52	nr	**42.66**
350 mm dia.	18.27	22.45	0.60	22.52	nr	**44.97**
400 mm dia.	20.42	25.08	0.60	22.52	nr	**47.60**
Valves						
15 mm dia.	22.95	28.19	0.78	29.27	nr	**57.46**
20 mm dia.	24.18	29.70	0.78	29.27	nr	**58.97**
25 mm dia.	25.63	31.49	0.78	29.27	nr	**60.76**
32 mm dia.	26.91	33.05	0.78	29.27	nr	**62.32**
40 mm dia.	28.19	34.63	0.78	29.27	nr	**63.90**
50 mm dia.	31.28	38.43	0.78	29.27	nr	**67.70**
65 mm dia.	33.79	41.51	0.78	29.27	nr	**70.78**
80 mm dia.	36.25	44.53	0.92	34.53	nr	**79.06**
100 mm dia.	45.32	55.67	0.92	34.53	nr	**90.20**
125 mm dia.	50.38	61.89	0.92	34.53	nr	**96.42**
150 mm dia.	57.99	71.24	0.92	34.53	nr	**105.77**
200 mm dia.	77.11	94.73	1.12	42.04	nr	**136.77**
250 mm dia.	89.09	109.45	1.12	42.04	nr	**151.49**
300 mm dia.	96.51	118.57	1.12	42.04	nr	**160.61**
350 mm dia.	106.87	131.29	1.12	42.04	nr	**173.33**
400 mm dia.	119.23	146.48	1.12	42.04	nr	**188.52**
Pumps						
15 mm dia.	42.49	52.20	2.34	87.82	nr	**140.02**
20 mm dia.	44.79	55.02	2.34	87.82	nr	**142.84**
25 mm dia.	47.45	58.30	2.34	87.82	nr	**146.12**
32 mm dia.	49.83	61.22	2.34	87.82	nr	**149.04**
40 mm dia.	52.21	64.14	2.34	87.82	nr	**151.96**
50 mm dia.	57.90	71.13	2.34	87.82	nr	**158.95**
65 mm dia.	82.79	101.70	2.34	87.82	nr	**189.52**
80 mm dia.	67.14	82.48	2.76	103.58	nr	**186.06**
100 mm dia.	83.92	103.10	2.76	103.58	nr	**206.68**
125 mm dia.	93.33	114.66	2.76	103.58	nr	**218.24**
150 mm dia.	107.39	131.93	2.76	103.58	nr	**235.51**
200 mm dia.	142.78	175.40	3.36	126.10	nr	**301.50**
250 mm dia.	164.98	202.68	3.36	126.10	nr	**328.78**
300 mm dia.	178.03	218.71	3.36	126.10	nr	**344.81**
350 mm dia.	197.90	243.12	3.36	126.10	nr	**369.22**
400 mm dia.	220.81	271.27	3.36	126.10	nr	**397.37**

38 MECHANICAL/COOLING/HEATING SYSTEMS

Item	Net Price £	Material £	Labour hours	Labour £	Unit	**Total rate £**
Expansion bellows						
15 mm dia.	33.98	41.75	1.05	39.41	nr	**81.16**
20 mm dia.	35.83	44.02	1.05	39.41	nr	**83.43**
25 mm dia.	37.96	46.63	1.05	39.41	nr	**86.04**
32 mm dia.	39.86	48.97	1.05	39.41	nr	**88.38**
40 mm dia.	41.75	51.29	1.05	39.41	nr	**90.70**
50 mm dia.	46.34	56.93	1.05	39.41	nr	**96.34**
65 mm dia.	50.06	61.50	1.05	39.41	nr	**100.91**
80 mm dia.	53.72	65.99	1.26	47.28	nr	**113.27**
100 mm dia.	67.13	82.47	1.26	47.28	nr	**129.75**
125 mm dia.	74.66	91.72	1.26	47.28	nr	**139.00**
150 mm dia.	85.93	105.57	1.26	47.28	nr	**152.85**
200 mm dia.	114.23	140.33	1.53	57.42	nr	**197.75**
250 mm dia.	131.98	162.14	1.53	57.42	nr	**219.56**
300 mm dia.	142.44	174.98	1.53	57.42	nr	**232.40**
350 mm dia.	158.32	194.49	1.53	57.42	nr	**251.91**
400 mm dia.	176.64	217.00	1.53	57.42	nr	**274.42**
Mineral fibre sectional insulation; bright class O foil faced; bright class O foil taped joints; 0.8 mm polyisobutylene sheeting; welded joints						
External pipework						
20 mm thick						
15 mm dia.	5.51	6.77	0.30	11.26	m	**18.03**
20 mm dia.	5.86	7.20	0.30	11.26	m	**18.46**
25 mm dia.	6.33	7.78	0.30	11.26	m	**19.04**
32 mm dia.	6.94	8.53	0.30	11.26	m	**19.79**
40 mm dia.	7.41	9.10	0.30	11.26	m	**20.36**
50 mm dia.	8.37	10.28	0.30	11.26	m	**21.54**
Extra over for fittings external insulation						
Flange/union						
15 mm dia.	8.43	10.35	0.75	28.14	nr	**38.49**
20 mm dia.	8.95	10.99	0.75	28.14	nr	**39.13**
25 mm dia.	9.61	11.80	0.75	28.14	nr	**39.94**
32 mm dia.	10.48	12.87	0.75	28.14	nr	**41.01**
40 mm dia.	11.06	13.59	0.75	28.14	nr	**41.73**
50 mm dia.	12.40	15.24	0.75	28.14	nr	**43.38**
Bends						
15 mm dia.	1.39	1.71	0.30	11.26	nr	**12.97**
20 mm dia.	1.47	1.81	0.30	11.26	nr	**13.07**
25 mm dia.	1.59	1.95	0.30	11.26	nr	**13.21**
32 mm dia.	1.74	2.14	0.30	11.26	nr	**13.40**
40 mm dia.	1.85	2.27	0.30	11.26	nr	**13.53**
50 mm dia.	2.09	2.57	0.30	11.26	nr	**13.83**
Tees						
15 mm dia.	1.39	1.71	0.30	11.26	nr	**12.97**
20 mm dia.	1.47	1.81	0.30	11.26	nr	**13.07**
25 mm dia.	1.59	1.95	0.30	11.26	nr	**13.21**

38 MECHANICAL/COOLING/HEATING SYSTEMS

Item	Net Price £	Material £	Labour hours	Labour £	Unit	Total rate £
THERMAL INSULATION – cont						
Extra over for fittings external insulation – cont						
Tees – cont						
32 mm dia.	1.74	2.14	0.30	11.26	nr	**13.40**
40 mm dia.	1.85	2.27	0.30	11.26	nr	**13.53**
50 mm dia.	2.09	2.57	0.30	11.26	nr	**13.83**
Valves						
15 mm dia.	13.39	16.45	1.03	38.66	nr	**55.11**
20 mm dia.	14.23	17.48	1.03	38.66	nr	**56.14**
25 mm dia.	15.27	18.76	1.03	38.66	nr	**57.42**
32 mm dia.	16.62	20.42	1.03	38.66	nr	**59.08**
40 mm dia.	17.55	21.56	1.03	38.66	nr	**60.22**
50 mm dia.	19.67	24.16	1.03	38.66	nr	**62.82**
Expansion bellows						
15 mm dia.	19.83	24.36	1.42	53.29	nr	**77.65**
20 mm dia.	21.08	25.89	1.42	53.29	nr	**79.18**
25 mm dia.	22.65	27.82	1.42	53.29	nr	**81.11**
32 mm dia.	24.65	30.28	1.42	53.29	nr	**83.57**
40 mm dia.	26.01	31.95	1.42	53.29	nr	**85.24**
50 mm dia.	29.14	35.79	1.42	53.29	nr	**89.08**
25 mm thick						
15 mm dia.	6.07	7.46	0.30	11.26	m	**18.72**
20 mm dia.	6.51	8.00	0.30	11.26	m	**19.26**
25 mm dia.	7.17	8.81	0.30	11.26	m	**20.07**
32 mm dia.	7.76	9.53	0.30	11.26	m	**20.79**
40 mm dia.	8.25	10.13	0.30	11.26	m	**21.39**
50 mm dia.	9.32	11.45	0.30	11.26	m	**22.71**
65 mm dia.	10.57	12.99	0.30	11.26	m	**24.25**
80 mm dia.	11.57	14.22	0.40	15.02	m	**29.24**
100 mm dia.	14.60	17.93	0.40	15.02	m	**32.95**
125 mm dia.	16.82	20.66	0.40	15.02	m	**35.68**
150 mm dia.	19.73	24.23	0.40	15.02	m	**39.25**
200 mm dia.	26.75	32.86	0.50	18.77	m	**51.63**
250 mm dia.	31.94	39.24	0.50	18.77	m	**58.01**
300 mm dia.	34.74	42.68	0.50	18.77	m	**61.45**
Extra over for fittings external insulation						
Flange/union						
15 mm dia.	9.27	11.39	0.75	28.14	nr	**39.53**
20 mm dia.	10.00	12.28	0.75	28.14	nr	**40.42**
25 mm dia.	10.97	13.47	0.75	28.14	nr	**41.61**
32 mm dia.	11.74	14.43	0.75	28.14	nr	**42.57**
40 mm dia.	12.56	15.43	0.75	28.14	nr	**43.57**
50 mm dia.	14.09	17.31	0.75	28.14	nr	**45.45**
65 mm dia.	15.93	19.57	0.75	28.14	nr	**47.71**
80 mm dia.	17.32	21.27	0.89	33.40	nr	**54.67**
100 mm dia.	21.53	26.45	0.89	33.40	nr	**59.85**
125 mm dia.	24.94	30.64	0.89	33.40	nr	**64.04**
150 mm dia.	29.02	35.65	0.89	33.40	nr	**69.05**

38 MECHANICAL/COOLING/HEATING SYSTEMS

Item	Net Price £	Material £	Labour hours	Labour £	Unit	**Total rate £**
200 mm dia.	39.26	48.23	1.15	43.16	nr	**91.39**
250 mm dia.	46.75	57.43	1.15	43.16	nr	**100.59**
300 mm dia.	51.92	63.79	1.15	43.16	nr	**106.95**
Bends						
15 mm dia.	1.51	1.86	0.30	11.26	nr	**13.12**
20 mm dia.	1.64	2.02	0.30	11.26	nr	**13.28**
25 mm dia.	1.78	2.18	0.30	11.26	nr	**13.44**
32 mm dia.	1.93	2.37	0.30	11.26	nr	**13.63**
40 mm dia.	2.07	2.54	0.30	11.26	nr	**13.80**
50 mm dia.	2.33	2.87	0.30	11.26	nr	**14.13**
65 mm dia.	2.66	3.27	0.30	11.26	nr	**14.53**
80 mm dia.	2.89	3.55	0.40	15.02	nr	**18.57**
100 mm dia.	3.65	4.48	0.40	15.02	nr	**19.50**
125 mm dia.	4.20	5.16	0.40	15.02	nr	**20.18**
150 mm dia.	4.94	6.07	0.40	15.02	nr	**21.09**
200 mm dia.	6.68	8.21	0.50	18.77	nr	**26.98**
250 mm dia.	7.99	9.82	0.50	18.77	nr	**28.59**
300 mm dia.	8.68	10.67	0.50	18.77	nr	**29.44**
Tees						
15 mm dia.	1.51	1.86	0.30	11.26	nr	**13.12**
20 mm dia.	1.64	2.02	0.30	11.26	nr	**13.28**
25 mm dia.	1.78	2.18	0.30	11.26	nr	**13.44**
32 mm dia.	1.93	2.37	0.30	11.26	nr	**13.63**
40 mm dia.	2.07	2.54	0.30	11.26	nr	**13.80**
50 mm dia.	2.33	2.87	0.30	11.26	nr	**14.13**
65 mm dia.	2.66	3.27	0.30	11.26	nr	**14.53**
80 mm dia.	2.89	3.55	0.40	15.02	nr	**18.57**
100 mm dia.	3.65	4.48	0.40	15.02	nr	**19.50**
125 mm dia.	4.20	5.16	0.40	15.02	nr	**20.18**
150 mm dia.	4.94	6.07	0.40	15.02	nr	**21.09**
200 mm dia.	6.68	8.21	0.50	18.77	nr	**26.98**
250 mm dia.	7.99	9.82	0.50	18.77	nr	**28.59**
300 mm dia.	8.68	10.67	0.50	18.77	nr	**29.44**
Valves						
15 mm dia.	14.74	18.11	1.03	38.66	nr	**56.77**
20 mm dia.	15.89	19.52	1.03	38.66	nr	**58.18**
25 mm dia.	17.43	21.41	1.03	38.66	nr	**60.07**
32 mm dia.	18.67	22.93	1.03	38.66	nr	**61.59**
40 mm dia.	19.95	24.51	1.03	38.66	nr	**63.17**
50 mm dia.	22.40	27.52	1.03	38.66	nr	**66.18**
65 mm dia.	25.33	31.12	1.03	38.66	nr	**69.78**
80 mm dia.	27.52	33.81	1.25	46.91	nr	**80.72**
100 mm dia.	34.22	42.04	1.25	46.91	nr	**88.95**
125 mm dia.	39.60	48.65	1.25	46.91	nr	**95.56**
150 mm dia.	46.10	56.64	1.25	46.91	nr	**103.55**
200 mm dia.	62.38	76.63	1.55	58.17	nr	**134.80**
250 mm dia.	74.24	91.20	1.55	58.17	nr	**149.37**
300 mm dia.	82.47	101.31	1.55	58.17	nr	**159.48**

38 MECHANICAL/COOLING/HEATING SYSTEMS

Item	Net Price £	Material £	Labour hours	Labour £	Unit	**Total rate £**
THERMAL INSULATION – cont						
Extra over for fittings external insulation – cont						
Expansion bellows						
15 mm dia.	21.85	26.84	1.42	53.29	nr	**80.13**
20 mm dia.	23.54	28.92	1.42	53.29	nr	**82.21**
25 mm dia.	25.82	31.72	1.42	53.29	nr	**85.01**
32 mm dia.	27.68	34.01	1.42	53.29	nr	**87.30**
40 mm dia.	29.57	36.33	1.42	53.29	nr	**89.62**
50 mm dia.	33.15	40.73	1.42	53.29	nr	**94.02**
65 mm dia.	37.51	46.08	1.42	53.29	nr	**99.37**
80 mm dia.	40.77	50.09	1.75	65.68	nr	**115.77**
100 mm dia.	50.69	62.28	1.75	65.68	nr	**127.96**
125 mm dia.	58.67	72.07	1.75	65.68	nr	**137.75**
150 mm dia.	68.29	83.90	1.75	65.68	nr	**149.58**
200 mm dia.	92.41	113.53	2.17	81.44	nr	**194.97**
250 mm dia.	109.97	135.09	2.17	81.44	nr	**216.53**
300 mm dia.	122.17	150.09	3.17	118.97	nr	**269.06**
30 mm thick						
15 mm dia.	8.79	10.79	0.30	11.26	m	**22.05**
20 mm dia.	7.95	9.77	0.30	11.26	m	**21.03**
25 mm dia.	8.43	10.35	0.30	11.26	m	**21.61**
32 mm dia.	9.15	11.25	0.30	11.26	m	**22.51**
40 mm dia.	9.64	11.84	0.30	11.26	m	**23.10**
50 mm dia.	10.85	13.32	0.30	11.26	m	**24.58**
65 mm dia.	12.20	14.98	0.30	11.26	m	**26.24**
80 mm dia.	13.27	16.31	0.40	15.02	m	**31.33**
100 mm dia.	16.59	20.38	0.40	15.02	m	**35.40**
125 mm dia.	18.99	23.33	0.40	15.02	m	**38.35**
150 mm dia.	22.05	27.09	0.40	15.02	m	**42.11**
200 mm dia.	29.70	36.49	0.50	18.77	m	**55.26**
250 mm dia.	35.24	43.29	0.50	18.77	m	**62.06**
300 mm dia.	38.01	46.69	0.50	18.77	m	**65.46**
350 mm dia.	41.57	51.07	0.50	18.77	m	**69.84**
Extra over for fittings external insulation						
Flange/union						
15 mm dia.	11.35	13.94	0.75	28.14	nr	**42.08**
20 mm dia.	12.02	14.76	0.75	28.14	nr	**42.90**
25 mm dia.	12.75	15.67	0.75	28.14	nr	**43.81**
32 mm dia.	13.91	17.08	0.75	28.14	nr	**45.22**
40 mm dia.	14.53	17.85	0.75	28.14	nr	**45.99**
50 mm dia.	16.22	19.93	0.75	28.14	nr	**48.07**
65 mm dia.	18.26	22.43	0.75	28.14	nr	**50.57**
80 mm dia.	19.90	24.44	0.89	33.40	nr	**57.84**
100 mm dia.	24.49	30.08	0.89	33.40	nr	**63.48**
125 mm dia.	27.98	34.38	0.89	33.40	nr	**67.78**
150 mm dia.	32.51	39.94	0.89	33.40	nr	**73.34**

38 MECHANICAL/COOLING/HEATING SYSTEMS

Item	Net Price £	Material £	Labour hours	Labour £	Unit	**Total rate £**
200 mm dia.	43.43	53.35	1.15	43.16	nr	**96.51**
250 mm dia.	51.43	63.18	1.15	43.16	nr	**106.34**
300 mm dia.	56.36	69.24	1.15	43.16	nr	**112.40**
350 mm dia.	62.26	76.48	1.15	43.16	nr	**119.64**
Bends						
15 mm dia.	1.86	2.29	0.30	11.26	nr	**13.55**
20 mm dia.	1.98	2.44	0.30	11.26	nr	**13.70**
25 mm dia.	2.12	2.60	0.30	11.26	nr	**13.86**
32 mm dia.	2.28	2.80	0.30	11.26	nr	**14.06**
40 mm dia.	2.41	2.96	0.30	11.26	nr	**14.22**
50 mm dia.	2.72	3.34	0.30	11.26	nr	**14.60**
65 mm dia.	3.06	3.76	0.30	11.26	nr	**15.02**
80 mm dia.	3.33	4.10	0.40	15.02	nr	**19.12**
100 mm dia.	4.16	5.11	0.40	15.02	nr	**20.13**
125 mm dia.	4.75	5.84	0.40	15.02	nr	**20.86**
150 mm dia.	5.52	6.78	0.40	15.02	nr	**21.80**
200 mm dia.	7.43	9.12	0.50	18.77	nr	**27.89**
250 mm dia.	8.82	10.84	0.50	18.77	nr	**29.61**
300 mm dia.	9.50	11.67	0.50	18.77	nr	**30.44**
350 mm dia.	10.38	12.75	0.50	18.77	nr	**31.52**
Tees						
15 mm dia.	1.86	2.29	0.30	11.26	nr	**13.55**
20 mm dia.	1.98	2.44	0.30	11.26	nr	**13.70**
25 mm dia.	2.12	2.60	0.30	11.26	nr	**13.86**
32 mm dia.	2.28	2.80	0.30	11.26	nr	**14.06**
40 mm dia.	2.41	2.96	0.30	11.26	nr	**14.22**
50 mm dia.	2.72	3.34	0.30	11.26	nr	**14.60**
65 mm dia.	3.06	3.76	0.30	11.26	nr	**15.02**
80 mm dia.	3.33	4.10	0.40	15.02	nr	**19.12**
100 mm dia.	4.16	5.11	0.40	15.02	nr	**20.13**
125 mm dia.	4.75	5.84	0.40	15.02	nr	**20.86**
150 mm dia.	5.52	6.78	0.40	15.02	nr	**21.80**
200 mm dia.	7.43	9.12	0.50	18.77	nr	**27.89**
250 mm dia.	8.82	10.84	0.50	18.77	nr	**29.61**
300 mm dia.	9.50	11.67	0.50	18.77	nr	**30.44**
350 mm dia.	10.38	12.75	0.50	18.77	nr	**31.52**
Valves						
15 mm dia.	18.03	22.16	1.03	38.66	nr	**60.82**
20 mm dia.	19.10	23.47	1.03	38.66	nr	**62.13**
25 mm dia.	20.26	24.89	1.03	38.66	nr	**63.55**
32 mm dia.	22.06	27.10	1.03	38.66	nr	**65.76**
40 mm dia.	23.10	28.38	1.03	38.66	nr	**67.04**
50 mm dia.	25.75	31.64	1.03	38.66	nr	**70.30**
65 mm dia.	28.99	35.62	1.03	38.66	nr	**74.28**
80 mm dia.	31.59	38.81	1.25	46.91	nr	**85.72**
100 mm dia.	38.87	47.75	1.25	46.91	nr	**94.66**
125 mm dia.	44.43	54.58	1.25	46.91	nr	**101.49**
150 mm dia.	51.61	63.40	1.25	46.91	nr	**110.31**
200 mm dia.	69.00	84.77	1.55	58.17	nr	**142.94**
250 mm dia.	81.70	100.37	1.55	58.17	nr	**158.54**
300 mm dia.	89.51	109.97	1.55	58.17	nr	**168.14**
350 mm dia.	98.88	121.47	1.55	58.17	nr	**179.64**

38 MECHANICAL/COOLING/HEATING SYSTEMS

Item	Net Price £	Material £	Labour hours	Labour £	Unit	Total rate £
THERMAL INSULATION – cont						
Extra over for fittings external insulation – cont						
Expansion bellows						
15 mm dia.	26.70	32.80	1.42	53.29	nr	**86.09**
20 mm dia.	28.31	34.78	1.42	53.29	nr	**88.07**
25 mm dia.	30.00	36.85	1.42	53.29	nr	**90.14**
32 mm dia.	32.71	40.18	1.42	53.29	nr	**93.47**
40 mm dia.	34.22	42.04	1.42	53.29	nr	**95.33**
50 mm dia.	38.14	46.85	1.42	53.29	nr	**100.14**
65 mm dia.	42.95	52.76	1.42	53.29	nr	**106.05**
80 mm dia.	46.82	57.52	1.75	65.68	nr	**123.20**
100 mm dia.	57.59	70.75	1.75	65.68	nr	**136.43**
125 mm dia.	65.82	80.86	1.75	65.68	nr	**146.54**
150 mm dia.	76.46	93.93	1.75	65.68	nr	**159.61**
200 mm dia.	102.21	125.57	2.17	81.44	nr	**207.01**
250 mm dia.	121.04	148.70	2.17	81.44	nr	**230.14**
300 mm dia.	132.60	162.90	2.17	81.44	nr	**244.34**
350 mm dia.	146.50	179.98	2.17	81.44	nr	**261.42**
40 mm thick						
15 mm dia.	9.35	11.49	0.30	11.26	m	**22.75**
20 mm dia.	9.68	11.90	0.30	11.26	m	**23.16**
25 mm dia.	10.33	12.69	0.30	11.26	m	**23.95**
32 mm dia.	11.00	13.51	0.30	11.26	m	**24.77**
40 mm dia.	11.50	14.13	0.30	11.26	m	**25.39**
50 mm dia.	12.83	15.76	0.30	11.26	m	**27.02**
65 mm dia.	14.31	17.58	0.30	11.26	m	**28.84**
80 mm dia.	15.51	19.06	0.40	15.02	m	**34.08**
100 mm dia.	19.37	23.79	0.40	15.02	m	**38.81**
125 mm dia.	21.82	26.81	0.40	15.02	m	**41.83**
150 mm dia.	25.19	30.94	0.40	15.02	m	**45.96**
200 mm dia.	33.60	41.28	0.50	18.77	m	**60.05**
250 mm dia.	39.35	48.34	0.50	18.77	m	**67.11**
300 mm dia.	42.60	52.33	0.50	18.77	m	**71.10**
350 mm dia.	46.82	57.52	0.50	18.77	m	**76.29**
400 mm dia.	52.11	64.02	0.50	18.77	m	**82.79**
Extra over for fittings external insulation						
Flange/union						
15 mm dia.	13.97	17.16	0.75	28.14	nr	**45.30**
20 mm dia.	14.59	17.92	0.75	28.14	nr	**46.06**
25 mm dia.	15.56	19.12	0.75	28.14	nr	**47.26**
32 mm dia.	16.39	20.14	0.75	28.14	nr	**48.28**
40 mm dia.	17.31	21.26	0.75	28.14	nr	**49.40**
50 mm dia.	19.18	23.56	0.75	28.14	nr	**51.70**
65 mm dia.	21.37	26.25	0.75	28.14	nr	**54.39**
80 mm dia.	23.03	28.30	0.89	33.40	nr	**61.70**
100 mm dia.	28.36	34.84	0.89	33.40	nr	**68.24**
125 mm dia.	31.73	38.98	0.89	33.40	nr	**72.38**
150 mm dia.	36.77	45.17	0.89	33.40	nr	**78.57**

38 MECHANICAL/COOLING/HEATING SYSTEMS

Item	Net Price £	Material £	Labour hours	Labour £	Unit	**Total rate £**
200 mm dia.	48.48	59.56	1.15	43.16	nr	**102.72**
250 mm dia.	56.83	69.81	1.15	43.16	nr	**112.97**
300 mm dia.	62.45	76.72	1.15	43.16	nr	**119.88**
350 mm dia.	69.26	85.08	1.15	43.16	nr	**128.24**
400 mm dia.	77.57	95.30	1.15	43.16	nr	**138.46**
Bends						
15 mm dia.	2.33	2.87	0.30	11.26	nr	**14.13**
20 mm dia.	2.42	2.97	0.30	11.26	nr	**14.23**
25 mm dia.	2.58	3.17	0.30	11.26	nr	**14.43**
32 mm dia.	2.75	3.38	0.30	11.26	nr	**14.64**
40 mm dia.	2.89	3.55	0.30	11.26	nr	**14.81**
50 mm dia.	3.23	3.97	0.30	11.26	nr	**15.23**
65 mm dia.	3.59	4.41	0.30	11.26	nr	**15.67**
80 mm dia.	3.89	4.78	0.40	15.02	nr	**19.80**
100 mm dia.	4.84	5.94	0.40	15.02	nr	**20.96**
125 mm dia.	5.46	6.71	0.40	15.02	nr	**21.73**
150 mm dia.	6.31	7.75	0.40	15.02	nr	**22.77**
200 mm dia.	8.40	10.32	0.50	18.77	nr	**29.09**
250 mm dia.	9.83	12.07	0.50	18.77	nr	**30.84**
300 mm dia.	10.65	13.08	0.50	18.77	nr	**31.85**
350 mm dia.	11.69	14.36	0.50	18.77	nr	**33.13**
400 mm dia.	13.03	16.01	0.50	18.77	nr	**34.78**
Tees						
15 mm dia.	2.33	2.87	0.30	11.26	nr	**14.13**
20 mm dia.	2.42	2.97	0.30	11.26	nr	**14.23**
25 mm dia.	2.58	3.17	0.30	11.26	nr	**14.43**
32 mm dia.	2.75	3.38	0.30	11.26	nr	**14.64**
40 mm dia.	2.89	3.55	0.30	11.26	nr	**14.81**
50 mm dia.	3.23	3.97	0.30	11.26	nr	**15.23**
65 mm dia.	3.59	4.41	0.30	11.26	nr	**15.67**
80 mm dia.	3.89	4.78	0.40	15.02	nr	**19.80**
100 mm dia.	4.84	5.94	0.40	15.02	nr	**20.96**
125 mm dia.	5.46	6.71	0.40	15.02	nr	**21.73**
150 mm dia.	6.31	7.75	0.40	15.02	nr	**22.77**
200 mm dia.	8.40	10.32	0.50	18.77	nr	**29.09**
250 mm dia.	9.83	12.07	0.50	18.77	nr	**30.84**
300 mm dia.	10.65	13.08	0.50	18.77	nr	**31.85**
350 mm dia.	11.69	14.36	0.50	18.77	nr	**33.13**
400 mm dia.	13.03	16.01	0.50	18.77	nr	**34.78**
Valves						
15 mm dia.	22.18	27.25	1.03	38.66	nr	**65.91**
20 mm dia.	23.17	28.47	1.03	38.66	nr	**67.13**
25 mm dia.	24.70	30.35	1.03	38.66	nr	**69.01**
32 mm dia.	26.02	31.96	1.03	38.66	nr	**70.62**
40 mm dia.	27.51	33.80	1.03	38.66	nr	**72.46**
50 mm dia.	30.48	37.44	1.03	38.66	nr	**76.10**
65 mm dia.	33.93	41.68	1.03	38.66	nr	**80.34**
80 mm dia.	36.58	44.94	1.25	46.91	nr	**91.85**
100 mm dia.	45.06	55.36	1.25	46.91	nr	**102.27**
125 mm dia.	50.39	61.91	1.25	46.91	nr	**108.82**
150 mm dia.	58.41	71.76	1.25	46.91	nr	**118.67**

38 MECHANICAL/COOLING/HEATING SYSTEMS

Item	Net Price £	Material £	Labour hours	Labour £	Unit	Total rate £
THERMAL INSULATION – cont						
Extra over for fittings external insulation – cont						
Valves – cont						
200 mm dia.	77.00	94.59	1.55	58.17	nr	**152.76**
250 mm dia.	90.22	110.84	1.55	58.17	nr	**169.01**
300 mm dia.	99.19	121.85	1.55	58.17	nr	**180.02**
350 mm dia.	110.01	135.15	1.55	58.17	nr	**193.32**
400 mm dia.	123.17	151.32	1.55	58.17	nr	**209.49**
Expansion bellows						
15 mm dia.	32.85	40.35	1.42	53.29	nr	**93.64**
20 mm dia.	34.34	42.19	1.42	53.29	nr	**95.48**
25 mm dia.	36.59	44.95	1.42	53.29	nr	**98.24**
32 mm dia.	38.54	47.34	1.42	53.29	nr	**100.63**
40 mm dia.	40.75	50.06	1.42	53.29	nr	**103.35**
50 mm dia.	45.17	55.49	1.42	53.29	nr	**108.78**
65 mm dia.	50.27	61.76	1.42	53.29	nr	**115.05**
80 mm dia.	54.19	66.57	1.75	65.68	nr	**132.25**
100 mm dia.	66.75	82.00	1.75	65.68	nr	**147.68**
125 mm dia.	74.66	91.72	1.75	65.68	nr	**157.40**
150 mm dia.	86.55	106.32	1.75	65.68	nr	**172.00**
200 mm dia.	114.08	140.14	2.17	81.44	nr	**221.58**
250 mm dia.	133.66	164.20	2.17	81.44	nr	**245.64**
300 mm dia.	146.94	180.52	2.17	81.44	nr	**261.96**
350 mm dia.	162.98	200.22	2.17	81.44	nr	**281.66**
400 mm dia.	182.49	224.19	2.17	81.44	nr	**305.63**
50 mm thick						
15 mm dia.	12.32	15.13	0.30	11.26	m	**26.39**
20 mm dia.	12.93	15.89	0.30	11.26	m	**27.15**
25 mm dia.	13.67	16.79	0.30	11.26	m	**28.05**
32 mm dia.	14.75	18.12	0.30	11.26	m	**29.38**
40 mm dia.	14.99	18.42	0.30	11.26	m	**29.68**
50 mm dia.	16.61	20.40	0.30	11.26	m	**31.66**
65 mm dia.	18.09	22.23	0.30	11.26	m	**33.49**
80 mm dia.	19.37	23.79	0.40	15.02	m	**38.81**
100 mm dia.	23.98	29.46	0.40	15.02	m	**44.48**
125 mm dia.	26.83	32.96	0.40	15.02	m	**47.98**
150 mm dia.	30.67	37.67	0.40	15.02	m	**52.69**
200 mm dia.	40.61	49.89	0.50	18.77	m	**68.66**
250 mm dia.	46.91	57.62	0.50	18.77	m	**76.39**
300 mm dia.	50.31	61.80	0.50	18.77	m	**80.57**
350 mm dia.	55.21	67.83	0.50	18.77	m	**86.60**
400 mm dia.	61.16	75.14	0.50	18.77	m	**93.91**

38 MECHANICAL/COOLING/HEATING SYSTEMS

Item	Net Price £	Material £	Labour hours	Labour £	Unit	Total rate £
Extra over for fittings external insulation						
Flange/union						
15 mm dia.	18.16	22.31	0.75	28.14	nr	**50.45**
20 mm dia.	19.10	23.47	0.75	28.14	nr	**51.61**
25 mm dia.	20.19	24.80	0.75	28.14	nr	**52.94**
32 mm dia.	21.19	26.03	0.75	28.14	nr	**54.17**
40 mm dia.	22.18	27.25	0.75	28.14	nr	**55.39**
50 mm dia.	24.44	30.02	0.75	28.14	nr	**58.16**
65 mm dia.	26.43	32.47	0.75	28.14	nr	**60.61**
80 mm dia.	28.34	34.82	0.89	33.40	nr	**68.22**
100 mm dia.	34.84	42.80	0.89	33.40	nr	**76.20**
125 mm dia.	38.72	47.57	0.89	33.40	nr	**80.97**
150 mm dia.	44.35	54.48	0.89	33.40	nr	**87.88**
200 mm dia.	58.01	71.26	1.15	43.16	nr	**114.42**
250 mm dia.	67.03	82.35	1.15	43.16	nr	**125.51**
300 mm dia.	72.74	89.37	1.15	43.16	nr	**132.53**
350 mm dia.	80.49	98.88	1.15	43.16	nr	**142.04**
400 mm dia.	89.69	110.19	1.15	43.16	nr	**153.35**
Bends						
15 mm dia.	3.09	3.80	0.30	11.26	nr	**15.06**
20 mm dia.	3.24	3.98	0.30	11.26	nr	**15.24**
25 mm dia.	3.41	4.19	0.30	11.26	nr	**15.45**
32 mm dia.	3.59	4.41	0.30	11.26	nr	**15.67**
40 mm dia.	3.75	4.61	0.30	11.26	nr	**15.87**
50 mm dia.	4.16	5.11	0.30	11.26	nr	**16.37**
65 mm dia.	4.52	5.55	0.30	11.26	nr	**16.81**
80 mm dia.	4.83	5.93	0.40	15.02	nr	**20.95**
100 mm dia.	6.00	7.37	0.40	15.02	nr	**22.39**
125 mm dia.	6.69	8.22	0.40	15.02	nr	**23.24**
150 mm dia.	7.67	9.42	0.40	15.02	nr	**24.44**
200 mm dia.	10.16	12.48	0.50	18.77	nr	**31.25**
250 mm dia.	11.73	14.41	0.50	18.77	nr	**33.18**
300 mm dia.	12.59	15.47	0.50	18.77	nr	**34.24**
350 mm dia.	13.81	16.97	0.50	18.77	nr	**35.74**
400 mm dia.	15.28	18.77	0.50	18.77	nr	**37.54**
Tees						
15 mm dia.	3.09	3.80	0.30	11.26	nr	**15.06**
20 mm dia.	3.24	3.98	0.30	11.26	nr	**15.24**
25 mm dia.	3.41	4.19	0.30	11.26	nr	**15.45**
32 mm dia.	3.59	4.41	0.30	11.26	nr	**15.67**
40 mm dia.	3.75	4.61	0.30	11.26	nr	**15.87**
50 mm dia.	4.16	5.11	0.30	11.26	nr	**16.37**
65 mm dia.	4.52	5.55	0.30	11.26	nr	**16.81**
80 mm dia.	4.83	5.93	0.40	15.02	nr	**20.95**
100 mm dia.	6.00	7.37	0.40	15.02	nr	**22.39**
125 mm dia.	6.69	8.22	0.40	15.02	nr	**23.24**
150 mm dia.	7.67	9.42	0.40	15.02	nr	**24.44**
200 mm dia.	10.16	12.48	0.50	18.77	nr	**31.25**
250 mm dia.	11.73	14.41	0.50	18.77	nr	**33.18**
300 mm dia.	12.59	15.47	0.50	18.77	nr	**34.24**
350 mm dia.	13.81	16.97	0.50	18.77	nr	**35.74**
400 mm dia.	15.28	18.77	0.50	18.77	nr	**37.54**

38 MECHANICAL/COOLING/HEATING SYSTEMS

Item	Net Price £	Material £	Labour hours	Labour £	Unit	**Total rate £**
THERMAL INSULATION – cont						
Extra over for fittings external insulation – cont						
Valves						
15 mm dia.	28.84	35.43	1.03	38.66	nr	**74.09**
20 mm dia.	30.34	37.27	1.03	38.66	nr	**75.93**
25 mm dia.	32.07	39.40	1.03	38.66	nr	**78.06**
32 mm dia.	33.67	41.36	1.03	38.66	nr	**80.02**
40 mm dia.	35.23	43.28	1.03	38.66	nr	**81.94**
50 mm dia.	38.85	47.72	1.03	38.66	nr	**86.38**
65 mm dia.	42.01	51.61	1.03	38.66	nr	**90.27**
80 mm dia.	45.01	55.29	1.25	46.91	nr	**102.20**
100 mm dia.	55.34	67.99	1.25	46.91	nr	**114.90**
125 mm dia.	61.50	75.56	1.25	46.91	nr	**122.47**
150 mm dia.	70.41	86.50	1.25	46.91	nr	**133.41**
200 mm dia.	92.13	113.18	1.55	58.17	nr	**171.35**
250 mm dia.	106.46	130.79	1.55	58.17	nr	**188.96**
300 mm dia.	115.52	141.92	1.55	58.17	nr	**200.09**
350 mm dia.	127.83	157.04	1.55	58.17	nr	**215.21**
400 mm dia.	142.43	174.97	1.55	58.17	nr	**233.14**
Expansion bellows						
15 mm dia.	42.73	52.49	1.42	53.29	nr	**105.78**
20 mm dia.	44.94	55.21	1.42	53.29	nr	**108.50**
25 mm dia.	47.51	58.37	1.42	53.29	nr	**111.66**
32 mm dia.	49.91	61.31	1.42	53.29	nr	**114.60**
40 mm dia.	52.18	64.10	1.42	53.29	nr	**117.39**
50 mm dia.	57.53	70.68	1.42	53.29	nr	**123.97**
65 mm dia.	62.23	76.45	1.42	53.29	nr	**129.74**
80 mm dia.	66.69	81.93	1.75	65.68	nr	**147.61**
100 mm dia.	81.99	100.73	1.75	65.68	nr	**166.41**
125 mm dia.	91.12	111.94	1.75	65.68	nr	**177.62**
150 mm dia.	104.33	128.17	1.75	65.68	nr	**193.85**
200 mm dia.	136.50	167.69	2.17	81.44	nr	**249.13**
250 mm dia.	157.73	193.77	2.17	81.44	nr	**275.21**
300 mm dia.	171.13	210.23	2.17	81.44	nr	**291.67**
350 mm dia.	189.38	232.65	2.17	81.44	nr	**314.09**
400 mm dia.	211.01	259.22	2.17	81.44	nr	**340.66**

38 MECHANICAL/COOLING/HEATING SYSTEMS

Item	Net Price £	Material £	Labour hours	Labour £	Unit	**Total rate £**
STEAM HEATING						
PIPELINES						
For pipework prices refer to Section – Low Temperature Hot Water Heating						
PIPELINE ANCILLARIES						
Steam traps and accessories						
Cast iron; inverted bucket type; steam trap pressure range up to 17 bar at 210°C; screwed ends						
½" dia.	101.84	125.11	0.85	31.90	nr	**157.01**
¾" dia.	101.84	125.11	1.13	42.41	nr	**167.52**
1" dia.	171.59	210.80	1.35	50.66	nr	**261.46**
1½" dia.	529.76	650.81	1.80	67.56	nr	**718.37**
2" dia.	592.99	728.49	2.18	81.82	nr	**810.31**
Cast iron; inverted bucket type; steam trap pressure range up to 17 bar at 210°C; flanged ends to BS 4504 PN16; bolted connections						
15 mm dia.	170.56	209.54	1.15	43.16	nr	**252.70**
20 mm dia.	206.13	253.23	1.25	46.91	nr	**300.14**
25 mm dia.	255.87	314.34	1.33	49.92	nr	**364.26**
40 mm dia.	542.83	666.87	1.46	54.79	nr	**721.66**
50 mm dia.	606.05	744.53	1.60	60.05	nr	**804.58**
Steam traps and strainers						
Stainless steel; thermodynamic trap with pressure range up to 42 bar; temperature range to 400°C; screwed ends to steel						
15 mm dia.	68.19	83.77	0.84	31.53	nr	**115.30**
20 mm dia.	77.17	94.80	1.14	42.80	nr	**137.60**
Stainless steel; thermodynamic trap with pressure range up to 24 bar; temperature range to 288°C; flanged ends to DIN 2456 PN64; bolted connections						
15 mm dia.	142.76	175.38	1.24	46.56	nr	**221.94**
20 mm dia.	156.49	192.24	1.34	50.31	nr	**242.55**
25 mm dia.	180.89	222.22	1.40	52.56	nr	**274.78**
Malleable iron pipeline strainer; max steam working pressure 14 bar and temperature range to 230°C; screwed ends to steel						
½" dia.	19.86	24.40	0.84	31.53	nr	**55.93**
¾" dia.	24.82	30.49	1.14	42.80	nr	**73.29**
1" dia.	38.72	47.57	1.30	48.80	nr	**96.37**
1½" dia.	79.52	97.69	1.50	56.35	nr	**154.04**
2" dia.	111.66	137.17	1.74	65.38	nr	**202.55**

38 MECHANICAL/COOLING/HEATING SYSTEMS

Item	Net Price £	Material £	Labour hours	Labour £	Unit	Total rate £
STEAM HEATING – cont						
Steam traps and strainers – cont						
Bronze pipeline strainer; max steam working pressure 25 bar; flanged ends to BS 4504 PN25; bolted connections						
15 mm dia.	170.55	209.52	1.24	46.56	nr	**256.08**
20 mm dia.	208.01	255.54	1.34	50.31	nr	**305.85**
25 mm dia.	238.51	293.01	1.40	52.56	nr	**345.57**
32 mm dia.	370.24	454.84	1.46	54.86	nr	**509.70**
40 mm dia.	420.17	516.18	1.54	57.82	nr	**574.00**
50 mm dia.	646.20	793.85	1.64	61.62	nr	**855.47**
65 mm dia.	715.53	879.03	2.50	93.83	nr	**972.86**
80 mm dia.	890.25	1093.67	2.91	109.11	nr	**1202.78**
100 mm dia.	1542.00	1894.35	3.51	131.68	nr	**2026.03**
Balanced pressure thermostatic steam trap and strainer; max working pressure up to 13 bar; screwed ends to steel						
½" dia.	141.32	173.61	1.26	47.32	nr	**220.93**
¾" dia.	141.69	174.07	1.71	64.26	nr	**238.33**
Bimetallic thermostatic steam trap and strainer; max working pressure up to 21 bar; flanged ends						
15 mm	183.84	225.84	1.24	46.56	nr	**272.40**
20 mm	201.96	248.10	1.34	50.31	nr	**298.41**
Sight glasses						
Pressed brass; straight; single window; screwed ends to steel						
15 mm dia.	42.98	52.80	0.84	31.53	nr	**84.33**
20 mm dia.	47.71	58.61	1.14	42.80	nr	**101.41**
25 mm dia.	59.61	73.23	1.30	48.80	nr	**122.03**
Gunmetal; straight; double window; screwed ends to steel						
15 mm dia.	69.33	85.18	0.84	31.53	nr	**116.71**
20 mm dia.	76.26	93.68	1.14	42.80	nr	**136.48**
25 mm dia.	94.28	115.83	1.30	48.80	nr	**164.63**
32 mm dia.	155.31	190.80	1.35	50.73	nr	**241.53**
40 mm dia.	155.31	190.80	1.74	65.38	nr	**256.18**
50 mm dia.	188.60	231.69	2.08	78.18	nr	**309.87**
SG iron flanged; BS 4504, PN 25						
15 mm dia.	141.44	173.75	1.00	37.53	nr	**211.28**
20 mm dia.	166.42	204.45	1.25	46.91	nr	**251.36**
25 mm dia.	212.18	260.66	1.50	56.35	nr	**317.01**
32 mm dia.	234.34	287.89	1.70	63.83	nr	**351.72**
40 mm dia.	306.47	376.50	2.00	75.06	nr	**451.56**
50 mm dia.	370.24	454.84	2.30	86.48	nr	**541.32**
Check valve and sight glass; gunmetal; screwed						
15 mm dia.	70.01	86.01	0.84	31.53	nr	**117.54**
20 mm dia.	74.17	91.12	1.14	42.80	nr	**133.92**
25 mm dia.	124.80	153.32	1.30	48.80	nr	**202.12**

38 MECHANICAL/COOLING/HEATING SYSTEMS

Item	Net Price £	Material £	Labour hours	Labour £	Unit	**Total rate £**
Pressure reducing valves						
Pressure reducing valve for steam; maximum range of 17 bar and 232°C; screwed ends to steel						
15 mm dia.	524.17	643.94	0.87	32.65	nr	**676.59**
20 mm dia.	567.15	696.75	0.91	34.16	nr	**730.91**
25 mm dia.	611.52	751.25	1.35	50.66	nr	**801.91**
Pressure reducing valve for steam; maximum range of 17 bar and 232°C; flanged ends to BS 4504 PN 25						
25 mm dia.	736.33	904.59	1.70	63.81	nr	**968.40**
32 mm dia.	836.19	1027.26	1.87	70.18	nr	**1097.44**
40 mm dia.	998.43	1226.57	2.12	79.57	nr	**1306.14**
50 mm dia.	1152.36	1415.67	2.57	96.45	nr	**1512.12**
Safety and relief valves						
Bronze safety valve; 'pop' type; side outlet; including easing lever; working pressure saturated steam up to 20.7 bar; screwed ends to steel						
15 mm dia.	178.49	219.27	0.32	12.01	nr	**231.28**
20 mm dia.	201.76	247.86	0.40	15.02	nr	**262.88**
Bronze safety valve; 'pop' type; side outlet; including easing lever; working pressure saturated steam up to 17.2 bar; screwed ends to steel						
25 mm dia.	287.12	352.73	0.47	17.64	nr	**370.37**
32 mm dia.	343.02	421.40	0.56	21.02	nr	**442.42**
Bronze safety valve; 'pop' type; side outlet; including easing lever; working pressure saturated steam up to 13.8 bar; screwed ends to steel						
40 mm dia.	443.90	545.33	0.64	24.01	nr	**569.34**
50 mm dia.	617.74	758.90	0.76	28.53	nr	**787.43**
65 mm dia.	878.50	1079.23	0.94	35.28	nr	**1114.51**
80 mm dia.	1139.26	1399.58	1.10	41.28	nr	**1440.86**

38 MECHANICAL/COOLING/HEATING SYSTEMS

Item	Net Price £	Material £	Labour hours	Labour £	Unit	**Total rate £**
STEAM HEATING – cont						
EQUIPMENT						
CALORIFIERS						
Non-storage calorifiers; mild steel shell construction with indirect steam heating for secondary LPHW at 82°C flow and 71°C return to BS 853; maximum test on shell 11 bar, tubes 26 bar						
Horizontal/vertical, for steam at 3 bar–5.5 bar						
88 kW capacity	658.39	808.84	8.00	300.25	nr	**1109.09**
176 kW capacity	958.46	1177.47	12.05	452.17	nr	**1629.64**
293 kW capacity	1044.19	1282.79	14.08	528.60	nr	**1811.39**
586 kW capacity	1524.52	1872.87	37.04	1390.02	nr	**3262.89**
879 kW capacity	1898.23	2331.98	40.00	1501.23	nr	**3833.21**
1465 kW capacity	2314.81	2843.75	45.45	1705.93	nr	**4549.68**

38 MECHANICAL/COOLING/HEATING SYSTEMS

Item	Net Price £	Material £	Labour hours	Labour £	Unit	Total rate £
LOCAL HEATING UNITS						
Warm air unit heater for connection to LTHW or steam supplies; suitable for heights upto 3 m; recirculating type; mild steel casing; heating coil; adjustable discharge louvre; axial fan; horizontal or vertical discharge; normal speed; entering air temperature 15°C; complete with enclosures; includes fixing in position; includes connections to primary heating supply; electrical work elsewhere						
Low pressure hot water						
7.5 kW, 265 l/s	385.74	473.89	6.53	245.07	nr	**718.96**
15.4 kW, 575 l/s	466.93	573.63	7.54	282.99	nr	**856.62**
26.9 kW, 1040 l/s	632.71	777.28	8.65	324.64	nr	**1101.92**
48.0 kW, 1620 l/s	834.03	1024.61	9.35	350.91	nr	**1375.52**
Steam, 2 bar						
9.2 kW, 265 l/s	551.50	677.51	6.53	245.07	nr	**922.58**
18.8 kW, 575 l/s	597.19	733.65	6.82	255.96	nr	**989.61**
34.8 kW, 1040 l/s	686.86	843.81	6.82	255.96	nr	**1099.77**
51.6 kW, 1625 l/s	930.49	1143.10	7.10	266.47	nr	**1409.57**

38 MECHANICAL/COOLING/HEATING SYSTEMS

Item	Net Price £	Material £	Labour hours	Labour £	Unit	Total rate £
CENTRAL REFRIGERATION PLANT						
CHILLERS; Air cooled						
Selection of air cooled chillers based on chilled water flow and return temperatures 6°C and 12°C, and an outdoor temperature of 35°C						
Air cooled liquid chiller; refrigerant 407C; scroll compressors; twin circuit; integral controls; includes placing in position; electrical work elsewhere						
Cooling load						
100 kW	17036.67	20929.54	8.00	300.25	nr	**21229.79**
150 kW	21470.48	26376.48	8.00	300.25	nr	**26676.73**
200 kW	26575.70	32648.25	8.00	300.25	nr	**32948.50**
Air cooled liquid chiller; refrigerant 407C; reciprocating compressors; twin circuit; integral controls; includes placing in position; electrical work elsewhere						
Cooling load						
400 kW	47408.74	58241.64	8.00	300.25	nr	**58541.89**
550 kW	67603.19	83050.52	8.00	300.25	nr	**83350.77**
700 kW	78537.13	96482.86	9.00	337.77	nr	**96820.63**
Air cooled liquid chiller; refrigerant R134a; screw compressors; twin circuit; integral controls; includes placing in position; electrical work elsewhere						
Cooling load						
250 kW	47802.75	58725.68	8.00	300.25	nr	**59025.93**
400 kW	57268.19	70353.97	8.00	300.25	nr	**70654.22**
600 kW	76878.42	94445.14	8.00	300.25	nr	**94745.39**
800 kW	92401.29	113514.99	9.00	337.77	nr	**113852.76**
1000 kW	108343.09	133099.49	9.00	337.77	nr	**133437.26**
1200 kW	137620.29	169066.53	10.00	375.31	nr	**169441.84**
Air cooled liquid chiller; ductable for indoor installation; refrigerant 407C; scroll compressors; integral controls; includes placing in position; electrical work elsewhere						
Cooling load						
40 kW	10132.33	12447.57	6.00	225.18	nr	**12672.75**
80 kW	15867.08	19492.70	6.00	225.18	nr	**19717.88**

38 MECHANICAL/COOLING/HEATING SYSTEMS

Item	Net Price £	Material £	Labour hours	Labour £	Unit	Total rate £
CHILLERS; Higher efficiency air cooled						
Selection of air cooled chillers based on chilled water flow and return temperatures of 6°C and 12°C and an outdoor temperature of 25°C						
These machines have significantly higher part load operating efficiencies than conventional air cooled machines						
Air cooled liquid chiller, refrigerant R410A; scroll compressors; complete with free cooling facility; integral controls; including placing in position; electrical work elsewhere						
Cooling load						
250 kW	43263.64	53149.38	8.00	300.25	nr	**53449.63**
300 kW	47734.82	58642.23	8.00	300.25	nr	**58942.48**
350 kW	55161.12	67765.44	8.00	300.25	nr	**68065.69**
400 kW	62254.54	76479.70	8.00	300.25	nr	**76779.95**
450 kW	64687.69	79468.83	8.00	300.25	nr	**79769.08**
500 kW	70321.68	86390.19	8.00	300.25	nr	**86690.44**
600 kW	72164.95	88654.64	8.00	300.25	nr	**88954.89**
650 kW	84286.60	103546.09	9.00	337.77	nr	**103883.86**
700 kW	89270.67	109669.01	9.00	337.77	nr	**110006.78**
CHILLERS; Water cooled						
Selection of water cooled chillers based on chilled water flow and return temperatures of 6°C and 12°C, and condenser entering and leaving temperatures of 27°C and 33°C						
Water cooled liquid chiller; refrigerant 407C; reciprocating compressors; twin circuit; integral controls; includes placing in position; electrical work elsewhere						
Cooling load						
200 kw	21899.59	26903.65	8.00	300.25	nr	**27203.90**
350 kW	38755.11	47610.65	8.00	300.25	nr	**47910.90**
500 kW	43973.55	54021.50	8.00	300.25	nr	**54321.75**
650 kW	50967.33	62613.37	9.00	337.77	nr	**62951.14**
750 kW	62081.30	76266.88	9.00	337.77	nr	**76604.65**
Water cooled condenserless liquid chiller; refrigerant 407C; reciprocating compressors; twin circuit; integral controls; includes placing in position; electrical work elsewhere						
Cooling load						
200 kW	18900.84	23219.68	8.00	300.25	nr	**23519.93**
350 kW	33179.27	40760.74	8.00	300.25	nr	**41060.99**
500 kW	41934.36	51516.36	8.00	300.25	nr	**51816.61**
650 kW	51286.72	63005.73	9.00	337.77	nr	**63343.50**
750 kW	58358.87	71693.87	9.00	337.77	nr	**72031.64**

38 MECHANICAL/COOLING/HEATING SYSTEMS

Item	Net Price £	Material £	Labour hours	Labour £	Unit	Total rate £
CENTRAL REFRIGERATION PLANT – cont						
CHILLERS – cont						
Selection of water cooled chillers – cont						
Water cooled liquid chiller; refrigerant R134a; screw compressors; twin circuit; integral controls; includes placing in position; electrical work elsewhere						
Cooling load						
300 kW	36517.81	44862.13	8.00	300.25	nr	**45162.38**
500 kW	46543.71	57178.95	8.00	300.25	nr	**57479.20**
700 kW	58691.41	72102.40	9.00	337.77	nr	**72440.17**
900 kW	82267.83	101066.03	9.00	337.77	nr	**101403.80**
1100 kW	90877.31	111642.77	10.00	375.31	nr	**112018.08**
1300 kW	105073.21	129082.44	10.00	375.31	nr	**129457.75**
Water cooled liquid chiller; refrigerant R134a; centrifugal compressors; twin circuit; integral controls; includes placing in position; electrical work elsewhere						
Cooling load						
1600 kW	148231.69	182102.63	11.00	412.84	nr	**182515.47**
1900 kW	180532.27	221783.90	11.00	412.84	nr	**222196.74**
2200 kW	215086.22	264233.42	13.00	487.90	nr	**264721.32**
2500 kW	222222.44	273000.26	13.00	487.90	nr	**273488.16**
3000 kW	270422.74	332214.34	15.00	562.96	nr	**332777.30**
3500 kW	368075.39	452180.62	15.00	562.96	nr	**452743.58**
4000 kW	390928.60	480255.78	20.00	750.61	nr	**481006.39**
4500 kW	417079.50	512382.16	20.00	750.61	nr	**513132.77**
5000 kW	508666.43	624896.71	25.00	938.27	nr	**625834.98**
CHILLERS; Absorption						
Absorption chiller, for operation using low pressure steam; selection based on chilled water flow and return temperatures of 6°C and 12°C, steam at 1 bar gauge and condenser entering and leaving temperatures of 27°C and 33°C; integral controls; includes placing in position; electrical work elsewhere						
Cooling load						
400 kW	63864.85	78457.96	8.00	300.25	nr	**78758.21**
700 kW	81093.03	99622.79	9.00	337.77	nr	**99960.56**
1000 kW	92515.88	113655.76	10.00	375.31	nr	**114031.07**
1300 kW	110407.61	135635.75	12.00	450.37	nr	**136086.12**
1600 kW	130666.79	160524.15	14.00	525.43	nr	**161049.58**
2000 kW	154864.65	190251.22	15.00	562.96	nr	**190814.18**

38 MECHANICAL/COOLING/HEATING SYSTEMS

Item	Net Price £	Material £	Labour hours	Labour £	Unit	Total rate £
Absorption chiller, for operation using low pressure hot water; selection based on chilled water flow and return temperatures of 6°C and 12°C, cooling water temperatures of 27°C and 33°C and hot water at 90°C; integral controls; includes placing in position; electrical work elsewhere						
Cooling load						
700 kW	92515.88	113655.76	9.00	337.77	nr	**113993.53**
1000 kW	115648.25	142073.87	10.00	375.31	nr	**142449.18**
1300 kW	134506.49	165241.22	12.00	450.37	nr	**165691.59**
1600 kW	154864.65	190251.22	14.00	525.43	nr	**190776.65**
HEAT REJECTION						
Dry air liquid coolers						
Flat coil configuration						
500 kW	19983.75	24550.04	15.00	562.96	nr	**25113.00**
Extra for inverter panels (factory wired and mounted on units)	7636.53	9381.48	15.00	562.96	nr	**9944.44**
800 kW	25732.50	31612.37	15.00	562.96	nr	**32175.33**
Extra for inverter panels (factory wired and mounted on units)	19983.75	24550.04	15.00	562.96	nr	**25113.00**
1100 kW	31210.78	38342.44	15.00	562.96	nr	**38905.40**
Extra for inverter panels (factory wired and mounted on units)	15274.16	18764.31	15.00	562.96	nr	**19327.27**
1400 kW	51469.38	63230.13	15.00	562.96	nr	**63793.09**
Extra for inverter panels (factory wired and mounted on units)	21053.56	25864.30	15.00	562.96	nr	**26427.26**
1700 kW	51469.38	63230.13	15.00	562.96	nr	**63793.09**
Extra for inverter panels (factory wired and mounted on units)	21406.15	26297.46	15.00	562.96	nr	**26860.42**
2000 kW	61320.00	75331.62	15.00	562.96	nr	**75894.58**
Extra for inverter panels (factory wired and mounted on units)	28174.35	34612.19	15.00	562.96	nr	**35175.15**
Note: heat rejection capacities above 500 kW require multiple units. Prices are therefore for total number of units.						
'Vee' type coil configuration						
500 kW	18342.13	22533.30	15.00	562.96	nr	**23096.26**
Extra for inverter panels (factory wired and mounted on units)	5554.72	6823.97	15.00	562.96	nr	**7386.93**
800 kW	26607.40	32687.19	15.00	562.96	nr	**33250.15**
Extra for inverter panels (factory wired and mounted on units)	12862.13	15801.12	15.00	562.96	nr	**16364.08**
1100 kW	36811.06	45222.39	15.00	562.96	nr	**45785.35**
Extra for inverter panels (factory wired and mounted on units)	15554.57	19108.79	15.00	562.96	nr	**19671.75**
1400 kW	52422.25	64400.73	15.00	562.96	nr	**64963.69**

38 MECHANICAL/COOLING/HEATING SYSTEMS

Item	Net Price £	Material £	Labour hours	Labour £	Unit	Total rate £
CENTRAL REFRIGERATION PLANT – cont						
HEAT REJECTION – cont						
Dry air liquid coolers – cont						
'Vee' type coil configuration – cont						
Extra for inverter panels (factory wired and mounted on units)	19341.88	23761.50	15.00	562.96	nr	**24324.46**
1700 kW	73696.85	90536.58	15.00	562.96	nr	**91099.54**
Extra for inverter panels (factory wired and mounted on units)	25727.66	31606.43	15.00	562.96	nr	**32169.39**
2000 kW	73594.95	90411.39	15.00	562.96	nr	**90974.35**
Extra for inverter panels (factory wired and mounted on units)	29012.27	35641.58	15.00	562.96	nr	**36204.54**
Note: Heat rejection capacities above 1100 kW require multiple units. Prices are for total number of units.						
Air cooled condensers						
Air cooled condenser; refrigerant 407C; selection based on condensing temperature of 45°C at 32°C dry bulb ambient; includes placing in position; electrical work elsewhere						
Flat coil configuration						
500 kW	16636.23	20437.61	15.00	562.96	nr	**21000.57**
Extra for inverter panels (factory wired and mounted on units)	7612.26	9351.66	15.00	562.96	nr	**9914.62**
800 kW	27293.40	33529.94	15.00	562.96	nr	**34092.90**
Extra for inverter panels (factory wired and mounted on units)	11653.67	14316.53	15.00	562.96	nr	**14879.49**
1100 kW	36538.85	44887.97	15.00	562.96	nr	**45450.93**
Extra for inverter panels (factory wired and mounted on units)	16567.04	20352.61	15.00	562.96	nr	**20915.57**
1400 kW	45618.19	56041.94	15.00	562.96	nr	**56604.90**
Extra for inverter panels (factory wired and mounted on units)	20373.18	25028.45	15.00	562.96	nr	**25591.41**
1700 kW	59403.30	72976.95	15.00	562.96	nr	**73539.91**
Extra for inverter panels (factory wired and mounted on units)	30559.76	37542.67	15.00	562.96	nr	**38105.63**
2000 kW	68427.29	84062.93	15.00	562.96	nr	**84625.89**
Extra for inverter panels (factory wired and mounted on units)	30559.76	37542.67	15.00	562.96	nr	**38105.63**
Note: Heat rejection capacities above 500 kW require multiple units. Prices are for total number of units						
'Vee' type coil configuration						
500 kW	16331.75	20063.56	15.00	562.96	nr	**20626.52**
Extra for inverter panels (factory wired and mounted on units)	5826.86	7158.30	15.00	562.96	nr	**7721.26**
800 kW	24566.83	30180.35	15.00	562.96	nr	**30743.31**

38 MECHANICAL/COOLING/HEATING SYSTEMS

Item	Net Price £	Material £	Labour hours	Labour £	Unit	Total rate £
Extra for inverter panels (factory wired and mounted on units)	11930.48	14656.59	15.00	562.96	nr	**15219.55**
1100 kW	36054.44	44292.87	15.00	562.96	nr	**44855.83**
Extra for inverter panels (factory wired and mounted on units)	20290.14	24926.43	15.00	562.96	nr	**25489.39**
1400 kW	41978.15	51570.16	15.00	562.96	nr	**52133.12**
Extra for inverter panels (factory wired and mounted on units)	23860.97	29313.20	15.00	562.96	nr	**29876.16**
1700 kW	54081.64	66439.30	15.00	562.96	nr	**67002.26**
Extra for inverter panels (factory wired and mounted on units)	30449.04	37406.65	15.00	562.96	nr	**37969.61**
2000 kW	62967.24	77355.25	15.00	562.96	nr	**77918.21**
Extra for inverter panels (factory wired and mounted on units)	35791.45	43969.80	15.00	562.96	nr	**44532.76**
Note: Heat rejection capacities above 1100 kW require multiple units. Prices are for total number of units.						
Cooling towers						
Cooling towers; forced draught, centrifugal fan, conterflow design; based on water temperatures of 35°C on and 29°C off at 21°C wet bulb ambient temperature; includes placing in position; electrical work elsewhere						
COOLING TOWERS: Open circuit type						
900 kW	11081.53	13613.66	20.00	750.61	nr	**14364.27**
Extra for stainless steel construction	4968.21	6103.45	–	–	nr	**6103.45**
Extra for intake and discharge sound attenuation	5347.09	6568.91	–	–	nr	**6568.91**
Extra for fan dampers for capacity control	1240.65	1524.14	–	–	nr	**1524.14**
1500 kW	17193.28	21121.95	20.00	750.61	nr	**21872.56**
Extra for stainless steel construction	7958.28	9776.75	–	–	nr	**9776.75**
Extra for intake and discharge sound attenuation	8012.77	9843.69	–	–	nr	**9843.69**
Extra for fan dampers for capacity control	1259.23	1546.97	–	–	nr	**1546.97**
2100 kW	24391.35	29964.77	20.00	750.61	nr	**30715.38**
Extra for stainless steel construction	11287.24	13866.37	–	–	nr	**13866.37**
Extra for intake and discharge sound attenuation	11897.68	14616.30	–	–	nr	**14616.30**
Extra for fan dampers for capacity control	1437.90	1766.46	–	–	nr	**1766.46**
2700 kW	29456.87	36187.77	20.00	750.61	nr	**36938.38**
Extra for stainless steel construction	13769.06	16915.29	–	–	nr	**16915.29**
Extra for intake and discharge sound attenuation	16028.40	19690.89	–	–	nr	**19690.89**
Extra for fan dampers for capacity control	2238.31	2749.76	–	–	nr	**2749.76**
3300 kW	36300.43	44595.07	20.00	750.61	nr	**45345.68**
Extra for stainless steel construction	16804.93	20644.86	–	–	nr	**20644.86**
Extra for intake and discharge sound attenuation	15116.49	18570.60	–	–	nr	**18570.60**

38 MECHANICAL/COOLING/HEATING SYSTEMS

Item	Net Price £	Material £	Labour hours	Labour £	Unit	Total rate £
CENTRAL REFRIGERATION PLANT – cont						
HEAT REJECTION – cont						
Cooling towers – cont						
COOLING TOWERS: Open circuit type – cont						
Extra for fan dampers for capacity control	1830.96	2249.33	–	–	nr	**2249.33**
3900 kW	40914.28	50263.20	20.00	750.61	nr	**51013.81**
Extra for stainless steel construction	18941.79	23269.98	–	–	nr	**23269.98**
Extra for intake and discharge sound attenuation	15406.64	18927.06	–	–	nr	**18927.06**
Extra for fan dampers for capacity control	1830.96	2249.33	–	–	nr	**2249.33**
4500 kW	46045.54	56566.94	23.00	863.20	nr	**57430.14**
Extra for stainless steel construction	21472.44	26378.89	–	–	nr	**26378.89**
Extra for intake and discharge sound attenuation	20433.57	25102.64	–	–	nr	**25102.64**
Extra for fan dampers for capacity control	2881.51	3539.94	–	–	nr	**3539.94**
5100 kW	53189.29	65343.04	23.00	863.20	nr	**66206.24**
Extra for stainless steel construction	24506.80	30106.61	–	–	nr	**30106.61**
Extra for intake and discharge sound attenuation	20379.24	25035.90	–	–	nr	**25035.90**
Extra for fan dampers for capacity control	2881.51	3539.94	–	–	nr	**3539.94**
5700 kW	56232.29	69081.37	30.00	1125.93	nr	**70207.30**
Extra for stainless steel construction	26257.51	32257.35	–	–	nr	**32257.35**
Extra for intake and discharge sound attenuation	30178.66	37074.48	–	–	nr	**37074.48**
Extra for fan dampers for capacity control	3168.81	3892.89	–	–	nr	**3892.89**
6300 kW	67801.22	83293.80	30.00	1125.93	nr	**84419.73**
Extra for stainless steel construction	31389.04	38561.44	–	–	nr	**38561.44**
Extra for intake and discharge sound attenuation	30614.59	37610.02	–	–	nr	**37610.02**
Extra for fan dampers for capacity control	3168.81	3892.89	–	–	nr	**3892.89**
COOLING TOWERS: Closed circuit type (includes 20% ethylene glycol)						
900 kW	30891.90	37950.70	20.00	750.61	nr	**38701.31**
Extra for stainless steel construction	25153.95	30901.63	–	–	nr	**30901.63**
Extra for intake and discharge sound attenuation	9295.76	11419.84	–	–	nr	**11419.84**
Extra for fan dampers for capacity control	1259.23	1546.97	–	–	nr	**1546.97**
1500 kW	57432.89	70556.30	20.00	750.61	nr	**71306.91**
Extra for stainless steel construction	45789.98	56252.99	–	–	nr	**56252.99**
Extra for intake and discharge sound attenuation	14837.78	18228.21	–	–	nr	**18228.21**
Extra for fan dampers for capacity control	1830.96	2249.33	–	–	nr	**2249.33**
2100 kW	69990.91	85983.83	20.00	750.61	nr	**86734.44**
Extra for stainless steel construction	62208.72	76423.41	–	–	nr	**76423.41**
Extra for intake and discharge sound attenuation	17860.79	21941.98	–	–	nr	**21941.98**

38 MECHANICAL/COOLING/HEATING SYSTEMS

Item	Net Price £	Material £	Labour hours	Labour £	Unit	**Total rate £**
Extra for fan dampers for capacity control	1830.96	2249.33	–	–	nr	**2249.33**
2700 kW	75462.36	92705.51	25.00	938.27	nr	**93643.78**
Extra for stainless steel construction	77079.79	94692.52	–	–	nr	**94692.52**
Extra for intake and discharge sound attenuation	26658.27	32749.69	–	–	nr	**32749.69**
Extra for fan dampers for capacity control	2770.02	3402.97	–	–	nr	**3402.97**
3300 kW	116268.01	142835.25	25.00	938.27	nr	**143773.52**
Extra for stainless steel construction	96158.48	118130.69	–	–	nr	**118130.69**
Extra for intake and discharge sound attenuation	35461.42	43564.35	–	–	nr	**43564.35**
Extra for fan dampers for capacity control	3168.81	3892.89	–	–	nr	**3892.89**
3900 kW	128017.00	157268.88	25.00	938.27	nr	**158207.15**
Extra for stainless steel construction	110117.28	135279.08	–	–	nr	**135279.08**
Extra for intake and discharge sound attenuation	36041.73	44277.26	–	–	nr	**44277.26**
Extra for fan dampers for capacity control	3168.81	3892.89	–	–	nr	**3892.89**
4500 kW	153261.62	188281.91	40.00	1501.23	nr	**189783.14**
Extra for stainless steel construction	114237.31	140340.53	–	–	nr	**140340.53**
Extra for intake and discharge sound attenuation	51514.11	63285.09	–	–	nr	**63285.09**
Extra for fan dampers for capacity control	5540.03	6805.93	–	–	nr	**6805.93**
5100 kW	170935.24	209993.94	40.00	1501.23	nr	**211495.17**
Extra for stainless steel construction	147627.40	181360.26	–	–	nr	**181360.26**
Extra for intake and discharge sound attenuation	51514.11	63285.09	–	–	nr	**63285.09**
Extra for fan dampers for capacity control	5540.03	6805.93	–	–	nr	**6805.93**
5700 kW	210773.23	258934.91	40.00	1501.23	nr	**260436.14**
Extra for stainless steel construction	152488.82	187332.52	–	–	nr	**187332.52**
Extra for intake and discharge sound attenuation	57588.72	70747.74	–	–	nr	**70747.74**
Extra for fan dampers for capacity control	6337.61	7785.75	–	–	nr	**7785.75**
6300 kW	225140.72	276585.37	40.00	1501.23	nr	**278086.60**
Extra for stainless steel construction	174422.52	214278.07	–	–	nr	**214278.07**
Extra for intake and discharge sound attenuation	57588.72	70747.74	–	–	nr	**70747.74**
Extra for fan dampers for capacity control	6337.61	7785.75	–	–	nr	**7785.75**

38 MECHANICAL/COOLING/HEATING SYSTEMS

Item	Net Price £	Material £	Labour hours	Labour £	Unit	Total rate £
CHILLED WATER						
SCREWED STEEL						
PIPELINES						
For pipework prices refer to Section – Low Temperature Hot Water Heating, with the exception of chilled water blocks within brackets as detailed hereafter. For minimum fixing distances, refer to the Tables and Memoranda to the rear of the book						
FIXINGS						
For steel pipes; black malleable iron						
Oversized pipe clip, to contain 30 mm insulation block for vapour barrier						
15 mm dia.	3.89	4.78	0.10	3.76	nr	**8.54**
20 mm dia.	3.92	4.82	0.11	4.13	nr	**8.95**
25 mm dia.	4.13	5.07	0.12	4.50	nr	**9.57**
32 mm dia.	4.24	5.21	0.14	5.26	nr	**10.47**
40 mm dia.	4.48	5.50	0.15	5.63	nr	**11.13**
50 mm dia.	6.40	7.86	0.16	6.01	nr	**13.87**
65 mm dia.	6.91	8.48	0.30	11.26	nr	**19.74**
80 mm dia.	7.33	9.01	0.35	13.14	nr	**22.15**
100 mm dia.	11.62	14.28	0.40	15.02	nr	**29.30**
125 mm dia.	13.48	16.55	0.55	20.60	nr	**37.15**
150 mm dia.	21.09	25.91	0.77	28.90	nr	**54.81**
200 mm dia.	26.92	33.08	0.90	33.78	nr	**66.86**
250 mm dia.	34.96	42.94	1.10	41.28	nr	**84.22**
300 mm dia.	41.61	51.11	1.25	46.91	nr	**98.02**
350 mm dia.	48.27	59.30	1.50	56.30	nr	**115.60**
400 mm dia.	54.05	66.40	1.75	65.68	nr	**132.08**
Screw on backplate (Male), black malleable iron; plugged and screwed						
M12	0.54	0.66	0.10	3.76	nr	**4.42**
Screw on backplate (Female), black malleable iron; plugged and screwed						
M12	0.54	1.32	0.10	3.76	nr	**5.08**
Extra over channel sections for fabricated hangers and brackets						
Galvanized steel; including inserts, bolts, nuts, washers; fixed to backgrounds						
41 × 21 mm	6.02	7.39	0.29	10.89	m	**18.28**
41 × 41 mm	7.22	8.87	0.29	10.89	m	**19.76**

38 MECHANICAL/COOLING/HEATING SYSTEMS

Item	Net Price £	Material £	Labour hours	Labour £	Unit	Total rate £
Threaded rods; metric thread; including nuts, washers etc.						
12 mm dia. × 600 mm long	3.04	3.74	0.18	6.76	nr	**10.50**
For plastic pipework suitable for chilled water systems, refer to ABS pipework details in Section – Cold Water, with the exception of chilled water blocks within brackets as detailed for the aforementioned steel pipe. For minimum fixing distances refer to the Tables and Memoranda to the rear of the book						
For copper pipework, refer to Section – Cold Water with the exception of chilled water blocks within brackets as detailed hereafter. For minimum fixing distances refer to the Tables and Memoranda to the rear of the book						
FIXINGS						
For copper pipework						
Oversized pipe clip, to contain 30 mm insulation block for vapour barrier						
15 mm dia.	3.89	4.78	0.10	3.76	nr	**8.54**
22 mm dia.	3.92	4.82	0.11	4.13	nr	**8.95**
28 mm dia.	4.13	5.07	0.12	4.50	nr	**9.57**
35 mm dia.	4.24	5.21	0.14	5.26	nr	**10.47**
42 mm dia.	4.48	5.50	0.15	5.63	nr	**11.13**
54 mm dia.	6.40	7.86	0.16	6.01	nr	**13.87**
67 mm dia.	6.91	8.48	0.30	11.26	nr	**19.74**
76 mm dia.	7.33	9.01	0.35	13.14	nr	**22.15**
108 mm dia.	3.01	17.98	0.40	15.02	nr	**33.00**
133 mm dia.	13.48	16.56	0.60	22.53	nr	**39.09**
159 mm dia.	21.09	25.91	0.77	28.90	nr	**54.81**
Screw on backplate, female						
All sizes 15 mm to 54 mm × 10 mm	1.56	1.92	0.10	3.76	nr	**5.68**
Screw on backplate, male						
All sizes 15 mm to 54 mm × 10 mm	2.07	2.54	0.10	3.76	nr	**6.30**

38 MECHANICAL/COOLING/HEATING SYSTEMS

Item	Net Price £	Material £	Labour hours	Labour £	Unit	**Total rate £**
CHILLED WATER – cont						
FIXINGS – cont						
Extra over channel sections for fabricated hangers and brackets						
Galvanized steel; including inserts, bolts, nuts, washers; fixed to backgrounds						
41 × 21 mm	6.02	7.39	0.29	10.89	m	**18.28**
41 × 41 mm	7.22	8.87	0.29	10.89	m	**19.76**
Threaded rods; metric thread; including nuts, washers etc.						
10 mm dia. × 600 mm long for ring clips up to 54 mm	1.97	2.42	0.18	6.76	nr	**9.18**
12 mm dia. × 600 mm long for ring clips from 54 mm	3.04	3.74	0.18	6.76	nr	**10.50**
PIPELINE ANCILLARIES						
For prices for ancillaries refer to Section: Low Temperature Hot Water Heating						

38 MECHANICAL/COOLING/HEATING SYSTEMS

Item	Net Price £	Material £	Labour hours	Labour £	Unit	Total rate £
HEAT EXCHANGERS						
Plate heat exchanger; for use in CHW systems; painted carbon steel frame, stainless steel plates, nitrile rubber gaskets, design pressure of 10 bar and operating temperature of 110/135°C						
Primary side; 13°C in, 8°C out; secondary side; 6°C in, 11°C out						
264 kW, 12.60 l/s	5460.23	6707.89	10.00	375.31	nr	**7083.20**
290 kW, 13.85 l/s	5755.59	7070.74	12.00	450.37	nr	**7521.11**
316 kW, 15.11 l/s	5991.62	7360.71	12.00	450.37	nr	**7811.08**
350 kW, 16.69 l/s	6443.49	7915.82	16.00	600.49	nr	**8516.31**
395 kW, 18.88 l/s	6895.35	8470.94	16.00	600.49	nr	**9071.43**
454 kW, 21.69 l/s	7489.84	9201.27	10.00	375.31	nr	**9576.58**
475 kW, 22.66 l/s	7634.99	9379.59	10.00	375.31	nr	**9754.90**
527 kW, 25.17 l/s	8273.66	10164.19	10.00	375.31	nr	**10539.50**
554 kW, 26.43 l/s	8493.28	10434.00	10.00	375.31	nr	**10809.31**
580 kW, 27.68 l/s	8784.30	10791.51	10.00	375.31	nr	**11166.82**
633 kW, 30.19 l/s	9312.45	11440.35	10.00	375.31	nr	**11815.66**
661 kW, 31.52 l/s	9537.11	11716.34	10.00	375.31	nr	**12091.65**
713 kW, 34.04 l/s	10180.83	12507.15	12.00	450.37	nr	**12957.52**
740 kW, 35.28 l/s	10402.97	12780.04	12.00	450.37	nr	**13230.41**
804 kW, 38.33 l/s	11077.00	13608.09	12.00	450.37	nr	**14058.46**
1925 kW, 91.82 l/s	18594.61	22843.47	15.00	562.96	nr	**23406.43**
2710 kW, 129.26 l/s	24698.56	30342.19	15.00	562.96	nr	**30905.15**
3100 kW, 147.87 l/s	27515.77	33803.12	15.00	562.96	nr	**34366.08**
Note: For temperature conditions different to those above, the cost of the units can vary significantly, and therefore the manufacturers advice should be sought.						

38 MECHANICAL/COOLING/HEATING SYSTEMS

Item	Net Price £	Material £	Labour hours	Labour £	Unit	Total rate £
TRACE HEATING						
Trace heating; for freeze protection or temperature maintenance of pipework; to BS 6351; including fixing to parent structures by plastic pull ties						
Straight laid						
15 mm	21.03	25.84	0.27	10.13	m	**35.97**
25 mm	21.03	25.84	0.27	10.13	m	**35.97**
28 mm	21.03	25.84	0.27	10.13	m	**35.97**
32 mm	21.03	25.84	0.30	11.26	m	**37.10**
35 mm	21.03	25.84	0.31	11.63	m	**37.47**
50 mm	21.03	25.84	0.34	12.77	m	**38.61**
100 mm	21.03	25.84	0.40	15.02	m	**40.86**
150 mm	21.03	25.84	0.40	15.02	m	**40.86**
Helically wound						
15 mm	26.79	32.91	1.00	37.53	m	**70.44**
25 mm	26.79	32.91	1.00	37.53	m	**70.44**
28 mm	26.79	32.91	1.00	37.53	m	**70.44**
32 mm	26.79	32.91	1.00	37.53	m	**70.44**
35 mm	26.79	32.91	1.00	37.53	m	**70.44**
50 mm	26.79	32.91	1.00	37.53	m	**70.44**
100 mm	26.79	32.91	1.00	37.53	m	**70.44**
150 mm	26.79	32.91	1.00	37.53	m	**70.44**
Accessories for trace heating; weatherproof; polycarbonate enclosure to IP standards; fully installed						
Connection junction box						
100 × 100 × 75 mm	63.36	77.84	1.40	52.54	nr	**130.38**
Single air thermostat						
150 × 150 × 75 mm	138.81	170.53	1.42	53.29	nr	**223.82**
Single capillary thermostat						
150 × 150 × 75 mm	199.71	245.34	1.46	54.79	nr	**300.13**
Twin capillary thermostat						
150 × 150 × 75 mm	358.29	440.16	1.46	54.79	nr	**494.95**

38 MECHANICAL/COOLING/HEATING SYSTEMS

Item	Net Price £	Material £	Labour hours	Labour £	Unit	**Total rate £**
PRESSURIZATION UNITS						
Chilled water packaged pressurization unit complete with expansion vessel(s), interconnecting pipework and necessary isolating and drain valves; includes placing in position; electrical work elsewhere. Selection based on a final working pressure of 4 bar, a 3 m static head and system operating temperatures of 6°/12°C						
System volume						
1800 litres	1078.48	1324.91	8.00	300.25	nr	**1625.16**
4500 litres	1174.16	1442.46	8.00	300.25	nr	**1742.71**
7200 litres	1236.56	1519.12	10.00	375.31	nr	**1894.43**
9900 litres	1657.76	2036.56	10.00	375.31	nr	**2411.87**
15300 litres	1733.68	2129.83	13.00	487.90	nr	**2617.73**
22500 litres	1811.68	2225.65	20.00	750.61	nr	**2976.26**
27000 litres	1811.68	2225.65	20.00	750.61	nr	**2976.26**
CHILLED BEAMS						
Static (passive) beams; cooling data based on 9K waterside cooling dt (e.g. water 14°C flow and 16°C return, room temperature 24°C)						
Static passive beam in perforated metal casing for exposed mounting from solid ceiling or roof soffit, 2 pipe cooling only	–	–	–	–	–	–
Length/Cooling Output						
1200 mm/300 W	395.00	485.26	3.00	112.59	m	**597.85**
1800 mm/450 W	515.00	632.68	3.50	131.35	m	**764.03**
2400 mm/600 W	615.00	755.53	4.00	150.12	m	**905.65**
3000 mm/750 W	730.00	896.80	4.60	172.64	m	**1069.44**
3600 mm/900 W	900.00	1105.65	5.00	187.66	m	**1293.31**
Static passive beam, chassis type for mounting above open grid or perforated metal ceiling or roof soffit, 2 pipe cooling only	–	–	–	–	–	–
Length/Cooling Output						
1200 mm/300 W	280.00	343.98	2.50	93.83	m	**437.81**
1800 mm/450 W	385.00	472.97	2.90	108.83	m	**581.80**
2400 mm/600 W	475.00	583.54	3.20	120.10	m	**703.64**
3000 mm/750 W	575.00	706.39	3.50	131.35	m	**837.74**
3600 mm/900 W	680.00	835.38	3.90	146.36	m	**981.74**

38 MECHANICAL/COOLING/HEATING SYSTEMS

Item	Net Price £	Material £	Labour hours	Labour £	Unit	Total rate £
PRESSURIZATION UNITS – cont						
CHILLED BEAMS – cont						
Multi-service beams; for exposed mounting on solid ceiling slabs, with Integrated lighting and lighting control, provision of fascia openings for other services (electrical work and other services elsewhere). Performance criteria equivalent to standard beams as listed above						
2 way discharge exposed linear multi-service climate beam for exposed mounting on solid ceiling; closed type with integrated secondary air circulation, 2 pipe cooling only, with primary air	–	–	–	–	–	–
Length/Cooling Output						
1200 mm/550 W	800.00	982.80	3.40	127.61	m	**1110.41**
1800 mm/900 W	1100.00	1351.35	3.90	146.36	m	**1497.71**
2400 mm/1200 W	1400.00	1719.90	4.40	165.13	m	**1885.03**
3000 mm/1550 W	1700.00	2088.45	4.90	183.90	m	**2272.35**
3600 mm/1850 W	2050.00	2518.42	5.40	202.66	m	**2721.08**
2 way discharge exposed linear multi-service climate beam for exposed mounting on solid ceiling; closed type with integrated secondary air circulation, 4 pipe cooling and heating, with primary air	–	–	–	–	–	–
Length/Cooling Output/Heating Output						
1200 mm/550 W/495W	830.00	1019.65	3.40	127.61	m	**1147.26**
1800 mm/900 W/810 W	1135.00	1394.35	3.90	146.36	m	**1540.71**
2400 mm/1200 W/1080 W	1440.00	1769.04	4.40	165.13	m	**1934.17**
3000 mm/1550 W/1395W	1745.00	2143.73	4.90	183.90	m	**2327.63**
3600 mm/1850 W/1665W	2100.00	2579.85	5.40	202.66	m	**2782.51**
Static passive multi-service beam for exposed mounting on solid ceiling, 2 pipe cooling only	–	–	–	–	–	–
Length/Cooling Output						
1200 mm/300 W	815.00	1001.23	3.40	127.61	m	**1128.84**
1800 mm/450 W	1115.00	1369.78	3.90	146.36	m	**1516.14**
2400 mm/600 W	1415.00	1738.33	4.40	165.13	m	**1903.46**
3000 mm/750 W	1710.00	2100.73	4.90	183.90	m	**2284.63**
3600 mm/900 W	2060.00	2530.71	5.40	202.66	m	**2733.37**

38 MECHANICAL/COOLING/HEATING SYSTEMS

Item	Net Price £	Material £	Labour hours	Labour £	Unit	Total rate £
Ventilated (active) beams; cooling data based on 9K airside and 9K waterside cooling dt (e.g. water 14°C flow and 16°C return, primary air 15°C, room temperature 24°C), heating data based on −4K airside and 20K waterside heating dt (e.g. water at 45°C flow and 35°C return, primary air 16°C, room temperature 20°C)						
1 way discharge recessed linear active climate beam mounted within a false ceiling; closed type with integrated secondary air circulation; 300 mm wide, 2 pipe cooling only, with 5 l/s/m primary air	–	–	–	–	–	–
Length/Cooling Output						
1200 mm/340 W	306.00	375.92	2.90	108.83	m	**484.75**
1800 mm/530 W	378.00	464.37	3.00	112.59	m	**576.96**
2400 mm/730 W	445.00	546.68	3.20	120.10	m	**666.78**
3000 mm/900 W	513.00	630.22	3.40	127.61	m	**757.83**
3600 mm/1100 W	554.00	680.59	3.50	131.35	m	**811.94**
1 way discharge recessed linear active climate beam mounted within a false ceiling; closed type with integrated secondary air circulation; 300 mm wide, 4 pipe cooling and heating, with 5 l/s/m primary air	–	–	–	–	–	–
Length/Cooling Output/Heating Output						
1200 mm/340 W/180 W	320.00	393.12	2.90	108.83	m	**501.95**
1800 mm/530 W/310 W	396.00	486.49	3.00	112.59	m	**599.08**
2400 mm/730 W/450 W	468.00	574.94	3.20	120.10	m	**695.04**
3000 mm/900 W/600 W	531.00	652.33	3.40	127.61	m	**779.94**
3600 mm/1100 W/750 W	573.00	703.93	3.50	131.35	m	**835.28**
2 way discharge recessed linear active climate beam mounted within a false ceiling; closed type with integrated secondary air circulation; 600 mm wide, 2 pipe cooling only, with 10 l/s/m primary air	–	–	–	–	–	–
Length/Cooling Output						
1200 mm/550 W	310.00	380.84	3.00	112.59	m	**493.43**
1800 mm/900 W	425.00	522.11	3.40	127.61	m	**649.72**
2400 mm/1200 W	530.00	651.10	3.70	138.86	m	**789.96**
3000 mm/1550 W	620.00	761.67	4.00	150.12	m	**911.79**
3600 mm/1850 W	710.00	872.23	4.40	165.13	m	**1037.36**
2 way discharge recessed linear active climate beam mounted within a false ceiling; closed type with integrated secondary air circulation; 600 mm wide, 4 pipe cooling and heating, with 10 l/s/m primary air	–	–	–	–	–	–
Length/Cooling Output/Heating Output						
1200 mm/550 W/495W	340.00	417.69	3.00	112.59	m	**530.28**
1800 mm/900 W/810 W	460.00	565.11	3.40	127.61	m	**692.72**
2400 mm/1200 W/1080 W	570.00	700.24	3.70	138.86	m	**839.10**

38 MECHANICAL/COOLING/HEATING SYSTEMS

Item	Net Price £	Material £	Labour hours	Labour £	Unit	Total rate £
PRESSURIZATION UNITS – cont						
CHILLED BEAMS – cont						
Ventilated (active) beams – cont						
Length/Cooling Output/Heating Output – cont						
3000 mm/1550 W/1395W	665.00	816.95	4.00	150.12	m	**967.07**
3600 mm/1850 W/1665W	760.00	933.66	4.40	165.13	m	**1098.79**
4 way discharge recessed modular active climate beam mounted within a false ceiling; closed type with integrated secondary air circulation; 600 mm wide, 2 pipe cooling only, with primary air at 10 l/s/m of beam discharge (600 × 600 unit 12 l/s, 1200 × 600 unit 18 l/s)	–	–	–	–	–	–
Length/Cooling Output						
600 × 600/500 W	325.00	399.26	2.90	108.83	m	**508.09**
1200 × 600/850 W	375.00	460.69	3.20	120.10	m	**580.79**
4 way discharge recessed modular active climate beam mounted within a false ceiling; closed type with integrated secondary air circulation; 600 mm wide, 4 pipe cooling and heating, with primary air at 10 l/s/m of beam discharge (600 × 600 unit 12 l/s, 1200 × 600 unit 18 l/s)	–	–	–	–	–	–
Length/Cooling Output/Heating Output						
600 × 600/500 W/450 W	350.00	429.98	2.90	108.83	m	**538.81**
1200 × 600/850 W/765W	405.00	497.54	3.20	120.10	m	**617.64**
1 way sidewall discharge active climate beam mounted within a ceiling bulkhead; closed type with integrated secondary air circulation, 2 pipe cooling only, with 20 l/s/m primary air	–	–	–	–	–	–
Length/Cooling Output						
900 mm/700 W	360.00	442.26	2.50	93.83	m	**536.09**
1100 mm/900 W	390.00	479.12	2.50	93.83	m	**572.95**
1300 mm/1100 W	460.00	565.11	2.70	101.33	m	**666.44**
1500 mm/1200 W	490.00	601.96	2.70	101.33	m	**703.29**
1 way sidewall discharge active climate beam mounted within a ceiling bulkhead; closed type with integrated secondary air circulation, 4 pipe cooling only, with 20 l/s/m primary air	–	–	–	–	–	–
Length/Cooling Output/Heating Output						
900 mm/700 W/400 W	390.00	479.12	2.50	93.83	m	**572.95**
1100 mm/900 W/600 W	420.00	515.97	2.50	93.83	m	**609.80**
1300 mm/1100 W/800 W	490.00	601.96	2.70	101.33	m	**703.29**
1500 mm/1200 W/900 W	520.00	638.82	2.70	101.33	m	**740.15**

38 MECHANICAL/COOLING/HEATING SYSTEMS

Item	Net Price £	Material £	Labour hours	Labour £	Unit	**Total rate £**
CHILLED CEILINGS						
Traditional (60% Radiant Absorption, 40% Convection) with bonded elements in perforated metal ceiling tiles (typically 30% free area and 2.5 mm perforated hole sizes)/tartan grid arrangement. Maximum waterside cooling effect approx. 65 W/m² at 9dTK.						
Radiant elements and metal ceilings combined	140.00	171.99	1.20	45.03	m	**217.02**
Hybrid (40% Radiant Absorption, 60% Convection) with radiant cooling chilled beam elements positioned above a perforated metal ceiling tiles (typically 30% free area and 2.5 mm perforated hole sizes)/tartan grid arrangement. Maximum waterside cooling effect approx. 100 W/m² at 9dTK.						
Radiant chilled beams	60.00	73.71	0.35	13.14	m	**86.85**
Metal ceilings	35.00	43.00	0.70	26.28	m	**69.28**
Convective beam (5% Radiant Absorption, 95% Convection) with convective cooling fin coil batteries positioned above a perforated metal ceiling tiles (typically 50% free area and 5.0 mm perforated hole sizes)/tartan grid arrangement. Maximum waterside cooling effect approx. 100 W/m² at 9dTK.						
Convective fin coil batteries	65.00	79.85	0.40	15.02	m	**94.87**
Metal ceilings	35.00	43.00	0.70	26.28	m	**69.28**
LEAK DETECTION						
Leak detection system consisting of a central control module connected by a leader cable to water sensing cables						
Control modules						
Alarm only	544.00	668.30	4.00	160.00	nr	**828.30**
Alarm and location	2393.00	2939.80	8.00	320.00	nr	**3259.80**
Cables						
Sensing – 3 m length	99.00	121.62	4.00	160.00	nr	**281.62**
Sensing – 5 m length	122.00	149.88	4.00	160.00	nr	**309.88**
Sensing – 7.5 m length	145.00	178.13	4.00	160.00	nr	**338.13**
Sensing – 15 m length	251.00	308.35	8.00	320.00	nr	**628.35**
Leader – 3.5 m length	44.00	54.05	2.00	80.00	nr	**134.05**
End terminal						
End terminal	18.00	22.11	0.05	2.01	nr	**24.12**

38 MECHANICAL/COOLING/HEATING SYSTEMS

Item	Net Price £	Material £	Labour hours	Labour £	Unit	Total rate £
ENERGY METERS						
Ultrasonic						
Energy meter for measuring energy use in chilled water systems; includes ultrasonic flow meter (with sensor and signal converter), energy calculator, pair of temperature sensors with brass pockets, and 3 m of interconnecting cable; includes fixing in position; electrical work elsewhere						
Pipe size (flanged connections to PN16); maximum flow rate						
50 mm, 36 m^3/hr	1278.34	1570.44	1.80	67.56	nr	**1638.00**
65 mm, 60 m^3/hr	1409.11	1731.09	2.32	87.08	nr	**1818.17**
80 mm, 100 m^3/hr	1579.51	1940.43	2.56	96.09	nr	**2036.52**
125 mm, 250 m^3/hr	1830.01	2248.17	3.60	135.11	nr	**2383.28**
150 mm, 360 m^3/hr	1986.58	2440.52	4.80	180.15	nr	**2620.67**
200 mm, 600 m^3/hr	2219.60	2726.78	6.24	234.19	nr	**2960.97**
250 mm, 1000 m^3/hr	2561.27	3146.52	9.60	360.30	nr	**3506.82**
300 mm, 1500 m^3/hr	3010.72	3698.67	10.80	405.33	nr	**4104.00**
350 mm, 2000 m^3/hr	3627.78	4456.73	13.20	495.40	nr	**4952.13**
400 mm, 2500 m^3/hr	4152.75	5101.66	15.60	585.48	nr	**5687.14**
500 mm, 3000 m^3/hr	4712.70	5789.55	24.00	900.73	nr	**6690.28**
600 mm, 3500 m^3/hr	5294.77	6504.62	28.00	1050.86	nr	**7555.48**
Electromagnetic						
Energy meter for measuring energy use in chilled water systems; includes electromagnetic flow meter (with sensor and signal converter), energy calculator, pair of temperature sensors with brass pockets, and 3 m of interconnecting cable; includes fixing in position; electrical work elsewhere						
Pipe size (flanged connections to PN40); maximum flow rate						
25 mm, 17.7 m^3/hr	916.48	1125.89	1.48	55.55	nr	**1181.44**
40 mm, 45 m^3/hr	925.25	1136.67	1.55	58.17	nr	**1194.84**
Pipe size (flanged connections to PN16); maximum flow rate						
50 mm, 70 m^3/hr	935.61	1149.39	1.80	67.56	nr	**1216.95**
65 mm, 120 m^3/hr	939.59	1154.29	2.32	87.08	nr	**1241.37**
80 mm, 180 m^3/hr	944.37	1160.16	2.56	96.09	nr	**1256.25**
125 mm, 450 m^3/hr	1028.05	1262.96	3.60	135.11	nr	**1398.07**
150 mm, 625 m^3/hr	1091.01	1340.30	4.80	180.15	nr	**1520.45**
200 mm, 1100 m^3/hr	1173.90	1442.13	6.24	234.19	nr	**1676.32**
250 mm, 1750 m^3/hr	1317.34	1618.35	9.60	360.30	nr	**1978.65**
300 mm, 2550 m^3/hr	1679.16	2062.85	10.80	405.33	nr	**2468.18**
350 mm, 3450 m^3/hr	2184.42	2683.56	13.20	495.40	nr	**3178.96**
400 mm, 4500 m^3/hr	2492.03	3061.46	15.60	585.48	nr	**3646.94**

38 MECHANICAL/COOLING/HEATING SYSTEMS

Item	Net Price £	Material £	Labour hours	Labour £	Unit	**Total rate £**
LOCAL COOLING UNITS						
Split system with ceiling void evaporator unit and external condensing unit						
Ceiling mounted 4 way blow cassette heat pump unit with remote fan speed and load control; refrigerant 407C; includes outdoor unit						
Cooling 3.6 kW, heating 4.1 kW	1749.96	2149.82	35.00	1313.57	nr	**3463.39**
Cooling 4.9 kW, heating 5.5 kW	1928.44	2369.08	35.00	1313.57	nr	**3682.65**
Cooling 7.1 kW, heating 8.2 kW	2334.19	2867.55	35.00	1313.57	nr	**4181.12**
Cooling 10 kW, heating 11.2 kW	2760.88	3391.74	35.00	1313.57	nr	**4705.31**
Cooling 12.20 kW, heating 14.60 kW	3020.23	3710.35	35.00	1313.57	nr	**5023.92**
Ceiling mounted 4 way blow cooling only unit with remote fan speed and load control; refrigerant 407C; includes outdoor unit						
Cooling 3.80 kW	1584.01	1945.95	35.00	1313.57	nr	**3259.52**
Cooling 5.20 kW	1769.46	2173.78	35.00	1313.57	nr	**3487.35**
Cooling 7.10 kW	2182.21	2680.85	35.00	1313.57	nr	**3994.42**
Cooling 10kw	2534.99	3114.24	35.00	1313.57	nr	**4427.81**
Cooling 12.2 kW	2686.98	3300.96	35.00	1313.57	nr	**4614.53**
In ceiling, ducted heat pump unit with remote fan speed and load control; refrigerant 407C; includes outdoor unit						
Cooling 3.60 kW, heating 4.10 kW	1282.84	1575.97	35.00	1313.57	nr	**2889.54**
Cooling 4.90 kW, heating 5.50 kW	1475.26	1812.35	35.00	1313.57	nr	**3125.92**
Cooling 7.10 kW, heating 8.20 kW	1475.26	1812.35	35.00	1313.57	nr	**3125.92**
Cooling 10 kW, heating 11.20 kW	1475.26	1812.35	35.00	1313.57	nr	**3125.92**
Cooling 12.20 kW, heating 14.50 kW	2974.23	3653.84	35.00	1313.57	nr	**4967.41**
In ceiling, ducted cooling only unit with remote fan speed and load control; refrigerant 407C; includes outdoor unit						
Cooling 3.70 kW	1157.34	1421.79	35.00	1313.57	nr	**2735.36**
Cooling 4.90 kW	1376.26	1690.73	35.00	1313.57	nr	**3004.30**
Cooling 7.10 kW	2077.64	2552.38	35.00	1313.57	nr	**3865.95**
Cooling 10 kW	2369.06	2910.39	35.00	1313.57	nr	**4223.96**
Cooling 12.3 kW	2640.95	3244.41	35.00	1313.57	nr	**4557.98**
Room units						
Ceiling mounted 4 way blow cassette heat pump unit with remote fan speed and load control; refrigerant 407C; excludes outdoor unit						
Cooling 3.6 kW, heating 4.1 kW	1104.35	1356.69	17.00	638.02	nr	**1994.71**
Cooling 4.9 kW, heating 5.5 kW	1130.84	1389.23	17.00	638.02	nr	**2027.25**
Cooling 7.1 kW, heating 8.2 kW	1224.27	1504.02	17.00	638.02	nr	**2142.04**
Cooling 10 kW, heating 11.2 kW	1326.06	1629.06	17.00	638.02	nr	**2267.08**
Cooling 12.20 kW, heating 14.60 kW	1437.61	1766.10	17.00	638.02	nr	**2404.12**

38 MECHANICAL/COOLING/HEATING SYSTEMS

Item	Net Price £	Material £	Labour hours	Labour £	Unit	Total rate £
LOCAL COOLING UNITS – cont						
Room units – cont						
Ceiling mounted 4 way blow cooling unit with remote fan speed and load control; refrigerant 407C; excludes outdoor unit						
Cooling 3.80 kW	1042.99	1281.32	17.00	638.02	nr	**1919.34**
Cooling 5.20 kW	1051.36	1291.59	17.00	638.02	nr	**1929.61**
Cooling 7.10 kW	1224.27	1504.02	17.00	638.02	nr	**2142.04**
Cooling 10 kW	1326.06	1629.06	17.00	638.02	nr	**2267.08**
Cooling 12.2 kW	1437.61	1766.10	17.00	638.02	nr	**2404.12**
In ceiling, ducted heat pump unit with remote fan speed and load control; refrigerant 407C; excludes outdoor unit						
Cooling 3.60 kW, heating 4.10 kW	637.24	782.85	17.00	638.02	nr	**1420.87**
Cooling 4.90 kW, heating 5.50 kW	677.68	832.53	17.00	638.02	nr	**1470.55**
Cooling 7.10 kW, heating 8.20 kW	1119.69	1375.54	17.00	638.02	nr	**2013.56**
Cooling 10 kW, heating 11.20 kW	1160.13	1425.22	17.00	638.02	nr	**2063.24**
Cooling 12.20 kW, heating 14.50 kW	1391.60	1709.58	17.00	638.02	nr	**2347.60**
In ceiling, ducted cooling unit only with remote fan speed and load control; refrigerant 407C; excludes outdoor unit						
Cooling 3.70 kW	616.32	757.14	17.00	638.02	nr	**1395.16**
Cooling 4.90 kW	658.15	808.54	17.00	638.02	nr	**1446.56**
Cooling 7.10 kW	1119.69	1375.54	17.00	638.02	nr	**2013.56**
Cooling 10 kW	1160.13	1425.22	17.00	638.02	nr	**2063.24**
Cooling 12.3 kW	1391.60	1709.58	17.00	638.02	nr	**2347.60**
External condensing units suitable for connection to multiple indoor units; inverter driven; refrigerant 407C						
Cooling only						
9 kW	2410.90	2961.79	17.00	638.02	nr	**3599.81**
Heat pump						
Cooling 5.20 kW, heating 6.10 kW	1717.88	2110.42	17.00	638.02	nr	**2748.44**
Cooling 6.80 kW, heating 2.50 kW	2190.57	2691.12	17.00	638.02	nr	**3329.14**
Cooling 8 kW, heating 9.60 kW	2536.37	3115.93	17.00	638.02	nr	**3753.95**
Cooling 14.50 kW, heating 16.50 kW	4209.64	5171.54	21.00	788.14	nr	**5959.68**

38 VENTILATION/AIR CONDITIONING SYSTEMS

Item	Net Price £	Material £	Labour hours	Labour £	Unit	**Total rate £**
DUCTWORK: CIRCULAR						
AIR DUCTLINES						
Galvanized sheet metal DW144 class B spirally wound circular section ductwork; including all necessary stiffeners, joints, couplers in the running length and duct supports						
Straight duct						
80 mm dia.	3.82	4.69	0.87	31.27	m	**35.96**
100 mm dia.	4.91	6.03	0.87	31.27	m	**37.30**
160 mm dia.	5.39	6.63	0.87	31.27	m	**37.90**
200 mm dia.	6.87	8.44	0.87	31.27	m	**39.71**
250 mm dia.	8.35	10.26	1.21	43.48	m	**53.74**
315 mm dia.	10.12	12.43	1.21	43.48	m	**55.91**
355 mm dia.	13.82	16.98	1.21	43.48	m	**60.46**
400 mm dia.	15.49	19.03	1.21	43.48	m	**62.51**
450 mm dia.	17.09	21.00	1.21	43.48	m	**64.48**
500 mm dia.	18.60	22.85	1.21	43.48	m	**66.33**
630 mm dia.	35.56	43.69	1.39	49.95	m	**93.64**
710 mm dia.	39.38	48.37	1.39	49.95	m	**98.32**
800 mm dia.	45.25	55.59	1.44	51.74	m	**107.33**
900 mm dia.	55.84	68.60	1.46	52.46	m	**121.06**
1000 mm dia.	69.75	85.69	1.65	59.28	m	**144.97**
1120 mm dia.	82.59	101.46	2.43	87.32	m	**188.78**
1250 mm dia.	90.27	110.90	2.43	87.32	m	**198.22**
1400 mm dia.	101.83	125.10	2.77	99.53	m	**224.63**
1600 mm dia.	131.38	161.40	3.06	109.95	m	**271.35**
Extra over fittings; circular duct class B						
End cap						
80 mm dia.	1.50	1.85	0.15	5.40	nr	**7.25**
100 mm dia.	1.64	2.02	0.15	5.40	nr	**7.42**
160 mm dia.	2.51	3.09	0.15	5.40	nr	**8.49**
200 mm dia.	2.97	3.64	0.20	7.18	nr	**10.82**
250 mm dia.	4.17	5.12	0.29	10.42	nr	**15.54**
315 mm dia.	5.29	6.50	0.29	10.42	nr	**16.92**
355 mm dia.	7.98	9.81	0.44	15.81	nr	**25.62**
400 mm dia.	8.27	10.16	0.44	15.81	nr	**25.97**
450 mm dia.	8.73	10.72	0.44	15.81	nr	**26.53**
500 mm dia.	9.16	11.26	0.44	15.81	nr	**27.07**
630 mm dia.	24.34	29.90	0.58	20.83	nr	**50.73**
710 mm dia.	28.45	34.95	0.69	24.79	nr	**59.74**
800 mm dia.	37.47	46.03	0.81	29.11	nr	**75.14**
900 mm dia.	42.31	51.98	0.92	33.05	nr	**85.03**
1000 mm dia.	58.34	71.67	1.04	37.37	nr	**109.04**
1120 mm dia.	64.52	79.26	1.16	41.68	nr	**120.94**
1250 mm dia.	71.09	87.34	1.16	41.68	nr	**129.02**
1400 mm dia.	90.18	110.79	1.16	41.68	nr	**152.47**
1600 mm dia.	101.93	125.22	1.16	41.68	nr	**166.90**

38 VENTILATION/AIR CONDITIONING SYSTEMS

Item	Net Price £	Material £	Labour hours	Labour £	Unit	**Total rate £**
DUCTWORK: CIRCULAR – cont						
Extra over fittings – cont						
Reducer						
80 mm dia.	4.86	5.97	0.29	10.42	nr	**16.39**
100 mm dia.	5.07	6.23	0.29	10.42	nr	**16.65**
160 mm dia.	6.65	8.17	0.29	10.42	nr	**18.59**
200 mm dia.	7.81	9.60	0.44	15.81	nr	**25.41**
250 mm dia.	9.29	11.41	0.58	20.83	nr	**32.24**
315 mm dia.	11.79	14.48	0.58	20.83	nr	**35.31**
355 mm dia.	15.47	19.01	0.87	31.27	nr	**50.28**
400 mm dia.	17.40	21.38	0.87	31.27	nr	**52.65**
450 mm dia.	18.70	22.97	0.87	31.27	nr	**54.24**
500 mm dia.	19.59	24.07	0.87	31.27	nr	**55.34**
630 mm dia.	54.38	66.80	0.87	31.27	nr	**98.07**
710 mm dia.	59.21	72.74	0.96	34.49	nr	**107.23**
800 mm dia.	75.84	93.17	1.06	38.08	nr	**131.25**
900 mm dia.	80.30	98.65	1.16	41.68	nr	**140.33**
1000 mm dia.	104.09	127.88	1.25	44.92	nr	**172.80**
1120 mm dia.	114.47	140.63	3.47	124.69	nr	**265.32**
1250 mm dia.	127.23	156.30	3.47	124.69	nr	**280.99**
1400 mm dia.	158.46	194.67	4.05	145.53	nr	**340.20**
1600 mm dia.	149.57	183.75	4.62	166.00	nr	**349.75**
90° segmented radius bend						
80 mm dia.	2.58	3.17	0.29	10.42	nr	**13.59**
100 mm dia.	2.96	3.63	0.29	10.42	nr	**14.05**
160 mm dia.	5.00	6.14	0.29	10.42	nr	**16.56**
200 mm dia.	6.72	8.25	0.44	15.81	nr	**24.06**
250 mm dia.	9.90	12.16	0.58	20.83	nr	**32.99**
315 mm dia.	11.53	14.16	0.58	20.83	nr	**34.99**
355 mm dia.	12.00	14.74	0.87	31.27	nr	**46.01**
400 mm dia.	14.14	17.37	0.87	31.27	nr	**48.64**
450 mm dia.	16.14	19.82	0.87	31.27	nr	**51.09**
500 mm dia.	16.91	20.77	0.87	31.27	nr	**52.04**
630 mm dia.	38.97	47.87	0.87	31.27	nr	**79.14**
710 mm dia.	54.27	66.68	0.96	34.49	nr	**101.17**
800 mm dia.	60.03	73.75	1.06	38.08	nr	**111.83**
900 mm dia.	65.38	80.31	1.16	41.68	nr	**121.99**
1000 mm dia.	121.59	149.37	1.25	44.92	nr	**194.29**
1120 mm dia.	135.12	165.99	3.47	124.69	nr	**290.68**
1250 mm dia.	173.97	213.72	3.47	124.69	nr	**338.41**
1400 mm dia.	310.50	381.44	4.05	145.53	nr	**526.97**
1600 mm dia.	321.33	394.76	4.62	166.00	nr	**560.76**
45° radius bend						
80 mm dia.	2.70	3.32	0.29	10.42	nr	**13.74**
100 mm dia.	3.04	3.74	0.29	10.42	nr	**14.16**
160 mm dia.	4.31	5.29	0.29	10.42	nr	**15.71**
200 mm dia.	5.59	6.87	0.40	14.37	nr	**21.24**
250 mm dia.	7.76	9.53	0.58	20.83	nr	**30.36**
315 mm dia.	10.42	12.80	0.58	20.83	nr	**33.63**
355 mm dia.	11.15	13.70	0.87	31.27	nr	**44.97**

38 VENTILATION/AIR CONDITIONING SYSTEMS

Item	Net Price £	Material £	Labour hours	Labour £	Unit	**Total rate £**
400 mm dia.	13.34	16.39	0.87	31.27	nr	**47.66**
450 mm dia.	13.51	16.60	0.87	31.27	nr	**47.87**
500 mm dia.	14.46	17.77	0.87	31.27	nr	**49.04**
630 mm dia.	41.84	51.40	0.87	31.27	nr	**82.67**
710 mm dia.	52.52	64.52	0.96	34.49	nr	**99.01**
800 mm dia.	62.02	76.19	1.06	38.08	nr	**114.27**
900 mm dia.	67.83	83.33	1.16	41.68	nr	**125.01**
1000 mm dia.	112.37	138.04	1.25	44.92	nr	**182.96**
1120 mm dia.	122.62	150.64	3.47	124.69	nr	**275.33**
1250 mm dia.	136.36	167.52	3.47	124.69	nr	**292.21**
1400 mm dia.	187.44	230.26	4.05	145.53	nr	**375.79**
1600 mm dia.	182.98	224.79	4.62	166.00	nr	**390.79**
90° equal twin bend						
80 mm dia.	7.70	9.46	0.58	20.83	nr	**30.29**
100 mm dia.	7.98	9.81	0.58	20.83	nr	**30.64**
160 mm dia.	14.12	17.35	0.58	20.83	nr	**38.18**
200 mm dia.	19.79	24.31	0.87	31.27	nr	**55.58**
250 mm dia.	30.50	37.46	1.16	41.68	nr	**79.14**
315 mm dia.	42.34	52.02	1.16	41.68	nr	**93.70**
355 mm dia.	47.17	57.95	1.73	62.16	nr	**120.11**
400 mm dia.	52.43	64.41	1.73	62.16	nr	**126.57**
450 mm dia.	56.74	69.71	1.73	62.16	nr	**131.87**
500 mm dia.	61.07	75.02	1.73	62.16	nr	**137.18**
630 mm dia.	117.99	144.95	1.73	62.16	nr	**207.11**
710 mm dia.	156.62	192.41	1.82	65.39	nr	**257.80**
800 mm dia.	191.36	235.08	1.93	69.35	nr	**304.43**
900 mm dia.	221.91	272.61	2.02	72.60	nr	**345.21**
1000 mm dia.	305.78	375.65	2.11	75.82	nr	**451.47**
1120 mm dia.	356.10	437.47	4.62	166.00	nr	**603.47**
1250 mm dia.	425.92	523.25	4.62	166.00	nr	**689.25**
1400 mm dia.	681.16	836.81	4.62	166.00	nr	**1002.81**
1600 mm dia.	681.12	836.76	4.62	166.00	nr	**1002.76**
Conical branch						
80 mm dia.	10.21	12.55	0.58	20.83	nr	**33.38**
100 mm dia.	10.43	12.81	0.58	20.83	nr	**33.64**
160 mm dia.	11.29	13.87	0.58	20.83	nr	**34.70**
200 mm dia.	12.01	14.75	0.87	31.27	nr	**46.02**
250 mm dia.	15.71	19.30	1.16	41.68	nr	**60.98**
315 mm dia.	17.53	21.54	1.16	41.68	nr	**63.22**
355 mm dia.	19.03	23.38	1.73	62.16	nr	**85.54**
400 mm dia.	19.63	24.12	1.73	62.16	nr	**86.28**
450 mm dia.	24.27	29.82	1.73	62.16	nr	**91.98**
500 mm dia.	25.00	30.71	1.73	62.16	nr	**92.87**
630 mm dia.	50.29	61.78	1.73	62.16	nr	**123.94**
710 mm dia.	56.02	68.82	1.82	65.39	nr	**134.21**
800 mm dia.	72.59	89.18	1.93	69.35	nr	**158.53**
900 mm dia.	76.95	94.53	2.02	72.60	nr	**167.13**
1000 mm dia.	97.55	119.84	2.11	75.82	nr	**195.66**
1120 mm dia.	135.93	166.99	4.62	166.00	nr	**332.99**
1250 mm dia.	135.93	166.99	5.20	186.86	nr	**353.85**
1400 mm dia.	163.76	201.18	5.20	186.86	nr	**388.04**
1600 mm dia.	192.89	236.96	5.20	186.86	nr	**423.82**

38 VENTILATION/AIR CONDITIONING SYSTEMS

Item	Net Price £	Material £	Labour hours	Labour £	Unit	**Total rate £**
DUCTWORK: CIRCULAR – cont						
Extra over fittings – cont						
45° branch						
80 mm dia.	8.10	9.95	0.58	20.83	nr	**30.78**
100 mm dia.	8.30	10.20	0.58	20.83	nr	**31.03**
160 mm dia.	12.63	15.52	0.58	20.83	nr	**36.35**
200 mm dia.	13.08	16.07	0.87	31.27	nr	**47.34**
250 mm dia.	13.71	16.84	1.16	41.68	nr	**58.52**
315 mm dia.	14.59	17.92	1.16	41.68	nr	**59.60**
355 mm dia.	18.80	23.10	1.73	62.16	nr	**85.26**
400 mm dia.	19.54	24.00	1.73	62.16	nr	**86.16**
450 mm dia.	20.36	25.01	1.73	62.16	nr	**87.17**
500 mm dia.	21.23	26.08	1.73	62.16	nr	**88.24**
630 mm dia.	42.22	51.87	1.73	62.16	nr	**114.03**
710 mm dia.	47.14	57.91	1.82	65.39	nr	**123.30**
800 mm dia.	67.26	82.62	2.13	76.53	nr	**159.15**
900 mm dia.	76.13	93.52	2.31	83.01	nr	**176.53**
1000 mm dia.	93.98	115.46	2.31	83.01	nr	**198.47**
1120 mm dia.	116.45	143.06	4.62	166.00	nr	**309.06**
1250 mm dia.	124.44	152.87	4.62	166.00	nr	**318.87**
1400 mm dia.	143.73	176.57	4.62	166.00	nr	**342.57**
1600 mm dia.	175.69	215.84	4.62	166.00	nr	**381.84**
For galvanized sheet metal DW144 class C rates, refer to galvanized sheet metal DW144 class B						

38 VENTILATION/AIR CONDITIONING SYSTEMS

Item	Net Price £	Material £	Labour hours	Labour £	Unit	**Total rate £**
DUCTWORK: FLAT OVAL						
AIR DUCTLINES						
Galvanized sheet metal DW144 class B spirally wound flat oval section ductwork; including all necessary stiffeners, joints, couplers in the running length and duct supports						
Straight duct						
345 × 102 mm	19.94	24.50	2.71	97.39	m	**121.89**
427 × 102 mm	22.83	28.05	2.99	107.45	m	**135.50**
508 × 102 mm	25.19	30.94	3.14	112.82	m	**143.76**
559 × 152 mm	34.01	41.78	3.43	123.26	m	**165.04**
531 × 203 mm	34.54	42.43	3.43	123.26	m	**165.69**
851 × 203 mm	44.95	55.22	5.72	205.54	m	**260.76**
582 × 254 mm	45.43	55.81	3.62	130.08	m	**185.89**
823 × 254 mm	45.59	56.01	5.80	208.41	m	**264.42**
1303 × 254 mm	108.05	132.74	8.13	292.14	m	**424.88**
632 × 305 mm	41.14	50.54	3.93	141.21	m	**191.75**
1275 × 305 mm	99.69	122.47	8.13	292.14	m	**414.61**
765 × 356 mm	47.67	58.56	5.72	205.54	m	**264.10**
1247 × 356 mm	100.78	123.81	8.13	292.14	m	**415.95**
1727 × 356 mm	135.75	166.77	10.41	374.06	m	**540.83**
737 × 406 mm	48.57	59.67	5.72	205.54	m	**265.21**
818 × 406 mm	51.30	63.02	6.21	223.15	m	**286.17**
978 × 406 mm	56.52	69.44	6.92	248.66	m	**318.10**
1379 × 406 mm	109.78	134.86	8.75	314.42	m	**449.28**
1699 × 406 mm	111.38	136.83	10.41	374.06	m	**510.89**
709 × 457 mm	48.63	59.75	5.72	205.54	m	**265.29**
1189 × 457 mm	92.69	113.87	8.80	316.22	m	**430.09**
1671 × 457 mm	133.80	164.38	10.31	370.48	m	**534.86**
678 × 508 mm	48.03	59.01	5.72	205.54	m	**264.55**
919 × 508 mm	56.52	69.44	7.30	262.32	m	**331.76**
1321 × 508 mm	109.59	134.63	8.75	314.42	m	**449.05**
Extra over fittings; flat oval duct class B						
End cap						
345 × 102 mm	17.49	21.48	0.20	7.18	nr	**28.66**
427 × 102 mm	17.97	22.07	0.20	7.18	nr	**29.25**
508 × 102 mm	18.44	22.65	0.20	7.18	nr	**29.83**
559 × 152 mm	26.70	32.80	0.29	10.42	nr	**43.22**
531 × 203 mm	26.82	32.95	0.29	10.42	nr	**43.37**
851 × 203 mm	29.77	36.57	0.44	15.81	nr	**52.38**
582 × 254 mm	27.54	33.83	0.44	15.81	nr	**49.64**
823 × 254 mm	29.28	35.97	0.44	15.81	nr	**51.78**
1303 × 254 mm	96.73	118.83	0.69	24.79	nr	**143.62**
632 × 305 mm	27.52	33.81	0.69	24.79	nr	**58.60**
1275 × 305 mm	85.33	104.83	0.69	24.79	nr	**129.62**

38 VENTILATION/AIR CONDITIONING SYSTEMS

Item	Net Price £	Material £	Labour hours	Labour £	Unit	Total rate £
DUCTWORK: FLAT OVAL – cont						
Extra over fittings – cont						
End cap – cont						
765 × 356 mm	28.88	35.48	0.69	24.79	nr	**60.27**
1247 × 356 mm	82.87	101.81	0.69	24.79	nr	**126.60**
1727 × 356 mm	123.12	151.25	0.69	24.79	nr	**176.04**
737 × 406 mm	29.71	36.50	1.04	37.37	nr	**73.87**
818 × 406 mm	30.47	37.43	0.69	24.79	nr	**62.22**
978 × 406 mm	31.95	39.25	0.69	24.79	nr	**64.04**
1379 × 406 mm	113.46	139.39	1.04	37.37	nr	**176.76**
1699 × 406 mm	123.85	152.15	1.04	37.37	nr	**189.52**
709 × 457 mm	28.90	35.50	1.04	37.37	nr	**72.87**
1189 × 457 mm	80.91	99.39	1.04	37.37	nr	**136.76**
1671 × 457 mm	120.59	148.14	1.04	37.37	nr	**185.51**
678 × 508 mm	38.99	47.90	1.04	37.37	nr	**85.27**
919 × 508 mm	42.06	51.67	1.04	37.37	nr	**89.04**
1321 × 508 mm	120.56	148.11	1.04	37.37	nr	**185.48**
Reducer						
345 × 102 mm	32.65	40.11	0.95	34.14	nr	**74.25**
427 × 102 mm	33.12	40.69	1.06	38.08	nr	**78.77**
508 × 102 mm	33.76	41.47	1.13	40.60	nr	**82.07**
559 × 152 mm	47.08	57.83	1.26	45.27	nr	**103.10**
531 × 203 mm	47.41	58.24	1.26	45.27	nr	**103.51**
851 × 203 mm	61.06	75.01	1.34	48.15	nr	**123.16**
582 × 254 mm	49.98	61.40	1.34	48.15	nr	**109.55**
823 × 254 mm	62.76	77.10	1.34	48.15	nr	**125.25**
1303 × 254 mm	142.93	175.59	1.34	48.15	nr	**223.74**
632 × 305 mm	54.96	67.51	0.70	25.16	nr	**92.67**
1275 × 305 mm	145.90	179.24	1.16	41.68	nr	**220.92**
765 × 356 mm	58.72	72.14	1.16	41.68	nr	**113.82**
1247 × 356 mm	148.42	182.33	1.16	41.68	nr	**224.01**
1727 × 356 mm	150.96	185.45	1.25	44.92	nr	**230.37**
737 × 406 mm	64.84	79.65	1.16	41.68	nr	**121.33**
818 × 406 mm	65.63	80.63	1.27	45.64	nr	**126.27**
978 × 406 mm	71.23	87.51	1.44	51.74	nr	**139.25**
1379 × 406 mm	150.65	185.07	1.44	51.74	nr	**236.81**
1699 × 406 mm	170.72	209.73	1.44	51.74	nr	**261.47**
709 × 457 mm	65.55	80.52	1.16	41.68	nr	**122.20**
1189 × 457 mm	132.86	163.22	1.34	48.15	nr	**211.37**
1671 × 457 mm	147.71	181.46	1.44	51.74	nr	**233.20**
678 × 508 mm	69.71	85.64	1.16	41.68	nr	**127.32**
919 × 508 mm	74.26	91.22	1.26	45.27	nr	**136.49**
1321 × 508 mm	148.74	182.73	1.44	51.74	nr	**234.47**
90° radius bend						
345 × 102 mm	42.65	52.40	0.29	10.42	nr	**62.82**
427 × 102 mm	40.78	50.10	0.58	20.83	nr	**70.93**
508 × 102 mm	39.31	48.29	0.58	20.83	nr	**69.12**
559 × 152 mm	54.30	66.71	0.58	20.83	nr	**87.54**
531 × 203 mm	54.90	67.44	0.87	31.27	nr	**98.71**
851 × 203 mm	74.18	91.13	0.87	31.27	nr	**122.40**
582 × 254 mm	52.04	63.93	0.87	31.27	nr	**95.20**

38 VENTILATION/AIR CONDITIONING SYSTEMS

Item	Net Price £	Material £	Labour hours	Labour £	Unit	**Total rate £**
823 × 254 mm	77.32	94.98	0.87	31.27	nr	**126.25**
1303 × 254 mm	173.98	213.74	0.96	34.49	nr	**248.23**
632 × 305 mm	49.12	60.34	0.87	31.27	nr	**91.61**
1275 × 305 mm	220.48	270.86	0.96	34.49	nr	**305.35**
765 × 356 mm	57.18	70.25	0.87	31.27	nr	**101.52**
1247 × 356 mm	221.69	272.35	0.96	34.49	nr	**306.84**
1727 × 356 mm	270.03	331.74	1.25	44.92	nr	**376.66**
737 × 406 mm	59.02	72.50	0.96	34.49	nr	**106.99**
818 × 406 mm	85.09	104.54	0.87	31.27	nr	**135.81**
978 × 406 mm	75.54	92.80	0.96	34.49	nr	**127.29**
1379 × 406 mm	198.30	243.61	1.16	41.68	nr	**285.29**
1699 × 406 mm	339.20	416.70	1.25	44.92	nr	**461.62**
709 × 457 mm	67.99	83.53	0.87	31.27	nr	**114.80**
1189 × 457 mm	164.17	201.68	0.96	34.49	nr	**236.17**
1671 × 457 mm	338.32	415.62	1.25	44.92	nr	**460.54**
678 × 508 mm	65.69	80.70	0.87	31.27	nr	**111.97**
919 × 508 mm	74.04	90.96	0.96	34.49	nr	**125.45**
1321 × 508 mm	205.28	252.19	1.16	41.68	nr	**293.87**
45° radius bend						
345 × 102 mm	30.18	37.08	0.79	28.39	nr	**65.47**
427 × 102 mm	29.59	36.35	0.85	30.54	nr	**66.89**
508 × 102 mm	29.00	35.63	0.95	34.14	nr	**69.77**
559 × 152 mm	40.27	49.48	0.79	28.39	nr	**77.87**
531 × 203 mm	40.53	49.79	0.85	30.54	nr	**80.33**
851 × 203 mm	56.62	69.56	0.98	35.21	nr	**104.77**
582 × 254 mm	39.50	48.52	0.76	27.31	nr	**75.83**
823 × 254 mm	58.84	72.28	0.95	34.14	nr	**106.42**
1303 × 254 mm	170.57	209.55	1.16	41.68	nr	**251.23**
632 × 305 mm	42.91	52.71	0.58	20.83	nr	**73.54**
1275 × 305 mm	196.45	241.34	1.16	41.68	nr	**283.02**
765 × 356 mm	44.04	54.11	0.87	31.27	nr	**85.38**
1247 × 356 mm	197.14	242.18	1.16	41.68	nr	**283.86**
1727 × 356 mm	191.73	235.54	1.26	45.27	nr	**280.81**
737 × 406 mm	46.46	57.08	0.69	24.79	nr	**81.87**
818 × 406 mm	64.49	79.22	0.78	28.03	nr	**107.25**
978 × 406 mm	60.44	74.25	0.87	31.27	nr	**105.52**
1379 × 406 mm	187.91	230.84	1.16	41.68	nr	**272.52**
1699 × 406 mm	226.63	278.42	1.27	45.64	nr	**324.06**
709 × 457 mm	52.34	64.30	0.81	29.11	nr	**93.41**
1189 × 457 mm	155.28	190.76	0.95	34.14	nr	**224.90**
1671 × 457 mm	223.28	274.30	1.26	45.27	nr	**319.57**
678 × 508 mm	51.04	62.71	0.92	33.05	nr	**95.76**
919 × 508 mm	59.34	72.90	1.10	39.52	nr	**112.42**
1321 × 508 mm	190.95	234.58	1.25	44.92	nr	**279.50**
90° hard bend with turning vanes						
345 × 102 mm	43.75	53.75	0.55	19.77	nr	**73.52**
427 × 102 mm	42.70	52.46	1.16	41.68	nr	**94.14**
508 × 102 mm	41.65	51.17	1.16	41.68	nr	**92.85**
559 × 152 mm	44.08	54.15	1.16	41.68	nr	**95.83**
531 × 203 mm	44.48	54.64	1.73	62.16	nr	**116.80**
851 × 203 mm	68.83	84.56	1.73	62.16	nr	**146.72**
582 × 254 mm	50.27	61.76	1.73	62.16	nr	**123.92**

38 VENTILATION/AIR CONDITIONING SYSTEMS

Item	Net Price £	Material £	Labour hours	Labour £	Unit	Total rate £
DUCTWORK: FLAT OVAL – cont						
Extra over fittings – cont						
90° hard bend with turning vanes – cont						
823 × 254 mm	68.41	84.04	1.73	62.16	nr	**146.20**
1303 × 254 mm	119.45	146.75	1.82	65.39	nr	**212.14**
632 × 305 mm	66.18	81.30	1.73	62.16	nr	**143.46**
1275 × 305 mm	121.63	149.43	1.82	65.39	nr	**214.82**
765 × 356 mm	64.95	79.79	1.73	62.16	nr	**141.95**
1247 × 356 mm	122.51	150.51	1.82	65.39	nr	**215.90**
1727 × 356 mm	205.92	252.98	1.82	65.39	nr	**318.37**
737 × 406 mm	71.34	87.64	1.73	62.16	nr	**149.80**
818 × 406 mm	68.86	84.60	1.73	62.16	nr	**146.76**
978 × 406 mm	75.19	92.37	1.73	62.16	nr	**154.53**
1379 × 406 mm	161.35	198.22	1.82	65.39	nr	**263.61**
1699 × 406 mm	252.54	310.24	2.11	75.82	nr	**386.06**
709 × 457 mm	79.74	97.97	1.73	62.16	nr	**160.13**
1189 × 457 mm	133.37	163.84	1.82	65.39	nr	**229.23**
1671 × 457 mm	209.45	257.31	2.11	75.82	nr	**333.13**
678 × 508 mm	77.09	94.71	1.82	65.39	nr	**160.10**
919 × 508 mm	84.60	103.93	1.82	65.39	nr	**169.32**
1321 × 508 mm	168.88	207.47	2.11	75.82	nr	**283.29**
90° branch						
345 × 102 mm	33.07	40.62	0.58	20.83	nr	**61.45**
427 × 102 mm	33.72	41.42	0.58	20.83	nr	**62.25**
508 × 102 mm	34.33	42.18	1.16	41.68	nr	**83.86**
559 × 152 mm	50.81	62.42	1.16	41.68	nr	**104.10**
531 × 203 mm	51.68	63.49	1.16	41.68	nr	**105.17**
851 × 203 mm	67.16	82.51	1.73	62.16	nr	**144.67**
582 × 254 mm	57.79	70.99	1.73	62.16	nr	**133.15**
823 × 254 mm	73.00	89.68	1.73	62.16	nr	**151.84**
1303 × 254 mm	155.39	190.90	1.82	65.39	nr	**256.29**
632 × 305 mm	64.45	79.18	1.73	62.16	nr	**141.34**
1275 × 305 mm	156.05	191.71	1.82	65.39	nr	**257.10**
765 × 356 mm	72.75	89.38	1.73	62.16	nr	**151.54**
1247 × 356 mm	160.52	197.20	1.82	65.39	nr	**262.59**
1727 × 356 mm	203.83	250.40	2.11	75.82	nr	**326.22**
737 × 406 mm	78.45	96.38	1.73	62.16	nr	**158.54**
818 × 406 mm	80.02	98.30	1.73	62.16	nr	**160.46**
978 × 406 mm	89.06	109.41	1.82	65.39	nr	**174.80**
1379 × 406 mm	172.40	211.80	1.93	69.35	nr	**281.15**
1699 × 406 mm	200.84	246.73	2.11	75.82	nr	**322.55**
709 × 457 mm	80.13	98.44	1.73	62.16	nr	**160.60**
1189 × 457 mm	141.55	173.89	1.82	65.39	nr	**239.28**
1671 × 457 mm	201.29	247.29	2.11	75.82	nr	**323.11**
678 × 508 mm	85.55	105.09	1.73	62.16	nr	**167.25**
919 × 508 mm	94.63	116.26	1.82	65.39	nr	**181.65**
1321 × 508 mm	173.12	212.68	2.11	75.82	nr	**288.50**

38 VENTILATION/AIR CONDITIONING SYSTEMS

Item	Net Price £	Material £	Labour hours	Labour £	Unit	**Total rate £**
45° branch						
345 × 102 mm	34.93	42.91	0.58	20.83	nr	**63.74**
427 × 102 mm	36.77	45.17	0.58	20.83	nr	**66.00**
508 × 102 mm	38.81	47.68	1.16	41.68	nr	**89.36**
559 × 152 mm	55.76	68.50	1.73	62.16	nr	**130.66**
531 × 203 mm	55.39	68.05	1.73	62.16	nr	**130.21**
851 × 203 mm	82.01	100.75	1.73	62.16	nr	**162.91**
582 × 254 mm	61.77	75.88	1.73	62.16	nr	**138.04**
823 × 254 mm	83.13	102.12	1.82	65.39	nr	**167.51**
1303 × 254 mm	192.17	236.08	1.92	69.00	nr	**305.08**
632 × 305 mm	68.64	84.33	1.73	62.16	nr	**146.49**
1275 × 305 mm	189.60	232.92	1.82	65.39	nr	**298.31**
765 × 356 mm	78.94	96.98	1.73	62.16	nr	**159.14**
1247 × 356 mm	190.35	233.85	1.82	65.39	nr	**299.24**
1727 × 356 mm	267.75	328.93	1.82	65.39	nr	**394.32**
737 × 406 mm	83.75	102.89	1.73	62.16	nr	**165.05**
818 × 406 mm	87.86	107.94	1.73	62.16	nr	**170.10**
978 × 406 mm	103.07	126.62	1.73	62.16	nr	**188.78**
1379 × 406 mm	207.50	254.91	1.93	69.35	nr	**324.26**
1699 × 406 mm	258.65	317.75	2.19	78.70	nr	**396.45**
709 × 457 mm	82.93	101.88	1.73	62.16	nr	**164.04**
1189 × 457 mm	163.51	200.88	1.82	65.39	nr	**266.27**
1671 × 457 mm	254.48	312.63	2.11	75.82	nr	**388.45**
678 × 508 mm	85.94	105.58	1.73	62.16	nr	**167.74**
919 × 508 mm	102.66	126.12	1.82	65.39	nr	**191.51**
1321 × 508 mm	200.24	245.99	1.93	69.35	nr	**315.34**
For rates for access doors refer to ancillaries in Ductwork Ancillaries Access Hatches						

38 VENTILATION/AIR CONDITIONING SYSTEMS

Item	Net Price £	Material £	Labour hours	Labour £	Unit	**Total rate £**
DUCTWORK: FLEXIBLE						
AIR DUCTLINES						
Aluminium foil flexible ductwork, DW 144 class B; multiply aluminium polyester laminate fabric, with high tensile steel wire helix						
Duct						
102 mm dia.	1.39	1.71	0.33	11.85	m	**13.56**
152 mm dia.	2.04	2.51	0.33	11.85	m	**14.36**
203 mm dia.	2.69	3.31	0.33	11.85	m	**15.16**
254 mm dia.	3.38	4.15	0.33	11.85	m	**16.00**
304 mm dia.	4.14	5.08	0.33	11.85	m	**16.93**
355 mm dia.	5.64	6.93	0.33	11.85	m	**18.78**
406 mm dia.	6.27	7.71	0.33	11.85	m	**19.56**
Insulated aluminium foil flexible ductwork, DW144 class B; laminate construction of aluminium and polyester multiply inner core with 25 mm insulation; outer layer of multiply aluminium polyester laminate, with high tensile steel wire helix						
Duct						
102 mm dia.	3.12	3.83	0.50	17.98	m	**21.81**
152 mm dia.	4.08	5.01	0.50	17.98	m	**22.99**
203 mm dia.	4.95	6.08	0.50	17.98	m	**24.06**
254 mm dia.	6.09	7.49	0.50	17.98	m	**25.47**
304 mm dia.	7.83	9.62	0.50	17.98	m	**27.60**
355 mm dia.	9.72	11.94	0.50	17.98	m	**29.92**
406 mm dia.	10.76	13.22	0.50	17.98	m	**31.20**

38 VENTILATION/AIR CONDITIONING SYSTEMS

Item	Net Price £	Material £	Labour hours	Labour £	Unit	Total rate £
DUCTWORK: PLASTIC						
AIR DUCTLINES						
Rigid grey PVC DW 154 circular section ductwork; solvent welded or filler rod welded joints; excludes couplers and supports (these are detailed separately); ductwork to conform to curent HSE regulations						
Straight duct (standard length 6 m)						
110 mm	3.70	4.55	1.74	62.53	m	**67.08**
160 mm	6.99	8.59	1.71	61.45	m	**70.04**
200 mm	9.19	11.29	1.79	64.32	m	**75.61**
225 mm	13.10	16.10	1.96	70.43	m	**86.53**
250 mm	12.88	15.82	1.98	71.14	m	**86.96**
315 mm	18.42	22.63	2.23	80.14	m	**102.77**
355 mm	24.67	30.30	2.25	80.85	m	**111.15**
400 mm	31.02	38.10	2.39	85.88	m	**123.98**
450 mm	38.33	47.09	2.95	106.01	m	**153.10**
500 mm	47.20	57.98	2.98	107.09	m	**165.07**
600 mm	72.97	89.64	3.12	112.11	m	**201.75**
Extra for supports (BZP finish); Horizontal – Maximum 2.4 m centres; Vertical – Maximum 4.0 m centres						
Duct size						
110 mm	22.87	28.10	0.59	21.20	m	**49.30**
160 mm	22.95	28.19	0.59	21.20	m	**49.39**
200 mm	23.03	28.30	0.59	21.20	m	**49.50**
225 mm	24.45	30.04	0.63	22.64	m	**52.68**
250 mm	24.46	30.05	0.63	22.64	m	**52.69**
315 mm	24.63	30.26	0.63	22.64	m	**52.90**
355 mm	23.56	28.95	0.81	29.11	m	**58.06**
400 mm	31.99	39.30	0.78	28.03	m	**67.33**
450 mm	32.89	40.40	0.80	28.75	m	**69.15**
500 mm	33.05	40.60	0.80	28.75	m	**69.35**
600 mm	35.09	43.11	0.85	30.54	m	**73.65**
Note: These are maximum figures and may be reduced subject to local conditions (i.e. a high number of changes of direction)						
Extra over fittings; Rigid grey PVC						
90° bend						
110 mm	18.93	23.26	1.11	39.89	m	**63.15**
160 mm	20.88	25.65	1.34	48.15	m	**73.80**
200 mm	17.48	21.47	2.00	71.87	m	**93.34**
225 mm	26.83	32.96	1.79	64.32	m	**97.28**
250 mm	28.57	35.10	2.01	72.22	m	**107.32**
315 mm	49.31	60.57	2.56	91.99	m	**152.56**
355 mm	58.72	72.14	2.71	97.39	m	**169.53**

38 VENTILATION/AIR CONDITIONING SYSTEMS

Item	Net Price £	Material £	Labour hours	Labour £	Unit	Total rate £
DUCTWORK: PLASTIC – cont						
Extra over fittings – cont						
90° bend – cont						
400 mm	74.58	91.62	3.58	128.65	m	**220.27**
450 mm	225.43	276.94	4.55	163.51	m	**440.45**
500 mm	261.66	321.45	5.01	180.02	m	**501.47**
600 mm	447.89	550.23	5.78	207.69	m	**757.92**
45° bend						
110 mm	7.80	9.59	0.71	25.51	m	**35.10**
160 mm	9.05	11.12	0.93	33.41	m	**44.53**
200 mm	9.94	12.21	1.14	40.96	m	**53.17**
225 mm	11.21	13.78	1.41	50.66	m	**64.44**
250 mm	12.01	14.75	1.62	58.20	m	**72.95**
315 mm	19.17	23.55	1.93	69.35	m	**92.90**
355 mm	24.94	30.64	2.22	79.78	m	**110.42**
400 mm	28.67	35.22	2.85	102.41	m	**137.63**
450 mm	110.43	135.66	3.77	135.47	m	**271.13**
500 mm	123.49	151.70	4.16	149.49	m	**301.19**
600 mm	201.18	247.15	4.81	172.84	m	**419.99**
Tee						
110 mm	16.54	20.32	1.04	37.37	m	**57.69**
160 mm	20.17	24.78	1.38	49.59	m	**74.37**
200 mm	26.83	32.96	1.79	64.32	m	**97.28**
225 mm	30.91	37.97	2.25	80.85	m	**118.82**
250 mm	37.84	46.48	2.62	94.15	m	**140.63**
315 mm	63.03	77.44	3.17	113.90	m	**191.34**
355 mm	88.16	108.31	3.73	134.03	m	**242.34**
400 mm	85.94	105.58	4.71	169.25	m	**274.83**
450 mm	168.99	207.61	5.72	205.54	m	**413.15**
500 mm	192.06	235.95	6.33	227.46	m	**463.41**
Coupler						
110 mm	4.42	5.43	0.70	25.16	m	**30.59**
160 mm	5.47	6.72	0.91	32.71	m	**39.43**
200 mm	6.80	8.36	1.12	40.25	m	**48.61**
225 mm	8.89	10.92	1.41	50.66	m	**61.58**
250 mm	9.90	12.16	1.60	57.50	m	**69.66**
315 mm	11.80	14.50	1.86	66.84	m	**81.34**
355 mm	12.40	15.24	2.43	87.32	m	**102.56**
400 mm	13.21	16.23	2.66	95.59	m	**111.82**
450 mm	31.79	39.05	3.20	114.99	m	**154.04**
500 mm	35.87	44.07	3.52	126.48	m	**170.55**
Damper						
110 mm	37.03	45.50	0.96	34.49	m	**79.99**
160 mm	37.59	46.18	1.26	45.27	m	**91.45**
200 mm	38.40	47.18	1.52	54.62	m	**101.80**
225 mm	39.69	48.76	1.88	67.56	m	**116.32**
250 mm	39.55	48.58	2.16	77.62	m	**126.20**
315 mm	45.36	55.72	2.94	105.65	m	**161.37**
355 mm	48.46	59.53	3.42	122.89	m	**182.42**
400 mm	52.03	63.92	3.79	136.20	m	**200.12**

38 VENTILATION/AIR CONDITIONING SYSTEMS

Item	Net Price £	Material £	Labour hours	Labour £	Unit	**Total rate £**
Reducer						
160 × 110 mm	9.68	11.90	0.85	30.54	m	**42.44**
200 × 110 mm	14.09	17.31	0.98	35.21	m	**52.52**
200 × 160 mm	12.20	14.98	1.07	38.45	m	**53.43**
225 × 200 mm	16.01	19.67	1.38	49.59	m	**69.26**
250 × 160 mm	16.93	20.80	1.38	49.59	m	**70.39**
250 × 200 mm	16.08	19.75	1.48	53.18	m	**72.93**
250 × 225 mm	15.78	19.38	1.58	56.78	m	**76.16**
315 × 200 mm	17.23	21.17	1.63	58.58	m	**79.75**
315 × 250 mm	17.71	21.76	1.84	66.12	m	**87.88**
355 × 200 mm	26.24	32.23	1.99	71.52	m	**103.75**
355 × 250 mm	19.57	24.05	1.97	70.79	m	**94.84**
355 × 315 mm	24.40	29.98	2.16	77.62	m	**107.60**
400 × 225 mm	29.62	36.39	2.47	88.76	m	**125.15**
400 × 315 mm	29.92	36.76	2.75	98.82	m	**135.58**
400 × 355 mm	31.51	38.71	2.73	98.09	m	**136.80**
450 × 315 mm	34.01	41.78	2.91	104.57	m	**146.35**
Flange						
110 mm	6.38	7.83	0.70	25.16	m	**32.99**
160 mm	6.46	7.94	0.91	32.71	m	**40.65**
200 mm	6.97	8.56	1.12	40.25	m	**48.81**
225 mm	6.77	8.32	1.35	48.51	m	**56.83**
250 mm	7.23	8.88	1.55	55.70	m	**64.58**
315 mm	10.30	12.65	1.85	66.48	m	**79.13**
355 mm	10.90	13.39	2.06	74.02	m	**87.41**
400 mm	10.93	13.43	2.62	94.15	m	**107.58**
Polypropylene (PPS) DW154 circular section ductwork; filler rod welded joints; excludes couplers and supports (these are detailed separately); ductwork to conform to current HSE regulations						
Straight duct (standard length 6 m)						
110 mm	17.32	21.27	0.65	23.35	m	**44.62**
160 mm	18.99	23.33	0.75	26.95	m	**50.28**
200 mm	20.16	24.77	0.85	30.54	m	**55.31**
225 mm	25.84	31.74	1.07	38.45	m	**70.19**
250 mm	26.32	32.33	1.18	42.41	m	**74.74**
315 mm	48.45	59.52	1.41	50.66	m	**110.18**
355 mm	51.46	63.22	1.53	54.98	m	**118.20**
400 mm	62.80	77.15	1.73	62.16	m	**139.31**

38 VENTILATION/AIR CONDITIONING SYSTEMS

Item	Net Price £	Material £	Labour hours	Labour £	Unit	Total rate £
DUCTWORK: PLASTIC – cont						
Extra for supports (BZP finish); Horizontal – Maximum 2.4 m centre; Vertical – Maximum 4.0 m centre						
Duct size						
110 mm	31.71	38.95	0.42	15.10	m	**54.05**
160 mm	31.71	38.95	0.42	15.10	m	**54.05**
200 mm	31.71	38.95	0.42	15.10	m	**54.05**
225 mm	31.71	38.95	0.42	15.10	m	**54.05**
250 mm	31.71	38.95	0.42	15.10	m	**54.05**
315 mm	31.71	38.95	0.42	15.10	m	**54.05**
355 mm	31.71	38.95	0.55	19.77	m	**58.72**
400 mm	44.88	55.14	0.55	19.77	m	**74.91**
Note: These are maximum figures and may be reduced subject to local conditions (i.e. a high number of changes of direction)						
Extra over fittings; Polypropylene (DW 154)						
90° bend						
110 mm	12.98	15.95	1.01	36.29	m	**52.24**
160 mm	16.17	19.87	1.30	46.70	m	**66.57**
200 mm	17.77	21.83	1.58	56.78	m	**78.61**
225 mm	24.77	30.43	2.10	75.45	m	**105.88**
250 mm	25.12	30.86	2.36	84.81	m	**115.67**
315 mm	74.89	92.00	3.14	112.82	m	**204.82**
355 mm	84.11	103.33	3.52	126.48	m	**229.81**
400 mm	82.42	101.25	4.39	157.75	m	**259.00**
45° bend						
110 mm	8.11	9.96	0.84	30.19	m	**40.15**
160 mm	11.81	14.51	1.13	40.60	m	**55.11**
200 mm	12.57	15.45	1.39	49.95	m	**65.40**
225 mm	15.41	18.93	1.73	62.16	m	**81.09**
250 mm	16.59	20.38	1.99	71.52	m	**91.90**
315 mm	47.01	57.75	2.57	92.35	m	**150.10**
355 mm	52.41	64.39	2.93	105.28	m	**169.67**
400 mm	50.29	61.78	3.60	129.36	m	**191.14**
Tee						
110 mm	36.23	44.51	1.55	55.70	m	**100.21**
160 mm	44.86	55.11	2.17	77.97	m	**133.08**
200 mm	51.87	63.72	2.68	96.31	m	**160.03**
225 mm	61.77	75.88	3.38	121.45	m	**197.33**
250 mm	71.69	88.07	3.95	141.94	m	**230.01**
315 mm	101.40	124.57	4.69	168.53	m	**293.10**
355 mm	126.76	155.73	5.44	195.48	m	**351.21**
400 mm	148.12	181.97	6.09	218.83	m	**400.80**
Coupler						
110 mm	7.98	9.81	0.84	30.19	m	**40.00**
160 mm	8.60	10.56	1.10	39.52	m	**50.08**
200 mm	9.64	11.84	1.36	48.87	m	**60.71**

38 VENTILATION/AIR CONDITIONING SYSTEMS

Item	Net Price £	Material £	Labour hours	Labour £	Unit	**Total rate £**
225 mm	10.38	12.75	1.67	60.01	m	**72.76**
250 mm	10.53	12.94	1.91	68.64	m	**81.58**
315 mm	14.55	17.87	2.24	80.49	m	**98.36**
355 mm	18.70	22.97	2.58	92.72	m	**115.69**
400 mm	19.10	23.47	3.30	118.59	m	**142.06**
Damper						
110 mm	75.66	92.95	0.79	28.39	m	**121.34**
160 mm	85.95	105.59	1.13	40.60	m	**146.19**
200 mm	92.59	113.75	1.44	51.74	m	**165.49**
225 mm	97.23	119.45	1.81	65.04	m	**184.49**
250 mm	99.45	122.18	2.08	74.74	m	**196.92**
315 mm	113.52	139.46	2.40	86.24	m	**225.70**
355 mm	122.77	150.82	2.75	98.82	m	**249.64**
400 mm	128.21	157.51	3.57	128.28	m	**285.79**
Reducer						
160 × 110 mm	27.27	33.51	0.87	31.27	m	**64.78**
200 × 160 mm	28.22	34.67	1.16	41.68	m	**76.35**
225 × 200 mm	32.90	40.41	1.46	52.46	m	**92.87**
250 × 200 mm	38.77	47.63	1.61	57.86	m	**105.49**
250 × 225 mm	47.52	58.38	1.70	61.09	m	**119.47**
315 × 200 mm	73.09	89.80	1.66	59.64	m	**149.44**
315 × 250 mm	66.09	81.20	2.03	72.94	m	**154.14**
355 × 200 mm	78.91	96.94	2.08	74.74	m	**171.68**
355 × 250 mm	83.98	103.17	2.05	73.66	m	**176.83**
355 × 315 mm	96.06	118.01	2.22	79.78	m	**197.79**
400 × 315 mm	114.17	140.26	2.88	103.49	m	**243.75**
400 × 355 mm	122.19	150.11	2.92	104.93	m	**255.04**
Flange						
110 mm	15.34	18.85	0.76	27.31	m	**46.16**
160 mm	17.24	21.18	1.02	36.66	m	**57.84**
200 mm	19.06	23.41	1.27	45.64	m	**69.05**
225 mm	19.46	23.91	1.55	55.70	m	**79.61**
250 mm	21.02	25.82	1.79	64.32	m	**90.14**
315 mm	23.76	29.19	2.08	74.74	m	**103.93**
355 mm	25.99	31.93	2.36	84.81	m	**116.74**
400 mm	27.61	33.91	3.07	110.32	m	**144.23**

38 VENTILATION/AIR CONDITIONING SYSTEMS

Item	Net Price £	Material £	Labour hours	Labour £	Unit	Total rate £
DUCTWORK: RECTANGULAR – CLASS B						
AIR DUCTLINES						
Galvanized sheet metal DW144 class B rectangular section ductwork; including all necessary stiffeners, joints, couplers in the running length and duct supports						
Ductwork up to 400 mm longest side						
Sum of two sides 200 mm	12.09	14.86	1.16	41.68	m	**56.54**
Sum of two sides 300 mm	12.94	15.90	1.16	41.68	m	**57.58**
Sum of two sides 400 mm	13.50	16.58	1.19	42.77	m	**59.35**
Sum of two sides 500 mm	14.29	17.56	1.19	42.77	m	**60.33**
Sum of two sides 600 mm	13.63	16.75	1.27	45.64	m	**62.39**
Sum of two sides 700 mm	14.35	17.63	1.27	45.64	m	**63.27**
Sum of two sides 800 mm	15.10	18.55	1.27	45.64	m	**64.19**
Extra over fittings; Rectangular ductwork class B; up to 400 mm longest side						
End cap						
Sum of two sides 200 mm	8.44	10.36	0.38	13.66	nr	**24.02**
Sum of two sides 300 mm	9.38	11.52	0.38	13.66	nr	**25.18**
Sum of two sides 400 mm	10.33	12.69	0.38	13.66	nr	**26.35**
Sum of two sides 500 mm	11.29	13.87	0.38	13.66	nr	**27.53**
Sum of two sides 600 mm	12.22	15.02	0.38	13.66	nr	**28.68**
Sum of two sides 700 mm	13.18	16.19	0.38	13.66	nr	**29.85**
Sum of two sides 800 mm	14.13	17.36	0.38	13.66	nr	**31.02**
Reducer						
Sum of two sides 200 mm	13.85	17.01	1.40	50.31	nr	**67.32**
Sum of two sides 300 mm	15.64	19.22	1.40	50.31	nr	**69.53**
Sum of two sides 400 mm	25.23	31.00	1.42	51.02	nr	**82.02**
Sum of two sides 500 mm	27.28	33.52	1.42	51.02	nr	**84.54**
Sum of two sides 600 mm	29.35	36.06	1.69	60.72	nr	**96.78**
Sum of two sides 700 mm	31.41	38.59	1.69	60.72	nr	**99.31**
Sum of two sides 800 mm	33.44	41.08	1.92	69.00	nr	**110.08**
Offset						
Sum of two sides 200 mm	20.45	25.13	1.63	58.58	nr	**83.71**
Sum of two sides 300 mm	23.12	28.40	1.63	58.58	nr	**86.98**
Sum of two sides 400 mm	33.85	41.58	1.65	59.28	nr	**100.86**
Sum of two sides 500 mm	36.79	45.19	1.65	59.28	nr	**104.47**
Sum of two sides 600 mm	39.21	48.17	1.92	69.00	nr	**117.17**
Sum of two sides 700 mm	41.88	51.45	1.92	69.00	nr	**120.45**
Sum of two sides 800 mm	44.23	54.34	1.92	69.00	nr	**123.34**
Square to round						
Sum of two sides 200 mm	18.00	22.11	1.63	58.58	nr	**80.69**
Sum of two sides 300 mm	20.22	24.84	1.63	58.58	nr	**83.42**
Sum of two sides 400 mm	26.75	32.86	1.65	59.28	nr	**92.14**
Sum of two sides 500 mm	29.11	35.76	1.65	59.28	nr	**95.04**
Sum of two sides 600 mm	31.46	38.65	1.92	69.00	nr	**107.65**
Sum of two sides 700 mm	33.81	41.54	1.92	69.00	nr	**110.54**
Sum of two sides 800 mm	36.15	44.41	1.92	69.00	nr	**113.41**

38 VENTILATION/AIR CONDITIONING SYSTEMS

Item	Net Price £	Material £	Labour hours	Labour £	Unit	**Total rate £**
90° radius bend						
Sum of two sides 200 mm	13.53	16.62	1.22	43.85	nr	**60.47**
Sum of two sides 300 mm	14.60	17.93	1.22	43.85	nr	**61.78**
Sum of two sides 400 mm	24.24	29.78	1.25	44.92	nr	**74.70**
Sum of two sides 500 mm	25.78	31.67	1.25	44.92	nr	**76.59**
Sum of two sides 600 mm	27.82	34.18	1.33	47.79	nr	**81.97**
Sum of two sides 700 mm	29.60	36.36	1.33	47.79	nr	**84.15**
Sum of two sides 800 mm	31.61	38.83	1.40	50.31	nr	**89.14**
45° radius bend						
Sum of two sides 200 mm	14.62	17.97	0.89	31.98	nr	**49.95**
Sum of two sides 300 mm	16.05	19.72	1.12	40.25	nr	**59.97**
Sum of two sides 400 mm	25.38	31.17	1.10	39.52	nr	**70.69**
Sum of two sides 500 mm	27.18	33.39	1.10	39.52	nr	**72.91**
Sum of two sides 600 mm	29.24	35.92	1.16	41.68	nr	**77.60**
Sum of two sides 700 mm	31.15	38.27	1.16	41.68	nr	**79.95**
Sum of two sides 800 mm	33.21	40.80	1.22	43.85	nr	**84.65**
90° mitre bend						
Sum of two sides 200 mm	22.56	27.72	1.29	46.35	nr	**74.07**
Sum of two sides 300 mm	24.70	30.35	1.29	46.35	nr	**76.70**
Sum of two sides 400 mm	35.54	43.66	1.29	46.35	nr	**90.01**
Sum of two sides 500 mm	38.32	47.07	1.29	46.35	nr	**93.42**
Sum of two sides 600 mm	41.86	51.43	1.39	49.95	nr	**101.38**
Sum of two sides 700 mm	45.14	55.45	1.39	49.95	nr	**105.40**
Sum of two sides 800 mm	48.73	59.86	1.46	52.46	nr	**112.32**
Branch						
Sum of two sides 200 mm	22.34	27.45	0.92	33.05	nr	**60.50**
Sum of two sides 300 mm	24.82	30.49	0.92	33.05	nr	**63.54**
Sum of two sides 400 mm	30.84	37.88	0.95	34.14	nr	**72.02**
Sum of two sides 500 mm	33.52	41.18	0.95	34.14	nr	**75.32**
Sum of two sides 600 mm	36.15	44.41	1.03	37.01	nr	**81.42**
Sum of two sides 700 mm	38.80	47.67	1.03	37.01	nr	**84.68**
Sum of two sides 800 mm	41.43	50.89	1.03	37.01	nr	**87.90**
Grille neck						
Sum of two sides 200 mm	25.49	31.31	1.10	39.52	nr	**70.83**
Sum of two sides 300 mm	28.45	34.95	1.10	39.52	nr	**74.47**
Sum of two sides 400 mm	31.41	38.59	1.16	41.68	nr	**80.27**
Sum of two sides 500 mm	34.36	42.21	1.16	41.68	nr	**83.89**
Sum of two sides 600 mm	37.33	45.86	1.18	42.41	nr	**88.27**
Sum of two sides 700 mm	40.27	49.48	1.18	42.41	nr	**91.89**
Sum of two sides 800 mm	43.26	53.14	1.18	42.41	nr	**95.55**
Ductwork 401 to 600 mm longest side						
Sum of two sides 600 mm	15.36	18.87	1.27	45.64	m	**64.51**
Sum of two sides 700 mm	16.39	20.14	1.27	45.64	m	**65.78**
Sum of two sides 800 mm	17.39	21.37	1.27	45.64	m	**67.01**
Sum of two sides 900 mm	18.31	22.49	1.27	45.64	m	**68.13**
Sum of two sides 1000 mm	19.24	23.64	1.37	49.22	m	**72.86**
Sum of two sides 1100 mm	20.34	24.99	1.37	49.22	m	**74.21**
Sum of two sides 1200 mm	21.26	26.11	1.37	49.22	m	**75.33**

38 VENTILATION/AIR CONDITIONING SYSTEMS

Item	Net Price £	Material £	Labour hours	Labour £	Unit	Total rate £
DUCTWORK: RECTANGULAR – CLASS B – cont						
Extra over fittings; Ductwork 401 to 600 mm longest side						
End cap						
Sum of two sides 600 mm	12.02	14.76	0.38	13.66	nr	**28.42**
Sum of two sides 700 mm	12.97	15.93	0.38	13.66	nr	**29.59**
Sum of two sides 800 mm	13.91	17.08	0.38	13.66	nr	**30.74**
Sum of two sides 900 mm	14.85	18.24	0.58	20.83	nr	**39.07**
Sum of two sides 1000 mm	15.78	19.38	0.58	20.83	nr	**40.21**
Sum of two sides 1100 mm	16.73	20.55	0.58	20.83	nr	**41.38**
Sum of two sides 1200 mm	17.67	21.70	0.58	20.83	nr	**42.53**
Reducer						
Sum of two sides 600 mm	27.72	34.05	1.69	60.72	nr	**94.77**
Sum of two sides 700 mm	29.72	36.51	1.69	60.72	nr	**97.23**
Sum of two sides 800 mm	31.68	38.92	1.92	69.00	nr	**107.92**
Sum of two sides 900 mm	33.68	41.38	1.92	69.00	nr	**110.38**
Sum of two sides 1000 mm	35.66	43.81	2.18	78.34	nr	**122.15**
Sum of two sides 1100 mm	37.80	46.44	2.18	78.34	nr	**124.78**
Sum of two sides 1200 mm	39.78	48.87	2.18	78.34	nr	**127.21**
Offset						
Sum of two sides 600 mm	38.77	47.63	1.92	69.00	nr	**116.63**
Sum of two sides 700 mm	41.63	51.15	1.92	69.00	nr	**120.15**
Sum of two sides 800 mm	43.91	53.94	1.92	69.00	nr	**122.94**
Sum of two sides 900 mm	46.22	56.78	1.92	69.00	nr	**125.78**
Sum of two sides 1000 mm	48.78	59.92	2.18	78.34	nr	**138.26**
Sum of two sides 1100 mm	51.16	62.85	2.18	78.34	nr	**141.19**
Sum of two sides 1200 mm	53.42	65.63	2.18	78.34	nr	**143.97**
Square to round						
Sum of two sides 600 mm	29.68	36.47	1.33	47.79	nr	**84.26**
Sum of two sides 700 mm	31.97	39.27	1.33	47.79	nr	**87.06**
Sum of two sides 800 mm	34.22	42.04	1.40	50.31	nr	**92.35**
Sum of two sides 900 mm	36.50	44.85	1.40	50.31	nr	**95.16**
Sum of two sides 1000 mm	38.78	47.64	1.82	65.39	nr	**113.03**
Sum of two sides 1100 mm	41.09	50.48	1.82	65.39	nr	**115.87**
Sum of two sides 1200 mm	43.37	53.28	1.82	65.39	nr	**118.67**
90° radius bend						
Sum of two sides 600 mm	26.94	33.10	1.16	41.68	nr	**74.78**
Sum of two sides 700 mm	28.43	34.92	1.16	41.68	nr	**76.60**
Sum of two sides 800 mm	30.57	37.56	1.22	43.85	nr	**81.41**
Sum of two sides 900 mm	32.76	40.25	1.22	43.85	nr	**84.10**
Sum of two sides 1000 mm	34.59	42.49	1.40	50.31	nr	**92.80**
Sum of two sides 1100 mm	36.88	45.31	1.40	50.31	nr	**95.62**
Sum of two sides 1200 mm	39.08	48.01	1.40	50.31	nr	**98.32**
45° bend						
Sum of two sides 600 mm	28.52	35.04	1.16	41.68	nr	**76.72**
Sum of two sides 700 mm	30.27	37.19	1.39	49.95	nr	**87.14**
Sum of two sides 800 mm	32.36	39.75	1.46	52.46	nr	**92.21**

38 VENTILATION/AIR CONDITIONING SYSTEMS

Item	Net Price £	Material £	Labour hours	Labour £	Unit	**Total rate £**
Sum of two sides 900 mm	34.49	42.37	1.46	52.46	nr	**94.83**
Sum of two sides 1000 mm	36.43	44.75	1.88	67.56	nr	**112.31**
Sum of two sides 1100 mm	38.70	47.54	1.88	67.56	nr	**115.10**
Sum of two sides 1200 mm	40.82	50.15	1.88	67.56	nr	**117.71**
90° mitre bend						
Sum of two sides 600 mm	45.39	55.77	1.39	49.95	nr	**105.72**
Sum of two sides 700 mm	48.04	59.02	2.16	77.62	nr	**136.64**
Sum of two sides 800 mm	51.78	63.61	2.26	81.22	nr	**144.83**
Sum of two sides 900 mm	55.59	68.29	2.26	81.22	nr	**149.51**
Sum of two sides 1000 mm	59.00	72.48	3.01	108.16	nr	**180.64**
Sum of two sides 1100 mm	62.97	77.35	3.01	108.16	nr	**185.51**
Sum of two sides 1200 mm	66.85	82.12	3.01	108.16	nr	**190.28**
Branch						
Sum of two sides 600 mm	35.68	43.84	1.03	37.01	nr	**80.85**
Sum of two sides 700 mm	38.31	47.06	1.03	37.01	nr	**84.07**
Sum of two sides 800 mm	40.95	50.31	1.03	37.01	nr	**87.32**
Sum of two sides 900 mm	43.58	53.54	1.03	37.01	nr	**90.55**
Sum of two sides 1000 mm	46.21	56.77	1.29	46.35	nr	**103.12**
Sum of two sides 1100 mm	48.93	60.11	1.29	46.35	nr	**106.46**
Sum of two sides 1200 mm	51.57	63.36	1.29	46.35	nr	**109.71**
Grille neck						
Sum of two sides 600 mm	36.91	45.34	1.18	42.41	nr	**87.75**
Sum of two sides 700 mm	39.90	49.01	1.18	42.41	nr	**91.42**
Sum of two sides 800 mm	42.85	52.64	1.18	42.41	nr	**95.05**
Sum of two sides 900 mm	45.84	56.31	1.18	42.41	nr	**98.72**
Sum of two sides 1000 mm	48.82	59.98	1.44	51.74	nr	**111.72**
Sum of two sides 1100 mm	51.79	63.62	1.44	51.74	nr	**115.36**
Sum of two sides 1200 mm	54.77	67.28	1.44	51.74	nr	**119.02**
Ductwork 601 to 800 mm longest side						
Sum of two sides 900 mm	20.87	25.64	1.27	45.64	m	**71.28**
Sum of two sides 1000 mm	21.79	26.76	1.37	49.22	m	**75.98**
Sum of two sides 1100 mm	22.73	27.92	1.37	49.22	m	**77.14**
Sum of two sides 1200 mm	23.82	29.26	1.37	49.22	m	**78.48**
Sum of two sides 1300 mm	24.74	30.40	1.40	50.31	m	**80.71**
Sum of two sides 1400 mm	25.67	31.53	1.40	50.31	m	**81.84**
Sum of two sides 1500 mm	26.60	32.68	1.48	53.18	m	**85.86**
Sum of two sides 1600 mm	27.54	33.83	1.55	55.70	m	**89.53**
Extra over fittings: Ductwork 601 to 800 mm longest side						
End cap						
Sum of two sides 900 mm	14.93	18.34	0.58	20.83	nr	**39.17**
Sum of two sides 1000 mm	15.87	19.50	0.58	20.83	nr	**40.33**
Sum of two sides 1100 mm	16.82	20.66	0.58	20.83	nr	**41.49**
Sum of two sides 1200 mm	17.76	21.82	0.58	20.83	nr	**42.65**
Sum of two sides 1300 mm	18.70	22.97	0.58	20.83	nr	**43.80**
Sum of two sides 1400 mm	19.52	23.98	0.58	20.83	nr	**44.81**
Sum of two sides 1500 mm	22.72	27.91	0.58	20.83	nr	**48.74**
Sum of two sides 1600 mm	25.91	31.83	0.58	20.83	nr	**52.66**

38 VENTILATION/AIR CONDITIONING SYSTEMS

Item	Net Price £	Material £	Labour hours	Labour £	Unit	Total rate £
DUCTWORK: RECTANGULAR – CLASS B – cont						
Extra over fittings: Ductwork 601 to 800 mm longest side – cont						
Reducer						
Sum of two sides 900 mm	34.18	41.99	1.92	69.00	nr	**110.99**
Sum of two sides 1000 mm	36.19	44.46	2.18	78.34	nr	**122.80**
Sum of two sides 1100 mm	38.18	46.90	2.18	78.34	nr	**125.24**
Sum of two sides 1200 mm	40.33	49.55	2.18	78.34	nr	**127.89**
Sum of two sides 1300 mm	42.34	52.02	2.30	82.66	nr	**134.68**
Sum of two sides 1400 mm	44.05	54.12	2.30	82.66	nr	**136.78**
Sum of two sides 1500 mm	50.57	62.13	2.47	88.76	nr	**150.89**
Sum of two sides 1600 mm	57.08	70.12	2.47	88.76	nr	**158.88**
Offset						
Sum of two sides 900 mm	47.72	58.62	1.92	69.00	nr	**127.62**
Sum of two sides 1000 mm	49.88	61.28	2.18	78.34	nr	**139.62**
Sum of two sides 1100 mm	52.02	63.90	2.18	78.34	nr	**142.24**
Sum of two sides 1200 mm	54.25	66.64	2.18	78.34	nr	**144.98**
Sum of two sides 1300 mm	56.36	69.24	2.30	82.66	nr	**151.90**
Sum of two sides 1400 mm	58.42	71.77	2.30	82.66	nr	**154.43**
Sum of two sides 1500 mm	67.28	82.66	2.47	88.76	nr	**171.42**
Sum of two sides 1600 mm	76.11	93.50	2.47	88.76	nr	**182.26**
Square to round						
Sum of two sides 900 mm	36.39	44.71	1.40	50.31	nr	**95.02**
Sum of two sides 1000 mm	38.69	47.53	1.82	65.39	nr	**112.92**
Sum of two sides 1100 mm	40.98	50.35	1.82	65.39	nr	**115.74**
Sum of two sides 1200 mm	43.31	53.20	1.82	65.39	nr	**118.59**
Sum of two sides 1300 mm	45.59	56.01	2.15	77.26	nr	**133.27**
Sum of two sides 1400 mm	47.53	58.39	2.15	77.26	nr	**135.65**
Sum of two sides 1500 mm	55.46	68.13	2.38	85.51	nr	**153.64**
Sum of two sides 1600 mm	63.39	77.88	2.38	85.51	nr	**163.39**
90° radius bend						
Sum of two sides 900 mm	30.99	38.07	1.22	43.85	nr	**81.92**
Sum of two sides 1000 mm	33.21	40.80	1.40	50.31	nr	**91.11**
Sum of two sides 1100 mm	35.44	43.53	1.40	50.31	nr	**93.84**
Sum of two sides 1200 mm	37.74	46.37	1.40	50.31	nr	**96.68**
Sum of two sides 1300 mm	39.96	49.09	1.91	68.64	nr	**117.73**
Sum of two sides 1400 mm	41.49	50.97	1.91	68.64	nr	**119.61**
Sum of two sides 1500 mm	48.22	59.24	2.11	75.82	nr	**135.06**
Sum of two sides 1600 mm	54.96	67.51	2.11	75.82	nr	**143.33**
45° bend						
Sum of two sides 900 mm	34.19	42.00	1.22	43.85	nr	**85.85**
Sum of two sides 1000 mm	36.34	44.65	1.40	50.31	nr	**94.96**
Sum of two sides 1100 mm	38.47	47.26	1.88	67.56	nr	**114.82**
Sum of two sides 1200 mm	40.75	50.06	1.88	67.56	nr	**117.62**
Sum of two sides 1300 mm	43.48	53.41	2.26	81.22	nr	**134.63**
Sum of two sides 1400 mm	44.54	54.72	2.26	81.22	nr	**135.94**
Sum of two sides 1500 mm	51.19	62.88	2.49	89.47	nr	**152.35**
Sum of two sides 1600 mm	57.85	71.06	2.49	89.47	nr	**160.53**

38 VENTILATION/AIR CONDITIONING SYSTEMS

Item	Net Price £	Material £	Labour hours	Labour £	Unit	**Total rate £**
90° mitre bend						
Sum of two sides 900 mm	52.81	64.88	1.22	43.85	nr	**108.73**
Sum of two sides 1000 mm	56.95	69.96	1.40	50.31	nr	**120.27**
Sum of two sides 1100 mm	61.09	75.05	3.01	108.16	nr	**183.21**
Sum of two sides 1200 mm	65.31	80.23	3.01	108.16	nr	**188.39**
Sum of two sides 1300 mm	69.45	85.32	3.67	131.88	nr	**217.20**
Sum of two sides 1400 mm	72.67	89.27	3.67	131.88	nr	**221.15**
Sum of two sides 1500 mm	83.66	102.77	4.07	146.25	nr	**249.02**
Sum of two sides 1600 mm	94.64	116.27	4.07	146.25	nr	**262.52**
Branch						
Sum of two sides 900 mm	44.93	55.20	1.22	43.85	nr	**99.05**
Sum of two sides 1000 mm	47.62	58.51	1.40	50.31	nr	**108.82**
Sum of two sides 1100 mm	50.31	61.80	1.29	46.35	nr	**108.15**
Sum of two sides 1200 mm	53.11	65.25	1.29	46.35	nr	**111.60**
Sum of two sides 1300 mm	55.81	68.57	1.39	49.95	nr	**118.52**
Sum of two sides 1400 mm	58.15	71.44	1.39	49.95	nr	**121.39**
Sum of two sides 1500 mm	66.85	82.12	1.64	58.93	nr	**141.05**
Sum of two sides 1600 mm	75.55	92.81	1.64	58.93	nr	**151.74**
Grille neck						
Sum of two sides 900 mm	46.09	56.63	1.22	43.85	nr	**100.48**
Sum of two sides 1000 mm	49.08	60.29	1.40	50.31	nr	**110.60**
Sum of two sides 1100 mm	52.07	63.97	1.44	51.74	nr	**115.71**
Sum of two sides 1200 mm	55.06	67.64	1.44	51.74	nr	**119.38**
Sum of two sides 1300 mm	58.04	71.31	1.69	60.72	nr	**132.03**
Sum of two sides 1400 mm	60.62	74.48	1.69	60.72	nr	**135.20**
Sum of two sides 1500 mm	70.38	86.46	1.79	64.32	nr	**150.78**
Sum of two sides 1600 mm	80.13	98.44	1.79	64.32	nr	**162.76**
Ductwork 801 to 1000 mm longest side						
Sum of two sides 1100 mm	33.13	40.70	1.37	49.22	m	**89.92**
Sum of two sides 1200 mm	34.96	42.94	1.37	49.22	m	**92.16**
Sum of two sides 1300 mm	36.78	45.18	1.40	50.31	m	**95.49**
Sum of two sides 1400 mm	38.77	47.63	1.40	50.31	m	**97.94**
Sum of two sides 1500 mm	40.60	49.88	1.48	53.18	m	**103.06**
Sum of two sides 1600 mm	42.43	52.12	1.55	55.70	m	**107.82**
Sum of two sides 1700 mm	44.25	54.36	1.55	55.70	m	**110.06**
Sum of two sides 1800 mm	46.23	56.79	1.61	57.86	m	**114.65**
Sum of two sides 1900 mm	48.07	59.05	1.61	57.86	m	**116.91**
Sum of two sides 2000 mm	49.89	61.29	1.61	57.86	m	**119.15**
Extra over fittings; Ductwork 801 to 1000 mm longest side						
End cap						
Sum of two sides 1100 mm	16.82	20.66	1.44	51.74	nr	**72.40**
Sum of two sides 1200 mm	17.76	21.82	1.44	51.74	nr	**73.56**
Sum of two sides 1300 mm	18.70	22.97	1.44	51.74	nr	**74.71**
Sum of two sides 1400 mm	19.52	23.98	1.44	51.74	nr	**75.72**
Sum of two sides 1500 mm	22.72	27.91	1.44	51.74	nr	**79.65**
Sum of two sides 1600 mm	25.91	31.83	1.44	51.74	nr	**83.57**
Sum of two sides 1700 mm	29.11	35.76	1.44	51.74	nr	**87.50**

38 VENTILATION/AIR CONDITIONING SYSTEMS

Item	Net Price £	Material £	Labour hours	Labour £	Unit	**Total rate £**
DUCTWORK: RECTANGULAR – CLASS B – cont						
Extra over fittings – cont						
End cap – cont						
Sum of two sides 1800 mm	32.32	39.70	1.44	51.74	nr	**91.44**
Sum of two sides 1900 mm	35.53	43.65	1.44	51.74	nr	**95.39**
Sum of two sides 2000 mm	38.73	47.58	1.44	51.74	nr	**99.32**
Reducer						
Sum of two sides 1100 mm	31.72	38.97	1.44	51.74	nr	**90.71**
Sum of two sides 1200 mm	33.22	40.81	1.44	51.74	nr	**92.55**
Sum of two sides 1300 mm	34.74	42.68	1.69	60.72	nr	**103.40**
Sum of two sides 1400 mm	36.10	44.35	1.69	60.72	nr	**105.07**
Sum of two sides 1500 mm	42.12	51.74	2.47	88.76	nr	**140.50**
Sum of two sides 1600 mm	48.14	59.14	2.47	88.76	nr	**147.90**
Sum of two sides 1700 mm	54.18	66.56	2.47	88.76	nr	**155.32**
Sum of two sides 1800 mm	60.31	74.09	2.59	93.07	nr	**167.16**
Sum of two sides 1900 mm	66.34	81.50	2.71	97.39	nr	**178.89**
Sum of two sides 2000 mm	72.36	88.89	2.71	97.39	nr	**186.28**
Offset						
Sum of two sides 1100 mm	49.01	60.21	1.44	51.74	nr	**111.95**
Sum of two sides 1200 mm	49.99	61.41	1.44	51.74	nr	**113.15**
Sum of two sides 1300 mm	50.82	62.43	1.69	60.72	nr	**123.15**
Sum of two sides 1400 mm	51.19	62.88	1.69	60.72	nr	**123.60**
Sum of two sides 1500 mm	58.51	71.88	2.47	88.76	nr	**160.64**
Sum of two sides 1600 mm	65.69	80.70	2.47	88.76	nr	**169.46**
Sum of two sides 1700 mm	72.75	89.38	2.59	93.07	nr	**182.45**
Sum of two sides 1800 mm	79.70	97.91	2.61	93.79	nr	**191.70**
Sum of two sides 1900 mm	87.44	107.42	2.71	97.39	nr	**204.81**
Sum of two sides 2000 mm	94.10	115.60	2.71	97.39	nr	**212.99**
Square to round						
Sum of two sides 1100 mm	34.39	42.25	1.44	51.74	nr	**93.99**
Sum of two sides 1200 mm	36.20	44.47	1.44	51.74	nr	**96.21**
Sum of two sides 1300 mm	38.00	46.68	1.69	60.72	nr	**107.40**
Sum of two sides 1400 mm	39.48	48.50	1.69	60.72	nr	**109.22**
Sum of two sides 1500 mm	46.91	57.62	2.38	85.51	nr	**143.13**
Sum of two sides 1600 mm	54.35	66.77	2.38	85.51	nr	**152.28**
Sum of two sides 1700 mm	61.79	75.90	2.55	91.63	nr	**167.53**
Sum of two sides 1800 mm	69.24	85.06	2.55	91.63	nr	**176.69**
Sum of two sides 1900 mm	76.68	94.21	2.83	101.69	nr	**195.90**
Sum of two sides 2000 mm	84.13	103.35	2.83	101.69	nr	**205.04**
90° radius bend						
Sum of two sides 1100 mm	25.38	31.17	1.44	51.74	nr	**82.91**
Sum of two sides 1200 mm	27.19	33.40	1.44	51.74	nr	**85.14**
Sum of two sides 1300 mm	28.98	35.61	1.69	60.72	nr	**96.33**
Sum of two sides 1400 mm	30.55	37.53	1.69	60.72	nr	**98.25**
Sum of two sides 1500 mm	36.84	45.26	2.11	75.82	nr	**121.08**
Sum of two sides 1600 mm	43.15	53.01	2.11	75.82	nr	**128.83**
Sum of two sides 1700 mm	49.45	60.75	2.26	81.22	nr	**141.97**
Sum of two sides 1800 mm	55.81	68.57	2.26	81.22	nr	**149.79**
Sum of two sides 1900 mm	61.02	74.96	2.48	89.11	nr	**164.07**
Sum of two sides 2000 mm	67.32	82.70	2.48	89.11	nr	**171.81**

38 VENTILATION/AIR CONDITIONING SYSTEMS

Item	Net Price £	Material £	Labour hours	Labour £	Unit	Total rate £
45° bend						
Sum of two sides 1100 mm	33.09	40.66	1.44	51.74	nr	**92.40**
Sum of two sides 1200 mm	35.01	43.01	1.44	51.74	nr	**94.75**
Sum of two sides 1300 mm	36.93	45.37	1.69	60.72	nr	**106.09**
Sum of two sides 1400 mm	38.69	47.53	1.69	60.72	nr	**108.25**
Sum of two sides 1500 mm	45.10	55.41	2.49	89.47	nr	**144.88**
Sum of two sides 1600 mm	51.52	63.29	2.49	89.47	nr	**152.76**
Sum of two sides 1700 mm	57.93	71.17	2.67	95.95	nr	**167.12**
Sum of two sides 1800 mm	64.48	79.21	2.67	95.95	nr	**175.16**
Sum of two sides 1900 mm	70.30	86.36	3.06	109.95	nr	**196.31**
Sum of two sides 2000 mm	76.73	94.26	3.06	109.95	nr	**204.21**
90° mitre bend						
Sum of two sides 1100 mm	52.98	65.09	1.44	51.74	nr	**116.83**
Sum of two sides 1200 mm	56.57	69.50	1.44	51.74	nr	**121.24**
Sum of two sides 1300 mm	60.16	73.91	1.69	60.72	nr	**134.63**
Sum of two sides 1400 mm	63.36	77.84	1.69	60.72	nr	**138.56**
Sum of two sides 1500 mm	73.71	90.55	2.07	74.38	nr	**164.93**
Sum of two sides 1600 mm	84.06	103.27	2.07	74.38	nr	**177.65**
Sum of two sides 1700 mm	94.40	115.97	2.80	100.61	nr	**216.58**
Sum of two sides 1800 mm	104.78	128.72	2.67	95.95	nr	**224.67**
Sum of two sides 1900 mm	113.91	139.93	2.95	106.01	nr	**245.94**
Sum of two sides 2000 mm	124.30	152.70	2.95	106.01	nr	**258.71**
Branch						
Sum of two sides 1100 mm	50.42	61.94	1.44	51.74	nr	**113.68**
Sum of two sides 1200 mm	53.11	65.25	1.44	51.74	nr	**116.99**
Sum of two sides 1300 mm	55.81	68.57	1.64	58.93	nr	**127.50**
Sum of two sides 1400 mm	58.24	71.55	1.64	58.93	nr	**130.48**
Sum of two sides 1500 mm	66.94	82.24	1.64	58.93	nr	**141.17**
Sum of two sides 1600 mm	75.65	92.94	1.64	58.93	nr	**151.87**
Sum of two sides 1700 mm	84.34	103.61	1.69	60.72	nr	**164.33**
Sum of two sides 1800 mm	93.16	114.45	1.69	60.72	nr	**175.17**
Sum of two sides 1900 mm	101.86	125.14	1.85	66.48	nr	**191.62**
Sum of two sides 2000 mm	110.56	135.83	1.85	66.48	nr	**202.31**
Grille neck						
Sum of two sides 1100 mm	52.07	63.97	1.44	51.74	nr	**115.71**
Sum of two sides 1200 mm	55.06	67.64	1.44	51.74	nr	**119.38**
Sum of two sides 1300 mm	58.04	71.31	1.69	60.72	nr	**132.03**
Sum of two sides 1400 mm	60.62	74.48	1.69	60.72	nr	**135.20**
Sum of two sides 1500 mm	70.38	86.46	1.79	64.32	nr	**150.78**
Sum of two sides 1600 mm	80.13	98.44	1.79	64.32	nr	**162.76**
Sum of two sides 1700 mm	89.89	110.43	1.86	66.84	nr	**177.27**
Sum of two sides 1800 mm	99.63	122.40	2.02	72.60	nr	**195.00**
Sum of two sides 1900 mm	109.39	134.39	2.02	72.60	nr	**206.99**
Sum of two sides 2000 mm	119.14	146.36	2.02	72.60	nr	**218.96**
Ductwork 1001 to 1250 mm longest side						
Sum of two sides 1300 mm	43.35	53.26	1.40	50.31	m	**103.57**
Sum of two sides 1400 mm	45.58	56.00	1.40	50.31	m	**106.31**
Sum of two sides 1500 mm	47.61	58.49	1.48	53.18	m	**111.67**

38 VENTILATION/AIR CONDITIONING SYSTEMS

Item	Net Price £	Material £	Labour hours	Labour £	Unit	Total rate £
DUCTWORK: RECTANGULAR – CLASS B – cont						
Ductwork 1001 to 1250 mm longest side – cont						
Sum of two sides 1600 mm	49.83	61.22	1.55	55.70	m	**116.92**
Sum of two sides 1700 mm	51.88	63.74	1.55	55.70	m	**119.44**
Sum of two sides 1800 mm	53.92	66.24	1.61	57.86	m	**124.10**
Sum of two sides 1900 mm	55.98	68.77	1.61	57.86	m	**126.63**
Sum of two sides 2000 mm	58.18	71.47	1.61	57.86	m	**129.33**
Sum of two sides 2100 mm	60.23	73.99	2.17	77.97	m	**151.96**
Sum of two sides 2200 mm	62.27	76.50	2.19	78.70	m	**155.20**
Sum of two sides 2300 mm	64.65	79.42	2.19	78.70	m	**158.12**
Sum of two sides 2400 mm	66.69	81.93	2.38	85.51	m	**167.44**
Sum of two sides 2500 mm	68.90	84.64	2.38	85.51	m	**170.15**
Extra over fittings; Ductwork 1001 to 1250 mm longest side						
End cap						
Sum of two sides 1300 mm	19.08	23.44	1.69	60.72	nr	**84.16**
Sum of two sides 1400 mm	19.92	24.48	1.69	60.72	nr	**85.20**
Sum of two sides 1500 mm	23.15	28.44	1.69	60.72	nr	**89.16**
Sum of two sides 1600 mm	26.40	32.43	1.69	60.72	nr	**93.15**
Sum of two sides 1700 mm	29.63	36.40	1.69	60.72	nr	**97.12**
Sum of two sides 1800 mm	32.87	40.38	1.69	60.72	nr	**101.10**
Sum of two sides 1900 mm	36.11	44.36	1.69	60.72	nr	**105.08**
Sum of two sides 2000 mm	39.35	48.34	1.69	60.72	nr	**109.06**
Sum of two sides 2100 mm	42.59	52.32	1.69	60.72	nr	**113.04**
Sum of two sides 2200 mm	45.83	56.30	1.69	60.72	nr	**117.02**
Sum of two sides 2300 mm	49.07	60.28	1.69	60.72	nr	**121.00**
Sum of two sides 2400 mm	52.30	64.25	1.69	60.72	nr	**124.97**
Sum of two sides 2500 mm	55.55	68.24	1.69	60.72	nr	**128.96**
Reducer						
Sum of two sides 1300 mm	33.89	41.63	1.69	60.72	nr	**102.35**
Sum of two sides 1400 mm	35.12	43.14	1.69	60.72	nr	**103.86**
Sum of two sides 1500 mm	41.18	50.59	2.47	88.76	nr	**139.35**
Sum of two sides 1600 mm	47.33	58.15	2.47	88.76	nr	**146.91**
Sum of two sides 1700 mm	53.38	65.57	2.47	88.76	nr	**154.33**
Sum of two sides 1800 mm	59.44	73.02	2.59	93.07	nr	**166.09**
Sum of two sides 1900 mm	65.48	80.44	2.71	97.39	nr	**177.83**
Sum of two sides 2000 mm	71.64	88.01	2.59	93.07	nr	**181.08**
Sum of two sides 2100 mm	77.68	95.43	2.92	104.93	nr	**200.36**
Sum of two sides 2200 mm	83.74	102.88	2.92	104.93	nr	**207.81**
Sum of two sides 2300 mm	89.91	110.45	2.92	104.93	nr	**215.38**
Sum of two sides 2400 mm	95.95	117.87	3.12	112.11	nr	**229.98**
Sum of two sides 2500 mm	102.12	125.45	3.12	112.11	nr	**237.56**
Offset						
Sum of two sides 1300 mm	66.96	82.26	1.69	60.72	nr	**142.98**
Sum of two sides 1400 mm	69.17	84.98	1.69	60.72	nr	**145.70**
Sum of two sides 1500 mm	77.56	95.29	2.47	88.76	nr	**184.05**

38 VENTILATION/AIR CONDITIONING SYSTEMS

Item	Net Price £	Material £	Labour hours	Labour £	Unit	**Total rate £**
Sum of two sides 1600 mm	85.90	105.53	2.47	88.76	nr	**194.29**
Sum of two sides 1700 mm	93.99	115.47	2.59	93.07	nr	**208.54**
Sum of two sides 1800 mm	101.94	125.23	2.61	93.79	nr	**219.02**
Sum of two sides 1900 mm	109.77	134.85	2.71	97.39	nr	**232.24**
Sum of two sides 2000 mm	117.49	144.33	2.71	97.39	nr	**241.72**
Sum of two sides 2100 mm	125.03	153.60	2.92	104.93	nr	**258.53**
Sum of two sides 2200 mm	133.63	164.17	3.26	117.15	nr	**281.32**
Sum of two sides 2300 mm	143.44	176.21	3.26	117.15	nr	**293.36**
Sum of two sides 2400 mm	153.22	188.23	3.48	125.04	nr	**313.27**
Sum of two sides 2500 mm	163.05	200.31	3.48	125.04	nr	**325.35**
Square to round						
Sum of two sides 1300 mm	36.84	45.26	1.69	60.72	nr	**105.98**
Sum of two sides 1400 mm	38.31	47.06	1.69	60.72	nr	**107.78**
Sum of two sides 1500 mm	45.78	56.24	2.38	85.51	nr	**141.75**
Sum of two sides 1600 mm	53.26	65.43	2.38	85.51	nr	**150.94**
Sum of two sides 1700 mm	60.75	74.63	2.55	91.63	nr	**166.26**
Sum of two sides 1800 mm	68.20	83.78	2.55	91.63	nr	**175.41**
Sum of two sides 1900 mm	75.66	92.95	2.83	101.69	nr	**194.64**
Sum of two sides 2000 mm	83.15	102.15	2.83	101.69	nr	**203.84**
Sum of two sides 2100 mm	90.61	111.31	3.85	138.34	nr	**249.65**
Sum of two sides 2200 mm	98.10	120.52	4.18	150.21	nr	**270.73**
Sum of two sides 2300 mm	105.57	129.70	4.22	151.64	nr	**281.34**
Sum of two sides 2400 mm	113.05	138.88	4.68	168.17	nr	**307.05**
Sum of two sides 2500 mm	120.53	148.07	4.70	168.88	nr	**316.95**
90° radius bend						
Sum of two sides 1300 mm	23.90	29.36	1.69	60.72	nr	**90.08**
Sum of two sides 1400 mm	24.26	29.80	1.69	60.72	nr	**90.52**
Sum of two sides 1500 mm	30.75	37.78	2.11	75.82	nr	**113.60**
Sum of two sides 1600 mm	37.24	45.75	2.11	75.82	nr	**121.57**
Sum of two sides 1700 mm	43.72	53.71	2.19	78.70	nr	**132.41**
Sum of two sides 1800 mm	50.19	61.66	2.19	78.70	nr	**140.36**
Sum of two sides 1900 mm	56.65	69.59	2.48	89.11	nr	**158.70**
Sum of two sides 2000 mm	63.17	77.61	2.26	81.22	nr	**158.83**
Sum of two sides 2100 mm	69.63	85.54	2.48	89.11	nr	**174.65**
Sum of two sides 2200 mm	76.10	93.49	2.48	89.11	nr	**182.60**
Sum of two sides 2300 mm	79.81	98.05	2.48	89.11	nr	**187.16**
Sum of two sides 2400 mm	86.26	105.97	3.90	140.14	nr	**246.11**
Sum of two sides 2500 mm	92.75	113.95	3.90	140.14	nr	**254.09**
45° bend						
Sum of two sides 1300 mm	34.63	42.55	1.69	60.72	nr	**103.27**
Sum of two sides 1400 mm	35.68	43.84	1.69	60.72	nr	**104.56**
Sum of two sides 1500 mm	42.22	51.87	2.49	89.47	nr	**141.34**
Sum of two sides 1600 mm	48.84	60.00	2.49	89.47	nr	**149.47**
Sum of two sides 1700 mm	55.38	68.03	2.67	95.95	nr	**163.98**
Sum of two sides 1800 mm	61.92	76.07	2.67	95.95	nr	**172.02**
Sum of two sides 1900 mm	68.45	84.09	3.06	109.95	nr	**194.04**
Sum of two sides 2000 mm	75.07	92.22	3.06	109.95	nr	**202.17**
Sum of two sides 2100 mm	81.60	100.24	4.05	145.53	nr	**245.77**
Sum of two sides 2200 mm	88.16	108.31	4.05	145.53	nr	**253.84**
Sum of two sides 2300 mm	93.33	114.66	4.39	157.75	nr	**272.41**
Sum of two sides 2400 mm	99.85	122.66	4.85	174.28	nr	**296.94**
Sum of two sides 2500 mm	106.48	130.81	4.85	174.28	nr	**305.09**

38 VENTILATION/AIR CONDITIONING SYSTEMS

Item	Net Price £	Material £	Labour hours	Labour £	Unit	Total rate £
DUCTWORK: RECTANGULAR – CLASS B – cont						
Extra over fittings – cont						
90° mitre bend						
Sum of two sides 1300 mm	72.40	88.95	1.69	60.72	nr	**149.67**
Sum of two sides 1400 mm	75.38	92.60	1.69	60.72	nr	**153.32**
Sum of two sides 1500 mm	87.43	107.40	2.80	100.61	nr	**208.01**
Sum of two sides 1600 mm	99.50	122.23	2.80	100.61	nr	**222.84**
Sum of two sides 1700 mm	111.55	137.04	2.95	106.01	nr	**243.05**
Sum of two sides 1800 mm	123.59	151.83	2.95	106.01	nr	**257.84**
Sum of two sides 1900 mm	135.66	166.66	4.05	145.53	nr	**312.19**
Sum of two sides 2000 mm	147.72	181.47	4.05	145.53	nr	**327.00**
Sum of two sides 2100 mm	159.76	196.27	4.07	146.25	nr	**342.52**
Sum of two sides 2200 mm	171.83	211.09	4.07	146.25	nr	**357.34**
Sum of two sides 2300 mm	180.69	221.98	4.39	157.75	nr	**379.73**
Sum of two sides 2400 mm	192.84	236.90	4.85	174.28	nr	**411.18**
Sum of two sides 2500 mm	204.97	251.80	4.85	174.28	nr	**426.08**
Branch						
Sum of two sides 1300 mm	57.27	70.36	1.44	51.74	nr	**122.10**
Sum of two sides 1400 mm	59.66	73.29	1.44	51.74	nr	**125.03**
Sum of two sides 1500 mm	68.48	84.13	1.64	58.93	nr	**143.06**
Sum of two sides 1600 mm	77.39	95.08	1.64	58.93	nr	**154.01**
Sum of two sides 1700 mm	86.22	105.92	1.64	58.93	nr	**164.85**
Sum of two sides 1800 mm	95.02	116.73	1.64	58.93	nr	**175.66**
Sum of two sides 1900 mm	103.83	127.55	1.69	60.72	nr	**188.27**
Sum of two sides 2000 mm	112.75	138.52	1.69	60.72	nr	**199.24**
Sum of two sides 2100 mm	121.56	149.34	1.85	66.48	nr	**215.82**
Sum of two sides 2200 mm	130.38	160.17	1.85	66.48	nr	**226.65**
Sum of two sides 2300 mm	139.30	171.13	2.61	93.79	nr	**264.92**
Sum of two sides 2400 mm	148.11	181.95	2.61	93.79	nr	**275.74**
Sum of two sides 2500 mm	157.02	192.90	2.61	93.79	nr	**286.69**
Grille neck						
Sum of two sides 1300 mm	59.74	73.40	1.79	64.32	nr	**137.72**
Sum of two sides 1400 mm	62.44	76.70	1.79	64.32	nr	**141.02**
Sum of two sides 1500 mm	72.36	88.89	1.79	64.32	nr	**153.21**
Sum of two sides 1600 mm	82.26	101.05	1.79	64.32	nr	**165.37**
Sum of two sides 1700 mm	92.19	113.25	1.86	66.84	nr	**180.09**
Sum of two sides 1800 mm	102.09	125.42	2.02	72.60	nr	**198.02**
Sum of two sides 1900 mm	112.00	137.59	2.02	72.60	nr	**210.19**
Sum of two sides 2000 mm	121.91	149.76	2.02	72.60	nr	**222.36**
Sum of two sides 2100 mm	131.83	161.95	2.61	93.79	nr	**255.74**
Sum of two sides 2200 mm	141.73	174.11	2.61	93.79	nr	**267.90**
Sum of two sides 2300 mm	151.66	186.31	2.61	93.79	nr	**280.10**
Sum of two sides 2400 mm	161.55	198.46	2.88	103.49	nr	**301.95**
Sum of two sides 2500 mm	171.47	210.65	2.88	103.49	nr	**314.14**
Ductwork 1251 to 1600 mm longest side						
Sum of two sides 1700 mm	68.37	83.99	1.55	55.70	m	**139.69**
Sum of two sides 1800 mm	70.89	87.09	1.61	57.86	m	**144.95**
Sum of two sides 1900 mm	73.43	90.21	1.61	57.86	m	**148.07**

38 VENTILATION/AIR CONDITIONING SYSTEMS

Item	Net Price £	Material £	Labour hours	Labour £	Unit	Total rate £
Sum of two sides 2000 mm	75.82	93.15	1.61	57.86	m	**151.01**
Sum of two sides 2100 mm	78.22	96.10	2.17	77.97	m	**174.07**
Sum of two sides 2200 mm	80.61	99.03	2.19	78.70	m	**177.73**
Sum of two sides 2300 mm	83.16	102.17	2.19	78.70	m	**180.87**
Sum of two sides 2400 mm	85.68	105.26	2.38	85.51	m	**190.77**
Sum of two sides 2500 mm	88.08	108.20	2.38	85.51	m	**193.71**
Sum of two sides 2600 mm	90.46	111.13	2.64	94.87	m	**206.00**
Sum of two sides 2700 mm	92.86	114.08	2.66	95.59	m	**209.67**
Sum of two sides 2800 mm	95.38	117.17	2.95	106.01	m	**223.18**
Sum of two sides 2900 mm	107.02	131.47	2.96	106.36	m	**237.83**
Sum of two sides 3000 mm	109.43	134.43	3.15	113.19	m	**247.62**
Sum of two sides 3100 mm	111.95	137.53	3.15	113.19	m	**250.72**
Sum of two sides 3200 mm	114.34	140.47	3.18	114.27	m	**254.74**
Extra over fittings; Ductwork 1251 to 1600 mm longest side						
End cap						
Sum of two sides 1700 mm	29.32	36.02	0.58	20.83	nr	**56.85**
Sum of two sides 1800 mm	32.52	39.95	0.58	20.83	nr	**60.78**
Sum of two sides 1900 mm	35.73	43.89	0.58	20.83	nr	**64.72**
Sum of two sides 2000 mm	38.93	47.83	0.58	20.83	nr	**68.66**
Sum of two sides 2100 mm	42.15	51.79	0.87	31.27	nr	**83.06**
Sum of two sides 2200 mm	45.35	55.71	0.87	31.27	nr	**86.98**
Sum of two sides 2300 mm	48.55	59.64	0.87	31.27	nr	**90.91**
Sum of two sides 2400 mm	51.75	63.58	0.87	31.27	nr	**94.85**
Sum of two sides 2500 mm	54.96	67.51	0.87	31.27	nr	**98.78**
Sum of two sides 2600 mm	58.17	71.46	0.87	31.27	nr	**102.73**
Sum of two sides 2700 mm	61.37	75.39	0.87	31.27	nr	**106.66**
Sum of two sides 2800 mm	64.57	79.33	1.16	41.68	nr	**121.01**
Sum of two sides 2900 mm	67.78	83.26	1.16	41.68	nr	**124.94**
Sum of two sides 3000 mm	74.95	92.07	1.30	46.70	nr	**138.77**
Sum of two sides 3100 mm	73.52	90.32	1.80	64.68	nr	**155.00**
Sum of two sides 3200 mm	75.83	93.16	1.80	64.68	nr	**157.84**
Reducer						
Sum of two sides 1700 mm	36.07	44.31	2.47	88.76	nr	**133.07**
Sum of two sides 1800 mm	41.67	51.19	2.59	93.07	nr	**144.26**
Sum of two sides 1900 mm	47.33	58.15	2.71	97.39	nr	**155.54**
Sum of two sides 2000 mm	52.90	64.98	2.71	97.39	nr	**162.37**
Sum of two sides 2100 mm	58.48	71.84	2.92	104.93	nr	**176.77**
Sum of two sides 2200 mm	64.04	78.68	2.92	104.93	nr	**183.61**
Sum of two sides 2300 mm	69.70	85.63	2.92	104.93	nr	**190.56**
Sum of two sides 2400 mm	75.30	92.51	3.12	112.11	nr	**204.62**
Sum of two sides 2500 mm	80.86	99.34	3.12	112.11	nr	**211.45**
Sum of two sides 2600 mm	86.45	106.21	3.12	112.11	nr	**218.32**
Sum of two sides 2700 mm	92.00	113.02	3.12	112.11	nr	**225.13**
Sum of two sides 2800 mm	97.61	119.91	3.95	141.94	nr	**261.85**
Sum of two sides 2900 mm	97.88	120.25	3.97	142.65	nr	**262.90**
Sum of two sides 3000 mm	102.29	125.66	4.52	162.42	nr	**288.08**
Sum of two sides 3100 mm	106.44	130.76	4.52	162.42	nr	**293.18**
Sum of two sides 3200 mm	110.20	135.38	4.52	162.42	nr	**297.80**

38 VENTILATION/AIR CONDITIONING SYSTEMS

Item	Net Price £	Material £	Labour hours	Labour £	Unit	Total rate £
DUCTWORK: RECTANGULAR – CLASS B – cont						
Extra over fittings – cont						
Offset						
Sum of two sides 1700 mm	91.81	112.79	2.59	93.07	nr	**205.86**
Sum of two sides 1800 mm	102.08	125.40	2.61	93.79	nr	**219.19**
Sum of two sides 1900 mm	109.04	133.96	2.71	97.39	nr	**231.35**
Sum of two sides 2000 mm	115.72	142.16	2.71	97.39	nr	**239.55**
Sum of two sides 2100 mm	122.23	150.16	2.92	104.93	nr	**255.09**
Sum of two sides 2200 mm	128.54	157.91	3.26	117.15	nr	**275.06**
Sum of two sides 2300 mm	134.70	165.48	3.26	117.15	nr	**282.63**
Sum of two sides 2400 mm	144.52	177.54	3.47	124.69	nr	**302.23**
Sum of two sides 2500 mm	150.38	184.74	3.48	125.04	nr	**309.78**
Sum of two sides 2600 mm	160.06	196.63	3.49	125.40	nr	**322.03**
Sum of two sides 2700 mm	169.74	208.53	3.50	125.77	nr	**334.30**
Sum of two sides 2800 mm	179.47	220.48	4.34	155.95	nr	**376.43**
Sum of two sides 2900 mm	179.45	220.46	4.76	171.04	nr	**391.50**
Sum of two sides 3000 mm	189.12	232.33	5.32	191.16	nr	**423.49**
Sum of two sides 3100 mm	196.65	241.58	5.35	192.24	nr	**433.82**
Sum of two sides 3200 mm	203.63	250.16	5.35	192.24	nr	**442.40**
Square to round						
Sum of two sides 1700 mm	43.35	53.26	2.55	91.63	nr	**144.89**
Sum of two sides 1800 mm	50.34	61.85	2.55	91.63	nr	**153.48**
Sum of two sides 1900 mm	57.31	70.40	2.83	101.69	nr	**172.09**
Sum of two sides 2000 mm	64.29	78.98	2.83	101.69	nr	**180.67**
Sum of two sides 2100 mm	71.26	87.54	3.85	138.34	nr	**225.88**
Sum of two sides 2200 mm	78.24	96.12	4.18	150.21	nr	**246.33**
Sum of two sides 2300 mm	85.20	104.66	4.22	151.64	nr	**256.30**
Sum of two sides 2400 mm	92.21	113.28	4.68	168.17	nr	**281.45**
Sum of two sides 2500 mm	99.21	121.88	4.70	168.88	nr	**290.76**
Sum of two sides 2600 mm	106.17	130.43	4.70	168.88	nr	**299.31**
Sum of two sides 2700 mm	113.14	138.99	4.71	169.25	nr	**308.24**
Sum of two sides 2800 mm	120.15	147.61	8.19	294.30	nr	**441.91**
Sum of two sides 2900 mm	120.58	148.13	8.62	309.74	nr	**457.87**
Sum of two sides 3000 mm	127.56	156.71	8.75	314.42	nr	**471.13**
Sum of two sides 3100 mm	132.77	163.11	8.75	314.42	nr	**477.53**
Sum of two sides 3200 mm	137.50	168.92	8.75	314.42	nr	**483.34**
90° radius bend						
Sum of two sides 1700 mm	89.14	109.50	2.19	78.70	nr	**188.20**
Sum of two sides 1800 mm	97.67	119.98	2.19	78.70	nr	**198.68**
Sum of two sides 1900 mm	110.35	135.57	2.26	81.22	nr	**216.79**
Sum of two sides 2000 mm	122.82	150.88	2.26	81.22	nr	**232.10**
Sum of two sides 2100 mm	135.30	166.22	2.48	89.11	nr	**255.33**
Sum of two sides 2200 mm	147.77	181.53	2.48	89.11	nr	**270.64**
Sum of two sides 2300 mm	160.46	197.13	2.48	89.11	nr	**286.24**
Sum of two sides 2400 mm	168.63	207.17	3.90	140.14	nr	**347.31**
Sum of two sides 2500 mm	181.05	222.42	3.90	140.14	nr	**362.56**
Sum of two sides 2600 mm	193.50	237.72	4.26	153.08	nr	**390.80**
Sum of two sides 2700 mm	205.95	253.01	4.55	163.51	nr	**416.52**

38 VENTILATION/AIR CONDITIONING SYSTEMS

Item	Net Price £	Material £	Labour hours	Labour £	Unit	Total rate £
Sum of two sides 2800 mm	213.82	262.68	4.55	163.51	nr	**426.19**
Sum of two sides 2900 mm	214.05	262.96	6.87	246.87	nr	**509.83**
Sum of two sides 3000 mm	226.44	278.18	7.00	251.54	nr	**529.72**
Sum of two sides 3100 mm	236.01	289.94	7.00	251.54	nr	**541.48**
Sum of two sides 3200 mm	244.80	300.74	7.00	251.54	nr	**552.28**
45° bend						
Sum of two sides 1700 mm	42.75	52.52	2.67	95.95	nr	**148.47**
Sum of two sides 1800 mm	46.94	57.67	2.67	95.95	nr	**153.62**
Sum of two sides 1900 mm	53.27	65.45	3.06	109.95	nr	**175.40**
Sum of two sides 2000 mm	59.50	73.10	3.06	109.95	nr	**183.05**
Sum of two sides 2100 mm	65.72	80.73	4.05	145.53	nr	**226.26**
Sum of two sides 2200 mm	71.96	88.40	4.05	145.53	nr	**233.93**
Sum of two sides 2300 mm	78.28	96.17	4.39	157.75	nr	**253.92**
Sum of two sides 2400 mm	82.30	101.10	4.85	174.28	nr	**275.38**
Sum of two sides 2500 mm	88.51	108.74	4.85	174.28	nr	**283.02**
Sum of two sides 2600 mm	94.73	116.37	4.87	175.00	nr	**291.37**
Sum of two sides 2700 mm	100.94	124.01	4.87	175.00	nr	**299.01**
Sum of two sides 2800 mm	104.81	128.76	8.81	316.57	nr	**445.33**
Sum of two sides 2900 mm	114.59	140.77	8.81	316.57	nr	**457.34**
Sum of two sides 3000 mm	110.70	136.00	9.31	334.54	nr	**470.54**
Sum of two sides 3100 mm	115.46	141.84	9.31	334.54	nr	**476.38**
Sum of two sides 3200 mm	119.84	147.22	9.39	337.42	nr	**484.64**
90° mitre bend						
Sum of two sides 1700 mm	96.86	119.00	2.67	95.95	nr	**214.95**
Sum of two sides 1800 mm	103.51	127.17	2.80	100.61	nr	**227.78**
Sum of two sides 1900 mm	115.64	142.06	2.95	106.01	nr	**248.07**
Sum of two sides 2000 mm	127.82	157.03	2.95	106.01	nr	**263.04**
Sum of two sides 2100 mm	139.97	171.95	4.05	145.53	nr	**317.48**
Sum of two sides 2200 mm	152.16	186.93	4.05	145.53	nr	**332.46**
Sum of two sides 2300 mm	164.30	201.84	4.39	157.75	nr	**359.59**
Sum of two sides 2400 mm	170.94	210.00	4.85	174.28	nr	**384.28**
Sum of two sides 2500 mm	183.16	225.01	4.85	174.28	nr	**399.29**
Sum of two sides 2600 mm	195.38	240.02	4.87	175.00	nr	**415.02**
Sum of two sides 2700 mm	207.60	255.03	4.87	175.00	nr	**430.03**
Sum of two sides 2800 mm	214.17	263.11	8.81	316.57	nr	**579.68**
Sum of two sides 2900 mm	215.17	264.34	14.81	532.17	nr	**796.51**
Sum of two sides 3000 mm	227.45	279.43	15.20	546.19	nr	**825.62**
Sum of two sides 3100 mm	237.50	291.77	15.60	560.56	nr	**852.33**
Sum of two sides 3200 mm	247.06	303.51	15.60	560.56	nr	**864.07**
Branch						
Sum of two sides 1700 mm	85.31	104.80	1.69	60.72	nr	**165.52**
Sum of two sides 1800 mm	94.02	115.50	1.69	60.72	nr	**176.22**
Sum of two sides 1900 mm	102.84	126.34	1.85	66.48	nr	**192.82**
Sum of two sides 2000 mm	111.56	137.06	1.85	66.48	nr	**203.54**
Sum of two sides 2100 mm	120.28	147.77	2.61	93.79	nr	**241.56**
Sum of two sides 2200 mm	129.01	158.49	2.61	93.79	nr	**252.28**
Sum of two sides 2300 mm	137.83	169.32	2.61	93.79	nr	**263.11**
Sum of two sides 2400 mm	147.18	180.81	2.88	103.49	nr	**284.30**
Sum of two sides 2500 mm	155.27	190.75	2.88	103.49	nr	**294.24**
Sum of two sides 2600 mm	164.01	201.48	2.88	103.49	nr	**304.97**
Sum of two sides 2700 mm	172.72	212.18	2.88	103.49	nr	**315.67**

38 VENTILATION/AIR CONDITIONING SYSTEMS

Item	Net Price £	Material £	Labour hours	Labour £	Unit	Total rate £
DUCTWORK: RECTANGULAR – CLASS B – cont						
Extra over fittings – cont						
Branch – cont						
Sum of two sides 2800 mm	181.44	222.89	3.94	141.58	nr	**364.47**
Sum of two sides 2900 mm	190.27	233.75	3.94	141.58	nr	**375.33**
Sum of two sides 3000 mm	198.99	244.46	4.83	173.57	nr	**418.03**
Sum of two sides 3100 mm	205.92	252.98	4.83	173.57	nr	**426.55**
Sum of two sides 3200 mm	212.24	260.74	4.83	173.57	nr	**434.31**
Grille neck						
Sum of two sides 1700 mm	91.21	112.06	1.86	66.84	nr	**178.90**
Sum of two sides 1800 mm	101.02	124.10	2.02	72.60	nr	**196.70**
Sum of two sides 1900 mm	110.82	136.14	2.02	72.60	nr	**208.74**
Sum of two sides 2000 mm	120.63	148.20	2.02	72.60	nr	**220.80**
Sum of two sides 2100 mm	130.45	160.26	2.61	93.79	nr	**254.05**
Sum of two sides 2200 mm	140.24	172.28	2.61	93.79	nr	**266.07**
Sum of two sides 2300 mm	150.07	184.36	2.61	93.79	nr	**278.15**
Sum of two sides 2400 mm	159.85	196.37	2.88	103.49	nr	**299.86**
Sum of two sides 2500 mm	169.67	208.44	2.88	103.49	nr	**311.93**
Sum of two sides 2600 mm	179.48	220.49	2.88	103.49	nr	**323.98**
Sum of two sides 2700 mm	189.29	232.54	2.88	103.49	nr	**336.03**
Sum of two sides 2800 mm	199.10	244.60	3.94	141.58	nr	**386.18**
Sum of two sides 2900 mm	208.90	256.63	4.12	148.05	nr	**404.68**
Sum of two sides 3000 mm	218.70	268.67	5.00	179.67	nr	**448.34**
Sum of two sides 3100 mm	226.50	278.26	5.00	179.67	nr	**457.93**
Sum of two sides 3200 mm	233.60	286.98	5.00	179.67	nr	**466.65**
Ductwork 1601 to 2000 mm longest side						
Sum of two sides 2100 mm	88.21	108.37	2.17	77.97	m	**186.34**
Sum of two sides 2200 mm	90.86	111.63	2.17	77.97	m	**189.60**
Sum of two sides 2300 mm	93.52	114.89	2.19	78.70	m	**193.59**
Sum of two sides 2400 mm	96.14	118.10	2.38	85.51	m	**203.61**
Sum of two sides 2500 mm	98.80	121.38	2.38	85.51	m	**206.89**
Sum of two sides 2600 mm	101.31	124.46	2.64	94.87	m	**219.33**
Sum of two sides 2700 mm	103.82	127.54	2.66	95.59	m	**223.13**
Sum of two sides 2800 mm	106.33	130.63	2.95	106.01	m	**236.64**
Sum of two sides 2900 mm	109.00	133.91	2.96	106.36	m	**240.27**
Sum of two sides 3000 mm	111.51	136.99	2.96	106.36	m	**243.35**
Sum of two sides 3100 mm	120.98	148.63	2.96	106.36	m	**254.99**
Sum of two sides 3200 mm	123.50	151.72	3.15	113.19	m	**264.91**
Sum of two sides 3300 mm	135.45	166.40	3.15	113.19	m	**279.59**
Sum of two sides 3400 mm	137.95	169.47	3.15	113.19	m	**282.66**
Sum of two sides 3500 mm	140.46	172.56	3.15	113.19	m	**285.75**
Sum of two sides 3600 mm	142.97	175.63	3.18	114.27	m	**289.90**
Sum of two sides 3700 mm	145.65	178.93	3.18	114.27	m	**293.20**
Sum of two sides 3800 mm	148.14	181.99	3.18	114.27	m	**296.26**
Sum of two sides 3900 mm	150.65	185.07	3.18	114.27	m	**299.34**
Sum of two sides 4000 mm	153.16	188.16	3.18	114.27	m	**302.43**

38 VENTILATION/AIR CONDITIONING SYSTEMS

Item	Net Price £	Material £	Labour hours	Labour £	Unit	Total rate £
Extra over fittings; Ductwork 1601 to 2000 mm longest side						
End cap						
Sum of two sides 2100 mm	42.15	51.79	0.87	31.27	nr	**83.06**
Sum of two sides 2200 mm	45.35	55.71	0.87	31.27	nr	**86.98**
Sum of two sides 2300 mm	48.55	59.64	0.87	31.27	nr	**90.91**
Sum of two sides 2400 mm	51.75	63.58	0.87	31.27	nr	**94.85**
Sum of two sides 2500 mm	54.96	67.51	0.87	31.27	nr	**98.78**
Sum of two sides 2600 mm	58.17	71.46	0.87	31.27	nr	**102.73**
Sum of two sides 2700 mm	61.37	75.39	0.87	31.27	nr	**106.66**
Sum of two sides 2800 mm	64.57	79.33	1.16	41.68	nr	**121.01**
Sum of two sides 2900 mm	67.78	83.26	1.16	41.68	nr	**124.94**
Sum of two sides 3000 mm	70.99	87.21	1.73	62.16	nr	**149.37**
Sum of two sides 3100 mm	73.52	90.32	1.80	64.68	nr	**155.00**
Sum of two sides 3200 mm	75.83	93.16	1.80	64.68	nr	**157.84**
Sum of two sides 3300 mm	78.13	95.98	1.80	64.68	nr	**160.66**
Sum of two sides 3400 mm	80.42	98.79	1.80	64.68	nr	**163.47**
Sum of two sides 3500 mm	82.73	101.63	1.80	64.68	nr	**166.31**
Sum of two sides 3600 mm	85.02	104.44	1.80	64.68	nr	**169.12**
Sum of two sides 3700 mm	87.32	107.27	1.80	64.68	nr	**171.95**
Sum of two sides 3800 mm	89.62	110.10	1.80	64.68	nr	**174.78**
Sum of two sides 3900 mm	91.91	112.91	1.80	64.68	nr	**177.59**
Sum of two sides 4000 mm	94.22	115.75	1.80	64.68	nr	**180.43**
Reducer						
Sum of two sides 2100 mm	58.31	71.63	2.61	93.79	nr	**165.42**
Sum of two sides 2200 mm	63.84	78.42	2.61	93.79	nr	**172.21**
Sum of two sides 2300 mm	69.42	85.28	2.61	93.79	nr	**179.07**
Sum of two sides 2400 mm	74.85	91.95	2.88	103.49	nr	**195.44**
Sum of two sides 2500 mm	80.47	98.86	3.12	112.11	nr	**210.97**
Sum of two sides 2600 mm	85.96	105.60	3.12	112.11	nr	**217.71**
Sum of two sides 2700 mm	91.48	112.38	3.12	112.11	nr	**224.49**
Sum of two sides 2800 mm	97.00	119.16	3.95	141.94	nr	**261.10**
Sum of two sides 2900 mm	102.58	126.02	3.97	142.65	nr	**268.67**
Sum of two sides 3000 mm	108.10	132.80	4.52	162.42	nr	**295.22**
Sum of two sides 3100 mm	108.42	133.19	4.52	162.42	nr	**295.61**
Sum of two sides 3200 mm	109.26	134.22	4.52	162.42	nr	**296.64**
Sum of two sides 3300 mm	110.99	136.35	4.52	162.42	nr	**298.77**
Sum of two sides 3400 mm	111.82	137.37	4.52	162.42	nr	**299.79**
Sum of two sides 3500 mm	115.52	141.92	4.52	162.42	nr	**304.34**
Sum of two sides 3600 mm	119.23	146.48	4.52	162.42	nr	**308.90**
Sum of two sides 3700 mm	128.83	158.27	4.52	162.42	nr	**320.69**
Sum of two sides 3800 mm	132.54	162.82	4.52	162.42	nr	**325.24**
Sum of two sides 3900 mm	136.23	167.36	4.52	162.42	nr	**329.78**
Sum of two sides 4000 mm	139.94	171.92	4.52	162.42	nr	**334.34**
Offset						
Sum of two sides 2100 mm	136.56	167.77	2.61	93.79	nr	**261.56**
Sum of two sides 2200 mm	146.91	180.47	2.61	93.79	nr	**274.26**
Sum of two sides 2300 mm	152.91	187.85	2.61	93.79	nr	**281.64**
Sum of two sides 2400 mm	167.57	205.86	2.88	103.49	nr	**309.35**
Sum of two sides 2500 mm	173.37	212.98	3.48	125.04	nr	**338.02**
Sum of two sides 2600 mm	178.90	219.78	3.49	125.40	nr	**345.18**
Sum of two sides 2700 mm	184.20	226.29	3.50	125.77	nr	**352.06**

38 VENTILATION/AIR CONDITIONING SYSTEMS

Item	Net Price £	Material £	Labour hours	Labour £	Unit	Total rate £
DUCTWORK: RECTANGULAR – CLASS B – cont						
Extra over fittings – cont						
Offset – cont						
Sum of two sides 2800 mm	189.32	232.57	4.34	155.95	nr	**388.52**
Sum of two sides 2900 mm	194.28	238.68	4.76	171.04	nr	**409.72**
Sum of two sides 3000 mm	198.96	244.42	5.32	191.16	nr	**435.58**
Sum of two sides 3100 mm	199.25	244.78	5.35	192.24	nr	**437.02**
Sum of two sides 3200 mm	203.39	249.87	5.35	192.24	nr	**442.11**
Sum of two sides 3300 mm	206.26	253.39	5.35	192.24	nr	**445.63**
Sum of two sides 3400 mm	210.41	258.49	5.35	192.24	nr	**450.73**
Sum of two sides 3500 mm	217.43	267.11	5.35	192.24	nr	**459.35**
Sum of two sides 3600 mm	224.45	275.74	5.35	192.24	nr	**467.98**
Sum of two sides 3700 mm	237.34	291.57	5.35	192.24	nr	**483.81**
Sum of two sides 3800 mm	244.35	300.18	5.35	192.24	nr	**492.42**
Sum of two sides 3900 mm	251.38	308.82	5.35	192.24	nr	**501.06**
Sum of two sides 4000 mm	258.37	317.40	5.35	192.24	nr	**509.64**
Square to round						
Sum of two sides 2100 mm	87.83	107.90	2.61	93.79	nr	**201.69**
Sum of two sides 2200 mm	96.29	118.29	2.61	93.79	nr	**212.08**
Sum of two sides 2300 mm	104.75	128.69	2.61	93.79	nr	**222.48**
Sum of two sides 2400 mm	113.26	139.14	2.88	103.49	nr	**242.63**
Sum of two sides 2500 mm	121.69	149.50	4.70	168.88	nr	**318.38**
Sum of two sides 2600 mm	130.16	159.90	4.70	168.88	nr	**328.78**
Sum of two sides 2700 mm	138.62	170.30	4.71	169.25	nr	**339.55**
Sum of two sides 2800 mm	147.08	180.68	8.19	294.30	nr	**474.98**
Sum of two sides 2900 mm	155.52	191.06	8.19	294.30	nr	**485.36**
Sum of two sides 3000 mm	163.98	201.45	8.19	294.30	nr	**495.75**
Sum of two sides 3100 mm	165.44	203.24	8.19	294.30	nr	**497.54**
Sum of two sides 3200 mm	170.26	209.16	8.19	294.30	nr	**503.46**
Sum of two sides 3300 mm	171.18	210.29	8.19	294.30	nr	**504.59**
Sum of two sides 3400 mm	176.00	216.22	8.62	309.74	nr	**525.96**
Sum of two sides 3500 mm	181.74	223.27	8.62	309.74	nr	**533.01**
Sum of two sides 3600 mm	187.49	230.33	8.62	309.74	nr	**540.07**
Sum of two sides 3700 mm	196.13	240.94	8.62	309.74	nr	**550.68**
Sum of two sides 3800 mm	201.88	248.01	8.75	314.42	nr	**562.43**
Sum of two sides 3900 mm	207.61	255.05	8.75	314.42	nr	**569.47**
Sum of two sides 4000 mm	213.38	262.13	8.75	314.42	nr	**576.55**
90° radius bend						
Sum of two sides 2100 mm	123.50	151.72	2.61	93.79	nr	**245.51**
Sum of two sides 2200 mm	238.31	292.76	2.61	93.79	nr	**386.55**
Sum of two sides 2300 mm	257.82	316.73	2.61	93.79	nr	**410.52**
Sum of two sides 2400 mm	264.78	325.28	2.88	103.49	nr	**428.77**
Sum of two sides 2500 mm	284.22	349.17	3.90	140.14	nr	**489.31**
Sum of two sides 2600 mm	303.33	372.64	4.26	153.08	nr	**525.72**
Sum of two sides 2700 mm	322.43	396.10	4.55	163.51	nr	**559.61**
Sum of two sides 2800 mm	341.53	419.57	4.55	163.51	nr	**583.08**
Sum of two sides 2900 mm	360.98	443.47	6.87	246.87	nr	**690.34**
Sum of two sides 3000 mm	380.10	466.96	6.87	246.87	nr	**713.83**
Sum of two sides 3100 mm	383.76	471.45	6.87	246.87	nr	**718.32**

38 VENTILATION/AIR CONDITIONING SYSTEMS

Item	Net Price £	Material £	Labour hours	Labour £	Unit	**Total rate £**
Sum of two sides 3200 mm	396.23	486.77	6.87	246.87	nr	**733.64**
Sum of two sides 3300 mm	397.42	488.23	6.87	246.87	nr	**735.10**
Sum of two sides 3400 mm	409.91	503.57	7.00	251.54	nr	**755.11**
Sum of two sides 3500 mm	423.59	520.38	7.00	251.54	nr	**771.92**
Sum of two sides 3600 mm	437.27	537.19	7.00	251.54	nr	**788.73**
Sum of two sides 3700 mm	468.76	575.87	7.00	251.54	nr	**827.41**
Sum of two sides 3800 mm	482.43	592.66	7.00	251.54	nr	**844.20**
Sum of two sides 3900 mm	496.12	609.48	7.00	251.54	nr	**861.02**
Sum of two sides 4000 mm	509.79	626.27	7.00	251.54	nr	**877.81**
45° bend						
Sum of two sides 2100 mm	59.36	72.92	2.61	93.79	nr	**166.71**
Sum of two sides 2200 mm	170.60	209.58	2.61	93.79	nr	**303.37**
Sum of two sides 2300 mm	183.73	225.71	2.61	93.79	nr	**319.50**
Sum of two sides 2400 mm	190.21	233.68	2.88	103.49	nr	**337.17**
Sum of two sides 2500 mm	203.29	249.74	4.85	174.28	nr	**424.02**
Sum of two sides 2600 mm	216.12	265.50	4.87	175.00	nr	**440.50**
Sum of two sides 2700 mm	228.96	281.27	4.87	175.00	nr	**456.27**
Sum of two sides 2800 mm	241.79	297.03	8.81	316.57	nr	**613.60**
Sum of two sides 2900 mm	254.89	313.13	8.81	316.57	nr	**629.70**
Sum of two sides 3000 mm	267.71	328.88	9.31	334.54	nr	**663.42**
Sum of two sides 3100 mm	281.06	345.28	9.31	334.54	nr	**679.82**
Sum of two sides 3200 mm	282.51	347.07	9.31	334.54	nr	**681.61**
Sum of two sides 3300 mm	283.26	347.98	9.31	334.54	nr	**682.52**
Sum of two sides 3400 mm	291.70	358.35	9.31	334.54	nr	**692.89**
Sum of two sides 3500 mm	300.92	369.68	9.39	337.42	nr	**707.10**
Sum of two sides 3600 mm	310.13	380.99	9.39	337.42	nr	**718.41**
Sum of two sides 3700 mm	331.25	406.94	9.39	337.42	nr	**744.36**
Sum of two sides 3800 mm	340.47	418.27	9.39	337.42	nr	**755.69**
Sum of two sides 3900 mm	349.67	429.57	9.39	337.42	nr	**766.99**
Sum of two sides 4000 mm	358.88	440.88	9.39	337.42	nr	**778.30**
90° mitre bend						
Sum of two sides 2100 mm	279.49	343.35	2.61	93.79	nr	**437.14**
Sum of two sides 2200 mm	295.44	362.94	2.61	93.79	nr	**456.73**
Sum of two sides 2300 mm	318.88	391.74	2.61	93.79	nr	**485.53**
Sum of two sides 2400 mm	326.57	401.19	2.88	103.49	nr	**504.68**
Sum of two sides 2500 mm	350.07	430.06	4.85	174.28	nr	**604.34**
Sum of two sides 2600 mm	373.68	459.07	4.87	175.00	nr	**634.07**
Sum of two sides 2700 mm	397.29	488.07	4.87	175.00	nr	**663.07**
Sum of two sides 2800 mm	420.90	517.07	8.81	316.57	nr	**833.64**
Sum of two sides 2900 mm	444.39	545.94	14.81	532.17	nr	**1078.11**
Sum of two sides 3000 mm	468.00	574.94	15.20	546.19	nr	**1121.13**
Sum of two sides 3100 mm	473.84	582.11	15.20	546.19	nr	**1128.30**
Sum of two sides 3200 mm	492.04	604.47	15.20	546.19	nr	**1150.66**
Sum of two sides 3300 mm	496.13	609.49	15.20	546.19	nr	**1155.68**
Sum of two sides 3400 mm	514.31	631.83	15.20	546.19	nr	**1178.02**
Sum of two sides 3500 mm	532.50	654.18	15.60	560.56	nr	**1214.74**
Sum of two sides 3600 mm	550.67	676.49	15.60	560.56	nr	**1237.05**
Sum of two sides 3700 mm	574.56	705.85	15.60	560.56	nr	**1266.41**
Sum of two sides 3800 mm	592.76	728.21	15.60	560.56	nr	**1288.77**
Sum of two sides 3900 mm	610.92	750.52	15.60	560.56	nr	**1311.08**
Sum of two sides 4000 mm	629.10	772.85	15.60	560.56	nr	**1333.41**

38 VENTILATION/AIR CONDITIONING SYSTEMS

Item	Net Price £	Material £	Labour hours	Labour £	Unit	Total rate £
DUCTWORK: RECTANGULAR – CLASS B – cont						
Extra over fittings – cont						
Branch						
Sum of two sides 2100 mm	123.61	151.85	2.61	93.79	nr	**245.64**
Sum of two sides 2200 mm	132.33	162.57	2.61	93.79	nr	**256.36**
Sum of two sides 2300 mm	141.17	173.43	2.61	93.79	nr	**267.22**
Sum of two sides 2400 mm	149.78	184.00	2.88	103.49	nr	**287.49**
Sum of two sides 2500 mm	158.62	194.87	2.88	103.49	nr	**298.36**
Sum of two sides 2600 mm	167.34	205.58	2.88	103.49	nr	**309.07**
Sum of two sides 2700 mm	176.07	216.30	2.88	103.49	nr	**319.79**
Sum of two sides 2800 mm	184.82	227.05	3.94	141.58	nr	**368.63**
Sum of two sides 2900 mm	193.64	237.89	3.94	141.58	nr	**379.47**
Sum of two sides 3000 mm	202.39	248.64	3.94	141.58	nr	**390.22**
Sum of two sides 3100 mm	209.33	257.17	3.94	141.58	nr	**398.75**
Sum of two sides 3200 mm	215.67	264.95	3.94	141.58	nr	**406.53**
Sum of two sides 3300 mm	222.09	272.84	3.94	141.58	nr	**414.42**
Sum of two sides 3400 mm	228.41	280.60	4.83	173.57	nr	**454.17**
Sum of two sides 3500 mm	234.73	288.36	4.83	173.57	nr	**461.93**
Sum of two sides 3600 mm	241.07	296.15	4.83	173.57	nr	**469.72**
Sum of two sides 3700 mm	250.40	307.62	4.83	173.57	nr	**481.19**
Sum of two sides 3800 mm	256.71	315.37	4.83	173.57	nr	**488.94**
Sum of two sides 3900 mm	263.05	323.16	4.83	173.57	nr	**496.73**
Sum of two sides 4000 mm	269.37	330.92	4.83	173.57	nr	**504.49**
Grille neck						
Sum of two sides 2100 mm	130.45	160.26	2.61	93.79	nr	**254.05**
Sum of two sides 2200 mm	140.24	172.28	2.61	93.79	nr	**266.07**
Sum of two sides 2300 mm	150.07	184.36	2.61	93.79	nr	**278.15**
Sum of two sides 2400 mm	159.85	196.37	2.88	103.49	nr	**299.86**
Sum of two sides 2500 mm	169.67	208.44	2.88	103.49	nr	**311.93**
Sum of two sides 2600 mm	179.48	220.49	2.88	103.49	nr	**323.98**
Sum of two sides 2700 mm	189.29	232.54	2.88	103.49	nr	**336.03**
Sum of two sides 2800 mm	199.10	244.60	3.94	141.58	nr	**386.18**
Sum of two sides 2900 mm	208.90	256.63	4.12	148.05	nr	**404.68**
Sum of two sides 3000 mm	218.70	268.67	4.12	148.05	nr	**416.72**
Sum of two sides 3100 mm	226.50	278.26	4.12	148.05	nr	**426.31**
Sum of two sides 3200 mm	233.60	286.98	4.12	148.05	nr	**435.03**
Sum of two sides 3300 mm	240.69	295.69	4.12	148.05	nr	**443.74**
Sum of two sides 3400 mm	247.80	304.43	5.00	179.67	nr	**484.10**
Sum of two sides 3500 mm	254.89	313.13	5.00	179.67	nr	**492.80**
Sum of two sides 3600 mm	261.97	321.83	5.00	179.67	nr	**501.50**
Sum of two sides 3700 mm	269.07	330.55	5.00	179.67	nr	**510.22**
Sum of two sides 3800 mm	276.17	339.28	5.00	179.67	nr	**518.95**
Sum of two sides 3900 mm	283.26	347.98	5.00	179.67	nr	**527.65**
Sum of two sides 4000 mm	290.35	356.70	5.00	179.67	nr	**536.37**
Ductwork 2001 to 2500 mm longest side						
Sum of two sides 2500 mm	125.35	153.99	2.38	85.51	m	**239.50**
Sum of two sides 2600 mm	128.47	157.83	2.64	94.87	m	**252.70**
Sum of two sides 2700 mm	131.05	161.00	2.66	95.59	m	**256.59**

38 VENTILATION/AIR CONDITIONING SYSTEMS

Item	Net Price £	Material £	Labour hours	Labour £	Unit	Total rate £
Sum of two sides 2800 mm	134.46	165.19	2.95	106.01	m	**271.20**
Sum of two sides 2900 mm	137.02	168.33	2.96	106.36	m	**274.69**
Sum of two sides 3000 mm	139.44	171.30	3.15	113.19	m	**284.49**
Sum of two sides 3100 mm	141.98	174.43	3.15	113.19	m	**287.62**
Sum of two sides 3200 mm	144.38	177.37	3.15	113.19	m	**290.56**
Sum of two sides 3300 mm	146.96	180.54	3.15	113.19	m	**293.73**
Sum of two sides 3400 mm	159.89	196.42	2.66	95.59	m	**292.01**
Sum of two sides 3500 mm	149.39	183.53	3.15	113.19	m	**296.72**
Sum of two sides 3600 mm	162.30	199.38	3.18	114.27	m	**313.65**
Sum of two sides 3700 mm	173.91	213.64	3.18	114.27	m	**327.91**
Sum of two sides 3800 mm	176.34	216.64	3.18	114.27	m	**330.91**
Sum of two sides 3900 mm	178.76	219.61	3.18	114.27	m	**333.88**
Sum of two sides 4000 mm	182.07	223.67	3.18	114.27	m	**337.94**
Extra over fittings; Ductwork 2001 to 2500 mm longest side						
End cap						
Sum of two sides 2500 mm	56.13	68.95	0.87	31.27	nr	**100.22**
Sum of two sides 2600 mm	59.40	72.98	0.87	31.27	nr	**104.25**
Sum of two sides 2700 mm	62.68	77.01	0.87	31.27	nr	**108.28**
Sum of two sides 2800 mm	65.95	81.02	1.16	41.68	nr	**122.70**
Sum of two sides 2900 mm	69.22	85.04	1.16	41.68	nr	**126.72**
Sum of two sides 3000 mm	72.50	89.06	1.73	62.16	nr	**151.22**
Sum of two sides 3100 mm	75.09	92.25	1.73	62.16	nr	**154.41**
Sum of two sides 3200 mm	77.44	95.13	1.73	62.16	nr	**157.29**
Sum of two sides 3300 mm	79.79	98.02	1.73	62.16	nr	**160.18**
Sum of two sides 3400 mm	82.13	100.89	1.80	64.68	nr	**165.57**
Sum of two sides 3500 mm	84.49	103.79	1.80	64.68	nr	**168.47**
Sum of two sides 3600 mm	86.83	106.67	1.80	64.68	nr	**171.35**
Sum of two sides 3700 mm	89.18	109.56	1.80	64.68	nr	**174.24**
Sum of two sides 3800 mm	91.52	112.43	1.80	64.68	nr	**177.11**
Sum of two sides 3900 mm	93.87	115.32	1.80	64.68	nr	**180.00**
Sum of two sides 4000 mm	96.22	118.21	1.80	64.68	nr	**182.89**
Reducer						
Sum of two sides 2500 mm	66.96	82.26	3.12	112.11	nr	**194.37**
Sum of two sides 2600 mm	72.56	89.15	3.12	112.11	nr	**201.26**
Sum of two sides 2700 mm	78.27	96.16	3.12	112.11	nr	**208.27**
Sum of two sides 2800 mm	83.75	102.89	3.95	141.94	nr	**244.83**
Sum of two sides 2900 mm	89.48	109.92	3.97	142.65	nr	**252.57**
Sum of two sides 3000 mm	95.09	116.82	3.97	142.65	nr	**259.47**
Sum of two sides 3100 mm	99.24	121.92	3.97	142.65	nr	**264.57**
Sum of two sides 3200 mm	103.02	126.56	3.97	142.65	nr	**269.21**
Sum of two sides 3300 mm	106.89	131.31	3.97	142.65	nr	**273.96**
Sum of two sides 3400 mm	108.89	133.77	4.52	162.42	nr	**296.19**
Sum of two sides 3500 mm	109.60	134.64	4.52	162.42	nr	**297.06**
Sum of two sides 3600 mm	110.67	135.95	4.52	162.42	nr	**298.37**
Sum of two sides 3700 mm	112.65	138.39	4.52	162.42	nr	**300.81**
Sum of two sides 3800 mm	113.37	139.27	4.52	162.42	nr	**301.69**
Sum of two sides 3900 mm	117.14	143.90	4.52	162.42	nr	**306.32**
Sum of two sides 4000 mm	120.95	148.59	4.52	162.42	nr	**311.01**

38 VENTILATION/AIR CONDITIONING SYSTEMS

Item	Net Price £	Material £	Labour hours	Labour £	Unit	Total rate £
DUCTWORK: RECTANGULAR – CLASS B – cont						
Extra over fittings – cont						
Offset						
Sum of two sides 2500 mm	168.13	206.55	3.48	125.04	nr	**331.59**
Sum of two sides 2600 mm	178.67	219.49	3.49	125.40	nr	**344.89**
Sum of two sides 2700 mm	182.82	224.60	3.50	125.77	nr	**350.37**
Sum of two sides 2800 mm	199.73	245.36	4.34	155.95	nr	**401.31**
Sum of two sides 2900 mm	203.68	250.23	4.76	171.04	nr	**421.27**
Sum of two sides 3000 mm	207.32	254.69	5.32	191.16	nr	**445.85**
Sum of two sides 3100 mm	208.60	256.26	5.35	192.24	nr	**448.50**
Sum of two sides 3200 mm	209.34	257.18	5.32	191.16	nr	**448.34**
Sum of two sides 3300 mm	209.36	257.20	5.32	191.16	nr	**448.36**
Sum of two sides 3400 mm	209.04	256.81	5.35	192.24	nr	**449.05**
Sum of two sides 3500 mm	207.94	255.45	5.32	191.16	nr	**446.61**
Sum of two sides 3600 mm	215.08	264.22	5.35	192.24	nr	**456.46**
Sum of two sides 3700 mm	217.57	267.29	5.35	192.24	nr	**459.53**
Sum of two sides 3800 mm	219.12	269.19	5.35	192.24	nr	**461.43**
Sum of two sides 3900 mm	226.27	277.98	5.35	192.24	nr	**470.22**
Sum of two sides 4000 mm	233.41	286.74	5.35	192.24	nr	**478.98**
Square to round						
Sum of two sides 2500 mm	108.76	133.61	4.70	168.88	nr	**302.49**
Sum of two sides 2600 mm	117.41	144.24	4.70	168.88	nr	**313.12**
Sum of two sides 2700 mm	126.01	154.80	4.71	169.25	nr	**324.05**
Sum of two sides 2800 mm	134.66	165.43	8.19	294.30	nr	**459.73**
Sum of two sides 2900 mm	143.28	176.02	8.19	294.30	nr	**470.32**
Sum of two sides 3000 mm	151.90	186.61	8.19	294.30	nr	**480.91**
Sum of two sides 3100 mm	158.38	194.57	8.19	294.30	nr	**488.87**
Sum of two sides 3200 mm	164.24	201.77	8.62	309.74	nr	**511.51**
Sum of two sides 3300 mm	170.09	208.96	8.62	309.74	nr	**518.70**
Sum of two sides 3400 mm	175.95	216.15	8.62	309.74	nr	**525.89**
Sum of two sides 3500 mm	175.96	216.16	8.62	309.74	nr	**525.90**
Sum of two sides 3600 mm	181.80	223.35	8.75	314.42	nr	**537.77**
Sum of two sides 3700 mm	184.22	226.32	8.75	314.42	nr	**540.74**
Sum of two sides 3800 mm	186.57	229.20	8.75	314.42	nr	**543.62**
Sum of two sides 3900 mm	192.44	236.41	8.75	314.42	nr	**550.83**
Sum of two sides 4000 mm	198.25	243.55	8.75	314.42	nr	**557.97**
90° radius bend						
Sum of two sides 2500 mm	248.06	304.74	3.90	140.14	nr	**444.88**
Sum of two sides 2600 mm	259.44	318.72	4.26	153.08	nr	**471.80**
Sum of two sides 2700 mm	279.21	343.01	4.55	163.51	nr	**506.52**
Sum of two sides 2800 mm	281.59	345.93	4.55	163.51	nr	**509.44**
Sum of two sides 2900 mm	301.29	370.14	6.87	246.87	nr	**617.01**
Sum of two sides 3000 mm	320.66	393.93	6.87	246.87	nr	**640.80**
Sum of two sides 3100 mm	335.67	412.37	6.87	246.87	nr	**659.24**
Sum of two sides 3200 mm	349.50	429.36	6.87	246.87	nr	**676.23**
Sum of two sides 3300 mm	363.65	446.74	6.87	246.87	nr	**693.61**
Sum of two sides 3400 mm	377.48	463.73	6.87	246.87	nr	**710.60**
Sum of two sides 3500 mm	376.66	462.72	7.00	251.54	nr	**714.26**

38 VENTILATION/AIR CONDITIONING SYSTEMS

Item	Net Price £	Material £	Labour hours	Labour £	Unit	Total rate £
Sum of two sides 3600 mm	386.09	474.32	7.00	251.54	nr	**725.86**
Sum of two sides 3700 mm	390.49	479.71	7.00	251.54	nr	**731.25**
Sum of two sides 3800 mm	399.92	491.31	7.00	251.54	nr	**742.85**
Sum of two sides 3900 mm	413.74	508.28	7.00	251.54	nr	**759.82**
Sum of two sides 4000 mm	417.55	512.96	7.00	251.54	nr	**764.50**
45° bend						
Sum of two sides 2500 mm	185.76	228.21	4.85	174.28	nr	**402.49**
Sum of two sides 2600 mm	194.72	239.21	4.87	175.00	nr	**414.21**
Sum of two sides 2700 mm	208.05	255.59	4.87	175.00	nr	**430.59**
Sum of two sides 2800 mm	212.30	260.81	8.81	316.57	nr	**577.38**
Sum of two sides 2900 mm	225.59	277.14	8.81	316.57	nr	**593.71**
Sum of two sides 3000 mm	238.62	293.15	9.31	334.54	nr	**627.69**
Sum of two sides 3100 mm	248.77	305.61	9.31	334.54	nr	**640.15**
Sum of two sides 3200 mm	258.13	317.11	9.31	334.54	nr	**651.65**
Sum of two sides 3300 mm	267.71	328.88	9.31	334.54	nr	**663.42**
Sum of two sides 3400 mm	277.05	340.36	9.31	334.54	nr	**674.90**
Sum of two sides 3500 mm	278.98	342.73	9.31	334.54	nr	**677.27**
Sum of two sides 3600 mm	288.21	354.07	9.39	337.42	nr	**691.49**
Sum of two sides 3700 mm	288.33	354.22	9.39	337.42	nr	**691.64**
Sum of two sides 3800 mm	297.56	365.56	9.39	337.42	nr	**702.98**
Sum of two sides 3900 mm	306.89	377.01	9.39	337.42	nr	**714.43**
Sum of two sides 4000 mm	311.21	382.33	9.39	337.42	nr	**719.75**
90° mitre bend						
Sum of two sides 2500 mm	303.96	373.41	4.85	174.28	nr	**547.69**
Sum of two sides 2600 mm	322.64	396.36	4.87	175.00	nr	**571.36**
Sum of two sides 2700 mm	346.72	425.94	4.87	175.00	nr	**600.94**
Sum of two sides 2800 mm	348.25	427.82	8.81	316.57	nr	**744.39**
Sum of two sides 2900 mm	372.36	457.44	14.81	532.17	nr	**989.61**
Sum of two sides 3000 mm	396.63	487.26	14.81	532.17	nr	**1019.43**
Sum of two sides 3100 mm	416.48	511.64	15.20	546.19	nr	**1057.83**
Sum of two sides 3200 mm	435.22	534.67	15.20	546.19	nr	**1080.86**
Sum of two sides 3300 mm	453.81	557.51	15.20	546.19	nr	**1103.70**
Sum of two sides 3400 mm	472.53	580.50	15.20	546.19	nr	**1126.69**
Sum of two sides 3500 mm	471.82	579.63	15.20	546.19	nr	**1125.82**
Sum of two sides 3600 mm	490.54	602.63	15.60	560.56	nr	**1163.19**
Sum of two sides 3700 mm	491.95	604.36	15.60	560.56	nr	**1164.92**
Sum of two sides 3800 mm	510.67	627.35	15.60	560.56	nr	**1187.91**
Sum of two sides 3900 mm	529.41	650.38	15.60	560.56	nr	**1210.94**
Sum of two sides 4000 mm	534.98	657.23	15.60	560.56	nr	**1217.79**
Branch						
Sum of two sides 2500 mm	162.29	199.37	2.88	103.49	nr	**302.86**
Sum of two sides 2600 mm	171.20	210.31	2.88	103.49	nr	**313.80**
Sum of two sides 2700 mm	180.21	221.39	2.88	103.49	nr	**324.88**
Sum of two sides 2800 mm	189.00	232.19	3.94	141.58	nr	**373.77**
Sum of two sides 2900 mm	198.03	243.29	3.94	141.58	nr	**384.87**
Sum of two sides 3000 mm	206.95	254.24	3.94	141.58	nr	**395.82**
Sum of two sides 3100 mm	214.05	262.96	3.94	141.58	nr	**404.54**
Sum of two sides 3200 mm	220.52	270.91	3.94	141.58	nr	**412.49**
Sum of two sides 3300 mm	227.07	278.95	3.94	141.58	nr	**420.53**
Sum of two sides 3400 mm	233.52	286.88	3.94	141.58	nr	**428.46**
Sum of two sides 3500 mm	240.31	295.22	4.83	173.57	nr	**468.79**

38 VENTILATION/AIR CONDITIONING SYSTEMS

Item	Net Price £	Material £	Labour hours	Labour £	Unit	Total rate £
DUCTWORK: RECTANGULAR – CLASS B – cont						
Extra over fittings – cont						
Branch – cont						
Sum of two sides 3600 mm	246.77	303.16	4.83	173.57	nr	**476.73**
Sum of two sides 3700 mm	253.32	311.20	4.83	173.57	nr	**484.77**
Sum of two sides 3800 mm	259.78	319.14	4.83	173.57	nr	**492.71**
Sum of two sides 3900 mm	266.23	327.06	4.83	173.57	nr	**500.63**
Sum of two sides 4000 mm	272.79	335.12	4.83	173.57	nr	**508.69**
Grille neck						
Sum of two sides 2500 mm	173.28	212.88	2.88	103.49	nr	**316.37**
Sum of two sides 2600 mm	183.30	225.18	2.88	103.49	nr	**328.67**
Sum of two sides 2700 mm	193.31	237.48	2.88	103.49	nr	**340.97**
Sum of two sides 2800 mm	203.34	249.81	3.94	141.58	nr	**391.39**
Sum of two sides 2900 mm	213.34	262.09	3.94	141.58	nr	**403.67**
Sum of two sides 3000 mm	223.36	274.40	3.94	141.58	nr	**415.98**
Sum of two sides 3100 mm	231.32	284.17	4.12	148.05	nr	**432.22**
Sum of two sides 3200 mm	238.57	293.09	4.12	148.05	nr	**441.14**
Sum of two sides 3300 mm	245.81	301.98	4.12	148.05	nr	**450.03**
Sum of two sides 3400 mm	253.07	310.89	4.12	148.05	nr	**458.94**
Sum of two sides 3500 mm	260.31	319.79	5.00	179.67	nr	**499.46**
Sum of two sides 3600 mm	267.55	328.68	5.00	179.67	nr	**508.35**
Sum of two sides 3700 mm	274.79	337.58	5.00	179.67	nr	**517.25**
Sum of two sides 3800 mm	282.05	346.50	5.00	179.67	nr	**526.17**
Sum of two sides 3900 mm	289.28	355.38	5.00	179.67	nr	**535.05**
Sum of two sides 4000 mm	296.53	364.29	5.00	179.67	nr	**543.96**
Ductwork 2501 to 4000 mm longest side						
Sum of two sides 3000 mm	205.87	252.91	2.38	85.51	m	**338.42**
Sum of two sides 3100 mm	210.69	258.84	2.38	85.51	m	**344.35**
Sum of two sides 3200 mm	215.23	264.41	2.38	85.51	m	**349.92**
Sum of two sides 3300 mm	219.75	269.97	2.38	85.51	m	**355.48**
Sum of two sides 3400 mm	225.60	277.15	2.64	94.87	m	**372.02**
Sum of two sides 3500 mm	230.13	282.71	2.66	95.59	m	**378.30**
Sum of two sides 3600 mm	234.66	288.28	2.95	106.01	m	**394.29**
Sum of two sides 3700 mm	240.63	295.62	2.96	106.36	m	**401.98**
Sum of two sides 3800 mm	245.16	301.18	3.15	113.19	m	**414.37**
Sum of two sides 3900 mm	258.65	317.75	3.15	113.19	m	**430.94**
Sum of two sides 4000 mm	263.18	323.32	3.15	113.19	m	**436.51**
Sum of two sides 4100 mm	267.87	329.08	3.35	120.38	m	**449.46**
Sum of two sides 4200 mm	272.40	334.65	3.35	120.38	m	**455.03**
Sum of two sides 4300 mm	276.93	340.21	3.60	129.36	m	**469.57**
Sum of two sides 4400 mm	283.54	348.33	3.60	129.36	m	**477.69**
Sum of two sides 4500 mm	288.10	353.93	3.60	129.36	m	**483.29**

38 VENTILATION/AIR CONDITIONING SYSTEMS

Item	Net Price £	Material £	Labour hours	Labour £	Unit	**Total rate £**
Extra over fittings; Ductwork 2501 to 4000 mm longest side						
End cap						
Sum of two sides 3000 mm	72.79	89.42	1.73	62.16	nr	**151.58**
Sum of two sides 3100 mm	75.39	92.62	1.73	62.16	nr	**154.78**
Sum of two sides 3200 mm	77.74	95.51	1.73	62.16	nr	**157.67**
Sum of two sides 3300 mm	80.10	98.41	1.73	62.16	nr	**160.57**
Sum of two sides 3400 mm	82.45	101.29	1.73	62.16	nr	**163.45**
Sum of two sides 3500 mm	84.83	104.21	1.73	62.16	nr	**166.37**
Sum of two sides 3600 mm	87.18	107.10	1.73	62.16	nr	**169.26**
Sum of two sides 3700 mm	89.55	110.01	1.73	62.16	nr	**172.17**
Sum of two sides 3800 mm	91.90	112.90	1.73	62.16	nr	**175.06**
Sum of two sides 3900 mm	94.26	115.79	1.80	64.68	nr	**180.47**
Sum of two sides 4000 mm	96.61	118.68	1.80	64.68	nr	**183.36**
Sum of two sides 4100 mm	98.97	121.58	1.80	64.68	nr	**186.26**
Sum of two sides 4200 mm	101.34	124.50	1.80	64.68	nr	**189.18**
Sum of two sides 4300 mm	103.70	127.40	1.88	67.56	nr	**194.96**
Sum of two sides 4400 mm	106.05	130.28	1.88	67.56	nr	**197.84**
Sum of two sides 4500 mm	108.41	133.18	1.88	67.56	nr	**200.74**
Reducer						
Sum of two sides 3000 mm	70.84	87.02	3.12	112.11	nr	**199.13**
Sum of two sides 3100 mm	74.25	91.21	3.12	112.11	nr	**203.32**
Sum of two sides 3200 mm	77.19	94.83	3.12	112.11	nr	**206.94**
Sum of two sides 3300 mm	80.12	98.43	3.12	112.11	nr	**210.54**
Sum of two sides 3400 mm	83.04	102.02	3.12	112.11	nr	**214.13**
Sum of two sides 3500 mm	85.95	105.59	3.12	112.11	nr	**217.70**
Sum of two sides 3600 mm	88.89	109.20	3.95	141.94	nr	**251.14**
Sum of two sides 3700 mm	91.90	112.90	3.97	142.65	nr	**255.55**
Sum of two sides 3800 mm	92.21	113.28	4.52	162.42	nr	**275.70**
Sum of two sides 3900 mm	94.83	116.50	4.52	162.42	nr	**278.92**
Sum of two sides 4000 mm	95.14	116.88	4.52	162.42	nr	**279.30**
Sum of two sides 4100 mm	98.18	120.61	4.52	162.42	nr	**283.03**
Sum of two sides 4200 mm	101.11	124.22	4.52	162.42	nr	**286.64**
Sum of two sides 4300 mm	104.05	127.83	4.92	176.79	nr	**304.62**
Sum of two sides 4400 mm	107.03	131.49	4.92	176.79	nr	**308.28**
Sum of two sides 4500 mm	109.96	135.08	5.12	183.98	nr	**319.06**
Offset						
Sum of two sides 3000 mm	193.63	237.88	3.48	125.04	nr	**362.92**
Sum of two sides 3100 mm	198.62	244.01	3.48	125.04	nr	**369.05**
Sum of two sides 3200 mm	203.15	249.57	3.48	125.04	nr	**374.61**
Sum of two sides 3300 mm	206.45	253.63	3.48	125.04	nr	**378.67**
Sum of two sides 3400 mm	206.67	253.89	3.50	125.77	nr	**379.66**
Sum of two sides 3500 mm	212.16	260.64	3.49	125.40	nr	**386.04**
Sum of two sides 3600 mm	216.98	266.56	4.25	152.71	nr	**419.27**
Sum of two sides 3700 mm	219.79	270.01	4.76	171.04	nr	**441.05**
Sum of two sides 3800 mm	213.33	262.08	5.32	191.16	nr	**453.24**
Sum of two sides 3900 mm	230.83	283.57	5.35	192.24	nr	**475.81**
Sum of two sides 4000 mm	218.56	268.51	5.35	192.24	nr	**460.75**
Sum of two sides 4100 mm	242.51	297.93	5.85	210.21	nr	**508.14**
Sum of two sides 4200 mm	244.43	300.28	5.85	210.21	nr	**510.49**
Sum of two sides 4300 mm	248.77	305.61	6.15	220.99	nr	**526.60**
Sum of two sides 4400 mm	249.19	306.13	6.30	226.39	nr	**532.52**

38 VENTILATION/AIR CONDITIONING SYSTEMS

Item	Net Price £	Material £	Labour hours	Labour £	Unit	Total rate £
DUCTWORK: RECTANGULAR – CLASS B – cont						
Extra over fittings – cont						
Offset – cont						
Sum of two sides 4500 mm	253.13	310.97	6.15	220.99	nr	**531.96**
Square to round						
Sum of two sides 3000 mm	152.06	186.81	4.70	168.88	nr	**355.69**
Sum of two sides 3100 mm	158.62	194.87	4.70	168.88	nr	**363.75**
Sum of two sides 3200 mm	164.45	202.03	4.70	168.88	nr	**370.91**
Sum of two sides 3300 mm	170.27	209.18	4.70	168.88	nr	**378.06**
Sum of two sides 3400 mm	176.10	216.34	4.71	169.25	nr	**385.59**
Sum of two sides 3500 mm	181.92	223.49	4.71	169.25	nr	**392.74**
Sum of two sides 3600 mm	187.77	230.67	8.19	294.30	nr	**524.97**
Sum of two sides 3700 mm	193.59	237.82	8.62	309.74	nr	**547.56**
Sum of two sides 3800 mm	199.41	244.98	8.62	309.74	nr	**554.72**
Sum of two sides 3900 mm	244.16	299.95	8.62	309.74	nr	**609.69**
Sum of two sides 4000 mm	251.45	308.91	8.75	314.42	nr	**623.33**
Sum of two sides 4100 mm	258.04	317.01	11.23	403.54	nr	**720.55**
Sum of two sides 4200 mm	264.99	325.54	11.23	403.54	nr	**729.08**
Sum of two sides 4300 mm	271.94	334.08	11.25	404.25	nr	**738.33**
Sum of two sides 4400 mm	278.88	342.60	11.25	404.25	nr	**746.85**
Sum of two sides 4500 mm	285.82	351.13	11.26	404.62	nr	**755.75**
90° radius bend						
Sum of two sides 3000 mm	421.27	517.53	3.90	140.14	nr	**657.67**
Sum of two sides 3100 mm	440.74	541.45	3.90	140.14	nr	**681.59**
Sum of two sides 3200 mm	458.09	562.77	4.26	153.08	nr	**715.85**
Sum of two sides 3300 mm	464.01	570.03	4.26	153.08	nr	**723.11**
Sum of two sides 3400 mm	475.44	584.07	4.26	153.08	nr	**737.15**
Sum of two sides 3500 mm	481.06	590.98	4.55	163.51	nr	**754.49**
Sum of two sides 3600 mm	498.11	611.93	4.55	163.51	nr	**775.44**
Sum of two sides 3700 mm	485.22	596.10	6.87	246.87	nr	**842.97**
Sum of two sides 3800 mm	501.96	616.65	6.87	246.87	nr	**863.52**
Sum of two sides 3900 mm	502.46	617.27	6.87	246.87	nr	**864.14**
Sum of two sides 4000 mm	503.16	618.14	7.00	251.54	nr	**869.68**
Sum of two sides 4100 mm	501.52	616.12	7.20	258.72	nr	**874.84**
Sum of two sides 4200 mm	517.95	636.30	7.20	258.72	nr	**895.02**
Sum of two sides 4300 mm	534.40	656.51	7.41	266.27	nr	**922.78**
Sum of two sides 4400 mm	535.95	658.41	7.41	266.27	nr	**924.68**
Sum of two sides 4500 mm	535.25	657.55	7.55	271.30	nr	**928.85**
45° bend						
Sum of two sides 3000 mm	203.97	250.57	4.85	174.28	nr	**424.85**
Sum of two sides 3100 mm	213.64	262.46	4.85	174.28	nr	**436.74**
Sum of two sides 3200 mm	222.24	273.02	4.85	174.28	nr	**447.30**
Sum of two sides 3300 mm	230.85	283.59	4.87	175.00	nr	**458.59**
Sum of two sides 3400 mm	250.20	307.37	4.87	175.00	nr	**482.37**
Sum of two sides 3500 mm	233.37	286.69	4.87	175.00	nr	**461.69**
Sum of two sides 3600 mm	252.67	310.40	8.81	316.57	nr	**626.97**
Sum of two sides 3700 mm	235.13	288.85	8.81	316.57	nr	**605.42**
Sum of two sides 3800 mm	243.44	299.06	9.31	334.54	nr	**633.60**
Sum of two sides 3900 mm	265.88	326.63	9.31	334.54	nr	**661.17**
Sum of two sides 4000 mm	273.98	336.59	9.31	334.54	nr	**671.13**

38 VENTILATION/AIR CONDITIONING SYSTEMS

Item	Net Price £	Material £	Labour hours	Labour £	Unit	**Total rate £**
Sum of two sides 4100 mm	242.44	297.83	10.01	359.69	nr	**657.52**
Sum of two sides 4200 mm	258.75	317.88	9.31	334.54	nr	**652.42**
Sum of two sides 4300 mm	250.60	307.86	10.01	359.69	nr	**667.55**
Sum of two sides 4400 mm	240.33	295.25	10.52	378.02	nr	**673.27**
Sum of two sides 4500 mm	248.25	304.97	10.52	378.02	nr	**682.99**
90° mitre bend						
Sum of two sides 3000 mm	524.27	644.07	4.85	174.28	nr	**818.35**
Sum of two sides 3100 mm	549.46	675.01	4.85	174.28	nr	**849.29**
Sum of two sides 3200 mm	572.99	703.92	4.87	175.00	nr	**878.92**
Sum of two sides 3300 mm	582.73	715.88	4.87	175.00	nr	**890.88**
Sum of two sides 3400 mm	596.52	732.83	4.87	175.00	nr	**907.83**
Sum of two sides 3500 mm	605.99	744.46	8.81	316.57	nr	**1061.03**
Sum of two sides 3600 mm	629.25	773.03	8.81	316.57	nr	**1089.60**
Sum of two sides 3700 mm	618.33	759.62	14.81	532.17	nr	**1291.79**
Sum of two sides 3800 mm	668.35	821.07	14.81	532.17	nr	**1353.24**
Sum of two sides 3900 mm	669.44	822.40	14.81	532.17	nr	**1354.57**
Sum of two sides 4000 mm	681.87	837.68	15.20	546.19	nr	**1383.87**
Sum of two sides 4100 mm	642.58	789.41	16.30	585.71	nr	**1375.12**
Sum of two sides 4200 mm	665.33	817.36	16.50	592.91	nr	**1410.27**
Sum of two sides 4300 mm	666.28	818.53	17.01	611.23	nr	**1429.76**
Sum of two sides 4400 mm	669.29	822.22	17.01	611.23	nr	**1433.45**
Sum of two sides 4500 mm	688.07	845.29	17.01	611.23	nr	**1456.52**
Branch						
Sum of two sides 3000 mm	217.75	267.51	2.88	103.49	nr	**371.00**
Sum of two sides 3100 mm	225.30	276.78	2.88	103.49	nr	**380.27**
Sum of two sides 3200 mm	232.08	285.11	2.88	103.49	nr	**388.60**
Sum of two sides 3300 mm	238.88	293.46	2.88	103.49	nr	**396.95**
Sum of two sides 3400 mm	245.63	301.76	2.88	103.49	nr	**405.25**
Sum of two sides 3500 mm	252.44	310.12	3.94	141.58	nr	**451.70**
Sum of two sides 3600 mm	259.22	318.45	3.94	141.58	nr	**460.03**
Sum of two sides 3700 mm	266.08	326.88	3.94	141.58	nr	**468.46**
Sum of two sides 3800 mm	272.87	335.22	4.83	173.57	nr	**508.79**
Sum of two sides 3900 mm	279.95	343.92	4.83	173.57	nr	**517.49**
Sum of two sides 4000 mm	286.74	352.26	4.83	173.57	nr	**525.83**
Sum of two sides 4100 mm	293.63	360.73	5.44	195.48	nr	**556.21**
Sum of two sides 4200 mm	300.43	369.07	5.44	195.48	nr	**564.55**
Sum of two sides 4300 mm	307.21	377.41	5.85	210.21	nr	**587.62**
Sum of two sides 4400 mm	314.07	385.83	5.85	210.21	nr	**596.04**
Sum of two sides 4500 mm	320.86	394.18	5.85	210.21	nr	**604.39**
Grille neck						
Sum of two sides 3000 mm	225.56	277.11	2.88	103.49	nr	**380.60**
Sum of two sides 3100 mm	233.60	286.98	2.88	103.49	nr	**390.47**
Sum of two sides 3200 mm	240.93	295.98	2.88	103.49	nr	**399.47**
Sum of two sides 3300 mm	248.23	304.95	2.88	103.49	nr	**408.44**
Sum of two sides 3400 mm	255.55	313.94	3.94	141.58	nr	**455.52**
Sum of two sides 3500 mm	262.88	322.95	3.94	141.58	nr	**464.53**
Sum of two sides 3600 mm	270.19	331.93	3.94	141.58	nr	**473.51**
Sum of two sides 3700 mm	277.51	340.92	4.12	148.05	nr	**488.97**
Sum of two sides 3800 mm	284.84	349.92	4.12	148.05	nr	**497.97**
Sum of two sides 3900 mm	292.16	358.92	4.12	148.05	nr	**506.97**
Sum of two sides 4000 mm	299.47	367.90	5.00	179.67	nr	**547.57**

38 VENTILATION/AIR CONDITIONING SYSTEMS

Item	Net Price £	Material £	Labour hours	Labour £	Unit	Total rate £
DUCTWORK: RECTANGULAR – CLASS B – cont						
Extra over fittings – cont						
Grille neck – cont						
Sum of two sides 4100 mm	306.78	376.88	5.00	179.67	nr	**556.55**
Sum of two sides 4200 mm	314.10	385.88	5.00	179.67	nr	**565.55**
Sum of two sides 4300 mm	321.44	394.88	5.23	187.94	nr	**582.82**
Sum of two sides 4400 mm	328.75	403.87	5.23	187.94	nr	**591.81**
Sum of two sides 4500 mm	336.07	412.86	5.39	193.68	nr	**606.54**

38 VENTILATION/AIR CONDITIONING SYSTEMS

Item	Net Price £	Material £	Labour hours	Labour £	Unit	Total rate £
DUCTWORK: RECTANGULAR – CLASS C						
AIR DUCTLINES						
Galvanized sheet metal DW144 class C rectangular section ductwork; including all necessary stiffeners, joints, couplers in the running length and duct supports						
Ductwork up to 400 mm longest side						
Sum of two sides 200 mm	12.98	15.95	1.17	42.04	m	**57.99**
Sum of two sides 300 mm	14.03	17.24	1.17	42.04	m	**59.28**
Sum of two sides 400 mm	13.93	17.11	1.19	42.77	m	**59.88**
Sum of two sides 500 mm	14.93	18.34	1.16	41.68	m	**60.02**
Sum of two sides 600 mm	14.96	18.38	1.19	42.77	m	**61.15**
Sum of two sides 700 mm	15.89	19.52	1.19	42.77	m	**62.29**
Sum of two sides 800 mm	16.87	20.73	1.19	42.77	m	**63.50**
Extra over fittings; Ductwork up to 400 mm longest side						
End cap						
Sum of two sides 200 mm	8.47	10.41	0.38	13.66	nr	**24.07**
Sum of two sides 300 mm	9.44	11.59	0.38	13.66	nr	**25.25**
Sum of two sides 400 mm	10.42	12.80	0.38	13.66	nr	**26.46**
Sum of two sides 500 mm	11.39	14.00	0.38	13.66	nr	**27.66**
Sum of two sides 600 mm	12.34	15.16	0.38	13.66	nr	**28.82**
Sum of two sides 700 mm	13.31	16.35	0.38	13.66	nr	**30.01**
Sum of two sides 800 mm	14.28	17.55	0.38	13.66	nr	**31.21**
Reducer						
Sum of two sides 200 mm	13.85	17.01	1.40	50.31	nr	**67.32**
Sum of two sides 300 mm	15.64	19.22	1.40	50.31	nr	**69.53**
Sum of two sides 400 mm	25.21	30.98	1.42	51.02	nr	**82.00**
Sum of two sides 500 mm	27.26	33.48	1.42	51.02	nr	**84.50**
Sum of two sides 600 mm	29.32	36.02	1.69	60.72	nr	**96.74**
Sum of two sides 700 mm	31.37	38.53	1.69	60.72	nr	**99.25**
Sum of two sides 800 mm	33.41	41.04	1.92	69.00	nr	**110.04**
Offset						
Sum of two sides 200 mm	20.45	25.13	1.63	58.58	nr	**83.71**
Sum of two sides 300 mm	23.12	28.40	1.63	58.58	nr	**86.98**
Sum of two sides 400 mm	33.81	41.54	1.65	59.28	nr	**100.82**
Sum of two sides 500 mm	36.74	45.14	1.65	59.28	nr	**104.42**
Sum of two sides 600 mm	39.16	48.11	1.92	69.00	nr	**117.11**
Sum of two sides 700 mm	41.81	51.37	1.92	69.00	nr	**120.37**
Sum of two sides 800 mm	44.16	54.25	1.92	69.00	nr	**123.25**
Square to round						
Sum of two sides 200 mm	18.00	22.11	1.22	43.85	nr	**65.96**
Sum of two sides 300 mm	20.22	24.84	1.22	43.85	nr	**68.69**
Sum of two sides 400 mm	26.74	32.85	1.25	44.92	nr	**77.77**
Sum of two sides 500 mm	29.09	35.74	1.25	44.92	nr	**80.66**
Sum of two sides 600 mm	31.43	38.61	1.33	47.79	nr	**86.40**
Sum of two sides 700 mm	33.78	41.50	1.33	47.79	nr	**89.29**
Sum of two sides 800 mm	36.12	44.37	1.40	50.31	nr	**94.68**

38 VENTILATION/AIR CONDITIONING SYSTEMS

Item	Net Price £	Material £	Labour hours	Labour £	Unit	Total rate £
DUCTWORK: RECTANGULAR – CLASS C – cont						
Extra over fittings – cont						
90° radius bend						
Sum of two sides 200 mm	13.53	16.62	1.10	39.52	nr	**56.14**
Sum of two sides 300 mm	14.60	17.93	1.10	39.52	nr	**57.45**
Sum of two sides 400 mm	24.21	29.75	1.12	40.25	nr	**70.00**
Sum of two sides 500 mm	25.74	31.63	1.12	40.25	nr	**71.88**
Sum of two sides 600 mm	27.78	34.13	1.16	41.68	nr	**75.81**
Sum of two sides 700 mm	29.56	36.32	1.16	41.68	nr	**78.00**
Sum of two sides 800 mm	31.55	38.76	1.22	43.85	nr	**82.61**
45° radius bend						
Sum of two sides 200 mm	14.62	17.97	1.29	46.35	nr	**64.32**
Sum of two sides 300 mm	16.05	19.72	1.29	46.35	nr	**66.07**
Sum of two sides 400 mm	25.37	31.16	1.29	46.35	nr	**77.51**
Sum of two sides 500 mm	27.16	33.37	1.29	46.35	nr	**79.72**
Sum of two sides 600 mm	29.22	35.90	1.39	49.95	nr	**85.85**
Sum of two sides 700 mm	31.13	38.24	1.39	49.95	nr	**88.19**
Sum of two sides 800 mm	33.17	40.75	1.46	52.46	nr	**93.21**
90° mitre bend						
Sum of two sides 200 mm	22.56	27.72	2.04	73.30	nr	**101.02**
Sum of two sides 300 mm	24.70	30.35	2.04	73.30	nr	**103.65**
Sum of two sides 400 mm	35.51	43.63	2.09	75.10	nr	**118.73**
Sum of two sides 500 mm	38.28	47.03	2.09	75.10	nr	**122.13**
Sum of two sides 600 mm	41.81	51.37	2.15	77.26	nr	**128.63**
Sum of two sides 700 mm	45.08	55.38	2.15	77.26	nr	**132.64**
Sum of two sides 800 mm	48.66	59.78	2.26	81.22	nr	**141.00**
Branch						
Sum of two sides 200 mm	22.41	27.53	0.92	33.05	nr	**60.58**
Sum of two sides 300 mm	24.90	30.59	0.92	33.05	nr	**63.64**
Sum of two sides 400 mm	30.98	38.06	0.95	34.14	nr	**72.20**
Sum of two sides 500 mm	33.71	41.41	0.95	34.14	nr	**75.55**
Sum of two sides 600 mm	36.39	44.71	1.03	37.01	nr	**81.72**
Sum of two sides 700 mm	39.06	47.99	1.03	37.01	nr	**85.00**
Sum of two sides 800 mm	41.73	51.26	1.03	37.01	nr	**88.27**
Grille neck						
Sum of two sides 200 mm	25.69	31.56	1.10	39.52	nr	**71.08**
Sum of two sides 300 mm	28.72	35.28	1.10	39.52	nr	**74.80**
Sum of two sides 400 mm	31.78	39.04	1.16	41.68	nr	**80.72**
Sum of two sides 500 mm	34.83	42.79	1.16	41.68	nr	**84.47**
Sum of two sides 600 mm	37.89	46.55	1.18	42.41	nr	**88.96**
Sum of two sides 700 mm	40.96	50.32	1.18	42.41	nr	**92.73**
Sum of two sides 800 mm	43.99	54.04	1.18	42.41	nr	**96.45**
Ductwork 401 to 600 mm longest side						
Sum of two sides 600 mm	20.66	25.38	1.17	42.04	m	**67.42**
Sum of two sides 700 mm	22.58	27.74	1.17	42.04	m	**69.78**
Sum of two sides 800 mm	24.44	30.02	1.17	42.04	m	**72.06**

38 VENTILATION/AIR CONDITIONING SYSTEMS

Item	Net Price £	Material £	Labour hours	Labour £	Unit	**Total rate £**
Sum of two sides 900 mm	26.26	32.26	1.27	45.64	m	**77.90**
Sum of two sides 1000 mm	28.08	34.49	1.48	53.18	m	**87.67**
Sum of two sides 1100 mm	30.05	36.92	1.49	53.54	m	**90.46**
Sum of two sides 1200 mm	31.85	39.12	1.49	53.54	m	**92.66**
Extra over fittings: Ductwork 401 to 600 mm longest side						
End cap						
Sum of two sides 600 mm	12.02	14.76	0.38	13.66	nr	**28.42**
Sum of two sides 700 mm	12.97	15.93	0.38	13.66	nr	**29.59**
Sum of two sides 800 mm	13.91	17.08	0.38	13.66	nr	**30.74**
Sum of two sides 900 mm	14.85	18.24	0.38	13.66	nr	**31.90**
Sum of two sides 1000 mm	15.78	19.38	0.38	13.66	nr	**33.04**
Sum of two sides 1100 mm	16.73	20.55	0.38	13.66	nr	**34.21**
Sum of two sides 1200 mm	17.67	21.70	0.38	13.66	nr	**35.36**
Reducer						
Sum of two sides 600 mm	25.60	31.45	1.69	60.72	nr	**92.17**
Sum of two sides 700 mm	27.24	33.46	1.69	60.72	nr	**94.18**
Sum of two sides 800 mm	28.85	35.44	1.92	69.00	nr	**104.44**
Sum of two sides 900 mm	30.50	37.46	1.92	69.00	nr	**106.46**
Sum of two sides 1000 mm	32.13	39.47	2.18	78.34	nr	**117.81**
Sum of two sides 1100 mm	33.89	41.63	2.18	78.34	nr	**119.97**
Sum of two sides 1200 mm	35.55	43.67	2.18	78.34	nr	**122.01**
Offset						
Sum of two sides 600 mm	36.30	44.59	1.92	69.00	nr	**113.59**
Sum of two sides 700 mm	38.74	47.60	1.92	69.00	nr	**116.60**
Sum of two sides 800 mm	40.11	49.28	1.92	69.00	nr	**118.28**
Sum of two sides 900 mm	41.39	50.85	1.92	69.00	nr	**119.85**
Sum of two sides 1000 mm	43.10	52.95	2.18	78.34	nr	**131.29**
Sum of two sides 1100 mm	44.22	54.33	2.18	78.34	nr	**132.67**
Sum of two sides 1200 mm	45.11	55.42	2.18	78.34	nr	**133.76**
Square to round						
Sum of two sides 600 mm	27.56	33.86	1.33	47.79	nr	**81.65**
Sum of two sides 700 mm	29.48	36.21	1.33	47.79	nr	**84.00**
Sum of two sides 800 mm	31.39	38.57	1.40	50.31	nr	**88.88**
Sum of two sides 900 mm	33.31	40.92	1.40	50.31	nr	**91.23**
Sum of two sides 1000 mm	35.22	43.27	1.82	65.39	nr	**108.66**
Sum of two sides 1100 mm	37.20	45.70	1.82	65.39	nr	**111.09**
Sum of two sides 1200 mm	39.11	48.05	1.82	65.39	nr	**113.44**
90° radius bend						
Sum of two sides 600 mm	24.05	29.55	1.16	41.68	nr	**71.23**
Sum of two sides 700 mm	24.63	30.26	1.16	41.68	nr	**71.94**
Sum of two sides 800 mm	26.22	32.21	1.22	43.85	nr	**76.06**
Sum of two sides 900 mm	27.87	34.24	1.22	43.85	nr	**78.09**
Sum of two sides 1000 mm	28.86	35.46	1.40	50.31	nr	**85.77**
Sum of two sides 1100 mm	30.57	37.56	1.40	50.31	nr	**87.87**
Sum of two sides 1200 mm	32.20	39.55	1.40	50.31	nr	**89.86**
45° bend						
Sum of two sides 600 mm	26.84	32.97	1.39	49.95	nr	**82.92**
Sum of two sides 700 mm	28.09	34.51	1.39	49.95	nr	**84.46**
Sum of two sides 800 mm	29.89	36.72	1.46	52.46	nr	**89.18**

38 VENTILATION/AIR CONDITIONING SYSTEMS

Item	Net Price £	Material £	Labour hours	Labour £	Unit	Total rate £
DUCTWORK: RECTANGULAR – CLASS C – cont						
Extra over fittings: Ductwork 401 to 600 mm longest side – cont						
45° bend – cont						
Sum of two sides 900 mm	31.69	38.93	1.46	52.46	nr	**91.39**
Sum of two sides 1000 mm	33.17	40.75	1.88	67.56	nr	**108.31**
Sum of two sides 1100 mm	35.11	43.13	1.88	67.56	nr	**110.69**
Sum of two sides 1200 mm	36.90	45.33	1.88	67.56	nr	**112.89**
90° mitre bend						
Sum of two sides 600 mm	42.32	51.99	2.15	77.26	nr	**129.25**
Sum of two sides 700 mm	43.91	53.94	2.15	77.26	nr	**131.20**
Sum of two sides 800 mm	47.08	57.83	2.26	81.22	nr	**139.05**
Sum of two sides 900 mm	50.27	61.76	2.26	81.22	nr	**142.98**
Sum of two sides 1000 mm	52.70	64.74	3.03	108.88	nr	**173.62**
Sum of two sides 1100 mm	56.03	68.84	3.03	108.88	nr	**177.72**
Sum of two sides 1200 mm	59.28	72.83	3.03	108.88	nr	**181.71**
Branch						
Sum of two sides 600 mm	35.68	43.84	1.03	37.01	nr	**80.85**
Sum of two sides 700 mm	38.31	47.06	1.03	37.01	nr	**84.07**
Sum of two sides 800 mm	40.95	50.31	1.03	37.01	nr	**87.32**
Sum of two sides 900 mm	43.58	53.54	1.03	37.01	nr	**90.55**
Sum of two sides 1000 mm	46.21	56.77	1.29	46.35	nr	**103.12**
Sum of two sides 1100 mm	48.93	60.11	1.29	46.35	nr	**106.46**
Sum of two sides 1200 mm	51.57	63.36	1.29	46.35	nr	**109.71**
Grille neck						
Sum of two sides 600 mm	36.91	45.34	1.18	42.41	nr	**87.75**
Sum of two sides 700 mm	39.90	49.01	1.18	42.41	nr	**91.42**
Sum of two sides 800 mm	42.85	52.64	1.18	42.41	nr	**95.05**
Sum of two sides 900 mm	45.84	56.31	1.18	42.41	nr	**98.72**
Sum of two sides 1000 mm	48.82	59.98	1.44	51.74	nr	**111.72**
Sum of two sides 1100 mm	51.79	63.62	1.44	51.74	nr	**115.36**
Sum of two sides 1200 mm	54.77	67.28	1.44	51.74	nr	**119.02**
Ductwork 601 to 800 mm longest side						
Sum of two sides 900 mm	28.80	35.39	1.27	45.64	m	**81.03**
Sum of two sides 1000 mm	30.62	37.62	1.48	53.18	m	**90.80**
Sum of two sides 1100 mm	32.42	39.83	1.49	53.54	m	**93.37**
Sum of two sides 1200 mm	33.64	41.33	1.49	53.54	m	**94.87**
Sum of two sides 1300 mm	36.20	44.47	1.51	54.26	m	**98.73**
Sum of two sides 1400 mm	38.02	46.70	1.55	55.70	m	**102.40**
Sum of two sides 1500 mm	39.84	48.94	1.61	57.86	m	**106.80**
Sum of two sides 1600 mm	41.65	51.17	1.62	58.20	m	**109.37**
Extra over fittings; Ductwork 601 to 800 mm longest side						
End cap						
Sum of two sides 900 mm	14.93	18.34	0.38	13.66	nr	**32.00**
Sum of two sides 1000 mm	15.87	19.50	0.38	13.66	nr	**33.16**
Sum of two sides 1100 mm	16.82	20.66	0.38	13.66	nr	**34.32**

38 VENTILATION/AIR CONDITIONING SYSTEMS

Item	Net Price £	Material £	Labour hours	Labour £	Unit	**Total rate £**
Sum of two sides 1200 mm	17.76	21.82	0.38	13.66	nr	**35.48**
Sum of two sides 1300 mm	18.70	22.97	0.38	13.66	nr	**36.63**
Sum of two sides 1400 mm	19.52	23.98	0.38	13.66	nr	**37.64**
Sum of two sides 1500 mm	22.72	27.91	0.38	13.66	nr	**41.57**
Sum of two sides 1600 mm	25.91	31.83	0.38	13.66	nr	**45.49**
Reducer						
Sum of two sides 900 mm	30.97	38.04	1.92	69.00	nr	**107.04**
Sum of two sides 1000 mm	32.62	40.08	2.18	78.34	nr	**118.42**
Sum of two sides 1100 mm	34.27	42.11	2.18	78.34	nr	**120.45**
Sum of two sides 1200 mm	36.06	44.30	2.18	78.34	nr	**122.64**
Sum of two sides 1300 mm	37.68	46.29	2.30	82.66	nr	**128.95**
Sum of two sides 1400 mm	39.06	47.99	2.30	82.66	nr	**130.65**
Sum of two sides 1500 mm	45.23	55.57	2.47	88.76	nr	**144.33**
Sum of two sides 1600 mm	51.38	63.12	2.47	88.76	nr	**151.88**
Offset						
Sum of two sides 900 mm	43.98	54.03	1.92	69.00	nr	**123.03**
Sum of two sides 1000 mm	45.10	55.41	2.18	78.34	nr	**133.75**
Sum of two sides 1100 mm	46.09	56.63	2.18	78.34	nr	**134.97**
Sum of two sides 1200 mm	47.02	57.76	2.18	78.34	nr	**136.10**
Sum of two sides 1300 mm	47.72	58.62	2.47	88.76	nr	**147.38**
Sum of two sides 1400 mm	48.69	59.82	2.47	88.76	nr	**148.58**
Sum of two sides 1500 mm	55.91	68.68	2.47	88.76	nr	**157.44**
Sum of two sides 1600 mm	63.00	77.40	2.47	88.76	nr	**166.16**
Square to round						
Sum of two sides 900 mm	33.18	40.76	1.40	50.31	nr	**91.07**
Sum of two sides 1000 mm	35.12	43.14	1.82	65.39	nr	**108.53**
Sum of two sides 1100 mm	37.06	45.53	1.82	65.39	nr	**110.92**
Sum of two sides 1200 mm	39.02	47.93	1.82	65.39	nr	**113.32**
Sum of two sides 1300 mm	40.96	50.32	2.32	83.37	nr	**133.69**
Sum of two sides 1400 mm	42.54	52.26	2.32	83.37	nr	**135.63**
Sum of two sides 1500 mm	50.12	61.57	2.56	91.99	nr	**153.56**
Sum of two sides 1600 mm	57.69	70.88	2.58	92.72	nr	**163.60**
90° radius bend						
Sum of two sides 900 mm	24.96	30.66	1.22	43.85	nr	**74.51**
Sum of two sides 1000 mm	26.51	32.57	1.40	50.31	nr	**82.88**
Sum of two sides 1100 mm	28.07	34.48	1.40	50.31	nr	**84.79**
Sum of two sides 1200 mm	29.71	36.50	1.40	50.31	nr	**86.81**
Sum of two sides 1300 mm	31.25	38.39	1.91	68.64	nr	**107.03**
Sum of two sides 1400 mm	31.66	38.89	1.91	68.64	nr	**107.53**
Sum of two sides 1500 mm	37.68	46.29	2.11	75.82	nr	**122.11**
Sum of two sides 1600 mm	43.74	53.74	2.11	75.82	nr	**129.56**
45° bend						
Sum of two sides 900 mm	30.81	37.85	1.46	52.46	nr	**90.31**
Sum of two sides 1000 mm	32.59	40.04	1.88	67.56	nr	**107.60**
Sum of two sides 1100 mm	34.37	42.22	1.88	67.56	nr	**109.78**
Sum of two sides 1200 mm	36.25	44.53	1.88	67.56	nr	**112.09**
Sum of two sides 1300 mm	38.03	46.73	2.26	81.22	nr	**127.95**
Sum of two sides 1400 mm	39.07	48.00	2.44	87.68	nr	**135.68**
Sum of two sides 1500 mm	45.34	55.70	2.44	87.68	nr	**143.38**
Sum of two sides 1600 mm	51.61	63.40	2.68	96.31	nr	**159.71**

38 VENTILATION/AIR CONDITIONING SYSTEMS

Item	Net Price £	Material £	Labour hours	Labour £	Unit	Total rate £
DUCTWORK: RECTANGULAR – CLASS C – cont						
Extra over fittings – cont						
90° mitre bend						
Sum of two sides 900 mm	46.03	56.55	2.26	81.22	nr	**137.77**
Sum of two sides 1000 mm	49.43	60.72	3.03	108.88	nr	**169.60**
Sum of two sides 1100 mm	52.82	64.89	3.03	108.88	nr	**173.77**
Sum of two sides 1200 mm	56.27	69.13	3.03	108.88	nr	**178.01**
Sum of two sides 1300 mm	59.67	73.30	3.85	138.34	nr	**211.64**
Sum of two sides 1400 mm	61.59	75.66	3.85	138.34	nr	**214.00**
Sum of two sides 1500 mm	71.78	88.18	4.25	152.71	nr	**240.89**
Sum of two sides 1600 mm	81.97	100.69	4.26	153.08	nr	**253.77**
Branch						
Sum of two sides 900 mm	44.93	55.20	1.03	37.01	nr	**92.21**
Sum of two sides 1000 mm	47.62	58.51	1.29	46.35	nr	**104.86**
Sum of two sides 1100 mm	50.31	61.80	1.29	46.35	nr	**108.15**
Sum of two sides 1200 mm	53.11	65.25	1.29	46.35	nr	**111.60**
Sum of two sides 1300 mm	55.81	68.57	1.39	49.95	nr	**118.52**
Sum of two sides 1400 mm	58.15	71.44	1.39	49.95	nr	**121.39**
Sum of two sides 1500 mm	66.85	82.12	1.64	58.93	nr	**141.05**
Sum of two sides 1600 mm	75.55	92.81	1.64	58.93	nr	**151.74**
Grille neck						
Sum of two sides 900 mm	46.09	56.63	1.18	42.41	nr	**99.04**
Sum of two sides 1000 mm	49.08	60.29	1.44	51.74	nr	**112.03**
Sum of two sides 1100 mm	52.07	63.97	1.44	51.74	nr	**115.71**
Sum of two sides 1200 mm	55.06	67.64	1.44	51.74	nr	**119.38**
Sum of two sides 1300 mm	58.04	71.31	1.69	60.72	nr	**132.03**
Sum of two sides 1400 mm	60.62	74.48	1.69	60.72	nr	**135.20**
Sum of two sides 1500 mm	70.38	86.46	1.79	64.32	nr	**150.78**
Sum of two sides 1600 mm	80.13	98.44	1.79	64.32	nr	**162.76**
Ductwork 801 to 1000 mm longest side						
Sum of two sides 1100 mm	32.34	39.73	1.49	53.54	m	**93.27**
Sum of two sides 1200 mm	34.12	41.92	1.49	53.54	m	**95.46**
Sum of two sides 1300 mm	35.91	44.11	1.51	54.26	m	**98.37**
Sum of two sides 1400 mm	37.85	46.49	1.55	55.70	m	**102.19**
Sum of two sides 1500 mm	39.63	48.69	1.61	57.86	m	**106.55**
Sum of two sides 1600 mm	41.42	50.88	1.62	58.20	m	**109.08**
Sum of two sides 1700 mm	43.20	53.07	1.74	62.53	m	**115.60**
Sum of two sides 1800 mm	45.13	55.44	1.76	63.24	m	**118.68**
Sum of two sides 1900 mm	46.92	57.65	1.81	65.04	m	**122.69**
Sum of two sides 2000 mm	48.70	59.83	1.82	65.39	m	**125.22**
Extra over fittings; Ductwork 801 to 1000 mm longest side						
End cap						
Sum of two sides 1100 mm	16.42	20.17	0.38	13.66	nr	**33.83**
Sum of two sides 1200 mm	17.34	21.30	0.38	13.66	nr	**34.96**
Sum of two sides 1300 mm	18.26	22.43	0.38	13.66	nr	**36.09**

38 VENTILATION/AIR CONDITIONING SYSTEMS

Item	Net Price £	Material £	Labour hours	Labour £	Unit	**Total rate £**
Sum of two sides 1400 mm	19.06	23.41	0.38	13.66	nr	**37.07**
Sum of two sides 1500 mm	22.18	27.25	0.38	13.66	nr	**40.91**
Sum of two sides 1600 mm	25.30	31.08	0.38	13.66	nr	**44.74**
Sum of two sides 1700 mm	28.42	34.91	0.58	20.83	nr	**55.74**
Sum of two sides 1800 mm	31.55	38.76	0.58	20.83	nr	**59.59**
Sum of two sides 1900 mm	34.68	42.61	0.58	20.83	nr	**63.44**
Sum of two sides 2000 mm	37.80	46.44	0.58	20.83	nr	**67.27**
Reducer						
Sum of two sides 1100 mm	30.96	38.03	2.18	78.34	nr	**116.37**
Sum of two sides 1200 mm	32.44	39.85	2.18	78.34	nr	**118.19**
Sum of two sides 1300 mm	33.91	41.65	2.30	82.66	nr	**124.31**
Sum of two sides 1400 mm	35.24	43.29	2.30	82.66	nr	**125.95**
Sum of two sides 1500 mm	41.12	50.52	2.47	88.76	nr	**139.28**
Sum of two sides 1600 mm	47.00	57.74	2.47	88.76	nr	**146.50**
Sum of two sides 1700 mm	52.88	64.96	2.59	93.07	nr	**158.03**
Sum of two sides 1800 mm	58.87	72.32	2.59	93.07	nr	**165.39**
Sum of two sides 1900 mm	64.76	79.56	2.71	97.39	nr	**176.95**
Sum of two sides 2000 mm	70.64	86.78	2.71	97.39	nr	**184.17**
Offset						
Sum of two sides 1100 mm	47.84	58.77	2.18	78.34	nr	**137.11**
Sum of two sides 1200 mm	48.79	59.93	2.18	78.34	nr	**138.27**
Sum of two sides 1300 mm	49.61	60.94	2.47	88.76	nr	**149.70**
Sum of two sides 1400 mm	49.98	61.40	2.47	88.76	nr	**150.16**
Sum of two sides 1500 mm	57.12	70.17	2.47	88.76	nr	**158.93**
Sum of two sides 1600 mm	64.13	78.78	2.47	88.76	nr	**167.54**
Sum of two sides 1700 mm	71.02	87.24	2.61	93.79	nr	**181.03**
Sum of two sides 1800 mm	77.80	95.58	2.61	93.79	nr	**189.37**
Sum of two sides 1900 mm	85.37	104.87	2.71	97.39	nr	**202.26**
Sum of two sides 2000 mm	91.86	112.85	2.71	97.39	nr	**210.24**
Square to round						
Sum of two sides 1100 mm	33.57	41.24	1.82	65.39	nr	**106.63**
Sum of two sides 1200 mm	35.34	43.42	1.82	65.39	nr	**108.81**
Sum of two sides 1300 mm	37.09	45.57	2.32	83.37	nr	**128.94**
Sum of two sides 1400 mm	38.55	47.35	2.32	83.37	nr	**130.72**
Sum of two sides 1500 mm	45.79	56.25	2.56	91.99	nr	**148.24**
Sum of two sides 1600 mm	53.06	65.18	2.58	92.72	nr	**157.90**
Sum of two sides 1700 mm	60.32	74.10	2.84	102.05	nr	**176.15**
Sum of two sides 1800 mm	67.59	83.03	2.84	102.05	nr	**185.08**
Sum of two sides 1900 mm	74.85	91.95	3.13	112.47	nr	**204.42**
Sum of two sides 2000 mm	82.12	100.88	3.13	112.47	nr	**213.35**
90° radius bend						
Sum of two sides 1100 mm	24.78	30.44	1.40	50.31	nr	**80.75**
Sum of two sides 1200 mm	26.54	32.60	1.40	50.31	nr	**82.91**
Sum of two sides 1300 mm	28.29	34.76	1.91	68.64	nr	**103.40**
Sum of two sides 1400 mm	29.82	36.63	1.91	68.64	nr	**105.27**
Sum of two sides 1500 mm	35.97	44.18	2.11	75.82	nr	**120.00**
Sum of two sides 1600 mm	42.12	51.74	2.11	75.82	nr	**127.56**
Sum of two sides 1700 mm	48.27	59.30	2.55	91.63	nr	**150.93**
Sum of two sides 1800 mm	54.49	66.94	2.55	91.63	nr	**158.57**
Sum of two sides 1900 mm	59.56	73.17	2.80	100.61	nr	**173.78**
Sum of two sides 2000 mm	65.72	80.73	2.80	100.61	nr	**181.34**

38 VENTILATION/AIR CONDITIONING SYSTEMS

Item	Net Price £	Material £	Labour hours	Labour £	Unit	Total rate £
DUCTWORK: RECTANGULAR – CLASS C – cont						
Extra over fittings – cont						
45° bend						
Sum of two sides 1100 mm	32.31	39.69	1.88	67.56	nr	**107.25**
Sum of two sides 1200 mm	34.18	41.99	1.88	67.56	nr	**109.55**
Sum of two sides 1300 mm	36.05	44.29	2.26	81.22	nr	**125.51**
Sum of two sides 1400 mm	37.76	46.39	2.44	87.68	nr	**134.07**
Sum of two sides 1500 mm	44.02	54.08	2.68	96.31	nr	**150.39**
Sum of two sides 1600 mm	50.29	61.78	2.69	96.66	nr	**158.44**
Sum of two sides 1700 mm	56.55	69.47	2.96	106.36	nr	**175.83**
Sum of two sides 1800 mm	62.94	77.32	2.96	106.36	nr	**183.68**
Sum of two sides 1900 mm	68.63	84.31	3.26	117.15	nr	**201.46**
Sum of two sides 2000 mm	74.89	92.00	3.26	117.15	nr	**209.15**
90° mitre bend						
Sum of two sides 1100 mm	51.72	63.54	3.03	108.88	nr	**172.42**
Sum of two sides 1200 mm	55.22	67.84	3.03	108.88	nr	**176.72**
Sum of two sides 1300 mm	58.72	72.14	3.85	138.34	nr	**210.48**
Sum of two sides 1400 mm	61.85	75.98	3.85	138.34	nr	**214.32**
Sum of two sides 1500 mm	71.95	88.39	4.25	152.71	nr	**241.10**
Sum of two sides 1600 mm	82.06	100.81	4.26	153.08	nr	**253.89**
Sum of two sides 1700 mm	92.15	113.21	4.68	168.17	nr	**281.38**
Sum of two sides 1800 mm	102.28	125.65	4.68	168.17	nr	**293.82**
Sum of two sides 1900 mm	111.20	136.60	4.87	175.00	nr	**311.60**
Sum of two sides 2000 mm	121.34	149.07	4.87	175.00	nr	**324.07**
Branch						
Sum of two sides 1100 mm	49.22	60.47	1.29	46.35	nr	**106.82**
Sum of two sides 1200 mm	51.85	63.69	1.29	46.35	nr	**110.04**
Sum of two sides 1300 mm	54.49	66.94	1.39	49.95	nr	**116.89**
Sum of two sides 1400 mm	56.85	69.84	1.39	49.95	nr	**119.79**
Sum of two sides 1500 mm	65.35	80.28	1.64	58.93	nr	**139.21**
Sum of two sides 1600 mm	73.85	90.72	1.64	58.93	nr	**149.65**
Sum of two sides 1700 mm	82.34	101.16	1.69	60.72	nr	**161.88**
Sum of two sides 1800 mm	90.95	111.73	1.69	60.72	nr	**172.45**
Sum of two sides 1900 mm	99.43	122.15	1.85	66.48	nr	**188.63**
Sum of two sides 2000 mm	107.93	132.59	1.85	66.48	nr	**199.07**
Grille neck						
Sum of two sides 1100 mm	50.83	62.44	1.44	51.74	nr	**114.18**
Sum of two sides 1200 mm	53.75	66.03	1.44	51.74	nr	**117.77**
Sum of two sides 1300 mm	56.67	69.61	1.69	60.72	nr	**130.33**
Sum of two sides 1400 mm	59.18	72.70	1.69	60.72	nr	**133.42**
Sum of two sides 1500 mm	68.70	84.40	1.79	64.32	nr	**148.72**
Sum of two sides 1600 mm	78.22	96.10	1.79	64.32	nr	**160.42**
Sum of two sides 1700 mm	87.75	107.80	1.86	66.84	nr	**174.64**
Sum of two sides 1800 mm	97.26	119.48	1.86	66.84	nr	**186.32**
Sum of two sides 1900 mm	106.78	131.18	2.02	72.60	nr	**203.78**
Sum of two sides 2000 mm	116.31	142.88	2.02	72.60	nr	**215.48**

38 VENTILATION/AIR CONDITIONING SYSTEMS

Item	Net Price £	Material £	Labour hours	Labour £	Unit	Total rate £
Ductwork 1001 to 1250 mm longest side						
Sum of two sides 1300 mm	55.57	68.27	1.51	54.26	m	**122.53**
Sum of two sides 1400 mm	58.11	71.39	1.55	55.70	m	**127.09**
Sum of two sides 1500 mm	60.50	74.33	1.61	57.86	m	**132.19**
Sum of two sides 1600 mm	63.03	77.44	1.62	58.20	m	**135.64**
Sum of two sides 1700 mm	65.41	80.36	1.74	62.53	m	**142.89**
Sum of two sides 1800 mm	67.81	83.31	1.76	63.24	m	**146.55**
Sum of two sides 1900 mm	70.16	86.19	1.81	65.04	m	**151.23**
Sum of two sides 2000 mm	72.72	89.33	1.82	65.39	m	**154.72**
Sum of two sides 2100 mm	75.10	92.26	2.53	90.91	m	**183.17**
Sum of two sides 2200 mm	86.36	106.09	2.55	91.63	m	**197.72**
Sum of two sides 2300 mm	89.07	109.42	2.56	91.99	m	**201.41**
Sum of two sides 2400 mm	91.45	112.35	2.76	99.17	m	**211.52**
Sum of two sides 2500 mm	94.00	115.48	2.77	99.53	m	**215.01**
Extra over fittings; Ductwork 1001 to 1250 mm longest side						
End cap						
Sum of two sides 1300 mm	19.08	23.44	0.38	13.66	nr	**37.10**
Sum of two sides 1400 mm	19.92	24.48	0.38	13.66	nr	**38.14**
Sum of two sides 1500 mm	23.15	28.44	0.38	13.66	nr	**42.10**
Sum of two sides 1600 mm	26.40	32.43	0.38	13.66	nr	**46.09**
Sum of two sides 1700 mm	29.63	36.40	0.58	20.83	nr	**57.23**
Sum of two sides 1800 mm	32.87	40.38	0.58	20.83	nr	**61.21**
Sum of two sides 1900 mm	36.11	44.36	0.58	20.83	nr	**65.19**
Sum of two sides 2000 mm	39.35	48.34	0.58	20.83	nr	**69.17**
Sum of two sides 2100 mm	42.59	52.32	0.87	31.27	nr	**83.59**
Sum of two sides 2200 mm	45.83	56.30	0.87	31.27	nr	**87.57**
Sum of two sides 2300 mm	49.07	60.28	0.87	31.27	nr	**91.55**
Sum of two sides 2400 mm	52.30	64.25	0.87	31.27	nr	**95.52**
Sum of two sides 2500 mm	55.55	68.24	0.87	31.27	nr	**99.51**
Reducer						
Sum of two sides 1300 mm	27.20	33.41	2.30	82.66	nr	**116.07**
Sum of two sides 1400 mm	28.28	34.74	2.30	82.66	nr	**117.40**
Sum of two sides 1500 mm	34.14	41.94	2.47	88.76	nr	**130.70**
Sum of two sides 1600 mm	40.11	49.28	2.47	88.76	nr	**138.04**
Sum of two sides 1700 mm	45.98	56.49	2.59	93.07	nr	**149.56**
Sum of two sides 1800 mm	51.85	63.69	2.59	93.07	nr	**156.76**
Sum of two sides 1900 mm	57.71	70.90	2.71	97.39	nr	**168.29**
Sum of two sides 2000 mm	63.70	78.26	2.71	97.39	nr	**175.65**
Sum of two sides 2100 mm	69.57	85.47	2.92	104.93	nr	**190.40**
Sum of two sides 2200 mm	70.59	86.72	2.92	104.93	nr	**191.65**
Sum of two sides 2300 mm	76.56	94.06	2.92	104.93	nr	**198.99**
Sum of two sides 2400 mm	82.44	101.27	3.12	112.11	nr	**213.38**
Sum of two sides 2500 mm	88.42	108.62	3.12	112.11	nr	**220.73**
Offset						
Sum of two sides 1300 mm	65.46	80.42	2.47	88.76	nr	**169.18**
Sum of two sides 1400 mm	67.41	82.81	2.47	88.76	nr	**171.57**
Sum of two sides 1500 mm	75.05	92.20	2.47	88.76	nr	**180.96**

38 VENTILATION/AIR CONDITIONING SYSTEMS

Item	Net Price £	Material £	Labour hours	Labour £	Unit	**Total rate £**
DUCTWORK: RECTANGULAR – CLASS C – cont						
Extra over fittings – cont						
Offset – cont						
Sum of two sides 1600 mm	82.61	101.48	2.47	88.76	nr	**190.24**
Sum of two sides 1700 mm	89.89	110.43	2.61	93.79	nr	**204.22**
Sum of two sides 1800 mm	96.95	119.10	2.61	93.79	nr	**212.89**
Sum of two sides 1900 mm	103.85	127.57	2.71	97.39	nr	**224.96**
Sum of two sides 2000 mm	110.60	135.87	2.71	97.39	nr	**233.26**
Sum of two sides 2100 mm	117.13	143.89	2.92	104.93	nr	**248.82**
Sum of two sides 2200 mm	115.57	141.98	3.26	117.15	nr	**259.13**
Sum of two sides 2300 mm	125.38	154.02	3.26	117.15	nr	**271.17**
Sum of two sides 2400 mm	135.17	166.06	3.47	124.69	nr	**290.75**
Sum of two sides 2500 mm	144.98	178.11	3.48	125.04	nr	**303.15**
Square to round						
Sum of two sides 1300 mm	30.17	37.07	2.32	83.37	nr	**120.44**
Sum of two sides 1400 mm	31.45	38.64	2.32	83.37	nr	**122.01**
Sum of two sides 1500 mm	38.74	47.60	2.56	91.99	nr	**139.59**
Sum of two sides 1600 mm	46.05	56.57	2.58	92.72	nr	**149.29**
Sum of two sides 1700 mm	53.33	65.52	2.84	102.05	nr	**167.57**
Sum of two sides 1800 mm	60.62	74.48	2.84	102.05	nr	**176.53**
Sum of two sides 1900 mm	75.66	92.95	3.13	112.47	nr	**205.42**
Sum of two sides 2000 mm	75.21	92.40	3.13	112.47	nr	**204.87**
Sum of two sides 2100 mm	82.49	101.34	4.26	153.08	nr	**254.42**
Sum of two sides 2200 mm	84.94	104.35	4.27	153.44	nr	**257.79**
Sum of two sides 2300 mm	92.25	113.33	4.30	154.52	nr	**267.85**
Sum of two sides 2400 mm	99.54	122.28	4.77	171.40	nr	**293.68**
Sum of two sides 2500 mm	106.83	131.24	4.79	172.13	nr	**303.37**
90° radius bend						
Sum of two sides 1300 mm	11.41	14.02	1.91	68.64	nr	**82.66**
Sum of two sides 1400 mm	14.40	17.69	1.91	68.64	nr	**86.33**
Sum of two sides 1500 mm	17.13	21.04	2.11	75.82	nr	**96.86**
Sum of two sides 1600 mm	23.28	28.60	2.11	75.82	nr	**104.42**
Sum of two sides 1700 mm	29.41	36.13	2.55	91.63	nr	**127.76**
Sum of two sides 1800 mm	35.53	43.65	2.55	91.63	nr	**135.28**
Sum of two sides 1900 mm	41.65	51.17	2.80	100.61	nr	**151.78**
Sum of two sides 2000 mm	47.80	58.73	2.80	100.61	nr	**159.34**
Sum of two sides 2100 mm	53.93	66.25	2.80	100.61	nr	**166.86**
Sum of two sides 2200 mm	50.67	62.24	2.80	100.61	nr	**162.85**
Sum of two sides 2300 mm	52.27	64.22	4.35	156.31	nr	**220.53**
Sum of two sides 2400 mm	58.35	71.68	4.35	156.31	nr	**227.99**
Sum of two sides 2500 mm	64.46	79.19	4.35	156.31	nr	**235.50**
45° bend						
Sum of two sides 1300 mm	27.84	34.20	2.26	81.22	nr	**115.42**
Sum of two sides 1400 mm	28.49	35.00	2.44	87.68	nr	**122.68**
Sum of two sides 1500 mm	34.84	42.80	2.68	96.31	nr	**139.11**
Sum of two sides 1600 mm	41.27	50.70	2.69	96.66	nr	**147.36**
Sum of two sides 1700 mm	47.62	58.51	2.96	106.36	nr	**164.87**
Sum of two sides 1800 mm	53.97	66.30	2.96	106.36	nr	**172.66**
Sum of two sides 1900 mm	60.30	74.08	3.26	117.15	nr	**191.23**

38 VENTILATION/AIR CONDITIONING SYSTEMS

Item	Net Price £	Material £	Labour hours	Labour £	Unit	Total rate £
Sum of two sides 2000 mm	66.75	82.00	3.26	117.15	nr	**199.15**
Sum of two sides 2100 mm	73.08	89.78	7.50	269.50	nr	**359.28**
Sum of two sides 2200 mm	74.33	91.32	7.50	269.50	nr	**360.82**
Sum of two sides 2300 mm	78.45	96.38	7.55	271.30	nr	**367.68**
Sum of two sides 2400 mm	84.77	104.14	8.13	292.14	nr	**396.28**
Sum of two sides 2500 mm	91.20	112.03	8.30	298.25	nr	**410.28**
90° mitre bend						
Sum of two sides 1300 mm	64.35	79.05	3.85	138.34	nr	**217.39**
Sum of two sides 1400 mm	66.89	82.17	3.85	138.34	nr	**220.51**
Sum of two sides 1500 mm	79.03	97.09	4.25	152.71	nr	**249.80**
Sum of two sides 1600 mm	91.18	112.01	4.26	153.08	nr	**265.09**
Sum of two sides 1700 mm	111.55	137.04	4.68	168.17	nr	**305.21**
Sum of two sides 1800 mm	123.59	151.83	4.68	168.17	nr	**320.00**
Sum of two sides 1900 mm	135.66	166.66	4.87	175.00	nr	**341.66**
Sum of two sides 2000 mm	139.73	171.65	4.87	175.00	nr	**346.65**
Sum of two sides 2100 mm	151.89	186.60	7.50	269.50	nr	**456.10**
Sum of two sides 2200 mm	153.13	188.12	7.50	269.50	nr	**457.62**
Sum of two sides 2300 mm	159.86	196.39	7.55	271.30	nr	**467.69**
Sum of two sides 2400 mm	172.03	211.34	8.13	292.14	nr	**503.48**
Sum of two sides 2500 mm	184.24	226.34	8.30	298.25	nr	**524.59**
Branch						
Sum of two sides 1300 mm	57.27	70.36	1.39	49.95	nr	**120.31**
Sum of two sides 1400 mm	59.66	73.29	1.39	49.95	nr	**123.24**
Sum of two sides 1500 mm	68.48	84.13	1.64	58.93	nr	**143.06**
Sum of two sides 1600 mm	77.39	95.08	1.64	58.93	nr	**154.01**
Sum of two sides 1700 mm	86.22	105.92	1.69	60.72	nr	**166.64**
Sum of two sides 1800 mm	95.02	116.73	1.69	60.72	nr	**177.45**
Sum of two sides 1900 mm	103.83	127.55	1.85	66.48	nr	**194.03**
Sum of two sides 2000 mm	112.75	138.52	1.85	66.48	nr	**205.00**
Sum of two sides 2100 mm	121.56	149.34	2.61	93.79	nr	**243.13**
Sum of two sides 2200 mm	130.38	160.17	2.61	93.79	nr	**253.96**
Sum of two sides 2300 mm	139.30	171.13	2.61	93.79	nr	**264.92**
Sum of two sides 2400 mm	148.11	181.95	2.88	103.49	nr	**285.44**
Sum of two sides 2500 mm	157.02	192.90	2.88	103.49	nr	**296.39**
Grille neck						
Sum of two sides 1300 mm	59.74	73.40	1.69	60.72	nr	**134.12**
Sum of two sides 1400 mm	62.44	76.70	1.69	60.72	nr	**137.42**
Sum of two sides 1500 mm	72.36	88.89	1.79	64.32	nr	**153.21**
Sum of two sides 1600 mm	82.26	101.05	1.79	64.32	nr	**165.37**
Sum of two sides 1700 mm	92.19	113.25	1.86	66.84	nr	**180.09**
Sum of two sides 1800 mm	102.09	125.42	1.86	66.84	nr	**192.26**
Sum of two sides 1900 mm	112.00	137.59	2.02	72.60	nr	**210.19**
Sum of two sides 2000 mm	121.91	149.76	2.02	72.60	nr	**222.36**
Sum of two sides 2100 mm	131.83	161.95	2.61	93.79	nr	**255.74**
Sum of two sides 2200 mm	141.73	174.11	2.80	100.61	nr	**274.72**
Sum of two sides 2300 mm	151.66	186.31	2.80	100.61	nr	**286.92**
Sum of two sides 2400 mm	161.55	198.46	3.06	109.95	nr	**308.41**
Sum of two sides 2500 mm	171.47	210.65	3.06	109.95	nr	**320.60**

38 VENTILATION/AIR CONDITIONING SYSTEMS

Item	Net Price £	Material £	Labour hours	Labour £	Unit	Total rate £
DUCTWORK: RECTANGULAR – CLASS C – cont						
Ductwork 1251 to 1600 mm longest side						
Sum of two sides 1700 mm	70.37	86.45	1.74	62.53	m	**148.98**
Sum of two sides 1800 mm	73.01	89.69	1.76	63.24	m	**152.93**
Sum of two sides 1900 mm	75.69	92.99	1.81	65.04	m	**158.03**
Sum of two sides 2000 mm	78.20	96.06	1.82	65.39	m	**161.45**
Sum of two sides 2100 mm	80.71	99.15	2.53	90.91	m	**190.06**
Sum of two sides 2200 mm	83.21	102.23	2.55	91.63	m	**193.86**
Sum of two sides 2300 mm	85.88	105.50	2.56	91.99	m	**197.49**
Sum of two sides 2400 mm	88.51	108.74	2.76	99.17	m	**207.91**
Sum of two sides 2500 mm	91.02	111.81	2.77	99.53	m	**211.34**
Sum of two sides 2600 mm	102.59	126.03	2.97	106.72	m	**232.75**
Sum of two sides 2700 mm	105.12	129.14	2.99	107.45	m	**236.59**
Sum of two sides 2800 mm	107.80	132.44	3.30	118.59	m	**251.03**
Sum of two sides 2900 mm	110.47	135.71	3.31	118.94	m	**254.65**
Sum of two sides 3000 mm	112.98	138.80	3.53	126.84	m	**265.64**
Sum of two sides 3100 mm	115.60	142.01	3.55	127.56	m	**269.57**
Sum of two sides 3200 mm	118.11	145.10	3.56	127.92	m	**273.02**
Extra over fittings; Ductwork 1251 to 1600 mm longest side						
End cap						
Sum of two sides 1700 mm	29.32	36.02	0.58	20.83	nr	**56.85**
Sum of two sides 1800 mm	32.52	39.95	0.58	20.83	nr	**60.78**
Sum of two sides 1900 mm	35.73	43.89	0.58	20.83	nr	**64.72**
Sum of two sides 2000 mm	38.93	47.83	0.58	20.83	nr	**68.66**
Sum of two sides 2100 mm	42.15	51.79	0.87	31.27	nr	**83.06**
Sum of two sides 2200 mm	45.35	55.71	0.87	31.27	nr	**86.98**
Sum of two sides 2300 mm	48.55	59.64	0.87	31.27	nr	**90.91**
Sum of two sides 2400 mm	51.75	63.58	0.87	31.27	nr	**94.85**
Sum of two sides 2500 mm	54.96	67.51	0.87	31.27	nr	**98.78**
Sum of two sides 2600 mm	58.17	71.46	0.87	31.27	nr	**102.73**
Sum of two sides 2700 mm	61.37	75.39	0.87	31.27	nr	**106.66**
Sum of two sides 2800 mm	64.57	79.33	1.16	41.68	nr	**121.01**
Sum of two sides 2900 mm	67.78	83.26	1.16	41.68	nr	**124.94**
Sum of two sides 3000 mm	77.66	95.40	1.73	62.16	nr	**157.56**
Sum of two sides 3100 mm	73.52	90.32	1.73	62.16	nr	**152.48**
Sum of two sides 3200 mm	75.83	93.16	1.73	62.16	nr	**155.32**
Reducer						
Sum of two sides 1700 mm	40.02	49.16	2.59	93.07	nr	**142.23**
Sum of two sides 1800 mm	45.54	55.94	2.59	93.07	nr	**149.01**
Sum of two sides 1900 mm	51.14	62.82	2.71	97.39	nr	**160.21**
Sum of two sides 2000 mm	66.34	81.50	2.71	97.39	nr	**178.89**
Sum of two sides 2100 mm	62.16	76.37	2.92	104.93	nr	**181.30**
Sum of two sides 2200 mm	67.67	83.13	2.92	104.93	nr	**188.06**
Sum of two sides 2300 mm	73.28	90.03	2.92	104.93	nr	**194.96**

38 VENTILATION/AIR CONDITIONING SYSTEMS

Item	Net Price £	Material £	Labour hours	Labour £	Unit	**Total rate £**
Sum of two sides 2400 mm	78.78	96.78	3.12	112.11	nr	**208.89**
Sum of two sides 2500 mm	84.29	103.55	3.12	112.11	nr	**215.66**
Sum of two sides 2600 mm	85.05	104.49	3.16	113.55	nr	**218.04**
Sum of two sides 2700 mm	88.81	109.11	3.16	113.55	nr	**222.66**
Sum of two sides 2800 mm	94.31	115.86	4.00	143.73	nr	**259.59**
Sum of two sides 2900 mm	105.63	129.77	4.01	144.09	nr	**273.86**
Sum of two sides 3000 mm	111.12	136.51	4.56	163.85	nr	**300.36**
Sum of two sides 3100 mm	115.22	141.55	4.56	163.85	nr	**305.40**
Sum of two sides 3200 mm	118.92	146.10	4.56	163.85	nr	**309.95**
Offset						
Sum of two sides 1700 mm	98.48	120.98	2.61	93.79	nr	**214.77**
Sum of two sides 1800 mm	108.81	133.68	2.61	93.79	nr	**227.47**
Sum of two sides 1900 mm	115.72	142.16	2.71	97.39	nr	**239.55**
Sum of two sides 2000 mm	122.33	150.29	2.71	97.39	nr	**247.68**
Sum of two sides 2100 mm	128.74	158.16	2.92	104.93	nr	**263.09**
Sum of two sides 2200 mm	134.96	165.80	3.26	117.15	nr	**282.95**
Sum of two sides 2300 mm	140.99	173.21	3.26	117.15	nr	**290.36**
Sum of two sides 2400 mm	150.85	185.31	3.47	124.69	nr	**310.00**
Sum of two sides 2500 mm	156.54	192.31	3.48	125.04	nr	**317.35**
Sum of two sides 2600 mm	156.59	192.37	3.49	125.40	nr	**317.77**
Sum of two sides 2700 mm	166.32	204.32	3.50	125.77	nr	**330.09**
Sum of two sides 2800 mm	176.03	216.26	4.33	155.59	nr	**371.85**
Sum of two sides 2900 mm	191.52	235.28	4.74	170.33	nr	**405.61**
Sum of two sides 3000 mm	201.24	247.22	5.31	190.81	nr	**438.03**
Sum of two sides 3100 mm	208.83	256.55	5.34	191.88	nr	**448.43**
Sum of two sides 3200 mm	215.85	265.17	5.35	192.24	nr	**457.41**
Square to round						
Sum of two sides 1700 mm	44.59	54.78	2.84	102.05	nr	**156.83**
Sum of two sides 1800 mm	51.53	63.30	2.84	102.05	nr	**165.35**
Sum of two sides 1900 mm	58.41	71.76	3.13	112.47	nr	**184.23**
Sum of two sides 2000 mm	65.31	80.23	3.13	112.47	nr	**192.70**
Sum of two sides 2100 mm	72.23	88.74	4.26	153.08	nr	**241.82**
Sum of two sides 2200 mm	79.13	97.21	4.27	153.44	nr	**250.65**
Sum of two sides 2300 mm	86.02	105.67	4.30	154.52	nr	**260.19**
Sum of two sides 2400 mm	92.95	114.19	4.77	171.40	nr	**285.59**
Sum of two sides 2500 mm	99.85	122.66	4.79	172.13	nr	**294.79**
Sum of two sides 2600 mm	100.25	123.15	4.95	177.87	nr	**301.02**
Sum of two sides 2700 mm	107.15	131.64	4.95	177.87	nr	**309.51**
Sum of two sides 2800 mm	114.04	140.10	8.49	305.08	nr	**445.18**
Sum of two sides 2900 mm	123.80	152.09	8.88	319.10	nr	**471.19**
Sum of two sides 3000 mm	130.70	160.57	9.02	324.11	nr	**484.68**
Sum of two sides 3100 mm	135.84	166.88	9.02	324.11	nr	**490.99**
Sum of two sides 3200 mm	140.48	172.58	9.09	326.63	nr	**499.21**
90° radius bend						
Sum of two sides 1700 mm	97.61	119.91	2.55	91.63	nr	**211.54**
Sum of two sides 1800 mm	105.85	130.03	2.55	91.63	nr	**221.66**
Sum of two sides 1900 mm	118.44	145.50	2.80	100.61	nr	**246.11**
Sum of two sides 2000 mm	130.83	160.72	2.80	100.61	nr	**261.33**
Sum of two sides 2100 mm	143.23	175.96	2.61	93.79	nr	**269.75**
Sum of two sides 2200 mm	155.61	191.16	2.62	94.15	nr	**285.31**
Sum of two sides 2300 mm	168.20	206.63	2.63	94.51	nr	**301.14**

38 VENTILATION/AIR CONDITIONING SYSTEMS

Item	Net Price £	Material £	Labour hours	Labour £	Unit	Total rate £
DUCTWORK: RECTANGULAR – CLASS C – cont						
Extra over fittings – cont						
90° radius bend – cont						
Sum of two sides 2400 mm	176.02	216.24	4.34	155.95	nr	**372.19**
Sum of two sides 2500 mm	188.37	231.41	4.35	156.31	nr	**387.72**
Sum of two sides 2600 mm	189.03	232.23	4.53	162.77	nr	**395.00**
Sum of two sides 2700 mm	201.39	247.41	4.53	162.77	nr	**410.18**
Sum of two sides 2800 mm	208.17	255.74	7.13	256.20	nr	**511.94**
Sum of two sides 2900 mm	232.09	285.13	7.17	257.64	nr	**542.77**
Sum of two sides 3000 mm	244.39	300.24	7.26	260.87	nr	**561.11**
Sum of two sides 3100 mm	253.83	311.83	7.26	260.87	nr	**572.70**
Sum of two sides 3200 mm	262.52	322.51	7.31	262.68	nr	**585.19**
45° bend						
Sum of two sides 1700 mm	46.88	57.59	2.96	106.36	nr	**163.95**
Sum of two sides 1800 mm	50.95	62.59	2.96	106.36	nr	**168.95**
Sum of two sides 1900 mm	57.23	70.31	3.26	117.15	nr	**187.46**
Sum of two sides 2000 mm	63.40	77.89	3.26	117.15	nr	**195.04**
Sum of two sides 2100 mm	69.58	85.48	7.50	269.50	nr	**354.98**
Sum of two sides 2200 mm	75.75	93.06	7.50	269.50	nr	**362.56**
Sum of two sides 2300 mm	82.02	100.76	7.55	271.30	nr	**372.06**
Sum of two sides 2400 mm	85.88	105.50	8.13	292.14	nr	**397.64**
Sum of two sides 2500 mm	92.03	113.06	8.30	298.25	nr	**411.31**
Sum of two sides 2600 mm	91.94	112.95	8.56	307.59	nr	**420.54**
Sum of two sides 2700 mm	98.11	120.53	8.62	309.74	nr	**430.27**
Sum of two sides 2800 mm	101.43	124.60	9.09	326.63	nr	**451.23**
Sum of two sides 2900 mm	113.38	139.28	9.09	326.63	nr	**465.91**
Sum of two sides 3000 mm	119.52	146.83	9.62	345.67	nr	**492.50**
Sum of two sides 3100 mm	124.21	152.60	9.62	345.67	nr	**498.27**
Sum of two sides 3200 mm	128.54	157.91	9.62	345.67	nr	**503.58**
90° mitre bend						
Sum of two sides 1700 mm	103.03	126.58	4.68	168.17	nr	**294.75**
Sum of two sides 1800 mm	109.52	134.55	4.68	168.17	nr	**302.72**
Sum of two sides 1900 mm	121.72	149.53	4.87	175.00	nr	**324.53**
Sum of two sides 2000 mm	133.95	164.56	4.87	175.00	nr	**339.56**
Sum of two sides 2100 mm	146.18	179.58	7.50	269.50	nr	**449.08**
Sum of two sides 2200 mm	158.41	194.61	7.50	269.50	nr	**464.11**
Sum of two sides 2300 mm	170.62	209.61	7.55	271.30	nr	**480.91**
Sum of two sides 2400 mm	177.03	217.49	8.13	292.14	nr	**509.63**
Sum of two sides 2500 mm	189.31	232.56	8.30	298.25	nr	**530.81**
Sum of two sides 2600 mm	187.75	230.65	8.56	307.59	nr	**538.24**
Sum of two sides 2700 mm	200.05	245.76	8.62	309.74	nr	**555.50**
Sum of two sides 2800 mm	205.46	252.41	15.20	546.19	nr	**798.60**
Sum of two sides 2900 mm	226.90	278.74	15.20	546.19	nr	**824.93**
Sum of two sides 3000 mm	239.21	293.87	15.60	560.56	nr	**854.43**
Sum of two sides 3100 mm	249.30	306.26	16.04	576.38	nr	**882.64**
Sum of two sides 3200 mm	258.90	318.06	16.04	576.38	nr	**894.44**

38 VENTILATION/AIR CONDITIONING SYSTEMS

Item	Net Price £	Material £	Labour hours	Labour £	Unit	**Total rate £**
Branch						
Sum of two sides 1700 mm	88.00	108.11	1.69	60.72	nr	**168.83**
Sum of two sides 1800 mm	96.72	118.82	1.69	60.72	nr	**179.54**
Sum of two sides 1900 mm	105.56	129.69	1.85	66.48	nr	**196.17**
Sum of two sides 2000 mm	114.29	140.41	1.85	66.48	nr	**206.89**
Sum of two sides 2100 mm	123.02	151.13	2.61	93.79	nr	**244.92**
Sum of two sides 2200 mm	131.76	161.87	2.61	93.79	nr	**255.66**
Sum of two sides 2300 mm	140.60	172.72	2.61	93.79	nr	**266.51**
Sum of two sides 2400 mm	149.32	183.43	2.88	103.49	nr	**286.92**
Sum of two sides 2500 mm	158.06	194.18	2.88	103.49	nr	**297.67**
Sum of two sides 2600 mm	166.79	204.90	2.88	103.49	nr	**308.39**
Sum of two sides 2700 mm	175.53	215.64	2.88	103.49	nr	**319.13**
Sum of two sides 2800 mm	184.25	226.35	3.94	141.58	nr	**367.93**
Sum of two sides 2900 mm	195.95	240.72	3.94	141.58	nr	**382.30**
Sum of two sides 3000 mm	204.68	251.45	4.83	173.57	nr	**425.02**
Sum of two sides 3100 mm	211.62	259.98	4.83	173.57	nr	**433.55**
Sum of two sides 3200 mm	217.95	267.75	4.83	173.57	nr	**441.32**
Grille neck						
Sum of two sides 1700 mm	91.21	112.06	1.86	66.84	nr	**178.90**
Sum of two sides 1800 mm	101.02	124.10	1.86	66.84	nr	**190.94**
Sum of two sides 1900 mm	110.82	136.14	2.02	72.60	nr	**208.74**
Sum of two sides 2000 mm	120.63	148.20	2.02	72.60	nr	**220.80**
Sum of two sides 2100 mm	130.45	160.26	2.80	100.61	nr	**260.87**
Sum of two sides 2200 mm	140.24	172.28	2.80	100.61	nr	**272.89**
Sum of two sides 2300 mm	150.07	184.36	2.80	100.61	nr	**284.97**
Sum of two sides 2400 mm	159.85	196.37	3.06	109.95	nr	**306.32**
Sum of two sides 2500 mm	169.67	208.44	3.06	109.95	nr	**318.39**
Sum of two sides 2600 mm	179.48	220.49	3.08	110.68	nr	**331.17**
Sum of two sides 2700 mm	189.29	232.54	3.08	110.68	nr	**343.22**
Sum of two sides 2800 mm	199.10	244.60	4.13	148.41	nr	**393.01**
Sum of two sides 2900 mm	208.90	256.63	4.13	148.41	nr	**405.04**
Sum of two sides 3000 mm	218.70	268.67	5.02	180.39	nr	**449.06**
Sum of two sides 3100 mm	226.50	278.26	5.02	180.39	nr	**458.65**
Sum of two sides 3200 mm	233.60	286.98	5.02	180.39	nr	**467.37**
Ductwork 1601 to 2000 mm longest side						
Sum of two sides 2100 mm	93.76	115.19	2.53	90.91	m	**206.10**
Sum of two sides 2200 mm	96.68	118.78	2.55	91.63	m	**210.41**
Sum of two sides 2300 mm	99.60	122.36	2.55	91.63	m	**213.99**
Sum of two sides 2400 mm	102.47	125.88	2.56	91.99	m	**217.87**
Sum of two sides 2500 mm	105.41	129.50	2.76	99.17	m	**228.67**
Sum of two sides 2600 mm	108.18	132.90	2.77	99.53	m	**232.43**
Sum of two sides 2700 mm	125.10	153.69	2.97	106.72	m	**260.41**
Sum of two sides 2800 mm	113.73	139.71	2.99	107.45	m	**247.16**
Sum of two sides 2900 mm	116.67	143.33	3.30	118.59	m	**261.92**
Sum of two sides 3000 mm	119.43	146.72	3.31	118.94	m	**265.66**
Sum of two sides 3100 mm	138.45	170.09	3.53	126.84	m	**296.93**
Sum of two sides 3200 mm	141.24	173.51	3.53	126.84	m	**300.35**
Sum of two sides 3300 mm	144.15	177.09	3.53	126.84	m	**303.93**
Sum of two sides 3400 mm	146.94	180.52	3.55	127.56	m	**308.08**
Sum of two sides 3500 mm	149.70	183.91	3.55	127.56	m	**311.47**

38 VENTILATION/AIR CONDITIONING SYSTEMS

Item	Net Price £	Material £	Labour hours	Labour £	Unit	Total rate £
DUCTWORK: RECTANGULAR – CLASS C – cont						
Ductwork 1601 to 2000 mm longest side – cont						
Sum of two sides 3600 mm	152.47	187.31	3.55	127.56	m	**314.87**
Sum of two sides 3700 mm	155.41	190.92	3.56	127.92	m	**318.84**
Sum of two sides 3800 mm	158.18	194.32	3.56	127.92	m	**322.24**
Sum of two sides 3900 mm	160.97	197.75	3.56	127.92	m	**325.67**
Sum of two sides 4000 mm	163.72	201.13	3.56	127.92	m	**329.05**
Extra over fittings; Ductwork 1601 to 2000 mm longest side						
End cap						
Sum of two sides 2100 mm	42.61	52.34	0.87	31.27	nr	**83.61**
Sum of two sides 2200 mm	45.84	56.31	0.87	31.27	nr	**87.58**
Sum of two sides 2300 mm	49.06	60.27	0.87	31.27	nr	**91.54**
Sum of two sides 2400 mm	52.28	64.23	0.87	31.27	nr	**95.50**
Sum of two sides 2500 mm	55.50	68.19	0.87	31.27	nr	**99.46**
Sum of two sides 2600 mm	58.75	72.18	0.87	31.27	nr	**103.45**
Sum of two sides 2700 mm	61.97	76.13	0.87	31.27	nr	**107.40**
Sum of two sides 2800 mm	65.20	80.09	1.16	41.68	nr	**121.77**
Sum of two sides 2900 mm	68.42	84.05	1.16	41.68	nr	**125.73**
Sum of two sides 3000 mm	71.65	88.02	1.73	62.16	nr	**150.18**
Sum of two sides 3100 mm	74.21	91.17	1.73	62.16	nr	**153.33**
Sum of two sides 3200 mm	76.52	94.01	1.73	62.16	nr	**156.17**
Sum of two sides 3300 mm	78.84	96.85	1.73	62.16	nr	**159.01**
Sum of two sides 3400 mm	81.15	99.70	1.73	62.16	nr	**161.86**
Sum of two sides 3500 mm	83.50	102.58	1.73	62.16	nr	**164.74**
Sum of two sides 3600 mm	85.81	105.42	1.73	62.16	nr	**167.58**
Sum of two sides 3700 mm	88.14	108.28	1.73	62.16	nr	**170.44**
Sum of two sides 3800 mm	90.45	111.12	1.73	62.16	nr	**173.28**
Sum of two sides 3900 mm	92.78	113.98	1.73	62.16	nr	**176.14**
Sum of two sides 4000 mm	95.09	116.82	1.73	62.16	nr	**178.98**
Reducer						
Sum of two sides 2100 mm	58.04	71.31	2.92	104.93	nr	**176.24**
Sum of two sides 2200 mm	63.55	78.07	2.92	104.93	nr	**183.00**
Sum of two sides 2300 mm	69.13	84.92	2.92	104.93	nr	**189.85**
Sum of two sides 2400 mm	74.55	91.58	3.12	112.11	nr	**203.69**
Sum of two sides 2500 mm	80.14	98.45	3.12	112.11	nr	**210.56**
Sum of two sides 2600 mm	85.63	105.20	3.16	113.55	nr	**218.75**
Sum of two sides 2700 mm	91.14	111.96	3.16	113.55	nr	**225.51**
Sum of two sides 2800 mm	96.65	118.73	4.00	143.73	nr	**262.46**
Sum of two sides 2900 mm	102.22	125.58	4.01	144.09	nr	**269.67**
Sum of two sides 3000 mm	107.72	132.33	4.01	144.09	nr	**276.42**
Sum of two sides 3100 mm	116.76	143.44	4.01	144.09	nr	**287.53**
Sum of two sides 3200 mm	119.53	146.84	4.56	163.85	nr	**310.69**
Sum of two sides 3300 mm	113.53	139.47	4.56	163.85	nr	**303.32**
Sum of two sides 3400 mm	117.22	144.01	4.56	163.85	nr	**307.86**
Sum of two sides 3500 mm	120.91	148.53	4.56	163.85	nr	**312.38**

38 VENTILATION/AIR CONDITIONING SYSTEMS

Item	Net Price £	Material £	Labour hours	Labour £	Unit	**Total rate £**
Sum of two sides 3600 mm	124.59	153.06	4.56	163.85	nr	**316.91**
Sum of two sides 3700 mm	128.37	157.70	4.56	163.85	nr	**321.55**
Sum of two sides 3800 mm	132.06	162.24	4.56	163.85	nr	**326.09**
Sum of two sides 3900 mm	135.75	166.77	4.56	163.85	nr	**330.62**
Sum of two sides 4000 mm	139.45	171.32	4.56	163.85	nr	**335.17**
Offset						
Sum of two sides 2100 mm	136.33	167.49	2.92	104.93	nr	**272.42**
Sum of two sides 2200 mm	146.65	180.16	3.26	117.15	nr	**297.31**
Sum of two sides 2300 mm	152.62	187.50	3.26	117.15	nr	**304.65**
Sum of two sides 2400 mm	167.30	205.53	3.47	124.69	nr	**330.22**
Sum of two sides 2500 mm	173.07	212.61	3.48	125.04	nr	**337.65**
Sum of two sides 2600 mm	178.55	219.35	3.49	125.40	nr	**344.75**
Sum of two sides 2700 mm	183.80	225.80	3.50	125.77	nr	**351.57**
Sum of two sides 2800 mm	188.87	232.03	4.33	155.59	nr	**387.62**
Sum of two sides 2900 mm	193.77	238.05	4.74	170.33	nr	**408.38**
Sum of two sides 3000 mm	188.74	231.87	5.34	191.88	nr	**423.75**
Sum of two sides 3100 mm	195.74	240.47	5.35	192.24	nr	**432.71**
Sum of two sides 3200 mm	198.41	243.75	5.31	190.81	nr	**434.56**
Sum of two sides 3300 mm	208.60	256.26	5.35	192.24	nr	**448.50**
Sum of two sides 3400 mm	215.58	264.84	5.35	192.24	nr	**457.08**
Sum of two sides 3500 mm	222.60	273.46	5.35	192.24	nr	**465.70**
Sum of two sides 3600 mm	229.59	282.05	5.35	192.24	nr	**474.29**
Sum of two sides 3700 mm	236.63	290.70	5.35	192.24	nr	**482.94**
Sum of two sides 3800 mm	243.64	299.31	5.35	192.24	nr	**491.55**
Sum of two sides 3900 mm	250.63	307.90	5.35	192.24	nr	**500.14**
Sum of two sides 4000 mm	257.64	316.51	5.35	192.24	nr	**508.75**
Square to round						
Sum of two sides 2100 mm	87.57	107.58	4.26	153.08	nr	**260.66**
Sum of two sides 2200 mm	96.02	117.96	4.27	153.44	nr	**271.40**
Sum of two sides 2300 mm	104.46	128.33	4.30	154.52	nr	**282.85**
Sum of two sides 2400 mm	178.47	219.25	4.77	171.40	nr	**390.65**
Sum of two sides 2500 mm	121.38	149.11	4.79	172.13	nr	**321.24**
Sum of two sides 2600 mm	129.84	159.51	4.95	177.87	nr	**337.38**
Sum of two sides 2700 mm	138.27	169.87	4.95	177.87	nr	**347.74**
Sum of two sides 2800 mm	146.72	180.24	8.49	305.08	nr	**485.32**
Sum of two sides 2900 mm	155.17	190.63	8.88	319.10	nr	**509.73**
Sum of two sides 3000 mm	158.39	194.59	9.02	324.11	nr	**518.70**
Sum of two sides 3100 mm	163.61	200.99	9.02	324.11	nr	**525.10**
Sum of two sides 3200 mm	164.13	201.63	9.09	326.63	nr	**528.26**
Sum of two sides 3300 mm	172.75	212.23	9.09	326.63	nr	**538.86**
Sum of two sides 3400 mm	178.47	219.25	9.09	326.63	nr	**545.88**
Sum of two sides 3500 mm	184.21	226.31	9.09	326.63	nr	**552.94**
Sum of two sides 3600 mm	189.95	233.35	9.09	326.63	nr	**559.98**
Sum of two sides 3700 mm	195.67	240.38	9.09	326.63	nr	**567.01**
Sum of two sides 3800 mm	201.41	247.43	9.09	326.63	nr	**574.06**
Sum of two sides 3900 mm	207.13	254.46	9.09	326.63	nr	**581.09**
Sum of two sides 4000 mm	212.87	261.51	9.09	326.63	nr	**588.14**
90° radius bend						
Sum of two sides 2100 mm	124.44	152.87	2.61	93.79	nr	**246.66**
Sum of two sides 2200 mm	239.27	293.95	2.62	94.15	nr	**388.10**
Sum of two sides 2300 mm	259.14	318.35	2.63	94.51	nr	**412.86**

38 VENTILATION/AIR CONDITIONING SYSTEMS

Item	Net Price £	Material £	Labour hours	Labour £	Unit	Total rate £
DUCTWORK: RECTANGULAR – CLASS C – cont						
Extra over fittings – cont						
90° radius bend – cont						
Sum of two sides 2400 mm	265.76	326.49	4.34	155.95	nr	**482.44**
Sum of two sides 2500 mm	285.59	350.85	4.35	156.31	nr	**507.16**
Sum of two sides 2600 mm	305.08	374.79	4.53	162.77	nr	**537.56**
Sum of two sides 2700 mm	324.57	398.74	4.53	162.77	nr	**561.51**
Sum of two sides 2800 mm	344.08	422.70	7.13	256.20	nr	**678.90**
Sum of two sides 2900 mm	363.90	447.05	7.17	257.64	nr	**704.69**
Sum of two sides 3000 mm	383.40	471.01	7.26	260.87	nr	**731.88**
Sum of two sides 3100 mm	372.24	457.30	7.26	260.87	nr	**718.17**
Sum of two sides 3200 mm	386.32	474.59	7.31	262.68	nr	**737.27**
Sum of two sides 3300 mm	418.19	513.74	7.31	262.68	nr	**776.42**
Sum of two sides 3400 mm	432.25	531.02	7.31	262.68	nr	**793.70**
Sum of two sides 3500 mm	446.32	548.30	7.31	262.68	nr	**810.98**
Sum of two sides 3600 mm	460.39	565.59	7.31	262.68	nr	**828.27**
Sum of two sides 3700 mm	474.79	583.28	7.31	262.68	nr	**845.96**
Sum of two sides 3800 mm	488.86	600.57	7.31	262.68	nr	**863.25**
Sum of two sides 3900 mm	502.92	617.84	7.31	262.68	nr	**880.52**
Sum of two sides 4000 mm	516.98	635.11	7.31	262.68	nr	**897.79**
45° bend						
Sum of two sides 2100 mm	161.09	197.90	7.50	269.50	nr	**467.40**
Sum of two sides 2200 mm	171.12	210.22	7.50	269.50	nr	**479.72**
Sum of two sides 2300 mm	184.44	226.58	7.55	271.30	nr	**497.88**
Sum of two sides 2400 mm	190.74	234.33	8.13	292.14	nr	**526.47**
Sum of two sides 2500 mm	204.05	250.68	8.30	298.25	nr	**548.93**
Sum of two sides 2600 mm	217.10	266.71	8.56	307.59	nr	**574.30**
Sum of two sides 2700 mm	230.12	282.70	8.62	309.74	nr	**592.44**
Sum of two sides 2800 mm	243.16	298.72	9.09	326.63	nr	**625.35**
Sum of two sides 2900 mm	256.48	315.08	9.09	326.63	nr	**641.71**
Sum of two sides 3000 mm	265.82	326.56	9.62	345.67	nr	**672.23**
Sum of two sides 3100 mm	269.51	331.10	9.62	345.67	nr	**676.77**
Sum of two sides 3200 mm	275.25	338.14	9.62	345.67	nr	**683.81**
Sum of two sides 3300 mm	296.58	364.35	9.62	345.67	nr	**710.02**
Sum of two sides 3400 mm	306.00	375.92	9.62	345.67	nr	**721.59**
Sum of two sides 3500 mm	315.41	387.48	9.62	345.67	nr	**733.15**
Sum of two sides 3600 mm	324.84	399.06	9.62	345.67	nr	**744.73**
Sum of two sides 3700 mm	334.53	410.97	9.62	345.67	nr	**756.64**
Sum of two sides 3800 mm	343.95	422.54	9.62	345.67	nr	**768.21**
Sum of two sides 3900 mm	353.36	434.10	9.62	345.67	nr	**779.77**
Sum of two sides 4000 mm	362.78	445.67	9.62	345.67	nr	**791.34**
90° mitre bend						
Sum of two sides 2100 mm	278.86	342.58	7.50	269.50	nr	**612.08**
Sum of two sides 2200 mm	294.75	362.10	7.50	269.50	nr	**631.60**
Sum of two sides 2300 mm	318.16	390.86	7.55	271.30	nr	**662.16**
Sum of two sides 2400 mm	325.75	400.19	8.13	292.14	nr	**692.33**
Sum of two sides 2500 mm	349.19	428.98	8.30	298.25	nr	**727.23**
Sum of two sides 2600 mm	372.78	457.96	8.56	307.59	nr	**765.55**
Sum of two sides 2700 mm	396.35	486.92	8.62	309.74	nr	**796.66**

38 VENTILATION/AIR CONDITIONING SYSTEMS

Item	Net Price £	Material £	Labour hours	Labour £	Unit	**Total rate £**
Sum of two sides 2800 mm	419.92	515.88	15.20	546.19	nr	**1062.07**
Sum of two sides 2900 mm	443.39	544.71	15.20	546.19	nr	**1090.90**
Sum of two sides 3000 mm	454.45	558.30	16.04	576.38	nr	**1134.68**
Sum of two sides 3100 mm	466.95	573.65	15.60	560.56	nr	**1134.21**
Sum of two sides 3200 mm	472.61	580.60	16.04	576.38	nr	**1156.98**
Sum of two sides 3300 mm	500.81	615.25	16.04	576.38	nr	**1191.63**
Sum of two sides 3400 mm	518.94	637.52	16.04	576.38	nr	**1213.90**
Sum of two sides 3500 mm	537.11	659.84	16.04	576.38	nr	**1236.22**
Sum of two sides 3600 mm	555.24	682.11	16.04	576.38	nr	**1258.49**
Sum of two sides 3700 mm	573.28	704.28	16.04	576.38	nr	**1280.66**
Sum of two sides 3800 mm	591.43	726.57	16.04	576.38	nr	**1302.95**
Sum of two sides 3900 mm	609.57	748.86	16.04	576.38	nr	**1325.24**
Sum of two sides 4000 mm	627.71	771.14	16.04	576.38	nr	**1347.52**
Branch						
Sum of two sides 2100 mm	125.00	153.56	2.61	93.79	nr	**247.35**
Sum of two sides 2200 mm	133.79	164.36	2.61	93.79	nr	**258.15**
Sum of two sides 2300 mm	142.69	175.30	2.61	93.79	nr	**269.09**
Sum of two sides 2400 mm	151.35	185.93	2.88	103.49	nr	**289.42**
Sum of two sides 2500 mm	160.27	196.90	2.88	103.49	nr	**300.39**
Sum of two sides 2600 mm	169.07	207.70	2.88	103.49	nr	**311.19**
Sum of two sides 2700 mm	177.86	218.51	2.88	103.49	nr	**322.00**
Sum of two sides 2800 mm	186.67	229.32	3.94	141.58	nr	**370.90**
Sum of two sides 2900 mm	195.58	240.27	3.94	141.58	nr	**381.85**
Sum of two sides 3000 mm	204.38	251.08	4.83	173.57	nr	**424.65**
Sum of two sides 3100 mm	211.39	259.70	4.83	173.57	nr	**433.27**
Sum of two sides 3200 mm	217.78	267.54	4.83	173.57	nr	**441.11**
Sum of two sides 3300 mm	227.18	279.09	4.83	173.57	nr	**452.66**
Sum of two sides 3400 mm	233.58	286.95	4.83	173.57	nr	**460.52**
Sum of two sides 3500 mm	239.96	294.79	4.83	173.57	nr	**468.36**
Sum of two sides 3600 mm	246.36	302.65	4.83	173.57	nr	**476.22**
Sum of two sides 3700 mm	252.84	310.61	4.83	173.57	nr	**484.18**
Sum of two sides 3800 mm	259.23	318.47	4.83	173.57	nr	**492.04**
Sum of two sides 3900 mm	265.63	326.33	4.83	173.57	nr	**499.90**
Sum of two sides 4000 mm	272.03	334.19	4.83	173.57	nr	**507.76**
Grille neck						
Sum of two sides 2100 mm	132.76	163.10	2.80	100.61	nr	**263.71**
Sum of two sides 2200 mm	142.68	175.29	2.80	100.61	nr	**275.90**
Sum of two sides 2300 mm	152.60	187.47	2.80	100.61	nr	**288.08**
Sum of two sides 2400 mm	162.51	199.65	3.06	109.95	nr	**309.60**
Sum of two sides 2500 mm	172.42	211.82	3.06	109.95	nr	**321.77**
Sum of two sides 2600 mm	182.34	224.01	3.08	110.68	nr	**334.69**
Sum of two sides 2700 mm	177.86	218.51	3.08	110.68	nr	**329.19**
Sum of two sides 2800 mm	186.67	229.32	4.13	148.41	nr	**377.73**
Sum of two sides 2900 mm	195.58	240.27	4.13	148.41	nr	**388.68**
Sum of two sides 3000 mm	222.01	272.74	5.02	180.39	nr	**453.13**
Sum of two sides 3100 mm	229.93	282.47	5.02	180.39	nr	**462.86**
Sum of two sides 3200 mm	237.15	291.34	5.02	180.39	nr	**471.73**
Sum of two sides 3300 mm	244.34	300.17	5.02	180.39	nr	**480.56**
Sum of two sides 3400 mm	251.54	309.01	5.02	180.39	nr	**489.40**
Sum of two sides 3500 mm	258.75	317.88	5.02	180.39	nr	**498.27**

38 VENTILATION/AIR CONDITIONING SYSTEMS

Item	Net Price £	Material £	Labour hours	Labour £	Unit	Total rate £
DUCTWORK: RECTANGULAR – CLASS C – cont						
Extra over fittings – cont						
Grille neck – cont						
Sum of two sides 3600 mm	265.95	326.72	5.02	180.39	nr	**507.11**
Sum of two sides 3700 mm	273.15	335.57	5.02	180.39	nr	**515.96**
Sum of two sides 3800 mm	280.36	344.42	5.02	180.39	nr	**524.81**
Sum of two sides 3900 mm	287.57	353.28	5.02	180.39	nr	**533.67**
Sum of two sides 4000 mm	294.76	362.11	5.02	180.39	nr	**542.50**
Ductwork 2001 to 2500 mm longest side						
Sum of two sides 2500 mm	179.12	220.05	2.77	99.53	m	**319.58**
Sum of two sides 2600 mm	184.37	226.50	2.97	106.72	m	**333.22**
Sum of two sides 2700 mm	189.02	232.21	2.99	107.45	m	**339.66**
Sum of two sides 2800 mm	194.54	238.99	3.30	118.59	m	**357.58**
Sum of two sides 2900 mm	199.18	244.69	3.31	118.94	m	**363.63**
Sum of two sides 3000 mm	203.72	250.27	3.53	126.84	m	**377.11**
Sum of two sides 3100 mm	208.34	255.95	3.55	127.56	m	**383.51**
Sum of two sides 3200 mm	212.88	261.52	3.56	127.92	m	**389.44**
Sum of two sides 3300 mm	217.54	267.25	3.56	127.92	m	**395.17**
Sum of two sides 3400 mm	222.05	272.79	3.56	127.92	m	**400.71**
Sum of two sides 3500 mm	243.71	299.40	3.56	127.92	m	**427.32**
Sum of two sides 3600 mm	248.22	304.94	3.56	127.92	m	**432.86**
Sum of two sides 3700 mm	252.88	310.66	3.56	127.92	m	**438.58**
Sum of two sides 3800 mm	257.40	316.22	3.56	127.92	m	**444.14**
Sum of two sides 3900 mm	261.93	321.78	3.56	127.92	m	**449.70**
Sum of two sides 4000 mm	267.33	328.42	3.56	127.92	m	**456.34**
Extra over fittings; Ductwork 2001 to 2500 mm longest side						
End cap						
Sum of two sides 2500 mm	56.68	69.64	0.87	31.27	nr	**100.91**
Sum of two sides 2600 mm	59.99	73.70	0.87	31.27	nr	**104.97**
Sum of two sides 2700 mm	63.29	77.75	0.87	31.27	nr	**109.02**
Sum of two sides 2800 mm	66.58	81.80	1.16	41.68	nr	**123.48**
Sum of two sides 2900 mm	69.88	85.85	1.16	41.68	nr	**127.53**
Sum of two sides 3000 mm	73.17	89.89	1.73	62.16	nr	**152.05**
Sum of two sides 3100 mm	75.79	93.10	1.73	62.16	nr	**155.26**
Sum of two sides 3200 mm	78.15	96.01	1.73	62.16	nr	**158.17**
Sum of two sides 3300 mm	80.52	98.92	1.73	62.16	nr	**161.08**
Sum of two sides 3400 mm	82.88	101.82	1.73	62.16	nr	**163.98**
Sum of two sides 3500 mm	85.28	104.77	1.73	62.16	nr	**166.93**
Sum of two sides 3600 mm	87.64	107.67	1.73	62.16	nr	**169.83**
Sum of two sides 3700 mm	90.01	110.58	1.73	62.16	nr	**172.74**
Sum of two sides 3800 mm	92.38	113.48	1.73	62.16	nr	**175.64**
Sum of two sides 3900 mm	94.75	116.40	1.73	62.16	nr	**178.56**
Sum of two sides 4000 mm	97.12	119.31	1.73	62.16	nr	**181.47**

38 VENTILATION/AIR CONDITIONING SYSTEMS

Item	Net Price £	Material £	Labour hours	Labour £	Unit	**Total rate £**
Reducer						
Sum of two sides 2500 mm	47.21	58.00	3.12	112.11	nr	**170.11**
Sum of two sides 2600 mm	51.99	63.87	3.16	113.55	nr	**177.42**
Sum of two sides 2700 mm	56.89	69.89	3.16	113.55	nr	**183.44**
Sum of two sides 2800 mm	61.56	75.63	4.00	143.73	nr	**219.36**
Sum of two sides 2900 mm	66.45	81.64	4.01	144.09	nr	**225.73**
Sum of two sides 3000 mm	71.25	87.53	4.56	163.85	nr	**251.38**
Sum of two sides 3100 mm	74.57	91.61	4.56	163.85	nr	**255.46**
Sum of two sides 3200 mm	77.53	95.25	4.56	163.85	nr	**259.10**
Sum of two sides 3300 mm	80.56	98.97	4.56	163.85	nr	**262.82**
Sum of two sides 3400 mm	83.52	102.61	4.56	163.85	nr	**266.46**
Sum of two sides 3500 mm	84.59	103.92	4.56	163.85	nr	**267.77**
Sum of two sides 3600 mm	84.89	104.29	4.56	163.85	nr	**268.14**
Sum of two sides 3700 mm	86.04	105.70	4.56	163.85	nr	**269.55**
Sum of two sides 3800 mm	88.99	109.33	4.56	163.85	nr	**273.18**
Sum of two sides 3900 mm	91.93	112.94	4.56	163.85	nr	**276.79**
Sum of two sides 4000 mm	94.97	116.67	4.56	163.85	nr	**280.52**
Offset						
Sum of two sides 2500 mm	157.14	193.04	3.48	125.04	nr	**318.08**
Sum of two sides 2600 mm	167.21	205.42	3.49	125.40	nr	**330.82**
Sum of two sides 2700 mm	167.01	205.17	3.50	125.77	nr	**330.94**
Sum of two sides 2800 mm	187.32	230.12	4.33	155.59	nr	**385.71**
Sum of two sides 2900 mm	179.58	220.62	4.74	170.33	nr	**390.95**
Sum of two sides 3000 mm	167.71	206.03	5.31	190.81	nr	**396.84**
Sum of two sides 3100 mm	167.73	206.05	5.34	191.88	nr	**397.93**
Sum of two sides 3200 mm	167.68	206.00	5.35	192.24	nr	**398.24**
Sum of two sides 3300 mm	168.25	206.69	5.35	192.24	nr	**398.93**
Sum of two sides 3400 mm	168.47	206.97	5.35	192.24	nr	**399.21**
Sum of two sides 3500 mm	168.93	207.53	5.35	192.24	nr	**399.77**
Sum of two sides 3600 mm	169.36	208.06	5.35	192.24	nr	**400.30**
Sum of two sides 3700 mm	169.66	208.43	5.35	192.24	nr	**400.67**
Sum of two sides 3800 mm	175.50	215.61	5.35	192.24	nr	**407.85**
Sum of two sides 3900 mm	181.34	222.78	5.35	192.24	nr	**415.02**
Sum of two sides 4000 mm	187.19	229.96	5.35	192.24	nr	**422.20**
Square to round						
Sum of two sides 2500 mm	80.60	99.01	4.79	172.13	nr	**271.14**
Sum of two sides 2600 mm	88.10	108.23	4.95	177.87	nr	**286.10**
Sum of two sides 2700 mm	95.60	117.44	4.95	177.87	nr	**295.31**
Sum of two sides 2800 mm	103.10	126.66	8.49	305.08	nr	**431.74**
Sum of two sides 2900 mm	110.61	135.88	8.88	319.10	nr	**454.98**
Sum of two sides 3000 mm	118.11	145.10	9.02	324.11	nr	**469.21**
Sum of two sides 3100 mm	123.48	151.69	9.02	324.11	nr	**475.80**
Sum of two sides 3200 mm	128.20	157.49	9.09	326.63	nr	**484.12**
Sum of two sides 3300 mm	129.61	159.22	9.09	326.63	nr	**485.85**
Sum of two sides 3400 mm	132.91	163.27	9.09	326.63	nr	**489.90**
Sum of two sides 3500 mm	137.66	169.11	9.09	326.63	nr	**495.74**
Sum of two sides 3600 mm	138.55	170.21	9.09	326.63	nr	**496.84**
Sum of two sides 3700 mm	142.09	174.56	9.09	326.63	nr	**501.19**
Sum of two sides 3800 mm	146.82	180.37	9.09	326.63	nr	**507.00**
Sum of two sides 3900 mm	151.56	186.20	9.09	326.63	nr	**512.83**
Sum of two sides 4000 mm	156.27	191.98	9.09	326.63	nr	**518.61**

38 VENTILATION/AIR CONDITIONING SYSTEMS

Item	Net Price £	Material £	Labour hours	Labour £	Unit	**Total rate £**
DUCTWORK: RECTANGULAR – CLASS C – cont						
Extra over fittings – cont						
90° radius bend						
Sum of two sides 2500 mm	213.21	261.93	4.35	156.31	nr	**418.24**
Sum of two sides 2600 mm	219.33	269.45	4.53	162.77	nr	**432.22**
Sum of two sides 2700 mm	237.85	292.19	4.53	162.77	nr	**454.96**
Sum of two sides 2800 mm	230.14	282.72	7.13	256.20	nr	**538.92**
Sum of two sides 2900 mm	248.31	305.05	7.17	257.64	nr	**562.69**
Sum of two sides 3000 mm	266.16	326.98	7.26	260.87	nr	**587.85**
Sum of two sides 3100 mm	279.66	343.56	7.26	260.87	nr	**604.43**
Sum of two sides 3200 mm	291.96	358.67	7.31	262.68	nr	**621.35**
Sum of two sides 3300 mm	296.04	363.69	7.31	262.68	nr	**626.37**
Sum of two sides 3400 mm	304.60	374.20	7.31	262.68	nr	**636.88**
Sum of two sides 3500 mm	308.35	378.81	7.31	262.68	nr	**641.49**
Sum of two sides 3600 mm	316.92	389.34	7.31	262.68	nr	**652.02**
Sum of two sides 3700 mm	339.15	416.65	7.31	262.68	nr	**679.33**
Sum of two sides 3800 mm	351.47	431.78	7.31	262.68	nr	**694.46**
Sum of two sides 3900 mm	360.21	442.52	7.31	262.68	nr	**705.20**
Sum of two sides 4000 mm	363.78	446.90	7.31	262.68	nr	**709.58**
45° bend						
Sum of two sides 2500 mm	174.57	214.46	8.30	298.25	nr	**512.71**
Sum of two sides 2600 mm	181.09	222.47	8.56	307.59	nr	**530.06**
Sum of two sides 2700 mm	194.02	238.35	8.62	309.74	nr	**548.09**
Sum of two sides 2800 mm	193.44	237.64	9.09	326.63	nr	**564.27**
Sum of two sides 2900 mm	206.19	253.30	9.09	326.63	nr	**579.93**
Sum of two sides 3000 mm	218.68	268.65	9.62	345.67	nr	**614.32**
Sum of two sides 3100 mm	228.31	280.48	9.62	345.67	nr	**626.15**
Sum of two sides 3200 mm	237.11	291.29	9.62	345.67	nr	**636.96**
Sum of two sides 3300 mm	246.20	302.45	9.62	345.67	nr	**648.12**
Sum of two sides 3400 mm	246.70	303.07	9.62	345.67	nr	**648.74**
Sum of two sides 3500 mm	255.00	313.27	9.62	345.67	nr	**658.94**
Sum of two sides 3600 mm	255.51	313.90	9.62	345.67	nr	**659.57**
Sum of two sides 3700 mm	276.69	339.92	9.62	345.67	nr	**685.59**
Sum of two sides 3800 mm	285.49	350.72	9.62	345.67	nr	**696.39**
Sum of two sides 3900 mm	294.31	361.56	9.62	345.67	nr	**707.23**
Sum of two sides 4000 mm	295.16	362.61	9.62	345.67	nr	**708.28**
90° mitre bend						
Sum of two sides 2500 mm	228.18	280.32	8.30	298.25	nr	**578.57**
Sum of two sides 2600 mm	239.12	293.76	8.56	307.59	nr	**601.35**
Sum of two sides 2700 mm	260.03	319.45	8.62	309.74	nr	**629.19**
Sum of two sides 2800 mm	248.16	304.87	15.20	546.19	nr	**851.06**
Sum of two sides 2900 mm	268.76	330.17	15.20	546.19	nr	**876.36**
Sum of two sides 3000 mm	289.50	355.66	15.20	546.19	nr	**901.85**
Sum of two sides 3100 mm	305.84	375.72	16.04	576.38	nr	**952.10**
Sum of two sides 3200 mm	321.04	394.40	16.04	576.38	nr	**970.78**
Sum of two sides 3300 mm	336.09	412.89	16.04	576.38	nr	**989.27**
Sum of two sides 3400 mm	351.30	431.57	16.04	576.38	nr	**1007.95**
Sum of two sides 3500 mm	352.80	433.42	16.04	576.38	nr	**1009.80**

38 VENTILATION/AIR CONDITIONING SYSTEMS

Item	Net Price £	Material £	Labour hours	Labour £	Unit	**Total rate £**
Sum of two sides 3600 mm	352.95	433.60	16.04	576.38	nr	**1009.98**
Sum of two sides 3700 mm	366.20	449.87	16.04	576.38	nr	**1026.25**
Sum of two sides 3800 mm	381.40	468.55	16.04	576.38	nr	**1044.93**
Sum of two sides 3900 mm	396.61	487.23	16.04	576.38	nr	**1063.61**
Sum of two sides 4000 mm	430.04	528.31	16.04	576.38	nr	**1104.69**
Branch						
Sum of two sides 2500 mm	172.42	211.82	2.88	103.49	nr	**315.31**
Sum of two sides 2600 mm	181.68	223.20	2.88	103.49	nr	**326.69**
Sum of two sides 2700 mm	191.09	234.76	2.88	103.49	nr	**338.25**
Sum of two sides 2800 mm	200.25	246.00	3.94	141.58	nr	**387.58**
Sum of two sides 2900 mm	209.63	257.53	3.94	141.58	nr	**399.11**
Sum of two sides 3000 mm	218.92	268.95	4.83	173.57	nr	**442.52**
Sum of two sides 3100 mm	226.40	278.13	4.83	173.57	nr	**451.70**
Sum of two sides 3200 mm	233.23	286.52	4.83	173.57	nr	**460.09**
Sum of two sides 3300 mm	240.16	295.04	4.83	173.57	nr	**468.61**
Sum of two sides 3400 mm	246.98	303.42	4.83	173.57	nr	**476.99**
Sum of two sides 3500 mm	254.12	312.19	4.83	173.57	nr	**485.76**
Sum of two sides 3600 mm	260.95	320.58	4.83	173.57	nr	**494.15**
Sum of two sides 3700 mm	270.91	332.81	4.83	173.57	nr	**506.38**
Sum of two sides 3800 mm	277.73	341.19	4.83	173.57	nr	**514.76**
Sum of two sides 3900 mm	284.55	349.57	4.83	173.57	nr	**523.14**
Sum of two sides 4000 mm	291.48	358.08	4.83	173.57	nr	**531.65**
Grille neck						
Sum of two sides 2500 mm	176.09	216.33	3.06	109.95	nr	**326.28**
Sum of two sides 2600 mm	186.22	228.77	3.08	110.68	nr	**339.45**
Sum of two sides 2700 mm	196.36	241.23	3.08	110.68	nr	**351.91**
Sum of two sides 2800 mm	206.48	253.66	4.13	148.41	nr	**402.07**
Sum of two sides 2900 mm	216.61	266.10	4.13	148.41	nr	**414.51**
Sum of two sides 3000 mm	226.74	278.55	5.02	180.39	nr	**458.94**
Sum of two sides 3100 mm	234.83	288.49	5.02	180.39	nr	**468.88**
Sum of two sides 3200 mm	242.19	297.53	5.02	180.39	nr	**477.92**
Sum of two sides 3300 mm	249.53	306.55	5.02	180.39	nr	**486.94**
Sum of two sides 3400 mm	256.89	315.59	5.02	180.39	nr	**495.98**
Sum of two sides 3500 mm	264.25	324.63	5.02	180.39	nr	**505.02**
Sum of two sides 3600 mm	271.61	333.67	5.02	180.39	nr	**514.06**
Sum of two sides 3700 mm	278.96	342.70	5.02	180.39	nr	**523.09**
Sum of two sides 3800 mm	286.33	351.76	5.02	180.39	nr	**532.15**
Sum of two sides 3900 mm	293.69	360.80	5.02	180.39	nr	**541.19**
Sum of two sides 4000 mm	301.04	369.83	5.02	180.39	nr	**550.22**
DUCTWORK ANCILLARIES: ACCESS DOORS						
Refer to ancillaries in Ductwork Ancillaries: Access Hatches						

38 VENTILATION/AIR CONDITIONING SYSTEMS

Item	Net Price £	Material £	Labour hours	Labour £	Unit	Total rate £
DUCTWORK ANCILLARIES: VOLUME CONTROL AND FIRE DAMPERS						
Volume control damper; opposed blade; galvanized steel casing; aluminium aerofoil blades; manually operated						
Rectangular						
Sum of two sides 200 mm	21.04	25.85	1.60	57.50	nr	**83.35**
Sum of two sides 300 mm	22.51	27.66	1.60	57.50	nr	**85.16**
Sum of two sides 400 mm	24.61	30.23	1.60	57.50	nr	**87.73**
Sum of two sides 500 mm	26.94	33.10	1.60	57.50	nr	**90.60**
Sum of two sides 600 mm	29.78	36.58	1.70	61.12	nr	**97.70**
Sum of two sides 700 mm	32.72	40.19	2.10	75.50	nr	**115.69**
Sum of two sides 800 mm	35.76	43.93	2.15	77.26	nr	**121.19**
Sum of two sides 900 mm	39.24	48.21	2.30	82.66	nr	**130.87**
Sum of two sides 1000 mm	42.61	52.34	2.40	86.24	nr	**138.58**
Sum of two sides 1100 mm	46.41	57.01	2.60	93.43	nr	**150.44**
Sum of two sides 1200 mm	52.50	64.50	2.80	100.61	nr	**165.11**
Sum of two sides 1300 mm	56.60	69.53	3.10	111.38	nr	**180.91**
Sum of two sides 1400 mm	61.03	74.98	3.25	116.78	nr	**191.76**
Sum of two sides 1500 mm	66.06	81.15	3.40	122.18	nr	**203.33**
Sum of two sides 1600 mm	71.02	87.24	3.45	123.96	nr	**211.20**
Sum of two sides 1700 mm	75.75	93.06	3.60	129.36	nr	**222.42**
Sum of two sides 1800 mm	81.42	100.02	3.90	140.14	nr	**240.16**
Sum of two sides 1900 mm	86.69	106.50	4.20	150.98	nr	**257.48**
Sum of two sides 2000 mm	93.00	114.25	4.33	155.59	nr	**269.84**
Circular						
100 mm dia.	28.09	34.51	0.80	28.75	nr	**63.26**
150 mm dia.	32.09	39.43	0.80	28.75	nr	**68.18**
160 mm dia.	33.45	41.10	0.90	32.33	nr	**73.43**
200 mm dia.	36.29	44.58	1.05	37.74	nr	**82.32**
250 mm dia.	40.40	49.63	1.20	43.13	nr	**92.76**
399 mm dia.	44.72	54.94	1.35	48.51	nr	**103.45**
315 mm dia.	46.71	57.38	1.35	48.56	nr	**105.94**
350 mm dia.	49.14	60.36	1.65	59.30	nr	**119.66**
400 mm dia.	53.65	65.91	1.90	68.31	nr	**134.22**
450 mm dia.	57.96	71.20	2.10	75.50	nr	**146.70**
500 mm dia.	63.14	77.56	2.95	106.00	nr	**183.56**
550 mm dia.	68.18	83.76	2.94	105.65	nr	**189.41**
600 mm dia.	73.85	90.72	2.94	105.65	nr	**196.37**
650 mm dia.	79.75	97.98	4.55	163.51	nr	**261.49**
700 mm dia.	86.17	105.86	5.20	186.86	nr	**292.72**
750 mm dia.	93.00	114.25	5.20	186.86	nr	**301.11**
800 mm dia.	100.06	122.92	5.80	208.41	nr	**331.33**
850 mm dia.	107.42	131.96	5.80	208.41	nr	**340.37**
900 mm dia.	115.10	141.40	6.40	229.97	nr	**371.37**
950 mm dia.	122.99	151.09	6.40	229.97	nr	**381.06**
1000 mm dia.	130.99	160.92	7.00	251.54	nr	**412.46**

38 VENTILATION/AIR CONDITIONING SYSTEMS

Item	Net Price £	Material £	Labour hours	Labour £	Unit	**Total rate £**
Flat oval						
345 × 102 mm	47.46	58.31	1.20	43.12	nr	**101.43**
508 × 102 mm	53.45	65.67	1.60	57.50	nr	**123.17**
559 × 152 mm	60.39	74.19	1.90	68.31	nr	**142.50**
531 × 203 mm	65.98	81.06	1.90	68.31	nr	**149.37**
851 × 203 mm	69.76	85.70	4.55	163.51	nr	**249.21**
582 × 254 mm	73.23	89.96	2.10	75.50	nr	**165.46**
823 × 254 mm	80.18	98.50	4.10	147.33	nr	**245.83**
632 × 305 mm	80.70	99.14	2.95	106.00	nr	**205.14**
765 × 356 mm	90.28	110.91	4.55	163.51	nr	**274.42**
737 × 406 mm	96.16	118.14	4.55	163.51	nr	**281.65**
818 × 406 mm	98.17	120.60	5.20	186.86	nr	**307.46**
978 × 406 mm	106.16	130.42	5.50	197.64	nr	**328.06**
709 × 457 mm	100.06	122.92	4.50	161.87	nr	**284.79**
678 × 508 mm	105.85	130.03	4.55	163.51	nr	**293.54**
919 × 508 mm	118.79	145.93	6.00	215.61	nr	**361.54**
Fire damper; galvanized steel casing; stainless steel folding shutter; fusible link with manual reset; BS 476 4 hour fire rated						
Rectangular						
Sum of two sides 200 mm	38.56	47.38	1.60	57.50	nr	**104.88**
Sum of two sides 300 mm	39.78	48.87	1.60	57.50	nr	**106.37**
Sum of two sides 400 mm	41.48	50.96	1.60	57.50	nr	**108.46**
Sum of two sides 500 mm	45.38	55.74	1.60	57.50	nr	**113.24**
Sum of two sides 600 mm	49.72	61.08	1.70	61.12	nr	**122.20**
Sum of two sides 700 mm	54.16	66.54	2.10	75.50	nr	**142.04**
Sum of two sides 800 mm	58.80	72.24	2.15	77.27	nr	**149.51**
Sum of two sides 900 mm	63.44	77.93	2.30	82.79	nr	**160.72**
Sum of two sides 1000 mm	68.18	83.76	2.40	86.37	nr	**170.13**
Sum of two sides 1100 mm	72.41	88.96	2.60	93.58	nr	**182.54**
Sum of two sides 1200 mm	77.46	95.16	2.80	100.65	nr	**195.81**
Sum of two sides 1300 mm	82.62	101.50	3.10	111.38	nr	**212.88**
Sum of two sides 1400 mm	88.18	108.33	3.25	116.78	nr	**225.11**
Sum of two sides 1500 mm	93.02	114.27	3.40	122.22	nr	**236.49**
Sum of two sides 1600 mm	98.37	120.84	3.45	123.96	nr	**244.80**
Sum of two sides 1700 mm	103.73	127.43	3.60	129.36	nr	**256.79**
Sum of two sides 1800 mm	109.39	134.39	3.90	140.14	nr	**274.53**
Sum of two sides 1900 mm	114.75	140.97	4.20	150.98	nr	**291.95**
Sum of two sides 2000 mm	119.79	147.16	4.33	155.59	nr	**302.75**
Sum of two sides 2100 mm	127.48	156.61	4.43	159.19	nr	**315.80**
Sum of two sides 2200 mm	134.87	165.69	4.55	163.51	nr	**329.20**
Circular						
100 mm dia.	46.96	57.69	0.80	28.75	nr	**86.44**
150 mm dia.	49.42	60.71	0.80	28.75	nr	**89.46**
160 mm dia.	49.72	61.08	0.90	32.33	nr	**93.41**
200 mm dia.	52.17	64.09	1.05	37.73	nr	**101.82**
250 mm dia.	58.15	71.44	1.20	43.13	nr	**114.57**
300 mm dia.	64.63	79.40	1.35	48.51	nr	**127.91**
315 mm dia.	67.44	82.84	1.35	48.56	nr	**131.40**

38 VENTILATION/AIR CONDITIONING SYSTEMS

Item	Net Price £	Material £	Labour hours	Labour £	Unit	Total rate £
DUCTWORK ANCILLARIES: VOLUME CONTROL AND FIRE DAMPERS – cont						
Fire damper – cont						
Circular – cont						
350 mm dia.	71.52	87.86	1.65	59.30	nr	**147.16**
400 mm dia.	79.02	97.07	1.90	68.31	nr	**165.38**
450 mm dia.	86.42	106.17	2.10	75.50	nr	**181.67**
500 mm dia.	94.53	116.13	2.95	106.00	nr	**222.13**
550 mm dia.	102.65	126.10	2.94	105.65	nr	**231.75**
600 mm dia.	111.27	136.70	4.55	163.51	nr	**300.21**
650 mm dia.	120.41	147.92	4.55	163.51	nr	**311.43**
700 mm dia.	129.76	159.41	5.20	186.86	nr	**346.27**
750 mm dia.	139.60	171.50	5.20	186.86	nr	**358.36**
800 mm dia.	149.64	183.83	5.80	208.41	nr	**392.24**
850 mm dia.	160.23	196.84	5.80	208.41	nr	**405.25**
900 mm dia.	171.10	210.20	6.40	229.97	nr	**440.17**
950 mm dia.	182.27	223.92	6.40	229.97	nr	**453.89**
1000 mm dia.	193.98	238.31	7.00	251.54	nr	**489.85**
Flat oval						
345 × 102 mm	64.58	79.34	1.20	43.13	nr	**122.47**
427 × 102 mm	70.14	86.16	1.35	48.56	nr	**134.72**
508 × 102 mm	72.50	89.06	1.60	57.50	nr	**146.56**
559 × 152 mm	79.05	97.11	1.90	68.31	nr	**165.42**
531 × 203 mm	83.14	102.13	1.90	68.31	nr	**170.44**
851 × 203 mm	104.48	128.35	4.55	163.51	nr	**291.86**
582 × 254 mm	110.25	135.44	2.10	75.50	nr	**210.94**
632 × 305 mm	115.69	142.13	2.95	106.01	nr	**248.14**
765 × 356 mm	132.94	163.32	4.55	163.51	nr	**326.83**
737 × 406 mm	138.67	170.35	4.55	163.51	nr	**333.86**
818 × 406 mm	147.19	180.82	5.20	186.86	nr	**367.68**
978 × 406 mm	160.02	196.58	5.50	197.64	nr	**394.22**
709 × 457 mm	139.72	171.64	4.50	161.87	nr	**333.51**
678 × 508 mm	145.36	178.57	4.55	163.51	nr	**342.08**
Smoke/fire damper; galvanized steel casing; stainless steel folding shutter; fusible link and 24V DC electromagnetic shutter release mechanism; spring operated; BS 476 4 hour fire rating						
Rectangular						
Sum of two sides 200 mm	295.28	362.75	1.60	57.50	nr	**420.25**
Sum of two sides 300 mm	296.14	363.80	1.60	57.50	nr	**421.30**
Sum of two sides 400 mm	296.99	364.85	1.60	57.50	nr	**422.35**
Sum of two sides 500 mm	304.11	373.60	1.60	57.50	nr	**431.10**
Sum of two sides 600 mm	311.55	382.74	1.70	61.12	nr	**443.86**
Sum of two sides 700 mm	319.31	392.27	2.10	75.50	nr	**467.77**
Sum of two sides 800 mm	327.28	402.07	2.15	77.27	nr	**479.34**
Sum of two sides 900 mm	335.59	412.27	2.30	82.79	nr	**495.06**
Sum of two sides 1000 mm	344.19	422.83	2.40	86.37	nr	**509.20**
Sum of two sides 1100 mm	353.02	433.68	2.60	93.58	nr	**527.26**
Sum of two sides 1200 mm	362.16	444.92	2.80	100.65	nr	**545.57**

38 VENTILATION/AIR CONDITIONING SYSTEMS

Item	Net Price £	Material £	Labour hours	Labour £	Unit	**Total rate £**
Sum of two sides 1300 mm	371.62	456.54	3.10	111.38	nr	**567.92**
Sum of two sides 1400 mm	381.28	468.40	3.25	116.78	nr	**585.18**
Sum of two sides 1500 mm	391.26	480.66	3.40	122.22	nr	**602.88**
Sum of two sides 1600 mm	401.58	493.34	3.45	123.96	nr	**617.30**
Sum of two sides 1700 mm	412.11	506.28	3.60	129.36	nr	**635.64**
Sum of two sides 1800 mm	422.95	519.59	3.90	140.14	nr	**659.73**
Sum of two sides 1900 mm	433.99	533.16	4.20	150.98	nr	**684.14**
Sum of two sides 2000 mm	445.36	547.12	4.33	155.59	nr	**702.71**
Circular						
100 mm dia.	144.04	176.96	0.80	28.75	nr	**205.71**
150 mm dia.	144.04	176.96	0.90	32.33	nr	**209.29**
200 mm dia.	144.04	176.96	1.05	37.73	nr	**214.69**
250 mm dia.	151.86	186.56	1.20	43.13	nr	**229.69**
300 mm dia.	159.99	196.55	1.35	48.56	nr	**245.11**
350 mm dia.	182.09	223.70	1.65	59.30	nr	**283.00**
400 mm dia.	200.24	245.99	1.90	68.31	nr	**314.30**
450 mm dia.	208.82	256.54	2.10	75.50	nr	**332.04**
500 mm dia.	227.42	279.38	2.95	106.00	nr	**385.38**
550 mm dia.	246.63	302.99	2.95	106.00	nr	**408.99**
600 mm dia.	258.25	317.26	4.55	163.51	nr	**480.77**
650 mm dia.	279.43	343.28	4.55	163.51	nr	**506.79**
700 mm dia.	302.46	371.57	5.20	186.86	nr	**558.43**
750 mm dia.	316.79	389.17	5.20	186.86	nr	**576.03**
800 mm dia.	344.35	423.03	5.80	208.41	nr	**631.44**
850 mm dia.	369.17	453.53	5.80	208.41	nr	**661.94**
900 mm dia.	383.07	470.60	6.40	229.97	nr	**700.57**
950 mm dia.	408.80	502.21	6.40	229.97	nr	**732.18**
1000 mm dia.	427.25	524.87	7.00	251.54	nr	**776.41**
Flat oval						
531 × 203 mm	255.54	313.93	1.90	68.31	nr	**382.24**
851 × 203 mm	285.26	350.44	4.55	163.51	nr	**513.95**
582 × 254 mm	278.16	341.72	2.10	75.50	nr	**417.22**
632 × 305 mm	296.54	364.30	2.95	106.01	nr	**470.31**
765 × 356 mm	319.91	393.00	4.55	163.51	nr	**556.51**
737 × 406 mm	334.35	410.75	4.55	163.51	nr	**574.26**
818 × 406 mm	339.03	416.50	5.20	186.86	nr	**603.36**
978 × 406 mm	358.14	439.97	5.50	197.64	nr	**637.61**
709 × 457 mm	343.94	422.53	4.50	161.87	nr	**584.40**
678 × 508 mm	358.05	439.87	4.55	163.51	nr	**603.38**

38 VENTILATION/AIR CONDITIONING SYSTEMS

Item	Net Price £	Material £	Labour hours	Labour £	Unit	**Total rate £**
DUCTWORK ANCILLARIES: ACCESS HATCHES						
Access doors, hollow steel construction; 25 mm mineral wool insulation; removeable or hinged; fixed with cams; including subframe and integral sealing gaskets						
Rectangular duct						
150 × 150 mm	16.80	20.64	1.25	44.92	nr	**65.56**
200 × 200 mm	18.49	22.71	1.25	44.92	nr	**67.63**
300 × 150 mm	19.00	23.34	1.25	44.92	nr	**68.26**
300 × 300 mm	21.38	26.26	1.25	44.92	nr	**71.18**
400 × 400 mm	24.38	29.95	1.35	48.56	nr	**78.51**
450 × 300 mm	24.38	29.95	1.50	53.96	nr	**83.91**
450 × 450 mm	26.88	33.02	1.50	53.96	nr	**86.98**
Access doors, hollow steel construction; 25 mm mineral wool insulation; removeable or hinged; fixed with cams; including subframe and integral sealing gaskets						
Flat oval duct						
235 × 90 mm	34.13	41.93	1.25	44.92	nr	**86.85**
235 × 140 mm	36.38	44.69	1.35	48.56	nr	**93.25**
335 × 235 mm	41.56	51.06	1.50	53.96	nr	**105.02**
535 × 235 mm	46.75	57.43	1.50	53.96	nr	**111.39**

38 VENTILATION/AIR CONDITIONING SYSTEMS

Item	Net Price £	Material £	Labour hours	Labour £	Unit	**Total rate £**
GRILLES/DIFFUSERS/LOUVRES						
Supply grilles; single deflection; extruded aluminium alloy frame and adjustable horizontal vanes; silver grey polyester powder coated; screw fixed						
Rectangular; for duct, ceiling and sidewall applications						
150 × 100 mm	11.47	14.09	0.60	21.56	nr	**35.65**
150 × 150 mm	14.02	17.22	0.60	21.56	nr	**38.78**
200 × 150 mm	16.66	20.46	0.65	23.35	nr	**43.81**
200 × 200 mm	14.42	17.71	0.72	25.87	nr	**43.58**
300 × 100 mm	12.38	15.20	0.72	25.87	nr	**41.07**
300 × 150 mm	14.35	17.63	0.80	28.75	nr	**46.38**
300 × 200 mm	16.32	20.04	0.88	31.63	nr	**51.67**
300 × 300 mm	20.25	24.87	1.04	37.37	nr	**62.24**
400 × 100 mm	13.86	17.03	0.88	31.63	nr	**48.66**
400 × 150 mm	16.03	19.70	0.94	33.78	nr	**53.48**
400 × 200 mm	18.21	22.38	1.04	37.37	nr	**59.75**
400 × 300 mm	22.56	27.72	1.12	40.25	nr	**67.97**
600 × 200 mm	27.77	34.11	1.26	45.27	nr	**79.38**
600 × 300 mm	44.93	55.20	1.40	50.31	nr	**105.51**
600 × 400 mm	53.51	65.74	1.61	57.86	nr	**123.60**
600 × 500 mm	62.12	76.31	1.76	63.24	nr	**139.55**
600 × 600 mm	70.72	86.88	2.17	77.97	nr	**164.85**
800 × 300 mm	52.59	64.61	1.76	63.24	nr	**127.85**
800 × 400 mm	62.61	76.91	2.17	77.97	nr	**154.88**
800 × 600 mm	82.63	101.51	3.00	107.80	nr	**209.31**
1000 × 300 mm	60.26	74.02	2.60	93.43	nr	**167.45**
1000 × 400 mm	71.69	88.07	3.00	107.80	nr	**195.87**
1000 × 600 mm	94.55	116.15	3.80	136.55	nr	**252.70**
1000 × 800 mm	117.40	144.23	3.80	136.55	nr	**280.78**
1200 × 600 mm	106.46	130.79	4.61	165.65	nr	**296.44**
1200 × 800 mm	132.14	162.33	4.61	165.65	nr	**327.98**
1200 × 1000 mm	157.82	193.88	4.61	165.65	nr	**359.53**
Rectangular; for duct, ceiling and sidewall applications; including opposed blade damper volume regulator						
150 × 100 mm	21.37	26.25	0.72	25.88	nr	**52.13**
150 × 150 mm	24.76	30.42	0.72	25.88	nr	**56.30**
200 × 150 mm	26.58	32.65	0.83	29.85	nr	**62.50**
200 × 200 mm	28.73	35.29	0.90	32.35	nr	**67.64**
300 × 100 mm	30.52	37.50	0.90	32.35	nr	**69.85**
300 × 150 mm	32.54	39.97	0.98	35.21	nr	**75.18**
300 × 200 mm	35.06	43.07	1.06	38.10	nr	**81.17**
300 × 300 mm	40.07	49.22	1.20	43.13	nr	**92.35**
400 × 100 mm	39.67	48.73	1.06	38.10	nr	**86.83**
400 × 150 mm	41.88	51.45	1.13	40.60	nr	**92.05**
400 × 200 mm	42.08	51.69	1.20	43.13	nr	**94.82**

38 VENTILATION/AIR CONDITIONING SYSTEMS

Item	Net Price £	Material £	Labour hours	Labour £	Unit	Total rate £
GRILLES/DIFFUSERS/LOUVRES – cont						
Rectangular – cont						
400 × 300 mm	47.82	58.75	1.34	48.17	nr	**106.92**
600 × 200 mm	58.28	71.60	1.50	53.96	nr	**125.56**
600 × 300 mm	77.41	95.10	1.66	59.69	nr	**154.79**
600 × 400 mm	88.01	108.12	1.80	64.74	nr	**172.86**
600 × 500 mm	99.66	122.43	2.00	71.87	nr	**194.30**
600 × 600 mm	110.26	135.45	2.60	93.58	nr	**229.03**
800 × 300 mm	106.17	130.43	2.00	71.87	nr	**202.30**
800 × 400 mm	119.05	146.25	2.60	93.58	nr	**239.83**
800 × 600 mm	146.25	179.67	3.61	129.73	nr	**309.40**
1000 × 300 mm	123.54	151.77	3.00	107.91	nr	**259.68**
1000 × 400 mm	138.41	170.04	3.61	129.73	nr	**299.77**
1000 × 600 mm	169.97	208.80	4.61	165.58	nr	**374.38**
1000 × 800 mm	251.37	308.81	4.61	165.58	nr	**474.39**
1200 × 600 mm	188.82	231.97	5.62	201.88	nr	**433.85**
1200 × 800 mm	277.05	340.36	6.10	219.20	nr	**559.56**
1200 × 1000 mm	315.03	387.02	6.50	233.57	nr	**620.59**
Supply grilles; double deflection; extruded aluminium alloy frame and adjustable horizontal and vertical vanes; white polyester powder coated; screw fixed						
Rectangular; for duct, ceiling and sidewall applications						
150 × 100 mm	11.47	14.09	0.88	31.63	nr	**45.72**
150 × 150 mm	14.80	18.19	0.88	31.63	nr	**49.82**
200 × 150 mm	16.66	20.46	1.08	38.81	nr	**59.27**
200 × 200 mm	18.58	22.83	1.25	44.92	nr	**67.75**
300 × 100 mm	18.09	22.23	1.25	44.92	nr	**67.15**
300 × 150 mm	20.42	25.08	1.50	53.91	nr	**78.99**
300 × 200 mm	22.77	27.97	1.75	62.88	nr	**90.85**
300 × 300 mm	27.46	33.74	2.15	77.26	nr	**111.00**
400 × 100 mm	21.39	26.28	1.75	62.88	nr	**89.16**
400 × 150 mm	24.18	29.70	1.95	70.08	nr	**99.78**
400 × 200 mm	26.98	33.15	2.15	77.26	nr	**110.41**
400 × 300 mm	32.56	40.01	2.55	91.63	nr	**131.64**
600 × 200 mm	60.15	73.90	3.01	108.16	nr	**182.06**
600 × 300 mm	72.68	89.29	3.36	120.74	nr	**210.03**
600 × 400 mm	85.22	104.70	3.80	136.55	nr	**241.25**
600 × 500 mm	97.75	120.09	4.20	150.92	nr	**271.01**
600 × 600 mm	110.29	135.49	4.51	162.07	nr	**297.56**
800 × 300 mm	90.01	110.58	4.20	150.92	nr	**261.50**
800 × 400 mm	105.59	129.72	4.51	162.07	nr	**291.79**
800 × 600 mm	136.74	167.99	5.10	183.27	nr	**351.26**
1000 × 300 mm	107.34	131.87	4.80	172.48	nr	**304.35**
1000 × 400 mm	125.96	154.74	5.10	183.27	nr	**338.01**
1000 × 600 mm	163.19	200.48	5.72	205.54	nr	**406.02**

38 VENTILATION/AIR CONDITIONING SYSTEMS

Item	Net Price £	Material £	Labour hours	Labour £	Unit	Total rate £
1000 × 800 mm	200.44	246.24	6.33	227.46	nr	**473.70**
1200 × 600 mm	189.64	232.97	5.72	205.54	nr	**438.51**
1200 × 800 mm	232.97	286.20	6.33	227.46	nr	**513.66**
1200 × 1000 mm	276.29	339.42	6.33	227.46	nr	**566.88**
Rectangular; for duct, ceiling and sidewall applications; including opposed blade damper volume regulator						
150 × 100 mm	23.38	28.72	1.00	35.93	nr	**64.65**
150 × 150 mm	27.73	34.06	1.00	35.93	nr	**69.99**
200 × 150 mm	30.59	37.58	1.26	45.27	nr	**82.85**
200 × 200 mm	32.88	40.39	1.43	51.39	nr	**91.78**
300 × 100 mm	36.23	44.51	1.43	51.39	nr	**95.90**
300 × 150 mm	38.62	47.45	1.68	60.36	nr	**107.81**
300 × 200 mm	41.49	50.97	1.93	69.35	nr	**120.32**
300 × 300 mm	47.29	58.10	2.31	83.01	nr	**141.11**
400 × 100 mm	47.21	58.00	1.93	69.35	nr	**127.35**
400 × 150 mm	50.04	61.48	2.14	76.89	nr	**138.37**
400 × 200 mm	50.84	62.45	2.31	83.01	nr	**145.46**
400 × 300 mm	57.82	71.03	2.77	99.53	nr	**170.56**
600 × 200 mm	90.66	111.37	3.25	116.78	nr	**228.15**
600 × 300 mm	105.18	129.21	3.62	130.08	nr	**259.29**
600 × 400 mm	119.72	147.07	3.99	143.38	nr	**290.45**
600 × 500 mm	135.30	166.22	4.44	159.55	nr	**325.77**
600 × 600 mm	149.83	184.07	4.94	177.52	nr	**361.59**
800 × 300 mm	143.60	176.41	4.44	159.55	nr	**335.96**
800 × 400 mm	162.02	199.04	4.94	177.52	nr	**376.56**
800 × 600 mm	200.35	246.13	5.71	205.18	nr	**451.31**
1000 × 300 mm	170.62	209.61	5.20	186.86	nr	**396.47**
1000 × 400 mm	192.68	236.71	5.71	205.18	nr	**441.89**
1000 × 600 mm	238.60	293.12	6.53	234.64	nr	**527.76**
1000 × 800 mm	334.41	410.82	6.53	234.64	nr	**645.46**
1200 × 600 mm	271.99	334.14	7.34	263.75	nr	**597.89**
1200 × 800 mm	377.88	464.23	8.80	316.22	nr	**780.45**
1200 × 1000 mm	433.49	532.54	8.80	316.22	nr	**848.76**
Floor grille suitable for mounting in raised access floors; heavy duty; extruded alumiinium; standard mill finish; complete with opposed blade volume control damper						
Diffuser						
600 mm × 600 mm	141.54	173.88	0.70	25.16	nr	**199.04**
Extra for nylon coated black finish	21.65	26.60	–	–	nr	**26.60**
Exhaust grilles; aluminium						
0° fixed blade core						
150 × 150 mm	14.66	18.01	0.60	21.65	nr	**39.66**
200 × 200 mm	17.24	21.18	0.72	25.88	nr	**47.06**
250 × 250 mm	20.07	24.65	0.80	28.75	nr	**53.40**
300 × 300 mm	23.14	28.42	1.00	35.93	nr	**64.35**
350 × 350 mm	26.45	32.50	1.20	43.12	nr	**75.62**

38 VENTILATION/AIR CONDITIONING SYSTEMS

Item	Net Price £	Material £	Labour hours	Labour £	Unit	**Total rate £**
GRILLES/DIFFUSERS/LOUVRES – cont						
Exhaust grilles – cont						
0° fixed blade core; including opposed blade damper volume regulator						
150 × 150 mm	27.59	33.89	0.62	22.28	nr	**56.17**
200 × 200 mm	31.55	38.76	0.72	25.88	nr	**64.64**
250 × 250 mm	35.90	44.10	0.80	28.75	nr	**72.85**
300 × 300 mm	42.96	52.77	1.00	35.93	nr	**88.70**
350 × 350 mm	48.09	59.08	1.20	43.12	nr	**102.20**
45° fixed blade core						
150 × 150 mm	14.66	18.01	0.62	22.28	nr	**40.29**
200 × 200 mm	17.24	21.18	0.72	25.88	nr	**47.06**
250 × 250 mm	20.07	24.65	0.80	28.75	nr	**53.40**
300 × 300 mm	23.14	28.42	1.00	35.93	nr	**64.35**
350 × 350 mm	26.45	32.50	1.20	43.12	nr	**75.62**
45° fixed blade core; including opposed blade damper volume regulator						
150 × 150 mm	27.59	33.89	0.62	22.28	nr	**56.17**
200 × 200 mm	31.55	38.76	0.72	25.88	nr	**64.64**
250 × 250 mm	35.90	44.10	0.80	28.75	nr	**72.85**
300 × 300 mm	42.96	52.77	1.00	35.93	nr	**88.70**
350 × 350 mm	48.09	59.08	1.20	43.12	nr	**102.20**
Eggcrate core						
150 × 150 mm	9.84	12.09	0.62	22.28	nr	**34.37**
200 × 200 mm	12.15	14.93	1.00	35.93	nr	**50.86**
250 × 250 mm	14.80	18.19	0.80	28.75	nr	**46.94**
300 × 300 mm	18.84	23.14	1.00	35.93	nr	**59.07**
350 × 350 mm	21.05	25.86	1.20	43.12	nr	**68.98**
Eggcrate core; including opposed blade damper volume regulator						
150 × 150 mm	22.77	27.97	0.62	22.28	nr	**50.25**
200 × 200 mm	26.45	32.50	0.72	25.88	nr	**58.38**
250 × 250 mm	30.62	37.62	0.80	28.75	nr	**66.37**
300 × 300 mm	38.66	47.49	1.00	35.93	nr	**83.42**
350 × 350 mm	42.70	52.46	1.20	43.12	nr	**95.58**
Mesh/perforated plate core						
150 × 150 mm	10.82	13.29	0.62	22.28	nr	**35.57**
200 × 200 mm	13.37	16.42	0.72	25.87	nr	**42.29**
250 × 250 mm	16.29	20.01	0.80	28.75	nr	**48.76**
300 × 300 mm	20.74	25.48	1.00	35.93	nr	**61.41**
350 × 350 mm	23.18	28.48	1.20	43.12	nr	**71.60**
Mesh/perforated plate core; including opposed blade damper volume regulator						
150 × 150 mm	23.76	29.19	0.62	22.28	nr	**51.47**
200 × 200 mm	27.67	33.99	0.72	25.87	nr	**59.86**
250 × 250 mm	32.11	39.45	0.80	28.75	nr	**68.20**
300 × 300 mm	40.56	49.83	0.80	28.75	nr	**78.58**
350 × 350 mm	44.81	55.05	1.20	43.12	nr	**98.17**

38 VENTILATION/AIR CONDITIONING SYSTEMS

Item	Net Price £	Material £	Labour hours	Labour £	Unit	Total rate £
Plastic air diffusion system						
Eggcrate grilles						
150 × 150 mm	4.41	5.42	0.62	23.28	nr	**28.70**
200 × 200 mm	7.71	9.47	0.72	27.04	nr	**36.51**
250 × 250 mm	8.81	10.83	0.80	30.02	nr	**40.85**
300 × 300 mm	8.81	10.83	1.00	37.53	nr	**48.36**
Single deflection grilles						
150 × 150 mm	6.94	8.53	0.62	23.28	nr	**31.81**
200 × 200 mm	8.60	10.56	0.72	27.04	nr	**37.60**
250 × 250 mm	9.47	11.63	0.80	30.02	nr	**41.65**
300 × 300 mm	11.90	14.62	1.00	37.53	nr	**52.15**
Double deflection grilles						
150 × 150 mm	8.65	10.63	0.62	23.28	nr	**33.91**
200 × 200 mm	11.48	14.10	0.72	27.04	nr	**41.14**
250 × 250 mm	13.11	16.11	0.80	30.02	nr	**46.13**
300 × 300 mm	18.51	22.74	1.00	37.53	nr	**60.27**
Door transfer grilles						
150 × 150 mm	13.00	15.97	0.62	23.28	nr	**39.25**
200 × 200 mm	15.09	18.54	0.72	27.04	nr	**45.58**
250 × 250 mm	15.64	19.22	0.80	30.02	nr	**49.24**
300 × 300 mm	16.97	20.84	1.00	37.53	nr	**58.37**
Opposed blade dampers						
150 × 150 mm	7.87	9.67	0.62	23.28	nr	**32.95**
200 × 200 mm	9.26	11.37	0.72	27.04	nr	**38.41**
250 × 250 mm	13.88	17.05	0.80	30.02	nr	**47.07**
300 × 300 mm	15.09	18.54	1.00	37.53	nr	**56.07**
Ceiling mounted diffusers; circular aluminium multi-cone diffuser						
Circular; for ceiling mounting						
141 mm dia. neck	50.24	61.72	0.80	28.75	nr	**90.47**
197 mm dia. neck	73.56	90.37	1.10	39.53	nr	**129.90**
309 mm dia. neck	99.64	122.41	1.40	50.33	nr	**172.74**
365 mm dia. neck	85.24	104.72	1.50	53.96	nr	**158.68**
457 mm dia. neck	134.40	165.11	2.00	71.87	nr	**236.98**
Circular; for ceiling mounting; including louvre damper volume control						
141 mm dia. neck	75.09	92.25	1.00	35.93	nr	**128.18**
197 mm dia. neck	98.42	120.91	1.20	43.13	nr	**164.04**
309 mm dia. neck	129.85	159.52	1.60	57.50	nr	**217.02**
365 mm dia. neck	117.72	144.62	1.90	68.31	nr	**212.93**
457 mm dia. neck	172.61	212.05	2.40	86.37	nr	**298.42**
Ceiling mounted diffusers; rectangular aluminium multi-cone diffuser; four way flow						
Rectangular; for ceiling mounting						
150 × 150 mm neck	20.67	25.39	1.80	64.74	nr	**90.13**
300 × 150 mm neck	35.27	43.33	2.30	82.66	nr	**125.99**
300 × 300 mm neck	31.50	38.70	2.80	100.61	nr	**139.31**

38 VENTILATION/AIR CONDITIONING SYSTEMS

Item	Net Price £	Material £	Labour hours	Labour £	Unit	Total rate £
GRILLES/DIFFUSERS/LOUVRES – cont						
Ceiling mounted diffusers – cont						
Rectangular – cont						
450 × 150 mm neck	44.94	55.21	2.80	100.61	nr	**155.82**
450 × 300 mm neck	43.99	54.04	3.20	114.99	nr	**169.03**
450 × 450 mm neck	43.93	53.97	3.40	122.22	nr	**176.19**
600 × 150 mm neck	55.05	67.63	3.20	114.99	nr	**182.62**
600 × 300 mm neck	54.75	67.26	3.50	125.77	nr	**193.03**
600 × 600 mm neck	61.10	75.06	4.00	143.73	nr	**218.79**
Rectangular; for ceiling mounting; including opposed blade damper volume regulator						
150 × 150 mm neck	26.55	32.61	1.80	64.74	nr	**97.35**
300 × 150 mm neck	53.46	65.68	2.30	82.66	nr	**148.34**
300 × 300 mm neck	39.60	48.65	2.80	100.65	nr	**149.30**
450 × 150 mm neck	71.68	88.06	2.80	100.61	nr	**188.67**
450 × 300 mm neck	70.37	86.45	3.30	118.59	nr	**205.04**
450 × 450 mm neck	55.67	68.39	3.51	126.08	nr	**194.47**
600 × 150 mm neck	87.90	107.98	3.30	118.59	nr	**226.57**
600 × 300 mm neck	87.24	107.17	4.00	143.73	nr	**250.90**
600 × 600 mm neck	89.93	110.48	5.62	201.88	nr	**312.36**
Slot diffusers; continuous aluminium slot diffuser with flanged frame (1500 mm sections)						
Diffuser						
1 slot	33.99	41.76	3.76	135.08	m	**176.84**
2 slot	41.37	50.82	3.76	135.05	m	**185.87**
3 slot	50.91	62.54	3.76	135.05	m	**197.59**
4 slot	57.30	70.39	4.50	161.87	m	**232.26**
6 slot	75.50	92.76	4.50	161.87	m	**254.63**
Diffuser; including equalizing deflector						
1 slot	70.35	86.43	5.26	189.13	m	**275.56**
2 slot	87.29	107.24	5.26	189.13	m	**296.37**
3 slot	106.76	131.16	5.26	189.13	m	**320.29**
4 slot	119.87	147.26	6.33	227.43	m	**374.69**
6 slot	157.69	193.72	6.33	227.43	m	**421.15**
Extra over for ends						
1 slot	3.79	4.65	1.00	35.93	nr	**40.58**
2 slot	4.00	4.91	1.00	35.93	nr	**40.84**
3 slot	4.24	5.21	1.00	35.93	nr	**41.14**
4 slot	4.45	5.47	1.30	46.70	nr	**52.17**
6 slot	5.08	6.24	1.40	50.31	nr	**56.55**
Plenum boxes; 1.0 m long; circular spigot; including cord operated flap damper						
1 slot	48.44	59.50	2.75	98.99	nr	**158.49**
2 slot	49.70	61.06	2.75	98.99	nr	**160.05**
3 slot	50.34	61.85	2.75	98.99	nr	**160.84**
4 slot	54.93	67.48	3.51	126.08	nr	**193.56**
6 slot	57.62	70.79	3.51	126.08	nr	**196.87**

38 VENTILATION/AIR CONDITIONING SYSTEMS

Item	Net Price £	Material £	Labour hours	Labour £	Unit	**Total rate £**
Plenum boxes; 2.0 m long; circular spigot; including cord operated flap damper						
1 slot	53.63	65.89	3.26	117.05	nr	**182.94**
2 slot	56.22	69.07	3.26	117.05	nr	**186.12**
3 slot	57.02	70.05	3.26	117.05	nr	**187.10**
4 slot	63.50	78.00	3.76	135.08	nr	**213.08**
6 slot	67.72	83.19	3.76	135.08	nr	**218.27**
Perforated diffusers; rectangular face aluminium perforated diffuser; quick release face plate; for integration with rectangular ceiling tiles						
Circular spigot; rectangular diffuser						
150 mm dia. spigot; 300 × 300 mm diffuser	49.51	60.83	1.00	35.93	nr	**96.76**
300 mm dia. spigot; 600 × 600 mm diffuser	73.87	90.75	1.40	50.33	nr	**141.08**
Circular spigot; rectangular diffuser; including louvre damper volume regulator						
150 mm dia. spigot; 300 × 300 mm diffuser	74.37	91.36	1.00	35.93	nr	**127.29**
300 mm dia. spigot; 600 × 600 mm diffuser	104.07	127.85	1.60	57.50	nr	**185.35**
Rectangular spigot; rectangular diffuser						
150 × 150 mm dia. spigot; 300 × 300 mm diffuser	49.51	60.83	1.00	37.53	nr	**98.36**
300 × 150 mm dia. spigot; 600 × 300 mm diffuser	66.64	81.87	1.20	45.03	nr	**126.90**
300 × 300 mm dia. spigot; 600 × 600 mm diffuser	73.87	90.75	1.40	52.54	nr	**143.29**
600 × 300 mm dia. spigot; 1200 × 600 mm diffuser	77.16	94.79	1.60	60.05	nr	**154.84**
Rectangular spigot; rectangular diffuser; including opposed blade damper volume regulator						
150 × 150 mm dia. spigot; 300 × 300 mm diffuser	74.37	91.36	1.20	45.03	nr	**136.39**
300 × 150 mm dia. spigot; 600 × 300 mm diffuser	99.12	121.77	1.40	52.56	nr	**174.33**
300 × 300 mm dia. spigot; 600 × 600 mm diffuser	104.07	127.85	1.60	60.05	nr	**187.90**
600 × 300 mm dia. spigot; 1200 × 600 mm diffuser	121.55	149.32	1.80	67.56	nr	**216.88**
Floor swirl diffuser; manual adjustment of air discharge direction; complete with damper and dirt trap						
Plastic diffuser						
150 mm dia.	27.43	33.69	0.50	17.98	nr	**51.67**
200 mm dia.	27.43	33.69	0.50	17.98	nr	**51.67**
Aluminium diffuser						
150 mm dia.	29.37	36.08	0.50	17.98	nr	**54.06**
200 mm dia.	42.93	52.74	0.50	17.98	nr	**70.72**

38 VENTILATION/AIR CONDITIONING SYSTEMS

Item	Net Price £	Material £	Labour hours	Labour £	Unit	**Total rate £**
GRILLES/DIFFUSERS/LOUVRES – cont						
Plastic air diffusion system						
Cellular diffusers						
300 × 300 mm	12.62	15.51	2.80	105.13	nr	**120.64**
600 × 600 mm	29.22	35.90	4.00	150.12	nr	**186.02**
Multi-cone diffusers						
300 × 300 mm	10.19	12.52	2.80	105.13	nr	**117.65**
450 × 450 mm	16.76	20.59	3.40	127.65	nr	**148.24**
500 × 500 mm	17.72	21.77	3.80	142.71	nr	**164.48**
600 × 600 mm	24.51	30.11	4.00	150.12	nr	**180.23**
625 × 625 mm	42.17	51.81	4.26	159.71	nr	**211.52**
Opposed blade dampers						
300 × 300 mm	13.59	16.70	1.20	45.07	nr	**61.77**
450 × 450 mm	14.94	18.35	1.50	56.35	nr	**74.70**
600 × 600 mm	30.91	37.97	2.60	97.74	nr	**135.71**
Plenum boxes						
300 mm	8.60	10.56	2.80	105.13	nr	**115.69**
450 mm	13.36	16.41	3.40	127.65	nr	**144.06**
600 mm	21.40	26.29	4.00	150.12	nr	**176.41**
Plenum spigot reducer						
600 mm	16.98	20.86	1.00	37.53	nr	**58.39**
Blanking kits for cellular diffusers						
300 mm	6.01	7.38	0.88	33.03	nr	**40.41**
600 mm	7.14	8.77	1.10	41.29	nr	**50.06**
Blanking kits for multi-cone diffusers						
300 mm	3.17	3.90	0.88	33.03	nr	**36.93**
450 mm	9.51	11.69	0.90	33.78	nr	**45.47**
600 mm	11.10	13.64	1.10	41.29	nr	**54.93**
Acoustic louvres; opening mounted; 300 mm deep steel louvres with blades packed with acoustic infill; 12 mm galvanized mesh birdscreen; screw fixing in opening						
Louvre units; self-finished galvanized steel						
900 mm high × 600 mm wide	124.23	152.62	3.00	112.59	nr	**265.21**
900 mm high × 900 mm wide	154.14	189.36	3.00	112.59	nr	**301.95**
900 mm high × 1200 mm wide	182.48	224.17	3.34	125.36	nr	**349.53**
900 mm high × 1500 mm wide	238.15	292.57	3.34	125.36	nr	**417.93**
900 mm high × 1800 mm wide	267.02	328.03	3.34	125.36	nr	**453.39**
900 mm high × 2100 mm wide	295.89	363.50	3.34	125.36	nr	**488.86**
900 mm high × 2400 mm wide	324.23	398.32	3.68	138.11	nr	**536.43**
900 mm high × 2700 mm wide	366.51	450.26	3.68	138.11	nr	**588.37**
900 mm high × 3000 mm wide	392.28	481.92	3.68	138.11	nr	**620.03**
1200 mm high × 600 mm wide	162.89	200.11	3.00	112.59	nr	**312.70**
1200 mm high × 900 mm wide	200.52	246.34	3.34	125.36	nr	**371.70**
1200 mm high × 1200 mm wide	237.63	291.93	3.34	125.36	nr	**417.29**
1200 mm high × 1500 mm wide	314.95	386.91	3.34	125.36	nr	**512.27**
1200 mm high × 1800 mm wide	352.07	432.52	3.68	138.11	nr	**570.63**
1200 mm high × 2100 mm wide	390.21	479.38	3.68	138.11	nr	**617.49**

38 VENTILATION/AIR CONDITIONING SYSTEMS

Item	Net Price £	Material £	Labour hours	Labour £	Unit	Total rate £
1200 mm high × 2400 mm wide	427.33	524.98	3.68	138.11	nr	**663.09**
1500 mm high × 600 mm wide	202.07	248.24	3.00	112.59	nr	**360.83**
1500 mm high × 900 mm wide	247.95	304.60	3.34	125.36	nr	**429.96**
1500 mm high × 1200 mm wide	293.82	360.96	3.34	125.36	nr	**486.32**
1500 mm high × 1500 mm wide	391.77	481.29	3.68	138.11	nr	**619.40**
1500 mm high × 1800 mm wide	437.65	537.65	3.68	138.11	nr	**675.76**
1500 mm high × 2100 mm wide	484.04	594.65	4.00	150.12	nr	**744.77**
1800 mm high × 600 mm wide	240.21	295.10	3.34	125.36	nr	**420.46**
1800 mm high × 900 mm wide	294.85	362.22	3.34	125.36	nr	**487.58**
1800 mm high × 1200 mm wide	350.00	429.98	3.68	138.11	nr	**568.09**
1800 mm high × 1500 mm wide	469.08	576.26	3.68	138.11	nr	**714.37**
Louvre units; polyester powder coated steel						
900 mm high × 600 mm wide	180.43	221.65	3.00	112.59	nr	**334.24**
900 mm high × 900 mm wide	237.63	291.93	3.00	112.59	nr	**404.52**
900 mm high × 1200 mm wide	293.82	360.96	3.34	125.36	nr	**486.32**
900 mm high × 1500 mm wide	377.33	463.55	3.34	125.36	nr	**588.91**
900 mm high × 1800 mm wide	434.03	533.21	3.34	125.36	nr	**658.57**
900 mm high × 2100 mm wide	491.25	603.50	3.34	125.36	nr	**728.86**
900 mm high × 2400 mm wide	547.43	672.51	3.68	138.11	nr	**810.62**
900 mm high × 2700 mm wide	617.02	758.01	3.68	138.11	nr	**896.12**
900 mm high × 3000 mm wide	670.63	823.87	3.68	138.11	nr	**961.98**
1200 mm high × 600 mm wide	237.12	291.30	3.00	112.59	nr	**403.89**
1200 mm high × 900 mm wide	311.86	383.12	3.34	125.36	nr	**508.48**
1200 mm high × 1200 mm wide	386.60	474.94	3.34	125.36	nr	**600.30**
1200 mm high × 1500 mm wide	500.52	614.89	3.34	125.36	nr	**740.25**
1200 mm high × 1800 mm wide	574.75	706.08	3.68	138.11	nr	**844.19**
1200 mm high × 2100 mm wide	650.02	798.55	3.68	138.11	nr	**936.66**
1200 mm high × 2400 mm wide	724.24	889.73	3.68	138.11	nr	**1027.84**
1500 mm high × 600 mm wide	294.85	362.22	3.00	112.59	nr	**474.81**
1500 mm high × 900 mm wide	387.11	475.57	3.34	125.36	nr	**600.93**
1500 mm high × 1200 mm wide	479.39	588.93	3.34	125.36	nr	**714.29**
1500 mm high × 1500 mm wide	623.73	766.25	3.68	138.11	nr	**904.36**
1500 mm high × 1800 mm wide	716.51	880.24	3.68	138.11	nr	**1018.35**
1500 mm high × 2100 mm wide	808.79	993.59	4.00	150.12	nr	**1143.71**
1800 mm high × 600 mm wide	351.54	431.87	3.34	125.36	nr	**557.23**
1800 mm high × 900 mm wide	461.86	567.40	3.34	125.36	nr	**692.76**
1800 mm high × 1200 mm wide	572.69	703.55	3.68	138.11	nr	**841.66**
1800 mm high × 1500 mm wide	747.44	918.23	3.68	138.11	nr	**1056.34**
Weather louvres; opening mounted; 300 mm deep galvanized steel louvres; screw fixing in position						
Louvre units; including 12 mm galvanized mesh birdscreen						
900 × 600 mm	113.50	139.44	2.25	84.54	nr	**223.98**
900 × 900 mm	163.99	201.46	2.25	84.54	nr	**286.00**
900 × 1200 mm	197.32	242.40	2.50	93.83	nr	**336.23**
900 × 1500 mm	240.13	295.00	2.50	93.83	nr	**388.83**
900 × 1800 mm	282.95	347.60	2.50	93.83	nr	**441.43**
900 × 2100 mm	354.67	435.71	2.50	93.83	nr	**529.54**
900 × 2400 mm	381.01	468.07	2.76	103.68	nr	**571.75**

38 VENTILATION/AIR CONDITIONING SYSTEMS

Item	Net Price £	Material £	Labour hours	Labour £	Unit	**Total rate £**
GRILLES/DIFFUSERS/LOUVRES – cont						
Weather louvres – cont						
Louvre units – cont						
900 × 2700 mm	430.14	528.42	2.76	103.68	nr	**632.10**
900 × 3000 mm	469.75	577.09	2.76	103.68	nr	**680.77**
1200 × 600 mm	153.71	188.83	2.25	84.54	nr	**273.37**
1200 × 900 mm	214.79	263.86	2.50	93.83	nr	**357.69**
1200 × 1200 mm	276.33	339.48	2.50	93.83	nr	**433.31**
1200 × 1500 mm	322.55	396.25	2.50	93.83	nr	**490.08**
1200 × 1800 mm	421.84	518.23	2.76	103.68	nr	**621.91**
1200 × 2100 mm	479.74	589.37	2.76	103.68	nr	**693.05**
1200 × 2400 mm	501.01	615.49	2.76	103.68	nr	**719.17**
1500 × 600 mm	181.68	223.20	2.25	84.54	nr	**307.74**
1500 × 900 mm	250.35	307.56	2.50	93.83	nr	**401.39**
1500 × 1200 mm	329.77	405.12	2.50	93.83	nr	**498.95**
1500 × 1500 mm	386.63	474.98	2.76	103.68	nr	**578.66**
1500 × 1800 mm	455.88	560.05	2.76	103.68	nr	**663.73**
1500 × 2100 mm	570.17	700.45	3.00	112.71	nr	**813.16**
1800 × 600 mm	203.22	249.66	2.50	93.83	nr	**343.49**
1800 × 900 mm	280.94	345.13	2.50	93.83	nr	**438.96**
1800 × 1200 mm	368.39	452.57	2.76	103.68	nr	**556.25**
1800 × 1500 mm	435.76	535.33	3.00	112.71	nr	**648.04**

38 VENTILATION/AIR CONDITIONING SYSTEMS

Item	Net Price £	Material £	Labour hours	Labour £	Unit	**Total rate £**
PLANT/EQUIPMENT: FANS						
Axial flow fan; including ancillaries, anti-vibration mountings, mounting feet, matching flanges, flexible connectors and clips; 415 V, 3 phase, 50 Hz motor; includes fixing in position; electrical work elsewhere						
Aerofoil blade fan unit; short duct case						
315 mm dia.; 0.47 m^3/s duty; 147 Pa	610.00	749.39	4.50	168.88	nr	**918.27**
500 mm dia.; 1.89 m^3/s duty; 500 Pa	785.00	964.37	5.00	187.66	nr	**1152.03**
560 mm dia.; 2.36 m^3/s duty; 147 Pa	890.00	1093.36	5.50	206.43	nr	**1299.79**
710 mm dia.; 5.67 m^3/s duty; 245 Pa	1149.00	1411.55	6.00	225.18	nr	**1636.73**
Aerofoil blade fan unit; long duct case						
315 mm dia.; 0.47 m^3/s duty; 147 Pa	900.00	1105.65	4.50	168.88	nr	**1274.53**
500 mm dia.; 1.89 m^3/s duty; 500 Pa	973.00	1195.33	5.00	187.66	nr	**1382.99**
560 mm dia.; 2.36 m^3/s duty; 147 Pa	912.00	1120.39	5.50	206.43	nr	**1326.82**
710 mm dia.; 5.67 m^3/s duty; 245 Pa	1571.00	1929.97	6.00	225.18	nr	**2155.15**
Aerofoil blade fan unit; two stage parallel fan arrangement; long duct case						
315 mm; 0.47 m^3/s @ 500 Pa	1730.00	2125.30	4.50	168.88	nr	**2294.18**
355 mm; 0.83 m^3/s @ 147 Pa	1791.00	2200.24	4.75	178.27	nr	**2378.51**
710 mm; 3.77 m^3/s @ 431 Pa	3164.00	3886.97	6.00	225.18	nr	**4112.15**
710 mm; 6.61 m^3/s @ 500 Pa	3164.00	3886.97	6.00	225.18	nr	**4112.15**
Axial flow fan; suitable for operation at 300°C for 90 minutes; including ancillaries, anti vibration mountings, mounting feet, matching flanges, flexible connectors and clips; 415 V, 3 phase, 50 Hz motor; includes fixing in position; price includes air operated damper; electrical work elsewhere						
450 mm; 2.0 m^3/s @ 300Pa	2874.00	3530.71	5.00	187.66	nr	**3718.37**
630 mm; 4.6 m^3/s @ 200Pa	2900.00	3562.65	5.50	206.43	nr	**3769.08**
900 mm; 9.0 m^3/s @ 300Pa	3015.00	3703.93	6.50	243.95	nr	**3947.88**
1000 mm; 15.0 m^3/s @ 400Pa	4890.00	6007.36	7.50	281.48	nr	**6288.84**
Bifurcated fan; suitable for temperature up to 200°C with motor protection to IP55; including ancillaries, anti-vibration mountings, mounting feet, matching flanges, flexible connectors and clips; 415 V, 3 phase, 50 Hz motor; includes fixing in position; electrical work elsewhere						
300 mm; 0.50 m^3/s @ 100Pa	1441.00	1770.27	4.50	168.88	nr	**1939.15**
400 mm; 1.97 m^3/s @ 200Pa	2139.00	2627.76	5.00	187.66	nr	**2815.42**
800 mm; 3.86 m^3/s @ 200Pa	2783.00	3418.92	5.50	206.43	nr	**3625.35**
1000 mm; 6.10 m^3/s @ 400Pa	5030.00	6179.35	6.50	243.95	nr	**6423.30**

38 VENTILATION/AIR CONDITIONING SYSTEMS

Item	Net Price £	Material £	Labour hours	Labour £	Unit	Total rate £
PLANT/EQUIPMENT: FANS – cont						
Duct mounted in line fan with backward curved centrifugal impellor; including ancillaries, matching flanges, flexible connectors and clips; 415 V, 3phase, 50 Hz motor; includes fixing in position; electrical work elsewhere						
0.5 m³/s @ 200Pa	1837.32	2257.14	4.50	168.88	nr	**2426.02**
1.0 m³/s @ 300Pa	2041.48	2507.96	5.00	187.66	nr	**2695.62**
3.0 m³/s @ 500Pa	3366.62	4135.90	5.50	206.43	nr	**4342.33**
5.0 m³/s @ 750Pa	5571.18	6844.19	6.50	243.95	nr	**7088.14**
7.0 m³/s @ 1000Pa	7003.84	8604.21	7.00	262.71	nr	**8866.92**
Twin fan extract unit; belt driven; located internally; complete with anti-vibration mounts and non-return shutter; including ancillaries, matching flanges, flexible connectors and clips; 3 phase, 50 Hz motor; includes fixing in position; electrical work elsewhere						
0.25 m³/s @ 150Pa	5865.00	7205.15	4.50	168.88	nr	**7374.03**
1.00 m³/s @ 200Pa	5865.00	7205.15	5.00	187.66	nr	**7392.81**
2.00 m³/s @ 250Pa	6874.00	8444.71	6.50	243.95	nr	**8688.66**
Roof mounted extract fan; including ancillaries, fibreglass cowling, fitted shutters and bird guard; 415 V, 3 phase, 50 Hz motor; includes fixing in position; electrical work elsewhere						
Flat roof installation, fixed to curb						
355 mm	1070.00	1314.49	4.50	168.88	nr	**1483.37**
400 mm	1178.00	1447.17	5.50	206.43	nr	**1653.60**
450 mm	1300.00	1597.05	7.00	262.71	nr	**1859.76**
500 mm	1784.00	2191.64	8.00	300.25	nr	**2491.89**
Centrifugal fan; single speed for internal domestic kitchens/utility rooms; fitted with standard overload protection; complete with housing; includes placing in position; electrical work elsewhere						
Window mounted						
245 m³/hr	221.09	271.61	0.50	18.77	nr	**290.38**
500 m³/hr	458.44	563.19	0.50	18.77	nr	**581.96**
Wall mounted						
245 m³/hr	287.60	353.31	0.83	31.28	nr	**384.59**
500 m³/hr	526.04	646.24	0.83	31.28	nr	**677.52**

38 VENTILATION/AIR CONDITIONING SYSTEMS

Item	Net Price £	Material £	Labour hours	Labour £	Unit	**Total rate £**
Centrifugal fan; various speeds, simultaneous ventilation from separate areas fitted with standard overload protection; complete with housing; includes placing in position; ducting and electrical work elsewhere						
Fan unit						
147–300 m^3/hr	109.38	134.37	1.00	37.53	nr	**171.90**
175–411 m^3/hr	136.43	167.60	1.00	37.53	nr	**205.13**
Toilet extract units; centrifugal fan; various speeds for internal domestic bathrooms/ W.Cs, complete with housing; includes placing in position; electrical work elsewhere						
Fan unit; fixed to wall; including shutter						
Single speed 85 m^3/hr	146.27	179.70	0.75	28.14	nr	**207.84**
Two speed 60-85 m^3/hr	178.21	218.94	0.83	31.28	nr	**250.22**
Humidity controlled; autospeed; fixed to wall; including shutter						
30-60-85 m^3/hr	233.65	287.04	1.00	37.53	nr	**324.57**

38 VENTILATION/AIR CONDITIONING SYSTEMS

Item	Net Price £	Material £	Labour hours	Labour £	Unit	Total rate £
PLANT/EQUIPMENT: AIR FILTRATION						
High efficiency duct mounted filters; 99.997% H13 (EU13); tested to BS 3928						
Standard; 1700 m³/hr air volume; continuous rating up to 80°C; sealed wood case, aluminium spacers, neoprene gaskets; water repellant filter media; includes placing in position						
610 × 610 × 292 mm	273.43	335.91	1.00	35.93	nr	**371.84**
Side withdrawal frame	103.59	127.26	2.50	89.84	nr	**217.10**
High capacity; 3400 m³/hr air volume; continuous rating up to 80°C; anti-corrosion coated mild steel frame, polyurethane sealant and neoprene gaskets; water repellant filter media; includes placing in position						
610 × 610 × 292 mm	297.30	365.23	1.00	35.93	nr	**401.16**
Side withdrawal frame	103.59	127.26	2.50	89.84	nr	**217.10**
Bag filters; 40/60% F5 (EU5); tested to BSEN 779						
Duct mounted bag filter; continuous rating up to 60°C; rigid filter assembly; sealed into one piece coated mild steel header with sealed pocket separators; includes placing in position						
6 pocket, 592 × 592 × 25 mm header; pockets 380 mm long; 1690 m³/hr	19.33	23.75	1.00	35.93	nr	**59.68**
Side withdrawal frame	17.93	22.03	2.00	71.87	nr	**93.90**
6 pocket, 592 × 592 × 25 mm header; pockets 500 mm long; 2550 m³/hr	19.62	24.11	1.50	53.91	nr	**78.02**
Side withdrawal frame	17.93	22.03	2.50	89.84	nr	**111.87**
6 pocket, 592 × 592 × 25 mm header; pockets 600 mm long; 3380 m³/hr	23.17	28.47	1.50	53.91	nr	**82.38**
Side withdrawal frame	17.93	22.03	3.00	107.80	nr	**129.83**
Bag filters; 80/90% F7, (EU7); tested to BSEN 779						
Duct mounted bag filter; continuous rating up to 60°C; rigid filter assembly; sealed into one piece coated mild steel header with sealed pocket separators; includes placing in position						
6 pocket, 592 × 592 × 25 mm header; pockets 500 mm long; 1688 m³/hr	19.62	24.11	1.00	35.93	nr	**60.04**
Side withdrawal frame	17.93	22.03	2.00	71.87	nr	**93.90**
6 pocket, 592 × 592 × 25 mm header; pockets 635 mm long; 2047 m³/hr	23.17	28.47	1.50	53.91	nr	**82.38**
Side withdrawal frame	17.93	22.03	2.50	89.84	nr	**111.87**
6 pocket, 592 × 592 × 25 mm header; pockets 762 mm long; 2729 m³/hr	26.71	32.81	1.50	53.91	nr	**86.72**
Side withdrawal frame	17.93	22.03	3.00	107.80	nr	**129.83**

38 VENTILATION/AIR CONDITIONING SYSTEMS

Item	Net Price £	Material £	Labour hours	Labour £	Unit	Total rate £
Grease filters, washable; minimum 65%						
Double sided extract unit; lightweight stainless steel construction; demountable composite filter media of woven metal mat and expanded metal mesh supports; for mounting on hood and extract systems (hood not included); includes placing in position						
500 × 686 × 565 mm, 4080 m³/hr	402.79	494.82	2.00	71.87	nr	**566.69**
1000 × 686 × 565 mm, 8160 m³/hr	614.25	754.60	3.00	107.80	nr	**862.40**
1500 × 686 × 565 mm, 12240 m³/hr	842.53	1035.05	3.50	125.77	nr	**1160.82**
Panel filters; 82% G3 (EU3); tested to BS EN779						
Modular duct mounted filter panels; continuous rating up to 100°C; graduated density media; rigid cardboard frame; includes placing in position						
596 × 596 × 47 mm, 2360 m³/hr	4.88	6.00	1.00	35.93	nr	**41.93**
Side withdrawal frame	34.98	42.98	2.50	89.84	nr	**132.82**
596 × 287 × 47 mm, 1140 m³/hr	3.37	4.14	1.00	35.93	nr	**40.07**
Side withdrawal frame	34.98	42.98	2.50	89.84	nr	**132.82**
Panel filters; 90% G4 (EU4); tested to BS EN779						
Modular duct mounted filter panels; continuous rating up to 100°C; pleated media with wire support; rigid cardboard frame; includes placing in position						
596 × 596 × 47 mm, 2560 m³/hr	7.38	9.06	1.00	35.93	nr	**44.99**
side withdrawal frame	44.11	54.19	3.00	107.80	nr	**161.99**
596 × 287 × 47 mm, 1230 m³/hr	5.10	6.27	1.00	35.93	nr	**42.20**
Side withdrawal frame	34.98	42.98	3.00	107.80	nr	**150.78**
Carbon filters; standard duty disposable carbon filters; steel frame with bonded carbon panels; for fixing to ductwork; including placing in position						
12 panels						
597 × 597 × 298 mm, 1460 m³/hr	343.29	421.73	0.33	11.85	nr	**433.58**
597 × 597 × 451 mm, 2200 m³/hr	387.00	475.43	0.33	11.85	nr	**487.28**
597 × 597 × 597 mm, 2930 m³/hr	431.57	530.19	0.33	11.85	nr	**542.04**
8 panels						
451 × 451 × 298 mm, 740 m³/hr	258.80	317.94	0.29	10.42	nr	**328.36**
451 × 451 × 451 mm, 1105 m³/hr	286.44	351.89	0.29	10.42	nr	**362.31**
451 × 451 × 597 mm, 1460 m³/hr	312.81	384.29	0.29	10.42	nr	**394.71**
6 panels						
298 × 298 × 298 mm, 365 m³/hr	183.38	225.28	0.25	8.98	nr	**234.26**
298 × 298 × 451 mm, 550 m³/hr	196.34	241.21	0.25	8.98	nr	**250.19**
298 × 298 × 597 mm, 780 m³/hr	208.74	256.44	0.25	8.98	nr	**265.42**

38 VENTILATION/AIR CONDITIONING SYSTEMS

Item	Net Price £	Material £	Labour hours	Labour £	Unit	Total rate £
PLANT/EQUIPMENT: AIR CURTAINS						
The selection of air curtains requires consideration of the particular conditions involved; climatic conditions, wind influence, construction and position all influence selection; consultation with a specialist manufacturer is therefore advisable.						
Commercial grade air curtains; recessed or exposed units with rigid sheet steel casing; aluminium grilles; high quality motor/centrifugal fan assembly; includes fixing in position; electrical work elsewhere						
Ambient temperature; 240 V single phase supply; mounting height 2.40 m						
1000 × 590 × 270 mm	2593.82	3186.51	12.05	452.17	nr	**3638.68**
1500 × 590 × 270 mm	3319.92	4078.53	12.05	452.17	nr	**4530.70**
2000 × 590 × 270 mm	4016.34	4934.08	12.05	452.25	nr	**5386.33**
2500 × 590 × 270 mm	4464.72	5484.91	13.00	487.90	nr	**5972.81**
Ambient temperature; 240 V single phase supply; mounting height 2.80 m						
1000 × 590 × 270 mm	2996.62	3681.35	16.13	605.34	nr	**4286.69**
1500 × 590 × 270 mm	3825.54	4699.67	16.13	605.34	nr	**5305.01**
2000 × 590 × 270 mm	4768.94	5858.64	16.13	605.34	nr	**6463.98**
2500 × 590 × 270 mm	5514.12	6774.10	17.10	641.77	nr	**7415.87**
Ambient temperature 240 V single phase supply; mounting height 3.30 m						
1000 × 774 × 370 mm	3914.58	4809.06	17.24	647.07	nr	**5456.13**
1500 × 774 × 370 mm	5293.64	6503.24	17.24	647.07	nr	**7150.31**
2000 × 774 × 370 mm	6547.62	8043.76	17.24	647.07	nr	**8690.83**
2500 × 774 × 370 mm	8276.48	10167.65	18.30	686.82	nr	**10854.47**
Ambient temperature; 240 V single phase supply; mounting height 4.00 m						
1000 × 774 × 370 mm	4323.74	5311.72	19.10	716.82	nr	**6028.54**
1500 × 774 × 370 mm	5672.06	6968.13	19.10	716.82	nr	**7684.95**
2000 × 774 × 370 mm	7061.72	8675.32	19.10	716.82	nr	**9392.14**
2500 × 774 × 370 mm	8276.48	10167.65	19.90	746.87	nr	**10914.52**
Water heated; 240 V single phase supply; mounting height 2.40 m						
1000 × 590 × 270 mm; 2.30–9.40 kW output	3051.74	3749.07	12.05	452.17	nr	**4201.24**
1500 × 590 × 270 mm; 3.50–14.20 kW output	3906.10	4798.65	12.05	452.17	nr	**5250.82**
2000 × 590 × 270 mm; 4.70–19.00 kW output	4725.48	5805.25	12.05	452.17	nr	**6257.42**
2500 × 590 × 270 mm; 5.90–23.70 kW output	5253.36	6453.75	13.00	487.90	nr	**6941.65**

38 VENTILATION/AIR CONDITIONING SYSTEMS

Item	Net Price £	Material £	Labour hours	Labour £	Unit	**Total rate £**
Water heated; 240 V single phase supply; mounting height 2.80 m						
1000 × 590 × 270 mm; 3.30–11.90 kW output	3526.62	4332.46	16.13	605.34	nr	**4937.80**
1500 × 590 × 270 mm; 5.00–17.90 kW output	4501.82	5530.49	16.13	605.34	nr	**6135.83**
2000 × 590 × 270 mm; 6.70–23.90 kW output	5611.64	6893.90	16.13	605.34	nr	**7499.24**
2500 × 590 × 270 mm; 8.30–29.80 kW output	6488.26	7970.82	17.10	641.77	nr	**8612.59**
Water heated; 240 V single phase supply; mounting height 3.30 m						
1000 × 774 × 370 mm; 6.10–21.80 kW output	4605.70	5658.10	17.24	647.07	nr	**6305.17**
1500 × 774 × 370 mm; 9.20–32.80 kW output	6228.56	7651.79	17.24	647.07	nr	**8298.86**
2000 × 774 × 370 mm; 12.30–43.70 kW output	7704.08	9464.46	17.24	647.07	nr	**10111.53**
2500 × 774 × 370 mm; 15.30–54.60 kW output	9119.18	11202.91	18.30	686.82	nr	**11889.73**
Water heated; 240 V single phase supply; mounting height 4.00 m						
1000 × 774 × 370 mm; 7.20–24.20 kW output	5086.94	6249.31	19.10	716.82	nr	**6966.13**
1500 × 774 × 370 mm; 10.90–36.30 kW output	6673.76	8198.72	19.10	716.82	nr	**8915.54**
2000 × 774 × 370 mm; 14.50–48.40 kW output	8308.28	10206.72	19.10	716.82	nr	**10923.54**
2500 × 774 × 370 mm; 18.10–60.60 kW output	9737.16	11962.10	19.90	746.87	nr	**12708.97**
Electrically heated; 415 V three phase supply; mounting height 2.40 m						
1000 × 590 × 270 mm; 2.30–9.40 kW output	3712.12	4560.34	12.05	452.17	nr	**5012.51**
1500 × 590 × 270 mm; 3.50–14.20 kW output	4618.42	5673.73	12.05	452.17	nr	**6125.90**
2000 × 590 × 270 mm; 4.70–19.00 kW output	5504.58	6762.38	12.05	481.93	nr	**7244.31**
2500 × 590 × 270 mm; 5.90–23.70 kW output	6258.24	7688.25	13.00	520.00	nr	**8208.25**
Electrically heated; 415 V three phase supply; mounting height 2.80 m						
1000 × 590 × 270 mm; 3.30–11.90 kW output	4370.38	5369.01	16.13	605.34	nr	**5974.35**
1500 × 590 × 270 mm; 5.00–17.90 kW output	5374.20	6602.20	16.13	605.34	nr	**7207.54**
2000 × 590 × 270 mm; 6.70–23.90 kW output	6786.12	8336.75	16.13	605.34	nr	**8942.09**
2500 × 590 × 270 mm; 8.30–29.80 kW output	7584.30	9317.31	17.10	641.77	nr	**9959.08**

38 VENTILATION/AIR CONDITIONING SYSTEMS

Item	Net Price £	Material £	Labour hours	Labour £	Unit	Total rate £
PLANT/EQUIPMENT: AIR CURTAINS – cont						
Commercial grade air curtains – cont						
Electrically heated; 415 V three phase supply; mounting height 3.30 m						
1000 × 774 × 370 mm; 6.10–21.80 kW output	6640.90	8158.34	17.24	647.03	nr	**8805.37**
1500 × 774 × 370 mm; 9.20–32.80 kW output	8897.64	10930.75	17.24	647.03	nr	**11577.78**
2000 × 774 × 370 mm; 12.30–43.70 kW output	10738.86	13192.69	17.24	647.03	nr	**13839.72**
2500 × 774 × 370 mm; 15.30–54.60 kW output	12712.58	15617.41	18.30	686.82	nr	**16304.23**
Electrically heated; 415 V three phase supply; mounting height 4.00 m						
1000 × 774 × 370 mm; 7.20–24.20 kW output	7328.84	9003.48	19.10	716.82	nr	**9720.30**
1500 × 774 × 370 mm; 10.90–36.30 kW output	9508.20	11680.82	19.10	716.82	nr	**12397.64**
2000 × 774 × 370 mm; 14.50–48.40 kW output	11499.94	14127.68	19.10	716.82	nr	**14844.50**
2500 × 774 × 370 mm; 18.10–60.60 kW output	13622.06	16734.70	19.90	746.87	nr	**17481.57**
Industrial grade air curtains; recessed or exposed units with rigid sheet steel casing; aluminium grilles; high quality motor/centrifugal fan assembly; includes fixing in position; electrical work elsewhere						
Ambient temperature; 230 V single phase supply; including wiring between multiple units; horizontally or vertically mounted; opening maximum 6.00 m						
1500 × 585 × 853 mm; 3.0 A supply	2425.28	2979.46	17.24	647.07	nr	**3626.53**
2000 × 585 × 853 mm; 4.0 A supply	2939.04	3610.61	17.24	647.07	nr	**4257.68**
Water heated; 230 V single phase supply; including wiring between multiple units; horizontally or vertically mounted; opening maximum 6.00 m						
1500 × 585 × 956 mm; 3.0 A supply; 34.80 kW output	2779.92	3415.14	17.24	647.07	nr	**4062.21**
2000 × 585 × 956 mm; 4.0 A supply; 50.70 kW output	3396.64	4172.77	17.24	647.07	nr	**4819.84**

38 VENTILATION/AIR CONDITIONING SYSTEMS

Item	Net Price £	Material £	Labour hours	Labour £	Unit	**Total rate £**
Water heated; 230 V single phase supply; including wiring between multiple units; vertically mounted in single bank for openings maximum 6.00 m wide or opposing twin banks for openings maximum 10.00 m wide						
1500 × 585 mm; 3.0 A supply; 41.1 kW output	2779.92	3415.14	17.24	647.07	nr	**4062.21**
2000 × 585 mm; 4.0 A supply; 57.7 kW output	3396.64	4172.77	17.24	647.07	nr	**4819.84**
Remote mounted electronic controller unit; excluding wiring to units						
0–10 V control	42.64	52.38	5.00	187.66	nr	**240.04**

38 VENTILATION/AIR CONDITIONING SYSTEMS

Item	Net Price £	Material £	Labour hours	Labour £	Unit	Total rate £
SILENCERS/ACOUSTIC TREATMENT						
Attenuators; DW144 galvanized construction c/w splitters; self-securing; fitted to ductwork						
To suit rectangular ducts; unit length 600 mm						
100 × 100 mm	102.97	126.49	0.75	26.95	nr	**153.44**
150 × 150 mm	107.49	132.05	0.75	26.95	nr	**159.00**
200 × 200 mm	111.99	137.58	0.75	26.95	nr	**164.53**
300 × 300 mm	125.29	153.92	0.75	26.95	nr	**180.87**
400 × 400 mm	140.65	172.79	1.00	35.93	nr	**208.72**
500 × 500 mm	164.15	201.66	1.25	44.92	nr	**246.58**
600 × 300 mm	158.12	194.25	1.25	44.92	nr	**239.17**
600 × 600 mm	221.30	271.87	1.25	44.92	nr	**316.79**
700 × 300 mm	167.65	205.96	1.50	53.91	nr	**259.87**
700 × 700 mm	249.64	306.68	1.50	53.91	nr	**360.59**
800 × 300 mm	173.66	213.34	2.00	71.87	nr	**285.21**
800 × 800 mm	288.02	353.83	2.00	71.87	nr	**425.70**
1000 × 1000 mm	368.67	452.91	3.00	107.80	nr	**560.71**
To suit rectangular ducts; unit length 1200 mm						
200 × 200 mm	123.49	151.70	1.00	35.93	nr	**187.63**
300 × 300 mm	152.68	187.57	1.00	35.93	nr	**223.50**
400 × 400 mm	183.57	225.52	1.33	47.79	nr	**273.31**
500 × 500 mm	221.57	272.20	1.66	59.64	nr	**331.84**
600 × 300 mm	211.27	259.55	1.66	59.64	nr	**319.19**
600 × 600 mm	309.11	379.74	1.66	59.64	nr	**439.38**
700 × 300 mm	226.47	278.22	2.00	71.87	nr	**350.09**
700 × 700 mm	353.73	434.55	2.00	71.87	nr	**506.42**
800 × 300 mm	236.77	290.87	2.66	95.59	nr	**386.46**
800 × 800 mm	408.16	501.43	2.66	95.59	nr	**597.02**
1000 × 1000 mm	552.52	678.77	4.00	143.73	nr	**822.50**
1300 × 1300 mm	948.43	1165.14	8.00	287.47	nr	**1452.61**
1500 × 1500 mm	1085.59	1333.65	8.00	287.47	nr	**1621.12**
1800 × 1800 mm	1485.33	1824.73	10.66	383.06	nr	**2207.79**
2000 × 2000 mm	1666.71	2047.55	13.33	478.99	nr	**2526.54**
To suit rectangular ducts; unit length 1800 mm						
200 × 200 mm	155.86	191.48	1.00	35.93	nr	**227.41**
300 × 300 mm	207.11	254.44	1.00	35.93	nr	**290.37**
400 × 400 mm	247.57	304.14	1.33	47.79	nr	**351.93**
500 × 500 mm	288.52	354.45	1.66	59.64	nr	**414.09**
600 × 300 mm	301.75	370.70	1.66	59.64	nr	**430.34**
600 × 600 mm	421.65	518.00	1.66	59.64	nr	**577.64**
700 × 300 mm	319.89	392.98	2.00	71.87	nr	**464.85**
700 × 700 mm	503.78	618.89	2.00	71.87	nr	**690.76**
800 × 300 mm	338.27	415.57	2.66	95.59	nr	**511.16**
800 × 800 mm	586.16	720.10	2.66	95.59	nr	**815.69**
1000 × 1000 mm	787.90	967.93	4.00	143.73	nr	**1111.66**
1300 × 1300 mm	1316.22	1616.98	8.00	287.47	nr	**1904.45**
1500 × 1500 mm	1491.62	1832.46	8.00	287.47	nr	**2119.93**
1800 × 1800 mm	2070.84	2544.02	13.33	478.99	nr	**3023.01**
2000 × 2000 mm	2344.93	2880.75	10.66	383.06	nr	**3263.81**

38 VENTILATION/AIR CONDITIONING SYSTEMS

Item	Net Price £	Material £	Labour hours	Labour £	Unit	**Total rate £**
2300 × 2300 mm	2931.12	3600.88	16.00	574.94	nr	**4175.82**
2500 × 2500 mm	4088.56	5022.80	18.66	670.52	nr	**5693.32**
To suit rectangular ducts; unit length 2400 mm						
500 × 500 mm	373.09	458.35	2.08	74.85	nr	**533.20**
600 × 300 mm	356.19	437.58	2.08	74.85	nr	**512.43**
600 × 600 mm	530.50	651.72	2.08	74.85	nr	**726.57**
700 × 300 mm	386.09	474.32	2.50	89.84	nr	**564.16**
700 × 700 mm	626.63	769.82	2.50	89.84	nr	**859.66**
800 × 300 mm	408.16	501.43	3.33	119.66	nr	**621.09**
800 × 800 mm	725.19	890.89	3.33	119.66	nr	**1010.55**
1000 × 1000 mm	978.40	1201.97	5.00	179.67	nr	**1381.64**
1300 × 1300 mm	1584.71	1946.82	10.00	359.34	nr	**2306.16**
1500 × 1500 mm	1808.65	2221.93	10.00	359.34	nr	**2581.27**
1800 × 1800 mm	2580.61	3170.28	13.33	478.92	nr	**3649.20**
2000 × 2000 mm	2895.87	3557.58	16.66	598.66	nr	**4156.24**
2300 × 2300 mm	3582.83	4401.51	20.00	718.67	nr	**5120.18**
2500 × 2500 mm	5030.83	6180.37	23.32	838.12	nr	**7018.49**
To suit circular ducts; unit length 600 mm						
100 mm dia.	89.42	109.85	0.75	26.95	nr	**136.80**
200 mm dia.	105.21	129.25	0.75	26.95	nr	**156.20**
250 mm dia.	119.27	146.53	0.75	26.95	nr	**173.48**
315 mm dia.	138.83	170.55	1.00	35.93	nr	**206.48**
355 mm dia.	160.65	197.36	1.00	35.93	nr	**233.29**
400 mm dia.	172.68	212.14	1.00	35.93	nr	**248.07**
450 mm dia.	208.52	256.17	1.00	35.93	nr	**292.10**
500 mm dia.	229.60	282.06	1.26	45.21	nr	**327.27**
630 mm dia.	270.70	332.56	1.50	53.91	nr	**386.47**
710 mm dia.	294.78	362.13	1.50	53.91	nr	**416.04**
800 mm dia.	327.87	402.79	2.00	71.87	nr	**474.66**
1000 mm dia.	440.13	540.70	3.00	107.80	nr	**648.50**
To suit circular ducts; unit length 1200 mm						
100 mm dia.	131.56	161.63	1.00	35.93	nr	**197.56**
200 mm dia.	144.33	177.31	1.00	35.93	nr	**213.24**
250 mm dia.	162.14	199.18	1.00	35.93	nr	**235.11**
315 mm dia.	185.46	227.84	1.33	47.79	nr	**275.63**
355 mm dia.	222.33	273.14	1.33	47.79	nr	**320.93**
400 mm dia.	235.86	289.76	1.33	47.79	nr	**337.55**
450 mm dia.	273.21	335.64	1.33	47.79	nr	**383.43**
500 mm dia.	280.00	343.98	1.66	59.64	nr	**403.62**
630 mm dia.	370.76	455.48	2.00	71.87	nr	**527.35**
710 mm dia.	409.86	503.52	2.00	71.87	nr	**575.39**
800 mm dia.	409.11	502.59	2.66	95.59	nr	**598.18**
1000 mm dia.	549.21	674.71	4.00	143.73	nr	**818.44**
1250 mm dia.	1034.58	1270.98	8.00	287.47	nr	**1558.45**
1400 mm dia.	1124.84	1381.86	8.00	287.47	nr	**1669.33**
1600 mm dia.	1247.03	1531.98	10.66	383.06	nr	**1915.04**

38 VENTILATION/AIR CONDITIONING SYSTEMS

Item	Net Price £	Material £	Labour hours	Labour £	Unit	**Total rate £**
SILENCERS/ACOUSTIC TREATMENT – cont						
To suit circular ducts; unit length 1800 mm						
200 mm dia.	202.26	248.47	1.25	44.92	nr	**293.39**
250 mm dia.	226.33	278.05	1.25	44.92	nr	**322.97**
315 mm dia.	268.97	330.42	1.66	59.64	nr	**390.06**
355 mm dia.	317.11	389.57	1.66	59.64	nr	**449.21**
400 mm dia.	328.13	403.11	1.66	59.64	nr	**462.75**
450 mm dia.	335.17	411.76	1.66	59.64	nr	**471.40**
500 mm dia.	368.46	452.65	2.08	74.74	nr	**527.39**
630 mm dia.	407.60	500.73	2.50	89.84	nr	**590.57**
710 mm dia.	476.82	585.77	2.50	89.84	nr	**675.61**
800 mm dia.	493.36	606.09	3.33	119.66	nr	**725.75**
1000 mm dia.	1288.08	1582.40	6.66	239.33	nr	**1821.73**
1250 mm dia.	1309.15	1608.30	6.66	239.33	nr	**1847.63**
1400 mm dia.	1446.38	1776.87	5.00	179.67	nr	**1956.54**
1600 mm dia.	1447.25	1777.94	9.90	355.74	nr	**2133.68**

38 VENTILATION/AIR CONDITIONING SYSTEMS

Item	Net Price £	Material £	Labour hours	Labour £	Unit	**Total rate £**
THERMAL INSULATION						
Concealed ductwork						
Flexible wrap; 20 kg–45 kg Bright Class O aluminium foil faced; Bright Class O foil taped joints; 62 mm metal pins and washers; aluminium bands						
40 mm thick insulation	12.78	15.70	0.40	15.02	m^2	**30.72**
Semi-rigid slab; 45 kg Bright Class O aluminium foil faced mineral fibre; Bright Class O foil taped joints; 62 mm metal pins and washers; aluminium bands						
40 mm thick insulation	16.72	20.54	0.65	24.40	m^2	**44.94**
Plantroom ductwork						
Semi-rigid slab; 45 kg Bright Class O aluminium foil faced mineral fibre; Bright Class O foil taped joints; 62 mm metal pins and washers; 22 swg plain/embossed aluminium cladding; pop rivited						
50 mm thick insulation	42.17	51.81	1.50	56.30	m^2	**108.11**
External ductwork						
Semi-rigid slab; 45 kg Bright Class O aluminium foil faced mineral fibre; Bright Class O foil taped joints; 62 mm metal pins and washers; 0.8 mm polyisobutylene sheeting; welded joints						
50 mm thick insulation	30.77	37.80	1.25	46.91	m^2	**84.71**

38 VENTILATION/AIR CONDITIONING SYSTEMS

Item	Net Price £	Material £	Labour hours	Labour £	Unit	Total rate £
DUCTWORK: FIRE RATED						
DUCTLINES						
The relevant BS requires that the fire rating of ductwork meets 3 criteria; stability (hours), integrity (hours) and insulation (hours). The least of the 3 periods defines the fire rating. The BS does however allow stability and integrity to be considered in isolation. Rates are therefore provided for both types of system.						
Care should to be taken when using the rates within this section to ensure thta the requiremements for stability, integrity and insulation are known and the appropriate rates are used						
High density single layer mineral wool fire rated ductwork slab, in accordance with BS476, Part 24 (ISO 6944: 1985), ducts 'Type A' and 'Type B'; 165 kg class O foil faced mineral fibre; 100 mm wide bright class O foil taped joints; welded pins; includes protection to all supports.						
½ hour stability, integrity and insulation						
25 mm thick, vertical and horizontal ductwork	45.91	56.40	1.25	46.91	m²	**103.31**
1 hour stability, integrity and insulation						
30 mm thick, vertical ductwork	53.90	66.21	1.50	56.30	m²	**122.51**
40 mm thick, horizontal ductwork	63.39	77.88	1.50	56.30	m²	**134.18**
1½ hour stability, integrity and insulation						
50 mm thick, vertical ductwork	75.73	93.03	1.75	65.68	m²	**158.71**
70 mm thick, horizontal ductwork	94.73	116.37	1.75	65.68	m²	**182.05**
2 hour stability, integrity and insulation						
70 mm, vertical ductwork	97.60	119.90	2.00	75.06	m²	**194.96**
90 mm horizontal ductwork	116.59	143.23	2.00	75.06	m²	**218.29**
Kitchen extract, 1 hour stability, integrity and insulation						
90 mm, vertical and horizontal	116.59	143.23	2.00	75.06	m²	**218.29**

38 VENTILATION/AIR CONDITIONING SYSTEMS

Item	Net Price £	Material £	Labour hours	Labour £	Unit	Total rate £
Galvanized sheet metal rectangular section ductwork to BS476 Part 24 (ISO 6944:1985), ducts 'Type A' and 'Type B'; provides 2 hours stability and 2 hours integrity at 1100°C (no rating for insulation); including all necessary stiffeners, joints and supports in the running length						
Ductwork up to 600 mm longest side						
Sum of two sides 200 mm	72.69	89.30	2.91	104.57	m	**193.87**
Sum of two sides 300 mm	78.92	96.96	2.99	107.45	m	**204.41**
Sum of two sides 400 mm	85.14	104.59	3.17	113.90	m	**218.49**
Sum of two sides 500 mm	91.36	112.23	3.37	121.10	m	**233.33**
Sum of two sides 600 mm	97.58	119.88	3.54	127.21	m	**247.09**
Sum of two sides 700 mm	103.81	127.53	3.72	133.68	m	**261.21**
Sum of two sides 800 mm	110.04	135.19	3.90	140.14	m	**275.33**
Sum of two sides 900 mm	116.25	142.81	5.04	181.10	m	**323.91**
Sum of two sides 1000 mm	122.48	150.47	5.58	200.52	m	**350.99**
Sum of two sides 1100 mm	128.69	158.10	5.84	209.85	m	**367.95**
Sum of two sides 1200 mm	134.90	165.72	6.11	219.56	m	**385.28**
Extra over fittings; Ductwork up to 600 mm longest side						
End cap						
Sum of two sides 200 mm	20.64	25.36	0.81	29.11	nr	**54.47**
Sum of two sides 300 mm	21.89	26.89	0.84	30.19	nr	**57.08**
Sum of two sides 400 mm	23.15	28.44	0.87	31.27	nr	**59.71**
Sum of two sides 500 mm	24.40	29.98	0.90	32.33	nr	**62.31**
Sum of two sides 600 mm	25.65	31.51	0.93	33.41	nr	**64.92**
Sum of two sides 700 mm	26.92	33.08	0.96	34.49	nr	**67.57**
Sum of two sides 800 mm	28.20	34.64	0.98	35.21	nr	**69.85**
Sum of two sides 900 mm	29.45	36.18	1.17	42.04	nr	**78.22**
Sum of two sides 1000 mm	30.69	37.71	1.22	43.85	nr	**81.56**
Sum of two sides 1100 mm	31.96	39.26	1.25	44.92	nr	**84.18**
Sum of two sides 1200 mm	33.21	40.80	1.28	45.99	nr	**86.79**
Reducer						
Sum of two sides 200 mm	81.87	100.58	2.23	80.14	nr	**180.72**
Sum of two sides 300 mm	85.64	105.21	2.37	85.15	nr	**190.36**
Sum of two sides 400 mm	89.38	109.80	2.51	90.19	nr	**199.99**
Sum of two sides 500 mm	93.15	114.44	2.65	95.22	nr	**209.66**
Sum of two sides 600 mm	96.90	119.04	2.79	100.25	nr	**219.29**
Sum of two sides 700 mm	100.66	123.66	2.87	103.13	nr	**226.79**
Sum of two sides 800 mm	104.43	128.29	3.01	108.16	nr	**236.45**
Sum of two sides 900 mm	108.19	132.91	3.06	109.95	nr	**242.86**
Sum of two sides 1000 mm	111.94	137.52	3.17	113.90	nr	**251.42**
Sum of two sides 1100 mm	115.68	142.12	3.22	115.71	nr	**257.83**
Sum of two sides 1200 mm	119.47	146.77	3.28	117.86	nr	**264.63**

38 VENTILATION/AIR CONDITIONING SYSTEMS

Item	Net Price £	Material £	Labour hours	Labour £	Unit	Total rate £
DUCTWORK: FIRE RATED – cont						
Extra over fittings – cont						
Offset						
Sum of two sides 200 mm	197.58	242.73	2.95	106.01	nr	**348.74**
Sum of two sides 300 mm	202.48	248.74	3.11	111.75	nr	**360.49**
Sum of two sides 400 mm	207.38	254.76	3.26	117.15	nr	**371.91**
Sum of two sides 500 mm	212.29	260.80	3.42	122.89	nr	**383.69**
Sum of two sides 600 mm	217.19	266.82	3.57	128.28	nr	**395.10**
Sum of two sides 700 mm	227.00	278.87	3.23	116.07	nr	**394.94**
Sum of two sides 800 mm	222.10	272.85	3.73	134.03	nr	**406.88**
Sum of two sides 900 mm	231.92	284.92	3.45	123.96	nr	**408.88**
Sum of two sides 1000 mm	236.82	290.93	3.67	131.88	nr	**422.81**
Sum of two sides 1100 mm	241.70	296.93	3.78	135.83	nr	**432.76**
Sum of two sides 1200 mm	246.63	302.99	3.89	139.78	nr	**442.77**
90° radius bend						
Sum of two sides 200 mm	80.11	98.42	2.06	74.02	nr	**172.44**
Sum of two sides 300 mm	86.18	105.87	2.21	79.41	nr	**185.28**
Sum of two sides 400 mm	92.24	113.32	2.36	84.81	nr	**198.13**
Sum of two sides 500 mm	98.33	120.80	2.51	90.19	nr	**210.99**
Sum of two sides 600 mm	104.40	128.26	2.66	95.59	nr	**223.85**
Sum of two sides 700 mm	110.44	135.67	2.81	100.97	nr	**236.64**
Sum of two sides 800 mm	116.53	143.16	2.97	106.72	nr	**249.88**
Sum of two sides 900 mm	122.57	150.58	2.99	107.45	nr	**258.03**
Sum of two sides 1000 mm	128.67	158.07	3.04	109.24	nr	**267.31**
Sum of two sides 1100 mm	134.72	165.50	3.07	110.32	nr	**275.82**
Sum of two sides 1200 mm	140.77	172.94	3.09	111.03	nr	**283.97**
45° radius bend						
Sum of two sides 200 mm	98.78	121.35	1.52	54.62	nr	**175.97**
Sum of two sides 300 mm	101.24	124.37	1.59	57.14	nr	**181.51**
Sum of two sides 400 mm	103.71	127.41	1.66	59.64	nr	**187.05**
Sum of two sides 500 mm	106.15	130.41	1.73	62.16	nr	**192.57**
Sum of two sides 600 mm	108.60	133.41	1.80	64.68	nr	**198.09**
Sum of two sides 700 mm	111.07	136.45	1.87	67.20	nr	**203.65**
Sum of two sides 800 mm	113.49	139.42	1.93	69.35	nr	**208.77**
Sum of two sides 900 mm	115.97	142.46	1.99	71.52	nr	**213.98**
Sum of two sides 1000 mm	118.41	145.47	2.05	73.66	nr	**219.13**
Sum of two sides 1100 mm	120.86	148.48	2.11	75.82	nr	**224.30**
Sum of two sides 1200 mm	123.31	151.48	2.17	77.97	nr	**229.45**
90° mitre bend						
Sum of two sides 200 mm	83.05	102.03	2.47	88.76	nr	**190.79**
Sum of two sides 300 mm	96.26	118.25	2.65	95.22	nr	**213.47**
Sum of two sides 400 mm	109.46	134.47	2.84	102.05	nr	**236.52**
Sum of two sides 500 mm	122.68	150.72	3.02	108.52	nr	**259.24**
Sum of two sides 600 mm	135.87	166.92	3.20	114.99	nr	**281.91**
Sum of two sides 700 mm	149.10	183.17	3.38	121.45	nr	**304.62**
Sum of two sides 800 mm	162.31	199.40	3.56	127.92	nr	**327.32**
Sum of two sides 900 mm	175.51	215.62	3.59	129.00	nr	**344.62**
Sum of two sides 1000 mm	188.74	231.87	3.66	131.52	nr	**363.39**
Sum of two sides 1100 mm	201.92	248.06	3.69	132.59	nr	**380.65**
Sum of two sides 1200 mm	215.17	264.34	3.72	133.68	nr	**398.02**

38 VENTILATION/AIR CONDITIONING SYSTEMS

Item	Net Price £	Material £	Labour hours	Labour £	Unit	**Total rate £**
Branch (side-on shoe)						
Sum of two sides 200 mm	30.83	37.87	0.98	35.21	nr	**73.08**
Sum of two sides 300 mm	31.00	38.08	1.06	38.08	nr	**76.16**
Sum of two sides 400 mm	31.17	38.29	1.14	40.96	nr	**79.25**
Sum of two sides 500 mm	31.35	38.51	1.22	43.85	nr	**82.36**
Sum of two sides 600 mm	31.50	38.70	1.30	46.70	nr	**85.40**
Sum of two sides 700 mm	31.68	38.92	1.38	49.59	nr	**88.51**
Sum of two sides 800 mm	32.08	39.41	1.37	49.22	nr	**88.63**
Sum of two sides 900 mm	31.83	39.10	1.46	52.46	nr	**91.56**
Sum of two sides 1000 mm	32.16	39.51	1.46	52.46	nr	**91.97**
Sum of two sides 1100 mm	32.33	39.72	1.50	53.91	nr	**93.63**
Sum of two sides 1200 mm	32.51	39.94	1.55	55.70	nr	**95.64**
Ductwork 601 to 800 mm longest side						
Sum of two sides 900 mm	116.23	142.79	5.04	181.10	m	**323.89**
Sum of two sides 1000 mm	122.46	150.44	5.58	200.52	m	**350.96**
Sum of two sides 1100 mm	128.69	158.10	5.84	209.85	m	**367.95**
Sum of two sides 1200 mm	134.90	165.72	6.11	219.56	m	**385.28**
Sum of two sides 1300 mm	141.16	173.42	6.38	229.27	m	**402.69**
Sum of two sides 1400 mm	147.38	181.05	6.65	238.96	m	**420.01**
Sum of two sides 1500 mm	153.62	188.73	6.91	248.30	m	**437.03**
Sum of two sides 1600 mm	159.85	196.37	7.18	258.00	m	**454.37**
Extra over fittings; Ductwork 601 to 800 mm longest side						
End cap						
Sum of two sides 900 mm	27.36	33.61	1.17	42.04	nr	**75.65**
Sum of two sides 1000 mm	29.31	36.00	1.22	43.85	nr	**79.85**
Sum of two sides 1100 mm	31.28	38.43	1.25	44.92	nr	**83.35**
Sum of two sides 1200 mm	33.22	40.81	1.28	45.99	nr	**86.80**
Sum of two sides 1300 mm	35.18	43.22	1.31	47.07	nr	**90.29**
Sum of two sides 1400 mm	37.14	45.62	1.34	48.15	nr	**93.77**
Sum of two sides 1500 mm	39.09	48.03	1.36	48.87	nr	**96.90**
Sum of two sides 1600 mm	41.05	50.43	1.39	49.95	nr	**100.38**
Reducer						
Sum of two sides 900 mm	102.40	125.80	3.06	109.95	nr	**235.75**
Sum of two sides 1000 mm	108.09	132.79	3.17	113.90	nr	**246.69**
Sum of two sides 1100 mm	113.78	139.78	3.22	115.71	nr	**255.49**
Sum of two sides 1200 mm	119.49	146.79	3.28	117.86	nr	**264.65**
Sum of two sides 1300 mm	125.18	153.78	3.33	119.66	nr	**273.44**
Sum of two sides 1400 mm	130.88	160.79	3.38	121.45	nr	**282.24**
Sum of two sides 1500 mm	136.58	167.79	3.44	123.62	nr	**291.41**
Sum of two sides 1600 mm	142.27	174.78	3.49	125.40	nr	**300.18**
Offset						
Sum of two sides 900 mm	217.77	267.53	3.45	123.96	nr	**391.49**
Sum of two sides 1000 mm	228.13	280.26	3.67	131.88	nr	**412.14**
Sum of two sides 1100 mm	238.51	293.01	3.78	135.83	nr	**428.84**
Sum of two sides 1200 mm	248.87	305.74	3.89	139.78	nr	**445.52**
Sum of two sides 1300 mm	259.23	318.47	4.00	143.73	nr	**462.20**
Sum of two sides 1400 mm	269.60	331.20	4.11	147.69	nr	**478.89**
Sum of two sides 1500 mm	279.96	343.93	4.23	152.00	nr	**495.93**
Sum of two sides 1600 mm	290.65	357.06	4.34	155.95	nr	**513.01**

38 VENTILATION/AIR CONDITIONING SYSTEMS

Item	Net Price £	Material £	Labour hours	Labour £	Unit	**Total rate £**
DUCTWORK: FIRE RATED – cont						
Extra over fittings – cont						
90° radius bend						
Sum of two sides 900 mm	108.88	133.76	2.99	107.45	nr	**241.21**
Sum of two sides 1000 mm	119.58	146.91	3.04	109.24	nr	**256.15**
Sum of two sides 1100 mm	130.24	160.00	3.07	110.32	nr	**270.32**
Sum of two sides 1200 mm	140.94	173.14	3.09	111.03	nr	**284.17**
Sum of two sides 1300 mm	151.62	186.27	3.12	112.11	nr	**298.38**
Sum of two sides 1400 mm	162.31	199.40	3.15	113.19	nr	**312.59**
Sum of two sides 1500 mm	173.00	212.53	3.17	113.90	nr	**326.43**
Sum of two sides 1600 mm	183.70	225.68	3.20	114.99	nr	**340.67**
45° bend						
Sum of two sides 900 mm	108.11	132.81	1.74	62.53	nr	**195.34**
Sum of two sides 1000 mm	113.18	139.04	1.85	66.48	nr	**205.52**
Sum of two sides 1100 mm	118.27	145.30	1.91	68.64	nr	**213.94**
Sum of two sides 1200 mm	123.32	151.49	1.96	70.43	nr	**221.92**
Sum of two sides 1300 mm	128.39	157.73	2.02	72.60	nr	**230.33**
Sum of two sides 1400 mm	133.46	163.96	2.07	74.38	nr	**238.34**
Sum of two sides 1500 mm	138.53	170.18	2.13	76.53	nr	**246.71**
Sum of two sides 1600 mm	143.60	176.41	2.18	78.34	nr	**254.75**
90° mitre bend						
Sum of two sides 900 mm	163.01	200.26	3.59	129.00	nr	**329.26**
Sum of two sides 1000 mm	180.40	221.62	3.66	131.52	nr	**353.14**
Sum of two sides 1100 mm	197.77	242.96	3.69	132.59	nr	**375.55**
Sum of two sides 1200 mm	215.17	264.34	3.72	133.68	nr	**398.02**
Sum of two sides 1300 mm	232.51	285.64	3.75	134.76	nr	**420.40**
Sum of two sides 1400 mm	249.90	307.00	3.78	135.83	nr	**442.83**
Sum of two sides 1500 mm	267.27	328.35	3.81	136.90	nr	**465.25**
Sum of two sides 1600 mm	284.66	349.70	3.84	137.98	nr	**487.68**
Branch (side-on shoe)						
Sum of two sides 900 mm	30.25	37.16	1.37	49.22	nr	**86.38**
Sum of two sides 1000 mm	31.00	38.08	1.46	52.46	nr	**90.54**
Sum of two sides 1100 mm	31.77	39.03	1.50	53.91	nr	**92.94**
Sum of two sides 1200 mm	32.52	39.95	1.55	55.70	nr	**95.65**
Sum of two sides 1300 mm	33.27	40.88	1.59	57.14	nr	**98.02**
Sum of two sides 1400 mm	32.69	40.16	1.68	60.36	nr	**100.52**
Sum of two sides 1500 mm	34.05	41.83	1.63	58.58	nr	**100.41**
Sum of two sides 1600 mm	35.56	43.69	1.72	61.80	nr	**105.49**
Ductwork 801 to 1000 mm longest side						
Sum of two sides 1100 mm	128.69	158.10	5.84	209.85	m	**367.95**
Sum of two sides 1200 mm	134.90	165.72	6.11	219.56	m	**385.28**
Sum of two sides 1300 mm	141.13	173.38	6.38	229.27	m	**402.65**
Sum of two sides 1400 mm	147.36	181.03	6.65	238.96	m	**419.99**
Sum of two sides 1500 mm	153.59	188.68	6.91	248.30	m	**436.98**
Sum of two sides 1600 mm	159.82	196.34	7.18	258.00	m	**454.34**
Sum of two sides 1700 mm	166.06	204.00	7.45	267.70	m	**471.70**
Sum of two sides 1800 mm	172.29	211.66	7.71	277.05	m	**488.71**
Sum of two sides 1900 mm	178.52	219.31	7.98	286.74	m	**506.05**
Sum of two sides 2000 mm	184.75	226.97	8.25	296.45	m	**523.42**

38 VENTILATION/AIR CONDITIONING SYSTEMS

Item	Net Price £	Material £	Labour hours	Labour £	Unit	Total rate £
Extra over fittings; Ductwork 801 to 1000 mm longest side						
End cap						
Sum of two sides 1100 mm	31.28	38.43	1.25	44.92	nr	**83.35**
Sum of two sides 1200 mm	33.22	40.81	1.28	45.99	nr	**86.80**
Sum of two sides 1300 mm	35.18	43.22	1.31	47.07	nr	**90.29**
Sum of two sides 1400 mm	37.14	45.62	1.34	48.15	nr	**93.77**
Sum of two sides 1500 mm	39.09	48.03	1.36	48.87	nr	**96.90**
Sum of two sides 1600 mm	41.03	50.41	1.39	49.95	nr	**100.36**
Sum of two sides 1700 mm	42.99	52.82	1.42	51.02	nr	**103.84**
Sum of two sides 1800 mm	44.95	55.22	1.45	52.10	nr	**107.32**
Sum of two sides 1900 mm	46.89	57.60	1.48	53.18	nr	**110.78**
Sum of two sides 2000 mm	48.84	60.00	1.51	54.26	nr	**114.26**
Reducer						
Sum of two sides 1100 mm	113.78	139.78	3.22	115.71	nr	**255.49**
Sum of two sides 1200 mm	119.50	146.81	3.28	117.86	nr	**264.67**
Sum of two sides 1300 mm	125.23	153.85	3.33	119.66	nr	**273.51**
Sum of two sides 1400 mm	130.97	160.89	3.38	121.45	nr	**282.34**
Sum of two sides 1500 mm	136.67	167.89	3.44	123.62	nr	**291.51**
Sum of two sides 1600 mm	142.43	174.97	3.49	125.40	nr	**300.37**
Sum of two sides 1700 mm	148.14	181.99	3.54	127.21	nr	**309.20**
Sum of two sides 1800 mm	153.87	189.03	3.60	129.36	nr	**318.39**
Sum of two sides 1900 mm	159.61	196.08	3.65	131.16	nr	**327.24**
Sum of two sides 2000 mm	168.19	206.62	3.70	132.95	nr	**339.57**
Offset						
Sum of two sides 1100 mm	236.63	290.70	3.78	135.83	nr	**426.53**
Sum of two sides 1200 mm	246.63	302.99	3.89	139.78	nr	**442.77**
Sum of two sides 1300 mm	256.63	315.27	4.00	143.73	nr	**459.00**
Sum of two sides 1400 mm	266.63	327.56	4.11	147.69	nr	**475.25**
Sum of two sides 1500 mm	276.63	339.84	4.23	152.00	nr	**491.84**
Sum of two sides 1600 mm	286.65	352.15	4.34	155.95	nr	**508.10**
Sum of two sides 1700 mm	296.65	364.43	4.45	159.90	nr	**524.33**
Sum of two sides 1800 mm	306.65	376.72	4.56	163.85	nr	**540.57**
Sum of two sides 1900 mm	316.63	388.98	4.67	167.81	nr	**556.79**
Sum of two sides 2000 mm	331.64	407.42	5.53	198.71	nr	**606.13**
90° radius bend						
Sum of two sides 1100 mm	130.05	159.77	3.07	110.32	nr	**270.09**
Sum of two sides 1200 mm	140.76	172.92	3.09	111.03	nr	**283.95**
Sum of two sides 1300 mm	151.48	186.09	3.12	112.11	nr	**298.20**
Sum of two sides 1400 mm	162.20	199.26	3.15	113.19	nr	**312.45**
Sum of two sides 1500 mm	172.90	212.40	3.17	113.90	nr	**326.30**
Sum of two sides 1600 mm	183.61	225.56	3.20	114.99	nr	**340.55**
Sum of two sides 1700 mm	194.31	238.71	3.22	115.71	nr	**354.42**
Sum of two sides 1800 mm	205.03	251.88	3.25	116.78	nr	**368.66**
Sum of two sides 1900 mm	215.74	265.04	3.28	117.86	nr	**382.90**
Sum of two sides 2000 mm	226.13	277.80	3.30	118.59	nr	**396.39**
45° bend						
Sum of two sides 1100 mm	119.03	146.23	1.91	68.64	nr	**214.87**
Sum of two sides 1200 mm	124.01	152.34	1.96	70.43	nr	**222.77**
Sum of two sides 1300 mm	128.99	158.47	2.02	72.60	nr	**231.07**

38 VENTILATION/AIR CONDITIONING SYSTEMS

Item	Net Price £	Material £	Labour hours	Labour £	Unit	Total rate £
DUCTWORK: FIRE RATED – cont						
Extra over fittings – cont						
45° bend – cont						
Sum of two sides 1400 mm	133.98	164.60	2.07	74.38	nr	**238.98**
Sum of two sides 1500 mm	138.96	170.71	2.13	76.53	nr	**247.24**
Sum of two sides 1600 mm	143.93	176.82	2.18	78.34	nr	**255.16**
Sum of two sides 1700 mm	148.91	182.93	2.24	80.49	nr	**263.42**
Sum of two sides 1800 mm	153.89	189.05	2.30	82.66	nr	**271.71**
Sum of two sides 1900 mm	158.85	195.14	2.35	84.45	nr	**279.59**
Sum of two sides 2000 mm	163.86	201.31	2.76	99.17	nr	**300.48**
90° mitre bend						
Sum of two sides 1100 mm	197.77	242.96	3.69	132.59	nr	**375.55**
Sum of two sides 1200 mm	215.17	264.34	3.72	133.68	nr	**398.02**
Sum of two sides 1300 mm	232.51	285.64	3.75	134.76	nr	**420.40**
Sum of two sides 1400 mm	249.90	307.00	3.78	135.83	nr	**442.83**
Sum of two sides 1500 mm	267.30	328.38	3.81	136.90	nr	**465.28**
Sum of two sides 1600 mm	284.66	349.70	3.84	137.98	nr	**487.68**
Sum of two sides 1700 mm	302.05	371.07	3.87	139.06	nr	**510.13**
Sum of two sides 1800 mm	319.43	392.42	3.90	140.14	nr	**532.56**
Sum of two sides 1900 mm	336.80	413.76	3.93	141.21	nr	**554.97**
Sum of two sides 2000 mm	354.19	435.12	3.96	142.30	nr	**577.42**
Branch (side-on shoe)						
Sum of two sides 1100 mm	31.74	39.00	1.50	53.91	nr	**92.91**
Sum of two sides 1200 mm	32.47	39.89	1.55	55.70	nr	**95.59**
Sum of two sides 1300 mm	33.26	40.86	1.59	57.14	nr	**98.00**
Sum of two sides 1400 mm	34.04	41.82	1.63	58.58	nr	**100.40**
Sum of two sides 1500 mm	34.81	42.77	1.68	60.36	nr	**103.13**
Sum of two sides 1600 mm	35.56	43.69	1.72	61.80	nr	**105.49**
Sum of two sides 1700 mm	36.34	44.65	1.77	63.60	nr	**108.25**
Sum of two sides 1800 mm	37.11	45.59	1.81	65.04	nr	**110.63**
Sum of two sides 1900 mm	37.87	46.53	1.86	66.84	nr	**113.37**
Sum of two sides 2000 mm	39.02	47.93	1.90	68.28	nr	**116.21**
Ductwork 1001 to 1250 mm longest side						
Sum of two sides 1300 mm	159.97	196.52	6.38	229.27	m	**425.79**
Sum of two sides 1400 mm	166.88	205.01	6.65	238.96	m	**443.97**
Sum of two sides 1500 mm	173.78	213.49	6.91	248.30	m	**461.79**
Sum of two sides 1600 mm	178.94	219.83	6.91	248.30	m	**468.13**
Sum of two sides 1700 mm	184.11	226.18	7.45	267.70	m	**493.88**
Sum of two sides 1800 mm	190.98	234.62	7.71	277.05	m	**511.67**
Sum of two sides 1900 mm	197.86	243.08	7.98	286.74	m	**529.82**
Sum of two sides 2000 mm	204.76	251.55	8.25	296.45	m	**548.00**
Sum of two sides 2100 mm	211.65	260.01	9.62	345.67	m	**605.68**
Sum of two sides 2200 mm	216.82	266.36	9.62	345.67	m	**612.03**
Sum of two sides 2300 mm	221.98	272.71	10.07	361.85	m	**634.56**
Sum of two sides 2400 mm	228.87	281.17	10.47	376.23	m	**657.40**
Sum of two sides 2500 mm	235.75	289.62	10.87	390.60	m	**680.22**

38 VENTILATION/AIR CONDITIONING SYSTEMS

Item	Net Price £	Material £	Labour hours	Labour £	Unit	Total rate £
Extra over fittings; Ductwork 1001 to 1250 mm longest side						
End cap						
Sum of two sides 1300 mm	35.20	43.24	1.31	47.07	nr	**90.31**
Sum of two sides 1400 mm	37.14	45.62	1.34	48.15	nr	**93.77**
Sum of two sides 1500 mm	39.06	47.99	1.36	48.87	nr	**96.86**
Sum of two sides 1600 mm	41.03	50.41	1.39	49.95	nr	**100.36**
Sum of two sides 1700 mm	42.99	52.82	1.42	51.02	nr	**103.84**
Sum of two sides 1800 mm	44.92	55.19	1.45	52.10	nr	**107.29**
Sum of two sides 1900 mm	46.89	57.60	1.48	53.18	nr	**110.78**
Sum of two sides 2000 mm	48.84	60.00	1.51	54.26	nr	**114.26**
Sum of two sides 2100 mm	50.80	62.41	2.66	95.59	nr	**158.00**
Sum of two sides 2200 mm	52.74	64.80	2.80	100.61	nr	**165.41**
Sum of two sides 2300 mm	54.70	67.20	2.95	106.01	nr	**173.21**
Sum of two sides 2400 mm	56.64	69.58	3.10	111.38	nr	**180.96**
Sum of two sides 2500 mm	58.59	71.98	3.24	116.42	nr	**188.40**
Reducer						
Sum of two sides 1300 mm	125.18	153.78	3.33	119.66	nr	**273.44**
Sum of two sides 1400 mm	130.82	160.71	3.38	121.45	nr	**282.16**
Sum of two sides 1500 mm	136.45	167.63	3.44	123.62	nr	**291.25**
Sum of two sides 1600 mm	142.12	174.59	3.49	125.40	nr	**299.99**
Sum of two sides 1700 mm	147.80	181.58	3.54	127.21	nr	**308.79**
Sum of two sides 1800 mm	153.46	188.53	3.60	129.36	nr	**317.89**
Sum of two sides 1900 mm	159.11	195.47	3.65	131.16	nr	**326.63**
Sum of two sides 2000 mm	164.78	202.43	3.70	132.95	nr	**335.38**
Sum of two sides 2100 mm	170.43	209.37	3.75	134.76	nr	**344.13**
Sum of two sides 2200 mm	176.08	216.31	3.80	136.55	nr	**352.86**
Sum of two sides 2300 mm	181.74	223.27	3.85	138.34	nr	**361.61**
Sum of two sides 2400 mm	187.41	230.23	3.90	140.14	nr	**370.37**
Sum of two sides 2500 mm	193.09	237.22	3.95	141.94	nr	**379.16**
Offset						
Sum of two sides 1300 mm	255.77	314.21	4.00	143.73	nr	**457.94**
Sum of two sides 1400 mm	265.92	326.69	4.11	147.69	nr	**474.38**
Sum of two sides 1500 mm	276.07	339.15	4.23	152.00	nr	**491.15**
Sum of two sides 1600 mm	286.21	351.61	4.34	155.95	nr	**507.56**
Sum of two sides 1700 mm	296.35	364.07	4.45	159.90	nr	**523.97**
Sum of two sides 1800 mm	306.49	376.52	4.56	163.85	nr	**540.37**
Sum of two sides 1900 mm	316.62	388.97	4.67	167.81	nr	**556.78**
Sum of two sides 2000 mm	326.76	401.43	5.53	198.71	nr	**600.14**
Sum of two sides 2100 mm	336.90	413.88	5.76	206.98	nr	**620.86**
Sum of two sides 2200 mm	346.87	426.13	5.99	215.25	nr	**641.38**
Sum of two sides 2300 mm	357.18	438.79	6.22	223.51	nr	**662.30**
Sum of two sides 2400 mm	367.31	451.24	6.45	231.77	nr	**683.01**
Sum of two sides 2500 mm	377.46	463.71	6.68	240.04	nr	**703.75**
90° radius bend						
Sum of two sides 1300 mm	151.27	185.84	3.12	112.11	nr	**297.95**
Sum of two sides 1400 mm	168.33	206.80	3.15	113.19	nr	**319.99**
Sum of two sides 1500 mm	172.78	212.26	3.17	113.90	nr	**326.16**
Sum of two sides 1600 mm	183.50	225.43	3.20	114.99	nr	**340.42**
Sum of two sides 1700 mm	194.25	238.63	3.22	115.71	nr	**354.34**
Sum of two sides 1800 mm	204.98	251.82	3.25	116.78	nr	**368.60**
Sum of two sides 1900 mm	215.72	265.01	3.28	117.86	nr	**382.87**

38 VENTILATION/AIR CONDITIONING SYSTEMS

Item	Net Price £	Material £	Labour hours	Labour £	Unit	Total rate £
DUCTWORK: FIRE RATED – cont						
Extra over fittings – cont						
90° radius bend – cont						
Sum of two sides 2000 mm	226.47	278.22	3.30	118.59	nr	**396.81**
Sum of two sides 2100 mm	237.21	291.42	3.32	119.30	nr	**410.72**
Sum of two sides 2200 mm	247.94	304.59	3.34	120.01	nr	**424.60**
Sum of two sides 2300 mm	258.69	317.80	3.36	120.74	nr	**438.54**
Sum of two sides 2400 mm	269.45	331.02	3.38	121.45	nr	**452.47**
Sum of two sides 2500 mm	280.18	344.20	3.40	122.18	nr	**466.38**
45° bend						
Sum of two sides 1300 mm	122.54	150.54	2.02	72.60	nr	**223.14**
Sum of two sides 1400 mm	128.05	157.31	2.07	74.38	nr	**231.69**
Sum of two sides 1500 mm	133.56	164.08	2.13	76.53	nr	**240.61**
Sum of two sides 1600 mm	139.08	170.86	2.18	78.34	nr	**249.20**
Sum of two sides 1700 mm	144.59	177.63	2.24	80.49	nr	**258.12**
Sum of two sides 1800 mm	150.11	184.41	2.30	82.66	nr	**267.07**
Sum of two sides 1900 mm	155.62	191.18	2.35	84.45	nr	**275.63**
Sum of two sides 2000 mm	161.15	197.98	2.76	99.17	nr	**297.15**
Sum of two sides 2100 mm	166.67	204.75	2.89	103.85	nr	**308.60**
Sum of two sides 2200 mm	172.20	211.54	3.01	108.16	nr	**319.70**
Sum of two sides 2300 mm	177.71	218.32	3.13	112.47	nr	**330.79**
Sum of two sides 2400 mm	183.23	225.10	3.26	117.15	nr	**342.25**
Sum of two sides 2500 mm	188.74	231.87	3.37	121.10	nr	**352.97**
90° mitre bend						
Sum of two sides 1300 mm	232.00	285.01	3.75	134.76	nr	**419.77**
Sum of two sides 1400 mm	249.21	306.16	3.78	135.83	nr	**441.99**
Sum of two sides 1500 mm	267.11	328.15	3.81	136.90	nr	**465.05**
Sum of two sides 1600 mm	284.64	349.68	3.84	137.98	nr	**487.66**
Sum of two sides 1700 mm	302.17	371.22	3.87	139.06	nr	**510.28**
Sum of two sides 1800 mm	319.70	392.75	3.90	140.14	nr	**532.89**
Sum of two sides 1900 mm	337.24	414.30	3.93	141.21	nr	**555.51**
Sum of two sides 2000 mm	354.80	435.88	3.96	142.30	nr	**578.18**
Sum of two sides 2100 mm	372.33	457.41	3.99	143.38	nr	**600.79**
Sum of two sides 2200 mm	389.87	478.96	4.02	144.45	nr	**623.41**
Sum of two sides 2300 mm	407.42	500.51	4.05	145.53	nr	**646.04**
Sum of two sides 2400 mm	424.93	522.03	4.08	146.61	nr	**668.64**
Sum of two sides 2500 mm	442.46	543.56	4.11	147.69	nr	**691.25**
Branch (side-on shoe)						
Sum of two sides 1300 mm	33.32	40.93	1.59	57.14	nr	**98.07**
Sum of two sides 1400 mm	34.09	41.88	1.63	58.58	nr	**100.46**
Sum of two sides 1500 mm	34.84	42.80	1.68	60.36	nr	**103.16**
Sum of two sides 1600 mm	35.59	43.72	1.72	61.80	nr	**105.52**
Sum of two sides 1700 mm	36.37	44.68	1.77	63.60	nr	**108.28**
Sum of two sides 1800 mm	37.14	45.62	1.81	65.04	nr	**110.66**
Sum of two sides 1900 mm	37.89	46.55	1.86	66.84	nr	**113.39**
Sum of two sides 2000 mm	38.67	47.50	1.90	68.28	nr	**115.78**
Sum of two sides 2100 mm	39.42	48.43	2.58	92.72	nr	**141.15**

38 VENTILATION/AIR CONDITIONING SYSTEMS

Item	Net Price £	Material £	Labour hours	Labour £	Unit	**Total rate £**
Sum of two sides 2200 mm	40.17	49.35	2.61	93.79	nr	**143.14**
Sum of two sides 2300 mm	40.96	50.32	2.64	94.87	nr	**145.19**
Sum of two sides 2400 mm	41.72	51.25	2.88	103.49	nr	**154.74**
Sum of two sides 2500 mm	42.47	52.17	2.91	104.57	nr	**156.74**
Ductwork 1251 to 2000 mm longest side						
Sum of two sides 1700 mm	175.42	215.50	7.45	267.70	m	**483.20**
Sum of two sides 1800 mm	186.67	229.32	7.71	277.05	m	**506.37**
Sum of two sides 1900 mm	197.91	243.13	7.98	286.74	m	**529.87**
Sum of two sides 2000 mm	209.15	256.95	8.25	296.45	m	**553.40**
Sum of two sides 2100 mm	220.40	270.76	9.62	345.67	m	**616.43**
Sum of two sides 2200 mm	231.64	284.57	9.66	347.13	m	**631.70**
Sum of two sides 2300 mm	242.86	298.36	10.07	361.85	m	**660.21**
Sum of two sides 2400 mm	254.09	312.15	10.47	376.23	m	**688.38**
Sum of two sides 2500 mm	265.33	325.96	10.87	390.60	m	**716.56**
Sum of two sides 2600 mm	276.60	339.80	11.27	404.97	m	**744.77**
Sum of two sides 2700 mm	287.82	353.59	11.67	419.35	m	**772.94**
Sum of two sides 2800 mm	299.07	367.41	12.08	434.08	m	**801.49**
Sum of two sides 2900 mm	310.29	381.19	12.48	448.46	m	**829.65**
Sum of two sides 3000 mm	321.53	395.00	12.88	462.83	m	**857.83**
Sum of two sides 3100 mm	332.78	408.82	13.26	476.49	m	**885.31**
Sum of two sides 3200 mm	344.02	422.63	13.69	491.93	m	**914.56**
Sum of two sides 3300 mm	365.49	449.00	14.09	506.30	m	**955.30**
Sum of two sides 3400 mm	377.03	463.19	14.49	520.67	m	**983.86**
Sum of two sides 3500 mm	388.59	477.38	14.89	535.05	m	**1012.43**
Sum of two sides 3600 mm	400.13	491.56	15.29	549.42	m	**1040.98**
Sum of two sides 3700 mm	411.73	505.81	15.69	563.80	m	**1069.61**
Sum of two sides 3800 mm	423.27	519.99	16.09	578.17	m	**1098.16**
Sum of two sides 3900 mm	434.84	534.20	16.49	592.55	m	**1126.75**
Sum of two sides 4000 mm	446.39	548.39	16.89	606.92	m	**1155.31**
Extra over fittings; Ductwork 1251 to 2000 mm longest side						
End cap						
Sum of two sides 1700 mm	63.85	78.43	1.42	51.02	nr	**129.45**
Sum of two sides 1800 mm	67.97	83.50	1.45	52.10	nr	**135.60**
Sum of two sides 1900 mm	72.07	88.54	1.48	53.18	nr	**141.72**
Sum of two sides 2000 mm	76.18	93.59	1.51	54.26	nr	**147.85**
Sum of two sides 2100 mm	80.28	98.63	2.66	95.59	nr	**194.22**
Sum of two sides 2200 mm	84.40	103.69	2.80	100.61	nr	**204.30**
Sum of two sides 2300 mm	88.49	108.71	2.95	106.01	nr	**214.72**
Sum of two sides 2400 mm	92.60	113.76	3.10	111.38	nr	**225.14**
Sum of two sides 2500 mm	96.70	118.80	3.24	116.42	nr	**235.22**
Sum of two sides 2600 mm	100.82	123.86	3.39	121.82	nr	**245.68**
Sum of two sides 2700 mm	104.90	128.87	3.54	127.21	nr	**256.08**
Sum of two sides 2800 mm	109.02	133.93	3.68	132.24	nr	**266.17**
Sum of two sides 2900 mm	113.15	139.01	3.83	137.63	nr	**276.64**
Sum of two sides 3000 mm	117.25	144.04	3.98	143.02	nr	**287.06**
Sum of two sides 3100 mm	121.35	149.08	4.12	148.05	nr	**297.13**

38 VENTILATION/AIR CONDITIONING SYSTEMS

Item	Net Price £	Material £	Labour hours	Labour £	Unit	Total rate £
DUCTWORK: FIRE RATED – cont						
Extra over fittings – cont						
End cap – cont						
Sum of two sides 3200 mm	125.47	154.14	4.27	153.44	nr	**307.58**
Sum of two sides 3300 mm	133.28	163.74	4.42	158.82	nr	**322.56**
Sum of two sides 3400 mm	137.51	168.93	4.57	164.21	nr	**333.14**
Sum of two sides 3500 mm	140.99	173.21	4.72	169.61	nr	**342.82**
Sum of two sides 3600 mm	143.82	176.68	4.87	175.00	nr	**351.68**
Sum of two sides 3700 mm	148.06	181.89	5.02	180.39	nr	**362.28**
Sum of two sides 3800 mm	222.29	273.08	5.17	185.78	nr	**458.86**
Sum of two sides 3900 mm	158.62	194.87	5.32	191.16	nr	**386.03**
Sum of two sides 4000 mm	162.85	200.06	5.47	196.56	nr	**396.62**
Reducer						
Sum of two sides 1700 mm	137.91	169.42	3.54	127.21	nr	**296.63**
Sum of two sides 1800 mm	154.70	190.05	3.60	129.36	nr	**319.41**
Sum of two sides 1900 mm	171.48	210.66	3.65	131.16	nr	**341.82**
Sum of two sides 2000 mm	205.06	251.92	2.92	104.93	nr	**356.85**
Sum of two sides 2100 mm	188.29	231.31	3.70	132.95	nr	**364.26**
Sum of two sides 2200 mm	221.85	272.54	3.10	111.38	nr	**383.92**
Sum of two sides 2300 mm	238.65	293.18	3.29	118.22	nr	**411.40**
Sum of two sides 2400 mm	255.44	313.80	3.48	125.04	nr	**438.84**
Sum of two sides 2500 mm	272.22	334.43	3.66	131.52	nr	**465.95**
Sum of two sides 2600 mm	289.02	355.06	3.85	138.34	nr	**493.40**
Sum of two sides 2700 mm	305.82	375.70	4.03	144.82	nr	**520.52**
Sum of two sides 2800 mm	322.61	396.32	4.22	151.64	nr	**547.96**
Sum of two sides 2900 mm	339.40	416.96	4.40	158.11	nr	**575.07**
Sum of two sides 3000 mm	364.55	447.85	4.59	164.93	nr	**612.78**
Sum of two sides 3100 mm	364.55	447.85	4.78	171.76	nr	**619.61**
Sum of two sides 3200 mm	381.36	468.50	4.96	178.23	nr	**646.73**
Sum of two sides 3300 mm	418.23	513.80	5.15	185.06	nr	**698.86**
Sum of two sides 3400 mm	435.50	535.01	5.34	191.88	nr	**726.89**
Sum of two sides 3500 mm	452.78	556.24	5.53	198.71	nr	**754.95**
Sum of two sides 3600 mm	470.04	577.45	5.72	205.54	nr	**782.99**
Sum of two sides 3700 mm	487.33	598.69	5.91	212.37	nr	**811.06**
Sum of two sides 3800 mm	504.59	619.89	6.10	219.20	nr	**839.09**
Sum of two sides 3900 mm	521.88	641.13	6.29	226.02	nr	**867.15**
Sum of two sides 4000 mm	539.16	662.36	6.48	232.85	nr	**895.21**
Offset						
Sum of two sides 1700 mm	137.91	169.42	4.45	159.90	nr	**329.32**
Sum of two sides 1800 mm	149.27	183.38	4.56	163.85	nr	**347.23**
Sum of two sides 1900 mm	160.65	197.36	4.67	167.81	nr	**365.17**
Sum of two sides 2000 mm	171.96	211.25	5.53	198.71	nr	**409.96**
Sum of two sides 2100 mm	183.34	225.24	5.76	206.98	nr	**432.22**
Sum of two sides 2200 mm	194.70	239.19	5.99	215.25	nr	**454.44**
Sum of two sides 2300 mm	206.08	253.17	6.22	223.51	nr	**476.68**
Sum of two sides 2400 mm	217.43	267.11	6.45	231.77	nr	**498.88**
Sum of two sides 2500 mm	228.79	281.06	6.68	240.04	nr	**521.10**
Sum of two sides 2600 mm	240.16	295.04	6.91	248.30	nr	**543.34**
Sum of two sides 2700 mm	251.53	309.00	7.14	256.56	nr	**565.56**

38 VENTILATION/AIR CONDITIONING SYSTEMS

Item	Net Price £	Material £	Labour hours	Labour £	Unit	**Total rate £**
Sum of two sides 2800 mm	262.90	322.97	7.37	264.83	nr	**587.80**
Sum of two sides 2900 mm	274.23	336.89	7.60	273.09	nr	**609.98**
Sum of two sides 3000 mm	285.61	350.87	7.83	281.37	nr	**632.24**
Sum of two sides 3100 mm	296.99	364.85	8.06	289.62	nr	**654.47**
Sum of two sides 3200 mm	308.35	378.81	8.29	297.88	nr	**676.69**
Sum of two sides 3300 mm	328.89	404.04	8.52	306.16	nr	**710.20**
Sum of two sides 3400 mm	340.56	418.38	8.75	314.42	nr	**732.80**
Sum of two sides 3500 mm	352.27	432.77	8.98	322.69	nr	**755.46**
Sum of two sides 3600 mm	363.96	447.12	9.21	330.95	nr	**778.07**
Sum of two sides 3700 mm	375.65	461.49	9.44	339.21	nr	**800.70**
Sum of two sides 3800 mm	387.33	475.84	9.67	347.48	nr	**823.32**
Sum of two sides 3900 mm	399.02	490.19	9.90	355.74	nr	**845.93**
Sum of two sides 4000 mm	410.73	504.58	10.13	364.00	nr	**868.58**
90° radius bend						
Sum of two sides 1700 mm	191.92	235.78	3.22	115.71	nr	**351.49**
Sum of two sides 1800 mm	209.26	257.07	3.25	116.78	nr	**373.85**
Sum of two sides 1900 mm	220.85	271.31	3.32	119.30	nr	**390.61**
Sum of two sides 2000 mm	226.61	278.39	3.28	117.86	nr	**396.25**
Sum of two sides 2100 mm	243.95	299.69	3.30	118.59	nr	**418.28**
Sum of two sides 2200 mm	279.12	342.90	3.32	119.30	nr	**462.20**
Sum of two sides 2300 mm	295.99	363.63	3.36	120.74	nr	**484.37**
Sum of two sides 2400 mm	313.30	384.89	3.38	121.45	nr	**506.34**
Sum of two sides 2500 mm	330.67	406.22	3.40	122.18	nr	**528.40**
Sum of two sides 2600 mm	348.02	427.54	3.42	122.89	nr	**550.43**
Sum of two sides 2700 mm	365.38	448.86	3.44	123.62	nr	**572.48**
Sum of two sides 2800 mm	382.72	470.17	3.46	124.32	nr	**594.49**
Sum of two sides 2900 mm	400.03	491.44	3.48	125.04	nr	**616.48**
Sum of two sides 3000 mm	417.39	512.77	3.50	125.77	nr	**638.54**
Sum of two sides 3100 mm	434.73	534.06	3.52	126.48	nr	**660.54**
Sum of two sides 3200 mm	437.14	537.02	3.54	127.21	nr	**664.23**
Sum of two sides 3300 mm	451.02	554.07	3.56	127.92	nr	**681.99**
Sum of two sides 3400 mm	664.37	816.18	3.58	128.65	nr	**944.83**
Sum of two sides 3500 mm	466.75	573.40	3.60	129.36	nr	**702.76**
Sum of two sides 3600 mm	482.81	593.13	3.62	130.08	nr	**723.21**
Sum of two sides 3700 mm	498.85	612.83	3.64	130.80	nr	**743.63**
Sum of two sides 3800 mm	514.92	632.58	3.66	131.52	nr	**764.10**
Sum of two sides 3900 mm	530.99	652.32	3.68	132.24	nr	**784.56**
Sum of two sides 4000 mm	547.04	672.04	3.70	132.95	nr	**804.99**
45° bend						
Sum of two sides 1700 mm	142.27	174.78	2.24	80.49	nr	**255.27**
Sum of two sides 1800 mm	158.00	194.10	2.30	82.66	nr	**276.76**
Sum of two sides 1900 mm	173.68	213.37	2.35	84.45	nr	**297.82**
Sum of two sides 2000 mm	189.43	232.71	2.76	99.17	nr	**331.88**
Sum of two sides 2100 mm	243.95	299.69	3.01	108.16	nr	**407.85**
Sum of two sides 2200 mm	236.60	290.66	3.13	112.47	nr	**403.13**
Sum of two sides 2300 mm	261.29	321.00	2.89	103.85	nr	**424.85**
Sum of two sides 2400 mm	252.29	309.94	3.26	117.15	nr	**427.09**
Sum of two sides 2500 mm	268.01	329.25	3.37	121.10	nr	**450.35**
Sum of two sides 2600 mm	283.70	348.53	3.49	125.40	nr	**473.93**
Sum of two sides 2700 mm	299.45	367.88	3.62	130.08	nr	**497.96**

38 VENTILATION/AIR CONDITIONING SYSTEMS

Item	Net Price £	Material £	Labour hours	Labour £	Unit	Total rate £
DUCTWORK: FIRE RATED – cont						
Extra over fittings – cont						
45° bend – cont						
Sum of two sides 2800 mm	315.16	387.18	3.74	134.39	nr	**521.57**
Sum of two sides 2900 mm	330.87	406.48	3.86	138.70	nr	**545.18**
Sum of two sides 3000 mm	346.58	425.77	3.98	143.02	nr	**568.79**
Sum of two sides 3100 mm	362.31	445.10	4.10	147.33	nr	**592.43**
Sum of two sides 3200 mm	378.01	464.38	4.22	151.64	nr	**616.02**
Sum of two sides 3300 mm	405.06	497.62	4.34	155.95	nr	**653.57**
Sum of two sides 3400 mm	421.22	517.47	4.46	160.27	nr	**677.74**
Sum of two sides 3500 mm	437.40	537.35	4.58	164.58	nr	**701.93**
Sum of two sides 3600 mm	453.59	557.24	4.70	168.88	nr	**726.12**
Sum of two sides 3700 mm	469.72	577.05	4.82	173.21	nr	**750.26**
Sum of two sides 3800 mm	485.90	596.92	4.94	177.52	nr	**774.44**
Sum of two sides 3900 mm	502.05	616.77	5.06	181.82	nr	**798.59**
Sum of two sides 4000 mm	518.24	636.66	5.18	186.13	nr	**822.79**
90° mitre bend						
Sum of two sides 1700 mm	287.44	353.12	3.87	139.06	nr	**492.18**
Sum of two sides 1800 mm	337.12	414.15	3.90	140.14	nr	**554.29**
Sum of two sides 1900 mm	386.81	475.20	3.93	141.21	nr	**616.41**
Sum of two sides 2000 mm	436.45	536.18	3.96	142.30	nr	**678.48**
Sum of two sides 2100 mm	486.14	597.22	3.82	137.26	nr	**734.48**
Sum of two sides 2200 mm	535.81	658.24	3.83	137.63	nr	**795.87**
Sum of two sides 2300 mm	585.50	719.28	3.83	137.63	nr	**856.91**
Sum of two sides 2400 mm	635.17	780.31	3.84	137.98	nr	**918.29**
Sum of two sides 2500 mm	684.84	841.32	3.85	138.34	nr	**979.66**
Sum of two sides 2600 mm	734.49	902.32	3.85	138.34	nr	**1040.66**
Sum of two sides 2700 mm	784.17	963.35	3.86	138.70	nr	**1102.05**
Sum of two sides 2800 mm	833.85	1024.38	3.86	138.70	nr	**1163.08**
Sum of two sides 2900 mm	883.52	1085.41	3.87	139.06	nr	**1224.47**
Sum of two sides 3000 mm	933.19	1146.42	3.87	139.06	nr	**1285.48**
Sum of two sides 3100 mm	982.88	1207.47	3.88	139.42	nr	**1346.89**
Sum of two sides 3200 mm	1032.54	1268.47	3.88	139.42	nr	**1407.89**
Sum of two sides 3300 mm	1113.35	1367.75	3.89	139.78	nr	**1507.53**
Sum of two sides 3400 mm	1164.45	1430.53	3.90	140.14	nr	**1570.67**
Sum of two sides 3500 mm	1215.55	1493.30	3.91	140.50	nr	**1633.80**
Sum of two sides 3600 mm	1266.65	1556.08	3.92	140.86	nr	**1696.94**
Sum of two sides 3700 mm	1317.77	1618.88	3.93	141.21	nr	**1760.09**
Sum of two sides 3800 mm	1368.85	1681.63	3.94	141.58	nr	**1823.21**
Sum of two sides 3900 mm	1419.97	1744.43	3.95	141.94	nr	**1886.37**
Sum of two sides 4000 mm	1471.07	1807.21	3.96	142.30	nr	**1949.51**
Branch (side-on shoe)						
Sum of two sides 1700 mm	32.58	40.03	1.77	63.60	nr	**103.63**
Sum of two sides 1800 mm	35.60	43.73	1.81	65.04	nr	**108.77**
Sum of two sides 1900 mm	38.66	47.49	1.86	66.84	nr	**114.33**
Sum of two sides 2000 mm	41.69	51.22	1.90	68.28	nr	**119.50**
Sum of two sides 2100 mm	44.71	54.93	2.58	92.72	nr	**147.65**
Sum of two sides 2200 mm	47.73	58.63	2.61	93.79	nr	**152.42**
Sum of two sides 2300 mm	50.78	62.38	2.65	95.22	nr	**157.60**

38 VENTILATION/AIR CONDITIONING SYSTEMS

Item	Net Price £	Material £	Labour hours	Labour £	Unit	**Total rate £**
Sum of two sides 2400 mm	53.80	66.10	2.68	96.31	nr	**162.41**
Sum of two sides 2500 mm	56.84	69.83	2.71	97.39	nr	**167.22**
Sum of two sides 2600 mm	59.88	73.56	2.75	98.82	nr	**172.38**
Sum of two sides 2700 mm	62.88	77.25	2.78	99.89	nr	**177.14**
Sum of two sides 2800 mm	65.93	81.00	2.81	100.97	nr	**181.97**
Sum of two sides 2900 mm	68.97	84.72	2.84	102.05	nr	**186.77**
Sum of two sides 3000 mm	71.99	88.44	2.87	103.13	nr	**191.57**
Sum of two sides 3100 mm	75.01	92.15	2.90	104.21	nr	**196.36**
Sum of two sides 3200 mm	78.06	95.90	2.93	105.28	nr	**201.18**
Sum of two sides 3300 mm	83.41	102.47	2.93	105.28	nr	**207.75**
Sum of two sides 3400 mm	86.54	106.31	3.00	107.80	nr	**214.11**
Sum of two sides 3500 mm	89.66	110.15	3.03	108.88	nr	**219.03**
Sum of two sides 3600 mm	92.75	113.95	3.06	109.95	nr	**223.90**
Sum of two sides 3700 mm	95.89	117.80	3.09	111.03	nr	**228.83**
Sum of two sides 3800 mm	99.02	121.64	3.12	112.11	nr	**233.75**
Sum of two sides 3900 mm	102.14	125.47	3.15	113.19	nr	**238.66**
Sum of two sides 4000 mm	105.23	129.28	3.18	114.27	nr	**243.55**
Rectangular section ductwork to BS476 Part 24 (ISO 6944:1985) (Duraduct LT), ducts 'Type A' and 'Type B'; manufactured from 6 mm thick laminate fire board consisting of steel circular hole punched facings pressed to a fibre cement core; provides upto 4 hours stability, 4 hours integrity and 32 minutes insulation; including all necessary stiffeners, joints and supports in the running length						
Ductwork up to 600 mm longest side						
Sum of two sides 200 mm	237.17	291.36	4.00	143.73	m	**435.09**
Sum of two sides 400 mm	242.96	298.47	4.00	143.73	m	**442.20**
Sum of two sides 600 mm	253.89	311.90	4.00	143.73	m	**455.63**
Sum of two sides 800 mm	335.03	411.59	5.50	197.64	m	**609.23**
Sum of two sides 1000 mm	451.35	554.48	5.50	197.64	m	**752.12**
Sum of two sides 1200 mm	466.53	573.13	6.00	215.61	m	**788.74**
Extra over fittings; Ductwork up to 600 mm longest side						
End cap						
Sum of two sides 200 mm	56.28	69.14	0.81	29.11	m	**98.25**
Sum of two sides 400 mm	56.28	69.14	0.87	31.27	m	**100.41**
Sum of two sides 600 mm	56.28	69.14	0.93	33.41	m	**102.55**
Sum of two sides 800 mm	90.92	111.70	0.98	35.21	m	**146.91**
Sum of two sides 1000 mm	90.92	111.70	1.22	43.85	m	**155.55**
Sum of two sides 1200 mm	138.52	170.17	1.28	45.99	m	**216.16**
Reducer						
Sum of two sides 200 mm	49.77	61.14	2.23	80.14	m	**141.28**
Sum of two sides 400 mm	49.77	61.14	2.51	90.19	m	**151.33**
Sum of two sides 600 mm	49.77	61.14	2.79	100.25	m	**161.39**

38 VENTILATION/AIR CONDITIONING SYSTEMS

Item	Net Price £	Material £	Labour hours	Labour £	Unit	Total rate £
DUCTWORK: FIRE RATED – cont						
Extra over fittings – cont						
Reducer – cont						
Sum of two sides 800 mm	97.40	119.66	3.01	108.16	m	**227.82**
Sum of two sides 1000 mm	97.40	119.66	3.17	113.90	m	**233.56**
Sum of two sides 1200 mm	123.38	151.57	3.28	117.86	m	**269.43**
Offset						
Sum of two sides 200 mm	49.77	61.14	2.95	106.01	m	**167.15**
Sum of two sides 400 mm	49.77	61.14	3.26	117.15	m	**178.29**
Sum of two sides 600 mm	49.77	61.14	3.57	128.28	m	**189.42**
Sum of two sides 800 mm	97.40	119.66	3.23	116.07	m	**235.73**
Sum of two sides 1000 mm	97.40	119.66	3.67	131.88	m	**251.54**
Sum of two sides 1200 mm	123.38	151.57	3.89	139.78	m	**291.35**
90° radius bend						
Sum of two sides 200 mm	166.67	204.75	2.06	74.02	m	**278.77**
Sum of two sides 400 mm	166.67	204.75	2.36	84.81	m	**289.56**
Sum of two sides 600 mm	166.67	204.75	2.66	95.59	m	**300.34**
Sum of two sides 800 mm	194.81	239.33	2.97	106.72	m	**346.05**
Sum of two sides 1000 mm	194.81	239.33	3.04	109.24	m	**348.57**
Sum of two sides 1200 mm	220.79	271.24	3.09	111.03	m	**382.27**
45° radius bend						
Sum of two sides 200 mm	166.67	204.75	1.52	54.62	m	**259.37**
Sum of two sides 400 mm	166.67	204.75	1.66	59.64	m	**264.39**
Sum of two sides 600 mm	166.67	204.75	1.80	64.68	m	**269.43**
Sum of two sides 800 mm	194.81	239.33	1.93	69.35	m	**308.68**
Sum of two sides 1000 mm	194.81	239.33	2.05	73.66	m	**312.99**
Sum of two sides 1200 mm	220.79	271.24	2.17	77.97	m	**349.21**
90° mitre bend						
Sum of two sides 200 mm	166.67	204.75	2.47	88.76	m	**293.51**
Sum of two sides 400 mm	166.67	204.75	2.84	102.05	m	**306.80**
Sum of two sides 600 mm	166.67	204.75	3.20	114.99	m	**319.74**
Sum of two sides 800 mm	194.81	239.33	3.56	127.92	m	**367.25**
Sum of two sides 1000 mm	194.81	239.33	3.66	131.52	m	**370.85**
Sum of two sides 1200 mm	220.79	271.24	3.72	133.68	m	**404.92**
Branch						
Sum of two sides 200 mm	60.61	74.46	0.98	35.21	m	**109.67**
Sum of two sides 400 mm	60.61	74.46	1.14	40.96	m	**115.42**
Sum of two sides 600 mm	60.61	74.46	1.30	46.70	m	**121.16**
Sum of two sides 800 mm	75.75	93.06	1.46	52.46	m	**145.52**
Sum of two sides 1000 mm	82.60	101.47	1.46	52.46	m	**153.93**
Sum of two sides 1200 mm	101.72	124.96	1.55	55.70	m	**180.66**
Ductwork 601 to 1000 mm longest side						
Sum of two sides 1000 mm	340.64	418.48	6.00	215.61	m	**634.09**
Sum of two sides 1100 mm	474.34	582.73	6.00	215.61	m	**798.34**
Sum of two sides 1300 mm	482.26	592.45	6.00	215.61	m	**808.06**
Sum of two sides 1500 mm	493.51	606.28	6.00	215.61	m	**821.89**
Sum of two sides 1700 mm	596.43	732.71	6.00	215.61	m	**948.32**
Sum of two sides 1900 mm	683.50	839.67	6.00	215.61	m	**1055.28**

38 VENTILATION/AIR CONDITIONING SYSTEMS

Item	Net Price £	Material £	Labour hours	Labour £	Unit	Total rate £
Extra over fittings; Dutcwork 601 to 1000 mm longest side						
End cap						
Sum of two sides 1000 mm	92.44	113.56	1.25	44.92	m	**158.48**
Sum of two sides 1100 mm	140.84	173.02	1.25	44.92	m	**217.94**
Sum of two sides 1300 mm	140.84	173.02	1.31	47.07	m	**220.09**
Sum of two sides 1500 mm	140.84	173.02	1.36	48.87	m	**221.89**
Sum of two sides 1700 mm	248.69	305.52	1.42	51.02	m	**356.54**
Sum of two sides 1900 mm	248.69	305.52	1.48	53.18	m	**358.70**
Reducer						
Sum of two sides 1000 mm	99.03	121.66	3.22	115.71	m	**237.37**
Sum of two sides 1100 mm	125.45	154.12	3.22	115.71	m	**269.83**
Sum of two sides 1300 mm	125.45	154.12	3.33	119.66	m	**273.78**
Sum of two sides 1500 mm	125.45	154.12	3.44	123.62	m	**277.74**
Sum of two sides 1700 mm	154.06	189.26	3.54	127.21	m	**316.47**
Sum of two sides 1900 mm	154.06	189.26	3.65	131.16	m	**320.42**
Offset						
Sum of two sides 1000 mm	99.03	121.66	3.78	135.83	m	**257.49**
Sum of two sides 1100 mm	125.45	154.12	3.78	135.83	m	**289.95**
Sum of two sides 1300 mm	125.45	154.12	4.00	143.73	m	**297.85**
Sum of two sides 1500 mm	125.45	154.12	4.23	152.00	m	**306.12**
Sum of two sides 1700 mm	154.06	189.26	4.45	159.90	m	**349.16**
Sum of two sides 1900 mm	154.06	189.26	4.67	167.81	m	**357.07**
90° radius bend						
Sum of two sides 1000 mm	198.07	243.33	3.78	135.83	m	**379.16**
Sum of two sides 1100 mm	224.48	275.77	3.07	110.32	m	**386.09**
Sum of two sides 1300 mm	224.48	275.77	3.12	112.11	m	**387.88**
Sum of two sides 1500 mm	224.48	275.77	3.17	113.90	m	**389.67**
Sum of two sides 1700 mm	294.90	362.28	3.22	115.71	m	**477.99**
Sum of two sides 1900 mm	294.90	362.28	3.28	117.86	m	**480.14**
45° bend						
Sum of two sides 1000 mm	224.48	275.77	1.91	68.64	m	**344.41**
Sum of two sides 1100 mm	224.48	275.77	2.00	71.87	m	**347.64**
Sum of two sides 1300 mm	224.48	275.77	2.02	72.60	m	**348.37**
Sum of two sides 1500 mm	224.48	275.77	2.13	76.53	m	**352.30**
Sum of two sides 1700 mm	294.90	362.28	2.24	80.49	m	**442.77**
Sum of two sides 1900 mm	294.90	362.28	2.35	84.45	m	**446.73**
90° mitre bend						
Sum of two sides 1000 mm	198.07	243.33	3.78	135.83	m	**379.16**
Sum of two sides 1100 mm	224.48	275.77	3.69	132.59	m	**408.36**
Sum of two sides 1300 mm	224.48	275.77	3.75	134.76	m	**410.53**
Sum of two sides 1500 mm	224.48	275.77	3.81	136.90	m	**412.67**
Sum of two sides 1700 mm	294.90	362.28	3.87	139.06	m	**501.34**
Sum of two sides 1900 mm	294.90	362.28	3.93	141.21	m	**503.49**
Branch						
Sum of two sides 1000 mm	80.08	98.37	2.38	85.51	m	**183.88**
Sum of two sides 1100 mm	103.42	127.05	1.50	53.91	m	**180.96**
Sum of two sides 1300 mm	103.42	127.05	1.59	57.14	m	**184.19**
Sum of two sides 1500 mm	103.42	127.05	1.68	60.36	m	**187.41**
Sum of two sides 1700 mm	132.05	162.23	1.77	63.60	m	**225.83**
Sum of two sides 1900 mm	132.05	162.23	1.86	66.84	m	**229.07**

38 VENTILATION/AIR CONDITIONING SYSTEMS

Item	Net Price £	Material £	Labour hours	Labour £	Unit	Total rate £
DUCTWORK: FIRE RATED – cont						
Ductwork 1001 to 1250 mm longest side						
Sum of two sides 1300 mm	474.34	582.73	6.00	215.61	m	**798.34**
Sum of two sides 1500 mm	504.16	619.36	6.00	215.61	m	**834.97**
Sum of two sides 1700 mm	596.43	732.71	6.00	215.61	m	**948.32**
Sum of two sides 1900 mm	606.37	744.92	6.00	215.61	m	**960.53**
Sum of two sides 2100 mm	794.61	976.17	6.50	233.57	m	**1209.74**
Sum of two sides 2300 mm	837.78	1029.21	6.50	233.57	m	**1262.78**
Sum of two sides 2500 mm	997.34	1225.23	6.50	233.57	m	**1458.80**
Extra over fittings; Ductwork 1001 to 1250 mm longest side						
End cap						
Sum of two sides 1300 mm	140.84	173.02	1.31	47.07	m	**220.09**
Sum of two sides 1500 mm	140.84	173.02	1.36	48.87	m	**221.89**
Sum of two sides 1700 mm	248.69	305.52	1.42	51.02	m	**356.54**
Sum of two sides 1900 mm	248.69	305.52	1.48	53.18	m	**358.70**
Sum of two sides 2100 mm	338.91	416.35	2.66	95.59	m	**511.94**
Sum of two sides 2300 mm	338.91	416.35	2.95	106.01	m	**522.36**
Sum of two sides 2500 mm	338.91	416.35	3.24	116.42	m	**532.77**
Reducer						
Sum of two sides 1300 mm	125.45	154.12	3.33	119.66	m	**273.78**
Sum of two sides 1500 mm	125.45	154.12	3.44	123.62	m	**277.74**
Sum of two sides 1700 mm	154.06	189.26	3.54	127.21	m	**316.47**
Sum of two sides 1900 mm	154.06	189.26	3.65	131.16	m	**320.42**
Sum of two sides 2100 mm	182.65	224.39	3.75	134.76	m	**359.15**
Sum of two sides 2300 mm	182.65	224.39	3.85	138.34	m	**362.73**
Sum of two sides 2500 mm	182.65	224.39	3.95	141.94	m	**366.33**
Offset						
Sum of two sides 1300 mm	125.45	154.12	4.00	143.73	m	**297.85**
Sum of two sides 1500 mm	125.45	154.12	4.23	152.00	m	**306.12**
Sum of two sides 1700 mm	154.06	189.26	4.45	159.90	m	**349.16**
Sum of two sides 1900 mm	154.06	189.26	4.67	167.81	m	**357.07**
Sum of two sides 2100 mm	182.65	224.39	5.76	206.98	m	**431.37**
Sum of two sides 2300 mm	182.65	224.39	6.22	223.51	m	**447.90**
Sum of two sides 2500 mm	182.65	224.39	6.68	240.04	m	**464.43**
90° radius bend						
Sum of two sides 1300 mm	224.48	275.77	3.12	112.11	m	**387.88**
Sum of two sides 1500 mm	224.48	275.77	3.17	113.90	m	**389.67**
Sum of two sides 1700 mm	294.90	362.28	3.22	115.71	m	**477.99**
Sum of two sides 1900 mm	294.90	362.28	3.28	117.86	m	**480.14**
Sum of two sides 2100 mm	301.50	370.40	3.32	119.30	m	**489.70**
Sum of two sides 2300 mm	301.50	370.40	3.36	120.74	m	**491.14**
Sum of two sides 2500 mm	301.50	370.40	3.40	122.18	m	**492.58**
45° bend						
Sum of two sides 1300 mm	224.48	275.77	2.02	72.60	m	**348.37**
Sum of two sides 1500 mm	224.48	275.77	2.13	76.53	m	**352.30**
Sum of two sides 1700 mm	294.90	362.28	2.24	80.49	m	**442.77**

38 VENTILATION/AIR CONDITIONING SYSTEMS

Item	Net Price £	Material £	Labour hours	Labour £	Unit	Total rate £
Sum of two sides 1900 mm	294.90	362.28	2.35	84.45	m	**446.73**
Sum of two sides 2100 mm	301.50	370.40	2.89	103.85	m	**474.25**
Sum of two sides 2300 mm	301.50	370.40	3.13	112.47	m	**482.87**
Sum of two sides 2500 mm	301.50	370.40	3.37	121.10	m	**491.50**
90° mitre bend						
Sum of two sides 1300 mm	224.48	275.77	3.75	134.76	m	**410.53**
Sum of two sides 1500 mm	224.48	275.77	3.81	136.90	m	**412.67**
Sum of two sides 1700 mm	294.90	362.28	3.87	139.06	m	**501.34**
Sum of two sides 1900 mm	294.90	362.28	3.93	141.21	m	**503.49**
Sum of two sides 2100 mm	301.50	370.40	3.99	143.38	m	**513.78**
Sum of two sides 2300 mm	301.50	370.40	4.05	145.53	m	**515.93**
Sum of two sides 2500 mm	301.50	370.40	4.11	147.69	m	**518.09**
Branch						
Sum of two sides 1300 mm	103.42	127.05	1.59	57.14	m	**184.19**
Sum of two sides 1500 mm	103.42	127.05	1.68	60.36	m	**187.41**
Sum of two sides 1700 mm	132.05	162.23	1.77	63.60	m	**225.83**
Sum of two sides 1900 mm	132.05	162.23	1.86	66.84	m	**229.07**
Sum of two sides 2100 mm	138.63	170.31	2.58	92.72	m	**263.03**
Sum of two sides 2300 mm	138.63	170.31	2.64	94.87	m	**265.18**
Sum of two sides 2500 mm	138.63	170.31	2.91	104.57	m	**274.88**
Ductwork 1251 to 2000 mm longest side						
Sum of two sides 1800 mm	596.43	732.71	6.00	215.61	m	**948.32**
Sum of two sides 2000 mm	695.82	854.82	6.00	215.61	m	**1070.43**
Sum of two sides 2200 mm	794.61	976.17	6.50	233.57	m	**1209.74**
Sum of two sides 2400 mm	893.58	1097.76	6.50	233.57	m	**1331.33**
Sum of two sides 2600 mm	941.41	1156.52	6.66	239.33	m	**1395.85**
Sum of two sides 2800 mm	1032.63	1268.59	6.66	239.33	m	**1507.92**
Sum of two sides 3000 mm	1123.29	1379.96	6.66	239.33	m	**1619.29**
Sum of two sides 3200 mm	1131.69	1390.28	9.00	323.40	m	**1713.68**
Sum of two sides 3400 mm	1244.85	1529.29	9.00	323.40	m	**1852.69**
Sum of two sides 3600 mm	1267.33	1556.92	11.70	420.43	m	**1977.35**
Sum of two sides 3800 mm	1380.09	1695.45	11.70	420.43	m	**2115.88**
Sum of two sides 4000 mm	1596.56	1961.38	11.70	420.43	m	**2381.81**
Extra over fittings; Ductwork 1251 to 2000 mm longest sides						
End cap						
Sum of two sides 1800 mm	248.69	305.52	1.45	52.10	m	**357.62**
Sum of two sides 2000 mm	248.69	305.52	1.51	54.26	m	**359.78**
Sum of two sides 2200 mm	338.91	416.35	2.80	100.61	m	**516.96**
Sum of two sides 2400 mm	338.91	416.35	3.10	111.38	m	**527.73**
Sum of two sides 2600 mm	464.34	570.44	3.39	121.82	m	**692.26**
Sum of two sides 2800 mm	464.34	570.44	3.68	132.24	m	**702.68**
Sum of two sides 3000 mm	464.34	570.44	3.98	143.02	m	**713.46**
Sum of two sides 3200 mm	627.20	770.51	4.27	153.44	m	**923.95**
Sum of two sides 3400 mm	627.20	770.51	4.57	164.21	m	**934.72**
Sum of two sides 3600 mm	772.42	948.92	4.87	175.00	m	**1123.92**
Sum of two sides 3800 mm	772.42	948.92	5.17	185.78	m	**1134.70**
Sum of two sides 4000 mm	772.42	948.92	5.47	196.56	m	**1145.48**

38 VENTILATION/AIR CONDITIONING SYSTEMS

Item	Net Price £	Material £	Labour hours	Labour £	Unit	Total rate £
DUCTWORK: FIRE RATED – cont						
Extra over fittings – cont						
Reducer						
Sum of two sides 1800 mm	154.06	189.26	3.60	129.36	m	**318.62**
Sum of two sides 2000 mm	154.06	189.26	3.70	132.95	m	**322.21**
Sum of two sides 2200 mm	182.65	224.39	3.10	111.38	m	**335.77**
Sum of two sides 2400 mm	182.65	224.39	3.48	125.04	m	**349.43**
Sum of two sides 2600 mm	253.07	310.89	3.85	138.34	m	**449.23**
Sum of two sides 2800 mm	253.07	310.89	4.22	151.64	m	**462.53**
Sum of two sides 3000 mm	253.07	310.89	4.59	164.93	m	**475.82**
Sum of two sides 3200 mm	281.69	346.06	4.96	178.23	m	**524.29**
Sum of two sides 3400 mm	281.69	346.06	5.34	191.88	m	**537.94**
Sum of two sides 3600 mm	308.10	378.50	5.72	205.54	m	**584.04**
Sum of two sides 3800 mm	308.10	378.50	6.10	219.20	m	**597.70**
Sum of two sides 4000 mm	308.10	378.50	6.48	232.85	m	**611.35**
Offset						
Sum of two sides 1800 mm	154.06	189.26	4.56	163.85	m	**353.11**
Sum of two sides 2000 mm	154.06	189.26	5.53	198.71	m	**387.97**
Sum of two sides 2200 mm	182.65	224.39	5.99	215.25	m	**439.64**
Sum of two sides 2400 mm	182.65	224.39	6.45	231.77	m	**456.16**
Sum of two sides 2600 mm	253.07	310.89	6.91	248.30	m	**559.19**
Sum of two sides 2800 mm	253.07	310.89	7.37	264.83	m	**575.72**
Sum of two sides 3000 mm	253.07	310.89	7.83	281.37	m	**592.26**
Sum of two sides 3200 mm	281.69	346.06	8.29	297.88	m	**643.94**
Sum of two sides 3400 mm	281.69	346.06	8.75	314.42	m	**660.48**
Sum of two sides 3600 mm	308.10	378.50	9.21	330.95	m	**709.45**
Sum of two sides 3800 mm	308.10	378.50	9.67	347.48	m	**725.98**
Sum of two sides 4000 mm	308.10	378.50	10.13	364.00	m	**742.50**
90° radius bend						
Sum of two sides 1800 mm	294.90	362.28	3.25	116.78	m	**479.06**
Sum of two sides 2000 mm	294.90	362.28	3.30	118.59	m	**480.87**
Sum of two sides 2200 mm	301.50	370.40	3.34	120.01	m	**490.41**
Sum of two sides 2400 mm	301.50	370.40	3.38	121.45	m	**491.85**
Sum of two sides 2600 mm	462.15	567.76	3.42	122.89	m	**690.65**
Sum of two sides 2800 mm	462.15	567.76	3.46	124.32	m	**692.08**
Sum of two sides 3000 mm	462.15	567.76	3.50	125.77	m	**693.53**
Sum of two sides 3200 mm	554.57	681.29	3.54	127.21	m	**808.50**
Sum of two sides 3400 mm	554.57	681.29	3.58	128.65	m	**809.94**
Sum of two sides 3600 mm	649.19	797.53	3.62	130.08	m	**927.61**
Sum of two sides 3800 mm	649.19	797.53	3.66	131.52	m	**929.05**
Sum of two sides 4000 mm	649.19	797.53	3.70	132.95	m	**930.48**
45° bend						
Sum of two sides 1800 mm	294.90	362.28	2.30	82.66	m	**444.94**
Sum of two sides 2000 mm	294.90	362.28	2.76	99.17	m	**461.45**
Sum of two sides 2200 mm	301.50	370.40	3.01	108.16	m	**478.56**
Sum of two sides 2400 mm	301.50	370.40	3.26	117.15	m	**487.55**
Sum of two sides 2600 mm	462.15	567.76	3.49	125.40	m	**693.16**
Sum of two sides 2800 mm	462.15	567.76	3.74	134.39	m	**702.15**
Sum of two sides 3000 mm	462.15	567.76	3.98	143.02	m	**710.78**

38 VENTILATION/AIR CONDITIONING SYSTEMS

Item	Net Price £	Material £	Labour hours	Labour £	Unit	**Total rate £**
Sum of two sides 3200 mm	554.57	681.29	4.22	151.64	m	**832.93**
Sum of two sides 3400 mm	554.57	681.29	4.46	160.27	m	**841.56**
Sum of two sides 3600 mm	649.19	797.53	4.70	168.88	m	**966.41**
Sum of two sides 3800 mm	649.19	797.53	4.94	177.52	m	**975.05**
Sum of two sides 4000 mm	649.19	797.53	5.18	186.13	m	**983.66**
90° mitre band						
Sum of two sides 1800 mm	294.90	362.28	3.90	140.14	m	**502.42**
Sum of two sides 2000 mm	294.90	362.28	3.96	142.30	m	**504.58**
Sum of two sides 2200 mm	301.50	370.40	3.83	137.63	m	**508.03**
Sum of two sides 2400 mm	400.89	492.49	3.84	137.98	m	**630.47**
Sum of two sides 2600 mm	462.15	567.76	3.85	138.34	m	**706.10**
Sum of two sides 2800 mm	462.15	567.76	3.86	138.70	m	**706.46**
Sum of two sides 3000 mm	462.15	567.76	3.87	139.06	m	**706.82**
Sum of two sides 3200 mm	554.57	681.29	3.88	139.42	m	**820.71**
Sum of two sides 3400 mm	554.57	681.29	3.90	140.14	m	**821.43**
Sum of two sides 3600 mm	649.19	797.53	3.92	140.86	m	**938.39**
Sum of two sides 3800 mm	649.19	797.53	3.94	141.58	m	**939.11**
Sum of two sides 4000 mm	649.19	797.53	3.96	142.30	m	**939.83**
Branch						
Sum of two sides 1800 mm	132.05	162.23	1.81	65.04	m	**227.27**
Sum of two sides 2000 mm	132.05	162.23	1.90	68.28	m	**230.51**
Sum of two sides 2200 mm	138.63	170.31	2.61	93.79	m	**264.10**
Sum of two sides 2400 mm	138.63	170.31	2.68	96.31	m	**266.62**
Sum of two sides 2600 mm	165.05	202.77	2.75	98.82	m	**301.59**
Sum of two sides 2800 mm	165.05	202.77	2.81	100.97	m	**303.74**
Sum of two sides 3000 mm	165.05	202.77	2.87	103.13	m	**305.90**
Sum of two sides 3200 mm	171.67	210.89	2.93	105.28	m	**316.17**
Sum of two sides 3400 mm	171.67	210.89	3.00	107.80	m	**318.69**
Sum of two sides 3600 mm	220.07	270.35	3.06	109.95	m	**380.30**
Sum of two sides 3800 mm	220.07	270.35	3.12	112.11	m	**382.46**
Sum of two sides 4000 mm	220.07	270.35	3.18	114.27	m	**384.62**

38 VENTILATION/AIR CONDITIONING SYSTEMS

Item	Net Price £	Material £	Labour hours	Labour £	Unit	Total rate £
LOW VELOCITY AIR CONDITIONING: AIR HANDLING UNITS						
Supply air handling unit; inlet with motorized damper, LTHW frost coil (at -5°C to +5°C), panel filter (EU4), bag filter (EU6), cooling coil (at 28°C db/20°C wb to 12°C db/11.5°C wb), LTHW heating coil (at 5°C to 21°C), supply fan, outlet plenum; includes access sections; all units located internally; Includes placing in position and fitting of sections together; electrical work elsewhere.						
Volume, external pressure						
2 m^3/s @ 350 Pa	6434.64	7904.96	40.00	1501.23	nr	**9406.19**
2 m^3/s @ 700 Pa	6874.49	8445.31	40.00	1501.23	nr	**9946.54**
5 m^3/s @ 350 Pa	10850.09	13329.34	65.00	2439.50	nr	**15768.84**
5 m^3/s @ 700 Pa	11229.65	13795.62	65.00	2439.50	nr	**16235.12**
8 m^3/s @ 350 Pa	16251.14	19964.52	77.00	2889.86	nr	**22854.38**
8 m^3/s @ 700 Pa	16504.47	20275.74	77.00	2889.86	nr	**23165.60**
10 m^3/s @ 350 Pa	17895.63	21984.78	100.00	3753.07	nr	**25737.85**
10 m^3/s @ 700 Pa	18750.02	23034.40	100.00	3753.07	nr	**26787.47**
13 m^3/s @ 350 Pa	22959.60	28205.87	108.00	4053.32	nr	**32259.19**
13 m^3/s @ 700 Pa	23668.71	29077.01	108.00	4053.32	nr	**33130.33**
15 m^3/s @ 350 Pa	26293.79	32301.92	120.00	4503.68	nr	**36805.60**
15 m^3/s @ 700 Pa	26772.22	32889.67	120.00	4503.68	nr	**37393.35**
18 m^3/s @ 350 Pa	29836.71	36654.40	133.00	4991.58	nr	**41645.98**
18 m^3/s @ 700 Pa	30385.20	37328.21	133.00	4991.58	nr	**42319.79**
20 m^3/s @ 350 Pa	34575.67	42476.21	142.00	5329.36	nr	**47805.57**
20 m^3/s @ 700 Pa	35072.91	43087.07	142.00	5329.36	nr	**48416.43**
Extra for inlet and discharge attenuators at 900 mm long						
2 m^3/s @ 350 Pa	1988.06	2442.33	5.00	187.66	nr	**2629.99**
5 m^3/s @ 350 Pa	4172.48	5125.89	10.00	375.31	nr	**5501.20**
10 m^3/s @ 700 Pa	5936.53	7293.03	13.00	487.90	nr	**7780.93**
15 m^3/s @ 700 Pa	9002.06	11059.03	16.00	600.49	nr	**11659.52**
20 m^3/s @ 700 Pa	11617.22	14271.76	20.00	750.61	nr	**15022.37**
Extra for locating units externally						
2 m^3/s @ 350 Pa	1473.05	1809.64	–	–	nr	**1809.64**
5 m^3/s @ 350 Pa	1869.02	2296.09	–	–	nr	**2296.09**
10 m^3/s @ 700 Pa	2990.07	3673.30	–	–	nr	**3673.30**
15 m^3/s @ 700 Pa	4321.10	5308.47	–	–	nr	**5308.47**
20 m^3/s @ 700 Pa	7489.86	9201.30	–	–	nr	**9201.30**

38 VENTILATION/AIR CONDITIONING SYSTEMS

Item	Net Price £	Material £	Labour hours	Labour £	Unit	Total rate £
Modular air handling unit with supply and extract sections. Supply side; inlet with motorized damper, LTHW frost coil (at -5°C to 5°C), panel filter (EU4), bag filter (EU6), cooling coil at 28°Cdb/20°Cwb to 12°Cdb/ 11.5°C wb), LTHW heating coil (at 5°C to 21°C), supply fan, outlet plenum. Extract side; inlet with motorized damper, extract fan; includes access sections; placing in position and fitting of sections together; electrical work elsewhere						
Volume, external pressure						
2 m^3/s @ 350 Pa	8988.96	11042.93	50.00	1876.54	nr	**12919.47**
2 m^3/s @ 700 Pa	9423.90	11577.26	50.00	1876.54	nr	**13453.80**
5 m^3/s @ 350 Pa	15492.01	19031.93	86.00	3227.64	nr	**22259.57**
5 m^3/s @ 700 Pa	15782.80	19389.17	86.00	3227.64	nr	**22616.81**
8 m^3/s @ 350 Pa	21512.89	26428.58	105.00	3940.72	nr	**30369.30**
8 m^3/s @ 700 Pa	22029.29	27062.98	105.00	3940.72	nr	**31003.70**
10 m^3/s @ 350 Pa	24840.97	30517.13	120.00	4503.68	nr	**35020.81**
10 m^3/s @ 700 Pa	25376.12	31174.56	120.00	4503.68	nr	**35678.24**
13 m^3/s @ 350 Pa	31966.65	39271.03	130.00	4878.98	nr	**44150.01**
13 m^3/s @ 700 Pa	32969.30	40502.78	130.00	4878.98	nr	**45381.76**
15 m^3/s @ 350 Pa	35720.49	43882.62	145.00	5441.95	nr	**49324.57**
15 m^3/s @ 700 Pa	37410.36	45958.63	145.00	5441.95	nr	**51400.58**
18 m^3/s @ 350 Pa	40832.81	50163.11	160.00	6004.91	nr	**56168.02**
18 m^3/s @ 700 Pa	41788.81	51337.56	160.00	6004.91	nr	**57342.47**
20 m^3/s @ 350 Pa	44628.62	54826.26	175.00	6567.87	nr	**61394.13**
20 m^3/s @ 700 Pa	46180.10	56732.26	175.00	6567.87	nr	**63300.13**
Extra for inlet and discharge attenuators at 900 mm long						
2 m^3/s @ 350 Pa	3624.34	4452.50	8.00	300.25	nr	**4752.75**
5 m^3/s @ 350 Pa	7157.03	8792.42	10.00	375.31	nr	**9167.73**
10 m^3/s @ 700 Pa	10241.11	12581.21	13.00	487.90	nr	**13069.11**
15 m^3/s @ 700 Pa	17517.87	21520.71	16.00	600.49	nr	**22121.20**
20 m^3/s @ 700 Pa	20449.31	25121.97	20.00	750.61	nr	**25872.58**
Extra for locating units externally						
2 m^3/s @ 350 Pa	2901.12	3564.03	–	–	nr	**3564.03**
5 m^3/s @ 350 Pa	4119.58	5060.91	–	–	nr	**5060.91**
10 m^3/s @ 700 Pa	5880.41	7224.08	–	–	nr	**7224.08**
15 m^3/s @ 700 Pa	11165.35	13716.63	–	–	nr	**13716.63**
20 m^3/s @ 700 Pa	17164.56	21086.67	–	–	nr	**21086.67**
Extra for humidifier, self-generating type						
2 m^3/s @ 350 Pa (10 kg/hr)	2528.13	3105.81	5.00	187.66	nr	**3293.47**
5 m^3/s @ 350 Pa (18 kg/hr)	3128.61	3843.49	5.00	187.66	nr	**4031.15**
10 m^3/s @ 700 Pa (30 kg/hr)	3560.19	4373.69	6.00	225.18	nr	**4598.87**
15 m^3/s @ 700 Pa (60 kg/hr)	6593.26	8099.82	8.00	300.25	nr	**8400.07**
20 m^3/s @ 700 Pa (90 kg/hr)	9912.92	12178.03	10.00	375.31	nr	**12553.34**

38 VENTILATION/AIR CONDITIONING SYSTEMS

Item	Net Price £	Material £	Labour hours	Labour £	Unit	Total rate £
LOW VELOCITY AIR CONDITIONING: AIR HANDLING UNITS – cont						
Modular air handling unit with supply and extract sections. – cont						
Extra for mixing box						
2 m^3/s @ 350 Pa	1213.95	1491.34	4.00	150.12	nr	**1641.46**
5 m^3/s @ 350 Pa	1711.13	2102.12	4.00	150.12	nr	**2252.24**
10 m^3/s @ 700 Pa	2397.28	2945.06	5.00	187.66	nr	**3132.72**
15 m^3/s @ 700 Pa	3199.48	3930.56	6.00	225.18	nr	**4155.74**
20 m^3/s @ 700 Pa	4877.52	5992.04	6.00	225.18	nr	**6217.22**
Extra for runaround coil, including pump and associated pipework; typical outputs in brackets (based on minimal distance between the supply and extract units)						
2 m^3/s @ 350 Pa (26 kW)	3634.19	4464.60	30.00	1125.93	nr	**5590.53**
5 m^3/s @ 350 Pa (37 kW)	6443.50	7915.83	30.00	1125.93	nr	**9041.76**
10 m^3/s @ 700 Pa (85 kW)	11155.94	13705.07	40.00	1501.23	nr	**15206.30**
15 m^3/s @ 700 Pa (151 kW)	15696.79	19283.50	50.00	1876.54	nr	**21160.04**
20 m^3/s @ 700 Pa (158 kW)	21396.96	26286.16	60.00	2251.84	nr	**28538.00**
Extra for thermal wheel (typical outputs in brackets)						
2 m^3/s @ 350 Pa (37 kW)	7516.84	9234.44	12.00	450.37	nr	**9684.81**
5 m^3/s @ 350 Pa (65 kW)	10796.00	13262.89	12.00	450.37	nr	**13713.26**
10 m^3/s @ 700 Pa (127 kW)	14397.26	17687.03	15.00	562.96	nr	**18249.99**
15 m^3/s @ 700 Pa (160 kW)	26974.35	33137.99	17.00	638.02	nr	**33776.01**
20 m^3/s @ 700 Pa (262 kW)	31509.18	38709.03	19.00	713.09	nr	**39422.12**
Extra for plate heat exchanger, including additional filtration in extract leg (typical outputs in brackets)						
2 m^3/s @ 350 Pa (25 kW)	4764.03	5852.62	12.00	450.37	nr	**6302.99**
5 m^3/s @ 350 Pa (51 kW)	9313.08	11441.12	12.00	450.37	nr	**11891.49**
10 m^3/s @ 700 Pa (98 kW)	13664.87	16787.30	15.00	562.96	nr	**17350.26**
15 m^3/s @ 700 Pa (160 kW)	23598.11	28990.28	17.00	638.02	nr	**29628.30**
20 m^3/s @ 700 Pa (190 kW)	30474.97	37438.50	19.00	713.09	nr	**38151.59**
Extra for electric heating in lieu of LTHW						
2 m^3/s @ 350 Pa	1651.15	2028.44	–	–	nr	**2028.44**
5 m^3/s @ 350 Pa	3093.63	3800.53	–	–	nr	**3800.53**
10 m^3/s @ 700 Pa	4176.88	5131.30	–	–	nr	**5131.30**
15 m^3/s @ 700 Pa	4379.85	5380.64	–	–	nr	**5380.64**
20 m^3/s @ 700 Pa	5168.85	6349.93	–	–	nr	**6349.93**

38 VENTILATION/AIR CONDITIONING SYSTEMS

Item	Net Price £	Material £	Labour hours	Labour £	Unit	Total rate £
VAV AIR CONDITIONING						
VAV TERMINAL BOXES						
VAV terminal box; integral acoustic silencer; factory installed and prewired control components (excluding electronic controller); selected at 200 Pa at entry to unit; includes fixing in position; electrical work elsewhere						
80–110 l/s	466.75	573.40	2.00	71.87	nr	**645.27**
Extra for secondary silencer	113.51	139.45	0.50	17.98	nr	**157.43**
Extra for 2 row LTHW heating coil	280.00	343.98	–	–	nr	**343.98**
150–190 l/s	491.15	603.38	2.00	71.87	nr	**675.25**
Extra for secondary silencer	124.64	153.12	0.50	17.98	nr	**171.10**
Extra for 2 row LTHW heating coil	300.00	368.55	–	–	nr	**368.55**
250–310 l/s	547.90	673.09	2.00	71.87	nr	**744.96**
Extra for secondary silencer	164.95	202.64	0.50	17.98	nr	**220.62**
Extra for 2 row LTHW heating coil	340.00	417.69	–	–	nr	**417.69**
420–520 l/s	593.52	729.14	2.00	71.87	nr	**801.01**
Extra for secondary silencer	187.76	230.66	0.50	17.98	nr	**248.64**
Extra for 2 row LTHW heating coil	414.00	508.60	–	–	nr	**508.60**
650–790 l/s	718.69	882.91	2.00	71.87	nr	**954.78**
Extra for secondary silencer	240.27	295.18	0.50	17.98	nr	**313.16**
Extra for 2 row LTHW heating coil	434.00	533.17	–	–	nr	**533.17**
1130–1370 l/s	834.85	1025.61	2.00	71.87	nr	**1097.48**
Extra for secondary silencer	335.74	412.46	0.50	17.98	nr	**430.44**
Extra for 2 row LTHW heating coil	454.00	557.74	–	–	nr	**557.74**
Extra for electric heater & thyristor controls, 3 kW/1ph (per box)	370.00	454.54	–	–	nr	**454.54**
Extra for fitting free issue box controller	–	–	–	–	nr	–
Fan assisted VAV terminal box; factory installed and prewired control components (excluding electronic controller); selected at 40 Pa external static pressure; includes fixing in position, electrical work elsewhere						
100–175 l/s	1033.22	1269.31	3.00	107.80	nr	**1377.11**
Extra for secondary silencer	123.05	151.17	0.50	17.98	nr	**169.15**
Extra for 1 row LTHW heating coil	330.00	405.40	–	–	nr	**405.40**
170–360 l/s	1127.10	1384.65	3.00	107.80	nr	**1492.45**
Extra for secondary silencer	155.94	191.57	0.50	17.98	nr	**209.55**
Extra for 1 row LTHW heating coil	350.00	429.98	–	–	nr	**429.98**
300–640 l/s	1292.05	1587.28	3.00	107.80	nr	**1695.08**
Extra for secondary silencer	238.68	293.22	0.50	17.98	nr	**311.20**
Extra for 1 row LTHW heating coil	380.00	466.83	–	–	nr	**466.83**
620–850 l/s	1292.05	1587.28	3.00	107.80	nr	**1695.08**
Extra for secondary silencer	238.68	293.22	0.50	17.98	nr	**311.20**
Extra for 1 row LTHW heating coil	410.00	503.68	–	–	nr	**503.68**
Extra for electric heater plus thyristor controls' 3 kW/lph (per box)	380.00	466.83	–	–	nr	**466.83**
Extra for fitting free issue controller	–	–	2.13	76.53	nr	**76.53**

38 VENTILATION/AIR CONDITIONING SYSTEMS

Item	Net Price £	Material £	Labour hours	Labour £	Unit	Total rate £
FAN COIL AIR CONDITIONING						
All selections based on summer return air condition of 23°C @ 50% RH, CHW @ 6°/12°C, LTHW @ 82°/71°C (where applicable), medium speed, external resistance of 30 Pa.						
All selections are based on heating and cooling units. For waterside control units there is no significant reduction in cost between 4 pipe heating and cooling and 2 pipe cooling only units (excluding controls). For airside control units, there is a marginal reduction (less than 5%) between 4 pipe heating and cooling units and 2 pipe cooling only units (excluding controls).						
Ceiling void mounted horizontal waterside control fan coil unit; cooling coil; LTHW heating coil; multi-tapped speed transformer; fine wire mesh filter; includes fixing in position; electrical work elsewhere.						
Total cooling load, heating load						
2800 W, 1000 W	462.00	567.57	4.00	150.12	nr	**717.69**
4000 W, 1700 W	486.00	597.05	4.00	150.12	nr	**747.17**
4500 W, 1900 W	682.00	837.84	4.00	150.12	nr	**987.96**
6000 W, 2600 W	892.00	1095.82	4.00	150.12	nr	**1245.94**
Ceiling void mounted horizontal waterside control fan coil unit; cooling coil; electric heating coil; multi-tapped speed transformer; fine wire mesh filter; includes fixing in position; electrical work elsewhere + thyristor and 2 No. HTCOs.						
Total cooling load, heating load						
2800 W, 1500 W	555.00	681.82	4.00	150.12	nr	**831.94**
4000 W, 2000 W	579.00	711.30	4.00	150.12	nr	**861.42**
4500 W, 2000 W	779.00	957.00	4.00	150.12	nr	**1107.12**
6000 W, 3000 W	989.00	1214.99	4.00	150.12	nr	**1365.11**
Ceiling void mounted horizontal airside control fan coil unit; cooling coil; LHTW heating coil; multi-tapped speed transformer; fine wire mesh filter, damper actuator & fixing kit; includes fixing in position; electrical work elsewhere.						
Total cooling load, heating load						
2600 W, 2200 W	726.00	891.89	4.00	150.12	nr	**1042.01**
3600 W, 3200 W	747.00	917.69	4.00	150.12	nr	**1067.81**
4000 W, 3600 W	836.00	1027.03	4.00	150.12	nr	**1177.15**
5400 W, 5000 W	998.00	1226.04	4.00	150.12	nr	**1376.16**

38 VENTILATION/AIR CONDITIONING SYSTEMS

Item	Net Price £	Material £	Labour hours	Labour £	Unit	**Total rate £**
Ceiling void mounted horizontal airside control fan coil unit; cooling coil; electric heating coil; multi-tapped speed transformer; fine wire mesh filter, damper actuator & fixing kit; includes fixing in position; electrical work elsewhere + thyristor and 2 No. HTCOs						
Total cooling load, heating load						
2600 W, 1500 W	773.00	949.63	4.00	150.12	nr	**1099.75**
3600 W, 2000 W	793.00	974.20	4.00	150.12	nr	**1124.32**
4000 W, 2000 W	877.00	1077.39	4.00	150.12	nr	**1227.51**
5400 W, 3000 W	1042.00	1280.10	4.00	150.12	nr	**1430.22**
Ceiling void mounted slimline horizontal waterside control fan coil unit, 170 mm deep; cooilng coil; LTHW heating coil; multi-tapped speed transformer; fine wire mesh filter; includes fixing in position; electrical work elsewhere						
Total cooling load, heating load						
1100 W, 1500 W	570.00	700.24	3.50	131.35	nr	**831.59**
3200 W, 3700 W	1089.00	1337.84	4.00	150.12	nr	**1487.96**
Ceiling void mounted slimline horizontal waterside control fan coil unit, 170 mm deep; cooilng coil; electric heating coil; multi-tapped speed transformer; fine wire mesh filter; includes fixing in position; electrical work elsewhere						
Total cooling load, heating load						
1100 W, 1000 W	421.00	517.20	3.50	131.35	nr	**648.55**
3400 W, 2000 W	1186.00	1457.00	4.00	150.12	nr	**1607.12**
Ceiling void mounted slimline horizontal airside control fan coil unit, 170 mm deep; cooling coil; multi-tapped speed transformer; fine wire mesh filter,damper actuator & fixing kit; includes fixing in position; electrical work elsewhere						
Total cooling load						
1000 W	740.00	909.09	3.50	131.35	nr	**1040.44**
3000 W	987.00	1212.53	4.00	150.12	nr	**1362.65**
Ceiling void mounted slimline horizontal airside control fan coil unit, 170 mm deep; cooilng coil; electric heating coil; multi-tapped speed transformer; fine wire mesh filter, damper actuator & fixing kit; includes fixing in position; electrical work elsewhere						
Total cooling load, heating load						
1000 W, 1000 W	548.00	673.22	3.50	131.35	nr	**804.57**
3000 W, 2000 W	1084.00	1331.69	4.00	150.12	nr	**1481.81**

38 VENTILATION/AIR CONDITIONING SYSTEMS

Item	Net Price £	Material £	Labour hours	Labour £	Unit	**Total rate £**
FAN COIL AIR CONDITIONING – cont						
Low level perimeter waterside control fan coil unit; cooling coil; LTHW heating coil; multi-tapped speed transformer; fine wire mesh filter; includes fixing in position; electrical work elsewhere						
Total cooling load, heating load						
1700 W, 1400 W	502.00	616.71	3.50	131.35	nr	**748.06**
Extra over for standard cabinet	198.71	244.11	1.00	37.53	nr	**281.64**
2200 W, 1900 W	649.00	797.30	3.50	131.35	nr	**928.65**
Extra over for standard cabinet	242.05	297.36	1.00	37.53	nr	**334.89**
2600 W, 2200 W	649.00	797.30	3.50	131.35	nr	**928.65**
Extra over for standard cabinet	248.02	304.69	1.00	37.53	nr	**342.22**
3900 W, 3200 W	827.00	1015.97	3.50	131.35	nr	**1147.32**
Extra over for standard cabinet	270.43	332.22	1.00	37.53	nr	**369.75**
Low level perimeter waterside control fan coil unit; cooling coil; electric heating coil; multi-tapped speed transformer; fine wire mesh filter; includes fixing in position; electrical work elsewhere						
Total cooling load, heating load						
1700 W, 1500 W	597.00	733.41	3.50	131.35	nr	**864.76**
Extra over for standard cabinet	198.71	244.11	1.00	37.53	nr	**281.64**
2200 W, 2000 W	746.00	916.46	3.50	131.35	nr	**1047.81**
Extra over for standard cabinet	242.05	297.36	1.00	37.53	nr	**334.89**
2600 W, 2000 W	746.00	916.46	3.50	131.35	nr	**1047.81**
Extra over for standard cabinet	248.02	304.69	1.00	37.53	nr	**342.22**
3800 W, 3000 W	926.00	1137.59	3.50	131.35	nr	**1268.94**
Extra over for standard cabinet	270.43	332.22	1.00	37.53	nr	**369.75**

ESSENTIAL READING FROM TAYLOR AND FRANCIS

Watts Pocket Handbook, 29th Edition

RUSHTON

Back in print for the first time in years, the Watts Pocket Handbook renews its commitment to share industry knowledge by providing technical and legal information across a comprehensive spread of property and construction topics.

Compiled by the Watts Technical Director, the Handbook provides specialist information and guidance on a vast selection of construction related subjects including:

- Contracts and procurement
- Insurance
- Materials and defects
- Environmental and sustainability issues

Watts Pocket Handbook remains the must-have reference book for professionals and students engaged in construction, building surveying, service engineering, property development and much more.

April 2016; 186 x 123 mm; 324 pp
Pb: 978-1-138-66595-8; £19.99

To Order: Tel: +44 (0) 1235 400524 Fax: +44 (0) 1235 400525
or Post: Taylor and Francis Customer Services,
Bookpoint Ltd, Unit T1, 200 Milton Park, Abingdon, Oxon, OX14 4TA UK
Email: book.orders@tandf.co.uk

For a complete listing of all our titles visit:
www.tandf.co.uk

Taylor & Francis
Taylor & Francis Group

ESSENTIAL READING FROM TAYLOR AND FRANCIS

Faber & Kell's Heating & Air-conditioning of Buildings

11th Edition

Doug Oughton & Ant Wilson

For over 70 years, Faber & Kell's has been the definitive reference text in its field. It provides an understanding of the principles of heating and air-conditioning of buildings in a concise manner, illustrating practical information with simple, easy-to-use diagrams, now in full-colour.

This new-look 11th edition has been re-organised for ease of use and includes fully updated chapters on sustainability and renewable energy sources, as well as information on the new Building Regulations Parts F and L. As well as extensive updates to regulations and codes, it now includes an introduction that explains the role of the building services engineer in the construction process. Its coverage of design calculations, advice on using the latest technologies, building management systems, operation and maintenance makes this an essential reference for all building services professionals.

January 2015: 246 x 189: 968pp
Pb: 978-0-415-52265-6: £105.00

To Order: Tel: +44 (0) 1235 400524 Fax: +44 (0) 1235 400525
or Post: Taylor and Francis Customer Services,
Bookpoint Ltd, Unit T1, 200 Milton Park, Abingdon, Oxon, OX14 4TA UK
Email: book.orders@tandf.co.uk

For a complete listing of all our titles visit:
www.tandf.co.uk

Taylor & Francis
Taylor & Francis Group

ESSENTIAL READING FROM TAYLOR AND FRANCIS

Cost Studies of Buildings

Allan Ashworth & Srinath Perera

This practical guide to cost studies of buildings has been updated and revised throughout for the 6th edition. New developments in RICS New Rules of Measurement (NRM) are incorporated throughout the book, in addition to new material on e-business, the internet, social media, building information modelling, sustainability, building resilience and carbon estimating.

This trusted and easy to use guide to the cost management role:

Focuses on the importance of costs of constructing projects during the different phases of the construction process

Features learning outcomes and self-assessment questions for each chapter

Addresses the requirements of international readers

From introductory data on the construction industry and the history of construction economics, to recommended methods for cost analysis and post-contract cost control, Cost Studies of Buildings is an ideal companion for anyone learning about cost management.

July 2015: 246x174: 544pp Pb:
978-1-138-01735-1: £42.99

To Order: Tel: +44 (0) 1235 400524 Fax: +44 (0) 1235 400525
or Post: Taylor and Francis Customer Services,
Bookpoint Ltd, Unit T1, 200 Milton Park, Abingdon, Oxon, OX14 4TA UK
Email: book.orders@tandf.co.uk

For a complete listing of all our titles visit:
www.tandf.co.uk

Taylor & Francis
Taylor & Francis Group

ESSENTIAL READING FROM TAYLOR AND FRANCIS

Quantity Surveyor's Pocket Book

2nd Edition

D Cartlidge

This second edition of the Quantity Surveyor's Pocket Book is fully updated in line with NRM1, NRM2 and JCT(11), and remains a must-have guide for students and qualified practitioners. Its focussed coverage of the data, techniques, and skills essential to the quantity surveying role make it an invaluable companion for everything from initial cost advice to the final account stage.

Key features include:

- the structure of the construction industry
- cost forecasting and feasibility studies
- measurement and quantification, with NRM2 and SMM7 examples
- estimating and bidding
- whole life costs
- contract selection
- final account procedure.

June 2012: 186x123: 440pp
Pb: 978-0-415-50110-1: £22.99

To Order: Tel: +44 (0) 1235 400524 Fax: +44 (0) 1235 400525
or Post: Taylor and Francis Customer Services,
Bookpoint Ltd, Unit T1, 200 Milton Park, Abingdon, Oxon, OX14 4TA UK
Email: book.orders@tandf.co.uk

For a complete listing of all our titles visit:
www.tandf.co.uk

Taylor & Francis
Taylor & Francis Group

Electrical Installations

Material Costs/Measured Work Prices

DIRECTIONS

The following explanations are given for each of the column headings and letter codes.

Unit	Prices for each unit are given as singular (i.e. 1 metre, 1 nr) unless stated otherwise.
Net price	Industry tender prices, plus nominal allowance for fixings, waste and applicable trade discounts.
Material cost	Net price plus percentage allowance for overheads (7%), profit (5%) and preliminaries (10%).
Labour norms	In man-hours for each operation.
Labour cost	Labour constant multiplied by the appropriate all-in man-hour cost based on gang rate (see also relevant Rates of Wages Section) plus percentage allowance for overheads, profit and preliminaries.
Measured work Price (total rate)	Material cost plus Labour cost.

MATERIAL COSTS

The Material Costs given are based at Second Quarter 2016 but exclude any charges in respect of VAT. The average rate of copper during this quarter is US$4,630/UK£3,305 per tonne. Users of the book are advised to register on the Spon's website www.pricebooks.co.uk/updates to receive the free quarterly updates – alerts will then be provided by e-mail as changes arise.

MEASURED WORK PRICES

These prices are intended to apply to new work in the London area. The prices are for reasonable quantities of work and the user should make suitable adjustments if the quantities are especially small or especially large. Adjustments may also be required for locality (e.g. outside London – refer to cost indices in approximate estimating section for details of adjustment factors) and for the market conditions (e.g. volume of work secured or being tendered) at the time of use.

ELECTRICAL INSTALLATIONS

The labour rate has been based on average gang rates per man hour effective from 4 January 2016 including allowances for all other emoluments and expenses. Future changes will be published in the free Spon's quarterly update by registering on their website. To this rate, has been added 10% and 7% to cover site and head office overheads and preliminary items together with a further 5% for profit, resulting in an inclusive rate of £32.56 per man hour. The rate has been calculated on a working year of 2025 hours; a detailed build-up of the rate is given at the end of these directions.

DIRECTIONS

In calculating the 'Measured Work Prices' the following assumptions have been made:

(a) That the work is carried out as a subcontract under the Standard Form of Building Contract.
(b) That, unless otherwise stated, the work is being carried out in open areas at a height which would not require more than simple scaffolding.
(c) That the building in which the work is being carried out is no more than six storey's high.

Where these assumptions are not valid, as for example where work is carried out in ducts and similar confined spaces or in multi-storey structures when additional time is needed to get to and from upper floors, then an appropriate adjustment must be made to the prices. Such adjustment will normally be to the labour element only.

DIRECTIONS

LABOUR RATE – ELECTRICAL

The annual cost of a notional eleven man gang of 11 men

	TECHNICIAN 1 NR	APPROVED ELECTRICIANS 4 NR	ELECTRICIANS 4 NR	LABOURERS 2 NR	SUB TOTALS
Hourly Rate from 4 January 2016	**19.68**	**17.48**	**16.13**	**12.94**	
Working hours per annum per man	1,687.50	1,687.50	1,687.50	1,687.50	
× Hourly rate × nr of men = £ per annum	**33,210.00**	**117,990.00**	**108,877.50**	**43,672.50**	**303,750.00**
Overtime Rate	29.52	26.22	24.20	19.41	
Overtime hours per annum per man	337.50	337.50	337.50	337.50	
× Hourly rate × nr of men = £ per annum	**9,963.00**	**35,397.00**	**32,663.25**	**13,101.75**	**91,125.00**
Total	**43,172.00**	**153,387.00**	**141,540.75**	**56,774.25**	**394,785.00**
Incentive schemes (insert percentage) 0.00%	**0.00**	**0.00**	**0.00**	**0.00**	**0.00**
Daily Travel Allowance (15–20 miles each way) effective from 04/01/16	5.47	5.47	5.47	5.47	
Days per annum per man	225.00	225.00	225.00	225.00	
× nr of men = £ per annum	**1,230.75**	**4,923.00**	**4,923.00**	**2,461.50**	**13,538.25**
Daily Travel Allowance (15–20 miles each way) effective from 04/01/16	3.75	3.75	3.75	3.75	
Days per annum per man	225.00	225.00	225.00	225.00	
× nr of men = £ per annum	**843.75**	**3,375.00**	**3,375.00**	**1,687.50**	**9,281.25**
JIB Pension Scheme @ 2.5%	**1,273.11**	**4,545.48**	**4,209.84**	**1,708.36**	**11,736.79**
JIB combined benefits scheme (nr of weeks per man)	52.00	52.00	52.00	52.00	
Benefit Credit effective from 4 January 2016	66.23	60.23	56.43	47.83	
× nr of men = £ per annum	**3,443.96**	**12,527.84**	**11,737.44**	**4,974.32**	**32,683.56**
Holiday Top-up Funding	**59.17**	**52.80**	**49.00**	**39.66**	
× nr of men @ 7.5 hrs per day = £ per annum	**3,076.84**	**10,981.36**	**10,192.26**	**4,124.38**	**28,374.84**
National Insurance Contributions:					
Annual gross pay (subject to NI) each	51,768.30	161,685.00	149,838.75	60,923.25	
% of NI Contributions	13.80	13.80	13.80	13.80	
£ Contributions/annum	**5,939.18**	**17,493.16**	**15,858.37**	**5,997.72**	**45,288.43**

	SUBTOTAL		**535,778.12**
	TRAINING (INCLUDING ANY TRADE REGISTRATIONS) – SAY	**1.00%**	**5,357.78**
	SEVERANCE PAY AND SUNDRY COSTS – SAY	**1.50%**	**8,117.04**
	EMPLOYER'S LIABILITY AND THIRD PARTY INSURANCE – SAY	**2.00%**	**10,985.06**
	ANNUAL COST OF NOTIONAL GANG		**560,238.00**
MEN ACTUALLY WORKING = 10.5	THEREFORE ANNUAL COST PERPRODUCTIVE MAN		**53,356.00**
AVERAGE NR OF HOURS WORKEDPER MAN = 2025	THEREFORE ALL-IN MAN HOUR		**26.35**
	PRELIMINARY ITEMS – SAY	**10%**	**2.63**
	SITE AND HEAD OFFICE OVERHEADS – SAY	**7%**	**2.03**
	PROFIT – SAY	**5%**	**1.55**
	THEREFORE INCLUSIVE MAN HOUR		**32.56**

DIRECTIONS

Notes:

(1) Hourly wage rates are those effective from 4 January 2016.
(2) The following assumptions have been made in the above calculations:
  (a) Hourly rates are based on London rate and job reporting own transport.
  (b) The working week of 37.5 hours is made up of 7.5 hours Monday to Friday.
  (c) Five days in the year are lost through sickness or similar reason.
  (d) A working year of 2025 hours.
(3) The incentive scheme addition of 0% is intended to reflect bonus schemes typically in use.
(4) National insurance contributions are those effective from 6 April 2016.
(5) The National Insurance concession will be withdrawn for the construction industry on 30 October 2012 therefore NI will be paid for 52 weeks.
(6) Weekly JIB Combined Benefit Credit Scheme are those effective from 5 January 2016.
(7) Overtime is paid after 37.5 hours.

ELECTRICAL SUPPLY/POWER/LIGHTING

Item	Net Price £	Material £	Labour hours	Labour £	Unit	Total rate £
ELECTRICAL GENERATION PLANT						
STANDBY GENERATORS						
Standby diesel generating sets; supply and installation; fixing to base; all supports and fixings; all necessary connections to equipment						
Three phase, 400 Volt, four wire 50 Hz packaged standby diesel generating set, complete with radio and television suppressors, daily service fuel tank and associated piping, 4 metres of exhaust pipe and primary exhaust silencer, control panel, mains failure relay, starting battery with charger, all internal wiring, interconnections, earthing and labels. Rated for standby duty; including UK delivery, installation and commissioning.						
60 kVA	17141.11	21057.85	100.00	4000.00	nr	**25057.85**
100 kVA	24030.09	29520.97	100.00	4000.00	nr	**33520.97**
150 kVA	27482.29	33761.99	100.00	4000.00	nr	**37761.99**
315 kVA	49548.95	60870.88	120.00	4799.99	nr	**65670.87**
500 kVA	58402.95	71748.02	120.00	4799.99	nr	**76548.01**
750 kVA	99386.94	122096.86	140.00	5600.00	nr	**127696.86**
1000 kVA	132093.76	162277.19	140.00	5600.00	nr	**167877.19**
1500 kVA	155461.59	190984.56	170.00	6799.99	nr	**197784.55**
2000 kVA	239468.97	294187.62	170.00	6799.99	nr	**300987.61**
2500 kVA	327865.38	402782.61	210.00	8399.99	nr	**411182.60**
Extra for residential silencer; peformance 75 dBA at 1 m; including connection to exhaust pipe						
60 kVA	920.31	1130.60	10.00	400.00	nr	**1530.60**
100 kVA	1099.69	1350.97	10.00	400.00	nr	**1750.97**
150 kVA	1261.11	1549.27	10.00	400.00	nr	**1949.27**
315 kVA	2453.24	3013.80	15.00	600.00	nr	**3613.80**
500 kVA	3137.61	3854.55	15.00	600.00	nr	**4454.55**
750 kVA	4674.65	5742.81	20.00	799.99	nr	**6542.80**
1000 kVA	6295.90	7734.51	20.00	799.99	nr	**8534.50**
1500 kVA	10029.55	12321.30	20.00	799.99	nr	**13121.29**
2000 kVA	10862.94	13345.12	30.00	1200.00	nr	**14545.12**
2500 kVA	12420.70	15258.83	30.00	1200.00	nr	**16458.83**

ELECTRICAL SUPPLY/POWER/LIGHTING

Item	Net Price £	Material £	Labour hours	Labour £	Unit	Total rate £
ELECTRICAL GENERATION PLANT – cont						
STANDBY GENERATORS – cont						
Synchronization panel for paralleling generators – not generators to mains; including interconnecting cables; commissioning and testing; fixing to backgrounds						
2 × 60 kVA	7666.00	9417.68	80.00	3200.00	nr	**12617.68**
2 × 100 kVA	8241.38	10124.53	80.00	3200.00	nr	**13324.53**
2 × 150 kVA	10367.59	12736.58	80.00	3200.00	nr	**15936.58**
2 × 315 kVA	17625.19	21652.54	80.00	3200.00	nr	**24852.54**
2 × 500 kVA	25445.74	31260.10	80.00	3200.00	nr	**34460.10**
2 × 750 kVA	28714.41	35275.65	80.00	3200.00	nr	**38475.65**
2 × 1000 kVA	30946.88	38018.24	120.00	4799.99	nr	**42818.23**
2 × 1500 kVA	36635.70	45006.96	120.00	4799.99	nr	**49806.95**
2 × 2000 kVA	39694.64	48764.87	120.00	4799.99	nr	**53564.86**
2 × 2500 kVA	46277.53	56851.95	120.00	4799.99	nr	**61651.94**
Prefabricated drop-over acoustic housing; performance 85 dBA at 1 m over the range from 60 kVA to 315 kVA, 75 dBA from 500 kVA to 2500 kVA.						
100 kVA	3425.96	4208.79	7.00	280.00	nr	**4488.79**
150 kVA	5563.24	6834.44	15.00	600.00	nr	**7434.44**
315 kVA	11870.16	14582.49	25.00	1000.00	nr	**15582.49**
500 kVA	19605.16	24084.94	40.00	1600.00	nr	**25684.94**
750 kVA	30895.84	37955.54	40.00	1600.00	nr	**39555.54**
1000 kVA	39087.54	48019.04	40.00	1600.00	nr	**49619.04**
1500 kVA	39087.54	48019.04	40.00	1600.00	nr	**49619.04**
2000 kVA	77533.47	95249.87	60.00	2400.00	nr	**97649.87**
2500 kVA	93193.87	114488.67	70.00	2799.99	nr	**117288.66**
COMBINED HEAT AND POWER (CHP) UNITS						
Gas fired engine; acoustic enclosure complete with exhaust fan and attenuators; exhaust gas attenuation; includes 6 m long pipe connections; dry air cooler for secondary water circuit to reject excess heat; controls and panel; commissioning						
Electrical output; Heat output						
82 kW; 132 kW	93268.41	114580.24	–	–	nr	**114580.24**
100 kW, 148 kW	99346.05	122046.62	–	–	nr	**122046.62**
118 kW, 181kW	104254.91	128077.15	–	–	nr	**128077.15**
130 kW, 201kW	113566.55	139516.50	–	–	nr	**139516.50**
140 kW, 207 kW	120851.55	148466.13	–	–	nr	**148466.13**
150 kW, 208 kW	121552.81	149327.63	–	–	nr	**149327.63**
160 kW, 216 kW	127279.82	156363.26	–	–	nr	**156363.26**

ELECTRICAL SUPPLY/POWER/LIGHTING

Item	Net Price £	Material £	Labour hours	Labour £	Unit	Total rate £
198 kW, 233 kW	150304.73	184649.36	–	–	nr	**184649.36**
210 kW, 319 kW	159012.68	195347.08	–	–	nr	**195347.08**
237 kW, 359 kW	165265.08	203028.15	–	–	nr	**203028.15**
307 kW, 435 kW	224113.56	275323.51	–	–	nr	**275323.51**
380 kW, 500 kW	259468.50	318757.05	–	–	nr	**318757.05**
490 kW, 679 kW	357178.27	438793.51	–	–	nr	**438793.51**
501kW, 518 kW	329595.13	404907.61	–	–	nr	**404907.61**
600 kW, 873 kW	392417.47	482084.86	–	–	nr	**482084.86**
725 kW, 1019 kW	530624.78	651872.54	–	–	nr	**651872.54**
975 kW, 1293 kW	658898.63	809456.97	–	–	nr	**809456.97**
1160 kW, 1442 kW	745972.52	916427.24	–	–	nr	**916427.24**
1379 kW, 1475 kW	890550.24	1094040.97	–	–	nr	**1094040.97**
1566 kW, 1647 kW	974175.67	1196774.81	–	–	nr	**1196774.81**
1600 kW, 1625 kW	971838.12	1193903.13	–	–	nr	**1193903.13**
1760 kW, 1821 kW	1067210.33	1311067.89	–	–	nr	**1311067.89**
2430 kW	1386073.50	1702791.29	–	–	nr	**1702791.29**
2931 kW	1675590.00	2058462.31	–	–	nr	**2058462.31**
3500 kW	1802850.00	2214801.23	–	–	nr	**2214801.23**
3900 kW	2121000.00	2605648.50	–	–	nr	**2605648.50**
Note: The costs detailed are based on a specialist subcontract package, as part of the M&E contract works, and include installation						

ELECTRICAL SUPPLY/POWER/LIGHTING

Item	Net Price £	Material £	Labour hours	Labour £	Unit	Total rate £
HV SUPPLY						
Cable; 6350/11000 volts, 3 core, XLPE; stranded copper conductors; steel wire armoured; LSOH to BS 7835						
Laid in trench/duct including marker tape (cable tiles measured elsewhere)						
95 mm²	32.34	39.73	0.23	9.20	m	**48.93**
120 mm²	29.38	36.09	0.23	9.20	m	**45.29**
150 mm²	36.74	45.14	0.25	10.00	m	**55.14**
185 mm²	38.86	47.74	0.25	10.00	m	**57.74**
240 mm²	50.42	61.94	0.27	10.79	m	**72.73**
300 mm²	72.86	89.51	0.29	11.59	m	**101.10**
Pile tape; ES 1-12-23; 1 m	2.47	3.03	0.01	0.41	m	**3.44**
Clipped direct to backgrounds including cleats						
95 mm²	50.21	61.69	0.47	18.80	m	**80.49**
120 mm²	54.19	66.57	0.50	20.00	m	**86.57**
150 mm²	62.44	76.70	0.53	21.20	m	**97.90**
185 mm²	68.42	84.05	0.55	22.00	m	**106.05**
240 mm²	79.11	97.19	0.60	24.00	m	**121.19**
300 mm²	89.92	110.47	0.68	27.19	m	**137.66**
Terminations for above cables, including heat-shrink kit and glanding off						
95–240 mm²	336.36	413.22	4.75	190.00	m	**603.22**
Cable tiles; single width; laid in trench above cables on prepared sand bed (cost of excavation excluded); reinforced concrete covers; concave/convex ends						
914 × 152 × 63/38 mm	2.47	3.03	0.11	4.40	m	**7.43**
914 × 229 × 63/38 mm	13.15	16.16	0.11	4.40	m	**20.56**
914 × 305 × 63/38 mm	14.64	17.99	0.11	4.40	m	**22.39**
Clipped direct to backgrounds including cleats						
95 mm²	53.36	65.55	0.55	22.00	m	**87.55**
120 mm²	56.66	69.60	0.63	25.20	m	**94.80**
150 mm²	66.38	81.54	0.66	26.40	m	**107.94**
185 mm²	70.31	86.37	0.72	28.79	m	**115.16**
240 mm²	81.04	99.56	0.82	32.80	m	**132.36**
Terminations for above cables, including compound joint and glanding off						
95 mm²	524.94	644.89	4.75	190.00	m	**834.89**
120 mm²	534.43	656.54	5.30	212.01	m	**868.55**
150 mm²	558.24	685.80	6.10	244.01	m	**929.81**
185 mm²	559.42	687.25	6.90	275.99	m	**963.24**
240 mm²	588.33	722.77	7.43	297.20	m	**1019.97**
300 mm²	623.91	766.47	8.75	350.00	m	**1116.47**

ELECTRICAL SUPPLY/POWER/LIGHTING

Item	Net Price £	Material £	Labour hours	Labour £	Unit	Total rate £
HV SWITCHGEAR AND TRANSFORMERS						
HV circuit breakers; installed on prepared foundations including all supports, fixings and inter panel connections where relevant. Excludes main and multicore cabling and heat shrink cable termination kits.						
Three phase 11kV, 630 amp, Air or SF6 insulated, with fixed pattern vacuum or SF6 circuit breaker panels; hand charged spring closing operation; prospective fault level up to 25 kA for 3 seconds. Feeders include ammeter with selector switch,VIP relays, overcurrent and earth fault relays with necessary current relays with necessary current transformers; incomers include 3 phase VT, voltmeter and phase selector switch; Includes IDMT overcurrent and earth fault relays/CTs.						
Single panel with cable chamber	29282.00	35972.94	31.70	1268.00	nr	**37240.94**
Three panel with one incomer and two feeders; with cable chambers	42620.00	52358.67	67.83	2713.19	nr	**55071.86**
Five panel with two incoming, two feeders and a bus section; with cable chambers	86270.00	105982.70	99.17	3966.81	nr	**109949.51**
Tripping Batteries						
Battery chargers; switchgear tripping and closing; double wound transfomer and earth screen; including fixing to background, commissioning and testing						
Valve regulated lead acid battery NGTS 3.12.2; BS 6290 Part 2; TPS 9/3; IEEE485						
30 volt; 19 Ah; 3A	2892.33	3553.23	6.50	260.00	nr	**3813.23**
40 volt; 29 Ah; 3A	4633.73	5692.53	8.50	340.00	nr	**6032.53**
110 volt; 19 Ah; 3A	4726.87	5806.96	6.50	260.00	nr	**6066.96**
110 volt; 29 Ah; 3A	5157.57	6336.08	8.50	340.00	nr	**6676.08**
100 volt; 38 Ah; 3A	5511.33	6770.67	10.00	400.00	nr	**7170.67**
Step down transformers; 11/0.415 kV, Dyn 11, 50 Hz. Complete with lifting lugs, mounting skids, provisions for wheels, undrilled gland plates to air-filled cable boxes, off load tapping facility, including UK delivery						
Oil-filled in free breathing ventialted steel tank						
500 kVA	10654.78	13089.39	30.00	1200.00	nr	**14289.39**
800 kVA	12134.11	14906.76	30.00	1200.00	nr	**16106.76**
1000 kVA	13752.44	16894.87	30.00	1200.00	nr	**18094.87**

ELECTRICAL SUPPLY/POWER/LIGHTING

Item	Net Price £	Material £	Labour hours	Labour £	Unit	Total rate £
HV SUPPLY – cont						
Step down transformers – cont						
1250 kVA	16711.06	20529.54	35.00	1400.00	nr	**21929.54**
1500 kVA	19802.74	24327.67	35.00	1400.00	nr	**25727.67**
2000 kVA	25719.97	31596.98	10.00	400.00	nr	**31996.98**
MIDEL – filled in gasket-sealed steel tank						
500 kVA	14112.41	17337.10	30.00	1200.00	nr	**18537.10**
800 kVA	16084.83	19760.21	30.00	1200.00	nr	**20960.21**
1000 kVA	18313.63	22498.30	30.00	1200.00	nr	**23698.30**
1250 kVA	22135.15	27193.04	35.00	1400.00	nr	**28593.04**
1500 kVA	26336.35	32354.21	35.00	1400.00	nr	**33754.21**
2000 kVA	34102.70	41895.17	40.00	1600.00	nr	**43495.17**
Extra for						
Fluid temperature indicator with 2 N/O contacts	395.89	486.35	2.00	80.00	nr	**566.35**
Winding temperature indicator with 2 N/O contacts	969.94	1191.57	2.00	80.00	nr	**1271.57**
Dehydrating breather	98.97	121.58	2.00	80.00	nr	**201.58**
Plain rollers	296.92	364.77	2.00	80.00	nr	**444.77**
Pressure relief device with 1 N/O contact	593.83	729.52	2.00	80.00	nr	**809.52**
Step down transformers; 11/0.415 kV, Dyn 11, 50 Hz. Complete with lifting lugs, mounting skids, provisions for wheels, undrilled gland plates to air-filled cable boxes, off load tapping facility, including UK delivery						
Cast Resin type in ventilated steel encloure, AN – Air Natural including winding temperture indicator with 2 N/O contacts						
500 kVA	15098.61	18548.64	40.00	1600.00	nr	**20148.64**
800 kVA	17564.12	21577.52	40.00	1600.00	nr	**23177.52**
1000 kVA	21247.58	26102.65	40.00	1600.00	nr	**27702.65**
1250 kVA	22998.09	28253.16	45.00	1799.99	nr	**30053.15**
1600 kVA	26089.81	32051.33	45.00	1799.99	nr	**33851.32**
2000 kVA	29541.55	36291.79	50.00	2000.00	nr	**38291.79**
Cast Resin type in ventilated steel enclosure with temperature controlled fans to achieve 40% increase to AN/AF rating. Includes winding temperature indicator with 2 N/O contacts						
500/700 kVA	16577.89	20365.94	42.00	1680.00	nr	**22045.94**
800/1120 kVA	19289.98	23697.74	42.00	1680.00	nr	**25377.74**
1000/1400 kVA	23866.92	29320.51	42.00	1680.00	nr	**31000.51**
12501750 kVA	24853.11	30532.05	47.00	1879.99	nr	**32412.04**
1600/2240 kVA	28077.94	34493.75	47.00	1879.99	nr	**36373.74**
2000/2800 kVA	32022.75	39339.95	52.00	2080.00	nr	**41419.95**

ELECTRICAL SUPPLY/POWER/LIGHTING

Item	Net Price £	Material £	Labour hours	Labour £	Unit	**Total rate £**
LV DISTRIBUTION: CONDUIT AND CABLE TRUNKING						
Heavy gauge, screwed drawn steel; surface fixed on saddles to backgrounds, with standard pattern boxes and fittings including all fixings and supports. (forming holes, conduit entry, draw wires etc. and components for earth continuity are included)						
Black enamelled						
20 mm dia.	3.70	4.55	0.49	19.59	m	**24.14**
25 mm dia.	5.37	6.59	0.56	22.40	m	**28.99**
32 mm dia.	10.98	13.49	0.64	25.60	m	**39.09**
38 mm dia.	13.06	16.04	0.73	29.20	m	**45.24**
50 mm dia.	20.73	25.46	1.04	41.60	m	**67.06**
Galvanized						
20 mm dia.	3.49	4.28	0.49	19.59	m	**23.87**
25 mm dia.	4.76	5.85	0.56	22.40	m	**28.25**
32 mm dia.	10.17	12.49	0.64	25.60	m	**38.09**
38 mm dia.	11.46	14.08	0.73	29.20	m	**43.28**
50 mm dia.	26.04	31.99	1.04	41.60	m	**73.59**
High impact PVC; surface fixed on saddles to backgrounds; with standard pattern boxes and fittings; including all fixings and supports						
Light gauge						
16 mm dia.	0.60	0.74	0.27	10.79	m	**11.53**
20 mm dia.	0.68	0.84	0.28	11.20	m	**12.04**
25 mm dia.	0.93	1.14	0.33	13.20	m	**14.34**
32 mm dia.	1.48	1.82	0.38	15.19	m	**17.01**
38 mm dia.	1.97	2.42	0.44	17.61	m	**20.03**
50 mm dia.	3.25	3.99	0.48	19.20	m	**23.19**
Heavy gauge						
16 mm dia.	0.71	0.87	0.27	10.79	m	**11.66**
20 mm dia.	0.80	0.99	0.28	11.20	m	**12.19**
25 mm dia.	1.10	1.35	0.33	13.20	m	**14.55**
32 mm dia.	1.75	2.15	0.38	15.19	m	**17.34**
38 mm dia.	2.31	2.84	0.44	17.61	m	**20.45**
50 mm dia.	3.82	4.69	0.48	19.20	m	**23.89**
Flexible conduits; including adaptors and locknuts (for connections to equipment)						
Metallic, PVC covered conduit; not exceeding 1 m long; including zinc plated mild steel adaptors, lock nuts and earth conductor						
16 mm dia.	9.51	11.69	0.42	16.81	nr	**28.50**
20 mm dia.	8.95	10.99	0.46	18.41	nr	**29.40**
25 mm dia.	14.10	17.32	0.43	17.20	nr	**34.52**

ELECTRICAL SUPPLY/POWER/LIGHTING

Item	Net Price £	Material £	Labour hours	Labour £	Unit	**Total rate £**
LV DISTRIBUTION: CONDUIT AND CABLE TRUNKING – cont						
Flexible conduits – cont						
32 mm dia.	21.84	26.83	0.51	20.40	nr	**47.23**
38 mm dia.	27.59	33.89	0.56	22.40	nr	**56.29**
50 mm dia.	96.29	118.29	0.82	32.80	nr	**151.09**
PVC conduit; not exceeding 1 m long; including nylon adaptors, lock nuts						
16 mm dia.	5.04	6.19	0.46	18.41	nr	**24.60**
20 mm dia.	5.04	6.19	0.48	19.20	nr	**25.39**
25 mm dia.	6.32	7.76	0.50	20.00	nr	**27.76**
32 mm.dia.	9.17	11.27	0.58	23.19	nr	**34.46**
PVC adaptable boxes; fixed to backgrounds; including all supports and fixings (cutting and connecting conduit to boxes is included)						
Square pattern						
75 × 75 × 53 mm	2.21	2.72	0.69	27.60	nr	**30.32**
100 × 100 × 75 mm	3.73	4.58	0.71	28.40	nr	**32.98**
150 × 150 × 75 mm	4.79	5.88	0.80	32.00	nr	**37.88**
Terminal strips to be fixed in metal or polythene adaptable boxes)						
20 amp high density polythene						
2 way	1.25	1.53	0.23	9.20	nr	**10.73**
3 way	1.38	1.69	0.23	9.20	nr	**10.89**
4 way	1.47	1.81	0.23	9.20	nr	**11.01**
5 way	1.54	1.89	0.23	9.20	nr	**11.09**
6 way	1.79	2.19	0.25	10.00	nr	**12.19**
7 way	1.83	2.25	0.25	10.00	nr	**12.25**
8 way	2.02	2.48	0.29	11.59	nr	**14.07**
9 way	2.14	2.63	0.30	12.00	nr	**14.63**
10 way	2.47	3.03	0.34	13.60	nr	**16.63**
11 way	2.51	3.09	0.34	13.60	nr	**16.69**
12 way	2.59	3.18	0.34	13.60	nr	**16.78**
13 way	3.00	3.69	0.37	14.81	nr	**18.50**
14 way	3.34	4.11	0.37	14.81	nr	**18.92**
15 way	3.42	4.20	0.39	15.60	nr	**19.80**
16 way	3.51	4.32	0.45	18.00	nr	**22.32**
18 way	3.76	4.62	0.45	18.00	nr	**22.62**

ELECTRICAL SUPPLY/POWER/LIGHTING

Item	Net Price £	Material £	Labour hours	Labour £	Unit	**Total rate £**
LV DISTRIBUTION: TRUNKING						
Galvanized steel trunking; fixed to backgrounds; jointed with standard connectors (including plates for air gap between trunking and background); earth continuity straps included						
Single compartment						
50 × 50 mm	5.85	7.18	0.39	15.60	m	**22.78**
75 × 50 mm	8.93	10.97	0.44	17.61	m	**28.58**
75 × 75 mm	9.55	11.73	0.47	18.80	m	**30.53**
100 × 50 mm	11.10	13.64	0.50	20.00	m	**33.64**
100 × 75 mm	11.98	14.72	0.57	22.81	m	**37.53**
100 × 100 mm	12.16	14.94	0.62	24.80	m	**39.74**
150 × 50 mm	16.16	19.86	0.78	31.21	m	**51.07**
150 × 100 mm	19.57	24.05	0.78	31.21	m	**55.26**
150 × 150 mm	20.91	25.68	0.86	34.40	m	**60.08**
225 × 75 mm	26.59	32.67	0.88	35.20	m	**67.87**
225 × 150 mm	33.23	40.82	0.84	33.60	m	**74.42**
225 × 225 mm	39.55	48.58	0.99	39.60	m	**88.18**
300 × 75 mm	27.34	33.59	0.96	38.40	m	**71.99**
300 × 100 mm	31.48	38.67	0.99	39.60	m	**78.27**
300 × 150 mm	24.08	29.58	0.99	39.60	m	**69.18**
300 × 225 mm	20.49	25.17	1.09	43.60	m	**68.77**
300 × 300 mm	52.08	63.98	1.16	46.40	m	**110.38**
Double compartment						
100 × 50 mm	7.84	9.63	0.54	21.60	m	**31.23**
100 × 75 mm	8.45	10.38	0.62	24.80	m	**35.18**
100 × 100 mm	8.77	10.77	0.66	26.40	m	**37.17**
150 × 50 mm	11.90	14.62	0.70	27.99	m	**42.61**
150 × 100 mm	14.76	18.13	0.83	33.19	m	**51.32**
150 × 150 mm	15.98	19.64	0.92	36.80	m	**56.44**
Triple compartment						
150 × 50 mm	13.71	16.84	0.79	31.59	m	**48.43**
150 × 100 mm	14.08	17.29	0.78	31.21	m	**48.50**
150 × 150 mm	17.24	21.18	1.01	40.40	m	**61.58**
Galvanized steel trunking fittings; cutting and jointing trunking to fittings is included						
Stop end						
50 × 50 mm	1.88	2.31	0.19	7.60	nr	**9.91**
75 × 50 mm	1.99	2.45	0.20	8.00	nr	**10.45**
75 × 75 mm	1.99	2.45	0.21	8.40	nr	**10.85**
100 × 50 mm	2.06	2.53	0.31	12.40	nr	**14.93**
100 × 75 mm	2.23	2.74	0.27	10.79	nr	**13.53**
100 × 100 mm	2.18	2.68	0.27	10.79	nr	**13.47**
150 × 50 mm	2.44	2.99	0.28	11.20	nr	**14.19**
150 × 100 mm	3.43	4.21	0.30	12.00	nr	**16.21**
150 × 150 mm	2.73	3.35	0.32	12.80	nr	**16.15**
225 × 75 mm	2.84	3.49	0.35	14.01	nr	**17.50**
225 × 150 mm	3.10	3.81	0.37	14.81	nr	**18.62**

ELECTRICAL SUPPLY/POWER/LIGHTING

Item	Net Price £	Material £	Labour hours	Labour £	Unit	Total rate £
LV DISTRIBUTION: TRUNKING – cont						
Galvanized steel trunking fittings – cont						
225 × 225 mm	3.37	4.14	0.38	15.19	nr	**19.33**
300 × 75 mm	4.92	6.05	0.42	16.81	nr	**22.86**
300 × 100 mm	4.10	5.04	0.42	16.81	nr	**21.85**
300 × 150 mm	4.92	6.05	0.43	17.20	nr	**23.25**
300 × 225 mm	5.91	7.26	0.45	18.00	nr	**25.26**
300 × 300 mm	6.17	7.58	0.48	19.20	nr	**26.78**
Flanged connector						
50 × 50 mm	1.99	2.45	0.19	7.60	nr	**10.05**
75 × 50 mm	3.05	3.75	0.20	8.00	nr	**11.75**
75 × 75 mm	2.47	3.03	0.21	8.40	nr	**11.43**
100 × 50 mm	3.17	3.90	0.26	10.41	nr	**14.31**
100 × 75 mm	3.27	4.02	0.27	10.79	nr	**14.81**
100 × 100 mm	3.17	3.90	0.27	10.79	nr	**14.69**
150 × 50 mm	2.66	3.27	0.28	11.20	nr	**14.47**
150 × 100 mm	3.43	4.21	0.30	12.00	nr	**16.21**
150 × 150 mm	3.47	4.26	0.32	12.80	nr	**17.06**
225 × 75 mm	2.84	3.49	0.35	14.01	nr	**17.50**
225 × 150 mm	3.37	4.14	0.37	14.81	nr	**18.95**
225 × 225 mm	4.87	5.99	0.38	15.19	nr	**21.18**
300 × 75 mm	4.10	5.04	0.42	16.81	nr	**21.85**
300 × 100 mm	4.92	6.05	0.42	16.81	nr	**22.86**
300 × 150 mm	5.41	6.65	0.43	17.20	nr	**23.85**
300 × 225 mm	5.95	7.31	0.45	18.00	nr	**25.31**
300 × 300 mm	6.17	7.58	0.48	19.20	nr	**26.78**
Bends 90°; single compartment						
50 × 50 mm	8.27	10.16	0.42	16.81	nr	**26.97**
75 × 50 mm	10.47	12.86	0.45	18.00	nr	**30.86**
75 × 75 mm	10.02	12.31	0.48	19.20	nr	**31.51**
100 × 50 mm	10.70	13.15	0.53	21.20	nr	**34.35**
100 × 75 mm	10.98	13.49	0.56	22.40	nr	**35.89**
100 × 100 mm	10.98	13.49	0.58	23.19	nr	**36.68**
150 × 50 mm	13.57	16.67	0.64	25.60	nr	**42.27**
150 × 100 mm	17.31	21.26	0.91	36.40	nr	**57.66**
150 × 150 mm	16.51	20.29	0.89	35.61	nr	**55.90**
225 × 75 mm	25.00	30.71	0.76	30.41	nr	**61.12**
225 × 150 mm	32.06	39.39	0.82	32.80	nr	**72.19**
225 × 225 mm	36.91	45.34	0.83	33.19	nr	**78.53**
300 × 75 mm	34.44	42.30	0.85	34.01	nr	**76.31**
300 × 100 mm	34.84	42.80	0.90	35.99	nr	**78.79**
300 × 150 mm	38.33	47.09	0.96	38.40	nr	**85.49**
300 × 225 mm	42.16	51.80	0.98	39.20	nr	**91.00**
300 × 300 mm	42.98	52.80	1.06	42.40	nr	**95.20**
Bends 90°; double compartment						
100 × 50 mm	9.96	12.23	0.53	21.20	nr	**33.43**
100 × 75 mm	10.48	12.87	0.56	22.40	nr	**35.27**
100 × 100 mm	11.03	13.56	0.58	23.19	nr	**36.75**
150 × 50 mm	10.94	13.44	0.65	26.00	nr	**39.44**
150 × 100 mm	11.53	14.16	0.69	27.60	nr	**41.76**
150 × 150 mm	12.13	14.90	0.73	29.20	nr	**44.10**

ELECTRICAL SUPPLY/POWER/LIGHTING

Item	Net Price £	Material £	Labour hours	Labour £	Unit	**Total rate £**
Bends 90°; triple compartment						
150 × 50 mm	14.38	17.66	0.68	27.19	nr	**44.85**
150 × 100 mm	20.10	24.70	0.73	29.20	nr	**53.90**
150 × 150 mm	21.16	26.00	0.77	30.80	nr	**56.80**
Tees; single compartment						
50 × 50 mm	7.52	9.24	0.56	22.40	nr	**31.64**
75 × 50 mm	11.73	14.41	0.57	22.81	nr	**37.22**
75 × 75 mm	11.36	13.95	0.60	24.00	nr	**37.95**
100 × 50 mm	13.73	16.86	0.65	26.00	nr	**42.86**
100 × 75 mm	13.81	16.97	0.72	28.79	nr	**45.76**
100 × 100 mm	13.25	16.28	0.71	28.40	nr	**44.68**
150 × 50 mm	15.71	19.30	0.82	32.80	nr	**52.10**
150 × 100 mm	19.51	23.97	0.84	33.60	nr	**57.57**
150 × 150 mm	18.92	23.25	0.91	36.40	nr	**59.65**
225 × 75 mm	33.67	41.36	0.94	37.60	nr	**78.96**
225 × 150 mm	45.48	55.87	1.01	40.40	nr	**96.27**
225 × 225 mm	52.05	63.94	1.02	40.80	nr	**104.74**
300 × 75 mm	48.60	59.70	1.07	42.80	nr	**102.50**
300 × 100 mm	50.36	61.87	1.07	42.80	nr	**104.67**
300 × 150 mm	54.26	66.65	1.14	45.60	nr	**112.25**
300 × 225 mm	58.75	72.18	1.19	47.61	nr	**119.79**
300 × 300 mm	62.14	76.34	1.26	50.41	nr	**126.75**
Tees; double compartment						
100 × 50 mm	9.67	11.88	0.65	26.00	nr	**37.88**
100 × 75 mm	12.22	15.02	0.71	28.40	nr	**43.42**
100 × 100 mm	11.96	14.69	0.72	28.79	nr	**43.48**
150 × 50 mm	12.70	15.60	0.82	32.80	nr	**48.40**
150 × 100 mm	13.30	16.34	0.85	34.01	nr	**50.35**
150 × 150 mm	14.92	18.33	0.91	36.40	nr	**54.73**
Tees; triple compartment						
150 × 50 mm	14.52	17.84	0.87	34.81	nr	**52.65**
150 × 100 mm	15.97	19.61	0.89	35.61	nr	**55.22**
150 × 150 mm	21.50	26.41	0.96	38.40	nr	**64.81**
Crossovers; single compartment						
50 × 50 mm	12.77	15.69	0.65	26.00	nr	**41.69**
75 × 50 mm	17.31	21.26	0.66	26.40	nr	**47.66**
75 × 75 mm	17.71	21.76	0.69	27.60	nr	**49.36**
100 × 50 mm	21.87	26.87	0.74	29.60	nr	**56.47**
100 × 75 mm	16.13	19.81	0.80	32.00	nr	**51.81**
100 × 100 mm	22.04	27.08	0.81	32.39	nr	**59.47**
150 × 50 mm	26.10	32.07	0.91	36.40	nr	**68.47**
150 × 100 mm	26.49	32.54	0.94	37.60	nr	**70.14**
150 × 150 mm	25.38	31.17	0.99	39.60	nr	**70.77**
225 × 75 mm	46.48	57.10	1.01	40.40	nr	**97.50**
225 × 150 mm	50.73	62.32	1.08	43.20	nr	**105.52**
225 × 225 mm	69.02	84.79	1.09	43.60	nr	**128.39**
300 × 75 mm	64.99	79.84	1.14	45.60	nr	**125.44**
300 × 100 mm	66.94	82.24	1.16	46.40	nr	**128.64**
300 × 150 mm	72.29	88.81	1.19	47.61	nr	**136.42**
300 × 225 mm	76.48	93.95	1.21	48.40	nr	**142.35**
300 × 300 mm	81.53	100.16	1.29	51.60	nr	**151.76**

ELECTRICAL SUPPLY/POWER/LIGHTING

Item	Net Price £	Material £	Labour hours	Labour £	Unit	Total rate £
LV DISTRIBUTION: TRUNKING – cont						
Galvanized steel trunking fittings – cont						
Crossovers; double compartment						
100 × 50 mm	12.83	15.76	0.74	29.60	nr	**45.36**
100 × 75 mm	13.46	16.54	0.80	32.00	nr	**48.54**
100 × 100 mm	13.76	16.90	0.81	32.39	nr	**49.29**
150 × 50 mm	15.21	18.69	0.86	34.40	nr	**53.09**
150 × 100 mm	15.25	18.73	0.94	37.60	nr	**56.33**
150 × 150 mm	23.19	28.49	1.00	40.01	nr	**68.50**
Crossovers; triple compartment						
150 × 50 mm	16.64	20.44	0.97	38.80	nr	**59.24**
150 × 100 mm	18.30	22.48	0.99	39.60	nr	**62.08**
150 × 150 mm	25.49	31.31	1.06	42.40	nr	**73.71**
Galvanized steel flush floor trunking; fixed to backgrounds; supports and fixings; standard coupling joints; earth continuity straps included						
Triple compartment						
350 × 60 mm	43.79	53.79	1.32	52.80	m	**106.59**
Four compartment						
350 × 60 mm	45.59	56.01	1.32	52.80	m	**108.81**
Galvanized steel flush floor trunking; fittings (cutting and jointing trunking to fittings is included)						
Stop end; triple compartment						
350 × 60 mm	4.59	5.64	0.53	21.20	nr	**26.84**
Stop end; four compartment						
350 × 60 mm	10.65	13.08	0.53	21.20	nr	**34.28**
Rising bend; standard; triple compartment						
350 × 60 mm	37.88	46.54	1.30	52.01	nr	**98.55**
Rising bend; standard; four compartment						
350 × 60 mm	39.69	48.76	1.30	52.01	nr	**100.77**
Rising bend; skirting; triple compartment						
350 × 60 mm	74.31	91.29	1.33	53.19	nr	**144.48**
Rising bend; skirting; four compartment						
350 × 60 mm	84.34	103.61	1.33	53.19	nr	**156.80**
Junction box; triple compartment						
350 × 60 mm	49.39	60.68	1.16	46.40	nr	**107.08**
Junction box; four compartment						
350 × 60 mm	51.37	63.10	1.16	46.40	nr	**109.50**
Body coupler (pair)						
3 and 4 compartment	2.55	3.13	0.16	6.40	nr	**9.53**
Service outlet module comprising flat lid with flanged carpet trim; twin 13 A outlet and drilled plate for mounting 2 telephone outlets; one blank plate; triple compartment						
3 compartment	61.26	75.25	0.47	18.80	nr	**94.05**

ELECTRICAL SUPPLY/POWER/LIGHTING

Item	Net Price £	Material £	Labour hours	Labour £	Unit	Total rate £
Service outlet module comprising flat lid with flanged carpet trim; twin 13 A outlet and drilled plate for mounting 2 telephone outlets; two blank plates; four compartment						
4 compartment	68.24	83.83	0.47	18.80	nr	**102.63**
Single compartment PVC trunking; grey finish; clip on lid; fixed to backgrounds; including supports and fixings (standard coupling joints)						
Dimensions						
50 × 50 mm	10.55	12.96	0.27	10.79	m	**23.75**
75 × 50 mm	11.43	14.04	0.28	11.20	m	**25.24**
75 × 75 mm	12.95	15.91	0.29	11.59	m	**27.50**
100 × 50 mm	14.67	18.02	0.34	13.60	m	**31.62**
100 × 75 mm	16.09	19.77	0.37	14.81	m	**34.58**
100 × 100 mm	16.99	20.87	0.37	14.81	m	**35.68**
150 × 50 mm	14.75	18.12	0.41	16.40	m	**34.52**
150 × 75 mm	26.28	32.29	0.44	17.61	m	**49.90**
150 × 100 mm	31.63	38.86	0.44	17.61	m	**56.47**
150 × 150 mm	32.36	39.75	0.48	19.20	m	**58.95**
Single compartment PVC trunking; fittings (cutting and jointing trunking to fittings is included)						
Crossover						
50 × 50 mm	17.28	21.23	0.29	11.59	nr	**32.82**
75 × 50 mm	19.16	23.54	0.30	12.00	nr	**35.54**
75 × 75 mm	20.79	25.54	0.31	12.40	nr	**37.94**
100 × 50 mm	27.72	34.05	0.35	14.01	nr	**48.06**
100 × 75 mm	34.96	42.94	0.36	14.40	nr	**57.34**
100 × 100 mm	30.92	37.99	0.40	15.99	nr	**53.98**
150 × 75 mm	39.61	48.66	0.45	18.00	nr	**66.66**
150 × 100 mm	47.52	58.38	0.46	18.41	nr	**76.79**
150 × 150 mm	73.76	90.62	0.47	18.80	nr	**109.42**
Stop end						
50 × 50 mm	0.72	0.88	0.12	4.80	nr	**5.68**
75 × 50 mm	1.35	1.66	0.12	4.80	nr	**6.46**
75 × 75 mm	1.03	1.27	0.13	5.20	nr	**6.47**
100 × 50 mm	1.84	2.26	0.16	6.40	nr	**8.66**
100 × 75 mm	2.88	3.54	0.16	6.40	nr	**9.94**
100 × 100 mm	2.89	3.55	0.18	7.20	nr	**10.75**
150 × 75 mm	7.52	9.24	0.20	8.00	nr	**17.24**
150 × 100 mm	9.31	11.43	0.21	8.40	nr	**19.83**
150 × 150 mm	9.50	11.67	0.22	8.80	nr	**20.47**
Flanged coupling						
50 × 50 mm	4.33	5.32	0.32	12.80	nr	**18.12**
75 × 50 mm	4.95	6.08	0.33	13.20	nr	**19.28**
75 × 75 mm	5.94	7.30	0.34	13.60	nr	**20.90**

ELECTRICAL SUPPLY/POWER/LIGHTING

Item	Net Price £	Material £	Labour hours	Labour £	Unit	Total rate £
LV DISTRIBUTION: TRUNKING – cont						
Single compartment PVC trunking – cont						
100 × 50 mm	6.69	8.22	0.44	17.61	nr	**25.83**
100 × 75 mm	7.58	9.31	0.45	18.00	nr	**27.31**
100 × 100 mm	8.07	9.91	0.46	18.41	nr	**28.32**
150 × 75 mm	8.55	10.50	0.57	22.81	nr	**33.31**
150 × 100 mm	9.01	11.07	0.57	22.81	nr	**33.88**
150 × 150 mm	9.45	11.61	0.59	23.60	nr	**35.21**
Internal coupling						
50 × 50 mm	1.53	1.88	0.07	2.80	nr	**4.68**
75 × 50 mm	1.81	2.23	0.07	2.80	nr	**5.03**
75 × 75 mm	1.82	2.24	0.07	2.80	nr	**5.04**
100 × 50 mm	2.43	2.98	0.08	3.19	nr	**6.17**
100 × 75 mm	2.73	3.35	0.08	3.19	nr	**6.54**
100 × 100 mm	3.01	3.70	0.08	3.19	nr	**6.89**
External coupling						
50 × 50 mm	1.67	2.05	0.09	3.60	nr	**5.65**
75 × 50 mm	2.00	2.46	0.09	3.60	nr	**6.06**
75 × 75 mm	1.98	2.44	0.09	3.60	nr	**6.04**
100 × 50 mm	2.66	3.27	0.10	4.00	nr	**7.27**
100 × 75 mm	3.03	3.73	0.10	4.00	nr	**7.73**
100 × 100 mm	3.30	4.05	0.10	4.00	nr	**8.05**
150 × 75 mm	3.78	4.64	0.11	4.40	nr	**9.04**
150 × 100 mm	3.94	4.84	0.11	4.40	nr	**9.24**
150 × 150 mm	4.08	5.01	0.11	4.40	nr	**9.41**
Angle; flat cover						
50 × 50 mm	6.13	7.53	0.18	7.20	nr	**14.73**
75 × 50 mm	8.17	10.04	0.19	7.60	nr	**17.64**
75 × 75 mm	9.40	11.55	0.20	8.00	nr	**19.55**
100 × 50 mm	14.04	17.25	0.23	9.20	nr	**26.45**
100 × 75 mm	21.38	26.26	0.26	10.41	nr	**36.67**
100 × 100 mm	19.50	23.95	0.26	10.41	nr	**34.36**
150 × 75 mm	26.58	32.65	0.30	12.00	nr	**44.65**
150 × 100 mm	31.61	38.83	0.33	13.20	nr	**52.03**
150 × 150 mm	47.91	58.85	0.34	13.60	nr	**72.45**
Angle; internal or external cover						
50 × 50 mm	7.46	9.17	0.18	7.20	nr	**16.37**
75 × 50 mm	10.28	12.63	0.19	7.60	nr	**20.23**
75 × 75 mm	13.21	16.23	0.20	8.00	nr	**24.23**
100 × 50 mm	14.23	17.48	0.23	9.20	nr	**26.68**
100 × 75 mm	22.88	28.11	0.26	10.41	nr	**38.52**
100 × 100 mm	22.98	28.23	0.26	10.41	nr	**38.64**
150 × 75 mm	28.09	34.51	0.30	12.00	nr	**46.51**
150 × 100 mm	33.11	40.68	0.33	13.20	nr	**53.88**
150 × 150 mm	46.51	57.14	0.34	13.60	nr	**70.74**

ELECTRICAL SUPPLY/POWER/LIGHTING

Item	Net Price £	Material £	Labour hours	Labour £	Unit	**Total rate £**
Tee; flat cover						
50 × 50 mm	5.13	6.30	0.24	9.60	nr	**15.90**
75 × 50 mm	7.83	9.62	0.25	10.00	nr	**19.62**
75 × 75 mm	8.05	9.89	0.26	10.41	nr	**20.30**
100 × 50 mm	17.39	21.37	0.32	12.80	nr	**34.17**
100 × 75 mm	18.47	22.69	0.33	13.20	nr	**35.89**
100 × 100 mm	24.10	29.61	0.34	13.60	nr	**43.21**
150 × 75 mm	31.88	39.16	0.41	16.40	nr	**55.56**
150 × 100 mm	40.89	50.23	0.42	16.81	nr	**67.04**
150 × 150 mm	55.62	68.33	0.44	17.61	nr	**85.94**
Tee; internal or external cover						
50 × 50 mm	14.55	17.87	0.24	9.60	nr	**27.47**
75 × 50 mm	15.98	19.64	0.25	10.00	nr	**29.64**
75 × 75 mm	17.77	21.83	0.26	10.41	nr	**32.24**
100 × 50 mm	22.81	28.02	0.32	12.80	nr	**40.82**
100 × 75 mm	26.06	32.01	0.33	13.20	nr	**45.21**
100 × 100 mm	29.24	35.92	0.34	13.60	nr	**49.52**
150 × 75 mm	37.88	46.54	0.41	16.40	nr	**62.94**
150 × 100 mm	45.65	56.08	0.42	16.81	nr	**72.89**
150 × 150 mm	60.95	74.88	0.44	17.61	nr	**92.49**
Division Strip (1.8 m long)						
50 mm	7.89	9.69	0.07	2.80	nr	**12.49**
75 mm	10.10	12.41	0.07	2.80	nr	**15.21**
100 mm	12.84	15.77	0.08	3.19	nr	**18.96**
PVC miniature trunking; white finish; fixed to backgrounds; including supports and fixing; standard coupling joints						
Single compartment						
16 × 16 mm	1.42	1.74	0.20	8.00	m	**9.74**
25 × 16 mm	1.74	2.14	0.21	8.40	m	**10.54**
38 × 16 mm	2.17	2.67	0.24	9.60	m	**12.27**
38 × 25 mm	2.59	3.18	0.25	10.00	m	**13.18**
Compartmented						
38 × 16 mm	2.54	3.12	0.24	9.60	m	**12.72**
38 × 25 mm	3.03	3.73	0.25	10.00	m	**13.73**
PVC miniature trunking fittings; single compartment; white finish; cutting and jointing trunking to fittings is included						
Coupling						
16 × 16 mm	0.44	0.54	0.10	4.00	nr	**4.54**
25 × 16 mm	0.44	0.54	0.12	4.60	nr	**5.14**
38 × 16 mm	0.44	0.54	0.12	4.80	nr	**5.34**
38 × 25 mm	1.04	1.28	0.14	5.61	nr	**6.89**
Stop end						
16 × 16 mm	0.44	0.54	0.12	4.80	nr	**5.34**
25 × 16 mm	0.44	0.54	0.13	5.08	nr	**5.62**
38 × 16 mm	0.44	0.54	0.15	6.00	nr	**6.54**
38 × 25 mm	1.04	1.28	0.17	6.80	nr	**8.08**

ELECTRICAL SUPPLY/POWER/LIGHTING

Item	Net Price £	Material £	Labour hours	Labour £	Unit	Total rate £
LV DISTRIBUTION: TRUNKING – cont						
PVC miniature trunking fittings – cont						
Bend; flat, internal or external						
16 × 16 mm	0.44	0.54	0.18	7.20	nr	**7.74**
25 × 16 mm	0.44	0.54	0.20	7.93	nr	**8.47**
38 × 16 mm	0.44	0.54	0.21	8.40	nr	**8.94**
38 × 25 mm	1.04	1.28	0.23	9.20	nr	**10.48**
Tee						
16 × 16 mm	0.74	0.91	0.19	7.60	nr	**8.51**
25 × 16 mm	0.74	0.91	0.23	9.20	nr	**10.11**
38 × 16 mm	0.74	0.91	0.26	10.41	nr	**11.32**
38 × 25 mm	1.03	1.27	0.29	11.59	nr	**12.86**
PVC bench trunking; white or grey finish; fixed to backgrounds; including supports and fixings; standard coupling joints						
Trunking						
90 × 90 mm	23.89	29.35	0.33	13.20	m	**42.55**
PVC bench trunking fittings; white or grey finish; cutting and jointing trunking to fittings is included						
Stop end						
90 × 90 mm	5.06	6.22	0.09	3.60	nr	**9.82**
Coupling						
90 × 90 mm	3.13	3.84	0.09	3.60	nr	**7.44**
Internal or external bend						
90 × 90 mm	17.18	21.10	0.28	11.20	nr	**32.30**
Socket plate						
90 × 90 mm – 1 gang	0.97	1.19	0.10	4.00	nr	**5.19**
90 × 90 mm – 2 gang	1.16	1.43	0.10	4.00	nr	**5.43**
PVC underfloor trunking; single compartment; fitted in floor screed; standard coupling joints						
Trunking						
60 × 25 mm	11.72	14.40	0.22	8.80	m	**23.20**
90 × 35 mm	16.88	20.74	0.27	10.79	m	**31.53**
PVC underfloor trunking fittings; single compartment; fitted in floor screed; (cutting and jointing trunking to fittings is included)						
Jointing sleeve						
60 × 25 mm	0.78	0.96	0.08	3.19	nr	**4.15**
90 × 35 mm	1.30	1.60	0.10	4.00	nr	**5.60**

ELECTRICAL SUPPLY/POWER/LIGHTING

Item	Net Price £	Material £	Labour hours	Labour £	Unit	**Total rate £**
Duct connector						
90 × 35 mm	0.60	0.74	0.17	6.80	nr	**7.54**
Socket reducer						
90 × 35 mm	1.08	1.32	0.12	4.80	nr	**6.12**
Vertical access box; 2 compartment						
Shallow	60.61	74.46	0.37	14.81	nr	**89.27**
Duct bend; vertical						
60 × 25 mm	11.82	14.52	0.27	10.79	nr	**25.31**
90 × 35 mm	13.34	16.39	0.35	14.01	nr	**30.40**
Duct bend; horizontal						
60 × 25 mm	13.94	17.13	0.30	12.00	nr	**29.13**
90 × 35 mm	14.15	17.39	0.37	14.81	nr	**32.20**
Zinc coated steel underfloor ducting; fixed to backgrounds; standard coupling joints; earth continuity straps; (Including supports and fixing, packing shims where required)						
Double compartment						
150 × 25 mm	9.54	11.72	0.57	22.81	m	**34.53**
Triple compartment						
225 × 25 mm	16.90	20.76	0.93	37.20	m	**57.96**
Zinc coated steel underfloor ducting fittings; (cutting and jointing to fittings is included)						
Stop end; double compartment						
150 × 25 mm	3.00	3.69	0.31	12.40	nr	**16.09**
Stop end; triple compartment						
225 × 25 mm	3.43	4.21	0.37	14.81	nr	**19.02**
Rising bend; double compartment; standard trunking						
150 × 25 mm	19.31	23.72	0.71	28.40	nr	**52.12**
Rising bend; triple compartment; standard trunking						
225 × 25 mm	34.70	42.63	0.85	34.01	nr	**76.64**
Rising bend; double compartment; to skirting						
150 × 25 mm	42.98	52.80	0.90	36.00	nr	**88.80**
Rising bend; triple compartment; to skirting						
225 × 25 mm	50.68	62.26	0.95	38.00	nr	**100.26**
horizontal bend; double compartment						
150 × 25 mm	29.36	36.07	0.64	25.60	nr	**61.67**
Horizontal bend; triple compartment						
225 × 25 mm	33.86	41.60	0.77	30.80	nr	**72.40**
Junction or service outlet boxes; terminal; double compartment						
150 mm	30.38	37.32	0.91	36.40	nr	**73.72**
Junction or service outlet boxes; terminal; triple compartment						
225 mm	34.72	42.65	1.11	44.39	nr	**87.04**

ELECTRICAL SUPPLY/POWER/LIGHTING

Item	Net Price £	Material £	Labour hours	Labour £	Unit	Total rate £
LV DISTRIBUTION: TRUNKING – cont						
Zinc coated steel underfloor ducting fittings – cont						
Junction or service outlet boxes; through or angle; double compartment						
150 mm	40.34	49.56	0.97	38.80	nr	**88.36**
Junction or service outlet boxes; through or angle; triple compartment						
225 mm	44.81	55.05	1.17	46.81	nr	**101.86**
Junction or service outlet boxes; tee; double compartment						
150 mm	40.34	49.56	1.02	40.80	nr	**90.36**
Junction or service outlet boxes; tee; triple compartment						
225 mm	44.81	55.05	1.22	48.79	nr	**103.84**
Junction or service outlet boxes; cross; double compartment						
up to 150 mm	40.34	49.56	1.03	41.20	nr	**90.76**
Junction or service outlet boxes; cross;triple compartment						
225 mm	44.81	55.05	1.23	49.20	nr	**104.25**
Plates for junction/inspection boxes; double and triple compartment						
Blank plate	8.81	10.83	0.92	36.80	nr	**47.63**
Conduit entry plate	11.06	13.59	0.86	34.40	nr	**47.99**
Trunking entry plate	11.06	13.59	0.86	34.40	nr	**47.99**
Service outlet box comprising flat lid with flanged carpet trim; twin 13A outlet and drilled plate for mounting 2 telephone outlets and terminal blocks; terminal outlet box; double compartment						
150 × 25 mm trunking	62.91	77.28	1.68	67.20	nr	**144.48**
Service outlet box comprising flat lid with flanged carpet trim; twin 13A outlet and drilled plate for mounting 2 telephone outlets and terminal blocks; terminal outlet box; triple compartment						
225 × 25 mm trunking	70.02	86.02	1.93	77.20	nr	**163.22**
PVC skirting/dado modular trunking; white (cutting and jointing trunking to fittings and backplates for fixing to walls is included)						
Main carrier/backplate						
50 × 170 mm	18.73	23.01	0.22	8.80	m	**31.81**
62 × 190 mm	20.79	25.54	0.22	8.80	m	**34.34**
Extension carrier/backplate						
50 × 42 mm	11.38	13.98	0.58	23.19	m	**37.17**

ELECTRICAL SUPPLY/POWER/LIGHTING

Item	Net Price £	Material £	Labour hours	Labour £	Unit	**Total rate £**
Carrier/backplate						
Including cover seal	6.64	8.16	0.53	21.20	m	**29.36**
Chamfered covers for fixing to backplates						
50 × 42 mm	3.86	4.75	0.33	13.20	m	**17.95**
Square covers for fixing to backplates						
50 × 42 mm	7.74	9.51	0.33	13.20	m	**22.71**
Plain covers for fixing to backplates						
85 mm	3.86	4.75	0.34	13.60	m	**18.35**
Retainers-clip to backplates to hold cables						
For chamfered covers	1.07	1.31	0.07	2.80	m	**4.11**
For square-recessed covers	0.92	1.13	0.07	2.80	m	**3.93**
For plain covers	3.56	4.38	0.07	2.80	m	**7.18**
Pre-packaged corner assemblies						
Internal; for 170 × 50 Assy	7.48	9.19	0.51	20.40	nr	**29.59**
Internal; for 190 × 62 Assy	8.31	10.21	0.51	20.40	nr	**30.61**
Internal; for 215 × 50 Assy	9.30	11.42	0.53	21.20	nr	**32.62**
Internal; for 254 × 50 Assy	11.08	13.61	0.53	21.20	nr	**34.81**
External; for 170 × 50 Assy	7.48	9.19	0.56	22.40	nr	**31.59**
External; for 190 × 62 Assy	8.31	10.21	0.56	22.40	nr	**32.61**
External; for 215 × 50 Assy	9.30	11.42	0.58	23.19	nr	**34.61**
External; for 254 × 50 Assy	11.08	13.61	0.58	23.19	nr	**36.80**
Clip on end caps						
170 × 50 Assy	4.48	5.50	0.11	4.40	nr	**9.90**
215 × 50 Assy	5.30	6.51	0.11	4.40	nr	**10.91**
254 × 50 Assy	6.22	7.64	0.11	4.40	nr	**12.04**
190 × 62 Assy	6.26	7.69	0.11	4.40	nr	**12.09**
Outlet box						
1 Gang; in horizontal trunking; clip in	4.00	4.91	0.34	13.60	nr	**18.51**
2 Gang; in horizontal trunking; clip in	4.99	6.13	0.34	13.60	nr	**19.73**
1 Gang; in vertical trunking; clip in	4.00	4.91	0.34	13.60	nr	**18.51**
Sheet steel adaptable boxes; with plain or knockout sides; fixed to backgrounds; including supports and fixings (cutting and connecting conduit to boxes is included)						
Square pattern – black						
75 × 75 × 37 mm	2.21	2.72	0.69	27.60	nr	**30.32**
75 × 75 × 50 mm	2.10	2.58	0.69	27.60	nr	**30.18**
75 × 75 × 75 mm	2.40	2.95	0.69	27.60	nr	**30.55**
100 × 100 × 50 mm	2.21	2.72	0.71	28.40	nr	**31.12**
150 × 150 × 50 mm	3.35	4.12	0.79	31.59	nr	**35.71**
150 × 150 × 75 mm	3.94	4.84	0.80	32.00	nr	**36.84**
150 × 150 × 100 mm	6.55	8.04	0.80	32.00	nr	**40.04**
200 × 200 × 50 mm	5.25	6.45	0.80	32.00	nr	**38.45**
225 × 225 × 50 mm	6.71	8.24	0.93	37.20	nr	**45.44**
225 × 225 × 100 mm	8.90	10.93	0.94	37.60	nr	**48.53**
300 × 300 × 100 mm	9.55	11.73	0.99	39.60	nr	**51.33**

ELECTRICAL SUPPLY/POWER/LIGHTING

Item	Net Price £	Material £	Labour hours	Labour £	Unit	Total rate £
LV DISTRIBUTION: TRUNKING – cont						
Sheet steel adaptable boxes – cont						
Square pattern – galvanized						
75 × 75 × 37 mm	2.03	2.50	0.69	27.60	nr	**30.10**
75 × 75 × 50 mm	2.61	3.20	0.70	27.99	nr	**31.19**
75 × 75 × 75 mm	2.36	2.90	0.69	27.60	nr	**30.50**
100 × 100 × 50 mm	2.85	3.50	0.71	28.40	nr	**31.90**
150 × 150 × 50 mm	3.43	4.21	0.84	33.60	nr	**37.81**
150 × 150 × 75 mm	4.45	5.47	0.80	32.00	nr	**37.47**
150 × 150 × 100 mm	5.42	6.66	0.80	32.00	nr	**38.66**
225 × 225 × 50 mm	6.28	7.72	0.93	37.20	nr	**44.92**
225 × 225 × 100 mm	6.60	8.11	0.94	37.60	nr	**45.71**
300 × 300 × 100 mm	9.20	11.20	1.10	38.88	nr	**50.08**
Rectangular pattern – black						
100 × 75 × 50 mm	3.53	4.34	0.69	27.60	nr	**31.94**
150 × 75 × 50 mm	3.70	4.55	0.70	27.99	nr	**32.54**
150 × 75 × 75 mm	4.01	4.92	0.71	28.40	nr	**33.32**
150 × 100 × 75 mm	8.93	10.97	0.71	28.40	nr	**39.37**
225 × 75 × 50 mm	7.72	9.48	0.78	31.21	nr	**40.69**
225 × 150 × 75 mm	12.33	15.15	0.81	32.39	nr	**47.54**
225 × 150 × 100 mm	23.14	28.42	0.81	32.39	nr	**60.81**
300 × 150 × 50 mm	23.14	28.42	0.93	37.20	nr	**65.62**
300 × 150 × 75 mm	23.14	28.42	0.94	37.60	nr	**66.02**
300 × 150 × 100 mm	23.14	28.42	0.96	38.40	nr	**66.82**
Rectangular pattern – galvanized						
100 × 75 × 50 mm	3.35	4.12	0.69	27.60	nr	**31.72**
150 × 75 × 50 mm	3.51	4.32	0.70	27.99	nr	**32.31**
150 × 75 × 75 mm	3.81	4.68	0.71	28.40	nr	**33.08**
150 × 100 × 75 mm	8.47	10.41	0.71	28.40	nr	**38.81**
225 × 75 × 50 mm	7.34	9.02	0.89	35.61	nr	**44.63**
225 × 150 × 75 mm	11.71	14.39	0.81	32.39	nr	**46.78**
225 × 150 × 100 mm	21.99	27.02	0.81	32.39	nr	**59.41**
300 × 150 × 50 mm	21.99	27.02	0.93	37.20	nr	**64.22**
300 × 150 × 75 mm	21.99	27.02	0.94	37.60	nr	**64.62**
300 × 150 × 100 mm	21.99	27.02	0.96	38.40	nr	**65.42**

ELECTRICAL SUPPLY/POWER/LIGHTING

Item	Net Price £	Material £	Labour hours	Labour £	Unit	**Total rate £**
LV DISTRIBUTION: CABLES AND WIRING						
ARMOURED CABLE						
Cable; XLPE insulated; PVC sheathed; copper stranded conductors to BS 5467; laid in trench/duct including marker tape; (cable tiles measured elsewhere)						
600/1000 Volt grade; single core (aluminium wire armour)						
25 mm²	0.42	0.51	0.15	6.00	m	**6.51**
35 mm²	0.47	0.58	0.15	6.00	m	**6.58**
50 mm²	0.69	0.85	0.17	6.80	m	**7.65**
70 mm²	4.91	6.03	0.18	7.20	m	**13.23**
95 mm²	6.40	7.86	0.20	8.00	m	**15.86**
120 mm²	7.71	9.47	0.22	8.80	m	**18.27**
150 mm²	9.33	11.47	0.24	9.60	m	**21.07**
185 mm²	11.35	13.94	0.26	10.41	m	**24.35**
240 mm²	14.61	17.94	0.30	12.00	m	**29.94**
300 mm²	18.04	22.17	0.31	12.40	m	**34.57**
400 mm²	22.86	28.09	0.38	15.19	m	**43.28**
500 mm²	39.54	48.57	0.44	17.61	m	**66.18**
630 mm²	48.22	59.24	0.52	20.80	m	**80.04**
800 mm²	29.90	36.73	0.62	24.80	m	**61.53**
1000 mm²	34.63	42.55	0.65	26.00	m	**68.55**
600/1000 Volt grade; two core (galvanized steel wire armour)						
1.5 mm²	0.75	0.92	0.06	2.39	m	**3.31**
2.5 mm²	0.93	1.14	0.06	2.39	m	**3.53**
4 mm²	1.06	1.30	0.08	3.19	m	**4.49**
6 mm²	1.55	1.90	0.08	3.19	m	**5.09**
10 mm²	2.01	2.47	0.10	4.00	m	**6.47**
16 mm²	2.93	3.60	0.10	4.00	m	**7.60**
25 mm²	4.18	5.13	0.15	6.00	m	**11.13**
35 mm²	5.29	6.50	0.15	6.00	m	**12.50**
50 mm²	6.82	8.38	0.17	6.80	m	**15.18**
70 mm²	9.36	11.50	0.18	7.20	m	**18.70**
95 mm²	13.04	16.02	0.20	8.00	m	**24.02**
120 mm²	15.86	19.49	0.22	8.80	m	**28.29**
150 mm²	19.55	24.01	0.24	9.60	m	**33.61**
185 mm²	24.96	30.66	0.26	10.41	m	**41.07**
240 mm²	32.19	39.54	0.30	12.00	m	**51.54**
300 mm²	39.74	48.83	0.31	12.40	m	**61.23**
400 mm²	46.28	56.86	0.35	14.01	m	**70.87**

ELECTRICAL SUPPLY/POWER/LIGHTING

Item	Net Price £	Material £	Labour hours	Labour £	Unit	Total rate £
LV DISTRIBUTION: CABLES AND WIRING – cont						
Cable – cont						
600/1000 Volt grade; three core (galvanized steel wire armour)						
1.5 mm²	0.74	0.91	0.07	2.80	m	**3.71**
2.5 mm²	1.01	1.24	0.07	2.80	m	**4.04**
4 mm²	1.33	1.64	0.09	3.60	m	**5.24**
6 mm²	1.69	2.08	0.10	4.00	m	**6.08**
10 mm²	2.62	3.22	0.11	4.40	m	**7.62**
16 mm²	3.78	4.64	0.11	4.40	m	**9.04**
25 mm²	5.41	6.65	0.16	6.40	m	**13.05**
35 mm²	7.15	8.79	0.16	6.40	m	**15.19**
50 mm²	9.33	11.47	0.19	7.60	m	**19.07**
70 mm²	13.22	16.24	0.21	8.40	m	**24.64**
95 mm²	18.23	22.40	0.23	9.20	m	**31.60**
120 mm²	22.52	27.67	0.24	9.60	m	**37.27**
150 mm²	27.83	34.19	0.27	10.79	m	**44.98**
185 mm²	34.46	42.34	0.30	12.00	m	**54.34**
240 mm²	44.73	54.95	0.33	13.20	m	**68.15**
300 mm²	55.07	67.65	0.35	14.01	m	**81.66**
400 mm²	64.33	79.03	0.41	16.40	m	**95.43**
600/1000 Volt grade; four core (galvanized steel wire armour)						
1.5 mm²	0.85	1.04	0.08	3.19	m	**4.23**
2.5 mm²	1.17	1.44	0.09	3.60	m	**5.04**
4 mm²	1.57	1.93	0.10	4.00	m	**5.93**
6 mm²	2.20	2.70	0.10	4.00	m	**6.70**
10 mm²	3.29	4.04	0.12	4.80	m	**8.84**
16 mm²	4.79	5.88	0.12	4.80	m	**10.68**
25 mm²	7.25	8.90	0.18	7.20	m	**16.10**
35 mm²	9.67	11.88	0.19	7.60	m	**19.48**
50 mm²	12.88	15.82	0.21	8.40	m	**24.22**
70 mm²	17.30	21.25	0.23	9.20	m	**30.45**
95 mm²	23.22	28.53	0.26	10.41	m	**38.94**
120 mm²	29.19	35.86	0.28	11.20	m	**47.06**
150 mm²	35.68	43.84	0.32	12.80	m	**56.64**
185 mm²	44.59	54.78	0.35	14.01	m	**68.79**
240 mm²	57.46	70.59	0.36	14.40	m	**84.99**
300 mm²	71.45	87.78	0.40	15.99	m	**103.77**
400 mm²	93.53	114.90	0.45	18.00	m	**132.90**
600/1000 Volt grade; seven core (galvanized steel wire armour)						
1.5 mm²	1.28	1.58	0.10	4.00	m	**5.58**
2.5 mm²	1.78	2.18	0.10	4.00	m	**6.18**
4 mm²	2.12	2.60	0.11	4.40	m	**7.00**
600/1000 Volt grade; twelve core (galvanized steel wire armour)						
1.5 mm²	2.07	2.54	0.11	4.40	m	**6.94**
2.5 mm²	2.92	3.59	0.11	4.40	m	**7.99**

ELECTRICAL SUPPLY/POWER/LIGHTING

Item	Net Price £	Material £	Labour hours	Labour £	Unit	**Total rate £**
600/1000 Volt grade; nineteen core (galvanized steel wire armour)						
1.5 mm²	2.98	3.66	0.13	5.20	m	**8.86**
2.5 mm²	4.46	5.48	0.14	5.61	m	**11.09**
600/1000 Volt grade; twenty seven core (galvanized steel wire armour)						
1.5 mm²	4.25	5.22	0.14	5.61	m	**10.83**
2.5 mm²	6.18	7.59	0.16	6.40	m	**13.99**
600/1000 Volt grade; thirty seven core (galvanized steel wire armour)						
1.5 mm²	5.46	6.71	0.15	6.00	m	**12.71**
2.5 mm²	8.05	9.89	0.17	6.80	m	**16.69**
Cable; XLPE insulated; PVC sheathed copper stranded conductors to BS 5467; clipped direct to backgrounds including cleat						
600/1000 Volt grade; single core (aluminium wire armour)						
25 mm²	1.92	2.36	0.35	14.01	m	**16.37**
35 mm²	2.10	2.58	0.36	14.40	m	**16.98**
50 mm²	2.31	2.84	0.37	14.81	m	**17.65**
70 mm²	5.27	6.48	0.39	15.60	m	**22.08**
95 mm²	6.94	8.53	0.42	16.81	m	**25.34**
120 mm²	8.32	10.22	0.47	18.80	m	**29.02**
150 mm²	9.94	12.21	0.51	20.40	m	**32.61**
185 mm²	11.96	14.69	0.59	23.60	m	**38.29**
240 mm²	15.98	19.64	0.68	27.19	m	**46.83**
300 mm²	19.42	23.86	0.74	29.60	m	**53.46**
400 mm²	25.38	31.17	0.88	35.20	m	**66.37**
500 mm²	32.10	39.44	0.88	35.20	m	**74.64**
630 mm²	42.65	52.40	1.05	42.00	m	**94.40**
800 mm²	36.26	44.54	1.33	53.19	m	**97.73**
1000 mm²	40.55	49.81	1.40	56.00	m	**105.81**
600/1000 Volt grade; two core (galvanized steel wire armour)						
1.5 mm²	0.78	0.96	0.20	8.00	m	**8.96**
2.5 mm²	0.94	1.16	0.20	8.00	m	**9.16**
4.0 mm²	1.23	1.51	0.21	8.40	m	**9.91**
6.0 mm²	1.69	2.08	0.22	8.80	m	**10.88**
10.0 mm²	2.50	3.07	0.24	9.60	m	**12.67**
16.0 mm²	3.42	4.20	0.25	10.00	m	**14.20**
25 mm²	4.59	5.64	0.35	14.01	m	**19.65**
35 mm²	5.80	7.13	0.36	14.40	m	**21.53**
50 mm²	7.28	8.95	0.37	14.81	m	**23.76**
70 mm²	9.93	12.20	0.39	15.60	m	**27.80**
95 mm²	13.01	15.98	0.42	16.81	m	**32.79**
120 mm²	16.53	20.31	0.47	18.80	m	**39.11**
150 mm²	19.98	24.55	0.51	20.40	m	**44.95**
185 mm²	24.97	30.67	0.59	23.60	m	**54.27**
240 mm²	31.77	39.03	0.68	27.19	m	**66.22**

ELECTRICAL SUPPLY/POWER/LIGHTING

Item	Net Price £	Material £	Labour hours	Labour £	Unit	Total rate £
LV DISTRIBUTION: CABLES AND WIRING – cont						
Cable – cont						
300 mm²	29.68	36.47	0.74	29.60	m	**66.07**
400 mm²	36.90	45.33	0.88	35.20	m	**80.53**
600/1000 Volt grade; three core (galvanized steel wire armour)						
1.5 mm²	0.90	1.10	0.20	8.00	m	**9.10**
2.5 mm²	1.17	1.44	0.21	8.40	m	**9.84**
4.0 mm²	1.52	1.87	0.22	8.80	m	**10.67**
6.0 mm²	2.05	2.52	0.22	8.80	m	**11.32**
10.0 mm²	3.16	3.89	0.25	10.00	m	**13.89**
16.0 mm²	4.32	5.30	0.26	10.41	m	**15.71**
25 mm²	6.02	7.39	0.37	14.81	m	**22.20**
35 mm²	7.76	9.53	0.39	15.60	m	**25.13**
50 mm²	10.02	12.31	0.40	15.99	m	**28.30**
70 mm²	13.50	16.58	0.42	16.81	m	**33.39**
95 mm²	18.74	23.03	0.45	18.00	m	**41.03**
120 mm²	23.04	28.31	0.52	20.80	m	**49.11**
150 mm²	28.47	34.98	0.55	22.00	m	**56.98**
185 mm²	35.10	43.12	0.63	25.20	m	**68.32**
240 mm²	46.06	56.58	0.71	28.40	m	**84.98**
300 mm²	56.40	69.29	0.78	31.21	m	**100.50**
400 mm²	72.51	89.08	0.87	34.81	m	**123.89**
600/1000 Volt grade; four core (galvanized steel wire armour)						
1.5 mm²	1.01	1.24	0.21	8.40	m	**9.64**
2.5 mm²	1.36	1.67	0.22	8.80	m	**10.47**
4.0 mm²	1.93	2.37	0.22	8.80	m	**11.17**
6.0 mm²	2.74	3.37	0.23	9.20	m	**12.57**
10.0 mm²	3.83	4.70	0.26	10.41	m	**15.11**
16.0 mm²	5.39	6.63	0.26	10.41	m	**17.04**
25 mm²	7.86	9.66	0.39	15.60	m	**25.26**
35 mm²	10.37	12.74	0.40	15.99	m	**28.73**
50 mm²	13.57	16.67	0.41	16.40	m	**33.07**
70 mm²	17.81	21.88	0.45	18.00	m	**39.88**
95 mm²	23.86	29.32	0.50	20.00	m	**49.32**
120 mm²	29.82	36.63	0.54	21.60	m	**58.23**
150 mm²	40.51	49.77	0.60	24.00	m	**73.77**
185 mm²	50.04	61.48	0.67	26.81	m	**88.29**
240 mm²	64.73	79.52	0.75	30.00	m	**109.52**
300 mm²	80.45	98.84	0.83	33.19	m	**132.03**
400 mm²	85.98	105.63	0.91	36.40	m	**142.03**
600/1000 Volt grade; seven core (galvanized steel wire armour)						
1.5 mm²	1.47	1.81	0.20	8.00	m	**9.81**
2.5 mm²	1.97	2.42	0.20	8.00	m	**10.42**
4.0 mm²	3.58	4.40	0.23	9.20	m	**13.60**

ELECTRICAL SUPPLY/POWER/LIGHTING

Item	Net Price £	Material £	Labour hours	Labour £	Unit	Total rate £
600/1000 Volt grade; twelve core (galvanized steel wire armour)						
1.5 mm²	2.44	7.32	0.23	9.20	m	**16.52**
2.5 mm²	3.46	4.25	0.24	9.60	m	**13.85**
600/1000 Volt grade; nineteen core (galvanized steel wire armour)						
1.5 mm²	3.52	4.33	0.26	10.41	m	**14.74**
2.5 mm²	5.07	6.23	0.28	11.20	m	**17.43**
600/1000 Volt grade; twenty- seven core (galvanized steel wire armour)						
1.5 mm²	4.85	5.95	0.29	11.59	m	**17.54**
2.5 mm²	6.87	8.44	0.30	12.00	m	**20.44**
600/1000 Volt grade; thirty-seven core (galvanized steel wire armour)						
1.5 mm²	6.16	7.57	0.32	12.80	m	**20.37**
2.5 mm²	8.74	10.74	0.33	13.20	m	**23.94**
Cable termination; brass weatherproof gland with inner and outer seal, shroud, brass locknut and earth ring (including drilling and cutting mild steel gland plate)						
600/1000 Volt grade; single core (aluminium wire armour)						
25 mm²	7.17	8.81	1.70	68.00	nr	**76.81**
35 mm²	7.17	8.81	1.79	71.60	nr	**80.41**
50 mm²	7.85	9.64	2.06	82.39	nr	**92.03**
70 mm²	13.42	16.49	2.12	84.81	nr	**101.30**
95 mm²	13.47	16.55	2.39	95.60	nr	**112.15**
120 mm²	13.50	16.58	2.47	98.79	nr	**115.37**
150 mm²	17.45	21.44	2.73	109.20	nr	**130.64**
185 mm²	17.68	21.72	3.05	122.00	nr	**143.72**
240 mm²	22.55	27.70	3.45	138.00	nr	**165.70**
300 mm²	22.86	28.09	3.84	153.60	nr	**181.69**
400 mm²	29.08	35.72	4.21	168.40	nr	**204.12**
500 mm²	30.53	37.51	5.70	228.00	m	**265.51**
630 mm²	41.24	50.66	6.20	248.00	m	**298.66**
800 mm²	70.34	86.42	7.50	300.00	m	**386.42**
1000 mm²	84.89	104.29	10.00	400.00	m	**504.29**
600/1000 Volt grade; two core (galvanized steel wire armour)						
1.5 mm²	4.44	5.45	0.58	23.19	nr	**28.64**
2.5 mm²	4.44	5.45	0.58	23.25	nr	**28.70**
4 mm²	4.44	5.45	0.58	23.19	nr	**28.64**
6 mm²	4.45	5.47	0.67	26.81	nr	**32.28**
10 mm²	5.01	6.15	1.00	40.01	nr	**46.16**
16 mm²	5.17	6.35	1.11	44.39	nr	**50.74**
25 mm²	5.88	7.22	1.70	68.00	nr	**75.22**
35 mm²	5.82	7.15	1.79	71.60	nr	**78.75**
50 mm²	10.32	12.67	2.06	82.39	nr	**95.06**
70 mm²	10.41	12.79	2.12	84.81	nr	**97.60**
95 mm²	12.47	15.32	2.39	95.60	nr	**110.92**

ELECTRICAL SUPPLY/POWER/LIGHTING

Item	Net Price £	Material £	Labour hours	Labour £	Unit	Total rate £
LV DISTRIBUTION: CABLES AND WIRING – cont						
Cable termination – cont						
120 mm²	12.52	15.38	2.47	98.79	nr	**114.17**
150 mm²	18.30	22.48	2.73	109.20	nr	**131.68**
185 mm²	18.73	23.01	3.05	122.00	nr	**145.01**
240 mm²	20.11	24.71	3.45	138.00	nr	**162.71**
300 mm²	33.01	40.55	3.84	153.60	nr	**194.15**
400 mm²	52.50	64.50	4.21	168.40	nr	**232.90**
600/1000 Volt grade; three core (galvanized steel wire armour)						
1.5 mm²	4.55	5.59	0.61	24.58	nr	**30.17**
2.5 mm²	4.55	5.59	0.62	24.80	nr	**30.39**
4 mm²	4.44	5.45	0.62	24.80	nr	**30.25**
6 mm²	5.13	6.30	0.71	28.40	nr	**34.70**
10 mm²	5.90	7.25	1.06	42.40	nr	**49.65**
16 mm²	6.13	7.53	1.19	47.61	nr	**55.14**
25 mm²	10.46	12.85	1.81	72.40	nr	**85.25**
35 mm²	10.48	12.87	1.99	79.59	nr	**92.46**
50 mm²	10.51	12.91	2.23	89.20	nr	**102.11**
70 mm²	12.60	15.48	2.40	95.99	nr	**111.47**
95 mm²	12.73	15.63	2.63	105.20	nr	**120.83**
120 mm²	12.83	15.76	2.83	113.19	nr	**128.95**
150 mm²	19.40	23.84	3.22	128.79	nr	**152.63**
185 mm²	20.08	24.66	3.44	137.60	nr	**162.26**
240 mm²	34.40	42.26	3.83	153.19	nr	**195.45**
300 mm²	35.34	43.42	4.28	171.20	nr	**214.62**
400 mm²	46.51	57.14	5.00	200.00	nr	**257.14**
600/1000 Volt grade; four core (galvanized steel wire armour)						
1.5 mm²	4.66	5.72	0.67	26.81	nr	**32.53**
2.5 mm²	4.66	5.72	0.67	26.81	nr	**32.53**
4 mm²	4.66	5.72	0.71	28.40	nr	**34.12**
6 mm²	6.02	7.39	0.76	30.41	nr	**37.80**
10 mm²	6.02	7.39	1.14	45.60	nr	**52.99**
16 mm²	6.33	7.78	1.29	51.60	nr	**59.38**
25 mm²	10.63	13.06	1.99	79.59	nr	**92.65**
35 mm²	10.67	13.10	2.16	86.40	nr	**99.50**
50 mm²	12.66	15.55	2.49	99.59	nr	**115.14**
70 mm²	12.83	15.76	2.65	106.00	nr	**121.76**
95 mm²	17.18	21.10	2.98	119.21	nr	**140.31**
120 mm²	17.29	21.24	3.15	126.00	nr	**147.24**
150 mm²	20.56	25.26	3.50	140.00	nr	**165.26**
185 mm²	33.69	41.39	3.72	148.80	nr	**190.19**
240 mm²	36.44	44.76	4.33	173.20	nr	**217.96**
300 mm²	52.70	64.74	4.86	194.40	nr	**259.14**
400 mm²	49.09	60.31	5.46	218.40	nr	**278.71**

ELECTRICAL SUPPLY/POWER/LIGHTING

Item	Net Price £	Material £	Labour hours	Labour £	Unit	**Total rate £**
600/1000 Volt grade; seven core (galvanized steel wire armour)						
1.5 mm²	5.17	6.35	0.81	32.39	nr	**38.74**
2.5 mm²	5.17	6.35	0.85	34.01	nr	**40.36**
4 mm²	8.27	10.16	0.93	37.20	nr	**47.36**
600/1000 Volt grade; twelve core (galvanized steel wire armour)						
1.5 mm²	6.32	7.76	1.14	45.60	nr	**53.36**
2.5 mm²	7.12	8.75	1.13	45.19	nr	**53.94**
600/1000 Volt grade; nineteen core (galvanized steel wire armour)						
1.5 mm²	7.93	9.74	1.54	61.59	nr	**71.33**
2.5 mm²	7.93	9.74	1.54	61.59	nr	**71.33**
600/1000 Volt grade; twenty-seven core (galvanized steel wire armour)						
1.5 mm²	8.87	10.90	1.94	77.61	nr	**88.51**
2.5 mm²	13.42	16.49	2.31	92.40	nr	**108.89**
600/1000 Volt grade; thirty-seven core (galvanized steel wire armour)						
1.5 mm²	14.58	17.91	2.53	101.20	nr	**119.11**
2.5 mm²	14.58	17.91	2.87	114.81	nr	**132.72**
Cable; XLPE insulated; LSOH sheathed (LSF); copper stranded conductors to BS 6724; laid in trench/duct including marker tape (cable tiles measured elsewhere)						
600/1000 Volt grade; single core (aluminium wire armour)						
50 mm²	3.96	4.86	0.17	6.80	m	**11.66**
70 mm²	5.45	6.70	0.18	7.20	m	**13.90**
95 mm²	7.17	8.81	0.20	8.00	m	**16.81**
120 mm²	8.59	10.55	0.22	8.80	m	**19.35**
150 mm²	10.26	12.60	0.24	9.60	m	**22.20**
185 mm²	12.35	15.17	0.26	10.41	m	**25.58**
240 mm²	16.50	20.28	0.30	12.00	m	**32.28**
300 mm²	20.05	24.63	0.31	12.40	m	**37.03**
400 mm²	26.20	32.18	0.35	14.01	m	**46.19**
500 mm²	33.15	40.73	0.44	17.61	m	**58.34**
630 mm²	44.03	54.10	0.52	20.80	m	**74.90**
800 mm²	55.38	68.03	0.62	24.80	m	**92.83**
1000 mm²	70.68	86.84	0.65	26.00	m	**112.84**
600/1000 Volt grade; two core (galvanized steel wire armour)						
1.5 mm²	0.81	1.00	0.06	2.39	m	**3.39**
2.5 mm²	0.97	1.19	0.06	2.39	m	**3.58**
4 mm²	1.27	1.56	0.08	3.19	m	**4.75**
6 mm²	1.74	2.14	0.08	3.19	m	**5.33**
10 mm²	2.58	3.17	0.10	4.00	m	**7.17**
16 mm²	3.53	4.34	0.10	4.00	m	**8.34**
25 mm²	4.74	5.83	0.15	6.00	m	**11.83**

ELECTRICAL SUPPLY/POWER/LIGHTING

Item	Net Price £	Material £	Labour hours	Labour £	Unit	Total rate £
LV DISTRIBUTION: CABLES AND WIRING – cont						
Cable – cont						
35 mm²	5.99	7.36	0.17	6.80	m	**14.16**
50 mm²	7.52	9.24	0.15	6.00	m	**15.24**
70 mm²	10.25	12.59	0.18	7.20	m	**19.79**
95 mm²	13.43	16.50	0.20	8.00	m	**24.50**
120 mm²	17.07	20.97	0.22	8.80	m	**29.77**
150 mm²	20.63	25.35	0.24	9.60	m	**34.95**
185 mm²	25.78	31.67	0.26	10.41	m	**42.08**
240 mm²	32.81	40.31	0.30	12.00	m	**52.31**
300 mm²	30.65	37.65	0.31	12.40	m	**50.05**
400 mm²	47.33	58.15	0.35	14.01	m	**72.16**
600/1000 Volt grade; three core (galvanized steel wire armour)						
1.5 mm²	0.92	1.13	0.07	2.80	m	**3.93**
2.5 mm²	1.20	1.47	0.07	2.80	m	**4.27**
4 mm²	1.57	1.93	0.09	3.60	m	**5.53**
6 mm²	2.12	2.60	0.10	4.00	m	**6.60**
10 mm²	3.26	4.00	0.11	4.40	m	**8.40**
16 mm²	4.46	5.48	0.11	4.40	m	**9.88**
25 mm²	6.21	7.63	0.16	6.40	m	**14.03**
35 mm²	8.01	9.84	0.16	6.40	m	**16.24**
50 mm²	10.34	12.70	0.19	7.60	m	**20.30**
70 mm²	13.94	17.13	0.21	8.40	m	**25.53**
95 mm²	19.35	23.77	0.23	9.20	m	**32.97**
120 mm²	23.79	29.22	0.24	9.60	m	**38.82**
150 mm²	29.39	36.11	0.27	10.79	m	**46.90**
185 mm²	36.24	44.52	0.30	12.00	m	**56.52**
240 mm²	47.55	58.41	0.33	13.20	m	**71.61**
300 mm²	58.24	71.55	0.35	14.01	m	**85.56**
400 mm²	81.91	100.62	0.41	16.40	m	**117.02**
600/1000 Volt grade; four core (galvanized steel wire armour)						
1.5 mm²	1.04	1.28	0.08	3.19	m	**4.47**
2.5 mm²	1.40	1.72	0.09	3.60	m	**5.32**
4 mm²	1.99	2.45	0.10	4.00	m	**6.45**
6 mm²	2.83	3.48	0.10	4.00	m	**7.48**
10 mm²	3.95	4.85	0.12	4.80	m	**9.65**
16 mm²	5.57	6.85	0.12	4.80	m	**11.65**
25 mm²	8.11	9.96	0.18	7.20	m	**17.16**
35 mm²	10.71	13.16	0.19	7.60	m	**20.76**
50 mm²	14.02	17.22	0.21	8.40	m	**25.62**
70 mm²	18.39	22.60	0.23	9.20	m	**31.80**
95 mm²	24.63	30.26	0.26	10.41	m	**40.67**
120 mm²	30.79	37.82	0.28	11.20	m	**49.02**
150 mm²	41.82	51.38	0.32	12.80	m	**64.18**
185 mm²	51.67	63.47	0.35	14.01	m	**77.48**
240 mm²	66.83	82.10	0.36	14.40	m	**96.50**
300 mm²	83.07	102.05	0.40	15.99	m	**118.04**
400 mm²	110.91	136.25	0.45	18.00	m	**154.25**

ELECTRICAL SUPPLY/POWER/LIGHTING

Item	Net Price £	Material £	Labour hours	Labour £	Unit	**Total rate £**
600/1000 Volt grade; seven core (galvanized steel wire armour)						
1.5 mm²	1.51	1.86	0.10	4.00	m	**5.86**
2.5 mm²	2.04	2.51	0.10	4.00	m	**6.51**
4 mm²	3.77	4.63	0.11	4.40	m	**9.03**
600/1000 Volt grade; twelve core (galvanized steel wire armour)						
1.5 mm²	2.51	3.09	0.11	4.40	m	**7.49**
2.5 mm²	3.58	4.40	0.11	4.40	m	**8.80**
600/1000 Volt grade; nineteen core (galvanized steel wire armour)						
1.5 mm²	3.64	4.47	0.13	5.20	m	**9.67**
2.5 mm²	5.24	6.44	0.14	5.61	m	**12.05**
600/1000 Volt grade; twenty-seven core (galvanized steel wire armour)						
1.5 mm²	5.01	6.15	0.14	5.61	m	**11.76**
2.5 mm²	7.10	8.73	0.16	6.40	m	**15.13**
600/1000 Volt grade; thirty-seven core (galvanized steel wire armour)						
1.5 mm²	6.36	7.81	0.15	6.00	m	**13.81**
2.5 mm²	9.03	11.10	0.17	6.80	m	**17.90**
Cable; XLPE insulated; LSOH sheathed (LSF) copper stranded conductors to BS 6724; clipped direct to backgrounds including cleat						
600/1000 Volt grade; single core (aluminium wire armour)						
50 mm²	3.14	3.85	0.37	14.81	m	**18.66**
70 mm²	4.77	5.86	0.39	15.60	m	**21.46**
95 mm²	6.41	7.88	0.42	16.81	m	**24.69**
120 mm²	7.64	9.39	0.47	18.80	m	**28.19**
150 mm²	9.44	11.59	0.51	20.40	m	**31.99**
185 mm²	11.34	13.93	0.59	23.60	m	**37.53**
240 mm²	14.62	17.97	0.68	27.19	m	**45.16**
300 mm²	17.81	21.88	0.74	29.60	m	**51.48**
400 mm²	23.15	28.44	0.81	32.39	m	**60.83**
500 mm²	29.70	36.49	0.88	35.20	m	**71.69**
630 mm²	37.24	45.75	1.05	42.00	m	**87.75**
800 mm²	47.99	58.96	1.33	53.19	m	**112.15**
1000 mm²	55.57	68.27	1.40	56.00	m	**124.27**
600/1000 Volt grade; two core (galvanized steel wire armour)						
1.5 mm²	0.68	0.84	0.20	8.00	m	**8.84**
2.5 mm²	0.78	0.96	0.20	8.00	m	**8.96**
4.0 mm²	1.02	1.25	0.21	8.40	m	**9.65**
6.0 mm²	1.24	1.52	0.22	8.80	m	**10.32**
10.0 mm²	1.66	2.04	0.24	9.60	m	**11.64**
16.0 mm²	2.50	3.07	0.25	10.00	m	**13.07**
25 mm²	2.50	3.07	0.35	14.01	m	**17.08**

ELECTRICAL SUPPLY/POWER/LIGHTING

Item	Net Price £	Material £	Labour hours	Labour £	Unit	Total rate £
LV DISTRIBUTION: CABLES AND WIRING – cont						
Cable; XLPE insulated – cont						
35 mm²	5.05	6.21	0.36	14.40	m	**20.61**
50 mm²	6.25	7.68	0.37	14.81	m	**22.49**
70 mm²	9.29	11.41	0.39	15.60	m	**27.01**
95 mm²	12.48	15.33	0.42	16.81	m	**32.14**
120 mm²	15.29	18.78	0.47	18.80	m	**37.58**
150 mm²	18.97	23.30	0.51	20.40	m	**43.70**
185 mm²	24.16	29.68	0.59	23.60	m	**53.28**
240 mm²	30.93	38.00	0.68	27.19	m	**65.19**
300 mm²	41.54	51.03	0.74	29.60	m	**80.63**
400 mm²	54.96	67.51	0.81	32.39	m	**99.90**
600/1000 Volt grade; three core (galvanized steel wire armour)						
1.5 mm²	0.75	0.92	0.20	8.00	m	**8.92**
2.5 mm²	0.95	1.17	0.21	8.40	m	**9.57**
4.0 mm²	1.17	1.44	0.22	8.80	m	**10.24**
6.0 mm²	1.51	1.86	0.22	8.80	m	**10.66**
10.0 mm²	2.31	2.84	0.25	10.00	m	**12.84**
16.0 mm²	3.23	3.97	0.26	10.41	m	**14.38**
25 mm²	4.36	5.36	0.37	14.81	m	**20.17**
35 mm²	6.07	7.46	0.39	15.60	m	**23.06**
50 mm²	8.66	10.64	0.40	15.99	m	**26.63**
70 mm²	12.86	15.80	0.42	16.81	m	**32.61**
95 mm²	17.70	21.75	0.45	18.00	m	**39.75**
120 mm²	22.01	27.04	0.52	20.80	m	**47.84**
150 mm²	27.34	33.59	0.55	22.00	m	**55.59**
185 mm²	33.69	41.39	0.63	25.20	m	**66.59**
240 mm²	43.11	52.96	0.71	28.40	m	**81.36**
300 mm²	54.63	67.12	0.78	31.21	m	**98.33**
400 mm²	64.88	79.71	0.87	34.81	m	**114.52**
600/1000 Volt grade; four core (galvanized steel wire armour)						
1.5 mm²	0.88	1.08	0.21	8.40	m	**9.48**
2.5 mm²	1.10	1.35	0.22	8.80	m	**10.15**
4.0 mm²	1.37	1.68	0.22	8.80	m	**10.48**
6.0 mm²	2.00	2.46	0.23	9.20	m	**11.66**
10.0 mm²	2.81	3.45	0.26	10.41	m	**13.86**
16.0 mm²	4.06	4.99	0.26	10.41	m	**15.40**
25 mm²	5.86	7.20	0.39	15.60	m	**22.80**
35 mm²	7.64	9.39	0.40	15.99	m	**25.38**
50 mm²	11.06	13.59	0.41	16.40	m	**29.99**
70 mm²	17.06	20.96	0.45	18.00	m	**38.96**
95 mm²	22.85	28.07	0.50	20.00	m	**48.07**
120 mm²	28.70	35.26	0.54	21.60	m	**56.86**
150 mm²	34.74	42.68	0.60	24.00	m	**66.68**
185 mm²	42.63	52.37	0.67	26.81	m	**79.18**
240 mm²	55.11	67.70	0.75	30.00	m	**97.70**
300 mm²	69.14	84.93	0.83	33.19	m	**118.12**
400 mm²	71.88	88.31	0.91	36.40	m	**124.71**

ELECTRICAL SUPPLY/POWER/LIGHTING

Item	Net Price £	Material £	Labour hours	Labour £	Unit	**Total rate £**
600/1000 Volt grade; seven core (galvanized steel wire armour)						
1.5 mm²	1.26	1.54	0.20	8.00	m	**9.54**
2.5 mm²	1.76	2.16	0.20	8.00	m	**10.16**
4.0 mm²	3.19	3.92	0.23	9.20	m	**13.12**
600/1000 Volt grade; twelve core (galvanized steel wire armour)						
1.5 mm²	2.00	2.46	0.23	9.20	m	**11.66**
2.5 mm²	3.01	3.70	0.24	9.60	m	**13.30**
600/1000 Volt grade; nineteen core (galvanized steel wire armour)						
1.5 mm²	3.03	3.73	0.26	10.41	m	**14.14**
2.5 mm²	4.54	5.58	0.28	11.20	m	**16.78**
600/1000 Volt grade; twenty-seven core (galvanized steel wire armour)						
1.5 mm²	4.16	5.11	0.29	11.59	m	**16.70**
2.5 mm²	5.80	7.13	0.30	12.00	m	**19.13**
600/1000 Volt grade; thirty-seven core (galvanized steel wire armour)						
1.5 mm²	5.19	6.37	0.32	12.80	m	**19.17**
2.5 mm²	7.35	9.03	0.33	13.20	m	**22.23**
Cable termination; brass weatherproof gland with inner and outer seal, shroud, brass locknut and earth ring (including drilling and cutting mild steel gland plate)						
600/1000 Volt grade; single core (aluminium wire armour)						
25 mm²	2.89	3.55	1.70	68.00	nr	**71.55**
35 mm²	3.15	3.87	1.79	71.60	nr	**75.47**
50 mm²	3.49	4.28	2.06	82.39	nr	**86.67**
70 mm²	4.99	6.13	2.12	84.81	nr	**90.94**
95 mm²	6.67	8.19	2.39	95.60	nr	**103.79**
120 mm²	7.90	9.70	2.47	98.79	nr	**108.49**
150 mm²	9.71	11.93	2.73	109.20	nr	**121.13**
185 mm²	11.57	14.22	3.05	122.00	nr	**136.22**
240 mm²	15.16	18.63	3.45	138.00	nr	**156.63**
300 mm²	18.32	22.50	3.84	153.60	nr	**176.10**
400 mm²	26.74	32.85	4.21	168.40	nr	**201.25**
500 mm²	33.70	41.40	5.70	228.00	m	**269.40**
630 mm²	43.83	53.84	6.20	248.00	m	**301.84**
800 mm²	54.50	66.96	7.50	300.00	m	**366.96**
1000 mm²	60.95	74.88	10.00	400.00	m	**474.88**
600/1000 Volt grade; two core (galvanized steel wire armour)						
1.5 mm²	0.85	1.04	0.52	20.67	nr	**21.71**
2.5 mm²	0.98	1.21	0.58	23.19	nr	**24.40**
4 mm²	1.24	1.52	0.58	23.19	nr	**24.71**
6 mm²	1.44	1.76	0.67	26.81	nr	**28.57**
10 mm²	1.87	2.30	1.00	40.01	nr	**42.31**
16 mm²	2.71	3.33	1.11	44.39	nr	**47.72**
25 mm²	2.36	2.90	1.70	68.00	nr	**70.90**

ELECTRICAL SUPPLY/POWER/LIGHTING

Item	Net Price £	Material £	Labour hours	Labour £	Unit	Total rate £
LV DISTRIBUTION: CABLES AND WIRING – cont						
Cable termination – cont						
35 mm²	5.27	6.48	1.79	71.60	nr	**78.08**
50 mm²	6.48	7.96	2.06	82.39	nr	**90.35**
70 mm²	9.63	11.83	2.12	84.81	nr	**96.64**
95 mm²	12.55	15.41	2.39	95.60	nr	**111.01**
120 mm²	15.43	18.95	2.47	98.79	nr	**117.74**
150 mm²	19.11	23.48	2.73	109.20	nr	**132.68**
185 mm²	24.33	29.89	3.05	122.00	nr	**151.89**
240 mm²	30.28	37.20	3.45	138.00	nr	**175.20**
300 mm²	42.40	52.09	3.84	153.60	nr	**205.69**
400 mm²	52.79	64.85	4.21	168.40	nr	**233.25**
600/1000 Volt grade; three core (galvanized steel wire armour)						
1.5 mm²	0.91	1.11	0.62	24.80	nr	**25.91**
2.5 mm²	1.09	1.34	0.62	24.80	nr	**26.14**
4 mm²	1.38	1.69	0.62	24.80	nr	**26.49**
6 mm²	1.72	2.11	0.71	28.40	nr	**30.51**
10 mm²	2.51	3.09	1.06	42.40	nr	**45.49**
16 mm²	3.45	4.24	1.19	47.61	nr	**51.85**
25 mm²	4.83	5.93	1.81	72.40	nr	**78.33**
35 mm²	6.28	7.72	1.99	79.59	nr	**87.31**
50 mm²	8.95	10.99	2.23	89.20	nr	**100.19**
70 mm²	12.99	15.96	2.40	95.99	nr	**111.95**
95 mm²	17.84	21.91	2.63	105.20	nr	**127.11**
120 mm²	22.15	27.22	2.83	113.19	nr	**140.41**
150 mm²	27.57	33.87	3.22	128.79	nr	**162.66**
185 mm²	33.90	41.64	3.44	137.60	nr	**179.24**
240 mm²	45.01	55.29	3.83	153.19	nr	**208.48**
300 mm²	56.66	69.60	4.28	171.20	nr	**240.80**
400 mm²	67.26	82.62	5.00	200.00	nr	**282.62**
600/1000 Volt grade; four core (galvanized steel wire armour)						
1.5 mm²	1.05	1.29	0.67	26.81	nr	**28.10**
2.5 mm²	1.30	1.60	0.67	26.81	nr	**28.41**
4 mm²	1.59	1.95	0.71	28.40	nr	**30.35**
6 mm²	2.22	2.73	0.76	30.41	nr	**33.14**
10 mm²	3.04	3.74	1.14	45.60	nr	**49.34**
16 mm²	4.27	5.25	1.29	51.60	nr	**56.85**
25 mm²	6.08	7.47	1.99	79.59	nr	**87.06**
35 mm²	7.93	9.74	2.16	86.40	nr	**96.14**
50 mm²	11.36	13.95	2.49	99.59	nr	**113.54**
70 mm²	17.20	21.13	2.65	106.00	nr	**127.13**
95 mm²	23.05	28.32	2.98	119.21	nr	**147.53**
120 mm²	28.88	35.48	3.15	126.00	nr	**161.48**
150 mm²	36.45	44.78	3.50	140.00	nr	**184.78**
185 mm²	44.54	54.72	3.72	148.80	nr	**203.52**
240 mm²	57.67	70.84	4.33	173.20	nr	**244.04**
300 mm²	72.37	88.90	4.86	194.40	nr	**283.30**
400 mm²	75.07	92.22	5.46	218.40	nr	**310.62**

ELECTRICAL SUPPLY/POWER/LIGHTING

Item	Net Price £	Material £	Labour hours	Labour £	Unit	Total rate £
600/1000 Volt grade; seven core (galvanized steel wire armour)						
1.5 mm²	1.47	1.81	0.81	32.39	nr	**34.20**
2.5 mm²	2.01	2.47	0.85	34.01	nr	**36.48**
4 mm²	3.64	4.47	0.93	37.20	nr	**41.67**
600/1000 Volt grade; twelve core (galvanized steel wire armour)						
1.5 mm²	2.14	2.63	1.14	45.60	nr	**48.23**
2.5 mm²	3.22	3.96	1.13	45.19	nr	**49.15**
600/1000 Volt grade; nineteen core (galvanized steel wire armour)						
1.5 mm²	3.24	3.98	1.54	61.59	nr	**65.57**
2.5 mm²	4.76	5.85	1.54	61.59	nr	**67.44**
600/1000 Volt grade; twenty-seven core (galvanized steel wire armour)						
1.5 mm²	4.37	5.37	1.94	77.61	nr	**82.98**
2.5 mm²	6.08	7.47	2.31	92.40	nr	**99.87**
600/1000 Volt grade; thirty-seven core (galvanized steel wire armour)						
1.5 mm²	5.46	6.71	2.53	101.20	nr	**107.91**
2.5 mm²	7.63	9.38	2.87	114.81	nr	**124.19**
UN-ARMOURED CABLE						
Cable: XLPE insulated; PVC sheathed 90c copper to CMA Code 6181e; for internal wiring; clipped to backgrounds; (Supports and fixings included)						
300/500 Volt grade; single core						
6.0 mm²	0.51	0.63	0.09	3.60	m	**4.23**
10 mm²	0.72	0.88	0.10	4.00	m	**4.88**
16 mm²	1.12	1.38	0.12	4.80	m	**6.18**
Cable; LSF insulated to CMA Code 6491B; non-sheathed copper; laid/drawn in trunking/conduit						
450/750 Volt grade; single core						
1.5 mm²	0.13	0.16	0.03	1.21	m	**1.37**
2.5 mm²	0.18	0.22	0.03	1.21	m	**1.43**
4.0 mm²	0.30	0.37	0.03	1.21	m	**1.58**
6.0 mm²	0.45	0.56	0.04	1.60	m	**2.16**
10.0 mm²	0.75	0.92	0.04	1.60	m	**2.52**
16.0 mm²	1.20	1.47	0.05	2.01	m	**3.48**
25.0 mm²	1.74	2.14	0.06	2.39	m	**4.53**
35.0 mm²	2.40	2.95	0.06	2.39	m	**5.34**
50.0 mm²	3.21	3.95	0.07	2.80	m	**6.75**
70.0 mm²	4.57	5.62	0.08	3.19	m	**8.81**
95.0 mm²	6.36	7.81	0.08	3.19	m	**11.00**
120.0 mm²	8.28	10.17	0.10	4.00	m	**14.17**
150.0 mm²	9.37	11.51	0.13	5.20	m	**16.71**
185.0 mm²	12.41	15.25	0.13	5.20	m	**20.45**
240.0 mm²	15.89	19.52	0.13	5.20	m	**24.72**

ELECTRICAL SUPPLY/POWER/LIGHTING

Item	Net Price £	Material £	Labour hours	Labour £	Unit	Total rate £
LV DISTRIBUTION: CABLES AND WIRING – cont						
Cable – cont						
300.0 mm²	20.00	24.57	0.13	5.20	m	**29.77**
Cable; twin & earth to CMA code 6242Y; clipped to backgrounds						
300/500 Volt grade; PVC/PVC						
1.5 mm² 2C+E	0.28	0.35	0.01	0.41	m	**0.76**
1.5 mm² 3C+E	0.34	0.42	0.02	0.80	m	**1.22**
2.5 mm² 2C+E	0.52	0.64	0.02	0.80	m	**1.44**
4.0 mm² 2C+E	0.79	0.97	0.02	0.80	m	**1.77**
6.0 mm² 2C+E	1.13	1.39	0.02	0.80	m	**2.19**
10.0 mm² 2C+E	1.88	2.31	0.03	1.21	m	**3.52**
16.0 mm² 2C+E	2.83	3.48	0.03	1.21	m	**4.69**
300/500 Volt grade; LSF/LSF						
1.5 mm² 2C+E	0.38	0.46	0.01	0.41	m	**0.87**
1.5 mm² 3C+E	0.55	0.67	0.02	0.80	m	**1.47**
2.5 mm² 2C+E	0.68	0.84	0.02	0.80	m	**1.64**
4.0 mm² 2C+E	0.81	1.00	0.02	0.80	m	**1.80**
6.0 mm² 2C+E	1.18	1.45	0.02	0.80	m	**2.25**
10.0 mm² 2C+E	2.70	3.32	0.03	1.21	m	**4.53**
16.0 mm² 2C+E	4.32	5.30	0.03	1.21	m	**6.51**
EARTH CABLE						
Cable; LSF insulated to CMA Code 6491B; non-sheathed copper; laid/drawn in trunking/conduit						
450/750 Volt grade; single core						
1.5 mm²	0.13	0.16	0.03	1.21	m	**1.37**
2.5 mm²	0.18	0.22	0.03	1.21	m	**1.43**
4.0 mm²	0.30	0.37	0.03	1.21	m	**1.58**
6.0 mm²	0.46	0.57	0.04	1.60	m	**2.17**
10.0 mm²	0.75	0.92	0.04	1.60	m	**2.52**
16.0 mm²	1.20	1.47	0.05	2.01	m	**3.48**
25.0 mm²	1.75	2.15	0.06	2.39	m	**4.54**
35.0 mm²	2.41	2.96	0.06	2.39	m	**5.35**
50.0 mm²	3.21	3.95	0.07	2.80	m	**6.75**
70.0 mm²	4.58	5.63	0.08	3.19	m	**8.82**
95.0 mm²	6.37	7.82	0.08	3.19	m	**11.01**
120.0 mm²	8.31	10.21	0.10	4.00	m	**14.21**
150.0 mm²	9.39	11.54	0.13	5.20	m	**16.74**
185.0 mm²	12.45	15.30	0.16	6.40	m	**21.70**
240.0 mm²	15.94	19.58	0.20	8.00	m	**27.58**

ELECTRICAL SUPPLY/POWER/LIGHTING

Item	Net Price £	Material £	Labour hours	Labour £	Unit	**Total rate £**
FLEXIBLE CABLE						
Flexible cord; PVC insulated; PVC sheathed; copper stranded to CMA Code 218*Y (laid loose)						
300 Volt grade; two core						
0.50 mm²	0.15	0.19	0.07	2.80	m	**2.99**
0.75 mm²	0.20	0.24	0.07	2.80	m	**3.04**
300 Volt grade; three core						
0.50 mm²	0.22	0.27	0.07	2.80	m	**3.07**
0.75 mm²	0.31	0.38	0.07	2.80	m	**3.18**
1.0 mm²	0.33	0.41	0.07	2.80	m	**3.21**
1.5 mm²	0.51	0.63	0.07	2.80	m	**3.43**
2.5 mm²	0.69	0.85	0.08	3.19	m	**4.04**
Flexible cord; PVC insulated; PVC sheathed; copper stranded to CMA Code 318Y (laid loose)						
300/500 Volt grade; two core						
0.75 mm²	0.22	0.27	0.07	2.80	m	**3.07**
1.0 mm²	0.27	0.34	0.07	2.80	m	**3.14**
1.5 mm²	0.36	0.44	0.07	2.80	m	**3.24**
2.5 mm²	0.62	0.77	0.07	2.80	m	**3.57**
300/500 Volt grade; three core						
0.75 mm²	0.29	0.36	0.07	2.80	m	**3.16**
1.0 mm²	0.36	0.44	0.07	2.80	m	**3.24**
1.5 mm²	0.51	0.63	0.07	2.80	m	**3.43**
2.5 mm²	0.81	1.00	0.08	3.19	m	**4.19**
300/500 Volt grade; four core						
0.75 mm²	0.38	0.46	0.08	3.19	m	**3.65**
1.0 mm²	0.48	0.59	0.08	3.19	m	**3.78**
1.5 mm²	0.69	0.85	0.08	3.19	m	**4.04**
2.5 mm²	1.12	1.38	0.09	3.60	m	**4.98**
Flexible cord; PVC insulated; PVC sheathed for use in high temperature zones; copper stranded to CMA Code 309Y (laid loose)						
300/500 Volt grade; two core						
0.50 mm²	0.23	0.28	0.07	2.80	m	**3.08**
0.75 mm²	0.28	0.35	0.07	2.80	m	**3.15**
1.0 mm²	0.33	0.41	0.07	2.80	m	**3.21**
1.5 mm²	0.45	0.56	0.07	2.80	m	**3.36**
2.5 mm²	0.66	0.81	0.07	2.80	m	**3.61**
300/500 Volt grade; three core						
0.50 mm²	0.33	0.41	0.07	2.80	m	**3.21**
0.75 mm²	0.38	0.46	0.07	2.80	m	**3.26**
1.0 mm²	0.44	0.54	0.07	2.80	m	**3.34**
1.5 mm²	0.61	0.75	0.07	2.80	m	**3.55**
2.5 mm²	0.84	1.03	0.07	2.80	m	**3.83**

ELECTRICAL SUPPLY/POWER/LIGHTING

Item	Net Price £	Material £	Labour hours	Labour £	Unit	Total rate £
LV DISTRIBUTION: CABLES AND WIRING – cont						
Flexible cord; rubber insulated; rubber sheathed; copper stranded to CMA code 318 (laid loose)						
300/500 Volt grade; two core						
0.50 mm²	0.38	0.46	0.07	2.80	m	**3.26**
0.75 mm²	0.48	0.59	0.07	2.80	m	**3.39**
1.0 mm²	0.58	0.71	0.07	2.80	m	**3.51**
1.5 mm²	0.73	0.89	0.07	2.80	m	**3.69**
2.5 mm²	1.08	1.32	0.07	2.80	m	**4.12**
300/500 Volt grade; three core						
0.50 mm²	0.48	0.59	0.07	2.80	m	**3.39**
0.75 mm²	0.58	0.71	0.07	2.80	m	**3.51**
1.0 mm²	0.68	0.84	0.07	2.80	m	**3.64**
1.5 mm²	0.94	1.16	0.07	2.80	m	**3.96**
2.5 mm²	1.39	1.71	0.07	2.80	m	**4.51**
300/500 Volt grade; four core						
0.50 mm²	0.54	0.66	0.08	3.19	m	**3.85**
0.75 mm²	0.65	0.80	0.08	3.19	m	**3.99**
1.0 mm²	0.84	1.03	0.08	3.19	m	**4.22**
1.5 mm²	1.30	1.60	0.08	3.19	m	**4.79**
2.5 mm²	1.82	2.24	0.08	3.19	m	**5.43**
Flexible cord; rubber insulated; rubber sheathed; for 90C operation; copper stranded to CMA Code 318 (laid loose)						
450/750 Volt grade; two core						
0.50 mm²	0.34	0.42	0.07	2.80	m	**3.22**
0.75 mm²	0.52	0.64	0.07	2.80	m	**3.44**
1.0 mm²	0.52	0.64	0.07	2.80	m	**3.44**
1.5 mm²	0.63	0.78	0.07	2.80	m	**3.58**
2.5 mm²	0.85	1.04	0.07	2.80	m	**3.84**
450/750 Volt grade; three core						
0.50 mm²	0.54	0.66	0.07	2.80	m	**3.46**
0.75 mm²	0.62	0.77	0.07	2.80	m	**3.57**
1.0 mm²	0.68	0.84	0.07	2.80	m	**3.64**
1.5 mm²	0.94	1.16	0.07	2.80	m	**3.96**
2.5 mm²	1.39	1.71	0.07	2.80	m	**4.51**
450/750 Volt grade; four core						
1.0 mm²	0.84	1.03	0.08	3.19	m	**4.22**
1.5 mm²	1.30	1.60	0.08	3.19	m	**4.79**
2.5 mm²	1.82	2.24	0.08	3.19	m	**5.43**

ELECTRICAL SUPPLY/POWER/LIGHTING

Item	Net Price £	Material £	Labour hours	Labour £	Unit	**Total rate £**
Heavy flexible cable; rubber insulated; rubber sheathed; copper stranded to CMA Code 638P (laid loose)						
450/750 Volt grade; two core						
1.0 mm²	0.52	0.64	0.08	3.19	m	**3.83**
1.5 mm²	0.63	0.78	0.08	3.19	m	**3.97**
2.5 mm²	0.85	1.04	0.08	3.19	m	**4.23**
450/750 Volt grade; three core						
1.0 mm²	0.63	0.78	0.08	3.19	m	**3.97**
1.5 mm²	0.73	0.89	0.08	3.19	m	**4.08**
2.5 mm²	0.84	1.03	0.08	3.19	m	**4.22**
450/750 Volt grade; four core						
1.0 mm²	0.85	1.04	0.08	3.19	m	**4.23**
1.5 mm²	0.93	1.14	0.08	3.19	m	**4.33**
2.5 mm²	1.28	1.58	0.08	3.19	m	**4.77**
FIRE RATED CABLE						
Cable, mineral insulated; copper sheathed with copper conductors; fixed with clips to backgrounds. BASEC approval to BS 6207 Part 1 1995; complies with BS 6387 Category CWZ						
Light duty 500 Volt grade; bare						
2L 1.0	4.29	5.27	0.23	9.20	m	**14.47**
2L 1.5	4.90	6.02	0.23	9.20	m	**15.22**
2L 2.5	5.90	7.25	0.25	10.00	m	**17.25**
2L 4.0	8.34	10.25	0.25	10.00	m	**20.25**
3L 1.0	5.07	6.23	0.24	9.60	m	**15.83**
3L 1.5	6.03	7.41	0.25	10.00	m	**17.41**
3L 2.5	8.61	10.57	0.25	10.00	m	**20.57**
4L 1.0	5.80	7.13	0.25	10.00	m	**17.13**
4L 1.5	6.50	7.98	0.25	10.00	m	**17.98**
4L 2.5	10.21	12.55	0.26	10.41	m	**22.96**
7L 1.5	10.28	12.63	0.28	11.20	m	**23.83**
7L 2.5	13.05	16.03	0.27	10.79	m	**26.82**
Light duty 500 Volt grade; LSF sheathed						
2L 1.0	3.35	4.12	0.23	9.20	m	**13.32**
2L 1.5	3.68	4.53	0.23	9.20	m	**13.73**
2L 2.5	4.44	5.45	0.25	10.00	m	**15.45**
2L 4.0	6.21	7.63	0.25	10.00	m	**17.63**
3L 1.0	3.94	4.84	0.24	9.60	m	**14.44**
3L 1.5	4.63	5.69	0.25	10.00	m	**15.69**
3L 2.5	6.45	7.93	0.25	10.00	m	**17.93**
4L 1.0	4.50	5.53	0.25	10.00	m	**15.53**
4L 1.5	5.30	6.51	0.25	10.00	m	**16.51**
4L 2.5	7.54	9.26	0.26	10.41	m	**19.67**
7L 1.5	6.56	8.06	0.28	11.20	m	**19.26**
7L 2.5	10.16	12.48	0.27	10.79	m	**23.27**

ELECTRICAL SUPPLY/POWER/LIGHTING

Item	Net Price £	Material £	Labour hours	Labour £	Unit	Total rate £
LV DISTRIBUTION: CABLES AND WIRING – cont						
Cable, mineral insulated – cont						
Heavy duty 750 Volt grade; bare						
1H 10	8.63	10.61	0.25	10.00	m	**20.61**
1H 16	11.69	14.36	0.26	10.41	m	**24.77**
1H 25	16.24	19.95	0.27	10.79	m	**30.74**
1H 35	23.44	28.79	0.32	12.80	m	**41.59**
1H 50	25.66	31.52	0.35	14.01	m	**45.53**
1H 70	34.24	42.06	0.38	15.19	m	**57.25**
1H 95	50.78	62.38	0.41	16.40	m	**78.78**
1H 120	53.42	65.63	0.46	18.41	m	**84.04**
1H 150	65.53	80.50	0.50	20.00	m	**100.50**
1H 185	79.91	98.16	0.56	22.40	m	**120.56**
1H 240	103.96	127.71	0.69	27.60	m	**155.31**
2H 1.5	7.96	9.78	0.25	10.00	m	**19.78**
2H 2.5	9.61	11.80	0.26	10.41	m	**22.21**
2H 4.0	11.77	14.46	0.26	10.41	m	**24.87**
2H 6.0	15.13	18.59	0.29	11.59	m	**30.18**
2H 10.0	19.34	23.76	0.34	13.60	m	**37.36**
2H 16.0	27.67	33.99	0.40	15.99	m	**49.98**
2H 25.0	38.15	46.87	0.44	17.61	m	**64.48**
3H 1.5	8.84	10.86	0.25	10.00	m	**20.86**
3H 2.5	10.96	13.46	0.25	10.00	m	**23.46**
3H 4.0	13.35	16.40	0.27	10.79	m	**27.19**
3H 6.0	16.50	20.28	0.30	12.00	m	**32.28**
3H 10.0	24.46	30.05	0.35	14.01	m	**44.06**
3H 16.0	32.94	40.47	0.41	16.40	m	**56.87**
3H 25.0	48.33	59.38	0.47	18.80	m	**78.18**
4H 1.5	10.75	13.21	0.24	9.60	m	**22.81**
4H 2.5	13.03	16.01	0.26	10.41	m	**26.42**
4H 4.0	16.09	19.77	0.29	11.59	m	**31.36**
4H 6.0	20.35	25.00	0.31	12.40	m	**37.40**
4H 10.0	28.90	35.50	0.37	14.81	m	**50.31**
4H 16.0	41.42	50.88	0.44	17.61	m	**68.49**
4H 25.0	59.14	72.65	0.52	20.80	m	**93.45**
7H 1.5	14.49	17.80	0.30	12.00	m	**29.80**
7H 2.5	19.31	23.72	0.32	12.80	m	**36.52**
12H 2.5	32.94	40.47	0.39	15.60	m	**56.07**
19H 1.5	48.29	59.33	0.42	16.81	m	**76.14**

ELECTRICAL SUPPLY/POWER/LIGHTING

Item	Net Price £	Material £	Labour hours	Labour £	Unit	**Total rate £**
Heavy duty 750 Volt grade; LSF sheathed						
1H 10	6.85	8.41	0.25	10.00	m	**18.41**
1H 16	9.04	11.11	0.26	10.41	m	**21.52**
1H 25	12.47	15.32	0.27	10.79	m	**26.11**
1H 35	17.25	21.19	0.32	12.80	m	**33.99**
1H 50	20.58	25.28	0.35	14.01	m	**39.29**
1H 70	26.20	32.18	0.38	15.19	m	**47.37**
1H 95	37.53	46.11	0.41	16.40	m	**62.51**
1H 120	41.36	50.81	0.46	18.41	m	**69.22**
1H 150	50.32	61.81	0.50	20.00	m	**81.81**
1H 185	61.08	75.03	0.56	22.40	m	**97.43**
1H 240	79.15	97.24	0.68	27.19	m	**124.43**
2H 1.5	6.16	7.57	0.25	10.00	m	**17.57**
2H 2.5	7.36	9.04	0.26	10.41	m	**19.45**
2H 4.0	9.12	11.20	0.26	10.41	m	**21.61**
2H 6.0	11.70	14.37	0.29	11.59	m	**25.96**
2H 10.0	15.49	19.03	0.34	13.60	m	**32.63**
2H 16.0	21.32	26.19	0.40	15.99	m	**42.18**
2H 25.0	29.69	36.48	0.44	17.61	m	**54.09**
3H 1.5	6.72	8.25	0.25	10.00	m	**18.25**
3H 2.5	8.42	10.34	0.25	10.00	m	**20.34**
3H 4.0	10.37	12.74	0.27	10.79	m	**23.53**
3H 6.0	13.31	16.35	0.30	12.00	m	**28.35**
3H 10.0	18.51	22.74	0.35	14.01	m	**36.75**
3H 16.0	25.96	31.89	0.41	16.40	m	**48.29**
3H 25.0	37.46	46.02	0.47	18.80	m	**64.82**
4H 1.5	8.29	10.18	0.24	9.60	m	**19.78**
4H 2.5	10.09	12.40	0.26	10.41	m	**22.81**
4H 4.0	12.82	15.75	0.29	11.59	m	**27.34**
4H 6.0	16.34	20.08	0.31	12.40	m	**32.48**
4H 10.0	22.52	27.67	0.37	14.81	m	**42.48**
4H 16.0	32.23	39.60	0.44	17.61	m	**57.21**
4H 25.0	45.69	56.13	0.52	20.80	m	**76.93**
7H 1.5	11.17	13.72	0.30	12.00	m	**25.72**
7H 2.5	15.29	18.78	0.32	12.80	m	**31.58**
12H 2.5	25.71	31.58	0.39	15.60	m	**47.18**
19H 1.5	37.40	45.95	0.42	16.81	m	**62.76**

ELECTRICAL SUPPLY/POWER/LIGHTING

Item	Net Price £	Material £	Labour hours	Labour £	Unit	Total rate £
LV DISTRIBUTION: CABLES AND WIRING – cont						
Cable terminations for MI Cable; Polymeric one piece moulding; containing grey sealing compound; testing; phase marking and connection						
Light Duty 500 Volt grade; brass gland; polymeric one moulding containing grey sealing compound; coloured conductor sleeving; Earth tag; plastic gland shroud						
2L 1.5	10.83	13.30	0.27	10.79	m	**24.09**
2L 2.5	10.83	14.66	0.27	10.79	m	**25.45**
3L 1.5	10.86	14.24	0.27	10.79	m	**25.03**
4L 1.5	10.86	14.24	0.27	10.79	m	**25.03**
Cable terminations; for MI copper sheathed cable. Certified for installation in potentially explosive atmospheres; testing; phase marking and connection; BS 6207 Part 2 1995						
Light duty 500 Volt grade; brass gland; brass pot with earth tail; pot closure; sealing compound; conductor sleving; plastic gland shroud; identification markers						
2L 1.0	10.83	13.30	0.39	15.60	nr	**28.90**
2L 1.5	10.83	13.30	0.41	16.40	nr	**29.70**
2L 2.5	10.83	13.30	0.44	17.44	nr	**30.74**
2L 4.0	10.83	13.30	0.46	18.41	nr	**31.71**
3L 1.0	10.86	13.35	0.43	17.20	nr	**30.55**
3L 1.5	10.86	13.35	0.44	17.49	nr	**30.84**
3L 2.5	10.86	13.35	0.44	17.61	nr	**30.96**
4L 1.0	10.86	13.35	0.47	18.80	nr	**32.15**
4L 1.5	10.86	13.35	0.47	18.98	nr	**32.33**
4L 2.5	10.86	13.35	0.50	20.00	nr	**33.35**
7L 1.0	10.51	12.91	0.69	27.60	nr	**40.51**
7L 1.5	26.31	32.32	0.70	27.99	nr	**60.31**
7L 2.5	26.31	32.32	0.74	29.60	nr	**61.92**
Heavy duty 750 Volt grade; brass gland; brass pot with earth tail; pot closure; sealing compound; conductor sleeving; plastic gland shroud; identification markers						
1H 10	10.89	13.38	0.37	14.81	nr	**28.19**
1H 16	10.89	13.38	0.39	15.60	nr	**28.98**
1H 25	10.89	13.38	0.56	22.40	nr	**35.78**
1H 35	10.89	13.38	0.57	22.81	nr	**36.19**
1H 50	26.39	32.42	0.60	24.00	nr	**56.42**
1H 70	26.14	32.11	0.67	26.81	nr	**58.92**
1H 95	26.14	32.11	0.75	30.00	nr	**62.11**
1H 120	42.34	52.02	0.94	37.60	nr	**89.62**
1H 150	42.34	52.02	0.99	39.60	nr	**91.62**
1H 185	42.34	52.02	1.26	50.41	nr	**102.43**
1H 240	66.79	82.05	1.37	54.80	nr	**136.85**

ELECTRICAL SUPPLY/POWER/LIGHTING

Item	Net Price £	Material £	Labour hours	Labour £	Unit	**Total rate £**
2H 1.5	11.53	14.16	0.42	16.81	nr	**30.97**
2H 2.5	11.53	14.16	0.44	17.71	nr	**31.87**
2H 4	11.53	14.16	0.47	18.80	nr	**32.96**
2H 6	11.53	14.16	0.54	21.60	nr	**35.76**
2H 10	27.97	34.36	0.58	23.19	nr	**57.55**
2H 16	27.97	34.36	0.69	27.60	nr	**61.96**
2H 25	45.31	55.66	0.77	30.80	nr	**86.46**
3H 1.5	11.54	14.17	0.44	17.61	nr	**31.78**
3H 2.5	11.54	14.17	0.47	18.92	nr	**33.09**
3H 4	11.54	14.17	0.57	22.81	nr	**36.98**
3H 6	28.07	34.48	0.61	24.40	nr	**58.88**
3H 10	28.07	34.48	0.65	26.00	nr	**60.48**
3H 16	28.07	34.48	0.78	31.21	nr	**65.69**
3H 25	71.53	87.87	0.85	34.01	nr	**121.88**
4H 1.5	11.54	14.17	0.52	20.80	nr	**34.97**
4H 2.5	11.54	14.17	0.53	21.20	nr	**35.37**
4H 4	28.07	34.48	0.60	24.00	nr	**58.48**
4H 6	28.07	34.48	0.65	26.00	nr	**60.48**
4H 10	28.07	34.48	0.69	27.60	nr	**62.08**
4H 16	45.31	55.66	0.88	35.20	nr	**90.86**
4H 25	73.62	90.45	0.93	37.20	nr	**127.65**
7H 1.5	27.97	34.36	0.71	28.40	nr	**62.76**
7H 2.5	27.97	34.36	0.74	29.60	nr	**63.96**
12H 1.5	45.31	55.66	0.85	34.01	nr	**89.67**
12H 2.5	45.31	55.66	1.00	40.01	nr	**95.67**
19H 2.5	71.53	87.87	1.11	44.39	nr	**132.26**
Cable; FP100; LOSH insulated; non-sheathed fire resistant to LPCB Approved to BS 6387 Catergory CWZ; in conduit or trunking including terminations						
450/750 Volt grade; single core						
1.0 mm²	0.46	0.57	0.13	5.20	m	**5.77**
1.5 mm²	0.23	0.28	0.13	5.20	m	**5.48**
2.5 mm²	0.34	0.42	0.13	5.20	m	**5.62**
4.0 mm²	0.56	0.69	0.14	5.48	m	**6.17**
6.0 mm²	1.34	1.65	0.13	5.20	m	**6.85**
10 mm²	2.27	2.79	0.16	6.40	m	**9.19**
16 mm²	3.52	4.33	0.16	6.40	m	**10.73**
Cable; FP200; Insudite insulated; LSOH sheathed screened fire resistant BASEC Approved to BS 7629; fixed with clips to backgrounds						
300/500 Volt grade; two core						
1.0 mm²	2.85	3.50	0.20	8.00	m	**11.50**
1.5 mm²	2.36	2.90	0.20	8.00	m	**10.90**
2.5 mm²	3.67	4.50	0.20	8.00	m	**12.50**
4.0 mm²	4.61	5.66	0.23	9.20	m	**14.86**

ELECTRICAL SUPPLY/POWER/LIGHTING

Item	Net Price £	Material £	Labour hours	Labour £	Unit	Total rate £
LV DISTRIBUTION: CABLES AND WIRING – cont						
Cable; FP200; Insudite insulated – cont						
300/500 Volt grade; three core						
1.0 mm²	3.14	3.85	0.23	9.20	m	**13.05**
1.5 mm²	3.74	4.60	0.23	9.20	m	**13.80**
2.5 mm²	4.51	5.54	0.23	9.20	m	**14.74**
4.0 mm²	7.42	9.11	0.25	10.00	m	**19.11**
300/500 Volt grade; four core						
1.0 mm²	3.66	4.49	0.25	10.00	m	**14.49**
1.5 mm²	4.42	5.43	0.25	10.00	m	**15.43**
2.5 mm²	6.12	7.52	0.25	10.00	m	**17.52**
4.0 mm	9.26	11.37	0.28	11.20	m	**22.57**
Terminations; including glanding-off, connection to equipment						
Two core						
1.0 mm²	0.68	0.84	0.35	14.01	nr	**14.85**
1.5 mm²	0.68	0.84	0.35	14.01	nr	**14.85**
2.5 mm²	0.68	0.84	0.35	14.01	nr	**14.85**
4.0 mm²	0.68	0.84	0.35	14.01	nr	**14.85**
Three core						
1.0 mm²	0.68	0.84	0.35	14.01	nr	**14.85**
1.5 mm²	0.68	0.84	0.35	14.01	nr	**14.85**
2.5 mm²	0.68	0.84	0.35	14.14	nr	**14.98**
4.0 mm²	0.68	0.84	0.35	14.01	nr	**14.85**
Four core						
1.0 mm²	0.68	0.84	0.35	14.01	nr	**14.85**
1.5 mm²	0.68	0.84	0.35	14.01	nr	**14.85**
2.5 mm²	0.68	0.84	0.35	14.01	nr	**14.85**
4.0 mm²	0.68	0.84	0.35	14.01	nr	**14.85**
Cable; FP400; polymeric insulated; LSOH sheathed fire resistant; armoured; with copper stranded copper conductors; BASEC Approved to BS 7846; fixed with clips to backgrounds						
600/1000 Volt grade; two core						
1.5 mm²	3.17	3.90	0.20	8.00	m	**11.90**
2.5 mm²	3.67	4.50	0.20	8.00	m	**12.50**
4.0 mm²	4.07	5.00	0.21	8.40	m	**13.40**
6.0 mm²	5.01	6.15	0.22	8.80	m	**14.95**
10 mm²	5.40	6.64	0.24	9.60	m	**16.24**
16 mm²	8.22	10.10	0.25	10.00	m	**20.10**
25 mm²	11.36	13.95	0.35	14.01	m	**27.96**
600/1000 Volt grade; three core						
1.5 mm²	3.51	4.32	0.20	8.00	m	**12.32**
2.5 mm²	4.10	5.04	0.21	8.40	m	**13.44**
4.0 mm²	4.83	5.93	0.22	8.80	m	**14.73**

ELECTRICAL SUPPLY/POWER/LIGHTING

Item	Net Price £	Material £	Labour hours	Labour £	Unit	**Total rate £**
6.0 mm²	5.12	6.29	0.22	8.80	m	**15.09**
10 mm²	6.22	7.64	0.25	10.00	m	**17.64**
16 mm²	10.18	12.51	0.26	10.41	m	**22.92**
25 mm²	12.71	15.61	0.37	14.81	m	**30.42**
600/1000 Volt grade; four core						
1.5 mm²	4.69	5.76	0.21	8.40	m	**14.16**
2.5 mm²	5.46	6.71	0.22	8.80	m	**15.51**
4.0 mm²	6.43	7.90	0.22	8.80	m	**16.70**
6.0 mm²	6.81	8.37	0.23	9.20	m	**17.57**
10 mm²	8.30	10.20	0.26	10.41	m	**20.61**
16 mm²	13.56	16.66	0.26	10.41	m	**27.07**
25 mm²	16.94	20.81	0.39	15.60	m	**36.41**
Terminations; including glanding-off, connection to equipment						
Two core						
1.5 mm²	2.03	2.50	0.58	23.19	nr	**25.69**
2.5 mm²	2.15	2.65	0.58	23.19	nr	**25.84**
4.0 mm²	2.35	2.89	0.61	24.52	nr	**27.41**
6.0 mm²	2.60	3.19	0.67	26.81	nr	**30.00**
10 mm²	2.86	3.52	1.00	40.01	nr	**43.53**
16 mm²	3.14	3.85	1.11	44.39	nr	**48.24**
25 mm²	3.45	4.24	1.70	68.00	nr	**72.24**
Three core						
1.5 mm²	3.06	3.76	0.62	24.80	nr	**28.56**
2.5 mm²	3.21	3.95	0.62	24.80	nr	**28.75**
4.0 mm²	3.54	4.35	0.66	26.57	nr	**30.92**
6.0 mm²	3.89	4.78	0.71	28.40	nr	**33.18**
10 mm²	4.27	5.25	1.06	42.40	nr	**47.65**
16 mm²	4.71	5.79	1.19	47.61	nr	**53.40**
25 mm²	5.17	6.35	1.81	72.40	nr	**78.75**
Four core						
1.5 mm²	4.07	5.00	0.67	26.81	nr	**31.81**
2.5 mm²	4.28	5.26	0.69	27.39	nr	**32.65**
4.0 mm²	4.50	5.53	0.71	28.40	nr	**33.93**
6.0 mm²	4.72	5.80	0.76	30.41	nr	**36.21**
10 mm²	4.85	5.95	1.14	45.60	nr	**51.55**
16 mm²	5.23	6.43	1.29	51.60	nr	**58.03**
25 mm²	5.83	7.16	1.35	54.00	nr	**61.16**
Cable; FP600; polymeric insulated; LSOH sheathed fire resistant; armoured; with copper stranded copper conductors; BASEC Approved to BS 7846; fixed with clips to backgrounds						
600/1000 Volt grade; two core						
4.0 mm²	5.55	6.81	0.21	8.40	m	**15.21**
10 mm²	6.45	7.93	0.24	9.60	m	**17.53**
16 mm²	7.63	9.38	0.25	10.00	m	**19.38**

ELECTRICAL SUPPLY/POWER/LIGHTING

Item	Net Price £	Material £	Labour hours	Labour £	Unit	Total rate £
LV DISTRIBUTION: CABLES AND WIRING – cont						
Cable; FP600; polymeric insulated – cont						
600/1000 Volt grade; three core						
4.0 mm²	6.45	7.93	0.22	8.80	m	**16.73**
6.0 mm²	6.36	7.81	0.22	8.80	m	**16.61**
10 mm²	8.08	9.92	0.25	10.00	m	**19.92**
16 mm²	9.71	11.93	0.26	10.41	m	**22.34**
25 mm²	12.92	15.88	0.37	14.81	m	**30.69**
35 mm²	15.58	19.14	0.38	15.19	m	**34.33**
50 mm²	20.71	25.44	0.40	15.99	m	**41.43**
70 mm²	25.32	31.10	0.42	16.81	m	**47.91**
95 mm²	37.42	45.97	0.45	18.00	m	**63.97**
600/1000 Volt grade; four core						
4.0 mm²	6.33	7.78	0.22	8.80	nr	**16.58**
6.0 mm²	6.98	8.58	0.23	9.20	nr	**17.78**
10 mm²	7.79	9.57	0.26	10.41	nr	**19.98**
16 mm²	8.95	10.99	0.26	10.41	nr	**21.40**
25 mm²	10.92	13.42	0.39	15.60	nr	**29.02**
35 mm²	13.98	17.18	0.40	15.99	m	**33.17**
50 mm²	17.36	21.33	0.42	16.81	m	**38.14**
70 mm²	22.43	27.55	0.45	18.00	m	**45.55**
95 mm²	29.69	36.48	0.47	18.80	m	**55.28**
120 mm²	36.86	45.29	0.48	19.20	m	**64.49**
150 mm²	44.30	54.42	0.51	20.40	m	**74.82**
185 mm²	54.48	66.93	0.52	20.80	m	**87.73**
240 mm²	70.04	86.05	0.55	22.00	m	**108.05**
300 mm²	85.81	105.42	0.59	23.60	m	**129.02**
400 mm²	107.49	132.05	0.63	25.20	m	**157.25**
Terminations; including glanding-off, connection to equipment						
Two core						
4.0 mm²	3.44	4.22	0.35	14.01	nr	**18.23**
10 mm²	4.00	4.91	0.35	14.01	nr	**18.92**
16 mm²	4.73	5.81	0.35	14.01	nr	**19.82**
Three core						
4.0 mm²	4.00	4.91	0.35	14.01	nr	**18.92**
6.0 mm²	3.94	4.84	0.35	14.01	nr	**18.85**
10 mm²	5.01	6.15	0.35	14.01	nr	**20.16**
16 mm²	6.02	7.39	0.35	14.01	nr	**21.40**
25 mm²	8.01	9.84	0.35	14.01	nr	**23.85**
35 mm²	9.66	11.86	0.35	14.01	m	**25.87**
50 mm²	12.84	15.77	0.35	14.01	m	**29.78**
70 mm²	15.70	19.29	0.35	14.01	m	**33.30**
95 mm²	23.20	28.50	0.35	14.01	m	**42.51**
Four core						
4.0 mm²	3.92	4.82	0.35	14.01	nr	**18.83**
6.0 mm²	4.33	5.32	0.35	14.01	nr	**19.33**
10 mm²	4.83	5.93	0.35	14.01	nr	**19.94**

ELECTRICAL SUPPLY/POWER/LIGHTING

Item	Net Price £	Material £	Labour hours	Labour £	Unit	**Total rate £**
16 mm²	5.55	6.81	0.35	14.01	nr	**20.82**
25 mm²	6.77	8.32	0.35	14.01	nr	**22.33**
35 mm²	8.67	10.65	0.35	14.01	m	**24.66**
50 mm²	10.76	13.22	0.35	14.01	m	**27.23**
70 mm²	13.91	17.08	0.35	14.01	m	**31.09**
95 mm²	18.41	22.62	0.35	14.01	m	**36.63**
120 mm²	22.85	28.07	0.35	14.01	m	**42.08**
150 mm²	27.47	33.75	0.35	14.01	m	**47.76**
185 mm²	33.78	41.50	0.35	14.01	m	**55.51**
240 mm²	43.42	53.34	0.35	14.01	m	**67.35**
300 mm²	53.20	65.35	0.35	14.01	m	**79.36**
400 mm²	66.64	81.87	0.35	14.01	m	**95.88**
Cable; Firetuff fire resistant to BS 6387; fixed with clips to backgrounds						
Two core						
1.5 mm²	2.39	2.94	0.20	8.00	m	**10.94**
2.5 mm²	2.89	3.55	0.20	8.00	m	**11.55**
4.0 mm²	3.75	4.61	0.21	8.40	m	**13.01**
Three core						
1.5 mm²	2.92	3.59	0.20	8.00	m	**11.59**
2.5 mm²	3.30	4.05	0.21	8.40	m	**12.45**
4.0 mm²	4.77	5.86	0.22	8.80	m	**14.66**
Four core						
1.5 mm²	3.26	4.00	0.21	8.40	m	**12.40**
2.5 mm²	4.00	4.91	0.22	8.80	m	**13.71**
4.0 mm²	5.45	6.70	0.22	8.80	m	**15.50**

ELECTRICAL SUPPLY/POWER/LIGHTING

Item	Net Price £	Material £	Labour hours	Labour £	Unit	Total rate £
LV DISTRUBITION: MODULAR WIRING						
Modular wiring systems; including commissioning						
Master distribution box; steel; fixed to backgrounds; 6 Port						
4.0 mm 18 core armoured home run cable	163.03	200.29	0.90	35.99	nr	**236.28**
4.0 mm 24 core armoured cable home run cable	163.03	200.29	0.95	38.00	nr	**238.29**
4.0 mm 18 core armoured home run cable & data cable	173.20	212.77	0.95	38.00	nr	**250.77**
6.0 mm 18 core armoured home run cable	163.03	200.29	1.00	40.01	nr	**240.30**
6.0 mm 24 core armoured home run cable	163.03	200.29	1.10	44.01	nr	**244.30**
6.0 mm 18 core armoured home run cable & data cable	173.20	212.77	1.10	44.01	nr	**256.78**
Master distribution box; steel; fixed to backgrounds; 9 Port						
4.0 mm 27 core armoured home run cable	203.77	250.33	1.30	52.01	nr	**302.34**
4.0 mm 27 core armoured home run cable & data cable	203.77	250.33	1.45	58.00	nr	**308.33**
6.0 mm 27 core armoured home run cable	213.96	262.85	1.45	58.00	nr	**320.85**
6.0 mm 27 core armoured home run cable & data cable	213.96	262.85	1.55	62.00	nr	**324.85**
Metal clad cable; BSEN 60439 Part 2 1993; BASEC approved						
4.0 mm 18 core	14.23	17.48	0.30	12.00	m	**29.48**
4.0 mm 24 core	21.90	26.90	0.32	12.80	m	**39.70**
4.0 mm 27 core	22.07	27.11	0.35	14.01	m	**41.12**
6.0 mm 18 core	18.95	23.28	0.32	12.80	m	**36.08**
6.0 mm 27 core	26.40	32.43	0.35	14.01	m	**46.44**
Metal clad data cable						
Single twisted pair	2.77	3.40	0.18	7.20	m	**10.60**
Twin twisted pair	4.47	5.49	0.18	7.20	m	**12.69**
Distribution cables; armoured; BSEN 60439 Part 2 1993; BASEC approved						
3 wire; 6.1 m long	52.56	64.58	0.92	36.80	nr	**101.38**
4 wire; 6.1 m long	59.28	72.83	0.96	38.40	nr	**111.23**
Extender cables; armoured; BSEN 60439 Part 2 1993; BASEC approved						
3 wire						
0.9 m long	32.98	40.52	0.13	5.20	nr	**45.72**
1.5 m long	36.31	44.60	0.23	9.20	nr	**53.80**
2.1 m long	39.65	48.71	0.31	12.40	nr	**61.11**
2.7 m long	42.98	52.80	0.40	15.99	nr	**68.79**
3.4 m long	46.86	57.57	0.51	20.40	nr	**77.97**
4.6 m long	53.53	65.76	0.69	27.60	nr	**93.36**
6.1 m long	61.85	75.98	0.92	36.80	nr	**112.78**

ELECTRICAL SUPPLY/POWER/LIGHTING

Item	Net Price £	Material £	Labour hours	Labour £	Unit	**Total rate £**
7.6 m long	70.18	86.22	1.14	45.60	nr	**131.82**
9.1 m long	78.50	96.44	1.37	54.80	nr	**151.24**
10.7 m long	90.22	110.84	1.61	64.40	nr	**175.24**
4 wire						
0.9 m long	34.97	42.96	0.14	5.61	nr	**48.57**
1.5 m long	38.92	47.82	0.24	9.60	nr	**57.42**
2.1 m long	42.86	52.66	0.32	12.80	nr	**65.46**
2.7 m long	46.81	57.51	0.43	17.20	nr	**74.71**
3.4 m long	51.40	63.15	0.51	20.40	nr	**83.55**
4.6 m long	59.30	72.85	0.67	26.81	nr	**99.66**
6.1 m long	69.38	85.23	0.92	36.80	nr	**122.03**
7.6 m long	79.02	97.07	1.22	48.79	nr	**145.86**
9.1 m long	88.87	109.18	1.46	58.40	nr	**167.58**
10.7 m long	99.41	122.13	1.71	68.41	nr	**190.54**
3 wire; including twisted pair						
0.9 m long	43.38	53.29	0.13	5.20	nr	**58.49**
1.5 m long	48.73	59.86	0.23	9.20	nr	**69.06**
2.1 m long	54.08	66.43	0.31	12.40	nr	**78.83**
2.7 m long	59.44	73.02	0.40	15.99	nr	**89.01**
3.4 m long	65.69	80.70	0.51	20.40	nr	**101.10**
4.6 m long	76.39	93.85	0.69	27.60	nr	**121.45**
6.1 m long	89.78	110.29	0.92	36.80	nr	**147.09**
7.6 m long	103.15	126.72	1.14	45.60	nr	**172.32**
9.1 m long	116.54	143.17	1.37	54.80	nr	**197.97**
10.7 m long	130.81	160.70	1.61	64.40	nr	**225.10**
Extender whip ended cables; armoured; BSEN 60439 Part 2 1993; BASEC approved						
3 wire; 3.0 m long	34.41	42.27	0.30	12.00	nr	**54.27**
4 wire; 3.0 m long	38.15	46.87	0.30	12.00	nr	**58.87**
T Connectors						
3 wire						
6 Pole 20a Tee	38.71	47.55	0.10	4.00	nr	**51.55**
Snap fix	19.73	24.23	0.10	4.00	nr	**28.23**
1.5 m flexible cable	22.87	28.10	0.10	4.00	nr	**32.10**
1.5 m armoured cable	25.83	31.73	0.15	6.00	nr	**37.73**
1.5 m armoured cable with twisted pair	32.92	40.45	0.15	6.00	nr	**46.45**
4 wire						
Snap fix	23.81	29.25	0.10	4.00	nr	**33.25**
1.5 m flexible cable	20.28	24.92	0.10	4.00	nr	**28.92**
1.5 m armoured cable	26.78	32.90	0.18	7.20	nr	**40.10**
Splitters						
5 wire	19.53	23.99	0.20	8.00	nr	**31.99**
5 wire converter	25.18	30.93	0.20	8.00	nr	**38.93**
Switch modules						
3 wire; 6.1 m long armoured cable	53.44	65.65	0.75	30.00	nr	**95.65**
4 wire; 6.1 m long armoured cable	60.14	73.88	0.80	32.00	nr	**105.88**

ELECTRICAL SUPPLY/POWER/LIGHTING

Item	Net Price £	Material £	Labour hours	Labour £	Unit	Total rate £
LV DISTRUBITION: MODULAR WIRING – cont						
Distribution cables; unarmoured; IEC 998 DIN/VDE 0628						
3 wire; 6.1 m long	20.31	24.95	0.70	27.99	nr	**52.94**
4 wire; 6.1 m long	25.41	31.22	0.75	30.00	nr	**61.22**
Extender cables; unarmoured; IEC 998 DIN/VDE 0628						
3 wire						
0.9 m long	12.41	15.25	0.07	2.80	nr	**18.05**
1.5 m long	13.58	16.68	0.12	4.80	nr	**21.48**
2.1 m long	14.76	18.13	0.17	6.80	nr	**24.93**
2.7 m long	19.55	24.01	0.27	10.79	nr	**34.80**
3.4 m long	17.32	21.27	0.22	8.80	nr	**30.07**
4.6 m long	19.69	24.19	0.37	14.81	nr	**39.00**
6.1 m long	22.64	27.81	0.49	19.59	nr	**47.40**
7.6 m long	25.59	31.44	0.61	24.40	nr	**55.84**
9.1 m long	28.54	35.06	0.73	29.20	nr	**64.26**
10.7 m long	31.69	38.93	0.86	34.40	nr	**73.33**
4 wire						
0.9 m long	14.94	18.35	0.08	3.19	nr	**21.54**
1.5 m long	16.48	20.24	0.14	5.61	nr	**25.85**
2.1 m long	18.02	22.13	0.19	7.60	nr	**29.73**
2.7 m long	19.55	24.01	0.24	9.60	nr	**33.61**
3.4 m long	21.34	26.22	0.31	12.40	nr	**38.62**
4.6 m long	24.41	29.99	0.41	16.40	nr	**46.39**
6.1 m long	28.25	34.70	0.55	22.00	nr	**56.70**
7.6 m long	32.09	39.43	0.68	27.19	nr	**66.62**
9.1 m long	35.92	44.13	0.82	32.80	nr	**76.93**
10.7 m long	40.02	49.16	0.96	38.40	nr	**87.56**
5 wire						
0.9 m long	22.66	27.84	0.09	3.60	nr	**31.44**
1.5 m long	24.65	30.28	0.15	6.00	nr	**36.28**
2.1 m long	26.64	32.73	0.21	8.40	nr	**41.13**
2.7 m long	28.62	35.16	0.27	10.79	nr	**45.95**
3.4 m long	30.93	38.00	0.34	13.60	nr	**51.60**
4.6 m long	34.90	42.87	0.46	18.41	nr	**61.28**
6.1 m long	39.86	48.97	0.61	24.40	nr	**73.37**
7.6 m long	44.82	55.06	0.76	30.41	nr	**85.47**
9.1 m long	49.77	61.14	0.91	36.40	nr	**97.54**
10.7 m long	55.08	67.66	1.07	42.80	nr	**110.46**
Extender whip ended cables; armoured; IEC 998 DIN/VDE 0628						
3 wire; 2.5 mm; 3.0 m long	13.02	15.99	0.30	12.00	nr	**27.99**
4 wire; 2.5 mm; 3.0 m long	15.29	18.78	0.30	12.00	nr	**30.78**

ELECTRICAL SUPPLY/POWER/LIGHTING

Item	Net Price £	Material £	Labour hours	Labour £	Unit	**Total rate £**
T Connectors						
3 wire						
5 pin; direct fix	12.01	14.75	0.10	4.00	nr	**18.75**
5 pin; 1.5 mm flexible cable; 0.3 m long	16.82	20.66	0.15	6.00	nr	**26.66**
4 wire						
5 pin; direct fix	15.31	18.81	0.20	8.00	nr	**26.81**
5 pin; 1.5 mm flexible cable; 0.3 m long	19.06	23.41	0.20	8.00	nr	**31.41**
5 wire						
5 pin; direct fix	17.43	21.41	0.20	8.00	nr	**29.41**
Splitters						
3 way; 5 pin	10.04	12.34	0.25	10.00	nr	**22.34**
Switch modules						
3 wire	28.62	35.16	0.20	8.00	nr	**43.16**
4 wire	29.84	36.66	0.22	8.80	nr	**45.46**

ELECTRICAL SUPPLY/POWER/LIGHTING

Item	Net Price £	Material £	Labour hours	Labour £	Unit	Total rate £
LV DISTRUBITION: BUSBAR TRUNKING						
MAINS BUSBAR						
Low impedance busbar trunking; fixed to backgrounds including supports, fixings and connections/jointing to equipment						
Straight copper busbar						
1000 amp TP&N	515.74	633.59	3.41	136.41	m	**770.00**
1350 amp TP&N	633.92	778.77	3.58	143.20	m	**921.97**
2000 amp TP&N	795.10	976.78	5.00	200.00	m	**1176.78**
2500 amp TP&N	1201.26	1475.74	5.90	236.00	m	**1711.74**
Extra for fittings mains bus bar						
IP54 protection						
1000 amp TP&N	35.17	43.21	2.16	86.40	m	**129.61**
1350 amp TP&N	37.79	46.42	2.61	104.40	m	**150.82**
2000 amp TP&N	52.43	64.41	3.51	140.41	m	**204.82**
2500 amp TP&N	62.25	76.47	3.96	158.40	m	**234.87**
End cover						
1000 amp TP&N	40.45	49.70	0.56	22.40	nr	**72.10**
1350 amp TP&N	42.19	51.83	0.56	22.40	nr	**74.23**
2000 amp TP&N	67.70	83.17	0.66	26.40	nr	**109.57**
2500 amp TP&N	68.57	84.24	0.66	26.40	nr	**110.64**
Edge elbow						
1000 amp TP&N	579.37	711.75	2.01	80.41	nr	**792.16**
1350 amp TP&N	652.04	801.03	2.01	80.41	nr	**881.44**
2000 amp TP&N	1011.32	1242.40	2.40	95.99	nr	**1338.39**
2500 amp TP&N	1328.30	1631.82	2.40	95.99	nr	**1727.81**
Flat elbow						
1000 amp TP&N	502.66	617.52	2.01	80.41	nr	**697.93**
1350 amp TP&N	545.04	669.58	2.01	80.41	nr	**749.99**
2000 amp TP&N	773.14	949.80	2.40	95.99	nr	**1045.79**
2500 amp TP&N	962.90	1182.92	2.40	95.99	nr	**1278.91**
Offset						
1000 amp TP&N	1009.34	1239.98	3.00	120.00	nr	**1359.98**
1350 amp TP&N	1241.47	1525.15	3.00	120.00	nr	**1645.15**
2000 amp TP&N	1956.10	2403.07	3.50	140.00	nr	**2543.07**
2500 amp TP&N	2226.57	2735.34	3.50	140.00	nr	**2875.34**
Edge Z unit						
1000 amp TP&N	1511.98	1857.47	3.00	120.00	nr	**1977.47**
1350 amp TP&N	1935.91	2378.26	3.00	120.00	nr	**2498.26**
2000 amp TP&N	2900.84	3563.68	3.50	140.00	nr	**3703.68**
2500 amp TP&N	3312.65	4069.59	3.50	140.00	nr	**4209.59**
Flat Z unit						
1000 amp TP&N	1634.76	2008.30	3.00	120.00	nr	**2128.30**
1350 amp TP&N	1635.13	2008.76	3.00	120.00	nr	**2128.76**
2000 amp TP&N	2434.52	2990.81	3.50	140.00	nr	**3130.81**
2500 amp TP&N	2927.07	3595.90	3.50	140.00	nr	**3735.90**

ELECTRICAL SUPPLY/POWER/LIGHTING

Item	Net Price £	Material £	Labour hours	Labour £	Unit	**Total rate £**
Edge tee						
1000 amp TP&N	1511.98	1857.47	2.20	88.00	nr	**1945.47**
1350 amp TP&N	1935.91	2378.26	2.20	88.00	nr	**2466.26**
2000 amp TP&N	2900.84	3563.68	2.60	104.00	nr	**3667.68**
2500 amp TP&N	3313.63	4070.80	2.60	104.00	nr	**4174.80**
Tap off; TP&N integral contactor/breaker						
18 amp	245.93	302.13	0.82	32.80	nr	**334.93**
Tap off; TP&N fusable with on-load switch; excludes fuses						
32 amp	695.83	854.83	0.82	32.80	nr	**887.63**
63 amp	710.98	873.44	0.88	35.20	nr	**908.64**
100 amp	869.49	1068.17	1.18	47.20	nr	**1115.37**
160 amp	988.38	1214.22	1.41	56.40	nr	**1270.62**
250 amp	1276.26	1567.88	1.76	70.40	nr	**1638.28**
315 amp	1507.04	1851.40	2.06	82.39	nr	**1933.79**
Tap off; TP&N MCCB						
63 amp	879.99	1081.07	0.88	35.20	nr	**1116.27**
125 amp	1054.82	1295.85	1.18	47.20	nr	**1343.05**
160 amp	1151.56	1414.70	1.41	56.40	nr	**1471.10**
250 amp	1481.40	1819.90	1.76	70.40	nr	**1890.30**
400 amp	1888.18	2319.63	2.06	82.39	nr	**2402.02**
RISING MAINS BUSBAR						
Rising mains busbar; insulated supports, earth continuity bar; including couplers; fixed to backgrounds						
Straight aluminium bar						
200 amp TP&N	179.79	220.87	2.13	85.20	m	**306.07**
315 amp TP&N	201.42	247.44	2.15	85.99	m	**333.43**
400 amp TP&N	233.27	286.58	2.15	85.99	m	**372.57**
630 amp TP&N	287.89	353.67	2.47	98.79	m	**452.46**
800 amp TP&N	432.41	531.22	2.88	115.20	m	**646.42**
Extra for fittings rising busbar						
End feed unit						
200 amp TP&N	359.58	441.75	2.57	102.81	nr	**544.56**
315 amp TP&N	360.72	443.14	2.76	110.41	nr	**553.55**
400 amp TP&N	402.82	494.87	2.76	110.41	nr	**605.28**
630 amp TP&N	403.96	496.26	3.64	145.60	nr	**641.86**
800 amp TP&N	448.34	550.79	4.54	181.60	nr	**732.39**
Top feeder unit						
200 amp TP&N	359.58	441.75	2.57	102.81	nr	**544.56**
315 amp TP&N	360.72	443.14	2.76	110.41	nr	**553.55**
400 amp TP&N	402.82	494.87	2.76	110.41	nr	**605.28**
630 amp TP&N	403.96	496.26	3.64	145.60	nr	**641.86**
800 amp TP&N	448.34	550.79	4.54	181.60	nr	**732.39**

ELECTRICAL SUPPLY/POWER/LIGHTING

Item	Net Price £	Material £	Labour hours	Labour £	Unit	Total rate £
LV DISTRUBITION: BUSBAR TRUNKING – cont						
Extra for fittings rising busbar – cont						
End cap						
200 amp TP&N	30.72	37.74	0.18	7.20	nr	**44.94**
315 amp TP&N	30.72	37.74	0.27	10.79	nr	**48.53**
400 amp TP&N	34.13	41.93	0.27	10.79	nr	**52.72**
630 amp TP&N	34.13	41.93	0.41	16.40	nr	**58.33**
800 amp TP&N	99.00	121.62	0.41	16.40	nr	**138.02**
Edge elbow						
200 amp TP&N	42.10	51.72	0.55	22.00	nr	**73.72**
315 amp TP&N	43.24	53.12	0.94	37.60	nr	**90.72**
400 amp TP&N	322.03	395.62	0.94	37.60	nr	**433.22**
630 amp TP&N	323.17	397.02	1.45	58.00	nr	**455.02**
800 amp TP&N	306.09	376.04	1.45	58.00	nr	**434.04**
Flat elbow						
200 amp TP&N	138.82	170.54	0.55	22.00	nr	**192.54**
315 amp TP&N	138.82	170.54	0.94	37.60	nr	**208.14**
400 amp TP&N	187.75	230.65	0.94	37.60	nr	**268.25**
630 amp TP&N	188.90	232.06	1.45	58.00	nr	**290.06**
800 amp TP&N	260.59	320.13	1.45	58.00	nr	**378.13**
Edge tee						
200 amp TP&N	194.58	239.04	0.61	24.40	nr	**263.44**
315 amp TP&N	195.72	240.44	1.02	40.80	nr	**281.24**
400 amp TP&N	274.24	336.90	1.02	40.80	nr	**377.70**
630 amp TP&N	274.24	336.90	1.57	62.80	nr	**399.70**
800 amp TP&N	448.34	550.79	1.57	62.80	nr	**613.59**
Flat tee						
200 amp TP&N	250.34	307.54	0.61	24.40	nr	**331.94**
315 amp TP&N	195.72	240.44	1.02	40.80	nr	**281.24**
400 amp TP&N	393.72	483.68	1.02	40.80	nr	**524.48**
630 amp TP&N	394.86	485.09	1.57	62.80	nr	**547.89**
800 amp TP&N	551.90	678.01	1.57	62.80	nr	**740.81**
Tap off units						
TP&N fusable with on-load switch; excludes fuses						
32 amp	179.78	220.86	0.82	32.80	nr	**253.66**
63 amp	237.85	292.19	0.88	35.20	nr	**327.39**
100 amp	319.36	392.33	1.18	47.20	nr	**439.53**
250 amp	479.04	588.50	1.41	56.40	nr	**644.90**
400 amp	697.90	857.37	2.06	82.39	nr	**939.76**
TP&N MCCB						
32 amp	183.13	224.97	0.82	32.80	nr	**257.77**
63 amp	253.48	311.40	0.88	35.20	nr	**346.60**
100 amp	406.45	499.33	1.18	47.20	nr	**546.53**
250 amp	688.97	846.39	1.41	56.40	nr	**902.79**
400 amp	1210.45	1487.04	2.06	82.39	nr	**1569.43**

ELECTRICAL SUPPLY/POWER/LIGHTING

Item	Net Price £	Material £	Labour hours	Labour £	Unit	**Total rate £**
LIGHTING BUSBAR						
Prewired busbar, plug-in trunking for lighting; galvanized sheet steel housing (PE); tin-plated copper conductors with tap-off units at 1 m intervals						
Straight lengths – 25 amp						
2 Pole & PE	29.86	36.69	0.16	6.40	m	**43.09**
4 Pole & PE	33.17	40.75	0.16	6.40	m	**47.15**
Straight lengths – 40 amp						
2 Pole & PE	29.86	36.69	0.16	6.40	m	**43.09**
4 Pole & PE	39.81	48.91	0.16	6.40	m	**55.31**
Components for prewired busbars, plug-in trunking for lighting						
Plug-in tap off units						
10 amp 4 Pole & PE; 3 m of cable	25.43	31.24	0.10	4.00	nr	**35.24**
16 amp 4 Pole & PE; 3 m of cable	27.65	33.97	0.10	4.00	nr	**37.97**
16 amp with phase selection, 2P & PE; no cable	21.01	25.81	0.10	4.00	nr	**29.81**
Trunking components						
End feed unit & cover; 4P & PE	32.07	39.40	0.23	9.20	nr	**48.60**
Centre feed unit	160.35	196.99	0.29	11.59	nr	**208.58**
Right hand, intermediate terminal box feed unit	34.28	42.12	0.23	9.20	nr	**51.32**
End cover (for R/hand feed)	8.85	10.87	0.06	2.39	nr	**13.26**
Flexible elbow unit	76.30	93.73	0.12	4.80	nr	**98.53**
Fixing bracket – universal	5.53	6.79	0.10	4.00	nr	**10.79**
Suspension Bracket – Flat	4.42	5.43	0.10	4.00	nr	**9.43**
UNDERFLOOR BUSBAR						
Prewired busbar, plug-in trunking for underfloor power distribution; galvanized sheet steel housing (PE); copper conductors with tap-off units at 300 mm intervals						
Straight lengths – 63 amp						
2 pole & PE	18.80	23.10	0.28	11.20	m	**34.30**
3 pole & PE; Clean Earth System	24.33	29.89	0.28	11.20	m	**41.09**
Components for prewired busbars, plug-in trunking for underfloor power distribution						
Plug-in tap off units						
32 amp 2P & PE; 3 m metal flexible pre-wired conduit	30.96	38.03	0.25	10.00	nr	**48.03**
32 amp 3P & PE; clean earth; 3 m metal flexible prewired conduit	37.60	46.19	0.28	11.20	nr	**57.39**

ELECTRICAL SUPPLY/POWER/LIGHTING

Item	Net Price £	Material £	Labour hours	Labour £	Unit	**Total rate £**
LV DISTRUBITION: BUSBAR TRUNKING – cont						
Components for prewired busbars, plug-in trunking for underfloor power distribution – cont						
Trunking components						
End feed unit & cover; 2P & PE	33.17	40.75	0.35	14.01	nr	**54.76**
End feed unit & cover; 3P & PE; clean earth	36.49	44.82	0.38	15.19	nr	**60.01**
End cover; 2P & PE	9.95	12.22	0.11	4.40	nr	**16.62**
End cover; 3P & PE	11.06	13.59	0.11	4.40	nr	**17.99**
Flexible interlink/corner; 2P&PE; 1 m long	57.50	70.64	0.34	13.60	nr	**84.24**
Flexible interlink/corner; 3P&PE; 1 m long	65.25	80.16	0.35	14.01	nr	**94.17**
Flexible interlink/corner; 2P&PE; 2 m long	70.77	86.94	0.37	14.81	nr	**101.75**
Flexible interlink/corner; 3P&PE; 2 m long	77.41	95.10	0.37	14.81	nr	**109.91**

ELECTRICAL SUPPLY/POWER/LIGHTING

Item	Net Price £	Material £	Labour hours	Labour £	Unit	Total rate £
LV DISTRUBITION: CABLE SUPPORTS						
LADDER RACK						
Light duty galvanized steel ladder rack; fixed to backgrounds; including supports, fixings and brackets; earth continuity straps						
Straight lengths						
150 mm wide ladder	20.95	25.74	0.69	27.60	m	**53.34**
300 mm wide ladder	22.83	28.05	0.88	35.20	m	**63.25**
450 mm wide ladder	16.15	19.84	1.26	50.41	m	**70.25**
600 mm wide ladder	26.08	32.04	1.51	60.41	m	**92.45**
Extra over (cutting and jointing racking to fittings is included)						
Inside riser bend						
150 mm wide ladder	74.66	94.64	0.33	13.20	nr	**107.84**
300 mm wide ladder	77.63	94.64	0.56	22.40	nr	**100.03**
450 mm wide ladder	87.00	91.72	0.85	34.01	nr	**125.73**
600 mm wide ladder	107.86	132.51	0.99	39.60	nr	**226.11**
Outside riser bend						
150 mm wide ladder	55.53	68.22	0.43	17.20	nr	**85.22**
300 mm wide ladder	71.49	87.82	0.43	17.20	nr	**105.02**
450 mm wide ladder	77.63	94.64	0.73	29.20	nr	**123.84**
600 mm wide ladder	107.86	132.51	0.86	34.40	nr	**166.91**
Equal tee						
150 mm wide ladder	64.89	79.72	0.62	24.80	nr	**104.52**
300 mm wide ladder	71.49	87.82	0.62	24.80	nr	**112.62**
450 mm wide ladder	76.93	94.51	1.09	43.60	nr	**138.11**
600 mm wide ladder	88.99	109.33	1.12	44.80	nr	**154.13**
Unequal tee						
150 mm wide ladder	64.89	79.72	0.57	22.81	nr	**102.53**
300 mm wide ladder	71.31	87.60	0.57	22.81	nr	**110.41**
450 mm wide ladder	80.95	99.45	1.17	46.81	nr	**146.26**
600 mm wide ladder	84.54	103.86	1.17	46.81	nr	**150.67**
4 way crossovers						
150 mm wide ladder	133.12	163.54	0.61	21.61	nr	**185.15**
300 mm wide ladder	166.67	204.75	0.72	28.79	nr	**233.54**
450 mm wide ladder	173.54	230.09	1.13	45.19	nr	**234.28**
600 mm wide ladder	187.29	332.23	1.29	51.60	nr	**383.83**
Heavy duty galvanized steel ladder rack; fixed to backgrounds; including supports, fixings and brackets; earth continuity straps						
Straight lengths						
150 mm wide ladder	26.30	32.31	0.68	27.19	m	**59.50**
300 mm wide ladder	28.57	35.10	0.79	31.59	m	**66.69**
450 mm wide ladder	30.73	37.75	1.07	42.80	m	**80.55**

ELECTRICAL SUPPLY/POWER/LIGHTING

Item	Net Price £	Material £	Labour hours	Labour £	Unit	**Total rate £**
LV DISTRUBITION: CABLE SUPPORTS – cont						
Heavy duty galvanized steel ladder rack – cont						
600 mm wide ladder	33.47	41.12	1.24	49.59	m	**90.71**
750 mm wide ladder	39.19	48.14	1.49	59.60	m	**107.74**
900 mm wide ladder	43.23	53.11	1.67	66.80	m	**119.91**
Extra over (cutting and jointing racking to fittings is included)						
Flat bend						
150 mm wide ladder	62.25	76.47	0.34	13.60	nr	**90.07**
300 mm wide ladder	66.58	81.80	0.39	15.60	nr	**97.40**
450 mm wide ladder	75.82	93.15	0.43	17.20	nr	**110.35**
600 mm wide ladder	86.28	106.00	0.61	24.40	nr	**130.40**
750 mm wide ladder	99.17	121.83	0.82	32.80	nr	**154.63**
900 mm wide ladder	109.51	134.54	0.97	38.80	nr	**173.34**
Inside riser bend						
150 mm wide ladder	92.18	113.24	0.27	10.79	nr	**124.03**
300 mm wide ladder	93.68	115.09	0.45	18.00	nr	**133.09**
450 mm wide ladder	101.49	124.68	0.65	26.00	nr	**150.68**
600 mm wide ladder	109.42	134.42	0.81	32.39	nr	**166.81**
750 mm wide ladder	114.34	140.47	0.92	36.80	nr	**177.27**
900 mm wide ladder	127.69	156.87	1.06	42.40	nr	**199.27**
Outside riser bend						
150 mm wide ladder	92.18	113.24	0.27	10.79	nr	**124.03**
300 mm wide ladder	93.78	115.21	0.33	13.20	nr	**128.41**
450 mm wide ladder	101.49	124.68	0.61	24.40	nr	**149.08**
600 mm wide ladder	109.42	134.42	0.76	30.41	nr	**164.83**
750 mm wide ladder	114.34	140.47	0.94	37.60	nr	**178.07**
900 mm wide ladder	127.69	156.87	1.05	42.00	nr	**198.87**
Equal tee						
150 mm wide ladder	98.55	121.06	0.37	14.81	nr	**135.87**
300 mm wide ladder	116.36	142.95	0.57	22.81	nr	**165.76**
450 mm wide ladder	127.38	156.48	0.83	33.19	nr	**189.67**
600 mm wide ladder	143.02	175.70	0.92	36.80	nr	**212.50**
750 mm wide ladder	185.07	227.36	1.13	45.19	nr	**272.55**
900 mm wide ladder	203.96	250.56	1.20	48.00	nr	**298.56**
Unequal tee						
300 mm wide ladder	111.07	136.45	0.57	22.81	nr	**159.26**
450 mm wide ladder	108.28	133.02	1.17	46.81	nr	**179.83**
600 mm wide ladder	117.33	144.14	1.17	46.81	nr	**190.95**
750 mm wide ladder	125.35	153.99	1.25	50.00	nr	**203.99**
900 mm wide ladder	161.55	198.46	1.33	53.19	nr	**251.65**
4 way crossovers						
150 mm wide ladder	159.05	195.39	0.50	20.00	nr	**215.39**
300 mm wide ladder	168.57	207.09	0.67	26.81	nr	**233.90**
450 mm wide ladder	215.91	265.24	0.92	36.80	nr	**302.04**
600 mm wide ladder	231.82	284.79	1.07	42.80	nr	**327.59**
750 mm wide ladder	244.23	300.04	1.25	50.00	nr	**350.04**
900 mm wide ladder	268.65	330.04	1.36	54.40	nr	**384.44**

ELECTRICAL SUPPLY/POWER/LIGHTING

Item	Net Price £	Material £	Labour hours	Labour £	Unit	**Total rate £**
Extra heavy duty galvanized steel ladder rack; fixed to backgrounds; including supports, fixings and brackets; earth continuity straps						
Straight lengths						
150 mm wide ladder	35.41	43.50	0.63	25.20	m	**68.70**
300 mm wide ladder	37.16	45.65	0.70	27.99	m	**73.64**
450 mm wide ladder	39.00	47.91	0.83	33.19	m	**81.10**
600 mm wide ladder	39.52	48.55	0.89	35.61	m	**84.16**
750 mm wide ladder	39.63	48.69	1.22	48.79	m	**97.48**
900 mm wide ladder	41.36	50.81	1.44	57.60	m	**108.41**
Extra over (cutting and jointing racking to fittings is included)						
Flat bend						
150 mm wide ladder	76.88	94.45	0.36	14.40	nr	**108.85**
300 mm wide ladder	80.00	98.28	0.39	15.60	nr	**113.88**
450 mm wide ladder	86.39	106.13	0.43	17.20	nr	**123.33**
600 mm wide ladder	97.72	120.05	0.61	24.40	nr	**144.45**
750 mm wide ladder	105.95	130.16	0.82	32.80	nr	**162.96**
900 mm wide ladder	114.74	140.96	0.97	38.80	nr	**179.76**
Inside riser bend						
150 mm wide ladder	94.86	116.54	0.36	14.40	nr	**130.94**
300 mm wide ladder	96.39	118.42	0.39	15.60	nr	**134.02**
450 mm wide ladder	101.90	125.18	0.43	17.20	nr	**142.38**
600 mm wide ladder	112.38	138.05	0.61	24.40	nr	**162.45**
750 mm wide ladder	115.30	141.65	0.82	32.80	nr	**174.45**
900 mm wide ladder	124.32	152.72	0.97	38.80	nr	**191.52**
Outside riser bend						
150 mm wide ladder	95.11	116.84	0.36	14.40	nr	**131.24**
300 mm wide ladder	96.39	118.42	0.39	15.60	nr	**134.02**
450 mm wide ladder	101.90	125.18	0.41	16.40	nr	**141.58**
600 mm wide ladder	112.38	138.05	0.57	22.81	nr	**160.86**
750 mm wide ladder	115.30	141.65	0.82	32.80	nr	**174.45**
900 mm wide ladder	124.32	152.72	0.93	37.20	nr	**189.92**
Equal tee						
150 mm wide ladder	114.65	140.85	0.37	14.81	nr	**155.66**
300 mm wide ladder	126.57	155.49	0.57	22.81	nr	**178.30**
450 mm wide ladder	134.07	164.70	0.83	33.19	nr	**197.89**
600 mm wide ladder	151.10	185.63	0.92	36.80	nr	**222.43**
750 mm wide ladder	182.64	224.37	1.13	45.19	nr	**269.56**
900 mm wide ladder	182.94	224.74	1.20	48.00	nr	**272.74**
Unequal tee						
150 mm wide ladder	118.04	145.02	0.37	14.81	nr	**159.83**
300 mm wide ladder	130.00	159.71	0.57	22.81	nr	**182.52**
450 mm wide ladder	137.50	168.92	1.17	46.81	nr	**215.73**
600 mm wide ladder	154.69	190.04	1.17	46.81	nr	**236.85**
750 mm wide ladder	182.64	673.10	1.25	50.00	nr	**723.10**
900 mm wide ladder	190.15	233.60	1.33	53.19	nr	**286.79**

ELECTRICAL SUPPLY/POWER/LIGHTING

Item	Net Price £	Material £	Labour hours	Labour £	Unit	Total rate £
LV DISTRUBITION: CABLE SUPPORTS – cont						
Extra over (cutting and jointing racking to fittings is included) – cont						
4 way crossovers						
150 mm wide ladder	159.42	195.85	0.50	20.00	nr	**215.85**
300 mm wide ladder	165.67	203.52	0.67	26.81	nr	**230.33**
450 mm wide ladder	193.81	238.10	0.92	36.80	nr	**274.90**
600 mm wide ladder	215.45	264.68	1.07	42.80	nr	**307.48**
750 mm wide ladder	223.97	275.14	1.25	50.00	nr	**325.14**
900 mm wide ladder	234.03	287.51	1.36	54.40	nr	**341.91**

ELECTRICAL SUPPLY/POWER/LIGHTING

Item	Net Price £	Material £	Labour hours	Labour £	Unit	**Total rate £**
LV DISTRUBITION: CABLE TRAYS						
Galvanized steel cable tray to BS 729; including standard coupling joints, fixings and earth continuity straps (supports and hangers are excluded)						
Light duty tray						
Straight lengths						
50 mm wide	2.85	3.50	0.19	7.60	m	**11.10**
75 mm wide	3.59	4.41	0.23	9.20	m	**13.61**
100 mm wide	4.43	5.44	0.31	12.40	m	**17.84**
150 mm wide	5.81	7.14	0.33	13.20	m	**20.34**
225 mm wide	11.13	13.67	0.39	15.60	m	**29.27**
300 mm wide	15.67	19.25	0.49	19.59	m	**38.84**
450 mm wide	10.76	13.22	0.60	24.00	m	**37.22**
600 mm wide	22.20	27.27	0.79	31.59	m	**58.86**
750 mm wide	28.20	34.64	1.04	41.60	m	**76.24**
900 mm wide	35.13	43.16	1.26	50.41	m	**93.57**
Extra over (cutting and jointing tray to fittings is included)						
Straight reducer						
75 mm wide	11.53	14.16	0.22	8.80	nr	**22.96**
100 mm wide	11.77	14.46	0.25	10.00	nr	**24.46**
150 mm wide	15.38	18.89	0.27	10.79	nr	**29.68**
225 mm wide	20.19	24.80	0.34	13.60	nr	**38.40**
300 mm wide	25.70	31.57	0.39	15.60	nr	**47.17**
450 mm wide	20.13	24.73	0.49	19.59	nr	**44.32**
600 mm wide	32.50	39.93	0.54	21.60	nr	**61.53**
750 mm wide	42.69	52.45	0.61	24.40	nr	**76.85**
900 mm wide	49.60	60.93	0.69	27.60	nr	**88.53**
Flat bend; 90°						
50 mm wide	6.87	8.44	0.19	7.60	nr	**16.04**
75 mm wide	7.15	8.79	0.24	9.60	nr	**18.39**
100 mm wide	7.44	9.14	0.28	11.20	nr	**20.34**
150 mm wide	8.74	10.74	0.30	12.00	nr	**22.74**
225 mm wide	11.38	13.98	0.36	14.40	nr	**28.38**
300 mm wide	16.72	20.54	0.44	17.61	nr	**38.15**
450 mm wide	14.43	17.72	0.57	22.81	nr	**40.53**
600 mm wide	26.84	32.97	0.69	27.60	nr	**60.57**
750 mm wide	38.00	46.68	0.81	32.39	nr	**79.07**
900 mm wide	55.88	68.65	0.94	37.60	nr	**106.25**
Adjustable riser						
50 mm wide	13.91	17.08	0.26	10.41	nr	**27.49**
75 mm wide	15.05	18.49	0.29	11.59	nr	**30.08**
100 mm wide	14.56	17.89	0.32	12.80	nr	**30.69**
150 mm wide	16.87	20.73	0.36	14.40	nr	**35.13**
225 mm wide	22.21	27.29	0.44	17.61	nr	**44.90**
300 mm wide	23.81	29.25	0.52	20.80	nr	**50.05**
450 mm wide	20.56	25.26	0.66	26.40	nr	**51.66**

ELECTRICAL SUPPLY/POWER/LIGHTING

Item	Net Price £	Material £	Labour hours	Labour £	Unit	Total rate £
LV DISTRUBITION: CABLE TRAYS – cont						
Extra over (cutting and jointing tray to fittings is included) – cont						
600 mm wide	31.78	39.04	0.79	31.59	nr	**70.63**
750 mm wide	40.91	50.25	1.03	41.20	nr	**91.45**
900 mm wide	48.14	59.14	1.10	44.01	nr	**103.15**
Inside riser; 90°						
50 mm wide	4.37	5.37	0.28	11.20	nr	**16.57**
75 mm wide	4.57	5.62	0.31	12.40	nr	**18.02**
100 mm wide	5.02	11.77	0.33	13.20	nr	**24.97**
150 mm wide	5.97	7.33	0.37	14.81	nr	**22.14**
225 mm wide	7.33	9.01	0.44	17.61	nr	**26.62**
300 mm wide	7.76	9.53	0.53	21.20	nr	**30.73**
450 mm wide	12.41	15.25	0.67	26.81	nr	**42.06**
600 mm wide	31.71	38.95	0.79	31.59	nr	**70.54**
750 mm wide	39.12	48.06	0.95	38.00	nr	**86.06**
900 mm wide	46.47	57.09	1.11	44.39	nr	**101.48**
Outside riser; 90°						
50 mm wide	4.37	5.37	0.28	11.20	nr	**16.57**
75 mm wide	4.57	5.62	0.31	12.40	nr	**18.02**
100 mm wide	5.02	6.16	0.33	13.20	nr	**19.36**
150 mm wide	5.97	7.33	0.37	14.81	nr	**22.14**
225 mm wide	7.33	9.01	0.44	17.61	nr	**26.62**
300 mm wide	7.76	9.53	0.53	21.20	nr	**30.73**
450 mm wide	12.41	15.25	0.67	26.81	nr	**42.06**
600 mm wide	17.63	21.66	0.79	31.59	nr	**53.25**
750 mm wide	23.95	29.42	0.95	38.00	nr	**67.42**
900 mm wide	23.77	29.20	1.11	44.39	nr	**73.59**
Equal tee						
50 mm wide	10.33	12.69	0.30	12.00	nr	**24.69**
75 mm wide	10.77	13.23	0.31	12.40	nr	**25.63**
100 mm wide	11.11	13.65	0.35	14.01	nr	**27.66**
150 mm wide	13.50	16.58	0.36	14.40	nr	**30.98**
225 mm wide	17.98	22.09	0.44	17.61	nr	**39.70**
300 mm wide	25.07	30.80	0.54	21.60	nr	**52.40**
450 mm wide	21.10	25.92	0.71	28.40	nr	**54.32**
600 mm wide	35.40	43.49	0.92	36.80	nr	**80.29**
750 mm wide	53.07	65.19	1.19	47.61	nr	**112.80**
900 mm wide	75.48	92.73	1.44	57.60	nr	**150.33**
Unequal tee						
75 mm wide	6.74	8.28	0.38	15.19	nr	**23.47**
100 mm wide	6.74	8.28	0.39	15.60	nr	**23.88**
150 mm wide	6.94	8.53	0.43	17.20	nr	**25.73**
225 mm wide	8.05	9.89	0.50	20.00	nr	**29.89**
300 mm wide	10.14	12.45	0.63	25.20	nr	**37.65**
450 mm wide	13.45	16.53	0.80	32.00	nr	**48.53**
600 mm wide	21.10	25.92	1.02	40.80	nr	**66.72**
750 mm wide	60.27	74.05	1.12	44.80	nr	**118.85**
900 mm wide	79.44	97.59	1.35	54.00	nr	**151.59**

ELECTRICAL SUPPLY/POWER/LIGHTING

Item	Net Price £	Material £	Labour hours	Labour £	Unit	**Total rate £**
4 way crossovers						
50 mm wide	15.35	18.86	0.38	15.19	nr	**34.05**
75 mm wide	15.63	19.20	0.40	15.99	nr	**35.19**
100 mm wide	16.45	20.21	0.40	15.99	nr	**36.20**
150 mm wide	19.49	23.94	0.44	17.61	nr	**41.55**
225 mm wide	26.26	32.26	0.53	21.20	nr	**53.46**
300 mm wide	36.15	44.41	0.64	25.60	nr	**70.01**
450 mm wide	31.43	38.61	0.84	33.60	nr	**72.21**
600 mm wide	35.40	43.49	1.03	41.20	nr	**84.69**
750 mm wide	53.07	65.19	1.13	45.19	nr	**110.38**
900 mm wide	75.48	92.73	1.36	54.40	nr	**147.13**
Medium duty tray with return flange						
Straight lengths						
50 mm wide	7.03	8.64	0.33	13.20	m	**21.84**
75 mm wide	8.20	10.07	0.33	13.20	m	**23.27**
100 mm wide	8.77	10.77	0.35	14.01	m	**24.78**
150 mm wide	10.72	13.17	0.39	15.60	m	**28.77**
225 mm wide	12.97	15.93	0.45	18.00	m	**33.93**
300 mm wide	17.95	22.05	0.57	22.81	m	**44.86**
450 mm wide	27.89	34.26	0.69	27.60	m	**61.86**
600 mm wide	38.94	47.84	0.91	36.40	m	**84.24**
Extra over (cutting and jointing tray to fittings is included)						
Straight reducer						
100 mm wide	17.71	21.76	0.25	10.00	nr	**31.76**
150 mm wide	19.34	23.76	0.27	10.79	nr	**34.55**
225 mm wide	22.29	27.38	0.34	13.60	nr	**40.98**
300 mm wide	25.81	31.71	0.39	15.60	nr	**47.31**
450 mm wide	33.35	40.97	0.49	19.59	nr	**60.56**
600 mm wide	41.32	50.76	0.54	21.60	nr	**72.36**
Flat bend; 90°						
75 mm wide	23.61	29.00	0.24	9.60	nr	**38.60**
100 mm wide	26.28	32.29	0.28	11.20	nr	**43.49**
150 mm wide	27.89	34.26	0.30	12.00	nr	**46.26**
225 mm wide	32.19	39.54	0.36	14.40	nr	**53.94**
300 mm wide	39.70	48.77	0.44	17.61	nr	**66.38**
450 mm wide	59.64	73.27	0.57	22.81	nr	**96.08**
600 mm wide	72.11	88.59	0.69	27.60	nr	**116.19**
Adjustable bend						
75 mm wide	27.00	33.17	0.29	11.59	nr	**44.76**
100 mm wide	29.38	36.09	0.32	12.80	nr	**48.89**
150 mm wide	33.59	41.27	0.36	14.40	nr	**55.67**
225 mm wide	37.65	46.25	0.44	17.61	nr	**63.86**
300 mm wide	42.55	52.27	0.52	20.80	nr	**73.07**
Adjustable riser						
75 mm wide	24.08	29.58	0.29	11.59	nr	**41.17**
100 mm wide	24.48	30.07	0.32	12.80	nr	**42.87**
150 mm wide	26.98	33.15	0.36	14.40	nr	**47.55**

ELECTRICAL SUPPLY/POWER/LIGHTING

Item	Net Price £	Material £	Labour hours	Labour £	Unit	Total rate £
LV DISTRUBITION: CABLE TRAYS – cont						
Extra over (cutting and jointing tray to fittings is included) – cont						
225 mm wide	28.61	35.14	0.44	17.61	nr	**52.75**
300 mm wide	30.51	37.49	0.52	20.80	nr	**58.29**
450 mm wide	40.02	49.16	0.66	26.40	nr	**75.56**
600 mm wide	49.09	60.31	0.79	31.59	nr	**91.90**
Inside riser; 90°						
75 mm wide	14.46	17.77	0.31	12.40	nr	**30.17**
100 mm wide	14.60	17.93	0.33	13.20	nr	**31.13**
150 mm wide	16.63	20.43	0.37	14.81	nr	**35.24**
225 mm wide	20.32	24.96	0.44	17.61	nr	**42.57**
300 mm wide	24.68	30.32	0.53	21.20	nr	**51.52**
450 mm wide	35.63	43.77	0.67	26.81	nr	**70.58**
600 mm wide	57.19	70.26	0.79	31.59	nr	**101.85**
Outside riser; 90°						
75 mm wide	14.46	17.77	0.31	12.40	nr	**30.17**
100 mm wide	14.60	17.93	0.33	13.20	nr	**31.13**
150 mm wide	16.63	20.43	0.37	14.81	nr	**35.24**
225 mm wide	20.32	24.96	0.44	17.61	nr	**42.57**
300 mm wide	24.68	30.32	0.53	21.20	nr	**51.52**
450 mm wide	35.63	43.77	0.67	26.81	nr	**70.58**
600 mm wide	57.19	70.26	0.79	31.59	nr	**101.85**
Equal tee						
75 mm wide	32.15	39.50	0.31	12.40	nr	**51.90**
100 mm wide	34.28	42.12	0.35	14.01	nr	**56.13**
150 mm wide	36.91	45.34	0.36	14.40	nr	**59.74**
225 mm wide	40.11	49.28	0.74	29.60	nr	**78.88**
300 mm wide	48.68	59.81	0.54	21.60	nr	**81.41**
450 mm wide	64.12	78.77	0.71	28.40	nr	**107.17**
600 mm wide	92.12	113.17	0.92	36.80	nr	**149.97**
Unequal tee						
100 mm wide	34.28	42.12	0.39	15.60	nr	**57.72**
150 mm wide	34.28	42.12	0.43	17.20	nr	**59.32**
225 mm wide	40.11	49.28	0.50	20.00	nr	**69.28**
300 mm wide	48.68	59.81	0.63	25.20	nr	**85.01**
450 mm wide	64.12	78.77	0.80	32.00	nr	**110.77**
600 mm wide	92.12	113.17	1.02	40.80	nr	**153.97**
4 way crossovers						
75 mm wide	45.59	56.01	0.40	15.99	nr	**72.00**
100 mm wide	49.14	60.36	0.40	15.99	nr	**76.35**
150 mm wide	42.70	52.46	0.44	17.61	nr	**70.07**
225 mm wide	61.74	75.85	0.53	21.20	nr	**97.05**
300 mm wide	71.84	88.25	0.64	25.60	nr	**113.85**
450 mm wide	91.44	112.33	0.84	33.60	nr	**145.93**
600 mm wide	133.29	163.75	1.03	41.20	nr	**204.95**

ELECTRICAL SUPPLY/POWER/LIGHTING

Item	Net Price £	Material £	Labour hours	Labour £	Unit	**Total rate £**
Heavy duty tray with return flange						
Straight lengths						
75 mm	11.02	13.53	0.34	13.60	m	**27.13**
100 mm	11.74	14.43	0.36	14.40	m	**28.83**
150 mm	13.57	16.67	0.40	15.99	m	**32.66**
225 mm	15.21	18.69	0.46	18.41	m	**37.10**
300 mm	18.55	22.78	0.58	23.19	m	**45.97**
450 mm	28.04	34.45	0.70	27.99	m	**62.44**
600 mm	34.00	41.77	0.92	36.80	m	**78.57**
750 mm	42.87	52.67	1.01	40.40	m	**93.07**
900 mm	47.43	58.26	1.14	45.60	m	**103.86**
Extra over (cutting and jointing tray to fittings is included)						
Straight reducer						
100 mm wide	27.77	34.11	0.25	10.00	nr	**44.11**
150 mm wide	28.84	35.43	0.27	10.79	nr	**46.22**
225 mm wide	32.74	40.23	0.34	13.60	nr	**53.83**
300 mm wide	36.50	44.85	0.39	15.60	nr	**60.45**
450 mm wide	52.21	64.14	0.49	19.59	nr	**83.73**
600 mm wide	58.04	71.31	0.54	21.60	nr	**92.91**
750 mm wide	73.62	90.45	0.60	24.00	nr	**114.45**
900 mm wide	80.75	99.20	0.66	26.40	nr	**125.60**
Flat bend; 90°						
75 mm wide	31.07	38.17	0.24	9.60	nr	**47.77**
100 mm wide	35.32	43.39	0.28	11.20	nr	**54.59**
150 mm wide	37.63	46.23	0.30	12.00	nr	**58.23**
225 mm wide	42.57	52.30	0.36	14.40	nr	**66.70**
300 mm wide	48.50	59.58	0.44	17.61	nr	**77.19**
450 mm wide	71.11	87.36	0.57	22.81	nr	**110.17**
600 mm wide	95.04	116.76	0.69	27.60	nr	**144.36**
750 mm wide	127.37	156.47	0.83	33.19	nr	**189.66**
900 mm wide	144.43	177.43	1.01	40.40	nr	**217.83**
Adjustable bend						
75 mm wide	32.31	39.69	0.29	11.59	nr	**51.28**
100 mm wide	35.72	43.88	0.32	12.80	nr	**56.68**
150 mm wide	37.74	46.37	0.36	14.40	nr	**60.77**
225 mm wide	41.53	51.02	0.44	17.61	nr	**68.63**
300 mm wide	49.28	60.54	0.52	20.80	nr	**81.34**
Adjustable riser						
75 mm wide	28.47	34.98	0.29	11.59	nr	**46.57**
100 mm wide	27.51	33.80	0.32	12.80	nr	**46.60**
150 mm wide	32.77	40.26	0.36	14.40	nr	**54.66**
225 mm wide	35.10	43.12	0.44	17.61	nr	**60.73**
300 mm wide	38.02	46.70	0.52	20.80	nr	**67.50**
450 mm wide	46.27	56.85	0.66	26.40	nr	**83.25**
600 mm wide	55.13	67.73	0.79	31.59	nr	**99.32**
750 mm wide	66.75	82.00	1.03	41.20	nr	**123.20**
900 mm wide	77.52	95.24	1.10	44.01	nr	**139.25**

ELECTRICAL SUPPLY/POWER/LIGHTING

Item	Net Price £	Material £	Labour hours	Labour £	Unit	Total rate £
LV DISTRUBITION: CABLE TRAYS – cont						
Extra over (cutting and jointing tray to fittings is included) – cont						
Inside riser; 90°						
75 mm wide	23.81	29.25	0.31	12.40	nr	**41.65**
100 mm wide	24.12	29.63	0.33	13.20	nr	**42.83**
150 mm wide	26.13	32.10	0.37	14.81	nr	**46.91**
225 mm wide	27.31	33.55	0.44	17.61	nr	**51.16**
300 mm wide	28.18	34.62	0.53	21.20	nr	**55.82**
450 mm wide	48.34	59.39	0.67	26.81	nr	**86.20**
600 mm wide	59.31	72.86	0.79	31.59	nr	**104.45**
750 mm wide	73.40	90.17	0.95	38.00	nr	**128.17**
900 mm wide	86.54	106.31	1.11	44.39	nr	**150.70**
Outside riser; 90°						
75 mm wide	23.81	29.25	0.31	12.40	nr	**41.65**
100 mm wide	24.12	29.63	0.33	13.20	nr	**42.83**
150 mm wide	26.13	32.10	0.37	14.81	nr	**46.91**
225 mm wide	27.31	33.55	0.44	17.61	nr	**51.16**
300 mm wide	28.18	34.62	0.53	21.20	nr	**55.82**
450 mm wide	48.34	59.39	0.67	26.81	nr	**86.20**
600 mm wide	59.31	72.86	0.79	31.59	nr	**104.45**
750 mm wide	73.40	90.17	0.95	38.00	nr	**128.17**
900 mm wide	86.54	106.31	1.11	44.39	nr	**150.70**
Equal tee						
75 mm wide	41.47	50.95	0.31	12.40	nr	**63.35**
100 mm wide	46.36	56.95	0.35	14.01	nr	**70.96**
150 mm wide	50.77	62.37	0.36	14.40	nr	**76.77**
225 mm wide	58.43	71.78	0.44	17.61	nr	**89.39**
300 mm wide	63.18	77.62	0.54	21.60	nr	**99.22**
450 mm wide	91.73	112.69	0.71	28.40	nr	**141.09**
600 mm wide	121.11	148.78	0.92	36.80	nr	**185.58**
750 mm wide	159.73	196.22	1.19	47.61	nr	**243.83**
900 mm wide	182.41	224.09	1.45	58.00	nr	**282.09**
Unequal tee						
75 mm wide	41.47	50.95	0.38	15.19	nr	**66.14**
100 mm wide	46.32	56.90	0.39	15.60	nr	**72.50**
150 mm wide	50.77	62.37	0.43	17.20	nr	**79.57**
225 mm wide	58.43	71.78	0.50	20.00	nr	**91.78**
300 mm wide	63.18	77.62	0.63	25.20	nr	**102.82**
450 mm wide	91.73	112.69	0.80	32.00	nr	**144.69**
600 mm wide	121.11	148.78	1.02	40.80	nr	**189.58**
750 mm wide	159.73	196.22	1.12	44.80	nr	**241.02**
900 mm wide	193.21	237.36	1.35	54.00	nr	**291.36**
4 way crossovers						
75 mm wide	42.09	51.71	0.40	15.99	nr	**67.70**
100 mm wide	60.33	74.12	0.40	15.99	nr	**90.11**
150 mm wide	60.33	74.12	0.44	17.61	nr	**91.73**
225 mm wide	83.88	103.05	0.53	21.20	nr	**124.25**
300 mm wide	92.50	113.63	0.64	25.60	nr	**139.23**
450 mm wide	133.05	163.45	0.84	33.60	nr	**197.05**
600 mm wide	180.83	222.15	1.03	41.20	nr	**263.35**

ELECTRICAL SUPPLY/POWER/LIGHTING

Item	Net Price £	Material £	Labour hours	Labour £	Unit	**Total rate £**
750 mm wide	220.55	270.94	1.13	45.19	nr	**316.13**
900 mm wide	266.04	326.83	1.36	54.40	nr	**381.23**
GRP cable tray including standard coupling joints and fixings (supports and hangers excluded)						
Tray						
100 mm wide	31.24	38.38	0.34	13.60	m	**51.98**
200 mm wide	40.01	49.15	0.39	15.60	m	**64.75**
400 mm wide	62.74	77.08	0.53	21.20	m	**98.28**
Cover						
100 mm wide	17.60	21.62	0.10	4.00	m	**25.62**
200 mm wide	23.32	28.64	0.11	4.40	m	**33.04**
400 mm wide	39.81	48.91	0.14	5.61	m	**54.52**
Extra for (cutting and jointing to fittings included)						
Reducer						
200 mm wide	83.03	102.01	0.23	9.20	nr	**111.21**
400 mm wide	108.47	133.26	0.30	12.00	nr	**145.26**
Reducer cover						
200 mm wide	51.96	63.83	0.25	10.00	nr	**73.83**
400 mm wide	75.95	93.30	0.28	11.20	nr	**104.50**
Bend						
100 mm wide	69.43	85.29	0.34	13.60	nr	**98.89**
200 mm wide	80.59	99.00	0.40	15.99	nr	**114.99**
400 mm wide	103.58	127.25	0.32	12.80	nr	**140.05**
Bend cover						
100 mm wide	34.50	42.39	0.10	4.00	nr	**46.39**
200 mm wide	46.08	56.61	0.10	4.00	nr	**60.61**
400 mm wide	63.66	78.20	0.13	5.20	nr	**83.40**
Tee						
100 mm wide	88.36	108.55	0.37	14.81	nr	**123.36**
200 mm wide	97.42	119.68	0.43	17.20	nr	**136.88**
400 mm wide	120.86	148.48	0.56	22.40	nr	**170.88**
Tee cover						
100 mm wide	44.52	54.69	0.27	10.79	nr	**65.48**
200 mm wide	53.59	65.84	0.31	12.40	nr	**78.24**
400 mm wide	75.59	92.86	0.37	14.81	nr	**107.67**

ELECTRICAL SUPPLY/POWER/LIGHTING

Item	Net Price £	Material £	Labour hours	Labour £	Unit	Total rate £
LV DISTRUBITION: BASKET TRAY						
Mild steel cable basket; zinc plated including standard coupling joints, fixings and earth continuity straps (supports and hangers are excluded)						
Basket 54 mm deep						
100 mm wide	4.59	5.64	0.22	8.80	m	**14.44**
150 mm wide	5.11	6.28	0.25	10.00	m	**16.28**
200 mm wide	5.59	6.87	0.28	11.20	m	**18.07**
300 mm wide	6.55	8.04	0.34	13.60	m	**21.64**
450 mm wide	7.95	9.77	0.44	17.61	m	**27.38**
600 mm wide	9.72	11.94	0.70	27.99	m	**39.93**
Extra for (cutting and jointing to fittings is included)						
Reducer						
150 mm wide	15.00	18.43	0.25	10.00	nr	**28.43**
200 mm wide	17.59	21.61	0.28	11.20	nr	**32.81**
300 mm wide	17.89	21.98	0.38	15.19	nr	**37.17**
450 mm wide	19.62	24.11	0.48	19.20	nr	**43.31**
600 mm wide	24.45	30.04	0.48	19.20	nr	**49.24**
Bend						
100 mm wide	14.32	17.59	0.23	9.20	nr	**26.79**
150 mm wide	16.78	20.61	0.26	10.41	nr	**31.02**
200 mm wide	17.07	20.97	0.30	12.00	nr	**32.97**
300 mm wide	18.68	22.95	0.35	14.01	nr	**36.96**
450 mm wide	23.28	28.60	0.50	20.00	nr	**48.60**
600 mm wide	27.38	33.63	0.58	23.19	nr	**56.82**
Tee						
100 mm wide	18.00	22.11	0.28	11.20	nr	**33.31**
150 mm wide	19.10	23.47	0.30	12.00	nr	**35.47**
200 mm wide	19.49	23.94	0.33	13.20	nr	**37.14**
300 mm wide	24.51	30.11	0.39	15.60	nr	**45.71**
450 mm wide	32.11	39.45	0.56	22.40	nr	**61.85**
600 mm wide	33.20	40.78	0.65	26.00	nr	**66.78**
Cross over						
100 mm wide	24.56	30.18	0.40	15.99	nr	**46.17**
150 mm wide	24.98	30.69	0.42	16.81	nr	**47.50**
200 mm wide	27.13	33.33	0.46	18.41	nr	**51.74**
300 mm wide	30.82	37.86	0.51	20.40	nr	**58.26**
450 mm wide	36.13	44.38	0.74	29.60	nr	**73.98**
600 mm wide	37.30	45.82	0.82	32.80	nr	**78.62**

ELECTRICAL SUPPLY/POWER/LIGHTING

Item	Net Price £	Material £	Labour hours	Labour £	Unit	**Total rate £**
Mild steel cable basket; expoxy coated including standard coupling joints, fixings and earth continuity straps (supports and hangers are excluded)						
Basket 54 mm deep						
100 mm wide	10.26	12.60	0.22	8.80	m	**21.40**
150 mm wide	11.60	14.25	0.25	10.00	m	**24.25**
200 mm wide	12.92	15.88	0.28	11.20	m	**27.08**
300 mm wide	14.68	18.04	0.34	13.60	m	**31.64**
450 mm wide	18.11	22.25	0.44	17.61	m	**39.86**
600 mm wide	20.62	25.34	0.70	27.99	m	**53.33**
Extra for (cutting and jointing to fittings is included)						
Reducer						
150 mm wide	21.46	26.37	0.28	11.20	nr	**37.57**
200 mm wide	24.05	29.55	0.28	11.20	nr	**40.75**
300 mm wide	25.65	31.51	0.38	15.19	nr	**46.70**
450 mm wide	29.99	36.84	0.48	19.20	nr	**56.04**
600 mm wide	37.37	45.91	0.48	19.20	nr	**65.11**
Bend						
100 mm wide	20.81	25.57	0.23	9.20	nr	**34.77**
150 mm wide	23.25	28.56	0.26	10.41	nr	**38.97**
200 mm wide	23.54	28.92	0.30	12.00	nr	**40.92**
300 mm wide	26.49	32.54	0.35	14.01	nr	**46.55**
450 mm wide	33.61	41.29	0.50	20.00	nr	**61.29**
600 mm wide	40.36	49.58	0.58	23.19	nr	**72.77**
Tee						
100 mm wide	24.46	30.05	0.28	11.20	nr	**41.25**
150 mm wide	25.58	31.43	0.30	12.00	nr	**43.43**
200 mm wide	25.96	31.89	0.33	13.20	nr	**45.09**
300 mm wide	32.28	39.66	0.39	15.60	nr	**55.26**
450 mm wide	41.14	50.54	0.56	22.40	nr	**72.94**
600 mm wide	46.14	56.68	0.65	26.00	nr	**82.68**
Cross over						
100 mm wide	31.03	38.13	0.40	15.99	nr	**54.12**
150 mm wide	31.38	38.55	0.42	16.81	nr	**55.36**
200 mm wide	33.61	41.29	0.46	18.41	nr	**59.70**
300 mm wide	38.58	47.40	0.51	20.40	nr	**67.80**
450 mm wide	46.47	57.09	0.74	29.60	nr	**86.69**
600 mm wide	50.21	61.69	0.82	32.80	nr	**94.49**

ELECTRICAL SUPPLY/POWER/LIGHTING

Item	Net Price £	Material £	Labour hours	Labour £	Unit	Total rate £
LV DISTRIBUTION: SWITCHGEAR AND DISTRIBUTION BOARDS						
LV switchboard components, factory-assembled modular construction to IP41; form 4, type 5; 2400 mm high, with front and rear access; top cable entry/exit; includes delivery, offloading, positioning and commissioning (hence separate labour costs are not detailed below); excludes cabling and cable terminations						
Air circuit breakers (ACBs) to BSEN 60947-2, withdrawable type, fitted with adjustable instantaneous and overload protection. Includes enclosure and copper links, assembled into LV switchboard.						
ACB-100 kA fault rated.						
4 pole, 6300 A (1600 mm wide)	35327.25	43399.52	–	–	nr	**43399.52**
4 pole, 5000 A (1600 mm wide)	28037.50	34444.07	–	–	nr	**34444.07**
4 pole, 4000 A (1600 mm wide)	16065.77	19736.80	–	–	nr	**19736.80**
4 pole, 3200 A (1600 mm wide)	12852.61	15789.43	–	–	nr	**15789.43**
4 pole, 2500 A (1600 mm wide)	9339.11	11473.10	–	–	nr	**11473.10**
4 pole, 2000 A (1600 mm wide) 85ka	7471.28	9178.47	–	–	nr	**9178.47**
4 pole, 1600 A (1600 mm wide) 85ka	6789.84	8341.32	–	–	nr	**8341.32**
4 pole, 1250 A (1600 mm wide)	5304.56	6516.66	–	–	nr	**6516.66**
4 pole, 1000 A (1600 mm wide)	6207.07	7625.38	–	–	nr	**7625.38**
4 pole, 800 A (1600 mm wide) 85ka	4965.65	6100.30	–	–	nr	**6100.30**
3 pole, 6300 A (1600 mm wide)	29240.37	35921.79	–	–	nr	**35921.79**
3 pole, 5000 A (1600 mm wide)	23206.64	28509.36	–	–	nr	**28509.36**
3 pole, 4000 A (1600 mm wide)	12347.94	15169.44	–	–	nr	**15169.44**
3 pole, 3200 A (1600 mm wide)	9878.36	12135.56	–	–	nr	**12135.56**
3 pole, 2500 A (1600 mm wide)	7717.46	9480.90	–	–	nr	**9480.90**
3 pole, 2000 A (1600 mm wide)	6838.71	8401.35	–	–	nr	**8401.35**
3 pole, 1600 A (1600 mm wide)	5470.97	6721.08	–	–	nr	**6721.08**
3 pole, 1250 A (1600 mm wide)	4274.20	5250.85	–	–	nr	**5250.85**
3 pole, 1000 A (1600 mm wide)	4211.21	5173.48	–	–	nr	**5173.48**
3 pole, 800 A (1600 mm wide)	3368.97	4138.77	–	–	nr	**4138.77**
ACB-65 kA fault rated.						
4 pole, 4000 A (1600 mm wide)	35327.25	43399.52	–	–	nr	**43399.52**
4 pole, 3200 A (1600 mm wide)	11856.65	14565.89	–	–	nr	**14565.89**
4 pole, 2500 A (1600 mm wide)	8880.44	10909.62	–	–	nr	**10909.62**
4 pole, 2000 A (1600 mm wide)	7104.35	8727.69	–	–	nr	**8727.69**
4 pole, 1600 A (1600 mm wide)	6422.90	7890.53	–	–	nr	**7890.53**
4 pole, 1250 A (1600 mm wide)	6057.57	7441.73	–	–	nr	**7441.73**
4 pole, 1000 A (1600 mm wide)	5886.01	7230.96	–	–	nr	**7230.96**
4 pole, 800 A (1600 mm wide)	4708.80	5784.77	–	–	nr	**5784.77**
3 pole, 6300 A (1600 mm wide)	29240.37	35921.79	–	–	nr	**35961.80**
3 pole, 4000 A (1600 mm wide)	18565.30	22807.47	–	–	nr	**22807.47**
3 pole, 3200 A (1600 mm wide)	9144.49	11234.00	–	–	nr	**11234.00**

ELECTRICAL SUPPLY/POWER/LIGHTING

Item	Net Price £	Material £	Labour hours	Labour £	Unit	**Total rate £**
3 pole, 2500 A (1600 mm wide)	7144.12	8776.55	–	–	nr	**8776.55**
3 pole, 2000 A (1600 mm wide)	6458.69	7934.50	–	–	nr	**7934.50**
3 pole, 1600 A (1600 mm wide)	5166.94	6347.59	–	–	nr	**6347.59**
3 pole, 1250 A (1600 mm wide)	4036.68	4959.07	–	–	nr	**4959.07**
3 pole, 1000 A (1600 mm wide)	4021.19	4940.03	–	–	nr	**4940.03**
3 pole, 800 A (1600 mm wide)	3216.95	3952.02	–	–	nr	**3952.02**
Extra for						
Cable box (one per ACB for form 4, types 6 & 7)	309.99	380.82	–	–	nr	**380.82**
Opening coil (shunt trip)	141.01	173.23	–	–	nr	**173.23**
Closing coil	93.41	114.75	–	–	nr	**114.75**
Undervoltage release	167.22	205.43	–	–	nr	**205.43**
Motor operator	562.65	691.21	–	–	nr	**691.21**
Mechnical interlock (per ACB)	388.95	477.82	–	–	nr	**477.82**
ACB fortress/Castell adaptor kit (one per ACB)	83.87	103.04	–	–	nr	**103.04**
Fortress/Castell ACB lock (one per ACB)	264.34	324.74	–	–	nr	**324.74**
Fortress/Castell key	65.82	80.86	–	–	nr	**80.86**
Moulded case circuit breakers (MCCBs) to BS EN 60947-2; plug-in type, fitted with electronic trip unit. Includes metalwork section and copper links, assembled into LV switchboard						
MCCB-150 kA fault rated						
4 pole, 630 A (800 mm wide, 600 mm high)	2381.74	2925.97	–	–	nr	**2925.97**
4 pole, 400 A (800 mm wide, 400 mm high)	1594.85	1959.27	–	–	nr	**1959.27**
4 pole, 250 A (800 mm wide, 400 mm high)	1472.65	1809.15	–	–	nr	**1809.15**
4 pole, 160 A (800 mm wide, 300 mm high)	1008.76	1239.26	–	–	nr	**1239.26**
4 pole, 100 A (800 mm wide, 200 mm high)	1590.15	1953.50	–	–	nr	**1953.50**
3 pole, 630 A (800 mm wide, 600 mm high)	1882.00	2312.04	–	–	nr	**2312.04**
3 pole, 400 A (800 mm wide, 400 mm high)	1280.99	1573.70	–	–	nr	**1573.70**
3 pole, 250 A (800 mm wide, 400 mm high)	837.73	1029.15	–	–	nr	**1029.15**
3 pole, 160 A (800 mm wide, 300 mm high)	601.92	739.46	–	–	nr	**739.46**
3 pole, 100 A (800 mm wide, 200 mm high)	489.81	601.73	–	–	nr	**601.73**
MCCB-70kA fault rated						
4 pole, 630 A (800 mm wide, 600 mm high)	1989.12	2443.63	–	–	nr	**2443.63**
4 pole, 400 A (800 mm wide, 400 mm high)	1367.31	1679.74	–	–	nr	**1679.74**
4 pole, 250 A (800 mm wide, 400 mm high)	995.72	1223.24	–	–	nr	**1223.24**
4 pole, 160 A (800 mm wide, 300 mm high)	750.48	921.96	–	–	nr	**921.96**
4 pole, 100 A (800 mm wide, 200 mm high)	618.43	759.74	–	–	nr	**759.74**
3 pole, 630 A (800 mm wide, 600 mm high)	1457.81	1790.92	–	–	nr	**1790.92**
3 pole, 400 A (800 mm wide, 400 mm high)	983.92	1208.75	–	–	nr	**1208.75**
3 pole, 250 A (800 mm wide, 400 mm high)	686.17	842.96	–	–	nr	**842.96**
3 pole, 160 A (800 mm wide, 300 mm high)	509.36	625.75	–	–	nr	**625.75**
3 pole, 100 A (800 mm wide, 200 mm high)	403.40	495.58	–	–	nr	**495.58**

ELECTRICAL SUPPLY/POWER/LIGHTING

Item	Net Price £	Material £	Labour hours	Labour £	Unit	Total rate £
LV DISTRIBUTION: SWITCHGEAR AND DISTRIBUTION BOARDS – cont						
Moulded case circuit breakers (MCCBs) to BS EN 60947-2 – cont						
MCCB-45kA fault rated						
4 pole, 630 A (800 mm wide, 600 mm, LI, SSKA high)	1831.94	2250.54	–	–	nr	**2250.54**
4 pole, 400 A (800 mm wide, 400 mm, LI, SSKA high)	1271.75	1562.35	–	–	nr	**1562.35**
3 pole, 630 A (800 mm wide, 600 mm, LI, SSKA high)	1365.67	1677.72	–	–	nr	**1677.72**
3 pole, 400 A (800 mm wide, 400 mm, LI, SSKA high)	937.75	1152.03	–	–	nr	**1152.03**
MCCB-36kA fault rated						
4 pole, 250 A (800 mm wide, 400 mm high)	1525.11	1873.60	–	–	nr	**1873.60**
4 pole, 160 A (800 mm wide, 300 mm high)	1155.30	1419.28	–	–	nr	**1419.28**
3 pole, 250 A (800 mm wide, 400 mm high)	866.47	1064.46	–	–	nr	**1064.46**
3 pole, 160 A (800 mm wide, 300 mm high)	763.09	937.46	–	–	nr	**937.46**
Extra for						
Cable box (one per MCCB for form 4, types 6 & 7)	131.10	161.06	–	–	nr	**161.06**
Shunt trip (for ratings 100 A to 630 A)	40.41	49.64	–	–	nr	**49.64**
Undervoltage release (for ratings 100 A to 630 A)	52.42	64.40	–	–	nr	**64.40**
Motor operator for 630 A MCCB	96.45	118.49	–	–	nr	**118.49**
Motor operator for 400 A MCCB	96.45	118.49	–	–	nr	**118.49**
Motor operator for 250 A MCCB	46.81	57.51	–	–	nr	**57.51**
Motor operator for 160/100 A MCCB	41.62	51.14	–	–	nr	**51.14**
Door handle for 630/400 A MCCB	82.06	100.81	–	–	nr	**100.81**
Door handle for 250/160/100 A MCCB	65.01	79.86	–	–	nr	**79.86**
MCCB earth fault protection	494.52	607.52	–	–	nr	**607.52**
LV switchboard busbar						
Copper busbar assembled into LV switchboard, ASTA type tested to appropriate fault level. Busbar Length may be estimated by adding the widths of the ACB sections to the width of the MCCB sections. ACBs up to 2000 A rating may be stacked two high; larger ratings are one per section. To determine the number of MCCB sections, add together all the MCCB heights and divide by 1800 mm, rounding up as necessary						
6000 A (6 × 10 mm × 100 mm)	2573.22	3161.20	–	–	nr	**3161.20**
5000 A (4 × 10 mm × 100 mm)	2115.02	2598.30	–	–	nr	**2598.30**
4000 A (4 × 10 mm × 100 mm)	2115.02	2598.30	–	–	nr	**2598.30**
3200 A (3 × 10 mm × 100 mm)	1469.02	1804.69	–	–	nr	**1804.69**
2500 A (2 × 10 mm × 100 mm)	1239.40	1522.61	–	–	nr	**1522.61**
2000 A (2 × 10 mm × 80 mm)	911.03	1119.21	–	–	nr	**1119.21**
1600 A (2 × 10 mm × 50 mm)	661.00	812.04	–	–	nr	**812.04**

ELECTRICAL SUPPLY/POWER/LIGHTING

Item	Net Price £	Material £	Labour hours	Labour £	Unit	Total rate £
1250 A (2 × 10 mm × 40 mm)	518.29	636.72	–	–	nr	**636.72**
1000 A (2 × 10 mm × 30 mm)	518.29	636.72	–	–	nr	**636.72**
800 A (2 × 10 mm × 20 mm)	403.46	495.65	–	–	nr	**495.65**
630 A (2 × 10 mm × 20 mm)	403.46	495.65	–	–	nr	**495.65**
400 A (2 × 10 mm × 10 mm)	345.53	424.48	–	–	nr	**424.48**
Automatic power factor correction (PFC); floor standing steel enclosure to IP 42, complete with microprocessor based relay and status indication; includes delivery, offloading, positioning and commissioning; excludes cabling and cable terminations						
Standard PFC (no de-tuning)						
100 kVAr	5045.57	6198.49	–	–	nr	**6198.49**
200 kVAr	7056.49	8668.89	–	–	nr	**8668.89**
400 kVAr	12046.25	14798.82	–	–	nr	**14798.82**
600 kVAr	16500.55	20270.92	–	–	nr	**20270.92**
PFC with de-tuning reactors						
100 kVAr	8208.97	10084.71	–	–	nr	**10084.71**
200 kVAr	11472.16	14093.55	–	–	nr	**14093.55**
400 kVAr	19772.33	24290.31	–	–	nr	**24290.31**
600 kVAr	29077.96	35722.27	–	–	nr	**35722.27**

ELECTRICAL SUPPLY/POWER/LIGHTING

Item	Net Price £	Material £	Labour hours	Labour £	Unit	Total rate £
LV DISTRIBUTION: AUTOMATIC TRANSFER SWITCHES						
Automatic transfer switches; steel enclosure; solenoid operating; programmable controller, keypad and LCD display; fixed to backgrounds; including commissioning and testing						
Panel mounting type 4 pole M6s; non-BMS connection						
40 amp	1775.93	2181.73	2.40	95.99	nr	**2277.72**
63 amp	1884.70	2315.35	2.50	100.00	nr	**2415.35**
80 amp	1928.00	2368.55	2.60	104.00	nr	**2472.55**
100 amp	1956.25	2403.25	2.60	104.00	nr	**2507.25**
125 amp	2008.96	2468.00	2.70	107.99	nr	**2575.99**
160 amp	2112.85	2595.63	2.80	112.00	nr	**2707.63**
Panel mounting type 4 pole M6e; BMS connection						
40 amp	2450.68	3010.66	2.50	100.00	nr	**3110.66**
63 amp	2595.71	3188.83	2.60	104.00	nr	**3292.83**
80 amp	2653.44	3259.75	2.70	107.99	nr	**3367.74**
100 amp	2691.08	3305.99	2.80	112.00	nr	**3417.99**
125 amp	2761.39	3392.37	2.90	115.99	nr	**3508.36**
160 amp	2899.88	3562.50	3.00	120.00	nr	**3682.50**
Enclosed type 3 pole or 4 pole						
63 amp	3297.17	4050.57	2.60	104.00	nr	**4154.57**
100 amp	3411.76	4191.35	2.90	115.99	nr	**4307.34**
125 amp	3637.63	4468.83	2.90	115.99	nr	**4584.82**
160 amp	3810.85	4681.62	2.60	104.00	nr	**4785.62**
250amp	4742.16	5825.75	3.30	132.01	nr	**5957.76**
400amp	7457.14	9161.09	4.30	172.00	nr	**9333.09**
630amp	10039.84	12333.94	4.84	193.60	nr	**12527.54**
800amp	13857.64	17024.11	5.12	204.80	nr	**17228.91**
1000amp	15361.99	18872.21	5.50	220.00	nr	**19092.21**
1250amp	19202.48	23590.25	6.00	240.00	nr	**23830.25**
1600amp	33048.12	40599.62	6.20	248.00	nr	**40847.62**
2000amp	45516.96	55917.58	6.90	275.99	nr	**56193.57**
2500 amp	51206.59	62907.30	2.60	104.00	nr	**63011.30**
3000amp	34536.73	42428.37	6.90	275.99	nr	**42704.36**
3200 amp	56896.59	69897.46	2.60	104.00	nr	**70001.46**
4000amp	42274.74	51934.52	6.90	275.99	nr	**52210.51**
Enclosed type 3 pole or 4 pole; with single by-pass						
40 amp	8449.09	10379.71	2.60	104.00	nr	**10483.71**
63 amp	8613.13	10581.23	2.60	104.00	nr	**10685.23**
80 amp	8722.18	10715.20	2.60	104.00	nr	**10819.20**
100 amp	8841.78	10862.12	2.60	104.00	nr	**10966.12**
125 amp	9299.11	11423.96	2.90	115.99	nr	**11539.95**
160 amp	10061.35	12360.37	2.90	115.99	nr	**12476.36**
250 amp	13094.93	16087.12	3.30	132.01	nr	**16219.13**

ELECTRICAL SUPPLY/POWER/LIGHTING

Item	Net Price £	Material £	Labour hours	Labour £	Unit	**Total rate £**
400 amp	13759.44	16903.47	4.30	172.00	nr	**17075.47**
630 amp	18873.13	23185.64	4.84	193.60	nr	**23379.24**
800 amp	20275.47	24908.42	5.12	204.80	nr	**25113.22**
1000 amp	28662.88	35212.35	5.50	220.00	nr	**35432.35**
1200 amp	30379.47	37321.18	2.60	104.00	nr	**37425.18**
1250 amp	28073.42	34488.19	6.00	240.00	nr	**34728.19**
1600 amp	31196.45	38324.84	6.20	248.00	nr	**38572.84**
Enclosed type 3 pole or 4 pole; with dual by-pass						
40 amp	11062.80	13590.65	2.60	104.00	nr	**13694.65**
63 amp	11484.16	14108.29	2.60	104.00	nr	**14212.29**
80 amp	11626.18	14282.76	2.60	104.00	nr	**14386.76**
100 amp	11789.05	14482.85	2.60	104.00	nr	**14586.85**
125 amp	12398.82	15231.95	2.90	115.99	nr	**15347.94**
160 amp	13415.12	16480.47	2.90	115.99	nr	**16596.46**
250 amp	17459.91	21449.49	3.30	132.01	nr	**21581.50**
400 amp	18345.93	22537.98	4.30	172.00	nr	**22709.98**
630 amp	25164.16	30914.17	4.84	193.60	nr	**31107.77**
800 amp	27033.98	33211.25	5.12	204.80	nr	**33416.05**
1000 amp	38217.19	46949.82	5.50	220.00	nr	**47169.82**
1200 amp	40505.96	49761.57	2.60	104.00	nr	**49865.57**
1250 amp	37431.23	45984.27	6.00	240.00	nr	**46224.27**
1600 amp	41595.26	51099.77	6.20	248.00	nr	**51347.77**

ELECTRICAL SUPPLY/POWER/LIGHTING

Item	Net Price £	Material £	Labour hours	Labour £	Unit	Total rate £
LV DISTRIBUTION: BREAKERS AND FUSES						
MCCB panelboards; IP4X construction, 50 kA busbars and fully-rated neutral; fitted with doorlock, removable glandplate; form 3b Type2; BSEN 60439-1; including fixing to backgrounds						
Panelboards cubicle with MCCB incomer						
Up to 250 A						
4 way TPN	959.81	1179.13	1.00	40.01	nr	**1219.14**
Extra over for integral incomer metering	983.04	1207.67	1.50	60.00	nr	**1267.67**
Up to 630 A						
6 way TPN	1726.56	2121.08	2.00	80.00	nr	**2201.08**
12 way TPN	1949.98	2395.55	2.50	100.00	nr	**2495.55**
18 Way TPN	2269.91	2788.58	3.00	120.00	nr	**2908.58**
Extra over for integral incomer metering	1161.76	1427.22	1.50	60.00	nr	**1487.22**
Up to 800 A						
6 way TPN	2954.46	3629.56	2.00	80.00	nr	**3709.56**
12 way TPN	3544.30	4354.17	2.50	100.00	nr	**4454.17**
18 way TPN	3773.07	4635.21	3.00	120.00	nr	**4755.21**
Extra over for integral incomer metering	1161.76	1427.22	1.50	60.00	nr	**1487.22**
Up to 1200 A						
20 way TPN	7905.40	9711.79	3.50	140.00	nr	**9851.79**
Up to 1600 A						
28 way TPN	10237.85	12577.19	3.50	140.00	nr	**12717.19**
Up to 2000 A						
28 way TPN	11144.04	13690.46	4.00	160.00	nr	**13850.46**
Feeder MCCBs						
Single pole						
32 A	81.07	99.59	0.75	30.00	nr	**129.59**
63 A	82.92	101.87	0.75	30.00	nr	**131.87**
100 A	84.75	104.12	0.75	30.00	nr	**134.12**
160 A	90.29	110.92	1.00	40.01	nr	**150.93**
Double pole						
32 A	121.61	149.39	0.75	30.00	nr	**179.39**
63 A	123.44	151.64	0.75	30.00	nr	**181.64**
100 A	180.57	221.83	0.75	30.00	nr	**251.83**
160 A	224.79	276.15	1.00	40.01	nr	**316.16**
Triple pole						
32 A	162.16	199.22	0.75	30.00	nr	**229.22**
63 A	165.82	203.71	0.75	30.00	nr	**233.71**
100 A	215.57	264.83	0.75	30.00	nr	**294.83**
160 A	278.23	341.81	1.00	40.01	nr	**381.82**
250 A	418.27	513.85	1.00	40.01	nr	**553.86**
400 A	565.68	694.94	1.25	50.00	nr	**744.94**
630 A	928.69	1140.90	1.50	60.00	nr	**1200.90**

ELECTRICAL SUPPLY/POWER/LIGHTING

Item	Net Price £	Material £	Labour hours	Labour £	Unit	Total rate £
MCB distribution boards; IP3X external protection enclosure; removable earth and neutral bars and DIN rail; 125/250 amp incomers; including fixing to backgrounds						
SP&N						
6 way	78.84	96.85	2.00	80.00	nr	**176.85**
8 way	93.50	114.86	2.50	100.00	nr	**214.86**
12 way	107.31	131.83	3.00	120.00	nr	**251.83**
16 way	127.56	156.71	4.00	160.00	nr	**316.71**
24 way	268.70	330.10	5.00	200.00	nr	**530.10**
TP&N						
4 way	546.33	671.17	3.00	120.00	nr	**791.17**
6 way	565.54	694.76	3.50	140.00	nr	**834.76**
8 way	592.13	727.43	4.00	160.00	nr	**887.43**
12 way	631.36	775.62	4.00	160.00	nr	**935.62**
16 way	722.45	887.53	5.00	200.00	nr	**1087.53**
24 way	909.81	1117.70	6.40	255.99	nr	**1373.69**
Miniature circuit breakers for distribution boards; BS EN 60 898; DIN rail mounting; including connecting to circuit						
SP&N; including connecting of wiring						
6 amp	9.53	11.71	0.10	4.00	nr	**15.71**
10–40 amp	9.92	12.19	0.10	4.00	nr	**16.19**
50–63 amp	10.38	12.75	0.14	5.61	nr	**18.36**
TP&N; including connecting of wiring						
6 amp	40.39	49.62	0.30	12.00	nr	**61.62**
10–40 amp	41.98	51.58	0.45	18.00	nr	**69.58**
50–63 amp	43.98	54.03	0.45	18.00	nr	**72.03**
Residual current circuit breakers for distribution boards; DIN rail mounting; including connecting to circuit						
SP&N						
10 mA						
6 amp	62.55	76.84	0.21	8.40	nr	**85.24**
10–32 amp	61.36	75.38	0.26	10.41	nr	**85.79**
45 amp	62.37	76.62	0.26	10.41	nr	**87.03**
30 mA						
6 amp	62.55	76.84	0.21	8.40	nr	**85.24**
10–40 amp	61.36	75.38	0.21	8.40	nr	**83.78**
50–63 amp	63.36	77.84	0.26	10.41	nr	**88.25**
100 mA						
6 amp	115.65	142.08	0.21	8.40	nr	**150.48**
10–40 amp	115.65	142.08	0.23	9.32	nr	**151.40**
50–63 amp	115.65	142.08	0.26	10.41	nr	**152.49**

ELECTRICAL SUPPLY/POWER/LIGHTING

Item	Net Price £	Material £	Labour hours	Labour £	Unit	Total rate £
LV DISTRIBUTION: BREAKERS AND FUSES – cont						
HRC fused distribution boards; IP4X external protection enclosure; including earth and neutral bars; fixing to backgrounds						
SP&N						
20 amp incomer						
4 way	148.93	182.96	1.00	40.01	nr	**222.97**
6 way	179.80	220.89	1.20	48.00	nr	**268.89**
8 way	210.77	258.93	1.40	56.00	nr	**314.93**
12 way	272.75	335.08	1.80	72.00	nr	**407.08**
32 amp incomer						
4 way	179.31	220.28	1.00	40.01	nr	**260.29**
6 way	235.80	289.68	1.20	48.00	nr	**337.68**
8 way	277.42	340.81	1.40	56.00	nr	**396.81**
12 way	357.18	438.79	1.80	72.00	nr	**510.79**
TP&N						
20 amp incomer						
4 way	280.99	345.20	1.50	60.00	nr	**405.20**
6 way	355.38	436.58	2.10	84.00	nr	**520.58**
8 way	420.19	516.20	2.70	107.99	nr	**624.19**
12 way	589.51	724.22	3.90	156.00	nr	**880.22**
32 amp incomer						
4 way	336.16	412.98	1.50	60.00	nr	**472.98**
6 way	451.31	554.43	2.10	84.00	nr	**638.43**
8 way	551.30	677.27	2.70	107.99	nr	**785.26**
12 way	764.64	939.36	3.90	156.00	nr	**1095.36**
63 amp incomer						
4 way	714.56	877.84	2.17	86.80	nr	**964.64**
6 way	916.42	1125.82	2.83	113.19	nr	**1239.01**
8 way	1103.58	1355.75	2.57	102.81	nr	**1458.56**
100 amp incomer						
4 way	1130.07	1388.29	2.40	95.99	nr	**1484.28**
6 way	1477.11	1814.63	2.73	109.20	nr	**1923.83**
8 way	1806.13	2218.83	3.87	154.80	nr	**2373.63**
200 amp incomer						
4 way	2799.08	3438.67	5.36	214.40	nr	**3653.07**
6 way	3699.58	4544.94	6.17	246.80	nr	**4791.74**
HRC fuse; includes fixing to fuse holder						
2–30 amp	3.03	3.73	0.10	4.00	nr	**7.73**
35–63 amp	6.56	8.06	0.12	4.80	nr	**12.86**
80 amp	9.67	11.88	0.15	6.00	nr	**17.88**
100 amp	11.63	14.29	0.15	6.00	nr	**20.29**
125 amp	17.58	21.60	0.15	6.00	nr	**27.60**
160 amp	18.44	22.65	0.15	6.00	nr	**28.65**
200 amp	19.10	23.47	0.15	6.00	nr	**29.47**

ELECTRICAL SUPPLY/POWER/LIGHTING

Item	Net Price £	Material £	Labour hours	Labour £	Unit	**Total rate £**
Consumer units; fixed to backgrounds; including supports, fixings, connections/ jointing to equipment.						
Switched and insulated; moulded plastic case, 63 amp 230 Volt SP&N; earth and neutral bars; 30 mA RCCB protection; fitted MCBs						
2 way	112.33	138.00	1.67	66.80	nr	**204.80**
4 way	126.05	154.85	1.59	63.60	nr	**218.45**
6 way	137.02	168.33	2.50	100.00	nr	**268.33**
8 way	147.85	181.63	3.00	120.00	nr	**301.63**
12 way	171.76	211.01	4.00	160.00	nr	**371.01**
16 way	207.54	254.96	5.50	220.00	nr	**474.96**
Switched and insulated; moulded plastic case, 100 amp 230 Volt SP&N; earth and neutral bars; 30 mA RCCB protection; fitted MCBs						
2 way	112.33	138.00	1.67	66.80	nr	**204.80**
4 way	126.05	154.85	1.59	63.60	nr	**218.45**
6 way	137.02	168.33	2.50	100.00	nr	**268.33**
8 way	147.85	181.63	3.00	120.00	nr	**301.63**
12 way	171.76	211.01	4.00	160.00	nr	**371.01**
16 way	207.54	254.96	5.50	220.00	nr	**474.96**
Extra for						
Residual current device; double pole; 230 Volt/ 30 mA tripping current						
16 amp	61.42	75.45	0.22	8.80	nr	**84.25**
30 amp	62.39	76.65	0.22	8.80	nr	**85.45**
40 amp	63.37	77.85	0.22	8.80	nr	**86.65**
63 amp	78.45	96.38	0.22	8.80	nr	**105.18**
80 amp	87.26	107.19	0.22	8.80	nr	**115.99**
100 amp	107.40	131.94	0.25	10.00	nr	**141.94**
Residual current device; double pole; 230 Volt/ 100 mA tripping current						
63 amp	71.72	88.11	0.22	8.80	nr	**96.91**
80 amp	82.97	101.92	0.22	8.80	nr	**110.72**
100 amp	107.43	131.97	0.25	10.00	nr	**141.97**
Heavy duty fuse switches; with HRC fuses BS 5419; short circuit rating 65 kA, 500 Volt; including retractable operating switches						
SP&N						
63 amp	277.47	340.87	1.30	52.01	nr	**392.88**
100 amp	405.59	498.27	1.95	77.99	nr	**576.26**
TP&N						
63 amp	349.56	429.44	1.83	73.20	nr	**502.64**
100 amp	491.21	603.46	2.48	99.20	nr	**702.66**
200 amp	757.72	930.86	3.13	125.19	nr	**1056.05**
300 amp	1316.37	1617.16	4.45	178.00	nr	**1795.16**
400 amp	1445.13	1775.34	4.45	178.00	nr	**1953.34**
600 amp	2181.35	2679.79	5.72	228.79	nr	**2908.58**
800 amp	3407.51	4186.13	7.88	315.20	nr	**4501.33**

ELECTRICAL SUPPLY/POWER/LIGHTING

Item	Net Price £	Material £	Labour hours	Labour £	Unit	Total rate £
LV DISTRIBUTION: BREAKERS AND FUSES – cont						
Switch disconnectors to BSEN 60947-3; in sheet steel case; IP41 with door interlock fixed to backgrounds						
Double pole						
20 amp	56.14	68.96	1.02	40.80	nr	**109.76**
32 amp	67.68	83.15	1.02	40.80	nr	**123.95**
63 amp	246.95	303.38	1.21	48.40	nr	**351.78**
100 amp	226.61	278.39	1.86	74.40	nr	**352.79**
TP&N						
20 amp	70.48	86.58	1.29	51.60	nr	**138.18**
32 amp	82.02	100.76	1.83	73.20	nr	**173.96**
63 amp	278.32	341.91	2.48	99.20	nr	**441.11**
100 amp	280.62	344.75	2.48	99.20	nr	**443.95**
125 amp	292.70	359.58	2.48	99.20	nr	**458.78**
160 amp	673.45	827.34	2.48	99.20	nr	**926.54**
Enclosed switch disconnector to BSEN 60947-3; enclosure minimum IP55 rating; complete with earth connection bar; fixed to backgrounds.						
TP						
20 amp	56.14	68.96	1.02	40.80	nr	**109.76**
32 amp	67.68	83.15	1.02	40.80	nr	**123.95**
63 amp	246.95	303.38	1.21	48.40	nr	**351.78**
TP&N						
20 amp	70.48	86.58	1.29	51.60	nr	**138.18**
32 amp	82.02	100.76	1.83	73.20	nr	**173.96**
63 amp	278.32	341.91	2.48	99.20	nr	**441.11**
Busbar chambers; fixed to background including all supports, fixings, connections/jointing to equipment.						
Sheet steel case enclosing 4 pole 550 Volt copper bars, detachable metal end plates						
600 mm long						
200 amp	511.49	628.36	2.62	104.80	nr	**733.16**
300 amp	657.28	807.47	3.03	121.20	nr	**928.67**
500 amp	1126.23	1383.57	4.48	179.20	nr	**1562.77**
900 mm long						
200 amp	736.75	905.10	3.04	121.60	nr	**1026.70**
300 amp	868.30	1066.71	3.59	143.60	nr	**1210.31**
500 amp	1284.37	1577.85	4.42	176.81	nr	**1754.66**
1350 mm long						
200 amp	1006.35	1236.30	3.38	135.20	nr	**1371.50**
300 amp	1184.46	1455.11	3.94	157.60	nr	**1612.71**
500 amp	1895.98	2329.22	4.82	192.80	nr	**2522.02**

ELECTRICAL SUPPLY/POWER/LIGHTING

Item	Net Price £	Material £	Labour hours	Labour £	Unit	Total rate £
Contactor relays; pressed steel enclosure; fixed to backgrounds including supports, fixings, connections/jointing to equipment						
Relays						
6 amp, 415/240 Volt, 4 pole N/O	59.24	72.78	0.52	20.80	nr	**93.58**
6 amp, 415/240 Volt, 8 pole N/O	72.44	88.99	0.85	34.01	nr	**123.00**
Push button stations; heavy gauge pressed steel enclosure; polycarbonate cover; IP65; fixed to backgrounds including supports, fixings,connections/ joining to equipment						
Standard units						
One button (start or stop)	74.97	92.10	0.39	15.60	nr	**107.70**
Two button (start or stop)	79.58	97.77	0.47	18.80	nr	**116.57**
Three button (forward-reverse-stop)	112.78	138.55	0.57	22.81	nr	**161.36**
Weatherproof junction boxes; enclosures with rail mounted terminal blocks; side hung door to receive padlock; fixed to backgrounds, including all supports and fixings (Suitable for cable up to 2.5 mm²; including glandplates and gaskets.)						
Sheet steel with zinc spray finish enclosure						
Overall size 229 × 152; suitable to receive 3 × 20(A) glands per gland plate	75.92	93.27	1.43	57.20	nr	**150.47**
Overall size 306 × 306; suitable to receive 14 × 20(A) glands per gland plate	101.86	125.14	2.17	86.80	nr	**211.94**
Overall size 458 × 382; suitable to receive 18 × 20(A) glands per gland plate	148.32	182.21	3.51	140.41	nr	**322.62**
Overall size 762 x508; suitable to receive 26 × 20(A) glands per gland plate	156.87	192.72	4.85	194.01	nr	**386.73**
Overall size 914 × 610; suitable to receive 45 × 20(A) glands per gland plate	173.38	212.99	7.01	280.40	nr	**493.39**
Weatherproof junction boxes; enclosures with rail mounted terminal blocks; screw fixed lid; fixed to backgrounds, including all supports and fixings (suitable for cable up to 2.5 mm²; including glandplates and gaskets)						
Glassfibre reinforced polycarbonate enclosure						
Overall size 190 × 190 × 130	103.70	127.40	1.43	57.20	nr	**184.60**
Overall size 190 × 190 × 180	151.83	186.52	1.53	61.20	nr	**247.72**
Overall size 280 × 190 × 130	171.47	210.65	2.17	86.80	nr	**297.45**
Overall size 280 × 190 × 180	192.11	236.01	2.37	94.80	nr	**330.81**
Overall size 380 × 190 × 130	214.34	263.32	3.33	133.19	nr	**396.51**
Overall size 380 × 190 × 180	230.20	282.80	3.30	132.01	nr	**414.81**
Overall size 380 × 280 × 130	246.09	302.33	4.66	186.40	nr	**488.73**
Overall size 380 × 280 × 180	265.13	325.71	5.36	214.40	nr	**540.11**
Overall size 560 × 280 × 130	319.11	392.03	7.01	280.40	nr	**672.43**
Overall size 560 × 380 × 180	328.65	403.75	7.67	306.81	nr	**710.56**

ELECTRICAL SUPPLY/POWER/LIGHTING

Item	Net Price £	Material £	Labour hours	Labour £	Unit	Total rate £
GENERAL LIGHTING						
Fluorescent Luminaires; surface fixed to backgrounds						
Batten type; surface mounted						
600 mm Single – 18 W	8.94	10.98	0.58	23.19	nr	**34.17**
600 mm Twin – 18 W	15.67	19.25	0.59	23.60	nr	**42.85**
1200 mm Single – 36 W	11.83	14.53	0.76	30.41	nr	**44.94**
1200 mm Twin – 36 W	13.26	16.29	0.84	33.60	nr	**49.89**
1500 mm Single – 58 W	22.66	27.84	0.77	30.80	nr	**58.64**
1500 mm Twin – 58 W	26.45	32.50	0.85	34.01	nr	**66.51**
1800 mm Single – 70 W	16.15	19.84	1.05	42.00	nr	**61.84**
1800 mm Twin – 70 W	29.24	35.92	1.06	42.40	nr	**78.32**
2400 mm Single – 100 W	22.11	27.16	1.25	50.00	nr	**77.16**
2400 mm Twin – 100 W	38.30	47.05	1.27	50.80	nr	**97.85**
Surface mounted, opal diffuser						
600 mm Twin – 18 W	24.32	29.87	0.62	24.80	nr	**54.67**
1200 mm Single – 36 W	20.99	25.79	0.79	31.59	nr	**57.38**
1200 mm Twin – 36 W	32.35	39.74	0.80	32.00	nr	**71.74**
1500 mm Single – 58 W	24.13	29.64	0.88	35.20	nr	**64.84**
1500 mm Twin – 58 W	37.83	46.47	0.90	35.99	nr	**82.46**
1800 mm Single – 70 W	30.10	36.98	1.09	43.60	nr	**80.58**
1800 mm Twin – 70 W	40.90	50.24	1.10	44.01	nr	**94.25**
2400 mm Single – 100 W	39.50	48.52	1.30	52.01	nr	**100.53**
2400 mm Twin – 100 W	56.51	69.43	1.31	52.40	nr	**121.83**
Surface mounted linear fluorescent; T8 lamp; high frequency control gear; low brightness; 65° cut-off; including wedge style louvre						
1200 mm, 1 × 36 watt	57.85	71.06	1.09	43.60	nr	**114.66**
1200 mm 2 × 36 watt	61.24	75.23	1.09	43.60	nr	**118.83**
Extra for emergency pack	52.71	64.75	0.25	10.00	nr	**74.75**
1500 mm, 1 × 58 watt	67.42	82.82	0.90	35.99	nr	**118.81**
1500 mm 2 × 58 watt	71.32	87.61	0.90	35.99	nr	**123.60**
Extra for emergency pack	52.61	64.63	0.25	10.00	nr	**74.63**
1800 mm, 1 × 70 watt	101.53	124.73	0.90	35.99	nr	**160.72**
1800 mm, 2 × 70 watt	115.36	141.72	0.90	35.99	nr	**177.71**
Extra for emergency pack	87.08	106.97	0.25	10.00	nr	**116.97**
Modular recessed linear fluorescent; high frequency control gear; low brightness; 65° cut off; including wedge style louvre; fitted to exposed T grid ceiling						
600 × 600 mm, 3 × 18 watt T8	45.49	55.88	0.84	33.60	nr	**89.48**
600 × 600 mm, 4 × 18 watt T8	46.53	57.16	0.87	34.81	nr	**91.97**
Extra for emergency pack	53.14	65.28	0.25	10.00	nr	**75.28**
300 × 1200 mm, 2 × 36 watt T8	63.49	77.99	0.87	34.81	nr	**112.80**
Extra for emergency pack	67.23	82.59	0.25	10.00	nr	**92.59**
600 × 1200 mm, 3 × 36 watt T8	69.28	85.11	0.89	35.61	nr	**120.72**
600 × 1200 mm, 4 × 36 watt T8	78.87	96.89	0.91	36.40	nr	**133.29**
Extra for emergency pack	62.26	76.48	0.25	10.00	nr	**86.48**
600 × 600 mm, 3 × 14 watt T5	64.63	79.40	0.84	33.60	nr	**113.00**
600 × 600 mm, 4 × 14 watt T5	66.57	81.78	0.87	34.81	nr	**116.59**
Extra for emergency pack	44.55	54.73	0.25	10.00	nr	**64.73**

ELECTRICAL SUPPLY/POWER/LIGHTING

Item	Net Price £	Material £	Labour hours	Labour £	Unit	**Total rate £**
Modular recessed; T8 lamp; high frequency control gear; cross-blade louvre; fitted to exposed T grid ceiling						
600 × 600 mm, 3 × 18 watt	49.67	61.02	0.84	33.60	nr	**94.62**
600 × 600 mm, 4 × 18 watt	66.53	81.73	0.87	34.81	nr	**116.54**
Extra for emergency pack	60.42	74.22	0.25	10.00	nr	**84.22**
Modular recessed compact fluorescent; TCL lamp; high frequency control gear; low brightness; 65° cut-off; including wedge style louvre; fitted to exposed T grid ceiling						
300 × 300 mm, 2 × 18 watt	100.61	123.60	0.75	30.00	nr	**153.60**
Extra for emergency pack	93.85	115.29	0.25	10.00	nr	**125.29**
500 × 500 mm, 2 × 36 watt	122.24	150.17	0.82	32.80	nr	**182.97**
600 × 600 mm, 2 × 36 watt	113.53	139.47	0.82	32.80	nr	**172.27**
600 × 600 mm, 2 × 40 watt	110.24	135.43	0.82	32.80	nr	**168.23**
Extra for emergency pack	44.55	54.73	0.25	10.00	nr	**64.73**
Ceiling recessed asymetric compact fluorescent downlighter; high frequency control gear; TCD lamp in 200 mm diameter luminaire; for wall-washing application						
1 × 18 watt	159.34	195.75	0.75	30.00	nr	**225.75**
1 × 26 watt	159.34	195.75	0.75	30.00	nr	**225.75**
2 × 18 watt	175.69	215.84	0.75	30.00	nr	**245.84**
2 × 26 watt	175.69	215.84	0.75	30.00	nr	**245.84**
Ceiling recessed asymetric compact fluorescesnt downlights; high frequency control gear; linear 200 mm × 600 mm luminaire with low glare louvre; for wall washing applications						
1 × 55 watt TCL	90.01	110.58	0.75	30.00	nr	**140.58**
Wall mounted compact fluorescent uplighter; high frequency control gear; TCL lamp in 300 mm × 600 mm luminaire						
2 × 36 watt	332.60	408.60	0.84	33.60	nr	**442.20**
2 × 40 watt	334.68	411.16	0.84	33.60	nr	**444.76**
2 × 55 watt	334.68	411.16	0.84	33.60	nr	**444.76**
Suspended linear fluorescent; T5 lamp; high frequency control gear; low brightness; 65° cut-off; 30% uplight, 70% downlight; including wedge style louvre						
1 × 49 watt	104.52	128.40	0.75	30.00	nr	**158.40**
Extra for emergency pack	44.55	54.73	0.25	10.00	nr	**64.73**
Semi-recessed 'architectural' linear fluorescent; T5 lamp; high frequency control gear; low brightness, delivers direct, ceiling and graduated wall washing illumination						
600 × 600 mm, 2 × 24 watt	140.38	172.45	0.87	34.81	nr	**207.26**
600 × 600 mm, 4 × 14 watt	151.01	185.51	0.87	34.81	nr	**220.32**
500 × 500 mm, 2 × 24 watt	137.97	169.49	0.87	34.81	nr	**204.30**
Extra for emergency pack	44.55	54.73	0.25	10.00	nr	**64.73**

ELECTRICAL SUPPLY/POWER/LIGHTING

Item	Net Price £	Material £	Labour hours	Labour £	Unit	Total rate £
GENERAL LIGHTING – cont						
Downlighter, recessed; low voltage; mirror reflector with white/chrome bezel; dimmable transformer; for dichroic lamps						
85 mm dia. × 20/50 watt	16.24	19.95	0.66	26.40	nr	**46.35**
118 mm dia. × 50 watt	21.26	26.11	0.66	26.40	nr	**52.51**
165 mm dia. × 100 watt	97.10	119.29	0.66	26.40	nr	**145.69**
LUMINAIRES						
High/Low Bay						
Compact discharge; aluminium reflector						
150 watt	178.59	219.40	1.50	60.00	nr	**279.40**
250 watt	133.55	164.06	1.50	60.00	nr	**224.06**
400 watt	133.55	164.06	1.50	60.00	nr	**224.06**
Sealed discharge; aluminium reflector						
150 watt	187.17	229.94	1.50	60.00	nr	**289.94**
250 watt	202.02	248.18	1.50	60.00	nr	**308.18**
400 watt	264.13	324.48	1.50	60.00	nr	**384.48**
Corrosion resistant GRP body; gasket sealed; acrylic diffuser						
600 mm Single – 18 W	30.48	37.44	0.49	19.59	nr	**57.03**
600 mm Twin – 18 W	39.68	48.75	0.49	19.59	nr	**68.34**
1200 mm Single – 36 W	34.42	42.28	0.64	25.60	nr	**67.88**
1200 mm Twin – 36 W	43.99	54.04	0.64	25.60	nr	**79.64**
1500 mm Single – 58 W	38.24	46.98	0.72	28.79	nr	**75.77**
1500 mm Twin – 58 W	47.54	58.40	0.72	28.79	nr	**87.19**
1800 mm Single – 70 W	56.60	69.53	0.94	37.60	nr	**107.13**
1800 mm Twin – 70 W	70.24	86.29	0.94	37.60	nr	**123.89**
Flameproof to IIA/IIB,I.P. 64; aluminium body; BS 229 and 899						
600 mm Single – 18 W	340.83	418.71	1.04	41.60	nr	**460.31**
600 mm Twin – 18 W	419.30	515.11	1.04	41.60	nr	**556.71**
1200 mm Single – 36 W	372.46	457.57	1.31	52.40	nr	**509.97**
1200 mm Twin – 36 W	455.84	560.00	1.18	47.20	nr	**607.20**
1500 mm Single – 58 W	398.75	489.87	1.64	65.60	nr	**555.47**
1500 mm Twin – 58 W	476.63	585.54	1.64	65.60	nr	**651.14**
1800 mm Single – 70 W	437.13	537.01	1.97	78.79	nr	**615.80**
1800 mm Twin – 70 W	499.20	613.26	1.97	78.79	nr	**692.05**
LED Luminaires						
Modular LED luminaire; recessed into ceiling; including driver, polycarbonate optical lenses/ diffuser, heat sink and 4000K colour temperature						
300 × 300 mm	83.30	102.33	0.75	30.00	nr	**132.33**
Extra for emergency pack	79.20	97.29	0.25	10.00	nr	**107.29**
Extra for DALI lighting control	25.55	31.38	0.25	10.00	nr	**41.38**
600 × 600 mm	196.00	240.79	0.82	32.80	nr	**273.59**
Extra for emergency pack	89.10	109.46	0.25	10.00	nr	**119.46**
Extra for DALI lighting control	24.75	30.41	0.25	10.00	nr	**40.41**

ELECTRICAL SUPPLY/POWER/LIGHTING

Item	Net Price £	Material £	Labour hours	Labour £	Unit	**Total rate £**
Modular linear LED luminaire; recessed into ceiling; including driver, polycarbonate optical lenses/diffuser, heat sink and 4000K colour temperature						
600 × 100 mm	303.56	372.93	0.84	33.60	nr	**406.53**
Extra for emergency pack	79.20	97.29	0.25	10.00	nr	**107.29**
Extra for DALI lighting control	25.55	31.38	0.25	10.00	nr	**41.38**
1200 × 100 mm	333.20	409.33	0.89	35.61	nr	**444.94**
Extra for emergency pack	79.20	97.29	0.25	10.00	nr	**107.29**
Extra for DALI lighting control	25.55	31.38	0.25	10.00	nr	**41.38**
LED downlight/spotlight; surface mounted; including driver, polycarbonate optical lenses/ diffuser, heat sink and 4000K colour temperature						
190 mm diameter	174.60	214.49	0.75	30.00	nr	**244.49**
Extra for emergency pack (remote)	94.05	115.54	0.25	10.00	nr	**125.54**
Extra for DALI lighting control (remote)	48.00	58.97	0.25	10.00	nr	**68.97**
LED downlight/spotlight; track mounted; adjustable fixing; including driver, polycarbonate optical lenses/diffuser, heat sink and 4000K colour temperature						
120 mm diameter	203.70	250.25	0.75	30.00	nr	**280.25**
Extra for emergency pack (track mounted remote)	287.10	352.71	0.25	10.00	nr	**362.71**
Extra for DALI lighting control (with DALI track)	48.00	58.97	0.25	10.00	nr	**68.97**
LED downlight/spotlight; recessed into ceiling; including driver, polycarbonate optical lenses/ diffuser, heat sink and 4000K colour temperature						
1 × spotlight	83.30	102.33	0.66	26.40	nr	**128.73**
Extra for emergency pack (remote)	94.05	115.54	0.25	10.00	nr	**125.54**
Extra for DALI lighting control	42.57	52.30	0.25	10.00	nr	**62.30**
2 × spotlight	93.10	114.38	0.66	26.40	nr	**140.78**
Extra for emergency pack (remote)	103.95	127.70	0.25	10.00	nr	**137.70**
Extra for DALI lighting control	24.75	30.41	0.25	10.00	nr	**40.41**
3 × spotlight	343.00	421.38	0.66	26.40	nr	**447.78**
Extra for emergency pack (remote)	108.90	133.78	0.25	10.00	nr	**143.78**
Extra for DALI lighting control	59.40	72.98	0.25	10.00	nr	**82.98**
LED downlight/spotlight; track mounted; adjustable fixing; including driver, polycarbonate optical lenses/diffuser, heat sink, 4000K colour temperature						
120 mm diameter	176.40	216.71	4.00	160.00	nr	**376.71**
Extra for emergency pack (remote)	203.70	250.25	4.00	160.00	nr	**410.25**
Extra for DALI lighting control (with DALI track)	49.50	60.81	0.25	10.00	nr	**70.81**

ELECTRICAL SUPPLY/POWER/LIGHTING

Item	Net Price £	Material £	Labour hours	Labour £	Unit	Total rate £
GENERAL LIGHTING – cont						
LED Luminaires – cont						
LED downlight/spotlight; track mounted; adjustable fixing; including driver, polycarbonate optical lenses/diffuser, heat sink, RGB Colour						
120 mm diameter	235.20	288.94	0.25	10.00	nr	**298.94**
Extra for emergency pack (remote)	203.70	250.25	0.25	10.00	nr	**260.25**
Extra for DALI lighting control (with DALI track)	49.50	60.81	0.25	10.00	nr	**70.81**
Internal flexible linear LED, IP20, including driver and controller, optical lense/diffuser, white 4000K colour temperature						
10 mm width (remote)	76.00	93.37	0.25	10.00	nr	**103.37**
Extra for emergency pack (remote)	152.00	186.73	0.25	10.00	nr	**196.73**
Extra for DALI lighting control (remote)	24.75	30.41	0.25	10.00	nr	**40.41**
Internal flexible linear LED, IP20, including driver and controller, optical lense/diffuser, RGB						
10 mm width	114.00	140.05	0.25	10.00	nr	**150.05**
Extra for emergency pack (per unit)	152.00	186.73	0.25	10.00	nr	**196.73**
Extra for DALI lighting control	24.75	30.41	0.25	10.00	nr	**40.41**
LED in-ground performance uplighters, IP68, including driver, polycarbonate optical lenses/ diffuser, heat sink, 4000K colour temperature						
50 mm diameter	361.00	443.49	0.25	10.00	nr	**453.49**
Extra for emergency pack (per unit)	176.40	216.71	0.25	10.00	nr	**226.71**
Extra for DALI lighting control	15.00	18.43	0.25	10.00	nr	**28.43**
LED in-ground performance uplighters, IP68, including driver, polycarbonate optical lenses/ diffuser, heat sink, RGB						
50 mm diameter	427.50	525.18	0.25	10.00	nr	**535.18**
Extra for emergency pack (per unit)	176.40	216.71	0.25	10.00	nr	**226.71**
Extra for DALI lighting control	15.00	18.43	0.25	10.00	nr	**28.43**
LED in-ground decorative uplighters, IP68, including driver, polycarbonate optical lenses/ diffuser, heat sink, 4000K colour						
50 mm diameter	171.00	210.07	0.25	10.00	nr	**220.07**
Extra for emergency pack (per unit)	176.40	216.71	0.25	10.00	nr	**226.71**
Extra for DALI lighting control	20.00	24.57	0.25	10.00	nr	**34.57**
LED in-ground decorative uplighters, IP68, including driver, polycarbonate optical lenses/ diffuser, heat sink, RGB						
50 mm diameter	190.00	233.42	0.25	10.00	nr	**243.42**
Extra for emergency pack (per unit)	176.40	216.71	0.25	10.00	nr	**226.71**
Extra for DALI lighting control	20.00	24.57	0.25	10.00	nr	**34.57**
External flexible linear LED, IP66, including driver and controller, optical lenses/diffuser, 4000K colour temperature						
10 mm width	114.00	140.05	0.25	10.00	nr	**150.05**
Extra for emergency pack (per unit)	156.80	192.63	0.25	10.00	nr	**202.63**
Extra for DALI lighting control	25.00	30.71	0.25	10.00	nr	**40.71**

ELECTRICAL SUPPLY/POWER/LIGHTING

Item	Net Price £	Material £	Labour hours	Labour £	Unit	Total rate £
External flexible linear LED, IP66, including driver and controller, optical lenses/diffuser, RGB						
10 mm width	152.00	186.73	0.25	10.00	nr	**196.73**
Extra for emergency pack (per unit)	156.80	192.63	0.25	10.00	nr	**202.63**
Extra for DALI lighting control	25.00	30.71	0.25	10.00	nr	**40.71**
Handrail lighting, LED, IP66, including driver and controller, optical lenses/diffuser 4000K colour temperature						
10 mm width	138.60	170.27	0.25	10.00	nr	**180.27**
Extra for emergency pack (per unit)	156.80	192.63	0.25	10.00	nr	**202.63**
Extra for DALI lighting control	25.00	30.71	0.25	10.00	nr	**40.71**
Handrail lighting, LED, IP66, including driver and controller, optical lenses/diffuser, RGB						
10 mm width	158.40	194.60	0.25	10.00	nr	**204.60**
Extra for emergency pack (per unit)	156.80	192.63	0.25	10.00	nr	**202.63**
Extra for DALI lighting control	25.00	30.71	0.25	10.00	nr	**40.71**
External lighting						
Flameproof						
Ground mounted 50 W	378.30	464.74	2.25	90.00	nr	**554.74**
Ceiling mounted 50 W	170.20	209.09	2.25	90.00	nr	**299.09**
Bulkhead; aluminium body and polycarbonate bowl; vandal resistant; IP65						
60 W	33.34	40.96	0.75	30.00	nr	**70.96**
Extra for						
Emergency version	90.54	111.23	0.25	10.00	nr	**121.23**
2D 2 pin 16 W	28.36	34.84	0.66	26.40	nr	**61.24**
2D 2 pin 28 W	56.82	69.80	0.66	26.40	nr	**96.20**
Photocell	20.18	24.79	0.75	30.00	nr	**54.79**
1500 mm high circular bollard; polycarbonate visor; vandal resistant; IP54						
50 W	213.05	261.73	1.75	70.00	nr	**331.73**
70 W	216.46	265.92	1.75	70.00	nr	**335.92**
80 W	259.04	318.23	1.75	70.00	nr	**388.23**
Floodlight; enclosed high performance dischargelight; integeral control gear; reflector; toughened glass; IP65						
70 W	94.38	115.94	1.25	50.00	nr	**165.94**
100 W	98.77	121.34	1.25	50.00	nr	**171.34**
150 W	104.71	128.64	1.25	50.00	nr	**178.64**
250 W	183.31	225.19	1.25	50.00	nr	**275.19**
400 W	189.89	233.28	1.25	50.00	nr	**283.28**
Extra for Photocell	20.16	24.77	0.75	30.00	nr	**54.77**
Lighting track						
Single circuit; 25A 2 P&E steel trunking; low voltage with copper conductors; including couplers and supports; fixed to backgrounds						
Straight track	13.62	16.74	0.50	20.00	m	**36.74**
Live end feed unit complete with end stop	23.71	29.13	0.33	13.20	nr	**42.33**
Flexible couplers 0.5 m	47.97	58.93	0.33	13.20	nr	**72.13**
Tap off complete with 0.8 m of cable	7.92	9.73	0.25	10.00	nr	**19.73**

ELECTRICAL SUPPLY/POWER/LIGHTING

Item	Net Price £	Material £	Labour hours	Labour £	Unit	Total rate £
GENERAL LIGHTING – cont						
Lighting track – cont						
Three circuit; 25A 2 P&E steel trunking; low voltage with copper conductors; including couplers and supports incorporating integral twisted pair comms bus bracket; fixed to backgrounds						
Straight track	22.20	27.27	0.75	30.00	nr	**57.27**
Live end feed unit complete with end stop	40.38	49.60	0.50	20.00	nr	**69.60**
Flexible couplers 0.5 m	57.53	70.68	0.45	18.00	nr	**88.68**
Tap off complete with 0.8 m of cable	10.40	12.78	0.25	10.00	nr	**22.78**
LIGHTING ACCESSORIES						
Switches						
6 amp metal clad surface mounted switch, gridswitch; one way						
1 gang	9.15	11.25	0.43	17.20	nr	**28.45**
2 gang	12.56	15.43	0.55	22.00	nr	**37.43**
3 gang	19.90	24.44	0.77	30.80	nr	**55.24**
4 gang	23.56	28.95	0.88	35.20	nr	**64.15**
6 gang	40.47	49.72	1.00	40.01	nr	**89.73**
8 gang	47.40	58.23	1.28	51.21	nr	**109.44**
10 gang	71.51	87.85	1.67	66.80	nr	**154.65**
Extra for						
10 amp – Two way switch	3.25	3.99	0.03	1.21	nr	**5.20**
20 amp – Two way switch	4.36	5.36	0.04	1.60	nr	**6.96**
20 amp – Intermediate	8.18	10.05	0.08	3.19	nr	**13.24**
20 amp – One way SP switch	3.36	4.13	0.08	3.19	nr	**7.32**
Steel blank plate; 1 gang	2.26	2.77	0.07	2.80	nr	**5.57**
Steel blank plate; 2 gang	3.82	4.69	0.08	3.19	nr	**7.88**
6 amp modular type switch; galvanized steel box, bronze or satin chrome coverplate; metalclad switches; flush mounting; one way						
1 gang	22.49	27.63	0.43	17.20	nr	**44.83**
2 gang	25.90	31.81	0.55	22.00	nr	**53.81**
3 gang	37.43	45.98	0.77	30.80	nr	**76.78**
4 gang	53.01	65.12	0.88	35.20	nr	**100.32**
6 gang	89.38	109.80	1.18	47.20	nr	**157.00**
8 gang	106.63	131.00	1.63	65.19	nr	**196.19**
9 gang	133.74	164.30	1.83	73.20	nr	**237.50**
12 gang	159.28	195.68	2.29	91.60	nr	**287.28**
6 amp modular type swtich; galvanized steel box; bronze or satin chrome coverplate; flush mounting; two way						
1 gang	23.39	28.74	0.43	17.20	nr	**45.94**
2 gang	32.05	39.37	0.55	22.00	nr	**61.37**
3 gang	46.31	56.89	0.77	30.80	nr	**87.69**

ELECTRICAL SUPPLY/POWER/LIGHTING

Item	Net Price £	Material £	Labour hours	Labour £	Unit	**Total rate £**
4 gang	55.13	67.73	0.88	35.20	nr	**102.93**
6 gang	92.95	114.19	1.18	47.20	nr	**161.39**
8 gang	110.88	136.22	1.63	65.19	nr	**201.41**
9 gang	139.10	170.89	1.83	73.20	nr	**244.09**
12 gang	165.66	203.51	2.22	88.80	nr	**292.31**
Plate switches; 10 amp flush mounted, white plastic fronted; 16 mm metal box; fitted brass earth terminal						
1 gang 1 way, single pole	8.59	10.55	0.28	11.20	nr	**21.75**
1 gang 2 way, single pole	9.62	11.82	0.33	13.20	nr	**25.02**
2 gang 2 way, single pole	12.02	14.76	0.44	17.61	nr	**32.37**
3 gang 2 way, single pole	14.10	17.32	0.56	22.40	nr	**39.72**
1 gang Intermediate	19.02	23.36	0.43	17.20	nr	**40.56**
1 gang 1 Way, Double Pole	16.44	20.19	0.33	13.20	nr	**33.39**
1 gang Single Pole with bell symbol	13.15	16.16	0.23	9.20	nr	**25.36**
1 gang Single Pole marked PRESS	10.51	12.91	0.23	9.20	nr	**22.11**
Time delay switch, suppressed	58.25	71.56	0.49	19.59	nr	**91.15**
Plate switches; 6 amp flush mounted white plastic fronted; 25 mm metal box; fitted brass earth terminal						
4 gang 2 way, single pole	19.44	23.88	0.42	16.81	nr	**40.69**
6 gang 2 aay, single pole	31.08	38.18	0.47	18.80	nr	**56.98**
Architrave plate switches; 6 amp flush mounted, white plastic fronted; 27 mm metal box; brass earth terminal						
1 gang 2 way, single pole	3.52	4.33	0.30	12.00	nr	**16.33**
2 gang 2 way, single pole	6.97	8.56	0.36	14.40	nr	**22.96**
Ceiling switches, white moulded plastic, pull cord; standard unit						
6 amp, 1 way, single pole	5.05	6.21	0.32	12.80	nr	**19.01**
6 amp, 2 way, single pole	6.07	7.46	0.34	13.60	nr	**21.06**
16 amp, 1 way, double pole	9.09	11.17	0.37	14.81	nr	**25.98**
45 amp, 1 way, double pole with neon indicator	14.11	17.34	0.47	18.80	nr	**36.14**
10 amp splash proof moulded switch with plain, threaded or PVC entry						
1 gang, 2 way single pole	21.66	26.61	0.34	13.60	nr	**40.21**
2 gang, 1 way single pole	24.48	30.07	0.36	14.40	nr	**44.47**
2 gang, 2 way single pole	31.01	38.09	0.40	15.99	nr	**54.08**
6 amp watertight switch; metalclad; BS 3676; ingress protected to IP65 surface mounted						
1 gang, 2 way; terminal entry	20.90	25.67	0.41	16.40	nr	**42.07**
1 gang, 2 way; through entry	20.90	25.67	0.42	16.81	nr	**42.48**
2 gang, 2 way; terminal entry	61.63	75.72	0.54	21.60	nr	**97.32**
2 gang, 2 way; through entry	61.63	75.72	0.53	21.20	nr	**96.92**
2 way replacement switch	16.60	20.39	0.10	4.00	nr	**24.39**
15 amp watertight switch; metalclad; BS 3676; ingress protected to IP65; surface mounted						
1 gang 2 way, terminal entry	27.99	34.39	0.42	16.81	nr	**51.20**
1 gang 2 way, through entry	27.99	34.39	0.43	17.20	nr	**51.59**
2 gang 2 way, terminal entry	65.70	80.71	0.55	22.00	nr	**102.71**

ELECTRICAL SUPPLY/POWER/LIGHTING

Item	Net Price £	Material £	Labour hours	Labour £	Unit	Total rate £
GENERAL LIGHTING – cont						
Switches – cont						
2 gang 2 way, through entry	65.70	80.71	0.54	21.60	nr	**102.31**
Intermediate interior only	16.60	20.39	0.11	4.40	nr	**24.79**
2 way interior only	16.60	20.39	0.11	4.40	nr	**24.79**
Double pole interior only	16.60	20.39	0.11	4.40	nr	**24.79**
Electrical accessories; fixed to backgrounds (Including fixings)						
Dimmer switches; rotary action; for individual lights; moulded plastic case; metal backbox; flush mounted						
1 gang, 1 way; 250 Watt	17.42	21.40	0.28	11.20	nr	**32.60**
1 gang, 1 way; 400 Watt	22.94	28.18	0.28	11.20	nr	**39.38**
Dimmer switches; push on/off action; for individual lights; moulded plastic case; metal backbox; flush mounted						
1 gang, 2 way; 250 Watt	26.62	32.71	0.34	13.60	nr	**46.31**
3 gang, 2 way; 250 Watt	39.23	48.19	0.48	19.20	nr	**67.39**
4 gang, 2 way; 250 Watt	51.62	63.42	0.57	22.81	nr	**86.23**
Dimmer switches; rotary action; metal cald; metal backbox; BS 5518 and BS 800; flush mounted						
1 gang, 1 way; 400 Watt	48.35	59.40	0.33	13.20	nr	**72.60**
Ceiling roses						
Ceiling rose: white moulded plastic; flush fixed to conduit box						
Plug in type; ceiling socket with 2 terminals, loop-in and ceiling plug with 3 terminals and cover	9.28	11.40	0.34	13.60	nr	**25.00**
BC lampholder; white moulded plastic; heat resistent PVC insulated and sheathed cable; flush fixed						
2 core; 0.75 mm²	2.94	3.61	0.33	13.20	nr	**16.81**
Batten holder: white moulded plastic; 3 terminals; BS 5042; fixed to conduit						
Straight pattern; 2 terminals with loop-in and earth	4.30	5.28	0.29	11.59	nr	**16.87**
Angled pattern; looped in terminal	–	–	0.29	11.59	nr	**11.59**
LIGHTING CONTROL MODULES						
SPV.92 Lighting control module; plug in; 9 output, 9 channel – switching						
Base and lid assembly	156.47	192.22	2.05	82.00	nr	**274.22**
SPV.92 Lighting control module; plug in; 9 output, 9 channel – dimming (DALI, DSI, 1–10V)						
Base and lid assembly	191.36	235.08	2.05	82.00	nr	**317.08**

ELECTRICAL SUPPLY/POWER/LIGHTING

Item	Net Price £	Material £	Labour hours	Labour £	Unit	**Total rate £**
SPH.12+ Lighting control module; hard wired; 4 circuit switching						
Base and lid assembly	168.85	207.43	1.85	74.00	nr	**281.43**
SPH.27r Lighting control module; hard wired; 2 circuit 40 ballast drive DALI (with relays)						
Base and lid assembly	242.03	297.34	1.85	74.00	nr	**371.34**
SPH.27 Lighting control module; hard wired; 2 circuit 40 ballast drive DALI (without relays)						
Base and lid assembly	230.77	283.50	1.85	74.00	nr	**357.50**
SPV.56 Compact lighting control module; 3 output 18 ballast drive; dimmable (DALI)						
Base and lid assembly	213.87	262.74	1.85	74.00	nr	**336.74**
SPB.82 Blind control module; 8 outputs						
Base and lid assembly	168.85	207.43	1.85	74.00	nr	**281.43**
Interfaces						
SPO.20 DALI to DMX converter	225.13	276.57	1.85	74.00	nr	**350.57**
Dual bus presence detectors (DALI & E-Bus)						
SPU.7-S Dual-bus presence detector and infra-red sensor	56.28	69.14	0.60	24.00	nr	**93.14**
SPU.6-S Dual-bus universal sensor	61.92	76.07	0.60	24.00	nr	**100.07**
SPU.6-C Compact Dual-bus Universal Sensor	62.92	153.36	0.60	24.00	nr	**177.36**
Scene switch plate; anodized aluminium finish						
SPK.9-4+4 Eight button intelligent switch plate	112.56	138.28	2.00	80.00	nr	**218.28**
SPK.9-4-S Four button intelligent switch plate	67.54	82.97	1.60	64.01	nr	**146.98**
SPK.9-2-S Two button intelligent switch plate	56.28	69.14	1.20	48.00	nr	**117.14**
Input Devices/Interfaces						
SPF.4d Compact DALI interface for 4 gang retractive switch plate	28.15	34.59	1.20	48.00	nr	**82.59**
SPL.20 External Photocell 20,000 Lux range	135.08	165.94	1.50	60.00	nr	**225.94**
SPF.6-T AV interface; RS232 interface	146.34	179.78	1.85	74.00	nr	**253.78**
Sub-addressing for DALI strings						
Sub addressing charge per SPH.27R (40 Sub-addresses)	110.00	135.13	3.00	120.00	nr	**255.13**
Sub addressing charge per SPV.56 (20 Sub-addresses)	55.00	202.70	3.00	120.00	nr	**322.70**
Commissioning						
Not included – worked out based on complexity of project						

ELECTRICAL SUPPLY/POWER/LIGHTING

Item	Net Price £	Material £	Labour hours	Labour £	Unit	**Total rate £**
GENERAL LV POWER						
ACCESSORIES						
OUTLETS						
Socket outlet: unswitched; 13 amp metal clad; BS 1363; galvanized steel box and coverplate with white plastic inserts; fixed surface mounted						
1 gang	7.04	8.65	0.41	16.40	nr	**25.05**
2 gang	12.98	15.95	0.41	16.40	nr	**32.35**
Socket outlet: switched; 13 amp metal clad; BS 1363; galvanized steel box and coverplate with white plastic inserts; fixed surface mounted						
1 gang	12.96	15.92	0.43	17.20	nr	**33.12**
2 gang	24.00	29.48	0.45	18.00	nr	**47.48**
Socket outlet: switched with neon indicator; 13 amp metal clad; BS 1363; galvanized steel box and coverplate withwhite plastic inserts; fixed surface mounted						
1 gang	15.33	18.84	0.43	17.20	nr	**36.04**
2 gang	27.87	34.24	0.45	18.00	nr	**52.24**
Socket outlet: unswitched; 13 amp; BS 1363; white moulded plastic box and coverplate; fixed surface mounted						
1 gang	5.30	6.51	0.41	16.40	nr	**22.91**
2 gang	10.46	12.85	0.41	16.40	nr	**29.25**
Socket outlet; switched; 13 amp; BS 1363; white moulded plastic box and coverplate; fixed surface mounted						
1 gang	6.40	7.86	0.43	17.20	nr	**25.06**
2 gang	10.26	12.60	0.45	18.00	nr	**30.60**
Socket outlet: switched with neon indicator; 13 amp; BS 1363; white moulded plastic box and coverplate; fixed surface mounted						
1 gang	13.08	16.07	0.43	17.20	nr	**33.27**
2 gang	17.63	21.66	0.45	18.00	nr	**39.66**
Socket outlet: switched; 13 amp; BS 1363; galvanized steel box, white moulded coverplate; flush fitted						
1 gang	6.40	7.86	0.43	17.20	nr	**25.06**
2 gang	12.37	15.19	0.45	18.00	nr	**33.19**
Socket outlet: switched with neon indicator; 13 amp; BS 1363; galvanized steel box, white moulded coverplate; flush fixed						
1 gang	13.08	32.15	0.43	17.20	nr	**49.35**
2 gang	22.62	55.58	0.45	18.00	nr	**73.58**

ELECTRICAL SUPPLY/POWER/LIGHTING

Item	Net Price £	Material £	Labour hours	Labour £	Unit	**Total rate £**
Socket outlet: switched; 13 amp; BS 1363; galvanized steel box, satin chrome coverplate; BS 4662; flush fixed						
1 gang	21.35	26.23	0.43	17.20	nr	**43.43**
2 gang	30.28	37.20	0.45	18.00	nr	**55.20**
Socket outlet: switched with neon indicator; 13 amp; BS 1363; steel backbox, satin chrome coverplate; BS 4662; flush fixed						
1 gang	16.79	20.62	0.43	17.20	nr	**37.82**
2 gang	30.28	37.20	0.45	18.00	nr	**55.20**
RCD protected socket outlets, 13 amp, to BS 1363; galvanized steel box, white moulded cover plate; flush fitted						
2 gang, 10 mA tripping (active control)	78.01	95.83	0.45	18.00	nr	**113.83**
2 gang, 30 mA tripping (active control)	69.50	85.38	0.45	18.00	nr	**103.38**
2 gang, 30 mA tripping (passive control)	69.50	85.38	0.45	18.00	nr	**103.38**
Filtered socket outlets, 13 amp, to BS 1363, with separate 'clean earth' terminal; galvanized steel box, white moulded cover plate; flush fitted						
2 gang (spike protected)	58.74	72.17	0.50	20.00	nr	**92.17**
2 gang (spike and RFI protected)	74.36	91.35	0.55	22.00	nr	**113.35**
Replacement filter cassette	20.60	25.31	0.15	6.00	nr	**31.31**
Non-standard socket outlets, 13 amp, to BS 1363, with separate 'clean earth' terminal; for plugs with T-shaped earth pin; galvanized steel box, white moulded cover plate; flush fitted						
1 gang	10.90	13.39	0.43	17.20	nr	**30.59**
2 gang	19.62	24.11	0.43	17.20	nr	**41.31**
2 gang coloured RED	27.25	33.47	0.43	17.20	nr	**50.67**
Weatherproof socket outlet: 40 amp; switched; single gang; RCD protected; water and dust protected to I.P.66; surface mounted						
40 A 30 mA tripping current protecting 1 socket	88.77	109.05	0.52	20.80	nr	**129.85**
40 A 30 mA tripping current protecting 2 sockets	94.67	116.30	0.64	25.60	nr	**141.90**
Plug for weatherproof socket outlet: protected to I.P.66						
13 amp plug	3.89	4.78	0.21	8.40	nr	**13.18**
Floor service outlet box; comprising flat lid with flanged carpet trim; twin 13 A switched socket outlets; punched plate for mounting 2 telephone outlets; one blank plate; triple compartment						
3 compartment	39.96	49.09	0.88	35.20	nr	**84.29**

ELECTRICAL SUPPLY/POWER/LIGHTING

Item	Net Price £	Material £	Labour hours	Labour £	Unit	Total rate £
GENERAL LV POWER – cont						
Floor service outlet box; comprising flat lid with flanged carpet trim; 2 twin 13A switched socket outlets; punched plate for mounting 1 telephone outlet; one blank plate; triple compartment						
3 compartment	41.44	50.90	0.88	35.20	nr	**86.10**
Floor service outlet box; comprising flat lid with flanged carpet trim; twin 13A switched socket outlets; punched plate for mounting 2 telephone outlets; two blank plates; four compartment						
4 compartment	54.50	66.96	0.88	35.20	nr	**102.16**
Floor service outlet box; comprising flat lid with flanged carpet trim; single 13A unswitched socket outlet; single compartment; circular						
1 compartment	58.13	71.41	0.79	31.59	nr	**103.00**
Floor service grommet, comprising flat lid with flanged carpet trim; circular						
Floor Grommet	26.29	32.30	0.49	19.59	nr	**51.89**
POWER POSTS/POLES/PILLARS						
Power post						
Power post; aluminium painted body; PVC-u cover; 5 nr outlets	313.86	385.58	4.00	160.00	nr	**545.58**
Power pole						
Power pole; 3.6 m high; aluminium painted body; PVC-u cover; 6 nr outlets	402.45	494.41	4.00	160.00	nr	**654.41**
Extra for						
Power pole extension bar; 900 mm long	43.98	54.03	1.50	60.00	nr	**114.03**
Vertical multi-compartment pillar; PVC-u; BS 4678 Part4 EN60529; excludes accessories						
Single						
630 mm long	140.18	172.21	2.00	80.00	nr	**252.21**
3000 mm long	403.96	496.26	2.00	80.00	nr	**576.26**
Double						
630 mm long	140.18	172.21	3.00	120.00	nr	**292.21**
3000 mm long	428.74	526.71	3.00	120.00	nr	**646.71**
CONNECTION UNITS						
Connection units: moulded pattern; BS 5733; moulded plastic box; white coverplate; knockout for flex outlet; surface mounted – standard fused						
DP Switched	8.15	10.02	0.49	19.59	nr	**29.61**
Unswitched	7.49	9.20	0.49	19.59	nr	**28.79**
DP Switched with neon indicator	10.34	12.70	0.49	19.59	nr	**32.29**

ELECTRICAL SUPPLY/POWER/LIGHTING

Item	Net Price £	Material £	Labour hours	Labour £	Unit	**Total rate £**
Connection units: moulded pattern; BS 5733; galvanized steel box; white coverplate; knockout for flex outlet; surface mounted						
DP Switched	10.27	22.64	0.49	19.59	nr	**42.23**
DP Unswitched	9.61	11.80	0.49	19.59	nr	**31.39**
DP Switched with neon indicator	12.47	15.32	0.49	19.59	nr	**34.91**
Connection units: galvanized pressed steel pattern; galvanized steel box; satin chrome or satin brass finish; white moulded plastic inserts; flush mounted – standard fused						
DP Switched	15.04	18.48	0.49	19.59	nr	**38.07**
Unswitched	14.15	17.39	0.49	19.59	nr	**36.98**
DP Switched with neon indicator	20.31	24.95	0.49	19.59	nr	**44.54**
Connection units: galvanized steel box; satin chrome or satin brass finish; white moulded plastic inserts; flex outlet; flush mounted – standard fused						
Switched	14.44	17.73	0.49	19.59	nr	**37.32**
Unswitched	13.70	16.83	0.49	19.59	nr	**36.42**
Switched with neon indicator	18.57	22.82	0.49	19.59	nr	**42.41**
SHAVER SOCKETS						
Shaver unit: self-setting overload device; 200/250 voltage supply; white moulded plastic faceplate; unswitched						
Surface type with moulded plastic box	27.13	33.33	0.55	22.00	nr	**55.33**
Flush type with galvanized steel box	28.46	34.97	0.57	22.81	nr	**57.78**
Shaver unit: dual voltage supply unit; white moulded plastic faceplate; unswitched						
Surface type with moulded plastic box	32.71	40.18	0.62	24.80	nr	**64.98**
Flush type with galvanized steel box	34.01	41.78	0.64	25.60	nr	**67.38**
COOKER CONTROL UNITS						
Cooker control unit: BS 4177; 45 amp D.P. main switch; 13 amp switched socket outlet; metal coverplate; plastic inserts; neon indicators						
Surface mounted with mounting box	38.41	47.19	0.61	24.40	nr	**71.59**
Flush mounted with galvanized steel box	36.71	45.10	0.61	24.40	nr	**69.50**
Cooker control unit: BS 4177; 45 amp D.P. main switch; 13 amp switched socket outlet; moulded plastic box and coverplate; surface mounted						
Standard	26.65	32.74	0.61	24.40	nr	**57.14**
With neon indicators	31.28	38.43	0.61	24.40	nr	**62.83**

ELECTRICAL SUPPLY/POWER/LIGHTING

Item	Net Price £	Material £	Labour hours	Labour £	Unit	Total rate £
GENERAL LV POWER – cont						
CONTROL COMPONENTS						
Connector unit: moulded white plastic cover and block; galvanized steel back box; to immersion heaters						
3 kW up to 915 mm long; fitted to thermostat	28.56	35.09	0.75	30.01	nr	**65.10**
Water heater switch: 20 amp; switched with neon indicator						
DP Switched with neon indicator	13.94	36.21	0.45	18.00	nr	**54.21**
SWITCH DISCONNECTORS						
Switch disconnectors; moulded plastic enclosure; fixed to backgrounds						
3 pole; IP54; Grey						
16 amp	25.17	30.92	0.80	32.00	nr	**62.92**
25 amp	29.78	36.58	0.80	32.00	nr	**68.58**
40 amp	48.51	59.60	0.80	32.00	nr	**91.60**
63 amp	75.51	92.77	1.00	40.01	nr	**132.78**
80 amp	130.91	160.82	1.25	50.00	nr	**210.82**
6 pole; IP54; Grey						
25 amp	41.85	51.41	1.00	40.01	nr	**91.42**
63 amp	70.69	86.85	1.25	50.00	nr	**136.85**
80 amp	133.97	164.58	1.80	72.00	nr	**236.58**
3 pole; IP54; Yellow						
16 amp	27.66	33.98	0.80	32.00	nr	**65.98**
25 amp	32.70	40.17	0.80	32.00	nr	**72.17**
40 amp	53.04	65.16	0.80	32.00	nr	**97.16**
63 amp	82.53	101.39	1.00	40.01	nr	**141.40**
6 pole; IP54; Yellow						
25 amp	41.85	51.41	1.00	40.01	nr	**91.42**
INDUSTRIAL SOCKETS/PLUGS						
Plugs; Splashproof; 100–130 volts, 50–60 Hz; IP 44 (Yellow)						
2 pole and earth						
16 amp	4.48	5.50	0.55	22.00	nr	**27.50**
32 amp	12.11	14.88	0.60	24.00	nr	**38.88**
3 pole and earth						
16 amp	13.27	16.31	0.65	26.00	nr	**42.31**
32 amp	16.42	20.17	0.72	28.79	nr	**48.96**
3 pole; neutral and earth						
16 amp	13.35	16.40	0.72	28.79	nr	**45.19**
32 amp	19.44	23.88	0.78	31.21	nr	**55.09**

ELECTRICAL SUPPLY/POWER/LIGHTING

Item	Net Price £	Material £	Labour hours	Labour £	Unit	**Total rate £**
Connectors; Splashproof; 100–130 volts, 50–60 Hz; IP 44 (Yellow)						
2 pole and earth						
16 amp	6.00	7.37	0.42	16.81	nr	**24.18**
32 amp	10.38	12.75	0.50	20.00	nr	**32.75**
3 pole and earth						
16 amp	11.27	13.85	0.48	19.20	nr	**33.05**
32 amp	16.80	20.64	0.58	23.19	nr	**43.83**
3 pole; neutral and earth						
16 amp	15.77	19.37	0.52	20.80	nr	**40.17**
32 amp	22.13	27.18	0.73	29.20	nr	**56.38**
Angled sockets; surface mounted; Splashproof; 100–130 volts, 50–60 Hz; IP 44 (Yellow)						
2 pole and earth						
16 amp	5.06	6.22	0.55	22.00	nr	**28.22**
32 amp	12.93	15.89	0.60	24.00	nr	**39.89**
3 pole and earth						
16 amp	12.32	15.13	0.65	26.00	nr	**41.13**
32 amp	23.14	28.42	0.72	28.79	nr	**57.21**
3 pole; neutral and earth						
16 amp	17.09	21.00	0.72	28.79	nr	**49.79**
32 amp	22.39	27.51	0.78	31.21	nr	**58.72**
Plugs; Watertight; 100–130 volts, 50–60 Hz; IP67 (Yellow)						
2 pole and earth						
16 amp	9.69	11.91	0.55	22.00	nr	**33.91**
32 amp	16.88	20.74	0.60	24.00	nr	**44.74**
63 amp	46.29	56.87	0.75	30.00	nr	**86.87**
Connectors; Watertight; 100–130 volts, 50–60 Hz; IP 67 (Yellow)						
2 pole and earth						
16 amp	19.79	24.31	0.42	16.81	nr	**41.12**
32 amp	33.53	41.19	0.50	20.00	nr	**61.19**
63 amp	81.16	99.71	0.67	26.81	nr	**126.52**
Angled sockets; surface mounted; Watertight; 100–130 volts, 50–60 Hz; IP 67 (Yellow)						
2 pole and earth						
16 amp	16.36	20.10	0.55	22.00	nr	**42.10**
32 amp	31.97	39.27	0.60	24.00	nr	**63.27**

ELECTRICAL SUPPLY/POWER/LIGHTING

Item	Net Price £	Material £	Labour hours	Labour £	Unit	Total rate £
GENERAL LV POWER – cont						
Plugs; Splashproof; 200–250 volts, 50–60 Hz; IP 44 (Blue)						
2 pole and earth						
16 amp	2.33	2.87	0.55	22.00	nr	**24.87**
32 amp	8.38	10.29	0.60	24.00	nr	**34.29**
63 amp	40.36	49.58	0.75	30.00	nr	**79.58**
3 pole and earth						
16 amp	9.27	11.39	0.65	26.00	nr	**37.39**
32 amp	12.46	15.31	0.72	28.79	nr	**44.10**
63 amp	40.51	49.77	0.83	33.19	nr	**82.96**
3 pole; neutral and earth						
16 amp	9.89	12.15	0.72	28.79	nr	**40.94**
32 amp	15.00	18.43	0.78	31.21	nr	**49.64**
Connectors; Splashproof; 200–250 volts, 50–60 Hz; IP 44 (Blue)						
2 pole and earth						
16 amp	5.76	7.08	0.42	16.81	nr	**23.89**
32 amp	14.32	17.59	0.50	20.00	nr	**37.59**
63 amp	50.59	62.15	0.67	26.81	nr	**88.96**
3 pole and earth						
16 amp	15.26	18.74	0.48	19.20	nr	**37.94**
32 amp	20.18	24.79	0.58	23.19	nr	**47.98**
63 amp	41.62	51.14	0.75	30.00	nr	**81.14**
3 pole; neutral and earth						
16 amp	17.74	21.80	0.52	20.80	nr	**42.60**
32 amp	60.76	74.64	0.73	29.20	nr	**103.84**
Angled sockets; surface mounted; Splashproof; 200–250 volts, 50–60 Hz; IP 44 (Blue)						
2 pole and earth						
16 amp	7.58	9.31	0.55	22.00	nr	**31.31**
32 amp	13.12	16.12	0.60	24.00	nr	**40.12**
63 amp	57.66	70.83	0.75	30.00	nr	**100.83**
3 pole and earth						
16 amp	14.74	18.11	0.65	26.00	nr	**44.11**
32 amp	25.51	31.34	0.72	28.79	nr	**60.13**
63 amp	56.17	69.01	0.83	33.19	nr	**102.20**
3 pole; neutral and earth						
16 amp	12.83	15.76	0.72	28.79	nr	**44.55**
32 amp	22.44	27.56	0.78	31.21	nr	**58.77**
Plugs; Watertight; 200–250 volts, 50–60 Hz; IP67 (Blue)						
2 pole and earth						
16 amp	9.88	12.14	0.41	16.40	nr	**28.54**
32 amp	17.19	21.12	0.50	20.00	nr	**41.12**
63 amp	47.21	58.00	0.66	26.40	nr	**84.40**
125 amp	121.03	148.69	0.86	34.40	nr	**183.09**

ELECTRICAL SUPPLY/POWER/LIGHTING

Item	Net Price £	Material £	Labour hours	Labour £	Unit	Total rate £
Connectors; Watertight; 200–250 volts, 50–60 Hz; IP 67 (Blue)						
2 pole and earth						
16 amp	9.88	12.14	0.41	16.40	nr	**28.54**
32 amp	16.85	20.70	0.50	20.00	nr	**40.70**
63 amp	65.34	80.27	0.67	26.81	nr	**107.08**
125 amp	171.39	210.56	0.87	34.81	nr	**245.37**
Angled sockets; surface mounted; Watertight; 200–250 volts, 50–60 Hz; IP 67 (Blue)						
2 pole and earth						
16 amp	16.36	20.10	0.55	22.00	nr	**42.10**
32 amp	31.96	39.26	0.60	24.00	nr	**63.26**
125 amp	232.83	286.03	1.00	40.01	nr	**326.04**

ELECTRICAL SUPPLY/POWER/LIGHTING

Item	Net Price £	Material £	Labour hours	Labour £	Unit	Total rate £
UNINTERRUPTIBLE POWER SUPPLY						
Uninterruptible power supply; sheet steel enclosure; self contained battery pack; including installation, testing and commissioning						
Single phase input and output; 5 year battery life; standard 13 A socket outlet connection						
1.0 kVA (10 minute supply)	1009.80	1240.54	0.30	12.00	nr	**1252.54**
1.0 kVA (30 minute supply)	1439.13	1767.97	0.50	20.00	nr	**1787.97**
2.0 kVA (10 minute supply)	1801.58	2213.24	0.50	20.00	nr	**2233.24**
2.0 kVA (60 minute supply)	3023.57	3714.46	0.50	20.00	nr	**3734.46**
3.0 kVA (10 minute supply)	2380.62	2924.60	0.50	20.00	nr	**2944.60**
3.0 kVA (40 minute supply)	3073.95	3776.35	1.00	40.01	nr	**3816.36**
5.0 kVA (30 minute supply)	3896.64	4787.02	1.00	40.01	nr	**4827.03**
8.0 kVA (10 minute supply)	4687.21	5758.24	2.00	80.00	nr	**5838.24**
8.0 kVA (30 minute supply)	6338.20	7786.47	2.00	80.00	nr	**7866.47**
Uninterruptible power supply; including final connections and testing and commissioning						
Medium size static; single phase input and output; 10 year battery life; in cubicle						
10.0 kVA (10 minute supply)	4687.21	5758.24	10.00	400.00	nr	**6158.24**
10.0 kVA (30 minute supply)	6338.20	7786.47	15.00	600.00	nr	**8386.47**
15.0 kVA (10 minute supply)	7656.32	9405.78	10.00	400.00	nr	**9805.78**
15.0 kVA (30 minute supply)	10582.88	13001.07	15.00	600.00	nr	**13601.07**
20.0 kVA (10 minute supply)	7656.32	9405.78	10.00	400.00	nr	**9805.78**
20.0 kVA (30 minute supply)	13622.56	16735.32	15.00	600.00	nr	**17335.32**
Medium size static; three phase input and output; 10 year battery life; in cubicle						
10.0 kVA (10 minute supply)	6326.27	7771.83	10.00	400.00	nr	**8171.83**
10.0 kVA (30 minute supply)	7925.01	9735.87	15.00	600.00	nr	**10335.87**
15.0 kVA (10 minute supply)	7215.01	8863.64	15.00	600.00	nr	**9463.64**
15.0 kVA (30 minute supply)	8916.00	10953.31	20.00	799.99	nr	**11753.30**
20.0 kVA (10 minute supply)	7284.36	8948.83	20.00	799.99	nr	**9748.82**
20.0 kVA (30 minute supply)	11021.56	13539.99	25.00	1000.00	nr	**14539.99**
30.0 kVA (10 minute supply)	10900.94	13391.81	25.00	1000.00	nr	**14391.81**
30.0 kVA (30 minute supply)	13703.64	16834.92	30.00	1200.00	nr	**18034.92**
Large size static; three phase input and output; 10 year battery life; in cubicle						
40 kVA (10 minute supply)	10900.94	13391.81	30.00	1200.00	nr	**14591.81**
40 kVA (30 minute supply)	16465.63	20228.03	30.00	1200.00	nr	**21428.03**
60 kVA (10 minute supply)	17534.44	21541.05	35.00	1400.00	nr	**22941.05**
60 kVA (30 minute supply)	20422.88	25089.51	35.00	1400.00	nr	**26489.51**
100 kVA (10 minute supply)	25769.04	31657.27	40.00	1600.00	nr	**33257.27**
200 kVA (10 minute supply)	42527.60	52245.15	40.00	1600.00	nr	**53845.15**
300 kVA (10 minute supply)	69264.16	85091.02	40.00	1600.00	nr	**86691.02**
400 kVA (10 minute supply)	81269.44	99839.50	50.00	2000.00	nr	**101839.50**
500 kVA (10 minute supply)	95375.84	117169.22	60.00	2400.00	nr	**119569.22**
600 kVA (10 minute supply)	128313.92	157633.65	70.00	2799.99	nr	**160433.64**
800 kVA (10 minute supply)	160248.48	196865.26	80.00	3200.00	nr	**200065.26**

ELECTRICAL SUPPLY/POWER/LIGHTING

Item	Net Price £	Material £	Labour hours	Labour £	Unit	**Total rate £**
Integral diesel rotary; 400 V three phase input and output; no break supply; including ventilation and acoustic attenuation oil day tank and interconnecting pipework						
300 kVA	1226775.00	1507093.09	120.00	4799.99	nr	**1511893.08**
500 kVA	482084.90	592241.30	120.00	4799.99	nr	**597041.29**
800 kVA	526373.10	646649.36	140.00	5600.00	nr	**652249.36**
1000 kVA	583976.80	717415.50	140.00	5600.00	nr	**723015.50**
1670 kVA	695608.10	854554.55	170.00	6799.99	nr	**861354.54**
2000 kVA	847419.10	1041054.37	200.00	7999.99	nr	**1049054.36**
2500 kVA	979336.60	1203115.01	230.00	9200.00	nr	**1212315.01**

ELECTRICAL SUPPLY/POWER/LIGHTING

Item	Net Price £	Material £	Labour hours	Labour £	Unit	Total rate £
EMERGENCY LIGHTING						
24 Volt/50 Volt/110 Volt fluorescent slave luminaires						
For use with DC central battery systems						
Indoor, 8 Watt	40.36	49.58	0.80	32.00	nr	**81.58**
Indoor, exit sign box	50.27	61.76	0.80	32.00	nr	**93.76**
Outdoor, 8 Watt weatherproof	44.97	55.24	0.80	32.00	nr	**87.24**
Conversion module AC/DC	50.27	61.76	0.25	10.00	nr	**71.76**
Self-contained; polycarbonate base and diffuser; LED charging light to European sign directive; 3 hour standby						
Non-maintained						
Indoor, 8 Watt	44.56	54.75	1.00	40.01	nr	**94.76**
Outdoor, 8 Watt weatherproof, vandal resistant IP65	63.97	78.58	1.00	40.01	nr	**118.59**
Maintained						
Indoor, 8 Watt	31.89	39.18	1.00	40.01	nr	**79.19**
Outdoor, 8 Watt weatherproof, vandal resistant IP65	94.37	115.93	1.00	40.01	nr	**155.94**
Exit signage						
Exit sign; gold effect, pendular including brackets						
Non-maintained, 8 Watt	121.17	148.86	1.00	40.01	nr	**188.87**
Maintained, 8 Watt	130.43	160.23	1.00	40.01	nr	**200.24**
Modification kit						
Module & battery for 58 W fluorescent modification from mains fitting to emergency; 3 hour standby	41.77	51.31	0.50	20.00	nr	**71.31**
Extra for remote box (when fitting is too small for modification)	20.73	25.46	0.50	20.00	nr	**45.46**
12 Volt low voltage lighting; non-maintained; 3 hour standby						
2 × 20 Watt lamp load	143.91	176.79	1.20	48.00	nr	**224.79**
1 × 50 Watt lamp load	157.10	193.00	1.00	40.01	nr	**233.01**
Maintained 3 hour standby						
2 × 20 Watt lamp load	136.23	167.36	1.20	48.00	nr	**215.36**
1 × 50 Watt lamp load	149.41	183.55	1.00	40.01	nr	**223.56**
DC central battery systems BS5266 compliant 24/50/110 Volt						
DC supply to luminaires on mains failure; metal cubicle with battery charger, changeover device and battery as integral unit; ICEL 1001 compliant; 10 year design life valve regulated lead acid battery; 24 hour recharge; LCD display & LED indication; ICEL alarm pack; Includes on-site commissioning on 110 Volt systems only						

ELECTRICAL SUPPLY/POWER/LIGHTING

Item	Net Price £	Material £	Labour hours	Labour £	Unit	Total rate £
24 Volt, wall mounted						
300 W maintained, 1 hour	2294.82	2819.19	4.00	160.00	nr	**2979.19**
635 W maintained, 3 hour	2784.04	3420.20	6.00	240.00	nr	**3660.20**
470 W non-maintained, 1 hour	2045.92	2513.42	4.00	160.00	nr	**2673.42**
780 W non-maintained, 3 hour	2417.12	2969.43	6.00	240.00	nr	**3209.43**
50 Volt						
935 W maintained, 3 hour	2198.26	2700.56	8.00	320.00	nr	**3020.56**
1965 W maintained, 3 hour	3059.76	3758.92	8.00	320.00	nr	**4078.92**
1311 W non-maintained, 3 hour	2995.39	3679.84	8.00	320.00	nr	**3999.84**
2510 W non-maintained 3, hour	4574.62	5619.93	8.00	320.00	nr	**5939.93**
110 Volt						
1603 W maintained, 3 hour	4311.77	5297.01	8.00	320.00	nr	**5617.01**
4446 W maintained, 3 hour	4924.36	6049.57	10.00	400.00	nr	**6449.57**
2492 W non-maintained, 3 hour	5470.44	6720.43	10.00	400.00	nr	**7120.43**
5429 W non-maintained, 3 hour	7467.01	9173.22	12.00	480.00	nr	**9653.22**
DC central battery systems; BS EN 50171 compliant; 24/50/110 Volt						
Central power systems						
DC supply to luminaires on mains failure; metal cubicle with battery charger, changeover device and battery as integral unit; 10 year design life valve regulated lead acid battery; 12 hour recharge to 80% of specified duty; low volts discount; LCD display & LED indication; includes on-site commissioning for CPS systems only; battery sized for 'end of life' @ 20°C test pushbutton						
24 Volt, floor standing						
400 W non-maintained, 1 hour	2155.35	2647.85	4.00	160.00	nr	**2807.85**
600 W maintained, 3 hour	2771.16	3404.37	6.00	240.00	nr	**3644.37**
50 Volt						
2133 W non-maintained, 3 hour	4844.97	5952.04	8.00	320.00	nr	**6272.04**
1900 W maintained, 3 hour	4630.40	5688.45	8.00	320.00	nr	**6008.45**
110 Volt						
2200 W non-maintained, 3 hour	4481.28	5505.25	8.00	320.00	nr	**5825.25**
4000 W maintained, 3 hour	5001.61	6144.47	12.00	480.00	nr	**6624.47**

ELECTRICAL SUPPLY/POWER/LIGHTING

Item	Net Price £	Material £	Labour hours	Labour £	Unit	Total rate £
EMERGENCY LIGHTING – cont						
Low power systems						
DC supply to luminaires on mains failure; metal cubicle with battery charger, changeover device and battery as integral unit; 5 year design life valve regulated lead acid battery; low volts discount; LED display & LED indication; battery sized for 'end of life' @ 20°C test pushbutton						
24 Volt, floor standing						
300 W non-maintained, 1 hour	2363.48	2903.53	4.00	160.00	nr	**3063.53**
600 W maintained, 3 hour	2616.67	3214.57	6.00	240.00	nr	**3454.57**
AC static inverter system; BS5266 compliant; one hour standby						
Central system supplying AC power on mains failure to mains luminaires; ICEL 1001 compliant metal cubicle(s) with changeover device, battery charger, battery & static inverter; 10 year design life valve regulated lead acid battery; 24 hour recharge; LED indication and LCD display; pure sinewave output						
One hour						
750 kVA, 600 kW single phase I/P & O/P	2792.62	3430.74	6.00	240.00	nr	**3670.74**
3 kVA, 2.55 kW single phase I/P & O/P	4581.05	5627.82	8.00	320.00	nr	**5947.82**
5 kVA, 4.25 kW single phase I/P & O/P	6049.78	7432.15	10.00	400.00	nr	**7832.15**
8 kVA, 6.80 kW single phase I/P & O/P	7126.92	8755.43	12.00	480.00	nr	**9235.43**
10 kVA, 8.5 kW single phase I/P & O/P	9200.73	11303.09	14.00	560.00	nr	**11863.09**
13 kVA, 11.05 kW single phase I/P & O/P	12308.76	15121.31	16.00	640.00	nr	**15761.31**
15 kVA, 12.75 kW single phase I/P & O/P	12757.21	15672.24	30.00	1200.00	nr	**16872.24**
20 kVA, 17.0 kW 3 phase I/P & single phase O/P	17619.35	21645.37	40.00	1600.00	nr	**23245.37**
30 kVA, 25.5 kW 3 phase I/P & O/P	23810.75	29251.51	60.00	2400.00	nr	**31651.51**
40 kVA, 34.0 kW 3 phase I/P & O/P	29684.58	36467.51	80.00	3200.00	nr	**39667.51**
50 kVA, 42.5 kW 3 phase I/P & O/P	38715.80	47562.36	90.00	3600.00	nr	**51162.36**
65 kVA, 55.25 kW 3 phase I/P & O/P	47493.83	58346.17	100.00	4000.00	nr	**62346.17**
90 kVA, 68.85 kW 3 phase I/P & O/P	62761.50	77102.50	120.00	4799.99	nr	**81902.49**
120 kVA, 102 kW 3 phase I/P & O/P	70056.86	86064.86	150.00	5999.99	nr	**92064.85**
Three hour						
750 VA, 600 W single phase I/P & O/P	3267.89	4014.60	6.00	240.00	nr	**4254.60**
3 kVA, 2.55 kW single phase I/P & O/P	5499.41	6756.03	8.00	320.00	nr	**7076.03**
5 kVA, 4.25 kW single phase I/P & O/P	8717.95	10710.00	10.00	400.00	nr	**11110.00**
8 kVA, 6.80 kW single phase I/P & O/P	10846.48	13324.90	12.00	480.00	nr	**13804.90**
10 kVA, 8.5 kW single phase I/P & O/P	14307.48	17576.74	14.00	560.00	nr	**18136.74**
13 kVA, 11.05 kW single phase I/P & O/P	17641.88	21673.05	16.00	640.00	nr	**22313.05**
15 kVA, 12.75 kW single phase I/P & O/P	17931.55	22028.91	30.00	1200.00	nr	**23228.91**

ELECTRICAL SUPPLY/POWER/LIGHTING

Item	Net Price £	Material £	Labour hours	Labour £	Unit	Total rate £
20 kVA, 17.0 kW 3 phase I/P & single phase O/P	25967.17	31900.67	40.00	1600.00	nr	**33500.67**
30 kVA, 25.5 kW 3 phase I/P & O/P	39020.49	47936.67	60.00	2400.00	nr	**50336.67**
40 kVA, 34.0 kW 3 phase I/P & O/P	46651.64	57311.54	80.00	3200.00	nr	**60511.54**
50 kVA, 42.5 kW 3 phase I/P & O/P	62922.43	77300.20	90.00	3600.00	nr	**80900.20**
65 kVA, 55.25 kW 3 phase I/P & O/P	76079.82	93464.06	100.00	4000.00	nr	**97464.06**
90 kVA, 68.85 kW 3 phase I/P & O/P	90040.76	110615.07	120.00	4799.99	nr	**115415.06**
120 kVA, 102 kW 3 phase I/P & O/P	110977.36	136335.69	150.00	5999.99	nr	**142335.68**
AC static inverter system; BS EN 50171 compliant; one hour standby; Low power system (typically wall mounted)						
Central system supplying AC power on mains failure to mains luminaires; metal cubicle(s) with changeover device, battery charger, battery & static inverter; 5 year design life valve regulated lead acid battery; LED indication and LCD display; 12 hour recharge to 80% duty; inverter rated for 120% of load for 100% of duty; battery sized for 'end of life' @ 20°C test pushbutton						
One hour						
300 VA, 240 W single phase I/P & O/P	1346.74	1654.47	3.00	120.00	nr	**1774.47**
600 VA, 480 W single phase I/P & O/P	1623.26	1994.17	4.00	160.00	nr	**2154.17**
750 VA, 600 W single phase I/P & O/P	2995.45	3679.91	6.00	240.00	nr	**3919.91**
Three hour						
150 VA, 120 W single phase I/P & O/P	1484.42	1823.61	3.00	120.00	nr	**1943.61**
450 VA, 360 W single phase I/P & O/P	1760.94	2163.32	4.00	160.00	nr	**2323.32**
750 VA, 600 W single phase I/P & O/P	3505.68	4306.73	6.00	240.00	nr	**4546.73**
AC static inverter system central power system; CPS BS EN 50171 compliant; one hour standby						
Central system supplying AC power on mains failure to mains luminaires; metal cubicle(s) with changeover device, battery charger, battery & static inverter; LED indication and LCD display; pure sinewave output; 10 year design life valve regulated lead acid battery; 12 hour recharge to 80% duty specified; unverter rated for 120% of load for 100% of duty; battery sized for 'end of life' @ 20°C test push button; includes on-site commissioning						
One hour						
750 VA, 600 W single phase I/P & O/P	3351.80	4117.69	6.00	240.00	nr	**4357.69**
3 kVA, 2.55 kW single phase I/P & O/P	5340.67	6561.01	8.00	320.00	nr	**6881.01**
5 kVA, 4.25 kW single phase I/P & O/P	6915.34	8495.50	10.00	400.00	nr	**8895.50**
8 kVA, 6.80 kW single phase I/P & O/P	8071.17	9915.43	12.00	480.00	nr	**10395.43**
10 kVA, 8.5 kW single phase I/P & O/P	10364.33	12732.58	14.00	560.00	nr	**13292.58**
13 kVA, 11.05 kW single phase I/P & O/P	13698.77	16828.94	16.00	640.00	nr	**17468.94**
15 kVA, 12.75 kW single phase I/P & O/P	14178.92	17418.81	30.00	1200.00	nr	**18618.81**

ELECTRICAL SUPPLY/POWER/LIGHTING

Item	Net Price £	Material £	Labour hours	Labour £	Unit	Total rate £
EMERGENCY LIGHTING – cont						
AC static inverter system central power system – cont						
20 kVA, 17.0 kW 3 phase I/P & single phase O/P	19532.32	23995.45	40.00	1600.00	nr	**25595.45**
30 kVA, 25.5 kW 3 phase I/P & O/P	26311.13	32323.22	60.00	2400.00	nr	**34723.22**
40 kVA, 34.0 kW 3 phase I/P & O/P	32610.95	40062.55	80.00	3200.00	nr	**43262.55**
50 kVA, 42.5 kW 3 phase I/P & O/P	42470.83	52175.41	90.00	3600.00	nr	**55775.41**
65 kVA, 55.25 kW 3 phase I/P & O/P	52456.81	64443.19	100.00	4000.00	nr	**68443.19**
90 kVA, 68.85 kW 3 phase I/P & O/P	68834.02	84562.59	120.00	4799.99	nr	**89362.58**
120 kVA, 102 kW 3 phase I/P & O/P	76659.91	94176.69	150.00	5999.99	nr	**100176.68**
Three hour						
750 VA, 600 W single phase I/P & O/P	3878.23	4764.41	6.00	240.00	nr	**5004.41**
3 kVA, 2.55 kW single phase I/P & O/P	6325.27	7770.60	8.00	320.00	nr	**8090.60**
5 kVA, 4.25 kW single phase I/P & O/P	9800.87	12040.37	10.00	400.00	nr	**12440.37**
8 kVA, 6.80 kW single phase I/P & O/P	12061.63	14817.72	12.00	480.00	nr	**15297.72**
10 kVA, 8.5 kW single phase I/P & O/P	15843.83	19464.14	14.00	560.00	nr	**20024.14**
13 kVA, 11.05 kW single phase I/P & O/P	19420.09	23857.59	16.00	640.00	nr	**24497.59**
15 kVA, 12.75 kW single phase I/P & O/P	19729.01	24237.09	30.00	1200.00	nr	**25437.09**
20 kVA, 17.0 kW 3 phase I/P & single phase O/P	28750.07	35319.46	40.00	1600.00	nr	**36919.46**
30 kVA, 25.5 kW 3 phase I/P & O/P	42624.71	52364.46	60.00	2400.00	nr	**54764.46**
40 kVA, 34.0 kW 3 phase I/P & O/P	50811.57	62422.02	80.00	3200.00	nr	**65622.02**
50 kVA, 42.5 kW 3 phase I/P & O/P	68436.02	84073.65	90.00	3600.00	nr	**87673.65**
65 kVA, 55.25 kW 3 phase I/P & O/P	83120.55	102113.59	100.00	4000.00	nr	**106113.59**
90 kVA, 68.85 kW 3 phase I/P & O/P	103337.81	126950.50	120.00	4799.99	nr	**131750.49**
120 kVA, 102 kW 3 phase I/P & O/P	120551.52	148097.54	150.00	5999.99	nr	**154097.53**

COMMUNICATIONS/SECURITY/CONTROL

Item	Net Price £	Material £	Labour hours	Labour £	Unit	**Total rate £**
TELECOMMUNICATIONS						
CABLES						
Multipair internal telephone cable; BS 6746; loose laid on tray/basket						
0.5 mm diameter conductor LSZH insulated and sheathed multipair cables; BT specification CW 1308						
3 pair	0.18	0.22	0.03	1.21	m	**1.43**
4 pair	0.21	0.26	0.03	1.21	m	**1.47**
6 pair	0.30	0.37	0.03	1.21	m	**1.58**
10 pair	0.54	0.66	0.05	2.01	m	**2.67**
15 pair	0.81	1.00	0.06	2.39	m	**3.39**
20 pair + 1 wire	1.14	1.40	0.06	2.39	m	**3.79**
25 pair	1.45	1.78	0.08	3.19	m	**4.97**
40 pair + earth	2.05	2.52	0.08	3.19	m	**5.71**
50 pair + earth	2.47	3.03	0.10	4.00	m	**7.03**
80 pair + earth	3.31	4.06	0.10	4.00	m	**8.06**
100 pair + earth	5.25	6.45	0.12	4.80	m	**11.25**
Multipair internal telephone cable; BS 6746; installed in conduit/trunking						
0.5 mm diameter conductor LSZH insulated and sheathed multipair cables; BT specification CW 1308						
3 pair	0.18	0.22	0.05	2.01	m	**2.23**
4 pair	0.21	0.26	0.06	2.39	m	**2.65**
6 pair	0.30	0.37	0.06	2.39	m	**2.76**
10 pair	0.54	0.66	0.07	2.80	m	**3.46**
15 pair	0.81	1.00	0.07	2.80	m	**3.80**
20 pair + 1 wire	1.14	1.40	0.09	3.60	m	**5.00**
25 pair	1.45	1.78	0.10	4.00	m	**5.78**
40 pair + earth	2.05	2.52	0.12	4.80	m	**7.32**
50 pair + earth	2.47	3.03	0.14	5.61	m	**8.64**
80 pair + earth	3.31	4.06	0.14	5.61	m	**9.67**
100 pair + earth	5.25	6.45	0.15	6.00	m	**12.45**
Low speed data; unshielded twisted pair; solid copper conductors; LSOH sheath; nominal impedance 100 Ohm; Category 3 to ISO IS 1801/EIA/TIA 568B and EN50173/50174 standards to current revisions						
Installed in riser						
25 pair 24 AWG	1.60	1.96	0.03	1.21	m	**3.17**
50 pair 24 AWG	2.68	3.30	0.06	2.39	m	**5.69**
100 pair 24 AWG	5.99	7.36	0.10	4.00	m	**11.36**
Installed below floor						
25 pair 24 AWG	1.60	1.96	0.02	0.80	m	**2.76**
50 pair 24 AWG	2.68	3.30	0.05	2.01	m	**5.31**
100 pair 24 AWG	5.99	7.36	0.08	3.19	m	**10.55**

COMMUNICATIONS/SECURITY/CONTROL

Item	Net Price £	Material £	Labour hours	Labour £	Unit	Total rate £
TELECOMMUNICATIONS – cont						
ACCESSORIES						
Telephone outlet: moulded plastic plate with box; fitted and connected; flush or surface mounted						
Single master outlet	6.79	8.34	0.35	14.01	nr	**22.35**
Single secondary outlet	5.02	6.16	0.35	14.01	nr	**20.17**
Telephone outlet: bronze or satin chromeplate; with box; fitted and connected; flush or surface mounted						
Single master outlet	11.06	13.59	0.35	14.01	nr	**27.60**
Single secondary outlet	12.09	14.86	0.35	14.01	nr	**28.87**
Frames and Box Connections						
Provision and Installation of a Dual Vertical Krone 108A Voice Distribution Frame that can accommodate a total of 138 × Krone 237A Strips	229.84	282.36	1.50	60.00	nr	**342.36**
Label Frame (Traffolyte style)	1.45	1.78	0.27	10.79	nr	**12.57**
Provision and Installation of a Box Connection 301A Voice Termination Unit that can accommodate a total of 10 × Krone 237A Strips	19.93	24.49	0.25	10.00	nr	**34.49**
Label Frame (Traffolyte style)	0.47	0.58	0.08	3.32	nr	**3.90**
Provision and Installation of a Box Connection 201 Voice Termination Unit that can accommodate 20 pairs	10.61	13.03	0.17	6.65	nr	**19.68**
Label Frame (Traffolyte style)	0.47	0.58	0.08	3.32	nr	**3.90**
Terminate, Test and Label Voice Multicore System						
Patch Panels						
Voice; 19" wide fully loaded, finished in black including termination and forming of cables (assuming 2 pairs per port)						
25 port – RJ45 UTP – Krone	49.88	61.28	2.60	104.00	nr	**165.28**
50 port – RJ45 UTP – Krone	67.99	83.53	4.65	186.00	nr	**269.53**
900 pair fully loaded Systimax style of frame including forming and termination of 9 × 100 pair cables	561.75	690.11	25.00	1000.00	nr	**1690.11**
Installation and termination of Krone Strip (10 pair block – 237A) including designation label strip	3.35	4.12	0.50	20.00	nr	**24.12**
Patch panel and Outlet labelling per port (Traffolyte style)	0.17	0.21	0.02	0.80	nr	**1.01**
Provision and installation of a voice jumper, for cross termination on Krone Termination Strips	0.05	0.07	0.06	2.39	nr	**2.46**
CW1308/Cat 3 cable circuit test per pair	–	–	0.04	1.60	nr	**1.60**

COMMUNICATIONS/SECURITY/CONTROL

Item	Net Price £	Material £	Labour hours	Labour £	Unit	Total rate £
RADIO AND TELEVISION						
RADIO						
CABLES						
Radio frequency cable; BS 2316; PVC sheathed; laid loose						
7/0.41 mm tinned copper inner conductor; solid polyethylene dielectric insulation; bare copper wire braid; PVC sheath; 75 ohm impedance						
Cable	1.02	1.25	0.05	2.01	m	**3.26**
Twin 1/0.58 mm copper covered steel solid core wire conductor; solid polyethylene dielectric insulation; barecopper wire braid; PVC sheath; 75 ohm impedance						
Cable	0.23	0.28	0.05	2.01	m	**2.29**
TELEVISION						
CABLES						
Television aerial cable; coaxial; PVC sheathed; fixed to backgrounds						
General purpose TV aerial downlead; copper stranded inner conductor; cellular polythene insulation; copper braid outer conductor; 75 ohm impedance						
7/0.25 mm	0.22	0.27	0.06	2.39	m	**2.66**
Low loss TV aerial downlead; solid copper inner conductor; cellular polythene insulation; copper braid outer; conductor; 75 ohm impedance						
1/1.12 mm	0.23	0.28	0.06	2.39	m	**2.67**
Low loss air spaced; solid copper inner conductor; air spaced polythene insulation; copper braid outer conductor; 75 ohm impedance						
1/1.00 mm	0.30	0.37	0.06	2.39	m	**2.76**
Satelite aerial downlead; solid copper inner conductor; air spaced polythene insulation; copper tape and braid outer conductor; 75 ohm impedance						
1/1.00 mm	0.27	0.34	0.06	2.39	m	**2.73**
Satelite TV coaxial; solid copper inner conductor; semi-air spaced polyethylene dielectric insulation; plain annealed copper foil and copper braid screen in outer conductor; PVC sheath; 75 ohm impedance						
1/1.25 mm	0.45	0.56	0.08	3.19	m	**3.75**

COMMUNICATIONS/SECURITY/CONTROL

Item	Net Price £	Material £	Labour hours	Labour £	Unit	Total rate £
RADIO AND TELEVISION – cont						
Television aerial cable – cont						
Satelite TV coaxial; solid copper inner conductor; air spaced polyethylene dielectric insulation; plain annealed copper foil and copper braid screen in outer conductor; PVC sheath; 75 ohm impedance						
1/1.67 mm	0.73	0.89	0.09	3.60	m	**4.49**
Video cable; PVC flame retardant sheath; laid loose						
7/0.1 mm silver coated copper covered annealed steel wire conductor; polyethylene dielectric insulation with tin coated copper wire braid; 75 ohm impedance						
Cable	0.17	0.21	0.05	2.01	m	**2.22**
ACCESSORIES						
TV co-axial socket outlet: moulded plastic box; flush or surface mounted						
One way Direct Connection	7.75	9.52	0.35	14.01	nr	**23.53**
Two way Direct Connection	10.80	13.27	0.35	14.01	nr	**27.28**
One way Isolated UHF/VHF	13.64	16.76	0.35	14.01	nr	**30.77**
Two way Isolated UHF/VHF	18.58	22.83	0.35	14.01	nr	**36.84**

COMMUNICATIONS/SECURITY/CONTROL

Item	Net Price £	Material £	Labour hours	Labour £	Unit	**Total rate £**
CLOCKS						
Master Clock						
Master clock module 230 V with programming circuits, NTP client and server for synchronization over Ethernet network, sounders Harmonys or Melodys, control of DHF and NTP clock systems						
Rack mounted	995.00	1222.36	3.00	120.00	nr	**1342.36**
Wall mounted	895.00	1099.51	5.00	200.00	nr	**1299.51**
Master clock module 230 V with programming circuits for control of wired and DHF clock systems, relays or sounders, NTP client and server for synchronization over Ethernet network						
Rack mounted	1344.00	1651.10	3.00	120.00	nr	**1771.10**
Wall mounted	1091.00	1340.29	5.00	200.00	nr	**1540.29**
Clocks and bells, master clock module 230 V with 3 programming circuits for control of wired and DHF clock systems, relays and sounders						
Rack mounted	867.00	1065.11	3.00	120.00	nr	**1185.11**
Wall mounted	686.00	842.75	5.00	200.00	nr	**1042.75**
Clocks only, master clock module 230 V for control of wired and DHF clock systems						
Rack mounted	656.00	805.90	3.00	120.00	nr	**925.90**
Wall mounted	510.00	626.54	5.00	200.00	nr	**826.54**
Extra over for						
GPS antenna with 20 m cable	258.00	316.95	2.50	100.00	nr	**416.95**
DHF transmitter	294.00	361.18	1.50	60.00	nr	**421.18**
DHF repeater, signal booster up to 200 m	337.00	414.00	3.00	120.00	nr	**534.00**
UPS	871.00	1070.02	5.00	200.00	nr	**1270.02**
Rack mounted external AC/DC power supply	501.00	615.48	3.00	120.00	nr	**735.48**
Indoor clock, 30 cm diameter						
Stand alone						
HMS display, quartz movement, battery operated	48.00	58.97	0.80	32.00	nr	**90.97**
DHF wireless						
Hour and minute display, battery operated	108.00	132.68	0.80	32.00	nr	**164.68**
Hour, minute and second display, battery operated	119.00	146.19	0.80	32.00	nr	**178.19**
HM display, TBT	151.00	185.50	0.80	32.00	nr	**217.50**
HMS display, TBT	121.00	148.65	0.80	32.00	nr	**180.65**
Slave wired						
Hour and minute display, 24V	65.00	79.85	1.00	40.01	nr	**119.86**
Hour, minute and second display, 24V	78.00	95.82	1.00	40.01	nr	**135.83**
Hour, minute display, TBT, AFNOR	107.00	131.45	1.00	40.01	nr	**171.46**

COMMUNICATIONS/SECURITY/CONTROL

Item	Net Price £	Material £	Labour hours	Labour £	Unit	Total rate £
CLOCKS – cont						
Hour, minute and second display, TBT AFNOR	119.00	146.19	1.00	40.01	nr	**186.20**
Hour, minute display, POE, NTP	151.00	185.50	1.00	40.01	nr	**225.51**
Hour, minute and second display, POE, NTP	155.00	190.42	1.00	40.01	nr	**230.43**
Indoor clock, 40 cm diameter						
Stand alone						
HMS display, quartz movement, battery operated	81.00	99.51	0.80	32.00	nr	**131.51**
DHF wireless						
Hour and minute display, battery operated	159.00	195.33	0.80	32.00	nr	**227.33**
Hour, minute and second display, battery operated	164.00	201.47	0.80	32.00	nr	**233.47**
HM display, TBT	159.00	195.33	0.80	32.00	nr	**227.33**
HMS display, TBT	170.00	208.85	0.80	32.00	nr	**240.85**
Slave wired						
Hour and minute display, 24 V	100.00	122.85	1.00	40.01	nr	**162.86**
Hour, minute and second display, 24 V	116.00	142.51	1.00	40.01	nr	**182.52**
Hour, minute display, TBT, AFNOR	159.00	195.33	1.00	40.01	nr	**235.34**
Hour, minute and second display, TBT AFNOR	174.00	213.76	1.00	40.01	nr	**253.77**
Hour, minute display, POE, NTP	222.00	272.73	1.00	40.01	nr	**312.74**
Hour, minute and second display, POE, NTP	226.00	277.64	1.00	40.01	nr	**317.65**
Digital clock, 7 cm digit height, 230 V						
Independent quartz movement						
Red display	196.00	240.79	0.80	32.00	nr	**272.79**
Green display	197.00	242.01	0.80	32.00	nr	**274.01**
Yellow display	197.00	242.01	0.80	32.00	nr	**274.01**
Blue display	328.00	402.95	0.80	32.00	nr	**434.95**
AFNOR, 230 V						
Red display	235.00	288.70	0.80	32.00	nr	**320.70**
Green display	235.00	288.70	0.80	32.00	nr	**320.70**
Yellow display	235.00	288.70	0.80	32.00	nr	**320.70**
Blue display	286.00	351.35	0.80	32.00	nr	**383.35**
DHF wireless, 230 V						
Red display	247.00	303.44	0.80	32.00	nr	**335.44**
Green display	247.00	303.44	0.80	32.00	nr	**335.44**
Yellow display	247.00	303.44	0.80	32.00	nr	**335.44**
Blue display	291.00	357.49	0.80	32.00	nr	**389.49**
NTP POE						
Red display	346.00	425.06	0.80	32.00	nr	**457.06**
Green display	346.00	425.06	0.80	32.00	nr	**457.06**
Yellow display	346.00	425.06	0.80	32.00	nr	**457.06**
Blue display	392.00	481.57	0.80	32.00	nr	**513.57**
Digital clock, 12 cm digit height, 230 V						
Independent quartz movement						
Red display	593.00	728.50	0.80	32.00	nr	**760.50**
Green display	593.00	728.50	0.80	32.00	nr	**760.50**
Yellow display	593.00	728.50	0.80	32.00	nr	**760.50**

COMMUNICATIONS/SECURITY/CONTROL

Item	Net Price £	Material £	Labour hours	Labour £	Unit	**Total rate £**
AFNOR, 230 V						
Red display	645.00	792.38	0.80	32.00	nr	**824.38**
Green display	634.00	778.87	0.80	32.00	nr	**810.87**
Yellow display	645.00	792.38	0.80	32.00	nr	**824.38**
DHF wireless, 230 V						
Red display	678.00	832.92	0.80	32.00	nr	**864.92**
Yellow display	677.00	831.69	0.80	32.00	nr	**863.69**
LCD digital clock, hour, minute and date						
Independent battery quartz movement	162.00	199.02	1.00	40.01	nr	**239.03**
DHS wireless, battery	227.00	278.87	1.00	40.01	nr	**318.88**
NTP, POE	309.00	379.61	1.00	40.01	nr	**419.62**

COMMUNICATIONS/SECURITY/CONTROL

Item	Net Price £	Material £	Labour hours	Labour £	Unit	Total rate £
DATA TRANSMISSION						
Cabinets						
Floor standing; suitable for 19” patch panels with glass lockable doors, metal rear doors, side panels, vertical cable management, 2 × 4 way PDUs, 4 way fan, earth bonding kit; installed on raised floor						
600 mm wide × 800 mm deep – 18U	681.41	837.11	3.00	120.00	nr	**957.11**
600 mm wide × 800 mm deep – 24U	709.80	871.99	3.00	120.00	nr	**991.99**
600 mm wide × 800 mm deep – 33U	762.32	936.51	4.00	160.00	nr	**1096.51**
600 mm wide × 800 mm deep – 42U	817.70	1004.55	4.00	160.00	nr	**1164.55**
600 mm wide × 800 mm deep – 47U	863.20	1060.44	4.00	160.00	nr	**1220.44**
800 mm wide × 800 mm deep – 42U	925.60	1137.10	4.00	160.00	nr	**1297.10**
800 mm wide × 800 mm deep – 47U	965.38	1185.96	4.00	160.00	nr	**1345.96**
Label cabinet	2.18	2.68	0.25	10.00	nr	**12.68**
Wall mounted; suitable for 19” patch panels with glass lockable doors, side panels, vertical cable management, 2 × 4 way PDUs, 4 way fan, earth bonding kit; fixed to wall						
19 mm wide × 500 mm deep – 9U	454.48	558.33	3.00	120.00	nr	**678.33**
19 mm wide × 500 mm deep – 12U	465.66	572.06	3.00	120.00	nr	**692.06**
19 mm wide × 500 mm deep – 15U	482.56	592.83	3.00	120.00	nr	**712.83**
19 mm wide × 500 mm deep – 18U	567.84	697.59	3.00	120.00	nr	**817.59**
19 mm wide × 500 mm deep – 21U	590.72	725.70	3.00	120.00	nr	**845.70**
Label cabinet	2.18	2.68	0.25	10.00	nr	**12.68**
Frames						
Floor standing; suitable for 19” patch panels with supports, vertical cable management, earth bonding kit; installed on raised floor						
19 mm wide × 500 mm deep – 25U	511.16	627.96	2.50	100.00	nr	**727.96**
19 mm wide × 500 mm deep – 39U	624.52	767.22	2.50	100.00	nr	**867.22**
19 mm wide × 500 mm deep – 42U	681.46	837.18	2.50	100.00	nr	**937.18**
19 mm wide × 500 mm deep – 47U	738.40	907.13	2.50	100.00	nr	**1007.13**
Label frame	2.18	2.68	0.25	10.00	nr	**12.68**
Copper Data Cabling						
Unshielded twisted pair; solid copper conductors; PVC insulation; nominal impedance 100 Ohm; Cat 5e to ISO 11801, EIA/TIA 568B and EN 50173/50174 standards to the current revisions						
4 pair 24 AWG; nominal outside diameter 5.6 mm; installed above ceiling	0.19	0.23	0.02	0.80	m	**1.03**
4 pair 24 AWG; nominal outside diameter 5.6 mm; installed in riser	0.19	0.23	0.02	0.80	m	**1.03**
4 pair 24 AWG; nominal outside diameter 5.6 mm; installed below floor	0.19	0.23	0.01	0.41	m	**0.64**
4 pair 24 AWG; nominal outside diameter 5.6 mm; installed in trunking	0.19	0.23	0.02	0.80	m	**1.03**

COMMUNICATIONS/SECURITY/CONTROL

Item	Net Price £	Material £	Labour hours	Labour £	Unit	**Total rate £**
Unshielded twisted pair; solid copper conductors; LSOH sheathed; nominal impedance 100 Ohm; Cat 5e to ISO 11801, EIA/TIA 568B and EN 50173/50174 standards to the current revisions						
4 pair 24 AWG; nominal outside diameter 5.6 mm; installed above ceiling	0.22	0.27	0.02	0.80	m	**1.07**
4 pair 24 AWG; nominal outside diameter 5.6 mm; installed in riser	0.22	0.27	0.02	0.80	m	**1.07**
4 pair 24 AWG; nominal outside diameter 5.6 mm; installed below floor	0.22	0.27	0.01	0.41	m	**0.68**
4 pair 24 AWG; nominal outside diameter 5.6 mm; installed in trunking	0.22	0.27	0.02	0.80	m	**1.07**
Unshielded twisted pair; solid copper conductors; PVC insulation; nominal impedance 100 Ohm; Cat 6 to ISO 11801, EIA/TIA 568B and EN 50173/50174 standards to the current revisions						
4 pair 24 AWG; nominal outside diameter 5.6 mm; installed above ceiling	0.29	0.36	0.02	0.60	m	**0.96**
4 pair 24 AWG; nominal outside diameter 5.6 mm; installed in riser	0.29	0.36	0.02	0.60	m	**0.96**
4 pair 24 AWG; nominal outside diameter 5.6 mm; installed below floor	0.29	0.36	0.01	0.32	m	**0.68**
4 pair 24 AWG; nominal outside diameter 5.6 mm; installed in trunking	0.29	0.36	0.02	0.80	m	**1.16**
Unshielded twisted pair; solid copper conductors; LSOH sheathed; nominal impedance 100 Ohm; Cat 6 to ISO 11801, EIA/TIA 568B and EN 50173/50174 standards to the current revisions						
4 pair 24 AWG; nominal outside diameter 5.6 mm; installed above ceiling	0.30	0.37	0.02	0.88	m	**1.25**
4 pair 24 AWG; nominal outside diameter 5.6 mm; installed in riser	0.30	0.37	0.02	0.88	m	**1.25**
4 pair 24 AWG; nominal outside diameter 5.6 mm; installed below floor	0.30	0.37	0.01	0.44	m	**0.81**
4 pair 24 AWG; nominal outside diameter 5.6 mm; installed in trunking	0.30	0.37	0.02	0.88	m	**1.25**
Patch Panels						
Category 5e; 19" wide fully loaded, finished in black including termination and forming of cables						
24 port – RJ45 UTP – Krone/110	50.73	62.32	4.75	190.00	nr	**252.32**
48 port – RJ45 UTP – Krone/110	95.21	116.97	9.35	374.00	nr	**490.97**
Patch panel labelling per port	0.22	0.27	0.02	0.80	nr	**1.07**

COMMUNICATIONS/SECURITY/CONTROL

Item	Net Price £	Material £	Labour hours	Labour £	Unit	Total rate £
DATA TRANSMISSION – cont						
Patch Panels – cont						
Category 6; 19" wide fully loaded, finished in black including termination and forming of cables						
24 port – RJ45 UTP – Krone/110	96.95	119.10	5.00	200.00	nr	**319.10**
48 port – RJ45 UTP – Krone/110	176.83	217.23	9.80	392.01	nr	**609.24**
Patch panel labelling per port	0.22	0.27	0.02	0.80	nr	**1.07**
Work Station						
Category 5e RJ45 data outlet plate and multiway outlet boxes for wall, ceiling and below floor installations including label to ISO 11801 standards						
Wall mounted; fully loaded						
One gang LSOH PVC plate	4.65	5.71	0.15	6.00	nr	**11.71**
Two gang LSOH PVC plate	6.64	8.16	0.20	8.00	nr	**16.16**
Four gang LSOH PVC plate	11.38	13.98	0.40	15.99	nr	**29.97**
One gang satin brass plate	15.08	18.52	0.25	10.00	nr	**28.52**
Two gang satin brass plate	17.00	20.88	0.33	13.20	nr	**34.08**
Ceiling mounted; fully loaded						
One gang metal clad plate	8.01	9.84	0.33	13.20	nr	**23.04**
Two gang metal clad plate	10.87	13.36	0.45	18.00	nr	**31.36**
Below floor; fully loaded						
Four way outlet box, 5 m length 20 mm flexible conduit with glands and starin relief bracket	27.06	33.24	0.80	32.00	nr	**65.24**
Six way outlet box, 5 m length 25 mm flexible conduit with glands and strain relief bracket	34.99	42.99	1.20	48.00	nr	**90.99**
Eight way outlet box, 5 m length 25 mm flexible conduit with glands and strain relief bracket	44.04	54.11	1.40	56.00	nr	**110.11**
Installation of outlet boxes to desks	–	–	0.40	15.99	nr	**15.99**
Category 6 RJ45 data outlet plate and multiway outlet boxes for wall, ceiling and below floor installations including label to ISO 11801 standards						
Wall mounted; fully loaded						
One gang LSOH PVC plate	5.92	7.28	0.17	6.59	nr	**13.87**
Two gang LSOH PVC plate	8.69	10.68	0.22	8.80	nr	**19.48**
Four gang LSOH PVC plate	15.71	19.30	0.44	17.61	nr	**36.91**
One gang satin brass plate	14.49	17.80	0.28	10.99	nr	**28.79**
Two gang satin brass plate	18.13	22.27	0.36	14.52	nr	**36.79**
Ceiling mounted; fully loaded						
One gang metal clad plate	9.61	11.80	0.36	14.52	nr	**26.32**
Two gang metal clad plate	13.56	16.66	0.50	19.80	nr	**36.46**

COMMUNICATIONS/SECURITY/CONTROL

Item	Net Price £	Material £	Labour hours	Labour £	Unit	Total rate £
Below floor; fully loaded						
Four way outlet box, 5 m length 20 mm flexible conduit with glands and starin relief bracket	33.17	40.75	0.88	35.20	nr	**75.95**
Six way outlet box, 5 m length 25 mm flexible conduit with glands and strain relief bracket	44.12	54.20	1.32	52.80	nr	**107.00**
Eight way outlet box, 5 m length 25 mm flexible conduit with glands and strain relief bracket	56.19	69.03	1.54	61.59	nr	**130.62**
Installation of outlet boxes to desks	–	–	0.40	15.99	nr	**15.99**
Category 5e cable test	–	–	0.10	4.00	nr	**4.00**
Category 6 cable test	–	–	0.11	4.40	nr	**4.40**
Copper Patch Leads						
Category 5e; straight through booted RJ45 UTP – RJ45 UTP						
Patch lead 1 m length	1.66	2.04	0.09	3.60	nr	**5.64**
Patch lead 3 m length	1.98	2.44	0.09	3.60	nr	**6.04**
Patch lead 5 m length	2.81	3.45	0.10	4.00	nr	**7.45**
Patch lead 7 m length	4.11	5.05	0.10	4.00	nr	**9.05**
Category 6; straight through booted RJ45 UTP – RJ45 UTP						
Patch lead 1 m length	4.42	5.43	0.09	3.60	nr	**9.03**
Patch lead 3 m length	6.66	8.18	0.09	3.60	nr	**11.78**
Patch lead 5 m length	7.80	9.59	0.10	4.00	nr	**13.59**
Patch lead 7 m length	9.88	12.14	0.10	4.00	nr	**16.14**
Note 1 – With an intelligent Patching System, add a 25% uplift to the Cat 5e and Cat 6 System price. This would be dependant on the System and IMS solution chosen, i.e. iPatch, RIT or iTRACs.						
Note 2 – For a Cat 6 augmented solution (10Gbps capable), add 25% to a Cat 6 System price.						
Fibre Infrastructure						
Fibre-optic cable, tight buffered, internal/external application, single mode, LSOH sheathed						
4 core fibre-optic cable	1.32	1.62	0.10	4.00	m	**5.62**
8 core fibre-optic cable	2.19	2.69	0.10	4.00	m	**6.69**
12 core fibre-optic cable	2.72	3.34	0.10	4.00	m	**7.34**
16 core fibre-optic cable	3.31	4.06	0.10	4.00	m	**8.06**
24 core fibre-optic cable	3.99	4.90	0.10	4.00	m	**8.90**
Fibre-optic cable OM1 and OM2, tight buffered, internal/external application, 62.5/125 multimode fibre, LSOH sheathed						
4 core fibre-optic cable	1.41	1.73	0.10	4.00	m	**5.73**
8 core fibre-optic cable	2.34	2.88	0.10	4.00	m	**6.88**
12 core fibre-optic cable	2.97	3.64	0.10	4.00	m	**7.64**

COMMUNICATIONS/SECURITY/CONTROL

Item	Net Price £	Material £	Labour hours	Labour £	Unit	Total rate £
DATA TRANSMISSION – cont						
Fibre Infrastructure – cont						
16 core fibre-optic cable	3.70	4.55	0.10	4.00	m	**8.55**
24 core fibre-optic cable	4.28	5.26	0.10	4.00	m	**9.26**
Fibre-optic cable OM3, tight buffered, internal only application, 50/125 multimode fibre, LSOH sheathed						
4 core fibre-optic cable	1.75	2.15	0.10	4.00	nr	**6.15**
8 core fibre-optic cable	2.72	3.34	0.10	4.00	nr	**7.34**
12 core fibre-optic cable	3.95	4.85	0.10	4.00	nr	**8.85**
24 core fibre cable	7.55	9.27	0.10	4.00	nr	**13.27**
Fibre-optic single and multimode connectors and couplers including termination						
ST singlemode booted connector	6.65	8.17	0.25	10.00	nr	**18.17**
ST multimode booted connector	3.21	3.95	0.25	10.00	nr	**13.95**
SC simplex singlemode booted connector	9.42	11.57	0.25	10.00	nr	**21.57**
SC simplex multimode booted connector	3.43	4.21	0.25	10.00	nr	**14.21**
SC duplex multimode booted connector	6.85	8.41	0.25	10.00	nr	**18.41**
ST – SC duplex adaptor	16.66	20.46	0.01	0.41	nr	**20.87**
ST inline bulkhead coupler	3.57	4.39	0.01	0.41	nr	**4.80**
SC duplex coupler	6.67	8.19	0.01	0.41	nr	**8.60**
MTRJ small form factor duplex connector	6.96	8.55	0.25	10.00	nr	**18.55**
LC simplex multimode booted connector	–	–	0.25	10.00	nr	**10.00**
Singlemode core test per core	–	–	0.20	8.00	nr	**8.00**
Multimode core test per core	–	–	0.20	8.00	nr	**8.00**
Fibre; 19” wide fully loaded, labled, alluminium alloy c/w couplers, fibre management and glands (excludes termination of fibre cores)						
8 way ST; fixed drawer	71.53	87.87	0.50	20.00	nr	**107.87**
16 way ST; fixed drawer	107.29	131.81	0.50	20.00	nr	**151.81**
24 way ST; fixed drawer	149.05	183.11	0.50	20.00	nr	**203.11**
8 way ST; sliding drawer	91.00	111.79	0.50	20.00	nr	**131.79**
16 way ST; sliding drawer	125.06	153.64	0.50	20.00	nr	**173.64**
24 way ST; sliding drawer	164.58	202.19	0.50	20.00	nr	**222.19**
8 way (4 duplex) SC; fixed drawer	85.28	104.77	0.50	20.00	nr	**124.77**
16 way (8 duplex) SC; fixed drawer	124.80	153.32	0.50	20.00	nr	**173.32**
24 way (12 duplex) SC; fixed drawer	153.40	188.45	0.50	20.00	nr	**208.45**
8 way (4 duplex) SC; sliding drawer	107.90	132.55	0.50	20.00	nr	**152.55**
16 way (8 duplex) SC; sliding drawer	147.68	181.43	0.50	20.00	nr	**201.43**
24 way (12 duplex) SC; sliding drawer	176.02	216.24	0.50	20.00	nr	**236.24**
8 way (4 duplex) MTRJ; fixed drawer	96.72	118.82	0.50	20.00	nr	**138.82**
16 way (8 duplex) MTRJ; fixed drawer	147.68	181.43	0.50	20.00	nr	**201.43**
24 way (12 duplex) MTRJ; fixed drawer	176.02	216.24	0.50	20.00	nr	**236.24**
8 way (4 duplex) FC/PC; fixed drawer	101.40	124.57	0.50	20.00	nr	**144.57**
16 way (8 duplex) FC/PC; fixed drawer	154.96	190.37	0.50	20.00	nr	**210.37**
24 way (12 duplex) FC/PC; fixed drawer	184.86	227.10	0.50	20.00	nr	**247.10**
Patch panel label per way	0.22	0.27	0.02	0.80	nr	**1.07**

COMMUNICATIONS/SECURITY/CONTROL

Item	Net Price £	Material £	Labour hours	Labour £	Unit	**Total rate £**
Fibre Patch Leads						
Single Mode						
Duplex OS1 LC – LC						
Fibre patch lead 1 m length	21.93	26.94	0.08	3.19	nr	**30.13**
Fibre patch lead 3 m length	20.52	25.21	0.08	3.19	nr	**28.40**
Fibre patch lead 5 m length	21.83	26.82	0.10	4.00	nr	**30.82**
Multimode						
Duplex 50/125 OM3 MTRJ – MTRJ						
Fibre patch lead 1 m length	19.25	23.65	0.08	3.19	nr	**26.84**
Fibre patch lead 3 m length	20.92	25.70	0.08	3.19	nr	**28.89**
Fibre patch lead 5 m length	21.99	27.02	0.10	4.00	nr	**31.02**
Duplex 50/125 OM3 ST – ST						
Fibre patch lead 1 m length	24.00	29.48	0.08	3.19	nr	**32.67**
Fibre patch lead 3 m length	26.11	32.08	0.08	3.19	nr	**35.27**
Fibre patch lead 5 m length	27.43	33.69	0.10	4.00	nr	**37.69**
Duplex 50/125 OM3 SC – SC						
Fibre patch lead 1 m length	21.55	26.47	0.08	3.19	nr	**29.66**
Fibre patch lead 3 m length	23.44	28.79	0.08	3.19	nr	**31.98**
Fibre patch lead 5 m length	24.62	30.25	0.10	4.00	nr	**34.25**
Duplex 50/125 OM3 LC – LC						
Fibre patch lead 1 m length	23.08	28.35	0.08	3.19	nr	**31.54**
Fibre patch lead 3 m length	25.12	30.86	0.08	3.19	nr	**34.05**
Fibre patch lead 5 m length	26.38	32.40	0.10	4.00	nr	**36.40**

COMMUNICATIONS/SECURITY/CONTROL

Item	Net Price £	Material £	Labour hours	Labour £	Unit	Total rate £
WI-FI						
Zone director controller, 1000 Mbps, RJ 45, 2 ports, auto MDX and auto sensing						
Up to 2,000 clients	812.00	997.54	4.00	160.00	nr	**1157.54**
Up to 5,000 clients, 500 access points	5620.00	6904.17	4.00	160.00	nr	**7064.17**
Up to 20,000 clients, 1,000 access points	6667.67	8191.23	4.00	160.00	nr	**8351.23**
Wi-Fi access points, suitable for large developments (commercial buildings, stadiums, hotels, education)						
Indoor access points, 802.11ac wi-fi, dual band, 2.4 GHz and 5 GHz						
Up to 100 connected devices	271.00	332.92	1.20	48.00	nr	**380.92**
Up to 100 connected devices, 867 Mbps	367.00	450.86	1.20	48.00	nr	**498.86**
Up to 300 connected devices, 867 Mbps	478.00	587.22	1.20	48.00	nr	**635.22**
Up to 400 connected devices, 1300 Mbps	589.00	723.59	1.20	48.00	nr	**771.59**
Up to 500 connected devices, 1300 Mbps	737.00	905.40	1.20	48.00	nr	**953.40**
Up to 500 connected devices, 1733 Mbps	970.00	1191.64	1.20	48.00	nr	**1239.64**
Outdoor access points, 802.11ac wi-fi, dual band, 2.4 GHz and 5 GHz, IP67, plastic enclosure, with flexible wall or pole mounting						
Standard	959.00	1178.13	1.80	72.00	nr	**1250.13**
Up to 500 Mbps	1770.00	2174.45	1.80	72.00	nr	**2246.45**
Wi-Fi access points, suitable for small/medium sized and residential						
Indoor access points, 802.11ac wi-fi, dual band, 2.4 GHz and 5 GHz						
Up to 100 connected devices, 867 Mbps	367.00	450.86	1.20	48.00	nr	**498.86**
Up to 300 connected devices, 567 Mbps	478.00	587.22	1.20	48.00	nr	**635.22**
Up to 400 connected devices, 1300 Mbps	589.00	723.59	1.20	48.00	nr	**771.59**
Outdoor access points, 802.11ac wi-fi, dual band, 2.4 GHz and 5 GHz, IP67, plastic enclosure, with flexible wall or pole mounting						
Standard	959.00	1178.13	1.80	72.00	nr	**1250.13**

COMMUNICATIONS/SECURITY/CONTROL

Item	Net Price £	Material £	Labour hours	Labour £	Unit	**Total rate £**
ACCESS CONTROL EQUIPMENT						
Equipment to control the movement of personnel into defined spaces; includes fixing to backgrounds, termination of power and data cables; excludes cable containment and cable installation						
Access control						
Magnetic swipe reader	163.20	200.49	1.50	60.00	nr	**260.49**
Proximity reader	112.20	137.83	1.50	60.00	nr	**197.83**
Exit button	20.40	25.06	1.50	60.00	nr	**85.06**
Exit PIR	134.64	165.41	2.00	80.00	nr	**245.41**
Emergency break glass double pole	30.60	37.59	1.00	40.01	nr	**77.60**
Alarm contact flush	5.10	6.27	1.00	40.01	nr	**46.28**
Alarm contact surface	5.10	6.27	1.00	40.01	nr	**46.28**
Reader controller 16 door	4375.80	5375.67	4.00	160.00	nr	**5535.67**
Reader controller 8 door	2903.94	3567.49	4.00	160.00	nr	**3727.49**
Reader controller 2 door	1016.94	1249.31	4.00	160.00	nr	**1409.31**
Reader interface	412.44	506.68	1.50	60.00	nr	**566.68**
Lock power supply 12 volt 3 amp	245.82	301.99	2.00	80.00	nr	**381.99**
Lock power supply 24 volt 3 amp	429.42	527.54	2.00	80.00	nr	**607.54**
Rechargeable battery 12 volt 7 AHr	12.24	15.04	0.25	10.00	nr	**25.04**
Lock Equipment						
Single slimline magnetic lock, monitored	106.08	130.32	2.00	80.00	nr	**210.32**
Single slimline magnetic lock, unmonitored	91.80	112.78	1.75	70.00	nr	**182.78**
Double slimline magnetic lock, monitored	210.12	258.13	4.00	160.00	nr	**418.13**
Double slimline magnetic lock, unmonitored	180.54	221.79	4.50	180.00	nr	**401.79**
Standard single magnetic lock, monitored	134.64	165.41	1.25	50.00	nr	**215.41**
Standard single magnetic lock, unmonitored	108.12	132.83	1.00	40.01	nr	**172.84**
Standard single magnetic lock, double monitored	180.54	221.79	1.50	60.00	nr	**281.79**
Standard double magnetic lock, monitored	269.28	330.81	1.50	60.00	nr	**390.81**
Standard double magnetic lock, unmonitored	211.14	259.38	1.25	50.00	nr	**309.38**
12 V electric release fail safe, monitored	140.76	172.92	1.25	50.00	nr	**222.92**
12 V electric release fail secure, monitored	140.76	172.92	1.00	40.01	nr	**212.93**
Solenoid bolt	266.22	327.05	1.50	60.00	nr	**387.05**
Electric mortice lock	99.96	122.80	1.00	40.01	nr	**162.81**

COMMUNICATIONS/SECURITY/CONTROL

Item	Net Price £	Material £	Labour hours	Labour £	Unit	**Total rate £**
SECURITY DETECTION AND ALARM						
Detection and alarm systems for the protection of property and persons; includes fixing of equipment to backgrounds and termination of power and data cabling; excludes cable containment and cable installation						
Detection, alarm equipment						
Alarm contact flush	5.10	6.27	1.00	40.01	nr	**46.28**
Alarm contact surface	5.10	6.27	1.00	40.01	nr	**46.28**
Roller shutter contact	16.32	20.04	1.00	40.01	nr	**60.05**
Personal attack button	17.34	21.30	1.00	40.01	nr	**61.31**
Acoustic break glass detectors	23.46	28.82	2.00	80.00	nr	**108.82**
Vibration detectors	30.60	37.59	2.00	80.00	nr	**117.59**
12 metre PIR detector	35.70	43.86	1.50	60.00	nr	**103.86**
15 metre dual detector	59.16	72.68	1.50	60.00	nr	**132.68**
8 zone alarm panel	314.16	385.95	3.00	120.00	nr	**505.95**
8–24 zone end station	308.71	379.25	4.00	160.00	nr	**539.25**
Remote keypad	104.04	127.82	2.00	80.00	nr	**207.82**
8 zone expansion	79.56	97.74	1.50	60.00	nr	**157.74**
Final exit set button	8.48	10.42	1.00	40.01	nr	**50.43**
Self-contained external sounder	76.50	93.99	2.00	80.00	nr	**173.99**
Internal loudspeaker	15.30	18.80	2.00	80.00	nr	**98.80**
Rechargeable battery 12 volt 7 AHr (ampere hours)	12.24	15.04	0.25	10.00	nr	**25.04**
Surveillance Equipment						
Vandal resistant camera, colour	150.00	184.28	3.00	120.00	nr	**304.28**
External camera, colour	300.00	368.55	5.00	200.00	nr	**568.55**
Auto dome external, colour	750.00	921.38	5.00	200.00	nr	**1121.38**
Auto dome external, colour/monochrome	750.00	921.38	5.00	200.00	nr	**1121.38**
Auto dome internal, colour	300.00	368.55	4.00	160.00	nr	**528.55**
Auto dome internal colour/monochrome	300.00	368.55	4.00	160.00	nr	**528.55**
Mini internal domes	499.80	614.01	3.00	120.00	nr	**734.01**
Camera switcher, 32 inputs	1084.68	1332.53	4.00	160.00	nr	**1492.53**
Full function keyboard	204.00	250.61	1.00	40.01	nr	**290.62**
16 CH multiplexers, duplex	601.15	738.52	2.00	80.00	nr	**818.52**
16 way DVR, 250 GB harddrive	750.00	921.38	3.00	120.00	nr	**1041.38**
Additional 250 GB harddrive	200.00	245.70	3.00	120.00	nr	**365.70**
16 way DVR, 500 GB harddrive	1000.00	2149.88	3.00	120.00	nr	**2269.88**
16 way DVR, 750 GB harddrive	1000.00	3685.50	1.00	40.01	nr	**3725.51**
10” colour monitor	130.00	159.71	2.00	80.00	nr	**239.71**
15” colour monitor	150.00	184.28	2.00	80.00	nr	**264.28**
17” colour monitor, high resolution	180.00	221.13	2.00	80.00	nr	**301.13**
21” colour monitor, high resolution	200.00	245.70	2.00	80.00	nr	**325.70**

COMMUNICATIONS/SECURITY/CONTROL

Item	Net Price £	Material £	Labour hours	Labour £	Unit	**Total rate £**
IP CCTV						
Internal fixed dome IP camera h.264	400.00	491.40	–	–	nr	**491.40**
Internal fixed dome IP dome camera 1MP	400.00	491.40	–	–	nr	**491.40**
External fixed dome IP	450.00	552.82	–	–	nr	**552.82**
External PTZ IP	1600.00	1965.60	–	–	nr	**1965.60**
24 Port PoE switch	1750.00	2149.88	–	–	nr	**2149.88**
16 Base channel network video recorder with 4TB storage	2000.00	2457.00	–	–	nr	**2457.00**
IP camera channel licence	80.00	98.28	–	–	nr	**98.28**

COMMUNICATIONS/SECURITY/CONTROL

Item	Net Price £	Material £	Labour hours	Labour £	Unit	Total rate £
FIRE DETECTION AND ALARM						
STANDARD FIRE DETECTION						
CONTROL PANEL						
Zone control panel; 16 zone, mild steel case						
Single loop	391.30	480.71	3.00	120.00	nr	**600.71**
Single loop (flush mounted)	412.30	506.51	0.25	10.00	nr	**516.51**
Loop extension card (1 loop)	208.60	256.26	3.51	140.35	nr	**396.61**
Two loop	584.50	718.05	4.00	160.00	nr	**878.05**
Two loop (flush mounted)	605.50	743.86	5.00	200.00	nr	**943.86**
Repeater panels						
Standard network card	339.50	417.08	4.00	160.00	nr	**577.08**
FR – focus repeater	339.50	417.08	4.00	160.00	nr	**577.08**
Focus repeater with controls	374.50	460.07	0.25	10.00	nr	**470.07**
EQUIPMENT						
Manual call point units: plastic covered						
Surface mounted						
Call point	37.37	45.91	0.50	20.00	nr	**65.91**
Call point; Weatherproof	135.35	166.28	0.80	32.00	nr	**198.28**
Flush mounted						
Call point	37.37	45.91	0.56	22.41	nr	**68.32**
Call point; Weatherproof	135.35	166.28	0.86	34.42	nr	**200.70**
Detectors						
Smoke, ionization type with mounting base	15.29	18.78	0.75	30.01	nr	**48.79**
Smoke, optical type with mounting base	14.70	18.06	0.75	30.01	nr	**48.07**
Fixed temperature heat detector with mounting base (60°C)	17.51	21.51	0.75	30.01	nr	**51.52**
Rate of Rise heat detector with mounting base (90°C)	19.92	24.48	0.75	30.01	nr	**54.49**
Duct detector including optical smoke detector and base	211.18	259.43	2.00	80.00	nr	**339.43**
Remote smoke detector LED indicator with base	5.01	6.15	0.50	20.00	nr	**26.15**
Sounders						
Intelligent wall sounder beacon	77.77	95.54	0.75	30.01	nr	**125.55**
Waterproof intelligent sounder beacon	131.72	161.82	0.75	30.00	nr	**191.82**
Siren; 230 V	64.69	79.47	1.25	50.00	nr	**129.47**
Magnetic Door Holder; 230 V; surface fixed	61.00	74.94	1.50	60.06	nr	**135.00**

COMMUNICATIONS/SECURITY/CONTROL

Item	Net Price £	Material £	Labour hours	Labour £	Unit	**Total rate £**
ADDRESSABLE FIRE DETECTION						
CONTROL PANEL						
Analogue addressable panel; BS EN54 Part 2 and 4 1998; incorporating 120 addresses per loop (maximum 1–2 km length); sounders wired on loop; sealed lead acid integral battery standby providing 48 hour standby; 24 volt DC; mild steel case; surface fixed						
1 loop; 4 × 12 volt batteries	1287.61	1581.82	6.00	240.00	nr	**1821.82**
Extra for 1 loop panel						
Loop expander card	77.00	94.59	1.00	40.01	nr	**134.60**
Repeater panel	893.91	1098.16	6.00	240.00	nr	**1338.16**
Network nodes	1529.68	1879.22	6.00	240.00	nr	**2119.22**
Interface unit; for other systems						
Mains powered	458.08	562.75	1.50	60.00	nr	**622.75**
Loop powered	201.08	247.02	1.00	40.01	nr	**287.03**
Single channel I/O	104.49	128.36	1.00	40.01	nr	**168.37**
Zone module	129.96	159.65	1.50	60.00	nr	**219.65**
4 loop; 4 × 12 volt batteries; 24 hour standby; 30 minute alarm	2158.64	2651.89	8.00	320.00	nr	**2971.89**
Extra for 4 loop panel						
Loop card	452.66	556.09	1.00	40.01	nr	**596.10**
Repeater panel	1281.80	1574.70	6.00	240.00	nr	**1814.70**
Mimic panel	3146.97	3866.05	5.00	200.00	nr	**4066.05**
Network nodes	1528.48	1877.74	6.00	240.00	nr	**2117.74**
Interface unit; for other systems						
Mains powered	458.08	562.75	1.50	60.00	nr	**622.75**
Loop powered	201.08	247.02	1.00	40.01	nr	**287.03**
Single channel I/O	104.49	128.36	1.00	40.01	nr	**168.37**
Zone module	129.96	159.65	1.50	60.00	nr	**219.65**
Line modules	23.96	29.43	1.00	40.01	nr	**69.44**
8 loop; 4 × 12 volt batteries; 24 hour standby; 30 minute alarm	4400.49	5406.00	12.00	480.00	nr	**5886.00**
Extra for 8 loop panel						
Loop card	452.66	556.09	1.00	40.01	nr	**596.10**
Repeater panel	1281.80	1574.70	6.00	240.00	nr	**1814.70**
Mimic panel	3146.97	3866.05	5.00	200.00	nr	**4066.05**
Network nodes	1528.48	1877.74	6.00	240.00	nr	**2117.74**
Interface unit; for other systems						
Mains powered	458.08	562.75	1.50	60.00	nr	**622.75**
Loop powered	201.10	247.05	1.00	40.01	nr	**287.06**
Single channel I/O	43.86	53.89	1.00	40.01	nr	**93.90**
Zone module	129.96	159.65	1.50	60.00	nr	**219.65**
Line modules	23.96	29.43	1.00	40.01	nr	**69.44**

COMMUNICATIONS/SECURITY/CONTROL

Item	Net Price £	Material £	Labour hours	Labour £	Unit	Total rate £
FIRE DETECTION AND ALARM – cont						
EQUIPMENT						
Manual Call Point						
Surface mounted						
Call point	60.36	74.15	1.00	40.01	nr	**114.16**
Call point; Weatherproof	221.16	271.70	1.25	50.00	nr	**321.70**
Flush mounted						
Call point	60.36	74.15	1.00	40.01	nr	**114.16**
Call point; Weatherproof	221.16	271.70	1.25	50.00	nr	**321.70**
Detectors						
Smoke, ionization type with mounting base	61.47	75.52	0.75	30.01	nr	**105.53**
Smoke, optical type with mounting base	60.84	74.74	0.75	30.01	nr	**104.75**
Fixed temperature heat detector with mounting base (60°C)	61.15	75.13	0.75	30.01	nr	**105.14**
Rate of rise heat detector with mounting base (90°C)	61.15	75.13	0.75	30.01	nr	**105.14**
Duct detector including optical smoke detector and addressable base	389.71	478.76	2.00	80.00	nr	**558.76**
Beam smoke detector with transmitter and receiver unit	600.80	738.09	2.00	80.00	nr	**818.09**
Zone short circuit isolator	38.64	47.47	0.75	30.00	nr	**77.47**
Plant interface unit	28.47	34.98	0.50	20.00	nr	**54.98**
Sounders						
Xenon flasher, 24 volt, conduit box	64.89	79.72	0.50	20.00	nr	**99.72**
Xenon flasher, 24 volt, conduit box; weatherproof	96.60	118.67	0.50	20.00	nr	**138.67**
6” bell, conduit box	77.77	255.76	0.75	30.01	nr	**285.77**
6” bell, conduit box; weatherproof	52.65	64.68	0.75	30.00	nr	**94.68**
Siren; 24 V polarized	65.72	80.73	1.00	40.01	nr	**120.74**
Siren; 240 V	64.69	79.47	1.25	50.00	nr	**129.47**
Magnetic door holder; 240 V; surface fixed	61.00	74.94	1.50	60.06	nr	**135.00**

COMMUNICATIONS/SECURITY/CONTROL

Item	Net Price £	Material £	Labour hours	Labour £	Unit	**Total rate £**
Addressable wireless fire detection system, BS 5839 and EN54 Part 25 compliant, 868 Mhz operating frequency						
CONTROL PANEL						
Wireless fire alarm panel, complete with back-up battery						
8 zone	430.00	528.26	4.00	160.00	nr	**688.26**
20 zone	580.00	712.53	5.00	200.00	nr	**912.53**
EQUIPMENT						
Smoke detector	115.00	141.28	0.50	20.00	nr	**161.28**
Combined smoke detector and sounder	220.00	270.27	0.50	20.00	nr	**290.27**
Combined smoke detector, sounder and beacon	260.00	319.41	0.50	20.00	nr	**339.41**
Heat detector	110.00	135.13	0.50	20.00	nr	**155.13**
Combined heat detector and sounder	215.00	264.13	0.50	20.00	nr	**284.13**
Combined heat detector, sounder and beacon	250.00	307.13	0.50	20.00	nr	**327.13**
Manual call point	115.00	141.28	0.50	20.00	nr	**161.28**
Manual call point (weatherproof)	270.00	331.69	0.75	30.00	nr	**361.69**
Sounder	155.00	190.42	0.50	20.00	nr	**210.42**
Beacon	190.00	233.42	0.50	20.00	nr	**253.42**
Combined sounder and beacon	195.00	239.56	0.50	20.00	nr	**259.56**
Remote silence button	110.00	135.13	0.50	20.00	nr	**155.13**
Output unit	205.00	251.84	0.50	20.00	nr	**271.84**
Signal booster panel	380.00	466.83	0.50	20.00	nr	**486.83**
Extra over for additional wired antenna	200.00	245.70	1.00	40.01	nr	**285.71**

COMMUNICATIONS/SECURITY/CONTROL

Item	Net Price £	Material £	Labour hours	Labour £	Unit	Total rate £
EARTHING AND BONDING						
Earth bar; polymer insulators and base mounting; including connections						
Non-disconnect link						
6 way	175.97	216.17	0.81	32.39	nr	**248.56**
8 way	194.85	239.37	0.81	32.39	nr	**271.76**
10 way	226.17	277.85	0.81	32.39	nr	**310.24**
Disconnect link						
6 way	198.24	243.54	1.01	40.40	nr	**283.94**
8 way	219.18	269.26	1.01	40.40	nr	**309.66**
10 way	243.06	298.60	1.01	40.40	nr	**339.00**
Soild earth bar; including connections						
150 × 50 × 6 mm	53.63	65.89	1.01	40.40	nr	**106.29**
Extra for earthing						
Disconnecting link						
300 × 50 × 6 mm	60.79	74.68	1.16	46.40	nr	**121.08**
500 × 50 × 6 mm	66.88	82.16	1.16	46.40	nr	**128.56**
Crimp lugs; including screws and connections to cable						
25 mm	0.50	0.62	0.31	12.40	nr	**13.02**
35 mm	0.68	0.84	0.31	12.40	nr	**13.24**
50 mm	0.78	0.96	0.32	12.80	nr	**13.76**
70 mm	1.40	1.72	0.32	12.80	nr	**14.52**
95 mm	2.33	2.87	0.46	18.41	nr	**21.28**
120 mm	1.65	2.03	1.25	50.00	nr	**52.03**
Earth clamps; connection to pipework						
15 mm to 32 mm dia.	1.37	1.68	0.15	6.00	nr	**7.68**
32 mm to 50 mm dia.	1.71	2.10	0.18	7.20	nr	**9.30**
50 mm to 75 mm dia.	2.00	2.46	0.20	8.00	nr	**10.46**

COMMUNICATIONS/SECURITY/CONTROL

Item	Net Price £	Material £	Labour hours	Labour £	Unit	**Total rate £**
LIGHTNING PROTECTION						
CONDUCTOR TAPE						
PVC sheathed copper tape						
25 × 3 mm	22.23	27.31	0.30	12.00	m	**39.31**
25 × 6 mm	42.04	51.65	0.30	12.00	m	**63.65**
50 × 6 mm	85.93	105.57	0.30	12.00	m	**117.57**
PVC sheathed copper solid circular conductor						
8 mm	12.97	15.93	0.50	20.00	m	**35.93**
Bare copper tape						
20 × 3 mm	18.24	22.41	0.30	12.00	m	**34.41**
25 × 3 mm	20.90	25.67	0.30	12.00	m	**37.67**
25 × 6 mm	38.31	47.06	0.40	15.99	m	**63.05**
50 × 6 mm	74.11	91.05	0.50	20.00	m	**111.05**
Bare copper solid circular conductor						
8 mm	11.79	14.48	0.50	20.00	m	**34.48**
Tape fixings; flat; metalic						
PVC sheathed copper						
25 × 3 mm	22.23	27.31	0.33	13.20	nr	**40.51**
25 × 6 mm	42.04	51.65	0.33	13.20	nr	**64.85**
50 × 6 mm	85.93	105.57	0.33	13.20	nr	**118.77**
8 mm	7.45	9.16	0.50	20.00	nr	**29.16**
Bare copper						
20 × 3 mm	19.16	23.54	0.30	12.00	nr	**35.54**
25 × 3 mm	20.90	25.67	0.30	12.00	nr	**37.67**
25 × 6 mm	38.31	47.06	0.40	15.99	nr	**63.05**
50 × 6 mm	74.11	91.05	0.50	20.00	nr	**111.05**
8 mm	12.38	15.20	0.50	20.00	nr	**35.20**
Tape fixings; flat; non-metalic; PVC sheathed copper						
25 × 3 mm	2.60	3.19	0.30	12.00	nr	**15.19**
Tape fixings; flat; non-metalic; Bare copper						
20 × 3 mm	1.02	1.25	0.30	12.00	nr	**13.25**
25 × 3 mm	1.02	1.25	0.30	12.00	nr	**13.25**
50 × 6 mm	5.41	6.65	0.30	12.00	nr	**18.65**
Puddle flanges; copper						
600 mm long	118.45	145.52	0.93	37.20	nr	**182.72**
AIR RODS						
Pointed air rod fixed to structure; copper						
10 mm diameter						
500 mm long	19.72	24.22	1.00	40.01	nr	**64.23**
1000 mm long	33.24	40.83	1.50	60.00	nr	**100.83**
Extra for						
Air terminal base	24.61	30.23	0.35	14.01	nr	**44.24**
Strike pad	26.14	32.11	0.35	14.01	nr	**46.12**
16 mm diameter						
500 mm long	31.25	38.39	0.91	36.40	nr	**74.79**
1000 mm long	57.10	70.15	1.75	70.00	nr	**140.15**
2000 mm long	104.26	128.08	2.50	100.00	nr	**228.08**

COMMUNICATIONS/SECURITY/CONTROL

Item	Net Price £	Material £	Labour hours	Labour £	Unit	**Total rate £**
LIGHTNING PROTECTION – cont						
AIR RODS – cont						
Extra for						
Multiple point	52.50	64.50	0.35	14.01	nr	**78.51**
Air terminal base	28.85	35.44	0.35	14.01	nr	**49.45**
Ridge saddle	54.17	66.55	0.35	14.01	nr	**80.56**
Side mounting bracket	53.91	66.22	0.50	20.00	nr	**86.22**
Rod to tape coupling	21.75	26.72	0.50	20.00	nr	**46.72**
Strike pad	26.14	32.11	0.35	14.01	nr	**46.12**
AIR TERMINALS						
16 mm diameter						
500 mm long	31.25	38.39	0.65	26.00	nr	**64.39**
1000 mm long	57.10	70.15	0.78	31.21	nr	**101.36**
2000 mm long	104.26	128.08	1.50	60.00	nr	**188.08**
Extra for						
Multiple point	52.50	64.50	0.35	14.01	nr	**78.51**
Flat saddle	53.91	66.22	0.35	14.01	nr	**80.23**
Side bracket	26.94	33.10	0.50	20.00	nr	**53.10**
Rod to cable coupling	22.37	27.48	0.50	20.00	nr	**47.48**
BONDS AND CLAMPS						
Bond to flat surface; copper						
26 mm	4.69	5.76	0.45	18.00	nr	**23.76**
8 mm diameter	19.32	23.73	0.33	13.20	nr	**36.93**
Pipe bond						
26 mm	4.69	5.76	0.45	18.00	nr	**23.76**
8 mm diameter	48.04	59.02	0.33	13.20	nr	**72.22**
Rod to tape clamp						
26 mm	8.67	10.65	0.45	18.00	nr	**28.65**
Square clamp; copper						
25 × 3 mm	9.84	12.09	0.33	13.20	nr	**25.29**
50 × 6 mm	50.39	61.91	0.50	20.00	nr	**81.91**
8 mm diameter	10.52	12.93	0.33	13.20	nr	**26.13**
Test clamp; copper						
26 × 8 mm; oblong	14.67	18.02	0.50	20.00	nr	**38.02**
26 × 8 mm; plate type	44.65	54.85	0.50	20.00	nr	**74.85**
26 × 8 mm; screw down	40.13	49.30	0.50	20.00	nr	**69.30**
Cast in earth points						
2 hole	32.61	40.06	0.75	30.00	nr	**70.06**
4 hole	48.40	59.46	1.00	40.01	nr	**99.47**
Extra for cast in earth points						
Cover plate; 25 × 3 mm	36.60	44.96	0.25	10.00	nr	**54.96**
Cover plate; 8 mm	36.60	44.96	0.25	10.00	nr	**54.96**
Rebar clamp; 8 mm	25.21	30.98	0.25	10.00	nr	**40.98**
Static earth receptacle	171.50	210.69	0.50	20.00	nr	**230.69**

COMMUNICATIONS/SECURITY/CONTROL

Item	Net Price £	Material £	Labour hours	Labour £	Unit	**Total rate £**
Copper braided bonds						
25 × 3 mm						
200 mm hole centres	16.93	20.80	0.33	13.20	nr	**34.00**
400 mm holes centres	25.95	31.88	0.40	15.99	nr	**47.87**
U bolt clamps						
16 mm	11.18	13.73	0.33	13.20	nr	**26.93**
20 mm	12.68	15.58	0.33	13.20	nr	**28.78**
25 mm	15.54	19.09	0.33	13.20	nr	**32.29**
EARTH PITS/MATS						
Earth inspection pit; hand to others for fixing						
Concrete	50.29	61.78	1.00	40.01	nr	**101.79**
Polypropylene	49.15	60.39	1.00	40.01	nr	**100.40**
Extra for						
5 hole copper earth bar; concrete pit	48.76	119.81	0.35	14.01	nr	**133.82**
5 hole earth bar; polypropylene	40.85	50.18	0.35	14.01	nr	**64.19**
Waterproof electrode seal						
Single flange	306.51	376.55	0.93	37.20	nr	**413.75**
Double flange	515.75	633.60	0.93	37.20	nr	**670.80**
Earth electrode mat; laid in ground and connected						
Copper tape lattice						
600 × 600 × 3 mm	138.34	169.95	0.93	37.20	nr	**207.15**
900 × 900 × 3 mm	248.08	304.76	0.93	37.20	nr	**341.96**
Copper tape plate						
600 × 600 × 1.5 mm	119.85	147.23	0.93	37.20	nr	**184.43**
600 × 600 × 3 mm	237.67	291.97	0.93	37.20	nr	**329.17**
900 × 900 × 1.5 mm	266.48	327.37	0.93	37.20	nr	**364.57**
900 × 900 × 3 mm	516.83	634.92	0.93	37.20	nr	**672.12**
EARTH RODS						
Solid cored copper earth electrodes driven into ground and connected						
15 mm diameter						
1200 mm long	48.69	59.82	0.93	37.20	nr	**97.02**
Extra for						
Coupling	1.47	1.81	0.06	2.39	nr	**4.20**
Driving stud	1.81	2.23	0.06	2.39	nr	**4.62**
Spike	1.67	2.05	0.06	2.39	nr	**4.44**
Rod clamp; flat tape	8.67	10.65	0.25	10.00	nr	**20.65**
Rod clamp; solid conductor	3.88	4.77	0.25	10.00	nr	**14.77**
20 mm diameter						
1200 mm long	94.44	116.01	0.98	39.20	nr	**155.21**
Extra for						
Coupling	1.47	1.81	0.06	2.39	nr	**4.20**
Driving stud	2.96	3.63	0.06	2.39	nr	**6.02**
Spike	2.73	3.35	0.06	2.39	nr	**5.74**
Rod clamp; flat tape	8.67	10.65	0.25	10.00	nr	**20.65**
Rod clamp; solid conductor	4.29	5.27	0.25	10.00	nr	**15.27**

COMMUNICATIONS/SECURITY/CONTROL

Item	Net Price £	Material £	Labour hours	Labour £	Unit	Total rate £
LIGHTNING PROTECTION – cont						
EARTH RODS – cont						
Stainless steel earth electrodes driven into ground and connected						
16 mm diameter						
1200 mm long	62.20	76.41	0.93	37.20	nr	**113.61**
Extra for						
Coupling	1.90	2.33	0.06	2.39	nr	**4.72**
Driving head	1.81	2.23	0.06	2.39	nr	**4.62**
Spike	1.67	2.05	0.06	2.39	nr	**4.44**
Rod clamp; flat tape	8.67	10.65	0.25	10.00	nr	**20.65**
Rod clamp; solid conductor	3.88	4.77	0.25	10.00	nr	**14.77**
SURGE PROTECTION						
Single phase; including connection to equipment						
90–150 V	355.49	436.72	5.00	200.00	nr	**636.72**
200–280 V	337.01	414.01	5.00	200.00	nr	**614.01**
Three phase; including connection to equipment						
156–260 V	717.50	881.44	10.00	400.00	nr	**1281.44**
346–484 V	668.58	821.35	10.00	400.00	nr	**1221.35**
349–484 V; remote display	743.59	913.50	10.00	400.00	nr	**1313.50**
346–484 V; 60 kA	1234.97	1517.16	10.00	400.00	nr	**1917.16**
346–484 V; 120 kA	2362.31	2902.09	10.00	400.00	nr	**3302.09**

COMMUNICATIONS/SECURITY/CONTROL

Item	Net Price £	Material £	Labour hours	Labour £	Unit	**Total rate £**
CENTRAL CONTROL/BUILDING MANAGEMENT						
EQUIPMENT						
Switches/sensors; includes fixing in position; electrical work elsewhere. **Note**: these are normally free issued to the mechanical contractor for fitting. The labour times applied assume the installation has been prepared for the fitting of the component.						
Pressure devices						
Liquid differential pressure sensor	148.11	181.95	0.50	20.00	nr	**201.95**
Liquid differential pressure switch	81.16	99.71	0.50	20.00	nr	**119.71**
Air differential pressure transmitter	107.12	131.60	0.50	20.00	nr	**151.60**
Air differential pressure switch	14.65	18.00	0.50	20.00	nr	**38.00**
Liquid level switch	73.91	90.79	0.50	20.00	nr	**110.79**
Static pressure sensor	83.69	102.82	0.50	20.00	nr	**122.82**
High pressure switch	73.91	90.79	0.50	20.00	nr	**110.79**
Low pressure switch	73.91	90.79	0.50	20.00	nr	**110.79**
Water pressure switch	113.28	139.17	0.50	20.00	nr	**159.17**
Duct averaging temperature sensor	49.28	60.54	1.00	40.01	nr	**100.55**
Temperature devices						
Return air sensor (fan coils)	4.82	5.92	1.00	40.01	nr	**45.93**
Frost thermostat	31.60	38.82	0.50	20.00	nr	**58.82**
Immersion thermostat	51.15	62.84	0.50	20.00	nr	**82.84**
Temperature high limit	51.15	62.84	0.50	20.00	nr	**82.84**
Temperature sensor with averaging element	49.28	60.54	0.50	20.00	nr	**80.54**
Immersion temperature sensor	20.93	25.71	0.50	20.00	nr	**45.71**
Space temperature sensor	6.07	7.46	1.00	40.01	nr	**47.47**
Combined space temperature & humidity sensor	83.69	102.82	1.00	40.01	nr	**142.83**
Outside air temperature sensor	11.72	14.40	2.00	80.00	nr	**94.40**
Outside air temperature & humidity sensor	274.50	337.22	2.00	80.00	nr	**417.22**
Duct humidity sensor	107.95	132.62	0.50	20.00	nr	**152.62**
Space humidity sensor	83.69	102.82	1.00	40.01	nr	**142.83**
Immersion water flow sensor	20.93	25.71	0.50	20.00	nr	**45.71**
Rain sensor	195.83	240.58	2.00	80.00	nr	**320.58**
Wind speed and direction sensor	619.29	760.80	2.00	80.00	nr	**840.80**
Controllers; includes fixing in position; electrical work elsewhere						
Zone						
Fan coil controller	158.17	194.31	2.00	80.00	nr	**274.31**
VAV controller	206.83	254.09	2.00	80.00	nr	**334.09**

COMMUNICATIONS/SECURITY/CONTROL

Item	Net Price £	Material £	Labour hours	Labour £	Unit	**Total rate £**
CENTRAL CONTROL/BUILDING MANAGEMENT – cont						
Controllers – cont						
Plant						
Controller, 96 I/O points (exact configuration is dependent upon the number of I/O boards added)	2903.12	3566.48	0.50	20.00	nr	**3586.48**
Controller, 48 I/O points (exact configuration is dependent upon the number of I/O boards added)	2423.59	2977.38	0.50	20.00	nr	**2997.38**
Controller, 32 I/O points (exact configuration is dependent upon the number of I/O boards added)	2012.62	2472.51	0.50	20.00	nr	**2492.51**
Additional Digital Input Boards (12 DI)	479.53	589.10	0.20	8.00	nr	**597.10**
Additional Digital Output Boards (6 DO)	410.07	503.77	0.20	8.00	nr	**511.77**
Additional analogue Input Boards (8 AI)	410.07	503.77	0.20	8.00	nr	**511.77**
Additional analogue Output Boards (8 AO)	410.07	503.77	0.20	8.00	nr	**511.77**
Outstation Enclosure (fitted in riser with space allowance for controller and network device)	410.07	503.77	5.00	200.00	nr	**703.77**
Damper actuator; electrical work elsewhere						
Damper actuator 0–10 V	87.03	106.92	–	–	nr	**106.92**
Damper actuator with auxiliary switches	57.74	70.94	–	–	nr	**70.94**
Frequency inverters: not mounted within MCC; includes fixing in position; electrical work elsewhere						
2.2 kW	662.29	813.62	2.00	80.00	nr	**893.62**
3 kW	732.33	899.67	2.00	80.00	nr	**979.67**
7.5 kW	1083.56	1331.16	2.00	80.00	nr	**1411.16**
11 kW	1324.58	1627.25	2.00	80.00	nr	**1707.25**
15 kW	1600.62	1966.37	2.00	80.00	nr	**2046.37**
18.5 kW	1870.48	2297.88	2.50	100.00	nr	**2397.88**
20 kW	2111.50	2593.97	2.50	100.00	nr	**2693.97**
30 kW	2558.52	3143.14	2.50	100.00	nr	**3243.14**
55 kW	4620.58	5676.38	3.00	120.00	nr	**5796.38**
Miscellaneous; includes fixing in position; electrical work elsewhere						
1 kW Thyristor	58.97	72.44	2.00	80.00	nr	**152.44**
10 kW Thyristor	113.56	139.51	2.00	80.00	nr	**219.51**
Front end and networking; electrical work elsewhere						
PC/monitor	875.50	1075.56	2.00	80.00	nr	**1155.56**
Printer	257.50	316.33	2.00	80.00	nr	**396.33**
PC software	–	–	–	–	nr	**-**
Network server software	2105.32	2586.38	–	–	nr	**2586.38**
Router (allows connection to a network)	106.09	130.34	–	–	nr	**130.34**

PART 4

Rates of Wages

ESSENTIAL READING FROM TAYLOR AND FRANCIS

Spon's Asia Pacific Construction Costs Handbook, Fifth Edition

LANGDON & SEAH

In the last few years, the global economic outlook has continued to be shrouded in uncertainty and volatility following the financial crisis in the Euro zone. While the US and Europe are going through a difficult period, investors are focusing more keenly on Asia. This fifth edition provides overarching construction cost data for 16 countries: Brunei, Cambodia, China, Hong Kong, India, Indonesia, Japan, Malaysia, Myanmar, Philippines, Singapore, South Korea, Sri Lanka, Taiwan, Thailand and Vietnam.

May 2015: 234X156 mm: 452 pp
Hb: 978-1-4822-4358-1: £160.00

To Order: Tel: +44 (0) 1235 400524 Fax: +44 (0) 1235 400525
or Post: Taylor and Francis Customer Services,
Bookpoint Ltd, Unit T1, 200 Milton Park, Abingdon, Oxon, OX14 4TA UK
Email: book.orders@tandf.co.uk

For a complete listing of all our titles visit:
www.tandf.co.uk

Taylor & Francis
Taylor & Francis Group

Mechanical Installations

Rates of Wages

HEATING, VENTILATING, AIR CONDITIONING, PIPING AND DOMESTIC ENGINEERING INDUSTRY

For full details of the wage agreement and the Heating Ventilating Air Conditioning Piping and Domestic Engineering Industry's National Working Rule Agreement, contact:

Note – At the time of printing, the October 2016 rates had not be been published.

Building & Engineering Services Association
ESCA House,
34 Palace Court,
Bayswater,
London W2 4JG

Telephone: 020 7313 4900
Internet: www.b-es.org

WAGE RATES, ALLOWANCES AND OTHER PROVISIONS

Hourly rates of wages

All districts of the United Kingdom

Main Grades	*From 5 October 2015 p/hr*
Foreman	16.08
Senior Craftsman (+2nd welding skill)	13.85
Senior Craftsman	13.30
Craftsman (+2nd welding skill)	12.75
Craftsman	12.20
Operative	11.04
Adult Trainee	9.31
Mate (18 and over)	9.31
Mate (16–17)	4.32
Modern Apprentices	
Junior	6.04
Intermediate	8.57
Senior	11.04

Note: Ductwork Erection Operatives are entitled to the same rates and allowances as the parallel Fitter grades shown.

HEATING, VENTILATING, AIR CONDITIONING, PIPING AND DOMESTIC ENGINEERING INDUSTRY

Trainee *Rates of Pay*

Junior Ductwork Trainees (Probationary)

Age at entry	*From 5 October 2015 p/hr*
17	5.48
18	5.48
19	5.48
20	5.48

Junior Ductwork Erectors (Year of Training)

	From 5 October 2015		
Age at entry	*1 yr p/h*	*2 yr p/h*	*3 yr p/h*
17	6.82	8.49	9.63
18	6.82	8.49	9.63
19	6.82	8.49	9.63
20	6.82	8.49	9.63

Responsibility Allowance (Craftsmen)	*From 5 October 2015 p/hr*
Second welding skill or supervisory responsibility (one unit)	0.55
Second welding skill and supervisory responsibility (two units)	0.55

Responsibility Allowance (Senior Craftsmen)	*From 5 October 2015 p/hr*
Second welding skill	0.55
Supervising responsibility	1.10
Second welding skill and supervisory responsibility	0.55

Daily travelling allowance – Scale 2

C: Craftsmen including Installers

M&A: Mates, Apprentices and Adult Trainees

Direct distance from centre to job in miles

		From 5 October 2015	
Over	*Not exceeding*	*C p/hr*	*M&A p/hr*
15	20	2.67	2.29
20	30	6.87	5.94
30	40	9.88	8.55
40	50	13.02	11.13

HEATING, VENTILATING, AIR CONDITIONING, PIPING AND DOMESTIC ENGINEERING INDUSTRY

Daily travelling allowance – Scale 1

C:	Craftsmen including Installers
M&A:	Mates, Apprentices and Adult Trainees

Direct distance from centre to job in miles

		From 5 October 2015	
Over	*Not exceeding*	*C p/hr*	*M&A p/hr*
0	15	7.30	7.30
15	20	9.98	9.60
20	30	14.18	13.25
30	40	17.19	15.86
40	50	20.31	18.46

Weekly Holiday Credit and Welfare Contributions

	From 5 October 2015						
	£	£	£	£	£	£	£
	a	b	c	d	E	f	g
Weekly Holiday Credit	74.73	69.40	66.84	64.29	61.75	59.25	56.69
Combined Weekly/Welfare Holiday Credit and Contribution	84.38	79.05	76.49	73.94	71.40	68.90	66.34

	From 5 October 2015			
	£	£	£	£
	h	i	j	k
Weekly Holiday Credit	51.34	43.27	39.90	28.11
Combined Weekly/Welfare Holiday Credit and Contribution	60.99	52.92	49.55	37.76

HEATING, VENTILATING, AIR CONDITIONING, PIPING AND DOMESTIC ENGINEERING INDUSTRY

The grades of H&V Operatives entitled to the different rates of Weekly Holiday Credit and Welfare Contribution are as follows:

a Foreman	*b* Senior Craftsman (RAS & RAW)	*c* Senior Craftsman (RAS)	*d* Senior Craftsman (RAW)
e Senior Craftsman Craftsman (+2 RA)	*f* Craftsman (+ 1RA)	*g* Craftsman	*h* Installer Senior Modern Apprentice
i Adult Trainee Mate (over 18)	*j* Intermediate Modern Apprentice	*k* Junior Modern Apprentice	*l* No grade allocated to this Credit Value Category

Daily abnormal conditions money	*From 5 October 2015*
Per day	3.28

Lodging allowance	*From 5 October 2015*
Per night	36.00

Explanatory Notes

1. Working Hours

   The normal working week (Monday to Friday) shall be 37.5 hours.

2. Overtime

   Time worked in excess of 37.5 hours during the normal working week shall be paid at time and a half until 12 hours have been worked since the actual starting time. Thereafter double time shall be paid until normal starting time the following morning. Weekend overtime shall be paid at time and a half for the first 5 hours worked on a Saturday and at double time thereafter until normal starting time on Monday morning.

PLUMBING MECHANICAL ENGINEERING SERVICES INDUSTRY

The Joint Industry Board for Plumbing Mechanical Engineering Services has agreed a two year wage agreement for 2014–2015 with effect from 6 January 2014.

For full details of this wage agreement and the JIB PMES National Working Rules, contact:

The Joint Industry Board for Plumbing Mechanical Engineering Services in England and Wales

JIB-PMES
PO Box 267
PE19 9DN

Telephone: 01480 476925
E-mail: info@jib-pmes.org.uk

WAGE RATES, ALLOWANCES AND OTHER PROVISIONS

EFFECTIVE FROM **5 January 2015**

Basic Rates of Hourly Pay

Applicable in England and Wales

	Hourly rate £
Operatives	
Technical plumber and gas service technician	15.61
Advanced plumber and gas service engineer	14.06
Trained plumber and gas service fitter	12.06
Apprentices	
4th year of training with NVQ level 3	11.67
4th year of training with NVQ level 2	10.57
4th year of training	9.31
3rd year of training with NVQ level 2	9.19
3rd year of training	7.56
2nd year of training	6.70
1st year of training	5.84
Adult Trainees	
3rd 6 months of employment	10.51
2nd 6 months of employment	10.10
1st 6 months of employment	9.41

PLUMBING MECHANICAL ENGINEERING SERVICES INDUSTRY

Major Projects Agreement

Where a job is designated as being a Major Project then the following Major Project Performance Payment hourly rate supplement shall be payable:

Employee Category

	National Payment £	London* Payment £
Technical Plumber and Gas Service Technician	2.20	3.57
Advanced Plumber and Gas Service Engineer	2.20	3.57
Trained Plumber and Gas Service Fitter	2.20	3.57
All 4th year apprentices	1.76	2.86
All 3rd year apprentices	1.32	2.68
2nd year apprentice	1.21	1.96
1st year apprentice	0.88	1.43
All adult trainees	1.76	2.86

* The London Payment Supplement applies only to designated Major Projects that are within the M25 London orbital motorway and are effective from 1 February 2007.
* National payment hourly rates are unchanged for 2007 and will continue to be at the rates shown in Promulgation 138A issued 14 October 2003.

Allowances

Daily travel time allowance plus return fares

All daily travel allowances are to be paid at the daily rate as follows:

Over	*Not exceeding*	*All Operatives*	*3rd & 4th Year Apprentices*	*1st & 2nd Year Apprentices*
20	30	£4.41	£2.85	£1.77
30	40	£10.30	£6.62	£4.24
40	50	£11.77	£7.03	£4.41

PLUMBING MECHANICAL ENGINEERING SERVICES INDUSTRY

Responsibility/Incentive Pay Allowance

As from 5 January 2015, Employers may, in consultation with the employees concerned, enhance the basic graded rates of pay by the payment of an additional amount, as per the bands shown below, where it is agreed that their work involves extra responsibility, productivity or flexibility.

Band 1	an additional rate of £0.27 per hour
Band 2	an additional rate of £0.48 per hour
Band 3	an additional rate of £0.71 per hour
Band 4	an additional rate of £0.93 per hour

This allowance forms part of an operative's basic rate of pay and shall be used to calculate premium payments.

Mileage allowance	£0.42 per mile
Lodging allowance	£34.00 per night
Subsistence Allowance (London Only)	£5.07 per night

Plumbers welding supplement

Possession of Gas or Arc Certificate	£0.30 per hour
Possession of Gas and Arc Certificate	£0.50 per hour

Weekly Holiday Credit Contributions (Holiday Credit Period 2014–2015)

	Public & Annual Gross Value, £	Combined Holiday Credit*, £
Technical Plumber and Gas Service Technician	66.75	63.20
Advance Plumber and Gas Service Engineer	60.05	56.90
Trained Plumber and Gas Service Fitter	51.40	48.80
Adult Trainee	39.35	38.65
Apprentice in last year of training	39.35	38.65
Apprentice 3rd year	28.25	27.65
Apprentice 2nd year	25.00	24.50
Apprentice 1st year	21.75	21.35
Working Principal	27.80	26.25
Ancillary Employee	34.10	32.55

Public and Annual gross value is the Combined Holiday credit value less JIB administration value.

PLUMBING MECHANICAL ENGINEERING SERVICES INDUSTRY

Explanatory Notes

1. Working Hours

   The normal working week (Monday to Friday) shall be 37½ hours, with 45 hours to be worked in the same period before overtime rates become applicable.

2. Overtime

   Overtime shall be paid at time and a half up to 8.00pm (Monday to Friday) and up to 1.00pm (Saturday). Overtime worked after these times shall be paid at double time.

3. Major Projects Agreement

   Under the Major Projects Agreement the normal working week shall be 38 hours (Monday to Friday) with overtime rates payable for all hours worked in excess of 38 hours in accordance with 2 above. However, it should be noted that the hourly rate supplement shall be paid for each hour worked but does not attract premium time enhancement.

4. Pension

   In addition to their hourly rates of pay, plumbing employees are entitled to inclusion within the Industry Pension Scheme (or one providing equivalent benefits). The current levels of industry scheme contributions are 6½% (employers) and 3¼% (employees).

5. Weekly Holiday Credit Contributions

   There has been a fundamental change in the way the JIB-PMES Holiday Pay Schemes shall apply for the Holiday Credit Period 2014–2015, in that 60 credits will be paid over a period of 53 weeks.

   For full details please refer to the Holiday Credit Period 2014–2015 Values as published by the JIB for Plumbing Mechanical Engineering Services in England and Wales.

Electrical Installations

Rates of Wages

ELECTRICAL CONTRACTING INDUSTRY

For full details of this wage agreement and the Joint Industry Board for the Electrical Contracting Industry's National Working Rules, contact:

Joint Industry Board
PO Box 127
Swanley
Kent
BR8 9BH

Telephone: 0333 321 230
Internet: www.jib.org.uk

WAGES (Graded Operatives)

Rates

Since 7 January 2002 two different wage rates have applied to JIB Graded Operatives working on site, depending on whether the Employer transports them to site or whether they provide their own transport. The two categories are:

Job Employed (Transport Provided)

Payable to an Operative who is transported to and from the job by his Employer. The Operative shall also be entitled to payment for Travel Time, when travelling in his own time, as detailed in the appropriate scale.

Job Employed (Own Transport)

Payable to an Operative who travels by his own means to and from the job. The Operative shall be entitled to payment for Travel Allowance and also Travel Time, when travelling in his own time, as detailed in the appropriate scale.

ELECTRICAL CONTRACTING INDUSTRY

The JIB rates of wages are set out below:

From and including 4 January 2016, the JIB hourly rates of wages shall be as set out below:

(i) **National Standard Rate:**

Grade	Transport Provided	Own Transport
Technician (or equivalent specialist grade)	£16.73	£17.57
Approved Electrician (or equivalent specialist grade)	£14.78	£15.61
Electrician (or equivalent specialist grade)	£13.54	£14.39
Senior Graded Electrical Trainee	£12.88	£13.68
Electrical Improver	£12.18	£12.97
Labourer	£10.75	£11.56
Adult Trainee	£10.68	£11.53

(ii) **London Rate:**

Grade	Transport Provided	Own Transport
Technician (or equivalent specialist grade)	£18.75	£19.68
Approved Electrician (or equivalent specialist grade)	£16.56	£17.48
Electrician (or equivalent specialist grade)	£15.16	£16.13
Senior Graded Electrical Trainee	£14.40	£15.33
Electrical Improver	£13.64	£14.54
Labourer	£12.06	£12.94
Adult Trainee	£11.97	£12.91

From and including 4 January 2016, the JIB hourly rates for Job Employed apprentices shall be:

(i) **National Standard Rates**

	Transport Provided	Own Transport
Stage 1	£4.74	£5.54
Stage 2	£6.97	£7.80
Stage 3	£10.09	£10.93
Stage 4	£10.68	£11.53

(ii) **London Rate**

	Transport Provided	Own Transport
Stage 1	£5.29	£6.21
Stage 2	£7.82	£8.72
Stage 3	£11.31	£12.25
Stage 4	£11.97	£12.91

ELECTRICAL CONTRACTING INDUSTRY

Travelling Time and Travel Allowances

From and including 4 January 2016

Operatives required to start/finish at the normal starting and finishing time on jobs which are 15 miles and over from the shop – in a straight line – receive payment for Travelling Time and where transport is not provided by the Employer, Travel Allowance, as follows:

	Distance	*Total Daily Travel Allowance*	*Total Daily Travelling Time*
(a)	**National Standard Rate**		
	Up to 15 miles	Nil	Nil
	Over 15 & up to 20 miles each way	£3.73	£5.10
	Over 20 & up to 25 miles each way	£4.95	£6.45
	Over 25 & up to 35 miles each way	£6.53	£7.89
	Over 35 & up to 55 miles each way	£10.40	£10.40
	Over 55 & up to 75 miles each way	£12.74	£12.74

For each additional 10 mile band over 75 miles, additional payment of £2.24 for Daily Travel Allowance and £1.51 for Daily Travelling Time will be made.

	Distance	*Total Daily Travel Allowance*	*Total Daily Travelling Time*
(b)	**London Rate**		
	Up to 15 miles	Nil	Nil
	Over 15 & up to 20 miles each way	£3.75	£5.47
	Over 20 & up to 25 miles each way	£5.00	£7.17
	Over 25 & up to 35 miles each way	£6.58	£8.60
	Over 35 & up to 55 miles each way	£10.50	£11.62
	Over 55 & up to 75 miles each way	£12.82	£13.60

For each additional 10 mile band over 75 miles, additional payments of £2.26 for Daily Travel Allowance and £1.52 for Daily Travelling Time will be made.

ELECTRICAL CONTRACTING INDUSTRY

Travelling time and Travel allowances

Apprentices required to start/finish at the normal starting and finishing time on jobs which are 15 miles and over from the shop – in a straight line – receive payment for Travelling Time and where transport is not provided by the Employer, Travel Allowance, as follows:

From and including 4 January 2016

	Distance	*Total Daily Travel Allowance*	*Total Daily Travelling Time*
(a)	**National Standard Rate**		
	Up to 15 miles	Nil	Nil
	Over 15 & up to 20 miles each way	£3.73	£2.75
	Over 20 & up to 25 miles each way	£4.95	£4.13
	Over 25 & up to 35 miles each way	£6.53	£5.50
	Over 35 & up to 55 miles each way	£10.40	£6.89
	Over 55 & up to 75 miles each way	£12.74	£8.28

For each additional 10 mile band over 75 miles, additional payments of £2.24 for Daily Travel Allowance and £1.39 for Daily Travelling Time will be made.

	Distance	*Total Daily Travel Allowance*	*Total Daily Travelling Time*
(b)	**London Rate**		
	Up to 15 miles	Nil	Nil
	Over 15 & up to 20 miles each way	£3.75	£3.09
	Over 20 & up to 25 miles each way	£5.00	£4.62
	Over 25 & up to 35 miles each way	£6.58	£6.18
	Over 35 & up to 55 miles each way	£10.50	£7.71
	Over 55 & up to 75 miles each way	£12.82	£9.28

For each additional 10 mile band over 75 miles, additional payments of £2.26 for Daily Travel Allowance and £1.57 for Daily Travelling Time will be made.

Lodging Allowances

£37.32 from and including 5 January 2015

Lodgings weekend retention fee, maximum reimbursement

£37.32 from and including 5 January 2015

Annual Holiday Lodging Allowance Retention

Maximum £12.28 per night (£85.96 per week) from and including 5 January 2015

ELECTRICAL CONTRACTING INDUSTRY

Responsibility money

From and including 30 March 1998 the minimum payment increased to 10p per hour and the maximum to £1.00 per hour (no change)

From and including 4 January 1992 responsibility payments are enhanced by overtime and shift premiums where appropriate (no change)

Combined JIB Benefits Stamp Value (from week commencing 4 January 2016)

JIB Grade	*Weekly JIB combined credit value* £	*Holiday Value* £
Technician	£59.17	£51.53
Approved Electrician	£52.80	£45.53
Electrician	£49.00	£41.73
Labourer & Adult Trainee	£39.66	£33.13

Explanatory Notes

1. Working Hours

   The normal working week (Monday to Friday) shall be 37½ hours, with 38 hours to be worked in the same period before overtime rates become applicable.

2. Overtime

   Overtime shall be paid at time and a half for all weekday overtime. Saturday overtime shall be paid at time and a half for the first 6 hours, or up to 3.00pm (whichever comes first). Thereafter double time shall be paid until normal starting time on Monday.

ESSENTIAL READING FROM TAYLOR AND FRANCIS

BIM and Quantity Surveying

PITTARD & SELL

The sudden arrival of Building Information Modelling (BIM) as a key part of the building industry is redefining the roles and working practices of its stakeholders. Many clients, designers, contractors, quantity surveyors, and building managers are still finding their feet in an industry where BIM compliance can bring great rewards.

This guide is designed to help quantity surveying practitioners and students understand what BIM means for them, and how they should prepare to work successfully on BIM compliant projects. The case studies show how firms at the forefront of this technology have integrated core quantity surveying responsibilities like cost estimating, tendering, and development appraisal into high profile BIM projects. In addition to this, the implications for project management, facilities management, contract administration and dispute resolution are also explored through case studies, making this a highly valuable guide for those in a range of construction project management roles.

Featuring a chapter describing how the role of the quantity surveyor is likely to permanently shift as a result of this development, as well as descriptions of tools used, this covers both the organisational and practical aspects of a crucial topic.

December 2015: 234 x 156 mm: 258 pp
Pb: 978-0-415-87043-6; £24.99

To Order: Tel: +44 (0) 1235 400524 Fax: +44 (0) 1235 400525
or Post: Taylor and Francis Customer Services,
Bookpoint Ltd, Unit T1, 200 Milton Park, Abingdon, Oxon, OX14 4TA UK
Email: book.orders@tandf.co.uk

For a complete listing of all our titles visit:
www.tandf.co.uk

Taylor & Francis
Taylor & Francis Group

PART 5

Daywork

When work is carried out in connection with a contract that cannot be valued in any other way, it is usual to assess the value on a cost basis with suitable allowances to cover overheads and profit. The basis of costing is a matter for agreement between the parties concerned but definitions of prime cost for the Heating and Ventilating and Electrical Industries have been published jointly by the Royal Institution of Chartered Surveyors and the appropriate bodies of the industries concerned, for those who wish to use them.

These, together with a schedule of basic plant hire charges are reproduced on the following pages, with the kind permission of the Royal Institution of Chartered Surveyors, who own the copyright.

ESSENTIAL READING FROM TAYLOR AND FRANCIS

NRM1 Cost Management Handbook

David P Benge

The definitive guide to measurement and estimating using NRM1, written by the author of NRM1

The 'RICS New rules of measurement: Order of cost estimating and cost planning of capital building works' (referred to as NRM1) is the cornerstone of good cost management of capital building works projects - enabling more effective and accurate cost advice to be given to clients and other project team members, while facilitating better cost control.

The NRM1 Cost Management Handbook is the essential guide to how to successfully interpret and apply these rules, including explanations of how to:

- quantify building works and prepare order of cost estimates and cost plans
- use the rules as a toolkit for risk management and procurement
- analyse actual costs for the purpose of collecting benchmark data and preparing cost analyses
- capture historical cost data for future order of cost estimates and elemental cost plans
- employ the rules to aid communication
- manage the complete 'cost management cycle'
- use the elemental breakdown and cost structures, together with the coding system developed for NRM1, to effectively integrate cost management with Building Information Modelling (BIM).

March 2014: 246 x 174: 640pp
Pb: 978-0-415-72077-9: £41.99

To Order: Tel: +44 (0) 1235 400524 Fax: +44 (0) 1235 400525
or Post: Taylor and Francis Customer Services,
Bookpoint Ltd, Unit T1, 200 Milton Park, Abingdon, Oxon, OX14 4TA UK
Email: book.orders@tandf.co.uk

For a complete listing of all our titles visit:
www.tandf.co.uk

Taylor & Francis
Taylor & Francis Group

HEATING AND VENTILATING INDUSTRY

DEFINITION OF PRIME COST OF DAYWORK CARRIED OUT UNDER A HEATING, VENTILATING, AIR CONDITIONING, REFRIGERATION, PIPEWORK AND/OR DOMESTIC ENGINEERING CONTRACT (JULY 1980 EDITION)

This Definition of Prime Cost is published by the Royal Institution of Chartered Surveyors and the Heating and Ventilating Contractors Association for convenience, and for use by people who choose to use it. Members of the Heating and Ventilating Contractors Association are not in any way debarred from defining Prime Cost and rendering accounts for work carried out on that basis in any way they choose. Building owners are advised to reach agreement with contractors on the Definition of Prime Cost to be used prior to entering into a contract or subcontract.

SECTION 1: APPLICATION

1.1 This Definition provides a basis for the valuation of daywork executed under such heating, ventilating, air conditioning, refrigeration, pipework and or domestic engineering contracts as provide for its use.

1.2 It is not applicable in any other circumstances, such as jobbing or other work carried out as a separate or main contract nor in the case of daywork executed after a date of practical completion.

1.3 The terms 'contract' and 'contractor' herein shall be read as 'subcontract' and 'subcontractor' as applicable.

SECTION 2: COMPOSITION OF TOTAL CHARGES

2.1 The Prime Cost of daywork comprises the sum of the following costs:
- (a) Labour as defined in Section 3.
- (b) Materials and goods as defined in Section 4.
- (c) Plant as defined in Section 5.

2.2 Incidental costs, overheads and profit as defined in Section 6, as provided in the contract and expressed therein as percentage adjustments, are applicable to each of 2.1 (a)–(c).

SECTION 3: LABOUR

3.1 The standard wage rates, emoluments and expenses referred to below and the standard working hours referred to in 3.2 are those laid down for the time being in the rules or decisions or agreements of the Joint Conciliation Committee of the Heating, Ventilating and Domestic Engineering Industry applicable to the works (or those of such other body as may be appropriate) and to the grade of operative concerned at the time when and the area where the daywork is executed.

3.2 Hourly base rates for labour are computed by dividing the annual prime cost of labour, based upon the standard working hours and as defined in 3.4, by the number of standard working hours per annum. See example.

3.3 The hourly rates computed in accordance with 3.2 shall be applied in respect of the time spent by operatives directly engaged on daywork, including those operating mechanical plant and transport and erecting and dismantling other plant (unless otherwise expressly provided in the contract) and handling and distributing the materials and goods used in the daywork.

3.4 The annual prime cost of labour comprises the following:
- (a) Standard weekly earnings (i.e. the standard working week as determined at the appropriate rate for the operative concerned).
- (b) Any supplemental payments.
- (c) Any guaranteed minimum payments (unless included in Section 6.1(a)–(p)).
- (d) Merit money.
- (e) Differentials or extra payments in respect of skill, responsibility, discomfort, inconvenience or risk (excluding those in respect of supervisory responsibility – see 3.5)
- (f) Payments in respect of public holidays.
- (g) Any amounts which may become payable by the contractor to or in respect of operatives arising from the rules etc. referred to in 3.1 which are not provided for in 3.4 (a) –(f) nor in Section 6.1 (a)–(p).
- (h) Employers contributions to the WELPLAN, the HVACR Welfare and Holiday Scheme or payments in lieu thereof.
- (i) Employers National Insurance contributions as applicable to 3.4 (a)–(h).
- (j) Any contribution, levy or tax imposed by Statute, payable by the contractor in his capacity as an employer.

HEATING AND VENTILATING INDUSTRY

3.5 Differentials or extra payments in respect of supervisory responsibility are excluded from the annual prime cost (see Section 6). The time of principals, staff, foremen, chargehands and the like when working manually is admissible under this Section at the rates for the appropriate grades.

SECTION 4: MATERIALS AND GOODS

4.1 The prime cost of materials and goods obtained specifically for the daywork is the invoice cost after deducting all trade discounts and any portion of cash discounts in excess of 5%.
4.2 The prime cost of all other materials and goods used in the daywork is based upon the current market prices plus any appropriate handling charges.
4.3 The prime cost referred to in 4.1 and 4.2 includes the cost of delivery to site.
4.4 Any Value Added Tax which is treated, or is capable of being treated, as input tax (as defined by the Finance Act 1972, or any re-enactment or amendment thereof or substitution therefore) by the contractor is excluded.

SECTION 5: PLANT

5.1 Unless otherwise stated in the contract, the prime cost of plant comprises the cost of the following:
  (a) Use or hire of mechanically-operated plant and transport for the time employed on and/or provided or retained for the daywork;
  (b) Use of non-mechanical plant (excluding non-mechanical hand tools) for the time employed on and/or provided or retained for the daywork;
  (c) Transport to and from the site and erection and dismantling where applicable.
5.2 The use of non-mechanical hand tools and of erected scaffolding, staging, trestles or the like is excluded (see Section 6), unless specifically retained for the daywork.

SECTION 6: INCIDENTAL COSTS, OVERHEADS AND PROFIT

6.1 The percentage adjustments provided in the contract which are applicable to each of the totals of Sections 3, 4 and 5 comprise the following:
  (a) Head office charges.
  (b) Site staff including site supervision.
  (c) The additional cost of overtime (other than that referred to in 6.2).
  (d) Time lost due to inclement weather.
  (e) The additional cost of bonuses and all other incentive payments in excess of any included in 3.4.
  (f) Apprentices' study time.
  (g) Fares and travelling allowances.
  (h) Country, lodging and periodic allowances.
  (i) Sick pay or insurances in respect thereof, other than as included in 3.4.
  (j) Third party and employers' liability insurance.
  (k) Liability in respect of redundancy payments to employees.
  (l) Employer's National Insurance contributions not included in 3.4.
  (m) Use and maintenance of non-mechanical hand tools.
  (n) Use of erected scaffolding, staging, trestles or the like (but see 5.2).
  (o) Use of tarpaulins, protective clothing, artificial lighting, safety and welfare facilities, storage and the like that may be available on site.
  (p) Any variation to basic rates required by the contractor in cases where the contract provides for the use of a specified schedule of basic plant charges (to the extent that no other provision is made for such variation – see 5.1).
  (q) In the case of a subcontract which provides that the subcontractor shall allow a cash discount, such provision as is necessary for the allowance of the prescribed rate of discount.
  (r) All other liabilities and obligations whatsoever not specifically referred to in this Section nor chargeable under any other Section.
  (s) Profit.
6.2 The additional cost of overtime where specifically ordered by the Architect/Supervising Officer shall only be chargeable in the terms of a prior written agreement between the parties.

HEATING AND VENTILATING INDUSTRY

MECHANICAL INSTALLATIONS

Calculation of Hourly Base Rate of Labour for Typical Main Grades applicable from 5 October 2016, refer to notes within Section Three – Rates of Wages.

	Foreman	Senior Craftsman (+ 2nd Welding Skill)	Senior Craftsman	Craftsman	Installer	Mate over 18
Hourly Rate from 5 October 2015	16.64	14.33	13.77	12.63	11.43	9.64
Annual standard earnings excluding all holidays, 45.6 weeks × 38 hours	28,584.01	24,619.93	23,642.25	21,686.87	19,624.84	16,549.57
Employers national insurance contributions from 6 April 2015	3,430.79	2,827.15	2,656.51	2,350.13	2,027.06	1,545.22
Weekly holiday credit and welfare contributions (52 weeks) from 5 October 2015	4,541.33	4,254.64	3,842.75	3,570.32	3,282.48	2,848.15
Annual prime cost of labour	36,403.62	31,425.75	30,018.60	27,492.04	24,827.90	20,854.49
Hourly base rate	21.20	18.30	17.48	16.01	14.46	12.14

Notes:

(1) Annual industry holiday (4.6 weeks × 37.5 hours) and public holidays (1.6 weeks × 37.5 hours) are paid through weekly holiday credit and welfare stamp scheme.
(2) Where applicable, Merit money and other variables (e.g. daily abnormal conditions money), which attract Employer's National Insurance contribution, should be included.
(3) Contractors in Northern Ireland should add the appropriate amount of CITB Levy to the annual prime cost of labour prior to calculating the hourly base rate.
(4) Hourly rate based on 1717.5 hours per annum and calculated as follows

52 Weeks @ 37.5 hrs/wk	=		1,950.00
Less			
Sickness = 1 week @ 37.5 hrs/wk	=	37.5	
Public Holiday = 9/5 = 1.6 weeks @ 37.5 hrs/wk	=	60	
Annual holidays = 4.6 weeks @ 3.75 hrs/wk	=	172.5	232.50
Hours	=		1,717.50

(5) For calculation of Holiday Credits and ENI refer to detailed labour rate evaluation.
(6) National Insurance contributions are those effective from 6 April 2016.
(7) Weekly holiday credit/welfare stamp values are those effective from 5 October 2016
(8) Hourly rates of wages are those effective from 5 October 2016.

ELECTRICAL INDUSTRY

DEFINITION OF PRIME COST OF DAYWORK CARRIED OUT UNDER AN ELECTRICAL CONTRACT (MARCH 1981 EDITION)

This Definition of Prime Cost is published by The Royal Institution of Chartered Surveyors and The Electrical Contractors' Associations for convenience and for use by people who choose to use it. Members of The Electrical Contractors' Association are not in any way debarred from defining Prime Cost and rendering accounts for work carried out on that basis in any way they choose. Building owners are advised to reach agreement with contractors on the Definition of Prime Cost to be used prior to entering into a contract or subcontract.

SECTION 1: APPLICATION

1.1 This Definition provides a basis for the valuation of daywork executed under such electrical contracts as provide for its use.

1.2 It is not applicable in any other circumstances, such as jobbing, or other work carried out as a separate or main contract, nor in the case of daywork executed after the date of practical completion.

1.3 The terms 'contract' and 'contractor' herein shall be read as 'subcontract' and 'subcontractor' as the context may require.

SECTION 2: COMPOSITION OF TOTAL CHARGES

2.1 The Prime Cost of daywork comprises the sum of the following costs:
  (a) Labour as defined in Section 3.
  (b) Materials and goods as defined in Section 4.
  (c) Plant as defined in Section 5.

2.2 Incidental costs, overheads and profit as defined in Section 6, as provided in the contract and expressed therein as percentage adjustments, are applicable to each of 2.1 (a)–(c).

SECTION 3: LABOUR

3.1 The standard wage rates, emoluments and expenses referred to below and the standard working hours referred to in 3.2 are those laid down for the time being in the rules and determinations or decisions of the Joint Industry Board or the Scottish Joint Industry Board for the Electrical Contracting Industry (or those of such other body as may be appropriate) applicable to the works and relating to the grade of operative concerned at the time when and in the area where daywork is executed.

3.2 Hourly base rates for labour are computed by dividing the annual prime cost of labour, based upon the standard working hours and as defined in 3.4 by the number of standard working hours per annum. See examples.

3.3 The hourly rates computed in accordance with 3.2 shall be applied in respect of the time spent by operatives directly engaged on daywork, including those operating mechanical plant and transport and erecting and dismantling other plant (unless otherwise expressly provided in the contract) and handling and distributing the materials and goods used in the daywork.

3.4 The annual prime cost of labour comprises the following:
  (a) Standard weekly earnings (i.e. the standard working week as determined at the appropriate rate for the operative concerned).
  (b) Payments in respect of public holidays.
  (c) Any amounts which may become payable by the Contractor to or in respect of operatives arising from operation of the rules etc. referred to in 3.1 which are not provided for in 3.4(a) and (b) nor in Section 6.
  (d) Employer's National Insurance Contributions as applicable to 3.4 (a)–(c).
  (e) Employer's contributions to the Joint Industry Board Combined Benefits Scheme or Scottish Joint Industry Board Holiday and Welfare Stamp Scheme, and holiday payments made to apprentices in compliance with the Joint Industry Board National Working Rules and Industrial Determinations as an employer.
  (f) Any contribution, levy or tax imposed by Statute, payable by the Contractor in his capacity as an employer.

3.5 Differentials or extra payments in respect of supervisory responsibility are excluded from the annual prime cost (see Section 6). The time of principals and similar categories, when working manually, is admissible under this Section at the rates for the appropriate grades.

ELECTRICAL INDUSTRY

SECTION 4: MATERIALS AND GOODS

4.1 The prime cost of materials and goods obtained specifically for the daywork is the invoice cost after deducting all trade discounts and any portion of cash discounts in excess of 5%.
4.2 The prime cost of all other materials and goods used in the daywork is based upon the current market prices plus any appropriate handling charges.
4.3 The prime cost referred to in 4.1 and 4.2 includes the cost of delivery to site.
4.4 Any Value Added Tax which is treated, or is capable of being treated, as input tax (as defined by the Finance Act 1972, or any re-enactment or amendment thereof or substitution therefore) by the Contractor is excluded.

SECTION 5: PLANT

5.1 Unless otherwise stated in the contract, the prime cost of plant comprises the cost of the following:
   (a) Use or hire of mechanically-operated plant and transport for the time employed on and/or provided or retained for the daywork.
   (b) Use of non-mechanical plant (excluding non-mechanical hand tools) for the time employed on and/or provided or retained for the daywork.
   (c) Transport to and from the site and erection and dismantling where applicable.
5.2 The use of non-mechanical hand tools and of erected scaffolding, staging, trestles or the likes is excluded (see Section 6), unless specifically retained for daywork.
5.3 Note: Where hired or other plant is operated by the Electrical Contractor's operatives, such time is to be included under Section 3 unless otherwise provided in the contract.

SECTION 6: INCIDENTAL COSTS, OVERHEADS AND PROFIT

6.1 The percentage adjustments provided in the contract which are applicable to each of the totals of Sections 3, 4 and 5, compromise the following:
   (a) Head Office charges.
   (b) Site staff including site supervision.
   (c) The additional cost of overtime (other than that referred to in 6.2).
   (d) Time lost due to inclement weather.
   (e) The additional cost of bonuses and other incentive payments.
   (f) Apprentices' study time.
   (g) Travelling time and fares.
   (h) Country and lodging allowances.
   (i) Sick pay or insurance in lieu thereof, in respect of apprentices.
   (j) Third party and employers' liability insurance.
   (k) Liability in respect of redundancy payments to employees.
   (l) Employers' National Insurance Contributions not included in 3.4.
   (m) Use and maintenance of non-mechanical hand tools.
   (n) Use of erected scaffolding, staging, trestles or the like (but see 5.2).
   (o) Use of tarpaulins, protective clothing, artificial lighting, safety and welfare facilities, storage and the like that may be available on site.
   (p) Any variation to basic rates required by the Contractor in cases where the contract provides for the use of a specified schedule of basic plant charges (to the extent that no other provision is made for such variation – see 5.1).
   (q) All other liabilities and obligations whatsoever not specifically referred to in this Section nor chargeable under any other Section.
   (r) Profit.
   (s) In the case of a subcontract which provides that the subcontractor shall allow a cash discount, such provision as is necessary for the allowance of the prescribed rate of discount.
6.2 The additional cost of overtime where specifically ordered by the Architect/Supervising Officer shall only be chargeable in the terms of a prior written agreement between the parties.

ELECTRICAL INDUSTRY

ELECTRICAL INSTALLATIONS

Calculation of Hourly Base Rate of Labour for Typical Main Grades applicable from 4 January 2016.

	Technician	Approved Electrician	Electrician	Labourer
Hourly Rate from 5 January 2015 (London Rates)	19.68	17.48	16.13	12.94
Annual standard earnings excluding all holidays, 45.8 weeks × 37.5 hours	33,948.00	30,153.00	27,824.25	22,321.50
Employers national insurance contributions from 6 April 2015	4,281.92	3,689.90	3,326.61	2,468.19
JIB Combined benefits from 5 January 2015	3,443.96	3,131.96	2,934.36	2,487.16
Holiday top up funding	1,748.44	1,565.44	1,459.29	1,188.74
Annual prime cost of labour	43,422.32	38,540.30	35,544.51	28,465.59
Hourly base rate	25.17	22.34	20.61	16.50

Notes:

(1) Annual industry holiday (4.4 weeks × 37.5 hours) and public holidays (1.6 weeks × 37.5 hours).
(2) It should be noted that all labour costs incurred by the Contractor in his capacity as an Employer, other than those contained in the hourly rate above, must be taken into account under Section 6.
(3) Public Holidays are paid through weekly holiday credit and welfare stamp scheme.
(4) Contractors in Northern Ireland should add the appropriate amount of CITB Levy to the annual prime cost of labour prior to calculating the hourly base rate.
(5) Hourly rate based on 1,725 hours per annum and calculated as follows:

52 Weeks @ 37.5 hrs/wk	=		1,950.00
Less			
Public Holiday = 9/5 = 1.6 weeks @ 37.5 hrs/wk	=	60.00	
Annual holidays = 4.4 weeks @ 37.5 hrs/wk	=	165.00	225
Hours	=		1,725.00

(6) For calculation of holiday credits and ENI refer to detailed labour rate evaluation.
(7) Hourly wage rates are those effective from 4 January 2016.
(8) National Insurance contributions are those effective from 6 April 2016.
(9) JIB Combined Benefits Values are those effective from 4 January 2016.

BUILDING INDUSTRY PLANT HIRE COSTS

SCHEDULE OF BASIC PLANT CHARGES (MAY 2001)

This Schedule is published by the Royal Institution of Chartered Surveyors and is for use in connection with Dayworks under a Building Contract.

EXPLANATORY NOTES

1 The rates in the Schedule are intended to apply solely to daywork carried out under and incidental to a Building Contract. They are NOT intended to apply to:
(i) jobbing or any other work carried out as a main or separate contract; or
(ii) work carried out after the date of commencement of the Defects Liability Period.

2 The rates apply to plant and machinery already on site, whether hired or owned by the Contractor.

3 The rates, unless otherwise stated, include the cost of fuel and power of every description, lubricating oils, grease, maintenance, sharpening of tools, replacement of spare parts, all consumable stores and for licences and insurances applicable to items of plant.

4 The rates, unless otherwise stated, do not include the costs of drivers and attendants (unless otherwise stated).

5 The rates in the Schedule are base costs and may be subject to an overall adjustment for price movement, overheads and profit, quoted by the Contractor prior to the placing of the Contract.

6 The rates should be applied to the time during which the plant is actually engaged in daywork.

7 Whether or not plant is chargeable on daywork depends on the daywork agreement in use and the inclusion of an item of plant in this schedule does not necessarily indicate that item is chargeable.

8 Rates for plant not included in the Schedule or which is not already on site and is specifically provided or hired for daywork shall be settled at prices which are reasonably related to the rates in the Schedule having regard to any overall adjustment quoted by the Contractor in the Conditions of Contract.

NOTE: All rates in the schedule were calculated during the first quarter of 2001.

BUILDING INDUSTRY PLANT HIRE COSTS

MECHANICAL PLANT AND TOOLS

Item of plant	Size/Rating	Unit	Rate per hour (£)
PUMPS			
Mobile Pumps			
Including pump hoses, values and strainers, etc.			
Diaphragm	50 mm dia.	Each	0.87
Diaphragm	76 mm dia.	Each	1.29
Submersible	50 mm dia.	Each	1.18
Induced flow	50 mm dia.	Each	1.54
Induced flow	76 mm dia.	Each	2.05
Centrifugal, self-priming	50 mm dia.	Each	1.96
Centrifugal, self-priming	102 mm dia.	Each	2.52
Centrifugal, self-priming	152 mm dia.	Each	3.87
SCAFFOLDING, SHORING, FENCING			
Complete Scaffolding			
Mobile working towers, single width	1.8 m × 0.8 m base × 7 m high	Each	2.00
Mobile working towers, single width	1.8 m × 0.8 m base × 9 m high	Each	2.80
Mobile working towers, double width	1.8 m × 1.4 m base × 7 m high	Each	2.15
Mobile working towers, double width	1.8 m × 1.4 m base × 15 m high	Each	5.10
Chimney scaffold, single unit		Each	1.79
Chimney scaffold, twin unit		Each	2.05
Chimney scaffold, four unit		Each	3.59
Trestles			
Trestle, adjustable	Any height	Pair	0.10
Trestle, painters	1.8 m high	Pair	0.21
Trestle, painters	2.4 m high	Pair	0.26
Shoring, Planking and Strutting			
'Acrow' adjustable prop	Sizes up to 4.9 m (open)	Each	0.10
'Strong boy' support attachment		Each	0.15
Adjustable trench strut	Sizes up to 1.67 m (open)	Each	0.10
Trench sheet		Metre	0.01
Backhoe trench box		Each	1.00
Temporary Fencing			
Including block and coupler			
Site fencing steel grid panel	3.5 m × 2.0 m	Each	0.08
Anti-climb site steel grid fence panel	3.5 m × 2.0 m	Each	0.08

BUILDING INDUSTRY PLANT HIRE COSTS

Item of plant	Size/Rating		Unit	Rate per hour (£)
LIFTING APPLIANCES AND CONVEYORS				
Cranes				
Mobile Cranes				
Rates are inclusive of drivers				
Lorry mounted, telescopic jib				
Two wheel drive	6 tonnes		Each	24.40
Two wheel drive	7 tonnes		Each	25.00
Two wheel drive	8 tonnes		Each	25.62
Two wheel drive	10 tonnes		Each	26.90
Two wheel drive	12 tonnes		Each	28.25
Two wheel drive	15 tonnes		Each	29.66
Two wheel drive	18 tonnes		Each	31.14
Two wheel drive	20 tonnes		Each	32.70
Two wheel drive	25 tonnes		Each	34.33
Four wheel drive	10 tonnes		Each	27.44
Four wheel drive	12 tonnes		Each	28.81
Four wheel drive	15 tonnes		Each	30.25
Four wheel drive	20 tonnes		Each	33.35
Four wheel drive	25 tonnes		Each	35.19
Four wheel drive	30 tonnes		Each	37.12
Four wheel drive	45 tonnes		Each	39.16
Four wheel drive	50 tonnes		Each	41.32
Track-mounted tower crane				
Rates inclusive of driver				
Note: Capacity equals maximum lift in tonnes times maximum radius at which it can be lifted				
	Capacity (metre/tonnes) Up to	Height under hook above ground (m) Up to		
Tower crane	10	17	Each	7.99
Tower crane	15	18	Each	8.59
Tower crane	20	20	Each	9.18
Tower crane	25	22	Each	11.56
Tower crane	30	22	Each	13.78
Tower crane	40	22	Each	18.09
Tower crane	50	22	Each	22.20
Tower crane	60	22	Each	24.32
Tower crane	70	22	Each	23.00

BUILDING INDUSTRY PLANT HIRE COSTS

Item of plant	Size/Rating		Unit	Rate per hour (£)
LIFTING APPLIANCES AND CONVEYORS – cont				
Tower crane	80	22	Each	25.91
Tower crane	110	22	Each	26.45
Tower crane	125	30	Each	29.38
Tower crane	150	30	Each	32.35
Static tower cranes				
Rates inclusive of driver				
To be charged at 90% of the above rates for track mounted tower cranes				
Crane Equipment				
Muck tipping skip	Up to 0.25 m³		Each	0.56
Muck tipping skip	0.5 m³		Each	0.67
Muck tipping skip	0.75 m³		Each	0.82
Muck tipping skip	1.0 m³		Each	1.03
Muck tipping skip	1.5 m³		Each	1.18
Muck tipping skip	2.0 m³		Each	1.38
Mortar skip	Up to 0.38 m³		Each	0.41
Boat skip	1.0 m³		Each	1.08
Boat skip	1.5 m³		Each	1.33
Boat skip	2.0 m³		Each	1.59
Concrete skips, hand levered	0.5 m³		Each	1.00
Concrete skips, hand levered	0.75 m³		Each	1.10
Concrete skips, hand levered	1.0 m³		Each	1.25
Concrete skips, hand levered	1.5 m³		Each	1.50
Concrete skips, hand levered	2.0 m³		Each	1.65
Concrete skips, geared	0.5 m³		Each	1.30
Concrete skips, geared	0.75 m³		Each	1.40
Concrete skips, geared	1.0 m³		Each	1.55
Concrete skips, geared	1.5 m³		Each	1.80
Concrete skips, geared	2.0 m³		Each	2.05
Hoists				
Scaffold hoists	200 kg		Each	1.92
Rack and pinion (goods only)	500 kg		Each	3.31
Rack and pinion (goods only)	1100 kg		Each	4.28
Rack and pinion (goods and passenger)	15 person, 1200 kg		Each	5.62
Wheelbarrow chain sling			Each	031

BUILDING INDUSTRY PLANT HIRE COSTS

Item of plant	Size/Rating		Unit	Rate per hour (£)
LIFTING APPLIANCES AND CONVEYORS – cont				
Conveyors				
Belt conveyors				
Conveyor	7.5 m long × 400 mm wide		Each	6.41
Miniveyor, control box and loading hopper	3 m unit		Each	3.59
Other conveying equipment				
Wheelbarrow			Each	0.21
Hydraulic superlift			Each	2.95
Pavac slab lifter			Each	1.03
Hand pad and hose attachment			Each	0.26
Lifting Trucks	Payload	Maximum Lift		
Fork lift, two wheel drive	1100 kg	up to 3.0 m	Each	4.87
Fork lift, two wheel drive	2540 kg	up to 3.7 m	Each	5.12
Fork lift, two wheel drive	1524 kg	up to 6.0 m	Each	6.04
Fork lift, four wheel drive	2600 kg	up to 5.4 m	Each	7.69
Lifting Platforms				
Hydraulic platform (Cherry picker)	7.5 m		Each	4.23
Hydraulic platform (Cherry picker)	13 m		Each	9.23
Scissor lift	7.8 m		Each	7.56
Telescopic handler	7 m, 2 tonnes		Each	7.18
Telescopic handler	13 m, 3 tonnes		Each	8.72
Lifting and Jacking Gear				
Pipe winch including gantry	1 tonne		Sets	1.92
Pipe winch including gantry	3 tonnes		Sets	3.21
Chain block	1 tonne		Each	0.45
Chain block	2 tonnes		Each	0.71
Chain block	5 tonnes		Each	1.22
Pull lift (Tirfor winch)	1 tonne		Each	0.64
Pull lift (Tirfor winch)	1.6 tonnes		Each	0.90
Pull lift (Tirfor winch)	3.2 tonnes		Each	1.15
Brother or chain slings, two legs	not exceeding 4.2 tonnes		Set	0.35
Brother or chain slings, two legs	not exceeding 7.5 tonnes		Set	0.45
Brother or chain slings, four legs	not exceeding 3.1 tonnes		Set	0.41
Brother or chain slings, four legs	not exceeding 11.2 tonnes		Set	1.28

BUILDING INDUSTRY PLANT HIRE COSTS

Item of plant	Size/Rating	Unit	Rate per hour (£)
CONSTRUCTION VEHICLES			
Lorries			
Plated lorries			
Rates are inclusive of driver			
Platform lorries	7.5 tonnes	Each	19.00
Platform lorries	17 tonnes	Each	21.00
Platform lorries	24 tonnes	Each	26.00
Platform lorries with winch and skids	7.5 tonnes	Each	21.40
Platform lorries with crane	17 tonnes	Each	27.50
Platform lorries with crane	24 tonnes	Each	32.10
Tipper Lorries			
Rates are inclusive of driver			
Tipper lorries	15/17 tonnes	Each	19.50
Tipper lorries	24 tonnes	Each	21.40
Tipper lorries	30 tonnes	Each	27.10
Dumpers			
Site use only (excluding tax, insurance and extra cost of DERV etc. when operating on highway)	Makers Capacity		
Two wheel drive	0.8 tonnes	Each	1.20
Two wheel drive	1 tonne	Each	1.30
Two wheel drive	1.2 tonnes	Each	1.60
Four wheel drive	2 tonnes	Each	2.50
Four wheel drive	3 tonnes	Each	3.00
Four wheel drive	4 tonnes	Each	3.50
Four wheel drive	5 tonnes	Each	4.00
Four wheel drive	6 tonnes	Each	4.50
Dumper Trucks			
Rates are inclusive of drivers			
Dumper trucks	10/13 tonnes	Each	20.00
Dumper trucks	18/20 tonnes	Each	20.40
Dumper trucks	22/25 tonnes	Each	26.30
Dumper trucks	35/40 tonnes	Each	36.60

BUILDING INDUSTRY PLANT HIRE COSTS

Item of plant	Size/Rating	Unit	Rate per hour (£)
CONSTRUCTION VEHICLES – cont			
Tractors			
Agricultural type			
Wheeled, rubber-clad tyred			
Light	48 h.p.	Each	4.65
Heavy	65 h.p.	Each	5.15
Crawler tractors			
With bull or angle dozer	80/90 h.p.	Each	21.40
With bull or angle dozer	115/130 h.p.	Each	25.10
With bull or angle dozer	130/150 h.p.	Each	26.00
With bull or angle dozer	155/175 h.p.	Each	27.74
With bull or angle dozer	210/230 h.p.	Each	28.00
With bull or angle dozer	300/340 h.p.	Each	31.10
With bull or angle dozer	400/440 h.p.	Each	46.90
With loading shovel	0.8 m^3	Each	25.00
With loading shovel	1.0 m^3	Each	28.00
With loading shovel	1.2 m^3	Each	32.00
With loading shovel	1.4 m^3	Each	36.00
With loading shovel	1.8 m^3	Each	45.00
Light Vans			
Ford Escort or the like		Each	4.74
Ford Transit or the like	1.0 tonne	Each	6.79
Luton Box Van or the like	1.8 tonnes	Each	8.33
Water/Fuel Storage			
Mobile water container	110 litres	Each	0.26
Water bowser	1100 litres	Each	0.55
Water bowser	3000 litres	Each	0.74
Mobile fuel container	110 litres	Each	0.28
Fuel bowser	1100 litres	Each	0.65
Fuel bowser	3000 litres	Each	1.02
EXCAVATIONS AND LOADERS			
Excavators			
Wheeled, hydraulic	7/10 tonnes	Each	12.00
Wheeled, hydraulic	11/13 tonnes	Each	12.70
Wheeled, hydraulic	15/16 tonnes	Each	14.80
Wheeled, hydraulic	17/18 tonnes	Each	16.70

BUILDING INDUSTRY PLANT HIRE COSTS

Item of plant	Size/Rating	Unit	Rate per hour (£)
EXCAVATIONS AND LOADERS – cont			
Wheeled, hydraulic	20/23 tonnes	Each	16.70
Crawler, hydraulic	12/14 tonnes	Each	12.00
Crawler, hydraulic	15/17.5 tonnes	Each	14.00
Crawler, hydraulic	20/23 tonnes	Each	16.00
Crawler, hydraulic	25/30 tonnes	Each	21.00
Crawler, hydraulic	30/35 tonnes	Each	30.00
Mini excavators	1000/1500 kg	Each	4.50
Mini excavators	2150/2400 kg	Each	5.50
Mini excavators	2700/3500 kg	Each	6.50
Mini excavators	3500/4500 kg	Each	8.50
Mini excavators	4500/6000 kg	Each	9.50
Loaders			
Wheeled skip loader		Each	4.50
Shovel loader, four wheel drive	1.6 m^3	Each	12.00
Shovel loader, four wheel drive	2.4 m^3	Each	19.00
Shovel loader, four wheel drive	3.6 m^3	Each	22.00
Shovel loader, four wheel drive	4.4 m^3	Each	23.00
Shovel loaders, crawlers	0.8 m^3	Each	11.00
Shovel loaders, crawlers	1.2 m^3	Each	14.00
Shovel loaders, crawlers	1.6 m^3	Each	16.00
Shovel loaders, crawlers	2 m^3	Each	17.00
Skid steer loader wheeled	300/400 kg payload	Each	6.00
Excavator Loaders			
Wheeled tractor type with black-hoe excavator			
Four wheel drive	2.5/3.5 tonnes	Each	7.00
Four wheel drive, 2 wheel steer	7/8 tonnes	Each	9.00
Four wheel drive, 4 wheel steer	7/8 tonnes	Each	10.00
Crawler, hydraulic	12 tonnes	Each	20.00
Crawler, hydraulic	20 tonnes	Each	26.00
Crawler, hydraulic	30 tonnes	Each	35.00
Crawler, hydraulic	40 tonnes	Each	38.00
COMPACTION EQUIPMENT			
Rollers			
Vibrating roller	368 kg–420 kg	Each	1.68

BUILDING INDUSTRY PLANT HIRE COSTS

Item of plant	Size/Rating	Unit	Rate per hour (£)
COMPACTION EQUIPMENT – cont			
Single roller	533 kg	Each	1.92
Single roller	750 kg	Each	2.41
Twin roller	698 kg	Each	1.93
Twin roller	851 kg	Each	2.41
Twin roller with seat end steering wheel	1067 kg	Each	3.03
Twin roller with seat end steering wheel	1397 kg	Each	3.17
Pavement roller	3–4 tonnes dead weight	Each	3.18
Pavement roller	4–6 tonnes	Each	4.13
Pavement roller	6–10 tonnes	Each	4.84
Rammers			
Tamper rammer 2 stroke – petrol	225 mm–275 mm	Each	1.59
Soil Compactors			
Plate compactor	375 mm–400 mm	Each	1.20
Plate compactor rubber pad	375 mm–1400 mm	Each	0.33
Plate compactor reversible plate – petrol	400 mm	Each	2.20
CONCRETE EQUIPMENT			
Concrete/Mortar Mixers			
Open drum without hopper	0.09/0.06 m³	Each	0.62
Open drum without hopper	0.12/0.09 m³	Each	0.68
Open drum without hopper	0.15/0.10 m³	Each	0.72
Open drum with hopper	0.20/0.15 m³	Each	0.80
Concrete/Mortar Transport Equipment			
Concrete pump incl. hose, valve and couplers			
Lorry mounted concrete pump	23 m max. distance	Each	36.00
Lorry mounted concrete pump	50 m max. distance	Each	46.00
Concrete Equipment			
Vibrator, poker, petrol type	up to 75 mm dia.	Each	1.62
Air vibrator (*excluding compressor and hose*)	up to 75 mm dia.	Each	0.79
Extra poker heads	5 m	Each	0.77
Vibrating screed unit with beam	3–5 m	Each	1.77
Vibrating screed unit with adjustable beam	725 mm–900 mm	Each	2.18
Power float		Each	1.72
Power grouter		Each	0.64

BUILDING INDUSTRY PLANT HIRE COSTS

Item of plant	Size/Rating	Unit	Rate per hour (£)
TESTING EQUIPMENT			
Pipe Testing Equipment			
Pressure testing pump, electric		Sets	1.87
Pressure testing pump, hydraulic		Sets	2.46
Pressure test pump		Sets	0.64
SITE ACCOMODATION AND TEMPORARY SERVICES			
Heating equipment			
Space heaters – propane	80,000 Btu/hr	Each	0.77
Space heaters – propane/electric	125,000 Btu/hr	Each	1.56
Space heaters – propane/electric	250,000 Btu/hr	Each	1.79
Space heaters, propane	125,000 Btu/hr	Each	1.33
Space heaters, propane	260,000 Btu/hr	Each	1.64
Cabinet heaters		Each	0.41
Cabinet heater, catalytic		Each	0.46
Electric halogen heaters		Each	1.28
Ceramic heaters	3 kW	Each	0.79
Fan heaters	3 kW	Each	0.41
Cooling fan		Each	1.15
Mobile cooling unit – small		Each	1.38
Mobile cooling unit – large		Each	1.54
Air conditioning unit		Each	2.62
Site Lighting and Equipment			
Tripod floodlight	500 W	Each	0.36
Tripod floodlight	1000 W	Each	0.34
Towable floodlight	4 × 1000 W	Each	2.00
Hand held floodlight	500 W	Each	0.22
Rechargeable light		Each	0.62
Inspection light		Each	0.15
Plasterer's light		Each	0.56
Lighting mast		Each	0.92
Festoon light string	33 m	Each	0.31
Site Electrical Equipment			
Extension leads	240 V/14 m	Each	0.20
Extension leads	110 V/14 m	Each	0.20
Cable reel	25 m 110 V/240 V	Each	0.28
Cable reel	50 m 110 V/240 V	Each	0.33

BUILDING INDUSTRY PLANT HIRE COSTS

Item of plant	Size/Rating	Unit	Rate per hour (£)
SITE ACCOMODATION AND TEMPORARY SERVICES – cont			
4 way junction box	110 V	Each	0.17
Power Generating Units			
Generator – petrol	2 kVA	Each	1.08
Generator – silenced petrol	2 kVA	Each	1.54
Generator – petrol	3 kVA	Each	1.38
Generator – diesel	5 kVA	Each	1.92
Generator – silenced diesel	8 kVA	Each	3.59
Generator – silenced diesel	15 kVA	Each	7.69
Tail adaptor	240 V	Each	0.20
Transformers			
Transformer	3 kVA	Each	0.36
Transformer	5 kVA	Each	0.51
Transformer	7.5 kVA	Each	0.82
Transformer	10 kVA	Each	0.87
Rubbish Collection and Disposal Equipment			
Rubbish chutes			
Standard plastic module	1 m section	Each	0.18
Steel liner insert		Each	0.26
Steel top hopper		Each	0.20
Plastic side entry hopper		Each	0.20
Plastic side entry hopper liner		Each	0.20
SITE EQUIPMENT			
Arc-(Electric) Complete With Leads			
Welder generator – petrol	200 amp	Each	2.26
Welder generator – diesel	300/350 amp	Each	3.33
Welder generator – diesel	4000 amp	Each	4..74
Extra welding lead sets		Each	0.29
Gas-Oxy Welder			
Welding and cutting set (including oxygen and acetylene, excluding underwater equipment and thermic boring)			
Small		Each	1.41
Large		Each	2.00
Mig welder		Each	1.00

BUILDING INDUSTRY PLANT HIRE COSTS

Item of plant	Size/Rating	Unit	Rate per hour (£)
SITE EQUIPMENT – cont			
Fume extractor		Each	0.92
Road Works Equipment			
Traffic lights, mains/generator	2-way	Set	4.01
Traffic lights, mains/generator	3-way	Set	7.92
Traffic lights, mains/generator	4-way	Set	9.81
Traffic lights, mains/generator – trailer mounted	2-way	Each	3.98
Flashing lights		10	0.20
Road safety cone	450 mm	10	0.26
Safety cone	750 mm	Each	0.38
Safety barrier plank	1.25 m	Each	0.03
Safety barrier plank	2 m	Each	0.04
Road sign		Each	0.26
DPC Equipment			
Damp-proofing injection machine		Each	1.49
Cleaning Equipment			
Vacuum cleaner (industrial wet) single motor		Each	0.62
Vacuum cleaner (industrial wet) twin motor		Each	1.23
Vacuum cleaner (industrial wet) triple motor		Each	1.44
Vacuum cleaner (industrial wet) back pack		Each	0.97
Pressure washer, light duty electric	1450 PSI	Each	0.97
Pressure washer, heavy duty, diesel	2500 PSI	Each	2.69
Cold pressure washer, electric		Each	1.79
Hot pressure washer, petrol		Each	2.92
Cold pressure washer, petrol		Each	2.00
Sandblast attachment to last washer		Each	0.54
Drain cleaning attachment to last washer		Each	0.31
Surface Preparation Equipment			
Rotavators	5 h.p.	Each	1.67
Scabbler, up to three heads		Each	1.15
Scabbler, pole		Each	1.50
Scrabbler, multi-headed floor		Each	4.00
Floor preparation machine		Each	2.82

BUILDING INDUSTRY PLANT HIRE COSTS

Item of plant	Size/Rating	Unit	Rate per hour (£)
SITE EQUIPMENT – cont			
Compressors and Equipment			
Portable Compressors			
Compressor – electric	0.23 m³/min	Each	1.59
Compressor – petrol	0.28 m³/min	Each	1.74
Compressor – petrol	0.71 m³/min	Each	2.00
Compressor – diesel	up to 2.83 m³/min	Each	1.24
Compressor – diesel	up to 3.68 m³/min	Each	1.49
Compressor – diesel	up to 4.25 m³/min	Each	1.60
Compressor – diesel	up to 4.81 m³/min	Each	1.92
Compressor – diesel	up to 7.64 m³/min	Each	3.08
Compressor – diesel	up to 11.32 m³/min	Each	4.23
Compressor – diesel	up to 18.40 m³/min	Each	5.73
Mobile Compressors			
Lorry mounted compressors *(machine plus lorry only)*	2.86–4.24 m³/min	Each	12.50
Tractor mounted compressors *(machine plus rubber tyred tractor)*	2..86–3.40 m³/min	Each	13.50
Accessories (Pneumatic Tools)			
(with and including up to 15 m of air hose)			
Demolition pick		Each	1.03
Breakers (with six steels) light	up to 150 kg	Each	0.79
Breakers (with six steels) medium	295 kg	Each	1.08
Breakers (with six steels) heavy	386 kg	Each	1.44
Rock drill (for use with compressor) hand held		Each	0.90
Additional hoses	15 m	Each	0.16
Muffler, tool silencer		Each	0.14
Breakers			
Demolition hammer drill, heavy duty, electric		Each	1.00
Road breaker, electric		Each	1.65
Road breaker, 2 stroke, petrol		Each	2.05
Hydraulic breaker unit, light duty, petrol		Each	2.05
Hydraulic breaker unit, heavy duty, petrol		Each	2.60
Hydraulic breaker unit, heavy duty, diesel		Each	2.95

BUILDING INDUSTRY PLANT HIRE COSTS

Item of plant	Size/Rating	Unit	Rate per hour (£)
SITE EQUIPMENT – cont			
Quarrying and Tooling Equipment			
Block and stone splitter, hydraulic	600 mm × 600 mm	Each	1.35
Block and stone splitter, manual		Each	1.10
Steel Reinforcement Equipment			
Bar bending machine – manual	up to 13 mm dia. rods	Each	0.90
Bar bending machine – manual	up to 20 mm dia. rods	Each	1.28
Bar shearing machine – electric	up to 38 mm dia. rods	Each	2.82
Bar shearing machine – electric	up to 40 mm dia. rods	Each	3.85
Bar cropper machine – electric	up to 13 mm dia. rods	Each	1.54
Bar cropper machine – electric	up to 20 mm dia. rods	Each	2.05
Bar cropper machine – electric	up to 40 mm dia. rods	Each	2.82
Bar cropper machine – 3 phase	up to 40 mm dia. rods	Each	3.85
Dehumidifiers			
110/240 V Water	68 litres extraction per 24 hours	Each	1.28
110/240 V Water	90 litres extraction per 24 hours	Each	1.85
SMALL TOOLS			
Saws			
Masonry bench saw	350 mm–500 mm dia.	Each	2.80
Floor saw	350 mm dia., 125 mm max. cut	Each	1.90
Floor saw	350 mm dia., 150 mm max. cut	Each	2.60
Floor saw, reversible	300 mm max. cut	Each	13.00
Chop/cut off saw, electric	350 mm dia.	Each	1.33
Circular saw, electric	230 mm dia.	Each	0.60
Tyrannosaw		Each	1.20
Reciprocating saw		Each	0.60
Door trimmer		Each	0.90
Chainsaw, petrol	500 mm	Each	2.13
Full chainsaw safety kit		Each	0.50
Worktop jig		Each	0.60
Pipework Equipment			
Pipe bender	15 mm–22 mm	Each	0.33
Pipe bender, hydraulic	50 mm	Each	0.60
Pipe bender, electric	50 mm–150 mm dia.	Each	1.35
Pipe cutter, hydraulic		Each	1.84
Tripod pipe vice		Set	0.40

BUILDING INDUSTRY PLANT HIRE COSTS

Item of plant	Size/Rating	Unit	Rate per hour (£)
SMALL TOOLS – cont			
Ratchet threader	12 mm–32 mm	Each	0.55
Pipe threading machine, electric	12 mm–75 mm	Each	2.40
Pipe threading machine, electric	12 mm–100 mm	Each	3.00
Impact wrench, electric		Each	0.54
Impact wrench, two stroke, petrol		Each	4.49
Impact wrench, heavy duty, electric		Each	1.13
Plumber's furnace, calor gas or similar		Each	2.16
Hand-Held Drills and Equipment			
Impact or hammer drill	up to 25 mm dia.	Each	0.50
Impact or hammer drill	35 mm dia.	Each	0.90
Angle head drill		Each	0.70
Stirrer, mixer drill		Each	0.70
Paint, Insulation Application Equipment			
Airless spray unit		Each	4.20
Portaspray unit		Each	1.65
HPVL turbine spray unit		Each	1.65
Compressor and spray gun		Each	2.20
Other Handtools			
Screwing machine	13 mm–50 mm dia.	Each	0.77
Screwing machine	25 mm–100 mm dia.	Each	1.57
Staple gun		Each	0.33
Air nail gun	110 V	Each	3.33
Cartridge hammer		Each	1.00
Tongue and groove nailer complete with mallet		Each	0.93
Chasing machine	152 mm	Each	1.72
Chasing machine	76 mm–203 mm	Each	5.99
Floor grinder		Each	3.00
Floor plane		Each	3.67
Diamond concrete planer		Each	2.05
Autofeed screwdriver, electric		Each	1.13
Laminate trimmer		Each	0.64
Biscuit jointer		Each	0.87
Random orbital sander		Each	0.72
Floor sander		Each	1.33
Palm, delta, flap or belt sander	300 mm	Each	0.38

BUILDING INDUSTRY PLANT HIRE COSTS

Item of plant	Size/Rating	Unit	Rate per hour (£)
SMALL TOOLS – cont			
Saw cutter, 2 stroke, petrol	up to 225 mm	Each	1.26
Grinder, angle or cutter	300 mm	Each	0.60
Grinder, angle or cutter		Each	1.10
Mortar raking tool attachment	325 mm	Each	0.15
Floor/polisher scrubber		Each	1.03
Floor tile stripper		Each	1.74
Wallpaper stripper, electric		Each	0.56
Electric scraper		Each	0.51
Hot air paint stripper	All sizes	Each	0.38
Electric diamond tile cutter		Each	1.38
Hand tile cutter		Each	0.36
Electric needle gun		Each	1.08
Needle chipping gun	1.2 m wide	Each	0.72
Pedestrian floor sweeper		Each	0.87

PART 6

Tables and Memoranda

This part contains the following sections:

ESSENTIAL READING FROM TAYLOR AND FRANCIS

Estimator's Pocket Book

Duncan Cartlidge

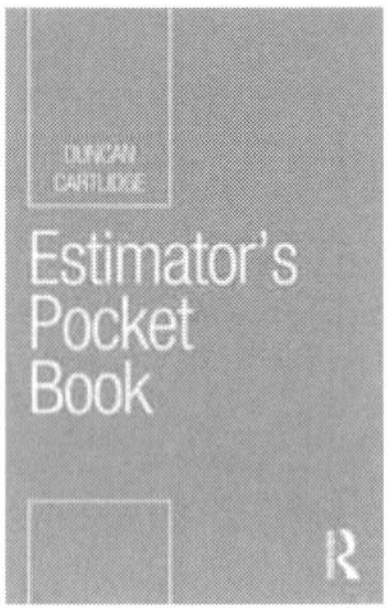

The Estimator's Pocket Book is a concise and practical reference covering the main pricing approaches, as well as useful information such as how to process sub-contractor quotations, tender settlement and adjudication. It is fully up-to-date with NRM2 throughout, features a look ahead to NRM3 and describes the implications of BIM for estimators.

It includes instructions on how to handle:

- the NRM order of cost estimate;
- unit-rate pricing for different trades;
- pro rata pricing and dayworks
- builders' quantities;
- approximate quantities.

Worked examples show how each of these techniques should be carried out in clear, easy-to-follow steps. This is the indispensible estimating reference for all quantity surveyors, cost managers, project managers and anybody else with estimating responsibilities. Particular attention is given to NRM2, but the overall focus is on the core estimating skills needed in practice.

May 2013 186x123: 310pp
Pb: 978-0-415-52711-8: £21.99

To Order: Tel: +44 (0) 1235 400524 Fax: +44 (0) 1235 400525
or Post: Taylor and Francis Customer Services,
Bookpoint Ltd, Unit T1, 200 Milton Park, Abingdon, Oxon, OX14 4TA UK
Email: book.orders@tandf.co.uk

For a complete listing of all our titles visit:
www.tandf.co.uk

Taylor & Francis
Taylor & Francis Group

CONVERSION TABLES

CONVERSION TABLES

Length	Unit	Conversion factors			
Millimetre	mm	1 in	= 25.4 mm	1 mm	= 0.0394 in
Centimetre	cm	1 in	= 2.54 cm	1 cm	= 0.3937 in
Metre	m	1 ft	= 0.3048 m	1 m	= 3.2808 ft
		1 yd	= 0.9144 m		= 1.0936 yd
Kilometre	km	1 mile	= 1.6093 km	1 km	= 0.6214 mile

Note: 1 cm = 10 mm; 1 m = 1 000 mm; 1 km = 1 000 m; 1 ft = 12 in; 1 yd = 3 ft; 1 mile = 1 760 yd

Area	Unit	Conversion factors			
Square Millimetre	mm^2	1 in^2	= 645.2 mm^2	1 mm^2	= 0.0016 in^2
Square Centimetre	cm^2	1 in^2	= 6.4516 cm^2	1 cm^2	= 1.1550 in^2
Square Metre	m^2	1 ft^2	= 0.0929 m^2	1 m^2	= 10.764 ft^2
		1 yd^2	= 0.8361 m^2	1 m^2	= 1.1960 yd^2
Square Kilometre	km^2	1 $mile^2$	= 2.590 km^2	1 km^2	= 0.3861 $mile^2$

Note: 1 cm^2 = 100 mm^2; 1 m^2 = 10 000 cm^2; 1 km^2 = 100 hectares; 1 ft^2 = 144 in^2; 1 yd^2 = 9 ft^2; 1 acre = 4 840 yd^2; 1 $mile^2$ = 640 acres

Volume	Unit	Conversion factors			
Cubic Centimetre	cm^3	1 cm^3	= 0.0610 in^3	1 in^3	= 16.387 cm^3
Cubic Decimetre	dm^3	1 dm^3	= 0.0353 ft^3	1 ft^3	= 28.329 dm^3
Cubic Metre	m^3	1 m^3	= 35.3147 ft^3	1 ft^3	= 0.0283 m^3
		1 m^3	= 1.3080 yd^3	1 yd^3	= 0.7646 m^3
Litre	l	1 l	= 1.76 pint	1 pint	= 0.5683 l
			= 2.113 US pt		= 0.4733 US l

Note: 1 dm^3 = 1 000 cm^3; 1 m^3 = 1 000 dm^3; 1 l = 1 dm^3; 1 ft^3 = 1 728 in^3; 1 yd^3 = 27 ft^3; 1 pint = 20 fl oz; 1 gal = 8 pints

Neither the Centimetre nor Decimetre are SI units, and as such their use, particularly that of the Decimetre, is not widespread outside educational circles.

Mass	Unit	Conversion factors			
Milligram	mg	1 mg	= 0.0154 grain	1 grain	= 64.935 mg
Gram	g	1 g	= 0.0353 oz	1 oz	= 28.35 g
Kilogram	kg	1 kg	= 2.2046 lb	1 lb	= 0.4536 kg
Tonne	t	1 t	= 0.9842 ton	1 ton	= 1.016 t

Note: 1 g = 1000 mg; 1 kg = 1000 g; 1 t = 1000 kg; 1 oz = 437.5 grains; 1 lb = 16 oz; 1 stone = 14 lb; 1 cwt = 112 lb; 1 ton = 20 cwt

Force	Unit	Conversion factors			
Newton	N	1 lbf	= 4.448 N	1 kgf	= 9.807 N
Kilonewton	kN	1 lbf	= 0.004448 kN	1 ton f	= 9.964 kN
Meganewton	MN	100 tonf	= 0.9964 MN		

CONVERSION TABLES

Pressure and stress	Unit	Conversion factors	
Kilonewton per square metre	kN/m^2	1 lbf/in^2	= 6.895 kN/m^2
		1 bar	= 100 kN/m^2
Meganewton per square metre	MN/m^2	1 $tonf/ft^2$	= 107.3 kN/m^2 = 0.1073 MN/m^2
		1 kgf/cm^2	= 98.07 kN/m^2
		1 lbf/ft^2	= 0.04788 kN/m^2

Coefficient of consolidation (Cv) or swelling	Unit	Conversion factors	
Square metre per year	m^2/year	1 cm^2/s	= 3 154 m^2/year
		1 ft^2/year	= 0.0929 m^2/year

Coefficient of permeability	Unit	Conversion factors	
Metre per second	m/s	1 cm/s	= 0.01 m/s
Metre per year	m/year	1 ft/year	= 0.3048 m/year
			= 0.9651 × $(10)^8$ m/s

Temperature	Unit	Conversion factors	
Degree Celsius	°C	°C = 5/9 × (°F − 32)	°F = (9 × °C)/ 5 + 32

CONVERSION TABLES

SPEED CONVERSION

km/h	m/min	mph	fpm
1	16.7	0.6	54.7
2	33.3	1.2	109.4
3	50.0	1.9	164.0
4	66.7	2.5	218.7
5	83.3	3.1	273.4
6	100.0	3.7	328.1
7	116.7	4.3	382.8
8	133.3	5.0	437.4
9	150.0	5.6	492.1
10	166.7	6.2	546.8
11	183.3	6.8	601.5
12	200.0	7.5	656.2
13	216.7	8.1	710.8
14	233.3	8.7	765.5
15	250.0	9.3	820.2
16	266.7	9.9	874.9
17	283.3	10.6	929.6
18	300.0	11.2	984.3
19	316.7	11.8	1038.9
20	333.3	12.4	1093.6
21	350.0	13.0	1148.3
22	366.7	13.7	1203.0
23	383.3	14.3	1257.7
24	400.0	14.9	1312.3
25	416.7	15.5	1367.0
26	433.3	16.2	1421.7
27	450.0	16.8	1476.4
28	466.7	17.4	1531.1
29	483.3	18.0	1585.7
30	500.0	18.6	1640.4
31	516.7	19.3	1695.1
32	533.3	19.9	1749.8
33	550.0	20.5	1804.5
34	566.7	21.1	1859.1
35	583.3	21.7	1913.8
36	600.0	22.4	1968.5
37	616.7	23.0	2023.2
38	633.3	23.6	2077.9
39	650.0	24.2	2132.5
40	666.7	24.9	2187.2
41	683.3	25.5	2241.9
42	700.0	26.1	2296.6
43	716.7	26.7	2351.3
44	733.3	27.3	2405.9
45	750.0	28.0	2460.6

CONVERSION TABLES

km/h	m/min	mph	fpm
46	766.7	28.6	2515.3
47	783.3	29.2	2570.0
48	800.0	29.8	2624.7
49	816.7	30.4	2679.4
50	833.3	31.1	2734.0

GEOMETRY

Two dimensional figures

Figure	Diagram of figure	Surface area	Perimeter
Square	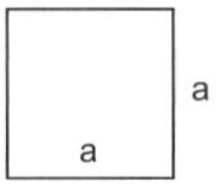	a^2	4a
Rectangle	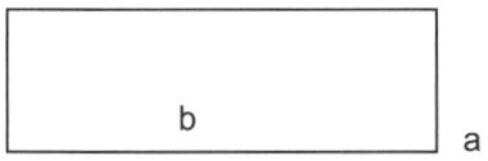	ab	$2(a + b)$
Triangle	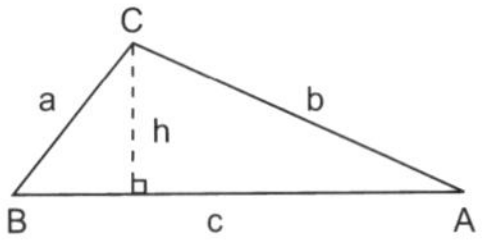	$\frac{1}{2}ch$	$a + b + c$
Circle	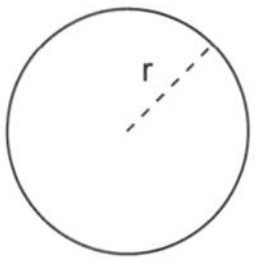	πr^2 $\frac{1}{4}\pi d^2$ where $2r = d$	$2\pi r$ πd
Parallelogram	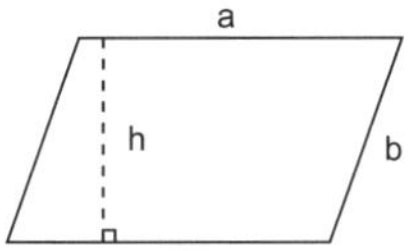	ah	$2(a + b)$
Trapezium	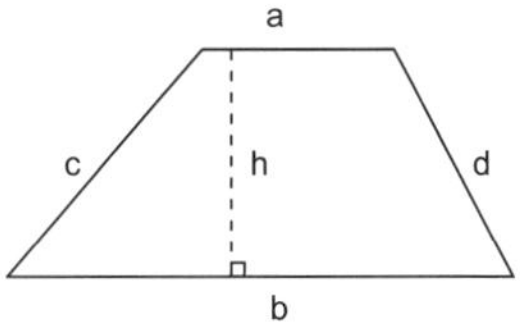	$\frac{1}{2}h(a + b)$	$a + b + c + d$
Ellipse	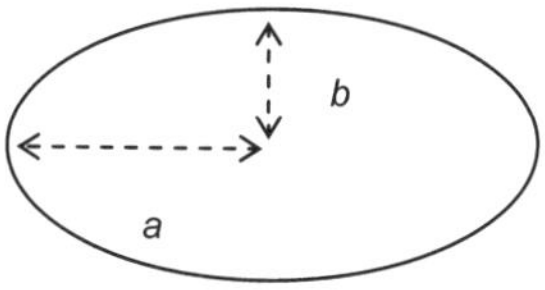	Approximately πab	$\pi(a + b)$
Hexagon	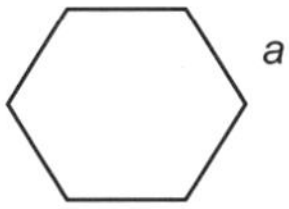	$2.6 \times a^2$	

GEOMETRY

Figure	Diagram of figure	Surface area	Perimeter
Octagon		$4.83 \times a^2$	6a
Sector of a circle	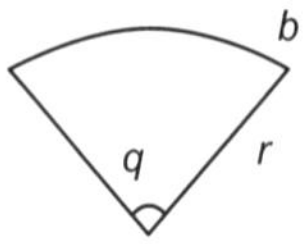	$\frac{1}{2}rb$ or $\frac{q}{360}\pi r^2$ note $b =$ angle $\frac{q}{360} \times \pi 2r$	
Segment of a circle	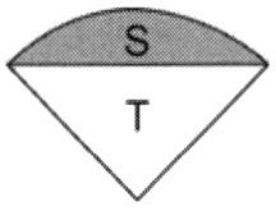	$S - T$ where S = area of sector, T = area of triangle	
Bellmouth	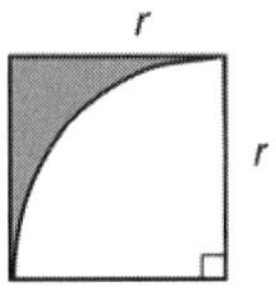	$\frac{3}{14} \times r^2$	

GEOMETRY

Three dimensional figures

Figure	Diagram of figure	Surface area	Volume
Cube	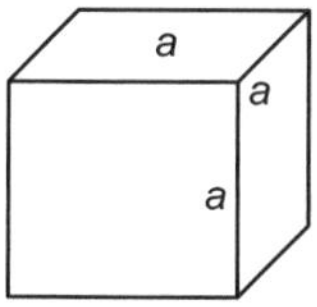	$6a^2$	a^3
Cuboid/ rectangular block	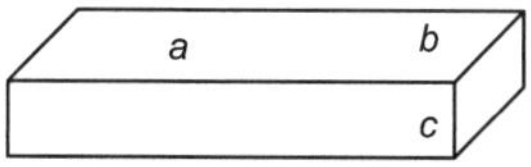	$2(ab + ac + bc)$	abc
Prism/ triangular block	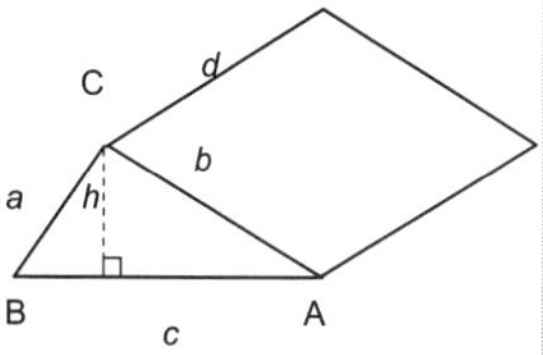	$bd + hc + dc + ad$	$\frac{1}{2}hcd$
Cylinder	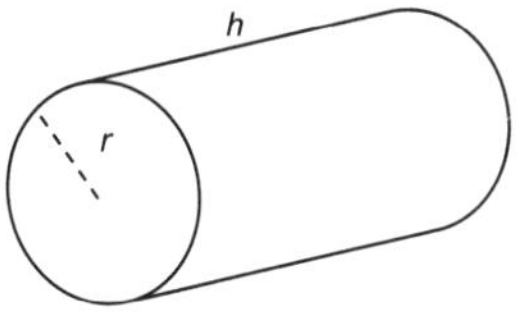	$2\pi r^2 + 2\pi h$	$\pi r^2 h$
Sphere	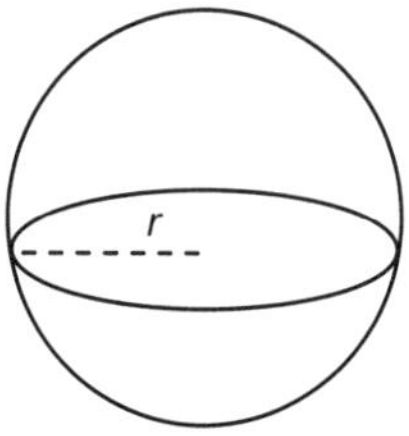	$4\pi r^2$	$\frac{4}{3}\pi r^3$
Segment of sphere	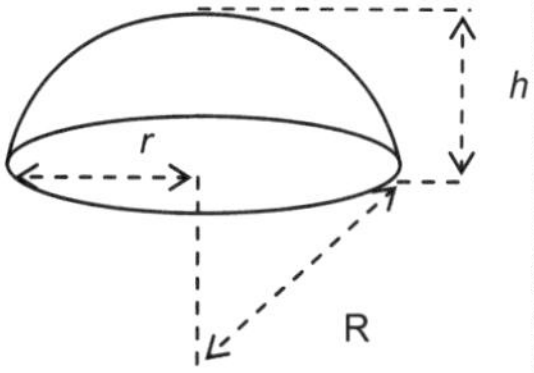	$2\pi Rh$	$\frac{1}{6}\pi h(3r^2 + h^2)$ $\frac{1}{3}\pi h^2(3R - H)$
Pyramid	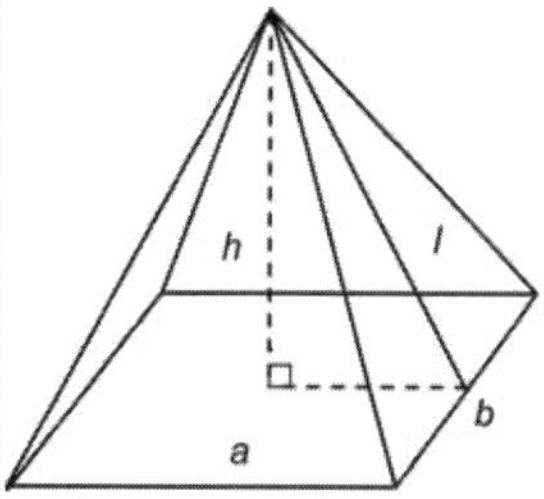	$(a + b)l + ab$	$\frac{1}{3}abh$

GEOMETRY

Figure	Diagram of figure	Surface area	Volume
Frustum of a pyramid	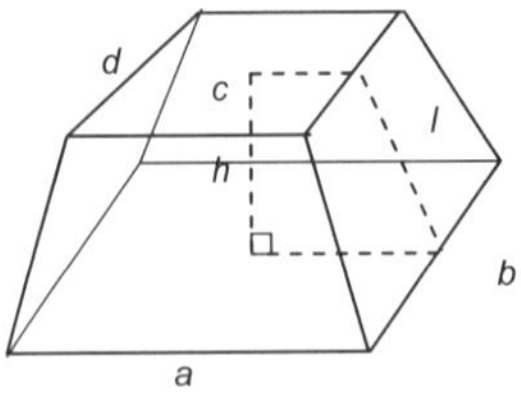	$l(a+b+c+d)+\sqrt{(ab+cd)}$ [rectangular figure only]	$\frac{h}{3}(ab+cd+\sqrt{abcd})$
Cone	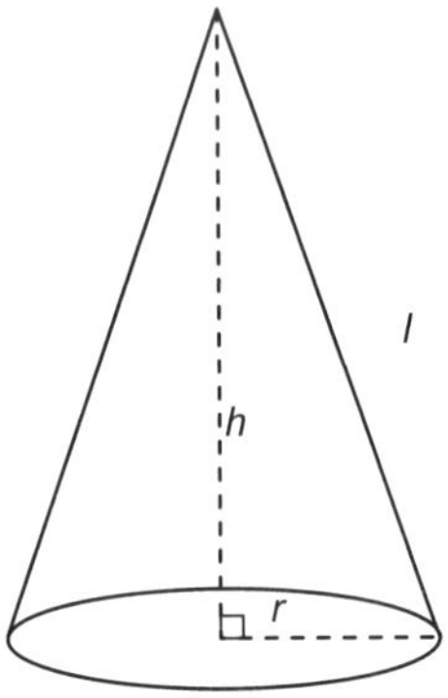	πrl (excluding base) $\pi rl+\pi r^2$ (including base)	$\frac{1}{3}\pi r^2 h$ $\frac{1}{12}\pi d^2 h$
Frustum of a cone	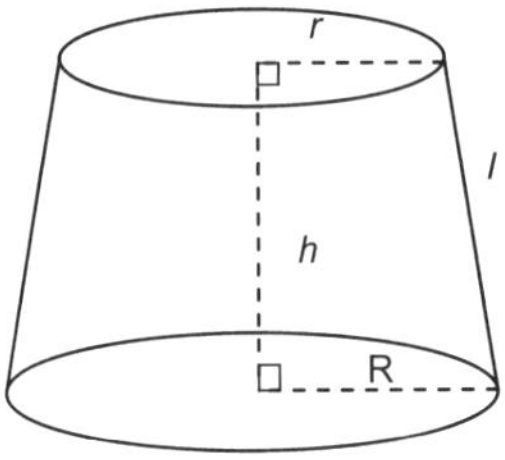	$\pi r^2+\pi R^2+\pi l(R+r)$	$\frac{1}{3}\pi(R^2+Rr+r^2)$

FORMULAE

Formulae

Formula	Description
Pythagoras Theorem	$A^2 = B^2 + C^2$ where A is the hypotenuse of a right-angled triangle and B and C are the two adjacent sides
Simpsons Rule	The Area is divided into an even number of strips of equal width, and therefore has an odd number of ordinates at the division points area = $\frac{S(A + 2B + 4C)}{3}$ where S = common interval (strip width) A = sum of first and last ordinates B = sum of remaining odd ordinates C = sum of the even ordinates The Volume can be calculated by the same formula, but by substituting the area of each coordinate rather than its length
Trapezoidal Rule	A given trench is divided into two equal sections, giving three ordinates, the first, the middle and the last volume = $\frac{S \times (A + B + 2C)}{2}$ where S = width of the strips A = area of the first section B = area of the last section C = area of the rest of the sections
Prismoidal Rule	A given trench is divided into two equal sections, giving three ordinates, the first, the middle and the last volume = $\frac{L \times (A + 4B + C)}{6}$ where L = total length of trench A = area of the first section B = area of the middle section C = area of the last section

TYPICAL THERMAL CONDUCTIVITY OF BUILDING MATERIALS

TYPICAL THERMAL CONDUCTIVITY OF BUILDING MATERIALS

(Always check manufacturer's details – variation will occur depending on product and nature of materials)

	Thermal conductivity (W/mK)		Thermal conductivity (W/mK)
Acoustic plasterboard	0.25	Oriented strand board	0.13
Aerated concrete slab (500 kg/m^3)	0.16	Outer leaf brick	0.77
Aluminium	237	Plasterboard	0.22
Asphalt (1700 kg/m^3)	0.5	Plaster dense (1300 kg/m^3)	0.5
Bitumen-impregnated fibreboard	0.05	Plaster lightweight (600 kg/m^3)	0.16
Blocks (standard grade 600 kg/m^3)	0.15	Plywood (950 kg/m^3)	0.16
Blocks (solar grade 460 kg/m^3)	0.11	Prefabricated timber wall panels (check manufacturer)	0.12
Brickwork (outer leaf 1700 kg/m^3)	0.84	Screed (1200 kg/m^3)	0.41
Brickwork (inner leaf 1700 kg/m^3)	0.62	Stone chippings (1800 kg/m^3)	0.96
Dense aggregate concrete block 1800 kg/m^3 (exposed)	1.21	Tile hanging (1900 kg/m^3)	0.84
Dense aggregate concrete block 1800 kg/m^3 (protected)	1.13	Timber (650 kg/m^3)	0.14
Calcium silicate board (600 kg/m^3)	0.17	Timber flooring (650 kg/m^3)	0.14
Concrete general	1.28	Timber rafters	0.13
Concrete (heavyweight 2300 kg/m^3)	1.63	Timber roof or floor joists	0.13
Concrete (dense 2100 kg/m^3 typical floor)	1.4	Roof tile (1900 kg/m^3)	0.84
Concrete (dense 2000 kg/m^3 typical floor)	1.13	Timber blocks (650 kg/m^3)	0.14
Concrete (medium 1400 kg/m^3)	0.51	Cellular glass	0.045
Concrete (lightweight 1200 kg/m^3)	0.38	Expanded polystyrene	0.034
Concrete (lightweight 600 kg/m^3)	0.19	Expanded polystyrene slab (25 kg/m^3)	0.035
Concrete slab (aerated 500 kg/m^3)	0.16	Extruded polystyrene	0.035
Copper	390	Glass mineral wool	0.04
External render sand/cement finish	1	Mineral quilt (12 kg/m^3)	0.04
External render (1300 kg/m^3)	0.5	Mineral wool slab (25 kg/m^3)	0.035
Felt – Bitumen layers (1700 kg/m^3)	0.5	Phenolic foam	0.022
Fibreboard (300 kg/m^3)	0.06	Polyisocyanurate	0.025
Glass	0.93	Polyurethane	0.025
Marble	3	Rigid polyurethane	0.025
Metal tray used in wriggly tin concrete floors (7800 kg/m^3)	50	Rock mineral wool	0.038
Mortar (1750 kg/m^3)	0.8		

EARTHWORK

Weights of Typical Materials Handled by Excavators

The weight of the material is that of the state in its natural bed and includes moisture. Adjustments should be made to allow for loose or compacted states

Material	Mass (kg/m³)	Mass (lb/cu yd)
Ashes, dry	610	1028
Ashes, wet	810	1365
Basalt, broken	1954	3293
Basalt, solid	2933	4943
Bauxite, crushed	1281	2159
Borax, fine	849	1431
Caliche	1440	2427
Cement, clinker	1415	2385
Chalk, fine	1221	2058
Chalk, solid	2406	4055
Cinders, coal, ash	641	1080
Cinders, furnace	913	1538
Clay, compacted	1746	2942
Clay, dry	1073	1808
Clay, wet	1602	2700
Coal, anthracite, solid	1506	2538
Coal, bituminous	1351	2277
Coke	610	1028
Dolomite, lumpy	1522	2565
Dolomite, solid	2886	4864
Earth, dense	2002	3374
Earth, dry, loam	1249	2105
Earth, Fullers, raw	673	1134
Earth, moist	1442	2430
Earth, wet	1602	2700
Felsite	2495	4205
Fieldspar, solid	2613	4404
Fluorite	3093	5213
Gabbro	3093	5213
Gneiss	2696	4544
Granite	2690	4534
Gravel, dry ¼ to 2 inch	1682	2835
Gravel, dry, loose	1522	2565
Gravel, wet ¼ to 2 inch	2002	3374
Gypsum, broken	1450	2444
Gypsum, solid	2787	4697
Hardcore (consolidated)	1928	3249
Lignite, dry	801	1350
Limestone, broken	1554	2619
Limestone, solid	2596	4375
Magnesite, magnesium ore	2993	5044
Marble	2679	4515
Marl, wet	2216	3735
Mica, broken	1602	2700
Mica, solid	2883	4859
Peat, dry	400	674
Peat, moist	700	1179
Peat, wet	1121	1889

EARTHWORK

Material	Mass (kg/m^3)	Mass (lb/cu yd)
Potash	1281	2159
Pumice, stone	640	1078
Quarry waste	1438	2423
Quartz sand	1201	2024
Quartz, solid	2584	4355
Rhyolite	2400	4045
Sand and gravel, dry	1650	2781
Sand and gravel, wet	2020	3404
Sand, dry	1602	2700
Sand, wet	1831	3086
Sandstone, solid	2412	4065
Shale, solid	2637	4444
Slag, broken	2114	3563
Slag, furnace granulated	961	1619
Slate, broken	1370	2309
Slate, solid	2667	4495
Snow, compacted	481	810
Snow, freshly fallen	160	269
Taconite	2803	4724
Trachyte	2400	4045
Trap rock, solid	2791	4704
Turf	400	674
Water	1000	1685

Transport Capacities

Type of vehicle	Capacity of vehicle	
	Payload	Heaped capacity
Wheelbarrow	150	0.10
1 tonne dumper	1250	1.00
2.5 tonne dumper	4000	2.50
Articulated dump truck (Volvo A20 6 × 4)	18500	11.00
Articulated dump truck (Volvo A35 6 × 6)	32000	19.00
Large capacity rear dumper (Euclid R35)	35000	22.00
Large capacity rear dumper (Euclid R85)	85000	50.00

EARTHWORK

Machine Volumes for Excavating and Filling

Machine type	Cycles per minute	Volume per minute (m^3)
1.5 tonne excavator	1	0.04
	2	0.08
	3	0.12
3 tonne excavator	1	0.13
	2	0.26
	3	0.39
5 tonne excavator	1	0.28
	2	0.56
	3	0.84
7 tonne excavator	1	0.28
	2	0.56
	3	0.84
21 tonne excavator	1	1.21
	2	2.42
	3	3.63
Backhoe loader JCB3CX excavator	1	0.28
Rear bucket capacity 0.28 m^3	2	0.56
	3	0.84
Backhoe loader JCB3CX loading	1	1.00
Front bucket capacity 1.00 m^3	2	2.00

Machine Volumes for Excavating and Filling

Machine type	Loads per hour	Volume per hour (m^3)
1 tonne high tip skip loader	5	2.43
Volume 0.485 m^3	7	3.40
	10	4.85
3 tonne dumper	4	7.60
Max volume 2.40 m^3	5	9.50
Available volume 1.9 m^3	7	13.30
	10	19.00
6 tonne dumper	4	15.08
Max volume 3.40 m^3	5	18.85
Available volume 3.77 m^3	7	26.39
	10	37.70

EARTHWORK

Bulkage of Soils (after excavation)

Type of soil	Approximate bulking of 1 m^3 after excavation
Vegetable soil and loam	25–30%
Soft clay	30–40%
Stiff clay	10–20%
Gravel	20–25%
Sand	40–50%
Chalk	40–50%
Rock, weathered	30–40%
Rock, unweathered	50–60%

Shrinkage of Materials (on being deposited)

Type of soil	Approximate bulking of 1 m^3 after excavation
Clay	10%
Gravel	8%
Gravel and sand	9%
Loam and light sandy soils	12%
Loose vegetable soils	15%

Voids in Material Used as Subbases or Beddings

Material	m^3 of voids/m^3
Alluvium	0.37
River grit	0.29
Quarry sand	0.24
Shingle	0.37
Gravel	0.39
Broken stone	0.45
Broken bricks	0.42

Angles of Repose

Type of soil		Degrees
Clay	– dry	30
	– damp, well drained	45
	– wet	15–20
Earth	– dry	30
	– damp	45
Gravel	– moist	48
Sand	– dry or moist	35
	– wet	25
Loam		40

EARTHWORK

Slopes and Angles

Ratio of base to height	Angle in degrees
5:1	11
4:1	14
3:1	18
2:1	27
1½:1	34
1:1	45
1:1½	56
1:2	63
1:3	72
1:4	76
1:5	79

Grades (in Degrees and Percents)

Degrees	Percent	Degrees	Percent
1	1.8	24	44.5
2	3.5	25	46.6
3	5.2	26	48.8
4	7.0	27	51.0
5	8.8	28	53.2
6	10.5	29	55.4
7	12.3	30	57.7
8	14.0	31	60.0
9	15.8	32	62.5
10	17.6	33	64.9
11	19.4	34	67.4
12	21.3	35	70.0
13	23.1	36	72.7
14	24.9	37	75.4
15	26.8	38	78.1
16	28.7	39	81.0
17	30.6	40	83.9
18	32.5	41	86.9
19	34.4	42	90.0
20	36.4	43	93.3
21	38.4	44	96.6
22	40.4	45	100.0
23	42.4		

EARTHWORK

Bearing Powers

Ground conditions		Bearing power		
		kg/m²	lb/in²	Metric t/m²
Rock,	broken	483	70	50
	solid	2415	350	240
Clay,	dry or hard	380	55	40
	medium dry	190	27	20
	soft or wet	100	14	10
Gravel,	cemented	760	110	80
Sand,	compacted	380	55	40
	clean dry	190	27	20
Swamp and alluvial soils		48	7	5

Earthwork Support

Maximum depth of excavation in various soils without the use of earthwork support

Ground conditions	Feet (ft)	Metres (m)
Compact soil	12	3.66
Drained loam	6	1.83
Dry sand	1	0.3
Gravelly earth	2	0.61
Ordinary earth	3	0.91
Stiff clay	10	3.05

It is important to note that the above table should only be used as a guide. Each case must be taken on its merits and, as the limited distances given above are approached, careful watch must be kept for the slightest signs of caving in

CONCRETE WORK

Weights of Concrete and Concrete Elements

Type of material		kg/m³	lb/cu ft
Ordinary concrete (dense aggregates)			
Non-reinforced plain or mass concrete			
Nominal weight		2305	144
Aggregate	– limestone	2162 to 2407	135 to 150
	– gravel	2244 to 2407	140 to 150
	– broken brick	2000 (av)	125 (av)
	– other crushed stone	2326 to 2489	145 to 155
Reinforced concrete			
Nominal weight		2407	150
Reinforcement	– 1%	2305 to 2468	144 to 154
	– 2%	2356 to 2519	147 to 157
	– 4%	2448 to 2703	153 to 163
Special concretes			
Heavy concrete			
Aggregates	– barytes, magnetite	3210 (min)	200 (min)
	– steel shot, punchings	5280	330
Lean mixes			
Dry-lean (gravel aggregate)		2244	140
Soil-cement (normal mix)		1601	100

CONCRETE WORK

Type of material		kg/m² per mm thick	lb/sq ft per inch thick
Ordinary concrete (dense aggregates)			
Solid slabs (floors, walls etc.)			
Thickness:	75 mm or 3 in	184	37.5
	100 mm or 4 in	245	50
	150 mm or 6 in	378	75
	250 mm or 10 in	612	125
	300 mm or 12 in	734	150
Ribbed slabs			
Thickness:	125 mm or 5 in	204	42
	150 mm or 6 in	219	45
	225 mm or 9 in	281	57
	300 mm or 12 in	342	70
Special concretes			
Finishes etc.			
	Rendering, screed etc. Granolithic, terrazzo	1928 to 2401	10 to 12.5
	Glass-block (hollow) concrete	1734 (approx)	9 (approx)
Prestressed concrete		Weights as for reinforced concrete (upper limits)	
Air-entrained concrete		Weights as for plain or reinforced concrete	

CONCRETE WORK

Average Weight of Aggregates

Materials	Voids %	Weight kg/m³
Sand	39	1660
Gravel 10–20 mm	45	1440
Gravel 35–75 mm	42	1555
Crushed stone	50	1330
Crushed granite (over 15 mm)	50	1345
(n.e. 15 mm)	47	1440
'All-in' ballast	32	1800–2000

Material	kg/m³	lb/cu yd
Vermiculite (aggregate)	64–80	108–135
All-in aggregate	1999	125

Applications and Mix Design

Site mixed concrete

Recommended mix	Class of work suitable for	Cement (kg)	Sand (kg)	Coarse aggregate (kg)	Nr 25 kg bags cement per m^3 of combined aggregate
1:3:6	Roughest type of mass concrete such as footings, road haunching over 300 mm thick	208	905	1509	8.30
1:2.5:5	Mass concrete of better class than 1:3:6 such as bases for machinery, walls below ground etc.	249	881	1474	10.00
1:2:4	Most ordinary uses of concrete, such as mass walls above ground, road slabs etc. and general reinforced concrete work	304	889	1431	12.20
1:1.5:3	Watertight floors, pavements and walls, tanks, pits, steps, paths, surface of 2 course roads, reinforced concrete where extra strength is required	371	801	1336	14.90
1:1:2	Works of thin section such as fence posts and small precast work	511	720	1206	20.40

CONCRETE WORK

Ready mixed concrete

Application	Designated concrete	Standardized prescribed concrete	Recommended consistence (nominal slump class)
Foundations			
Mass concrete fill or blinding	GEN 1	ST2	S3
Strip footings	GEN 1	ST2	S3
Mass concrete foundations			
Single storey buildings	GEN 1	ST2	S3
Double storey buildings	GEN 3	ST4	S3
Trench fill foundations			
Single storey buildings	GEN 1	ST2	S4
Double storey buildings	GEN 3	ST4	S4
General applications			
Kerb bedding and haunching	GEN 0	ST1	S1
Drainage works – immediate support	GEN 1	ST2	S1
Other drainage works	GEN 1	ST2	S3
Oversite below suspended slabs	GEN 1	ST2	S3
Floors			
Garage and house floors with no embedded steel	GEN 3	ST4	S2
Wearing surface: Light foot and trolley traffic	RC30	ST4	S2
Wearing surface: General industrial	RC40	N/A	S2
Wearing surface: Heavy industrial	RC50	N/A	S2
Paving			
House drives, domestic parking and external parking	PAV 1	N/A	S2
Heavy-duty external paving	PAV 2	N/A	S2

CONCRETE WORK

Prescribed Mixes for Ordinary Structural Concrete

Weights of cement and total dry aggregates in kg to produce approximately one cubic metre of fully compacted concrete together with the percentages by weight of fine aggregate in total dry aggregates

Conc. grade	Nominal max size of aggregate (mm)	40		20		14		10	
	Workability	**Med.**	**High**	**Med.**	**High**	**Med.**	**High**	**Med.**	**High**
	Limits to slump that may be expected (mm)	**50–100**	**100–150**	**25–75**	**75–125**	**10–50**	**50–100**	**10–25**	**25–50**
7	Cement (kg)	180	200	210	230	–	–	–	–
	Total aggregate (kg)	1950	1850	1900	1800	–	–	–	–
	Fine aggregate (%)	30–45	30–45	35–50	35–50	–	–	–	–
10	Cement (kg)	210	230	240	260	–	–	–	–
	Total aggregate (kg)	1900	1850	1850	1800	–	–	–	–
	Fine aggregate (%)	30–45	30–45	35–50	35–50	–	–	–	–
15	Cement (kg)	250	270	280	310	–	–	–	–
	Total aggregate (kg)	1850	1800	1800	1750	–	–	–	–
	Fine aggregate (%)	30–45	30–45	35–50	35–50	–	–	–	–
20	Cement (kg)	300	320	320	350	340	380	360	410
	Total aggregate (kg)	1850	1750	1800	1750	1750	1700	1750	1650
	Sand								
	Zone 1 (%)	35	40	40	45	45	50	50	55
	Zone 2 (%)	30	35	35	40	40	45	45	50
	Zone 3 (%)	30	30	30	35	35	40	40	45
25	Cement (kg)	340	360	360	390	380	420	400	450
	Total aggregate (kg)	1800	1750	1750	1700	1700	1650	1700	1600
	Sand								
	Zone 1 (%)	35	40	40	45	45	50	50	55
	Zone 2 (%)	30	35	35	40	40	45	45	50
	Zone 3 (%)	30	30	30	35	35	40	40	45
30	Cement (kg)	370	390	400	430	430	470	460	510
	Total aggregate (kg)	1750	1700	1700	1650	1700	1600	1650	1550
	Sand								
	Zone 1 (%)	35	40	40	45	45	50	50	55
	Zone 2 (%)	30	35	35	40	40	45	45	50
	Zone 3 (%)	30	30	30	35	35	40	40	45

REINFORCEMENT

Weights of Bar Reinforcement

Nominal sizes (mm)	Cross-sectional area (mm^2)	Mass (kg/m)	Length of bar (m/tonne)
6	28.27	0.222	4505
8	50.27	0.395	2534
10	78.54	0.617	1622
12	113.10	0.888	1126
16	201.06	1.578	634
20	314.16	2.466	405
25	490.87	3.853	260
32	804.25	6.313	158
40	1265.64	9.865	101
50	1963.50	15.413	65

Weights of Bars (at specific spacings)

Weights of metric bars in kilogrammes per square metre

Size	Spacing of bars in millimetres									
(mm)	75	100	125	150	175	200	225	250	275	300
6	2.96	2.220	1.776	1.480	1.27	1.110	0.99	0.89	0.81	0.74
8	5.26	3.95	3.16	2.63	2.26	1.97	1.75	1.58	1.44	1.32
10	8.22	6.17	4.93	4.11	3.52	3.08	2.74	2.47	2.24	2.06
12	11.84	8.88	7.10	5.92	5.07	4.44	3.95	3.55	3.23	2.96
16	21.04	15.78	12.63	10.52	9.02	7.89	7.02	6.31	5.74	5.26
20	32.88	24.66	19.73	16.44	14.09	12.33	10.96	9.87	8.97	8.22
25	51.38	38.53	30.83	25.69	22.02	19.27	17.13	15.41	14.01	12.84
32	84.18	63.13	50.51	42.09	36.08	31.57	28.06	25.25	22.96	21.04
40	131.53	98.65	78.92	65.76	56.37	49.32	43.84	39.46	35.87	32.88
50	205.51	154.13	123.31	102.76	88.08	77.07	68.50	61.65	56.05	51.38

Basic weight of steelwork taken as 7850 kg/m^3
Basic weight of bar reinforcement per metre run = 0.00785 kg/mm^2
The value of π has been taken as 3.141592654

REINFORCEMENT

Fabric Reinforcement

Preferred range of designated fabric types and stock sheet sizes

Fabric reference	Longitudinal wires			Cross wires			
	Nominal wire size (mm)	Pitch (mm)	Area (mm²/m)	Nominal wire size (mm)	Pitch (mm)	Area (mm²/m)	Mass (kg/m²)
Square mesh							
A393	10	200	393	10	200	393	6.16
A252	8	200	252	8	200	252	3.95
A193	7	200	193	7	200	193	3.02
A142	6	200	142	6	200	142	2.22
A98	5	200	98	5	200	98	1.54
Structural mesh							
B1131	12	100	1131	8	200	252	10.90
B785	10	100	785	8	200	252	8.14
B503	8	100	503	8	200	252	5.93
B385	7	100	385	7	200	193	4.53
B283	6	100	283	7	200	193	3.73
B196	5	100	196	7	200	193	3.05
Long mesh							
C785	10	100	785	6	400	70.8	6.72
C636	9	100	636	6	400	70.8	5.55
C503	8	100	503	5	400	49.0	4.34
C385	7	100	385	5	400	49.0	3.41
C283	6	100	283	5	400	49.0	2.61
Wrapping mesh							
D98	5	200	98	5	200	98	1.54
D49	2.5	100	49	2.5	100	49	0.77

Stock sheet size 4.8 m × 2.4 m, Area 11.52 m²

Average weight kg/m³ of steelwork reinforcement in concrete for various building elements

Substructure	kg/m³ concrete	Substructure	kg/m³ concrete
Pile caps	110–150	Plate slab	150–220
Tie beams	130–170	Cant slab	145–210
Ground beams	230–330	Ribbed floors	130–200
Bases	125–180	Topping to block floor	30–40
Footings	100–150	Columns	210–310
Retaining walls	150–210	Beams	250–350
Raft	60–70	Stairs	130–170
Slabs – one way	120–200	Walls – normal	40–100
Slabs – two way	110–220	Walls – wind	70–125

Note: For exposed elements add the following %:
Walls 50%, Beams 100%, Columns 15%

FORMWORK

Formwork Stripping Times – Normal Curing Periods

Conditions under which concrete is maturing	Minimum periods of protection for different types of cement					
	Number of days (where the average surface temperature of the concrete exceeds 10°C during the whole period)			Equivalent maturity (degree hours) calculated as the age of the concrete in hours multiplied by the number of degrees Celsius by which the average surface temperature of the concrete exceeds 10°C		
	Other	SRPC	OPC or RHPC	Other	SRPC	OPC or RHPC
1. Hot weather or drying winds	7	4	3	3500	2000	1500
2. Conditions not covered by 1	4	3	2	2000	1500	1000

KEY
OPC – Ordinary Portland Cement
RHPC – Rapid-hardening Portland Cement
SRPC – Sulphate-resisting Portland Cement

Minimum Period before Striking Formwork

	Minimum period before striking		
	Surface temperature of concrete		
	16°C	17°C	t°C (0–25)
Vertical formwork to columns, walls and large beams	12 hours	18 hours	300 hours t+10
Soffit formwork to slabs	4 days	6 days	100 days t+10
Props to slabs	10 days	15 days	250 days t+10
Soffit formwork to beams	9 days	14 days	230 days t+10
Props to beams	14 days	21 days	360 days t+10

MASONRY

Number of Bricks Required for Various Types of Work per m² of Walling

Description	Brick size	
	215 × 102.5 × 50 mm	215 × 102.5 × 65 mm
Half brick thick		
Stretcher bond	74	59
English bond	108	86
English garden wall bond	90	72
Flemish bond	96	79
Flemish garden wall bond	83	66
One brick thick and cavity wall of two half brick skins		
Stretcher bond	148	119

Quantities of Bricks and Mortar Required per m² of Walling

	Unit	No of bricks required	Mortar required (cubic metres)		
Standard bricks			No frogs	Single frogs	Double frogs
Brick size 215 × 102.5 × 50 mm					
half brick wall (103 mm)	m^2	72	0.022	0.027	0.032
2 × half brick cavity wall (270 mm)	m^2	144	0.044	0.054	0.064
one brick wall (215 mm)	m^2	144	0.052	0.064	0.076
one and a half brick wall (322 mm)	m^2	216	0.073	0.091	0.108
Mass brickwork	m^3	576	0.347	0.413	0.480
Brick size 215 × 102.5 × 65 mm					
half brick wall (103 mm)	m^2	58	0.019	0.022	0.026
2 × half brick cavity wall (270 mm)	m^2	116	0.038	0.045	0.055
one brick wall (215 mm)	m^2	116	0.046	0.055	0.064
one and a half brick wall (322 mm)	m^2	174	0.063	0.074	0.088
Mass brickwork	m^3	464	0.307	0.360	0.413
Metric modular bricks			Perforated		
Brick size 200 × 100 × 75 mm					
90 mm thick	m^2	67	0.016	0.019	
190 mm thick	m^2	133	0.042	0.048	
290 mm thick	m^2	200	0.068	0.078	
Brick size 200 × 100 × 100 mm					
90 mm thick	m^2	50	0.013	0.016	
190 mm thick	m^2	100	0.036	0.041	
290 mm thick	m^2	150	0.059	0.067	
Brick size 300 × 100 × 75 mm					
90 mm thick	m^2	33	–	0.015	
Brick size 300 × 100 × 100 mm					
90 mm thick	m^2	44	0.015	0.018	

Note: Assuming 10 mm thick joints

MASONRY

Mortar Required per m² Blockwork (9.88 blocks/m²)

Wall thickness Mortar m^3/m^2	75 0.005	90 0.006	100 0.007	125 0.008	140 0.009	190 0.013	215 0.014

Mortar Group	Cement: lime: sand	Masonry cement: sand	Cement: sand with plasticizer
1	1:0–0.25:3		
2	1:0.5:4–4.5	1:2.5-3.5	1:3–4
3	1:1:5–6	1:4–5	1:5–6
4	1:2:8–9	1:5.5–6.5	1:7–8
5	1:3:10–12	1:6.5–7	1:8

Group 1: strong inflexible mortar
Group 5: weak but flexible

All mixes within a group are of approximately similar strength
Frost resistance increases with the use of plasticizers
Cement: lime: sand mixes give the strongest bond and greatest resistance to rain penetration
Masonry cement equals ordinary Portland cement plus a fine neutral mineral filler and an air entraining agent

Calcium Silicate Bricks

Type	Strength	Location
Class 2 crushing strength	14.0 N/mm²	not suitable for walls
Class 3	20.5 N/mm²	walls above dpc
Class 4	27.5 N/mm²	cappings and copings
Class 5	34.5 N/mm²	retaining walls
Class 6	41.5 N/mm²	walls below ground
Class 7	48.5 N/mm²	walls below ground

The Class 7 calcium silicate bricks are therefore equal in strength to Class B bricks
Calcium silicate bricks are not suitable for DPCs

Durability of Bricks	
FL	Frost resistant with low salt content
FN	Frost resistant with normal salt content
ML	Moderately frost resistant with low salt content
MN	Moderately frost resistant with normal salt content

MASONRY

Brickwork Dimensions

No. of horizontal bricks	Dimensions (mm)	No. of vertical courses	Height of vertical courses (mm)
½	112.5	1	75
1	225.0	2	150
1½	337.5	3	225
2	450.0	4	300
2½	562.5	5	375
3	675.0	6	450
3½	787.5	7	525
4	900.0	8	600
4½	1012.5	9	675
5	1125.0	10	750
5½	1237.5	11	825
6	1350.0	12	900
6½	1462.5	13	975
7	1575.0	14	1050
7½	1687.5	15	1125
8	1800.0	16	1200
8½	1912.5	17	1275
9	2025.0	18	1350
9½	2137.5	19	1425
10	2250.0	20	1500
20	4500.0	24	1575
40	9000.0	28	2100
50	11250.0	32	2400
60	13500.0	36	2700
75	16875.0	40	3000

TIMBER

Weights of Timber

Material	kg/m^3	lb/cu ft
General	806 (avg)	50 (avg)
Douglas fir	479	30
Yellow pine, spruce	479	30
Pitch pine	673	42
Larch, elm	561	35
Oak (English)	724 to 959	45 to 60
Teak	643 to 877	40 to 55
Jarrah	959	60
Greenheart	1040 to 1204	65 to 75
Quebracho	1285	80
Material	**kg/m^2 per mm thickness**	**lb/sq ft per inch thickness**
Wooden boarding and blocks		
Softwood	0.48	2.5
Hardwood	0.76	4
Hardboard	1.06	5.5
Chipboard	0.76	4
Plywood	0.62	3.25
Blockboard	0.48	2.5
Fibreboard	0.29	1.5
Wood-wool	0.58	3
Plasterboard	0.96	5
Weather boarding	0.35	1.8

TIMBER

Conversion Tables (for timber only)

Inches	Millimetres	Feet	Metres
1	25	1	0.300
2	50	2	0.600
3	75	3	0.900
4	100	4	1.200
5	125	5	1.500
6	150	6	1.800
7	175	7	2.100
8	200	8	2.400
9	225	9	2.700
10	250	10	3.000
11	275	11	3.300
12	300	12	3.600
13	325	13	3.900
14	350	14	4.200
15	375	15	4.500
16	400	16	4.800
17	425	17	5.100
18	450	18	5.400
19	475	19	5.700
20	500	20	6.000
21	525	21	6.300
22	550	22	6.600
23	575	23	6.900
24	600	24	7.200

Planed Softwood

The finished end section size of planed timber is usually 3/16" less than the original size from which it is produced. This however varies slightly depending upon availability of material and origin of the species used.

Standards (timber) to cubic metres and cubic metres to standards (timber)

Cubic metres	Cubic metres standards	Standards
4.672	1	0.214
9.344	2	0.428
14.017	3	0.642
18.689	4	0.856
23.361	5	1.070
28.033	6	1.284
32.706	7	1.498
37.378	8	1.712
42.050	9	1.926
46.722	10	2.140
93.445	20	4.281
140.167	30	6.421
186.890	40	8.561
233.612	50	10.702
280.335	60	12.842
327.057	70	14.982
373.779	80	17.122

TIMBER

1 cu metre = 35.3148 cu ft = 0.21403 std

1 cu ft = 0.028317 cu metres

1 std = 4.67227 cu metres

Basic sizes of sawn softwood available (cross-sectional areas)

Thickness (mm)	Width (mm)								
	75	100	125	150	175	200	225	250	300
16	X	X	X	X					
19	X	X	X	X					
22	X	X	X	X					
25	X	X	X	X	X	X	X	X	X
32	X	X	X	X	X	X	X	X	X
36	X	X	X	X					
38	X	X	X	X	X	X	X		
44	X	X	X	X	X	X	X	X	X
47*	X	X	X	X	X	X	X	X	X
50	X	X	X	X	X	X	X	X	X
63	X	X	X	X	X	X	X		
75	X	X	X	X	X	X	X	X	
100		X		X		X		X	X
150				X		X			X
200						X			
250								X	
300									X

* This range of widths for 47 mm thickness will usually be found to be available in construction quality only

Note: The smaller sizes below 100 mm thick and 250 mm width are normally but not exclusively of European origin. Sizes beyond this are usually of North and South American origin

Basic lengths of sawn softwood available (metres)

1.80	2.10 2.40 2.70	3.00 3.30 3.60 3.90	4.20 4.50 4.80	5.10 5.40 5.70	6.00 6.30 6.60 6.90	7.20

Note: Lengths of 6.00 m and over will generally only be available from North American species and may have to be recut from larger sizes

TIMBER

Reductions from basic size to finished size by planning of two opposed faces

Purpose	Reductions from basic sizes for timber			
	15–35 mm	36–100 mm	101–150 mm	over 150 mm
a) Constructional timber	3 mm	3 mm	5 mm	6 mm
b) Matching interlocking boards	4 mm	4 mm	6 mm	6 mm
c) Wood trim not specified in BS 584	5 mm	7 mm	7 mm	9 mm
d) Joinery and cabinet work	7 mm	9 mm	11 mm	13 mm

Note: The reduction of width or depth is overall the extreme size and is exclusive of any reduction of the face by the machining of a tongue or lap joints

Maximum Spans for Various Roof Trusses

Maximum permissible spans for rafters for Fink trussed rafters

Basic size (mm)	Actual size (mm)	Pitch (degrees)								
		15 (m)	17.5 (m)	20 (m)	22.5 (m)	25 (m)	27.5 (m)	30 (m)	32.5 (m)	35 (m)
38 × 75	35 × 72	6.03	6.16	6.29	6.41	6.51	6.60	6.70	6.80	6.90
38 × 100	35 × 97	7.48	7.67	7.83	7.97	8.10	8.22	8.34	8.47	8.61
38 × 125	35 × 120	8.80	9.00	9.20	9.37	9.54	9.68	9.82	9.98	10.16
44 × 75	41 × 72	6.45	6.59	6.71	6.83	6.93	7.03	7.14	7.24	7.35
44 × 100	41 × 97	8.05	8.23	8.40	8.55	8.68	8.81	8.93	9.09	9.22
44 × 125	41 × 120	9.38	9.60	9.81	9.99	10.15	10.31	10.45	10.64	10.81
50 × 75	47 × 72	6.87	7.01	7.13	7.25	7.35	7.45	7.53	7.67	7.78
50 × 100	47 × 97	8.62	8.80	8.97	9.12	9.25	9.38	9.50	9.66	9.80
50 × 125	47 × 120	10.01	10.24	10.44	10.62	10.77	10.94	11.00	11.00	11.00

TIMBER

Sizes of Internal and External Doorsets

Description	Internal size (mm)	Permissible deviation	External size (mm)	Permissible deviation
Coordinating dimension: height of door leaf height sets	2100		2100	
Coordinating dimension: height of ceiling height set	2300 2350 2400 2700 3000		2300 2350 2400 2700 3000	
Coordinating dimension: width of all doorsets S = Single leaf set D = Double leaf set	600 S 700 S 800 S&D 900 S&D 1000 S&D 1200 D 1500 D 1800 D 2100 D		900 S 1000 S 1200 D 1800 D 2100 D	
Work size: height of door leaf height set	2090	± 2.0	2095	± 2.0
Work size: height of ceiling height set	2285 2335 2385 2685 2985	± 2.0	2295 2345 2395 2695 2995	± 2.0
Work size: width of all doorsets S = Single leaf set D = Double leaf set	590 S 690 S 790 S&D 890 S&D 990 S&D 1190 D 1490 D 1790 D 2090 D	± 2.0	895 S 995 S 1195 D 1495 D 1795 D 2095 D	± 2.0
Width of door leaf in single leaf sets F = Flush leaf P = Panel leaf	526 F 626 F 726 F&P 826 F&P 926 F&P	± 1.5	806 F&P 906 F&P	± 1.5
Width of door leaf in double leaf sets F = Flush leaf P = Panel leaf	362 F 412 F 426 F 562 F&P 712 F&P 826 F&P 1012 F&P	± 1.5	552 F&P 702 F&P 852 F&P 1002 F&P	± 1.5
Door leaf height for all doorsets	2040	± 1.5	1994	± 1.5

ROOFING

Total Roof Loadings for Various Types of Tiles/Slates

		Roof load (slope) kg/m^2		
		Slate/Tile	Roofing underlay and battens^2	Total dead load kg/m
Asbestos cement slate (600 × 300)		21.50	3.14	24.64
Clay tile	interlocking	67.00	5.50	72.50
	plain	43.50	2.87	46.37
Concrete tile	interlocking	47.20	2.69	49.89
	plain	78.20	5.50	83.70
Natural slate (18" × 10")		35.40	3.40	38.80
		Roof load (plan) kg/m^2		
Asbestos cement slate (600 × 300)		28.45	76.50	104.95
Clay tile	interlocking	53.54	76.50	130.04
	plain	83.71	76.50	60.21
Concrete tile	interlocking	57.60	76.50	134.10
	plain	96.64	76.50	173.14

ROOFING

Tiling Data

Product		Lap (mm)	Gauge of battens	No. slates per m²	Battens (m/m²)	Weight as laid (kg/m²)
CEMENT SLATES						
Eternit slates	600 × 300 mm	100	250	13.4	4.00	19.50
(Duracem)		90	255	13.1	3.92	19.20
		80	260	12.9	3.85	19.00
		70	265	12.7	3.77	18.60
	600 × 350 mm	100	250	11.5	4.00	19.50
		90	255	11.2	3.92	19.20
	500 × 250 mm	100	200	20.0	5.00	20.00
		90	205	19.5	4.88	19.50
		80	210	19.1	4.76	19.00
		70	215	18.6	4.65	18.60
	400 × 200 mm	90	155	32.3	6.45	20.80
		80	160	31.3	6.25	20.20
		70	165	30.3	6.06	19.60
CONCRETE TILES/SLATES						
Redland Roofing						
Stonewold slate	430 × 380 mm	75	355	8.2	2.82	51.20
Double Roman tile	418 × 330 mm	75	355	8.2	2.91	45.50
Grovebury pantile	418 × 332 mm	75	343	9.7	2.91	47.90
Norfolk pantile	381 × 227 mm	75	306	16.3	3.26	44.01
		100	281	17.8	3.56	48.06
Renown interlocking tile	418 × 330 mm	75	343	9.7	2.91	46.40
'49' tile	381 × 227 mm	75	306	16.3	3.26	44.80
		100	281	17.8	3.56	48.95
Plain, vertical tiling	265 × 165 mm	35	115	52.7	8.70	62.20
Marley Roofing						
Bold roll tile	420 × 330 mm	75	344	9.7	2.90	47.00
		100	–	10.5	3.20	51.00
Modern roof tile	420 × 330 mm	75	338	10.2	3.00	54.00
		100	–	11.0	3.20	58.00
Ludlow major	420 × 330 mm	75	338	10.2	3.00	45.00
		100	–	11.0	3.20	49.00
Ludlow plus	387 × 229 mm	75	305	16.1	3.30	47.00
		100	–	17.5	3.60	51.00
Mendip tile	420 × 330 mm	75	338	10.2	3.00	47.00
		100	–	11.0	3.20	51.00
Wessex	413 × 330 mm	75	338	10.2	3.00	54.00
		100	–	11.0	3.20	58.00
Plain tile	267 × 165 mm	65	100	60.0	10.00	76.00
		75	95	64.0	10.50	81.00
		85	90	68.0	11.30	86.00
Plain vertical tiles (feature)	267 × 165 mm	35	110	53.0	8.70	67.00
		34	115	56.0	9.10	71.00

ROOFING

Slate Nails, Quantity per Kilogram

Length	Type			
	Plain wire	Galvanized wire	Copper nail	Zinc nail
28.5 mm	325	305	325	415
34.4 mm	286	256	254	292
50.8 mm	242	224	194	200

Metal Sheet Coverings

Thicknesses and weights of sheet metal coverings								
Lead to BS 1178								
BS Code No	3	4	5	6	7	8		
Colour code	Green	Blue	Red	Black	White	Orange		
Thickness (mm)	1.25	1.80	2.24	2.50	3.15	3.55		
Density kg/m^2	14.18	20.41	25.40	30.05	35.72	40.26		
Copper to BS 2870								
Thickness (mm)		0.60	0.70					
Bay width								
Roll (mm)		500	650					
Seam (mm)		525	600					
Standard width to form bay	600	750						
Normal length of sheet	1.80	1.80						
Zinc to BS 849								
Zinc Gauge (Nr)	9	10	11	12	13	14	15	16
Thickness (mm)	0.43	0.48	0.56	0.64	0.71	0.79	0.91	1.04
Density (kg/m^2)	3.1	3.2	3.8	4.3	4.8	5.3	6.2	7.0
Aluminium to BS 4868								
Thickness (mm)	0.5	0.6	0.7	0.8	0.9	1.0	1.2	
Density (kg/m^2)	12.8	15.4	17.9	20.5	23.0	25.6	30.7	

ROOFING

Type of felt	Nominal mass per unit area (kg/10 m)	Nominal mass per unit area of fibre base (g/m^2)	Nominal length of roll (m)
Class 1			
1B fine granule	14	220	10 or 20
surfaced bitumen	18	330	10 or 20
	25	470	10
1E mineral surfaced bitumen	38	470	10
1F reinforced bitumen	15	160 (fibre) 110 (hessian)	15
1F reinforced bitumen, aluminium faced	13	160 (fibre) 110 (hessian)	15
Class 2			
2B fine granule surfaced bitumen asbestos	18	500	10 or 20
2E mineral surfaced bitumen asbestos	38	600	10
Class 3			
3B fine granule surfaced bitumen glass fibre	18	60	20
3E mineral surfaced bitumen glass fibre	28	60	10
3E venting base layer bitumen glass fibre	32	60*	10
3H venting base layer bitumen glass fibre	17	60*	20

* Excluding effect of perforations

GLAZING

Nominal thickness (mm)	Tolerance on thickness (mm)	Approximate weight (kg/m²)	Normal maximum size (mm)
Float and polished plate glass			
3	+ 0.2	7.50	2140 × 1220
4	+ 0.2	10.00	2760 × 1220
5	+ 0.2	12.50	3180 × 2100
6	+ 0.2	15.00	4600 × 3180
10	+ 0.3	25.00)	6000 × 3300
12	+ 0.3	30.00)	
15	+ 0.5	37.50	3050 × 3000
19	+ 1.0	47.50)	3000 × 2900
25	+ 1.0	63.50)	
Clear sheet glass			
2 *	+ 0.2	5.00	1920 × 1220
3	+ 0.3	7.50	2130 × 1320
4	+ 0.3	10.00	2760 × 1220
5 *	+ 0.3	12.50)	2130 × 2400
6 *	+ 0.3	15.00)	
Cast glass			
3	+ 0.4		
	− 0.2	6.00)	2140 × 1280
4	+ 0.5	7.50)	
5	+ 0.5	9.50	2140 × 1320
6	+ 0.5	11.50)	3700 × 1280
10	+ 0.8	21.50)	
Wired glass (Cast wired glass)			
6	+ 0.3	−)	
	− 0.7	)	3700 × 1840
7	+ 0.7	−)	
(Polished wire glass)			
6	+ 1.0	−	330 × 1830

* The 5 mm and 6 mm thickness are known as *thick drawn sheet*. Although 2 mm sheet glass is available it is not recommended for general glazing purposes

METAL

Weights of Metals

Material	kg/m³	lb/cu ft
Metals, steel construction, etc.		
Iron		
– cast	7207	450
– wrought	7687	480
– ore – general	2407	150
– (crushed) Swedish	3682	230
Steel	7854	490
Copper		
– cast	8731	545
– wrought	8945	558
Brass	8497	530
Bronze	8945	558
Aluminium	2774	173
Lead	11322	707
Zinc (rolled)	7140	446
	g/mm² per metre	**lb/sq ft per foot**
Steel bars	7.85	3.4
Structural steelwork	Net weight of member @ 7854 kg/m³	
riveted	+ 10% for cleats, rivets, bolts, etc.	
welded	+ 1.25% to 2.5% for welds, etc.	
Rolled sections		
beams	+ 2.5%	
stanchions	+ 5% (extra for caps and bases)	
Plate		
web girders	+ 10% for rivets or welds, stiffeners, etc.	
	kg/m	**lb/ft**
Steel stairs: industrial type		
1 m or 3 ft wide	84	56
Steel tubes		
50 mm or 2 in bore	5 to 6	3 to 4
Gas piping		
20 mm or ¾ in	2	1¼

METAL

Universal Beams BS 4: Part 1: 2005

Designation	Mass (kg/m)	Depth of section (mm)	Width of section (mm)	Thickness		Surface area (m²/m)
				Web (mm)	Flange (mm)	
1016 × 305 × 487	487.0	1036.1	308.5	30.0	54.1	3.20
1016 × 305 × 438	438.0	1025.9	305.4	26.9	49.0	3.17
1016 × 305 × 393	393.0	1016.0	303.0	24.4	43.9	3.15
1016 × 305 × 349	349.0	1008.1	302.0	21.1	40.0	3.13
1016 × 305 × 314	314.0	1000.0	300.0	19.1	35.9	3.11
1016 × 305 × 272	272.0	990.1	300.0	16.5	31.0	3.10
1016 × 305 × 249	249.0	980.2	300.0	16.5	26.0	3.08
1016 × 305 × 222	222.0	970.3	300.0	16.0	21.1	3.06
914 × 419 × 388	388.0	921.0	420.5	21.4	36.6	3.44
914 × 419 × 343	343.3	911.8	418.5	19.4	32.0	3.42
914 × 305 × 289	289.1	926.6	307.7	19.5	32.0	3.01
914 × 305 × 253	253.4	918.4	305.5	17.3	27.9	2.99
914 × 305 × 224	224.2	910.4	304.1	15.9	23.9	2.97
914 × 305 × 201	200.9	903.0	303.3	15.1	20.2	2.96
838 × 292 × 226	226.5	850.9	293.8	16.1	26.8	2.81
838 × 292 × 194	193.8	840.7	292.4	14.7	21.7	2.79
838 × 292 × 176	175.9	834.9	291.7	14.0	18.8	2.78
762 × 267 × 197	196.8	769.8	268.0	15.6	25.4	2.55
762 × 267 × 173	173.0	762.2	266.7	14.3	21.6	2.53
762 × 267 × 147	146.9	754.0	265.2	12.8	17.5	2.51
762 × 267 × 134	133.9	750.0	264.4	12.0	15.5	2.51
686 × 254 × 170	170.2	692.9	255.8	14.5	23.7	2.35
686 × 254 × 152	152.4	687.5	254.5	13.2	21.0	2.34
686 × 254 × 140	140.1	383.5	253.7	12.4	19.0	2.33
686 × 254 × 125	125.2	677.9	253.0	11.7	16.2	2.32
610 × 305 × 238	238.1	635.8	311.4	18.4	31.4	2.45
610 × 305 × 179	179.0	620.2	307.1	14.1	23.6	2.41
610 × 305 × 149	149.1	612.4	304.8	11.8	19.7	2.39
610 × 229 × 140	139.9	617.2	230.2	13.1	22.1	2.11
610 × 229 × 125	125.1	612.2	229.0	11.9	19.6	2.09
610 × 229 × 113	113.0	607.6	228.2	11.1	17.3	2.08
610 × 229 × 101	101.2	602.6	227.6	10.5	14.8	2.07
533 × 210 × 122	122.0	544.5	211.9	12.7	21.3	1.89
533 × 210 × 109	109.0	539.5	210.8	11.6	18.8	1.88
533 × 210 × 101	101.0	536.7	210.0	10.8	17.4	1.87
533 × 210 × 92	92.1	533.1	209.3	10.1	15.6	1.86
533 × 210 × 82	82.2	528.3	208.8	9.6	13.2	1.85
457 × 191 × 98	98.3	467.2	192.8	11.4	19.6	1.67
457 × 191 × 89	89.3	463.4	191.9	10.5	17.7	1.66
457 × 191 × 82	82.0	460.0	191.3	9.9	16.0	1.65
457 × 191 × 74	74.3	457.0	190.4	9.0	14.5	1.64
457 × 191 × 67	67.1	453.4	189.9	8.5	12.7	1.63
457 × 152 × 82	82.1	465.8	155.3	10.5	18.9	1.51
457 × 152 × 74	74.2	462.0	154.4	9.6	17.0	1.50
457 × 152 × 67	67.2	458.0	153.8	9.0	15.0	1.50
457 × 152 × 60	59.8	454.6	152.9	8.1	13.3	1.50
457 × 152 × 52	52.3	449.8	152.4	7.6	10.9	1.48
406 × 178 × 74	74.2	412.8	179.5	9.5	16.0	1.51
406 × 178 × 67	67.1	409.4	178.8	8.8	14.3	1.50
406 × 178 × 60	60.1	406.4	177.9	7.9	12.8	1.49

METAL

Designation	Mass (kg/m)	Depth of section (mm)	Width of section (mm)	Thickness		Surface area (m^2/m)
				Web (mm)	Flange (mm)	
406 × 178 × 50	54.1	402.6	177.7	7.7	10.9	1.48
406 × 140 × 46	46.0	403.2	142.2	6.8	11.2	1.34
406 × 140 × 39	39.0	398.0	141.8	6.4	8.6	1.33
356 × 171 × 67	67.1	363.4	173.2	9.1	15.7	1.38
356 × 171 × 57	57.0	358.0	172.2	8.1	13.0	1.37
356 × 171 × 51	51.0	355.0	171.5	7.4	11.5	1.36
356 × 171 × 45	45.0	351.4	171.1	7.0	9.7	1.36
356 × 127 × 39	39.1	353.4	126.0	6.6	10.7	1.18
356 × 127 × 33	33.1	349.0	125.4	6.0	8.5	1.17
305 × 165 × 54	54.0	310.4	166.9	7.9	13.7	1.26
305 × 165 × 46	46.1	306.6	165.7	6.7	11.8	1.25
305 × 165 × 40	40.3	303.4	165.0	6.0	10.2	1.24
305 × 127 × 48	48.1	311.0	125.3	9.0	14.0	1.09
305 × 127 × 42	41.9	307.2	124.3	8.0	12.1	1.08
305 × 127 × 37	37.0	304.4	123.3	7.1	10.7	1.07
305 × 102 × 33	32.8	312.7	102.4	6.6	10.8	1.01
305 × 102 × 28	28.2	308.7	101.8	6.0	8.8	1.00
305 × 102 × 25	24.8	305.1	101.6	5.8	7.0	0.992
254 × 146 × 43	43.0	259.6	147.3	7.2	12.7	1.08
254 × 146 × 37	37.0	256.0	146.4	6.3	10.9	1.07
254 × 146 × 31	31.1	251.4	146.1	6.0	8.6	1.06
254 × 102 × 28	28.3	260.4	102.2	6.3	10.0	0.904
254 × 102 × 25	25.2	257.2	101.9	6.0	8.4	0.897
254 × 102 × 22	22.0	254.0	101.6	5.7	6.8	0.890
203 × 133 × 30	30.0	206.8	133.9	6.4	9.6	0.923
203 × 133 × 25	25.1	203.2	133.2	5.7	7.8	0.915
203 × 102 × 23	23.1	203.2	101.8	5.4	9.3	0.790
178 × 102 × 19	19.0	177.8	101.2	4.8	7.9	0.738
152 × 89 × 16	16.0	152.4	88.7	4.5	7.7	0.638
127 × 76 × 13	13.0	127.0	76.0	4.0	7.6	0.537

METAL

Universal Columns BS 4: Part 1: 2005

Designation	Mass (kg/m)	Depth of section (mm)	Width of section (mm)	Thickness		Surface area (m^2/m)
				Web (mm)	Flange (mm)	
356 × 406 × 634	633.9	474.7	424.0	47.6	77.0	2.52
356 × 406 × 551	551.0	455.6	418.5	42.1	67.5	2.47
356 × 406 × 467	467.0	436.6	412.2	35.8	58.0	2.42
356 × 406 × 393	393.0	419.0	407.0	30.6	49.2	2.38
356 × 406 × 340	339.9	406.4	403.0	26.6	42.9	2.35
356 × 406 × 287	287.1	393.6	399.0	22.6	36.5	2.31
356 × 406 × 235	235.1	381.0	384.8	18.4	30.2	2.28
356 × 368 × 202	201.9	374.6	374.7	16.5	27.0	2.19
356 × 368 × 177	177.0	368.2	372.6	14.4	23.8	2.17
356 × 368 × 153	152.9	362.0	370.5	12.3	20.7	2.16
356 × 368 × 129	129.0	355.6	368.6	10.4	17.5	2.14
305 × 305 × 283	282.9	365.3	322.2	26.8	44.1	1.94
305 × 305 × 240	240.0	352.5	318.4	23.0	37.7	1.91
305 × 305 × 198	198.1	339.9	314.5	19.1	31.4	1.87
305 × 305 × 158	158.1	327.1	311.2	15.8	25.0	1.84
305 × 305 × 137	136.9	320.5	309.2	13.8	21.7	1.82
305 × 305 × 118	117.9	314.5	307.4	12.0	18.7	1.81
305 × 305 × 97	96.9	307.9	305.3	9.9	15.4	1.79
254 × 254 × 167	167.1	289.1	265.2	19.2	31.7	1.58
254 × 254 × 132	132.0	276.3	261.3	15.3	25.3	1.55
254 × 254 × 107	107.1	266.7	258.8	12.8	20.5	1.52
254 × 254 × 89	88.9	260.3	256.3	10.3	17.3	1.50
254 × 254 × 73	73.1	254.1	254.6	8.6	14.2	1.49
203 × 203 × 86	86.1	222.2	209.1	12.7	20.5	1.24
203 × 203 × 71	71.0	215.8	206.4	10.0	17.3	1.22
203 × 203 × 60	60.0	209.6	205.8	9.4	14.2	1.21
203 × 203 × 52	52.0	206.2	204.3	7.9	12.5	1.20
203 × 203 × 46	46.1	203.2	203.6	7.2	11.0	1.19
152 × 152 × 37	37.0	161.8	154.4	8.0	11.5	0.912
152 × 152 × 30	30.0	157.6	152.9	6.5	9.4	0.901
152 × 152 × 23	23.0	152.4	152.2	5.8	6.8	0.889

METAL

Joists BS 4: Part 1: 2005 (retained for reference, Corus have ceased manufacture in UK)

Designation	Mass (kg/m)	Depth of section (mm)	Width of section (mm)	Thickness		Surface area (m^2/m)
				Web (mm)	Flange (mm)	
254 × 203 × 82	82.0	254.0	203.2	10.2	19.9	1.210
203 × 152 × 52	52.3	203.2	152.4	8.9	16.5	0.932
152 × 127 × 37	37.3	152.4	127.0	10.4	13.2	0.737
127 × 114 × 29	29.3	127.0	114.3	10.2	11.5	0.646
127 × 114 × 27	26.9	127.0	114.3	7.4	11.4	0.650
102 × 102 × 23	23.0	101.6	101.6	9.5	10.3	0.549
102 × 44 × 7	7.5	101.6	44.5	4.3	6.1	0.350
89 × 89 × 19	19.5	88.9	88.9	9.5	9.9	0.476
76 × 76 × 13	12.8	76.2	76.2	5.1	8.4	0.411

Parallel Flange Channels

Designation	Mass (kg/m)	Depth of section (mm)	Width of section (mm)	Thickness		Surface area (m^2/m)
				Web (mm)	Flange (mm)	
430 × 100 × 64	64.4	430	100	11.0	19.0	1.23
380 × 100 × 54	54.0	380	100	9.5	17.5	1.13
300 × 100 × 46	45.5	300	100	9.0	16.5	0.969
300 × 90 × 41	41.4	300	90	9.0	15.5	0.932
260 × 90 × 35	34.8	260	90	8.0	14.0	0.854
260 × 75 × 28	27.6	260	75	7.0	12.0	0.79
230 × 90 × 32	32.2	230	90	7.5	14.0	0.795
230 × 75 × 26	25.7	230	75	6.5	12.5	0.737
200 × 90 × 30	29.7	200	90	7.0	14.0	0.736
200 × 75 × 23	23.4	200	75	6.0	12.5	0.678
180 × 90 × 26	26.1	180	90	6.5	12.5	0.697
180 × 75 × 20	20.3	180	75	6.0	10.5	0.638
150 × 90 × 24	23.9	150	90	6.5	12.0	0.637
150 × 75 × 18	17.9	150	75	5.5	10.0	0.579
125 × 65 × 15	14.8	125	65	5.5	9.5	0.489
100 × 50 × 10	10.2	100	50	5.0	8.5	0.382

METAL

Equal Angles BS EN 10056-1

Designation	Mass (kg/m)	Surface area (m²/m)
200 × 200 × 24	71.1	0.790
200 × 200 × 20	59.9	0.790
200 × 200 × 18	54.2	0.790
200 × 200 × 16	48.5	0.790
150 × 150 × 18	40.1	0.59
150 × 150 × 15	33.8	0.59
150 × 150 × 12	27.3	0.59
150 × 150 × 10	23.0	0.59
120 × 120 × 15	26.6	0.47
120 × 120 × 12	21.6	0.47
120 × 120 × 10	18.2	0.47
120 × 120 × 8	14.7	0.47
100 × 100 × 15	21.9	0.39
100 × 100 × 12	17.8	0.39
100 × 100 × 10	15.0	0.39
100 × 100 × 8	12.2	0.39
90 × 90 × 12	15.9	0.35
90 × 90 × 10	13.4	0.35
90 × 90 × 8	10.9	0.35
90 × 90 × 7	9.61	0.35
90 × 90 × 6	8.30	0.35

Unequal Angles BS EN 10056-1

Designation	Mass (kg/m)	Surface area (m²/m)
200 × 150 × 18	47.1	0.69
200 × 150 × 15	39.6	0.69
200 × 150 × 12	32.0	0.69
200 × 100 × 15	33.7	0.59
200 × 100 × 12	27.3	0.59
200 × 100 × 10	23.0	0.59
150 × 90 × 15	26.6	0.47
150 × 90 × 12	21.6	0.47
150 × 90 × 10	18.2	0.47
150 × 75 × 15	24.8	0.44
150 × 75 × 12	20.2	0.44
150 × 75 × 10	17.0	0.44
125 × 75 × 12	17.8	0.40
125 × 75 × 10	15.0	0.40
125 × 75 × 8	12.2	0.40
100 × 75 × 12	15.4	0.34
100 × 75 × 10	13.0	0.34
100 × 75 × 8	10.6	0.34
100 × 65 × 10	12.3	0.32
100 × 65 × 8	9.94	0.32
100 × 65 × 7	8.77	0.32

METAL

Structural Tees Split from Universal Beams BS 4: Part 1: 2005

Designation	Mass (kg/m)	Surface area (m²/m)
305 × 305 × 90	89.5	1.22
305 × 305 × 75	74.6	1.22
254 × 343 × 63	62.6	1.19
229 × 305 × 70	69.9	1.07
229 × 305 × 63	62.5	1.07
229 × 305 × 57	56.5	1.07
229 × 305 × 51	50.6	1.07
210 × 267 × 61	61.0	0.95
210 × 267 × 55	54.5	0.95
210 × 267 × 51	50.5	0.95
210 × 267 × 46	46.1	0.95
210 × 267 × 41	41.1	0.95
191 × 229 × 49	49.2	0.84
191 × 229 × 45	44.6	0.84
191 × 229 × 41	41.0	0.84
191 × 229 × 37	37.1	0.84
191 × 229 × 34	33.6	0.84
152 × 229 × 41	41.0	0.76
152 × 229 × 37	37.1	0.76
152 × 229 × 34	33.6	0.76
152 × 229 × 30	29.9	0.76
152 × 229 × 26	26.2	0.76

Universal Bearing Piles BS 4: Part 1: 2005

Designation	Mass (kg/m)	Depth of Section (mm)	Width of section (mm)	Thickness	
				Web (mm)	Flange (mm)
356 × 368 × 174	173.9	361.4	378.5	20.3	20.4
356 × 368 × 152	152.0	356.4	376.0	17.8	17.9
356 × 368 × 133	133.0	352.0	373.8	15.6	15.7
356 × 368 × 109	108.9	346.4	371.0	12.8	12.9
305 × 305 × 223	222.9	337.9	325.7	30.3	30.4
305 × 305 × 186	186.0	328.3	320.9	25.5	25.6
305 × 305 × 149	149.1	318.5	316.0	20.6	20.7
305 × 305 × 126	126.1	312.3	312.9	17.5	17.6
305 × 305 × 110	110.0	307.9	310.7	15.3	15.4
305 × 305 × 95	94.9	303.7	308.7	13.3	13.3
305 × 305 × 88	88.0	301.7	307.8	12.4	12.3
305 × 305 × 79	78.9	299.3	306.4	11.0	11.1
254 × 254 × 85	85.1	254.3	260.4	14.4	14.3
254 × 254 × 71	71.0	249.7	258.0	12.0	12.0
254 × 254 × 63	63.0	247.1	256.6	10.6	10.7
203 × 203 × 54	53.9	204.0	207.7	11.3	11.4
203 × 203 × 45	44.9	200.2	205.9	9.5	9.5

METAL

Hot Formed Square Hollow Sections EN 10210 S275J2H & S355J2H

Size (mm)	Wall thickness (mm)	Mass (kg/m)	Superficial area (m^2/m)
40 × 40	2.5	2.89	0.154
	3.0	3.41	0.152
	3.2	3.61	0.152
	3.6	4.01	0.151
	4.0	4.39	0.150
	5.0	5.28	0.147
50 × 50	2.5	3.68	0.194
	3.0	4.35	0.192
	3.2	4.62	0.192
	3.6	5.14	0.191
	4.0	5.64	0.190
	5.0	6.85	0.187
	6.0	7.99	0.185
	6.3	8.31	0.184
60 × 60	3.0	5.29	0.232
	3.2	5.62	0.232
	3.6	6.27	0.231
	4.0	6.90	0.230
	5.0	8.42	0.227
	6.0	9.87	0.225
	6.3	10.30	0.224
	8.0	12.50	0.219
70 × 70	3.0	6.24	0.272
	3.2	6.63	0.272
	3.6	7.40	0.271
	4.0	8.15	0.270
	5.0	9.99	0.267
	6.0	11.80	0.265
	6.3	12.30	0.264
	8.0	15.00	0.259
80 × 80	3.2	7.63	0.312
	3.6	8.53	0.311
	4.0	9.41	0.310
	5.0	11.60	0.307
	6.0	13.60	0.305
	6.3	14.20	0.304
	8.0	17.50	0.299
90 × 90	3.6	9.66	0.351
	4.0	10.70	0.350
	5.0	13.10	0.347
	6.0	15.50	0.345
	6.3	16.20	0.344
	8.0	20.10	0.339
100 × 100	3.6	10.80	0.391
	4.0	11.90	0.390
	5.0	14.70	0.387
	6.0	17.40	0.385
	6.3	18.20	0.384
	8.0	22.60	0.379
	10.0	27.40	0.374
120 × 120	4.0	14.40	0.470
	5.0	17.80	0.467
	6.0	21.20	0.465

METAL

Size (mm)	Wall thickness (mm)	Mass (kg/m)	Superficial area (m²/m)
	6.3	22.20	0.464
	8.0	27.60	0.459
	10.0	33.70	0.454
	12.0	39.50	0.449
	12.5	40.90	0.448
140 × 140	5.0	21.00	0.547
	6.0	24.90	0.545
	6.3	26.10	0.544
	8.0	32.60	0.539
	10.0	40.00	0.534
	12.0	47.00	0.529
	12.5	48.70	0.528
150 × 150	5.0	22.60	0.587
	6.0	26.80	0.585
	6.3	28.10	0.584
	8.0	35.10	0.579
	10.0	43.10	0.574
	12.0	50.80	0.569
	12.5	52.70	0.568
Hot formed from seamless hollow	16.0	65.2	0.559
160 × 160	5.0	24.10	0.627
	6.0	28.70	0.625
	6.3	30.10	0.624
	8.0	37.60	0.619
	10.0	46.30	0.614
	12.0	54.60	0.609
	12.5	56.60	0.608
	16.0	70.20	0.599
180 × 180	5.0	27.30	0.707
	6.0	32.50	0.705
	6.3	34.00	0.704
	8.0	42.70	0.699
	10.0	52.50	0.694
	12.0	62.10	0.689
	12.5	64.40	0.688
	16.0	80.20	0.679
200 × 200	5.0	30.40	0.787
	6.0	36.20	0.785
	6.3	38.00	0.784
	8.0	47.70	0.779
	10.0	58.80	0.774
	12.0	69.60	0.769
	12.5	72.30	0.768
	16.0	90.30	0.759
250 × 250	5.0	38.30	0.987
	6.0	45.70	0.985
	6.3	47.90	0.984
	8.0	60.30	0.979
	10.0	74.50	0.974
	12.0	88.50	0.969
	12.5	91.90	0.968
	16.0	115.00	0.959

METAL

Size (mm)	Wall thickness (mm)	Mass (kg/m)	Superficial area (m^2/m)
300 × 300	6.0	55.10	1.18
	6.3	57.80	1.18
	8.0	72.80	1.18
	10.0	90.20	1.17
	12.0	107.00	1.17
	12.5	112.00	1.17
	16.0	141.00	1.16
350 × 350	8.0	85.40	1.38
	10.0	106.00	1.37
	12.0	126.00	1.37
	12.5	131.00	1.37
	16.0	166.00	1.36
400 × 400	8.0	97.90	1.58
	10.0	122.00	1.57
	12.0	145.00	1.57
	12.5	151.00	1.57
	16.0	191.00	1.56
(Grade S355J2H only)	20.00*	235.00	1.55

Note: * SAW process

METAL

Hot Formed Square Hollow Sections JUMBO RHS: JIS G3136

Size (mm)	Wall thickness (mm)	Mass (kg/m)	Superficial area (m²/m)
350 × 350	19.0	190.00	1.33
	22.0	217.00	1.32
	25.0	242.00	1.31
400 × 400	22.0	251.00	1.52
	25.0	282.00	1.51
450 × 450	12.0	162.00	1.76
	16.0	213.00	1.75
	19.0	250.00	1.73
	22.0	286.00	1.72
	25.0	321.00	1.71
	28.0 *	355.00	1.70
	32.0 *	399.00	1.69
500 × 500	12.0	181.00	1.96
	16.0	238.00	1.95
	19.0	280.00	1.93
	22.0	320.00	1.92
	25.0	360.00	1.91
	28.0 *	399.00	1.90
	32.0 *	450.00	1.89
	36.0 *	498.00	1.88
550 × 550	16.0	263.00	2.15
	19.0	309.00	2.13
	22.0	355.00	2.12
	25.0	399.00	2.11
	28.0 *	443.00	2.10
	32.0 *	500.00	2.09
	36.0 *	555.00	2.08
	40.0 *	608.00	2.06
600 × 600	25.0 *	439.00	2.31
	28.0 *	487.00	2.30
	32.0 *	550.00	2.29
	36.0 *	611.00	2.28
	40.0 *	671.00	2.26
700 × 700	25.0 *	517.00	2.71
	28.0 *	575.00	2.70
	32.0 *	651.00	2.69
	36.0 *	724.00	2.68
	40.0 *	797.00	2.68

Note: * SAW process

METAL

Hot Formed Rectangular Hollow Sections: EN10210 S275J2h & S355J2H

Size (mm)	Wall thickness (mm)	Mass (kg/m)	Superficial area (m²/m)
50 × 30	2.5	2.89	0.154
	3.0	3.41	0.152
	3.2	3.61	0.152
	3.6	4.01	0.151
	4.0	4.39	0.150
	5.0	5.28	0.147
60 × 40	2.5	3.68	0.194
	3.0	4.35	0.192
	3.2	4.62	0.192
	3.6	5.14	0.191
	4.0	5.64	0.190
	5.0	6.85	0.187
	6.0	7.99	0.185
	6.3	8.31	0.184
80 × 40	3.0	5.29	0.232
	3.2	5.62	0.232
	3.6	6.27	0.231
	4.0	6.90	0.230
	5.0	8.42	0.227
	6.0	9.87	0.225
	6.3	10.30	0.224
	8.0	12.50	0.219
76.2 × 50.8	3.0	5.62	0.246
	3.2	5.97	0.246
	3.6	6.66	0.245
	4.0	7.34	0.244
	5.0	8.97	0.241
	6.0	10.50	0.239
	6.3	11.00	0.238
	8.0	13.40	0.233
90 × 50	3.0	6.24	0.272
	3.2	6.63	0.272
	3.6	7.40	0.271
	4.0	8.15	0.270
	5.0	9.99	0.267
	6.0	11.80	0.265
	6.3	12.30	0.264
	8.0	15.00	0.259
100 × 50	3.0	6.71	0.292
	3.2	7.13	0.292
	3.6	7.96	0.291
	4.0	8.78	0.290
	5.0	10.80	0.287
	6.0	12.70	0.285
	6.3	13.30	0.284
	8.0	16.30	0.279

METAL

Size (mm)	Wall thickness (mm)	Mass (kg/m)	Superficial area (m²/m)
100 × 60	3.0	7.18	0.312
	3.2	7.63	0.312
	3.6	8.53	0.311
	4.0	9.41	0.310
	5.0	11.60	0.307
	6.0	13.60	0.305
	6.3	14.20	0.304
	8.0	17.50	0.299
120 × 60	3.6	9.70	0.351
	4.0	10.70	0.350
	5.0	13.10	0.347
	6.0	15.50	0.345
	6.3	16.20	0.344
	8.0	20.10	0.339
120 × 80	3.6	10.80	0.391
	4.0	11.90	0.390
	5.0	14.70	0.387
	6.0	17.40	0.385
	6.3	18.20	0.384
	8.0	22.60	0.379
	10.0	27.40	0.374
150 × 100	4.0	15.10	0.490
	5.0	18.60	0.487
	6.0	22.10	0.485
	6.3	23.10	0.484
	8.0	28.90	0.479
	10.0	35.30	0.474
	12.0	41.40	0.469
	12.5	42.80	0.468
160 × 80	4.0	14.40	0.470
	5.0	17.80	0.467
	6.0	21.20	0.465
	6.3	22.20	0.464
	8.0	27.60	0.459
	10.0	33.70	0.454
	12.0	39.50	0.449
	12.5	40.90	0.448
200 × 100	5.0	22.60	0.587
	6.0	26.80	0.585
	6.3	28.10	0.584
	8.0	35.10	0.579
	10.0	43.10	0.574
	12.0	50.80	0.569
	12.5	52.70	0.568
	16.0	65.20	0.559
250 × 150	5.0	30.40	0.787
	6.0	36.20	0.785
	6.3	38.00	0.784
	8.0	47.70	0.779
	10.0	58.80	0.774
	12.0	69.60	0.769
	12.5	72.30	0.768
	16.0	90.30	0.759

METAL

Size (mm)	Wall thickness (mm)	Mass (kg/m)	Superficial area (m^2/m)
300 × 200	5.0	38.30	0.987
	6.0	45.70	0.985
	6.3	47.90	0.984
	8.0	60.30	0.979
	10.0	74.50	0.974
	12.0	88.50	0.969
	12.5	91.90	0.968
	16.0	115.00	0.959
400 × 200	6.0	55.10	1.18
	6.3	57.80	1.18
	8.0	72.80	1.18
	10.0	90.20	1.17
	12.0	107.00	1.17
	12.5	112.00	1.17
	16.0	141.00	1.16
450 × 250	8.0	85.40	1.38
	10.0	106.00	1.37
	12.0	126.00	1.37
	12.5	131.00	1.37
	16.0	166.00	1.36
500 × 300	8.0	98.00	1.58
	10.0	122.00	1.57
	12.0	145.00	1.57
	12.5	151.00	1.57
	16.0	191.00	1.56
	20.0	235.00	1.55

METAL

Hot Formed Circular Hollow Sections EN 10210 S275J2H & S355J2H

Outside diameter (mm)	Wall thickness (mm)	Mass (kg/m)	Superficial area (m²/m)
21.3	3.2	1.43	0.067
26.9	3.2	1.87	0.085
33.7	3.0	2.27	0.106
	3.2	2.41	0.106
	3.6	2.67	0.106
	4.0	2.93	0.106
42.4	3.0	2.91	0.133
	3.2	3.09	0.133
	3.6	3.44	0.133
	4.0	3.79	0.133
48.3	2.5	2.82	0.152
	3.0	3.35	0.152
	3.2	3.56	0.152
	3.6	3.97	0.152
	4.0	4.37	0.152
	5.0	5.34	0.152
60.3	2.5	3.56	0.189
	3.0	4.24	0.189
	3.2	4.51	0.189
	3.6	5.03	0.189
	4.0	5.55	0.189
	5.0	6.82	0.189
76.1	2.5	4.54	0.239
	3.0	5.41	0.239
	3.2	5.75	0.239
	3.6	6.44	0.239
	4.0	7.11	0.239
	5.0	8.77	0.239
	6.0	10.40	0.239
	6.3	10.80	0.239
88.9	2.5	5.33	0.279
	3.0	6.36	0.279
	3.2	6.76	0.27
	3.6	7.57	0.279
	4.0	8.38	0.279
	5.0	10.30	0.279
	6.0	12.30	0.279
	6.3	12.80	0.279
114.3	3.0	8.23	0.359
	3.2	8.77	0.359
	3.6	9.83	0.359
	4.0	10.09	0.359
	5.0	13.50	0.359
	6.0	16.00	0.359
	6.3	16.80	0.359

METAL

Outside diameter (mm)	Wall thickness (mm)	Mass (kg/m)	Superficial area (m^2/m)
139.7	3.2	10.80	0.439
	3.6	12.10	0.439
	4.0	13.40	0.439
	5.0	16.60	0.439
	6.0	19.80	0.439
	6.3	20.70	0.439
	8.0	26.00	0.439
	10.0	32.00	0.439
168.3	3.2	13.00	0.529
	3.6	14.60	0.529
	4.0	16.20	0.529
	5.0	20.10	0.529
	6.0	24.00	0.529
	6.3	25.20	0.529
	8.0	31.60	0.529
	10.0	39.00	0.529
	12.0	46.30	0.529
	12.5	48.00	0.529
193.7	5.0	23.30	0.609
	6.0	27.80	0.609
	6.3	29.10	0.609
	8.0	36.60	0.609
	10.0	45.30	0.609
	12.0	53.80	0.609
	12.5	55.90	0.609
219.1	5.0	26.40	0.688
	6.0	31.50	0.688
	6.3	33.10	0.688
	8.0	41.60	0.688
	10.0	51.60	0.688
	12.0	61.30	0.688
	12.5	63.70	0.688
	16.0	80.10	0.688
244.5	5.0	29.50	0.768
	6.0	35.30	0.768
	6.3	37.00	0.768
	8.0	46.70	0.768
	10.0	57.80	0.768
	12.0	68.80	0.768
	12.5	71.50	0.768
	16.0	90.20	0.768
273.0	5.0	33.00	0.858
	6.0	39.50	0.858
	6.3	41.40	0.858
	8.0	52.30	0.858
	10.0	64.90	0.858
	12.0	77.20	0.858
	12.5	80.30	0.858
	16.0	101.00	0.858

METAL

Outside diameter (mm)	Wall thickness (mm)	Mass (kg/m)	Superficial area (m^2/m)
323.9	5.0	39.30	1.02
	6.0	47.00	1.02
	6.3	49.30	1.02
	8.0	62.30	1.02
	10.0	77.40	1.02
	12.0	92.30	1.02
	12.5	96.00	1.02
	16.0	121.00	1.02
355.6	6.3	54.30	1.12
	8.0	68.60	1.12
	10.0	85.30	1.12
	12.0	102.00	1.12
	12.5	106.00	1.12
	16.0	134.00	1.12
406.4	6.3	62.20	1.28
	8.0	79.60	1.28
	10.0	97.80	1.28
	12.0	117.00	1.28
	12.5	121.00	1.28
	16.0	154.00	1.28
457.0	6.3	70.00	1.44
	8.0	88.60	1.44
	10.0	110.00	1.44
	12.0	132.00	1.44
	12.5	137.00	1.44
	16.0	174.00	1.44
508.0	6.3	77.90	1.60
	8.0	98.60	1.60
	10.0	123.00	1.60
	12.0	147.00	1.60
	12.5	153.00	1.60
	16.0	194.00	1.60

METAL

Spacing of Holes in Angles

Nominal leg length (mm)	Spacing of holes						Maximum diameter of bolt or rivet		
	A	B	C	D	E	F	A	B and C	D, E and F
200		75	75	55	55	55		30	20
150		55	55					20	
125		45	60					20	
120									
100	55						24		
90	50						24		
80	45						20		
75	45						20		
70	40						20		
65	35						20		
60	35						16		
50	28						12		
45	25								
40	23								
30	20								
25	15								

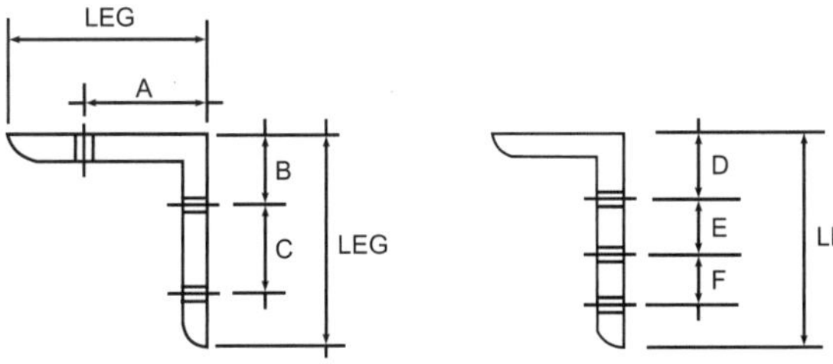

KERBS, PAVING, ETC.

KERBS/EDGINGS/CHANNELS

Precast Concrete Kerbs to BS 7263

Straight kerb units: length from 450 to 915 mm

150 mm high × 125 mm thick		
bullnosed	type BN	
half battered	type HB3	
255 mm high × 125 mm thick		
45° splayed	type SP	
half battered	type HB2	
305 mm high × 150 mm thick		
half battered	type HB1	
Quadrant kerb units		
150 mm high × 305 and 455 mm radius to match	type BN	type QBN
150 mm high × 305 and 455 mm radius to match	type HB2, HB3	type QHB
150 mm high × 305 and 455 mm radius to match	type SP	type QSP
255 mm high × 305 and 455 mm radius to match	type BN	type QBN
255 mm high × 305 and 455 mm radius to match	type HB2, HB3	type QHB
225 mm high × 305 and 455 mm radius to match	type SP	type QSP
Angle kerb units		
305 × 305 × 225 mm high × 125 mm thick		
bullnosed external angle	type XA	
splayed external angle to match type SP	type XA	
bullnosed internal angle	type IA	
splayed internal angle to match type SP	type IA	
Channels		
255 mm wide × 125 mm high flat	type CS1	
150 mm wide × 125 mm high flat type	CS2	
255 mm wide × 125 mm high dished	type CD	

KERBS, PAVING, ETC.

Transition kerb units			
from kerb type SP to HB	left handed	type TL	
	right handed	type TR	
from kerb type BN to HB	left handed	type DL1	
	right handed	type DR1	
from kerb type BN to SP	left handed	type DL2	
	right handed	type DR2	

Number of kerbs required per quarter circle (780 mm kerb lengths)

Radius (m)	Number in quarter circle
12	24
10	20
8	16
6	12
5	10
4	8
3	6
2	4
1	2

Precast Concrete Edgings

Round top type ER	Flat top type EF	Bullnosed top type EBN
150 × 50 mm	150 × 50 mm	150 × 50 mm
200 × 50 mm	200 × 50 mm	200 × 50 mm
250 × 50 mm	250 × 50 mm	250 × 50 mm

KERBS, PAVING, ETC.

BASES

Cement Bound Material for Bases and Subbases

CBM1:	very carefully graded aggregate from 37.5–75 mm, with a 7-day strength of 4.5 N/mm^2
CBM2:	same range of aggregate as CBM1 but with more tolerance in each size of aggregate with a 7-day strength of 7.0 N/mm^2
CBM3:	crushed natural aggregate or blast furnace slag, graded from 37.5–150 mm for 40 mm aggregate, and from 20–75 mm for 20 mm aggregate, with a 7-day strength of 10 N/mm^2
CBM4:	crushed natural aggregate or blast furnace slag, graded from 37.5–150 mm for 40 mm aggregate, and from 20–75 mm for 20 mm aggregate, with a 7-day strength of 15 N/mm^2

INTERLOCKING BRICK/BLOCK ROADS/PAVINGS

Sizes of Precast Concrete Paving Blocks

Type R blocks	**Type S**
200 × 100 × 60 mm	Any shape within a 295 mm space
200 × 100 × 65 mm	
200 × 100 × 80 mm	
200 × 100 × 100 mm	
Sizes of clay brick pavers	
200 × 100 × 50 mm	
200 × 100 × 65 mm	
210 × 105 × 50 mm	
210 × 105 × 65 mm	
215 × 102.5 × 50 mm	
215 × 102.5 × 65 mm	
Type PA: 3 kN	
Footpaths and pedestrian areas, private driveways, car parks, light vehicle traffic and over-run	
Type PB: 7 kN	
Residential roads, lorry parks, factory yards, docks, petrol station forecourts, hardstandings, bus stations	

KERBS, PAVING, ETC.

PAVING AND SURFACING

Weights and Sizes of Paving and Surfacing

Description of item	Size	Quantity per tonne
Paving 50 mm thick	900 × 600 mm	15
Paving 50 mm thick	750 × 600 mm	18
Paving 50 mm thick	600 × 600 mm	23
Paving 50 mm thick	450 × 600 mm	30
Paving 38 mm thick	600 × 600 mm	30
Path edging	914 × 50 × 150 mm	60
Kerb (including radius and tapers)	125 × 254 × 914 mm	15
Kerb (including radius and tapers)	125 × 150 × 914 mm	25
Square channel	125 × 254 × 914 mm	15
Dished channel	125 × 254 × 914 mm	15
Quadrants	300 × 300 × 254 mm	19
Quadrants	450 × 450 × 254 mm	12
Quadrants	300 × 300 × 150 mm	30
Internal angles	300 × 300 × 254 mm	30
Fluted pavement channel	255 × 75 × 914 mm	25
Corner stones	300 × 300 mm	80
Corner stones	360 × 360 mm	60
Cable covers	914 × 175 mm	55
Gulley kerbs	220 × 220 × 150 mm	60
Gulley kerbs	220 × 200 × 75 mm	120

KERBS, PAVING, ETC.

Weights and Sizes of Paving and Surfacing

Material	kg/m³	lb/cu yd
Tarmacadam	2306	3891
Macadam (waterbound)	2563	4325
Vermiculite (aggregate)	64–80	108–135
Terracotta	2114	3568
Cork – compressed	388	24
	kg/m²	**lb/sq ft**
Clay floor tiles, 12.7 mm	27.3	5.6
Pavement lights	122	25
Damp-proof course	5	1
	kg/m² per mm thickness	**lb/sq ft per inch thickness**
Paving slabs (stone)	2.3	12
Granite setts	2.88	15
Asphalt	2.30	12
Rubber flooring	1.68	9
Polyvinyl chloride	1.94 (avg)	10 (avg)

Coverage (m²) Per Cubic Metre of Materials Used as Subbases or Capping Layers

Consolidated thickness laid in (mm)	Square metre coverage		
	Gravel	Sand	Hardcore
50	15.80	16.50	–
75	10.50	11.00	–
100	7.92	8.20	7.42
125	6.34	6.60	5.90
150	5.28	5.50	4.95
175	–	–	4.23
200	–	–	3.71
225	–	–	3.30
300	–	–	2.47

KERBS, PAVING, ETC.

Approximate Rate of Spreads

Average thickness of course (mm)	Description	Approximate rate of spread			
		Open Textured		Dense, Medium & Fine Textured	
		(kg/m²)	(m²/t)	(kg/m²)	(m²/t)
35	14 mm open textured or dense wearing course	60–75	13–17	70–85	12–14
40	20 mm open textured or dense base course	70–85	12–14	80–100	10–12
45	20 mm open textured or dense base course	80–100	10–12	95–100	9–10
50	20 mm open textured or dense, or 28 mm dense base course	85–110	9–12	110–120	8–9
60	28 mm dense base course, 40 mm open textured of dense base course or 40 mm single course as base course		8–10	130–150	7–8
65	28 mm dense base course, 40 mm open textured or dense base course or 40 mm single course	100–135	7–10	140–160	6–7
75	40 mm single course, 40 mm open textured or dense base course, 40 mm dense roadbase	120–150	7–8	165–185	5–6
100	40 mm dense base course or roadbase	–	–	220–240	4–4.5

KERBS, PAVING, ETC.

Surface Dressing Roads: Coverage (m^2) per Tonne of Material

Size in mm	Sand	Granite chips	Gravel	Limestone chips
Sand	168	–	–	–
3	–	148	152	165
6	–	130	133	144
9	–	111	114	123
13	–	85	87	95
19	–	68	71	78

Sizes of Flags

Reference	Nominal size (mm)	Thickness (mm)
A	600 × 450	50 and 63
B	600 × 600	50 and 63
C	600 × 750	50 and 63
D	600 × 900	50 and 63
E	450 × 450	50 and 70 chamfered top surface
F	400 × 400	50 and 65 chamfered top surface
G	300 × 300	50 and 60 chamfered top surface

Sizes of Natural Stone Setts

Width (mm)		Length (mm)		Depth (mm)
100	×	100	×	100
75	×	150 to 250	×	125
75	×	150 to 250	×	150
100	×	150 to 250	×	100
100	×	150 to 250	×	150

SEEDING/TURFING AND PLANTING

Topsoil Quality

Topsoil grade	Properties
Premium	Natural topsoil, high fertility, loamy texture, good soil structure, suitable for intensive cultivation.
General purpose	Natural or manufactured topsoil of lesser quality than Premium, suitable for agriculture or amenity landscape, may need fertilizer or soil structure improvement.
Economy	Selected subsoil, natural mineral deposit such as river silt or greensand. The grade comprises two subgrades; 'Low clay' and 'High clay' which is more liable to compaction in handling. This grade is suitable for low-production agricultural land and amenity woodland or conservation planting areas.

Forms of Trees

Standards:	Shall be clear with substantially straight stems. Grafted and budded trees shall have no more than a slight bend at the union. Standards shall be designated as Half, Extra light, Light, Standard, Selected standard, Heavy, and Extra heavy.
Sizes of Standards	
Heavy standard	12–14 cm girth × 3.50 to 5.00 m high
Extra Heavy standard	14–16 cm girth × 4.25 to 5.00 m high
Extra Heavy standard	16–18 cm girth × 4.25 to 6.00 m high
Extra Heavy standard	18–20 cm girth × 5.00 to 6.00 m high
Semi-mature trees:	Between 6.0 m and 12.0 m tall with a girth of 20 to 75 cm at 1.0 m above ground.
Feathered trees:	Shall have a defined upright central leader, with stem furnished with evenly spread and balanced lateral shoots down to or near the ground.
Whips:	Shall be without significant feather growth as determined by visual inspection.
Multi-stemmed trees:	Shall have two or more main stems at, near, above or below ground.

Seedlings grown from seed and not transplanted shall be specified when ordered for sale as:

1+0	one year old seedling
2+0	two year old seedling
1+1	one year seed bed, one year transplanted = two year old seedling
1+2	one year seed bed, two years transplanted = three year old seedling
2+1	two years seed bed, one year transplanted = three year old seedling
1u1	two years seed bed, undercut after 1 year = two year old seedling
2u2	four years seed bed, undercut after 2 years = four year old seedling

SEEDING/TURFING AND PLANTING

Cuttings

The age of cuttings (plants grown from shoots, stems, or roots of the mother plant) shall be specified when ordered for sale. The height of transplants and undercut seedlings/cuttings (which have been transplanted or undercut at least once) shall be stated in centimetres. The number of growing seasons before and after transplanting or undercutting shall be stated.

0 + 1	one year cutting
0 + 2	two year cutting
0 + 1 + 1	one year cutting bed, one year transplanted = two year old seedling
0 + 1 + 2	one year cutting bed, two years transplanted = three year old seedling

Grass Cutting Capacities in m^2 per hour

Speed mph	Width of cut in metres												
	0.5	0.7	1.0	1.2	1.5	1.7	2.0	2.0	2.1	2.5	2.8	3.0	3.4
1.0	724	1127	1529	1931	2334	2736	3138	3219	3380	4023	4506	4828	5472
1.5	1086	1690	2293	2897	3500	4104	4707	4828	5069	6035	6759	7242	8208
2.0	1448	2253	3058	3862	4667	5472	6276	6437	6759	8047	9012	9656	10944
2.5	1811	2816	3822	4828	5834	6840	7846	8047	8449	10058	11265	12070	13679
3.0	2173	3380	4587	5794	7001	8208	9415	9656	10139	12070	13518	14484	16415
3.5	2535	3943	5351	6759	8167	9576	10984	11265	11829	14082	15772	16898	19151
4.0	2897	4506	6115	7725	9334	10944	12553	12875	13518	16093	18025	19312	21887
4.5	3259	5069	6880	8690	10501	12311	14122	14484	15208	18105	20278	21726	24623
5.0	3621	5633	7644	9656	11668	13679	15691	16093	16898	20117	22531	24140	27359
5.5	3983	6196	8409	10622	12834	15047	17260	17703	18588	22128	24784	26554	30095
6.0	4345	6759	9173	11587	14001	16415	18829	19312	20278	24140	27037	28968	32831
6.5	4707	7322	9938	12553	15168	17783	20398	20921	21967	26152	29290	31382	35566
7.0	5069	7886	10702	13518	16335	19151	21967	22531	23657	28163	31543	33796	38302

Number of Plants per m^2: For Plants Planted on an Evenly Spaced Grid

Planting distances

mm	0.10	0.15	0.20	0.25	0.35	0.40	0.45	0.50	0.60	0.75	0.90	1.00	1.20	1.50
0.10	100.00	66.67	50.00	40.00	28.57	25.00	22.22	20.00	16.67	13.33	11.11	10.00	8.33	6.67
0.15	66.67	44.44	33.33	26.67	19.05	16.67	14.81	13.33	11.11	8.89	7.41	6.67	5.56	4.44
0.20	50.00	33.33	25.00	20.00	14.29	12.50	11.11	10.00	8.33	6.67	5.56	5.00	4.17	3.33
0.25	40.00	26.67	20.00	16.00	11.43	10.00	8.89	8.00	6.67	5.33	4.44	4.00	3.33	2.67
0.35	28.57	19.05	14.29	11.43	8.16	7.14	6.35	5.71	4.76	3.81	3.17	2.86	2.38	1.90
0.40	25.00	16.67	12.50	10.00	7.14	6.25	5.56	5.00	4.17	3.33	2.78	2.50	2.08	1.67
0.45	22.22	14.81	11.11	8.89	6.35	5.56	4.94	4.44	3.70	2.96	2.47	2.22	1.85	1.48
0.50	20.00	13.33	10.00	8.00	5.71	5.00	4.44	4.00	3.33	2.67	2.22	2.00	1.67	1.33
0.60	16.67	11.11	8.33	6.67	4.76	4.17	3.70	3.33	2.78	2.22	1.85	1.67	1.39	1.11
0.75	13.33	8.89	6.67	5.33	3.81	3.33	2.96	2.67	2.22	1.78	1.48	1.33	1.11	0.89
0.90	11.11	7.41	5.56	4.44	3.17	2.78	2.47	2.22	1.85	1.48	1.23	1.11	0.93	0.74
1.00	10.00	6.67	5.00	4.00	2.86	2.50	2.22	2.00	1.67	1.33	1.11	1.00	0.83	0.67
1.20	8.33	5.56	4.17	3.33	2.38	2.08	1.85	1.67	1.39	1.11	0.93	0.83	0.69	0.56
1.50	6.67	4.44	3.33	2.67	1.90	1.67	1.48	1.33	1.11	0.89	0.74	0.67	0.56	0.44

SEEDING/TURFING AND PLANTING

Grass Clippings Wet: Based on 3.5 m³/tonne

Annual kg/100 m²	Average 20 cuts kg/100 m²	m²/tonne	m²/m³
32.0	1.6	61162.1	214067.3

Nr of cuts	22	20	18	16	12	4
kg/cut	1.45	1.60	1.78	2.00	2.67	8.00
	Area capacity of 3 tonne vehicle per load					
m²	206250	187500	168750	150000	112500	37500
Load m³	100 m² units/m³ of vehicle space					
1	196.4	178.6	160.7	142.9	107.1	35.7
2	392.9	357.1	321.4	285.7	214.3	71.4
3	589.3	535.7	482.1	428.6	321.4	107.1
4	785.7	714.3	642.9	571.4	428.6	142.9
5	982.1	892.9	803.6	714.3	535.7	178.6

Transportation of Trees

To unload large trees a machine with the necessary lifting strength is required. The weight of the trees must therefore be known in advance. The following table gives a rough overview. The additional columns with root ball dimensions and the number of plants per trailer provide additional information, for example about preparing planting holes and calculating unloading times.

Girth in cm	Rootball diameter in cm	Ball height in cm	Weight in kg	Numbers of trees per trailer
16–18	50–60	40	150	100–120
18–20	60–70	40–50	200	80–100
20–25	60–70	40–50	270	50–70
25–30	80	50–60	350	50
30–35	90–100	60–70	500	12–18
35–40	100–110	60–70	650	10–15
40–45	110–120	60–70	850	8–12
45–50	110–120	60–70	1100	5–7
50–60	130–140	60–70	1600	1–3
60–70	150–160	60–70	2500	1
70–80	180–200	70	4000	1
80–90	200–220	70–80	5500	1
90–100	230–250	80–90	7500	1
100–120	250–270	80–90	9500	1

Data supplied by Lorenz von Ehren GmbH
The information in the table is approximate; deviations depend on soil type, genus and weather

FENCING AND GATES

Types of Preservative

Creosote (tar oil) can be 'factory' applied	by pressure to BS 144: pts 1&2 by immersion to BS 144: pt 1 by hot and cold open tank to BS 144: pts 1&2
Copper/chromium/arsenic (CCA)	by full cell process to BS 4072 pts 1&2
Organic solvent (OS)	by double vacuum (vacvac) to BS 5707 pts 1&3 by immersion to BS 5057 pts 1&3
Pentachlorophenol (PCP)	by heavy oil double vacuum to BS 5705 pts 2&3
Boron diffusion process (treated with disodium octaborate to BWPA Manual 1986)	

Note: Boron is used on green timber at source and the timber is supplied dry

Cleft Chestnut Pale Fences

Pales	Pale spacing	Wire lines	
900 mm	75 mm	2	temporary protection
1050 mm	75 or 100 mm	2	light protective fences
1200 mm	75 mm	3	perimeter fences
1350 mm	75 mm	3	perimeter fences
1500 mm	50 mm	3	narrow perimeter fences
1800 mm	50 mm	3	light security fences

Close-Boarded Fences

Close-boarded fences 1.05 to 1.8 m high
Type BCR (recessed) or BCM (morticed) with concrete posts 140 × 115 mm tapered and Type BW with timber posts

Palisade Fences

Wooden palisade fences
Type WPC with concrete posts 140 × 115 mm tapered and Type WPW with timber posts

For both types of fence:
Height of fence 1050 mm: two rails
Height of fence 1200 mm: two rails
Height of fence 1500 mm: three rails
Height of fence 1650 mm: three rails
Height of fence 1800 mm: three rails

FENCING AND GATES

Post and Rail Fences

Wooden post and rail fences Type MPR 11/3 morticed rails and Type SPR 11/3 nailed rails Height to top of rail 1100 mm Rails: three rails 87 mm, 38 mm Type MPR 11/4 morticed rails and Type SPR 11/4 nailed rails Height to top of rail 1100 mm Rails: four rails 87 mm, 38 mm Type MPR 13/4 morticed rails and Type SPR 13/4 nailed rails Height to top of rail 1300 mm Rail spacing 250 mm, 250 mm, and 225 mm from top Rails: four rails 87 mm, 38 mm

Steel Posts

Rolled steel angle iron posts for chain link fencing

Posts	Fence height	Strut	Straining post
1500 × 40 × 40 × 5 mm	900 mm	1500 × 40 × 40 × 5 mm	1500 × 50 × 50 × 6 mm
1800 × 40 × 40 × 5 mm	1200 mm	1800 × 40 × 40 × 5 mm	1800 × 50 × 50 × 6 mm
2000 × 45 × 45 × 5 mm	1400 mm	2000 × 45 × 45 × 5 mm	2000 × 60 × 60 × 6 mm
2600 × 45 × 45 × 5 mm	1800 mm	2600 × 45 × 45 × 5 mm	2600 × 60 × 60 × 6 mm
3000 × 50 × 50 × 6 mm with arms	1800 mm	2600 × 45 × 45 × 5 mm	3000 × 60 × 60 × 6 mm

Concrete Posts

Concrete posts for chain link fencing

Posts and straining posts	Fence height	Strut
1570 mm 100 × 100 mm	900 mm	1500 mm × 75 × 75 mm
1870 mm 125 × 125 mm	1200 mm	1830 mm × 100 × 75 mm
2070 mm 125 × 125 mm	1400 mm	1980 mm × 100 × 75 mm
2620 mm 125 × 125 mm	1800 mm	2590 mm × 100 × 85 mm
3040 mm 125 × 125 mm	1800 mm	2590 mm × 100 × 85 mm (with arms)

FENCING AND GATES

Rolled Steel Angle Posts

Rolled steel angle posts for rectangular wire mesh (field) fencing

Posts	Fence height	Strut	Straining post
1200 × 40 × 40 × 5 mm	600 mm	1200 × 75 × 75 mm	1350 × 100 × 100 mm
1400 × 40 × 40 × 5 mm	800 mm	1400 × 75 × 75 mm	1550 × 100 × 100 mm
1500 × 40 × 40 × 5 mm	900 mm	1500 × 75 × 75 mm	1650 × 100 × 100 mm
1600 × 40 × 40 × 5 mm	1000 mm	1600 × 75 × 75 mm	1750 × 100 × 100 mm
1750 × 40 × 40 × 5 mm	1150 mm	1750 × 75 × 100 mm	1900 × 125 × 125 mm

Concrete Posts

Concrete posts for rectangular wire mesh (field) fencing

Posts	Fence height	Strut	Straining post
1270 × 100 × 100 mm	600 mm	1200 × 75 × 75 mm	1420 × 100 × 100 mm
1470 × 100 × 100 mm	800 mm	1350 × 75 × 75 mm	1620 × 100 × 100 mm
1570 × 100 × 100 mm	900 mm	1500 × 75 × 75 mm	1720 × 100 × 100 mm
1670 × 100 × 100 mm	600 mm	1650 × 75 × 75 mm	1820 × 100 × 100 mm
1820 × 125 × 125 mm	1150 mm	1830 × 75 × 100 mm	1970 × 125 × 125 mm

Cleft Chestnut Pale Fences

Timber Posts

Timber posts for wire mesh and hexagonal wire netting fences

Round timber for general fences

Posts	Fence height	Strut	Straining post
1300 × 65 mm dia.	600 mm	1200 × 80 mm dia.	1450 × 100 mm dia.
1500 × 65 mm dia.	800 mm	1400 × 80 mm dia.	1650 × 100 mm dia.
1600 × 65 mm dia.	900 mm	1500 × 80 mm dia.	1750 × 100 mm dia.
1700 × 65 mm dia.	1050 mm	1600 × 80 mm dia.	1850 × 100 mm dia.
1800 × 65 mm dia.	1150 mm	1750 × 80 mm dia.	2000 × 120 mm dia.

Squared timber for general fences

Posts	Fence height	Strut	Straining post
1300 × 75 × 75 mm	600 mm	1200 × 75 × 75 mm	1450 × 100 × 100 mm
1500 × 75 × 75 mm	800 mm	1400 × 75 × 75 mm	1650 × 100 × 100 mm
1600 × 75 × 75 mm	900 mm	1500 × 75 × 75 mm	1750 × 100 × 100 mm
1700 × 75 × 75 mm	1050 mm	1600 × 75 × 75 mm	1850 × 100 × 100 mm
1800 × 75 × 75 mm	1150 mm	1750 × 75 × 75 mm	2000 × 125 × 100 mm

FENCING AND GATES

Steel Fences to BS 1722: Part 9: 1992

	Fence height	Top/bottom rails and flat posts	Vertical bars
Light	1000 mm	40 × 10 mm 450 mm in ground	12 mm dia. at 115 mm cs
	1200 mm	40 × 10 mm 550 mm in ground	12 mm dia. at 115 mm cs
	1400 mm	40 × 10 mm 550 mm in ground	12 mm dia. at 115 mm cs
Light	1000 mm	40 × 10 mm 450 mm in ground	16 mm dia. at 120 mm cs
	1200 mm	40 × 10 mm 550 mm in ground	16 mm dia. at 120 mm cs
	1400 mm	40 × 10 mm 550 mm in ground	16 mm dia. at 120 mm cs
Medium	1200 mm	50 × 10 mm 550 mm in ground	20 mm dia. at 125 mm cs
	1400 mm	50 × 10 mm 550 mm in ground	20 mm dia. at 125 mm cs
	1600 mm	50 × 10 mm 600 mm in ground	22 mm dia. at 145 mm cs
	1800 mm	50 × 10 mm 600 mm in ground	22 mm dia. at 145 mm cs
Heavy	1600 mm	50 × 10 mm 600 mm in ground	22 mm dia. at 145 mm cs
	1800 mm	50 × 10 mm 600 mm in ground	22 mm dia. at 145 mm cs
	2000 mm	50 × 10 mm 600 mm in ground	22 mm dia. at 145 mm cs
	2200 mm	50 × 10 mm 600 mm in ground	22 mm dia. at 145 mm cs

Notes: Mild steel fences: round or square verticals; flat standards and horizontals. Tops of vertical bars may be bow-top, blunt, or pointed. Round or square bar railings

Timber Field Gates to BS 3470: 1975

Gates made to this standard are designed to open one way only
All timber gates are 1100 mm high
Width over stiles 2400, 2700, 3000, 3300, 3600, and 4200 mm
Gates over 4200 mm should be made in two leaves

Steel Field Gates to BS 3470: 1975

All steel gates are 1100 mm high

Heavy duty:	width over stiles 2400, 3000, 3600 and 4500 mm
Light duty:	width over stiles 2400, 3000, and 3600 mm

FENCING AND GATES

Domestic Front Entrance Gates to BS 4092: Part 1: 1966

Metal gates:	Single gates are 900 mm high minimum, 900 mm, 1000 mm and 1100 mm wide

Domestic Front Entrance Gates to BS 4092: Part 2: 1966

Wooden gates:	All rails shall be tenoned into the stiles Single gates are 840 mm high minimum, 801 mm and 1020 mm wide Double gates are 840 mm high minimum, 2130, 2340 and 2640 mm wide

Timber Bridle Gates to BS 5709:1979 (Horse or Hunting Gates)

Gates open one way only	
Minimum width between posts	1525 mm
Minimum height	1100 mm

Timber Kissing Gates to BS 5709:1979

Minimum width	700 mm
Minimum height	1000 mm
Minimum distance between shutting posts	600 mm
Minimum clearance at mid-point	600 mm

Metal Kissing Gates to BS 5709:1979

Sizes are the same as those for timber kissing gates Maximum gaps between rails 120 mm

Categories of Pedestrian Guard Rail to BS 3049:1976

Class A for normal use Class B where vandalism is expected Class C where crowd pressure is likely

DRAINAGE

Width Required for Trenches for Various Diameters of Pipes

Pipe diameter (mm)	Trench n.e. 1.50 m deep	Trench over 1.50 m deep
n.e. 100 mm	450 mm	600 mm
100–150 mm	500 mm	650 mm
150–225 mm	600 mm	750 mm
225–300 mm	650 mm	800 mm
300–400 mm	750 mm	900 mm
400–450 mm	900 mm	1050 mm
450–600 mm	1100 mm	1300 mm

Weights and Dimensions – Vitrified Clay Pipes

Product	Nominal diameter (mm)	Effective length (mm)	BS 65 limits of tolerance		Crushing strength (kN/m)	Weight	
			min (mm)	max (mm)		(kg/pipe)	(kg/m)
Supersleve	100	1600	96	105	35.00	14.71	9.19
	150	1750	146	158	35.00	29.24	16.71
Hepsleve	225	1850	221	236	28.00	84.03	45.42
	300	2500	295	313	34.00	193.05	77.22
	150	1500	146	158	22.00	37.04	24.69
Hepseal	225	1750	221	236	28.00	85.47	48.84
	300	2500	295	313	34.00	204.08	81.63
	400	2500	394	414	44.00	357.14	142.86
	450	2500	444	464	44.00	454.55	181.63
	500	2500	494	514	48.00	555.56	222.22
	600	2500	591	615	57.00	796.23	307.69
	700	3000	689	719	67.00	1111.11	370.45
	800	3000	788	822	72.00	1351.35	450.45
Hepline	100	1600	95	107	22.00	14.71	9.19
	150	1750	145	160	22.00	29.24	16.71
	225	1850	219	239	28.00	84.03	45.42
	300	1850	292	317	34.00	142.86	77.22
Hepduct (conduit)	90	1500	–	–	28.00	12.05	8.03
	100	1600	–	–	28.00	14.71	9.19
	125	1750	–	–	28.00	20.73	11.84
	150	1750	–	–	28.00	29.24	16.71
	225	1850	–	–	28.00	84.03	45.42
	300	1850	–	–	34.00	142.86	77.22

DRAINAGE

Weights and Dimensions – Vitrified Clay Pipes

Nominal internal diameter (mm)	Nominal wall thickness (mm)	Approximate weight (kg/m)
150	25	45
225	29	71
300	32	122
375	35	162
450	38	191
600	48	317
750	54	454
900	60	616
1200	76	912
1500	89	1458
1800	102	1884
2100	127	2619

Wall thickness, weights and pipe lengths vary, depending on type of pipe required

The particulars shown above represent a selection of available diameters and are applicable to strength class 1 pipes with flexible rubber ring joints

Tubes with Ogee joints are also available

DRAINAGE

Weights and Dimensions – PVC-u Pipes

	Nominal size	Mean outside diameter (mm)		Wall thickness	Weight
		min	max	(mm)	(kg/m)
Standard pipes	82.4	82.4	82.7	3.2	1.2
	110.0	110.0	110.4	3.2	1.6
	160.0	160.0	160.6	4.1	3.0
	200.0	200.0	200.6	4.9	4.6
	250.0	250.0	250.7	6.1	7.2
Perforated pipes heavy grade	As above	As above	As above	As above	As above
thin wall	82.4	82.4	82.7	1.7	–
	110.0	110.0	110.4	2.2	–
	160.0	160.0	160.6	3.2	–

Width of Trenches Required for Various Diameters of Pipes

Pipe diameter (mm)	Trench n.e. 1.5 m deep (mm)	Trench over 1.5 m deep (mm)
n.e. 100	450	600
100–150	500	650
150–225	600	750
225–300	650	800
300–400	750	900
400–450	900	1050
450–600	1100	1300

DRAINAGE

DRAINAGE BELOW GROUND AND LAND DRAINAGE

Flow of Water Which Can Be Carried by Various Sizes of Pipe

Clay or concrete pipes

	Gradient of pipeline							
	1:10	1:20	1:30	1:40	1:50	1:60	1:80	1:100
Pipe size	**Flow in litres per second**							
DN 100 15.0	8.5	6.8	5.8	5.2	4.7	4.0	3.5	
DN 150 28.0	19.0	16.0	14.0	12.0	11.0	9.1	8.0	
DN 225 140.0	95.0	76.0	66.0	58.0	53.0	46.0	40.0	

Plastic pipes

	Gradient of pipeline							
	1:10	1:20	1:30	1:40	1:50	1:60	1:80	1:100
Pipe size	**Flow in litres per second**							
82.4 mm i/dia.	12.0	8.5	6.8	5.8	5.2	4.7	4.0	3.5
110 mm i/dia.	28.0	19.0	16.0	14.0	12.0	11.0	9.1	8.0
160 mm i/dia.	76.0	53.0	43.0	37.0	33.0	29.0	25.0	22.0
200 mm i/dia.	140.0	95.0	76.0	66.0	58.0	53.0	46.0	40.0

Vitrified (Perforated) Clay Pipes and Fittings to BS En 295-5 1994

Length not specified		
75 mm bore	**250 mm bore**	**600 mm bore**
100	300	700
125	350	800
150	400	1000
200	450	1200
225	500	

Precast Concrete Pipes: Prestressed Non-pressure Pipes and Fittings: Flexible Joints to BS 5911: Pt. 103: 1994

Rationalized metric nominal sizes: 450, 500	
Length:	500–1000 by 100 increments 1000–2200 by 200 increments 2200–2800 by 300 increments
Angles: length:	450–600 angles 45, 22.5,11.25° 600 or more angles 22.5, 11.25°

DRAINAGE

Precast Concrete Pipes: Unreinforced and Circular Manholes and Soakaways to BS 5911: Pt. 200: 1994

Nominal sizes:	
Shafts:	675, 900 mm
Chambers:	900, 1050, 1200, 1350, 1500, 1800, 2100, 2400, 2700, 3000 mm
Large chambers:	To have either tapered reducing rings or a flat reducing slab in order to accept the standard cover
Ring depths:	1. 300–1200 mm by 300 mm increments except for bottom slab and rings below cover slab, these are by 150 mm increments 2. 250–1000 mm by 250 mm increments except for bottom slab and rings below cover slab, these are by 125 mm increments
Access hole:	750 × 750 mm for DN 1050 chamber 1200 × 675 mm for DN 1350 chamber

Calculation of Soakaway Depth

The following formula determines the depth of concrete ring soakaway that would be required for draining given amounts of water.

$$\mathbf{h} = \frac{\mathbf{4ar}}{\mathbf{3\pi D^2}}$$

h = depth of the chamber below the invert pipe
a = the area to be drained
r = the hourly rate of rainfall (50 mm per hour)
π = pi
D = internal diameter of the soakaway

This table shows the depth of chambers in each ring size which would be required to contain the volume of water specified. These allow a recommended storage capacity of ⅓ (one third of the hourly rainfall figure).

Table Showing Required Depth of Concrete Ring Chambers in Metres

Area m²	**50**	**100**	**150**	**200**	**300**	**400**	**500**
Ring size							
0.9	1.31	2.62	3.93	5.24	7.86	10.48	13.10
1.1	0.96	1.92	2.89	3.85	5.77	7.70	9.62
1.2	0.74	1.47	2.21	2.95	4.42	5.89	7.37
1.4	0.58	1.16	1.75	2.33	3.49	4.66	5.82
1.5	0.47	0.94	1.41	1.89	2.83	3.77	4.72
1.8	0.33	0.65	0.98	1.31	1.96	2.62	3.27
2.1	0.24	0.48	0.72	0.96	1.44	1.92	2.41
2.4	0.18	0.37	0.55	0.74	1.11	1.47	1.84
2.7	0.15	0.29	0.44	0.58	0.87	1.16	1.46
3.0	0.12	0.24	0.35	0.47	0.71	0.94	1.18

DRAINAGE

Precast Concrete Inspection Chambers and Gullies to BS 5911: Part 230: 1994

Nominal sizes:	375 diameter, 750, 900 mm deep 450 diameter, 750, 900, 1050, 1200 mm deep
Depths:	from the top for trapped or untrapped units: centre of outlet 300 mm invert (bottom) of the outlet pipe 400 mm
Depth of water seal for trapped gullies:	85 mm, rodding eye int. dia. 100 mm
Cover slab:	65 mm min

Bedding Flexible Pipes: PVC-u Or Ductile Iron

Type 1 =	100 mm fill below pipe, 300 mm above pipe: single size material
Type 2 =	100 mm fill below pipe, 300 mm above pipe: single size or graded material
Type 3 =	100 mm fill below pipe, 75 mm above pipe with concrete protective slab over
Type 4 =	100 mm fill below pipe, fill laid level with top of pipe
Type 5 =	200 mm fill below pipe, fill laid level with top of pipe
Concrete =	25 mm sand blinding to bottom of trench, pipe supported on chocks,100 mm concrete under the pipe, 150 mm concrete over the pipe

DRAINAGE

Bedding Rigid Pipes: Clay or Concrete
(for vitrified clay pipes the manufacturer should be consulted)

Class D:	Pipe laid on natural ground with cut-outs for joints, soil screened to remove stones over 40 mm and returned over pipe to 150 m min depth. Suitable for firm ground with trenches trimmed by hand.
Class N:	Pipe laid on 50 mm granular material of graded aggregate to Table 4 of BS 882, or 10 mm aggregate to Table 6 of BS 882, or as dug light soil (not clay) screened to remove stones over 10 mm. Suitable for machine dug trenches.
Class B:	As Class N, but with granular bedding extending half way up the pipe diameter.
Class F:	Pipe laid on 100 mm granular fill to BS 882 below pipe, minimum 150 mm granular fill above pipe: single size material. Suitable for machine dug trenches.
Class A:	Concrete 100 mm thick under the pipe extending half way up the pipe, backfilled with the appropriate class of fill. Used where there is only a very shallow fall to the drain. Class A bedding allows the pipes to be laid to an exact gradient.
Concrete surround:	25 mm sand blinding to bottom of trench, pipe supported on chocks, 100 mm concrete under the pipe, 150 mm concrete over the pipe. It is preferable to bed pipes under slabs or wall in granular material.

PIPED SUPPLY SYSTEMS

Identification of Service Tubes From Utility to Dwellings

Utility	Colour	Size	Depth
British Telecom	grey	54 mm od	450 mm
Electricity	black	38 mm od	450 mm
Gas	yellow	42 mm od rigid 60 mm od convoluted	450 mm
Water	may be blue	(normally untubed)	750 mm

ELECTRICAL SUPPLY/POWER/LIGHTING SYSTEMS

Electrical Insulation Class En 60.598 BS 4533

Class 1:	luminaires comply with class 1 (I) earthed electrical requirements
Class 2:	luminaires comply with class 2 (II) double insulated electrical requirements
Class 3:	luminaires comply with class 3 (III) electrical requirements

Protection to Light Fittings

BS EN 60529:1992 Classification for degrees of protection provided by enclosures.
(IP Code – International or ingress Protection)

1st characteristic: against ingress of solid foreign objects		
The figure	2	indicates that fingers cannot enter
	3	that a 2.5 mm diameter probe cannot enter
	4	that a 1.0 mm diameter probe cannot enter
	5	the fitting is dust proof (no dust around live parts)
	6	the fitting is dust tight (no dust entry)
2nd characteristic: ingress of water with harmful effects		
The figure	0	indicates unprotected
	1	vertically dripping water cannot enter
	2	water dripping 15° (tilt) cannot enter
	3	spraying water cannot enter
	4	splashing water cannot enter
	5	jetting water cannot enter
	6	powerful jetting water cannot enter
	7	proof against temporary immersion
	8	proof against continuous immersion
Optional additional codes:		A–D protects against access to hazardous parts
	H	high voltage apparatus
	M	fitting was in motion during water test
	S	fitting was static during water test
	W	protects against weather
Marking code arrangement:		(example) IPX5S = IP (International or Ingress Protection) X (denotes omission of first characteristic) 5 = jetting S = static during water test

RAIL TRACKS

RAIL TRACKS

	kg/m of track	lb/ft of track
Standard gauge Bull-head rails, chairs, transverse timber (softwood) sleepers etc.	245	165
Main lines		
Flat-bottom rails, transverse prestressed concrete sleepers, etc.	418	280
Add for electric third rail	51	35
Add for crushed stone ballast	2600	1750
	kg/m^2	**lb/sq ft**
Overall average weight – rails connections, sleepers, ballast, etc.	733	150
	kg/m of track	**lb/ft of track**
Bridge rails, longitudinal timber sleepers, etc.	112	75

RAIL TRACKS

Heavy Rails

British Standard Section No.	Rail height (mm)	Foot width (mm)	Head width (mm)	Min web thickness (mm)	Section weight (kg/m)
Flat Bottom Rails					
60 A	114.30	109.54	57.15	11.11	30.62
70 A	123.82	111.12	60.32	12.30	34.81
75 A	128.59	114.30	61.91	12.70	37.45
80 A	133.35	117.47	63.50	13.10	39.76
90 A	142.88	127.00	66.67	13.89	45.10
95 A	147.64	130.17	69.85	14.68	47.31
100 A	152.40	133.35	69.85	15.08	50.18
110 A	158.75	139.70	69.85	15.87	54.52
113 A	158.75	139.70	69.85	20.00	56.22
50 'O'	100.01	100.01	52.39	10.32	24.82
80 'O'	127.00	127.00	63.50	13.89	39.74
60R	114.30	109.54	57.15	11.11	29.85
75R	128.59	122.24	61.91	13.10	37.09
80R	133.35	127.00	63.50	13.49	39.72
90R	142.88	136.53	66.67	13.89	44.58
95R	147.64	141.29	68.26	14.29	47.21
100R	152.40	146.05	69.85	14.29	49.60
95N	147.64	139.70	69.85	13.89	47.27
Bull Head Rails					
95R BH	145.26	69.85	69.85	19.05	47.07

Light Rails

British Standard Section No.	Rail height (mm)	Foot width (mm)	Head width (mm)	Min web thickness (mm)	Section weight (kg/m)
Flat Bottom Rails					
20M	65.09	55.56	30.96	6.75	9.88
30M	75.41	69.85	38.10	9.13	14.79
35M	80.96	76.20	42.86	9.13	17.39
35R	85.73	82.55	44.45	8.33	17.40
40	88.11	80.57	45.64	12.3	19.89
Bridge Rails					
13	48.00	92	36.00	18.0	13.31
16	54.00	108	44.50	16.0	16.06
20	55.50	127	50.00	20.5	19.86
28	67.00	152	50.00	31.0	28.62
35	76.00	160	58.00	34.5	35.38
50	76.00	165	58.50	–	50.18
Crane Rails					
A65	75.00	175.00	65.00	38.0	43.10
A75	85.00	200.00	75.00	45.0	56.20
A100	95.00	200.00	100.00	60.0	74.30
A120	105.00	220.00	120.00	72.0	100.00
175CR	152.40	152.40	107.95	38.1	86.92

RAIL TRACKS

Fish Plates

British Standard Section No.	Overall plate length		Hole diameter	Finished weight per pair	
	4 Hole (mm)	6 Hole (mm)	(mm)	4 Hole (kg/pair)	6 Hole (kg/pair)
For British Standard Heavy Rails: Flat Bottom Rails					
60 A	406.40	609.60	20.64	9.87	14.76
70 A	406.40	609.60	22.22	11.15	16.65
75 A	406.40	–	23.81	11.82	17.73
80 A	406.40	609.60	23.81	13.15	19.72
90 A	457.20	685.80	25.40	17.49	26.23
100 A	508.00	–	pear	25.02	–
110 A (shallow)	507.00	–	27.00	30.11	54.64
113 A (heavy)	507.00	–	27.00	30.11	54.64
50 'O' (shallow)	406.40	–	–	6.68	10.14
80 'O' (shallow)	495.30	–	23.81	14.72	22.69
60R (shallow)	406.40	609.60	20.64	8.76	13.13
60R (angled)	406.40	609.60	20.64	11.27	16.90
75R (shallow)	406.40	–	23.81	10.94	16.42
75R (angled)	406.40	–	23.81	13.67	–
80R (shallow)	406.40	609.60	23.81	11.93	17.89
80R (angled)	406.40	609.60	23.81	14.90	22.33
For British Standard Heavy Rails: Bull head rails					
95R BH (shallow)	–	457.20	27.00	14.59	14.61
For British Standard Light Rails: Flat Bottom Rails					
30M	355.6	–	–	–	2.72
35M	355.6	–	–	–	2.83
40	355.6	–	–	3.76	–

FRACTIONS, DECIMALS AND MILLIMETRE EQUIVALENTS

Fractions	Decimals	(mm)		Fractions	Decimals	(mm)
1/64	0.015625	0.396875		33/64	0.515625	13.096875
1/32	0.03125	0.79375		17/32	0.53125	13.49375
3/64	0.046875	1.190625		35/64	0.546875	13.890625
1/16	0.0625	1.5875		9/16	0.5625	14.2875
5/64	0.078125	1.984375		37/64	0.578125	14.684375
3/32	0.09375	2.38125		19/32	0.59375	15.08125
7/64	0.109375	2.778125		39/64	0.609375	15.478125
1/8	0.125	3.175		5/8	0.625	15.875
9/64	0.140625	3.571875		41/64	0.640625	16.271875
5/32	0.15625	3.96875		21/32	0.65625	16.66875
11/64	0.171875	4.365625		43/64	0.671875	17.065625
3/16	0.1875	4.7625		11/16	0.6875	17.4625
13/64	0.203125	5.159375		45/64	0.703125	17.859375
7/32	0.21875	5.55625		23/32	0.71875	18.25625
15/64	0.234375	5.953125		47/64	0.734375	18.653125
1/4	0.25	6.35		3/4	0.75	19.05
17/64	0.265625	6.746875		49/64	0.765625	19.446875
9/32	0.28125	7.14375		25/32	0.78125	19.84375
19/64	0.296875	7.540625		51/64	0.796875	20.240625
5/16	0.3125	7.9375		13/16	0.8125	20.6375
21/64	0.328125	8.334375		53/64	0.828125	21.034375
11/32	0.34375	8.73125		27/32	0.84375	21.43125
23/64	0.359375	9.128125		55/64	0.859375	21.828125
3/8	0.375	9.525		7/8	0.875	22.225
25/64	0.390625	9.921875		57/64	0.890625	22.621875
13/32	0.40625	10.31875		29/32	0.90625	23.01875
27/64	0.421875	10.71563		59/64	0.921875	23.415625
7/16	0.4375	11.1125		15/16	0.9375	23.8125
29/64	0.453125	11.50938		61/64	0.953125	24.209375
15/32	0.46875	11.90625		31/32	0.96875	24.60625
31/64	0.484375	12.30313		63/64	0.984375	25.003125
1/2	0.5	12.7		1.0	1	25.4

IMPERIAL STANDARD WIRE GAUGE (SWG)

IMPERIAL STANDARD WIRE GAUGE (SWG)

SWG No.	Diameter (inches)	Diameter (mm)	SWG No.	Diameter (inches)	Diameter (mm)
7/0	0.5	12.7	23	0.024	0.61
6/0	0.464	11.79	24	0.022	0.559
5/0	0.432	10.97	25	0.02	0.508
4/0	0.4	10.16	26	0.018	0.457
3/0	0.372	9.45	27	0.0164	0.417
2/0	0.348	8.84	28	0.0148	0.376
1/0	0.324	8.23	29	0.0136	0.345
1	0.3	7.62	30	0.0124	0.315
2	0.276	7.01	31	0.0116	0.295
3	0.252	6.4	32	0.0108	0.274
4	0.232	5.89	33	0.01	0.254
5	0.212	5.38	34	0.009	0.234
6	0.192	4.88	35	0.008	0.213
7	0.176	4.47	36	0.008	0.193
8	0.16	4.06	37	0.007	0.173
9	0.144	3.66	38	0.006	0.152
10	0.128	3.25	39	0.005	0.132
11	0.116	2.95	40	0.005	0.122
12	0.104	2.64	41	0.004	0.112
13	0.092	2.34	42	0.004	0.102
14	0.08	2.03	43	0.004	0.091
15	0.072	1.83	44	0.003	0.081
16	0.064	1.63	45	0.003	0.071
17	0.056	1.42	46	0.002	0.061
18	0.048	1.22	47	0.002	0.051
19	0.04	1.016	48	0.002	0.041
20	0.036	0.914	49	0.001	0.031
21	0.032	0.813	50	0.001	0.025
22	0.028	0.711			

PIPES, WATER, STORAGE, INSULATION

WATER PRESSURE DUE TO HEIGHT

Imperial

Head (Feet)	Pressure (lb/in^2)		Head (Feet)	Pressure (lb/in^2)
1	0.43		70	30.35
5	2.17		75	32.51
10	4.34		80	34.68
15	6.5		85	36.85
20	8.67		90	39.02
25	10.84		95	41.18
30	13.01		100	43.35
35	15.17		105	45.52
40	17.34		110	47.69
45	19.51		120	52.02
50	21.68		130	56.36
55	23.84		140	60.69
60	26.01		150	65.03
65	28.18			

Metric

Head (m)	Pressure (bar)		Head (m)	Pressure (bar)
0.5	0.049		18.0	1.766
1.0	0.098		19.0	1.864
1.5	0.147		20.0	1.962
2.0	0.196		21.0	2.06
3.0	0.294		22.0	2.158
4.0	0.392		23.0	2.256
5.0	0.491		24.0	2.354
6.0	0.589		25.0	2.453
7.0	0.687		26.0	2.551
8.0	0.785		27.0	2.649
9.0	0.883		28.0	2.747
10.0	0.981		29.0	2.845
11.0	1.079		30.0	2.943
12.0	1.177		32.5	3.188
13.0	1.275		35.0	3.434
14.0	1.373		37.5	3.679
15.0	1.472		40.0	3.924
16.0	1.57		42.5	4.169
17.0	1.668		45.0	4.415

1 bar	=	14.5038 lbf/in^2
1 lbf/in^2	=	0.06895 bar
1 metre	=	3.2808 ft or 39.3701 in
1 foot	=	0.3048 metres
1 in wg	=	2.5 mbar (249.1 N/m^2)

PIPES, WATER, STORAGE, INSULATION

Dimensions and Weights of Copper Pipes to BSEN 1057, BSEN 12499, BSEN 14251

Outside diameter (mm)	Internal diameter (mm)	Weight per metre (kg)	Internal diameter (mm)	Weight per metre (kg)	Internal siameter (mm)	Weight per metre (kg)
	Formerly Table X		Formerly Table Y		Formerly Table Z	
6	4.80	0.0911	4.40	0.1170	5.00	0.0774
8	6.80	0.1246	6.40	0.1617	7.00	0.1054
10	8.80	0.1580	8.40	0.2064	9.00	0.1334
12	10.80	0.1914	10.40	0.2511	11.00	0.1612
15	13.60	0.2796	13.00	0.3923	14.00	0.2031
18	16.40	0.3852	16.00	0.4760	16.80	0.2918
22	20.22	0.5308	19.62	0.6974	20.82	0.3589
28	26.22	0.6814	25.62	0.8985	26.82	0.4594
35	32.63	1.1334	32.03	1.4085	33.63	0.6701
42	39.63	1.3675	39.03	1.6996	40.43	0.9216
54	51.63	1.7691	50.03	2.9052	52.23	1.3343
76.1	73.22	3.1287	72.22	4.1437	73.82	2.5131
108	105.12	4.4666	103.12	7.3745	105.72	3.5834
133	130.38	5.5151	–	–	130.38	5.5151
159	155.38	8.7795	–	–	156.38	6.6056

Dimensions of Stainless Steel Pipes to BS 4127

Outside siameter (mm)	Maximum outside siameter (mm)	Minimum outside diameter (mm)	Wall thickness (mm)	Working pressure (bar)
6	6.045	5.940	0.6	330
8	8.045	7.940	0.6	260
10	10.045	9.940	0.6	210
12	12.045	11.940	0.6	170
15	15.045	14.940	0.6	140
18	18.045	17.940	0.7	135
22	22.055	21.950	0.7	110
28	28.055	27.950	0.8	121
35	35.070	34.965	1.0	100
42	42.070	41.965	1.1	91
54	54.090	53.940	1.2	77

PIPES, WATER, STORAGE, INSULATION

Dimensions of Steel Pipes to BS 1387

Nominal Size (mm)	Approx. Outside Diameter (mm)	Outside diameter				Thickness		
		Light		Medium & Heavy		Light (mm)	Medium (mm)	Heavy (mm)
		Max (mm)	Min (mm)	Max (mm)	Min (mm)			
6	10.20	10.10	9.70	10.40	9.80	1.80	2.00	2.65
8	13.50	13.60	13.20	13.90	13.30	1.80	2.35	2.90
10	17.20	17.10	16.70	17.40	16.80	1.80	2.35	2.90
15	21.30	21.40	21.00	21.70	21.10	2.00	2.65	3.25
20	26.90	26.90	26.40	27.20	26.60	2.35	2.65	3.25
25	33.70	33.80	33.20	34.20	33.40	2.65	3.25	4.05
32	42.40	42.50	41.90	42.90	42.10	2.65	3.25	4.05
40	48.30	48.40	47.80	48.80	48.00	2.90	3.25	4.05
50	60.30	60.20	59.60	60.80	59.80	2.90	3.65	4.50
65	76.10	76.00	75.20	76.60	75.40	3.25	3.65	4.50
80	88.90	88.70	87.90	89.50	88.10	3.25	4.05	4.85
100	114.30	113.90	113.00	114.90	113.30	3.65	4.50	5.40
125	139.70	–	–	140.60	138.70	–	4.85	5.40
150	165.1*	–	–	166.10	164.10	–	4.85	5.40

* 165.1 mm (6.5in) outside diameter is not generally recommended except where screwing to BS 21 is necessary
All dimensions are in accordance with ISO R65 except approximate outside diameters which are in accordance with ISO R64
Light quality is equivalent to ISO R65 Light Series II

Approximate Metres Per Tonne of Tubes to BS 1387

Nom. size (mm)	BLACK						GALVANIZED					
	Plain/screwed ends			Screwed & socketed			Plain/screwed ends			Screwed & socketed		
	L (m)	M (m)	H (m)	L (m)	M (m)	H (m)	L (m)	M (m)	H (m)	L (m)	M (m)	H (m)
6	2765	2461	2030	2743	2443	2018	2604	2333	1948	2584	2317	1937
8	1936	1538	1300	1920	1527	1292	1826	1467	1254	1811	1458	1247
10	1483	1173	979	1471	1165	974	1400	1120	944	1386	1113	939
15	1050	817	688	1040	811	684	996	785	665	987	779	661
20	712	634	529	704	628	525	679	609	512	673	603	508
25	498	410	336	494	407	334	478	396	327	474	394	325
32	388	319	260	384	316	259	373	308	254	369	305	252
40	307	277	226	303	273	223	296	268	220	292	264	217
50	244	196	162	239	194	160	235	191	158	231	188	157
65	172	153	127	169	151	125	167	149	124	163	146	122
80	147	118	99	143	116	98	142	115	97	139	113	96
100	101	82	69	98	81	68	98	81	68	95	79	67
125	–	62	56	–	60	55	–	60	55	–	59	54
150	–	52	47	–	50	46	–	51	46	–	49	45

The figures for 'plain or screwed ends' apply also to tubes to BS 1775 of equivalent size and thickness
Key:
L – Light
M – Medium
H – Heavy

PIPES, WATER, STORAGE, INSULATION

Flange Dimension Chart to BS 4504 & BS 10

Normal Pressure Rating (PN 6) 6 Bar

Nom. size	Flange outside dia.	Table 6/2 Forged Welding Neck	Table 6/3 Plate Slip on	Table 6/4 Forged Bossed Screwed	Table 6/5 Forged Bossed Slip on	Table 6/8 Plate Blank	Raised face		Nr. bolt hole	Size of bolt
							Dia.	T'ness		
15	80	12	12	12	12	12	40	2	4	M10 × 40
20	90	14	14	14	14	14	50	2	4	M10 × 45
25	100	14	14	14	14	14	60	2	4	M10 × 45
32	120	14	16	14	14	14	70	2	4	M12 × 45
40	130	14	16	14	14	14	80	3	4	M12 × 45
50	140	14	16	14	14	14	90	3	4	M12 × 45
65	160	14	16	14	14	14	110	3	4	M12 × 45
80	190	16	18	16	16	16	128	3	4	M16 × 55
100	210	16	18	16	16	16	148	3	4	M16 × 55
125	240	18	20	18	18	18	178	3	8	M16 × 60
150	265	18	20	18	18	18	202	3	8	M16 × 60
200	320	20	22	–	20	20	258	3	8	M16 × 60
250	375	22	24	–	22	22	312	3	12	M16 × 65
300	440	22	24	–	22	22	365	4	12	M20 × 70

Normal Pressure Rating (PN 16) 16 Bar

Nom. size	Flange outside dia.	Table 6/2 Forged Welding Neck	Table 6/3 Plate Slip on	Table 6/4 Forged Bossed Screwed	Table 6/5 Forged Bossed Slip on	Table 6/8 Plate Blank	Raised face		Nr. bolt hole	Size of bolt
							Dia.	T'ness		
15	95	14	14	14	14	14	45	2	4	M12 × 45
20	105	16	16	16	16	16	58	2	4	M12 × 50
25	115	16	16	16	16	16	68	2	4	M12 × 50
32	140	16	16	16	16	16	78	2	4	M16 × 55
40	150	16	16	16	16	16	88	3	4	M16 × 55
50	165	18	18	18	18	18	102	3	4	M16 × 60
65	185	18	18	18	18	18	122	3	4	M16 × 60
80	200	20	20	20	20	20	138	3	8	M16 × 60
100	220	20	20	20	20	20	158	3	8	M16 × 65
125	250	22	22	22	22	22	188	3	8	M16 × 70
150	285	22	22	22	22	22	212	3	8	M20 × 70
200	340	24	24	–	24	24	268	3	12	M20 × 75
250	405	26	26	–	26	26	320	3	12	M24 × 90
300	460	28	28	–	28	28	378	4	12	M24 × 90

PIPES, WATER, STORAGE, INSULATION

Minimum Distances Between Supports/Fixings

Material	BS Nominal pipe size		Pipes – Vertical	Pipes – Horizontal on to low gradients
	(inch)	(mm)	Support distance in metres	Support distance in metres
Copper	0.50	15.00	1.90	1.30
	0.75	22.00	2.50	1.90
	1.00	28.00	2.50	1.90
	1.25	35.00	2.80	2.50
	1.50	42.00	2.80	2.50
	2.00	54.00	3.90	2.50
	2.50	67.00	3.90	2.80
	3.00	76.10	3.90	2.80
	4.00	108.00	3.90	2.80
	5.00	133.00	3.90	2.80
	6.00	159.00	3.90	2.80
muPVC	1.25	32.00	1.20	0.50
	1.50	40.00	1.20	0.50
	2.00	50.00	1.20	0.60
Polypropylene	1.25	32.00	1.20	0.50
	1.50	40.00	1.20	0.50
uPVC	–	82.40	1.20	0.50
	–	110.00	1.80	0.90
	–	160.00	1.80	1.20
Steel	0.50	15.00	2.40	1.80
	0.75	20.00	3.00	2.40
	1.00	25.00	3.00	2.40
	1.25	32.00	3.00	2.40
	1.50	40.00	3.70	2.40
	2.00	50.00	3.70	2.40
	2.50	65.00	4.60	3.00
	3.00	80.40	4.60	3.00
	4.00	100.00	4.60	3.00
	5.00	125.00	5.50	3.70
	6.00	150.00	5.50	4.50
	8.00	200.00	8.50	6.00
	10.00	250.00	9.00	6.50
	12.00	300.00	10.00	7.00
	16.00	400.00	10.00	8.25

PIPES, WATER, STORAGE, INSULATION

Litres of Water Storage Required Per Person Per Building Type

Type of building	Storage (litres)
Houses and flats (up to 4 bedrooms)	120/bedroom
Houses and flats (more than 4 bedrooms)	100/bedroom
Hostels	90/bed
Hotels	200/bed
Nurses homes and medical quarters	120/bed
Offices with canteen	45/person
Offices without canteen	40/person
Restaurants	7/meal
Boarding schools	90/person
Day schools – Primary	15/person
Day schools – Secondary	20/person

Recommended Air Conditioning Design Loads

Building type	Design loading
Computer rooms	500 W/m^2 of floor area
Restaurants	150 W/m^2 of floor area
Banks (main area)	100 W/m^2 of floor area
Supermarkets	25 W/m^2 of floor area
Large office block (exterior zone)	100 W/m^2 of floor area
Large office block (interior zone)	80 W/m^2 of floor area
Small office block (interior zone)	80 W/m^2 of floor area

PIPES, WATER, STORAGE, INSULATION

Capacity and Dimensions of Galvanized Mild Steel Cisterns – BS 417

Capacity (litres)	BS type (SCM)	Dimensions		
		Length (mm)	Width (mm)	Depth (mm)
18	45	457	305	305
36	70	610	305	371
54	90	610	406	371
68	110	610	432	432
86	135	610	457	482
114	180	686	508	508
159	230	736	559	559
191	270	762	584	610
227	320	914	610	584
264	360	914	660	610
327	450/1	1220	610	610
336	450/2	965	686	686
423	570	965	762	787
491	680	1090	864	736
709	910	1070	889	889

Capacity of Cold Water Polypropylene Storage Cisterns – BS 4213

Capacity (litres)	BS type (PC)	Maximum height (mm)
18	4	310
36	8	380
68	15	430
91	20	510
114	25	530
182	40	610
227	50	660
273	60	660
318	70	660
455	100	760

PIPES, WATER, STORAGE, INSULATION

Minimum Insulation Thickness to Protect Against Freezing for Domestic Cold Water Systems (8 Hour Evaluation Period)

Pipe size (mm)	Insulation thickness (mm)					
	Condition 1			Condition 2		
	λ = 0.020	λ = 0.030	λ = 0.040	λ = 0.020	λ = 0.030	λ = 0.040
Copper pipes						
15	11	20	34	12	23	41
22	6	9	13	6	10	15
28	4	6	9	4	7	10
35	3	5	7	4	5	7
42	3	4	5	8	4	6
54	2	3	4	2	3	4
76	2	2	3	2	2	3
Steel pipes						
15	9	15	24	10	18	29
20	6	9	13	6	10	15
25	4	7	9	5	7	10
32	3	5	6	3	5	7
40	3	4	5	3	4	6
50	2	3	4	2	3	4
65	2	2	3	2	3	3

Condition 1: water temperature 7°C; ambient temperature –6°C; evaluation period 8 h; permitted ice formation 50%; normal installation, i.e. inside the building and inside the envelope of the structural insulation

Condition 2: water temperature 2°C; ambient temperature –6°C; evaluation period 8 h; permitted ice formation 50%; extreme installation, i.e. inside the building but outside the envelope of the structural insulation

λ = thermal conductivity [W/(mK)]

Insulation Thickness for Chilled And Cold Water Supplies to Prevent Condensation

On a Low Emissivity Outer Surface (0.05, i.e. Bright Reinforced Aluminium Foil) with an Ambient Temperature of +25°C and a Relative Humidity of 80%

Steel pipe size (mm)	*t* = +10			*t* = +5			*t* = 0		
	Insulation thickness (mm)			Insulation thickness (mm)			Insulation thickness (mm)		
	λ = 0.030	λ = 0.040	λ = 0.050	λ = 0.030	λ = 0.040	λ = 0.050	λ = 0.030	λ = 0.040	λ = 0.050
15	16	20	25	22	28	34	28	36	43
25	18	24	29	25	32	39	32	41	50
50	22	28	34	30	39	47	38	49	60
100	26	34	41	36	47	57	46	60	73
150	29	38	46	40	52	64	51	67	82
250	33	43	53	46	60	74	59	77	94
Flat surfaces	39	52	65	56	75	93	73	97	122

t = temperature of contents (°C)

λ = thermal conductivity at mean temperature of insulation [W/(mK)]

PIPES, WATER, STORAGE, INSULATION

Insulation Thickness for Non-domestic Heating Installations to Control Heat Loss

Steel pipe size (mm)	*t* = 75			*t* = 100			*t* = 150		
	Insulation thickness (mm)			Insulation thickness (mm)			Insulation thickness (mm)		
	λ = 0.030	λ = 0.040	λ = 0.050	λ = 0.030	λ = 0.040	λ = 0.050	λ = 0.030	λ = 0.040	λ = 0.050
10	18	32	55	20	36	62	23	44	77
15	19	34	56	21	38	64	26	47	80
20	21	36	57	23	40	65	28	50	83
25	23	38	58	26	43	68	31	53	85
32	24	39	59	28	45	69	33	55	87
40	25	40	60	29	47	70	35	57	88
50	27	42	61	31	49	72	37	59	90
65	29	43	62	33	51	74	40	63	92
80	30	44	62	35	52	75	42	65	94
100	31	46	63	37	54	76	45	68	96
150	33	48	64	40	57	77	50	73	100
200	35	49	65	42	59	79	53	76	103
250	36	50	66	43	61	80	55	78	105

t = hot face temperature (°C)
λ = thermal conductivity at mean temperature of insulation [W/(mK)]

Index

Ebook Single-User Licence Agreement

We welcome you as a user of this Spon Price Book ebook and hope that you find it a useful and valuable tool. Please read this document carefully. **This is a legal agreement** between you (hereinafter referred to as the "Licensee") and Taylor and Francis Books Ltd. (the "Publisher"), which defines the terms under which you may use the Product. **By accessing and retrieving the access code on the label inside the front cover of this book you agree to these terms and conditions outlined herein. If you do not agree to these terms you must return the Product to your supplier intact, with the seal on the label unbroken and with the access code not accessed.**

1. Definition of the Product

The product which is the subject of this Agreement, (the "Product") consists of online and offline access to the VitalSource ebook edition of *Spon's Mechanical & Electrical Services Price Book 2017..*

2. Commencement and licence

2.1 This Agreement commences upon the breaking open of the document containing the access code by the Licensee (the "Commencement Date").

2.2 This is a licence agreement (the "Agreement") for the use of the Product by the Licensee, and not an agreement for sale.

2.3 The Publisher licenses the Licensee on a non-exclusive and non-transferable basis to use the Product on condition that the Licensee complies with this Agreement. The Licensee acknowledges that it is only permitted to use the Product in accordance with this Agreement.

3. Multiple use

Use of the Product is not provided or allowed for more than one user or for a wide area network or consortium.

4. Installation and Use

4.1 The Licensee may provide access to the Product for individual study in the following manner: The Licensee may install the Product on a secure local area network on a single site for use by one user.

4.2 The Licensee shall be responsible for installing the Product and for the effectiveness of such installation.

4.3 Text from the Product may be incorporated in a coursepack. Such use is only permissible with the express permission of the Publisher in writing and requires the payment of the appropriate fee as specified by the Publisher and signature of a separate licence agreement.

4.4 The Product is a free addition to the book and the Publisher is under no obligation to provide any technical support.

5. Permitted Activities

5.1 The Licensee shall be entitled to use the Product for its own internal purposes;

5.2 The Licensee acknowledges that its rights to use the Product are strictly set out in this Agreement, and all other uses (whether expressly mentioned in Clause 6 below or not) are prohibited.

6. Prohibited Activities

The following are prohibited without the express permission of the Publisher:

6.1 The commercial exploitation of any part of the Product.

6.2 The rental, loan, (free or for money or money's worth) or hire purchase of this product, save with the express consent of the Publisher.

6.3 Any activity which raises the reasonable prospect of impeding the Publisher's ability or opportunities to market the Product.

6.4 Any networking, physical or electronic distribution or dissemination of the product save as expressly permitted by this Agreement.

6.5 Any reverse engineering, decompilation, disassembly or other alteration of the Product save in accordance with applicable national laws.

6.6 The right to create any derivative product or service from the Product save as expressly provided for in this Agreement.

6.7 Any alteration, amendment, modification or deletion from the Product, whether for the purposes of error correction or otherwise.

7. **General Responsibilities of the License**

7.1 The Licensee will take all reasonable steps to ensure that the Product is used in accordance with the terms and conditions of this Agreement.

7.2 The Licensee acknowledges that damages may not be a sufficient remedy for the Publisher in the event of breach of this Agreement by the Licensee, and that an injunction may be appropriate.

7.3 The Licensee undertakes to keep the Product safe and to use its best endeavours to ensure that the product does not fall into the hands of third parties, whether as a result of theft or otherwise.

7.4 Where information of a confidential nature relating to the product of the business affairs of the Publisher comes into the possession of the Licensee pursuant to this Agreement (or otherwise), the Licensee agrees to use such information solely for the purposes of this Agreement, and under no circumstances to disclose any element of the information to any third party save strictly as permitted under this Agreement. For the avoidance of doubt, the Licensee's obligations under this sub-clause 7.4 shall survive the termination of this Agreement.

8. **Warrant and Liability**

8.1 The Publisher warrants that it has the authority to enter into this agreement and that it has secured all rights and permissions necessary to enable the Licensee to use the Product in accordance with this Agreement.

8.2 The Publisher warrants that the Product as supplied on the Commencement Date shall be free of defects in materials and workmanship, and undertakes to replace any defective Product within 28 days of notice of such defect being received provided such notice is received within 30 days of such supply. As an alternative to replacement, the Publisher agrees fully to refund the Licensee in such circumstances, if the Licensee so requests, provided that the Licensee returns this copy of *Spon's Mechanical & Electrical Services Price Book 2017* to the Publisher. The provisions of this sub-clause 8.2 do not apply where the defect results from an accident or from misuse of the product by the Licensee.

8.3 Sub-clause 8.2 sets out the sole and exclusive remedy of the Licensee in relation to defects in the Product.

8.4 The Publisher and the Licensee acknowledge that the Publisher supplies the Product on an "as is" basis. The Publisher gives no warranties:

8.4.1 that the Product satisfies the individual requirements of the Licensee; or

8.4.2 that the Product is otherwise fit for the Licensee's purpose; or

8.4.3 that the Product is compatible with the Licensee's hardware equipment and software operating environment.

8.5 The Publisher hereby disclaims all warranties and conditions, express or implied, which are not stated above.

8.6 Nothing in this Clause 8 limits the Publisher's liability to the Licensee in the event of death or personal injury resulting from the Publisher's negligence.

8.7 The Publisher hereby excludes liability for loss of revenue, reputation, business, profits, or for indirect or consequential losses, irrespective of whether the Publisher was advised by the Licensee of the potential of such losses.

8.8 The Licensee acknowledges the merit of independently verifying the price book data prior to taking any decisions of material significance (commercial or otherwise) based on such data. It is agreed that the Publisher shall not be liable for any losses which result from the Licensee placing reliance on the data under any circumstances.

8.9 Subject to sub-clause 8.6 above, the Publisher's liability under this Agreement shall be limited to the purchase price.

9. **Intellectual Property Rights**

9.1 Nothing in this Agreement affects the ownership of copyright or other intellectual property rights in the Product.

9.2 The Licensee agrees to display the Publishers' copyright notice in the manner described in the Product.

9.3 The Licensee hereby agrees to abide by copyright and similar notice requirements required by the Publisher, details of which are as follows:

"© 2017 Taylor & Francis. All rights reserved. All materials in *Spon's Mechanical & Electrical Services Price Book 2017* are copyright protected. All rights reserved. No such materials may be used, displayed, modified, adapted, distributed, transmitted, transferred, published or otherwise reproduced in any form or by any means now or hereafter developed other than strictly in accordance with the terms of the licence agreement enclosed with *Spon's Civil Engineering & Highway Works Price Book 2017*. However, text and images may be printed and copied for research and private study within the preset program limitations. Please note the copyright notice above, and that any text or images printed or copied must credit the source."

9.4 This Product contains material proprietary to and copyedited by the Publisher and others. Except for the licence granted herein, all rights, title and interest in the Product, in all languages, formats and media throughout the world, including copyrights therein, are and remain the property of the Publisher or other copyright holders identified in the Product.

10. Non-assignment

This Agreement and the licence contained within it may not be assigned to any other person or entity without the written consent of the Publisher.

11. Termination and Consequences of Termination.

11.1 The Publisher shall have the right to terminate this Agreement if:

11.1.1 the Licensee is in material breach of this Agreement and fails to remedy such breach (where capable of remedy) within 14 days of a written notice from the Publisher requiring it to do so; or

11.1.2 the Licensee becomes insolvent, becomes subject to receivership, liquidation or similar external administration; or

11.1.3 the Licensee ceases to operate in business.

11.2 The Licensee shall have the right to terminate this Agreement for any reason upon two month's written notice. The Licensee shall not be entitled to any refund for payments made under this Agreement prior to termination under this sub-clause 11.2.

11.3 Termination by either of the parties is without prejudice to any other rights or remedies under the general law to which they may be entitled, or which survive such termination (including rights of the Publisher under sub-clause 7.4 above).

11.4 Upon termination of this Agreement, or expiry of its terms, the Licensee must destroy all copies and any back up copies of the product or part thereof.

12. General

12.1 *Compliance with export provisions*

The Publisher hereby agrees to comply fully with all relevant export laws and regulations of the United Kingdom to ensure that the Product is not exported, directly or indirectly, in violation of English law.

12.2 *Force majeure*

The parties accept no responsibility for breaches of this Agreement occurring as a result of circumstances beyond their control.

12.3 *No waiver*

Any failure or delay by either party to exercise or enforce any right conferred by this Agreement shall not be deemed to be a waiver of such right.

12.4 *Entire agreement*

This Agreement represents the entire agreement between the Publisher and the Licensee concerning the Product. The terms of this Agreement supersede all prior purchase orders, written terms and conditions, written or verbal representations, advertising or statements relating in any way to the Product.

12.5 *Severability*

If any provision of this Agreement is found to be invalid or unenforceable by a court of law of competent jurisdiction, such a finding shall not affect the other provisions of this Agreement and all provisions of this Agreement unaffected by such a finding shall remain in full force and effect.

12.6 *Variations*

This agreement may only be varied in writing by means of variation signed in writing by both parties.

12.7 *Notices*

All notices to be delivered to: Spon's Price Books, Taylor & Francis Books Ltd., 3 Park Square, Milton Park, Abingdon, Oxfordshire, OX14 4RN, UK.

12.8 *Governing law*

This Agreement is governed by English law and the parties hereby agree that any dispute arising under this Agreement shall be subject to the jurisdiction of the English courts.

If you have any queries about the terms of this licence, please contact:

Spon's Price Books
Taylor & Francis Books Ltd.
3 Park Square, Milton Park, Abingdon, Oxfordshire, OX14 4RN
Tel: +44 (0) 20 7017 6000
www.sponpress.com

Spon Press
an imprint of Taylor & Francis

Ebook options

Your print copy comes with a free ebook on the VitalSource® Bookshelf platform. Further copies of the ebook are available from https://www.crcpress.com/search/results?kw=spon+2017
Pick your price book and select the ebook under the 'Select Format' option.

To buy Spon's ebooks for five or more users of in your organisation please contact:

Spon's Price Books
eBooks & Online Sales
Taylor & Francis Books Ltd.
3 Park Square, Milton Park, Oxfordshire, OX14 4RN
Tel: (020) 337 73480
onlinesales@informa.com